CAMBRIDGE STUDIES IN ADVANCED MATHEMATICS 183

DERIVED CATEGORIES

There have been remarkably few systematic expositions of the theory of derived categories since its inception in the work of Grothendieck and Verdier in the 1960s. This book is the first in-depth treatment of this important component of homological algebra. It carefully explains the foundations in detail before moving on to key applications in commutative and noncommutative algebra, many otherwise unavailable outside of research articles. These include commutative and noncommutative dualizing complexes, perfect DG modules and tilting DG bimodules.

Written with graduate students in mind, the emphasis here is on explicit constructions (with many examples and exercises) as opposed to axiomatics, with the goal of demystifying this difficult subject. Beyond serving as a thorough introduction for students, it will serve as an important reference for researchers in algebra, geometry and mathematical physics.

Amnon Yekutieli is Professor of Mathematics at Ben-Gurion University of the Negev, Israel. His research interests are in algebraic geometry, ring theory, derived categories and deformation quantization. He has taught several graduate-level courses on derived categories. This is his fourth book.

CAMBRIDGE STUDIES IN ADVANCED MATHEMATICS

All the titles listed below can be obtained from good booksellers or from Cambridge University Press. For a complete series listing, visit www.cambridge.org/mathematics.

Already Published

145 M. Viana *Lectures on Lyapunov Exponents*
146 J.-H. Evertse & K. Győry *Unit Equations in Diophantine Number Theory*
147 A. Prasad *Representation Theory*
148 S. R. Garcia, J. Mashreghi & W. T. Ross *Introduction to Model Spaces and Their Operators*
149 C. Godsil & K. Meagher *Erdös–Ko–Rado Theorems: Algebraic Approaches*
150 P. Mattila *Fourier Analysis and Hausdorff Dimension*
151 M. Viana & K. Oliveira *Foundations of Ergodic Theory*
152 V. I. Paulsen & M. Raghupathi *An Introduction to the Theory of Reproducing Kernel Hilbert Spaces*
153 R. Beals & R. Wong *Special Functions and Orthogonal Polynomials*
154 V. Jurdjevic *Optimal Control and Geometry: Integrable Systems*
155 G. Pisier *Martingales in Banach Spaces*
156 C. T. C. Wall *Differential Topology*
157 J. C. Robinson, J. L. Rodrigo & W. Sadowski *The Three-Dimensional NavierStokes Equations*
158 D. Huybrechts *Lectures on K3 Surfaces*
159 H. Matsumoto & S. Taniguchi *Stochastic Analysis*
160 A. Borodin & G. Olshanski *Representations of the Infinite Symmetric Group*
161 P. Webb *Finite Group Representations for the Pure Mathematician*
162 C. J. Bishop & Y. Peres *Fractals in Probability and Analysis*
163 A. Bovier *Gaussian Processes on Trees*
164 P. Schneider *Galois Representations and* (φ, Γ)*-Modules*
165 P. Gille & T. Szamuely *Central Simple Algebras and Galois Cohomology* (2nd Edition)
166 D. Li & H. Queffelec *Introduction to Banach Spaces, I*
167 D. Li & H. Queffelec *Introduction to Banach Spaces, II*
168 J. Carlson, S. Müller-Stach & C. Peters *Period Mappings and Period Domains* (2nd Edition)
169 J. M. Landsberg *Geometry and Complexity Theory*
170 J. S. Milne *Algebraic Groups*
171 J. Gough & J. Kupsch *Quantum Fields and Processes*
172 T. Ceccherini-Silberstein, F. Scarabotti & F. Tolli *Discrete Harmonic Analysis*
173 P. Garrett *Modern Analysis of Automorphic Forms by Example, I*
174 P. Garrett *Modern Analysis of Automorphic Forms by Example, II*
175 G. Navarro *Character Theory and the McKay Conjecture*
176 P. Fleig, H. P. A. Gustafsson, A. Kleinschmidt & D. Persson *Eisenstein Series and Automorphic Representations*
177 E. Peterson *Formal Geometry and Bordism Operators*
178 A. Ogus *Lectures on Logarithmic Algebraic Geometry*
179 N. Nikolski *Hardy Spaces*
180 D.-C. Cisinski *Higher Categories and Homotopical Algebra*
181 A. Agrachev, D. Barilari & U. Boscain *A Comprehensive Introduction to Sub-Riemannian Geometry*
182 N. Nikolski *Toeplitz Matrices and Operators*
183 A. Yekutieli *Derived Categories*
184 C. Demeter *Fourier Restriction, Decoupling and Applications*

Derived Categories

AMNON YEKUTIELI
Ben-Gurion University of the Negev, Israel

CAMBRIDGE
UNIVERSITY PRESS

CAMBRIDGE
UNIVERSITY PRESS

University Printing House, Cambridge CB2 8BS, United Kingdom

One Liberty Plaza, 20th Floor, New York, NY 10006, USA

477 Williamstown Road, Port Melbourne, VIC 3207, Australia

314–321, 3rd Floor, Plot 3, Splendor Forum, Jasola District Centre, New Delhi – 110025, India

79 Anson Road, #06—04/06, Singapore 079906

Cambridge University Press is part of the University of Cambridge.

It furthers the University's mission by disseminating knowledge in the pursuit of education, learning, and research at the highest international levels of excellence.

www.cambridge.org
Information on this title: www.cambridge.org/9781108419338
DOI: 10.1017/9781108292825

First published 2020

A catalogue record for this publication is available from the British Library.

ISBN 978-1-108-41933-8 Hardback

Dedicated to Alexander Grothendieck, in memoriam

Contents

0

Introduction

0.1 On the Subject

Derived categories were introduced by A. Grothendieck and J.-L. Verdier around 1960 and were first published in the book [62] by R. Hartshorne. The basic idea was as follows. They had realized that the derived functors of classical homological algebra, namely the functors $\mathrm{R}^q F, \mathrm{L}_q F : \mathsf{M} \to \mathsf{N}$ derived from an additive functor $F : \mathsf{M} \to \mathsf{N}$ between abelian categories, are too limited to allow several rather natural manipulations. Perhaps the most important operation that was lacking was the composition of derived functors; the best approximation of it was a spectral sequence.

The solution to the problem was to invent a new category, starting from a given abelian category M. The objects of this new category are the complexes of objects of M. These are the same complexes that play an auxiliary role in classical homological algebra, as resolutions of objects of M. The complexes form a category $\mathbf{C}(\mathsf{M})$, but this category is not sufficiently intricate to carry in it the information of derived functors. So it must be modified.

A morphism $\phi : M \to N$ in $\mathbf{C}(\mathsf{M})$ is called a *quasi-isomorphism* if in each degree q the cohomology morphism $\mathrm{H}^q(\phi) : \mathrm{H}^q(M) \to \mathrm{H}^q(N)$ in M is an isomorphism. The modification that is needed is to make the quasi-isomorphisms invertible. This is done by a formal localization procedure, and the resulting category (with the same objects as $\mathbf{C}(\mathsf{M})$) is the derived category $\mathbf{D}(\mathsf{M})$. There is a functor $\mathrm{Q} : \mathbf{C}(\mathsf{M}) \to \mathbf{D}(\mathsf{M})$, which is the identity on objects, and it has a universal property (it is initial among the functors that send the quasi-isomorphisms to isomorphisms). A theorem (analogous to Ore localization in noncommutative ring theory) says that every morphism θ in $\mathbf{D}(\mathsf{M})$ can be written as a simple left or right fraction:

$$\theta = \mathrm{Q}(\psi_0)^{-1} \circ \mathrm{Q}(\phi_0) = \mathrm{Q}(\phi_1) \circ \mathrm{Q}(\psi_1)^{-1}, \tag{0.1.1}$$

where ϕ_i and ψ_i are morphisms in $\mathbf{C}(\mathsf{M})$ and ψ_i are quasi-isomorphisms.

The cohomology functors $\mathrm{H}^q : \mathbf{D}(\mathsf{M}) \to \mathsf{M}$, for all $q \in \mathbb{Z}$, are still defined. It turns out that the functor $\mathsf{M} \to \mathbf{D}(\mathsf{M})$, which sends an object M to the complex M concentrated in degree 0, is fully faithful.

The next step is to say what is a left or a right derived functor of an additive functor $F : \mathsf{M} \to \mathsf{N}$. The functor F can be extended in an obvious manner to a functor on complexes $F : \mathbf{C}(\mathsf{M}) \to \mathbf{C}(\mathsf{N})$. A *right derived functor* of F is a functor

$$\mathrm{R}F : \mathbf{D}(\mathsf{M}) \to \mathbf{D}(\mathsf{N}), \tag{0.1.2}$$

together with a morphism of functors $\eta^{\mathrm{R}} : \mathrm{Q}_{\mathsf{N}} \circ F \to \mathrm{R}F \circ \mathrm{Q}_{\mathsf{M}}$. The pair $(\mathrm{R}F, \eta^{\mathrm{R}})$ has to be *initial among all such pairs*. The uniqueness of such a functor $\mathrm{R}F$, up to a unique isomorphism, is relatively easy to prove (using the language of 2-categories). As for existence of $\mathrm{R}F$, it relies on the existence of suitable resolutions (similar to the injective resolutions in the classical situation). If these resolutions exist, and if the original functor F is left exact, then there is a canonical isomorphism of functors

$$\mathrm{R}^q F \cong \mathrm{H}^q \circ \mathrm{R}F : \mathsf{M} \to \mathsf{N} \tag{0.1.3}$$

for every $q \geq 0$.

The left derived functor

$$\mathrm{L}F : \mathbf{D}(\mathsf{M}) \to \mathbf{D}(\mathsf{N}) \tag{0.1.4}$$

is defined similarly. When suitable resolutions exist, and when F is right exact, there is a canonical isomorphism of functors

$$\mathrm{L}_q F \cong \mathrm{H}^{-q} \circ \mathrm{L}F : \mathsf{M} \to \mathsf{N} \tag{0.1.5}$$

for every $q \geq 0$.

There are several variations: F could be a contravariant additive functor, or it could be an additive bifunctor, contravariant in one or two of its arguments. In all these situations the derived (bi)functors $\mathrm{R}F$ and $\mathrm{L}F$ can be defined.

The derived category $\mathbf{D}(\mathsf{M})$ is additive, but it is not abelian. The notion of short exact sequence (in M and in $\mathbf{C}(\mathsf{M})$) is replaced by that of *distinguished triangle*, and thus $\mathbf{D}(\mathsf{M})$ is a *triangulated category*. The derived functors $\mathrm{R}F$ and $\mathrm{L}F$ are *triangulated functors*, which means that they send distinguished triangles in $\mathbf{D}(\mathsf{M})$ to distinguished triangles in $\mathbf{D}(\mathsf{N})$.

Already in classical homological algebra, we are interested in the *bifunctors* $\mathrm{Hom}(-,-)$ and $(- \otimes -)$. These bifunctors can also be derived. To simplify matters, let's assume that A is a commutative ring and $\mathsf{M} = \mathsf{N} = \mathsf{Mod}\,A$, the category of A-modules. We then have bifunctors

$$\mathrm{Hom}_A(-,-) : (\mathsf{Mod}\,A)^{\mathrm{op}} \times \mathsf{Mod}\,A \to \mathsf{Mod}\,A$$

and

$$(- \otimes_A -) : \mathsf{Mod}\,A \times \mathsf{Mod}\,A \to \mathsf{Mod}\,A,$$

where the superscript "op" denotes the opposite category, which encodes the contravariance in the first argument of Hom. In this situation, all resolutions exist, and we have the right derived bifunctor

$$\mathrm{RHom}_A(-,-) : \mathbf{D}(\mathsf{Mod}\,A)^{\mathrm{op}} \times \mathbf{D}(\mathsf{Mod}\,A) \to \mathbf{D}(\mathsf{Mod}\,A) \tag{0.1.6}$$

and the left derived bifunctor

$$(- \otimes^{\mathrm{L}}_A -) : \mathbf{D}(\mathsf{Mod}\,A) \times \mathbf{D}(\mathsf{Mod}\,A) \to \mathbf{D}(\mathsf{Mod}\,A). \tag{0.1.7}$$

The compatibility with the classical derived bifunctors is this: there are canonical isomorphisms

$$\mathrm{Ext}^q_A(M,N) \cong \mathrm{H}^q(\mathrm{RHom}_A(M,N)) \tag{0.1.8}$$

and

$$\mathrm{Tor}^A_q(M,N) \cong \mathrm{H}^{-q}(M \otimes^{\mathrm{L}}_A N) \tag{0.1.9}$$

for all $M, N \in \mathsf{Mod}\,A$ and $q \geq 0$.

This is what derived categories and derived functors are. As to what can be done with them, here are some of the things we will explore in this book:

- *Dualizing complexes* and *residue complexes* over noetherian commutative rings. Besides the original treatment from [62], which we present in detail here, we also include *Van den Bergh rigidity* in the commutative setting, which gives rise to *rigid residue complexes.*
- *Perfect DG modules* and *tilting DG bimodules* over noncommutative DG rings, and a few variants of *derived Morita Theory*, including the *Rickard–Keller Theorem.*
- *Derived torsion* and *balanced dualizing complexes* over connected graded NC rings, and *rigid dualizing complexes* over NC rings, including a full proof of the *Van den Bergh Existence Theorem* for NC rigid dualizing complexes.

A topic that is beyond the scope of this book, but of which we provide an outline here, is

- The *rigid approach to Grothendieck Duality* on noetherian schemes and Deligne–Mumford stacks.

Derived categories have important roles in several areas of mathematics; below is a partial list. We will not be able to talk about any of these topics in this book, so instead we give some references alongside each topic.

- $\mathcal{D}$-modules, perverse sheaves and representations of algebraic groups and Lie algebras. See [16] and [27]. More recently, the focus in this area is on the *Geometric Langlands Correspondence*, which can only be stated in terms of derived categories (see the survey [50]).
- Algebraic analysis, including differential, microdifferential and DQ modules (see [74], [134], [77]) and microlocal sheaf theory (see [75]), with its application to symplectic topology (see [149], [110]).
- Representations of finite groups and quivers, including *cluster algebras* and the *Broué Conjecture*. See [59], [84], [43].
- Birational algebraic geometry. This includes *Fourier–Mukai transforms* and *semi-orthogonal decompositions*. See the surveys [64] and [103], and the book [70].
- Homological mirror symmetry. It relates the derived category of coherent sheaves on a complex algebraic variety X to the *derived Fukaya category* of the mirror partner Y, which is a symplectic manifold. See Remark 3.8.22 and the online reference [88].
- Derived algebraic geometry. Here not only is the category of sheaves derived but also the underlying geometric objects (schemes or stacks). See Example 6.2.35, Remark 6.2.38 and the references [99] and [152].

0.2 A Motivating Discussion: Duality

Let us now approach derived categories from another perspective, very different from the one taken in the previous section, by considering the idea of *duality in algebra*.

We begin with something elementary: linear algebra. Take a field $\mathbb{K}$. Given a $\mathbb{K}$-module M (i.e. a vector space), let $D(M) := \mathrm{Hom}_{\mathbb{K}}(M, \mathbb{K})$, the dual module. There is a canonical homomorphism

$$\mathrm{ev}_M : M \to D(D(M)), \tag{0.2.1}$$

called *Hom-evaluation*, whose formula is $\mathrm{ev}_M(m)(\phi) := \phi(m)$ for $m \in M$ and $\phi \in D(M)$. If M is finitely generated, then ev_M is an isomorphism (actually this is "if and only if").

To formalize this situation, let $\mathsf{Mod}\,\mathbb{K}$ denote the category of $\mathbb{K}$-modules. Then $D : \mathsf{Mod}\,\mathbb{K} \to \mathsf{Mod}\,\mathbb{K}$ is a contravariant functor, and $\mathrm{ev} : \mathrm{Id} \to D \circ D$ is a morphism of functors (i.e. a natural transformation). Here Id is the identity functor of $\mathsf{Mod}\,\mathbb{K}$.

Now let us replace $\mathbb{K}$ by some nonzero commutative ring A. Again we can define a contravariant functor

$$D : \mathsf{Mod}\,A \to \mathsf{Mod}\,A, \quad D(M) := \mathrm{Hom}_A(M, A), \tag{0.2.2}$$

and a morphism of functors $\mathrm{ev} : \mathrm{Id} \to D \circ D$. It is easy to see that $\mathrm{ev}_M : M \to D(D(M))$ is an isomorphism if M is a finitely generated free A-module. Of course, we can't expect reflexivity (i.e. ev_M being an isomorphism) if M is not finitely generated, but what about a finitely generated module that is not free?

In order to understand this better, let us concentrate on the ring $A = \mathbb{Z}$. Since $\mathbb{Z}$-modules are just abelian groups, the category $\mathsf{Mod}\,\mathbb{Z}$ is often denoted by Ab. Let Ab_{f} be the full subcategory of finitely generated abelian groups. Every finitely generated abelian group is of the form $M \cong T \oplus F$, with T finite and F free. (The letters "T" and "F" stand for "torsion" and "free," respectively.) It is important to note that this is *not a canonical isomorphism*. There is a canonical short exact sequence

$$0 \to T \xrightarrow{\phi} M \xrightarrow{\psi} F \to 0 \tag{0.2.3}$$

in Ab_{f}, but the decomposition $M \cong T \oplus F$ comes from *choosing a splitting* $\sigma : F \to M$ of this sequence.

Exercise 0.2.4 Prove that the exact sequence (0.2.3) is functorial, namely that there are functors $T, F : \mathsf{Ab}_{\mathrm{f}} \to \mathsf{Ab}_{\mathrm{f}}$, and natural transformations $\phi : T \to \mathrm{Id}$ and $\psi : \mathrm{Id} \to F$, such that for each $M \in \mathsf{Ab}_{\mathrm{f}}$ the group $T(M)$ is finite, the group $F(M)$ is free and the sequence of homomorphisms

$$0 \to T(M) \xrightarrow{\phi_M} M \xrightarrow{\psi_M} F(M) \to 0 \tag{0.2.5}$$

is exact.

Next, prove that there does not exist a *functorial decomposition* of a finitely generated abelian group into a free part and a finite part. Namely, there is no natural transformation $\sigma : F \to \mathrm{Id}$, such that for every M the homomorphism $\sigma_M : F(M) \to M$ splits the sequence (0.2.5). (Hint: find a counterexample.)

We know that for a free finitely generated abelian group F, there is reflexivity, i.e. $\mathrm{ev}_F : F \to D(D(F))$ is an isomorphism. But for a finite abelian group T we have $D(T) = \mathrm{Hom}_{\mathbb{Z}}(T, \mathbb{Z}) = 0$. Thus, for a group $M \in \mathsf{Ab}_{\mathrm{f}}$ with a nonzero torsion subgroup T, reflexivity fails: $\mathrm{ev}_M : M \to D(D(M))$ is not an isomorphism.

On the other hand, for an abelian group M we can define another sort of dual: $D'(M) := \mathrm{Hom}_{\mathbb{Z}}(M, \mathbb{Q}/\mathbb{Z})$. There is a morphism of functors $\mathrm{ev}' : \mathrm{Id} \to$

$D' \circ D'$. For a finite abelian group T the homomorphism $\mathrm{ev}'_T : T \to D'(D'(T))$ is an isomorphism; this can be seen by decomposing T into cyclic groups, and for a finite cyclic group it is clear. So D' is a duality for finite abelian groups. (We may view the abelian group $\mathbb{Q}/\mathbb{Z}$ as the group of roots of 1 in $\mathbb{C}$, via the exponential function; then D' becomes *Pontryagin Duality*.)

But for a finitely generated free abelian group F we get $D'(D'(F)) = \widehat{F}$, the profinite completion of F. So once more, this is not a good duality for all finitely generated abelian groups.

This is where the *derived category* enters. For every commutative ring A, there is the derived category $\mathbf{D}(\mathsf{Mod}\, A)$. Here is a very quick explanation of it, in concrete terms – as opposed to the abstract point of view taken in the previous section.

Recall that a *complex* of A-modules is a diagram

$$M = (\cdots \to M^{-1} \xrightarrow{\mathrm{d}_M^{-1}} M^0 \xrightarrow{\mathrm{d}_M^0} M^1 \to \cdots) \tag{0.2.6}$$

in the category $\mathsf{Mod}\, A$. Namely the M^i are A-modules, and the d_M^i are homomorphisms. The condition is that $\mathrm{d}_M^{i+1} \circ \mathrm{d}_M^i = 0$. We sometimes write $M = \{M^i\}_{i \in \mathbb{Z}}$. The collection $\mathrm{d}_M = \{\mathrm{d}_M^i\}_{i \in \mathbb{Z}}$ is called the *differential* of M.

Given a second complex

$$N = (\cdots \to N^{-1} \xrightarrow{\mathrm{d}_N^{-1}} N^0 \xrightarrow{\mathrm{d}_N^0} N^1 \to \cdots),$$

a *homomorphism of complexes* $\phi : M \to N$ is a collection $\phi = \{\phi^i\}_{i \in \mathbb{Z}}$ of homomorphisms $\phi^i : M^i \to N^i$ in $\mathsf{Mod}\, A$ satisfying $\phi^{i+1} \circ \mathrm{d}_M^i = \mathrm{d}_N^i \circ \phi^i$. The resulting category is denoted by $\mathbf{C}(\mathsf{Mod}\, A)$.

The ith *cohomology* of the complex M is

$$\mathrm{H}^i(M) := \mathrm{Ker}(\mathrm{d}_M^i) \,/\, \mathrm{Im}(\mathrm{d}_M^{i-1}) \in \mathsf{Mod}\, A. \tag{0.2.7}$$

A homomorphism $\phi : M \to N$ in $\mathbf{C}(\mathsf{Mod}\, A)$ induces homomorphisms $\mathrm{H}^i(\phi) : \mathrm{H}^i(M) \to \mathrm{H}^i(N)$ in $\mathsf{Mod}\, A$. We call ϕ a *quasi-isomorphism* if all the homomorphisms $\mathrm{H}^i(\phi)$ are isomorphisms.

The derived category $\mathbf{D}(\mathsf{Mod}\, A)$ is the localization of $\mathbf{C}(\mathsf{Mod}\, A)$ with respect to the quasi-isomorphisms. This means that $\mathbf{D}(\mathsf{Mod}\, A)$ has the same objects as $\mathbf{C}(\mathsf{Mod}\, A)$. There is a functor

$$\mathrm{Q} : \mathbf{C}(\mathsf{Mod}\, A) \to \mathbf{D}(\mathsf{Mod}\, A), \tag{0.2.8}$$

which is the identity on objects it sends quasi-isomorphisms to isomorphisms, and it is universal for this property.

A single A-module M^0 can be viewed as a complex M concentrated in degree 0:

$$M = (\cdots \to 0 \xrightarrow{0} M^0 \xrightarrow{0} 0 \to \cdots).$$

This turns out to be a fully faithful embedding

$$\mathsf{Mod}\, A \to \mathbf{D}(\mathsf{Mod}\, A). \tag{0.2.9}$$

The essential image of this embedding is the full subcategory of $\mathbf{D}(\mathsf{Mod}\, A)$ on the complexes M whose cohomology is concentrated in degree 0. In this way we have *enlarged* the category of A-modules. All this is explained in Chapters 6 and 7 of the book.

Here is a very important kind of quasi-isomorphism. Suppose M is an A-module and

$$\cdots \to P^{-2} \xrightarrow{\mathrm{d}_P^{-2}} P^{-1} \xrightarrow{\mathrm{d}_P^{-1}} P^0 \xrightarrow{\rho} M \to 0 \tag{0.2.10}$$

is a projective resolution of it. We can view M as a complex concentrated in degree 0, by the embedding (0.2.9). Define the complex

$$P := (\cdots \to P^{-2} \xrightarrow{\mathrm{d}_P^{-2}} P^{-1} \xrightarrow{\mathrm{d}_P^{-1}} P^0 \to 0 \to \cdots), \tag{0.2.11}$$

concentrated in nonpositive degrees. Then ρ becomes a morphism of complexes $\rho : P \to M$. The exactness of the sequence (0.2.10) says that ρ is actually a quasi-isomorphism. Thus $\mathrm{Q}(\rho) : P \to M$ is an isomorphism in $\mathbf{D}(\mathsf{Mod}\, A)$.

Let us fix a complex $R \in \mathbf{C}(\mathsf{Mod}\, A)$. For every complex $M \in \mathbf{C}(\mathsf{Mod}\, A)$ we can form the complex

$$D(M) := \mathrm{Hom}_A(M, R) \in \mathbf{C}(\mathsf{Mod}\, A).$$

This is the usual Hom complex (we recall it in Section 3.6). As M changes, we get a contravariant functor

$$D : \mathbf{C}(\mathsf{Mod}\, A) \to \mathbf{C}(\mathsf{Mod}\, A).$$

The functor D has a *contravariant right derived functor*

$$\mathrm{R}D : \mathbf{D}(\mathsf{Mod}\, A) \to \mathbf{D}(\mathsf{Mod}\, A). \tag{0.2.12}$$

If P is a bounded above complex of projective modules (as in formula (0.2.11)), or more generally a *K-projective complex* (see Section 10.2), then there is a canonical isomorphism

$$\mathrm{R}D(P) \cong D(P) = \mathrm{Hom}_A(P, R). \tag{0.2.13}$$

Every complex M admits a K-projective resolution $\rho : P \to M$, and this allows us to calculate $\mathrm{R}D(M)$. Indeed, because the morphism $\mathrm{Q}(\rho) : P \to M$ is an isomorphism in $\mathbf{D}(\mathsf{Mod}\, A)$, it follows that $\mathrm{R}D(\mathrm{Q}(\rho)) : \mathrm{R}D(M) \to \mathrm{R}D(P)$ is an isomorphism in $\mathbf{D}(\mathsf{Mod}\, A)$. And the complex $\mathrm{R}D(P)$ is known by the canonical isomorphism (0.2.13). All this is explained in Chapters 8, 10 and 11 of the book.

It turns out that there is a canonical morphism

$$\mathrm{ev}^{\mathrm{R}} : \mathrm{Id} \to \mathrm{R}D \circ \mathrm{R}D \tag{0.2.14}$$

of functors from $\mathsf{D}(\operatorname{Mod} A)$ to itself, called *derived Hom-evaluation*. See Section 13.1.

Let us now return to the ring $A = \mathbb{Z}$ and the complex $R = \mathbb{Z}$. So the functor D is the same one we had in (0.2.2). Given a finitely generated abelian group M, we want to calculate the complexes $\mathrm{R}D(M)$ and $\mathrm{R}D(\mathrm{R}D(M))$ and the morphism

$$\mathrm{ev}^{\mathrm{R}}_{M} : M \to \mathrm{R}D(\mathrm{R}D(M)) \tag{0.2.15}$$

in $\mathsf{D}(\operatorname{Mod} A)$. As explained above, for this we choose a projective resolution $\rho : P \to M$, and then we calculate the complexes $\mathrm{R}D(P)$ and $\mathrm{R}D(\mathrm{R}D(P))$ and the morphism $\mathrm{ev}^{\mathrm{R}}_{P}$. For convenience we choose a projective resolution P of the shape

$$\begin{aligned} P &= (\cdots \to 0 \to P^{-1} \xrightarrow{\mathrm{d}_P^{-1}} P^0 \to 0 \to \cdots) \\ &= (\cdots \to 0 \to \mathbb{Z}^{r_1} \xrightarrow{\boldsymbol{a}\cdot(-)} \mathbb{Z}^{r_0} \to 0 \cdots), \end{aligned}$$

where $r_0, r_1 \in \mathbb{N}$ and $\boldsymbol{a}$ is a matrix of integers. The complex $\mathrm{R}D(P)$ is this:

$$\mathrm{R}D(P) \cong D(P) = \operatorname{Hom}_{\mathbb{Z}}(P, \mathbb{Z}) = (\cdots \to 0 \to \mathbb{Z}^{r_0} \xrightarrow{\boldsymbol{a}^{\mathrm{t}}\cdot(-)} \mathbb{Z}^{r_1} \to 0 \cdots),$$

a complex of free modules concentrated in degrees 0 and 1, with the transpose matrix $\boldsymbol{a}^{\mathrm{t}}$ as its differential. (We are deliberately ignoring signs here; the correct signs are shown in formulas (3.6.4) and (13.1.15).)

Because $\mathrm{R}D(P) \cong D(P)$ is itself a bounded complex of free modules, its derived dual is

$$\mathrm{R}D(\mathrm{R}D(P)) \cong D(D(P)) = \operatorname{Hom}_{\mathbb{Z}}(\operatorname{Hom}_{\mathbb{Z}}(P, \mathbb{Z}), \mathbb{Z}). \tag{0.2.16}$$

Under the isomorphism (0.2.16), the derived Hom-evaluation morphism $\mathrm{ev}^{\mathrm{R}}_{P}$ in this case is just the naive Hom-evaluation homomorphism $\mathrm{ev}_P : P \to D(D(P))$ in $\mathsf{C}(\operatorname{Mod} \mathbb{Z})$ from (0.2.1); see Exercise 13.1.17. Because P^0 and P^{-1} are finite rank free modules, it follows that ev_P is an isomorphism in $\mathsf{C}(\operatorname{Mod} \mathbb{Z})$. Therefore the morphism $\mathrm{ev}^{\mathrm{R}}_{M}$ in $\mathsf{D}(\operatorname{Mod} \mathbb{Z})$ is an isomorphism. We see that $\mathrm{R}D$ *is a duality that holds for all finitely generated* $\mathbb{Z}$*-modules* M!

Actually, much more is true. Let us denote by $\mathsf{D}_{\mathrm{f}}(\operatorname{Mod} \mathbb{Z})$ the full subcategory of $\mathsf{D}(\operatorname{Mod} \mathbb{Z})$ on the complexes M such that $\mathrm{H}^i(M)$ is finitely generated for all i. Then, according to Theorem 13.1.18, $\mathrm{ev}^{\mathrm{R}}_{M}$ is an isomorphism for every $M \in \mathsf{D}_{\mathrm{f}}(\operatorname{Mod} \mathbb{Z})$. It follows that

$$\mathrm{R}D : \mathsf{D}_{\mathrm{f}}(\operatorname{Mod} \mathbb{Z}) \to \mathsf{D}_{\mathrm{f}}(\operatorname{Mod} \mathbb{Z}) \tag{0.2.17}$$

is a duality (a contravariant equivalence). This is the celebrated *Grothendieck Duality*.

Here is the connection between the derived duality $\mathrm{R}D$ and the classical dualities D and D'. Take a finitely generated abelian group M, with short exact sequence (0.2.3). There are canonical isomorphisms

$$\mathrm{H}^0(\mathrm{R}D(M)) \cong \mathrm{Ext}^0_{\mathbb{Z}}(M,\mathbb{Z}) \cong \mathrm{Hom}_{\mathbb{Z}}(M,\mathbb{Z}) \cong \mathrm{Hom}_{\mathbb{Z}}(F,\mathbb{Z}) = D(F)$$

and

$$\mathrm{H}^1(\mathrm{R}D(M)) \cong \mathrm{Ext}^1_{\mathbb{Z}}(M,\mathbb{Z}) \cong \mathrm{Ext}^1_{\mathbb{Z}}(T,\mathbb{Z}) \cong D'(T).$$

The cohomologies $\mathrm{H}^i(\mathrm{R}D(M))$ vanish for $i \neq 0, 1$. We see that if M is neither free nor finite, then $\mathrm{H}^0(\mathrm{R}D(M))$ and $\mathrm{H}^1(\mathrm{R}D(M))$ are both nonzero, so that the complex $\mathrm{R}D(M)$ is not isomorphic in $\mathbf{D}(\mathsf{Mod}\,\mathbb{Z})$ to an object of $\mathsf{Mod}\,\mathbb{Z}$, under the embedding (0.2.9).

Grothendieck Duality holds for many noetherian commutative rings A. A sufficient condition is that A is a finitely generated ring over a regular noetherian ring $\mathbb{K}$ (e.g. $\mathbb{K} = \mathbb{Z}$ or a field). A complex $R \in \mathbf{D}(\mathsf{Mod}\,A)$ for which the contravariant functor

$$\mathrm{R}D = \mathrm{RHom}_A(-, R) : \mathbf{D}_{\mathrm{f}}(\mathsf{Mod}\,A) \to \mathbf{D}_{\mathrm{f}}(\mathsf{Mod}\,A) \tag{0.2.18}$$

is an equivalence is called a *dualizing complex*. (This is not quite accurate – see Definition 13.1.9 for the precise technical conditions on R.) A dualizing complex R over A is unique (up to a degree translation and tensoring with an invertible module). See Theorems 13.1.18, 13.1.34 and 13.1.35.

Interestingly, the structure of the dualizing complex R depends on the geometry of the ring A (i.e. of the affine scheme $\mathrm{Spec}(A)$). If A is a regular ring (like $\mathbb{Z}$) then $R = A$ is dualizing. If A is a Cohen–Macaulay ring (and $\mathrm{Spec}(A)$ is connected), then R is a single A-module (up to a shift in degrees). But if A is a more singular ring, then R must live in several degrees, as the next example shows.

Example 0.2.19 Consider the affine algebraic variety $X \subseteq \mathbf{A}^3_{\mathbb{R}}$, which is the union of a plane and a line that meet at a point, with coordinate ring

$$A = \mathbb{R}[t_1, t_2, t_3]/(t_3 \cdot t_1,\ t_3 \cdot t_2).$$

See Figure 1. A dualizing complex R over A must live in two adjacent degrees; namely there is some i such that both $\mathrm{H}^i(R)$ and $\mathrm{H}^{i+1}(R)$ are nonzero. This calculation is worked out in full in Example 13.3.12.

One can also talk about dualizing complexes over *noncommutative rings*. We will do this in Chapters 17 and 18.

0.3 On the Book

This book develops the theory of derived categories, starting from the foundations, and going all the way to applications in commutative and noncommutative

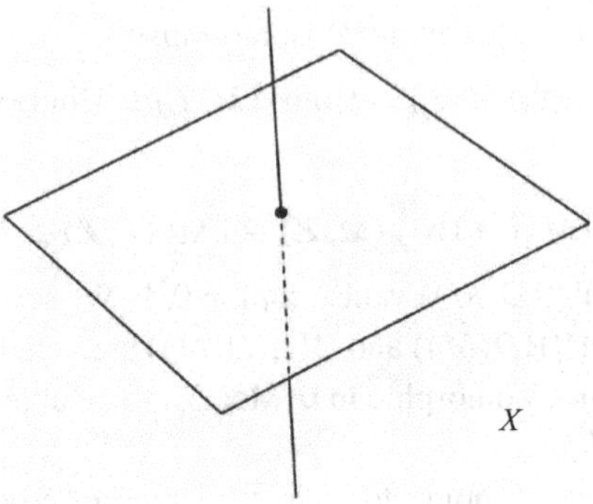

Figure 1. An algebraic variety X that is connected but not equidimensional, and hence not Cohen–Macaulay.

algebra. The emphasis is on explicit constructions (with examples), as opposed to axiomatics. The most abstract concept we use is probably that of an abelian category (which seems indispensable).

A special feature of this book is that most of the theory deals with the category $\mathsf{C}(A, \mathsf{M})$ of *DG A-modules in* M, where A is a DG ring and M is an abelian category. Here "DG" is short for "differential graded," and our DG rings are more commonly known as unital associative DG algebras. See Sections 3.3 and 3.8 for the definitions. The notion $\mathsf{C}(A, \mathsf{M})$ covers most important examples that arise in algebra and geometry:

- The category $\mathsf{C}(A)$ of DG A-modules, for any DG ring A. This includes unbounded complexes of modules over an ordinary ring A.
- The category $\mathsf{C}(\mathsf{M})$ of unbounded complexes in any abelian category M. This includes $\mathsf{M} = \mathsf{Mod}\,\mathcal{A}$, the category of sheaves of $\mathcal{A}$-modules on a ringed space $(X, \mathcal{A})$.

The category $\mathsf{C}(A, \mathsf{M})$ is a *DG category*, and its DG structure determines the *homotopy category* $\mathsf{K}(A, \mathsf{M})$ with its *triangulated structure*. We prove that every *DG functor* $F : \mathsf{C}(A, \mathsf{M}) \to \mathsf{C}(B, \mathsf{N})$ induces a *triangulated functor*

$$F : \mathsf{K}(A, \mathsf{M}) \to \mathsf{K}(B, \mathsf{N}) \tag{0.3.1}$$

between the homotopy categories.

We can now reveal that in the previous sections we were a bit imprecise (for the sake of simplifying the exposition): what we referred to there as $\mathsf{C}(\mathsf{M})$ was actually the *strict subcategory* $\mathsf{C}_{\mathrm{str}}(\mathsf{M})$, whose morphisms are the degree 0 cocycles in the DG category $\mathsf{C}(\mathsf{M})$. For the same reason the homotopy category $\mathsf{K}(\mathsf{M})$ was suppressed there.

The *derived category* $\mathsf{D}(A, \mathsf{M})$ is obtained from $\mathsf{K}(A, \mathsf{M})$ by inverting the quasi-isomorphisms. If suitable resolutions exist, the triangulated functor (0.3.1) can be derived, on the left or on the right, to yield triangulated functors

$$\mathrm{L}F,\ \mathrm{R}F : \mathsf{D}(A, \mathsf{M}) \to \mathsf{D}(B, \mathsf{N}). \tag{0.3.2}$$

We prove existence of *K-injective*, *K-projective* and *K-flat* resolutions in $\mathsf{K}(A, \mathsf{M})$ in several important contexts, and explain their roles in deriving functors and in presenting morphisms in $\mathsf{D}(A, \mathsf{M})$.

In the last six chapters of the book we discuss a few key applications of derived categories to commutative and noncommutative algebra.

There is a chapter-by-chapter synopsis of the book in Section 0.4 below.

The book is based on notes for advanced courses given at Ben-Gurion University, in the academic years 2011–12, 2015–16 and 2016–17. The main sources for Chapters 1–12 of the book are [62] and [75], but the DG theory component is absent from those earlier texts and is pretty much our own interpretation of folklore results. The material covered in Chapters 13–18 is adapted from research papers.

0.4 Synopsis of the Book

Here is a chapter-by-chapter description of the material in the book.

Chapters 1–2 These chapters are pretty much a review of the standard material on categories and functors, especially abelian categories and additive functors, that is needed for the book.

There is one new topic here – it is our *sheaf tricks*, in Section 2.4, which are a "cheap substitute" for the Freyd–Mitchell Theorem, in the sense that they facilitate proofs of various results in abstract abelian categories, yet they are themselves easy to prove (and we provide these proofs).

A reader who is familiar with this material can skip these chapters; yet we do recommend looking at our notational conventions, which are spelled out in Conventions 1.2.4 and 1.2.5.

Chapter 3 A good understanding of *DG algebra* ("DG" is short for "differential graded") is essential in our approach to derived categories. By DG algebra we mean DG rings, DG modules, DG categories and DG functors. There do not exist (to our knowledge) detailed textbook references for DG algebra. Therefore we have included a lot of basic material in this chapter.

We work over a fixed nonzero commutative base ring $\mathbb{K}$ (e.g. a field or the ring of integers $\mathbb{Z}$). All linear operations (rings, categories, functors, etc.) are assumed to be $\mathbb{K}$-linear. A *DG module* is a module M with a direct sum decomposition $M = \bigoplus_{i \in \mathbb{Z}} M^i$ into submodules, with a differential d_M of degree 1 satisfying $\mathrm{d}_M \circ \mathrm{d}_M = 0$. The grading on M is called *cohomological grading*. Tensor products of DG modules have *signed braiding isomorphisms*: for DG modules M and N, and for homogeneous elements $m \in M^i$ and $n \in N^j$, we define

$$\mathrm{br}_{M,N}(m \otimes n) := (-1)^{i \cdot j} \cdot n \otimes m \in N \otimes_{\mathbb{K}} M.$$

This braiding is usually referred to as the *Koszul sign rule*.

A *DG category* is a $\mathbb{K}$-linear category C in which the morphism sets $\mathrm{Hom}_{\mathsf{C}}(M, N)$ are endowed with DG module structures, and the composition functions are DG bilinear. There are two important sources of DG categories: the category $\mathbf{C}(\mathsf{M})$ of complexes in a $\mathbb{K}$-linear abelian category M, and the category of DG modules over a central DG $\mathbb{K}$-ring A (traditionally called a unital associative DG $\mathbb{K}$-algebra). Since we want to consider both setups, but we wish to avoid repetition, we have devised a new concept that combines both: the DG category $\mathbf{C}(A, \mathsf{M})$ of *DG A-modules in* M. By definition, a DG A-module in M is a pair (M, f), consisting of a complex $M \in \mathbf{C}(\mathsf{M})$, together with a DG $\mathbb{K}$-ring homomorphism $f : A \to \mathrm{End}_{\mathsf{M}}(M)$. The morphisms in $\mathbf{C}(A, \mathsf{M})$ are the morphism in $\mathbf{C}(\mathsf{M})$ that respect the action of A. When $A = \mathbb{K}$ we are in the case $\mathbf{C}(A, \mathsf{M}) = \mathbf{C}(\mathsf{M})$, and when $\mathsf{M} = \mathsf{Mod}\,\mathbb{K}$ we are in the case $\mathbf{C}(A, \mathsf{M}) = \mathbf{C}(A)$.

A morphism $\phi : M \to N$ in $\mathbf{C}(A, \mathsf{M})$ is called a *strict morphism* if it is a degree 0 cocycle. The *strict subcategory* $\mathbf{C}_{\mathrm{str}}(A, \mathsf{M})$ is the subcategory of $\mathbf{C}(A, \mathsf{M})$ on all the objects, and its morphisms are the strict morphisms. The strict category is abelian.

It is important to mention that $\mathbf{C}(A, \mathsf{M})$ has more structure than just a DG category. Here the objects have DG structures too, and there is the cohomology functor $\mathrm{H} : \mathbf{C}_{\mathrm{str}}(A, \mathsf{M}) \to \mathbf{G}_{\mathrm{str}}(\mathsf{M})$, where the latter is the category of graded objects in M. The cohomology functor determines the *quasi-isomorphisms*: these are the morphisms ψ in $\mathbf{C}_{\mathrm{str}}(A, \mathsf{M})$ s.t. $\mathrm{H}(\psi)$ is an isomorphism. The set of quasi-isomorphisms is denoted by $\mathbf{S}(A, \mathsf{M})$.

Chapter 4 We talk about the *translation functor* and *standard triangles* in $\mathbf{C}(A, \mathsf{M})$. This chapter consists mostly of new material, some of it implicit in the paper [25] on *pretriangulated DG categories*.

The translation $\mathrm{T}(M)$ of a DG module M is the usual one (a shift by 1 in degree, and differential $-\mathrm{d}_M$). A calculation shows that $\mathrm{T} : \mathbf{C}(A, \mathsf{M}) \to \mathbf{C}(A, \mathsf{M})$ is a DG functor. We introduce the "little t operator," which is an invertible degree -1 morphism $\mathrm{t} : \mathrm{Id} \to \mathrm{T}$ of DG functors from $\mathbf{C}(A, \mathsf{M})$ to itself. The operator t facilitates many calculations.

A morphism $\phi : M \to N$ in $\mathbf{C}_{\mathrm{str}}(A, \mathsf{M})$ gives rise to the standard triangle

$$M \xrightarrow{\phi} N \xrightarrow{e_\phi} \mathrm{Cone}(\phi) \xrightarrow{p_\phi} \mathrm{T}(M)$$

in $\mathbf{C}_{\mathrm{str}}(A, \mathsf{M})$. As a graded module, the standard cone is

$$\mathrm{Cone}(\phi) := N \oplus \mathrm{T}(M) = \begin{bmatrix} N \\ \mathrm{T}(M) \end{bmatrix},$$

written as a column module. The differential of $\mathrm{Cone}(\phi)$ is left multiplication by the matrix of degree 1 operators

$$\mathrm{d}_{\mathrm{Cone}} := \begin{bmatrix} \mathrm{d}_N & \phi \circ \mathrm{t}_M^{-1} \\ 0 & \mathrm{d}_{\mathrm{T}(M)} \end{bmatrix}.$$

Consider a DG functor

$$F : \mathbf{C}(A, \mathsf{M}) \to \mathbf{C}(B, \mathsf{N}), \tag{0.4.1}$$

where B is another DG ring and N is another abelian category. In Theorem 4.4.3 we show that there is a canonical isomorphism

$$\tau_F : F \circ \mathrm{T} \xrightarrow{\simeq} \mathrm{T} \circ F \tag{0.4.2}$$

of functors $\mathbf{C}_{\mathrm{str}}(A, \mathsf{M}) \to \mathbf{C}_{\mathrm{str}}(B, \mathsf{N})$, called the *translation isomorphism.* The pair (F, τ_F) is called a *T-additive functor.* Then, in Theorem 4.5.5, we prove that F sends standard triangles in $\mathbf{C}_{\mathrm{str}}(A, \mathsf{M})$ to standard triangles in $\mathbf{C}_{\mathrm{str}}(B, \mathsf{N})$.

We end this chapter with several examples of DG functors. These examples are prototypes – they can be easily extended to other setups.

Chapter 5 We start with the theory of *triangulated categories* and *triangulated functors*, mainly following [62]. Because the octahedral axiom plays no role in our book, we give it minimal attention.

The *homotopy category* $\mathbf{K}(A, \mathsf{M})$ has the same objects as $\mathbf{C}(A, \mathsf{M})$, and its morphisms are

$$\mathrm{Hom}_{\mathbf{K}(A,\mathsf{M})}(M, N) := \mathrm{H}^0(\mathrm{Hom}_{\mathbf{C}(A,\mathsf{M})}(M, N)).$$

There is a functor $\mathrm{P} : \mathbf{C}_{\mathrm{str}}(A, \mathsf{M}) \to \mathbf{K}(A, \mathsf{M})$, which is the identity on objects and surjective on morphisms.

In Section 5.4 we prove that the homotopy category $\mathbf{K}(A, \mathsf{M})$ is triangulated. Its *distinguished triangles* are the triangles that are isomorphic to the images under the functor P of the standard triangles in $\mathbf{C}_{\mathrm{str}}(A, \mathsf{M})$.

Theorem 5.5.1 says that given a DG functor F as in (0.4.1), with translation isomorphism τ_F from (0.4.2), the T-additive functor

$$(F, \tau_F) : \mathbf{K}(A, \mathsf{M}) \to \mathbf{K}(B, \mathsf{N}) \tag{0.4.3}$$

is triangulated. This seems to be a new result, unifying well-known yet disparate examples.

In Section 5.6 we put a triangulated structure on the opposite homotopy category $\mathbf{K}(A, \mathsf{M})^{\mathrm{op}}$. We prove that a contravariant DG functor F, like (0.4.1) but with source $\mathbf{C}(A, \mathsf{M})^{\mathrm{op}}$, induces a contravariant triangulated functor (F, τ_F), like (0.4.3) but with source $\mathbf{K}(A, \mathsf{M})^{\mathrm{op}}$.

Chapter 6 In this chapter we take a close look at *localization of categories.* We give a detailed proof of the theorem on *Ore localization* (also known as noncommutative localization). We then prove that the localization K_S of a linear category K at a denominator set S is a linear category too, and the localization functor $\mathrm{Q} : \mathsf{K} \to \mathsf{K}_\mathsf{S}$ is linear.

Chapter 7 We begin by proving that if K is a triangulated category and $\mathsf{S} \subseteq \mathsf{K}$ is a *denominator set of cohomological origin*, then the localized category K_S is triangulated, and the localization functor $\mathrm{Q} : \mathsf{K} \to \mathsf{K}_\mathsf{S}$ is triangulated.

In the case of the triangulated category $\mathbf{K}(A, \mathsf{M})$, and the quasi-isomorphisms $\mathbf{S}(A, \mathsf{M})$ in it, we get the *derived category*

$$\mathbf{D}(A, \mathsf{M}) := \mathbf{K}(A, \mathsf{M})_{\mathbf{S}(A,\mathsf{M})}, \tag{0.4.4}$$

and the triangulated localization functor

$$\mathrm{Q} : \mathbf{K}(A, \mathsf{M}) \to \mathbf{D}(A, \mathsf{M}). \tag{0.4.5}$$

We look at the full subcategory $\mathbf{K}^\star(A, \mathsf{M})$ of $\mathbf{K}(A, \mathsf{M})$ corresponding to a boundedness condition $\star \in \{+, -, \mathrm{b}\}$. We prove that when the DG ring A is nonpositive, the localization $\mathbf{D}^\star(A, \mathsf{M})$ of $\mathbf{K}^\star(A, \mathsf{M})$ with respect to the quasi-isomorphisms in it embeds fully faithfully in $\mathbf{D}(A, \mathsf{M})$. We also prove that the obvious functor $\mathsf{M} \to \mathbf{D}(\mathsf{M})$ is fully faithful.

The chapter ends with a study of the triangulated structure of the opposite derived category $\mathbf{D}(A, \mathsf{M})^{\mathrm{op}}$.

Chapter 8 In this chapter we talk about *derived functors.* To make the definitions of derived functors precise, we introduce some 2*-categorical notation* here (in Sections 8.1–8.2).

In Section 8.3 we look at *abstract derived functors.* Namely, K and E are categories (without any extra structure), $F : \mathsf{K} \to \mathsf{E}$ is a functor, and $\mathsf{S} \subseteq \mathsf{K}$ is a multiplicatively closed set of morphisms. The localization of K w.r.t. S is $\mathrm{Q} : \mathsf{K} \to \mathsf{K}_\mathsf{S}$. A *right derived functor of F w.r.t.* S is a pair $(\mathrm{R}F, \eta^{\mathrm{R}})$, where $\mathrm{R}F : \mathsf{K}_\mathsf{S} \to \mathsf{E}$ is a functor, and $\eta^{\mathrm{R}} : F \to \mathrm{R}F \circ \mathrm{Q}$ is a morphism of functors. The pair $(\mathrm{R}F, \eta^{\mathrm{R}})$ has a universal property (it is initial among all such pairs), making it unique up to a unique isomorphism. The *left derived functor* $(\mathrm{L}F, \eta^{\mathrm{L}})$ is defined similarly.

We provide a general existence theorem for derived functors. For the right derived functor $\mathrm{R}F$, existence is proved when S is a left denominator set, and there exists a full subcategory $\mathsf{J} \subseteq \mathsf{K}$ which is F-acyclic and right-resolves all objects of K. Likewise, for the left derived functor $\mathrm{L}F$ we prove existence when S is a right denominator set, and there exists a full subcategory $\mathsf{P} \subseteq \mathsf{K}$ which is F-acyclic and left-resolves all objects of K.

In Section 8.4 we specialize to *triangulated derived functors*. Here K and E are triangulated categories, $F : \mathsf{K} \to \mathsf{E}$ is a triangulated functor and $\mathsf{S} \subseteq \mathsf{K}$ is a multiplicatively closed set of cohomological origin (i.e. S consists of the quasi-isomorphisms w.r.t. some cohomological functor $H : \mathsf{K} \to \mathsf{M}$). The right and left derived functors $\mathrm{R}F, \mathrm{L}F : \mathsf{K}_{\mathsf{S}} \to \mathsf{E}$ are defined as in the abstract setting, and their uniqueness is also proved the same way. Existence requires resolving subcategories $\mathsf{J}, \mathsf{P} \subseteq \mathsf{K}$ as above, which are also full triangulated subcategories.

The chapter is concluded with a discussion of *contravariant triangulated derived functors*.

Chapter 9 This chapter is devoted to *DG and triangulated bifunctors*. We prove that a DG bifunctor

$$F : \mathbf{C}(A_1, \mathsf{M}_1) \times \mathbf{C}(A_2, \mathsf{M}_2) \to \mathbf{C}(B, \mathsf{N})$$

induces a triangulated bifunctor

$$(F, \tau_1, \tau_2) : \mathbf{K}(A_1, \mathsf{M}_1) \times \mathbf{K}(A_2, \mathsf{M}_2) \to \mathbf{K}(B, \mathsf{N}).$$

Then we define left and right derived bifunctors, and prove their uniqueness and existence, under suitable conditions. Bifunctors that are contravariant in one or two of the arguments are also studied.

Chapter 10 Here we define *K-injective* and *K-projective* objects in $\mathbf{K}(A, \mathsf{M})$, and also *K-flat* objects in $\mathbf{K}(A)$. These constitute full triangulated subcategories of $\mathbf{K}(A, \mathsf{M})$, and we refer to them as *resolving subcategories*. The category $\mathbf{K}^{\star}(A, \mathsf{M})_{\mathrm{inj}}$ of K-injectives in $\mathbf{K}^{\star}(A, \mathsf{M})$, for a boundedness condition $\star$, plays the role of the category J in Chapter 8 above, and the category of K-projectives $\mathbf{K}^{\star}(A, \mathsf{M})_{\mathrm{prj}}$ plays the role of the category P there. The K-flat DG modules are acyclic for the tensor functor.

Furthermore, we prove that the functors $\mathrm{Q} : \mathbf{K}^{\star}(A, \mathsf{M})_{\mathrm{inj}} \to \mathbf{D}^{\star}(A, \mathsf{M})$ and $\mathrm{Q} : \mathbf{K}^{\star}(A, \mathsf{M})_{\mathrm{prj}} \to \mathbf{D}^{\star}(A, \mathsf{M})$ (see equation (0.4.5)) are fully faithful.

Chapter 11 In this chapter we prove *existence* of K-injective, K-projective and K-flat resolutions in several important cases of $\mathbf{C}^{\star}(A, \mathsf{M})$:

- K-projective resolutions in $\mathbf{C}^{-}(\mathsf{M})$, where M is an abelian category with enough projectives. This is classical (i.e. it is already in [62]).
- K-projective resolutions in $\mathbf{C}(A)$, where A is any DG ring. This includes the category of unbounded complexes of modules over a ring A.

- K-injective resolutions in $\mathbf{C}^+(\mathsf{M})$, where M is an abelian category with enough injectives. This is classical too.
- K-injective resolutions in $\mathbf{C}(A)$, where A is any DG ring.

Our proofs are explicit, and we use limits of complexes cautiously (since this is known to be a pitfall). We follow several sources: [62], [143], [79], [75] and private notes provided by B. Keller.

Chapter 12 This chapter is quite varied. In Sections 12.2–12.3 there is a detailed look at the derived bifunctors $\mathrm{RHom}(-,-)$ and $(- \otimes^{\mathrm{L}} -)$.

Next, in Section 12.4, we study *cohomological dimensions of functors*. This is a refinement of the notion of *way-out functor* from [62]. It is used in Section 12.5 to prove a few theorems about triangulated functors, such as a sufficient condition for a morphism $\zeta : F \to G$ of triangulated functors to be an isomorphism.

In Sections 12.6–12.7 we study several adjunction formulas that involve the bifunctors $\mathrm{RHom}(-,-)$ and $(- \otimes^{\mathrm{L}} -)$. We define *derived forward adjunction* and *derived backward adjunction*. These adjunction formulas hold for arbitrary DG rings (without any commutativity or boundedness conditions). We prove that if $A \to B$ is a *quasi-isomorphism of DG rings*, then the restriction functor $\mathrm{Rest}_{B/A} : \mathbf{D}(B) \to \mathbf{D}(A)$ is an equivalence, and it respects the derived bifunctors $\mathrm{RHom}(-,-)$ and $(- \otimes^{\mathrm{L}} -)$.

Resolutions of DG rings are important in several contexts. In Section 12.8 we provide a full proof that given a DG $\mathbb{K}$-ring A, there exists a *noncommutative semi-free DG ring resolution* $\tilde{A} \to A$ over $\mathbb{K}$.

In Section 12.9 there is a very general theorem providing sufficient conditions for the *derived tensor-evaluation morphism*

$$\mathrm{ev}^{\mathrm{R,L}}_{L,M,N} : \mathrm{RHom}_A(L, M) \otimes^{\mathrm{L}}_B N \to \mathrm{RHom}_A(L, M \otimes^{\mathrm{L}}_B N) \qquad (0.4.6)$$

in $\mathbf{D}(\mathbb{K})$ to be an isomorphism. Here A and B are DG $\mathbb{K}$-rings, $L \in \mathbf{D}(A)$, $M \in \mathbf{D}(A \otimes_{\mathbb{K}} B^{\mathrm{op}})$ and $N \in \mathbf{D}(B)$.

Lastly, in Section 12.10 we present some adjunction formulas that pertain only to weakly commutative DG rings.

Chapter 13 This chapter starts (in the first three sections) by retelling the material in [62] on *dualizing complexes* and *residue complexes* over noetherian commutative rings.

A complex $R \in \mathbf{D}(A)$ is called *dualizing* if it has finitely generated cohomology modules, finite injective dimension, and the *derived Morita property*, which says that the *derived homothety morphism*

$$\mathrm{hm}^{\mathrm{R}}_R : A \to \mathrm{RHom}_A(R, R) \qquad (0.4.7)$$

in $\mathbf{D}(A)$ is an isomorphism. (This is a variant of the derived Hom-evaluation morphism $\mathrm{ev}^{\mathrm{R}}_{M}$ from equation (0.2.15), with $M = A$.)

To a dualizing complex R, we associate the *duality functor*

$$D := \mathrm{RHom}_A(-, R) : \mathbf{D}(A)^{\mathrm{op}} \to \mathbf{D}(A).$$

The duality functor induces an equivalence of triangulated categories $D : \mathbf{D}_{\mathrm{f}}(A)^{\mathrm{op}} \to \mathbf{D}_{\mathrm{f}}(A)$. Here $\mathbf{D}_{\mathrm{f}}(A)$ is the full subcategory of $\mathbf{D}(A)$ on the complexes with finitely generated cohomology modules.

A residue complex is a dualizing complex $\mathcal{K}$ that consists of injective modules of the correct multiplicity in each degree. If $R = \mathcal{K}$ is a residue complex, then the associated duality functor is $D = \mathrm{Hom}_A(-, \mathcal{K})$.

We prove uniqueness of dualizing complexes over a noetherian commutative ring A (up to the obvious twists), and existence when A is *essentially finite type over a regular noetherian ring* $\mathbb{K}$. In this book the adjective "regular noetherian" includes the condition that $\mathbb{K}$ has finite Krull dimension.

There is a stronger uniqueness property for residue complexes. Residue complexes exist whenever dualizing complexes exist: given a dualizing complex R, its minimal injective resolution $\mathcal{K}$ is a residue complex. To understand residue complexes, we review the *Matlis classification of injective A-modules.*

In remarks we provide sketches of Matlis Duality, Local Duality and the interpretation of Cohen–Macaulay complexes as perverse modules.

In the last two sections we talk about *Van den Bergh rigidity*. Let $\mathbb{K}$ be a regular noetherian ring, and let A be a flat essentially finite type (FEFT) $\mathbb{K}$-ring. (The flatness condition is just to simplify matters; see Remark 13.4.24.) Given a complex $M \in \mathbf{D}(A)$, its *square relative to* $\mathbb{K}$ is the complex

$$\mathrm{Sq}_{A/\mathbb{K}}(M) := \mathrm{RHom}_{A \otimes_{\mathbb{K}} A}(A, M \otimes^{\mathrm{L}}_{\mathbb{K}} M) \in \mathbf{D}(A).$$

We prove that $\mathrm{Sq}_{A/\mathbb{K}}$ is a *quadratic functor*. A *rigid complex over A relative to* $\mathbb{K}$ is a pair (M, ρ), where $M \in \mathbf{D}(A)$ and $\rho : M \xrightarrow{\simeq} \mathrm{Sq}_{A/\mathbb{K}}(M)$ is an isomorphism in $\mathbf{D}(A)$, called a *rigidifying isomorphism*. Given another rigid complex (N, σ), a *rigid morphism* between them is a morphism $\phi : M \to N$ in $\mathbf{D}(A)$ such that the diagram

$$\begin{array}{ccc} M & \xrightarrow{\ \rho\ } & \mathrm{Sq}_{A/\mathbb{K}}(M) \\ {\scriptstyle \phi}\downarrow & & \downarrow{\scriptstyle \mathrm{Sq}_{A/\mathbb{K}}(\phi)} \\ N & \xrightarrow{\ \sigma\ } & \mathrm{Sq}_{A/\mathbb{K}}(N) \end{array} \tag{0.4.8}$$

in $\mathbf{D}(A)$ is commutative. We prove that if (M, ρ) is a rigid complex such that M has the derived Morita property, then the only rigid automorphism of (M, ρ) is the identity.

A *rigid dualizing complex over A relative to* $\mathbb{K}$ is a rigid complex (R, ρ) such that R is a dualizing complex. We prove that if A has a rigid dualizing complex, then it is unique up to a unique rigid isomorphism. Existence of a rigid dualizing complex is harder to prove, and we just give a reference to it.

A *rigid residue complex over A relative to* $\mathbb{K}$ is a rigid complex $(\mathcal{K}, \rho)$ such that $\mathcal{K}$ is a residue complex. These always exist, and they are unique in the following very strong sense: if $(\mathcal{K}', \rho')$ is another rigid residue complex, then there is a unique isomorphism $\phi : \mathcal{K} \xrightarrow{\simeq} \mathcal{K}'$ in $\mathbf{C}_{\mathrm{str}}(A)$, such that $\mathrm{Q}(\phi)$ is a rigid isomorphism in $\mathbf{D}(A)$.

We end this chapter with two remarks (Remark 13.5.18 and 13.5.19) that explain how rigid residue complexes allow a new approach to *residues and duality on schemes and Deligne–Mumford stacks*, with references.

Chapter 14 We begin (in Section 14.1) with a systematic study of *algebraically perfect DG modules* over a NC DG ring A. The abbreviation "NC" stands for "noncommutative." By definition, a DG module $L \in \mathbf{D}(A)$ is called algebraically perfect if it belongs to the saturated full triangulated subcategory of $\mathbf{D}(A)$ generated by the DG module A. We give several characterizations of algebraically perfect DG modules; one of them is that L is a *compact object* of $\mathbf{D}(A)$. When A is a ring, we prove that L is algebraically perfect if and only if it is isomorphic, in $\mathbf{D}(A)$, to a bounded complex of finitely generated projective A-modules.

In Section 14.2 we prove the following *Derived Morita Theorem*, Theorem 14.2.29: Let $\mathsf{E} \subseteq \mathbf{D}(A)$ be a full triangulated subcategory which is closed under infinite direct sums, let P be a compact generator of E which is either K-projective or K-injective, and let $B := \mathrm{End}_A(P)^{\mathrm{op}}$. Then the functor $\mathrm{RHom}_A(P, -) : \mathsf{E} \to \mathbf{D}(B)$ is an equivalence of triangulated categories, with quasi-inverse $P \otimes^{\mathrm{L}}_B (-)$.

From Section 14.3 to the end of this chapter we assume that the DG rings in question are K-flat over the base ring $\mathbb{K}$. Note that every DG $\mathbb{K}$-ring A admits a NC semi-free DG ring resolution $\tilde{A} \to A$. The DG ring $\tilde{A}$ is K-flat over $\mathbb{K}$, and the restriction functor $\mathbf{D}(A) \to \mathbf{D}(\tilde{A})$ is an equivalence of triangulated categories; so the flatness restriction can be circumvented. Section 14.3 contains some basic constructions of derived functors between categories of DG bimodules (that require the flatness over the base ring).

Next, in Section 14.4, we define *tilting DG bimodules*. A DG bimodule $T \in \mathbf{D}(B \otimes_{\mathbb{K}} A^{\mathrm{op}})$ is called a tilting DG B-A-bimodule if there exists some $S \in \mathbf{D}(A \otimes_{\mathbb{K}} B^{\mathrm{op}})$ with isomorphisms $S \otimes^{\mathrm{L}}_B T \cong A$ in $\mathbf{D}(A^{\mathrm{en}})$ and $T \otimes^{\mathrm{L}}_A S \cong B$ in $\mathbf{D}(B^{\mathrm{en}})$. Here $A^{\mathrm{en}} := A \otimes_{\mathbb{K}} A^{\mathrm{op}}$ is the *enveloping DG ring of* A, and likewise B^{en}. Among other results, we prove that $T \in \mathbf{D}(B \otimes_{\mathbb{K}} A^{\mathrm{op}})$ is tilting if and only if T is a compact generator on the B side (i.e. it is a compact generator of $\mathsf{D}(B)$), and it has the *NC derived Morita property* on the B side, namely the canonical morphism $A \to \mathrm{RHom}_B(T, T)$ in $\mathsf{D}(A^{\mathrm{en}})$ is an isomorphism. We also prove the *Rickard–Keller Theorem*, asserting that if A and B are rings, and there exists a $\mathbb{K}$-linear equivalence of triangulated categories $\mathsf{D}(A) \to \mathsf{D}(B)$, then there exists a tilting DG B-A-bimodule.

Lastly, in Section 14.5, we introduce the *NC derived Picard group* $\mathrm{DPic}_{\mathbb{K}}(A)$ of a flat $\mathbb{K}$-ring A. The structure of this group is calculated when A is either local or commutative.

Chapter 15 This chapter, as well as Chapters 16 and 17, are on *algebraically graded rings*, which is our name for $\mathbb{Z}$-graded rings that have lower indices and do not involve the Koszul sign rule. (This is in contrast with the cohomologically graded rings mentioned above, which underly DG rings). Simply put, these are the usual graded rings that one encounters in textbooks on commutative and noncommutative algebra. With few exceptions, the base ring $\mathbb{K}$ in the four final chapters of the book is a field.

Let $A = \bigoplus_{i \in \mathbb{Z}} A_i$ be an algebraically graded central $\mathbb{K}$-ring. In Chapter 15 we define the category of algebraically graded A-modules $\mathsf{M}(A, \mathrm{gr})$. Its objects are the algebraically graded (left) A-modules $M = \bigoplus_{i \in \mathbb{Z}} M_i$, and the homomorphisms are the degree 0 A-linear homomorphisms. It is a $\mathbb{K}$-linear abelian category. We talk about finiteness in the algebraically graded context, and about various kinds of homological properties, such as *graded-injectivity*.

The category of complexes with entries in $\mathsf{M}(A, \mathrm{gr})$ is the DG category $\mathsf{C}(A, \mathrm{gr}) := \mathsf{C}(\mathsf{M}(A, \mathrm{gr}))$. Its objects are bigraded: a complex $M \in \mathsf{C}(A, \mathrm{gr})$ has a direct sum decomposition $M = \bigoplus_{i,j} M^i_j$ into $\mathbb{K}$-modules. Here i is the cohomological degree and j is the algebraic degree. The differential goes like this: $\mathrm{d}_M : M^i_j \to M^{i+1}_j$.

Just as for any other abelian category, we have the derived category $\mathsf{D}(A, \mathrm{gr}) := \mathsf{D}(\mathsf{M}(A, \mathrm{gr}))$. This is a triangulated category. We present (quickly) the algebraically graded variants of K-projective resolutions, etc., and the relevant derived functors.

Special emphasis is given to *connected graded rings*. An algebraically graded ring A is called connected if $A = \bigoplus_{i \in \mathbb{N}} A_i$, $A_0 = \mathbb{K}$, and each A_i is a finite $\mathbb{K}$-module. In a connected graded ring A there is the *augmentation ideal* $\mathfrak{m} := \bigoplus_{i>0} A_i$. We view $A/\mathfrak{m} \cong \mathbb{K}$ as a graded A-bimodule.

Among the connected graded rings we single out the *Artin–Schelter regular graded rings*. A noetherian connected graded ring A is called AS regular if it has finite graded global dimension and if there are isomorphisms

$$\mathrm{RHom}_A(\mathbb{K}, A) \cong \mathrm{RHom}_{A^{\mathrm{op}}}(\mathbb{K}, A) \cong \mathbb{K}(l)[-n] \qquad (0.4.9)$$

in $\mathsf{D}(\mathbb{K})$ for some integers l and n. One of the results is that if A is a noetherian connected graded ring, $a \in A$ is a regular central homogeneous element of positive degree, and the ring $B := A/(a)$ is AS regular, then A is also AS regular.

Chapter 16 Let A be a connected graded ring over the base field $\mathbb{K}$, with augmentation ideal $\mathfrak{m}$. In this chapter we study *derived* $\mathfrak{m}$*-torsion*, both for complexes of graded A-modules and for complexes of graded bimodules. By this we mean that, taking a second graded ring B, we look at the triangulated functor

$$\mathrm{R}\Gamma_{\mathfrak{m}} : \mathsf{D}(A \otimes_{\mathbb{K}} B^{\mathrm{op}}, \mathrm{gr}) \to \mathsf{D}(A \otimes_{\mathbb{K}} B^{\mathrm{op}}, \mathrm{gr}).$$

One of the main results (Theorem 16.4.3) says that if A is a left noetherian connected graded ring, and if the functor $\mathrm{R}\Gamma_{\mathfrak{m}}$ has finite cohomological dimension, then there is a functorial isomorphism

$$\mathrm{ev}^{\mathrm{R,L}}_{\mathfrak{m},M} : P \otimes^{\mathrm{L}}_{A} M \xrightarrow{\simeq} \mathrm{R}\Gamma_{\mathfrak{m}}(M) \tag{0.4.10}$$

for $M \in \mathsf{D}(A \otimes_{\mathbb{K}} B^{\mathrm{op}}, \mathrm{gr})$, where

$$P := \mathrm{R}\Gamma_{\mathfrak{m}}(A) \in \mathsf{D}(A^{\mathrm{en}}, \mathrm{gr}). \tag{0.4.11}$$

We also prove the *NC MGM Equivalence* in the connected graded context (see Theorem 16.6.32).

The χ *condition* of M. Artin and J. J. Zhang is introduced in Section 16.5. We study how this condition interacts with derived torsion. To a complex of graded A-bimodules, namely an object of $\mathsf{D}(A^{\mathrm{en}}, \mathrm{gr})$, we can also apply derived $\mathfrak{m}$-torsion from the right side, thus obtaining a complex $\mathrm{R}\Gamma_{\mathfrak{m}^{\mathrm{op}}}(M) \in \mathsf{D}(A^{\mathrm{en}}, \mathrm{gr})$. Theorem 16.5.27 says that if A is a noetherian connected graded ring of finite local cohomological dimension (i.e. both functors $\mathrm{R}\Gamma_{\mathfrak{m}}$ and $\mathrm{R}\Gamma_{\mathfrak{m}^{\mathrm{op}}}$ have finite cohomological dimension), that satisfies the χ condition, then there is a functorial isomorphism

$$\epsilon_M : \mathrm{R}\Gamma_{\mathfrak{m}}(M) \xrightarrow{\simeq} \mathrm{R}\Gamma_{\mathfrak{m}^{\mathrm{op}}}(M) \tag{0.4.12}$$

in $\mathsf{D}(A^{\mathrm{en}}, \mathrm{gr})$ for all complexes $M \in \mathsf{D}(A^{\mathrm{en}}, \mathrm{gr})$ whose cohomologies are finite A-modules on both sides. We call this phenomenon *symmetric derived* $\mathfrak{m}$*-torsion*.

Chapter 17 The focus of this chapter is on *balanced NC dualizing complexes*, following [161]. Let A be a noetherian connected graded ring over the base field $\mathbb{K}$. A complex $R \in \mathsf{D}^{\mathrm{b}}(A^{\mathrm{en}}, \mathrm{gr})$ is called a *graded NC dualizing complex* if it satisfies these three conditions:

(i) The bimodules $\mathrm{H}^q(R)$ are finitely generated A-modules on both sides.
(ii) The complex R has finite graded-injective dimension on both sides.

(iii) The complex R has the NC derived Morita property on both sides; namely the canonical morphisms $A \to \mathrm{RHom}_A(R, R)$ and $A \to \mathrm{RHom}_{A^{\mathrm{op}}}(R, R)$ in $\mathbf{D}(A^{\mathrm{en}}, \mathrm{gr})$ are isomorphisms.

A balanced NC dualizing complex over A is a pair (R, β), where R is a graded NC dualizing complex over A with symmetric derived $\mathfrak{m}$-torsion, and $\beta : \mathrm{R}\Gamma_{\mathfrak{m}}(R) \xrightarrow{\simeq} A^*$ is an isomorphism in $\mathbf{D}(A^{\mathrm{en}}, \mathrm{gr})$, called a *balancing isomorphism*. Here and later we write $M^* := \mathrm{Hom}_{\mathbb{K}}(M, \mathbb{K})$, the graded $\mathbb{K}$-linear dual of a graded module M. The graded bimodule A^* is a graded-injective A-module on both sides.

To a balanced dualizing complex R, we associate the duality functors

$$D_A := \mathrm{RHom}_A(-, R) : \mathbf{D}(A, \mathrm{gr})^{\mathrm{op}} \to \mathbf{D}(A^{\mathrm{op}}, \mathrm{gr})$$

and

$$D_{A^{\mathrm{op}}} := \mathrm{RHom}_{A^{\mathrm{op}}}(-, R) : \mathbf{D}(A^{\mathrm{op}}, \mathrm{gr})^{\mathrm{op}} \to \mathbf{D}(A, \mathrm{gr}).$$

When restricted to the subcategories of complexes with finitely generated cohomology modules, this gives an equivalence

$$D_A : \mathbf{D}_{\mathrm{f}}(A, \mathrm{gr})^{\mathrm{op}} \to \mathbf{D}_{\mathrm{f}}(A^{\mathrm{op}}, \mathrm{gr}) \tag{0.4.13}$$

with quasi-inverse $D_{A^{\mathrm{op}}}$.

A balanced dualizing complex (R, β) is unique up to a unique isomorphism (Theorem 17.2.4), and it satisfies the NC Local Duality Theorem 17.2.7. Results of Yekutieli, Zhang and M. Van den Bergh say that a noetherian connected graded ring A has a balanced dualizing complex if and only if A satisfies the χ condition and has finite local cohomological dimension. The formula for the balanced dualizing complex is $R := P^*$, where P is the complex from (0.4.11). See Corollary 17.3.24.

If A is an AS regular (or more generally AS Gorenstein) graded ring, then it has a balanced dualizing complex $R = A(\phi, -l)[n]$. Here l and n are the integers from formula (0.4.9), and ϕ is a graded ring automorphism of A. See Corollary 17.3.14.

Chapter 18 The final chapter of the book deals with *NC rigid dualizing complexes*. Let A be a NC central $\mathbb{K}$-ring (where $\mathbb{K}$ is the base field). A *NC dualizing complex* over A is a complex $R \in \mathbf{D}^{\mathrm{b}}(A^{\mathrm{en}})$ whose cohomology bimodules are finitely generated on both sides, it has finite injective dimension on both sides, and it has the NC derived Morita property on both sides. This is a modification of the graded definition given above.

Let $M \in \mathbf{D}(A^{\mathrm{en}})$. The *NC square* of M is the complex

$$\mathrm{Sq}_{A/\mathbb{K}}(M) := \mathrm{RHom}_{A^{\mathrm{en}}}(A, M \otimes_{\mathbb{K}} M) \in \mathbf{D}(A^{\mathrm{en}}). \tag{0.4.14}$$

This formula is ambiguous, because the complex of $\mathbb{K}$-modules $M \otimes_{\mathbb{K}} M$ has on it four distinct commuting actions by the ring A; but formula (0.4.14) is made precise at the beginning of Section 18.2. A *rigidifying isomorphism* for M is an isomorphism $\rho : M \xrightarrow{\simeq} \mathrm{Sq}_{A/\mathbb{K}}(M)$ in $\mathbf{D}(A^{\mathrm{en}})$, and the pair (M, ρ) is a *NC rigid complex.* A NC rigid dualizing complex is a NC rigid complex (R, ρ) such that R is a NC dualizing complex.

The definition of NC rigid dualizing complex was introduced by Van den Bergh in his paper [153]. He also proved that a rigid NC dualizing complex is unique up to isomorphism. Later, in [165], it was proved that this isomorphism is itself unique. We reproduce these results in Section 18.2.

We also give Van den Bergh's theorem on the existence of NC rigid dualizing complexes from [153], with a full proof. This is Theorem 18.5.4. Here is what it says: suppose the ring A admits a filtration $F = \{F_j(A)\}_{j \geq -1}$ such that the associated graded ring $\mathrm{Gr}^F(A)$ is noetherian connected and has a balanced dualizing complex. Then A has a rigid dualizing complex.

An important special case of the Van den Bergh Existence Theorem is Theorem 18.6.11. It says that if the graded ring $\mathrm{Gr}^F(A)$ from the previous paragraph is AS regular, then the rigid NC dualizing complex of A is $R = A(\mu)[n]$, where μ is a $\mathbb{K}$-ring automorphism of A that respects the filtration F, and n is an integer. The automorphism $\nu := \mu^{-1}$ is called the *Nakayama automorphism* of A. In modern terminology the ring A is called an *n-dimensional twisted Calabi–Yau ring*.

Finally we state and prove the *Van den Bergh Duality Theorem* for Hochschild (co)homology, and give an example of a *Calabi–Yau category* of fractional dimension.

0.5 What Is Not in the Book

A very important aspect of the theory of derived categories is *geometric*. Unfortunately our book does not discuss this aspect (except in passing). We had hoped to include the geometric aspect, but as the book grew longer this became impractical.

Geometric derived categories come in two kinds. The first kind is the derived category $\mathbf{D}(X) = \mathbf{D}(\mathsf{Mod}\, \mathcal{O}_X)$, where $(X, \mathcal{O}_X)$ is a scheme, and $\mathsf{Mod}\, \mathcal{O}_X$ is the abelian category of sheaves of $\mathcal{O}_X$-modules. This kind of derived category is the subject of the original book [62]. For recent treatments see [70], [144] or [116]. The last two references also treat derived category $\mathbf{D}(\mathfrak{X})$ of sheaves of modules on an algebraic stack $\mathfrak{X}$. We address this aspect of derived categories in Remarks 11.4.45, 11.6.32, 13.5.18 and 13.5.19.

The second kind of geometric derived category is $\mathbf{D}(\mathbb{K}_X) = \mathbf{D}(\mathsf{Mod}\,\mathbb{K}_X)$, where X is a topological space (or a site, such as the étale site of a scheme), $\mathbb{K}_X$ is the constant sheaf of rings $\mathbb{K}$ on X, and $\mathsf{Mod}\,\mathbb{K}_X$ is the abelian category of sheaves of $\mathbb{K}_X$-modules on X. For this kind of derived category we recommend the book [75].

One should also mention, in this context, the new and important theories of *derived algebraic geometry* (DAG), in which schemes are replaced by *derived stacks*. These new geometric objects carry derived categories of modules. See the references [99], [152] and [115], Remark 3.8.22 and Example 6.2.35. The *commutative DG rings* that we discuss briefly in Section 3.3 are *affine derived schemes* from the point of view of DAG.

0.6 Prerequisites and Recommended Bibliography

In preparing the book, the assumptions were that the reader is already familiar with these topics:

- Categories and functors, and classical homological algebra, namely the derived functors $\mathrm{R}^q F$ and $\mathrm{L}_q F$ of an additive functor $F : \mathsf{M} \to \mathsf{N}$ between abelian categories. For these topics we recommend the books [101], [65] and [129].
- Commutative and noncommutative ring theory, e.g. from the books [46], [105], [2], [130] and [129].

For the topics above we merely review the material, and point to references when necessary.

There are a few earlier books that deal with derived categories, in varying degrees of detail and depth. Some of them – [62] (the original reference), [75] and [76] – served as sources for us when writing the present book. Other books, such as [51], [158] and [70], are somewhat sketchy in their treatment of derived categories, but they might be useful for a reader who wants another perspective on the subject.

Finally, we want to mention the evolving online reference [144], which contains a huge amount of information on all the topics listed above.

0.7 Credo, Writing Style and Goals

Since its inception around 1960, there has been very little literature on the theory of derived categories. In some respect, the only detailed account for many years was the original book [62], written by R. Hartshorne, following notes by J.-L. Verdier (for Chapter I) and by A. Grothendieck (for Chapters II–VII). Several accounts appeared later as parts of the books [158], [51], [75], [76], [70], and maybe a few others – but none of these accounts provided enough detailed content to make it possible for a mathematician to learn how to work with derived categories, beyond a rather superficial level. *The theory thus remained mysterious.*

A personal belief of mine is that *mathematics should not be mysterious.* Some mathematics is very easy to explain. However, a few branches of mathematics are truly hard; among them are algebraic geometry and derived categories. My moral goal in this book is to demonstrate that *the theory of derived categories is difficult, but not mysterious.*

The series of books [55] by Grothendieck and J. Dieudonné, and then the book [63] by Hartshorne, have shown us that *algebraic geometry is difficult but not mysterious.* The definitions and the statements are precise, and the proofs are available (to be read or to be taken on trust, as the reader prefers). I hope that the present book will do the same for derived categories. (Although I doubt I can match the excellent writing talent of the aforementioned authors!)

In more practical terms, the goal of this book is to develop the theory of derived categories in a systematic fashion, with full details, and with several important applications. The expectation is that our book will open the doors for researchers in algebra and geometry (and related disciplines such as mathematical physics) to productive work using derived categories – doors that have been to a large extent locked until now (or at least hidden by the shrouds of mystery, to use the prior metaphor).

This book is far too lengthy for a one-semester graduate course. (In my rather sluggish lecturing style, Chapters 1–13 of the book took about four semesters.) The book is intended to be used as a reference, or for personal learning. Perhaps a lecturer who has the ability to concentrate the material sufficiently, or to choose only certain key aspects, might extract a one-semester course from this book; if so, please let me know!

0.8 Acknowledgments

As already mentioned, the book originated in a course on derived categories that was held at Ben-Gurion University in Spring 2012. I want to thank the participants of this course for correcting many of my mistakes (both in real time during the lectures and afterward when writing the notes [170]). Thanks also to Joseph Lipman, Pierre Schapira, Amnon Neeman and Charles Weibel for helpful discussions during the preparation of that course. Vincent Beck, Yang Han and Lucas Simon sent me corrections and useful comments on the material in [170].

I started writing the book itself while teaching a four-semester course on the subject at Ben-Gurion University, spanning the academic years 2015–16 and 2016–17. I wish to thank the participants of this course, and especially Rishi Vyas, Stephan Snigerov, Asaf Yekutieli, S. D. Banerjee, Alberto Boix and William Woods, who contributed material and corrected numerous mistakes.

Ben-Gurion University was generous enough to permit this long project. The project was supported by the Israel Science Foundation grant no. 253/13.

Stephan Snigerov, William Woods and Rishi Vyas helped me prepare the manuscript for publication, and I wish to thank them for that. Bernhard Keller kindly sent me private notes containing clean proofs of several theorems on the existence of resolutions of DG modules, and he also helped me with numerous suggestions.

In the process of writing the book I have also benefited from the advice of Pierre Schapira, Robin Hartshorne, Rodney Sharp, Manuel Saorin, Suresh Nayak, Liran Shaul, Johan de Jong, Steven Kleiman, Louis Rowen, Amnon Neeman, Amiram Braun, Jesse Wolfson, Bjarn de Jong, James Zhang, Jun-Ichi Miyachi, Vladimir Hinich, Brooke Shipley, Jeremy Rickard, Peter Jørgensen, Michael Sharpe (on LaTeX), Roland Berger, Mattia Ornaghi, Georges Maltsiniotis, Harry Gindi, Goncalo Tabuada, Sefi Ladkani, Saurabh Singh and the anonymous referees for Cambridge University Press.

Special thanks to Thomas Harris, my editor at Cambridge University Press, whose initiative made the publication of this book possible, to Clare Dennison and Libby Haynes at CUP, for their assistance, and to the copy editor Holly Monteith, who polished the final manuscript.

1

Basic Facts on Categories

It is assumed that the reader has a working knowledge of categories and functors. References for this material are [101], [102], [129], [65], [158], [75] and [76]. In this chapter we review the relevant material and establish notation.

1.1 Set Theory

In this book we will not try to be precise about issues of set theory. The blanket assumption is that we are given a *Grothendieck universe* U. This is a "large" infinite set. A *small set*, or a U-small set, is a set S that is an element of U. We want all the products $\prod_{i \in I} S_i$ and disjoint unions $\coprod_{i \in I} S_i$, with I and S_i small sets, to be small sets too. (This requirement is not crucial for us, and it is more a matter of convenience. When dealing with higher categories, one usually needs a hierarchy of universes anyhow.) We assume that the axiom of choice holds in U.

A U-*category* is a category C whose set of objects Ob(C) is a subset of U, and for every $C, D \in \mathrm{Ob}(\mathsf{C})$ the set of morphisms $\mathrm{Hom}_{\mathsf{C}}(C, D)$ is small. If Ob(C) is also small, then C is called a *small category*. See [5], [102] or [76, Section 1.1]. Another approach, involving "sets" versus "classes," can be found in [113].

We denote by Set the category of all small sets. So Ob(Set) = U, and Set is a U-category. A group (or a ring, etc.) is called small if its underlying set is small. We denote by Grp, Ab, Rng and $\mathsf{Rng}_{\mathrm{c}}$ the categories of small groups, small abelian groups, small rings and small commutative rings, respectively. For a small ring A we denote by Mod A the category of small left A-modules.

By default we work with U-categories, and from now on, U will remain implicit. There are several places in which we shall encounter set theoretical issues (regarding functor categories and localization of categories), but these problems can be solved by introducing a bigger universe V such that $\mathsf{U} \in \mathsf{V}$.

1.2 Notation and Conventions

Let C be a category. We often write $C \in \mathsf{C}$ as an abbreviation for $C \in \mathrm{Ob}(\mathsf{C})$. For an object C, its identity automorphism is denoted by id_C. The identity functor of C is denoted by Id_{C}.

The opposite category of C is C^{op}. It has the same objects as C, but the morphism sets are $\mathrm{Hom}_{\mathsf{C}^{\mathrm{op}}}(C_0, C_1) := \mathrm{Hom}_{\mathsf{C}}(C_1, C_0)$, and composition is reversed. Of course, $(\mathsf{C}^{\mathrm{op}})^{\mathrm{op}} = \mathsf{C}$. The identity functor of C can be viewed as a contravariant functor $\mathrm{Op} : \mathsf{C} \to \mathsf{C}^{\mathrm{op}}$. To be explicit, on objects, we take $\mathrm{Op}(C) := C$. As for morphisms, given a morphism $\phi : C_0 \to C_1$ in C, we let $\mathrm{Op}(\phi) : \mathrm{Op}(C_1) \to \mathrm{Op}(C_0)$ be the morphism $\mathrm{Op}(\phi) := \phi$ in C^{op}. The inverse functor $\mathsf{C}^{\mathrm{op}} \to \mathsf{C}$ is also denoted by Op. Thus $\mathrm{Op} \circ \mathrm{Op} = \mathrm{Id}$.

A contravariant functor $F : \mathsf{C} \to \mathsf{D}$ is the same as a covariant functor $F \circ \mathrm{Op} : \mathsf{C}^{\mathrm{op}} \to \mathsf{D}$. By default, all functors will be covariant, unless explicitly mentioned otherwise. Contravariant functors will almost always be dealt with by replacing the source category with its opposite.

Definition 1.2.1 Let $\mathbb{K}$ be a commutative ring. By a *central $\mathbb{K}$-ring*, we mean a ring A, together with a ring homomorphism $\mathbb{K} \to A$, called the structural homomorphism, such that the image of $\mathbb{K}$ is inside the center of A.

The category of central $\mathbb{K}$-rings, whose morphisms are the ring homomorphisms $f : A \to B$ that respect the structural homomorphisms from $\mathbb{K}$, is denoted by $\mathsf{Rng}/_{\mathrm{c}}\,\mathbb{K}$.

Traditionally, a central $\mathbb{K}$-ring was called a "unital associative $\mathbb{K}$-algebra." Of course, all rings and ring homomorphisms are unital. When $\mathbb{K} = \mathbb{Z}$, a central $\mathbb{K}$-ring is just a ring, and then we sometimes use the notation Rng.

Example 1.2.2 Let $\mathbb{K}$ be a nonzero commutative ring, and let n be a positive integer. Then $\mathrm{Mat}_{n \times n}(\mathbb{K})$, the ring of $n \times n$ matrices with entries in $\mathbb{K}$, is a central $\mathbb{K}$-ring.

Definition 1.2.3 Let A be a ring. We denote by $\mathsf{Mod}\,A$, or by the abbreviated notation $\mathbf{M}(A)$, the category of left A-modules.

Rings and modules are very important for us, so let us also put forth the next convention.

Convention 1.2.4 Below are the default implicit assumptions for linear structures and operations.

(1) There is a nonzero commutative base ring $\mathbb{K}$ (e.g. the ring of integers $\mathbb{Z}$ or a field).
(2) The unadorned tensor symbol $\otimes$ means $\otimes_{\mathbb{K}}$.
(3) All rings are central $\mathbb{K}$-rings (see Definition 1.2.1), all ring homomorphisms are over $\mathbb{K}$, and all bimodules are $\mathbb{K}$-central.
(4) Generalizing (3), all linear categories are $\mathbb{K}$-linear (see Definition 2.1.1), and all linear functors are $\mathbb{K}$-linear (see Definition 2.5.1).
(5) For a ring A, all A-modules are left A-modules, unless explicitly stated otherwise.

Right A-modules are left modules over the opposite ring A^{op}, and this is the way we shall most often deal with them. Morphisms in the categories of groups, rings, A-modules, etc. will usually be called group homomorphism, ring homomorphisms, A-module homomorphisms, etc., respectively.

Convention 1.2.5 We will try to keep the following font and letter conventions:

- $f : C \to D$ is a morphism between objects in a category.
- $F : \mathsf{C} \to \mathsf{D}$ is a functor between categories.
- $\eta : F \to G$ is morphism of functors (i.e. a natural transformation) between functors $F, G : \mathsf{C} \to \mathsf{D}$.
- $\phi, \psi, \phi_i : M \to N$ are morphisms between objects in an abelian category M.
- $F : \mathsf{M} \to \mathsf{N}$ is a linear functor between abelian categories.
- The category of complexes in an abelian category M is $\mathbf{C}(\mathsf{M})$.
- If M is a module category, and $M \in \mathrm{Ob}(\mathsf{M})$, then elements of M will be denoted by $m, n, m_i, \ldots$.

1.3 Epimorphisms and Monomorphisms

Let C be a category. Recall that a morphism $f : C \to D$ in C is called an *isomorphism* if there is a morphism $g : D \to C$ such that $f \circ g = \mathrm{id}_D$ and $g \circ f = \mathrm{id}_C$. The morphism g is called the *inverse* of f, it is unique (if it exists), and it is denoted by f^{-1}. An isomorphism is often denoted by this shape of arrow: $f : C \xrightarrow{\simeq} D$.

A morphism $f : C \to D$ in C is called an *epimorphism* if it has the right cancellation property: for every $g, g' : D \to E$, $g \circ f = g' \circ f$ implies $g = g'$. An epimorphism is often denoted by this shape of arrow: $f : C \twoheadrightarrow D$.

A morphism $f : C \to D$ is called a *monomorphism* if it has the left cancellation property: for every $g, g' : E \to C$, $f \circ g = f \circ g'$ implies $g = g'$. A monomorphism is often denoted by this shape of arrow: $f : C \rightarrowtail D$.

Example 1.3.1 In Set the monomorphisms are the injections, and the epimorphisms are the surjections. A morphism $f : C \to D$ in Set that is both a monomorphism and an epimorphism is an isomorphism. The same holds in the category $\mathsf{Mod}\, A$ of left modules over a ring A.

This example could be misleading, because the property of being an epimorphism is often not preserved by forgetful functors, as the next exercise shows.

Exercise 1.3.2 Consider the category of rings Rng. Show that the forgetful functor $\mathsf{Rng} \to \mathsf{Set}$ respects monomorphisms, but it does not respect epimorphisms. (Hint: show that the inclusion $\mathbb{Z} \to \mathbb{Q}$ is an epimorphism in Rng.)

By a *subobject* of an object $C \in \mathsf{C}$ we mean a monomorphism $f : C' \rightarrowtail C$ in C. We sometimes write $C' \subseteq C$ in this situation, but this is only notational (and does not mean inclusion of sets). We say that two subobjects $f_0 : C'_0 \rightarrowtail C$ and $f_1 : C'_1 \rightarrowtail C$ of C are *isomorphic* if there is an isomorphism $g : C'_0 \xrightarrow{\simeq} C'_1$ such that $f_1 \circ g = f_0$.

Likewise, by a *quotient* of C we mean an epimorphism $g : C \twoheadrightarrow C''$ in C. There is an analogous notion of isomorphic quotients.

Exercise 1.3.3 Let C be a category, and let C be an object of C.

(1) Suppose $f_0 : C'_0 \rightarrowtail C$ and $f_1 : C'_1 \rightarrowtail C$ are subobjects of C. Show that there is at most one morphism $g : C'_0 \to C'_1$ such that $f_1 \circ g = f_0$; and if g exists, then it is a monomorphism.

(2) Show that isomorphism is an equivalence relation on the set of subobjects of C. Show that the set of equivalence classes of subobjects of C is partially ordered by "inclusion." (Ignore set theoretical issues.)

(3) Formulate and prove the analogous statements for quotient objects.

An *initial object* in a category C is an object $C_0 \in \mathsf{C}$, such that for every object $C \in \mathsf{C}$ there is exactly one morphism $C_0 \to C$. Thus the set $\mathrm{Hom}_{\mathsf{C}}(C_0, C)$ is a singleton. A *terminal object* in C is an object $C_\infty \in \mathsf{C}$, such that for every object $C \in \mathsf{C}$ there is exactly one morphism $C \to C_\infty$.

Definition 1.3.4 A *zero object* in a category C is an object which is both initial and terminal.

Initial, terminal and zero objects are unique up to unique isomorphisms (but they need not exist).

Example 1.3.5 In Set, $\varnothing$ is an initial object, and every singleton is a terminal object. There is no zero object.

Example 1.3.6 In $\mathsf{Mod}\,A$, every trivial module (with only the zero element) is a zero object, and we denote this module by 0. This is allowed, since any other zero module is uniquely isomorphic to it.

1.4 Products and Coproducts

Let C be a category. By a *collection of objects of* C indexed by a (small) set I, we mean a function $I \to \mathrm{Ob}(\mathsf{C})$, $i \mapsto C_i$. We usually denote this collection using the curly brackets notation: $\{C_i\}_{i\in I}$.

Given a collection $\{C_i\}_{i\in I}$ of objects of C, its *product* is a pair $(C, \{p_i\}_{i\in I})$ consisting of an object $C \in \mathsf{C}$, and a collection $\{p_i\}_{i\in I}$ of morphisms $p_i : C \to C_i$, called *projections*. The pair $(C, \{p_i\}_{i\in I})$ must have this universal property: given an object D and morphisms $f_i : D \to C_i$, there is a unique morphism $f : D \to C$ s.t. $f_i = p_i \circ f$. Of course, if a product $(C, \{p_i\}_{i\in I})$ exists, then it is unique up to a unique isomorphism; and we usually write $\prod_{i\in I} C_i := C$, leaving the projection morphisms implicit.

Example 1.4.1 In Set and $\mathsf{Mod}\,A$ all products exist, and they are the usual cartesian products.

For a collection $\{C_i\}_{i\in I}$ of objects of C, its *coproduct* is a pair $(C, \{e_i\}_{i\in I})$, consisting of an object C and a collection $\{e_i\}_{i\in I}$ of morphisms $e_i : C_i \to C$, called *embeddings*. The pair $(C, \{e_i\}_{i\in I})$ must have this universal property: given an object D and morphisms $f_i : C_i \to D$, there is a unique morphism $f : C \to D$ s.t. $f_i = f \circ e_i$. If a coproduct $(C, \{e_i\}_{i\in I})$ exists, then it is unique up to a unique isomorphism; and we write $\coprod_{i\in I} C_i := C$, leaving the embeddings implicit.

Example 1.4.2 In Set the coproduct is the disjoint union. In $\mathsf{Mod}\,A$ the coproduct is the direct sum.

Products and coproducts are very degenerate cases of *limits* and *colimits*, respectively. In this book we will not need to use limits and colimits in their most general form. All we shall need is inverse limits and direct limits indexed by $\mathbb{N}$; and these will be recalled in Section 1.8 below.

We do need to talk about *fibered products*. Let C be some category. Recall that a commutative diagram

$$\begin{array}{ccc} E & \xrightarrow{g_2} & D_2 \\ {\scriptstyle g_1}\downarrow & & \downarrow{\scriptstyle f_2} \\ D_1 & \xrightarrow{f_1} & C \end{array} \tag{1.4.3}$$

in C is called *cartesian* if for every object $E' \in$ C, with morphisms $g_1' : E' \to D_1$ and $g_2' : E' \to D_2$ that satisfy $f_1 \circ g_1' = f_2 \circ g_2'$, there exists a unique morphism $h : E' \to E$ such that $g_i' = g_i \circ h$. See the following commutative diagram.

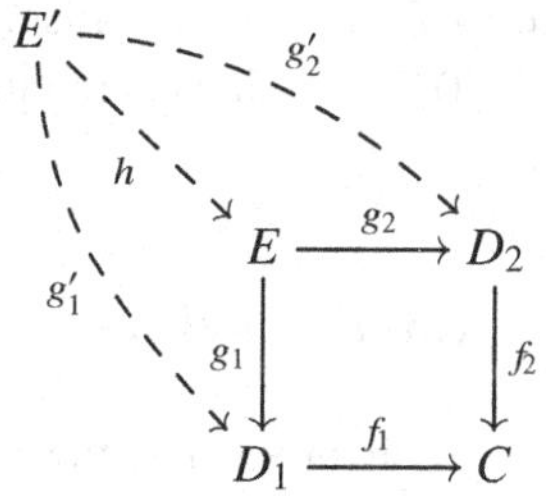

A cartesian diagram is also called a *pullback diagram*, and the object E is called the *fibered product* of D_1 and D_2 over C, with notation $D_1 \times_C D_2 := E$. This notation leaves the morphisms implicit. Of course, if a fibered product exists, than it is unique up to a unique isomorphism that commutes with the given arrows.

There is a dual notion: *fibered coproduct*. The input is morphisms $C \to D_1$ and $C \to D_2$ in C, and the fibered coproduct $D_1 \sqcup_C D_1$ in C is just the fibered product in the opposite category C^{op}.

1.5 Equivalence of Categories

Recall that a functor $F : \mathsf{C} \to \mathsf{D}$ is an *equivalence* if there exist a functor $G : \mathsf{D} \to \mathsf{C}$, and isomorphisms of functors (i.e. natural isomorphisms) $G \circ F \xrightarrow{\simeq} \mathrm{Id}_{\mathsf{C}}$ and $F \circ G \xrightarrow{\simeq} \mathrm{Id}_{\mathsf{D}}$. Such a functor G is called a *quasi-inverse* of F, and it is unique up to isomorphism (if it exists).

The functor $F : \mathsf{C} \to \mathsf{D}$ is *full* (resp. *faithful*) if for every $C_0, C_1 \in \mathsf{C}$ the function

$$F : \mathrm{Hom}_{\mathsf{C}}(C_0, C_1) \to \mathrm{Hom}_{\mathsf{D}}(F(C_0), F(C_1))$$

is surjective (resp. injective).

We know that $F : \mathsf{C} \to \mathsf{D}$ is an equivalence if and only if these two conditions hold:

(i) F is essentially surjective on objects. This means that for every $D \in \mathsf{D}$ there is an isomorphism $F(C) \xrightarrow{\simeq} D$ for some $C \in \mathsf{C}$.
(ii) F is fully faithful (i.e. full and faithful).

Exercise 1.5.1 If you are not sure about the last claim (characterization of equivalences), then prove it. (Hint: use the axiom of choice to construct a quasi-inverse of F.)

A functor $F : \mathsf{C} \to \mathsf{D}$ is called an *isomorphism of categories* if it is bijective on sets of objects and on sets of morphisms. It is clear that an isomorphism of categories is an equivalence. If F is an isomorphism of categories, then it has an inverse isomorphism $F^{-1} : \mathsf{D} \to \mathsf{C}$, which is unique.

1.6 Bifunctors

Let C and D be categories. Their product is the category $\mathsf{C} \times \mathsf{D}$ defined as follows: the set of objects is

$$\mathrm{Ob}(\mathsf{C} \times \mathsf{D}) := \mathrm{Ob}(\mathsf{C}) \times \mathrm{Ob}(\mathsf{D}).$$

The sets of morphisms are

$$\mathrm{Hom}_{\mathsf{C} \times \mathsf{D}}\big((C_0, D_0), (C_1, D_1)\big) := \mathrm{Hom}_{\mathsf{C}}(C_0, C_1) \times \mathrm{Hom}_{\mathsf{D}}(D_0, D_1).$$

The composition is

$$(f_1, g_1) \circ (f_0, g_0) := (f_1 \circ f_0, g_1 \circ g_0),$$

and the identity morphism of an object (C, D) is $(\mathrm{id}_C, \mathrm{id}_D)$.

A *bifunctor* from (C, D) to E is a functor $F : \mathsf{C} \times \mathsf{D} \to \mathsf{E}$. The extra information that is given when we call F a bifunctor is that the source category $\mathsf{C} \times \mathsf{D}$ is a product.

1.7 Representable Functors

Let C be a category. An object $C \in \mathsf{C}$ gives rise to a functor

$$\mathrm{Y}_{\mathsf{C}}(C) : \mathsf{C}^{\mathrm{op}} \to \mathsf{Set}, \quad \mathrm{Y}_{\mathsf{C}}(C) := \mathrm{Hom}_{\mathsf{C}}(-, C). \tag{1.7.1}$$

Explicitly, the functor $\mathrm{Y}_{\mathsf{C}}(C)$ sends an object $D \in \mathsf{C}$ to the set $\mathrm{Y}_{\mathsf{C}}(C)(D) := \mathrm{Hom}_{\mathsf{C}}(D, C)$, and a morphism $\psi : D_0 \to D_1$ in C goes to the function

$$\mathrm{Y}_{\mathsf{C}}(C)(\psi) := \mathrm{Hom}_{\mathsf{C}}(\psi, \mathrm{id}_C) : \mathrm{Hom}_{\mathsf{C}}(D_1, C) \to \mathrm{Hom}_{\mathsf{C}}(D_0, C).$$

Now suppose we are given a morphism $\phi : C_0 \to C_1$ in C. There is a morphism of functors (a natural transformation)

$$\mathrm{Y}_{\mathsf{C}}(\phi) := \mathrm{Hom}_{\mathsf{C}}(-, \phi) : \mathrm{Y}_{\mathsf{C}}(C_0) \to \mathrm{Y}_{\mathsf{C}}(C_1). \tag{1.7.2}$$

Consider the category $\mathsf{Fun}(\mathsf{C}^{\mathrm{op}}, \mathsf{Set})$, whose objects are the functors $F : \mathsf{C}^{\mathrm{op}} \to \mathsf{Set}$, and whose morphisms are the morphisms of functors $\eta : F_0 \to F_1$. There is a set theoretic difficulty here: the sets of objects and morphisms of $\mathsf{Fun}(\mathsf{C}^{\mathrm{op}}, \mathsf{Set})$ are too big (unless C is a small category), and this is not a U-category. Hence we must enlarge the universe, as mentioned in Section 1.1.

Definition 1.7.3 The *Yoneda functor* of the category C is the functor

$$\mathrm{Y}_{\mathsf{C}} : \mathsf{C} \to \mathsf{Fun}(\mathsf{C}^{\mathrm{op}}, \mathsf{Set})$$

described by formulas (1.7.1) and (1.7.2).

Theorem 1.7.4 (Yoneda Lemma) *The Yoneda functor* Y_{C} *is fully faithful.*

See [102, Section III.2] or [76, Section 1.4] for a proof. The proof is not hard, but it is very confusing.

A functor $F : \mathsf{C}^{\mathrm{op}} \to \mathsf{Set}$ is called *representable* if there is an isomorphism of functors $\eta : F \xrightarrow{\simeq} \mathrm{Y}_{\mathsf{C}}(C)$ for some object $C \in \mathsf{C}$. Such an object C is said to *represent* the functor F. The Yoneda Lemma says that Y_{C} is an equivalence from C to the category of representable functors. Thus the pair (C, η) is unique up to a unique isomorphism (if it exists). Note that the isomorphism of sets $\eta_C : F(C) \xrightarrow{\simeq} \mathrm{Y}_{\mathsf{C}}(C)(C)$ gives a special element $\tilde{\eta} \in F(C)$ such that $\eta_C(\tilde{\eta}) = \mathrm{id}_C$.

Dually, any object $C \in \mathsf{C}$ gives rise to a functor

$$\mathrm{Y}^{\vee}_{\mathsf{C}}(C) : \mathsf{C} \to \mathsf{Set}, \quad \mathrm{Y}^{\vee}_{\mathsf{C}}(C) := \mathrm{Hom}_{\mathsf{C}}(C, -). \tag{1.7.5}$$

A morphism $\phi : C_0 \to C_1$ in C induces a morphism of functors

$$\mathrm{Y}^{\vee}_{\mathsf{C}}(\phi) := \mathrm{Hom}_{\mathsf{C}}(\phi, -) : \mathrm{Y}^{\vee}_{\mathsf{C}}(C_1) \to \mathrm{Y}^{\vee}_{\mathsf{C}}(C_0). \tag{1.7.6}$$

Definition 1.7.7 The *dual Yoneda functor* of the category C is the functor

$$\mathrm{Y}^{\vee}_{\mathsf{C}} : \mathsf{C}^{\mathrm{op}} \to \mathsf{Fun}(\mathsf{C}, \mathsf{Set})$$

described by formulas (1.7.5) and (1.7.6).

Theorem 1.7.8 (Dual Yoneda Lemma) *The dual Yoneda functor* $\mathrm{Y}^{\vee}_{\mathsf{C}}$ *is fully faithful.*

This is also proved in [102, Section III.2] and [76, Section 1.4].

A functor $F : \mathsf{C} \to \mathsf{Set}$ is called *corepresentable* if there is an isomorphism of functors $\eta : F \xrightarrow{\simeq} \mathrm{Y}^{\vee}_{\mathsf{C}}(C)$ for some object $C \in \mathsf{C}$. The object C is said to *corepresent* the functor F. The dual Yoneda Lemma says that the functor $\mathrm{Y}_{\mathsf{C}}^{\vee}$ is an equivalence from C^{op} to the category of corepresentable functors. The identity automorphism id_C corresponds to a special element $\tilde{\eta} \in F(C)$.

1.8 Inverse and Direct Limits

We are only interested in direct and inverse limits indexed by the ordered set $\mathbb{N}$. For a general discussion see [101] or [75].

Let C be a category. Recall that an $\mathbb{N}$-*indexed direct system* in C is data

$$(\{C_k\}_{k\in\mathbb{N}}, \{\mu_k\}_{k\in\mathbb{N}}),$$

where C_k are objects of C, and $\mu_k : C_k \to C_{k+1}$ are morphisms, which we call *transitions*. A *direct limit* of this system is data $(C, \{\epsilon_k\}_{k\in\mathbb{N}})$, where $C \in \mathsf{C}$, and $\epsilon_k : C_k \to C$ are morphisms, which we call *abutments*, such that $\epsilon_{k+1} \circ \mu_k = \epsilon_k$ for all k. The universal property required is this: if $(C', \{\epsilon'_k\}_{k\in\mathbb{N}})$ is another pair such that $\epsilon'_{k+1} \circ \mu_k = \epsilon'_k$, then there is a unique morphism $\epsilon : C \to C'$ such that $\epsilon'_k = \epsilon \circ \epsilon_k$. If a direct limit C exists, then of course it is unique, up to a unique isomorphism. We then write $\lim_{k\to} C_k := C$ and call this the direct limit of the system $\{C_k\}_{k\in\mathbb{N}}$, keeping the transitions and the abutments implicit. Sometimes we look at the morphisms

$$\mu_{k_0,k_1} := \mu_{k_1-1} \circ \cdots \circ \mu_{k_0} : C_{k_0} \to C_{k_1}$$

for $k_0 < k_1$, and $\mu_{k,k} := \mathrm{id}_{C_k}$.

By an $\mathbb{N}$-*indexed inverse system* in the category C we mean data

$$(\{C_k\}_{k\in\mathbb{N}}, \{\mu_k\}_{k\in\mathbb{N}}),$$

where $\{C_k\}_{k\in\mathbb{N}}$ is a collection of objects, and $\mu_k : C_{k+1} \to C_k$ are morphisms, also called transitions. An *inverse limit* of this system is data $(C, \{\epsilon_k\}_{k\in\mathbb{N}})$, where $C \in \mathsf{C}$, and $\epsilon_k : C \to C_k$ are morphisms, which we also call abutments, such that $\mu_k \circ \epsilon_{k+1} = \epsilon_k$. These satisfy an analogous universal property. If an inverse limit C exists, then it is unique, up to a unique isomorphism. We then write $\lim_{\leftarrow k} C_k := C$, and we call this the inverse limit of the system $\{C_k\}_{k\in\mathbb{N}}$. We define the morphisms

$$\mu_{k_0,k_1} := \mu_{k_0} \circ \cdots \circ \mu_{k_1-1} : C_{k_1} \to C_{k_0}$$

for $k_0 < k_1$, and $\mu_{k,k} := \mathrm{id}_{C_k}$.

Exercise 1.8.1 We can view the ordered set $\mathbb{N}$ as a category, with a single morphism $k \to l$ when $k \leq l$, and no morphisms otherwise.

(1) Interpret $\mathbb{N}$-indexed direct and inverse systems in C as functors $F : \mathbb{N} \to \mathsf{C}$ and $G : \mathbb{N}^{\mathrm{op}} \to \mathsf{C}$, respectively.

(2) Let $\bar{\mathbb{N}} := \mathbb{N} \cup \{\infty\}$. Interpret the direct and inverse limits of F and G, respectively, as functors $\bar{F} : \bar{\mathbb{N}} \to \mathsf{C}$ and $\bar{G} : \bar{\mathbb{N}}^{\mathrm{op}} \to \mathsf{C}$, extending F and G, with suitable universal properties.

Exercise 1.8.2 Prove that $\mathbb{N}$-indexed direct and inverse limits exist in the categories Set and $\mathsf{Mod}\, A$, for any ring A. Give explicit formulas.

Example 1.8.3 Let M be the category of finite abelian groups. The inverse system $\{M_k\}_{k\in\mathbb{N}}$, where $M_k := \mathbb{Z}/(2^k)$, and the transition $\mu_k : M_{k+1} \to M_k$ is the canonical surjection, does not have an inverse limit in M. We can also make $\{M_k\}_{k\in\mathbb{N}}$ into a direct system, in which the transition $\nu_k : M_k \to M_{k+1}$ is multiplication by 2. The direct limit does not exist in M.

If $\{C_k\}_{k\in\mathbb{N}}$ is a direct system in C, and $D \in \mathsf{C}$ is any object, then there is an induced inverse system $\{\mathrm{Hom}_{\mathsf{C}}(C_k, D)\}_{k\in\mathbb{N}}$ in Set, and it has a limit. If $C := \lim_{k\to} C_k$ exists, then the abutments $\epsilon_k : C_k \to C$ induce a morphism

$$\mathrm{Hom}_{\mathsf{C}}(C, D) \to \lim_{\leftarrow k} \mathrm{Hom}_{\mathsf{C}}(C_k, D) \tag{1.8.4}$$

in Set.

Similarly, if $\{C_k\}_{k\in\mathbb{N}}$ is an inverse system in C, and $D \in \mathsf{C}$ is any object, then there is an induced inverse system $\{\mathrm{Hom}_{\mathsf{C}}(D, C_k)\}_{k\in\mathbb{N}}$ in Set, and it has a limit. If $C := \lim_{\leftarrow k} C_k$ exists, then the abutments $\epsilon_k : C \to C_k$ induce a morphism

$$\mathrm{Hom}_{\mathsf{C}}(D, C) \to \lim_{\leftarrow k} \mathrm{Hom}_{\mathsf{C}}(D, C_k), \tag{1.8.5}$$

in Set.

Proposition 1.8.6 *Let C be a category.*

(1) *Let $\{C_k\}_{k\in\mathbb{N}}$ be a direct system in C, and let $(C, \{\epsilon_k\}_{k\in\mathbb{N}})$ be data as in the definition of a direct limit. Then $C = \lim_{k\to} C_k$ if and only if for every object $D \in \mathsf{C}$, the function* (1.8.4) *is bijective.*

(2) *Let $\{C_k\}_{k\in\mathbb{N}}$ be an inverse system in C, and let $(C, \{\epsilon_k\}_{k\in\mathbb{N}})$ be data as in the definition of an inverse limit. Then $C = \lim_{\leftarrow k} C_k$ if and only if for every object $D \in \mathsf{C}$, the function* (1.8.5) *is bijective.*

Exercise 1.8.7 Prove Proposition 1.8.6.

Remark 1.8.8 In many cases inverse and direct limits do not exist in C because "its objects are too small." This happens in the category $\mathsf{Set}_{\mathrm{fin}}$ of finite sets, and also in the category $\mathsf{Ab}_{\mathrm{fin}}$ of finite abelian groups (see Example 1.8.3 above).

There is a very effective method to enlarge C just enough so that the bigger category will have the desired limits. This is done by means of the categories

$\mathsf{Ind}(\mathsf{C})$ and $\mathsf{Pro}(\mathsf{C})$ of *ind-objects* and *pro-objects* of C, respectively. See [75, Section 1.11] and [76, Section 6.1] for detailed discussions.

Here are two examples. For $\mathsf{C} = \mathsf{Set}_{\mathrm{fin}}$, the category $\mathsf{Ind}(\mathsf{Set}_{\mathrm{fin}})$ is (canonically equivalent to) Set. The category $\mathsf{Pro}(\mathsf{Set}_{\mathrm{fin}})$ is the category of totally disconnected compact Hausdorff topological spaces, whose morphisms are the continuous functions.

For $\mathsf{C} = \mathsf{Ab}_{\mathrm{fin}}$, the category $\mathsf{Ind}(\mathsf{Ab}_{\mathrm{fin}})$ of ind-objects is the category Ab of abelian groups. The category $\mathsf{Pro}(\mathsf{Ab}_{\mathrm{fin}})$ of pro-objects is the category of profinite abelian groups, whose morphisms are the continuous group homomorphisms.

2

Abelian Categories and Additive Functors

The concept of *abelian category* is an extremely useful abstraction of a category of modules. It was introduced by A. Grothendieck in his foundational paper [53] from 1957. References for this material are [101], [102], [129], [65], [158], [75] and [76].

2.1 Linear Categories

Definition 2.1.1 Let $\mathbb{K}$ be a commutative ring. A $\mathbb{K}$-*linear category* is a category M, endowed with a $\mathbb{K}$-module structure on each of the sets of morphisms $\mathrm{Hom}_{\mathsf{M}}(M_0, M_1)$. The condition is this:

- For all $M_0, M_1, M_2 \in \mathsf{M}$ the composition function

$$\mathrm{Hom}_{\mathsf{M}}(M_1, M_2) \times \mathrm{Hom}_{\mathsf{M}}(M_0, M_1) \to \mathrm{Hom}_{\mathsf{M}}(M_0, M_2)$$
$$(\phi_1, \phi_0) \mapsto \phi_1 \circ \phi_0$$

  is $\mathbb{K}$ bilinear.

If $\mathbb{K} = \mathbb{Z}$, we say that M is a *linear category*.

As already mentioned in Convention 1.2.4, in our book all linear categories are $\mathbb{K}$-linear.

Proposition 2.1.2 *Let* M *be a* $\mathbb{K}$*-linear category.*

(1) *For every object* $M \in \mathsf{M}$*, the set* $\mathrm{End}_{\mathsf{M}}(M) := \mathrm{Hom}_{\mathsf{M}}(M, M)$*, with its given addition operation, and with the operation of composition, is a central* $\mathbb{K}$*-ring.*

(2) *For every two objects* $M_0, M_1 \in \mathsf{M}$, *the set* $\mathrm{Hom}_{\mathsf{M}}(M_0, M_1)$, *with its given addition operation, and with the operations of composition, is a left module over the ring* $\mathrm{End}_{\mathsf{M}}(M_1)$, *and a right module over the ring* $\mathrm{End}_{\mathsf{M}}(M_0)$. *Furthermore, these left and right actions commute with each other.*

Exercise 2.1.3 Prove Proposition 2.1.2.

This result can be reversed:

Example 2.1.4 Let A be a central $\mathbb{K}$-ring. Define a category M like this: there is a single object M, and its set of morphisms is $\mathrm{Hom}_{\mathsf{M}}(M, M) := A$. Composition in M is the multiplication of A. Then M is a $\mathbb{K}$-linear category.

For a central $\mathbb{K}$-ring A, the opposite ring A^{op} has the same $\mathbb{K}$-module structure as A, but the multiplication is reversed.

Exercise 2.1.5 Let A be a nonzero ring. Let $P, Q \in \mathsf{Mod}\, A$ be distinct free A-modules of rank 1.

(1) Prove that there is a ring isomorphism $\mathrm{End}_{\mathsf{Mod}\, A}(P) \cong A^{\mathrm{op}}$. Is this ring isomorphism canonical?
(2) Let M be the full subcategory of $\mathsf{Mod}\, A$ on the set of objects $\{P, Q\}$. Compare the linear category M to the ring of matrices $\mathrm{Mat}_{2\times 2}(A^{\mathrm{op}})$.

2.2 Additive Categories

Definition 2.2.1 An *additive category* is a linear category M satisfying these two conditions:

(i) M has a zero object 0.
(ii) M has finite coproducts.

Observe that $\mathrm{Hom}_{\mathsf{M}}(M, N) \neq \varnothing$ for every $M, N \in \mathsf{M}$, since this is an abelian group. For the zero object $0 \in \mathsf{M}$ we have $\mathrm{Hom}_{\mathsf{M}}(M, 0) = \mathrm{Hom}_{\mathsf{M}}(0, M) = 0$, the zero abelian group. We denote the unique arrows $0 \to M$ and $M \to 0$ also by 0. So the numeral 0 has a lot of meanings; but they are always (hopefully) clear from the context. The coproduct in a linear category M is usually denoted by $\bigoplus$, and is called the *direct sum*; cf. Example 1.4.2.

Example 2.2.2 Let A be a central $\mathbb{K}$-ring. The category $\mathsf{Mod}\, A$ is a $\mathbb{K}$-linear additive category. The full subcategory $\mathsf{F} \subseteq \mathsf{Mod}\, A$ on the free modules is also additive.

Proposition 2.2.3 *Let* M *be a linear category. Let* $\{M_i\}_{i \in I}$ *be a finite collection of objects of* M*, and assume the coproduct* $M = \bigoplus_{i \in I} M_i$ *exists, with embeddings* $e_i : M_i \to M$.

(1) *For each i let* $p_i : M \to M_i$ *be the unique morphism such that* $p_i \circ e_i = \mathrm{id}_{M_i}$*, and* $p_i \circ e_j = 0$ *for* $j \neq i$*. Then* $(M, \{p_i\}_{i \in I})$ *is a product of the collection* $\{M_i\}_{i \in I}$.

(2) $\sum_{i \in I} e_i \circ p_i = \mathrm{id}_M$.

Exercise 2.2.4 Prove this proposition.

Part (1) of Proposition 2.2.3 directly implies:

Corollary 2.2.5 *An additive category has finite products.*

Definition 2.2.6 Let M be an additive category, and let N be a full subcategory of M. We say that N is a *full additive subcategory* of M if N contains the zero object, and is closed under finite direct sums.

Exercise 2.2.7 In the situation of Definition 2.2.6, show that the category N is itself additive.

Example 2.2.8 Consider the linear category M from Example 2.1.4, built from a ring A. It does not have a zero object (unless the ring A is the zero ring), so it is not additive.

A more puzzling question is this: Does M have finite direct sums? This turns out to be equivalent to whether or not $A \cong A \oplus A$ as right A-modules. One can show that when A is nonzero and commutative, or nonzero and noetherian, then $A \not\cong A \oplus A$ in $\mathsf{Mod}\, A^{\mathrm{op}}$. On the other hand, if we take a field $\mathbb{K}$, and a countable rank $\mathbb{K}$-module N, then $A := \mathrm{End}_{\mathbb{K}}(N)$ will satisfy $A \cong A \oplus A$.

Proposition 2.2.9 *Let* M *be a linear category, and* $N \in$ M*. The following conditions are equivalent:*

(i) *The ring* $\mathrm{End}_{\mathsf{M}}(N)$ *is trivial.*

(ii) N *is a zero object of* M.

Proof

(ii) ⇒ (i): Since the set $\mathrm{End}_{\mathsf{M}}(N)$ is a singleton, it must be the trivial ring.

(i) ⇒ (ii): If the ring $\mathrm{End}_{\mathsf{M}}(N)$ is trivial, then all left and right modules over it must be trivial. Now use Proposition 2.1.2(2). □

2.3 Abelian Categories

Definition 2.3.1 Let M be an additive category, and let $f : M \to N$ be a morphism in M. A *kernel* of f is a pair (K, k), consisting of an object $K \in \mathsf{M}$ and a morphism $k : K \to M$, with these two properties:

(i) $f \circ k = 0$.

(ii) If $k' : K' \to M$ is a morphism in M such that $f \circ k' = 0$, then there is a unique morphism $g : K' \to K$ such that $k' = k \circ g$.

In other words, the object K represents the functor $\mathsf{M}^{\mathrm{op}} \to \mathsf{Ab}$,

$$K' \mapsto \{k' \in \mathrm{Hom}_{\mathsf{M}}(K', M) \mid f \circ k' = 0\}.$$

The kernel of f is of course unique up to a unique isomorphism (if it exists), and we denote if by $\mathrm{Ker}(f)$. Sometimes $\mathrm{Ker}(f)$ refers only to the object K, and other times it refers only to the morphism k; as usual, this should be clear from the context.

Definition 2.3.2 Let M be an additive category, and let $f : M \to N$ be a morphism in M. A *cokernel* of f is a pair (C, c), consisting of an object $C \in \mathsf{M}$ and a morphism $c : N \to C$, with these two properties:

(i) $c \circ f = 0$.

(ii) If $c' : N \to C'$ is a morphism in M such that $c' \circ f = 0$, then there is a unique morphism $g : C \to C'$ such that $c' = g \circ c$.

In other words, the object C corepresents the functor $\mathsf{M} \to \mathsf{Ab}$,

$$C' \mapsto \{c' \in \mathrm{Hom}_{\mathsf{M}}(N, C') \mid c' \circ f = 0\}.$$

The cokernel of f is of course unique up to a unique isomorphism (if it exists), and we denote if by $\mathrm{Coker}(f)$. Sometimes $\mathrm{Coker}(f)$ refers only to the object C, and other times it refers only to the morphism c; as usual, this should be clear from the context.

Example 2.3.3 In $\mathsf{Mod}\,A$ all kernels and cokernels exist. Given $f : M \to N$, the kernel is $k : K \to M$, where $K := \{m \in M \mid f(m) = 0\}$, and the k is the inclusion. The cokernel is $c : N \to C$, where $C := N/f(M)$, and c is the canonical projection.

Proposition 2.3.4 *Let M be an additive category, and let $f : M \to N$ be a morphism in M.*

(1) *If $k : K \to M$ is a kernel of f, then k is a monomorphism.*

(2) *If $c : N \to C$ is a cokernel of f, then c is an epimorphism.*

Exercise 2.3.5 Prove the proposition.

Definition 2.3.6 Assume the additive category M has kernels and cokernels. Let $f : M \to N$ be a morphism in M.

(1) Define the *image* of f to be $\operatorname{Im}(f) := \operatorname{Ker}(\operatorname{Coker}(f))$.

(2) Define the *coimage* of f to be $\operatorname{Coim}(f) := \operatorname{Coker}(\operatorname{Ker}(f))$.

The image is familiar, but the coimage is probably not. The next diagram should help. We start with a morphism $f : M \to N$ in M. The kernel and cokernel of f fit into this diagram: $K \xrightarrow{k} M \xrightarrow{f} N \xrightarrow{c} C$. Inserting $\alpha := \operatorname{Coker}(k) = \operatorname{Coim}(f)$ and $\beta := \operatorname{Ker}(c) = \operatorname{Im}(f)$ we get the following commutative diagram (solid arrows):

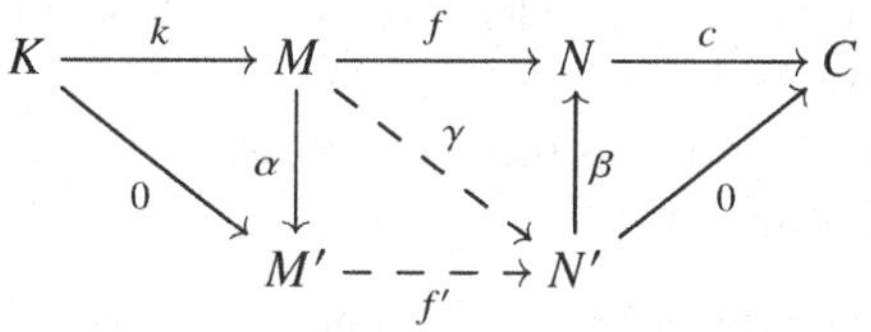

(2.3.7)

Since $c \circ f = 0$, there is a unique morphism γ (the dashed arrow) making the diagram commutative. Now $\beta \circ \gamma \circ k = f \circ k = 0$; and β is a monomorphism; so $\gamma \circ k = 0$. Hence there is a unique morphism $f' : M' \to N'$ making the diagram commutative. We conclude that $f : M \to N$ induces a morphism

$$f' : \operatorname{Coim}(f) \to \operatorname{Im}(f). \tag{2.3.8}$$

Definition 2.3.9 An *abelian category* is an additive category M with these two extra properties:

(i) All morphisms in M admit kernels and cokernels.

(ii) For every morphism $f : M \to N$ in M, the induced morphism f' in equation (2.3.8) is an isomorphism.

Here is a less precise but (maybe) easier way to remember to state property (ii). Because $M' = \operatorname{Coker}(\operatorname{Ker}(f))$ and $N' = \operatorname{Ker}(\operatorname{Coker}(f))$, we see that

$$\operatorname{Coker}(\operatorname{Ker}(f)) = \operatorname{Ker}(\operatorname{Coker}(f)). \tag{2.3.10}$$

From now on we forget all about the coimage.

Exercise 2.3.11 For any ring A, prove that the category Mod A is abelian.

This includes the category Ab = Mod $\mathbb{Z}$, from which the name derives.

Definition 2.3.12 Let M be an abelian category, and let N be a full subcategory of M. We say that N is a *full abelian subcategory* of M if the zero object belongs to N, and N is closed in M under taking finite direct sums, kernels and cokernels.

Exercise 2.3.13 In the situation of Definition 2.3.12, show that the category N is itself abelian.

Example 2.3.14 Let M_1 be the category of finitely generated abelian groups, and let M_0 be the category of finite abelian groups. Then M_1 is a full abelian subcategory of Ab, and M_0 is a full abelian subcategory of M_1.

Exercise 2.3.15 Let N be the full subcategory of Ab whose objects are the finitely generated free abelian groups. It is an additive subcategory of Ab (since it is closed under direct sums).

(1) Show that N is closed under kernels in Ab.
(2) Show that N is not closed under cokernels in Ab, so it is not a full abelian subcategory of Ab.
(3) Show that N has cokernels (not the same as those of Ab). Still, it fails to be an abelian category.

Exercise 2.3.16 The category Grp is not linear of course. Still, it does have a zero object (the trivial group). Show that Grp has kernels and cokernels, but condition (ii) of Definition 2.3.9 fails.

Exercise 2.3.17 Let Hilb be the category of Hilbert spaces over $\mathbb{C}$. The morphisms are the continuous $\mathbb{C}$-linear homomorphisms. Show that Hilb is a $\mathbb{C}$-linear additive category with kernels and cokernels, but it is not an abelian category.

Exercise 2.3.18 Let A be a ring. Show that A is *left noetherian* iff the category $\mathsf{Mod}_{\mathrm{f}}\, A$ of finitely generated left modules is a full abelian subcategory of $\mathsf{Mod}\, A$.

Example 2.3.19 Let $(X, \mathcal{A})$ be a *ringed space*; namely X is a topological space and $\mathcal{A}$ is a sheaf of rings on X. There is a very detailed discussion of sheaves of modules in [63, Section II.1] and [75, Sections 2.1–2.2]. Let us mention only one aspect of it.

We denote by $\mathsf{Mod}^{\mathrm{pr}}\, \mathcal{A}$ the category of *presheaves* of left $\mathcal{A}$-modules on X. This is an abelian category. Given a morphism $\phi : \mathcal{M} \to \mathcal{N}$ in $\mathsf{Mod}^{\mathrm{pr}}\, \mathcal{A}$, its kernel is the presheaf $\mathrm{Ker}^{\mathrm{pr}}(\phi)$ defined by

$$\Gamma(U, \mathrm{Ker}^{\mathrm{pr}}(\phi)) := \mathrm{Ker}\,(\phi : \Gamma(U, \mathcal{M}) \to \Gamma(U, \mathcal{N}))$$

on every open set $U \subseteq X$. The cokernel of ϕ is the presheaf $\mathrm{Coker}^{\mathrm{pr}}(\phi)$ defined by

$$\Gamma(U, \mathrm{Coker}^{\mathrm{pr}}(\phi)) : \mathrm{Coker}\,(\phi : \Gamma(U, \mathcal{M}) \to \Gamma(U, \mathcal{N})).$$

Now let $\mathsf{Mod}\, \mathcal{A}$ be the full subcategory of $\mathsf{Mod}^{\mathrm{pr}}\, \mathcal{A}$ consisting of *sheaves*. It is a full additive subcategory of $\mathsf{Mod}^{\mathrm{pr}}\, \mathcal{A}$, closed under kernels. However,

$\mathsf{Mod}\,\mathcal{A}$ is not closed under cokernels inside $\mathsf{Mod}^{\mathrm{pr}}\,\mathcal{A}$, and hence it is not a full abelian subcategory.

Nonetheless, $\mathsf{Mod}\,\mathcal{A}$ is itself an abelian category, but with different cokernels. Given a morphism $\phi : \mathcal{M} \to \mathcal{N}$ in $\mathsf{Mod}\,\mathcal{A}$, its cokernel in $\mathsf{Mod}\,\mathcal{A}$ is the sheaf $\mathrm{Coker}(\phi)$ that's associated to the presheaf $\mathrm{Coker}^{\mathrm{pr}}(\phi)$.

Proposition 2.3.20 *Let* M *be a linear category.*

(1) *The opposite category* M^{op} *has a canonical structure of a linear category.*
(2) *If* M *is additive, then* M^{op} *is also additive.*
(3) *If* M *is abelian, then* M^{op} *is also abelian.*

Proof (1) Since $\mathrm{Hom}_{\mathsf{M}^{\mathrm{op}}}(M, N) = \mathrm{Hom}_{\mathsf{M}}(N, M)$, this is an abelian group. The bilinearity of the composition in M^{op} is clear.

(2) The zero objects in M and M^{op} are the same. Existence of finite coproducts in M^{op} is because of the existence of finite products in M; see Proposition 2.2.3(1).

(3) M^{op} has kernels and cokernels, since $\mathrm{Ker}_{\mathsf{M}^{\mathrm{op}}}(\mathrm{Op}(\phi)) = \mathrm{Coker}_{\mathsf{M}}(\phi)$ and vice versa. Also the symmetric condition (ii) of Definition 2.3.9 holds. □

Proposition 2.3.21 *Let* $\phi : M \to N$ *be a morphism in an abelian category* M.

(1) ϕ *is a monomorphism if and only if* $\mathrm{Ker}(\phi) = 0$.
(2) ϕ *is an epimorphism if and only if* $\mathrm{Coker}(\phi) = 0$.
(3) ϕ *is an isomorphism if and only if it is both a monomorphism and an epimorphism.*

Exercise 2.3.22 Prove this proposition.

Consider a diagram

$$S = (\cdots M_{-1} \xrightarrow{\phi_{-1}} M_0 \xrightarrow{\phi_0} M_1 \xrightarrow{\phi_1} M_2 \cdots) \tag{2.3.23}$$

in an abelian category M, extending finitely or infinitely to either side. Such a diagram is called a *sequence in* M. An object M_i appearing in S is called *interior in* S if there is an object M_{i-1} appearing to the left of it, and an object M_{i+1} appearing to the right of it.

Definition 2.3.24 Let S be a sequence in the abelian category M, with notation as in (2.3.23).

(1) Suppose M_i is an interior object in S. We say that the sequence S is *exact at* M_i if $\mathrm{Im}(\phi_{i-1}) = \mathrm{Ker}(\phi_i)$, as subobjects of M_i.
(2) The sequence S is said to be *exact* if it is exact at all of its interior objects.

Example 2.3.25 A morphism $\phi : M \to N$ in an abelian category M is a monomorphism iff $0 \to M \xrightarrow{\phi} N$ is an exact sequence. The morphism ϕ is an epimorphism iff the sequence $M \xrightarrow{\phi} N \to 0$ is exact.

Definition 2.3.26 A *short exact sequence* in an abelian category M is an exact sequence of the form

$$S = (0 \to M_0 \xrightarrow{\phi_0} M_1 \xrightarrow{\phi_1} M_2 \to 0).$$

Proposition 2.3.27 *Let* M *be a* $\mathbb{K}$*-linear abelian category.*

(1) *Let* $0 \to M' \xrightarrow{\phi} M \xrightarrow{\psi} M''$ *be an exact sequence in* M*. Then for every* $L \in \mathsf{M}$ *the sequence*

$$0 \to \operatorname{Hom}_{\mathsf{M}}(L, M') \xrightarrow{\operatorname{Hom}(\mathrm{id},\phi)} \operatorname{Hom}_{\mathsf{M}}(L, M) \xrightarrow{\operatorname{Hom}(\mathrm{id},\psi)} \operatorname{Hom}_{\mathsf{M}}(L, M'')$$

in $\mathsf{Mod}\,\mathbb{K}$ *is exact.*

(2) *Let* $M' \xrightarrow{\phi} M \xrightarrow{\psi} M'' \to 0$ *be an exact sequence in* M*. Then for every* $N \in \mathsf{M}$ *the sequence*

$$0 \to \operatorname{Hom}_{\mathsf{M}}(M'', N) \xrightarrow{\operatorname{Hom}(\psi,\mathrm{id})} \operatorname{Hom}_{\mathsf{M}}(M, N) \xrightarrow{\operatorname{Hom}(\phi,\mathrm{id})} \operatorname{Hom}_{\mathsf{M}}(M', N)$$

in $\mathsf{Mod}\,\mathbb{K}$ *is exact.*

Exercise 2.3.28 Prove Proposition 2.3.27. (Hint: use the definitions of kernel, cokernel and image.)

2.4 A Method for Producing Proofs in Abelian Categories

A well-known difficulty in the theory of abelian categories is this: formulas that are easy to prove for a module category $\mathsf{M} = \mathsf{Mod}\,A$, using *elements*, are often very hard to prove in an abstract abelian category M (directly from the axioms).

A neat solution to this difficulty was found by P. Freyd and B. Mitchell:

Theorem 2.4.1 (Freyd–Mitchell) *Let* M *be a small abelian category. Then* M *is equivalent to a full abelian subcategory of* $\mathsf{Mod}\,A$*, for a suitable ring* A*.*

Remark 2.4.2 This is a deep and difficult result. See [47], a book that's basically devoted to proving this theorem.

A modern proof can be found in [76, Theorem 9.6.10] – but it too is very involved. Roughly speaking, they show that the abelian category $\mathsf{Ind}(\mathsf{M}^{\mathrm{op}})$ of ind-objects of M^{op} has an injective cogenerator. This implies that the abelian category $\mathsf{Pro}(\mathsf{M})$ of pro-objects, which is equivalent to $\mathsf{Ind}(\mathsf{M}^{\mathrm{op}})^{\mathrm{op}}$, has a projective generator, say P. Defining the ring $A := \operatorname{End}_{\mathsf{Pro}(\mathsf{M})}(P)^{\mathrm{op}}$, there is an equivalence of abelian categories $\mathsf{Pro}(\mathsf{M}) \approx \mathsf{Mod}\,A$. On the other hand, the Yoneda functor is a fully faithful embedding $\mathsf{M} \to \mathsf{Pro}(\mathsf{M})$.

The Freyd–Mitchell Theorem implies that for purposes of finitary calculations in the abelian category M (e.g. checking whether a sequence is exact, see Definition 2.3.24) we can assume that objects of M have elements. This often simplifies the work.

Since we do not give a proof of the Freyd–Mitchell Theorem in our book, we feel it is improper to use it. As a substitute, we provide the two "sheaf tricks" below, namely Propositions 2.4.9 and 2.4.11, with full proofs. Later in the book these sheaf tricks are used to give relatively easy proofs of several results on abstract abelian categories, most notably Theorem 3.7.10 on the existence of the long exact cohomology sequence. The method of proof using these tricks is explained in Remark 2.4.6. It is not as slick as the method that the Freyd–Mitchell Theorem offers; but at least we have self-contained proofs. See Remark 2.4.12 for some background (influence and history) on the sheaf tricks.

First an important technical lemma. In an abelian category M we have finite products, and they are also coproducts (see Proposition 2.2.3).

Lemma 2.4.3 *Let* M *be an abelian category. Consider a diagram*

$$\begin{array}{ccc} M_1 \times M_2 & \xrightarrow{p_2} & M_2 \\ {\scriptstyle p_1}\downarrow & & \downarrow{\scriptstyle \phi_2} \\ M_1 & \xrightarrow{\phi_1} & N \end{array} \tag{D}$$

in M*, where p_i are the projections. Define the object*

$$L := \operatorname{Ker}(\phi_1 \circ p_1 - \phi_2 \circ p_2) \subseteq M_1 \times M_2,$$

with inclusion morphism $e : L \to M_1 \times M_2$. Next define the morphisms $\psi_i := p_i \circ e : L \to M_i$.

(1) *The diagram*

$$\begin{array}{ccc} L & \xrightarrow{\psi_2} & M_2 \\ {\scriptstyle \psi_1}\downarrow & & \downarrow{\scriptstyle \phi_2} \\ M_1 & \xrightarrow{\phi_1} & N \end{array}$$

is cartesian, and $L = M_1 \times_N M_2$.

(2) *If ϕ_1 is an epimorphism, then ψ_2 is an epimorphism.*

Note that the diagram (D) is not assumed to be commutative. In case (D) does happen to be commutative, then $M_1 \times_N M_2 = M_1 \times M_2$.

Proof (1) The fact that L (with the morphisms ψ_i) is the fibered product is immediate from the definitions of product and kernel.

(2) Here we follow [102]. Let ρ be a morphism such that

$$\rho \circ (\phi_1 \circ p_1 - \phi_2 \circ p_2) = 0.$$

Consider the embedding

$$e_1 : M_1 \to M_1 \oplus M_2 = M_1 \times M_2.$$

Then

$$0 = \rho \circ (\phi_1 \circ p_1 - \phi_2 \circ p_2) \circ e_1 = \rho \circ \phi_1.$$

Because ϕ_1 is an epimorphism, it follows that $\rho = 0$. We conclude that

$$\phi_1 \circ p_1 - \phi_2 \circ p_2 : M_1 \times M_2 \to N \tag{2.4.4}$$

is an epimorphism. Thus (2.4.4) is the cokernel of e.

Next let $\sigma : M_2 \to P$ be a morphism such that $\sigma \circ \psi_2 = 0$. Since $\psi_2 = p_2 \circ e$, we get $\sigma \circ p_2 \circ e = 0$. Hence $\sigma \circ p_2$ factors through $\operatorname{Coker}(e)$. Namely there is a morphism $\sigma' : N \to P$ such that

$$\sigma \circ p_2 = \sigma' \circ (\phi_1 \circ p_1 - \phi_2 \circ p_2) : M_1 \times M_2 \to P. \tag{2.4.5}$$

But $p_2 \circ e_1 = 0$, and therefore

$$0 = \sigma \circ p_2 \circ e_1 = \sigma' \circ (\phi_1 \circ p_1 - \phi_2 \circ p_2) \circ e_1 = \sigma' \circ \phi_1.$$

As ϕ_1 is an epimorphism, it follows that $\sigma' = 0$. Finally, using (2.4.5), we get

$$\sigma = \sigma \circ p_2 \circ e_2 = \sigma' \circ (\phi_1 \circ p_1 - \phi_2 \circ p_2) \circ e_2 = -\sigma' \circ \phi_2 = 0.$$

We conclude that ψ_2 is an epimorphism. □

Before giving the precise statements, here is a heuristic.

Remark 2.4.6 The sheaf tricks (Propositions 2.4.9 and 2.4.11) work like this: we pretend that the objects of our $\mathbb{K}$-linear abelian category M are "sheaves on an imaginary topological space X"; objects playing this role are denoted by letters $M, N, \ldots$. The objects of M are also "open sets in the topological space X"; and objects playing this role are denoted by letters $U, V, \ldots$. Given a sheaf M and an open set U, there is a $\mathbb{K}$-module $\Gamma(U, M)$ of "sections of M over U." These "sections" can be added and subtracted, being elements of a $\mathbb{K}$-module. Sometimes we require "refinement," i.e. replacing an "open set U" by a "covering $V \twoheadrightarrow U$," giving rise to an embedding $\mathbb{K}$-modules $\Gamma(U, M) \rightarrowtail \Gamma(V, M)$. This allegory is made precise in Definition 2.4.7 below.

Definition 2.4.7 (Sheaf Metaphor) Let M be a $\mathbb{K}$-linear abelian category.

(1) Objects of M are called sheaves or open sets, depending on the role they play in each context.

(2) For an open set $U \in \mathsf{M}$ and a sheaf $M \in \mathsf{M}$ we write $\Gamma(U, M) := \operatorname{Hom}_{\mathsf{M}}(U, M)$. This is a $\mathbb{K}$-module, and we call it the *module of sections of the sheaf M over the open set U*.

(3) Given a morphism $\rho : V \to U$ of open sets in M, and a sheaf $M \in \mathsf{M}$, we use this notation for the resulting $\mathbb{K}$-module homomorphism:

$$\rho^* := \operatorname{Hom}_{\mathsf{M}}(\rho, \mathrm{id}_M) : \Gamma(U, M) \to \Gamma(V, M).$$

We call ρ^* the *pullback along* ρ.

(4) Given a morphism $\phi : M \to N$ of sheaves in M, and an open set $U \in \mathsf{M}$, we use this notation for the resulting $\mathbb{K}$-module homomorphism:

$$\Gamma(U, \phi) := \operatorname{Hom}_{\mathsf{M}}(\mathrm{id}_U, \phi) : \Gamma(U, M) \to \Gamma(U, N).$$

(5) If $\rho : V \to U$ is a morphism of open sets in M that is an epimorphism, then we call ρ a *covering of U*. The homomorphism ρ^* in this case is called the *restriction of M from U to V*.

Note that given a covering $\rho : V \twoheadrightarrow U$, the restriction homomorphism ρ^* is injective (see Proposition 2.3.27(2)). By slight abuse of notation, when the covering ρ is clear from the context, we will often identify the $\mathbb{K}$-module $\Gamma(U, M)$ with its image in $\Gamma(V, M)$ under ρ^*. Another convenient abuse of notation is writing ϕ instead of $\Gamma(U, \phi)$, for a morphism $\phi : M \to N$ in M and open set U. Combined, these notations give sense to such a commutative diagram

$$\begin{array}{ccc} \Gamma(U, M) & \xrightarrow{\ \phi\ } & \Gamma(U, N) \\ \big\downarrow{\scriptstyle \subseteq} & & \big\downarrow{\scriptstyle \subseteq} \\ \Gamma(V, M) & \xrightarrow{\ \phi\ } & \Gamma(V, N) \end{array} \tag{2.4.8}$$

in $\mathsf{Mod}\,\mathbb{K}$. This will be used in item (3) of the next proposition.

Proposition 2.4.9 (First Sheaf Trick) *Let $\phi : M \to N$ be a morphism in an abelian category* M.

(1) *The morphism ϕ is zero if and only if for every $U \in \mathsf{M}$ the homomorphism $\Gamma(U, \phi)$ in $\mathsf{Mod}\,\mathbb{K}$ is zero.*

(2) *The morphism ϕ is a monomorphism if and only if for every $U \in \mathsf{M}$ the homomorphism $\Gamma(U, \phi)$ in $\mathsf{Mod}\,\mathbb{K}$ is injective.*

(3) *The morphism ϕ is an epimorphism if and only if for every $U \in \mathsf{M}$ and every section $n \in \Gamma(U, N)$, there exists a covering $V \twoheadrightarrow U$ and a section $m \in \Gamma(V, M)$, such that $\phi(m) = n$ in $\Gamma(V, N)$.*

The heuristic interpretation of item (3) of the proposition is this: $\phi : M \to N$ is an epimorphism in M if and only if it is "locally surjective."

Proof (1) If $\phi = 0$ then for every $m \in \Gamma(U, M)$ we have $\Gamma(U, \phi)(m) = \phi \circ m = 0$ in $\Gamma(U, N)$; thus $\Gamma(U, \phi) = 0$. Conversely, if $\Gamma(U, \phi) = 0$ for all U, then take $U := M$ and $m := \mathrm{id}_M \in \Gamma(U, M)$. We obtain $\phi = \Gamma(U, \phi)(m) = 0$.

(2) First assume ϕ is a monomorphism. Take an arbitrary object U and a section $m \in \Gamma(U, M)$. So $m : U \to M$ is a morphism in M. If $\phi \circ m = 0$, then, by the definition of a monomorphism, we must have $m = 0$. Thus $\Gamma(U, \phi)$ is injective.

Conversely, assume that $\Gamma(U, \phi)$ is injective for every U. Take $U := \mathrm{Ker}(\phi)$ and let $m : U \to M$ be the inclusion. Then $\Gamma(U, \phi)(m) = \phi \circ m = 0$. But then $m = 0$ and $U = 0$. Thus ϕ is a monomorphism.

(3) First assume ϕ is an epimorphism. Consider a section $n \in \Gamma(U, N)$. So we have morphisms $\phi : M \twoheadrightarrow N$ and $n : U \to N$. Let $V := M \times_N U$, the fibered product. By Lemma 2.4.3(2) the projection $\rho : V \to U$ is an epimorphism, i.e. a covering in our terminology. Now the other projection $m : V \to M$ satisfies $\phi \circ m = n \circ \rho$. This means that $\phi(m) = n$ in $\Gamma(V, N)$.

Conversely, let us take $U := N$ and $n := \mathrm{id}_N \in \Gamma(U, N)$. There exists an epimorphism $\rho : V \twoheadrightarrow U$ and a morphism $m : V \to M$ such that $\phi \circ m = n \circ \rho = \rho$. We see that $\phi \circ m$ is an epimorphism, and hence ϕ is an epimorphism. □

Example 2.4.10 Suppose

$$S = (0 \to M' \xrightarrow{\phi} M \xrightarrow{\psi} M'' \to 0)$$

is a sequence in M. Here is how exactness of S is tested using the first sheaf trick.

By item (1) of the first sheaf trick, the condition $\psi \circ \phi = 0$ is same as $\Gamma(U, \psi \circ \phi) = 0$ for every $U \in$ M.

Exactness at M' is the same as this condition: for every $U \in$ M and every section $m' \in \Gamma(U, M')$, if $\phi(m') = 0$ then $m' = 0$. This is by item (2).

Exactness at M'' is the same as this condition: for every $U \in$ M and every section $m'' \in \Gamma(U, M'')$, there exists a covering $V \twoheadrightarrow U$ and a section $m \in \Gamma(V, M)$, such that $\psi(m) = m''$ in $\Gamma(V, M'')$.

Finally, exactness at M is the same as this condition (given that $\psi \circ \phi = 0$): for every $U \in$ M and every section $m \in \Gamma(U, M)$ such that $\psi(m) = 0$, there exists a covering $V \twoheadrightarrow U$ and a section $m' \in \Gamma(V, M')$, such that $\phi(m') = m$ in $\Gamma(V, M'')$.

Proposition 2.4.11 (Second Sheaf Trick) *Let M be an object in an abelian category* M. *Suppose $\rho_1 : V_1 \to U$ and $\rho_2 : V_2 \to U$ are morphisms in* M, *such that $(\rho_1, \rho_2) : V_1 \oplus V_2 \to U$ is an epimorphism. Let $W := V_1 \times_U V_2$, with projection morphisms $\sigma_1 : W \to V_1$ and $\sigma_2 : W \to V_2$. Then the sequence*

$$0 \to \Gamma(U, M) \xrightarrow{(\rho_1^*, \rho_2^*)} \Gamma(V_1, M) \times \Gamma(V_2, M) \xrightarrow{(\sigma_1^*, -\sigma_2^*)} \Gamma(W, M)$$

in $\mathsf{Mod}\,\mathbb{K}$ *is exact.*

Proof Note that $V_1 \times V_2 = V_1 \oplus V_2$. By Lemma 2.4.3(1) there is an exact sequence

$$0 \to W \xrightarrow{(\sigma_1,-\sigma_2)} V_1 \oplus V_2 \xrightarrow{(\rho_1,\rho_2)} U \to 0$$

in M. By Proposition 2.3.27(2) we get an exact sequence

$$0 \to \Gamma(U, M) \xrightarrow{(\rho_1^*,\rho_2^*)} \Gamma(V_1 \oplus V_2, M) \xrightarrow{(\sigma_1,-\sigma_2)^*} \Gamma(W, M)$$

in Ab. But

$$\Gamma(V_1 \oplus V_2, M) \cong \Gamma(V_1, M) \times \Gamma(V_2, M). \qquad \square$$

Remark 2.4.12 As can be seen in the practical application of our sheaf tricks (e.g. in the proofs of Proposition 2.5.18 and Theorem 3.7.10) these tricks reduce proofs about an abstract abelian category M, to proofs that are just like in the concrete case of the abelian category $\mathsf{M} = \mathsf{Ab}\,X$ of sheaves of abelian groups on a topological space X. This is not as easy as the case $\mathsf{M} = \mathsf{Mod}\,A$ of modules over a ring A, which the Freyd–Mitchell Theorem permits, but – at least for someone with experience in algebraic geometry – our method is quite straightforward.

Our sheaf tricks are inspired by [144, tag 05PL]. There it is shown that the geometric allegory is genuine, except that instead of a topological space X, there is a *site* X (in the sense of Grothendieck). The underlying category of the site X is the abelian category M itself, and the coverings of X are the epimorphisms in M. It is proved in [144] that the category M embeds as a full abelian subcategory of $\mathsf{Ab}\,X$, the category of sheaves of abelian groups on X.

Proposition 2.4.9 is very similar to [102, Theorem VIII.3], where what we call “sections” are called “members.” But S. MacLane’s method loses the abelian group structure (the members are not elements of abelian groups). An improvement of MacLane’s method can be found in an unpublished note of G. Bergman [18].

2.5 Additive Functors

Definition 2.5.1 Let M and N be $\mathbb{K}$-linear categories. A functor $F : \mathsf{M} \to \mathsf{N}$ is called a $\mathbb{K}$-*linear functor* if for every $M_0, M_1 \in \mathsf{M}$ the function

$$F : \mathrm{Hom}_{\mathsf{M}}(M_0, M_1) \to \mathrm{Hom}_{\mathsf{N}}(F(M_0), F(M_1))$$

is a $\mathbb{K}$-linear homomorphism.

When the base ring $\mathbb{K}$ is implicit (cf. Convention 1.2.4), we sometimes say that F is an *additive functor*, with the same meaning as $\mathbb{K}$-linear functor.

Additive functors commute with finite direct sums. More precisely:

Proposition 2.5.2 *Let $F : \mathsf{M} \to \mathsf{N}$ be an additive functor between linear categories, let $\{M_i\}_{i \in I}$ be a finite collection of objects of* M*, and assume that the direct sum $(M, \{e_i\}_{i \in I})$ of the collection $\{M_i\}_{i \in I}$ exists in* M*. Then $(F(M), \{F(e_i)\}_{i \in I})$ is a direct sum of the collection $\{F(M_i)\}_{i \in I}$ in* N.

Exercise 2.5.3 Prove Proposition 2.5.2. (Hint: use Proposition 2.2.3.)

Note that the proposition above also talks about finite products, because of Proposition 2.2.3.

Proposition 2.5.4 *Suppose $F, G : \mathsf{K} \to \mathsf{L}$ are additive functors between linear categories, and $\eta : F \to G$ is a morphism of functors. Let M, M', N be objects of* K*, and assume that $N \cong M \oplus M'$. Then the following two conditions are equivalent.*

(i) *$\eta_N : F(N) \to G(N)$ is an isomorphism.*

(ii) *$\eta_M : F(M) \to G(M)$ and $\eta_{M'} : F(M') \to G(M')$ are isomorphisms.*

Exercise 2.5.5 Prove Proposition 2.5.4.

Example 2.5.6 Let $f : A \to B$ be a ring homomorphism. There are two additive functors associated to f : the *restriction functor* $\operatorname{Rest}_f : \mathsf{Mod}\, B \to \mathsf{Mod}\, A$ and the *induction functor* $\operatorname{Ind}_f : \mathsf{Mod}\, A \to \mathsf{Mod}\, B$. Given a B-module N, the A-module $\operatorname{Rest}_f(N)$ has the same underlying $\mathbb{K}$-module, and A acts on it through f. For an A-module M, the induced B-module is $\operatorname{Ind}_f(M) := B \otimes_A M$.

Proposition 2.5.7 *Let $F : \mathsf{M} \to \mathsf{N}$ be an additive functor between linear categories. Then:*

(1) *For every $M \in \mathsf{M}$ the function $F : \operatorname{End}_{\mathsf{M}}(M) \to \operatorname{End}_{\mathsf{N}}(F(M))$ is a ring homomorphism.*

(2) *For every $M_0, M_1 \in \mathsf{M}$ the function*

$$F : \operatorname{Hom}_{\mathsf{M}}(M_0, M_1) \to \operatorname{Hom}_{\mathsf{N}}(F(M_0), F(M_1))$$

is a homomorphism of left $\operatorname{End}_{\mathsf{M}}(M_1)$-modules, and of right $\operatorname{End}_{\mathsf{M}}(M_0)$-modules.

(3) *If M is a zero object of* M*, then $F(M)$ is a zero object of* N.

Proof

(1) By Definition 2.5.1 the function F respects addition. By the definition of a functor, it respects multiplication and units.

(2) Immediate from the definitions, such as (1).

(3) Combine part (1) with Proposition 2.2.9. □

Definition 2.5.8 Let $F : \mathsf{M} \to \mathsf{N}$ be an additive functor between abelian categories.

(1) F is called *left exact* if it commutes with kernels. Namely for every morphism $\phi : M_0 \to M_1$ in M, with kernel $k : K \to M_0$, the morphism $F(k) : F(K) \to F(M_0)$ is a kernel of $F(\phi) : F(M_0) \to F(M_1)$.

(2) F is called *right exact* if it commutes with cokernels. Namely for every morphism $\phi : M_0 \to M_1$ in M, with cokernel $c : M_1 \to C$, the morphism $F(c) : F(M_1) \to F(C)$ is a cokernel of $F(\phi) : F(M_0) \to F(M_1)$.

(3) F is called *exact* if it is both left exact and right exact.

Let us illustrate this. Suppose $\phi : M_0 \to M_1$ is a morphism in M, with kernel K and cokernel C. Applying F to the sequence $K \xrightarrow{k} M_0 \xrightarrow{\phi} M_1 \xrightarrow{c} C$ in M we get the solid arrows in this diagram

$$\begin{array}{ccccccc} F(K) & \xrightarrow{F(k)} & F(M_0) & \xrightarrow{F(\phi)} & F(M_1) & \xrightarrow{F(c)} & F(C) \\ & \overset{\psi}{\searrow} & \uparrow & & \downarrow & \overset{\chi}{\nearrow} & \\ & & \mathrm{Ker}_{\mathsf{N}}(F(\phi)) & & \mathrm{Coker}_{\mathsf{N}}(F(\phi)) & & \end{array}$$

in N. Because N is abelian, we get the vertical dashed arrows: the kernel and cokernel of $F(\phi)$. The slanted dashed arrows exist and are unique because $F(\phi) \circ F(k) = 0$ and $F(c) \circ F(\phi) = 0$. Left exactness of F requires ψ to be an isomorphism, and right exactness requires χ to be an isomorphism.

Recall that a short exact sequence in M is an exact sequence of the form

$$S = (0 \to M_0 \xrightarrow{\phi_0} M_1 \xrightarrow{\phi_1} M_2 \to 0). \tag{2.5.9}$$

Proposition 2.5.10 *Let $F : \mathsf{M} \to \mathsf{N}$ be an additive functor between abelian categories.*

(1) *The functor F is left exact if and only if for every short exact sequence* $\mathbf{S}$ *in* M, *with notation* (2.5.9), *the sequence*

$$0 \to F(M_0) \xrightarrow{F(\phi_0)} F(M_1) \xrightarrow{F(\phi_0)} F(M_2)$$

is exact in N.

(2) *The functor F is right exact if and only if for every short exact sequence* $\mathbf{S}$ *in* M, *with the notation with notation* (2.5.9), *the sequence*

$$F(M_0) \xrightarrow{F(\phi_0)} F(M_1) \xrightarrow{F(\phi_1)} F(M_2) \to 0$$

is exact in N.

Exercise 2.5.11 Prove Proposition 2.5.10. (Hint: $M_0 \cong \mathrm{Ker}(M_1 \to M_2)$, etc.)

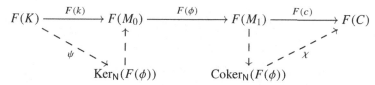

Example 2.5.12 Let A be a commutative ring, and let M be a fixed A-module. Define functors $F, G : \mathsf{Mod}\, A \to \mathsf{Mod}\, A$ like this: $F(N) := M \otimes_A N$ and $G(N) := \mathrm{Hom}_A(M, N)$. Then F is right exact and G is left exact.

Example 2.5.13 Let $f : A \to B$ be a ring homomorphism. In the notation of Example 2.5.6, the restriction functor Rest_f is exact, and the induction functor Ind_f is right exact.

Proposition 2.5.14 *Let $F : \mathsf{M} \to \mathsf{N}$ be an additive functor between abelian categories. If F is an equivalence then it is exact.*

Proof We will prove that F respects kernels; the proof for cokernels is similar. Take a morphism $\phi : M_0 \to M_1$ in M, with kernel K. We have this diagram (solid arrows):

$$\begin{array}{ccccc} M & & & & \\ {\scriptstyle \psi}\downarrow & \searrow^{\theta} & & & \\ K & \xrightarrow{\ k\ } & M_0 & \xrightarrow{\ \phi\ } & M_1 \end{array}$$

Applying F we obtain this diagram (solid arrows):

$$\begin{array}{ccccc} N = F(M) & & & & \\ {\scriptstyle F(\psi)}\downarrow & \searrow^{\bar{\theta}} & & & \\ F(K) & \xrightarrow{\ F(k)\ } & F(M_0) & \xrightarrow{\ F(\phi)\ } & F(M_1) \end{array}$$

in N. Suppose $\bar{\theta} : N \to F(M_0)$ is a morphism in N s.t. $F(\phi) \circ \bar{\theta} = 0$. Since F is essentially surjective on objects, there is some $M \in \mathsf{M}$ with an isomorphism $\alpha : F(M) \xrightarrow{\simeq} N$. After replacing N with $F(M)$ and $\bar{\theta}$ with $\bar{\theta} \circ \alpha$, we can assume that $N = F(M)$.

Now since F is fully faithful, there is a unique $\theta : M \to M_0$ s.t. $F(\theta) = \bar{\theta}$; and $\phi \circ \theta = 0$. So there is a unique $\psi : M \to K$ s.t. $\theta = k \circ \psi$. It follows that $F(\psi) : F(M) \to F(K)$ is the unique morphism s.t. $\bar{\theta} = F(k) \circ F(\psi)$. □

Here is a result that could afford another proof of the previous proposition.

Proposition 2.5.15 *Let $F : \mathsf{M} \to \mathsf{N}$ be an additive functor between linear categories. Assume F is an equivalence, with quasi-inverse G. Then $G : \mathsf{N} \to \mathsf{M}$ is an additive functor.*

Exercise 2.5.16 Prove Proposition 2.5.15.

Definition 2.5.17 Consider abelian categories M and N. Suppose we are given a sequence

$$\cdots F_{-1} \xrightarrow{\phi_{-1}} F_0 \xrightarrow{\phi_0} F_1 \xrightarrow{\phi_1} F_2 \cdots$$

(finite or infinite on either side), where each $F_i : \mathsf{M} \to \mathsf{N}$ is an additive functor, and each $\phi_i : F_i \to F_{i+1}$ is a morphism of functors. We say that this sequence is an *exact sequence of additive functors* if for every object $M \in \mathsf{M}$ the sequence

$$\cdots F_{-1}(M) \xrightarrow{\phi_{-1,M}} F_0(M) \xrightarrow{\phi_{0,M}} F_1(M) \xrightarrow{\phi_{1,M}} F_2(M) \cdots$$

in N is exact.

Proposition 2.5.18 *Let* M *and* N *be abelian categories, and let* $F_0 \xrightarrow{\phi_0} F_1 \xrightarrow{\phi_1} F_2$ *be a sequence of additive functors* $\mathsf{M} \to \mathsf{N}$.

(1) *If* $F_0 \xrightarrow{\phi_0} F_1 \xrightarrow{\phi_1} F_2 \to 0$ *is an exact sequence of additive functors, and if the functors* F_0 *and* F_1 *are both right exact, then the functor* F_2 *is right exact.*

(2) *If* $0 \to F_0 \xrightarrow{\phi_0} F_1 \xrightarrow{\phi_1} F_2$ *is an exact sequence of additive functors, and if the functors* F_1 *and* F_2 *are both left exact, then the functor* F_0 *is left exact.*

Proof

(1) Let $M' \xrightarrow{\sigma} M \xrightarrow{\tau} M'' \to 0$ be an exact sequence in M. We must show that $F_2(M') \to F_2(M) \to F_2(M'') \to 0$ is an exact sequence in N. Let us examine the commutative diagram

$$\begin{array}{ccccccc}
F_0(M') & \xrightarrow{\phi_{0,M'}} & F_1(M') & \xrightarrow{\phi_{1,M'}} & F_2(M') & \longrightarrow & 0 \\
\downarrow{\scriptstyle F_0(\sigma)} & & \downarrow{\scriptstyle F_1(\sigma)} & & \downarrow{\scriptstyle F_2(\sigma)} & & \\
F_0(M) & \xrightarrow{\phi_{0,M}} & F_1(M) & \xrightarrow{\phi_{1,M}} & F_2(M) & \longrightarrow & 0 \\
\downarrow{\scriptstyle F_0(\tau)} & & \downarrow{\scriptstyle F_1(\tau)} & & \downarrow{\scriptstyle F_2(\tau)} & & \\
F_0(M'') & \xrightarrow{\phi_{0,M''}} & F_1(M'') & \xrightarrow{\phi_{1,M''}} & F_2(M'') & \longrightarrow & 0 \\
\downarrow & & \downarrow & & \downarrow & & \\
0 & & 0 & & 0 & &
\end{array} \tag{2.5.19}$$

in N. It is known that the rows and the first two columns are exact. We must prove that the third column is exact.

First let us prove that $F_2(\tau)$ is an epimorphism. This is easy: we know that $F_1(\tau)$ and $\phi_{1,M''}$ are epimorphisms; hence $\phi_{1,M''} \circ F_1(\tau) = F_2(\tau) \circ \phi_{1,M}$ is an epimorphism; and thus $F_2(\tau)$ is an epimorphism.

More challenging is the proof that the third column is exact at $F_2(M)$, namely that $\operatorname{Im}(F_2(\sigma)) \to \operatorname{Ker}(F_2(\tau))$ is an epimorphism. For this we shall use the first sheaf trick (Proposition 2.4.9) and a diagram chase. Consider a section $m_2 \in \Gamma(U, \operatorname{Ker}(F_2(\tau)))$ on some "open set" U, i.e. for some object

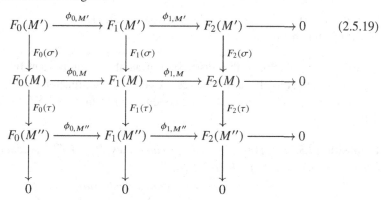

$U \in \mathsf{M}$. We shall prove that there exist a covering $V \twoheadrightarrow U$ and a section $m_2' \in \Gamma(V, F_2(M'))$ such that $F_2(\sigma)(m_2') = m_2$ in $\Gamma(V, F_2(M))$.

Because $\phi_{1,M}$ is an epimorphism, there is a covering $U_1 \twoheadrightarrow U$ and a section $m_1 \in \Gamma(U_1, F_1(M))$ such that $\phi_{1,M}(m_1) = m_2$ in $\Gamma(U_1, F_2(M))$. Let $m_1'' := F_1(\tau)(m_1) \in \Gamma(U_1, F_1(M''))$. We have $\phi_{1,M''}(m_1'') = F_2(\tau)(m_2) = 0$. This means that $m_1'' \in \Gamma(U_1, \operatorname{Ker}(\phi_{1,M''}))$.

The exactness of the third row says that for some covering $U_2 \twoheadrightarrow U_1$ there is a section $m_0'' \in \Gamma(U_2, F_0(M''))$ such that $\phi_{0,M''}(m_0'') = m_1''$ in $\Gamma(U_2, F_1(M''))$.

The exactness of the first column implies that for some covering $U_3 \twoheadrightarrow U_2$ there is a section $m_0 \in \Gamma(U_3, F_0(M))$ such that $m_0'' = F_0(\tau)(m_0)$ in $\Gamma(U_3, F_0(M''))$. Define $\tilde{m}_1 := m_1 - \phi_{0,M}(m_0)$ in $\Gamma(U_3, F_1(M))$. Note that $\phi_{1,M}(\tilde{m}_1) = \phi_{1,M}(m_1) = m_2$ in $\Gamma(U_3, F_2(M))$. Also $F_1(\tau)(\tilde{m}_1) = 0$, i.e. $\tilde{m}_1 \in \Gamma(U_3, \operatorname{Ker}(F_1(\tau)))$.

Due to the exactness of the second column, for some covering $V \twoheadrightarrow U_3$ there is a section $m_1' \in \Gamma(V, F_1(M'))$ such that $\tilde{m}_1 = F_1(\sigma)(m_1')$ in $\Gamma(V, F_1(M))$. Define $m_2' := \phi_{1,M'}(m_1') \in \Gamma(V, F_2(M'))$. Then $F_2(\sigma)(m_2') = \phi_{1,M}(\tilde{m}_1) = m_2$ in $\Gamma(V, F_2(M))$.

(2) See next exercise. □

Exercise 2.5.20 Prove part (2) of Proposition 2.5.17. (Hint: imitate the proof of part (1); but this is easier.)

We end this section with a discussion of additive contravariant functors. Suppose M and N are linear categories. A contravariant functor $F : \mathsf{M} \to \mathsf{N}$ is said to be additive if it satisfies the condition in Definition 2.5.1, with the obvious changes.

Proposition 2.5.21 *Let* M *and* N *be linear categories. Put on* M^{op} *the canonical linear structure* (*see Proposition* 2.3.20).

(1) *The functor* $\mathrm{Op} : \mathsf{M} \to \mathsf{M}^{\mathrm{op}}$ *is an additive contravariant functor.*
(2) *If* $F : \mathsf{M} \to \mathsf{N}$ *is an additive contravariant functor, then* $F \circ \mathrm{Op} : \mathsf{M}^{\mathrm{op}} \to \mathsf{N}$ *is an additive functor; and vice versa.*

Exercise 2.5.22 Prove Proposition 2.5.21.

In view of Proposition 2.5.10, we can give an unambiguous definition of left and right exact contravariant functors.

Definition 2.5.23 Let $F : \mathsf{M} \to \mathsf{N}$ be an additive contravariant functor between abelian categories.

(1) F a *left exact contravariant functor* if for every short exact sequence S in M, in the notation of (2.5.9), the sequence

$$0 \to F(M_2) \xrightarrow{F(\phi_1)} F(M_1) \xrightarrow{F(\phi_0)} F(M_0)$$

in N is exact.

(2) F is a *right exact contravariant functor* if for every short exact sequence S in M, the sequence

$$F(M_2) \xrightarrow{F(\phi_1)} F(M_1) \xrightarrow{F(\phi_0)} F(M_0) \to 0$$

in N is exact.

(3) F is an *exact contravariant functor* if it sends every short exact sequence S in M to a short exact sequence in N.

Proposition 2.5.24 *Let* M *and* N *be abelian categories. Recall that* M^{op} *is also an abelian category.*

(1) *The functor* $\mathrm{Op} : \mathsf{M} \to \mathsf{M}^{\mathrm{op}}$ *is an exact contravariant functor.*

(2) *If* $F : \mathsf{M} \to \mathsf{N}$ *is an exact contravariant functor, then* $F \circ \mathrm{Op} : \mathsf{M}^{\mathrm{op}} \to \mathsf{N}$ *is an exact functor; and vice versa. Likewise is true for left exactness and right exactness.*

Exercise 2.5.25 Prove Proposition 2.5.24.

Example 2.5.26 Let A be a commutative ring, and let M be a fixed A-module. Define the contravariant functor $F : \mathsf{Mod}\, A \to \mathsf{Mod}\, A$ to be $F(N) := \mathrm{Hom}_A(N, M)$. Then F is a left exact contravariant function.

Sometimes M and M^{op} are equivalent as abelian categories, as the next exercise shows. For a counterexample see Remark 2.7.20 below.

Exercise 2.5.27 Let $\mathbb{K}$ be a field, and consider the category $\mathsf{M} := \mathsf{Mod}_{\mathrm{f}}\, \mathbb{K}$ of finitely generated $\mathbb{K}$-modules (traditionally known as "finite dimensional vector spaces over $\mathbb{K}$"). This is a $\mathbb{K}$-linear abelian category. Find a $\mathbb{K}$-linear equivalence $F : \mathsf{M}^{\mathrm{op}} \to \mathsf{M}$.

2.6 Projective Objects

In this section M is an abelian category.

A *splitting* of an epimorphism $\psi : M \to M''$ in M is a morphism $\alpha : M'' \to M$ s.t. $\psi \circ \alpha = \mathrm{id}_{M''}$. A splitting of a monomorphism $\phi : M' \to M$ is a morphism $\beta : M \to M'$ s.t. $\beta \circ \phi = \mathrm{id}_{M'}$. A splitting of a short exact sequence

$$0 \to M' \xrightarrow{\phi} M \xrightarrow{\psi} M'' \to 0 \tag{2.6.1}$$

is a splitting of the epimorphism ψ, or equivalently a splitting of the monomorphism ϕ. The short exact sequence is said to be *split* if it has some splitting.

Exercise 2.6.2 Show how to get from a splitting of ϕ to a splitting of ψ, and vice versa. Show how any of those gives rise to an isomorphism $M \cong M' \oplus M''$.

Definition 2.6.3 An object $P \in \mathsf{M}$ is called a *projective object* if for every morphism $\gamma : P \to N$ and every *epimorphism* $\psi : M \twoheadrightarrow N$, there exists a morphism $\tilde{\gamma} : P \to N$ such that $\psi \circ \tilde{\gamma} = \gamma$.

This is described in the following commutative diagram in M :

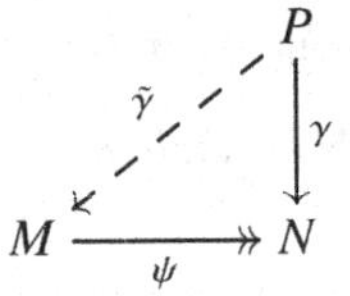

Proposition 2.6.4 *The following conditions are equivalent for* $P \in \mathsf{M}$:

(i) *P is projective.*

(ii) *The additive functor* $\mathrm{Hom}_{\mathsf{M}}(P, -) : \mathsf{M} \to \mathsf{Ab}$ *is exact.*

(iii) *Every short exact sequence* (2.6.1) *with* $M'' = P$ *is split.*

Proof Exercise. □

Definition 2.6.5 We say M *has enough projectives* if every $M \in \mathsf{M}$ admits an epimorphism $P \to M$ from a projective object P.

Exercise 2.6.6 Let A be a ring.

(1) Prove that an A-module P is projective iff it is a direct summand of a free module; i.e. $P \oplus P' \cong Q$ for some module P' and free module Q.

(2) Prove that the category $\mathsf{Mod}\, A$ has enough projectives.

Exercise 2.6.7 Let M be the category of finite abelian groups. Prove that the only projective object in M is 0. So M does not have enough projectives. (Hint: use Proposition 2.6.4.)

Example 2.6.8 Consider the scheme $X := \mathbf{P}^1_{\mathbb{K}}$, the projective line over a field $\mathbb{K}$, with structure sheaf $\mathcal{O}_X$. The category $\mathsf{Coh}\, \mathcal{O}_X$ of coherent $\mathcal{O}_X$-modules is abelian (it is a full abelian subcategory of $\mathsf{Mod}\, \mathcal{O}_X$, cf. Example 2.3.19). One can show that the only projective object of $\mathsf{Coh}\, \mathcal{O}_X$ is 0, but this is quite involved.

Let us only indicate why the sheaf $\mathcal{O}_X$ is not a projective object of $\mathsf{Coh}\, \mathcal{O}_X$. Denote by t_0, t_1 the homogeneous coordinates of X. These belong to

$\Gamma(X, \mathcal{O}_X(1))$, so each determines a homomorphism of sheaves $t_j : \mathcal{O}_X(i) \to \mathcal{O}_X(i+1)$. We get a sequence

$$0 \to \mathcal{O}_X(-2) \xrightarrow{[t_0\ t_1]} \mathcal{O}_X(-1)^{\oplus 2} \xrightarrow{\left[\begin{smallmatrix} -t_1 \\ t_0 \end{smallmatrix}\right]} \mathcal{O}_X \to 0 \tag{2.6.9}$$

in $\mathsf{Coh}\, \mathcal{O}_X$, which is known to be exact. It is also known that $\Gamma(X, \mathcal{O}_X) = \mathbb{K}$ and $\Gamma(X, \mathcal{O}_X(-1)) = 0$. Therefore the sequence (2.6.9) is not split.

2.7 Injective Objects

In this section M is an abelian category.

Definition 2.7.1 An object $I \in \mathsf{M}$ is called an *injective object* if for every morphism $\gamma : M \to I$ and every *monomorphism* $\psi : M \rightarrowtail N$, there exists a morphism $\tilde{\gamma} : N \to I$ such that $\tilde{\gamma} \circ \psi = \gamma$.

This is depicted in the following commutative diagram in M :

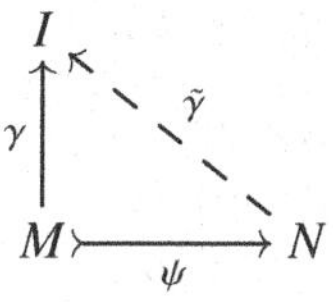

Proposition 2.7.2 *The following conditions are equivalent for $I \in \mathsf{M}$:*

(i) *I is injective.*

(ii) *The additive functor* $\mathrm{Hom}_{\mathsf{M}}(-, I) : \mathsf{M}^{\mathrm{op}} \to \mathsf{Ab}$ *is exact.*

(iii) *Every short exact sequence* (2.6.1) *with $M' = I$ is split.*

Exercise 2.7.3 Prove Proposition 2.7.2.

Recall that $\mathrm{Op} : \mathsf{M} \to \mathsf{M}^{\mathrm{op}}$ is an exact functor.

Proposition 2.7.4 *An object $J \in \mathsf{M}$ is injective if and only if the object* $\mathrm{Op}(J) \in \mathsf{M}^{\mathrm{op}}$ *is projective.*

Exercise 2.7.5 Prove Proposition 2.7.4.

Example 2.7.6 Let A be a ring. Unlike projectives, the structure of injective objects in $\mathsf{Mod}\, A$ is very complicated, and not much is known (except that they exist). However if A is a commutative noetherian ring then we know this: every injective module I is a direct sum of indecomposable injective modules; and the indecomposables are parameterized by $\mathrm{Spec}(A)$, the set of prime ideals of A. These facts are due to Matlis; see Section 13.2 in the book.

Definition 2.7.7 We say M *has enough injectives* if every $M \in \mathsf{M}$ admits a monomorphism $M \to I$ to an injective object I.

Here are a few results about injective objects. Recall that modules over a ring are always left modules by default.

Proposition 2.7.8 *Let $f : A \to B$ be a ring homomorphism, and let I be an injective A-module. Then $J := \mathrm{Hom}_A(B, I)$ is an injective B-module.*

Proof Note that B is a left A-module via f, and a right B-module. This makes J into a left B-module. In a formula: for $\phi \in J$ and $b, b' \in B$ we have $(b \cdot \phi)(b') = \phi(b' \cdot b)$.

Now given any $N \in \mathsf{Mod}\, B$ there is an isomorphism

$$\mathrm{Hom}_B(N, J) = \mathrm{Hom}_B(N, \mathrm{Hom}_A(B, I)) \cong \mathrm{Hom}_A(N, I). \tag{2.7.9}$$

This is a natural isomorphism (of functors in N). So the functor $\mathrm{Hom}_B(-, J)$ is exact, and hence J is injective. □

Theorem 2.7.10 (Baer Criterion) *Let A be a ring and I an A-module. Assume that every A-module homomorphism $\mathfrak{a} \to I$ from a left ideal $\mathfrak{a} \subseteq A$ extends to a homomorphism $A \to I$. Then the module I is injective.*

Proof Consider an A-module M, a submodule $N \subseteq M$, and a homomorphism $\gamma : N \to I$. We have to prove that γ extends to a homomorphism $M \to I$. Look at the pairs (N', γ') consisting of a submodule $N' \subseteq M$ that contains N, and a homomorphism $\gamma' : N' \to I$ that extends γ. The set of all such pairs is ordered by inclusion, and it satisfies the conditions of Zorn's Lemma. Therefore there exists a maximal pair (N', γ'). We claim that $N' = M$.

Otherwise, there is an element $m \in M$ that does not belong to N'. Define $N'' := N' + A \cdot m$, so $N' \subsetneq N'' \subseteq M$. Let $\mathfrak{a} := \{a \in A \mid a \cdot m \in N'\}$, which is a left ideal of A. There is a short exact sequence

$$0 \to \mathfrak{a} \xrightarrow{\alpha} N' \oplus A \to N'' \to 0$$

of A-modules, where $\alpha(a) := (a \cdot m, -a)$. Let $\phi : \mathfrak{a} \to I$ be the homomorphism $\phi(a) := \gamma'(a \cdot m)$. By assumption, it extends to a homomorphism $\tilde{\phi} : A \to I$. We get a homomorphism $\gamma' + \tilde{\phi} : N' \oplus A \to I$ that vanishes on the image of α. Thus there is an induced homomorphism $\gamma'' : N'' \to I$. This contradicts the maximality of (N', γ'). □

Lemma 2.7.11 *The $\mathbb{Z}$-module $\mathbb{Q}/\mathbb{Z}$ is injective.*

Proof By the Baer criterion, it is enough to consider a homomorphism $\gamma : \mathfrak{a} \to \mathbb{Q}/\mathbb{Z}$ for an ideal $\mathfrak{a} = n \cdot \mathbb{Z} \subseteq \mathbb{Z}$. We may assume that $n \neq 0$. Say $\gamma(n) = r + \mathbb{Z}$ with $r \in \mathbb{Q}$. Then we can extend γ to $\tilde{\gamma} : \mathbb{Z} \to \mathbb{Q}/\mathbb{Z}$ with $\tilde{\gamma}(1) := r/n + \mathbb{Z}$. □

Lemma 2.7.12 *Let $\{I_x\}_{x \in X}$ be a collection of injective objects of* M. *If the product $\prod_{x \in X} I_x$ exists in* M, *then it is an injective object.*

Proof Exercise. □

Theorem 2.7.13 *Let A be any ring. The category* Mod A *has enough injectives.*

Proof The proof is done in a few steps.

Step 1. Here $A = \mathbb{Z}$. Take any nonzero $\mathbb{Z}$-module M and any nonzero $m \in M$. Consider the cyclic submodule $M' := \mathbb{Z} \cdot m \subseteq M$. There is a homomorphism $\gamma' : M' \to \mathbb{Q}/\mathbb{Z}$ s.t. $\gamma'(m) \neq 0$. Indeed, if $M' \cong \mathbb{Z}$, then we take any $r \in \mathbb{Q} - \mathbb{Z}$; and if $M' \cong \mathbb{Z}/(n)$ for some $n > 0$, then we take $r := 1/n$. In either case, we define $\gamma'(m) := r + \mathbb{Z} \in \mathbb{Q}/\mathbb{Z}$. Since $\mathbb{Q}/\mathbb{Z}$ is an injective $\mathbb{Z}$-module, γ' extends to a homomorphism $\gamma : M \to \mathbb{Q}/\mathbb{Z}$. By construction we have $\gamma(m) \neq 0$.

Step 2. Now A is any ring, M is any nonzero A-module, and $m \in M$ a nonzero element. Define the A-module $I := \mathrm{Hom}_{\mathbb{Z}}(A, \mathbb{Q}/\mathbb{Z})$, which, according to Lemma 2.7.11 and Proposition 2.7.8, is an injective A-module. Let $\gamma : M \to \mathbb{Q}/\mathbb{Z}$ be a $\mathbb{Z}$-linear homomorphism such that $\gamma(m) \neq 0$. Such γ exists by step 1. Let $\theta : I \to \mathbb{Q}/\mathbb{Z}$ be the $\mathbb{Z}$-linear homomorphism that sends an element $\chi \in I$ to $\chi(1) \in \mathbb{Q}/\mathbb{Z}$. The adjunction formula (2.7.9) gives an A-module homomorphism $\psi : M \to I$ s.t. $\theta \circ \psi = \gamma$. We note that $(\theta \circ \psi)(m) = \gamma(m) \neq 0$, and hence $\psi(m) \neq 0$.

Step 3. Here A and M are arbitrary. Let I be as in step 2. For every nonzero $m \in M$ there is an A-linear homomorphism $\psi_m : M \to I$ such that $\psi_m(m) \neq 0$. For $m = 0$ let $\psi_0 : M \to I$ be an arbitrary homomorphism (e.g. $\psi_0 = 0$). Define the A-module $J := \prod_{m \in M} I$. There is a homomorphism $\psi := \prod_{m \in M} \psi_m : M \to J$, and it is easy to check that ψ is a monomorphism. By Lemma 2.7.12, J is an injective A-module. □

Exercise 2.7.14 At the price of getting a bigger injective module, we can make the construction of injective resolutions functorial. Let $I := \mathrm{Hom}_{\mathbb{Z}}(A, \mathbb{Q}/\mathbb{Z})$ as above. Given an A-module M, consider the set $X(M) := \mathrm{Hom}_A(M, I) \cong \mathrm{Hom}_{\mathbb{Z}}(M, \mathbb{Q}/\mathbb{Z})$. Let $J(M) := \prod_{\psi \in X(M)} I$. There is a "tautological" homomorphism $\phi_M : M \to J(M)$. Show that ϕ_M is a monomorphism, $J : M \mapsto J(M)$ is a functor, and $\phi : \mathrm{Id} \to J$ is a natural transformation.

Is the functor $J : \mathsf{Mod}\, A \to \mathsf{Mod}\, A$ additive?

Example 2.7.15 Let N be the category of torsion abelian groups, and M the category of finite abelian groups. Then N $\subseteq$ Ab and M $\subseteq$ N are full abelian subcategories. M has no projectives nor injectives except 0 (see Exercise 2.6.7

regarding projectives). The only projective in N is 0. However, it can be shown that N has enough injectives; see [63, Lemma III.3.2] or [161, Proposition 4.6].

Proposition 2.7.16 *If A is a left noetherian ring, then every direct sum of injective A-modules is an injective module.*

Exercise 2.7.17 Prove Proposition 2.7.16. (Hint: use the Baer criterion.)

Exercise 2.7.18 Here we study injectives in the category $\mathsf{Ab} = \mathsf{Mod}\,\mathbb{Z}$. By Lemma 2.7.11, the module $I := \mathbb{Q}/\mathbb{Z}$ is injective. For a (positive) prime number p, we denote by $\widehat{\mathbb{Z}}_p$ the ring of p-adic integers, and by $\widehat{\mathbb{Q}}_p$ its field of fractions (namely the p-adic completions of $\mathbb{Z}$ and $\mathbb{Q}$, respectively). Define the abelian group $I_p := \widehat{\mathbb{Q}}_p/\widehat{\mathbb{Z}}_p$.

(1) Show that I_p is an injective object of Ab.
(2) Show that I_p is indecomposable (i.e. it is not the direct sum of two nonzero objects).
(3) Show that $I \cong \bigoplus_p I_p$.
(4) The theory (see Section 13.2) tells us that there is another indecomposable injective object in Ab, besides the I_p. Try to identify it.

The abelian category $\mathsf{Mod}\,\mathcal{A}$ associated to a ringed space $(X, \mathcal{A})$ was introduced in Example 2.3.19.

Proposition 2.7.19 *Let $(X, \mathcal{A})$ be a ringed space. The category $\mathsf{Mod}\,\mathcal{A}$ has enough injectives.*

Proof Let $\mathcal{M}$ be an $\mathcal{A}$-module. Take a point $x \in X$. The stalk $\mathcal{M}_x$ is a module over the ring $\mathcal{A}_x$, and by Theorem 2.7.13 we can find an embedding $\phi_x : \mathcal{M}_x \to I_x$ into an injective $\mathcal{A}_x$-module. Let $g_x : \{x\} \to X$ be the inclusion, which we may view as a map of ringed spaces from $(\{x\}, \mathcal{A}_x)$ to $(X, \mathcal{A})$. Define $\mathcal{I}_x := g_{x*}(I_x)$, which is an $\mathcal{A}$-module (in fact it is a constant sheaf supported on the closed set $\overline{\{x\}} \subseteq X$). The adjunction formula gives rise to a sheaf homomorphism $\psi_x : \mathcal{M} \to \mathcal{I}_x$. Since the functor $g_x^* : \mathsf{Mod}\,\mathcal{A} \to \mathsf{Mod}\,\mathcal{A}_x$ is exact, the adjunction formula shows that $\mathcal{I}_x$ is an injective object of $\mathsf{Mod}\,\mathcal{A}$.

Finally let $\mathcal{J} := \prod_{x \in X} \mathcal{I}_x$. This is an injective $\mathcal{A}$-module. There is a homomorphism $\psi := \prod_{x \in X} \psi_x : \mathcal{M} \to \mathcal{J}$ in $\mathsf{Mod}\,\mathcal{A}$. This is a monomorphism, since for every point x, letting $\mathcal{J}_x$ be the stalk of the sheaf $\mathcal{J}$ at x, the composition $\mathcal{M}_x \xrightarrow{\psi_x} \mathcal{J}_x \xrightarrow{p_x} \mathcal{I}_x$ is the embedding $\phi_x : \mathcal{M}_x \to I_x$. □

Remark 2.7.20 Let A be a nonzero ring, and consider the abelian category $\mathsf{M} := \mathsf{Mod}\,A$, the category of A-modules. A reasonable question to ask is this: Are the abelian categories M and M^{op} equivalent? The answer is negative. In fact, P. Freyd, in [47, Exercise 5.B.3], shows that $(\mathsf{Mod}\, A)^{\mathrm{op}}$ is not equivalent, as an abelian category, to $\mathsf{Mod}\, B$ for any ring B. The argument involves a delicate study of countable coproducts and products, and properties of Grothendieck abelian categories.

Here is a special case of the previous remark, which might shed some more light on the issue.

Example 2.7.21 Consider the category $\mathsf{Mod}\, \mathbb{Z} = \mathsf{Ab}$ of abelian groups. Here is a proof that there does not exist an additive equivalence $F : \mathsf{Ab}^{\mathrm{op}} \to \mathsf{Ab}$. Suppose we had such an equivalence. Consider the object $P := \mathbb{Z} \in \mathsf{Ab}$, and let $I := F(P) \in \mathsf{Ab}$. Because P is an indecomposable projective object and $F : \mathsf{Ab} \to \mathsf{Ab}$ is a contravariant equivalence, the object I has to be an indecomposable injective. The endomorphism rings are $\mathrm{End}_{\mathsf{Ab}}(I) \cong \mathrm{End}_{\mathsf{Ab}}(P)^{\mathrm{op}} = \mathbb{Z}^{\mathrm{op}} = \mathbb{Z}$. However, the structure theorem for injective modules over commutative noetherian rings (Theorem 13.2.15) says that the only indecomposable injectives in Ab are $I = \widehat{\mathbb{Q}}_p / \widehat{\mathbb{Z}}_p$ and $I = \mathbb{Q}$; and their endomorphism rings are $\widehat{\mathbb{Z}}_p$ and $\mathbb{Q}$, respectively.

3

Differential Graded Algebra

Recall that according to Convention 1.2.4, there is a nonzero commutative base ring $\mathbb{K}$. By default, all rings are $\mathbb{K}$-central, all linear categories are $\mathbb{K}$-linear, all linear operations (such as ring homomorphisms and linear functors) are $\mathbb{K}$-linear and $\otimes$ means $\otimes_{\mathbb{K}}$.

Throughout, "DG" stands for "differential graded." There is some material about DG algebra in a few published references, such as the book [101] and the papers [79], [151], [133] and [12]. However, for our purposes we need a much more detailed understanding of this theory, and this is what the present chapter provides.

3.1 Graded Algebra

Before entering the DG world, it is good to understand the graded world.

Definition 3.1.1 A *cohomologically graded* $\mathbb{K}$*-module* is a $\mathbb{K}$-module M equipped with a direct sum decomposition $M = \bigoplus_{i \in \mathbb{Z}} M^i$ into $\mathbb{K}$-submodules. The $\mathbb{K}$-module M^i is called the *homogeneous component of cohomological degree* i *of* M. The nonzero elements of M^i are called *homogeneous elements of cohomological degree* i.

From here until Chapter 15 we are going to respect the following convention, which will simplify the discussion.

Convention 3.1.2 By "graded $\mathbb{K}$-module" we mean a *cohomologically graded* $\mathbb{K}$*-module*, as defined above.

In Chapter 15 we shall introduce *algebraically graded rings and modules*, and then we shall have to make a careful distinction between these notions. See Remark 3.1.25 regarding commutativity in the two settings.

Suppose M and N are graded $\mathbb{K}$-modules. For an integer i let

$$(M \otimes N)^i := \bigoplus_{j \in \mathbb{Z}} (M^j \otimes N^{i-j}).$$

Then

$$M \otimes N = \bigoplus_{i \in \mathbb{Z}} (M \otimes N)^i \tag{3.1.3}$$

is a graded $\mathbb{K}$-module.

A $\mathbb{K}$-linear homomorphism $\phi : M \to N$ is said to be *homogeneous of degree* i if $\phi(M^j) \subseteq N^{j+i}$ for all j. We denote by $\mathrm{Hom}_{\mathbb{K}}(M, N)^i$ the $\mathbb{K}$-module of degree i homomorphisms $M \to N$. In other words,

$$\mathrm{Hom}_{\mathbb{K}}(M, N)^i := \prod_{j \in \mathbb{Z}} \mathrm{Hom}_{\mathbb{K}}(M^j, N^{j+i}). \tag{3.1.4}$$

Definition 3.1.5 Let M and N be graded $\mathbb{K}$-modules.

(1) The *module of graded $\mathbb{K}$-linear homomorphisms* from M to N is the graded $\mathbb{K}$-module

$$\mathrm{Hom}_{\mathbb{K}}(M, N) := \bigoplus_{i \in \mathbb{Z}} \mathrm{Hom}_{\mathbb{K}}(M, N)^i.$$

(2) A degree 0 homomorphism $\phi : M \to N$ is called a *strict homomorphism of graded $\mathbb{K}$-modules*.

If M_0, M_1, M_2 are graded $\mathbb{K}$-modules, and $\phi_k : M_k \to M_{k+1}$ are $\mathbb{K}$-linear homomorphisms of degrees i_k, then $\phi_1 \circ \phi_0 : M_0 \to M_2$ is a $\mathbb{K}$-linear homomorphism of degree $i_0 + i_1$. The identity automorphism $\mathrm{id}_M : M \to M$ has degree 0.

Definition 3.1.6 The *strict category of graded $\mathbb{K}$-modules* is the category $\mathbf{G}_{\mathrm{str}}(\mathbb{K})$, whose objects are the cohomologically graded $\mathbb{K}$-modules, and whose morphisms are the strict homomorphisms of cohomologically graded $\mathbb{K}$-modules.

It is easy to see that $\mathbf{G}_{\mathrm{str}}(\mathbb{K})$ is a $\mathbb{K}$-linear abelian category – the kernels and cokernels are degreewise.

Remark 3.1.7 Let $\mathrm{Ungr} : \mathbf{G}_{\mathrm{str}}(\mathbb{K}) \to \mathbf{M}(\mathbb{K})$ be the functor that forgets the grading. It is faithful, but often not full. Namely the obvious homomorphism

$$\mathrm{Ungr}(\mathrm{Hom}_{\mathbb{K}}(M, N)) \to \mathrm{Hom}_{\mathbb{K}}(\mathrm{Ungr}(M), \mathrm{Ungr}(N))$$

is injective but not bijective. See Remark 15.1.11 for a discussion.

The tensor operation $(- \otimes -)$ from (3.1.3) makes $\mathbf{G}_{\mathrm{str}}(\mathbb{K})$ into a *monoidal* $\mathbb{K}$*-linear category*, with monoidal unit $\mathbb{K}$. This means that the bifunctor

$$(- \otimes -) : \mathbf{G}_{\mathrm{str}}(\mathbb{K}) \times \mathbf{G}_{\mathrm{str}}(\mathbb{K}) \to \mathbf{G}_{\mathrm{str}}(\mathbb{K})$$

satisfies the monoidal axioms (associativity up to a trifunctorial cocycle, etc., see [102, Chapter XI]), and it is $\mathbb{K}$-bilinear. Moreover, given $M, N \in \mathbf{G}_{\mathrm{str}}(\mathbb{K})$, let us define the *braiding isomorphism*

$$\mathrm{br}_{M,N} : M \otimes N \xrightarrow{\simeq} N \otimes M, \tag{3.1.8}$$

$$\mathrm{br}_{M,N}(m \otimes n) := (-1)^{i \cdot j} \cdot n \otimes m \tag{3.1.9}$$

for homogeneous elements $m \in M^i$ and $n \in N^j$. Because $\mathrm{br}_{N,M} \circ \mathrm{br}_{M,N} = \mathrm{id}_{M \otimes N}$, this makes $\mathbf{G}_{\mathrm{str}}(\mathbb{K})$ into a *symmetric monoidal* $\mathbb{K}$*-linear category*. The braiding isomorphism (3.1.9) is often called the *Koszul sign rule*. See Remark 3.1.25 for a background discussion.

Exercise 3.1.10 For an integer $l \geq 1$ let τ be a permutation of the set $\{1, \ldots, l\}$. Show that for every $M_1, \ldots, M_l \in \mathbf{G}_{\mathrm{str}}(\mathbb{K})$ there is an isomorphism

$$\mathrm{br}_{\tau} : M_1 \otimes \cdots \otimes M_l \xrightarrow{\simeq} M_{\tau(1)} \otimes \cdots \otimes M_{\tau(l)}$$

in $\mathbf{G}_{\mathrm{str}}(\mathbb{K})$, which is functorial in the sequence of objects $\{M_i\}_{i=1,\ldots,l}$, functorial in the permutation τ, monoidal (for a partition $l = l_1 + l_2$), and $\mathrm{br}_{\tau} = \mathrm{br}_{M_1,M_2}$ when $l = 2$ and τ is the transposition. See [102, Theorem XI.1.1] or [17] for the answer.

Definition 3.1.11 A *cohomologically graded central* $\mathbb{K}$*-ring* is a central $\mathbb{K}$-ring A, equipped with a direct sum decomposition $A = \bigoplus_{i \in \mathbb{Z}} A^i$ into $\mathbb{K}$-submodules, such that $1_A \in A^0$, and $A^i \cdot A^j \subseteq A^{i+j}$.

Definition 3.1.12 Let A and B be cohomologically graded central $\mathbb{K}$-rings. A *homomorphism of cohomologically graded central* $\mathbb{K}$*-rings* is a $\mathbb{K}$-ring homomorphism $f : A \to B$ that respects the gradings, namely $f(A^i) \subseteq B^i$.

The category of cohomologically graded central $\mathbb{K}$-rings is denoted by $\mathsf{GRng}/_{\mathrm{c}}\, \mathbb{K}$.

As always for ring homomorphisms, f must preserve units, i.e. $f(1_A) = 1_B$. Note that $\mathbb{K}$ itself is a graded ring, concentrated in degree 0; and it is the initial object of $\mathsf{GRng}/_{\mathrm{c}}\, \mathbb{K}$.

To simplify the discussion, we shall follow the next convention (until Chapter 15).

Convention 3.1.13 By "graded ring" we mean a *cohomologically graded central* $\mathbb{K}$*-ring*, as defined above.

Recall that by Convention 1.2.4, all ring homomorphisms, and that includes graded rings, are over $\mathbb{K}$.

Example 3.1.14 Let M be a graded $\mathbb{K}$-module. Then the graded module

$$\operatorname{End}_{\mathbb{K}}(M) := \operatorname{Hom}_{\mathbb{K}}(M, M) = \bigoplus_{i \in \mathbb{Z}} \operatorname{Hom}_{\mathbb{K}}(M, M)^i,$$

with the operation of composition, is a graded $\mathbb{K}$-ring.

The prototypical manifestation of the Koszul sign rule is the next definition.

Definition 3.1.15 Let $A = \bigoplus_{i \in \mathbb{Z}} A^i$ be a graded $\mathbb{K}$-ring. We say that A is a *weakly commutative graded ring* if $b \cdot a = (-1)^{i \cdot j} \cdot a \cdot b$ for all $a \in A^i$ and $b \in A^j$.

Remark 3.1.16 Let us give a categorical explanation of Definition 3.1.15, using the symmetric monoidal structure of $\mathbf{G}_{\mathrm{str}}(\mathbb{K})$.

The general categorical way to define a *ring object* in the symmetric monoidal linear category $(\mathbf{G}_{\mathrm{str}}(\mathbb{K}), \otimes, \mathbb{K}, \mathrm{br})$ is as an object $A \in \mathbf{G}_{\mathrm{str}}(\mathbb{K})$ equipped with a multiplication morphism $\mathrm{m} : A \otimes A \to A$ and a unit morphism $\mathrm{u} : \mathbb{K} \to A$, such that the triple $(A, \mathrm{m}, \mathrm{u})$ obeys the ring axioms. In our case it is precisely Definition 3.1.11.

The *categorical commutativity condition* for the ring object $(A, \mathrm{m}, \mathrm{u})$ is that the diagram

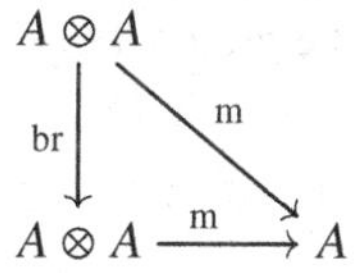

in $\mathbf{G}_{\mathrm{str}}(\mathbb{K})$ is commutative. This is what weak commutativity is, in Definition 3.1.15.

Here is a definition similar to Definition 3.1.15.

Definition 3.1.17 Let $A = \bigoplus_{i \in \mathbb{Z}} A^i$ be a graded ring.

(1) Homogeneous elements $a \in A^i$ and $b \in A^j$ are said to *graded-commute* with each other if $b \cdot a = (-1)^{i \cdot j} \cdot a \cdot b$.

(2) A homogeneous element $a \in A^i$ is called a *graded-central element* if it graded-commutes with all homogeneous elements of A.

(3) The *graded-center* of A is the $\mathbb{K}$-submodule $\operatorname{Cent}(A) \subseteq A$ generated by the homogeneous graded-central elements.

Exercise 3.1.18 Let A be a graded ring. Show that

(1) $\mathrm{Cent}(A)$ is a graded subring of A, it is weakly commutative, and it contains the image of the base ring $\mathbb{K}$.
(2) A is weakly commutative iff $\mathrm{Cent}(A) = A$.

Below are several sign formulas that are more subtle consequences of the Koszul sign rule. They can be traced – with effort – to the *biclosed monoidal structure* of $\mathbf{G}_{\mathrm{str}}(\mathbb{K})$, namely to the interaction between the bifunctors $(- \otimes -)$ and $\mathrm{Hom}_{\mathbb{K}}(-,-)$.

Suppose that for $k = 0, 1$ we are given graded $\mathbb{K}$-module homomorphisms $\phi_k : M_k \to N_k$ of degrees i_k. Then the homomorphism

$$\phi_0 \otimes \phi_1 \in \mathrm{Hom}_{\mathbb{K}}((M_0 \otimes M_1, N_0 \otimes N_1)^{i_0+i_1}$$

acts on a tensor $m_0 \otimes m_1 \in M_0 \otimes M_1$, with $m_k \in M_k^{j_k}$, like this:

$$(\phi_0 \otimes \phi_1)(m_0 \otimes m_1) := (-1)^{i_1 \cdot j_0} \cdot \phi_0(m_0) \otimes \phi_1(m_1) \in N_0 \otimes N_1. \quad (3.1.19)$$

The rule of thumb explaining this formula is that ϕ_1 and m_0 were transposed.

Suppose we are given graded $\mathbb{K}$-module homomorphisms $\phi_0 : N_0 \to M_0$ and $\phi_1 : M_1 \to N_1$ of degrees i_0 and i_1. Then the homomorphism

$$\mathrm{Hom}(\phi_0, \phi_1) \in \mathrm{Hom}_{\mathbb{K}}(\mathrm{Hom}_{\mathbb{K}}(M_0, M_1), \mathrm{Hom}_{\mathbb{K}}(N_0, N_1))^{i_0+i_1}$$

acts on $\gamma \in \mathrm{Hom}_{\mathbb{K}}(M_0, M_1)^j$ as follows: for an element $n_0 \in N_0^k$ we have

$$\mathrm{Hom}(\phi_0, \phi_1)(\gamma)(n_0) := (-1)^{i_0 \cdot (i_1+j)}(\phi_1 \circ \gamma \circ \phi_0)(n_0) \in N_1^{k+i_0+i_1+j}. \quad (3.1.20)$$

The sign is because ϕ_0 jumped across ϕ_1 and γ.

Definition 3.1.21 Let A and B be graded rings. Then $A \otimes B$ is a graded ring, with multiplication

$$(a_0 \otimes b_0) \cdot (a_1 \otimes b_1) := (-1)^{i_1 \cdot j_0} \cdot (a_0 \cdot a_1) \otimes (b_0 \cdot b_1)$$

for elements $a_k \in A^{i_k}$ and $b_k \in B^{j_k}$.

We shall require another notion of commutativity that is not of categorical nature.

Definition 3.1.22 Let $A = \bigoplus_{i \in \mathbb{Z}} A^i$ be a graded ring.
(1) The graded ring A is called *strongly commutative* if it is weakly commutative (Definition 3.1.15), and also $a^2 = 0$ if $a \in A^i$ and i is odd.
(2) The graded ring A is called *nonpositive* if $A^i = 0$ for all $i > 0$.
(3) The graded ring A is called a *commutative graded ring* if it is nonpositive and strongly commutative.

This definition is taken from [173]. In [188] the term *super-commutative* was used instead of *strongly commutative*. The name *strongly commutative* was suggested to us by J. Palmieri.

Example 3.1.23 By a *graded set* we mean a set X partitioned as $X = \coprod_{i \in \mathbb{Z}} X^i$. The elements of X^i are called variables of degree i. The *noncommutative graded polynomial ring* on the graded set X is the graded ring $\mathbb{K}\langle X \rangle$, which is the free graded $\mathbb{K}$-module spanned by the monomials (i.e. words) $x_1 \cdots x_n$ in the elements of X, and its multiplication is by concatenation of monomials.

The *strongly commutative graded polynomial ring* on the graded set X is the graded ring $\mathbb{K}[X] := \mathbb{K}\langle X \rangle / I$, where I is the two-sided ideal of $\mathbb{K}\langle X \rangle$ generated by the elements $y \cdot x - (-1)^{i \cdot j} \cdot x \cdot y$ for all variables $x \in X^i$ and $y \in X^j$, together with the elements $z \cdot z$ for all $z \in X^k$ and odd k. This is a strongly commutative graded ring.

Exercise 3.1.24 Show that if A and B are weakly (resp. strongly) commutative graded rings, then so is $A \otimes B$.

Remark 3.1.25 Weak commutativity is the obvious commutativity condition in the cohomologically graded setting, when the Koszul sign rule is imposed; see Remark 3.1.16. Of course, there are many instances of commutative graded rings that do not involve the Koszul sign rule; see e.g. [46] or [105]. In our book we call them *algebraically graded rings*, and they are studied in Chapters 15–17.

Strong commutativity has another reason. Its role is to guarantee that the strongly commutative graded polynomial ring $\mathbb{K}[X]$ from Example 3.1.23 above will be a graded-free $\mathbb{K}$-module. Without this condition, the square of an odd variable z would be a nonzero 2-torsion element.

Of course, if 2 is invertible in $\mathbb{K}$ (e.g. if $\mathbb{K}$ contains $\mathbb{Q}$), then weak and strong commutativity of a graded central $\mathbb{K}$-ring coincide. Since most texts dealing with DG rings assume that $\mathbb{Q} \subseteq \mathbb{K}$, the subtle distinction we make is absent from them.

Remark 3.1.26 Let $A = \bigoplus_{i \in \mathbb{Z}} A^i$ be a weakly commutative graded ring, and assume A has nonzero odd elements. Then, after we forget the grading, the ring A is no longer commutative (except in special cases, as in characteristic 2).

This phenomenon will reappear often in the study of DG algebra.

Remark 3.1.27 The origin of the Koszul sign rule, predating the work of J.L. Koszul, appears to be in the formula for the differential d of the tensor product of two complexes, which occurs in classical homological algebra (and is repeated here in Definition 3.2.3). Without the sign we won't have $\mathrm{d} \circ \mathrm{d} = 0$.

In his thesis [89], Koszul talks about the sign rule explicitly, when discussing the Chevalley–Eilenberg complex. See [17] for a detailed study of the Koszul sign rule and its group-theoretic interpretation.

It should be noted that there are some sign inconsistencies in the book [62]. The Koszul sign rule could have prevented them.

In our book we made a strenuous effort to have correct (namely consistent) signs. This did not always produce satisfactory outcomes – for instance, in the context of the triangulated structure of the opposite homotopy category, as explained in Remark 5.6.12.

Definition 3.1.28 Let A be a graded ring. A *graded left A-module* is a left A-module M, equipped with a $\mathbb{K}$-module decomposition $M = \bigoplus_{i \in \mathbb{Z}} M^i$, such that $A^i \cdot M^j \subseteq M^{i+j}$ for all i, j.

We can also talk about graded right A-modules, and graded bimodules. But our default option (see Convention 1.2.4) is that modules are left modules.

Exercise 3.1.29 Let M be a graded $\mathbb{K}$-module, A a graded central $\mathbb{K}$-ring, and $f : A \to \operatorname{End}_{\mathbb{K}}(M)$ a homomorphism in $\mathsf{GRng}/_{\mathrm{c}} \mathbb{K}$.

(1) Show that M becomes a graded A-module, with action $a \cdot m := f(a)(m)$.
(2) Show that every graded A-module structure on M that is consistent with the given graded $\mathbb{K}$-module structure arises this way.

Lemma 3.1.30 *Let A be a graded ring, let M be a graded right A-module, and let N be a graded left A-module. Then the $\mathbb{K}$-module $M \otimes_A N$ has a direct sum decomposition*

$$M \otimes_A N = \bigoplus_{i \in \mathbb{Z}} (M \otimes_A N)^i,$$

where $(M \otimes_A N)^i$ is the $\mathbb{K}$-linear span of the tensors $m \otimes n$, for all $j \in \mathbb{Z}$, $m \in M^j$ and $n \in N^{i-j}$.

Proof There is a canonical surjection of $\mathbb{K}$-modules $M \otimes N \to M \otimes_A N$. Its kernel is the $\mathbb{K}$-submodule $L \subseteq M \otimes N$ generated by the elements $(m \cdot a) \otimes n - m \otimes (a \cdot n)$, for $m \in M^j$, $n \in N^k$ and $a \in A^l$. So L is a graded submodule of $M \otimes N$, and therefore so is the quotient. Finally, by formula (3.1.3) the ith homogeneous component of $M \otimes_A N$ is precisely $(M \otimes_A N)^i$. □

Definition 3.1.31 Let A be a graded ring, and let M, N be graded A-modules. For each $i \in \mathbb{Z}$ define $\operatorname{Hom}_A(M, N)^i$ to be the subset of $\operatorname{Hom}_{\mathbb{K}}(M, N)^i$ consisting of the homomorphisms $\phi : M \to N$ such that $\phi(a \cdot m) = (-1)^{i \cdot k} \cdot a \cdot \phi(m)$ for all $a \in A^k$. Next define the graded $\mathbb{K}$-module

$$\operatorname{Hom}_A(M, N) := \bigoplus_{i \in \mathbb{Z}} \operatorname{Hom}_A(M, N)^i.$$

Suppose C is a $\mathbb{K}$-linear category (Definition 2.1.1). Since the composition of morphisms is $\mathbb{K}$-bilinear, for every triple of objects $M_0, M_1, M_2 \in \mathsf{C}$, composition can be expressed as a $\mathbb{K}$-linear homomorphism

$$\mathrm{Hom}_{\mathsf{C}}(M_1, M_2) \otimes \mathrm{Hom}_{\mathsf{C}}(M_0, M_1) \to \mathrm{Hom}_{\mathsf{C}}(M_0, M_2),$$

$\phi_1 \otimes \phi_0 \mapsto \phi_1 \circ \phi_0$. We refer to it as the composition homomorphism. It will be used in the following definition.

Definition 3.1.32 A *graded $\mathbb{K}$-linear category* is a $\mathbb{K}$-linear category C, endowed with a grading on each of the $\mathbb{K}$-modules $\mathrm{Hom}_{\mathsf{C}}(M_0, M_1)$. The conditions are these:

(a) For every object M, the identity automorphism id_M has degree 0.

(b) For every triple of objects $M_0, M_1, M_2 \in \mathsf{C}$, the composition homomorphism

$$\mathrm{Hom}_{\mathsf{C}}(M_1, M_2) \otimes \mathrm{Hom}_{\mathsf{C}}(M_0, M_1) \to \mathrm{Hom}_{\mathsf{C}}(M_0, M_2)$$

is a strict homomorphism of graded $\mathbb{K}$-modules.

In item (b) we use the graded module structure on a tensor product from equation (3.1.3). A morphism $\phi \in \mathrm{Hom}_{\mathsf{C}}(M_0, M_1)^i$ is called a morphism of degree i.

Definition 3.1.33 Let C be a graded $\mathbb{K}$-linear category. The *strict subcategory of* C is the subcategory $\mathrm{Str}(\mathsf{C})$ on all objects of C, whose morphisms are the degree 0 morphisms of C.

Example 3.1.34 Let A be a graded ring. Define $\mathsf{GMod}\, A$ to be the category whose objects are the graded A-modules. For $M, N \in \mathsf{GMod}\, A$, the set of morphisms is the graded $\mathbb{K}$-module $\mathrm{Hom}_{\mathsf{GMod}\, A}(M, N) := \mathrm{Hom}_A(M, N)$ from Definition 3.1.31. Then $\mathsf{GMod}\, A$ is a graded $\mathbb{K}$-linear category. The morphisms in the subcategory $\mathsf{GMod}_{\mathrm{str}}\, A := \mathrm{Str}(\mathsf{GMod}\, A)$ are the strict homomorphisms of graded A-modules. We often write $\mathbf{G}(A) := \mathsf{GMod}\, A$ and $\mathbf{G}_{\mathrm{str}}(A) := \mathsf{GMod}_{\mathrm{str}}\, A$. In the special case $A = \mathbb{K}$, the category $\mathbf{G}_{\mathrm{str}}(\mathbb{K})$ already appeared in Definition 3.1.6.

Remark 3.1.35 The name "strict homomorphism of graded modules," and the corresponding notations $\mathsf{GMod}_{\mathrm{str}}\, A = \mathbf{G}_{\mathrm{str}}(A)$, are new. We introduced them to distinguish the abelian category $\mathsf{GMod}_{\mathrm{str}}\, A$ from the graded category $\mathsf{GMod}\, A$ that contains it. See Definitions 3.4.1 and 3.4.6 for the DG versions of these notions.

Definition 3.1.36 Let C and D be graded $\mathbb{K}$-linear categories. A functor $F : \mathsf{C} \to \mathsf{D}$ is called a *graded $\mathbb{K}$-linear functor* if it satisfies this condition:

▷ For every pair of objects $M_0, M_1 \in \mathsf{C}$, the function
$$F : \mathrm{Hom}_{\mathsf{C}}(M_0, M_1) \to \mathrm{Hom}_{\mathsf{D}}(F(M_0), F(M_1))$$
is a strict homomorphism of graded $\mathbb{K}$-modules.

Convention 3.1.37 To simplify the terminology, we shall often use the expressions "graded category" and "graded functor" as abbreviations for "graded $\mathbb{K}$-linear category" and "graded $\mathbb{K}$-linear functor," respectively.

Example 3.1.38 Let A be a graded ring. We can view A as a category A with a single object, and it is a graded category. If $f : A \to B$ is a homomorphism of graded rings, then passing to single-object categories we get a graded functor $F : \mathsf{A} \to \mathsf{B}$.

Recall that "morphism of functors" is synonymous with "natural transformation."

Definition 3.1.39 Let $F, G : \mathsf{C} \to \mathsf{D}$ be graded functors between graded categories, and let $i \in \mathbb{Z}$. A *degree i morphism of graded functors* $\eta : F \to G$ is a collection $\eta = \{\eta_M\}_{M \in \mathsf{C}}$ of morphisms $\eta_M \in \mathrm{Hom}_{\mathsf{D}}(F(M), G(M))^i$, such that for every morphism $\phi \in \mathrm{Hom}_{\mathsf{C}}(M_0, M_1)^j$ there is equality
$$G(\phi) \circ \eta_{M_0} = (-1)^{i \cdot j} \cdot \eta_{M_1} \circ F(\phi)$$
inside $\mathrm{Hom}_{\mathsf{D}}(F(M_0), G(M_1))^{i+j}$.

If i is odd in the definition above, then after we forget the grading, $\eta : F \to G$ is usually no longer a morphism of functors; this is another instance of the phenomenon mentioned in Remark 3.1.26.

Definition 3.1.40 Let M be an abelian category. A *graded object in* M is a collection $\{M^i\}_{i \in \mathbb{Z}}$ of objects $M^i \in \mathsf{M}$.

Because we did not assume that M has countable direct sums, the graded objects are "external" to M; cf. Exercise 3.1.44.

Suppose $M = \{M^i\}_{i \in \mathbb{Z}}$ and $N = \{N^i\}_{i \in \mathbb{Z}}$ are graded objects in M. For an integer i we define the $\mathbb{K}$-module
$$\mathrm{Hom}_{\mathsf{M}}(M, N)^i := \prod_{j \in \mathbb{Z}} \mathrm{Hom}_{\mathsf{M}}(M^j, N^{j+i}). \tag{3.1.41}$$

We get a graded $\mathbb{K}$-module
$$\mathrm{Hom}_{\mathsf{M}}(M, N) := \bigoplus_{i \in \mathbb{Z}} \mathrm{Hom}_{\mathsf{M}}(M, N)^i. \tag{3.1.42}$$

Definition 3.1.43 Let M be an abelian category. The *category of graded objects in* M is the graded linear category $\mathbf{G}(\mathsf{M})$, whose objects are the graded objects in M, and the morphism sets are the graded modules

$$\mathrm{Hom}_{\mathbf{G}(\mathsf{M})}(M, N) := \mathrm{Hom}_{\mathsf{M}}(M, N)$$

from equation 3.1.42. The composition operation is the obvious one.

Exercise 3.1.44 Suppose $\mathsf{M} = \mathbf{M}(A)$, the category of modules over a central $\mathbb{K}$-ring A. For every $M = \{M^i\}_{i\in\mathbb{Z}} \in \mathbf{G}(\mathsf{M})$ let $F(M) := \bigoplus_{i\in\mathbb{Z}} M^i$. Then $F(M)$ is a graded A-module, as discussed earlier, so $F(M)$ is an object of the category $\mathbf{G}(A)$ from Example 3.1.34. Prove that $F : \mathbf{G}(\mathsf{M}) \to \mathbf{G}(A)$ is an isomorphism of graded categories.

In the next definition we combine graded rings and linear categories, to concoct a new hybrid.

Definition 3.1.45 Let M be an abelian category, and let A be a graded ring. A *graded A-module in* M is an object $M \in \mathbf{G}(\mathsf{M})$, together with a graded ring homomorphism $f : A \to \mathrm{End}_{\mathsf{M}}(M)$. The set of graded A-modules in M is denoted by $\mathbf{G}(A, \mathsf{M})$.

What the definition says is that an element $a \in A^i$ gives rise to a degree i endomorphism $f(a)$ of the graded object $M = \{M^j\}_{j\in\mathbb{Z}}$. In turn, this means that for every j, $f(a) : M^j \to M^{j+i}$ is a morphism in M. The operation f satisfies $f(1_A) = \mathrm{id}_M$ and $f(a_1 \cdot a_2) = f(a_1) \circ f(a_2)$.

Example 3.1.46 If $A = \mathbb{K}$, then $\mathbf{G}(A, \mathsf{M}) = \mathbf{G}(\mathsf{M})$; and if $\mathsf{M} = \mathsf{Mod}\,\mathbb{K}$, then $\mathbf{G}(A, \mathsf{M}) = \mathbf{G}(A)$.

The next definition is a variant of Definition 3.1.31.

Definition 3.1.47 Let M be an abelian category, and let A be a graded ring. For $M, N \in \mathbf{G}(A, \mathsf{M})$ and $i \in \mathbb{Z}$ we define $\mathrm{Hom}_{A,\mathsf{M}}(M, N)^i$ to be the subset of $\mathrm{Hom}_{\mathsf{M}}(M, N)^i$ consisting of the morphisms $\phi : M \to N$ such that

$$\phi \circ f_M(a) = (-1)^{i\cdot k} \cdot f_N(a) \circ \phi$$

for all $a \in A^k$. Next let

$$\mathrm{Hom}_{A,\mathsf{M}}(M, N) := \bigoplus_{i\in\mathbb{Z}} \mathrm{Hom}_{A,\mathsf{M}}(M, N)^i .$$

This is a graded $\mathbb{K}$-module.

Definition 3.1.48 Let M be an abelian category, and let A be a graded ring. The *category of graded A-modules in* M is the graded category $\mathbf{G}(A, \mathsf{M})$ whose objects are the graded A-modules in M, and the morphism graded $\mathbb{K}$-modules are

$$\mathrm{Hom}_{\mathbf{G}(A,\mathsf{M})}(M_0, M_1) := \mathrm{Hom}_{A,\mathsf{M}}(M_0, M_1)$$

from Definition 3.1.47. The compositions are those of $\mathbf{G}(\mathsf{M})$.

In the special case that A is a ring (i.e. a graded ring concentrated in degree 0), this is [76, Definition 8.5.1].

Notice that forgetting the action of A is a faithful graded functor $\mathbf{G}(A, \mathsf{M}) \to \mathbf{G}(\mathsf{M})$. As in every graded category, there is the subcategory

$$\mathbf{G}_{\mathrm{str}}(A, \mathsf{M}) := \mathrm{Str}(\mathbf{G}(A, \mathsf{M})) \subseteq \mathbf{G}(A, \mathsf{M}) \tag{3.1.49}$$

of strict (i.e. degree 0) morphisms.

Exercise 3.1.50 Show that $\mathbf{G}_{\mathrm{str}}(A, \mathsf{M})$ is an abelian category.

Remark 3.1.51 The reader may have noticed that we can talk about the graded category $\mathbf{G}(\mathsf{M})$ for every linear category M, regardless of whether it is abelian or not. We choose to restrict attention to the abelian case for a pedagogical reason: this will hopefully reduce confusion between the many sorts of graded (and later DG) categories that occur in our discussion.

3.2 DG $\mathbb{K}$-Modules

Recall that Conventions 1.2.4 and 3.1.2 are in force.

Definition 3.2.1 A *DG $\mathbb{K}$-module* is a graded $\mathbb{K}$-module $M = \bigoplus_{i\in\mathbb{Z}} M^i$, together with a $\mathbb{K}$-linear homomorphism $\mathrm{d}_M : M \to M$ of degree 1, called the *differential*, satisfying $\mathrm{d}_M \circ \mathrm{d}_M = 0$.

When there is no danger of confusion, we may write d instead of d_M.

Definition 3.2.2 Let M and N be DG $\mathbb{K}$-modules. A *strict homomorphism of DG $\mathbb{K}$-modules* is a $\mathbb{K}$-linear homomorphism $\phi : M \to N$ of degree 0 that commutes with the differentials. The resulting category is denoted by $\mathsf{DGMod}_{\mathrm{str}}\, \mathbb{K}$, or by the abbreviated notation $\mathbf{C}_{\mathrm{str}}(\mathbb{K})$.

It is easy to see that $\mathbf{C}_{\mathrm{str}}(\mathbb{K})$ is a $\mathbb{K}$-linear abelian category. There is a forgetful functor $\mathbf{C}_{\mathrm{str}}(\mathbb{K}) \to \mathbf{G}_{\mathrm{str}}(\mathbb{K})$, $(M, \mathrm{d}_M) \mapsto M$.

Definition 3.2.3 Suppose M and N are DG $\mathbb{K}$-modules.

(1) The graded $\mathbb{K}$-module structure on the tensor product $M \otimes N$ was given in equation (3.1.3). We put on it the differential

$$\mathrm{d}(m \otimes n) := \mathrm{d}_M(m) \otimes n + (-1)^i \cdot m \otimes \mathrm{d}_N(n)$$

for $m \in M^i$ and $n \in N^j$. In this way $M \otimes N$ becomes a DG $\mathbb{K}$-module. We sometimes write $\mathrm{d}_{M \otimes N}$ for this differential.

(2) The graded module $\mathrm{Hom}_{\mathbb{K}}(M, N)$ was introduced in Definition 3.1.5. We give it this differential:

$$\mathrm{d}(\phi) := \mathrm{d}_N \circ \phi - (-1)^i \cdot \phi \circ \mathrm{d}_M$$

for $\phi \in \mathrm{Hom}_{\mathbb{K}}(M, N)^i$. Thus $\mathrm{Hom}_{\mathbb{K}}(M, N)$ becomes a DG $\mathbb{K}$-module. We sometimes denote this differential by $\mathrm{d}_{\mathrm{Hom}_{\mathbb{K}}(M,N)}$.

Definition 3.2.4 Let M be a DG $\mathbb{K}$-module, and let i be an integer.

(1) The module of degree i *cocycles* of M is

$$\mathrm{Z}^i(M) := \mathrm{Ker}(M^i \xrightarrow{\mathrm{d}_M} M^{i+1}).$$

(2) The module of degree i *coboundaries* of M is

$$\mathrm{B}^i(M) := \mathrm{Im}(M^{i-1} \xrightarrow{\mathrm{d}_M} M^i).$$

(3) The module of degree i *decocycles* of M is

$$\mathrm{Y}^i(M) := \mathrm{Coker}(M^{i-1} \xrightarrow{\mathrm{d}_M} M^i).$$

(4) The ith *cohomology* module of M is

$$\mathrm{H}^i(M) := \mathrm{Z}^i(M) / \mathrm{B}^i(M) \cong \mathrm{Coker}(M^{i-1} \xrightarrow{\mathrm{d}_M} \mathrm{Z}^i(M)).$$

The definition above is standard, with the exception of item (3), which is a new contribution (both the notation $\mathrm{Y}^i(M)$ and the name "decocycles"). The fact that $\mathrm{d}_M \circ \mathrm{d}_M = 0$ implies that $\mathrm{B}^i(M) \subseteq \mathrm{Z}^i(M) \subseteq M^i$, so item (4) makes sense. On the other hand, since $\mathrm{Y}^i(M) = M^i / \mathrm{B}^i(M)$, there is a canonical isomorphism

$$\mathrm{H}^i(M) \cong \mathrm{Ker}(\mathrm{Y}^i(M) \xrightarrow{\mathrm{d}_M} M^{i+1}). \tag{3.2.5}$$

The modules defined above are functorial in M; to be precise, these are $\mathbb{K}$-linear functors

$$\mathrm{Z}^i, \mathrm{B}^i, \mathrm{Y}^i, \mathrm{H}^i : \mathbf{C}_{\mathrm{str}}(\mathbb{K}) \to \mathbf{M}(\mathbb{K}). \tag{3.2.6}$$

We can rephrase Definition 3.2.2 using the notion of cocycles: for DG $\mathbb{K}$-modules M and N there is equality

$$\mathrm{Hom}_{\mathbf{C}_{\mathrm{str}}(\mathbb{K})}(M, N) = \mathrm{Z}^0(\mathrm{Hom}_{\mathbb{K}}(M, N)) \tag{3.2.7}$$

of submodules of $\mathrm{Hom}_{\mathbb{K}}(M, N)$.

3.3 DG Rings and Modules

Recall that Conventions 1.2.4, 3.1.2 and 3.1.13 are in place.

Definition 3.3.1 A *DG central* $\mathbb{K}$*-ring* is a graded central $\mathbb{K}$-ring $A = \bigoplus_{i \in \mathbb{Z}} A^i$ (see Definition 3.1.11), together with a $\mathbb{K}$-linear homomorphism $\mathrm{d}_A : A \to A$ of degree 1, called the *differential*, satisfying the equation $\mathrm{d}_A \circ \mathrm{d}_A = 0$, and the graded Leibniz rule

$$\mathrm{d}_A(a \cdot b) = \mathrm{d}_A(a) \cdot b + (-1)^i \cdot a \cdot \mathrm{d}_A(b)$$

for all $a \in A^i$ and $b \in A^j$.

We sometimes write d instead of d_A.

Definition 3.3.2 Let A and B be DG central $\mathbb{K}$-rings. A *homomorphism of DG central* $\mathbb{K}$*-rings* $f : A \to B$ is a graded central $\mathbb{K}$-ring homomorphism that commutes with the differentials of A and B. The resulting category is denoted by $\mathsf{DGRng}/_{\mathrm{c}}\, \mathbb{K}$.

Central $\mathbb{K}$-rings are viewed as DG central $\mathbb{K}$-rings concentrated in degree 0 (and with trivial differentials). Thus the category $\mathsf{Rng}/_{\mathrm{c}}\, \mathbb{K}$ is a full subcategory of $\mathsf{DGRng}/_{\mathrm{c}}\, \mathbb{K}$.

Convention 3.3.3 To simplify terminology, we usually write "DG ring" instead of "DG central $\mathbb{K}$-ring."

Definition 3.3.4 A DG ring A is called *weakly commutative*, *strongly commutative*, *nonpositive* or *commutative* if it is so, respectively, as a graded ring (after forgetting the differential), in the sense of Definition 3.1.22.

The corresponding full subcategories of $\mathsf{DGRng}/_{\mathrm{c}}\, \mathbb{K}$ are $\mathsf{DGRng}_{\mathrm{wc}}/\mathbb{K}$, $\mathsf{DGRng}_{\mathrm{sc}}/\mathbb{K}$, $\mathsf{DGRng}^{\leq 0}/_{\mathrm{c}}\, \mathbb{K}$ and $\mathsf{DGRng}^{\leq 0}_{\mathrm{sc}}/\mathbb{K}$.

When $\mathbb{K} = \mathbb{Z}$ we can write $\mathsf{DGRng} := \mathsf{DGRng}/_{\mathrm{c}}\, \mathbb{Z}$, etc.

Here are a few examples of DG rings. First, a silly example.

Example 3.3.5 Let A be a graded ring. Then A, with the zero differential, is a DG ring.

Example 3.3.6 Let X be a differentiable (i.e. of type C^∞) manifold over $\mathbb{K} := \mathbb{R}$. The de Rham complex A of X is a DG $\mathbb{R}$-ring, with the wedge product and the exterior differential. See [75, Section 2.9.7] for details. This is a strongly commutative DG ring, in the sense of Definition 3.3.4. However, it is almost never nonpositive.

The next example is the algebraic analogue of the previous one.

Example 3.3.7 Let C be a commutative $\mathbb{K}$-ring. Then the algebraic de Rham complex $A := \Omega_{C/\mathbb{K}} = \bigoplus_{p \geq 0} \Omega^p_{C/\mathbb{K}}$ is a DG $\mathbb{K}$-ring. It is a strongly commutative DG ring, and it is almost never nonpositive. See [46, Exercise 16.15] or [105, Section 25] for details.

Example 3.3.8 Let M be a DG $\mathbb{K}$-module. Consider the DG $\mathbb{K}$-module $\mathrm{End}_{\mathbb{K}}(M) := \mathrm{Hom}_{\mathbb{K}}(M, M)$ from Definition 3.2.3(2). Composition of endomorphisms is an associative multiplication on $\mathrm{End}_{\mathbb{K}}(M)$ that respects the grading, and the graded Leibniz rule holds. We see that $\mathrm{End}_{\mathbb{K}}(M)$ is a DG $\mathbb{K}$-ring. It is usually neither weakly commutative nor nonpositive.

Example 3.3.9 Let C be a commutative ring and let $c \in C$ be an element. The *Koszul complex* of c is the DG C-module $\mathrm{K}(C; c)$ defined as follows. In degree 0 we let $\mathrm{K}^0(C; c) := C$. In degree -1, $\mathrm{K}^{-1}(C; c)$ is a free C-module of rank 1, with basis element x. All other homogeneous components are trivial. The differential d is determined by what it does to the basis element $x \in \mathrm{K}^{-1}(C; c)$, and we let $\mathrm{d}(x) := c \in \mathrm{K}^0(C; c)$.

We want to make $\mathrm{K}(C; c)$ into a strongly commutative DG ring (in the sense of Definition 3.3.4). Since x is an odd element, we must define the graded ring structure by $\mathrm{K}(C; c) := C \otimes \mathbb{K}[x]$, where $\mathbb{K}[x]$ is the strongly commutative graded polynomial ring from Example 3.1.23, on the graded set $X = X^{-1} := \{x\}$; i.e. $x^2 = 0$.

Example 3.3.10 Let A and B be DG rings. The graded ring $A \otimes B$ from Definition 3.1.21, with the differential from Definition 3.2.3(1), is a DG ring.

Example 3.3.11 Let C be a commutative ring and let $\boldsymbol{c} = (c_1, \ldots, c_n)$ be a sequence of elements in C. By combining Examples 3.3.9 and 3.3.10 we obtain the Koszul complex

$$\mathrm{K}(C; \boldsymbol{c}) := \mathrm{K}(C; c_1) \otimes_C \cdots \otimes_C \mathrm{K}(C; c_n).$$

This is a strongly commutative DG C-ring.

Remark 3.3.12 In the book [46, Section 17, Exercises] there are a few variants of the Koszul complex. Note that in most of the classical literature (including [46] and [105]), the multiplicative structure of $\mathrm{K}(C; \boldsymbol{c})$ has been ignored.

Definition 3.3.13 Let A be a DG ring. The *opposite DG ring* A^{op} is the same DG $\mathbb{K}$-module as A, but the multiplication $\cdot^{\mathrm{op}}$ of A^{op} is the multiplication $\cdot$ of A, reversed and twisted by signs: $a \cdot^{\mathrm{op}} b := (-1)^{i \cdot j} \cdot b \cdot a$ for $a \in A^i$ and $b \in A^j$.

Exercise 3.3.14 Verify that A^{op} is a DG ring.

Note that A is weakly commutative iff $A = A^{\mathrm{op}}$.

Remark 3.3.15 In algebraic topology and homotopy theory it is customary to use lower indices for DG rings (and to call them DG algebras). In other words, they use *homological grading*, as opposed to our cohomological grading. Thus our nonpositive DG ring $A = \bigoplus_{i \leq 0} A^i$ becomes a nonnegative graded algebra $A = \bigoplus_{i \geq 0} A_i$ in their language.

Another notion that is common in homotopy theory is that of a *connective DG algebra*; this is a DG ring $A = \bigoplus_{i \in \mathbb{Z}} A_i$ whose homology $\mathrm{H}(A) = \bigoplus_{i \in \mathbb{Z}} \mathrm{H}_i(A)$ is nonnegative. In our cohomological language, the analogue would be a *coconnective DG ring*, that is a DG ring A whose cohomology $\mathrm{H}(A) = \bigoplus_{i \in \mathbb{Z}} \mathrm{H}^i(A)$ is a nonpositive graded ring. Cf. Exercise 3.3.20 below.

If A is a coconnective DG ring, then its smart truncation $A' := \mathrm{smt}^{\leq 0}(A)$, in the sense of Definition 7.3.6, is a nonpositive DG ring, and the inclusion $A' \to A$ is a quasi-isomorphism.

Definition 3.3.16 Let A be a DG ring. A *DG left A-module* is a graded left A-module $M = \bigoplus_{i \in \mathbb{Z}} M^i$, together with a $\mathbb{K}$-linear homomorphism $\mathrm{d}_M : M \to M$ of degree 1, called the differential, satisfying $\mathrm{d}_M \circ \mathrm{d}_M = 0$ and

$$\mathrm{d}_M(a \cdot m) = \mathrm{d}_A(a) \cdot m + (-1)^i \cdot a \cdot \mathrm{d}_M(m)$$

for $a \in A^i$ and $m \in M^j$.

DG right A-modules are defined likewise, but we won't deal with them much. This is because DG right A-modules are DG left modules over the opposite DG ring A^{op}. More precisely, if M is a DG right A-module, then the formula $a \cdot m := (-1)^{i \cdot j} \cdot m \cdot a$, for $a \in A^i$ and $m \in M^j$, makes M into a DG left A^{op}-module.

As implied by Convention 1.2.4(5), all DG modules are by default DG left modules.

Proposition 3.3.17 *Let A be a DG $\mathbb{K}$-ring, and let M be a DG $\mathbb{K}$-module.*

(1) *Suppose $f : A \to \mathrm{End}_{\mathbb{K}}(M)$ is a DG $\mathbb{K}$-ring homomorphism. Then the formula $a \cdot m := f(a)(m)$, for $a \in A^i$ and $m \in M^j$, makes M into a DG A-module.*
(2) *Conversely, every DG A-module structure on M that's compatible with the DG $\mathbb{K}$-module structure arises in this way from a DG $\mathbb{K}$-ring homomorphism $f : A \to \mathrm{End}_{\mathbb{K}}(M)$.*

Exercise 3.3.18 Prove this proposition.

Definition 3.3.19 Let M and N be DG A-modules. A *strict homomorphism of DG A-modules* $\phi : M \to N$ is a strict homomorphism of DG $\mathbb{K}$-modules (Definition 3.2.2) that respects the action of A. The resulting category is denoted by $\mathsf{DGMod}_{\mathrm{str}} A$, or by the short notation $\mathbf{C}_{\mathrm{str}}(A)$.

The category $\mathbf{C}_{\mathrm{str}}(A)$ is abelian. See Proposition 3.8.6 for a more general statement.

Exercise 3.3.20 Let A be a DG ring. Show that the module of cocycles $\mathrm{Z}(A) := \bigoplus_{i \in \mathbb{Z}} \mathrm{Z}^i(A)$ is a graded subring of A, and the module of coboundaries $\mathrm{B}(A) := \bigoplus_{i \in \mathbb{Z}} \mathrm{B}^i(A)$ is a two-sided graded ideal of $\mathrm{Z}(A)$. Conclude that the cohomology module $\mathrm{H}(A) := \bigoplus_{i \in \mathbb{Z}} \mathrm{H}^i(A)$ is a graded ring.

Let $f : A \to B$ be a homomorphism of DG rings. Show that $\mathrm{H}(f) : \mathrm{H}(A) \to \mathrm{H}(B)$ is a graded ring homomorphism.

Exercise 3.3.21 Let A be a DG ring. Given a DG A-module M, show that its cohomology $\mathrm{H}(M)$ is a graded $\mathrm{H}(A)$-module. If $\phi : M \to N$ is a homomorphism in $\mathbf{C}_{\mathrm{str}}(A)$, then $\mathrm{H}(\phi) : \mathrm{H}(M) \to \mathrm{H}(N)$ is a homomorphism in $\mathbf{G}_{\mathrm{str}}(\mathrm{H}(A))$. Conclude that

$$\mathrm{H} : \mathbf{C}_{\mathrm{str}}(A) \to \mathbf{G}_{\mathrm{str}}(\mathrm{H}(A)) \tag{3.3.22}$$

is a linear functor.

Definition 3.3.23 Let A be a DG $\mathbb{K}$-ring, let M be a right DG A-module, and let N be a left DG A-module. By Lemma 3.1.30, $M \otimes_A N$ is a graded $\mathbb{K}$-module. We make it into a DG $\mathbb{K}$-module with the differential from Definition 3.2.3(1).

Definition 3.3.24 Let A be a DG $\mathbb{K}$-ring, and let M and N be left DG A-modules. The graded $\mathbb{K}$-module $\mathrm{Hom}_A(M, N)$ from Definition 3.1.31 is made into a DG $\mathbb{K}$-module with the differential from Definition 3.2.3(2).

Generalizing formula (3.2.7), for DG A-modules M and N there is equality

$$\mathrm{Hom}_{\mathbf{C}_{\mathrm{str}}(A)}(M, N) = \mathrm{Z}^0(\mathrm{Hom}_A(M, N)).$$

Proposition 3.3.25 *Let A be a DG ring.*

(1) *The category $\mathbf{C}_{\mathrm{str}}(A)$ has products. Given a collection $\{M_x\}_{x \in X}$ of DG A-modules, their product $M = \prod_{x \in X} M_x$ in $\mathbf{C}_{\mathrm{str}}(A)$ is $M := \bigoplus_{i \in \mathbb{Z}} M^i$, where $M^i := \prod_{x \in X} M^i_x$.*

(2) *The category $\mathbf{C}_{\mathrm{str}}(A)$ has direct sums. Given a collection $\{M_x\}_{x \in X}$ of DG A-modules, their direct sum $M = \bigoplus_{x \in X} M_x$ in $\mathbf{C}_{\mathrm{str}}(A)$ is $M := \bigoplus_{i \in \mathbb{Z}} M^i$, where $M^i := \bigoplus_{x \in X} M^i_x$.*

(3) *The functor* H *from* (3.3.22) *commutes with all products and direct sums.*

Exercise 3.3.26 Prove this proposition.

3.4 DG Categories

In Definition 3.1.32 we saw graded categories. Here is the DG version.

Definition 3.4.1 A *DG* $\mathbb{K}$*-linear category* is a $\mathbb{K}$-linear category C, endowed with a DG $\mathbb{K}$-module structure on each of the morphism $\mathbb{K}$-modules $\operatorname{Hom}_{\mathsf{C}}(M_0, M_1)$. The conditions are these:

(a) For every object M, the identity automorphism id_M is a degree 0 cocycle in $\operatorname{Hom}_{\mathsf{C}}(M, M)$.

(b) For every triple of objects $M_0, M_1, M_2 \in \mathsf{C}$, the composition homomorphism

$$\operatorname{Hom}_{\mathsf{C}}(M_1, M_2) \otimes \operatorname{Hom}_{\mathsf{C}}(M_0, M_1) \to \operatorname{Hom}_{\mathsf{C}}(M_0, M_2)$$

is a strict homomorphism of DG $\mathbb{K}$-modules.

The differential in a DG $\mathbb{K}$-linear category C is sometimes denoted by d_{C}; for instance

$$\mathrm{d}_{\mathsf{C}} : \operatorname{Hom}_{\mathsf{C}}(M_0, M_1)^i \to \operatorname{Hom}_{\mathsf{C}}(M_0, M_1)^{i+1}. \tag{3.4.2}$$

The next convention extends Convention 3.3.3.

Convention 3.4.3 We shall often write "DG category" instead of the longer expression "DG $\mathbb{K}$-linear category."

If C' is a full subcategory of a DG category C, then of course C' is itself a DG category.

Definition 3.4.4 Let C be a DG category.

(1) A morphism $\phi \in \operatorname{Hom}_{\mathsf{C}}(M, N)^i$ is called a *degree i morphism*.

(2) A morphism $\phi \in \operatorname{Hom}_{\mathsf{C}}(M, N)$ is called a *cocycle* if $\mathrm{d}_{\mathsf{C}}(\phi) = 0$.

(3) A morphism $\phi : M \to N$ in C is called a *strict morphism* if it is a degree 0 cocycle.

Lemma 3.4.5 *Let C be a DG category, and for $i = 0, 1, 2$ let $\phi_i : M_i \to M_{i+1}$ be a morphism in C of degree k_i.*

(1) *The morphism $\phi_1 \circ \phi_0$ has degree $k_0 + k_1$, and*

$$\mathrm{d}_{\mathsf{C}}(\phi_1 \circ \phi_0) = \mathrm{d}_{\mathsf{C}}(\phi_1) \circ \phi_0 + (-1)^{k_1} \cdot \phi_1 \circ \mathrm{d}_{\mathsf{C}}(\phi_0).$$

(2) *If ϕ_0 and ϕ_1 are cocycles, then so is $\phi_1 \circ \phi_0$.*

(3) *If ϕ_1 is a coboundary, and ϕ_0 and ϕ_2 are cocycles, then $\phi_2 \circ \phi_1 \circ \phi_0$ is a coboundary.*

Proof (1) This is just a rephrasing of item (b) in Definition 3.4.1.

(2) This is immediate from (1).

(3) Say $\phi_1 = \mathrm{d}_{\mathsf{C}}(\psi_1)$ for some degree $k_1 - 1$ morphism $\psi_1 : M_1 \to M_2$. Then

$$\phi_2 \circ \phi_1 \circ \phi_0 = \mathrm{d}_{\mathsf{C}}((-1)^{k_2} \cdot \phi_2 \circ \psi_1 \circ \phi_0). \qquad \square$$

The previous lemma makes the next definition possible.

Definition 3.4.6 Let C be a DG category.

(1) The *strict subcategory* of C is the category Str(C), with the same objects as C, but with strict morphisms only. Thus

$$\mathrm{Hom}_{\mathrm{Str}(\mathsf{C})}(M, N) = \mathrm{Z}^0(\mathrm{Hom}_{\mathsf{C}}(M, N)).$$

(2) The *homotopy category* of C is the category Ho(C), with the same objects as C, and with morphism sets

$$\mathrm{Hom}_{\mathrm{Ho}(\mathsf{C})}(M, N) := \mathrm{H}^0(\mathrm{Hom}_{\mathsf{C}}(M, N)).$$

(3) We denote by $\mathrm{P} : \mathrm{Str}(\mathsf{C}) \to \mathrm{Ho}(\mathsf{C})$ the functor which is the identity on objects, and sends a strict morphism to its homotopy class.

The categories Str(C) and Ho(C) are linear, and the inclusion functor $\mathrm{Str}(\mathsf{C}) \to \mathsf{C}$ and the projection functor $\mathrm{P} : \mathrm{Str}(\mathsf{C}) \to \mathrm{Ho}(\mathsf{C})$ are linear. The first is faithful, and the second is full.

Example 3.4.7 If A is a DG category, then for every object $x \in \mathsf{A}$, its set of endomorphisms $A := \mathrm{End}_{\mathsf{A}}(x)$ is a DG ring. Conversely, every DG ring A can be viewed as a DG category A with a single object.

Example 3.4.8 Let A be a DG ring. The set of DG A-modules forms a DG category $\mathsf{DGMod}\, A$, in which the morphism DG $\mathbb{K}$-modules are

$$\mathrm{Hom}_{\mathsf{DGMod}\, A}(M, N) := \mathrm{Hom}_A(M, N)$$

from Definition 3.3.24. We shall often write $\mathbf{C}(A) := \mathsf{DGMod}\, A$. The strict subcategory here is

$$\mathrm{Str}(\mathsf{DGMod}\, A) = \mathsf{DGMod}_{\mathrm{str}}\, A = \mathbf{C}_{\mathrm{str}}(A);$$

cf. Definition 3.3.19.

Here is a result that will be used later.

Proposition 3.4.9 *Let $\phi : M \to N$ be a degree i isomorphism in a DG category* C. *Assume ϕ is a cocycle, namely $\mathrm{d}_{\mathsf{C}}(\phi) = 0$. Then its inverse $\phi^{-1} : N \to M$ is also a cocycle.*

Proof According the Leibniz rule (Lemma 3.4.5(1)) and the fact that id_M is a cocycle, we have

$$\begin{aligned} 0 &= \mathrm{d}_{\mathsf{C}}(\mathrm{id}_M) = \mathrm{d}(\phi^{-1} \circ \phi) \\ &= \mathrm{d}_{\mathsf{C}}(\phi^{-1}) \circ \phi + (-1)^{-i} \cdot \phi^{-1} \circ \mathrm{d}_{\mathsf{C}}(\phi) = \mathrm{d}_{\mathsf{C}}(\phi^{-1}) \circ \phi. \end{aligned}$$

Because ϕ is an isomorphism, we conclude that $\mathrm{d}_{\mathsf{C}}(\phi^{-1}) = 0$. □

Remark 3.4.10 The fact that the concept "DG category" refers both to DG rings (Example 3.4.7) and to the categories of DG modules over them (Example 3.4.8) is a source of frequent confusion. See Remarks 3.1.51 and 3.8.21.

Remark 3.4.11 For other accounts of DG categories see the relatively old references [85], [79], [25], or the more recent [151]. An internet search can give plenty more information, including the relation to simplicial and infinity categories.

In this book we shall be exclusively concerned with the DG categories $\mathbf{C}(A, \mathsf{M})$, to be introduced in Section 3.8, which have a lot more structure than other DG categories.

See Remark 3.8.21 regarding the DG category $\mathbf{C}(\mathsf{A})$ of DG modules over a DG category A, in the sense of [79]; and Remark 3.8.22 regarding A_∞ categories.

3.5 DG Functors

Here C and D are DG $\mathbb{K}$-linear categories (see Definition 3.4.1). When we forget differentials, C and D become graded $\mathbb{K}$-linear categories. So we can talk about graded $\mathbb{K}$-linear functors $\mathsf{C} \to \mathsf{D}$, as in Definition 3.1.36.

Recall the meaning of a strict homomorphism of DG $\mathbb{K}$-modules: it has degree 0 and commutes with the differentials.

Definition 3.5.1 Let C and D be DG $\mathbb{K}$-linear categories. A functor $F : \mathsf{C} \to \mathsf{D}$ is called a *DG $\mathbb{K}$-linear functor* if it satisfies this condition:

▷ For every pair of objects $M_0, M_1 \in \mathsf{C}$, the function

$$F : \mathrm{Hom}_{\mathsf{C}}(M_0, M_1) \to \mathrm{Hom}_{\mathsf{D}}(F(M_0), F(M_1))$$

is a strict homomorphism of DG $\mathbb{K}$-modules.

In other words, F is a DG functor if it is a graded functor, and $\mathrm{d}_{\mathsf{D}} \circ F = F \circ \mathrm{d}_{\mathsf{C}}$ as degree 1 homomorphisms

$$\mathrm{Hom}_{\mathsf{C}}(M_0, M_1) \to \mathrm{Hom}_{\mathsf{D}}(F(M_0), F(M_1)).$$

To match Convention 3.4.3, we now give:

Convention 3.5.2 We shall usually write “DG functor” instead of the longer expression “DG $\mathbb{K}$-linear functor.”

Example 3.5.3 Let $f : A \to B$ be a homomorphism of DG rings, and let A and B be the corresponding single-object DG categories. Then f gives rise to a DG functor $F : \mathsf{A} \to \mathsf{B}$.

Other examples of DG functors, more relevant to our study, will be given in Section 4.6.

Definition 3.5.4 Let $F, G : \mathsf{C} \to \mathsf{D}$ be DG functors.

(1) A *degree i morphism of DG functors* $\eta : F \to G$ is a degree i morphism of graded functors, as in Definition 3.1.39.
(2) Let $\eta : F \to G$ be a degree i morphism of DG functors. For each object $M \in \mathsf{C}$ there is a degree $i + 1$ morphism $\mathrm{d}_{\mathsf{D}}(\eta_M) : F(M) \to G(M)$ in D. We define the collection of morphisms $\mathrm{d}_{\mathsf{D}}(\eta) := \{\mathrm{d}_{\mathsf{D}}(\eta_M)\}_{M \in \mathsf{C}}$.
(3) A *strict morphism of DG functors* is a degree 0 morphism of graded functors $\eta : F \to G$ such that $\mathrm{d}_{\mathsf{D}}(\eta) = 0$.

Explanation of item (2) of the definition: for an object $M \in \mathsf{C}$, we have

$$\eta_M \in \mathrm{Hom}_{\mathsf{D}}(F(M), G(M))^i.$$

Now $\mathrm{Hom}_{\mathsf{D}}(F(M), G(M))$ is a DG $\mathbb{K}$-module with differential d_{D}; see formula (3.4.2). Thus

$$\mathrm{d}_{\mathsf{D}}(\eta_M) \in \mathrm{Hom}_{\mathsf{D}}(F(M), G(M))^{i+1}.$$

Proposition 3.5.5 *In the situation of Definition* 3.5.4, *the collection of morphisms* $\mathrm{d}_{\mathsf{D}}(\eta)$ *is a degree* $i + 1$ *morphism of DG functors* $F \to G$.

Exercise 3.5.6 Prove this proposition.

The categories $\mathrm{Str}(\mathsf{C})$ and $\mathrm{Ho}(\mathsf{C})$ were introduced in Definition 3.4.6.

Proposition 3.5.7 *Let* $F : \mathsf{C} \to \mathsf{D}$ *be a DG functor. Then* F *induces linear functors* $\mathrm{Str}(F) : \mathrm{Str}(\mathsf{C}) \to \mathrm{Str}(\mathsf{D})$ *and* $\mathrm{Ho}(F) : \mathrm{Ho}(\mathsf{C}) \to \mathrm{Ho}(\mathsf{D})$.

Proof Because F is a DG functor, it sends 0-cocycles in $\mathrm{Hom}_{\mathsf{C}}(M_0, M_1)$ to 0-cocycles in $\mathrm{Hom}_{\mathsf{D}}(F(M_0), F(M_1))$. The same for 0-coboundaries. □

By abuse of notation, and when there is no danger of confusion, we will sometimes write F instead of $\mathrm{Str}(F)$ or $\mathrm{Ho}(F)$.

Exercise 3.5.8 Let A and C be DG categories, and assume A is small. Define $\mathsf{DGFun}(\mathsf{A}, \mathsf{C})$ to be the set of DG functors $F : \mathsf{A} \to \mathsf{C}$. Show that $\mathsf{DGFun}(\mathsf{A}, \mathsf{C})$ is a DG category, where the morphisms are from Definition 3.5.4(1), and their differentials are from Definition 3.5.4(2).

3.6 Complexes in Abelian Categories

Here we recall facts about complexes from the classical homological theory, and place them within our context. In this section M is a $\mathbb{K}$-linear abelian category.

A *complex of objects of* M, or a *complex in* M, is a diagram

$$(\cdots \to M^{-1} \xrightarrow{\mathrm{d}_M^{-1}} M^0 \xrightarrow{\mathrm{d}_M^{0}} M^1 \xrightarrow{\mathrm{d}_M^{1}} M^2 \to \cdots) \tag{3.6.1}$$

of objects and morphisms in M, such that $\mathrm{d}_M^{i+1} \circ \mathrm{d}_M^i = 0$. The collection of objects $M := \{M^i\}_{i\in\mathbb{Z}}$ is nothing but a graded object of M, as defined in Section 3.1. The collection of morphisms $\mathrm{d}_M := \{\mathrm{d}_M^i\}_{i\in\mathbb{Z}}$ is called a *differential*, or a *coboundary operator*. Thus a complex is a pair (M, d_M) made up of a graded object M and a differential d_M on it. We sometimes write d instead of d_M or d_M^i. At other times we leave the differential implicit, and just refer to the complex as M.

Let N be another complex in M. A *strict morphism of complexes* $\phi : M \to N$ is a collection $\phi = \{\phi^i\}_{i\in\mathbb{Z}}$ of morphisms $\phi^i : M^i \to N^i$ in M, such that

$$\mathrm{d}_N^i \circ \phi^i = \phi^{i+1} \circ \mathrm{d}_M^i. \tag{3.6.2}$$

Note that a strict morphism $\phi : M \to N$ can be viewed as a commutative diagram

$$\begin{array}{ccccccc}
\cdots & \longrightarrow & M^i & \xrightarrow{\mathrm{d}_M^i} & M^{i+1} & \longrightarrow & \cdots \\
 & & \Big\downarrow{\scriptstyle \phi^i} & & \Big\downarrow{\scriptstyle \phi^{i+1}} & & \\
\cdots & \longrightarrow & N^i & \xrightarrow{\mathrm{d}_N^i} & N^{i+1} & \longrightarrow & \cdots
\end{array}$$

in M. The identity automorphism id_M of the complex M is a strict morphism.

Remark 3.6.3 In most textbooks, what we call a "strict morphism of complexes" is simply called a "morphism of complexes." See Remark 3.1.35 for an explanation.

Let us denote by $\mathbf{C}_{\mathrm{str}}(\mathsf{M})$ the category of complexes in M, with strict morphisms. This is a $\mathbb{K}$-linear abelian category. Indeed, the direct sum of complexes is the degree-wise direct sum, i.e. $(M \oplus N)^i = M^i \oplus N^i$. The same for kernels and cokernels. If N is a full abelian subcategory of M, then $\mathbf{C}_{\mathrm{str}}(\mathsf{N})$ is a full abelian subcategory of $\mathbf{C}_{\mathrm{str}}(\mathsf{M})$.

A single object $M^0 \in \mathsf{M}$ can be viewed as a complex

$$M := (\cdots \to 0 \to M^0 \to 0 \to \cdots),$$

where M^0 is in degree 0; the differential of this complex is of course zero. The assignment $M^0 \mapsto M$ is a fully faithful $\mathbb{K}$-linear functor $\mathsf{M} \to \mathsf{C}_{\mathrm{str}}(\mathsf{M})$.

Let M and N be complexes in M. As in (3.1.42) there is a graded $\mathbb{K}$-module $\mathrm{Hom}_{\mathsf{M}}(M, N)$. It is a DG $\mathbb{K}$-module with differential d given by the formula

$$\mathrm{d}(\phi) := \mathrm{d}_N \circ \phi - (-1)^i \cdot \phi \circ \mathrm{d}_M \tag{3.6.4}$$

for $\phi \in \mathrm{Hom}_{\mathsf{M}}(M, N)^i$. It is easy to check that $\mathrm{d} \circ \mathrm{d} = 0$. We sometimes denote this differential by $\mathrm{d}_{\mathrm{Hom}}$ or $\mathrm{d}_{\mathrm{Hom}_{\mathsf{M}}(M,N)}$.

Thus, an element $\phi \in \mathrm{Hom}_{\mathsf{M}}(M, N)^i$ is a collection $\phi = \{\phi^j\}_{j \in \mathbb{Z}}$ of morphisms $\phi^j : M^j \to N^{j+i}$. In a diagram, for $i = 2$, it looks like this:

$$\cdots \longrightarrow M^j \xrightarrow{\ \mathrm{d}\ } M^{j+1} \xrightarrow{\ \mathrm{d}\ } M^{j+2} \xrightarrow{\ \mathrm{d}\ } M^{j+3} \longrightarrow \cdots$$

$$\cdots \longrightarrow N^j \xrightarrow{\ \mathrm{d}\ } N^{j+1} \xrightarrow{\ \mathrm{d}\ } N^{j+2} \xrightarrow{\ \mathrm{d}\ } N^{j+3} \longrightarrow \cdots$$

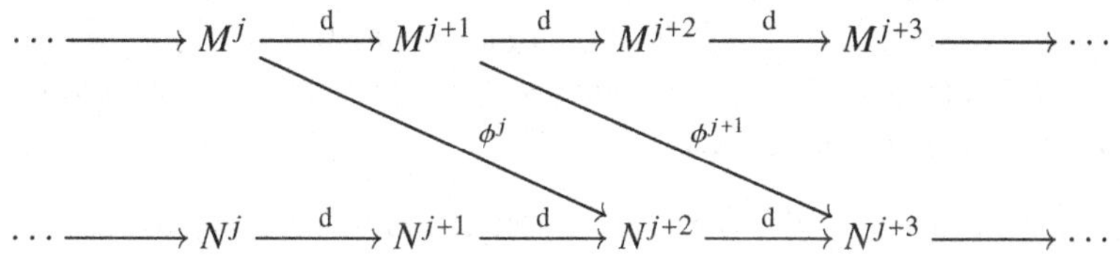

Warning: since ϕ does not have to commute with the differentials, this is usually not a commutative diagram!

For a triple of complexes M_0, M_1, M_2 and degrees i_0, i_1 there are $\mathbb{K}$-linear homomorphisms

$$\mathrm{Hom}_{\mathsf{M}}(M_1, M_2)^{i_1} \otimes \mathrm{Hom}_{\mathsf{M}}(M_0, M_1)^{i_0} \to \mathrm{Hom}_{\mathsf{M}}(M_0, M_2)^{i_0 + i_1},$$

$$\phi_1 \otimes \phi_0 \mapsto \phi_1 \circ \phi_0.$$

Lemma 3.6.5 *The composition homomorphism*

$$\mathrm{Hom}_{\mathsf{M}}(M_1, M_2) \otimes \mathrm{Hom}_{\mathsf{M}}(M_0, M_1) \to \mathrm{Hom}_{\mathsf{M}}(M_0, M_2)$$

is a strict homomorphism of DG $\mathbb{K}$-modules.

Exercise 3.6.6 Prove the lemma.

The lemma justifies the next definition.

Definition 3.6.7 Let $\mathsf{C}(\mathsf{M})$ be the DG $\mathbb{K}$-linear category whose objects are the complexes in M, and whose morphism DG $\mathbb{K}$-modules are $\mathrm{Hom}_{\mathsf{M}}(M, N)$, from formulas (3.1.42) and (3.6.4).

It is clear, from comparing formulas (3.6.4) and (3.6.2), that the strict morphisms of complexes defined at the top of this section are the same as those from Definition 3.4.6(1). In other words, $\mathrm{Str}(\mathsf{C}(\mathsf{M})) = \mathsf{C}_{\mathrm{str}}(\mathsf{M})$.

Remark 3.6.8 A possible ambiguity could arise in the meaning of $\mathrm{Hom}_{\mathsf{M}}(M, N)$ if $M, N \in \mathsf{M}$. Does it mean the $\mathbb{K}$-module of morphisms in the category M ? Or, if we view M and N as complexes by the canonical embedding $\mathsf{M} \subseteq \mathbf{C}(\mathsf{M})$, does $\operatorname{Hom}_{\mathsf{M}}(M, N)$ mean the complex of morphisms in the DG category $\mathbf{C}(\mathsf{M})$? It turns out that there is no actual difficulty: since the complex of $\mathbb{K}$-modules $\operatorname{Hom}_{\mathsf{M}}(M, N)$ is concentrated in degree 0, we may view it as a single $\mathbb{K}$-module, and this is precisely the $\mathbb{K}$-module of morphisms in the category M.

When $\mathsf{M} = \mathsf{Mod}\, A$ for a ring A, there is no essential distinction between complexes and DG modules. The next proposition is the DG version of Exercise 3.1.44.

Proposition 3.6.9 *Let A be a ring. Given a complex $M \in \mathbf{C}(\mathsf{Mod}\, A)$, with notation as in* (3.6.1), *define the DG A-module $F(M) := \bigoplus_{i \in \mathbb{Z}} M^i$, with differential* $\mathrm{d} := \sum_{i \in \mathbb{Z}} \mathrm{d}_M^i$. *Then the functor $F : \mathbf{C}(\mathsf{Mod}\, A) \to \mathsf{DGMod}\, A$ is an isomorphism of DG categories.*

Exercise 3.6.10 Prove this proposition. (Hint: choose good notation.)

3.7 The Long Exact Cohomology Sequence

As in the previous section, M is a $\mathbb{K}$-linear abelian category. Here we give a detailed proof of the existence and functoriality of the long exact cohomology sequence, using our *sheaf tricks* from Section 2.4. This allows us to avoid the Freyd–Mitchell Theorem.

Consider a short exact sequence

$$\boldsymbol{E} = (0 \to L \xrightarrow{\phi} M \xrightarrow{\psi} N \to 0) \tag{3.7.1}$$

in the abelian category $\mathbf{C}_{\mathrm{str}}(\mathsf{M})$. This means that ϕ and ψ are morphisms in $\mathbf{C}_{\mathrm{str}}(\mathsf{M})$, and in each degree i there is a short exact sequence

$$0 \to L^i \xrightarrow{\phi^i} M^i \xrightarrow{\psi^i} N^i \to 0$$

in M. The next definitions and lemmas refer to the short exact sequence $\boldsymbol{E}$. In them we shall make use of the notation from Section 2.4.

Let us denote by $\pi_N : \mathrm{Z}^i(N) \twoheadrightarrow \mathrm{H}^i(N)$ the canonical epimorphism in M; and likewise for L and M.

Definition 3.7.2 Let $\bar{n} \in \Gamma(U, \mathrm{H}^i(N))$ for some $U \in \mathsf{M}$, and let $V \twoheadrightarrow U$ be a covering of U. By a *connecting triple* for $\bar{n}$ over V we mean a triple (n, m, l) of sections $n \in \Gamma(V, \mathrm{Z}^i(N))$, $m \in \Gamma(V, M^i)$ and $l \in \Gamma(V, \mathrm{Z}^{i+1}(L))$, such that $\pi_N(n) = \bar{n} \in \Gamma(V, \mathrm{H}^i(N))$, $\psi(m) = n \in \Gamma(V, N^i)$ and $\phi(l) = \mathrm{d}(m) \in \Gamma(V, M^{i+1})$.

Here is the relevant diagram in M :

$$\begin{array}{ccccc}
\mathrm{H}^{i+1}(L) & \overset{\pi_L}{\twoheadleftarrow} & \mathrm{Z}^{i+1}(L) & \overset{\phi}{\longrightarrow} & M^{i+1} & & & & \\
 & & & & \uparrow \mathrm{d} & & & & \\
 & & & & M^{i} & \overset{\psi}{\longrightarrow} & \mathrm{Z}^{i}(N) & \overset{\pi_N}{\twoheadrightarrow} & \mathrm{H}^{i}(N)
\end{array}$$

It is possible that some coverings do not admit a connecting triple. But some do:

Lemma 3.7.3 *Let $\bar{n} \in \Gamma(U, \mathrm{H}^i(N))$. There exists a covering $\rho : V \twoheadrightarrow U$ that admits a connecting triple (n, m, l) for $\bar{n}$.*

Proof Because $\pi_N : \mathrm{Z}^i(N) \to \mathrm{H}^i(N)$ is an epimorphism, by the first sheaf trick, there is a covering $V' \twoheadrightarrow U$ with a section $n \in \Gamma(V', \mathrm{Z}^i(N))$ such that $\pi_N(n) = \bar{n}$.

Because $\psi : M^i \to N^i$ is an epimorphism, there is a covering $V'' \twoheadrightarrow V'$ with a section $m \in \Gamma(V'', M^i)$ such that $\psi(m) = n$.

Consider the section $\mathrm{d}(m) \in \Gamma(V'', M^{i+1})$. We have $\psi(\mathrm{d}(m)) = \mathrm{d}(\psi(m)) = \mathrm{d}(n) = 0$. By exactness at M^{i+1}, there is a covering $V \twoheadrightarrow V''$ with a section $l \in \Gamma(V, L^{i+1})$ such that $\phi(l) = \mathrm{d}(m)$ in $\Gamma(V, M^{i+1})$.

Now $\phi(\mathrm{d}(l)) = \mathrm{d}(\phi(l)) = \mathrm{d}(\mathrm{d}(m)) = 0$. Because ϕ is a monomorphism, it follows that $\mathrm{d}(l) = 0$. Thus $l \in \Gamma(V, \mathrm{Z}^{i+1}(L))$. □

Lemma 3.7.4 *Let $\bar{n} \in \Gamma(U, \mathrm{H}^i(N))$. Suppose $V \twoheadrightarrow U$ is a covering that admits connecting triples (n, m, l) and (n', m', l') for $\bar{n}$. Then $\pi_L(l') = \pi_L(l)$ in $\Gamma(V, \mathrm{H}^{i+1}(L))$.*

Proof We know that $\pi_N(n) = \bar{n} = \pi_N(n')$ in $\Gamma(V, \mathrm{H}^i(N))$. Thus $n - n' \in \mathrm{Ker}(\pi_N) \subseteq \Gamma(V, \mathrm{Z}^i(N))$. Looking at the exact sequence

$$N^{i-1} \xrightarrow{\mathrm{d}} \mathrm{Z}^i(N) \xrightarrow{\pi_N} \mathrm{H}^i(N) \to 0$$

and invoking the first sheaf trick, we see that there's a covering $W \twoheadrightarrow V$ with a section $\tilde{n} \in \Gamma(W, N^{i-1})$ such that $\mathrm{d}(\tilde{n}) = n - n'$.

Again invoking the first sheaf trick, there's a covering $W' \twoheadrightarrow W$ with a section $\tilde{m} \in \Gamma(W', M^{i-1})$ such that $\psi(\tilde{m}) = \tilde{n}$.

Consider the section $(m - m') - \mathrm{d}(\tilde{m}) \in \Gamma(W', M^i)$. It satisfies

$$\psi((m - m') - \mathrm{d}(\tilde{m})) = (n - n') - \mathrm{d}(\tilde{n}) = 0.$$

By the first sheaf trick, there's a covering $W'' \twoheadrightarrow W'$ with a section $\tilde{l} \in \Gamma(W'', L^i)$ such that $\phi(\tilde{l}) = (m - m') - \mathrm{d}(\tilde{m})$. Now

$$\phi(\mathrm{d}(\tilde{l}) - (l - l')) = \mathrm{d}(m - m') - \phi(l - l') = 0.$$

Because ϕ is a monomorphism, it follows that $l - l' = \mathrm{d}(\tilde{l})$. Therefore $\pi_L(l) = \pi_L(l')$ in $\Gamma(W'', \mathrm{H}^{i+1}(L))$. But $W'' \twoheadrightarrow V$ is a covering, so $\pi_L(l) = \pi_L(l')$ in $\Gamma(V, \mathrm{H}^{i+1}(L))$. □

Lemma 3.7.4 justifies the next definition.

Definition 3.7.5 Let $\bar{n} \in \Gamma(U, \mathrm{H}^i(N))$. Suppose $V \twoheadrightarrow U$ is a covering that admits a connecting triple. We define $\delta_V(\bar{n}) \in \Gamma(V, \mathrm{H}^{i+1}(L))$ to be the unique section such that $\delta_V(\bar{n}) = \pi_L(l)$ for every connecting triple (n, m, l) for $\bar{n}$ over V.

Lemma 3.7.6 *Let $\bar{n} \in \Gamma(U, \mathrm{H}^i(N))$, and let*

$$\begin{array}{ccc} V' & \xrightarrow{\rho'} & U' \\ \tau \downarrow & & \downarrow \sigma \\ V & \xrightarrow{\rho} & U \end{array}$$

be a commutative diagram in M*, in which the horizontal arrows are coverings. Suppose V admits a connecting triple for $\bar{n}$. Then V' admits a connecting triple for $\bar{n}' := \sigma^*(\bar{n}) \in \Gamma(U', \mathrm{H}^i(N))$, and there is equality $\delta_{V'}(\bar{n}') = \tau^*(\delta_V(\bar{n}))$ in $\Gamma(V', \mathrm{H}^{i+1}(L))$.*

Proof If (n, m, l) is a connecting triple for $\bar{n}$ over V, then $(\tau^*(n), \tau^*(m), \tau^*(l))$ is a connecting triple for $\bar{n}'$ over V'. Thus $\delta_{V'}(\bar{n}') = \pi_L(\tau^*(l)) = \tau^*(\pi_L(l)) = \tau^*(\delta_V(\bar{n}))$. □

Lemma 3.7.7 *Let $\bar{n} \in \Gamma(U, \mathrm{H}^i(N))$. Suppose $\rho : V \twoheadrightarrow U$ is a covering that admits a connecting triple for $\bar{n}$. Then $\delta_V(\bar{n})$ lies in the subgroup $\Gamma(U, \mathrm{H}^{i+1}(L))$ of $\Gamma(V, \mathrm{H}^{i+1}(L))$.*

Proof This is a "descent" argument, using the second sheaf trick. Let $V_1, V_2 := V$ and $\rho_1, \rho_2 := \rho$. In the notation of Proposition 2.4.11, applied to the object $Q := \mathrm{H}^{i+1}(L) \in$ M, there is an exact sequence

$$0 \to \Gamma(U, Q) \xrightarrow{(\rho_1^*, \rho_2^*)} \Gamma(V_1, Q) \times \Gamma(V_2, Q) \xrightarrow{(\sigma_1^*, -\sigma_2^*)} \Gamma(W, Q).$$

Here

$$\rho_1 \circ \sigma_1 = \rho_2 \circ \sigma_2 : W = V_1 \times_U V_2 \to U$$

is a covering too. According to Lemma 3.7.6, the section

$$(\delta_{V_1}(\bar{n}), \delta_{V_2}(\bar{n})) \in \Gamma(V_1, Q) \times \Gamma(V_2, Q)$$

satisfies

$$(\sigma_1^*, -\sigma_2^*)(\delta_{V_1}(\bar{n}), \delta_{V_2}(\bar{n})) = \delta_W(\bar{n}) - \delta_W(\bar{n}) = 0.$$

Hence $\delta_V(\bar{n})$ is in the image of ρ^*. □

Lemma 3.7.8 *There is a unique morphism* $\delta^i_{\boldsymbol{E}} : \mathrm{H}^i(N) \to \mathrm{H}^{i+1}(L)$ *in* M *with the following property*:

(†) *For every* $U \in \mathsf{M}$, *every section* $\bar{n} \in \Gamma(U, \mathrm{H}^i(N))$, *and every covering* $V \twoheadrightarrow U$ *that admits a connecting triple for* $\bar{n}$, *there is equality* $\delta^i_{\boldsymbol{E}}(\bar{n}) = \delta_V(\bar{n})$ *in* $\Gamma(U, \mathrm{H}^{i+1}(L))$.

Proof We look at the universal section of $\mathrm{H}^i(N)$, namely we take $U_0 := \mathrm{H}^i(N)$ and $\bar{n}_0 := \mathrm{id}_{U_0} \in \Gamma(U_0, \mathrm{H}^i(N))$. Let $\rho_0 : V_0 \twoheadrightarrow U_0$ be a covering that admits a connecting triple for $\bar{n}_0$; see Lemma 3.7.3. According to Lemma 3.7.7, we know that

$$\delta_{V_0}(\bar{n}_0) \in \Gamma(U_0, \mathrm{H}^{i+1}(L)) = \mathrm{Hom}_{\mathsf{M}}(\mathrm{H}^i(N), \mathrm{H}^{i+1}(L)).$$

Define $\delta^i_{\boldsymbol{E}} := \delta_{V_0}(\bar{n}_0)$.

We need to prove that $\delta^i_{\boldsymbol{E}}$ has property (†). So let $U \in \mathsf{M}$, let $\bar{n} \in \Gamma(U, \mathrm{H}^i(N))$, and let $\rho : V \twoheadrightarrow U$ a covering that admits a connecting triple for $\bar{n}$. Define the morphism $\sigma := \bar{n} : U \to U_0$. We get a morphism $\sigma \circ \rho : V \to U_0$. Define $V' := V \times_{U_0} V_0$. By Lemma 2.4.3(2) the induced morphism $V' \to V$ is a covering. Therefore, by composing it with ρ we get a covering $\rho' : V' \twoheadrightarrow U$.

Consider the commutative diagram

$$\begin{array}{ccc} V' & \xrightarrow{\ \rho'\ } & U \\ {\scriptstyle \tau}\downarrow & & \downarrow{\scriptstyle \sigma} \\ V_0 & \xrightarrow{\ \rho_0\ } & U_0 \end{array}$$

in M. The horizontal arrows are coverings. Now $\bar{n} = \sigma = \sigma \circ \mathrm{id}_{U_0} = \sigma^*(\bar{n}_0)$. So, according to Lemma 3.7.6, there is equality

$$\delta_{V'}(\bar{n}) = \tau^*(\delta_{V_0}(\bar{n}_0)) = \tau^*(\delta^i_{\boldsymbol{E}} \circ \rho_0) = \delta^i_{\boldsymbol{E}} \circ \rho_0 \circ \tau = \delta^i_{\boldsymbol{E}} \circ \sigma \circ \rho' = \delta^i_{\boldsymbol{E}}(\bar{n})$$

in $\Gamma(V', \mathrm{H}^{i+1}(L))$. But $\rho' : V' \twoheadrightarrow U$ is a covering, and hence there is equality $\delta_{V'}(\bar{n}) = \delta^i_{\boldsymbol{E}}(\bar{n})$ in $\Gamma(U, \mathrm{H}^{i+1}(L))$. □

Definition 3.7.9 The morphism $\delta^i_{\boldsymbol{E}} : \mathrm{H}^i(N) \to \mathrm{H}^{i+1}(L)$ is called the *ith connecting morphism* of the short exact sequence $\boldsymbol{E}$ from (3.7.1).

Theorem 3.7.10 (Long Exact Sequence in Cohomology) *Let* M *be an abelian category. Given a short exact sequence*

$$\boldsymbol{E} = (0 \to L \xrightarrow{\phi} M \xrightarrow{\psi} N \to 0)$$

in $\mathbf{C}_{\mathrm{str}}(\mathsf{M})$, *the sequence*

$$\cdots \to \mathrm{H}^i(L) \xrightarrow{\mathrm{H}^i(\phi)} \mathrm{H}^i(M) \xrightarrow{\mathrm{H}^i(\psi)} \mathrm{H}^i(N) \xrightarrow{\delta^i_{\boldsymbol{E}}} \mathrm{H}^{i+1}(L) \to \cdots$$

in M *is exact.*

Proof We need to check exactness at $\mathrm{H}^i(N)$, $\mathrm{H}^{i+1}(L)$ and $\mathrm{H}^i(M)$. This will be done using the first sheaf trick (Proposition 2.4.9) and the second sheaf trick (Proposition 2.4.11).

▹ Step 1. Exactness at $\mathrm{H}^i(N)$. This is done in two substeps. For simplification we shall write $\bar{\psi} := \mathrm{H}^i(\psi)$.

▹ ▹ Substep 1.a. We start by proving that $\delta^i_{\boldsymbol{E}} \circ \bar{\psi} = 0$. According to Proposition 2.4.9(1), it suffices to show that given an element $\bar{m} \in \Gamma(U, \mathrm{H}^i(N))$, with image $\bar{n} := \bar{\psi}(\bar{m}) \in \Gamma(U, \mathrm{H}^i(N))$, the element $\delta^i_{\boldsymbol{E}}(\bar{n})$ is zero.

By Proposition 2.4.9(3) there is a covering $V \twoheadrightarrow U$, and an element $m \in \Gamma(V, \mathrm{Z}^i(M))$, such that $\pi_M(m) = \bar{m} \in \Gamma(V, \mathrm{H}^i(M))$. Letting $n := \psi(m) \in \Gamma(V, \mathrm{Z}^i(N))$, we see that $\pi_N(n) = \bar{n} \in \Gamma(V, \mathrm{H}^i(N))$. Now $\mathrm{d}_M(m) = 0$, and hence, taking $l := 0$, the triple (n, m, l) is a connecting triple for $\bar{n}$ over V. Therefore $\delta^i_{\boldsymbol{E}}(\bar{n}) = \pi_L(l) = 0$.

▹ ▹ Substep 1.b. Now we prove that the morphism $\bar{\psi} : \mathrm{H}^i(M) \to \mathrm{Ker}(\delta^i_{\boldsymbol{E}})$ is an epimorphism. We shall use Proposition 2.4.9(3). Let $\bar{n} \in \Gamma(U, \mathrm{Ker}(\delta^i_{\boldsymbol{E}}))$; so $\bar{n} \in \Gamma(U, \mathrm{H}^i(N))$ and $\delta^i_{\boldsymbol{E}}(\bar{n}) = 0$. It suffices to find a covering $V' \twoheadrightarrow U$ and a section $\bar{m} \in \Gamma(V', \mathrm{H}^i(M))$ such that $\bar{\psi}(\bar{m}) = \bar{n}$ in $\Gamma(V', \mathrm{H}^i(N))$.

Take a covering $V \twoheadrightarrow U$ that admits a connecting triple (n, m, l) for $\bar{n}$. Then $\pi_L(l) = \delta_V(\bar{n}) = \delta^i_{\boldsymbol{E}}(\bar{n}) = 0$ in $\Gamma(V, \mathrm{H}^{i+1}(L))$. By Proposition 2.4.9(3) there is some covering $V' \twoheadrightarrow V$ and a section $l' \in \Gamma(V', L^i)$ s.t. $\mathrm{d}_L(l') = l$ in $\Gamma(V', L^{i+1})$. Define $m' := \phi(l') \in \Gamma(V', M^i)$. Now $\mathrm{d}_M(m - m') = \mathrm{d}_M(m) - \mathrm{d}_M(m') = \phi(l) - \phi(l) = 0$, so $m - m'$ belongs to $\Gamma(V', \mathrm{Z}^i(M))$. Because $\psi(m - m') = \psi(m) = n$, the cohomology class $\bar{m} := \pi_M(m - m')$ in $\Gamma(V', \mathrm{H}^i(M))$ satisfies $\bar{\psi}(\bar{m}) = \bar{n}$ in $\Gamma(V', \mathrm{H}^i(N))$.

▹ Step 2. Exactness at $\mathrm{H}^{i+1}(L)$. This is also done in two substeps. We will use the abbreviation $\bar{\phi} := \mathrm{H}^{i+1}(\phi)$.

▹ ▹ Substep 2.a. We start by proving that $\bar{\phi} \circ \delta^i_{\boldsymbol{E}} = 0$. Take $\bar{n} \in \Gamma(U, \mathrm{H}^i(N))$, and let $\bar{l} := \delta^i_{\boldsymbol{E}}(\bar{n})$. We need to prove that $\bar{\phi}(\bar{l}) = 0$. Let $V \twoheadrightarrow U$ be a covering that admits a connecting triple (n, m, l) for $\bar{n}$. This says that $\bar{l} = \pi_L(l)$, and hence $\bar{\phi}(\bar{l}) = \pi_L(\phi(l)) = \pi_M(\mathrm{d}_M(m)) = 0$.

▹ ▹ Substep 2.b. Now we prove that $\delta^i_{\boldsymbol{E}} : \mathrm{H}^i(N) \to \mathrm{Ker}(\bar{\phi})$ is an epimorphism. Let $\bar{l} \in \Gamma(U, \mathrm{H}^{i+1}(L))$ be such that $\bar{\phi}(\bar{l}) = 0$. Take a covering $V \twoheadrightarrow U$ for which there's a section $l \in \Gamma(V, \mathrm{Z}^{i+1}(L))$ s.t. $\pi_L(l) = \bar{l}$. The section $\phi(l) \in \Gamma(V, \mathrm{Z}^{i+1}(M))$ satisfies $\pi_M(\phi(l)) = \bar{\phi}(\bar{l}) = 0$ in $\Gamma(V, \mathrm{H}^{i+1}(M))$, and therefore there is some covering $V' \twoheadrightarrow V$ and a section $m \in \Gamma(V', M^i)$ s.t. $\mathrm{d}_M(m) = \phi(l)$ in $\Gamma(V', M^{i+1})$.

Define $n := \psi(m) \in \Gamma(V', N^i)$. Then $\mathrm{d}_N(n) = \mathrm{d}_N(\psi(m)) = \psi(\mathrm{d}_M(m)) = \psi(\phi(l)) = 0$, so in fact $n \in \Gamma(V', \mathrm{Z}^i(N))$. Let $\bar{n} := \pi_N(n) \in \Gamma(V', \mathrm{H}^i(N))$. We see that (n, m, l) is a connecting triple for $\bar{n}$ over V'. Therefore $\bar{l} = \pi_L(l) = \delta_{V'}(\bar{n}) = \delta^i_{\boldsymbol{E}}(\bar{n})$ in $\Gamma(V', \mathrm{H}^{i+1}(L))$.

▹ Step 3. Exactness at $\mathrm{H}^i(M)$: let's use the abbreviations $\bar{\phi} := \mathrm{H}^i(\phi)$ and $\bar{\psi} := \mathrm{H}^i(\psi)$. It is clear that $\bar{\psi} \circ \bar{\phi} = 0$. It remains to prove that $\bar{\phi} : \mathrm{H}^i(L) \to \mathrm{Ker}(\bar{\psi})$ is an epimorphism.

Let $\bar{m} \in \Gamma(U, \mathrm{H}^i(M))$ be such that $\bar{\psi}(\bar{m}) = 0$. We are going to find a covering $W \to U$ and a section $\bar{l} \in \Gamma(W, \mathrm{H}^i(L))$ such that $\bar{\phi}(\bar{l}) = \bar{m}$ in $\Gamma(W, \mathrm{H}^i(M))$.

Take a covering $V \twoheadrightarrow U$ for which there's a section $m \in \Gamma(V, \mathrm{Z}^i(M))$ s.t. $\pi_M(m) = \bar{m}$. The section $n := \psi(m) \in \Gamma(V, \mathrm{Z}^i(N))$ satisfies $\pi_N(n) = \bar{\psi}(\bar{m}) = 0$ in $\Gamma(V, \mathrm{H}^i(N))$, and therefore there is some covering $V' \twoheadrightarrow V$ and a section $n' \in \Gamma(V', N^{i-1})$ s.t. $\mathrm{d}(n') = n$ in $\Gamma(V', N^i)$.

Take a covering $V'' \twoheadrightarrow V'$ for which there is a section $m'' \in \Gamma(V'', M^{i-1})$ s.t. $\psi(m'') = n'$. The section $m - \mathrm{d}_M(m'') \in \Gamma(V'', \mathrm{Z}^i(M))$ satisfies $\pi_M(m - \mathrm{d}(m'')) = \bar{m}$ in $\Gamma(V'', \mathrm{H}^i(M)$, and also $\psi(m - \mathrm{d}_M(m'')) = \psi(m) - \psi(\mathrm{d}_M(m'')) = n - n = 0$.

There is a covering $W \twoheadrightarrow V''$ with a section $l \in \Gamma(W, L^i)$ s.t. $\phi(l) = m - \mathrm{d}_M(m'')$. Because ϕ is a monomorphism and $m - \mathrm{d}_M(m'')$ is a cocycle, it follows that the section l belongs to $\Gamma(W, \mathrm{Z}^i(L))$. Its cohomology class $\bar{l} := \pi_L(l)$ in $\Gamma(W, \mathrm{H}^i(L))$ satisfies $\bar{\phi}(\bar{l}) = \pi_M(\phi(l)) = \pi_M(m - \mathrm{d}_M(m'')) = \bar{m}$ in $\Gamma(W, \mathrm{H}^i(M))$. □

Proposition 3.7.11 *Let $\chi : \boldsymbol{E} \to \boldsymbol{E}'$ be a morphism of short exact sequences in $\mathbf{C}_{\mathrm{str}}(\mathsf{M})$. Namely $\chi = (\chi_L, \chi_M, \chi_N)$ in the commutative diagram with exact rows*

$$\begin{array}{ccccccccc}
0 & \longrightarrow & L & \xrightarrow{\phi} & M & \xrightarrow{\psi} & N & \longrightarrow & 0 \\
 & & \Big\downarrow{\scriptstyle \chi_L} & & \Big\downarrow{\scriptstyle \chi_M} & & \Big\downarrow{\scriptstyle \chi_N} & & \\
0 & \longrightarrow & L' & \xrightarrow{\phi'} & M' & \xrightarrow{\psi'} & N' & \longrightarrow & 0
\end{array}$$

in $\mathbf{C}_{\mathrm{str}}(\mathsf{M})$. Then, for every i, the diagram

$$\begin{array}{ccc}
\mathrm{H}^i(N) & \xrightarrow{\delta^i_{\boldsymbol{E}}} & \mathrm{H}^{i+1}(L) \\
\Big\downarrow{\scriptstyle \mathrm{H}^i(\chi_N)} & & \Big\downarrow{\scriptstyle \mathrm{H}^{i+1}(\chi_L)} \\
\mathrm{H}^i(N') & \xrightarrow{\delta^i_{\boldsymbol{E}'}} & \mathrm{H}^{i+1}(L')
\end{array}$$

is commutative.

Exercise 3.7.12 Prove Proposition 3.7.11. (Hint: study the proof of the previous theorem.)

3.8 The DG Category $\mathbf{C}(A, \mathsf{M})$

We now combine material from previous sections. The concept introduced in the definition below is new. It is the DG version of Definition 3.1.45.

Conventions 1.2.4, 3.3.3, 3.4.3 and 3.5.2 are in place. Thus all linear structures and operations are $\mathbb{K}$-linear, and rings and bimodules are central over $\mathbb{K}$.

Recall that given an abelian category M, the category of complexes $\mathbf{C}(\mathsf{M})$ is a DG category. For a complex $M \in \mathbf{C}(\mathsf{M})$ we have the DG ring $\mathrm{End}_{\mathsf{M}}(M) = \mathrm{Hom}_{\mathsf{M}}(M, M)$. The multiplication in this ring is composition.

Definition 3.8.1 Let M be an abelian category, and let A be a DG ring. A *DG A-module in* M is an object $M \in \mathbf{C}(\mathsf{M})$, together with a DG ring homomorphism $f : A \to \mathrm{End}_{\mathsf{M}}(M)$.

If M is a DG A-module in M, then, after forgetting the differentials, M becomes a graded A-module in M.

Definition 3.8.2 Let M be an abelian category, let A be a DG ring and let M, N be DG A-modules in M. In Definition 3.1.47 we introduced the graded $\mathbb{K}$-module $\mathrm{Hom}_{A,\mathsf{M}}(M, N)$. This is made into a DG $\mathbb{K}$-module with differential

$$\mathrm{d}(\phi) := \mathrm{d}_N \circ \phi - (-1)^i \cdot \phi \circ \mathrm{d}_M$$

for $\phi \in \mathrm{Hom}_{A,\mathsf{M}}(M, N)^i$.

When we have to be specific, we denote the differential of $\mathrm{Hom}_{A,\mathsf{M}}(M, N)$ by $\mathrm{d}_{\mathrm{Hom}}$, $\mathrm{d}_{A,\mathsf{M}}$, or some similar expression.

As we have seen before (in Lemmas 3.6.5 and 3.4.5), given morphisms

$$\phi_k \in \mathrm{Hom}_{A,\mathsf{M}}(M_k, M_{k+1})^{i_k}$$

for $k \in \{0, 1\}$, we have

$$\phi_1 \circ \phi_0 \in \mathrm{Hom}_{A,\mathsf{M}}(M_0, M_2)^{i_0+i_1},$$

and

$$\mathrm{d}(\phi_1 \circ \phi_0) = \mathrm{d}(\phi_1) \circ \phi_0 + (-1)^{i_1} \cdot \phi_1 \circ \mathrm{d}(\phi_0).$$

Also the identity automorphism id_M belongs to $\mathrm{Hom}_{A,\mathsf{M}}(M, M)^0$, and $\mathrm{d}(\mathrm{id}_M) = 0$. Therefore the next definition is legitimate.

Definition 3.8.3 Let M be an abelian category, and let A be a DG ring. The DG category of DG A-modules in M is denoted by $\mathbf{C}(A, \mathsf{M})$. The morphism DG $\mathbb{K}$-modules are

$$\mathrm{Hom}_{\mathbf{C}(A,\mathsf{M})}(M_0, M_1) := \mathrm{Hom}_{A,\mathsf{M}}(M_0, M_1)$$

from Definition 3.8.2. The composition is that of $\mathbf{C}(\mathsf{M})$.

Notice that forgetting the action of A is a faithful DG functor $\mathbf{C}(A, \mathsf{M}) \to \mathbf{C}(\mathsf{M})$.

Example 3.8.4 If $A = \mathbb{K}$, then $\mathbf{C}(A, \mathsf{M}) = \mathbf{C}(\mathsf{M})$; and if $\mathsf{M} = \mathsf{Mod}\,\mathbb{K}$, then $\mathbf{C}(A, \mathsf{M}) = \mathbf{C}(A) = \mathsf{DGMod}\,A$.

Definition 3.8.5 In the situation of Definition 3.8.3:

(1) The strict category of $\mathbf{C}(A, \mathsf{M})$ (see Definition 3.4.6(1)) is denoted by $\mathbf{C}_{\mathrm{str}}(A, \mathsf{M})$.

(2) The homotopy category of $\mathbf{C}(A, \mathsf{M})$ (see Definition 3.4.6(2)) is denoted by $\mathbf{K}(A, \mathsf{M})$.

To say this in words: a morphism $\phi : M \to N$ in $\mathbf{C}(A, \mathsf{M})$ is strict if and only if it has degree 0 and $\phi \circ \mathrm{d}_M = \mathrm{d}_N \circ \phi$. The morphisms in $\mathbf{K}(A, \mathsf{M})$ are the homotopy classes of strict morphisms.

An additive functor $F : \mathsf{M} \to \mathsf{N}$ between abelian categories is called *faithfully exact* if for every sequence $\boldsymbol{E}$ in $\mathbf{C}_{\mathrm{str}}(A, \mathsf{M})$, the sequence $\boldsymbol{E}$ is exact if and only if the sequence $F(\boldsymbol{E})$ in N is exact.

Proposition 3.8.6 *The category* $\mathbf{C}_{\mathrm{str}}(A, \mathsf{M})$ *is a* $\mathbb{K}$*-linear abelian category, and the forgetful functors*

$$\mathbf{C}_{\mathrm{str}}(A, \mathsf{M}) \to \mathbf{C}_{\mathrm{str}}(\mathsf{M}) \xrightarrow{\mathrm{Und}} \mathbf{G}_{\mathrm{str}}(\mathsf{M})$$

are faithfully exact.

Exercise 3.8.7 Prove the proposition above.

Like Definition 3.2.4, given a DG module $M \in \mathbf{C}(A, \mathsf{M})$ and an integer i, we can consider the objects of degree i cocycles $\mathrm{Z}^i(M)$, decocycles $\mathrm{Y}^i(M)$, coboundaries $\mathrm{B}^i(M)$ and cohomology $\mathrm{H}^i(M)$. These are all objects of M. As M varies we get linear functors

$$\mathrm{Z}, \mathrm{Y}, \mathrm{B}, \mathrm{H} : \mathbf{C}_{\mathrm{str}}(A, \mathsf{M}) \to \mathbf{G}_{\mathrm{str}}(\mathsf{M}). \tag{3.8.8}$$

These objects are related by the following exact sequences in M :

$$0 \to \mathrm{Z}^i(M) \to M^i \xrightarrow{\mathrm{d}_M^i} M^{i+1}, \tag{3.8.9}$$

$$M^{i-1} \xrightarrow{\mathrm{d}_M^{i-1}} M \to \mathrm{Y}^i(M) \to 0, \tag{3.8.10}$$

$$0 \to \mathrm{B}^i(M) \to \mathrm{Z}^i(M) \to \mathrm{H}^i(M) \to 0, \tag{3.8.11}$$

$$0 \to \mathrm{H}^i(M) \to \mathrm{Y}^i(M) \to \mathrm{B}^{i+1}(M) \to 0. \tag{3.8.12}$$

Remark 3.8.13 Actually, the cohomology functor H from equation (3.8.8) factors through the category $\mathbf{G}_{\mathrm{str}}(\mathrm{H}(A), \mathsf{M})$. See formula (3.1.49). We will not need this fact.

Proposition 3.8.14 *The functor* Z *in equation* (3.8.8) *is left exact, and the functor* Y *in equation* (3.8.8) *is right exact.*

Proof It is enough to consider each degree separately. And by Proposition 3.8.6, we may ignore the DG ring A. Let $F^i : \mathbf{G}_{\text{str}}(\mathsf{M}) \to \mathsf{M}$ be the functor that sends a graded module M to its degree i component M^i. This is an exact functor.

Formula (3.8.9) exhibits Z^i as the kernel of the homomorphism of functors $\mathrm{d}^i : F^i \to F^{i+1}$. According to Proposition 2.5.18(2), the functor Z^i is left exact.

Similarly, formula (3.8.10) exhibits Y^i as the cokernel of the homomorphism of functors $\mathrm{d}^{i-1} : F^{i-1} \to F^i$. According to Proposition 2.5.18(2), the functor Y^i is right exact. □

Proposition 3.8.15 *Let* $\{M_x\}_{x \in X}$ *be a collection of objects of* $\mathbf{C}(A, \mathsf{M})$, *indexed by a set* X. *Assume that for every* $i \in \mathbb{Z}$ *the direct sum* $M^i := \bigoplus_{x \in X} M^i_x$ *exists in* M. *Then the graded object* $M := \{M^i\}_{i \in \mathbb{Z}} \in \mathbf{G}(\mathsf{M})$ *has a canonical structure of DG A-module in* M, *and* M *is a direct sum of the collection* $\{M_x\}_{x \in X}$ *in* $\mathbf{C}_{\text{str}}(A, \mathsf{M})$.

Exercise 3.8.16 Prove Proposition 3.8.15. (Hint: look at the proof of Theorem 3.9.16.)

Example 3.8.17 Since $\mathbf{M}(\mathbb{K})$ has infinite direct sums, the proposition above shows that $\mathbf{C}_{\text{str}}(A)$ has infinite direct sums.

Proposition 3.8.18 *Given a DG ring* A, *let* $A^\natural$ *be the graded ring gotten by forgetting the differential. Consider the forgetful functor* $\mathrm{Und} : \mathbf{C}(A, \mathsf{M}) \to \mathbf{G}(A^\natural, \mathsf{M})$ *that forgets the differentials of DG modules, i.e.* $\mathrm{Und}(M, \mathrm{d}_M) := M$.

(1) Und *is a fully faithful graded functor.*

(2) *On the strict categories,* $\mathrm{Str}(\mathrm{Und}) : \mathbf{C}_{\text{str}}(A, \mathsf{M}) \to \mathbf{G}_{\text{str}}(A, \mathsf{M})$ *is a faithfully exact additive functor.*

Exercise 3.8.19 Prove Proposition 3.8.18.

A graded object $M = \{M^i\}_{i \in \mathbb{Z}}$ in M is said to be *bounded above* if the set $\{i \mid M^i \neq 0\}$ is bounded above. Likewise, we define *bounded below* and *bounded* graded objects.

Definition 3.8.20 We define $\mathbf{C}^-(A, \mathsf{M})$, $\mathbf{C}^+(A, \mathsf{M})$ and $\mathbf{C}^{\mathrm{b}}(A, \mathsf{M})$ to be the full subcategories of $\mathbf{C}(A, \mathsf{M})$ consisting of bounded above, bounded below and bounded DG modules, respectively.

The symbols "−," "+" and "b" and "⟨empty⟩" (the empty symbol) are called *boundedness indicators*. We shall usually use the symbol "⋆" to denote an unspecified boundedness indicator. Thus the expression $\mathsf{C}^{\star}(A, \mathsf{M})$ can refer to any of the four full subcategories of $\mathsf{C}(A, \mathsf{M})$ that are mentioned in the definition above.

Remark 3.8.21 Here is a generalization of Definition 3.8.3. Instead of a DG central $\mathbb{K}$-ring A we can take a small $\mathbb{K}$-linear DG category A. We then define the $\mathbb{K}$-linear DG category $\mathsf{C}(\mathsf{A}, \mathsf{M}) := \mathsf{DGFun}(\mathsf{A}, \mathsf{C}(\mathsf{M}))$ as in Exercise 3.5.8.

This is indeed a generalization of Definition 3.8.3: when A has a single object x, and we write $A := \mathrm{End}_{\mathsf{A}}(x)$, then the functor $M \mapsto M(x)$ is an isomorphism of DG categories $\mathsf{C}(\mathsf{A}, \mathsf{M}) \xrightarrow{\simeq} \mathsf{C}(A, \mathsf{M})$.

In the special case of $\mathsf{M} = \mathbf{M}(\mathbb{K}) = \mathsf{Mod}\,\mathbb{K}$, the DG category $\mathsf{C}(\mathsf{A}, \mathsf{M})$ is what B. Keller [79] calls the DG category of *DG* A*-modules*.

Almost everything we do in this book for $\mathsf{C}(A, \mathsf{M})$ holds in the more general context of $\mathsf{C}(\mathsf{A}, \mathsf{M})$. However, in the more general context a lot of the intuition is lost, and some aspects become pretty cumbersome. This is the reason we have decided to stick with the less general context.

Remark 3.8.22 The theory of DG categories has a generalization – this is the theory of A_∞ *categories*, which will be outlined in this remark. All set theoretical issues are going to be ignored.

The concept of an A_∞ *algebra*, namely a single-object A_∞ category, was introduced by J. Stasheff [146], [147] in the 1960s, in relation with his work on *higher operations on H-spaces* in topology.

A_∞ categories are at the core of *Homological Mirror Symmetry*, conceived by K. Fukaya and M. Kontsevich during the 1990s. Indeed, a symplectic manifold Y gives rise to the *Fukaya category* $\mathsf{F}(Y)$, which is an A_∞ category, and is a categorification of *Floer cohomology*. See [48], [86], [87], [141] and [88].

Further developments on A_∞ algebras and categories were made, among others, by T.V. Kadeishvili [73], K. Lefèvre–Hasegawa [94], B. Keller [80], [83], and Canonaco–Ornaghi–Stellari [37]. For the analogous theory of L_∞ *algebras*, generalizing DG Lie algebras, see [169].

We shall only discuss strictly unital A_∞ categories, giving very few details (full details can be found in [83] and [37]), and our notation deviates somewhat from the references (because we want it to be compatible with the rest of the book). The base ring $\mathbb{K}$ is assumed to be a field. An A_∞ category A over $\mathbb{K}$ has a set of objects $\mathrm{Ob}(\mathsf{A})$. For each pair of objects $x, y \in \mathrm{Ob}(\mathsf{A})$ there is a graded $\mathbb{K}$-module $\mathsf{A}(x, y) = \mathrm{Hom}_{\mathsf{A}}(x, y)$. For each $x \in \mathrm{Ob}(\mathsf{A})$ there is a degree 0 element $\mathrm{id}_x \in \mathsf{A}(x, x)$. For each $i \geq 1$ and each sequence $x_0, \ldots, x_i \in \mathrm{Ob}(\mathsf{A})$, there is a degree $2 - i$ homomorphism

$$\mathrm{m}_i : \mathsf{A}(x_{i-1}, x_i) \otimes \cdots \otimes \mathsf{A}(x_0, x_1) \to \mathsf{A}(x_0, x_i) \tag{3.8.23}$$

$\mathbf{G}(\mathbb{K})$. The operations m_i satisfy certain relations. Roughly speaking, the operation m_1 is a differential, the operation m_2 is a multiplication, the element id_x is a unit for the multiplication m_2, the multiplication m_2 is associative up to the hotomopy m_3, etc. The relations imply that the cohomology $\mathrm{H}(\mathsf{A})$, w.r.t. the differential m_1, is a graded category, in the sense of Definition 3.1.32. If the higher (i.e. $i > 2$) operations m_i vanish, then A is a DG category, in the sense of Definition 3.4.1.

Suppose A and B are A_∞ categories. An A_∞ *functor* $F : \mathsf{A} \to \mathsf{B}$ consists of a function $F_{\mathrm{ob}} : \mathrm{Ob}(\mathsf{A}) \to \mathrm{Ob}(\mathsf{B})$, and for each $i \geq 1$ and each sequence $x_0, \ldots, x_i \in \mathrm{Ob}(\mathsf{A})$, a degree $1 - i$ homomorphism

$$F_i : \mathsf{A}(x_{i-1}, x_i) \otimes \cdots \otimes \mathsf{A}(x_0, x_1) \to \mathsf{B}(F_{\mathrm{ob}}(x_0), F_{\mathrm{ob}}(x_i)) \tag{3.8.24}$$

in $\mathbf{G}(\mathbb{K})$. There are relations that the F_i have to satisfy. Given another A_∞ functor $G : \mathsf{B} \to \mathsf{C}$, there is a composed A_∞ functor $G \circ F : \mathsf{A} \to \mathsf{C}$, which is hard to describe directly (it is usually defined in terms of *cocategories*, via the *bar construction*). We denote by $\mathrm{A}_\infty\mathsf{Cat}$ the category whose objects are the A_∞ categories and whose morphisms are the A_∞ functors.

For a pair $\mathsf{A}, \mathsf{B} \in \mathrm{Ob}(\mathrm{A}_\infty\mathsf{Cat})$ we denote by $\mathrm{A}_\infty\mathsf{Fun}(\mathsf{A}, \mathsf{B})$ the set of A_∞ functors from A to B, so that

$$\mathrm{Hom}_{\mathrm{A}_\infty\mathsf{Cat}}(\mathsf{A}, \mathsf{B}) = \mathrm{A}_\infty\mathsf{Fun}(\mathsf{A}, \mathsf{B}). \tag{3.8.25}$$

An important fact is that $\mathrm{A}_\infty\mathsf{Fun}(\mathsf{A}, \mathsf{B})$ is itself the set of objects of an A_∞ category, i.e. $\mathrm{A}_\infty\mathsf{Fun}(\mathsf{A}, \mathsf{B}) \in \mathrm{Ob}(\mathrm{A}_\infty\mathsf{Cat})$. If B is a DG category (i.e. its higher operations vanish), then $\mathrm{A}_\infty\mathsf{Fun}(\mathsf{A}, \mathsf{B})$ is also a DG category.

We shall discuss more aspects of A_∞ categories in Remarks 6.2.38 and 7.2.15.

3.9 Contravariant DG Functors

In this section we address, in a systematic fashion, the issue of reversing arrows in DG categories. As always, we work over a commutative base ring $\mathbb{K}$.

Definition 3.9.1 Let C and D be DG categories. A *contravariant DG functor* $F : \mathsf{C} \to \mathsf{D}$ consists of a function $F : \mathrm{Ob}(\mathsf{C}) \to \mathrm{Ob}(\mathsf{D})$, and for each pair of objects $M_0, M_1 \in \mathrm{Ob}(\mathsf{C})$ a homomorphism

$$F : \mathrm{Hom}_{\mathsf{C}}(M_0, M_1) \to \mathrm{Hom}_{\mathsf{D}}(F(M_1), F(M_0))$$

in $\mathbf{C}_{\mathrm{str}}(\mathbb{K})$. The conditions are:

(a) Units: $F(\mathrm{id}_M) = \mathrm{id}_{F(M)}$.

(b) Graded reversed composition: suppose that for $k \in \{0, 1\}$ we are given morphisms $\phi_k \in \mathrm{Hom}_{\mathsf{C}}(M_k, M_{k+1})^{i_k}$. Then there is equality

$$F(\phi_1 \circ \phi_0) = (-1)^{i_0 \cdot i_1} \cdot F(\phi_0) \circ F(\phi_1)$$

inside $\mathrm{Hom}_{\mathsf{D}}(F(M_2), F(M_0))^{i_0+i_1}$.

Warning: a contravariant DG functor does not remain a contravariant functor after the grading is forgotten; cf. Remark 3.1.26.

Here is the categorical version of Definition 3.3.13.

Definition 3.9.2 Let C be a DG category. The *opposite DG category* C^{op} has the same set of objects. The morphism DG $\mathbb{K}$-modules are

$$\mathrm{Hom}_{\mathsf{C}^{\mathrm{op}}}(M_0, M_1) := \mathrm{Hom}_{\mathsf{C}}(M_1, M_0).$$

The composition $\circ^{\mathrm{op}}$ of C^{op} is the composition $\circ$ of C, reversed and multiplied by signs:

$$\phi_0 \circ^{\mathrm{op}} \phi_1 := (-1)^{i_0 \cdot i_1} \cdot \phi_1 \circ \phi_0$$

for morphisms $\phi_k \in \mathrm{Hom}_{\mathsf{C}}(M_k, M_{k+1})^{i_k}$.

One needs to verify that this is indeed a DG category. This is basically the same verification as in Exercise 3.3.14.

As before, we define the operation $\mathrm{Op} : \mathsf{C} \to \mathsf{C}^{\mathrm{op}}$ to be the identity on objects, and the identity on morphisms in reversed order, i.e.

$$\mathrm{Op} = \mathrm{id} : \mathrm{Hom}_{\mathsf{C}}(M_0, M_1) \xrightarrow{\simeq} \mathrm{Hom}_{\mathsf{C}^{\mathrm{op}}}(M_1, M_0).$$

Note that $(\mathsf{C}^{\mathrm{op}})^{\mathrm{op}} = \mathsf{C}$, and we denote the inverse operation $\mathsf{C}^{\mathrm{op}} \to \mathsf{C}$ also by Op.

Proposition 3.9.3 *Let* C, D *and* E *be DG categories.*

(1) *The operations* $\mathrm{Op} : \mathsf{C} \to \mathsf{C}^{\mathrm{op}}$ *and* $\mathrm{Op} : \mathsf{C}^{\mathrm{op}} \to \mathsf{C}$ *are contravariant DG functors.*
(2) *If* $F : \mathsf{C} \to \mathsf{D}$ *is a contravariant DG functor, then the composition* $F \circ \mathrm{Op} : \mathsf{C}^{\mathrm{op}} \to \mathsf{D}$ *is a DG functor; and vice versa.*
(3) *If* $F : \mathsf{C} \to \mathsf{D}$ *and* $G : \mathsf{D} \to \mathsf{E}$ *are contravariant DG functors, then the composition* $G \circ F : \mathsf{C} \to \mathsf{E}$ *is a DG functor.*

Exercise 3.9.4 Prove the previous proposition.

Definitions 3.9.2 and 3.9.1 make sense for graded categories, by forgetting differentials. Thus for graded categories C and D we can talk about contravariant graded functors $\mathsf{C} \to \mathsf{D}$, and about the graded category C^{op}.

We already met $\mathbf{G}(\mathsf{M})$, the category of graded objects in an abelian category M; see Definition 3.1.43. It is a graded category. Its objects are collections $M = \{M^i\}_{i \in \mathbb{Z}}$ of objects $M^i \in \mathsf{M}$.

Let M and N be abelian categories, and let $F : \mathsf{M} \to \mathsf{N}$ be a contravariant linear functor. For a graded object $M = \{M^i\}_{i \in \mathbb{Z}} \in \mathbf{G}(\mathsf{M})$ let us define the graded object

$$\mathbf{G}(F)(M) := \{N^i\}_{i \in \mathbb{Z}} \in \mathbf{G}(\mathsf{N}), \quad N^i := F(M^{-i}) \in \mathsf{N}. \tag{3.9.5}$$

Next consider a pair of objects $M_0, M_1 \in \mathbf{G}(\mathsf{M})$ and a degree i morphism $\phi : M_0 \to M_1$ in $\mathbf{G}(\mathsf{M})$. Thus

$$\phi = \{\phi^j\}_{j \in \mathbb{Z}} \in \mathrm{Hom}_{\mathbf{G}(\mathsf{M})}(M_0, M_1)^i,$$

where, as in formula (3.1.41), the morphism ϕ^j belongs to $\mathrm{Hom}_{\mathsf{M}}(M_0^j, M_1^{j+i})$. We have objects $N_k := \mathbf{G}(F)(M_k) \in \mathbf{G}(\mathsf{N})$, for $k \in \{0, 1\}$, defined by (3.9.5). Explicitly, $N_k = \{N_k^i\}_{i \in \mathbb{Z}}$ and $N_k^i = F(M_k^{-i})$. For each $j \in \mathbb{Z}$ define the morphism

$$\psi^j := (-1)^{i \cdot j} \cdot F(\phi^{-j-i}) \in \mathrm{Hom}_{\mathsf{N}}(N_1^j, N_0^{j+i}). \tag{3.9.6}$$

Collecting them we obtain a morphism

$$\mathbf{G}(F)(\phi) := \{\psi^j\}_{j \in \mathbb{Z}} \in \mathrm{Hom}_{\mathbf{G}(\mathsf{N})}(N_1, N_0)^i. \tag{3.9.7}$$

Lemma 3.9.8 *The assignments* (3.9.5) *and* (3.9.7) *produce a contravariant graded functor* $\mathbf{G}(F) : \mathbf{G}(\mathsf{M}) \to \mathbf{G}(\mathsf{N})$.

Proof Since for morphisms of degree 0 there is no sign twist, the identity automorphism $\mathrm{id}_M = \{\mathrm{id}_{M^i}\}_{i \in \mathbb{Z}}$ of $M = \{M^i\}_{i \in \mathbb{Z}}$ in $\mathbf{G}(\mathsf{M})$ is sent to the identity automorphism of $\mathbf{G}(F)(M)$ in $\mathbf{G}(\mathsf{N})$.

Next we look at morphisms $\phi_k = \{\phi_k^j\}_{j \in \mathbb{Z}}$ in $\mathrm{Hom}_{\mathbf{G}(\mathsf{M})}(M_k, M_{k+1})^{i_k}$ for $k = 0, 1$. The composition $\phi_1 \circ \phi_0$ has degree $i_0 + i_1$, and the jth component of $\phi_1 \circ \phi_0$ is $\phi_1^{j+i_0} \circ \phi_0^j$. Therefore the jth component of $\mathbf{G}(F)(\phi_1 \circ \phi_0)$ is

$$\begin{aligned} \mathbf{G}(F)(\phi_1 \circ \phi_0)^j &= (-1)^{j \cdot (i_0+i_1)} \cdot F(\phi_1^{-j-i_1} \circ \phi_0^{-j-(i_0+i_1)}) \\ &= (-1)^{j \cdot (i_0+i_1)} \cdot F(\phi_0^{-j-(i_0+i_1)}) \circ F(\phi_1^{-j-i_1}). \end{aligned} \tag{3.9.9}$$

The jth component of $\mathbf{G}(F)(\phi_k)$ is $\mathbf{G}(F)(\phi_k)^j = (-1)^{j \cdot i_k} \cdot F(\phi_k^{-j-i_k})$. So the jth component of $(-1)^{i_0 \cdot i_1} \cdot \mathbf{G}(F)(\phi_0) \circ \mathbf{G}(F)(\phi_1)$ is

$$\begin{aligned} &(-1)^{i_0 \cdot i_1} \cdot (\mathbf{G}(F)(\phi_0) \circ \mathbf{G}(F)(\phi_1))^j \\ &= (-1)^{i_0 \cdot i_1} \cdot (-1)^{(j+i_1) \cdot i_0} \cdot F(\phi_0^{-(j+i_1)-i_0}) \circ (-1)^{j \cdot i_1} \cdot F(\phi_1^{-j-i_1}). \end{aligned} \tag{3.9.10}$$

We see that the morphisms (3.9.9) and (3.9.10) are equal. □

Now we consider a complex $(M, \mathrm{d}_M) \in \mathbf{C}(\mathsf{M})$. This is made up of a graded object $M = \{M^i\}_{i \in \mathbb{Z}} \in \mathbf{G}(\mathsf{M})$ together with a differential $\mathrm{d}_M = \{\mathrm{d}_M^i\}_{i \in \mathbb{Z}}$, where $\mathrm{d}_M^i : M^i \to M^{i+1}$. We can view d_M as an element of $\mathrm{End}_{\mathbf{G}(\mathsf{M})}(M)^1 =$

$\mathrm{Hom}_{\mathbf{G}(\mathsf{M})}(M, M)^1$. We specify a differential $\mathrm{d}_{\mathbf{C}(F)(M)}$ on the graded object $\mathbf{G}(F)(M) \in \mathbf{G}(\mathsf{N})$ as follows:

$$\mathrm{d}_{\mathbf{C}(F)(M)} := -\mathbf{G}(F)(\mathrm{d}_M) \in \mathrm{End}_{\mathbf{G}(\mathsf{N})}(\mathbf{G}(F)(M))^1. \tag{3.9.11}$$

To be explicit, the component

$$\mathrm{d}^i_{\mathbf{C}(F)(M)} : \mathbf{G}(F)(M)^i = F(M^{-i}) \to F(M^{-i-1}) = \mathbf{G}(F)(M)^{i+1}$$

of $\mathrm{d}_{\mathbf{C}(F)(M)}$ is, by (3.9.6),

$$\mathrm{d}^i_{\mathbf{C}(F)(M)} = (-1)^{i+1} \cdot F(\mathrm{d}_M^{-i-1}) : F(M^{-i}) \to F(M^{-i-1}). \tag{3.9.12}$$

This shows that our formula coincides with the one in [75, Remark 1.8.11].

The full subcategory $\mathbf{C}^\star(\mathsf{N}) \subseteq \mathbf{C}(\mathsf{N})$, for a given boundedness condition $\star$, was introduced in Definition 3.8.20. Each boundedness condition $\star$ has a *reversed boundedness condition* $-\star$, with the obvious meaning; e.g. if $\star$ is $+$ then $-\star$ is $-$.

Theorem 3.9.13 *Let* M *and* N *be abelian categories, and let* $F : \mathsf{M} \to \mathsf{N}$ *be a contravariant linear functor. The assignments* (3.9.5), (3.9.7) *and* (3.9.11) *produce a contravariant DG functor*

$$\mathbf{C}(F) : \mathbf{C}(\mathsf{M}) \to \mathbf{C}(\mathsf{N}).$$

For every boundedness condition $\star$ *we have* $\mathbf{C}(F)(\mathbf{C}^\star(\mathsf{M})) \subseteq \mathbf{C}^{-\star}(\mathsf{N})$, *where* $-\star$ *is the reversed boundedness condition.*

Proof We already know, by Lemma 3.9.8, that $\mathbf{G}(F)$ is a graded functor. We need to prove that for a pair of DG modules (M_0, d_{M_0}) and (M_1, d_{M_1}) in $\mathbf{C}(\mathsf{M})$ the strict homomorphism of graded $\mathbb{K}$-modules

$$\mathbf{G}(F) : \mathrm{Hom}_{\mathbf{G}(\mathsf{M})}(M_0, M_1) \to \mathrm{Hom}_{\mathbf{G}(\mathsf{N})}(\mathbf{G}(F)(M_1), \mathbf{G}(F)(M_0))$$

respects differentials. Take any $\phi \in \mathrm{Hom}_{\mathbf{G}(\mathsf{M})}(M_0, M_1)^i$. By definition we have $\mathrm{d}(\phi) = \mathrm{d}_{M_1} \circ \phi - (-1)^i \cdot \phi \circ \mathrm{d}_{M_0}$. Using the fact that $\mathbf{G}(F)$ is a contravariant graded functor, we obtain these equalities:

$$\begin{aligned}
&\mathbf{G}(F)(\mathrm{d}(\phi)) \\
&\quad = (-1)^i \cdot \mathbf{G}(F)(\phi) \circ \mathbf{G}(F)(\mathrm{d}_{M_1}) - (-1)^i \cdot (-1)^i \cdot \mathbf{G}(F)(\mathrm{d}_{M_0}) \circ \mathbf{G}(F)(\phi) \\
&\quad = \mathrm{d}_{\mathbf{C}(F)(M_0)} \circ \mathbf{G}(F)(\phi) - (-1)^i \cdot \mathbf{G}(F)(\phi) \circ \mathrm{d}_{\mathbf{C}(F)(M_1)} = \mathrm{d}(\mathbf{G}(F)(\phi)).
\end{aligned}$$

The claim about the boundedness conditions is immediate from equation (3.9.5). □

The sign appearing in formula (3.9.11) might seem arbitrary. Besides being the only sign for which Theorem 3.9.13 holds, there is another explanation, which can be seen in the next exercise.

Exercise 3.9.14 Take $\mathsf{M} = \mathsf{N} := \operatorname{Mod} \mathbb{K}$, and consider the contravariant additive functor $F := \operatorname{Hom}_{\mathbb{K}}(-, \mathbb{K})$ from M to itself. Let $M \in \mathbf{C}(\mathsf{M})$; we can view M as a complex of $\mathbb{K}$-modules or as a DG $\mathbb{K}$-module, as done in Proposition 3.6.9. Show that $\mathbf{C}(F)(M) \cong \operatorname{Hom}_{\mathbb{K}}(M, \mathbb{K})$ in $\mathbf{C}_{\mathrm{str}}(\mathbb{K})$, where the second object is the graded module from formula (3.1.42), with the differential d from Definition 3.2.3(2).

The next definition and theorem will help us later when studying contravariant triangulated functors.

Definition 3.9.15 Let A be a DG ring and let M be an abelian category. The *flipped category* of $\mathbf{C}(A, \mathsf{M})$ is the DG category $\mathbf{C}(A, \mathsf{M})^{\mathrm{flip}} := \mathbf{C}(A^{\mathrm{op}}, \mathsf{M}^{\mathrm{op}})$.

Theorem 3.9.16 *Let A be a DG ring and let M be an abelian category. Then:*

(1) *There is a canonical isomorphism of DG categories*

$$\operatorname{Flip} : \mathbf{C}(A, \mathsf{M})^{\mathrm{op}} \xrightarrow{\simeq} \mathbf{C}(A^{\mathrm{op}}, \mathsf{M}^{\mathrm{op}}) = \mathbf{C}(A, \mathsf{M})^{\mathrm{flip}}.$$

(2) *For every boundedness condition $\star$ we have*

$$\operatorname{Flip}(\mathbf{C}^{\star}(A, \mathsf{M})^{\mathrm{op}}) = \mathbf{C}^{-\star}(A^{\mathrm{op}}, \mathsf{M}^{\mathrm{op}}),$$

where $-\star$ is the reversed boundedness condition.

(3) *The induced isomorphism on the strict categories*

$$\operatorname{Str}(\operatorname{Flip}) : \mathbf{C}_{\mathrm{str}}(A, \mathsf{M})^{\mathrm{op}} \xrightarrow{\simeq} \mathbf{C}_{\mathrm{str}}(A^{\mathrm{op}}, \mathsf{M}^{\mathrm{op}})$$

is an exact functor, for the respective abelian category structures.

(4) *For every integer i, the functor* Str(Flip) *makes the diagram*

$$\begin{array}{ccc} \mathbf{C}_{\mathrm{str}}(A, \mathsf{M})^{\mathrm{op}} & \xrightarrow[\cong]{\operatorname{Str}(\operatorname{Flip})} & \mathbf{C}_{\mathrm{str}}(A^{\mathrm{op}}, \mathsf{M}^{\mathrm{op}}) \\ & \searrow^{\mathrm{H}^{i}} & \downarrow{\scriptstyle \mathrm{H}^{-i}} \\ & & \mathsf{M}^{\mathrm{op}} \end{array}$$

commutative, up to an isomorphism of linear functors.

Proof

(1) According to Proposition 3.9.3, there is a contravariant DG functor $\operatorname{Op} : \mathbf{C}(A, \mathsf{M})^{\mathrm{op}} \to \mathbf{C}(A, \mathsf{M})$. It is bijective on objects and morphisms. We are going to construct a contravariant DG functor $E : \mathbf{C}(A, \mathsf{M}) \to \mathbf{C}(A^{\mathrm{op}}, \mathsf{M}^{\mathrm{op}})$ which is also bijective on objects and morphisms. The composed DG functor

$$\operatorname{Flip} := E \circ \operatorname{Op} : \mathbf{C}(A, \mathsf{M})^{\mathrm{op}} \to \mathbf{C}(A^{\mathrm{op}}, \mathsf{M}^{\mathrm{op}})$$

will have the desired properties.

Let us construct E. We start with the contravariant additive functor $F := \operatorname{Op} : \mathsf{M} \to \mathsf{M}^{\mathrm{op}}$. Theorem 3.9.13 says that $\mathbf{C}(F) : \mathbf{C}(\mathsf{M}) \to \mathbf{C}(\mathsf{M}^{\mathrm{op}})$

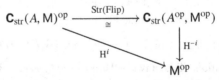

is a contravariant DG functor. Recall that an object of $\mathbf{C}(A, \mathsf{M})$ is a triple (M, d_M, f_M), where $M \in \mathbf{G}(\mathsf{M})$; d_M is a differential on the graded object M; and $f_M : A \to \mathrm{End}_{\mathbf{C}(\mathsf{M})}(M)$ is a DG ring homomorphism. See Definitions 3.1.47, 3.8.1 and 3.8.3. Define $(N, \mathrm{d}_N) := \mathbf{C}(F)(M, \mathrm{d}_M) \in \mathbf{C}(\mathsf{M}^{\mathrm{op}})$. Since $\mathbf{C}(F) : \mathrm{End}_{\mathbf{C}(\mathsf{M})}(M, \mathrm{d}_M) \to \mathrm{End}_{\mathbf{C}(\mathsf{M}^{\mathrm{op}})}(N, \mathrm{d}_N)$ is a DG ring antihomomorphism (by which we mean the single object version of a contravariant DG functor), and $\mathrm{Op} : A^{\mathrm{op}} \to A$ is also such an antihomomorphism, it follows that

$$f_N := \mathbf{C}(F) \circ f_M \circ \mathrm{Op} : A^{\mathrm{op}} \to \mathrm{End}_{\mathbf{C}(\mathsf{M}^{\mathrm{op}})}(N, \mathrm{d}_N)$$

is a DG ring homomorphism. Thus $E(M, \mathrm{d}_M, f_M) := (N, \mathrm{d}_N, f_N)$ is an object of $\mathbf{C}(A^{\mathrm{op}}, \mathsf{M}^{\mathrm{op}})$. In this way we have a function

$$E : \mathrm{Ob}(\mathbf{C}(A, \mathsf{M})) \to \mathrm{Ob}(\mathbf{C}(A^{\mathrm{op}}, \mathsf{M}^{\mathrm{op}})),$$

and it is clearly bijective.

The operation of E on morphisms is of course that of $\mathbf{C}(F)$. It remains to verify that the resulting morphisms in $\mathbf{C}(\mathsf{M}^{\mathrm{op}})$ respect the action of elements of A^{op}. Namely that the condition in Definition 3.1.47 is satisfied. Take any morphism

$$\phi \in \mathrm{Hom}_{\mathbf{C}(A,\mathsf{M})}\big((M_0, \mathrm{d}_{M_0}, f_{M_0}), (M_1, \mathrm{d}_{M_1}, f_{M_1})\big)^i$$

and any element $a \in (A^{\mathrm{op}})^j$; and write $(N_k, \mathrm{d}_{N_k}, f_{N_k}) := E(M_k, \mathrm{d}_{M_k}, f_{M_k})$ and

$$\psi := \mathbf{G}(F)(\phi) \in \mathrm{Hom}_{\mathbf{C}(\mathsf{M}^{\mathrm{op}})}\big((N_1, \mathrm{d}_{N_1}), (N_0, \mathrm{d}_{N_0})\big)^i.$$

We have to prove that

$$\psi \circ f_{N_1}(a) = (-1)^{i \cdot j} \cdot f_{N_0}(a) \circ \psi. \tag{3.9.17}$$

This is done using Lemma 3.9.8, like in the proof of Theorem 3.9.13; and we leave this to the reader.

(2) Clear from Theorem 3.9.13.

(3) Exactness in the categories $\mathbf{C}_{\mathrm{str}}(A, \mathsf{M})^{\mathrm{op}}$ and $\mathbf{C}_{\mathrm{str}}(A^{\mathrm{op}}, \mathsf{M}^{\mathrm{op}})$ is checked in each degree separately, and both are exactness in the abelian category M^{op}. The functor Str(Flip) is $\pm\,\mathrm{Id}_{\mathsf{M}^{\mathrm{op}}}$ in each degree, so it is exact.

(4) As complexes in M^{op}, M and $\mathrm{Flip}(M)$ are equal, up to the renumbering of degrees and the signs of the differentials in various degrees. So the cohomology objects satisfy $\mathrm{H}^i(M) \cong \mathrm{H}^{-i}(\mathrm{Flip}(M))$ in M^{op}. □

Exercise 3.9.18 Prove formula (3.9.17) above.

Remark 3.9.19 Theorem 3.9.16 will be used to introduce a triangulated structure on the category $\mathbf{K}(A, \mathsf{M})^{\mathrm{op}}$. This will be done in Section 5.6.

Combined with Proposition 3.9.3, Theorem 3.9.16 allows us to replace a contravariant DG functor $F : \mathbf{C}(A, \mathsf{M}) \to \mathsf{D}$ with a usual, covariant, DG functor $F \circ \mathrm{Flip}^{-1} : \mathbf{C}(A^{\mathrm{op}}, \mathsf{M}^{\mathrm{op}}) \to \mathsf{D}$. This replacement is going to be very useful when discussing formal properties, such as existence of derived functors, etc.

However, in practical terms (e.g. for producing resolutions of DG modules), the category $\mathbf{C}(A^{\mathrm{op}}, \mathsf{M}^{\mathrm{op}})$ is not very helpful. The reason is that the opposite abelian category M^{op} is almost always a synthetic construction (it does not "really exist in concrete terms"). See Remark 2.7.20 and Example 2.7.21.

We are going to maneuver between the two approaches for reversal of morphisms, each time choosing the more useful approach.

4

Translations and Standard Triangles

As before, we fix a $\mathbb{K}$-linear abelian category M and a DG central $\mathbb{K}$-ring A. In this chapter we study the translation functor and the standard cone of a strict morphism, all in the context of the DG category $\mathbf{C}(A, \mathsf{M})$.

We then study properties of DG functors $F : \mathbf{C}(A, \mathsf{M}) \to \mathbf{C}(B, \mathsf{N})$ between such DG categories. In view of Theorem 3.9.16 it suffices to look at covariant DG functors (and not to worry about contravariant DG functors).

4.1 The Translation Functor

The translation functor goes back to the beginnings of derived categories – see Remark 4.1.13. The treatment in this section, with the "little t operator," is taken from [173, Section 1]. Here and later, the word *operator* is used as a synonym for "morphism in a linear category."

Definition 4.1.1 Let $M = \{M^i\}_{i \in \mathbb{Z}}$ be a graded object in M, i.e. an object of $\mathbf{G}(\mathsf{M})$. The *translation* of M is the object

$$\mathrm{T}(M) = \{\mathrm{T}(M)^i\}_{i \in \mathbb{Z}} \in \mathbf{G}(\mathsf{M}),$$

which is defined as follows: the graded component of degree i of $\mathrm{T}(M)$ is $\mathrm{T}(M)^i := M^{i+1}$.

Definition 4.1.2 (The little t operator) Let $M = \{M^i\}_{i \in \mathbb{Z}}$ be an object of $\mathbf{G}(\mathsf{M})$. We define

$$\mathrm{t}_M : M \to \mathrm{T}(M)$$

to be the degree -1 morphism of graded objects of M, that for every degree i is the identity morphism $\mathrm{t}_M|_{M^i} := \mathrm{id}_{M^i} : M^i \xrightarrow{\simeq} \mathrm{T}(M)^{i-1}$ of the object M^i in M.

Note that the morphism $\mathrm{t}_M : M \to \mathrm{T}(M)$ is a degree -1 isomorphism in the graded category $\mathbf{G}(\mathsf{M})$.

Definition 4.1.3 For $M \in \mathbf{G}(\mathsf{M})$ we define the morphism $\mathrm{t}_M^{-1} : \mathrm{T}(M) \to M$ in $\mathbf{G}(\mathsf{M})$ to be the inverse of t_M.

Of course, the morphism t_M^{-1} has degree $+1$. The reason for stating this definition is to avoid the potential confusion between the morphism t_M^{-1} in $\mathbf{G}(\mathsf{M})$ and the degree -1 component of the morphism t_M, which we denote by $\mathrm{t}_M|_{M^{-1}}$, as in Definition 4.1.2 above.

Definition 4.1.4 Let $M = \{M^i\}_{i\in\mathbb{Z}}$ be a DG A-module in M, i.e. an object of $\mathbf{C}(A, \mathsf{M})$. The *translation* of M is the object

$$\mathrm{T}(M) \in \mathbf{C}(A, \mathsf{M})$$

defined as follows.

(1) As a graded object of M, it is as specified in Definition 4.1.1.

(2) The differential $\mathrm{d}_{\mathrm{T}(M)}$ is defined by the formula

$$\mathrm{d}_{\mathrm{T}(M)} := -\,\mathrm{t}_M \circ \mathrm{d}_M \circ \mathrm{t}_M^{-1}.$$

(3) Let $f_M : A \to \mathrm{End}_{\mathsf{M}}(M)$ be the DG ring homomorphism that determines the action of A on M. Then

$$f_{\mathrm{T}(M)} : A \to \mathrm{End}_{\mathsf{M}}(\mathrm{T}(M))$$

is defined by

$$f_{\mathrm{T}(M)}(a) := (-1)^j \cdot \mathrm{t}_M \circ f_M(a) \circ \mathrm{t}_M^{-1}$$

for $a \in A^j$.

Thus, the differential $\mathrm{d}_{\mathrm{T}(M)} = \{\mathrm{d}^i_{\mathrm{T}(M)}\}_{i\in\mathbb{Z}}$ makes this diagram in M commutative for every i:

$$\begin{array}{ccc} \mathrm{T}(M)^i & \xrightarrow{\mathrm{d}^i_{\mathrm{T}(M)}} & \mathrm{T}(M)^{i+1} \\ {\scriptstyle \mathrm{t}_M}\uparrow & & \uparrow{\scriptstyle \mathrm{t}_M} \\ M^{i+1} & \xrightarrow{-\mathrm{d}^{i+1}_M} & M^{i+2} \end{array}$$

And the left A-module structure makes this diagram in M commutative for every i and every $a \in A^j$:

$$\begin{array}{ccc} \mathrm{T}(M)^i & \xrightarrow{f_{\mathrm{T}(M)}(a)} & \mathrm{T}(M)^{i+j} \\ {\scriptstyle \mathrm{t}_M}\uparrow & & \uparrow{\scriptstyle \mathrm{t}_M} \\ M^{i+1} & \xrightarrow{(-1)^j \cdot f_M(a)} & M^{i+j+1} \end{array}$$

Proposition 4.1.5 *The morphisms* t_M *and* t_M^{-1} *are cocycles, in the DG* $\mathbb{K}$-*modules* $\mathrm{Hom}_{A,\mathsf{M}}(M, \mathrm{T}(M))$ *and* $\mathrm{Hom}_{A,\mathsf{M}}(\mathrm{T}(M), M)$, *respectively.*

Proof We use the notation $\mathrm{d}_{\mathrm{Hom}}$ for the differential in the DG module $\mathrm{Hom}_{A,\mathsf{M}}(M, \mathrm{T}(M))$. Let us calculate. Because t_M has degree -1, we have

$$\begin{aligned} \mathrm{d}_{\mathrm{Hom}}(\mathrm{t}_M) &= \mathrm{d}_{\mathrm{T}(M)} \circ \mathrm{t}_M + \mathrm{t}_M \circ \mathrm{d}_M \\ &= (-\mathrm{t}_M \circ \mathrm{d}_M \circ \mathrm{t}_M^{-1}) \circ \mathrm{t}_M + \mathrm{t}_M \circ \mathrm{d}_M = 0. \end{aligned}$$

As for t_M^{-1} : this is done using the graded Leibniz rule, just as in the proof of Proposition 3.4.9. □

Definition 4.1.6 Given a morphism $\phi \in \mathrm{Hom}_{A,\mathsf{M}}(M, N)^i$ we define the morphism

$$\mathrm{T}(\phi) \in \mathrm{Hom}_{A,\mathsf{M}}(\mathrm{T}(M), \mathrm{T}(N))^i$$

to be

$$\mathrm{T}(\phi) := (-1)^i \cdot \mathrm{t}_N \circ \phi \circ \mathrm{t}_M^{-1} .$$

To clarify this definition, let us write $\phi = \{\phi^j\}_{j \in \mathbb{Z}}$, so that $\phi^j : M^j \to N^{j+i}$ is a morphism in M. Then $\mathrm{T}(\phi)^j : \mathrm{T}(M)^j \to \mathrm{T}(N)^{j+i}$ is $\mathrm{T}(\phi)^j = (-1)^i \cdot \mathrm{t}_N \circ \phi^{j+1} \circ \mathrm{t}_M^{-1}$. The corresponding commutative diagram in M, for each i, j, is:

$$\begin{array}{ccc} \mathrm{T}(M)^j & \xrightarrow{\mathrm{T}(\phi)^j} & \mathrm{T}(N)^{j+i} \\ {\scriptstyle \mathrm{t}_M}\uparrow & & \uparrow{\scriptstyle \mathrm{t}_N} \\ M^{j+1} & \xrightarrow{(-1)^i \cdot \phi^{j+1}} & N^{j+i+1} \end{array} \qquad (4.1.7)$$

Theorem 4.1.8 *Let* M *be an abelian category and let* A *be a DG ring.*

(1) *The assignments* $M \mapsto \mathrm{T}(M)$ *and* $\phi \mapsto \mathrm{T}(\phi)$ *are a DG functor*

$$\mathrm{T} : \mathbf{C}(A, \mathsf{M}) \to \mathbf{C}(A, \mathsf{M}).$$

(2) *The collection* $\mathrm{t} := \{\mathrm{t}_M\}_{M \in \mathbf{C}(A,\mathsf{M})}$ *is a degree* -1 *isomorphism*

$$\mathrm{t} : \mathrm{Id} \to \mathrm{T}$$

of DG functors from $\mathbf{C}(A, \mathsf{M})$ *to itself.*

Proof

(1) Take morphisms $\phi_1 : M_0 \to M_1$ and $\phi_2 : M_1 \to M_2$, of degrees i_1 and i_2, respectively. Then

$$\begin{aligned}
\mathrm{T}(\phi_2 \circ \phi_1) &= (-1)^{i_1+i_2} \cdot \mathrm{t}_{M_2} \circ (\phi_2 \circ \phi_1) \circ \mathrm{t}_{M_0}^{-1} \\
&= (-1)^{i_1+i_2} \cdot \mathrm{t}_{M_2} \circ \phi_2 \circ (\mathrm{t}_{M_1}^{-1} \circ \mathrm{t}_{M_1}) \circ \phi_1 \circ \mathrm{t}_{M_0}^{-1} \\
&= ((-1)^{i_2} \cdot \mathrm{t}_{M_2} \circ \phi_2 \circ \mathrm{t}_{M_1}^{-1}) \circ ((-1)^{i_1} \cdot \mathrm{t}_{M_1} \circ \phi_1 \circ \mathrm{t}_{M_0}^{-1}) \\
&= \mathrm{T}(\phi_2) \circ \mathrm{T}(\phi_1).
\end{aligned}$$

Clearly, $\mathrm{T}(\mathrm{id}_M) = \mathrm{id}_{\mathrm{T}(M)}$, and $\mathrm{T}(\lambda \cdot \phi + \psi) = \lambda \cdot \mathrm{T}(\phi) + \mathrm{T}(\psi)$ for all $\lambda \in \mathbb{K}$ and $\phi, \psi \in \mathrm{Hom}_{A,\mathrm{M}}(M_0, M_1)^i$. So T is a $\mathbb{K}$-linear graded functor.

By Proposition 4.1.5 we know that $\mathrm{d} \circ \mathrm{t} = -\mathrm{t} \circ \mathrm{d}$ and $\mathrm{d} \circ \mathrm{t}^{-1} = -\mathrm{t}^{-1} \circ \mathrm{d}$. This implies that for every morphism ϕ in $\mathbf{C}(A, \mathrm{M})$, we have $\mathrm{T}(\mathrm{d}(\phi)) = \mathrm{d}(\mathrm{T}(\phi))$. So T is a DG functor.

(2) Take some $\phi \in \mathrm{Hom}_{A,\mathrm{M}}(M_0, M_1)^i$. We have to prove that $\mathrm{t}_{M_1} \circ \phi = (-1)^i \cdot \mathrm{T}(\phi) \circ \mathrm{t}_{M_0}$ as elements of $\mathrm{Hom}_{A,\mathrm{M}}(M_0, \mathrm{T}(M_1))^{i+1}$. But by Definition 4.1.6 we have

$$\mathrm{T}(\phi) \circ \mathrm{t}_{M_0} = ((-1)^i \cdot \mathrm{t}_{M_1} \circ \phi \circ \mathrm{t}_{M_0}^{-1}) \circ \mathrm{t}_{M_0} = (-1)^i \cdot \mathrm{t}_{M_1} \circ \phi. \qquad \square$$

Definition 4.1.9 We call T the *translation functor* of the DG category $\mathbf{C}(A, \mathrm{M})$.

Corollary 4.1.10

1. *The functor* T *is an automorphism of the DG category* $\mathbf{C}(A, \mathrm{M})$.
2. *For every* $k, l \in \mathbb{Z}$ *there is an equality of functors* $\mathrm{T}^l \circ \mathrm{T}^k = \mathrm{T}^{l+k}$.

Proof (1) By the theorem, T is a DG functor. By definition, T is bijective on the set of objects of $\mathbf{C}(A, \mathrm{M})$ and on the sets of morphisms.

(2) By part (1) of this corollary, the inverse T^{-1} is a uniquely defined functor (not just up to an isomorphism of functors). Thus we can define T^k, the kth power of T, and the equality stated holds. □

Proposition 4.1.11 *Consider any* $M \in \mathbf{C}(A, \mathrm{M})$.

1. *There is equality* $\mathrm{t}_{\mathrm{T}(M)} = -\mathrm{T}(\mathrm{t}_M)$ *of degree* -1 *morphisms* $\mathrm{T}(M) \to \mathrm{T}^2(M)$ *in* $\mathbf{C}(A, \mathrm{M})$.
2. *There is equality* $\mathrm{t}_{\mathrm{T}^{-1}(M)} = -\mathrm{T}^{-1}(\mathrm{t}_M)$ *between these degree* -1 *morphisms* $\mathrm{T}^{-1}(M) \to \mathrm{T}(\mathrm{T}^{-1}(M)) = M = \mathrm{T}^{-1}(\mathrm{T}(M))$ *in* $\mathbf{C}(A, \mathrm{M})$.

Proof (1) This is an easy calculation, using Definition 4.1.6:

$$\mathrm{T}(\mathrm{t}_M) = -\mathrm{t}_{\mathrm{T}(M)} \circ \mathrm{t}_M \circ \mathrm{t}_M^{-1} = -\mathrm{t}_{\mathrm{T}(M)}\,.$$

(2) A similar calculation. □

Proposition 4.1.12 *Let $M \in \mathbf{C}(A, \mathsf{M})$ and $i \in \mathbb{Z}$. There is equality* $\mathrm{d}_{\mathrm{T}^i(M)} = \mathrm{T}^i(\mathrm{d}_M)$ *in* $\mathrm{Hom}_{A,\mathsf{M}}(\mathrm{T}^i(M), \mathrm{T}^i(M))^1$.

Proof We start with $i = 1$. The differential d_M is an element of $\mathrm{Hom}_{\mathsf{M}}(M, M)^1$. By Definitions 4.1.4 and 4.1.6 we get $\mathrm{d}_{\mathrm{T}(M)} = -\mathrm{t}_M \circ \mathrm{d}_M \circ \mathrm{t}_M^{-1} = \mathrm{T}(\mathrm{d}_M)$. Using induction on i the assertion holds for all $i \geq 0$.

For $i \leq 0$ we use descending induction on i. We assume that the assertion holds for i. Let us define $N := \mathrm{T}^{i-1}(M)$, so $\mathrm{T}(N) = \mathrm{T}^i(M)$. By the previous paragraph, with the DG module N, we know that $\mathrm{d}_{\mathrm{T}^i(M)} = \mathrm{d}_{\mathrm{T}(N)} = \mathrm{T}(\mathrm{d}_N) = \mathrm{T}(\mathrm{d}_{\mathrm{T}^{i-1}(M)})$. Applying the functor T^{-1} to this equality we get $\mathrm{T}^{i-1}(\mathrm{d}_M) = \mathrm{T}^{-1}(\mathrm{T}^i(\mathrm{d}_M)) = \mathrm{T}^{-1}(\mathrm{d}_{\mathrm{T}^i(M)}) = \mathrm{d}_{\mathrm{T}^{i-1}(M)}$. □

Remark 4.1.13 There are several names in the literature for the translation functor T : *twist*, *shift* and *suspension*. There are also several notations: $\mathrm{T}(M) = M[1] = \Sigma M$. In the later part of this book we shall use the notation $M[k] := \mathrm{T}^k(M)$ for the kth translation.

4.2 The Standard Triangle of a Strict Morphism

As before, we fix an abelian category M and a DG ring A. Here is the cone construction in $\mathbf{C}(A, \mathsf{M})$, as it looks using the operator t.

Definition 4.2.1 Let $\phi : M \to N$ be a strict morphism in $\mathbf{C}(A, \mathsf{M})$. The *standard cone of* ϕ is the object $\mathrm{Cone}(\phi) \in \mathbf{C}(A, \mathsf{M})$ defined as follows. As a graded A-module in M we let

$$\mathrm{Cone}(\phi) := N \oplus \mathrm{T}(M).$$

The differential $\mathrm{d}_{\mathrm{Cone}}$ is this: if we express the graded module as a column

$$\mathrm{Cone}(\phi) = \begin{bmatrix} N \\ \mathrm{T}(M) \end{bmatrix},$$

then $\mathrm{d}_{\mathrm{Cone}}$ is left multiplication by the matrix

$$\mathrm{d}_{\mathrm{Cone}} := \begin{bmatrix} \mathrm{d}_N & \phi \circ \mathrm{t}_M^{-1} \\ 0 & \mathrm{d}_{\mathrm{T}(M)} \end{bmatrix}$$

of degree 1 morphisms of graded A-modules in M.

In other words, the morphism $\mathrm{d}^i_{\mathrm{Cone}} : \mathrm{Cone}(\phi)^i \to \mathrm{Cone}(\phi)^{i+1}$ is $\mathrm{d}^i_{\mathrm{Cone}} = \mathrm{d}^i_N + \mathrm{d}^i_{\mathrm{T}(M)} + \phi^{i+1} \circ \mathrm{t}_M^{-1}$, where $\phi^{i+1} \circ \mathrm{t}_M^{-1}$ is the composed morphism $\mathrm{T}(M)^i \xrightarrow{\mathrm{t}_M^{-1}} M^{i+1} \xrightarrow{\phi^{i+i}} N^{i+1}$.

In the situation of Definition 4.2.1, let us denote by

$$e_\phi : N \to N \oplus \mathrm{T}(M) \tag{4.2.2}$$

the embedding, and by

$$p_\phi : N \oplus \mathrm{T}(M) \to \mathrm{T}(M) \tag{4.2.3}$$

the projection. Thus, as matrices we have

$$e_\phi = \begin{bmatrix} \mathrm{id}_N \\ 0 \end{bmatrix} \quad \text{and} \quad p_\phi = \begin{bmatrix} 0 & \mathrm{id}_{\mathrm{T}(M)} \end{bmatrix}.$$

The standard cone of ϕ sits in the exact sequence

$$0 \to N \xrightarrow{e_\phi} \mathrm{Cone}(\phi) \xrightarrow{p_\phi} \mathrm{T}(M) \to 0 \tag{4.2.4}$$

in the abelian category $\mathbf{C}_{\mathrm{str}}(A, \mathsf{M})$.

Definition 4.2.5 Let $\phi : M \to N$ be a morphism in $\mathbf{C}_{\mathrm{str}}(A, \mathsf{M})$. The diagram

$$M \xrightarrow{\phi} N \xrightarrow{e_\phi} \mathrm{Cone}(\phi) \xrightarrow{p_\phi} \mathrm{T}(M)$$

in $\mathbf{C}_{\mathrm{str}}(A, \mathsf{M})$ is called the *standard triangle* associated to ϕ.

The cone construction is functorial, in the following sense.

Proposition 4.2.6 *Let*

$$\begin{array}{ccc} M_0 & \xrightarrow{\phi_0} & N_0 \\ \downarrow{\scriptstyle\psi} & & \downarrow{\scriptstyle\chi} \\ M_1 & \xrightarrow{\phi_1} & N_1 \end{array}$$

be a commutative diagram in $\mathbf{C}_{\mathrm{str}}(A, \mathsf{M})$. *Then*

$$(\chi, \mathrm{T}(\psi)) : \mathrm{Cone}(\phi_0) \to \mathrm{Cone}(\phi_1) \tag{4.2.7}$$

is a morphism in $\mathbf{C}_{\mathrm{str}}(A, \mathsf{M})$, *and the diagram*

$$\begin{array}{ccccccc} M_0 & \xrightarrow{\phi_0} & N_0 & \xrightarrow{e_{\phi_0}} & \mathrm{Cone}(\phi_0) & \xrightarrow{p_{\phi_0}} & \mathrm{T}(M_0) \\ \downarrow{\scriptstyle\psi} & & \downarrow{\scriptstyle\chi} & & \downarrow{\scriptstyle(\chi,\mathrm{T}(\psi))} & & \downarrow{\scriptstyle\mathrm{T}(\psi)} \\ M_1 & \xrightarrow{\phi_1} & N_1 & \xrightarrow{e_{\phi_1}} & \mathrm{Cone}(\phi_1) & \xrightarrow{p_{\phi_1}} & \mathrm{T}(M_1) \end{array}$$

in $\mathbf{C}_{\mathrm{str}}(A, \mathsf{M})$ *is commutative.*

Proof This is a simple consequence of the definitions. □

4.3 The Gauge of a Graded Functor

Recall that we have an abelian category M and a DG ring A. The next definition is new.

Definition 4.3.1 Let $F : \mathbf{C}(A, \mathsf{M}) \to \mathbf{C}(B, \mathsf{N})$ be a graded functor. For every object $M \in \mathbf{C}(A, \mathsf{M})$ let

$$\gamma_{F,M} := \mathrm{d}_{F(M)} - F(\mathrm{d}_M) \in \mathrm{Hom}_{B,\mathsf{N}}(F(M), F(M))^1.$$

The collection of morphisms

$$\gamma_F := \{\gamma_{F,M}\}_{M \in \mathbf{C}(A,\mathsf{M})}$$

is called the *gauge of* F.

The next theorem is due to R. Vyas.

Theorem 4.3.2 *The following two conditions are equivalent for a graded functor* $F : \mathbf{C}(A, \mathsf{M}) \to \mathbf{C}(B, \mathsf{N})$.

(i) F *is a DG functor.*

(ii) *The gauge* γ_F *is a degree* 1 *morphism of graded functors* $\gamma_F : F \to F$.

Proof Recall that F is a DG functor (condition (i)) iff

$$(F \circ \mathrm{d}_{A,\mathsf{M}})(\phi) = (\mathrm{d}_{B,\mathsf{N}} \circ F)(\phi) \tag{4.3.3}$$

for every $\phi \in \mathrm{Hom}_{A,\mathsf{M}}(M_0, M_1)^i$. And γ_F is a degree 1 morphism of graded functors (condition (ii)) iff

$$\gamma_{F,M_1} \circ F(\phi) = (-1)^i \cdot F(\phi) \circ \gamma_{F,M_0} \tag{4.3.4}$$

for every such ϕ.

Here is the calculation. Because F is a graded functor, we get

$$\begin{aligned} F(\mathrm{d}_{A,\mathsf{M}}(\phi)) &= F(\mathrm{d}_{M_1} \circ \phi - (-1)^i \cdot \phi \circ \mathrm{d}_{M_0}) \\ &= F(\mathrm{d}_{M_1}) \circ F(\phi) - (-1)^i \cdot F(\phi) \circ F(\mathrm{d}_{M_0}) \end{aligned} \tag{4.3.5}$$

and

$$\mathrm{d}_{B,\mathsf{N}}(F(\phi)) = \mathrm{d}_{F(M_1)} \circ F(\phi) - (-1)^i \cdot F(\phi) \circ \mathrm{d}_{F(M_0)}. \tag{4.3.6}$$

Using equations (4.3.5) and (4.3.6), and the definition of γ_F, we obtain

$$\begin{aligned} (F \circ \mathrm{d}_{A,\mathsf{M}} - \mathrm{d}_{B,\mathsf{N}} \circ F)(\phi) &= F(\mathrm{d}_{A,\mathsf{M}}(\phi)) - \mathrm{d}_{B,\mathsf{N}}(F(\phi)) \\ &= (F(\mathrm{d}_{M_1}) - \mathrm{d}_{F(M_1)}) \circ F(\phi) - (-1)^i \cdot F(\phi) \circ (F(\mathrm{d}_{M_0}) - \mathrm{d}_{F(M_0)}) \\ &= -\gamma_{F,M_1} \circ F(\phi) + (-1)^i \cdot F(\phi) \circ \gamma_{F,M_0}. \end{aligned} \tag{4.3.7}$$

Finally, the vanishing of the first expression in (4.3.7) is the same as the equality in (4.3.3); whereas the vanishing of the last expression in (4.3.7) is the same as the equality in (4.3.4). □

4.4 The Translation Isomorphism of a DG Functor

Here we consider abelian categories M and N, and DG rings A and B. The translation functor of the DG category $\mathbf{C}(A, \mathsf{M})$ will be denoted here by $\mathrm{T}_{A,\mathsf{M}}$. For an object $M \in \mathbf{C}(A, \mathsf{M})$, we have the little t operator

$$\mathrm{t}_M \in \mathrm{Hom}_{A,\mathsf{M}}(M, \mathrm{T}_{A,\mathsf{M}}(M))^{-1}.$$

This is an isomorphism in the DG category $\mathbf{C}(A, \mathsf{M})$, and likewise for the DG category $\mathbf{C}(B, \mathsf{N})$.

Definition 4.4.1 Let $F : \mathbf{C}(A, \mathsf{M}) \to \mathbf{C}(B, \mathsf{N})$ be a DG functor. For an object $M \in \mathbf{C}(A, \mathsf{M})$, let

$$\tau_{F,M} : F(\mathrm{T}_{A,\mathsf{M}}(M)) \to \mathrm{T}_{B,\mathsf{N}}(F(M))$$

be the degree 0 isomorphism

$$\tau_{F,M} := \mathrm{t}_{F(M)} \circ F(\mathrm{t}_M)^{-1}$$

in $\mathbf{C}(B, \mathsf{N})$, called the *translation isomorphism* of the functor F at the object M.

The isomorphism $\tau_{F,M}$ sits in the following commutative diagram

$$\begin{array}{ccc} F(\mathrm{T}_{A,\mathsf{M}}(M)) & \xrightarrow{\tau_{F,M}} & \mathrm{T}_{B,\mathsf{N}}(F(M)) \\ {\scriptstyle F(\mathrm{t}_M)} \uparrow & \nearrow {\scriptstyle \mathrm{t}_{F(M)}} & \\ F(M) & & \end{array}$$

of isomorphisms in the category $\mathbf{C}(B, \mathsf{N})$.

Proposition 4.4.2 *$\tau_{F,M}$ is an isomorphism in $\mathbf{C}_{\mathrm{str}}(B, \mathsf{N})$.*

Proof We know that $\tau_{F,M}$ is an isomorphism in $\mathbf{C}(B, \mathsf{N})$. It suffices to prove that both $\tau_{F,M}$ and its inverse $\tau_{F,M}^{-1}$ are strict morphisms. Now by Proposition 4.1.5, t_M and t_M^{-1} are cocycles. Therefore, $F(\mathrm{t}_M)$ and $F(\mathrm{t}_M)^{-1} = F(\mathrm{t}_M^{-1})$ are cocycles. For the same reason, $\mathrm{t}_{F(M)}$ and $\mathrm{t}_{F(M)}^{-1}$ are cocycles. But $\tau_{F,M} = \mathrm{t}_{F(M)} \circ F(\mathrm{t}_M)^{-1}$, and $\tau_{F,M}^{-1} = F(\mathrm{t}_M) \circ \mathrm{t}_{F(M)}^{-1}$. □

Theorem 4.4.3 *Let $F : \mathbf{C}(A, \mathsf{M}) \to \mathbf{C}(B, \mathsf{N})$ be a DG functor. Then the collection*

$$\tau_F := \{\tau_{F,M}\}_{M \in \mathbf{C}(A,\mathsf{M})}$$

is an isomorphism

$$\tau_F : F \circ \mathrm{T}_{A,\mathsf{M}} \xrightarrow{\simeq} \mathrm{T}_{B,\mathsf{N}} \circ F$$

of functors $\mathbf{C}_{\mathrm{str}}(A, \mathsf{M}) \to \mathbf{C}_{\mathrm{str}}(B, \mathsf{N})$.

The slogan summarizing this theorem is "A DG functor commutes with translations."

Proof In view of Proposition 4.4.2, all we need to prove is that τ_F is a morphism of functors (i.e. it is a natural transformation).

Let $\phi : M_0 \to M_1$ be a morphism in $\mathbf{C}_{\mathrm{str}}(A, \mathsf{M})$. We must prove that the diagram

$$\begin{array}{ccc}
(F \circ \mathrm{T}_{A,\mathsf{M}})(M_0) & \xrightarrow{\tau_{F,M_0}} & (\mathrm{T}_{B,\mathsf{N}} \circ F)(M_0) \\
{\scriptstyle (F\circ \mathrm{T}_{A,\mathsf{M}})(\phi)} \downarrow & & \downarrow {\scriptstyle (\mathrm{T}_{B,\mathsf{N}} \circ F)(\phi)} \\
(F \circ \mathrm{T}_{A,\mathsf{M}})(M_1) & \xrightarrow{\tau_{F,M_1}} & (\mathrm{T}_{B,\mathsf{N}} \circ F)(M_1)
\end{array}$$

in $\mathbf{C}_{\mathrm{str}}(B, \mathsf{N})$ is commutative. This will be true if the next diagram

$$\begin{array}{ccccc}
(F \circ \mathrm{T}_{A,\mathsf{M}})(M_0) & \xleftarrow{F(\mathrm{t}_{M_0})} & F(M_0) & \xrightarrow{\mathrm{t}_{F(M_0)}} & (\mathrm{T}_{B,\mathsf{N}} \circ F)(M_0) \\
{\scriptstyle (F\circ \mathrm{T}_{A,\mathsf{M}})(\phi)} \downarrow & & {\scriptstyle F(\phi)} \downarrow & & \downarrow {\scriptstyle (\mathrm{T}_{B,\mathsf{N}} \circ F)(\phi)} \\
(F \circ \mathrm{T}_{A,\mathsf{M}})(M_1) & \xleftarrow{F(\mathrm{t}_{M_1})} & F(M_1) & \xrightarrow{\mathrm{t}_{F(M_1)}} & (\mathrm{T}_{B,\mathsf{N}} \circ F)(M_1)
\end{array}$$

in $\mathbf{C}(B, \mathsf{N})$, whose horizontal arrows are isomorphisms, is commutative. For this to be true, it is enough to prove that both squares in this diagram are commutative. This is true by Theorem 4.1.8(2). □

Recall that the translation T and all its powers are DG functors. To finish this section, we calculate their translation isomorphisms.

Proposition 4.4.4 *For every integer k, the translation isomorphism of the DG functor T^k is $\tau_{\mathrm{T}^k} = (-1)^k \cdot \mathrm{id}_{\mathrm{T}^{k+1}}$, where $\mathrm{id}_{\mathrm{T}^{k+1}}$ is the identity automorphism of the functor T^{k+1}.*

Proof By Definition 4.4.1 and Proposition 4.1.11(1), for $k = 1$ the formula is $\tau_{\mathrm{T},M} = \mathrm{t}_{\mathrm{T}(M)} \circ \mathrm{T}(\mathrm{t}_M)^{-1} = -\,\mathrm{id}_{\mathrm{T}^2(M)}$, where $\mathrm{id}_{\mathrm{T}^2(M)}$ is the identity automorphism of the DG module $\mathrm{T}^2(M)$. Hence $\tau_{\mathrm{T}} = -\,\mathrm{id}_{\mathrm{T}^2}$. For other integers k the calculation is similar. □

4.5 Standard Triangles and DG Functors

In Section 4.2 we defined the standard triangle associated to a strict morphism; and in Section 3.5 we defined DG functors. Now we show how these notions interact with each other. As before, we consider abelian categories M and N, and DG rings A and B.

Let $F : \mathbf{C}(A, \mathsf{M}) \to \mathbf{C}(B, \mathsf{N})$ be a DG functor. Given a morphism $\phi : M_0 \to M_1$ in $\mathbf{C}_{\mathrm{str}}(A, \mathsf{M})$, we have a morphism $F(\phi) : F(M_0) \to F(M_1)$ in $\mathbf{C}_{\mathrm{str}}(B, \mathsf{N})$, and objects $F(\mathrm{Cone}_{A,\mathsf{M}}(\phi))$ and $\mathrm{Cone}_{B,\mathsf{N}}(F(\phi))$ in $\mathbf{C}(B, \mathsf{N})$. By definition, and using the fully faithful graded functor Und from Proposition 3.8.18, there is a canonical isomorphism

$$\mathrm{Cone}_{A,\mathsf{M}}(\phi) \cong M_1 \oplus \mathrm{T}_{A,\mathsf{M}}(M_0) \tag{4.5.1}$$

in $\mathbf{G}_{\mathrm{str}}(A, \mathsf{M})$. Since F is a linear functor, it commutes with finite direct sums, and therefore there is a canonical isomorphism

$$F(\mathrm{Cone}_{A,\mathsf{M}}(\phi)) \cong F(M_1) \oplus F(\mathrm{T}_{A,\mathsf{M}}(M_0)) \tag{4.5.2}$$

in $\mathbf{G}_{\mathrm{str}}(A, \mathsf{M})$. And by definition there is a canonical isomorphism

$$\mathrm{Cone}_{B,\mathsf{N}}(F(\phi)) \cong F(M_1) \oplus \mathrm{T}_{B,\mathsf{N}}(F(M_0)) \tag{4.5.3}$$

in $\mathbf{G}_{\mathrm{str}}(A, \mathsf{M})$. Warning: the isomorphisms (4.5.1), (4.5.2) and (4.5.3) might not commute with the differentials. The differentials on the right sides are diagonal matrices, but on the left sides they are upper-triangular matrices (see Definition 4.2.1).

Lemma 4.5.4 *Let $F, G : \mathbf{G}(A, \mathsf{M}) \to \mathbf{G}(B, \mathsf{N})$ be graded functors, and let $\eta : F \to G$ be a degree j morphism of graded functors. Suppose $M \cong M_0 \oplus M_1$ in $\mathbf{G}_{\mathrm{str}}(A, \mathsf{M})$, with embeddings $e_i : M_i \to M$ and projections $p_i : M \to M_i$. Then*

$$\eta_M = (G(e_0), G(e_1)) \circ (\eta_{M_0}, \eta_{M_1}) \circ (F(p_0), F(p_1)),$$

as degree j morphisms $F(M) \to G(M)$ in $\mathbf{G}(B, \mathsf{N})$.

The lemma says that the diagram

$$\begin{array}{ccc} F(M) & \xrightarrow{(F(p_0), F(p_1))} & F(M_0) \oplus F(M_1) \\ {\scriptstyle \eta_M} \downarrow & & \downarrow {\scriptstyle (\eta_{M_0}, \eta_{M_1})} \\ G(M) & \xleftarrow{(G(e_0), G(e_1))} & G(M_0) \oplus G(M_1) \end{array}$$

in $\mathbf{G}(B, \mathsf{N})$ is commutative.

Proof It suffices to prove that the diagram below is commutative for $i = 0, 1$:

$$\begin{array}{ccccc} & & \overset{\mathrm{id}}{\frown} & & \\ F(M_i) & \xrightarrow{F(e_i)} & F(M) & \xrightarrow{F(p_i)} & F(M_i) \\ \downarrow{\scriptstyle \eta_{M_i}} & & \downarrow{\scriptstyle \eta_M} & & \downarrow{\scriptstyle \eta_{M_i}} \\ G(M_i) & \xrightarrow{G(e_i)} & G(M) & \xrightarrow{G(p_i)} & G(M_i) \\ & & \underset{\mathrm{id}}{\smile} & & \end{array}$$

This is true because η is a morphism of functors (a natural transformation). □

Theorem 4.5.5 *Let* $F : \mathbf{C}(A, \mathsf{M}) \to \mathbf{C}(B, \mathsf{N})$ *be a DG functor, and let* $\phi : M_0 \to M_1$ *be a morphism in* $\mathbf{C}_{\mathrm{str}}(A, \mathsf{M})$. *Define the isomorphism*

$$\mathrm{cone}(F, \phi) : F(\mathrm{Cone}_{A,\mathsf{M}}(\phi)) \xrightarrow{\simeq} \mathrm{Cone}_{B,\mathsf{N}}(F(\phi))$$

in $\mathbf{G}_{\mathrm{str}}(A, \mathsf{M})$ *to be*

$$\mathrm{cone}(F, \phi) := (\mathrm{id}_{F(M_1)}, \tau_{F,M_0}).$$

Then:

(1) *The isomorphism* $\mathrm{cone}(F, \phi)$ *commutes with the differentials, so it is an isomorphism in* $\mathbf{C}_{\mathrm{str}}(A, \mathsf{M})$.

(2) *The diagram*

$$\begin{array}{ccccccc} F(M_0) & \xrightarrow{F(\phi)} & F(M_1) & \xrightarrow{F(e_\phi)} & F(\mathrm{Cone}_{A,\mathsf{M}}(\phi)) & \xrightarrow{F(p_\phi)} & F(\mathrm{T}_{A,\mathsf{M}}(M_0)) \\ \downarrow{\scriptstyle \mathrm{id}} & & \downarrow{\scriptstyle \mathrm{id}} & & \downarrow{\scriptstyle \mathrm{cone}(F,\phi)} & & \downarrow{\scriptstyle \tau_{F,M_0}} \\ F(M_0) & \xrightarrow{F(\phi)} & F(M_1) & \xrightarrow{e_{F(\phi)}} & \mathrm{Cone}_{B,\mathsf{N}}(F(\phi)) & \xrightarrow{p_{F(\phi)}} & \mathrm{T}_{B,\mathsf{N}}(F(M_0)) \end{array}$$

in $\mathbf{C}_{\mathrm{str}}(B, \mathsf{N})$ *is commutative.*

When defining $\mathrm{cone}(F, \phi)$ above, we are using the decompositions (4.5.2) and (4.5.3) in the category $\mathbf{G}_{\mathrm{str}}(A, \mathsf{M})$, and the isomorphism τ_{F,M_0} from Definition 4.4.1.

The slogan summarizing this theorem is "A DG functor sends standard triangles to standard triangles."

Proof

(1) To save space, let us write $\theta := \mathrm{cone}(F, \phi)$. We have to prove that $\mathrm{d}_{B,\mathsf{N}}(\theta) = 0$. Let's write $P := \mathrm{Cone}_{A,\mathsf{M}}(\phi)$ and $Q := \mathrm{Cone}_{B,\mathsf{N}}(F(\phi))$. Recall that $\mathrm{d}_{B,\mathsf{N}}(\theta) = \mathrm{d}_Q \circ \theta - \theta \circ \mathrm{d}_{F(P)}$. We have to prove that this is the zero element in $\mathrm{Hom}_{B,\mathsf{N}}(F(P), Q)^1$.

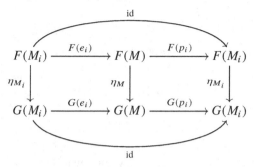

Writing the cones as column modules:

$$P = \begin{bmatrix} M_1 \\ \mathrm{T}_{A,\mathrm{M}}(M_0) \end{bmatrix} \quad \text{and} \quad Q = \begin{bmatrix} F(M_1) \\ \mathrm{T}_{B,\mathrm{N}}(F(M_0)) \end{bmatrix},$$

the matrices representing the morphisms in question are

$$\theta = \begin{bmatrix} \mathrm{id}_{F(M_1)} & 0 \\ 0 & \tau_{F,M_0} \end{bmatrix}, \quad \mathrm{d}_P = \begin{bmatrix} \mathrm{d}_{M_1} & \phi \circ \mathrm{t}_{M_0}^{-1} \\ 0 & \mathrm{d}_{\mathrm{T}_{A,\mathrm{M}}(M_0)} \end{bmatrix}$$

and

$$\mathrm{d}_Q = \begin{bmatrix} \mathrm{d}_{F(M_1)} & F(\phi) \circ \mathrm{t}_{F(M_0)}^{-1} \\ 0 & \mathrm{d}_{\mathrm{T}_{B,\mathrm{N}}(F(M_0))} \end{bmatrix}.$$

Let us write $\gamma := \gamma_F$ for simplicity. According to Theorem 4.3.2, the gauge $\gamma : F \to F$ is a degree 1 morphism of DG functors $\mathbf{C}(A, \mathrm{M}) \to \mathbf{C}(B, \mathrm{N})$. Because the decomposition (4.5.1) is in the category $\mathbf{G}_{\mathrm{str}}(A, \mathrm{M})$, Lemma 4.5.4 tells us that γ_P decomposes too, i.e.

$$\gamma_P = \begin{bmatrix} \gamma_{M_1} & 0 \\ 0 & \gamma_{\mathrm{T}_{A,\mathrm{M}}(M_0)} \end{bmatrix}.$$

By definition of γ_P we have $\mathrm{d}_{F(P)} = F(\mathrm{d}_P) + \gamma_P$ in $\mathrm{Hom}_{B,\mathrm{N}}(F(P), F(P))^1$. It follows that

$$\begin{aligned} \mathrm{d}_{F(P)} &= F(\mathrm{d}_P) + \gamma_P \\ &= \begin{bmatrix} F(\mathrm{d}_{M_1}) & F(\phi \circ \mathrm{t}_{M_0}^{-1}) \\ 0 & F(\mathrm{d}_{\mathrm{T}_{A,\mathrm{M}}(M_0)}) \end{bmatrix} + \begin{bmatrix} \gamma_{M_1} & 0 \\ 0 & \gamma_{\mathrm{T}_{A,\mathrm{M}}(M_0)} \end{bmatrix} \\ &= \begin{bmatrix} F(\mathrm{d}_{M_1}) + \gamma_{M_1} & F(\phi \circ \mathrm{t}_{M_0}^{-1}) \\ 0 & F(\mathrm{d}_{\mathrm{T}_{A,\mathrm{M}}(M_0)}) + \gamma_{\mathrm{T}_{A,\mathrm{M}}(M_0)} \end{bmatrix} \\ &= \begin{bmatrix} \mathrm{d}_{F(M_1)} & F(\phi \circ \mathrm{t}_{M_0}^{-1}) \\ 0 & \mathrm{d}_{F(\mathrm{T}_{A,\mathrm{M}}(M_0))} \end{bmatrix}. \end{aligned}$$

Finally we will check that $\theta \circ \mathrm{d}_{F(P)}$ and $\mathrm{d}_Q \circ \theta$ are equal as matrices of morphisms. We do that in each matrix entry separately. The two left entries in the matrices $\theta \circ \mathrm{d}_{F(P)}$ and $\mathrm{d}_Q \circ \theta$ agree trivially. The bottom right entries in these matrices are $\tau_{F,M_0} \circ \mathrm{d}_{F(\mathrm{T}_{A,\mathrm{M}}(M_0))}$ and $\mathrm{d}_{\mathrm{T}_{B,\mathrm{N}}(F(M_0))} \circ \tau_{F,M_0}$, respectively; they are equal by Proposition 4.4.2. And in the top right entries we have $F(\phi \circ \mathrm{t}_{M_0}^{-1})$ and $F(\phi) \circ \mathrm{t}_{F(M_0)}^{-1} \circ \tau_{F,M_0}$, respectively. Now $F(\phi \circ \mathrm{t}_{M_0}^{-1}) = F(\phi) \circ F(\mathrm{t}_{M_0}^{-1})$; so it suffices to prove that $F(\mathrm{t}_{M_0}^{-1}) = \mathrm{t}_{F(M_0)}^{-1} \circ \tau_{F,M_0}$. This is immediate from the definition of τ_{F,M_0}.

(2) By definition of $\theta = \mathrm{cone}(F, \phi)$, the diagram is commutative in $\mathbf{G}_{\mathrm{str}}(A, \mathrm{M})$. But by part (1) we know that all morphisms in it commute with the differentials, so they lie in $\mathbf{C}_{\mathrm{str}}(B, \mathrm{N})$. And the functor Und from Proposition 3.8.18 is faithful. □

Corollary 4.5.6 *In the situation of Theorem 4.5.5, the diagram*

$$\begin{array}{ccccccc} F(M_0) & \xrightarrow{F(\phi)} & F(M_1) & \xrightarrow{F(e_\phi)} & F(\mathrm{Cone}_{A,\mathsf{M}}(\phi)) & \xrightarrow{\tau_{F,M_0} \circ F(p_\phi)} & \mathrm{T}_{B,\mathsf{N}}(F(M_0)) \\ \downarrow{\scriptstyle \mathrm{id}} & & \downarrow{\scriptstyle \mathrm{id}} & & \downarrow{\scriptstyle \mathrm{cone}(F,\phi)} & & \downarrow{\scriptstyle \mathrm{id}} \\ F(M_0) & \xrightarrow{F(\phi)} & F(M_1) & \xrightarrow{e_{F(\phi)}} & \mathrm{Cone}_{B,\mathsf{N}}(F(\phi)) & \xrightarrow{p_{F(\phi)}} & \mathrm{T}_{B,\mathsf{N}}(F(M_0)) \end{array}$$

is an isomorphism of triangles in $\mathbf{C}_{\mathrm{str}}(B, \mathsf{N})$.

Proof Just rearrange the diagram in item (2) of the theorem. □

4.6 Examples of DG Functors

Recall that M and N are abelian categories, and A and B are DG rings. Here are four examples of DG functors, of various types. We work out in detail the translation isomorphism, the cone isomorphism and the gauge in each example. These examples should serve as templates for constructing other DG functors.

Example 4.6.1 Here $A = B = \mathbb{K}$, so $\mathbf{C}(A, \mathsf{M}) = \mathbf{C}(\mathsf{M})$ and $\mathbf{C}(B, \mathsf{N}) = \mathbf{C}(\mathsf{N})$. Let $F : \mathsf{M} \to \mathsf{N}$ be a $\mathbb{K}$-linear functor. It extends to a functor $\mathbf{C}(F) : \mathbf{C}(\mathsf{M}) \to \mathbf{C}(\mathsf{N})$ as follows: on objects, a complex

$$M = (\{M^i\}_{i\in\mathbb{Z}}, \{\mathrm{d}^i_M\}_{i\in\mathbb{Z}}) \in \mathbf{C}(\mathsf{M})$$

goes to the complex

$$\mathbf{C}(F)(M) := (\{F(M^i)\}, \{F(\mathrm{d}^i_M)\}) \in \mathbf{C}(\mathsf{N}).$$

A morphism $\phi = \{\phi^j\}_{j\in\mathbb{Z}}$ in $\mathbf{C}(\mathsf{M})$ goes to the morphism $\mathbf{C}(F)(\phi) := \{F(\phi^j)\}_{j\in\mathbb{Z}}$ in $\mathbf{C}(\mathsf{N})$. A slightly tedious calculation shows that $\mathbf{C}(F)$ is a graded functor.

Given a complex $M \in \mathbf{C}(\mathsf{M})$, let $N := \mathbf{C}(F)(M) \in \mathbf{C}(\mathsf{N})$. Then the translation of N is $\mathrm{T}_{\mathsf{N}}(N) = \mathbf{C}(F)(\mathrm{T}_{\mathsf{M}}(M))$; and the little t operator of N is $\mathrm{t}_N = \mathbf{C}(F)(\mathrm{t}_M)$. So the translation isomorphism

$$\tau_{\mathbf{C}(F)} : \mathbf{C}(F) \circ \mathrm{T}_{\mathsf{M}} \xrightarrow{\simeq} \mathrm{T}_{\mathsf{N}} \circ \mathbf{C}(F)$$

of functors $\mathbf{C}_{\mathrm{str}}(\mathsf{M}) \to \mathbf{C}_{\mathrm{str}}(\mathsf{N})$ is the identity automorphism of this functor.

Let $\phi : M_0 \to M_1$ be a morphism in $\mathbf{C}_{\mathrm{str}}(\mathsf{M})$, whose image under $\mathbf{C}(F)$ is the morphism $\psi : N_0 \to N_1$ in $\mathbf{C}_{\mathrm{str}}(\mathsf{N})$. Then

$$\mathrm{Cone}(\psi) = N_1 \oplus \mathrm{T}_{\mathsf{N}}(N_0) = \mathbf{C}(F)(\mathrm{Cone}(\phi))$$

as graded objects in N, with differential

$$\mathrm{d}_{\mathrm{Cone}(\psi)} = \begin{bmatrix} \mathrm{d}_{N_1} & \psi \circ \mathrm{t}^{-1}_{N_0} \\ 0 & \mathrm{d}_{\mathrm{T}(N_0)} \end{bmatrix} = \mathbf{C}(F)\left(\begin{bmatrix} \mathrm{d}_{M_1} & \phi \circ \mathrm{t}^{-1}_{M_0} \\ 0 & \mathrm{d}_{\mathrm{T}(M_0)} \end{bmatrix}\right) = \mathbf{C}(F)(\mathrm{d}_{\mathrm{Cone}(\phi)}).$$

We see that the cone isomorphism $\operatorname{cone}(F, \phi)$ is the identity automorphism of the DG module $\operatorname{Cone}(\psi)$. The gauge $\gamma_{\mathbf{C}(F)}$ of the graded functor $\mathbf{C}(F)$ is zero. Therefore, by Theorem 4.3.2, $\mathbf{C}(F)$ is a DG functor.

The next example is much more complicated, and we work out the full details (only once – later on, such details will be left to the reader).

Example 4.6.2 Let A and B be DG rings, and fix some $N \in \mathsf{DGMod}\, B \otimes A^{\mathrm{op}} = \mathbf{C}(B \otimes A^{\mathrm{op}})$. In other words, N is a DG B-A-bimodule. For every $M \in \mathbf{C}(A)$ we have a DG $\mathbb{K}$-module $F(M) := N \otimes_A M$, as in Definition 3.3.23. The differential of $F(M)$ is

$$\mathrm{d}_{F(M)} = \mathrm{d}_N \otimes \mathrm{id}_M + \mathrm{id}_N \otimes \mathrm{d}_M. \tag{4.6.3}$$

See formula (3.1.19) regarding the Koszul sign rule that's involved. But $F(M)$ has the structure of a DG B-module: for every $b \in B$, $n \in N$ and $m \in M$, the action is $b \cdot (n \otimes m) := (b \cdot n) \otimes m$. Clearly

$$F : \mathbf{C}(A) = \mathsf{DGMod}\, A \to \mathbf{C}(B) = \mathsf{DGMod}\, B$$

is a linear functor. We will show that it is actually a DG functor.

Let $M_0, M_1 \in \mathbf{C}(A)$, and consider the $\mathbb{K}$-linear homomorphism

$$F : \operatorname{Hom}_A(M_0, M_1) \to \operatorname{Hom}_B(N \otimes_A M_0, N \otimes_A M_1). \tag{4.6.4}$$

Take any $\phi \in \operatorname{Hom}_A(M_0, M_1)^i$. Then

$$F(\phi) \in \operatorname{Hom}_B(N \otimes_A M_0, N \otimes_A M_1)$$

is the homomorphism that on a homogeneous tensor $n \otimes m \in (N \otimes_A M_0)^{k+j}$, with $n \in N^k$ and $m \in M_0^j$, has the value

$$F(\phi)(n \otimes m) = (-1)^{i \cdot k} \cdot n \otimes \phi(m) \in (N \otimes_A M_1)^{k+j+i}.$$

In other words,

$$F(\phi) = \mathrm{id}_N \otimes \phi. \tag{4.6.5}$$

We see that the homomorphism $F(\phi)$ has degree i. So F is a graded functor.

Let us calculate γ_F, the gauge of F. From (4.6.5) and (4.6.3) we get $\gamma_{F,M} = \mathrm{d}_N \otimes \mathrm{id}_M$, which is often a nonzero endomorphism of $F(M)$. Still, take any degree i morphism $\phi : M_0 \to M_1$ in $\mathbf{C}(A)$. Then

$$\begin{aligned} \gamma_{M_1} \circ F(\phi) &= (\mathrm{d}_N \otimes \mathrm{id}_{M_1}) \circ (\mathrm{id}_N \otimes \phi) \\ &= \mathrm{d}_N \otimes \phi = (-1)^i \cdot (\mathrm{id}_N \otimes \phi) \circ (\mathrm{d}_N \otimes \mathrm{id}_{M_0}) = (-1)^i \cdot F(\phi) \circ \gamma_{M_0}. \end{aligned}$$

We see that γ_F satisfies the condition of Definition 3.5.4(1), which is really Definition 3.1.39. By Theorem 4.3.2, F is a DG functor. (It is also possible to calculate directly that F is a DG functor.)

Finally let us figure out what is the translation isomorphism τ_F of the functor F. Take $M \in \mathbf{C}(A)$. Then $\tau_{F,M} : F(\mathrm{T}_A(M)) \to \mathrm{T}_B(F(M))$ is an isomorphism in $\mathbf{C}_{\mathrm{str}}(B)$. By Definition 4.4.1 we have $\tau_{F,M} = \mathrm{t}_{F(M)} \circ F(\mathrm{t}_M)^{-1}$. Take any $n \in N^k$ and $m \in M^{j+1}$, so that $n \otimes \mathrm{t}_M(m) \in (N \otimes_A \mathrm{T}_A(M))^{k+j} = F(\mathrm{T}_A(M))^{k+j}$, a degree $k + j$ element of $F(\mathrm{T}_A(M))$. But

$$n \otimes \mathrm{t}_M(m) = (-1)^k \cdot (\mathrm{id}_N \otimes \mathrm{t}_M)(n \otimes m) = (-1)^k \cdot F(\mathrm{t}_M)(n \otimes m).$$

Therefore

$$\tau_{F,M}(n \otimes \mathrm{t}_M(m)) = (-1)^k \cdot \mathrm{t}_{F(M)}(n \otimes m) \in \mathrm{T}_B(F(M))^{k+j}.$$

Observe that when N is concentrated in degree 0, we are back in the situation of Example 4.6.1, in which there are no sign twists, and $\tau_{F,M}$ is the identity automorphism.

Example 4.6.6 Let A and B be DG rings, and fix some $N \in \mathsf{DGMod}\, A \otimes B^{\mathrm{op}} = \mathbf{C}(B \otimes A^{\mathrm{op}})$. For any $M \in \mathsf{DGMod}\, A$ we define $F(M) := \mathrm{Hom}_A(N, M)$. This is a DG B-module: for every $b \in B^i$ and $\phi \in \mathrm{Hom}_A(N, M)^j$, the homomorphism $b \cdot \phi \in \mathrm{Hom}_A(N, M)^{i+j}$ has value $(b \cdot \phi)(n) := (-1)^{i \cdot (j+k)} \cdot \phi(n \cdot b) \in M^{i+j+k}$ on $n \in N^k$. As in the previous example,

$$F : \mathbf{C}(A) = \mathsf{DGMod}\, A \to \mathbf{C}(B) = \mathsf{DGMod}\, B$$

is a $\mathbb{K}$-linear graded functor.

The value of the gauge γ_F at $M \in \mathbf{C}(A)$ is $\gamma_{F,M} = \mathrm{Hom}(\mathrm{d}_N, \mathrm{id}_M)$. See formula (3.1.20) regarding this notation. Namely for $\psi \in F(M)^j = \mathrm{Hom}_A(N, M)^j$ we have $\gamma_{F,M}(\psi) = (-1)^j \cdot \psi \circ \mathrm{d}_N$. It is not too hard to check that γ_F is a degree 1 morphism of graded functors. Hence, by Theorem 4.3.2, F is a DG functor.

The formula for the translation isomorphism τ_F is as follows. Take $M \in \mathbf{C}(A)$. Then

$$\tau_{F,M} : F(\mathrm{T}_A(M)) = \mathrm{Hom}_A(N, \mathrm{T}_A(M)) \to \mathrm{T}_B(F(M)) = \mathrm{T}_B(\mathrm{Hom}_A(N, M))$$

is, by definition, $\tau_{F,M} = \mathrm{t}_{F(M)} \circ F(\mathrm{t}_M)^{-1}$. Now $F(\mathrm{t}_M)^{-1} = \mathrm{Hom}(\mathrm{id}_N, \mathrm{t}_M^{-1})$. So given any $\psi \in F(\mathrm{T}_A(M))^k$, we have $\tau_{F,M}(\psi) = \mathrm{t}_{F(M)}(\mathrm{t}_M^{-1} \circ \psi) \in \mathrm{T}_B(F(M))^k$.

We end with a contravariant example.

Example 4.6.7 Let A be a commutative ring. Fix some complex $N \in \mathbf{C}(A)$. For any $M \in \mathbf{C}(A)$ let $F(M) := \mathrm{Hom}_A(M, N) \in \mathbf{C}(A)$. For every degree i morphism $\phi : M_0 \to M_1$ in $\mathbf{C}(A)$ let $F(\phi) : F(M_1) \to F(M_0)$ be the degree i morphism $F(\phi) := \mathrm{Hom}_A(\phi, \mathrm{id}_N)$.

A direct calculation, which we leave to the reader, shows that

$$F : \mathrm{Hom}_{\mathbf{C}(A)}(M_0, M_1) \to \mathrm{Hom}_{\mathbf{C}(A)}(F(M_1), F(M_0))$$

is a strict homomorphism of DG $\mathbb{K}$-modules. It also satisfies conditions (a) and (b) of Definition 3.9.1. Thus we have a contravariant DG functor $F : \mathbf{C}(A) \to \mathbf{C}(A)$. For contravariant DG functors we do not talk about translation isomorphisms or gauges.

We will return to the *DG bifunctors* $\mathrm{Hom}(-, -)$ and $(- \otimes -)$ later in the book, in Section 9.1.

5

Triangulated Categories and Functors

In this chapter we introduce triangulated categories and triangulated functors. We prove that for a DG ring A and an abelian category M, the homotopy category $\mathbf{K}(A, \mathsf{M})$ is triangulated. We also prove that a DG functor $F : \mathbf{C}(A, \mathsf{M}) \to \mathbf{C}(B, \mathsf{N})$ induces a triangulated functor $F : \mathbf{K}(A, \mathsf{M}) \to \mathbf{K}(B, \mathsf{N})$.

Recall that by Convention 1.2.4, there is a nonzero commutative base ring $\mathbb{K}$, which is implicit most of the time. All rings are central $\mathbb{K}$-rings, and all ring homomorphisms are over $\mathbb{K}$. All linear categories are $\mathbb{K}$-linear, and all linear (i.e. additive) functors between them are $\mathbb{K}$-linear.

5.1 Triangulated Categories

In this section we define triangulated categories and make some remarks regarding them.

Recall that a functor is called an isomorphism of categories if it is bijective on sets of objects and on sets of morphisms.

Definition 5.1.1 Let K be an additive category. A *translation functor* on K is an additive automorphism T of K. The pair (K, T) is called a *T-additive category*.

Remark 5.1.2 Some texts give a more relaxed definition: the translation functor T is only required to be an additive autoequivalence of K. The resulting theory is more complicated (it is 2-categorical, but most texts try to suppress this fact).

Later in the book (see Definition 12.1.7) we will write $M[k] := \mathrm{T}^k(M)$, the kth translation of an object M.

Definition 5.1.3 Suppose $(\mathsf{K}, \mathrm{T}_{\mathsf{K}})$ and $(\mathsf{L}, \mathrm{T}_{\mathsf{L}})$ are T-additive categories. A *T-additive functor* between them is a pair (F, τ), consisting of an additive functor $F : \mathsf{K} \to \mathsf{L}$, together with an isomorphism

$$\tau : F \circ \mathrm{T}_{\mathsf{K}} \xrightarrow{\simeq} \mathrm{T}_{\mathsf{L}} \circ F$$

of functors $\mathsf{K} \to \mathsf{L}$, called a *translation isomorphism.*

Definition 5.1.4 Let $(\mathsf{K}_i, \mathrm{T}_i)$ be T-additive categories, for $i = 0, 1, 2$, and let

$$(F_i, \tau_i) : (\mathsf{K}_{i-1}, \mathrm{T}_{i-1}) \to (\mathsf{K}_i, \mathrm{T}_i)$$

be T-additive functors. The composition

$$(F, \tau) = (F_2, \tau_2) \circ (F_1, \tau_1)$$

is the T-additive functor $(\mathsf{K}_0, \mathrm{T}_0) \to (\mathsf{K}_2, \mathrm{T}_2)$ defined as follows: the functor is $F := F_2 \circ F_1$, and the translation isomorphism $\tau : F \circ T_0 \xrightarrow{\simeq} T_2 \circ F$ is $\tau := \tau_2 \circ F_2(\tau_1)$.

Definition 5.1.5 Suppose $(\mathsf{K}, \mathrm{T}_{\mathsf{K}})$ and $(\mathsf{L}, \mathrm{T}_{\mathsf{L}})$ are T-additive categories, and

$$(F, \tau), (G, \nu) : (\mathsf{K}, \mathrm{T}_{\mathsf{K}}) \to (\mathsf{L}, \mathrm{T}_{\mathsf{L}})$$

are T-additive functors. A *morphism of T-additive functors*

$$\eta : (F, \tau) \to (G, \nu)$$

is a morphism of functors $\eta : F \to G$, such that for every object $M \in \mathsf{K}$ this diagram in L is commutative:

$$\begin{array}{ccc} F(\mathrm{T}_{\mathsf{K}}(M)) & \xrightarrow{\tau_M} & \mathrm{T}_{\mathsf{L}}(F(M)) \\ \Big\downarrow{\scriptstyle \eta_{\mathrm{T}_{\mathsf{K}}(M)}} & & \Big\downarrow{\scriptstyle \mathrm{T}_{\mathsf{L}}(\eta_M)} \\ G(\mathrm{T}_{\mathsf{K}}(M)) & \xrightarrow{\nu_M} & \mathrm{T}_{\mathsf{L}}(G(M)) \end{array}$$

Definition 5.1.6 Let (K, T) be a T-additive category. A *triangle* in (K, T) is a diagram

$$L \xrightarrow{\alpha} M \xrightarrow{\beta} N \xrightarrow{\gamma} \mathrm{T}(L)$$

in K.

Sometimes, when the names of the morphisms in the triangle above are not important, we may use the more compact notation

$$L \to M \to N \xrightarrow{\triangle} . \tag{5.1.7}$$

Definition 5.1.8 Let (K, T) be a T-additive category. Suppose

$$L \xrightarrow{\alpha} M \xrightarrow{\beta} N \xrightarrow{\gamma} \mathrm{T}(L)$$

and

$$L' \xrightarrow{\alpha'} M' \xrightarrow{\beta'} N' \xrightarrow{\gamma'} \mathrm{T}(L')$$

are triangles in (K, T). A *morphism of triangles* between them is a commutative diagram

$$\begin{array}{ccccccc} L & \xrightarrow{\alpha} & M & \xrightarrow{\beta} & N & \xrightarrow{\gamma} & \mathrm{T}(L) \\ \phi \downarrow & & \psi \downarrow & & \chi \downarrow & & \downarrow \mathrm{T}(\phi) \\ L' & \xrightarrow{\alpha'} & M' & \xrightarrow{\beta'} & N' & \xrightarrow{\gamma'} & \mathrm{T}(L') \end{array}$$

in K.

The morphism of triangles (ϕ, ψ, χ) is called an isomorphism if ϕ, ψ and χ are all isomorphisms.

Remark 5.1.9 Why "triangle"? This is because sometimes a triangle

$$L \xrightarrow{\alpha} M \xrightarrow{\beta} N \xrightarrow{\gamma} \mathrm{T}(L)$$

is written as a diagram

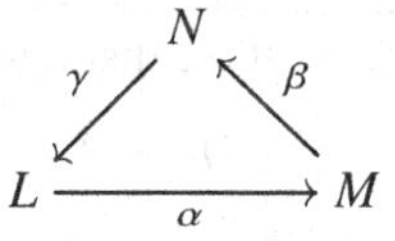

See [62, page 20].

Definition 5.1.10 A *triangulated category* is a T-additive category (K, T), equipped with a set of triangles called *distinguished triangles*. The following axioms have to be satisfied:

(TR1) (a) Every triangle that is isomorphic to a distinguished triangle is also a distinguished triangle.

(b) For every morphism $\alpha : L \to M$ in K there is a distinguished triangle

$$L \xrightarrow{\alpha} M \to N \to \mathrm{T}(L).$$

(c) For every object M the triangle

$$M \xrightarrow{\mathrm{id}_M} M \to 0 \to \mathrm{T}(M)$$

is distinguished.

(TR2) A triangle

$$L \xrightarrow{\alpha} M \xrightarrow{\beta} N \xrightarrow{\gamma} \mathrm{T}(L)$$

is distinguished iff the triangle

$$M \xrightarrow{\beta} N \xrightarrow{\gamma} \mathrm{T}(L) \xrightarrow{-\mathrm{T}(\alpha)} \mathrm{T}(M)$$

is distinguished.

(TR3) Suppose

$$\begin{array}{ccccccc} L & \xrightarrow{\alpha} & M & \xrightarrow{\beta} & N & \xrightarrow{\gamma} & \mathrm{T}(L) \\ \phi \downarrow & & \psi \downarrow & & & & \\ L' & \xrightarrow{\alpha'} & M' & \xrightarrow{\beta'} & N' & \xrightarrow{\gamma'} & \mathrm{T}(L') \end{array}$$

is a commutative diagram in K in which the rows are distinguished triangles. Then there exists a morphism $\chi : N \to N'$ such that the diagram

$$\begin{array}{ccccccc} L & \xrightarrow{\alpha} & M & \xrightarrow{\beta} & N & \xrightarrow{\gamma} & \mathrm{T}(L) \\ \phi \downarrow & & \psi \downarrow & & \chi \downarrow & & \downarrow \mathrm{T}(\phi) \\ L' & \xrightarrow{\alpha'} & M' & \xrightarrow{\beta'} & N' & \xrightarrow{\gamma'} & \mathrm{T}(L') \end{array}$$

is a morphism of triangles.

(TR4) Suppose we are given these three distinguished triangles:

$$L \xrightarrow{\alpha} M \xrightarrow{\gamma} P \to \mathrm{T}(L),$$

$$M \xrightarrow{\beta} N \xrightarrow{\epsilon} R \to \mathrm{T}(M),$$

$$L \xrightarrow{\beta\circ\alpha} N \xrightarrow{\delta} Q \to \mathrm{T}(L).$$

Then there is a distinguished triangle

$$P \xrightarrow{\phi} Q \xrightarrow{\psi} R \xrightarrow{\rho} \mathrm{T}(P)$$

making diagram (5.1.11) commutative.

Here are a few remarks on triangulated categories. See Remark 5.4.16 regarding triangulated categories arising from algebraic topology.

Remark 5.1.12 The numbering of the axioms we use is taken from [62], and it agrees with the numbering in [158]. The numbering in [75] and [76] is different.

In the situation that we care about, namely $\mathsf{K} = \mathbf{K}(A, \mathsf{M})$, the distinguished triangles will be those triangles that are isomorphic, in $\mathbf{K}(A, \mathsf{M})$, to the standard

$$
\begin{array}{ccccccc}
L & \xrightarrow{\alpha} & M & \xrightarrow{\gamma} & P & \longrightarrow & \mathrm{T}(L)\\
{\scriptstyle \mathrm{id}}\downarrow && {\scriptstyle \beta}\downarrow && {\scriptstyle \phi}\downarrow && {\scriptstyle \mathrm{id}}\downarrow\\
L & \xrightarrow{\beta\circ\alpha} & N & \xrightarrow{\delta} & Q & \longrightarrow & \mathrm{T}(L)\\
{\scriptstyle \alpha}\downarrow && {\scriptstyle \mathrm{id}}\downarrow && {\scriptstyle \psi}\downarrow && {\scriptstyle \mathrm{T}(\alpha)}\downarrow\\
M & \xrightarrow{\beta} & N & \xrightarrow{\epsilon} & R & \longrightarrow & \mathrm{T}(M)\\
{\scriptstyle \gamma}\downarrow && {\scriptstyle \delta}\downarrow && {\scriptstyle \mathrm{id}}\downarrow && {\scriptstyle \mathrm{T}(\gamma)}\downarrow\\
P & \xrightarrow{\phi} & Q & \xrightarrow{\psi} & R & \xrightarrow{\rho} & \mathrm{T}(P)
\end{array}
\tag{5.1.11}
$$

triangles in $\mathbf{C}(A, \mathsf{M})$ from Definition 4.2.5. See Definition 5.4.2 below for the precise statement.

The object N in item (b) of axiom (TR1) is referred to as a *cone* of the morphism α. We should think of the cone as something combining "the cokernel" and "the kernel" of α. Axiom (TR3) guarantees a sort of weak functoriality of the cone. See Section 5.3, and in particular Remark 5.3.9, for more properties of the cone.

Axiom (TR2) says that if we "turn" a distinguished triangle (cf. Remark 5.1.9) we remain with a distinguished triangle.

Remark 5.1.13 The axiom (TR4) is called the *octahedral axiom*. It is supposed to replace the isomorphism

$$(N/L)/(M/L) \cong N/M$$

for objects $L \subseteq M \subseteq N$ in an abelian category M. The octahedral axiom is needed for the theory of *t-structures*: it is used, in [16], to show that the heart of a t-structure is an abelian category (see Remark 5.3.26). This axiom is also needed to form Verdier quotients of triangulated categories (see Remark 7.1.14). The book [113] has detailed discussions of the octahedral axiom.

A T-additive category (K, T) that only satisfies axioms (TR1)–(TR3) is called a *pretriangulated category*. (The reader should not confuse "pretriangulated category," as used here, with the "pretriangulated DG category" from [25]; see Remark 5.5.4.) It is not known whether the octahedral axiom is a consequence of the other axioms; there was a recent paper by A. Maccioca (arxiv:1506.00887) claiming that, but it had a fatal error in it.

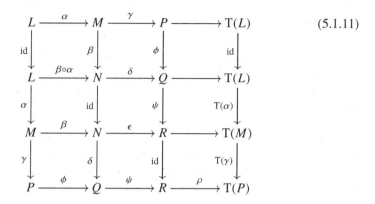

In our book the octahedral axiom does not play any role. For this reason we had excluded it from an earlier version of the book, in which we had discussed pretriangulated categories only. Our decision to include this axiom in the current version of the book, and thus to talk about triangulated categories (rather than about pretriangulated ones) is just to be more in line with the mainstream usage. With the exception of a longer proof of Theorem 5.4.3 – stating that $\mathbf{K}(A, \mathsf{M})$ is a triangulated category – there is virtually no change in the content of the book, and almost all definitions and results are valid for pretriangulated categories.

Remark 5.1.14 The structure of triangulated categories is poorly understood, and there is no classification of triangulated functors between them. This is true even for the derived category $\mathbf{D}(A)$ of a ring A. For instance, a famous open question of J. Rickard is whether every triangulated autoequivalence F of $\mathbf{D}(A)$ is isomorphic to $P \otimes^{\mathrm{L}}_{A} (-)$ for some tilting complex P. More on this issue in Chapter 14.

In this book the role of the triangulated structure on derived categories is secondary. It is mostly used for induction on the amplitude of the cohomology of complexes; cf. Chapter 12.

Furthermore, one of the most important functors studied here – the squaring operation $\mathrm{Sq}_{B/A}$ from Sections 13.4 and 18.2 – is a functor from $\mathbf{D}(B)$ to itself, which is not triangulated, nor even linear: it is a *quadratic functor*.

5.2 Triangulated and Cohomological Functors

Suppose K and L are T-additive categories, with translation functors T_{K} and T_{L}, respectively. A T-additive functor $F : \mathsf{K} \to \mathsf{L}$ was defined in Definition 5.1.3. A morphism of T-additive functors $\eta : F \to G$ was introduced in Definition 5.1.5.

Definition 5.2.1 Let K and L be triangulated categories.

(1) A *triangulated functor* from K to L is a T-additive functor $(F, \tau) : \mathsf{K} \to \mathsf{L}$ that satisfies this condition: for every distinguished triangle

$$L \xrightarrow{\alpha} M \xrightarrow{\beta} N \xrightarrow{\gamma} \mathrm{T}_{\mathsf{K}}(L)$$

in K, the triangle

$$F(L) \xrightarrow{F(\alpha)} F(M) \xrightarrow{F(\beta)} F(N) \xrightarrow{\tau_L \circ F(\gamma)} \mathrm{T}_{\mathsf{L}}(F(L))$$

is a distinguished triangle in L.

(2) Suppose $(G, \nu) : \mathsf{K} \to \mathsf{L}$ is another triangulated functor. A *morphism of triangulated functors* $\eta : (F, \tau) \to (G, \nu)$ is a morphism of T-additive functors, as in Definition 5.1.5.

Sometimes we keep the translation isomorphism τ implicit, and refer to F as a triangulated functor.

Definition 5.2.2 Let K be a triangulated category. A *full triangulated subcategory* of K is a subcategory $\mathsf{L} \subseteq \mathsf{K}$ satisfying these conditions:

(a) L is a full additive subcategory (see Definition 2.2.6).

(b) L is closed under translations, i.e. $L \in \mathsf{L}$ iff $\mathrm{T}(L) \in \mathsf{L}$.

(c) L is closed under distinguished triangles, i.e. if $L' \to L \to L'' \xrightarrow{\triangle}$ is a distinguished triangle in K s.t. $L', L \in \mathsf{L}$, then also $L'' \in \mathsf{L}$.

Observe that L itself is triangulated, and the inclusion $\mathsf{L} \to \mathsf{K}$ is a triangulated functor.

Proposition 5.2.3 *For* $i = 0, 1, 2$ *let* $(\mathsf{K}_i, \mathrm{T}_i)$ *be triangulated categories, and let* $(F_i, \tau_i) : (\mathsf{K}_{i-1}, \mathrm{T}_{i-1}) \to (\mathsf{K}_i, \mathrm{T}_i)$ *be triangulated functors. Define the T-additive functor* $(F, \tau) := (F_2, \tau_2) \circ (F_1, \tau_1)$ *as in Definition* 5.1.4. *Then* $(F, \tau) : (\mathsf{K}_0, \mathrm{T}_0) \to (\mathsf{K}_2, \mathrm{T}_2)$ *is a triangulated functor.*

Exercise 5.2.4 Prove Proposition 5.2.3.

Definition 5.2.5 Let K be a triangulated category, and let M be an abelian category.

(1) A *cohomological functor* $F : \mathsf{K} \to \mathsf{M}$ is an additive functor, such that for every distinguished triangle

$$L \xrightarrow{\alpha} M \xrightarrow{\beta} N \xrightarrow{\gamma} \mathrm{T}(L)$$

in K, the sequence

$$F(L) \xrightarrow{F(\alpha)} F(M) \xrightarrow{F(\beta)} F(N)$$

is exact in M.

(2) A *contravariant cohomological functor* $F : \mathsf{K} \to \mathsf{M}$ is a contravariant additive functor, such that for every distinguished triangle

$$L \xrightarrow{\alpha} M \xrightarrow{\beta} N \xrightarrow{\gamma} \mathrm{T}(L)$$

in K, the sequence

$$F(N) \xrightarrow{F(\beta)} F(M) \xrightarrow{F(\alpha)} F(L)$$

is exact in M.

Proposition 5.2.6 *Let* $F : \mathsf{K} \to \mathsf{M}$ *be a cohomological functor, and let*

$$L \xrightarrow{\alpha} M \xrightarrow{\beta} N \xrightarrow{\gamma} \mathrm{T}(L)$$

be a distinguished triangle in K*. Then the sequence*

$$\cdots \to F(\mathrm{T}^i(L)) \xrightarrow{F(\mathrm{T}^i(\alpha))} F(\mathrm{T}^i(M)) \xrightarrow{F(\mathrm{T}^i(\beta))} F(\mathrm{T}^i(N))$$
$$\xrightarrow{F(\mathrm{T}^i(\gamma))} F(\mathrm{T}^{i+1}(L)) \xrightarrow{F(\mathrm{T}^{i+1}(\alpha))} F(\mathrm{T}^{i+1}(M)) \to \cdots$$

in M *is exact.*

Proof By axiom (TR2) we have these distinguished triangles:

$$\mathrm{T}^i(L) \xrightarrow{(-1)^i \cdot \mathrm{T}^i(\alpha)} \mathrm{T}^i(M) \xrightarrow{(-1)^i \cdot \mathrm{T}^i(\beta)} \mathrm{T}^i(N) \xrightarrow{(-1)^i \cdot \mathrm{T}^i(\gamma)} \mathrm{T}^{i+1}(L),$$

$$\mathrm{T}^i(M) \xrightarrow{(-1)^i \cdot \mathrm{T}^i(\beta)} \mathrm{T}^i(N) \xrightarrow{(-1)^i \cdot \mathrm{T}^i(\gamma)} \mathrm{T}^{i+1}(L) \xrightarrow{(-1)^{i+1} \cdot \mathrm{T}^{i+1}(\alpha)} \mathrm{T}^{i+1}(M),$$

$$\mathrm{T}^i(N) \xrightarrow{(-1)^i \cdot \mathrm{T}^i(\gamma)} \mathrm{T}^{i+1}(L) \xrightarrow{(-1)^{i+1} \cdot \mathrm{T}^{i+1}(\alpha)} \mathrm{T}^{i+1}(M) \xrightarrow{(-1)^{i+1} \cdot \mathrm{T}^{i+1}(\beta)} \mathrm{T}^{i+1}(N).$$

Now use the definition of a cohomological functor, noting that multiplying morphisms in an exact sequence by -1 preserves exactness. □

Proposition 5.2.7 *Let* K *be a triangulated category. For every* $P \in \mathsf{K}$ *the functor*

$$\mathrm{Hom}_{\mathsf{K}}(P, -) : \mathsf{K} \to \mathsf{Mod}\,\mathbb{K}$$

is a cohomological functor, and the functor

$$\mathrm{Hom}_{\mathsf{K}}(-, P) : \mathsf{K} \to \mathsf{Mod}\,\mathbb{K}$$

is a contravariant cohomological functor.

Proof We will prove the covariant statement. The contravariant statement is proved similarly, and we leave this to the reader.

Consider a distinguished triangle $L \xrightarrow{\alpha} M \xrightarrow{\beta} N \xrightarrow{\gamma} \mathrm{T}(L)$ in K. We have to prove that the sequence

$$\mathrm{Hom}_{\mathsf{K}}(P, L) \xrightarrow{\mathrm{Hom}(\mathrm{id}_P, \alpha)} \mathrm{Hom}_{\mathsf{K}}(P, M) \xrightarrow{\mathrm{Hom}(\mathrm{id}_P, \beta)} \mathrm{Hom}_{\mathsf{K}}(P, N)$$

in Mod $\mathbb{K}$ is exact. In view of Proposition 5.3.1, all we need to show is that for every $\psi : P \to M$ s.t. $\beta \circ \psi = 0$, there is some $\phi : P \to L$ s.t. $\psi = \alpha \circ \phi$. In a picture, we must show that the diagram below (solid arrows)

$$\begin{array}{ccccccc}
P & \xrightarrow{\mathrm{id}} & P & \longrightarrow & 0 & \longrightarrow & \mathrm{T}(P) \\
{\scriptstyle \phi}\downarrow & & \downarrow{\scriptstyle \psi} & & \downarrow & & \downarrow{\scriptstyle \mathrm{T}(\phi)} \\
L & \xrightarrow{\alpha} & M & \xrightarrow{\beta} & N & \xrightarrow{\gamma} & \mathrm{T}(L)
\end{array}$$

can be completed (dashed arrow). This is true by (TR2) (= turning) and (TR3) (= extending). □

Exercise 5.2.8 Prove the contravariant statement in the proposition above.

Question 5.2.9 Let K and L be triangulated categories, and let $F : \mathsf{K} \to \mathsf{L}$ be an additive functor. Is it true that there is at most one isomorphism of functors $\tau : F \circ \mathrm{T}_{\mathsf{K}} \xrightarrow{\simeq} \mathrm{T}_{\mathsf{L}} \circ F$ such that the pair (F, τ) is a triangulated functor?

5.3 Some Properties of Triangulated Categories

In this section we prove a few general results on triangulated categories. Recall that all triangulated categories, and all triangulated functors between them, are (implicitly) $\mathbb{K}$-linear.

Proposition 5.3.1 *Let* K *be a triangulated category. If* $L \xrightarrow{\alpha} M \xrightarrow{\beta} N \xrightarrow{\gamma} \mathrm{T}(L)$ *is a distinguished triangle in* K*, then* $\beta \circ \alpha = 0$.

Proof By axioms (TR1) and (TR3) we have a commutative diagram

$$\begin{array}{ccccccc} L & \xrightarrow{\mathrm{id}_L} & L & \longrightarrow & 0 & \longrightarrow & \mathrm{T}(L) \\ {\scriptstyle \mathrm{id}_L}\downarrow & & {\scriptstyle \alpha}\downarrow & & \downarrow & & \downarrow{\scriptstyle \mathrm{T}(\mathrm{id}_L)} \\ L & \xrightarrow{\alpha} & M & \xrightarrow{\beta} & N & \xrightarrow{\gamma} & \mathrm{T}(L) \end{array}$$

We see that $\beta \circ \alpha$ factors through 0. □

Proposition 5.3.2 *Let* K *be a triangulated category, and let*

$$\begin{array}{ccccccc} L & \xrightarrow{\alpha} & M & \xrightarrow{\beta} & N & \xrightarrow{\gamma} & \mathrm{T}(L) \\ {\scriptstyle \phi}\downarrow & & {\scriptstyle \psi}\downarrow & & {\scriptstyle \chi}\downarrow & & \downarrow{\scriptstyle \mathrm{T}(\phi)} \\ L' & \xrightarrow{\alpha'} & M' & \xrightarrow{\beta'} & N' & \xrightarrow{\gamma'} & \mathrm{T}(L') \end{array}$$

be a morphism of distinguished triangles. If ϕ *and* ψ *are isomorphisms, then* χ *is also an isomorphism.*

Proof Take an arbitrary $P \in \mathsf{K}$, and let $F := \mathrm{Hom}_{\mathsf{K}}(P, -)$. We get a commutative diagram

$$\begin{array}{ccccccccc} F(L) & \xrightarrow{F(\alpha)} & F(M) & \xrightarrow{F(\beta)} & F(N) & \xrightarrow{F(\gamma)} & F(\mathrm{T}(L)) & \xrightarrow{F(\mathrm{T}(\alpha))} & F(\mathrm{T}(M)) \\ \downarrow{\scriptstyle F(\phi)} & & {\scriptstyle F(\psi)}\downarrow & & {\scriptstyle F(\chi)}\downarrow & & {\scriptstyle F(\mathrm{T}(\phi))}\downarrow & & {\scriptstyle F(\mathrm{T}(\psi))}\downarrow \\ F(L') & \xrightarrow{F(\alpha')} & F(M') & \xrightarrow{F(\beta')} & F(N') & \xrightarrow{F(\gamma')} & F(\mathrm{T}(L')) & \xrightarrow{F(\mathrm{T}(\alpha'))} & F(\mathrm{T}(M')) \end{array}$$

in $\mathsf{Mod}\,\mathbb{K}$. By Proposition 5.2.7 the rows in the diagram are exact sequences. Since the other vertical arrows are isomorphisms, it follows that

$$F(\chi) : \mathrm{Hom}_{\mathsf{K}}(P, N) \to \mathrm{Hom}_{\mathsf{K}}(P, N')$$

is an isomorphism of abelian groups. By forgetting structure, we see that $F(\chi)$ is an isomorphism of sets.

Consider the Yoneda functor $\mathrm{Y}_{\mathsf{K}} : \mathsf{K} \to \mathsf{Fun}(\mathsf{K}^{\mathrm{op}}, \mathsf{Set})$ from Section 1.7. There is a morphism $\mathrm{Y}_{\mathsf{K}}(\chi) : \mathrm{Y}_{\mathsf{K}}(N) \to \mathrm{Y}_{\mathsf{K}}(N')$ in $\mathsf{Fun}(\mathsf{K}^{\mathrm{op}}, \mathsf{Set})$. For every object $P \in \mathsf{K}$, letting $F := \mathrm{Hom}_{\mathsf{K}}(P, -)$ as above, we have $\mathrm{Y}_{\mathsf{K}}(N)(P) = F(N)$ and $\mathrm{Y}_{\mathsf{K}}(N')(P) = F(N')$. The calculation above shows that

$$F(\chi) = \mathrm{Y}_{\mathsf{K}}(\chi)(P) : \mathrm{Y}_{\mathsf{K}}(N)(P) \to \mathrm{Y}_{\mathsf{K}}(N')(P)$$

is an isomorphism in Set. Therefore $\mathrm{Y}_{\mathsf{K}}(\chi)$ is an isomorphism in $\mathsf{Fun}(\mathsf{K}^{\mathrm{op}}, \mathsf{Set})$. According to Yoneda Lemma (Theorem 1.7.4), the morphism $\chi : N \to N'$ in K is an isomorphism. □

Exercise 5.3.3 Try to prove the last proposition without the Yoneda Lemma.

Recall that if $\eta : F \to G$ is a morphism of functors $\mathsf{K} \to \mathsf{L}$, then for an object $M \in \mathsf{K}$ we denote by $\eta_M : F(M) \to G(M)$ the corresponding morphism in L.

Corollary 5.3.4 *Suppose $F, G : \mathsf{K} \to \mathsf{L}$ are triangulated functors between triangulated categories, and $\eta : F \to G$ is a morphism of triangulated functors. Let $L \xrightarrow{\alpha} M \xrightarrow{\beta} N \xrightarrow{\gamma} \mathrm{T}(L)$ be a distinguished triangle in K. If η_L and η_M are isomorphisms, then η_N is an isomorphism.*

Proof After applying the functors F and G, and the morphisms of functors τ_F, τ_G and η, we obtain a commutative diagram

$$\begin{array}{ccccccc}
F(L) & \xrightarrow{F(\alpha)} & F(M) & \xrightarrow{F(\beta)} & F(N) & \xrightarrow{\tau_{F,L} \circ F(\gamma)} & \mathrm{T}_{\mathsf{L}}(F(L)) \\
\downarrow{\scriptstyle \eta_L} & & \downarrow{\scriptstyle \eta_M} & & \downarrow{\scriptstyle \eta_N} & & \downarrow{\scriptstyle \mathrm{T}_{\mathsf{L}}(\eta_L)} \\
G(L) & \xrightarrow{G(\alpha)} & G(M) & \xrightarrow{G(\beta)} & G(N) & \xrightarrow{\tau_{G,L} \circ G(\gamma)} & \mathrm{T}_{\mathsf{L}}(G(L))
\end{array}$$

in L in which the rows are distinguished triangles. According to Proposition 5.3.2, the morphism η_N is an isomorphism. □

Corollary 5.3.5 *Let K be a triangulated category, and let $L \xrightarrow{\alpha} M \to N \xrightarrow{\triangle}$ be a distinguished triangle in it. The two conditions below are equivalent.*

(i) *$\alpha : L \to M$ is an isomorphism.*

(ii) *$N \cong 0$.*

Exercise 5.3.6 Prove Corollary 5.3.5. (Hint: use Proposition 5.3.2 and axiom (TR1)(c).)

Definition 5.3.7 Let K be a triangulated category, and let $\alpha : L \to M$ be a morphism in K. By axiom (TR1)(b) there exists a distinguished triangle

$$L \xrightarrow{\alpha} M \xrightarrow{\beta} N \xrightarrow{\gamma} \mathrm{T}(L) \tag{†}$$

in K. The object N is called a *cone of* α, and the distinguished triangle (†) is called a *distinguished triangle built on* α.

Corollary 5.3.8 *In the situation of Definition 5.3.7, the object N and the distinguished triangle (†) are unique up to isomorphism.*

Proof Suppose

$$L \xrightarrow{\alpha} M \xrightarrow{\beta'} N' \xrightarrow{\gamma'} \mathrm{T}(L) \tag{†'}$$

is another distinguished triangle built on α. By axiom (TR3) there is a commutative diagram

$$\begin{array}{ccccccc} L & \xrightarrow{\alpha} & M & \xrightarrow{\beta} & N & \xrightarrow{\gamma} & \mathrm{T}(L) \\ {\scriptstyle \mathrm{id}_L}\downarrow & & {\scriptstyle \mathrm{id}_M}\downarrow & & {\scriptstyle \chi}\downarrow & & \downarrow{\scriptstyle \mathrm{T}(\mathrm{id}_L)} \\ L & \xrightarrow{\alpha} & M & \xrightarrow{\beta'} & N' & \xrightarrow{\gamma'} & \mathrm{T}(L) \end{array}$$

in K. Proposition 5.3.2 says that χ is an isomorphism. □

Remark 5.3.9 In general the isomorphism χ in the corollary above is not unique, and thus the cone of α is not functorial in the morphism α. However, in some special cases χ is unique – see Remark 5.3.26 below.

Note also that the standard cones and triangles in the DG category $\mathbf{C}(A, \mathsf{M})$ are functorial, by Proposition 4.2.6. This implies that in the triangulated categories $\mathbf{K}(A, \mathsf{M})$ and $\mathbf{D}(A, \mathsf{M})$ we can sometimes arrange to have functorial cones.

Proposition 5.3.10 *Let K be a triangulated category, and let* $L \xrightarrow{\alpha} M \xrightarrow{\beta} N \xrightarrow{\gamma} \mathrm{T}(L)$ *be a distinguished triangle in it. The two conditions below are equivalent.*

(i) *The morphism* γ *is zero.*

(ii) *There exists a morphism* $\tau : M \to L$ *such that* $\tau \circ \alpha = \mathrm{id}_L$.

Proof

(i) ⇒ (ii): By Propositions 5.2.7 the contravariant functor $\mathrm{Hom}_{\mathsf{K}}(-, L)$ is cohomological. Applying it to the given distinguished triangle, and using Proposition 5.2.6, we get an exact sequence of $\mathbb{K}$-modules

$$\mathrm{Hom}_{\mathsf{K}}(M, L) \xrightarrow{\mathrm{Hom}(\alpha, \mathrm{id}_L)} \mathrm{Hom}_{\mathsf{K}}(L, L) \xrightarrow{\mathrm{Hom}(\mathrm{T}^{-1}(\gamma), \mathrm{id}_L)} \mathrm{Hom}_{\mathsf{K}}(\mathrm{T}^{-1}(N), L).$$

Since the homomorphism $\mathrm{Hom}_{\mathsf{K}}(\mathrm{T}^{-1}(\gamma), \mathrm{id}_L)$ is zero, there exists some morphism $\tau \in \mathrm{Hom}_{\mathsf{K}}(M, L)$ such that $\mathrm{id}_L = \mathrm{Hom}_{\mathsf{K}}(\alpha, \mathrm{id}_L)(\tau) = \tau \circ \alpha$.

(ii) ⇒ (i): Let us examine the commutative diagram

$$\begin{array}{ccc}
\mathrm{Hom}_{\mathsf{K}}(N, N) & & \\
\Big\downarrow {\scriptstyle \mathrm{Hom}(\mathrm{id},\gamma)} & & \\
\mathrm{Hom}_{\mathsf{K}}(N, \mathrm{T}(L)) & \xrightarrow{\mathrm{Hom}(\mathrm{id},\mathrm{id})} & \\
\Big\downarrow {\scriptstyle \mathrm{Hom}(\mathrm{id},\mathrm{T}(\alpha))} & & \searrow \\
\mathrm{Hom}_{\mathsf{K}}(N, \mathrm{T}(M)) & \xrightarrow{\mathrm{Hom}(\mathrm{id},\mathrm{T}(\tau))} & \mathrm{Hom}_{\mathsf{K}}(N, \mathrm{T}(L))
\end{array}$$

in $\mathsf{Mod}\,\mathbb{K}$. Because the homomorphism $\mathrm{Hom}_{\mathsf{K}}(\mathrm{id}, \mathrm{id})$ is bijective, it follows that $\mathrm{Hom}_{\mathsf{K}}(\mathrm{id}, \mathrm{T}(\alpha))$ is injective. But the column is an exact sequence (by Propositions 5.2.7 and 5.2.6), and therefore $\mathrm{Hom}_{\mathsf{K}}(\mathrm{id}_N, \gamma) = 0$. We conclude that $\gamma = \mathrm{Hom}_{\mathsf{K}}(\mathrm{id}_N, \gamma)(\mathrm{id}_N) = 0$. □

Lemma 5.3.11 *Let M, M' be objects in a triangulated category* K*. Consider the canonical morphisms*

$$M \xrightarrow{e} M \oplus M' \xrightarrow{p} M \quad \text{and} \quad M' \xrightarrow{e'} M \oplus M' \xrightarrow{p'} M'.$$

Then

$$M \xrightarrow{e} M \oplus M' \xrightarrow{p'} M' \xrightarrow{0} \mathrm{T}(M)$$

is a distinguished triangle in K.

Proof By axiom (TR1)(c) and axiom (TR2) there is a distinguished triangle $0 \xrightarrow{0} M' \xrightarrow{\mathrm{id}_{M'}} M' \xrightarrow{0} 0$. By axiom (TR1)(b) there is a distinguished triangle

$$M \xrightarrow{e} M \oplus M' \xrightarrow{\beta} N \xrightarrow{\gamma} \mathrm{T}(M) \tag{5.3.12}$$

in K, for some object N. Because $p \circ e = \mathrm{id}_M$, Proposition 5.3.10 says that $\gamma = 0$.

Next, since $p' \circ e = 0$, axiom (TR3) produces a morphism of triangles

$$\begin{array}{ccccccc}
M & \xrightarrow{e} & M \oplus M' & \xrightarrow{\beta} & N & \xrightarrow{\gamma=0} & \mathrm{T}(M) \\
\Big\downarrow {\scriptstyle 0} & & \Big\downarrow {\scriptstyle p'} & & \Big\downarrow {\scriptstyle \psi} & & \Big\downarrow {\scriptstyle 0} \\
0 & \xrightarrow{0} & M' & \xrightarrow{\mathrm{id}_{M'}} & M' & \xrightarrow{0} & 0
\end{array} \tag{5.3.13}$$

We claim that ψ is an isomorphism. This is proved indirectly. For every object

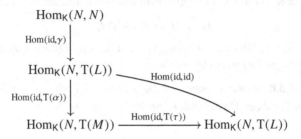

$L \in \mathsf{K}$ there is a commutative diagram

$$\begin{array}{ccccccc} \operatorname{Hom}_{\mathsf{K}}(L, M) & \longrightarrow & \operatorname{Hom}_{\mathsf{K}}(L, M \oplus M') & \longrightarrow & \operatorname{Hom}_{\mathsf{K}}(L, N) & \longrightarrow & 0 \\ & & \Big\downarrow {\scriptstyle \operatorname{Hom}(\mathrm{id}_L, p')} & & \Big\downarrow {\scriptstyle \operatorname{Hom}(\mathrm{id}_L, \psi)} & & \\ 0 & \longrightarrow & \operatorname{Hom}_{\mathsf{K}}(L, M') & \xrightarrow{\mathrm{id}} & \operatorname{Hom}_{\mathsf{K}}(L, M') & \longrightarrow & 0 \end{array} \tag{5.3.14}$$

in $\mathsf{Mod}\, \mathbb{K}$, which is gotten by applying the functor $\operatorname{Hom}_{\mathsf{K}}(L, -)$ to the left part of diagram (5.3.13). The rows in (5.3.14) are exact sequences. Because

$$\operatorname{Hom}_{\mathsf{K}}(L, M \oplus M') \cong \operatorname{Hom}_{\mathsf{K}}(L, M) \oplus \operatorname{Hom}_{\mathsf{K}}(L, M'),$$

we can replace (5.3.14) with the next commutative diagram

$$\begin{array}{ccccccc} 0 & \longrightarrow & \operatorname{Hom}_{\mathsf{K}}(L, M') & \longrightarrow & \operatorname{Hom}_{\mathsf{K}}(L, N) & \longrightarrow & 0 \\ & & \Big\downarrow {\scriptstyle \operatorname{Hom}(\mathrm{id}_L, \mathrm{id}_{M'})} & & \Big\downarrow {\scriptstyle \operatorname{Hom}(\mathrm{id}_L, \psi)} & & \\ 0 & \longrightarrow & \operatorname{Hom}_{\mathsf{K}}(L, M') & \xrightarrow{\mathrm{id}} & \operatorname{Hom}_{\mathsf{K}}(L, M') & \longrightarrow & 0 \end{array} \tag{5.3.15}$$

with exact rows. We see that $\operatorname{Hom}(\mathrm{id}_L, \psi)$ is an isomorphism of $\mathbb{K}$-modules. Hence it is a bijection of sets. Using the Yoneda Lemma, like in the proof of Proposition 5.3.2, we conclude that ψ is an isomorphism in K.

Finally, from the commutative diagram (5.3.13) we know that $\psi \circ \beta = p'$. So we have this isomorphism of triangles:

$$\begin{array}{ccccccc} M & \xrightarrow{e} & M \oplus M' & \xrightarrow{\beta} & N & \xrightarrow{\gamma=0} & \mathrm{T}(M) \\ \Big\downarrow {\scriptstyle \mathrm{id}} & & \Big\downarrow {\scriptstyle \mathrm{id}} & & \Big\downarrow {\scriptstyle \psi\ \cong} & & \Big\downarrow {\scriptstyle \mathrm{id}} \\ M & \xrightarrow{e} & M \oplus M' & \xrightarrow{p'} & M' & \xrightarrow{0} & \mathrm{T}(M) \end{array} \tag{5.3.16}$$

The first triangle is distinguished. By axiom (TR1)(a) the second triangle is also distinguished. □

Definition 5.3.17 Let K be a linear category, and let $M, N \in \mathsf{K}$.

(1) The object M is called a *retract* of N if there are morphisms $M \xrightarrow{e} N \xrightarrow{p} M$ such that $p \circ e = \mathrm{id}_M$. The morphism $e : M \to N$ is called an embedding.
(2) The object M is called a *direct summand* of N if there is an object $M' \in \mathsf{K}$ and an isomorphism $M \oplus M' \cong N$. The corresponding morphism $e : M \to N$ is called an embedding.

Note that in both items above the embedding $e : M \to N$ is a monomorphism in K.

Theorem 5.3.18 *Let K be a triangulated category, and let $e : M \to N$ be a morphism in K. The following two conditions are equivalent:*

(i) *There is a morphism $e' : M' \to N$ in* K *such that*

$$(e, e') : M \oplus M' \to N$$

is an isomorphism. In other words, M is a direct summand of N, with embedding $e : M \to N$.

(ii) *There is a morphism $p : N \to M$ in* K *such that $p \circ e = \mathrm{id}_M$. In other words, M is a retract of N, with embedding $e : M \to N$.*

Proof

(i) ⇒ (ii): This is trivial (and only requires K to be a linear category).

(ii) ⇒ (i): Let $M \xrightarrow{e} N \xrightarrow{\beta} M' \xrightarrow{\gamma} \mathrm{T}(M)$ be a distinguished triangle in K built on e. By Proposition 5.3.10 we know that $\gamma = 0$. According to Lemma 5.3.11, $M \xrightarrow{\epsilon} M \oplus M' \xrightarrow{\pi'} M' \xrightarrow{0} \mathrm{T}(M)$ is also the distinguished triangle in K; here ϵ is the embedding and π' is the projection. Turning these two distinguished triangles and using axiom (TR3) we get a morphism θ such that the diagram

$$\begin{array}{ccccccc}
\mathrm{T}^{-1}(M') & \xrightarrow{0} & M & \xrightarrow{\epsilon} & M \oplus M' & \xrightarrow{\pi'} & M' \\
\downarrow{\scriptstyle \mathrm{id}} & & \downarrow{\scriptstyle \mathrm{id}} & & \downarrow{\scriptstyle \theta} & & \downarrow{\scriptstyle \mathrm{id}} \\
\mathrm{T}^{-1}(M') & \xrightarrow{0} & M & \xrightarrow{e} & N & \xrightarrow{\beta} & M'
\end{array}$$

is commutative. By Proposition 5.3.2, θ is an isomorphism. Finally we define $e' := \theta \circ \epsilon'$, where $\epsilon' : M' \to M \oplus M'$ is the embedding. □

Definition 5.3.19 Let K be a triangulated category. A *saturated full triangulated subcategory* of K is a full triangulated subcategory $\mathsf{K}' \subseteq \mathsf{K}$ (Definition 5.2.2) which is closed in K under taking direct summands and under isomorphisms.

Definition 5.3.20 Let K be a triangulated category and let $\mathsf{Z} \subseteq \mathrm{Ob}(\mathsf{K})$ be a set of objects. The *saturated full triangulated subcategory of* K *generated by* Z is the smallest saturated full triangulated subcategory $\mathsf{K}' \subseteq \mathsf{K}$ such that $\mathsf{Z} \subseteq \mathrm{Ob}(\mathsf{K}')$.

Proposition 5.3.21 *Let* K *be a triangulated category, let* $\mathsf{Z} \subseteq \mathrm{Ob}(\mathsf{K})$ *be a set of objects, and let* $\mathsf{K}' \subseteq \mathsf{K}$ *be the saturated full triangulated subcategory of* K *generated by* Z*. An object $M \in \mathsf{K}$ belongs to K' if and only if there is a sequence $M_0, \ldots, M_r$ of objects of K, such that $M_r = M$, and for every $i \leq r$ at least one of the following conditions holds*:

(a) $M_i \in \mathsf{Z}$.

(b) *There is an isomorphism $M_i \cong \mathrm{T}^p(M_j)$ in K for some $j < i$ and $p \in \mathbb{Z}$.*

(c) *There is a distinguished triangle* $M_j \to M_k \to M_i \xrightarrow{\triangle}$ *in* K *for some* $j, k < i$.
(d) M_i *is a direct summand of* M_j *in* K *for some* $j < i$.

Proposition 5.3.22 *Let* K *and* L *be triangulated categories, let* $F, G : \mathsf{K} \to \mathsf{L}$ *be triangulated functors, and let* $\zeta : F \to G$ *be a morphism of triangulated functors. Denote by* K' *the full subcategory of* K *on the objects* M *such that* $\zeta_M : F(M) \to G(M)$ *is an isomorphism in* L. *Then* K' *is a saturated full triangulated subcategory of* K.

Corollary 5.3.23 *Let* K *and* L *be triangulated categories, let* $F : \mathsf{K} \to \mathsf{L}$ *be a triangulated functor, and let* K' *be the kernel of* F, *i.e.* $\mathsf{K}' \subseteq \mathsf{K}$ *is the full subcategory on the objects* $M \in \mathsf{K}$ *such that* $F(M) = 0$. *Then* K' *is a saturated full triangulated subcategory of* K.

Exercise 5.3.24 Prove Proposition 5.3.21, Proposition 5.3.22 and Corollary 5.3.23.

We end this section with two remarks.

Remark 5.3.25 The story of the concept "saturated full triangulated subcategory" is convoluted. In the text [155] from 1977, J.-L. Verdier introduced *épaisse full triangulated subcategories* (a rather messy definition). However, J. Rickard proved (see [125, Proposition 1.3], published in 1989) that épaisse is equivalent to saturated, as defined above. Rickard did not give a name for this new and useful property.

Verdier died in 1989. A few years later, in 1996, G. Maltsiniotis published Verdier's thesis [156]. In this thesis the word "épaisse" is absent, and instead we find the definition of saturated categories, identical to Definition 5.3.19. Maltsiniotis notes that Verdier was aware (although this is only implicit in his writing) of the equivalence between épaisse and saturated; indeed, these are precisely the categories that occur as kernels of triangulated functors (see Corollary 5.3.23 above, and Remark 7.1.14 on Verdier quotients and saturated denominator sets of morphisms).

Remark 5.3.26 As mentioned above, in Remark 5.3.9, the cone of a morphism $\alpha : L \to M$ in a triangulated category D is usually not functorial. However, if D is endowed with a *t-structure*, then some cones can be made functorial. Here is a quick explanation.

The concept of t-structure was introduced by A.A. Beilinson, J. Bernstein and P. Deligne in the book [16]. A t-structure on a triangulated category D consists of a pair $(\mathsf{D}^{\leq 0}, \mathsf{D}^{\geq 0})$ of full subcategories of D that satisfy a few axioms (see e.g. [75, Chapter X] or [187, Definition 4.1]). The prototypical example is the *standard t-structure* on $\mathsf{D} := \mathbf{D}(\mathsf{M})$, the derived category of an abelian category M (see Chapter 7 of the book). Here $\mathsf{D}^{\leq 0}$ is the full subcategory of $\mathbf{D}(\mathsf{M})$ on the complexes M with nonpositive cohomology, and $\mathsf{D}^{\geq 0}$ is the full subcategory of $\mathbf{D}(\mathsf{M})$ on the complexes M with nonnegative cohomology. Other t-structures on $\mathsf{D} = \mathbf{D}(\mathsf{M})$ are often called *perverse t-structures.*

The *heart* of the t-structure $(\mathsf{D}^{\leq 0}, \mathsf{D}^{\geq 0})$ is the category $\mathsf{D}^0 := \mathsf{D}^{\leq 0} \cap \mathsf{D}^{\geq 0}$. In the prototypical example above, the heart is the category of complexes M with cohomology concentrated in degree 0, so it is equivalent to M. Something similar happens in general: the heart D^0 is always an abelian category, and there is a cohomological functor $\mathrm{H}^0 : \mathsf{D} \to \mathsf{D}^0$. The short exact sequences $0 \to L \xrightarrow{\alpha} M \xrightarrow{\beta} N \to 0$ in D^0 are the distinguished triangles $L \xrightarrow{\alpha} M \xrightarrow{\beta} N \xrightarrow{\triangle}$ in D such that $L, M, N \in \mathsf{D}^0$.

Here is how a t-structure $(\mathsf{D}^{\leq 0}, \mathsf{D}^{\geq 0})$ affects functoriality of cones. Axiom (TR3) guarantees that for a pair of distinguished triangles and morphism ϕ, ψ as in the solid diagram below, there is *at least one* morphism χ (the dashed arrow) making the whole diagram commutative.

$$
\begin{array}{ccccccc}
L & \xrightarrow{\alpha} & M & \xrightarrow{\beta} & N & \xrightarrow{\gamma} & \mathrm{T}(L) \\
\downarrow{\scriptstyle \phi} & & \downarrow{\scriptstyle \psi} & & \downarrow{\scriptstyle \chi} & & \downarrow{\scriptstyle \mathrm{T}(\phi)} \\
L' & \xrightarrow{\alpha'} & M' & \xrightarrow{\beta'} & N' & \xrightarrow{\gamma'} & \mathrm{T}(L')
\end{array}
$$

According to [16, Proposition 1.1.9] (see also [187, Lemma 4.10]), if $L \in \mathsf{D}^{\leq 0}$ and $\mathrm{T}(N') \in \mathsf{D}^{\geq 0}$, then this morphism χ is *unique.*

The idea of a t-structure emerged in the study of algebraic topology, microlocal analysis and representations of Lie algebras. In that context the triangulated category D was the full subcategory of $\mathbf{D}(\mathbb{K}_X)$ on complexes with suitable finiteness and constructibility conditions. The base ring $\mathbb{K}$ was either a field or the l-adic integers $\widehat{\mathbb{Z}}_l$. The space (or site) X was either a topological space with a classical metric topology, or a scheme with its étale topology. The objects of the heart where called *perverse sheaves.* This theory is explained in the original book [16], and also in [27] and in [75, Chapter X]. In algebraic geometry one can also consider *perverse coherent sheaves*; see [187] and [4].

5.4 The Homotopy Category Is Triangulated

In this section we consider a $\mathbb{K}$-linear abelian category M and a DG central $\mathbb{K}$-ring A, where $\mathbb{K}$ is the commutative base ring. These ingredients give rise to the $\mathbb{K}$-linear DG category $\mathbf{C}(A, \mathsf{M})$ of DG A-modules in M, as in Section 3.8.

The strict category $\mathbf{C}_{\mathrm{str}}(A, \mathsf{M})$ and the homotopy category $\mathbf{K}(A, \mathsf{M})$ were introduced in Definition 3.8.5. Recall that these $\mathbb{K}$-linear categories have the same objects as $\mathbf{C}(A, \mathsf{M})$. The morphism $\mathbb{K}$-modules are

$$\mathrm{Hom}_{\mathbf{C}_{\mathrm{str}}(A,\mathsf{M})}(M_0, M_1) = \mathrm{Z}^0(\mathrm{Hom}_{\mathbf{C}(A,\mathsf{M})}(M_0, M_1))$$

and

$$\mathrm{Hom}_{\mathbf{K}(A,\mathsf{M})}(M_0, M_1) = \mathrm{H}^0(\mathrm{Hom}_{\mathbf{C}(A,\mathsf{M})}(M_0, M_1)).$$

There is a full additive functor $\mathrm{P} : \mathbf{C}_{\mathrm{str}}(A, \mathsf{M}) \to \mathbf{K}(A, \mathsf{M})$ which is the identity on objects, and on morphisms it sends a homomorphism to its homotopy class. Morphisms in $\mathbf{K}(A, \mathsf{M})$ will often by decorated with a bar, like $\bar{\phi}$.

Consider the translation functor T from Definition 4.1.9. Since T is a DG functor from $\mathbf{C}(A, \mathsf{M})$ to itself (see Corollary 4.1.10), it restricts to a linear functor from $\mathbf{C}_{\mathrm{str}}(A, \mathsf{M})$ to itself, and it induces a linear functor $\bar{\mathrm{T}}$ from $\mathbf{K}(A, \mathsf{M})$ to itself, such that $\mathrm{P} \circ \mathrm{T} = \bar{\mathrm{T}} \circ \mathrm{P}$.

Proposition 5.4.1

(1) *The category* $\mathbf{C}_{\mathrm{str}}(A, \mathsf{M})$*, equipped with the translation functor* T*, is a T-additive category.*

(2) *The category* $\mathbf{K}(A, \mathsf{M})$*, equipped with the translation functor* $\bar{\mathrm{T}}$*, is a T-additive category.*

(3) *Let* $\tau : \mathrm{P} \circ \mathrm{T} \xrightarrow{\simeq} \bar{\mathrm{T}} \circ \mathrm{P}$ *be the identity automorphism. Then the pair*

$$(\mathrm{P}, \tau) : \mathbf{C}_{\mathrm{str}}(A, \mathsf{M}) \to \mathbf{K}(A, \mathsf{M})$$

is a T-additive functor.

Proof (1) We need to prove that $\mathbf{C}_{\mathrm{str}}(A, \mathsf{M})$ is additive. Of course, the zero complex is a zero object. Next we consider finite direct sums. Let $M_1, \ldots, M_r$ be a finite collection of objects in $\mathbf{C}(A, \mathsf{M})$. Each M_i is a DG A-module in M, and we write it as $M_i = \{M_i^j\}_{j\in\mathbb{Z}}$. In each degree j the direct sum $M^j := \bigoplus_{i=1}^r M_i^j$ exists in M. Let $M := \{M^j\}_{j\in\mathbb{Z}}$ be the resulting graded object in M. The differential $\mathrm{d}_M : M^j \to M^{j+1}$ exists by the universal property of direct sums; so we obtain a complex $M \in \mathbf{C}(\mathsf{M})$. The DG A-module structure on M is defined similarly: for $a \in A^k$, there is an induced degree k morphism $f(a) : M \to M$ in $\mathbf{C}(\mathsf{M})$. Thus M becomes an object of $\mathbf{C}(A, \mathsf{M})$. But the embeddings $e_i : M_i \to M$ are strict morphisms, so $(M, \{e_i\})$ is a coproduct of the collection $\{M_i\}$ in $\mathbf{C}_{\mathrm{str}}(A, \mathsf{M})$.

(2) Now consider the category $\mathbf{K}(A, \mathsf{M})$. Because the functor $\mathrm{P} : \mathbf{C}_{\mathrm{str}}(A, \mathsf{M}) \to \mathbf{K}(A, \mathsf{M})$ is additive and is bijective on objects, part (1) above and Proposition 2.5.2 say that $\mathbf{K}(A, \mathsf{M})$ is an additive category.

(3) Clear. □

From now on we denote by T, instead of by $\bar{\mathrm{T}}$, the translation functor of $\mathbf{K}(A, \mathrm{M})$.

Definition 5.4.2 A triangle

$$L \xrightarrow{\bar{\alpha}} M \xrightarrow{\bar{\beta}} N \xrightarrow{\bar{\gamma}} \mathrm{T}(L)$$

in $\mathbf{K}(A, \mathrm{M})$ is said to be a *distinguished triangle* if there is a standard triangle

$$L' \xrightarrow{\alpha'} M' \xrightarrow{\beta'} N' \xrightarrow{\gamma'} \mathrm{T}(L')$$

in $\mathbf{C}_{\mathrm{str}}(A, \mathrm{M})$, as in Definition 4.2.5, and an isomorphism of triangles

$$\begin{array}{ccccccc} L' & \xrightarrow{\mathrm{P}(\alpha')} & M' & \xrightarrow{\mathrm{P}(\beta')} & N' & \xrightarrow{\mathrm{P}(\gamma')} & \mathrm{T}(L') \\ \downarrow{\scriptstyle \bar{\phi}} & & \downarrow{\scriptstyle \bar{\psi}} & & \downarrow{\scriptstyle \bar{\chi}} & & \downarrow{\scriptstyle \mathrm{T}(\bar{\phi})} \\ L & \xrightarrow{\bar{\alpha}} & M & \xrightarrow{\bar{\beta}} & N & \xrightarrow{\bar{\gamma}} & \mathrm{T}(L) \end{array}$$

in $\mathbf{K}(A, \mathrm{M})$.

Theorem 5.4.3 *The T-additive category* $\mathbf{K}(A, \mathrm{M})$, *with the set of distinguished triangles defined above, is a triangulated category.*

The proof is after three lemmas.

Lemma 5.4.4 *Let* $M \in \mathbf{C}(A, \mathrm{M})$, *and consider the standard cone* $N := \mathrm{Cone}(\mathrm{id}_M)$. *Then the DG module* N *is null-homotopic, i.e.* $0 \to N$ *is an isomorphism in* $\mathbf{K}(A, \mathrm{M})$.

Proof We shall exhibit a homotopy θ from 0_N to id_N. Recall from Section 4.2 that $N = \mathrm{Cone}(\mathrm{id}_M) = M \oplus \mathrm{T}(M) = \left[\begin{smallmatrix} M \\ \mathrm{T}(M) \end{smallmatrix}\right]$ as graded modules, with a differential whose matrix presentation is $\mathrm{d}_N = \left[\begin{smallmatrix} \mathrm{d}_M & \mathrm{t}_M^{-1} \\ 0 & \mathrm{d}_{\mathrm{T}(M)} \end{smallmatrix}\right]$. And by the definition in Section 4.1 we have $\mathrm{d}_{\mathrm{T}(M)} = -\mathrm{t}_M \circ \mathrm{d}_M \circ \mathrm{t}_M^{-1}$. Define $\theta : N \to N$ to be the degree -1 morphism with matrix presentation $\theta := \left[\begin{smallmatrix} 0 & 0 \\ \mathrm{t}_M & 0 \end{smallmatrix}\right]$. Then, using the formulas above for d_N and $\mathrm{d}_{\mathrm{T}(M)}$, we get

$$\mathrm{d}_N \circ \theta + \theta \circ \mathrm{d}_N = \begin{bmatrix} \mathrm{id}_M & 0 \\ 0 & \mathrm{id}_{\mathrm{T}(M)} \end{bmatrix} = \mathrm{id}_N . \qquad \square$$

Exercise 5.4.5 Here is a generalization of Lemma 5.4.4. Consider a morphism $\phi : M_0 \to M_1$ in $\mathbf{C}_{\mathrm{str}}(A, \mathrm{M})$. Show that the three conditions below are equivalent:

(i) ϕ is a homotopy equivalence.

(ii) $\bar{\phi}$ is an isomorphism in $\mathbf{K}(A, \mathsf{M})$.
(iii) The DG module $\mathrm{Cone}(\phi)$ is null-homotopic.

Try to do this directly, not using Proposition 5.2.7(2) and Theorem 5.4.3.

The next lemma is based on [75, Lemma 1.4.2].

Lemma 5.4.6 *Consider a morphism $\alpha : L \to M$ in $\mathbf{C}_{\mathrm{str}}(A, \mathsf{M})$, the standard triangle $L \xrightarrow{\alpha} M \xrightarrow{\beta} N \xrightarrow{\gamma} \mathrm{T}(L)$ built on α, and the standard triangle $M \xrightarrow{\beta} N \xrightarrow{\phi} P \xrightarrow{\psi} \mathrm{T}(M)$ built on β, all in $\mathbf{C}_{\mathrm{str}}(A, \mathsf{M})$. So $N = \mathrm{Cone}(\alpha)$ and $P = \mathrm{Cone}(\beta)$. There is a morphism $\rho : \mathrm{T}(L) \to P$ in $\mathbf{C}_{\mathrm{str}}(A, \mathsf{M})$ s.t. $\bar{\rho}$ is an isomorphism in $\mathbf{K}(A, \mathsf{M})$, and the diagram*

$$\begin{array}{ccccccc} M & \xrightarrow{\bar{\beta}} & N & \xrightarrow{\bar{\gamma}} & \mathrm{T}(L) & \xrightarrow{-\mathrm{T}(\bar{\alpha})} & \mathrm{T}(M) \\ \downarrow{\scriptstyle \mathrm{id}_M} & & \downarrow{\scriptstyle \mathrm{id}_N} & & \downarrow{\scriptstyle \bar{\rho}} & & \downarrow{\scriptstyle \mathrm{id}_{\mathrm{T}(M)}} \\ M & \xrightarrow{\bar{\beta}} & N & \xrightarrow{\bar{\phi}} & P & \xrightarrow{\bar{\psi}} & \mathrm{T}(M) \end{array}$$

commutes in $\mathbf{K}(A, \mathsf{M})$.

Proof Note that $N = M \oplus \mathrm{T}(L)$ and $P = N \oplus \mathrm{T}(M) = M \oplus \mathrm{T}(L) \oplus \mathrm{T}(M)$ as graded modules. Thus P and d_P have the following matrix presentations:

$$P = \begin{bmatrix} M \\ \mathrm{T}(L) \\ \mathrm{T}(M) \end{bmatrix}, \quad \mathrm{d}_P = \begin{bmatrix} \mathrm{d}_M & \alpha \circ \mathrm{t}_L^{-1} & \mathrm{t}_M^{-1} \\ 0 & \mathrm{d}_{\mathrm{T}(L)} & 0 \\ 0 & 0 & \mathrm{d}_{\mathrm{T}(M)} \end{bmatrix}.$$

Define morphisms $\rho : \mathrm{T}(L) \to P$ and $\chi : P \to \mathrm{T}(L)$ in $\mathbf{C}_{\mathrm{str}}(A, \mathsf{M})$ by the matrix presentations

$$\rho := \begin{bmatrix} 0 \\ \mathrm{id}_{\mathrm{T}(L)} \\ -\mathrm{T}(\alpha) \end{bmatrix}, \quad \chi := \begin{bmatrix} 0 & \mathrm{id}_{\mathrm{T}(L)} & 0 \end{bmatrix}.$$

Direct calculations show that $\chi \circ \rho = \mathrm{id}_{\mathrm{T}(L)}$, $\rho \circ \gamma = \rho \circ \chi \circ \phi$ and $\psi \circ \rho = -\mathrm{T}(\alpha)$.

It remains to prove that $\rho \circ \chi$ is homotopic to id_P. Define a degree -1 morphism $\theta : P \to P$ by the matrix

$$\theta := \begin{bmatrix} 0 & 0 & 0 \\ 0 & 0 & 0 \\ \mathrm{t}_M & 0 & 0 \end{bmatrix}.$$

Then a direct calculation, using the equalities $\mathrm{t}_M \circ \mathrm{d}_M + \mathrm{d}_{\mathrm{T}(M)} \circ \mathrm{t}_M = 0$ and $\mathrm{T}(\alpha) = \mathrm{t}_M \circ \alpha \circ \mathrm{t}_L^{-1}$ gives $\theta \circ \mathrm{d}_P + \mathrm{d}_P \circ \theta = \mathrm{id}_P - \rho \circ \chi$. □

Lemma 5.4.7 *Consider a standard triangle* $L \xrightarrow{\alpha} M \xrightarrow{\beta} N \xrightarrow{\gamma} \mathrm{T}(L)$ *in* $\mathbf{C}_{\mathrm{str}}(A, \mathsf{M})$. *For every integer k, the triangle*

$$\mathrm{T}^k(L) \xrightarrow{\mathrm{T}^k(\alpha)} \mathrm{T}^k(M) \xrightarrow{\mathrm{T}^k(\beta)} \mathrm{T}^k(N) \xrightarrow{(-1)^k \cdot \mathrm{T}^k(\gamma)} \mathrm{T}^{k+1}(L)$$

is isomorphic, in $\mathbf{C}_{\mathrm{str}}(A, \mathsf{M})$, *to a standard triangle.*

Proof Combine Corollary 4.1.10, Corollary 4.5.6 with $F = \mathrm{T}$, and Proposition 4.4.4. □

Proof of Theorem 5.4.3 We essentially follow the proof of [75, Proposition 1.4.4], adding some details.

(TR1): By definition the set of distinguished triangles in $\mathbf{K}(A, \mathsf{M})$ is closed under isomorphisms. This establishes item (a).

As for item (b): consider any morphism $\bar{\alpha} : L \to M$ in $\mathbf{K}(A, \mathsf{M})$. It is represented by a morphism $\alpha : L \to M$ in $\mathbf{C}_{\mathrm{str}}(A, \mathsf{M})$. Take the standard triangle on α in $\mathbf{C}_{\mathrm{str}}(A, \mathsf{M})$. Its image in $\mathbf{K}(A, \mathsf{M})$ has the desired property.

Finally, Lemma 5.4.4 shows that the triangle $M \xrightarrow{\mathrm{id}_M} M \to 0 \to \mathrm{T}(M)$ is isomorphic in $\mathbf{K}(A, \mathsf{M})$ to the triangle $M \xrightarrow{\mathrm{id}_M} M \xrightarrow{\bar{e}} \mathrm{Cone}(\mathrm{id}_M) \xrightarrow{\bar{p}} \mathrm{T}(M)$. The latter is the image of a standard triangle, and so it is distinguished.

(TR2): Consider the triangles

$$L \xrightarrow{\bar{\alpha}} M \xrightarrow{\bar{\beta}} N \xrightarrow{\bar{\gamma}} \mathrm{T}(L) \tag{5.4.8}$$

and

$$M \xrightarrow{\bar{\beta}} N \xrightarrow{\bar{\gamma}} \mathrm{T}(L) \xrightarrow{-\mathrm{T}(\bar{\alpha})} \mathrm{T}(M) \tag{5.4.9}$$

in $\mathbf{K}(A, \mathsf{M})$. If (5.4.8) is distinguished, then by Lemma 5.4.6 so is (5.4.9).

Conversely, if (5.4.9) is distinguished, then by turning it five times, and using the previous step (namely by Lemma 5.4.6), we see that the triangle

$$\mathrm{T}^2(L) \xrightarrow{\mathrm{T}^2(\bar{\alpha})} \mathrm{T}^2(M) \xrightarrow{\mathrm{T}^2(\bar{\beta})} \mathrm{T}^2(N) \xrightarrow{\mathrm{T}^2(\bar{\gamma})} \mathrm{T}^3(L)$$

is distinguished. According to Lemma 5.4.7 (with $k = -2$), the triangle gotten by applying T^{-2} to this is distinguished. But this is just the triangle (5.4.8).

(TR3): Consider a commutative diagram in $\mathbf{K}(A, \mathsf{M})$:

$$\begin{array}{ccccccc}
\bar{L} & \xrightarrow{\bar{\alpha}} & \bar{M} & \xrightarrow{\bar{\beta}} & \bar{N} & \xrightarrow{\bar{\gamma}} & \mathrm{T}(\bar{L}) \\
\downarrow{\scriptstyle \bar{\phi}} & & \downarrow{\scriptstyle \bar{\psi}} & & & & \\
\bar{L}' & \xrightarrow{\bar{\alpha}'} & \bar{M}' & \xrightarrow{\bar{\beta}'} & \bar{N}' & \xrightarrow{\bar{\gamma}'} & \mathrm{T}(\bar{L}')
\end{array} \tag{5.4.10}$$

where the horizontal triangles are distinguished. By definition the rows in (5.4.10) are isomorphic in $\mathbf{K}(A, \mathrm{M})$ to the images under the functor P of standard triangles in $\mathbf{C}_{\mathrm{str}}(A, \mathrm{M})$. These are the rows in diagram (5.4.11) below. The vertical morphisms in (5.4.10) are also induced from morphisms in $\mathbf{C}_{\mathrm{str}}(A, \mathrm{M})$, i.e. $\bar{\phi} = \mathrm{P}(\phi)$ and $\bar{\psi} = \mathrm{P}(\psi)$. Thus (5.4.10) is isomorphic to the image under P of the following diagram:

$$\begin{array}{ccccccc} L & \xrightarrow{\alpha} & M & \xrightarrow{\beta} & N & \xrightarrow{\gamma} & \mathrm{T}(L) \\ \downarrow{\scriptstyle \phi} & & \downarrow{\scriptstyle \psi} & & & & \\ L' & \xrightarrow{\alpha'} & M' & \xrightarrow{\beta'} & N' & \xrightarrow{\gamma'} & \mathrm{T}(L') \end{array} \tag{5.4.11}$$

Warning: the diagram (5.4.11) is only commutative up to homotopy in $\mathbf{C}_{\mathrm{str}}(A, \mathrm{M})$.

Since the rows in (5.4.11) are standard triangles (see Definition 4.2.5), the objects N and N' are cones: $N = \mathrm{Cone}(\alpha)$ and $N' = \mathrm{Cone}(\alpha')$. The commutativity up to homotopy of this diagram means that there is a degree -1 morphism $\theta : L \to M'$ in $\mathbf{C}(A, \mathrm{M})$ such that $\alpha' \circ \phi = \psi \circ \alpha + \mathrm{d}(\theta)$.

Define the morphism $\chi : N = \begin{bmatrix} M \\ \mathrm{T}(L) \end{bmatrix} \to N' = \begin{bmatrix} M' \\ \mathrm{T}(L') \end{bmatrix}$ by the matrix presentation

$$\chi := \begin{bmatrix} \psi & \theta \circ \mathrm{t}_L^{-1} \\ 0 & \mathrm{T}(\phi) \end{bmatrix}.$$

An easy calculation shows that χ is a morphism in $\mathbf{C}_{\mathrm{str}}(A, \mathrm{M})$, and that there are equalities $\mathrm{T}(\phi) \circ \gamma = \gamma' \circ \chi$ and $\chi \circ \beta = \beta' \circ \psi$. Therefore, when we apply the functor P, and conjugate by the original isomorphism between (5.4.10) and the image of (5.4.11), we obtain a commutative diagram

$$\begin{array}{ccccccc} \bar{L} & \xrightarrow{\bar{\alpha}} & \bar{M} & \xrightarrow{\bar{\beta}} & \bar{N} & \xrightarrow{\bar{\gamma}} & \mathrm{T}(\bar{L}) \\ \downarrow{\scriptstyle \bar{\phi}} & & \downarrow{\scriptstyle \bar{\psi}} & & \downarrow{\scriptstyle \bar{\chi}} & & \downarrow{\scriptstyle \mathrm{T}(\bar{\phi})} \\ \bar{L}' & \xrightarrow{\bar{\alpha}'} & \bar{M}' & \xrightarrow{\bar{\beta}'} & \bar{N}' & \xrightarrow{\bar{\gamma}'} & \mathrm{T}(\bar{L}') \end{array}$$

in $\mathbf{K}(A, \mathrm{M})$, where $\bar{\chi}$ is conjugate to $\mathrm{P}(\chi)$.

(TR4): We may assume that the three given distinguished triangles are standard triangles in $\mathbf{C}_{\mathrm{str}}(A, \mathrm{M})$. Namely, we can assume that $\alpha : L \to M$ and $\beta : M \to N$ are morphisms in $\mathbf{C}_{\mathrm{str}}(A, \mathrm{M})$; the DG modules P, Q, R are $P = \mathrm{Cone}(\alpha)$, $Q = \mathrm{Cone}(\beta \circ \alpha)$ and $R = \mathrm{Cone}(\beta)$; and the morphisms γ, δ, ϵ in $\mathbf{C}_{\mathrm{str}}(A, \mathrm{M})$ are $\gamma = e_\alpha$, $\delta = e_{\beta \circ \alpha}$ and $\epsilon = e_\beta$. All this is in the notation of Section 4.2.

In matrix notation we have

$$P = \begin{bmatrix} M \\ \mathrm{T}(L) \end{bmatrix}, \quad Q = \begin{bmatrix} N \\ \mathrm{T}(L) \end{bmatrix}, \quad R = \begin{bmatrix} N \\ \mathrm{T}(M) \end{bmatrix}.$$

We define the morphisms $\phi : P \to Q$ and $\psi : Q \to R$ in $\mathbf{C}_{\mathrm{str}}(A, \mathsf{M})$ by the matrix presentations

$$\phi := \begin{bmatrix} \beta & 0 \\ 0 & \mathrm{id}_{\mathrm{T}(L)} \end{bmatrix}, \quad \psi := \begin{bmatrix} \mathrm{id}_N & 0 \\ 0 & \mathrm{T}(\alpha) \end{bmatrix}.$$

(We leave to the reader to verify that ϕ and ψ commute with the differentials d_P, d_Q and d_R; this is just linear algebra, using the matrix presentations of the differentials of the cones from Definition 4.2.1.) Define the morphism $\rho : R \to \mathrm{T}(Q)$ in $\mathbf{C}_{\mathrm{str}}(A, \mathsf{M})$ to be the composition of the morphisms $R \to \mathrm{T}(M) \xrightarrow{\mathrm{T}(\gamma)} \mathrm{T}(Q)$. Then the diagram (5.1.11) in $\mathbf{C}_{\mathrm{str}}(A, \mathsf{M})$ is commutative.

It remains to prove that the triangle

$$P \xrightarrow{\bar{\phi}} Q \xrightarrow{\bar{\psi}} R \xrightarrow{\bar{\rho}} \mathrm{T}(P) \tag{5.4.12}$$

in $\mathbf{K}(A, \mathsf{M})$ is distinguished. Let $C := \mathrm{Cone}(\phi)$; so we have a standard triangle

$$P \xrightarrow{\phi} Q \xrightarrow{e_\phi} C \xrightarrow{p_\phi} \mathrm{T}(P) \tag{5.4.13}$$

in $\mathbf{C}_{\mathrm{str}}(A, \mathsf{M})$. We are going to prove that the triangles (5.4.12) and (5.4.13) are isomorphic in $\mathbf{K}(A, \mathsf{M})$, by producing an isomorphism $\bar{\chi} : C \xrightarrow{\simeq} R$ in $\mathbf{K}(A, \mathsf{M})$ that makes the diagram

$$\begin{array}{ccccccc} P & \xrightarrow{\bar{\phi}} & Q & \xrightarrow{\bar{e}_\phi} & C & \xrightarrow{\bar{p}_\phi} & \mathrm{T}(P) \\ \downarrow{\scriptstyle \mathrm{id}} & & \downarrow{\scriptstyle \mathrm{id}} & & \downarrow{\scriptstyle \bar{\chi}} & & \downarrow{\scriptstyle \mathrm{id}} \\ P & \xrightarrow{\bar{\phi}} & Q & \xrightarrow{\bar{\psi}} & R & \xrightarrow{\bar{\rho}} & \mathrm{T}(P) \end{array}$$

commutative.

Here are the matrices for the object C, the morphism $\chi : C \to R$, and another morphism $\omega : R \to C$, both in $\mathbf{C}_{\mathrm{str}}(A, \mathsf{M})$.

$$C = \begin{bmatrix} N \\ \mathrm{T}(L) \\ \mathrm{T}(M) \\ \mathrm{T}^2(L) \end{bmatrix}, \quad \chi := \begin{bmatrix} \mathrm{id}_N & 0 & 0 & 0 \\ 0 & \mathrm{T}(\alpha) & \mathrm{id}_{\mathrm{T}(M)} & 0 \end{bmatrix}, \quad \omega := \begin{bmatrix} \mathrm{id}_N & 0 \\ 0 & 0 \\ 0 & \mathrm{id}_{\mathrm{T}(M)} \\ 0 & 0 \end{bmatrix}.$$

Again, we leave it to the reader to check that χ and ω commute with the differentials. It is easy to see that $\omega \circ \psi = e_\phi$, $\rho \circ \chi = p_\phi$ and $\chi \circ \omega = \mathrm{id}_R$.

Finally we must find a homotopy between $\omega \circ \chi$ and id_C. Consider the degree -1 endomorphism θ of C :

$$\theta := \begin{bmatrix} 0 & 0 & 0 & 0 \\ 0 & 0 & 0 & 0 \\ 0 & 0 & 0 & 0 \\ 0 & \mathrm{t}_{\mathrm{T}(L)} & 0 & 0 \end{bmatrix}.$$

Then $\mathrm{d}_C \circ \theta + \theta \circ \mathrm{d}_C = \mathrm{id}_C - \omega \circ \chi$. □

A full subcategory of a DG category is of course a DG category itself. Full additive (resp. triangulated) subcategories of additive (resp. triangulated) categories were defined in Definition 2.2.6 (resp. 5.2.2).

Corollary 5.4.14 *Let* $\mathsf{C} \subseteq \mathbf{C}(A, \mathsf{M})$ *be a full subcategory satisfying these three conditions*:

(a) $\mathsf{C}_{\mathrm{str}}$ *is a full additive subcategory of* $\mathbf{C}_{\mathrm{str}}(A, \mathsf{M})$.
(b) C *is translation invariant, i.e.* $M \in \mathsf{C}$ *if and only* $\mathrm{T}(M) \in \mathsf{C}$.
(c) C *is closed under standard cones, i.e. for every morphism* ϕ *in* $\mathsf{C}_{\mathrm{str}}$ *the object* $\mathrm{Cone}(\phi)$ *belongs to* C.

Then $\mathsf{K} := \mathrm{Ho}(\mathsf{C})$ *is a full triangulated subcategory of* $\mathbf{K}(A, \mathsf{M})$.

Proof Each condition here implies the same numbered condition as in Definition 5.2.2. □

Here is a partial converse to the corollary.

Proposition 5.4.15 *Suppose* $\mathsf{K} \subseteq \mathbf{K}(A, \mathsf{M})$ *is a full triangulated subcategory. Let* $\mathsf{C} \subseteq \mathbf{C}(A, \mathsf{M})$ *be the full subcategory on the set of objects* $\mathrm{Ob}(\mathsf{K})$. *Then the DG category* C *satisfies conditions* (a)-(c) *of Corollary* 5.4.14, *and* $\mathsf{K} = \mathrm{Ho}(\mathsf{C})$.

Exercise 5.4.16 Prove Proposition 5.4.15.

Here are two remarks about other sorts of triangulated categories.

Remark 5.4.17 Let E be an exact category in the sense of D. Quillen [122]. Assume that E is a *Frobenius category*, i.e. E has enough injectives and enough projectives, and an object is injective if and only if it is projective. The *stable category* of E is the quotient category $\underline{\mathsf{E}} := \mathsf{E} / \mathsf{E}_0$, where $\mathsf{E}_0 \subseteq \mathsf{E}$ is the two-sided ideal of morphisms factoring through a projective object. The set of objects of $\underline{\mathsf{E}}$ is the same as that of E. D. Happel [58] has shown that $\underline{\mathsf{E}}$ is a triangulated category.

By definition, an *algebraic triangulated category* is a triangulated category K which is equivalent to the stable category $\underline{\mathsf{E}}$ of some Frobenius category E. It turns out that all the derived categories studied in this book (see Chapter 7) are in fact algebraic triangulated categories; see the paper [78] of B. Keller.

Here are a couple of other examples of algebraic triangulated categories. Let M be an additive category, and let $\mathsf{E} := \mathbf{C}_{\mathrm{str}}(\mathsf{M})$, the strict category of complexes in M. Put on E the exact structure in which the short exact sequences are the degreewise split sequences. Then E becomes a Frobenius category; the projective-injective objects in it are the contractible complexes. In this case the ideal E_0 coincides with the null-homotopic morphisms in $\mathsf{E} = \mathbf{C}_{\mathrm{str}}(\mathsf{M})$, and hence $\underline{\mathsf{E}} = \mathbf{K}(\mathsf{M})$, the homotopy category of M.

For the second example, suppose that $\mathbb{K}$ is a field and A a noetherian $\mathbb{K}$-ring which is Gorenstein, i.e. A has finite injective dimension as a left and as a right module over itself. The category of finitely generated A-modules is $\mathbf{M}_{\mathrm{f}}(A) = \mathsf{Mod}_{\mathrm{f}}\, A$. Let $\mathsf{E} \subseteq \mathbf{M}_{\mathrm{f}}(A)$ be the full subcategory on the modules M such that $\mathrm{Ext}^i_A(M, A) = 0$ for all $i > 0$. These modules are sometimes called *maximal Cohen–Macaulay modules*. Put on E the exact structure in which the exact sequences are the exact sequences in $\mathbf{M}_{\mathrm{f}}(A)$ with all terms belonging to E. Then E is a Frobenius category. The projective-injective objects of E are the modules in it that are projective A-modules. Moreover, according to R.-O. Buchweitz [34], the triangulated category $\underline{\mathsf{E}}$ in this case is equivalent to the *stable derived category* of A, which is the Verdier quotient $\mathbf{D}^{\mathrm{b}}_{\mathrm{f}}(A)/\mathbf{D}(A)_{\mathrm{perf}}$. Here $\mathbf{D}^{\mathrm{b}}_{\mathrm{f}}(A)$ is the derived category of bounded complexes with finitely generated cohomologies (see Section 7.4), $\mathbf{D}(A)_{\mathrm{perf}}$ is the full subcategory on the perfect complexes (see Chapter 14), and Verdier quotients are explained in Remark 7.1.14. Note that the triangulated category $\mathbf{D}^{\mathrm{b}}_{\mathrm{f}}(A)/\mathbf{D}(A)_{\mathrm{perf}}$ is called the *triangulated category of singularities* by D.O. Orlov [117], and this was later shortened to the *singularity category* of A.

Remark 5.4.18 In 1961, A. Dold and D. Puppe [44] tried to formalize the properties of the *stable homotopy category of topological spaces*. What they got is a theory almost identical to the theory of triangulated categories studied here (Definition 5.1.10), with the exception of the octahedral axiom (TR4).

In more modern terms we can consider the notion of a *stable model category*, which we now define. Let C be a pointed Quillen model category, meaning that it has a zero object. One defines a *suspension functor* Σ and a *loop functor* Ω on C using pushout and pullback diagrams involving the zero object. Let D be the *homotopy category* of C. (In the terminology of our book, D is actually the *derived category* of C, because it is gotten by inverting the weak equivalences in C; see Chapters 6–8.) If the functors Σ and Ω induce autoequivalences on D, quasi-inverse to each other, then C is called a stable model category. In this case it can be shown that D is a triangulated category, with translation functor $\mathrm{T} := \Sigma$. A more general notion is that of a *stable infinity category* (see [100] or [67]).

A *topological triangulated category* is, by definition, a triangulated category K which is equivalent to a full triangulated subcategory of the homotopy category D of a stable model category C, as above. For more information on these topics see [68], [135], [136], [100] and [38].

5.5 From DG Functors to Triangulated Functors

We now add a second DG ring B, and a second abelian category N. DG functors were introduced in Section 3.5.

Consider a DG functor $F : \mathbf{C}(A, \mathsf{M}) \to \mathbf{C}(B, \mathsf{N})$. From Theorem 4.4.3 we know that the translation isomorphism τ_F is an isomorphism of DG functors $\tau_F : F \circ \mathrm{T}_{A,\mathsf{M}} \xrightarrow{\simeq} \mathrm{T}_{B,\mathsf{N}} \circ F$. Therefore, when we pass to the homotopy categories, and writing $\bar{F} := \mathrm{Ho}(F)$ and $\bar{\tau}_F := \mathrm{Ho}(\tau_F)$, we get a T-additive functor $(\bar{F}, \bar{\tau}_F) : \mathbf{K}(A, \mathsf{M}) \to \mathbf{K}(B, \mathsf{N})$.

If $G : \mathbf{C}(A, \mathsf{M}) \to \mathbf{C}(B, \mathsf{N})$ is another DG functor, then we have another T-additive functor $(\bar{G}, \bar{\tau}_G) : \mathbf{K}(A, \mathsf{M}) \to \mathbf{K}(B, \mathsf{N})$. And if $\eta : F \to G$ is a strict morphism of DG functors, then there is a morphism of additive functors $\bar{\eta} := \mathrm{Ho}(\eta) : \bar{F} \to \bar{G}$. This notation will be used in the next theorem.

Theorem 5.5.1 *Let A and B be DG rings, let M and N be abelian categories, and let $F : \mathbf{C}(A, \mathsf{M}) \to \mathbf{C}(B, \mathsf{N})$ be a DG functor.*

(1) *The T-additive functor*

$$(\bar{F}, \bar{\tau}_F) : \mathbf{K}(A, \mathsf{M}) \to \mathbf{K}(B, \mathsf{N})$$

is a triangulated functor.

(2) *Suppose $G : \mathbf{C}(A, \mathsf{M}) \to \mathbf{C}(B, \mathsf{N})$ is another DG functor, and $\eta : F \to G$ is a strict morphism of DG functors. Then*

$$\bar{\eta} : (\bar{F}, \bar{\tau}_F) \to (\bar{G}, \bar{\tau}_G)$$

is a morphism of triangulated functors.

Proof

(1) Take a distinguished triangle $L \xrightarrow{\bar{\alpha}} M \xrightarrow{\bar{\beta}} N \xrightarrow{\bar{\gamma}} \mathrm{T}(L)$ in $\mathbf{K}(A, \mathsf{M})$. Since we are only interested in triangles up to isomorphism, we can assume that this is the image under the functor P of a standard triangle $L \xrightarrow{\alpha} M \xrightarrow{\beta} N \xrightarrow{\gamma} \mathrm{T}(L)$ in $\mathbf{C}_{\mathrm{str}}(A, \mathsf{M})$. According to Theorem 4.5.5 and Corollary 4.5.6, there is a standard triangle $L' \xrightarrow{\alpha'} M' \xrightarrow{\beta'} N' \xrightarrow{\gamma'} \mathrm{T}(L')$ in $\mathbf{C}_{\mathrm{str}}(B, \mathsf{N})$, and a commutative diagram

$$\begin{array}{ccccccc}
F(L) & \xrightarrow{F(\alpha)} & F(M) & \xrightarrow{F(\beta)} & F(N) & \xrightarrow{\tau_{F,L} \circ F(\gamma)} & \mathrm{T}(F(L)) \\
\downarrow{\scriptstyle \phi} & & \downarrow{\scriptstyle \psi} & & \downarrow{\scriptstyle \chi} & & \downarrow{\scriptstyle \mathrm{T}(\phi)} \\
L' & \xrightarrow{\alpha'} & M' & \xrightarrow{\beta'} & N' & \xrightarrow{\gamma'} & \mathrm{T}(L')
\end{array}$$

in $\mathbf{C}_{\mathrm{str}}(B, \mathsf{N})$, in which the vertical arrows are isomorphisms. (Actually, we can take $L' = F(L)$, $\phi = \mathrm{id}_{F(L)}$, etc.) After applying the functor P to this diagram, we see that the condition in Definition 5.2.1(1) is satisfied.

(2) By Definition 5.2.1(2), all we need to prove that $\bar{\eta}$ is a morphism of T-additive functors. Let's use these abbreviations: $\mathsf{K} := \mathbf{K}(A, \mathsf{M})$ and $\mathsf{L} := \mathbf{K}(B, \mathsf{N})$. Definition 5.1.5 requires that for every object $M \in \mathsf{K}$ the diagram

$$\begin{array}{ccc} \bar{F}(\mathrm{T}_{\mathsf{K}}(M)) & \xrightarrow{\bar{\tau}_{F,M}} & \mathrm{T}_{\mathsf{L}}(\bar{F}(M)) \\ \Big\downarrow{\scriptstyle \bar{\eta}_{\mathrm{T}_{\mathsf{K}}(M)}} & & \Big\downarrow{\scriptstyle \mathrm{T}_{\mathsf{L}}(\bar{\eta}_M)} \\ \bar{G}(\mathrm{T}_{\mathsf{K}}(M)) & \xrightarrow{\bar{\tau}_{G,M}} & \mathrm{T}_{\mathsf{L}}(\bar{G}(M)) \end{array} \tag{5.5.2}$$

in L is commutative. Going back to Definitions 4.1.6 and 4.4.1 we see that $\mathrm{T}_{\mathsf{L}}(\eta_M) = \mathrm{t}_{G(M)} \circ \eta_M \circ \mathrm{t}_{F(M)}^{-1}$, $\tau_{F,M} = \mathrm{t}_{F(M)} \circ F(\mathrm{t}_M^{-1})$ and $\tau_{G,M} = \mathrm{t}_{G(M)} \circ G(\mathrm{t}_M^{-1})$, as morphisms in $\mathbf{C}(B, \mathsf{N})$. Thus the path in (5.5.2) that starts by going right is represented by the morphism

$$\begin{aligned} \mathrm{T}_{\mathsf{L}}(\eta_M) \circ \tau_{F,M} &= \mathrm{t}_{G(M)} \circ \eta_M \circ \mathrm{t}_{F(M)}^{-1} \circ \mathrm{t}_{F(M)} \circ F(\mathrm{t}_M^{-1}) \\ &= \mathrm{t}_{G(M)} \circ \eta_M \circ F(\mathrm{t}_M^{-1}) \end{aligned}$$

in $\mathbf{C}_{\mathrm{str}}(B, \mathsf{N})$, and other path is represented by $\mathrm{t}_{G(M)} \circ G(\mathrm{t}_M^{-1}) \circ \eta_{\mathrm{T}_{\mathsf{K}}(M)}$. Because $\mathrm{t}_{G(M)}$ is an isomorphism (of degree -1) in $\mathbf{C}(B, \mathsf{N})$, it suffices to prove that $\eta_M \circ F(\mathrm{t}_M^{-1}) = G(\mathrm{t}_M^{-1}) \circ \eta_{\mathrm{T}_{\mathsf{K}}(M)}$ in $\mathbf{C}(B, \mathsf{N})$; namely that the diagram

$$\begin{array}{ccc} F(\mathrm{T}_{\mathsf{K}}(M)) & \xrightarrow{F(\mathrm{t}_M^{-1})} & F(M) \\ \Big\downarrow{\scriptstyle \eta_{\mathrm{T}_{\mathsf{K}}(M)}} & & \Big\downarrow{\scriptstyle \eta_M} \\ G(\mathrm{T}_{\mathsf{K}}(M)) & \xrightarrow{G(\mathrm{t}_M^{-1})} & G(M) \end{array}$$

is commutative. This is true because $\eta : F \to G$ is a strict morphism of DG functors, and here it acts on the degree 1 morphism $\mathrm{t}_M^{-1} : \mathrm{T}_{\mathsf{K}}(M) \to M$ in $\mathbf{C}(B, \mathsf{N})$. □

Corollary 5.5.3 *For every integer k, the pair* $(\mathrm{T}^k, (-1)^k \cdot \mathrm{id}_{\mathrm{T}^{k+1}})$ *is a triangulated functor from* $\mathbf{K}(A, \mathsf{M})$ *to itself.*

Proof Combine Theorem 5.5.1 and Proposition 4.4.4. □

Remark 5.5.4 In [25], A.I. Bondal and M.M. Kapranov introduced the concept of *pretriangulated DG category*. This is a DG category C for which the homotopy category $\mathrm{Ho}(\mathsf{C})$ is canonically triangulated (the details of the definition are too complicated to mention here). Our DG categories $\mathbf{C}(A, \mathsf{M})$ are pretriangulated in the sense of [25]; but they have a lot more structure (e.g. the objects have cohomologies too).

Suppose C and C' are pretriangulated DG categories. In [25] there is a (rather complicated) definition of *preexact DG functor* $F : \mathsf{C} \to \mathsf{C}'$. It is stated there that if F is a preexact DG functor, then $\mathrm{Ho}(F) : \mathrm{Ho}(\mathsf{C}) \to \mathrm{Ho}(\mathsf{C}')$ is a triangulated functor. This is analogous to our Theorem 5.5.1. Presumably, Theorems 4.4.3 and 4.5.5 imply that every DG functor $F : \mathbf{C}(A, \mathsf{M}) \to \mathbf{C}(A', \mathsf{M}')$ is preexact in the sense of [25]; but we did not verify this.

5.6 The Opposite Homotopy Category Is Triangulated

Here we introduce a canonical triangulated structure on the opposite homotopy category $\mathbf{K}(A, \mathsf{M})^{\mathrm{op}}$. This gives us a way to talk about contravariant triangulated functors whose source is a full subcategory of $\mathbf{K}(A, \mathsf{M})$. Our solution is precise, but it is not totally satisfactory – see Remark 5.6.12.

We already gave a thorough treatment of contravariant DG functors in Section 3.9. In the previous section we explained precisely how to pass from DG functors to triangulated functors. In this section we treat the contravariant case.

As before, A is a central DG $\mathbb{K}$-ring, and M is a $\mathbb{K}$-linear abelian category. The DG category of DG A-modules in M is $\mathbf{C}(A, \mathsf{M})$. In Section 3.9 we introduced the flipped DG category $\mathbf{C}(A^{\mathrm{op}}, \mathsf{M}^{\mathrm{op}})$. In Theorem 3.9.16 we had a canonical isomorphism of DG categories

$$\mathrm{Flip} : \mathbf{C}(A, \mathsf{M})^{\mathrm{op}} \xrightarrow{\simeq} \mathbf{C}(A^{\mathrm{op}}, \mathsf{M}^{\mathrm{op}}). \tag{5.6.1}$$

Definition 5.6.2 The *flipped category* of $\mathbf{K}(A, \mathsf{M})$ is the triangulated category

$$\mathbf{K}(A, \mathsf{M})^{\mathrm{flip}} := \mathbf{K}(A^{\mathrm{op}}, \mathsf{M}^{\mathrm{op}}) = \mathrm{Ho}(\mathbf{C}(A^{\mathrm{op}}, \mathsf{M}^{\mathrm{op}})).$$

See Theorem 5.4.3. Its translation functor is denoted by $\mathrm{T}^{\mathrm{flip}}$.

Since $\mathrm{Ho}(\mathbf{C}(A, \mathsf{M})^{\mathrm{op}}) = \mathbf{K}(A, \mathsf{M})^{\mathrm{op}}$, the isomorphism (5.6.1) induces a $\mathbb{K}$-linear isomorphism of categories

$$\overline{\mathrm{Flip}} := \mathrm{Ho}(\mathrm{Flip}) : \mathbf{K}(A, \mathsf{M})^{\mathrm{op}} \xrightarrow{\simeq} \mathbf{K}(A, \mathsf{M})^{\mathrm{flip}}. \tag{5.6.3}$$

Definition 5.6.4 The category $\mathbf{K}(A, \mathsf{M})^{\mathrm{op}}$ is given the triangulated category structure induced from the flipped category $\mathbf{K}(A, \mathsf{M})^{\mathrm{flip}}$ under the isomorphism (5.6.3). The translation functor of $\mathbf{K}(A, \mathsf{M})^{\mathrm{op}}$ is denoted by T^{op}.

Thus

$$\mathrm{T}^{\mathrm{op}} = \overline{\mathrm{Flip}}^{-1} \circ \mathrm{T}^{\mathrm{flip}} \circ \overline{\mathrm{Flip}}. \tag{5.6.5}$$

The distinguished triangles of $\mathbf{K}(A, \mathsf{M})^{\mathrm{op}}$ are the images under $\overline{\mathrm{Flip}}^{-1}$ of the distinguished triangles of $\mathbf{K}(A, \mathsf{M})^{\mathrm{flip}}$. And tautologically,

$$(\overline{\mathrm{Flip}}, \mathrm{id}) : \mathbf{K}(A, \mathsf{M})^{\mathrm{op}} \to \mathbf{K}(A, \mathsf{M})^{\mathrm{flip}} \tag{5.6.6}$$

is an isomorphism of triangulated categories.

Definition 5.6.7 Let $\mathsf{K} \subseteq \mathbf{K}(A, \mathsf{M})$ be a full additive subcategory, and assume that K^{op} is a triangulated subcategory of $\mathbf{K}(A, \mathsf{M})^{\mathrm{op}}$. Then we give K^{op} the triangulated structure induced from $\mathbf{K}(A, \mathsf{M})^{\mathrm{op}}$.

To say this a bit differently, the condition on K in the definition is that $\mathsf{K}^{\mathrm{flip}} := \overline{\mathrm{Flip}}(\mathsf{K}^{\mathrm{op}}) \subseteq \mathbf{K}(A, \mathsf{M})^{\mathrm{flip}}$ is a triangulated subcategory. The triangulated structure we put on K^{op} is such that $(\overline{\mathrm{Flip}}, \mathrm{id}) : (\mathsf{K}^{\mathrm{op}}, \mathrm{T}^{\mathrm{op}}) \to (\mathsf{K}^{\mathrm{flip}}, \mathrm{T}^{\mathrm{flip}})$ is an isomorphism of triangulated categories.

Remark 5.6.8 It could happen (though we have no example of it) that $\mathsf{K} \subseteq \mathbf{K}(A, \mathsf{M})$ is a full triangulated subcategory, yet $\mathsf{K}^{\mathrm{op}} \subseteq \mathbf{K}(A, \mathsf{M})^{\mathrm{op}}$ is not a triangulated subcategory; or vice versa. More on this issue in Remark 5.6.12.

The remark above notwithstanding, in many important cases, such as in Propositions 7.6.1 and 7.6.2, both $\mathsf{K} \subseteq \mathbf{K}(A, \mathsf{M})$ and $\mathsf{K}^{\mathrm{op}} \subseteq \mathbf{K}(A, \mathsf{M})^{\mathrm{op}}$ are triangulated.

Definition 5.6.9 Let L be a triangulated category, and let $\mathsf{K} \subseteq \mathbf{K}(A, \mathsf{M})$ be a full subcategory such that $\mathsf{K}^{\mathrm{op}} \subseteq \mathbf{K}(A, \mathsf{M})^{\mathrm{op}}$ is triangulated. A *contravariant triangulated functor* from K to L is, by definition, a triangulated functor $(F, \tau) : \mathsf{K}^{\mathrm{op}} \to \mathsf{L}$ in the sense of Definition 5.2.1, where K^{op} has the triangulated structure from Definition 5.6.7.

Theorem 5.6.10 *Let A and B be DG rings, and let M and N be abelian categories. Let $\mathsf{C} \subseteq \mathbf{C}(A, \mathsf{M})$ be a full subcategory, and let $F : \mathsf{C}^{\mathrm{op}} \to \mathbf{C}(B, \mathsf{N})$ be a DG functor. Consider the homotopy category $\mathsf{K} := \mathrm{Ho}(\mathsf{C}) \subseteq \mathrm{Ho}(\mathbf{C}(A, \mathsf{M})) = \mathbf{K}(A, \mathsf{M})$ and the induced additive functor $\bar{F} := \mathrm{Ho}(F) : \mathsf{K}^{\mathrm{op}} \to \mathbf{K}(B, \mathsf{N})$. Suppose that $\mathsf{K}^{\mathrm{op}} \subseteq \mathbf{K}(A, \mathsf{M})^{\mathrm{op}}$ is triangulated. Then there is a canonical translation isomorphism $\bar{\tau}$ such that*

$$(\bar{F}, \bar{\tau}) : \mathsf{K}^{\mathrm{op}} \to \mathbf{K}(B, \mathsf{N})$$

is a triangulated functor.

Proof Define

$$\mathsf{C}^{\mathrm{flip}} := \mathrm{Flip}(\mathsf{C}^{\mathrm{op}}) \subseteq \mathbf{C}(A^{\mathrm{op}}, \mathsf{M}^{\mathrm{op}}) = \mathbf{C}(A, \mathsf{M})^{\mathrm{flip}}.$$

This is a DG category, and $\mathrm{Flip} : \mathsf{C}^{\mathrm{op}} \to \mathsf{C}^{\mathrm{flip}}$ is an isomorphism of DG categories. Also $\mathrm{Ho}(\mathsf{C}^{\mathrm{flip}}) = \mathsf{K}^{\mathrm{flip}}$. There is a DG functor

$$F^{\mathrm{flip}} := F \circ \mathrm{Flip}^{-1} : \mathsf{C}^{\mathrm{flip}} \to \mathbf{C}(B, \mathsf{N}),$$

and it induces an additive functor

$$\bar{F}^{\mathrm{flip}} := \mathrm{Ho}(F^{\mathrm{flip}}) : \mathsf{K}^{\mathrm{flip}} \to \mathbf{K}(B, \mathsf{N}).$$

By Theorem 5.5.1 there is a translation isomorphism $\bar{\tau}^{\mathrm{flip}}$ such that

$$(\bar{F}^{\mathrm{flip}}, \bar{\tau}^{\mathrm{flip}}) : \mathsf{K}^{\mathrm{flip}} \to \mathbf{K}(B, \mathsf{N})$$

is a triangulated functor. Using formula (5.6.5) we obtain these equalities and isomorphisms:

$$\begin{aligned} \bar{F} \circ \mathrm{T}^{\mathrm{op}} &= \bar{F} \circ \overline{\mathrm{Flip}}^{-1} \circ \mathrm{T}^{\mathrm{flip}} \circ \overline{\mathrm{Flip}} = \bar{F}^{\mathrm{flip}} \circ \mathrm{T}^{\mathrm{flip}} \circ \overline{\mathrm{Flip}} \\ &\xrightarrow{\bar{\tau}^{\mathrm{flip}}} \mathrm{T}_{\mathbf{K}(B,\mathsf{N})} \circ \bar{F}^{\mathrm{flip}} \circ \overline{\mathrm{Flip}} = \mathrm{T}_{\mathbf{K}(B,\mathsf{N})} \circ \bar{F}. \end{aligned}$$

We define $\bar{\tau}$ to be the composed isomorphism. By construction the pair $(\bar{F}, \bar{\tau})$ is a triangulated functor. □

Exercise 5.6.11 Let $\mathsf{C} \subseteq \mathbf{C}(A, \mathsf{M})$ be a full subcategory s.t. $\mathsf{K} := \mathrm{Ho}(\mathsf{C})$ is a triangulated subcategory of $\mathbf{K}(A, \mathsf{M})$, and let $G : \mathsf{C} \to \mathbf{C}(B, \mathsf{N})^{\mathrm{op}}$ be a DG functor. Write $\bar{G} := \mathrm{Ho}(G)$. Show that there is a translation isomorphism $\bar{\nu}$ s.t. $(\bar{G}, \bar{\nu}) : \mathsf{K} \to \mathbf{K}(B, \mathsf{N})^{\mathrm{op}}$ is a triangulated functor. (Hint: study the proof of the theorem above.)

Remark 5.6.12 Let K be a triangulated category. There is a way to introduce a triangulated structure on the opposite category K^{op}. The translation functor is $\mathrm{T}^{\mathrm{op}} := \mathrm{Op} \circ \mathrm{T}^{-1} \circ \mathrm{Op}^{-1}$. The distinguished triangles are defined to be

$$N \xrightarrow{\mathrm{Op}(\beta)} M \xrightarrow{\mathrm{Op}(\alpha)} L \xrightarrow{\mathrm{Op}(-\mathrm{T}^{-1}(\gamma))} \mathrm{T}^{\mathrm{op}}(N),$$

where $L \xrightarrow{\alpha} M \xrightarrow{\beta} N \xrightarrow{\gamma} \mathrm{T}(L)$ goes over all distinguished triangles in K. See [76, Remark 10.1.10(ii)]. This approach has an advantage: it allows to talk about contravariant triangulated functors $\mathsf{K} \to \mathsf{L}$ without any restriction on the category K. This is in contrast to our quite restricted Definition 5.6.9. The disadvantage of the approach presented in this paragraph is that there is no easy way to tie it with the DG theory (as opposed to our Theorems 5.5.1 and 5.6.10).

It would have been very pleasing if in the case $\mathsf{K} = \mathbf{K}(A, \mathsf{M})$, the triangulated structure on $\mathbf{K}(A, \mathsf{M})^{\mathrm{op}}$ presented in the paragraph above would coincide with the triangulated structure from Definition 5.6.4. However, our calculations seem to indicate otherwise.

Since, in this book, we are only interested in triangulated functors that are of DG origin (covariant, as in Theorem 5.5.1, or contravariant, as in Theorem 5.6.10), we decided to adopt Definition 5.6.9. It is a reliable, yet somewhat awkward approach. For instance, it requires us to perform particular calculations for the composition of contravariant triangulated functors, as in Lemma 13.1.16.

6

Localization of Categories

This chapter is devoted to the general theory of *Ore localization of categories*. We are given a category A and a multiplicatively closed set of morphisms $\mathsf{S} \subseteq \mathsf{A}$. The localized category A_S, gotten by formally inverting the morphisms in S, always exists. The goal is to have a presentation of the morphisms of A_S as left or right fractions.

In Chapter 7 we shall apply the results of this chapter to triangulated categories.

6.1 The Formalism of Localization

We will start with a category A, without even assuming it is linear. Still we use the notation A, because it will be suggestive to think about a linear category A with a single object, which is just a ring A. The reason is that our localization procedure is the same as that in noncommutative ring theory – even when the category is not linear and it has multiple objects. In Section 6.3 we treat linear categories, and thus we recover the ring theoretic localization as a special case.

The emphasis will be on morphisms rather than on objects. Thus it will be convenient to write $\mathsf{A}(M, N) := \mathrm{Hom}_\mathsf{A}(M, N)$ for $M, N \in \mathrm{Ob}(\mathsf{A})$. We sometimes use the notation $a \in \mathsf{A}$ for a morphism $a \in \mathsf{A}(M, N)$, leaving the objects implicit. When we write $b \circ a$ for $a, b \in \mathsf{A}$, we implicitly mean that these morphisms are composable.

For heuristic purposes, we can think of A as a linear category (e.g. living inside some category of modules), with objects $M, N, \ldots$. For any given object

M, we then have a genuine ring $\mathsf{A}(M) := \mathsf{A}(M, M)$.

Definition 6.1.1 Let A be a category. A *multiplicatively closed set of morphisms* in A is a subcategory $\mathsf{S} \subseteq \mathsf{A}$ such that $\mathrm{Ob}(\mathsf{S}) = \mathrm{Ob}(\mathsf{A})$.

In other words, for any pair of objects $M, N \in \mathsf{A}$ there is a subset $\mathsf{S}(M, N) \subseteq \mathsf{A}(M, N)$, such that $\mathrm{id}_M \in \mathsf{S}(M, M)$, and such that for any $s \in \mathsf{S}(L, M)$ and $t \in \mathsf{S}(M, N)$, the composition $t \circ s \in \mathsf{S}(L, N)$.

Using our shorthand, we can write the definition like this: $\mathrm{id}_M \in \mathsf{S}$, and $s, t \in \mathsf{S}$ implies $t \circ s \in \mathsf{S}$.

If $A = \mathsf{A}$ is a single object linear category, namely a ring, then $S = \mathsf{S}$ is a multiplicatively closed set in the sense of ring theory.

There are various notions of localization in the literature. We restrict attention to two of them. Here is the first:

Definition 6.1.2 Let S be a multiplicatively closed set of morphisms in a category A. A *localization* of A with respect to S is a pair $(\mathsf{A}_\mathsf{S}, \mathrm{Q})$, consisting of a category A_S and a functor $\mathrm{Q} : \mathsf{A} \to \mathsf{A}_\mathsf{S}$, called the localization functor, having the following properties:

(Loc1) There is equality $\mathrm{Ob}(\mathsf{A}_\mathsf{S}) = \mathrm{Ob}(\mathsf{A})$, and Q is the identity on objects.

(Loc2) For every $s \in \mathsf{S}$, the morphism $\mathrm{Q}(s) \in \mathsf{A}_\mathsf{S}$ is invertible (i.e. it is an isomorphism).

(Loc3) Suppose B is a category, and $F : \mathsf{A} \to \mathsf{B}$ is a functor such that $F(s)$ is invertible for every $s \in \mathsf{S}$. Then there is a unique functor $F_\mathsf{S} : \mathsf{A}_\mathsf{S} \to \mathsf{B}$ such that $F_\mathsf{S} \circ \mathrm{Q} = F$ as functors $\mathsf{A} \to \mathsf{B}$.

In a commutative diagram:

$$\begin{array}{ccccc} \mathsf{S} & \xrightarrow{\text{inc}} & \mathsf{A} & \overset{F}{\dashrightarrow} & \mathsf{B} \\ & & {\scriptstyle \mathrm{Q}}\downarrow & \nearrow_{F_\mathsf{S}} & \\ & & \mathsf{A}_\mathsf{S} & & \end{array}$$

Example 6.1.3 Suppose A has a single object, say x. Here is what the definition above says in this case. We have a ring $A := \mathsf{A}(x, x)$ and a multiplicatively closed set $S := \mathsf{S}(x, x) \subseteq A$. The localization of A with respect to S is a ring $A_S := \mathsf{A}_\mathsf{S}(x, x)$, with a ring homomorphism $\mathrm{Q} : A \to A_S$, such that $\mathrm{Q}(S) \subseteq (A_S)^\times$, the group of invertible elements of A_S. Condition (Loc3) says that any ring homomorphism $F : A \to B$ such that $F(S) \subseteq B^\times$ factors uniquely through A_S.

Theorem 6.1.4 *Let* S *be a multiplicatively closed set of morphisms in a category* A. *A localization* $(\mathsf{A}_\mathsf{S}, \mathrm{Q})$ *of* A *with respect to* S*, in the sense of Definition* 6.1.2*, exists, and it is unique up to a unique isomorphism.*

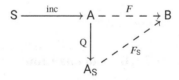

Proof Uniqueness: Suppose $(\mathsf{A}'_{\mathsf{S}}, \mathrm{Q}')$ is another localization. By condition (Loc3), for both localizations, there are unique functors $G : \mathsf{A}_{\mathsf{S}} \to \mathsf{A}'_{\mathsf{S}}$ and $G' : \mathsf{A}'_{\mathsf{S}} \to \mathsf{A}_{\mathsf{S}}$ that satisfy $\mathrm{Q}' = G \circ \mathrm{Q}$ and $\mathrm{Q} = G' \circ \mathrm{Q}'$. These functors must be the identity on objects, and are inverses to each other.

Existence: Here we encounter the set theoretic issue alluded to in Section 1.1. The category A_{S} that we will construct might have to belong to a bigger universe than that in which A lives; this is because of its large morphism sets. We proceed as follows. Let B be the free category on the set of objects $\mathrm{Ob}(\mathsf{A})$, with set of generating morphisms $\mathsf{A} \sqcup \mathsf{S}^{\mathrm{op}}$. Thus, the morphisms in B are the finite sequences of composable morphisms in $\mathsf{A} \sqcup \mathsf{S}^{\mathrm{op}}$, and composition is concatenation. Inside B we have the congruence (two-sided ideal) $\mathsf{N} \subseteq \mathsf{B}$ generated by the relations of A, the relations of S^{op}, and the relations $\mathrm{Op}(s) \circ s = \mathrm{id}$ and $s \circ \mathrm{Op}(s) = \mathrm{id}$ for all $s \in \mathsf{S}$. The quotient category $\mathsf{A}_{\mathsf{S}} := \mathsf{B} / \mathsf{N}$ has the desired properties. Cf. [49] for more details. □

Proposition 6.1.5 *Let* S *be a multiplicatively closed set in a category* A*, let* B *be a category, and let* $\mathrm{Q} : \mathsf{A} \to \mathsf{B}$ *be a functor. Consider the multiplicatively closed set* $\mathsf{S}^{\mathrm{op}} \subseteq \mathsf{A}^{\mathrm{op}}$ *and the functor* $\mathrm{Q}^{\mathrm{op}} : \mathsf{A}^{\mathrm{op}} \to \mathsf{B}^{\mathrm{op}}$*. The two conditions below are equivalent:*

(i) *The pair* (B, Q) *is a localization of* A *with respect to* S.
(ii) *The pair* $(\mathsf{B}^{\mathrm{op}}, \mathrm{Q}^{\mathrm{op}})$ *is a localization of* A^{op} *with respect to* S^{op}.

Exercise 6.1.6 Prove Proposition 6.1.5.

Often the localization A_{S} is of little value, because there is no practical way to describe the morphisms in it. This issue will be addressed in the next section.

6.2 Ore Localization

There is a better notion of localization. The references here are [62], [49], [158], [75], [148] and [130]. The first four references talk about localization of categories; and the last two talk about noncommutative rings. It seems that historically, this *noncommutative calculus of fractions* was discovered by Ore and Asano in ring theory, around 1930. There was progress in the categorical side, notably by Gabriel around 1960.

In this section we mostly follow the treatment of [148]; but we sometimes use diagrams instead of formulas.

Definition 6.2.1 Let S be a multiplicatively closed set of morphisms in a category A. A *right Ore localization* of A with respect to S is a pair $(\mathsf{A}_{\mathsf{S}}, \mathrm{Q})$, consisting of a category A_S and a functor $\mathrm{Q} : \mathsf{A} \to \mathsf{A}_\mathsf{S}$, having the following properties:

(RO1) There is equality $\mathrm{Ob}(\mathsf{A}_\mathsf{S}) = \mathrm{Ob}(\mathsf{A})$, and Q is the identity on objects.

(RO2) For every $s \in \mathsf{S}$, the morphism $\mathrm{Q}(s) \in \mathsf{A}_\mathsf{S}$ is an isomorphism.

(RO3) Every morphism $q \in \mathsf{A}_\mathsf{S}$ can be written as $q = \mathrm{Q}(a) \circ \mathrm{Q}(s)^{-1}$ for some $a \in \mathsf{A}$ and $s \in \mathsf{S}$.

(RO4) Suppose $a, b \in \mathsf{A}$ satisfy $\mathrm{Q}(a) = \mathrm{Q}(b)$. Then $a \circ s = b \circ s$ for some $s \in \mathsf{S}$.

The letters "RO" stand for "right Ore." We refer to the expression $q = \mathrm{Q}(a) \circ \mathrm{Q}(s)^{-1}$ as a *right fraction representation of q*. Here is the left sided version of this definition:

Definition 6.2.2 Let S be a multiplicatively closed set of morphisms in a category A. A *left Ore localization* of A with respect to S is a pair $(\mathsf{A}_\mathsf{S}, \mathrm{Q})$, consisting of a category A_S and a functor $\mathrm{Q} : \mathsf{A} \to \mathsf{A}_\mathsf{S}$, having the following properties:

(LO1) There is equality $\mathrm{Ob}(\mathsf{A}_\mathsf{S}) = \mathrm{Ob}(\mathsf{A})$, and Q is the identity on objects.

(LO2) For every $s \in \mathsf{S}$, the morphism $\mathrm{Q}(s) \in \mathsf{A}_\mathsf{S}$ is an isomorphism.

(LO3) Every morphism $q \in \mathsf{A}_\mathsf{S}$ can be written as $q = \mathrm{Q}(s)^{-1} \circ \mathrm{Q}(a)$ for some $a \in \mathsf{A}$ and $s \in \mathsf{S}$.

(LO4) Suppose $a, b \in \mathsf{A}$ satisfy $\mathrm{Q}(a) = \mathrm{Q}(b)$. Then $s \circ a = s \circ b$ for some $s \in \mathsf{S}$.

As in the right case, the letters "LO" stand for "left Ore"; and we refer to the expression $q = \mathrm{Q}(s)^{-1} \circ \mathrm{Q}(a)$ as a *left fraction representation of q*.

Remark 6.2.3 The results to follow in this section will be stated for right Ore localizations only. They all have "left" versions, with identical proofs (just a matter of reversing some arrows or compositions), and therefore they will be omitted.

To reinforce the last remark, we give:

Proposition 6.2.4 *Let* S *be a multiplicatively closed set in a category* A, *let* B *be a category, and let* $\mathrm{Q} : \mathsf{A} \to \mathsf{B}$ *be a functor. Consider the multiplicatively closed set* $\mathsf{S}^{\mathrm{op}} \subseteq \mathsf{A}^{\mathrm{op}}$ *and the functor* $\mathrm{Q}^{\mathrm{op}} : \mathsf{A}^{\mathrm{op}} \to \mathsf{B}^{\mathrm{op}}$. *The two conditions below are equivalent*:

(i) *The pair* (B, Q) *is a left Ore localization of* A *with respect to* S.

(ii) *The pair* $(\mathsf{B}^{\mathrm{op}}, \mathrm{Q}^{\mathrm{op}})$ *is a right Ore localization of* A^{op} *with respect to* S^{op}.

Exercise 6.2.5 Prove Proposition 6.2.4.

Lemma 6.2.6 *Let* (A_S, Q) *be a right Ore localization, let* $a_1, a_2 \in A$ *and* $s_1, s_2 \in S$. *The following conditions are equivalent*:

(i) $Q(a_1) \circ Q(s_1)^{-1} = Q(a_2) \circ Q(s_2)^{-1}$ *in* A_S.

(ii) *There are* $b_1, b_2 \in A$ *s.t.* $a_1 \circ b_1 = a_2 \circ b_2$, *and* $s_1 \circ b_1 = s_2 \circ b_2 \in S$.

Proof

(ii) ⇒ (i): Since $Q(s_i)$ and $Q(s_i \circ b_i)$ are invertible, it follows that $Q(b_i)$ are invertible. So

$$\begin{aligned} Q(a_1) \circ Q(s_1)^{-1} &= Q(a_1) \circ Q(b_1) \circ Q(b_1)^{-1} \circ Q(s_1)^{-1} \\ &= Q(a_2) \circ Q(b_2) \circ Q(b_2)^{-1} \circ Q(s_2)^{-1} = Q(a_2) \circ Q(s_2)^{-1}. \end{aligned}$$

(i) ⇒ (ii): By property (RO3) there are $c \in A$ and $u \in S$ s.t.

$$Q(s_2)^{-1} \circ Q(s_1) = Q(c) \circ Q(u)^{-1}. \tag{6.2.7}$$

Rewriting this equation we get

$$Q(s_1 \circ u) = Q(s_2 \circ c). \tag{6.2.8}$$

It is given that $Q(a_1) = Q(a_2) \circ Q(s_2)^{-1} \circ Q(s_1)$. Plugging (6.2.7) into it we obtain $Q(a_1) = Q(a_2) \circ Q(c) \circ Q(u)^{-1}$. Rearranging this equation we get $Q(a_1 \circ u) = Q(a_2 \circ c)$. By property (RO4) there is $v \in S$ s.t. $a_1 \circ u \circ v = a_2 \circ c \circ v$. Likewise, from equation (6.2.8) and property (RO4), there is $v' \in S$ s.t. $s_1 \circ u \circ v' = s_2 \circ c \circ v'$.

Again using property (RO3), there are $d \in A$ and $w \in S$ s.t. $Q(v)^{-1} \circ Q(v') = Q(d) \circ Q(w)^{-1}$. Rearranging we get $Q(v' \circ w) = Q(v \circ d)$. By property (RO4) there is $w' \in S$ s.t. $v' \circ w \circ w' = v \circ d \circ w'$. Define $b_1 := u \circ v \circ d \circ w'$ and $b_2 := c \circ v \circ d \circ w'$. Then

$$\begin{aligned} s_1 \circ b_1 &= s_1 \circ u \circ v \circ d \circ w' = s_1 \circ u \circ v' \circ w \circ w' \\ &= s_2 \circ c \circ v' \circ w \circ w' = s_2 \circ b_2, \end{aligned}$$

and it is in S. Also

$$a_1 \circ b_1 = a_1 \circ u \circ v \circ d \circ w' = a_2 \circ c \circ v \circ d \circ w' = a_2 \circ b_2. \qquad \square$$

Proposition 6.2.9 *A right Ore localization* (A_S, Q) *is a localization in the sense of Definition* 6.1.2.

Proof Say B is a category, and $F : A \to B$ is a functor such that $F(s)$ is an isomorphism for every $s \in S$.

The uniqueness of a functor $F_S : A_S \to B$ satisfying $F_S \circ Q = F$ is clear from property (RO3). We have to prove existence.

Define F_S to be F on objects. For a morphism q in A_S, property (RO3) says that there is a right fraction presentation $q = Q(a_1) \circ Q(s_1)^{-1}$, with $a_1 \in A$

and $s_1 \in \mathsf{S}$. Let $F_{\mathsf{S}}(q) := F(a_1) \circ F(s_1)^{-1}$ in B. We have to prove that this is well-defined, i.e. it does not depend on the presentation. So suppose that $q = \mathrm{Q}(a_2) \circ \mathrm{Q}(s_2)^{-1}$ is another right fraction presentation of q. Let $b_1, b_2 \in \mathsf{A}$ be as in Lemma 6.2.6. Since $F(s_i)$ and $F(s_i \circ b_i)$ are invertible, then so is $F(b_i)$. We get $F(a_2) = F(a_1) \circ F(b_1) \circ F(b_2)^{-1}$ and $F(s_2) = F(s_1) \circ F(b_1) \circ F(b_2)^{-1}$. Hence $F(a_2) \circ F(s_2)^{-1} = F(a_1) \circ F(s_1)^{-1}$, as required.

It remains to prove that F_{S} is a functor. Since the identity id_M of the object M in A_{S} can be presented as $\mathrm{id}_M = \mathrm{Q}(\mathrm{id}_M) \circ \mathrm{Q}(\mathrm{id}_M)^{-1}$, we see that $F_{\mathsf{S}}(\mathrm{id}_M) = \mathrm{id}_{F(M)}$.

Next let q_1 and q_2 be morphisms in A_{S}, such that the composition $q_2 \circ q_1$ exists (i.e. the target of q_1 is the source of q_2). We have to show that $F_{\mathsf{S}}(q_2 \circ q_1)$ equals $F_{\mathsf{S}}(q_2) \circ F_{\mathsf{S}}(q_1)$. Choose presentations $q_i = \mathrm{Q}(a_i) \circ \mathrm{Q}(s_i)^{-1}$, so that

$$F_{\mathsf{S}}(q_2) \circ F_{\mathsf{S}}(q_1) = F(a_2) \circ F(s_2)^{-1} \circ F(a_1) \circ F(s_1)^{-1}. \tag{6.2.10}$$

By property (RO3) there is a right fraction presentation

$$\mathrm{Q}(s_2)^{-1} \circ \mathrm{Q}(a_1) = \mathrm{Q}(b) \circ \mathrm{Q}(t)^{-1} \tag{6.2.11}$$

for some $b \in \mathsf{A}$ and $t \in \mathsf{S}$. Because $\mathrm{Q}(a_1 \circ t) = \mathrm{Q}(s_2 \circ b)$, by (RO4), there is some $r \in \mathsf{S}$ such that $a_1 \circ t \circ r = s_2 \circ b \circ r$. Therefore $F(a_1 \circ t \circ r) = F(s_2 \circ b \circ r)$, which implies, by canceling the invertible morphism $F(r)$ and rearranging, that

$$F(s_2)^{-1} \circ F(a_1) = F(b) \circ F(t)^{-1} \tag{6.2.12}$$

in B.

Let us continue. Using equation (6.2.11) we have

$$\begin{aligned} q_2 \circ q_1 &= \mathrm{Q}(a_2) \circ \mathrm{Q}(s_2)^{-1} \circ \mathrm{Q}(a_1) \circ \mathrm{Q}(s_1)^{-1} \\ &= \mathrm{Q}(a_2) \circ \mathrm{Q}(b) \circ \mathrm{Q}(t)^{-1} \circ \mathrm{Q}(s_1)^{-1} = \mathrm{Q}(a_2 \circ b) \circ \mathrm{Q}(s_1 \circ t)^{-1}. \end{aligned}$$

Using this presentation of $q_2 \circ q_1$, and the equality (6.2.12), we obtain

$$\begin{aligned} F_{\mathsf{S}}(q_2 \circ q_1) &= F(a_2 \circ b) \circ F(s_1 \circ t)^{-1} = F(a_2) \circ F(b) \circ F(t)^{-1} \circ F(s_1)^{-1} \\ &= F(a_2) \circ F(s_2)^{-1} \circ F(a_1) \circ F(s_1)^{-1}. \end{aligned}$$

This is the same as (6.2.10). □

Corollary 6.2.13 *Let* S *be a multiplicatively closed set of morphisms in a category* A. *Assume that these two assertions hold*:

- *The pair* (B, Q) *is either a right or a left Ore localization of* A *with respect to* S.
- *The pair* $(\mathsf{B}', \mathrm{Q}')$ *is either a right or a left Ore localization of* A *with respect to* S.

Then there is a unique isomorphism of localizations $(\mathsf{B}, \mathrm{Q}) \cong (\mathsf{B}', \mathrm{Q}')$.

Proof By Proposition 6.2.9 (in its right or left versions, as the case may be), both (B, Q) and $(\mathsf{B}', \mathrm{Q}')$ are localizations in the sense of Definition 6.1.2. Hence, by Theorem 6.1.4, there is a unique isomorphism $(\mathsf{B}, \mathrm{Q}) \cong (\mathsf{B}, \mathrm{Q}')$. □

Definition 6.2.14 Let S be multiplicatively closed set of morphisms in a category A. We say that S is a *right denominator set* if it satisfies these two conditions:

(RD1) (Right Ore condition) Given $a \in \mathsf{A}$ and $s \in \mathsf{S}$, there exist $b \in \mathsf{A}$ and $t \in \mathsf{S}$ such that $a \circ t = s \circ b$.

(RD2) (Right cancellation condition) Given $a, b \in \mathsf{A}$ and $s \in \mathsf{S}$ such that $s \circ a = s \circ b$, there exists $t \in \mathsf{S}$ such that $a \circ t = b \circ t$.

In the definition above the implicit assumption is that the sources and targets of the morphisms match; e.g. in (RD1) the morphisms a and s have the same target. Here are the diagrams illustrating the definition:

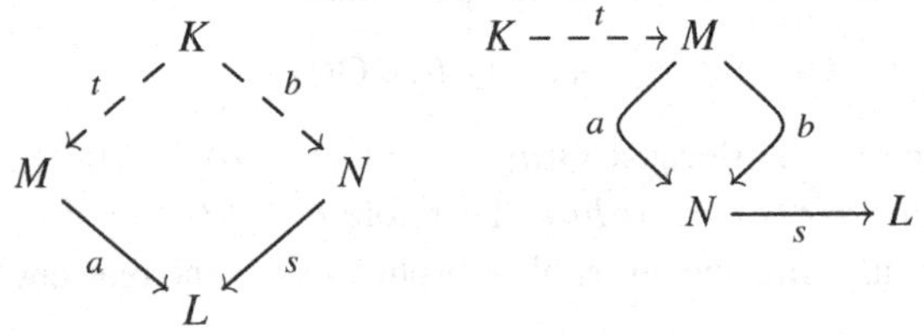

(6.2.15)

Now the left version of Definition 6.2.14:

Definition 6.2.16 Let S be multiplicatively closed set of morphisms in a category A. We say that S is a *left denominator set* if it satisfies these two conditions:

(LD1) (Left Ore condition) Given $a \in \mathsf{A}$ and $s \in \mathsf{S}$, there exist $b \in \mathsf{A}$ and $t \in \mathsf{S}$ such that $t \circ a = b \circ s$.

(LD2) (Left cancellation condition) Given $a, b \in \mathsf{A}$ and $s \in \mathsf{S}$ such that $a \circ s = b \circ s$, there exists $t \in \mathsf{S}$ such that $t \circ a = t \circ b$.

Proposition 6.2.17 *Let* S *be multiplicatively closed set of morphisms in a category* A. *The two conditions below are equivalent:*

(i) S *is a left denominator set in* A.

(ii) S^{op} *is a right denominator set in* A^{op}.

Exercise 6.2.18 Prove Proposition 6.2.17.

Here is the main theorem regarding Ore localization.

Theorem 6.2.19 *The following conditions are equivalent for a category* A *and a multiplicatively closed set of morphisms* $\mathsf{S} \subseteq \mathsf{A}$.

(i) *The right Ore localization* $(\mathsf{A}_{\mathsf{S}}, \mathrm{Q})$ *exists.*
(ii) S *is a right denominator set.*

The proof of Theorem 6.2.19 is after some preparation. The hard part is proving that (ii) $\Rightarrow$ (i). The general idea is the same as in commutative localization: we consider the set of pairs of morphisms $\mathsf{A} \times \mathsf{S}$, and define a relation $\sim$ on it, with the hope that this is an equivalence relation, and that the quotient set A_{S} will be a category, and it will have the desired properties.

Let's assume that S is a right denominator set. For any $M, N \in \mathrm{Ob}(A)$ consider the set

$$(\mathsf{A} \times \mathsf{S})(M, N) := \coprod_{L \in \mathrm{Ob}(A)} \mathsf{A}(L, N) \times \mathsf{S}(L, M).$$

Remark 6.2.20 The set $(\mathsf{A} \times \mathsf{S})(M, N)$ could be big, namely not an element of the initial universe U. This would require the introduction of a larger universe, say V, in which U is an element. And the resulting category A_{S} will be a V-category. This is the same issue we had in the proof of Theorem 6.1.4.

We will ignore this issue. Moreover, in many cases of interest (derived categories where there are DG enhancements, such as the K-injective enhancement), there will be an alternative presentation of A_{S} as a U-category. We will refer to this when we get to it.

An element $(a, s) \in (\mathsf{A} \times \mathsf{S})(M, N)$ can be pictured as a diagram

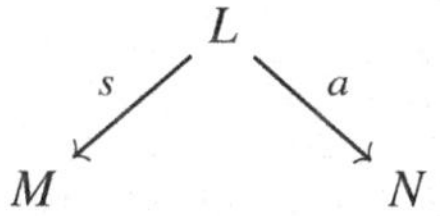

(6.2.21)

in A. This diagram will eventually represent the right fraction $\mathrm{Q}(a) \circ \mathrm{Q}(s)^{-1} : M \to N$.

Definition 6.2.22 We define a relation $\sim$ on the set $\mathsf{A} \times \mathsf{S}$ like this: $(a_1, s_1) \sim (a_2, s_2)$ if there exist $b_1, b_2 \in \mathsf{A}$ s.t. $a_1 \circ b_1 = a_2 \circ b_2$ and $s_1 \circ b_1 = s_2 \circ b_2 \in \mathsf{S}$.

Note that the relation $\sim$ imposes condition (ii) of Lemma 6.2.6. Here it is in a commutative diagram, in which we have made the objects explicit:

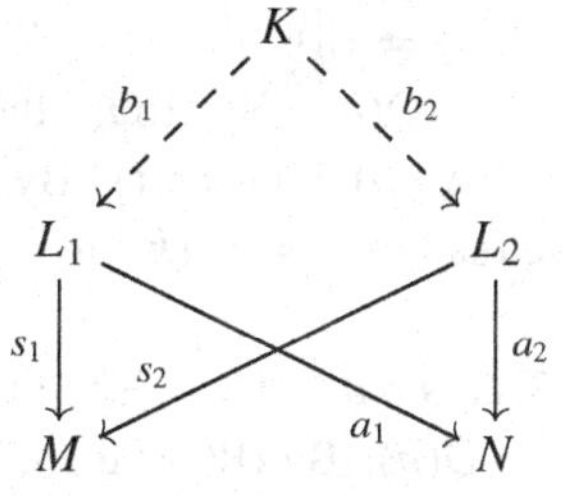

(6.2.23)

The paths (i.e. morphisms) ending at M are in S.

Lemma 6.2.24 *If* S *is a right denominator set, then the relation* $\sim$ *is an equivalence.*

Proof Reflexivity: take $K := L$ and $b_i := \mathrm{id}_L : L \to L$. Symmetry is trivial.

Now to prove transitivity. Suppose we are given $(a_1, s_1) \sim (a_2, s_2)$ and $(a_2, s_2) \sim (a_3, s_3)$. So we have the solid part of the first diagram in (6.2.25), and it is commutative. The morphisms ending at M are in S.

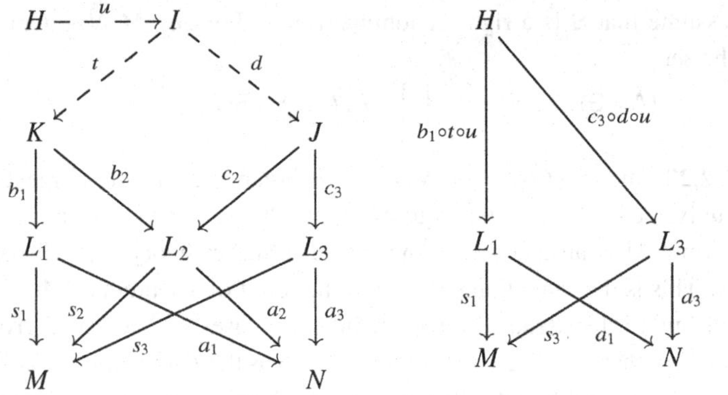

(6.2.25)

By condition (RD1) applied to the morphisms $K \to M \leftarrow J$ there are $t \in \mathrm{S}$ and $d \in \mathrm{A}$ s.t. $(s_3 \circ c_3) \circ d = (s_1 \circ b_1) \circ t$. Comparing the morphisms $I \to M$ in this diagram, we see that

$$s_2 \circ (b_2 \circ t) = s_1 \circ b_1 \circ t = s_3 \circ c_3 \circ d = s_2 \circ (c_2 \circ d).$$

By (RD2) there is $u \in \mathrm{S}$ s.t. $(b_2 \circ t) \circ u = (c_2 \circ d) \circ u$. So all morphisms $H \to M$ are equal and belong to S, and all morphisms $H \to N$ are equal. Now delete the object L_2 and the arrows going through it. Then delete the objects I, J, K, but keep the paths going through them. We get the second diagram in (6.2.25). It is commutative, and all morphisms ending at M are in S. This is evidence for $(a_1, s_1) \sim (a_3, s_3)$. □

Proof of Theorem 6.2.19

Step 1. In this step we prove (i) ⇒ (ii). Take $a \in \mathrm{A}$ and $s \in \mathrm{S}$ with the same target. Consider $q := \mathrm{Q}(s)^{-1} \circ \mathrm{Q}(a)$. By (RO3) there are $b \in \mathrm{A}$ and $t \in \mathrm{S}$ s.t. $q = \mathrm{Q}(b) \circ \mathrm{Q}(t)^{-1}$. So $\mathrm{Q}(s \circ b) = \mathrm{Q}(a \circ t)$. By (RO4) there is $u \in \mathrm{S}$ s.t. $(s \circ b) \circ u = (a \circ t) \circ u$. We read this as $s \circ (b \circ u) = a \circ (t \circ u)$, and note that $t \circ u \in \mathrm{S}$. So (RD1) holds.

Next $a, b \in \mathrm{A}$ and $s \in \mathrm{S}$ s.t. $s \circ a = s \circ b$. Then $\mathrm{Q}(s \circ a) = \mathrm{Q}(s \circ b)$. But $\mathrm{Q}(s)$ is invertible, so $\mathrm{Q}(a) = \mathrm{Q}(b)$. By (RO4) there is $t \in \mathrm{S}$ s.t. $a \circ t = b \circ t$. We have proved (RD2).

Step 2. Now we assume that condition (ii) holds, and we define the morphism sets $\mathsf{A}_{\mathsf{S}}(M, N)$, composition between them, and the identity morphisms.

For any $M, N \in \mathrm{Ob}(\mathsf{A})$ let

$$\mathsf{A}_{\mathsf{S}}(M, N) := (\mathsf{A} \times \mathsf{S})(M, N) / \sim,$$

the quotient set modulo the relation $\sim$ from Definition 6.2.22, which is an equivalence relation by Lemma 6.2.24.

We define composition like this. Given $q_1 \in \mathsf{A}_{\mathsf{S}}(M_0, M_1)$ and $q_2 \in \mathsf{A}_{\mathsf{S}}(M_1, M_2)$, choose representatives $(a_i, s_i) \in (\mathsf{A} \times \mathsf{S})(M_{i-1}, M_i)$. We use the notation $q_i = \overline{(a_i, s_i)}$ to indicate this. By (RD1) there are $c \in \mathsf{A}$ and $u \in \mathsf{S}$ s.t. $s_2 \circ c = a_1 \circ u$. The composition $q_2 \circ q_1 \in \mathsf{A}_{\mathsf{S}}(M_0, M_2)$ is defined to be

$$q_2 \circ q_1 := \overline{(a_2 \circ c, s_1 \circ u)} \in (\mathsf{A} \times \mathsf{S})(M_0, M_2)).$$

The idea behind the formula can be seen in diagram (6.2.26).

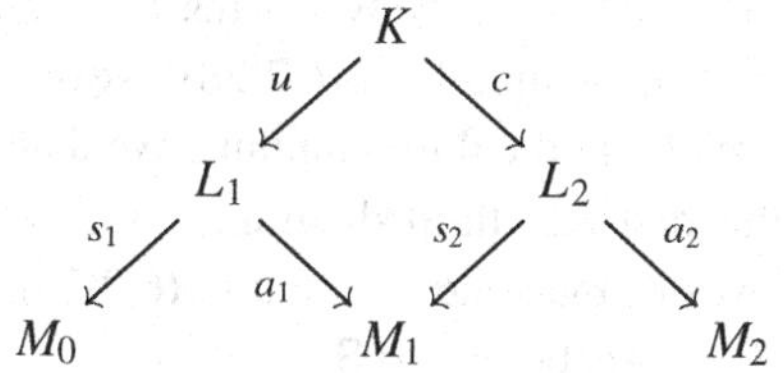

(6.2.26)

We have to verify that this definition is independent of the representatives. So suppose we take other representatives $q_i = \overline{(a'_i, s'_i)}$, and we choose morphisms u', c' to construct the composition. This is the solid part of the diagram (6.2.27), and it is a commutative diagram. We must prove that

$$\overline{(a_2 \circ c, s_1 \circ u)} = \overline{(a'_2 \circ c', s'_1 \circ u')}.$$

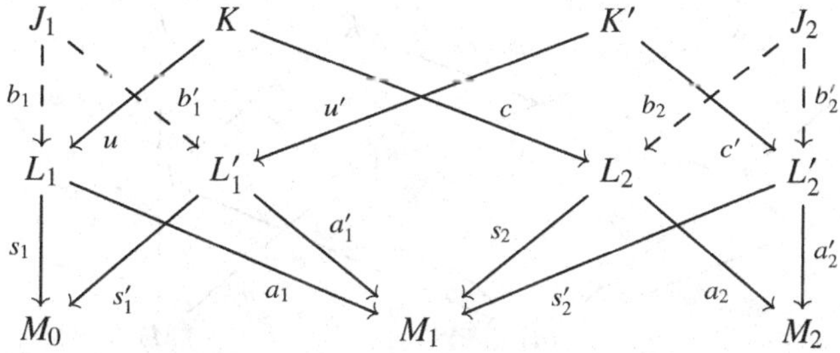

(6.2.27)

There are morphisms b_i, b'_i that are evidence for $(a_i, s_i) \sim (a'_i, s'_i)$. They are depicted as the dashed arrows in (6.2.27). That whole diagram is commutative. The morphisms $J_1 \to M_0$, $K \to M_0$, $K' \to M_0$ and $J_2 \to M_1$ are all in S.

Choose $v_1 \in \mathsf{S}$ and $d_1 \in \mathsf{A}$ s.t. the first diagram in (6.2.28) is commutative.

This can be done by (RD1).

(6.2.28)

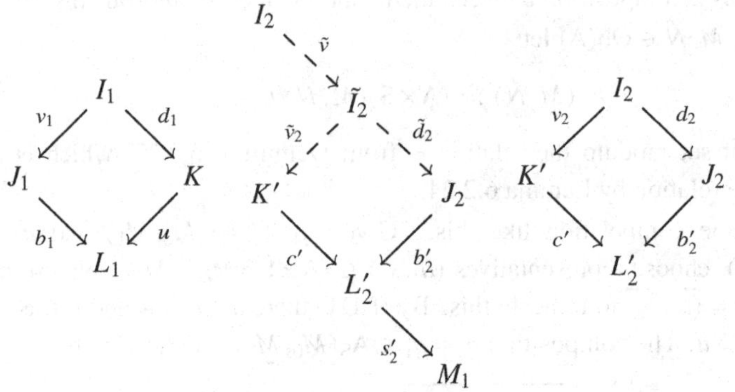

Consider the solid part of the middle diagram in (6.2.28). Since $J_2 \to M_1$ is in S, by (RD1), there are $\tilde{v}_2 \in$ S and $\tilde{d}_2 \in$ A s.t. the two paths $\tilde{I}_2 \to M_1$ are equal. By (RD2), there is $\tilde{v} \in$ S s.t. the two paths $I_2 \to L'_2$ are equal. We get the commutative diagram in the middle of (6.2.28). Next, defining $d_2 := \tilde{d}_2 \circ \tilde{v}$ and $v_2 := \tilde{v}_2 \circ \tilde{v} \in$ S, we obtain the third commutative diagram in (6.2.28).

We now embed the first and third diagrams from (6.2.28) into the diagram (6.2.27). This gives us the solid diagram in (6.2.29), and it is commutative. The morphisms $I_1 \to M_0$ belong to S.

(6.2.29)

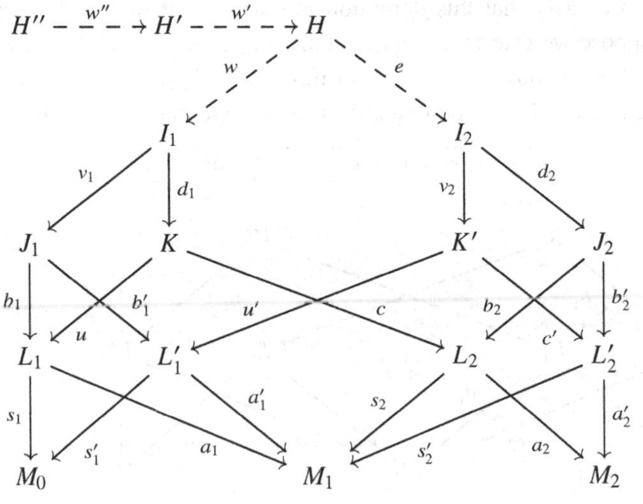

Choose $w \in$ S and $e \in$ A, starting at an object H, to fill the diagram $I_1 \to M_0 \leftarrow I_2$, using (RD1). The path $H \to I_1 \to M_0$ is in S, and all the paths $H \to M_0$ are equal. But we could have failure of commutativity in the paths $H \to L'_1$ and $H \to L_2$.

The two paths $H \to L'_1$ in (6.2.29) satisfy

$$s'_1 \circ (b'_1 \circ v_1 \circ w) = s'_1 \circ (u' \circ v_2 \circ e).$$

Therefore, by (RD2), there is $w' \in \mathsf{S}$ s.t.

$$(b'_1 \circ v_1 \circ w) \circ w' = (u' \circ v_2 \circ e) \circ w'.$$

Next, the two paths $H' \to L_2$ satisfy

$$s_2 \circ (c \circ d_1 \circ w \circ w') = s_2 \circ (b_2 \circ d_2 \circ e \circ w');$$

this is because we can take a detour through L'_1. Therefore, again by (RD2), there is $w'' \in \mathsf{S}$ s.t.

$$(c \circ d_1 \circ w \circ w') \circ w'' = (b_2 \circ d_2 \circ e \circ w') \circ w''.$$

Now all paths $H'' \to M_2$ in (6.2.29) are equal. All paths $H'' \to M_0$ are equal and are in S.

Erase the objects M_1, J_1, J_2 and all arrows touching them from (6.2.29). Then erase H, H', but keep the paths through them. We obtain the commutative diagram (6.2.30).

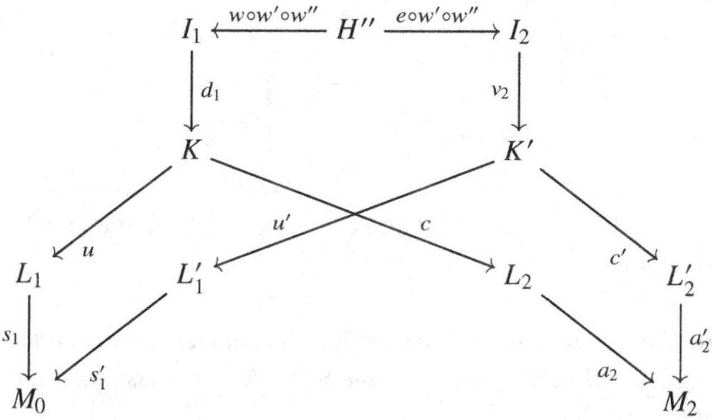

(6.2.30)

This is evidence for

$$(a_2 \circ c, s_1 \circ u) \sim (a'_2 \circ c', s'_1 \circ u').$$

The proof that composition is well-defined is done.

The identity morphism id_M in $\mathsf{A_S}$ of an object M is $\overline{(\mathrm{id}_M, \mathrm{id}_M)}$.

Step 3. We have to verify the associativity and the identity properties of composition in $\mathsf{A_S}$. Namely that $\mathsf{A_S}$ is a category. This seems to be not too hard, given Step 2, and we leave it as an exercise.

Step 4. The functor $\mathrm{Q} : \mathsf{A} \to \mathsf{A_S}$ is defined to be $\mathrm{Q}(M) := M$ on objects, and $\mathrm{Q}(a) := \overline{(a, \mathrm{id}_M)}$ for $a : M \to N$ in A. We have to verify this is a functor. Again, an exercise.

Step 5. Finally we verify properties (RO1)-(RO4). (RO1) is clear. The inverse of $\mathrm{Q}(s)$ is $\overline{(\mathrm{id}, s)}$, so (RO2) holds.

It is not hard to see that $\overline{(a, s)} = \overline{(a, \mathrm{id})} \circ \overline{(\mathrm{id}, s)}$; this is (RO3).

If $\mathrm{Q}(a_1) = \mathrm{Q}(a_2)$, then $(a_1, \mathrm{id}_M) \sim (a_2, \mathrm{id}_M)$; so there are $b_1, b_2 \in \mathsf{A}$ s.t. $a_1 \circ b_1 = a_2 \circ b_2$ and $\mathrm{id} \circ b_1 = \mathrm{id} \circ b_2 \in \mathsf{S}$. Writing $s := b_1 \in \mathsf{S}$, we get $a_1 \circ s = a_2 \circ s$. This proves (RO4). □

Exercise 6.2.31 Finish the details in the proof of Theorem 6.2.19.

Proposition 6.2.32 (Common Denominator) *Let* A *be a category, let* S *be a right denominator set in* A, *and let* $(\mathsf{A}_\mathsf{S}, \mathrm{Q})$ *be the right Ore localization. For every two morphisms* $q_1, q_2 : M \to N$ *in* A_S *there is a common right denominator. Namely we can write* $q_i = \mathrm{Q}(a_i) \circ \mathrm{Q}(s)^{-1}$ *for suitable* $a_i \in \mathsf{A}$ *and* $s \in \mathsf{S}$.

Proof Choose representatives $q_i = \mathrm{Q}(a'_i) \circ \mathrm{Q}(s'_i)^{-1}$. By (RD1) applied to $L_1 \to M \leftarrow L_2$, there are $b \in \mathsf{A}$ and $t \in \mathsf{S}$ such that the part of diagram (6.2.33) that lies above M is commutative:

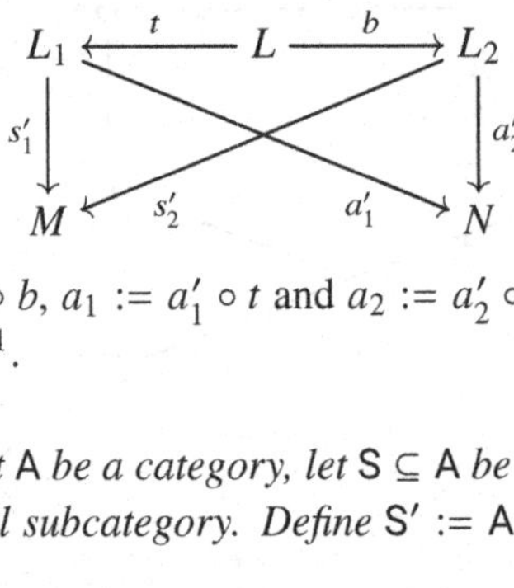

(6.2.33)

Write $s := s'_1 \circ t = s'_2 \circ b$, $a_1 := a'_1 \circ t$ and $a_2 := a'_2 \circ b$. By Lemma 6.2.6 we get $q_i = \mathrm{Q}(a_i) \circ \mathrm{Q}(s)^{-1}$. □

Proposition 6.2.34 *Let* A *be a category, let* $\mathsf{S} \subseteq \mathsf{A}$ *be a right denominator set, and let* $\mathsf{A}' \subseteq \mathsf{A}$ *be a full subcategory. Define* $\mathsf{S}' := \mathsf{A}' \cap \mathsf{S}$. *Assume these two conditions hold:*

(i) S' *is a right denominator set in* A'.

(ii) *Let* $M \in \mathrm{Ob}(\mathsf{A})$. *If there exists a morphism* $s : M \to L'$ *in* S *with* $L' \in \mathrm{Ob}(\mathsf{A}')$, *there exists a morphism* $t : K' \to M$ *in* S *with* $K' \in \mathrm{Ob}(\mathsf{A}')$.

Then the canonical functor $\mathsf{A}'_{\mathsf{S}'} \to \mathsf{A}_\mathsf{S}$ *is fully faithful.*

Proof Let's denote the inclusion functor by $F : \mathsf{A}' \to \mathsf{A}$. We want to prove that its localization $F_{\mathsf{S}'} : \mathsf{A}'_{\mathsf{S}'} \to \mathsf{A}_\mathsf{S}$ is fully faithful.

Step 1. Let $L'_1, L'_2 \in \mathrm{Ob}(\mathsf{A}')$, and let $q : L'_1 \to L'_2$ be a morphism in A_S. Choose a presentation $q = \mathrm{Q}(a) \circ \mathrm{Q}(s)^{-1}$ with $s : M \to L'_1$ a morphism in S and $a : M \to L'_2$ a morphism in A. This is possible because S is a right denominator set in A. By condition (ii) we can find a morphism $t : K' \to M$ in

S with $K' \in \mathrm{Ob}(\mathsf{A}')$. See next diagram.

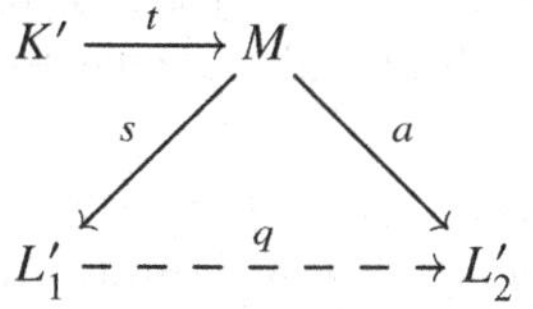

Then $q = \mathrm{Q}(a \circ t) \circ \mathrm{Q}(s \circ t)^{-1}$. But $s \circ t \in \mathsf{S}'$ and $a \circ t \in \mathsf{A}'$, so q is in the image of the functor $F_{\mathsf{S}'}$. We see that $F_{\mathsf{S}'}$ is full.

Step 2. Let $p', q' : L'_1 \to L'_2$ be morphisms in $\mathsf{A}'_{\mathsf{S}'}$ such that $F_{\mathsf{S}'}(p') = F_{\mathsf{S}'}(q')$. Let us denote the localization functor of A' by $\mathrm{Q}' : \mathsf{A}' \to \mathsf{A}'_{\mathsf{S}'}$. Because S' is a right denominator set in A', and using Proposition 6.2.32, we can find presentations $p' = \mathrm{Q}'(a') \circ \mathrm{Q}'(s')^{-1}$ and $q' = \mathrm{Q}'(b') \circ \mathrm{Q}'(s')^{-1}$ with morphisms $s' : N' \to L'_1$ in S' and $a', b' : N' \to L'_2$ in A'. Next, since $F_{\mathsf{S}'}(p') = F_{\mathsf{S}'}(q')$, Lemma 6.2.6 tells us that there are morphisms $u, v : M \to N'$ in A s.t. $a' \circ u = b' \circ v$, and $s' \circ u = s' \circ v \in \mathsf{S}$. Condition (ii), applied to the morphism $s' \circ u : M \to L'_1$, says that there is a morphism $t : K' \to M$ in S with source $K' \in \mathrm{Ob}(\mathsf{A}')$. See diagram below.

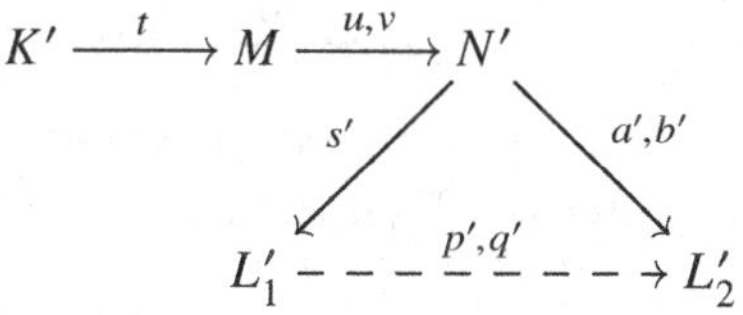

Then we have

$$\begin{aligned} p' &= \mathrm{Q}'(a' \circ u \circ t) \circ \mathrm{Q}'(s' \circ u \circ t)^{-1} \\ &= \mathrm{Q}'(b' \circ u \circ t) \circ \mathrm{Q}'(s' \circ u \circ t)^{-1} = q'. \end{aligned}$$

This proves that $F_{\mathsf{S}'}$ is faithful. □

In the next section we study Ore localization of linear categories, and give rings as examples (see Examples 6.3.6 and 6.3.5). And in Chapter 7 we study localization of triangulated categories. Here is a nonlinear example.

Example 6.2.35 Fix a base commutative ring $\mathbb{K}$ and a topological space (or, more generally, a site) X. Let $\mathbb{K}_X$ be the constant sheaf of rings $\mathbb{K}$ on X. Consider the category $\mathsf{DGRng}^{\leq 0}_{\mathrm{sc}} / \mathbb{K}_X$ of sheaves of commutative DG $\mathbb{K}_X$-rings on X (see Definition 3.3.4 and Remark 12.8.22). For the sake of brevity let us just write $\mathsf{DGRng}_X := \mathsf{DGRng}^{\leq 0}_{\mathrm{sc}} / \mathbb{K}_X$.

The localization of DGRng_X with respect to the quasi-isomorphisms in it is called the *derived category of commutative DG $\mathbb{K}_X$-rings*, and the notation is $\mathsf{D}(\mathsf{DGRng}_X)$. The categorical localization functor is $\mathrm{Q} : \mathsf{DGRng}_X \to \mathsf{D}(\mathsf{DGRng}_X)$, and it is the identity on objects. (The unfortunate tradition in homotopy theory is to call such a localization the "homotopy category," and to denote it by "$\mathrm{Ho}(\mathsf{DGRng}_X)$.")

As explained in the lecture notes [179], every DG $\mathbb{K}_X$-ring $\mathcal{A}$ admits a *semi-pseudo-free resolution* $\tilde{\mathcal{A}} \to \mathcal{A}$ in DGRng_X. The semi-pseudo-free DG $\mathbb{K}_X$-rings have good lifting properties. Indeed, they behave like cofibrant objects in a Quillen model structure (even though there does not seems to be such a structure on DGRng_X, except when X is a discrete space).

There is a congruence on the category DGRng_X, called the *quasi-homotopy relation*. (It turns out that this idea was already known to some homotopy theorists, see Remark 6.2.37 below.) Let us denote by $\mathsf{K}(\mathsf{DGRng}_X)$ the *homotopy category*, which is the quotient of DGRng_X modulo this congruence. So there is a functor $\mathrm{P} : \mathsf{DGRng}_X \to \mathsf{K}(\mathsf{DGRng}_X)$, which is the identity on objects, and it sends a DG ring homomorphism to its quasi-homotopy class.

The functors Q and P fit into this commutative diagram:

$$\mathsf{DGRng}_X \xrightarrow{\ \mathrm{P}\ } \mathsf{K}(\mathsf{DGRng}_X) \xrightarrow{\ \bar{\mathrm{Q}}\ } \mathsf{D}(\mathsf{DGRng}_X), \qquad \mathrm{Q} = \bar{\mathrm{Q}} \circ \mathrm{P} \tag{6.2.36}$$

The functor $\bar{\mathrm{Q}}$ is a *faithful right Ore localization*. The proof relies on the lifting properties of semi-pseudo-free DG $\mathbb{K}_X$-rings. See Remark 12.8.22 for related material.

Remark 6.2.37 Suppose C is a category, with a given multiplicatively closed set of morphisms $\mathsf{S} \subseteq \mathsf{C}$. If C admits a *Quillen model structure* for which S is the set of *weak equivalences*, then there is automatically a *calculus of fractions* for the localization functor $\mathrm{Q} : \mathsf{C} \to \mathsf{C}_{\mathsf{S}}$.

For instance, suppose that all objects of C are *fibrant*. In this case C is called a *category of fibrant objects*. Then there is a congruence relation H on C, and a commutative diagram

$$\mathsf{C} \xrightarrow{\ \mathrm{P}\ } \mathsf{C}/\mathsf{H} \xrightarrow{\ \bar{\mathrm{Q}}\ } \mathsf{C}_{\mathsf{S}}, \qquad \mathrm{Q} = \bar{\mathrm{Q}} \circ \mathrm{P}$$

in which, like in (6.2.36), the functor P is full, and the functor $\bar{\mathrm{Q}}$ is a right Ore localization. These facts are explained in [115] (search for the emphasized text in this remark).

Remark 6.2.38 This is a continuation of Remark 3.8.22 on strictly unital A_∞ categories, and it is related to Example 6.2.35 and Remark 6.2.37 above, and to

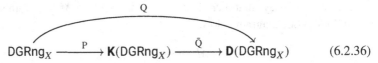

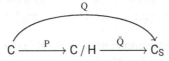

Remark 12.8.21 below. The references are [83] and [37]. Recall that the base ring $\mathbb{K}$ is a field.

Let A and B be A_∞ categories, i.e. objects of the category $\mathrm{A}_\infty\mathsf{Cat}$. An A_∞ functor $F : \mathsf{A} \to \mathsf{B}$ is called a *quasi-equivalence* if $\mathrm{H}(F) : \mathrm{H}(\mathsf{A}) \to \mathrm{H}(\mathsf{B})$ is an equivalence of graded categories. The *derived category* of $\mathrm{A}_\infty\mathsf{Cat}$ is the localization $\mathbf{D}(\mathrm{A}_\infty\mathsf{Cat})$ of $\mathrm{A}_\infty\mathsf{Cat}$ with respect to the quasi-equivalences. (As mentioned above, the tradition is to call $\mathbf{D}(\mathrm{A}_\infty\mathsf{Cat})$ the "homotopy category" of $\mathrm{A}_\infty\mathsf{Cat}$, and to denote it by "$\mathrm{Ho}(\mathrm{A}_\infty\mathsf{Cat})$.") There is a localization functor $\mathrm{Q} : \mathrm{A}_\infty\mathsf{Cat} \to \mathbf{D}(\mathrm{A}_\infty\mathsf{Cat})$, which is the identity on objects.

Suppose $F, F' : \mathsf{A} \to \mathsf{B}$ are A_∞ functors. So F and F' are objects of the A_∞ category $\mathrm{A}_\infty\mathsf{Fun}(\mathsf{A}, \mathsf{B})$. We say that F and F' are A_∞ *homotopic* if they are isomorphic in the category $\mathrm{H}^0(\mathrm{A}_\infty\mathsf{Fun}(\mathsf{A}, \mathsf{B}))$, in which case we write $F \approx F'$. Define $\mathbf{K}(\mathrm{A}_\infty\mathsf{Cat})$ to be the quotient of $\mathrm{A}_\infty\mathsf{Cat}$ modulo the relation $\approx$. There is a full functor $\mathrm{P} : \mathrm{A}_\infty\mathsf{Cat} \to \mathbf{K}(\mathrm{A}_\infty\mathsf{Cat})$, which is the identity on objects. An A_∞ functor $F : \mathsf{A} \to \mathsf{B}$ is called an A_∞ *equivalence* if it is an isomorphism in $\mathbf{K}(\mathrm{A}_\infty\mathsf{Cat})$. This means that there is an A_∞ functor $G : \mathsf{B} \to \mathsf{A}$, such that $G \circ F \approx \mathrm{Id}_\mathsf{A}$ and $F \circ G \approx \mathrm{Id}_\mathsf{B}$. There is a commutative diagram

$$\mathrm{A}_\infty\mathsf{Cat} \xrightarrow{\mathrm{P}} \mathbf{K}(\mathrm{A}_\infty\mathsf{Cat}) \xrightarrow{\bar{\mathrm{Q}}} \mathbf{D}(\mathrm{A}_\infty\mathsf{Cat}), \qquad \mathrm{Q} = \bar{\mathrm{Q}} \circ \mathrm{P} \tag{6.2.39}$$

An important theorem (see [94, Theorem 9.2.0.4]) says that an A_∞ functor F is a quasi-equivalence iff it is an A_∞ equivalence. This implies that the functor $\bar{\mathrm{Q}}$ in diagram (6.2.39) is an isomorphism of categories.

It is worthwhile to mention that the category $\mathrm{A}_\infty\mathsf{Cat}$ does not have a Quillen model structure (for which the weak equivalences are the quasi-equivalences).

Let DGCat be the category of $\mathbb{K}$-linear DG categories, where the morphisms are the DG functors (see Definition 3.5.1). There is a fully faithful embedding $\mathsf{DGRng} /_\mathrm{c} \mathbb{K} \to \mathsf{DGCat}$, since the DG rings (see Definition 3.3.2) are the single-object DG categories. Because an A_∞ category whose higher operations vanish is a DG category, there is a faithful (but not full) embedding $\mathsf{DGCat} \to \mathrm{A}_\infty\mathsf{Cat}$. The *derived category* of DGCat is the localization $\mathbf{D}(\mathsf{DGCat})$ of DGCat with respect to the quasi-equivalences in it. (Once more, the traditional notation for $\mathbf{D}(\mathsf{DGCat})$ is "$\mathrm{Ho}(\mathsf{DGCat})$.") The embedding of DGCat into $\mathrm{A}_\infty\mathsf{Cat}$ induces a functor $\mathbf{D}(\mathsf{DGCat}) \to \mathbf{D}(\mathrm{A}_\infty\mathsf{Cat})$, and this is an equivalence. Together with the equivalence $\bar{\mathrm{Q}}$ from (6.2.39), this implies that every morphism $F : \mathsf{A} \to \mathsf{B}$ in $\mathbf{D}(\mathsf{DGCat})$ can be represented by an A_∞ functor $\tilde{F} : \mathsf{A} \to \mathsf{B}$, which is unique up to a unique isomorphism in $\mathbf{K}(\mathrm{A}_\infty\mathsf{Cat})$.

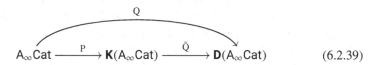

Moreover, the category $\mathbf{D}(\mathsf{DGCat})$ is symmetric monoidal, with tensor operation $\mathsf{A} \otimes \mathsf{B}$. It turns out that the DG category $\mathsf{A}_\infty\mathsf{Fun}(\mathsf{A}, \mathsf{B})$ is the *internal Hom object*, making $\mathbf{D}(\mathsf{DGCat})$ into a *closed monoidal category*.

6.3 Localization of Linear Categories

Until now in this chapter we dealt with arbitrary categories. In this section our categories will be linear over the commutative base ring $\mathbb{K}$ (which will be implicit most of the time). See Convention 1.2.4.

For convenience we only talk about right denominator sets here. All the statements hold equally for left denominator sets; cf. Remark 6.2.3 and Proposition 6.2.4.

Theorem 6.3.1 *Let* A *be a* $\mathbb{K}$*-linear category, let* S *be a right denominator set in* A*, and let* $(\mathsf{A}_\mathsf{S}, \mathrm{Q})$ *be the right Ore localization.*

(1) *The category* A_S *has a unique* $\mathbb{K}$*-linear structure, such that* $\mathrm{Q} : \mathsf{A} \to \mathsf{A}_\mathsf{S}$ *is a* $\mathbb{K}$*-linear functor.*
(2) *Suppose* B *is another* $\mathbb{K}$*-linear category, and* $F : \mathsf{A} \to \mathsf{B}$ *is a* $\mathbb{K}$*-linear functor s.t.* $F(s)$ *is invertible for every* $s \in \mathsf{S}$*. Let* $F_\mathsf{S} : \mathsf{A}_\mathsf{S} \to \mathsf{B}$ *be the localization of* F*. Then* F_S *is a* $\mathbb{K}$*-linear functor.*
(3) *If* A *is an additive category, then so is* A_S*.*

Proof

(1) Let $q_1, q_2 : M \to N$ be morphisms in A_S. Choose common denominator presentations $q_i = \mathrm{Q}(a_i) \circ \mathrm{Q}(s)^{-1}$. Since Q must be an additive functor, we have to define

$$\mathrm{Q}(a_1) + \mathrm{Q}(a_2) := \mathrm{Q}(a_1 + a_2). \tag{6.3.2}$$

By the distributive law (bilinearity of composition) we must define $q_1 + q_2 := \mathrm{Q}(a_1 + a_2) \circ \mathrm{Q}(s)^{-1}$. And for $\lambda \in \mathbb{K}$ we must define $\lambda \cdot q_i := \mathrm{Q}(\lambda \cdot a_i) \circ \mathrm{Q}(s)^{-1}$.

The usual tricks are then used to prove independence of representatives. For instance, to prove that (6.3.2) is independent of choices, suppose that $\mathrm{Q}(a_1) = \mathrm{Q}(a_1')$ and $\mathrm{Q}(a_2) = \mathrm{Q}(a_2')$. Then, by (RO4), there are $t_1, t_2 \in \mathsf{S}$ such that $a_1 \circ t_1 = a_1' \circ t_1$ and $a_2 \circ t_2 = a_2' \circ t_2$. By (RD1) there exist $b \in \mathsf{A}$ and $v \in \mathsf{S}$ s.t. $t_1 \circ b = t_2 \circ v$. Let $t_3 := t_2 \circ v \in \mathsf{S}$. Then $(a_1 + a_2) \circ t_3 = (a_1' + a_2') \circ t_3$, and hence $\mathrm{Q}(a_1 + a_2) = \mathrm{Q}(a_1' + a_2')$.

In this way A_S is a $\mathbb{K}$-linear category, and Q is a $\mathbb{K}$-linear functor.

(2) The only option for F_S is $F_\mathsf{S}(q_i) := F(a_i) \circ F(s)^{-1}$. The usual tricks are used to prove independence of representatives.

(3) Clear from Propositions 2.5.2 and 2.5.7. □

Exercise 6.3.3 Finish the proofs of items (1) and (2) of the theorem.

Remark 6.3.4 Let A be a ring, which we can think of as a one-object linear category A. In this context, Theorem 6.3.1 is one of the most important results in noncommutative ring theory. See the books [106], [130] or [148].

Here are somewhat concrete examples.

Example 6.3.5 Suppose A is a noncommutative ring (i.e. it is not necessarily commutative), and S is a central multiplicatively closed set in A (i.e. $S \subseteq \operatorname{Cent}(A)$). Because the elements of S commute with all elements of A, the denominator conditions hold automatically. We get a left and right Ore localization of rings $\mathrm{Q} : A \to A_S$.

Note that if S contains a nilpotent element, then the ring A_S is trivial.

The last observation should serve as a warning: localization can sometimes kill everything. Fortunately, the localization procedure that gives rise to the derived category does not cause any catastrophe, as we shall see in Proposition 7.1.10.

Example 6.3.6 Suppose A is a noncommutative ring that is both noetherian and an integral domain (i.e. the only zero-divisor in A is 0). Let S be the set of nonzero elements of A. Then S is both a left and a right denominator set. See [106, Theorem 2.1.15].

Remark 6.3.7 Suppose A is a ring and S is a right denominator set in it. Then the right Ore localization A_S is *flat* as left A-module. See [130, Theorem 3.1.20]. We have no idea if something like this is true for linear categories with more than one object.

Proposition 6.3.8 *Let* (K, T) *be a T-additive category, let* S *be a right denominator set in* K *such that* $\mathrm{T}(\mathsf{S}) = \mathsf{S}$, *and let* $\mathrm{Q} : \mathsf{K} \to \mathsf{K}_\mathsf{S}$ *be the localization functor.*

(1) *There is a unique additive automorphism* T_S *of the category* K_S, *such that* $\mathrm{T}_\mathsf{S} \circ \mathrm{Q} = \mathrm{Q} \circ \mathrm{T}$ *as functors* $\mathsf{K} \to \mathsf{K}_\mathsf{S}$.
(2) *Let* τ *be the identity automorphism of the functor* $\mathrm{Q} \circ \mathrm{T}$. *Then* $(\mathrm{Q}, \tau) : (\mathsf{K}, \mathrm{T}) \to (\mathsf{K}_\mathsf{S}, \mathrm{T}_\mathsf{S})$ *is a T-additive functor.*

Proof

(1) By the assumption the functor $\mathrm{Q} \circ \mathrm{T} : \mathsf{K} \to \mathsf{K}_\mathsf{S}$ sends the morphisms in S to isomorphisms. By the property (Loc3) of localization in Definition 6.1.2, the functor $\mathrm{T}_\mathsf{S} : \mathsf{K}_\mathsf{S} \to \mathsf{K}_\mathsf{S}$ satisfying $\mathrm{T}_\mathsf{S} \circ \mathrm{Q} = \mathrm{Q} \circ \mathrm{T}$ exists and is unique. Similarly, there is a unique functor $\mathrm{T}_\mathsf{S}^{-1} : \mathsf{K}_\mathsf{S} \to \mathsf{K}_\mathsf{S}$ satisfying $\mathrm{T}_\mathsf{S}^{-1} \circ \mathrm{Q} = \mathrm{Q} \circ \mathrm{T}^{-1}$. An easy

calculation shows that $\mathrm{T}_{\mathrm{S}}^{-1} \circ \mathrm{T}_{\mathrm{S}} = \mathrm{Id} = \mathrm{T}_{\mathrm{S}} \circ \mathrm{T}_{\mathrm{S}}^{-1}$. Hence T_{S} is an automorphism of K_{S}. By Theorem 6.3.1 the functor T_{S} is additive.

(2) This is clear. □

The composition of T-additive functors was defined in Definition 5.1.4.

Proposition 6.3.9 *In the situation of Proposition* 6.3.8, *suppose* $(\mathsf{K}', \mathrm{T}')$ *is another T-additive category, and* $(F, \nu) : (\mathsf{K}, \mathrm{T}) \to (\mathsf{K}', \mathrm{T}')$ *is a T-additive functor, such that* $F(s)$ *is invertible for every* $s \in \mathrm{S}$. *Let* $F_{\mathrm{S}} : \mathsf{K}_{\mathrm{S}} \to \mathsf{K}'$ *be the localized functor. Then there is a unique isomorphism* $\nu_{\mathrm{S}} : F_{\mathrm{S}} \circ \mathrm{T}_{\mathrm{S}} \xrightarrow{\simeq} \mathrm{T}' \circ F_{\mathrm{S}}$ *of functors* $\mathsf{K}_{\mathrm{S}} \to \mathsf{K}'$, *such that* $(F, \nu) = (F_{\mathrm{S}}, \nu_{\mathrm{S}}) \circ (\mathrm{Q}, \tau)$ *as T-additive functors* $(\mathsf{K}, \mathrm{T}) \to (\mathsf{K}', \mathrm{T}')$.

Exercise 6.3.10 Prove Proposition 6.3.9.

7

The Derived Category $\mathsf{D}(A, \mathsf{M})$

In this chapter we introduce the main mathematical concept of the book: the *derived category* $\mathsf{D}(A, \mathsf{M})$ of DG A-modules in M. Here A is a central DG $\mathbb{K}$-ring and M is a $\mathbb{K}$-linear abelian category. The base ring $\mathbb{K}$ can be any nonzero commutative ring, and it will remain implicit most of the time.

7.1 Localization of Triangulated Categories

Let K be a triangulated category, with translation functor T. As we did in Chapter 6, we shall often write $a \in \mathsf{K}$ for a morphism $a \in \mathsf{K}(M, N) = \mathrm{Hom}_{\mathsf{K}}(M, N)$, leaving the objects implicit.

Proposition 7.1.1 *Suppose $H : \mathsf{K} \to \mathsf{M}$ is a cohomological functor, where M is some abelian category. Let*

$$\mathsf{S} := \{s \in \mathsf{K} \mid H(\mathrm{T}^i(s)) \text{ is invertible for all } i \in \mathbb{Z}\}.$$

Then S is a left and right denominator set in K.

Proof It is clear that S is closed under composition and contains the identity morphisms. So it is a multiplicatively closed set.

Let's prove that condition (RD1) of Definition 6.2.14 holds. Suppose we are given morphisms $L \xrightarrow{a} N \xleftarrow{s} M$ with $s \in \mathsf{S}$. We need to find morphisms $L \xleftarrow{t} K \xrightarrow{b} M$ with $t \in \mathsf{S}$ and such that $a \circ t = s \circ b$.

Consider the solid commutative diagram

$$\begin{array}{ccccccc} K & \xrightarrow{t} & L & \xrightarrow{c \circ a} & P & \longrightarrow & \mathrm{T}(K) \\ {\scriptstyle b}\downarrow & & \downarrow{\scriptstyle a} & & \downarrow{\scriptstyle =} & & \downarrow{\scriptstyle \mathrm{T}(b)} \\ M & \xrightarrow{s} & N & \xrightarrow{c} & P & \longrightarrow & \mathrm{T}(M) \end{array}$$

where the bottom row is a distinguished triangle built on $M \xrightarrow{s} N$, and the top row is a distinguished triangle built on $L \xrightarrow{c \circ a} P$, then turned 120° to the right. By axiom (TR3) there is a morphism b making the diagram commutative. Thus $a \circ t = s \circ b$. Since $H(\mathrm{T}^i(s))$ are invertible for all $i \in \mathbb{Z}$, it follows that $H(\mathrm{T}^i(P)) = 0$. But then $H(\mathrm{T}^i(t))$ are invertible for all $i \in \mathbb{Z}$, so $t \in \mathsf{S}$.

Next we prove condition (RD2) of Definition 6.2.14. Because we are in an additive category, this condition is simplified: given $a \in \mathsf{K}$ and $s \in \mathsf{S}$ satisfying $s \circ a = 0$, we have to find $t \in \mathsf{S}$ satisfying $a \circ t = 0$.

Say the objects involved are $L \xrightarrow{a} M \xrightarrow{s} N$. Take a distinguished triangle built on s and then turned: $P \xrightarrow{b} M \xrightarrow{s} N \to \mathrm{T}(P)$. We get an exact sequence

$$\mathrm{Hom}_{\mathsf{K}}(L, P) \xrightarrow{b \circ (-)} \mathrm{Hom}_{\mathsf{K}}(L, M) \xrightarrow{s \circ (-)} \mathrm{Hom}_{\mathsf{K}}(L, N).$$

Since $s \circ a = 0$, there is $c : L \to P$ s.t. $a = b \circ c$. Now look at a distinguished triangle built on c, and then turned: $K \xrightarrow{t} L \xrightarrow{c} P \to \mathrm{T}(K)$. We know that $c \circ t = 0$; hence $a \circ t = b \circ c \circ t = 0$. But $(s \in \mathsf{S}) \Rightarrow (H(\mathrm{T}^i(P)) = 0$ for all $i)$ $\Rightarrow (t \in \mathsf{S})$.

The left versions (LD1) and (LD2) are proved the same way. □

Definition 7.1.2 A *denominator set of cohomological origin* in K is a denominator set $\mathsf{S} \subseteq \mathsf{K}$ that arises from a cohomological functor H, as in Proposition 7.1.1. The morphisms in S are called *quasi-isomorphisms relative to H*.

Theorem 7.1.3 *Let* (K, T) *be a triangulated category, let* S *be a denominator set of cohomological origin in* K*, and let* $(\mathrm{Q}, \tau) : (\mathsf{K}, \mathrm{T}) \to (\mathsf{K}_{\mathsf{S}}, \mathrm{T}_{\mathsf{S}})$ *be the T-additive functor from Proposition* 6.3.8. *Then the T-additive category* $(\mathsf{K}_{\mathsf{S}}, \mathrm{T}_{\mathsf{S}})$ *has a unique triangulated structure such that these two properties hold:*

(i) *The pair* (Q, τ) *is a triangulated functor.*

(ii) *Suppose* $(\mathsf{K}', \mathrm{T}')$ *is another triangulated category, and* $(F, \nu) : (\mathsf{K}, \mathrm{T}) \to (\mathsf{K}', \mathrm{T}')$ *is a triangulated functor, such that* $F(s)$ *is invertible for every* $s \in \mathsf{S}$. *Let* $(F_{\mathsf{S}}, \nu_{\mathsf{S}}) : (\mathsf{K}_{\mathsf{S}}, \mathrm{T}_{\mathsf{S}}) \to (\mathsf{K}', \mathrm{T}')$ *be the T-additive functor from Proposition* 6.3.9. *Then* $(F_{\mathsf{S}}, \nu_{\mathsf{S}})$ *is a triangulated functor.*

This result is stated as [62, Proposition I.3.2] and as [75, Proposition 1.6.9]. Both sources do not give proofs (just hints).

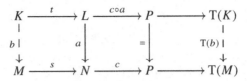

Proof Since S is of cohomological origin, we have $\mathrm{T}(\mathsf{S}) = \mathsf{S}$. Recall that the translation isomorphism τ is the identity automorphism of the functor $\mathrm{Q} \circ \mathrm{T}$; see Proposition 6.3.8. So we will ignore it.

Step 1. The distinguished triangles in K_S are defined to be those triangles that are isomorphic to the images under Q of distinguished triangles in K. Let us verify the axioms of triangulated category.

(TR1). By definition every triangle that's isomorphic to a distinguished triangle is distinguished; and the triangle $M \xrightarrow{\mathrm{id}_M} M \to 0 \to \mathrm{T}(M)$ in K_S is clearly distinguished.

Suppose we are given a morphism $\alpha : L \to M$ in K_S. We have to build a distinguished triangle on it. Choose a fraction presentation $\alpha = \mathrm{Q}(a) \circ \mathrm{Q}(s)^{-1}$. Using condition (LD1) we can find $b \in \mathsf{K}$ and $t \in \mathsf{S}$ such that $t \circ a = b \circ s$. These fit into the solid commutative diagram

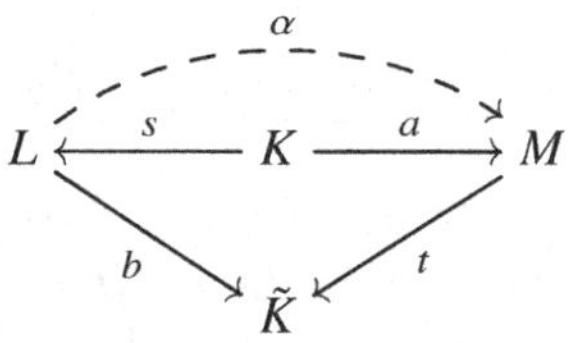

in K. (The dashed arrow α is in K_S.)

Consider the solid commutative diagram below, where the rows are distinguished triangles built on a and b, respectively.

$$\begin{array}{ccccccc} K & \xrightarrow{a} & M & \xrightarrow{e} & N & \xrightarrow{c} & \mathrm{T}(K) \\ \downarrow{\scriptstyle s} & & \downarrow{\scriptstyle t} & & \downarrow{\scriptstyle u} & & \downarrow{\scriptstyle \mathrm{T}(s)} \\ L & \xrightarrow{b} & \tilde{K} & \longrightarrow & P & \xrightarrow{d} & \mathrm{T}(L) \end{array} \tag{7.1.4}$$

By (TR3) there is a morphism u that makes the whole diagram commutative. Since $s, t \in \mathsf{S}$ and H is a cohomological functor, it follows that $u \in \mathsf{S}$. Applying the functor Q to (7.1.4), and using the isomorphism $\mathrm{Q}(t) : M \to \tilde{K}$ to replace $\tilde{K}$ with M, we get the commutative diagram

$$\begin{array}{ccccccc} K & \xrightarrow{\mathrm{Q}(a)} & M & \xrightarrow{\mathrm{Q}(e)} & N & \xrightarrow{\mathrm{Q}(c)} & \mathrm{T}(K) \\ \downarrow{\scriptstyle \mathrm{Q}(s)} & & \downarrow{\scriptstyle \mathrm{Q}(\mathrm{id}_M)} & & \downarrow{\scriptstyle \mathrm{Q}(u)} & & \downarrow{\scriptstyle \mathrm{T}(\mathrm{Q}(s))} \\ L & \xrightarrow{\alpha} & M & \xrightarrow{\mathrm{Q}(u \circ e)} & P & \xrightarrow{\mathrm{Q}(d)} & \mathrm{T}(L) \end{array}$$

in K_S. The top row is a distinguished triangle, and the vertical arrows are isomorphisms. So the bottom row is a distinguished triangle. This is the triangle we were looking for.

(TR2). Turning: this is trivial.

(TR3). We are given the solid commutative diagram in K_S, where the rows are distinguished triangles:

$$\begin{array}{ccccccc} L & \xrightarrow{\alpha} & M & \xrightarrow{\beta} & N & \xrightarrow{\gamma} & \mathrm{T}(L) \\ \downarrow{\scriptstyle \phi} & & \downarrow{\scriptstyle \psi} & & \downarrow{\scriptstyle \chi} & & \downarrow{\scriptstyle \mathrm{T}(\phi)} \\ L' & \xrightarrow{\alpha'} & M' & \xrightarrow{\beta'} & N' & \xrightarrow{\gamma'} & \mathrm{T}(L') \end{array} \tag{7.1.5}$$

and we have to find χ to complete the diagram.

By replacing the rows with isomorphic triangles, we can assume they come from K. Thus we can replace (7.1.5) with this diagram:

$$\begin{array}{ccccccc} L & \xrightarrow{\mathrm{Q}(\alpha)} & M & \xrightarrow{\mathrm{Q}(\beta)} & N & \xrightarrow{\mathrm{Q}(\gamma)} & \mathrm{T}(L) \\ \downarrow{\scriptstyle \phi} & & \downarrow{\scriptstyle \psi} & & \downarrow{\scriptstyle \chi} & & \downarrow{\scriptstyle \mathrm{T}(\phi)} \\ L' & \xrightarrow{\mathrm{Q}(\alpha')} & M' & \xrightarrow{\mathrm{Q}(\beta')} & N' & \xrightarrow{\mathrm{Q}(\gamma')} & \mathrm{T}(L') \end{array} \tag{7.1.6}$$

in which $\alpha, \beta, \gamma, \alpha', \beta', \gamma'$ are morphisms in K. It is a commutative diagram. Let us choose fraction presentations $\phi = \mathrm{Q}(a) \circ \mathrm{Q}(s)^{-1}$ and $\psi = \mathrm{Q}(b) \circ \mathrm{Q}(t)^{-1}$. Then the solid diagram (7.1.6) comes from applying Q to the diagram

$$\begin{array}{ccccccc} L & \xrightarrow{\alpha} & M & \xrightarrow{\beta} & N & \xrightarrow{\gamma} & \mathrm{T}(L) \\ \uparrow{\scriptstyle s} & & \uparrow{\scriptstyle t} & & & & \uparrow{\scriptstyle \mathrm{T}(s)} \\ \tilde{L} & & \tilde{M} & & & & \mathrm{T}(\tilde{L}) \\ \downarrow{\scriptstyle a} & & \downarrow{\scriptstyle b} & & & & \downarrow{\scriptstyle \mathrm{T}(a)} \\ L' & \xrightarrow{\alpha'} & M' & \xrightarrow{\beta'} & N' & \xrightarrow{\gamma'} & \mathrm{T}(L') \end{array} \tag{7.1.7}$$

in K. Here the rows are distinguished triangles in K; but the diagram might fail to be commutative.

By axiom (RO3) we can find $c \in \mathsf{K}$ and $u \in \mathsf{S}$ s.t. $\mathrm{Q}(t)^{-1} \circ \mathrm{Q}(\alpha) \circ \mathrm{Q}(s) = \mathrm{Q}(c) \circ \mathrm{Q}(u)^{-1}$. This is the solid diagram below:

$$\begin{array}{ccccccc} & & & & L & \xrightarrow{\alpha} & M \\ & & & & \uparrow{\scriptstyle s} & & \uparrow{\scriptstyle t} \\ \tilde{L}'' & \overset{u'}{\dashrightarrow} & \tilde{L}' & \xrightarrow{u} & \tilde{L} & & \tilde{M} \\ & & & & \downarrow{\scriptstyle a} & & \downarrow{\scriptstyle b} \\ & & & & L' & \xrightarrow{\alpha'} & M' \end{array} \qquad (c : \tilde{L}' \to \tilde{M})$$

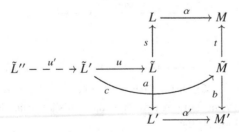

Thus $\mathrm{Q}(\alpha \circ s \circ u) = \mathrm{Q}(t \circ c)$. By (RO4) there is a morphism $u' \in \mathrm{S}$ s.t. $(\alpha \circ s \circ u) \circ u' = (t \circ c) \circ u'$. We get

$$\phi = \mathrm{Q}(a) \circ \mathrm{Q}(s)^{-1} = \mathrm{Q}(a \circ u \circ u') \circ \mathrm{Q}(s \circ u \circ u')^{-1}$$

in K_{S}. Thus, after substituting $\tilde{L} := \tilde{L}''$, $s := s \circ u \circ u'$, $a := a \circ u \circ u'$ and $c := c \circ u'$, we get a new diagram

$$\begin{array}{ccccccc} L & \xrightarrow{\alpha} & M & \xrightarrow{\beta} & N & \xrightarrow{\gamma} & \mathrm{T}(L) \\ {\scriptstyle s}\uparrow & & {\scriptstyle t}\uparrow & & & & \uparrow{\scriptstyle \mathrm{T}(s)} \\ \tilde{L} & \xrightarrow{c} & \tilde{M} & & & & \mathrm{T}(\tilde{L}) \\ {\scriptstyle a}\downarrow & & {\scriptstyle b}\downarrow & & & & \downarrow{\scriptstyle \mathrm{T}(a)} \\ L' & \xrightarrow{\alpha'} & M' & \xrightarrow{\beta'} & N' & \xrightarrow{\gamma'} & \mathrm{T}(L') \end{array} \tag{7.1.8}$$

in K instead of (7.1.7). In this new diagram the top left square is commutative; but maybe the bottom left square is not commutative.

When we apply Q to the diagram (7.1.8), the whole diagram, including the bottom left square, becomes commutative, since (7.1.6) is commutative. Again using condition (RO4), there is $v \in \mathrm{S}$ s.t. $(\alpha' \circ a) \circ v = (b \circ c) \circ v$. In a diagram:

$$\begin{array}{ccccc} & & L & \xrightarrow{\alpha} & M \\ & & {\scriptstyle s}\uparrow & & \uparrow{\scriptstyle t} \\ \tilde{L}' & \xrightarrow{v} & \tilde{L} & \xrightarrow{c} & \tilde{M} \\ & & {\scriptstyle a}\downarrow & & \downarrow{\scriptstyle b} \\ & & L' & \xrightarrow{\alpha'} & M' \end{array}$$

Performing the replacements $\tilde{L} := \tilde{L}'$, $s := s \circ v$, $c := c \circ v$ and $a := a \circ v$ we now have a commutative square also at the bottom left of (7.1.8). Since $\gamma \circ \beta = 0$ and $\gamma' \circ \beta' = 0$, in fact the whole diagram (7.1.8) in K is now commutative.

Now by (TR1) we can embed the morphism c in a distinguished triangle. We get the solid diagram

$$\begin{array}{ccccccc} L & \xrightarrow{\alpha} & M & \xrightarrow{\beta} & N & \xrightarrow{\gamma} & \mathrm{T}(L) \\ {\scriptstyle s}\uparrow & & {\scriptstyle t}\uparrow & & {\scriptstyle w}\uparrow & & \uparrow{\scriptstyle \mathrm{T}(s)} \\ \tilde{L} & \xrightarrow{c} & \tilde{M} & \xrightarrow{\tilde{\beta}} & \tilde{N} & \xrightarrow{\tilde{\gamma}} & \mathrm{T}(\tilde{L}) \\ {\scriptstyle a}\downarrow & & {\scriptstyle b}\downarrow & & {\scriptstyle d}\downarrow & & \downarrow{\scriptstyle \mathrm{T}(a)} \\ L' & \xrightarrow{\alpha'} & M' & \xrightarrow{\beta'} & N' & \xrightarrow{\gamma'} & \mathrm{T}(L') \end{array} \tag{7.1.9}$$

in K. The rows are distinguished triangles. Since $\tilde{\gamma} \circ \tilde{\beta} = 0$, the solid diagram is commutative. By (TR3) there are morphisms w and d that make the whole diagram commutative. Now by the long exact cohomology sequence (see Theorem 3.7.10) the morphism w belongs to S. The morphism $\chi := \mathrm{Q}(d) \circ \mathrm{Q}(w)^{-1} : N \to N'$ solves the problem.

(TR4). We will not give any details, as this axiom is not an important feature for us (see Remark 5.1.13). A proof can be found as part of the proof of [113, Theorem 2.1.8, page 97], where this axiom is called (TR4′).

Step 2. Suppose (F, ν) is a triangulated functor as in condition (ii). By Proposition 6.3.9 this extends uniquely to a T-additive functor $(F_{\mathsf{S}}, \nu_{\mathsf{S}})$. The construction of the triangulated structure on $(\mathsf{K}_{\mathsf{S}}, \mathrm{T}_{\mathsf{S}})$ in the previous steps, and the defining property of the translation isomorphism ν_{S} in Proposition 6.3.9, show that $(F_{\mathsf{S}}, \nu_{\mathsf{S}})$ is a triangulated functor.

Step 3. At this point $(\mathsf{K}_{\mathsf{S}}, \mathrm{T}_{\mathsf{S}})$ is a triangulated category, and conditions (i)-(ii) of the theorem are satisfied. We need to prove the uniqueness of the triangulated structure on $(\mathsf{K}_{\mathsf{S}}, \mathrm{T}_{\mathsf{S}})$. Condition (i) says that we can't have less distinguished triangles than those we declared. We can't have more distinguished triangles, because of condition (ii). □

Proposition 7.1.10 *Consider the situation of Proposition* 7.1.1 *and Theorem* 7.1.3.

(1) *The cohomological functor* $H : \mathsf{K} \to \mathsf{M}$ *factors into* $H = H_{\mathsf{S}} \circ \mathrm{Q}$, *where* $H_{\mathsf{S}} : \mathsf{K}_{\mathsf{S}} \to \mathsf{M}$ *is a cohomological functor.*
(2) *Let* M *be an object of* K. *The object* $\mathrm{Q}(M)$ *is zero in* K_{S} *iff the objects* $H(\mathrm{T}^i((M))$ *are zero in* M *for all* i.

Proof

(1) The existence and uniqueness of the functor H_{S} are by the universal property (Loc3) in Definition 6.1.2. We leave it as an exercise to show that H_{S} is a cohomological functor.

(2) Since H_{S} is an additive functor, if $\mathrm{Q}(M) = 0$, then so is $H(M) = H_{\mathsf{S}}(\mathrm{Q}(M))$. And of course $\mathrm{Q}(M) = 0$ iff $\mathrm{Q}(\mathrm{T}^i(M)) = 0$ for all i.

For the converse, let $\phi : 0 \to M$ be the zero morphism in K. If $H(\mathrm{T}^i(M)) = 0$ for all i, then $H(\mathrm{T}^i(\phi)) : 0 \to H(\mathrm{T}^i(M))$ are isomorphisms for all i. Then $\phi \in \mathsf{S}$, and so $\mathrm{Q}(\phi) : 0 \to \mathrm{Q}(M)$ is an isomorphism in K_{S}. □

Proposition 7.1.11 *Let* K *be a triangulated category, let* S *be a denominator set of cohomological origin in* K, *and let* K' *be a full triangulated subcategory of* K. *Then* $\mathsf{S}' := \mathsf{K}' \cap \mathsf{S}$ *is a denominator set of cohomological origin in* K', *the Ore localization* $\mathsf{K}'_{\mathsf{S}'}$ *exists, and* $\mathsf{K}'_{\mathsf{S}'}$ *is a triangulated category.*

Proof Let $H : \mathsf{K} \to \mathsf{M}$ be a cohomological functor that determines S. The functor $H|_{\mathsf{K}'} : \mathsf{K}' \to \mathsf{M}$ is also cohomological, and the set of morphisms S' satisfies

$$\mathsf{S}' = \{s \in \mathsf{K}' \mid H|_{\mathsf{K}'}(\mathrm{T}^i(s)) \text{ is an isomorphism for all } i\}.$$

Hence Proposition 7.1.1 and Theorem 7.1.3 apply. □

In the situation of the proposition, the localization functor is denoted by $\mathrm{Q}' : \mathsf{K}' \to \mathsf{K}'_{\mathsf{S}'}$.

Proposition 7.1.12 *In the situation of Proposition* 7.1.11, *let* $F : \mathsf{K}' \to \mathsf{E}$ *be a triangulated functor into some triangulated category* E. *Assume that for every* $s \in \mathsf{S}'$, *the morphism* $F(s)$ *is an isomorphism in* E. *Then there is a unique triangulated functor* $F_{\mathsf{S}'} : \mathsf{K}'_{\mathsf{S}'} \to \mathsf{E}$ *that extends* F; *Namely* $F_{\mathsf{S}'} \circ \mathrm{Q}' = F$ *as functors* $\mathsf{K}' \to \mathsf{E}$.

Proof This is part of Theorem 7.1.3. □

In particular we can look at the functor $F : \mathsf{K}' \xrightarrow{\text{inc}} \mathsf{K} \xrightarrow{\mathrm{Q}} \mathsf{K}_{\mathsf{S}}$, and its extension $F_{\mathsf{S}'} : \mathsf{K}'_{\mathsf{S}'} \to \mathsf{K}_{\mathsf{S}}$. We are interested in sufficient conditions for the functor $F_{\mathsf{S}'}$ to be fully faithful.

Proposition 7.1.13 *Let* K *be a triangulated category, let* S *be a denominator set of cohomological origin in* K, *and let* $\mathsf{K}' \subseteq \mathsf{K}$ *be a full triangulated subcategory. Define* $\mathsf{S}' := \mathsf{K}' \cap \mathsf{S}$. *Assume either of these conditions holds*:

(r) *Let* $M \in \mathrm{Ob}(\mathsf{K})$. *If there exists a morphism* $s : M \to L$ *in* S *with* $L \in \mathrm{Ob}(\mathsf{K}')$, *there exists a morphism* $t : K \to M$ *in* S *with* $K \in \mathrm{Ob}(\mathsf{K}')$.

(l) *Let* $M \in \mathrm{Ob}(\mathsf{K})$. *If there exists a morphism* $s : L \to M$ *in* S *with* $L \in \mathrm{Ob}(\mathsf{K}')$, *there exists a morphism* $t : M \to K$ *in* S *with* $K \in \mathrm{Ob}(\mathsf{K}')$.

Then the functor $F_{\mathsf{S}'} : \mathsf{K}'_{\mathsf{S}'} \to \mathsf{K}_{\mathsf{S}}$ *is fully faithful.*

Proof By Proposition 7.1.11, S' is a multiplicatively closed set of cohomological origin. So by Proposition 7.1.1 it is a right and left denominator set in K'. According to Proposition 6.2.34, under condition (r) the functor $F_{\mathsf{S}'}$ is fully faithful.

For condition (l) we pass to the opposite categories, using Proposition 6.2.17. □

Remark 7.1.14 Let K be a triangulated category, and let $\mathsf{N} \subseteq \mathsf{K}$ be a full triangulated subcategory. Define S to be the set of morphisms s in K such that the cone of s (see Definition 5.3.7) is isomorphic to an object of N. It turns out that S is a left and right denominator set in K. We call S the *denominator set associated to* N. The localization K_{S} is a triangulated category, and it is called the *Verdier quotient* of K by N, with notation $\mathsf{K}/\mathsf{N} := \mathsf{K}_{\mathsf{S}}$. See [156, Chapter 2] or [75, Section 1.6].

We want to relate this to Remark 5.3.25. Let $\tilde{\mathsf{S}}$ be a left and right denominator set in K. A morphism s in K is called a *divisor* of $\tilde{\mathsf{S}}$ if there exist morphisms $u, v \in \mathsf{K}$ such that $u \circ s$ and $s \circ v$ belong to $\tilde{\mathsf{S}}$. We call $\tilde{\mathsf{S}}$ a *saturated denominator set* if it contains all its divisors.

Let $\mathsf{N} \subseteq \mathsf{K}$ be a full triangulated subcategory, and let S be the associated denominator set, as in the first paragraph. Let $\tilde{\mathsf{N}}$ be the saturated full triangulated subcategory of K generated by N, in the sense of Definition 5.3.20. Let $\tilde{\mathsf{S}}$ be the set of morphisms gotten by adjoining to S all its divisors. Then $\tilde{\mathsf{S}}$ is a saturated left and right denominator set of morphisms in K, the canonical functor $\mathsf{K}_{\mathsf{S}} \to \mathsf{K}_{\tilde{\mathsf{S}}}$ is an isomorphism, the kernel of the localization functor $\mathrm{Q} : \mathsf{K} \to \mathsf{K}_{\mathsf{S}}$ (see Corollary 5.3.23) is $\tilde{\mathsf{N}}$, and $\tilde{\mathsf{S}}$ is the denominator set associated to $\tilde{\mathsf{N}}$. This is proved in [156, Chapter 2] and [76, Proposition 7.1.20].

An important example of a Verdier quotient is the passage from the homotopy category $\mathbf{K}(A, \mathsf{M})$ to the derived category $\mathbf{D}(A, \mathsf{M})$, which we shall study in Section 7.2 below. Indeed, the set of quasi-isomorphisms $\mathbf{S}(A, \mathsf{M})$ in $\mathbf{K}(A, \mathsf{M})$ can be described as the denominator set associated to the full subcategory $\mathbf{N}(A, \mathsf{M}) \subseteq \mathbf{K}(A, \mathsf{M})$ on the acyclic complexes, which is a saturated full triangulated subcategory.

7.2 Definition of the Derived Category

We now specialize to the triangulated category $\mathsf{K} = \mathbf{K}(A, \mathsf{M})$, for a central DG $\mathbb{K}$-ring A and a $\mathbb{K}$-linear abelian category M.

Recall the cohomology functor $\mathrm{H} : \mathbf{C}_{\mathrm{str}}(A, \mathsf{M}) \to \mathbf{G}_{\mathrm{str}}(\mathsf{M})$ from Definition 3.2.4.

Lemma 7.2.1 *Suppose $\phi, \psi : M \to N$ are morphisms in $\mathbf{C}_{\mathrm{str}}(A, \mathsf{M})$ such that $\psi - \phi$ is a coboundary in $\mathrm{Hom}_{A,\mathsf{M}}(M, N)$. Then $\mathrm{H}(\phi) = \mathrm{H}(\psi)$, as morphisms $\mathrm{H}(M) \to \mathrm{H}(N)$ in $\mathbf{G}_{\mathrm{str}}(\mathsf{M})$.*

Exercise 7.2.2 Prove the lemma.

It follows that there is an induced functor

$$\mathrm{H} : \mathbf{K}(A, \mathsf{M}) \to \mathbf{G}_{\mathrm{str}}(\mathsf{M}). \tag{7.2.3}$$

We shall be interested in its degree 0 component H^0. Of course, all other components can be recovered from H^0 by the translation functor: $\mathrm{H}^i = \mathrm{H}^0 \circ \mathrm{T}^i$.

Proposition 7.2.4 *Let* M *be an abelian category and let A be a DG ring. Then the functor* $\mathrm{H}^0 : \mathbf{K}(A, \mathsf{M}) \to \mathsf{M}$ *is cohomological.*

Proof Clearly H^0 is additive. Consider a distinguished triangle

$$L \xrightarrow{\alpha} M \xrightarrow{\beta} N \xrightarrow{\gamma} \mathrm{T}(L) \tag{7.2.5}$$

in $\mathbf{K}(A, \mathsf{M})$. We must prove that

$$\mathrm{H}^0(L) \xrightarrow{\mathrm{H}^0(\alpha)} \mathrm{H}^0(M) \xrightarrow{\mathrm{H}^0(\beta)} \mathrm{H}^0(N)$$

is an exact sequence in M.

By Definition 5.4.2 we can assume that the distinguished triangle (7.2.5) is the image of a standard triangle in $\mathbf{C}(A, \mathsf{M})$. Namely that $N = \mathrm{Cone}(\alpha)$, the standard cone associated to α, and the morphisms are $\beta = e_\alpha$ and $\gamma = p_\alpha$; see Definitions 4.2.1 and 4.2.5. To be explicit, in matrix notation we have $N = \left[\begin{smallmatrix} M \\ \mathrm{T}(L) \end{smallmatrix}\right]$, $\mathrm{d}_N = \left[\begin{smallmatrix} \mathrm{d}_M & \alpha \circ \mathrm{t}_L^{-1} \\ 0 & \mathrm{d}_{\mathrm{T}(L)} \end{smallmatrix}\right]$ and $\beta = \left[\begin{smallmatrix} \mathrm{id}_M \\ 0 \end{smallmatrix}\right]$.

Let us use the abbreviations $\bar{\beta} := \mathrm{H}^0(\beta)$ and $\bar{\alpha} := \mathrm{H}^0(\alpha)$. By Proposition 5.3.1 we know that $\beta \circ \alpha = 0$, and hence $\bar{\beta} \circ \bar{\alpha} = 0$. It remains to prove that the morphism $\bar{\alpha} : \mathrm{H}^0(L) \to \mathrm{Ker}(\bar{\beta})$ in M is an epimorphism. This will be done using the first sheaf trick (Proposition 2.4.9(3)): we will prove that given a section $\bar{m} \in \Gamma(U, \mathrm{Ker}(\bar{\beta}))$ on an open set $U \in \mathsf{M}$, there is a refinement $V' \twoheadrightarrow U$ and a section $\bar{l} \in \Gamma(V', \mathrm{H}^0(L))$ such that $\bar{\alpha}(\bar{l}) = \bar{m}$ in $\Gamma(V', \mathrm{Ker}(\bar{\beta}))$. See Section 2.4 for the "geometric" terminology that's used in the sheaf tricks.

Take some section $\bar{m} \in \Gamma(U, \mathrm{Ker}(\bar{\beta}))$ on an open set $U \in \mathsf{M}$. This means that $\bar{m} \in \Gamma(U, \mathrm{H}^0(M))$ and $\bar{\beta}(\bar{m}) = 0$. There is a covering $V \twoheadrightarrow U$ and a section $m \in \Gamma(V, \mathrm{Z}^0(M))$ such that $\pi_M(m) = \bar{m}$. Now

$$\beta(m) = \begin{bmatrix} m \\ 0 \end{bmatrix} \in \Gamma(V, \mathrm{Z}^0(N)) \subseteq \Gamma(V, N^0) = \begin{bmatrix} \Gamma(V, M^0) \\ \Gamma(V, L^1) \end{bmatrix}$$

in column notation. The vanishing of $\bar{\beta}(\bar{m}) = \pi_N(\beta(m))$ in $\Gamma(V, \mathrm{H}^0(N))$ means that there is a covering $V' \twoheadrightarrow V$ and a section $n \in \Gamma(V', N^{-1})$ such that $\mathrm{d}_N(n) = \beta(m)$. In column notation we have $n = \left[\begin{smallmatrix} m' \\ l \end{smallmatrix}\right]$, with $m' \in \Gamma(V', M^{-1})$ and $l \in \Gamma(V', L^0)$. The equality $\beta(m) = \mathrm{d}_N(n)$ becomes the matrix equality

$$\begin{bmatrix} m \\ 0 \end{bmatrix} = \begin{bmatrix} \mathrm{d}_M & \alpha \\ 0 & -\mathrm{d}_L \end{bmatrix} \cdot \begin{bmatrix} m' \\ l \end{bmatrix} = \begin{bmatrix} \mathrm{d}_M(m') + \alpha(l) \\ -\mathrm{d}_L(l) \end{bmatrix}.$$

We see that $l \in \Gamma(V', \mathrm{Z}^0(L))$. Then the cohomology class $\bar{l} := \pi_L(l) \in \Gamma(V', \mathrm{H}^0(L))$ satisfies

$$\bar{\alpha}(\bar{l}) = \pi_M(\alpha(l)) = \pi_M(\mathrm{d}_M(m') + \alpha(l)) = \pi_M(m) = \bar{m}$$

in $\Gamma(V', \mathrm{H}^0(M))$, as required. □

Definition 7.2.6 A morphism ϕ in $\mathbf{K}(A, \mathsf{M})$ is called a *quasi-isomorphism* if the morphisms $\mathrm{H}^i(\phi)$ in M are isomorphisms for all i. The set of quasi-isomorphisms in $\mathbf{K}(A, \mathsf{M})$ is denoted by $\mathbf{S}(A, \mathsf{M})$.

By Proposition 7.2.4 the functor H^0 is cohomological. According to Proposition 7.1.1, $\mathbf{S}(A, \mathsf{M})$ is a denominator set of cohomological origin; so Theorem 7.1.3 applies to it, and the next definition makes sense.

Definition 7.2.7 Let M be a $\mathbb{K}$-linear abelian category and A a central DG $\mathbb{K}$-ring. The *derived category of DG A-modules in* M is the $\mathbb{K}$-linear triangulated category

$$\mathbf{D}(A, \mathsf{M}) := \mathbf{K}(A, \mathsf{M})_{\mathbf{S}(A, \mathsf{M})}.$$

The corresponding triangulated localization functor is

$$\mathrm{Q} : \mathbf{K}(A, \mathsf{M}) \to \mathbf{D}(A, \mathsf{M}).$$

We also have the additive functor $\mathrm{P} : \mathbf{C}_{\mathrm{str}}(A, \mathsf{M}) \to \mathbf{K}(A, \mathsf{M})$ which sends a strict morphism of DG modules to its homotopy class.

Definition 7.2.8 Let M be an abelian category and let A be a DG ring. Define the functor

$$\tilde{\mathrm{Q}} := \mathrm{Q} \circ \mathrm{P} : \mathbf{C}_{\mathrm{str}}(A, \mathsf{M}) \to \mathbf{D}(A, \mathsf{M}).$$

We thus get a commutative diagram of additive functors

$$\mathbf{C}_{\mathrm{str}}(A, \mathsf{M}) \xrightarrow{\mathrm{P}} \mathbf{K}(A, \mathsf{M}) \xrightarrow{\mathrm{Q}} \mathbf{D}(A, \mathsf{M}), \qquad \tilde{\mathrm{Q}} : \mathbf{C}_{\mathrm{str}}(A, \mathsf{M}) \to \mathbf{D}(A, \mathsf{M}) \tag{7.2.9}$$

that are all the identity on objects.

It is sometimes convenient to describe morphisms in $\mathbf{D}(A, \mathsf{M})$ in terms of the functor $\tilde{\mathrm{Q}}$. A morphism $s \in \mathbf{C}_{\mathrm{str}}(A, \mathsf{M})$ is called a quasi-isomorphism if $\mathrm{P}(s)$ is a quasi-isomorphism in $\mathbf{K}(A, \mathsf{M})$; i.e. if $\mathrm{H}(s)$ is an isomorphism in $\mathbf{G}_{\mathrm{str}}(\mathsf{M})$.

Proposition 7.2.10

(1) *Every morphism ϕ in $\mathbf{D}(A, \mathsf{M})$ can be written as a right fraction $\phi = \tilde{\mathrm{Q}}(a) \circ \tilde{\mathrm{Q}}(s)^{-1}$ where $a, s \in \mathbf{C}_{\mathrm{str}}(A, \mathsf{M})$ and s is a quasi-isomorphism.*

(2) *Let $a \in \mathbf{C}_{\mathrm{str}}(A, \mathsf{M})$. Then $\tilde{\mathrm{Q}}(a) = 0$ in $\mathbf{D}(A, \mathsf{M})$ iff there exists a quasi-isomorphism s in $\mathbf{C}_{\mathrm{str}}(A, \mathsf{M})$ such that $a \circ s$ is a coboundary in $\mathbf{C}(A, \mathsf{M})$.*

Proof

(1) This is because of property (RO3) of Definition 6.2.1 and the fact that P is full.

(2) Let us write $\bar{a} := \mathrm{P}(a) \in \mathbf{K}(A, \mathsf{M})$. Since $\mathrm{Q}(\bar{a}) = 0$, by Property (RO4) of Definition 6.2.1, there is a quasi-isomorphism $\bar{s} \in \mathbf{K}(A, \mathsf{M})$ such that $\bar{a} \circ \bar{s} = 0$ in $\mathbf{K}(A, \mathsf{M})$. Choose a quasi-isomorphism $s \in \mathbf{C}_{\mathrm{str}}(A, \mathsf{M})$ such that $\bar{s} = \mathrm{P}(s)$. Then $\mathrm{P}(a \circ s) = 0$, and this means that $a \circ s$ is a coboundary. □

Of course, there is a left version of this proposition. The next definition is about arbitrary (not necessarily linear) categories.

Definition 7.2.11 A functor $F : \mathsf{C} \to \mathsf{D}$ between categories is called *conservative* if for every morphism ϕ in C, ϕ is an isomorphism if and only if $F(\phi)$ is an isomorphism.

By Proposition 7.1.10 the cohomology functor (7.2.3) extends uniquely to a functor

$$\mathrm{H} : \mathbf{D}(A, \mathsf{M}) \to \mathbf{G}_{\mathrm{str}}(\mathsf{M}). \tag{7.2.12}$$

Corollary 7.2.13 *The functor* H *in* (7.2.12) *is conservative.*

Proof One implication is trivial. For the other implication, assume that ϕ is a morphism in $\mathbf{D}(A, \mathsf{M})$ such that $\mathrm{H}(\phi)$ is an isomorphism. We can write ϕ as a right fraction: $\phi = \mathrm{Q}(a) \circ \mathrm{Q}(s)^{-1}$, where $a \in \mathbf{K}(A, \mathsf{M})$ and $s \in \mathbf{S}(A, \mathsf{M})$. Then $\mathrm{H}(\phi) = \mathrm{H}(\mathrm{Q}(a)) \circ \mathrm{H}(\mathrm{Q}(s))^{-1}$, and we see that $\mathrm{H}(\mathrm{Q}(a))$ is an isomorphism. But $\mathrm{H}(a) = \mathrm{H}(\mathrm{Q}(a))$, so in fact $a \in \mathbf{S}(A, \mathsf{M})$ too. Therefore $\mathrm{Q}(a)$ is an isomorphism in $\mathbf{D}(A, \mathsf{M})$. It follows that ϕ is an isomorphism in $\mathbf{D}(A, \mathsf{M})$. □

Exercise 7.2.14 Here $\mathsf{M} = \mathbf{M}(\mathbb{K})$, so $\mathbf{K}(A, \mathsf{M}) = \mathbf{K}(A)$. Show that the functor $\mathrm{H}^0 : \mathbf{K}(A) \to \mathbf{M}(\mathbb{K})$ is corepresentable by the object $A \in \mathbf{K}(A)$ (see Section 1.7).

Remark 7.2.15 This is a continuation of Remarks 3.8.22 and 6.2.38 on strictly unital A_∞ categories. The base ring $\mathbb{K}$ is a field. The references are [83] and [37]. Recall that if A is an A_∞ category and B is a DG category, then $\mathrm{A}_\infty\mathsf{Fun}(\mathsf{A}, \mathsf{B})$ is a DG category.

Consider the special case is when $\mathsf{B} = \mathbf{C}(\mathbb{K})$, the DG category of complexes of $\mathbb{K}$-modules. An A_∞ functor $M : \mathsf{A} \to \mathbf{C}(\mathbb{K})$ is called an A_∞ A-*module*. The A_∞ A-modules are the objects of the DG category $\mathbf{C}_\infty(\mathsf{A}) := \mathrm{A}_\infty\mathsf{Fun}(\mathsf{A}, \mathbf{C}(\mathbb{K}))$. The strict subcategory $\mathrm{Str}(\mathbf{C}_\infty(\mathsf{A}))$, in the sense of Definition 3.4.6(1), is what most texts call the *category of* A_∞ A-*modules*.

A morphism $\phi : M \to N$ in $\mathrm{Str}(\mathbf{C}_\infty(\mathsf{A}))$ is called a *quasi-isomorphism* if for every object $x \in \mathrm{Ob}(\mathsf{A})$ the morphism $\phi(x) : M(x) \to N(x)$ in $\mathrm{Str}(\mathbf{C}(\mathbb{K})) = \mathbf{C}_{\mathrm{str}}(\mathbb{K})$ is a quasi-isomorphism. The localization of $\mathrm{Str}(\mathbf{C}_\infty(\mathsf{A}))$ w.r.t. the quasi-isomorphisms is the *derived category of* A_∞ A-*modules*, denoted by

$\mathbf{D}_\infty(\mathsf{A})$. (As usual, the tradition is to call $\mathbf{D}_\infty(\mathsf{A})$ the "homotopy category" of $\mathbf{C}_\infty(\mathsf{A})$, and to denote it by $\mathrm{Ho}(\mathbf{C}_\infty(\mathsf{A}))$.) There is a localization functor $\mathrm{Q} : \mathrm{Str}(\mathbf{C}_\infty(\mathsf{A})) \to \mathbf{D}_\infty(\mathsf{A})$.

We can also consider the category $\mathbf{K}_\infty(\mathsf{A}) := \mathrm{H}^0(\mathbf{C}_\infty(\mathsf{A}))$, see Definition 3.4.6(2). There is a full functor $\mathrm{P} : \mathrm{Str}(\mathbf{C}_\infty(\mathsf{A})) \to \mathbf{K}_\infty(\mathsf{A})$, and a commutative diagram

$$\mathrm{Str}(\mathbf{C}_\infty(\mathsf{A})) \xrightarrow{\ \mathrm{P}\ } \mathbf{K}_\infty(\mathsf{A}) \xrightarrow{\ \bar{\mathrm{Q}}\ } \mathbf{D}_\infty(\mathsf{A}), \qquad \mathrm{Q} = \bar{\mathrm{Q}} \circ \mathrm{P} \tag{7.2.16}$$

All three functors are the identity on objects. The categories $\mathbf{K}_\infty(\mathsf{A})$ and $\mathbf{D}_\infty(\mathsf{A})$ are triangulated, and the functor $\bar{\mathrm{Q}}$ is an isomorphism of triangulated categories.

Finally, let A be a DG category. Then the DG category $\mathbf{C}(\mathsf{A})$ of DG A-modules embeds faithfully (but not fully) into the DG category $\mathbf{C}_\infty(\mathsf{A})$. The induced triangulated functor $\mathbf{D}(\mathsf{A}) \to \mathbf{D}_\infty(\mathsf{A})$ is an equivalence.

7.3 Boundedness Conditions in $\mathbf{K}(A, \mathsf{M})$

We continue with a central DG $\mathbb{K}$-ring A and a $\mathbb{K}$-linear abelian category M. Boundedness conditions in $\mathbf{C}(A, \mathsf{M})$ were introduced in Definition 3.8.20. Similarly:

Definition 7.3.1 We define $\mathbf{K}^-(A, \mathsf{M})$, $\mathbf{K}^+(A, \mathsf{M})$ and $\mathbf{K}^{\mathrm{b}}(A, \mathsf{M})$ to be the full subcategories of $\mathbf{K}(A, \mathsf{M})$ consisting of bounded above, bounded below and bounded DG modules, respectively.

Of course, $\mathbf{K}^{\mathrm{b}}(A, \mathsf{M}) = \mathbf{K}^-(A, \mathsf{M}) \cap \mathbf{K}^+(A, \mathsf{M})$. The subcategories $\mathbf{K}^\star(A, \mathsf{M})$, for $\star \in \{-, +, \mathrm{b}\}$, are full triangulated subcategories of $\mathbf{K}(A, \mathsf{M})$; this is because the operations of translation and cone preserve the various boundedness conditions. Note that $\mathbf{K}^\star(A, \mathsf{M}) = \mathrm{Ho}(\mathbf{C}^\star(A, \mathsf{M}))$. As the next example shows, sometimes the categories $\mathbf{C}^\star(A, \mathsf{M})$ and $\mathbf{K}^\star(A, \mathsf{M})$ can be very degenerate.

Example 7.3.2 Let A be the DG ring $\mathbb{K}[t, t^{-1}]$, the ring of Laurent polynomials in the variable t of degree 1, with the zero differential. If $M = \{M^i\}_{i \in \mathbb{Z}}$ is a nonzero object of $\mathbf{C}(A, \mathsf{M})$, then $M^i \neq 0$ for all i. Therefore the categories $\mathbf{C}^\star(A, \mathsf{M})$ and $\mathbf{K}^\star(A, \mathsf{M})$ are zero for $\star \in \{-, +, \mathrm{b}\}$.

Let $\mathbf{S}^\star(A, \mathsf{M}) := \mathbf{K}^\star(A, \mathsf{M}) \cap \mathbf{S}(A, \mathsf{M})$, the category of quasi-isomorphisms in $\mathbf{K}^\star(A, \mathsf{M})$. As already mentioned, Theorem 7.1.3 applies here, so we can localize.

Definition 7.3.3 For $\star \in \{-, +, \mathrm{b}\}$ we define

$$\mathbf{D}^{\star}(A, \mathsf{M}) := \mathbf{K}^{\star}(A, \mathsf{M})_{\mathbf{S}^{\star}(A, \mathsf{M})},$$

the Ore localization of $\mathbf{K}^{\star}(A, \mathsf{M})$ with respect to $\mathbf{S}^{\star}(A, \mathsf{M})$.

Here is another kind of boundedness condition.

Definition 7.3.4 For $\star \in \{-, +, \mathrm{b}\}$ we define $\mathbf{D}(A, \mathsf{M})^{\star}$ to be the full subcategory of $\mathbf{D}(A, \mathsf{M})$ on the complexes M whose cohomology $\mathrm{H}(M)$ is of boundedness type $\star$.

Of course, $\mathbf{D}(A, \mathsf{M})^{\star}$ is a full triangulated subcategory of $\mathbf{D}(A, \mathsf{M})$. See Remark 7.3.13 regarding Definitions 7.3.3 and 7.3.4.

Handling boundedness conditions requires the use of *truncations*. These will be introduced after the next proposition.

Proposition 7.3.5 *Let* $0 \to L \xrightarrow{\phi} M \xrightarrow{\psi} N \to 0$ *be a short exact sequence in* $\mathbf{C}_{\mathrm{str}}(A, \mathsf{M})$. *Then there is a morphism* $\theta : N \to \mathrm{T}(L)$ *in* $\mathbf{D}(A, \mathsf{M})$ *such that* $L \xrightarrow{\mathrm{Q}(\phi)} M \xrightarrow{\mathrm{Q}(\psi)} N \xrightarrow{\theta} \mathrm{T}(L)$ *is a distinguished triangle in* $\mathbf{D}(A, \mathsf{M})$.

Proof We are following the proof of [75, Proposition 1.7.5]. Let $\tilde{N}$ be the standard cone on ϕ. In matrix notation, as in Definition 4.2.1, we have $\tilde{N} = \left[\begin{smallmatrix} M \\ \mathrm{T}(L) \end{smallmatrix}\right]$ and $\mathrm{d}_{\tilde{N}} = \left[\begin{smallmatrix} \mathrm{d}_M & \phi \circ \mathrm{t}^{-1} \\ 0 & \mathrm{d}_{\mathrm{T}(L)} \end{smallmatrix}\right]$. The object $\tilde{N}$ sits inside the standard triangle $L \xrightarrow{\phi} M \xrightarrow{\tilde{\psi}} \tilde{N} \xrightarrow{\tilde{\chi}} \mathrm{T}(L)$ in $\mathbf{C}_{\mathrm{str}}(A, \mathsf{M})$, where $\tilde{\psi} := \left[\begin{smallmatrix} \mathrm{id} \\ 0 \end{smallmatrix}\right]$ and $\tilde{\chi} := [0 \ \mathrm{id}]$ in matrix notation. Define the morphism $\gamma = \tilde{N} \to N$ to be the matrix $\gamma := [\psi \ 0]$. We get a commutative diagram

$$\begin{array}{ccccccc} L & \xrightarrow{\phi} & M & \xrightarrow{\tilde{\psi}} & \tilde{N} & \xrightarrow{\tilde{\chi}} & \mathrm{T}(L) \\ & & & \searrow^{\psi} & \downarrow{\scriptstyle \gamma} & & \\ & & & & N & & \end{array}$$

in $\mathbf{C}_{\mathrm{str}}(A, \mathsf{M})$. We shall prove below that γ is a quasi-isomorphism. Then the morphism $\theta := \mathrm{Q}(\tilde{\chi}) \circ \mathrm{Q}(\gamma)^{-1} : N \to \mathrm{T}(L)$ will work.

Let $\tilde{K}$ be the standard cone on id_L, and let $\tilde{\beta} : \tilde{K} \to \tilde{N}$ be the matrix morphism $\left[\begin{smallmatrix} \phi & 0 \\ 0 & \mathrm{id} \end{smallmatrix}\right] : \left[\begin{smallmatrix} L \\ \mathrm{T}(L) \end{smallmatrix}\right] \to \left[\begin{smallmatrix} M \\ \mathrm{T}(L) \end{smallmatrix}\right]$. This fits into a short exact sequence $0 \to \tilde{K} \xrightarrow{\tilde{\beta}} \tilde{N} \xrightarrow{\gamma} N \to 0$ in $\mathbf{C}_{\mathrm{str}}(A, \mathsf{M})$. But the DG module $\tilde{K}$ is acyclic, and therefore γ is a quasi-isomorphism, as claimed. □

Definition 7.3.6 Let $M \in \mathbf{C}(\mathsf{M})$ and $i \in \mathbb{Z}$. The *smart truncation of M below i* is the complex

$$\mathrm{smt}^{\leq i}(M) := (\cdots \to M^{i-2} \xrightarrow{\mathrm{d}} M^{i-1} \xrightarrow{\mathrm{d}} \mathrm{Z}^i(M) \to 0 \to \cdots).$$

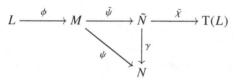

The *smart truncation of M above i* is the complex

$$\operatorname{smt}^{\geq i}(M) := (\cdots \to 0 \to \mathrm{Y}^i(M) \xrightarrow{\mathrm{d}} M^{i+1} \xrightarrow{\mathrm{d}} M^{i+2} \to \cdots).$$

In the definition we use the objects of cocycles $\mathrm{Z}^i(M)$ and decocycles $\mathrm{Y}^i(M)$ from Definition 3.2.4. Note that $\operatorname{smt}^{\leq i}(M)$ is a subcomplex of M, whereas $\operatorname{smt}^{\geq i}(M)$ a quotient complex of M. Denoting the inclusion and the projection by e and p, respectively, we get a diagram

$$\operatorname{smt}^{\leq i}(M) \xrightarrow{e} M \xrightarrow{p} \operatorname{smt}^{\geq i+1}(M) \tag{7.3.7}$$

in $\mathbf{C}_{\mathrm{str}}(\mathsf{M})$. This diagram is functorial in $M \in \mathbf{C}_{\mathrm{str}}(\mathsf{M})$. But diagram (7.3.7) is not an exact sequence in general – there is a defect in degree i.

Recall that a DG ring A is called nonpositive if $A^i = 0$ for all $i > 0$.

Proposition 7.3.8 *Assume A is a nonpositive DG ring.*

(1) *The differential of every $M \in \mathbf{C}_{\mathrm{str}}(A, \mathsf{M})$ is A^0-linear.*
(2) *The smart truncations from Definition* 7.3.6 *are functors from $\mathbf{C}_{\mathrm{str}}(A, \mathsf{M})$ to itself.*

Exercise 7.3.9 Prove this proposition.

Proposition 7.3.10 *Assume A is nonpositive. For every $M \in \mathbf{C}(A, \mathsf{M})$ there is a distinguished triangle*

$$\operatorname{smt}^{\leq i}(M) \xrightarrow{e} M \xrightarrow{p} \operatorname{smt}^{\geq i+1}(M) \xrightarrow{\theta} \mathrm{T}(\operatorname{smt}^{\leq i}(M))$$

in $\mathbf{D}(A, \mathsf{M})$. Also, $\mathrm{H}^j(e) : \mathrm{H}^j(\operatorname{smt}^{\leq i}(M)) \to \mathrm{H}^j(M)$ is an isomorphism in M for all $j \leq i$, and $\mathrm{H}^j(p) : \mathrm{H}^j(M) \to \mathrm{H}^j(\operatorname{smt}^{\geq i+1}(M))$ is an isomorphism in M for all $j \geq i + 1$.

Proof The claims about $\mathrm{H}^j(e)$ and $\mathrm{H}^j(p)$ are trivial to verify. Now there is a short exact sequence

$$0 \to \operatorname{smt}^{\leq i}(M) \xrightarrow{e} M \xrightarrow{p'} N \to 0 \tag{7.3.11}$$

in $\mathbf{C}_{\mathrm{str}}(A, \mathsf{M})$, where

$$N := (\cdots \to 0 \to M^i / \mathrm{Z}^i(M) \xrightarrow{\mathrm{d}} M^{i+1} \xrightarrow{\mathrm{d}} M^{i+2} \to \cdots).$$

According to Proposition 7.3.5, we get a distinguished triangle

$$\operatorname{smt}^{\leq i}(M) \xrightarrow{e} M \xrightarrow{p'} N \xrightarrow{\theta'} \mathrm{T}(\operatorname{smt}^{\leq i}(M))$$

in $\mathbf{D}(A, \mathsf{M})$. Next, there is an obvious quasi-isomorphism $\phi : N \to \operatorname{smt}^{\geq i+1}(M)$ in $\mathbf{C}_{\mathrm{str}}(A, \mathsf{M})$ such that $p = \phi \circ p'$. We define the morphism

$$\theta := \theta' \circ \mathrm{Q}(\phi)^{-1} : \operatorname{smt}^{\geq i+1}(M) \to \mathrm{T}(\operatorname{smt}^{\leq i}(M))$$

in $\mathbf{D}(A, \mathsf{M})$. □

Proposition 7.3.12 *Assume A is nonpositive. For $\star \in \{-, +, \mathrm{b}\}$ the canonical functor $\mathsf{D}^{\star}(A, \mathsf{M}) \to \mathsf{D}(A, \mathsf{M})^{\star}$ is an equivalence of triangulated categories.*

Proof It is done in four steps.

Step 1. Here we prove that the functor $F^- : \mathsf{D}^-(A, \mathsf{M}) \to \mathsf{D}(A, \mathsf{M})$ is fully faithful. Let $s : M \to L$ be a quasi-isomorphism in $\mathsf{K}(A, \mathsf{M})$ with $L \in \mathsf{K}^-(A, \mathsf{M})$. Say L is concentrated in degrees $\leq i$. Then $\mathrm{H}^j(M) = \mathrm{H}^j(L) = 0$ for all $j > i$. The smart truncation $\mathrm{smt}^{\leq i}(M)$ belongs to $\mathsf{K}^-(A, \mathsf{M})$, and the inclusion $t : \mathrm{smt}^{\leq i}(M) \to M$ is a quasi-isomorphism. According to Proposition 7.1.13, with $\mathsf{K} = \mathsf{K}(A, \mathsf{M})$ and $\mathsf{K}' = \mathsf{K}^-(A, \mathsf{M})$, and with condition (r), we see that F^- is fully faithful.

Step 2. Here we prove that the functor $F^+ : \mathsf{D}^+(\mathsf{M}) \to \mathsf{D}(\mathsf{M})$ is fully faithful. Let $s : L \to M$ be a quasi-isomorphism in $\mathsf{K}(A, \mathsf{M})$ with $L \in \mathsf{K}^+(A, \mathsf{M})$. Say L is concentrated in degrees $\geq i$. Then $\mathrm{H}^j(M) = \mathrm{H}^j(L) = 0$ for all $j < i$. The smart truncation $\mathrm{smt}^{\geq i}(M)$ belongs to $\mathsf{K}^+(A, \mathsf{M})$, and the projection $t : M \to \mathrm{smt}^{\geq i}(M)$ is a quasi-isomorphism. According to Proposition 7.1.13, with condition (l), we see that F^+ is fully faithful.

Step 3. The arguments in step 1 we show that $\mathsf{D}^{\mathrm{b}}(A, \mathsf{M}) \to \mathsf{D}^+(A, \mathsf{M})$ is fully faithful. And by step 2, $\mathsf{D}^+(A, \mathsf{M}) \to \mathsf{D}(A, \mathsf{M})$ is fully faithful. Therefore $\mathsf{D}^{\mathrm{b}}(A, \mathsf{M}) \to \mathsf{D}(A, \mathsf{M})$ is fully faithful.

Step 4. Smart truncation shows that the functor $\mathsf{D}^{\star}(A, \mathsf{M}) \to \mathsf{D}(A, \mathsf{M})^{\star}$ is essentially surjective on objects. □

Remark 7.3.13 Definition 7.3.3 with $A = \mathbb{K}$, namely $\mathsf{D}^{\star}(\mathsf{M})$, is standard usage of this notation – starting from [62, Section I.4], all the way to the recent book [76, Definition 13.1.2]. An exception is [144, Definition tag=05RU].

On the other hand, definition 7.3.4 with $A = \mathbb{K}$, namely $\mathsf{D}(\mathsf{M})^{\star}$, is almost never used (an exception is [120]).

We are going to require both distinct concepts, $\mathsf{D}^{\star}(A, \mathsf{M})$ and $\mathsf{D}(A, \mathsf{M})^{\star}$, until the end of Chapter 11. Beginning with Chapter 12, in Definition 12.1.6, the notation $\mathsf{D}(A, \mathsf{M})^{\star}$ will be eliminated, and $\mathsf{D}^{\star}(A, \mathsf{M})$ will be redefined to replace it.

7.4 Thick Subcategories of M

We begin by recalling a definition about abelian categories.

Definition 7.4.1 Let M be an abelian category. A *thick abelian subcategory* of M is a full abelian subcategory N that is closed under extensions. Namely if

$$0 \to M' \to M \to M'' \to 0$$

is a short exact sequence in M with $M', M'' \in \mathsf{N}$, then $M \in \mathsf{N}$ too.

Proposition 7.4.2 *Let M be an abelian category, and let $\mathsf{M}' \subseteq \mathsf{M}$ be a thick abelian subcategory. Suppose $M_1 \to M_2 \to N \to M_3 \to M_4$ is an exact sequence in M, and the objects M_i belong M'. Then $N \in \mathsf{M}'$ too.*

Exercise 7.4.3 Prove this proposition.

Definition 7.4.4 Let M be an abelian category and $\mathsf{N} \subseteq \mathsf{M}$ a thick abelian subcategory. We denote by $\mathsf{D}_{\mathsf{N}}(\mathsf{M})$ the full subcategory of $\mathsf{D}(\mathsf{M})$ consisting of complexes M such that $\mathrm{H}^i(M) \in \mathsf{N}$ for every i.

Given a boundedness condition $\star$, we write $\mathsf{D}^{\star}_{\mathsf{N}}(\mathsf{M}) := \mathsf{D}_{\mathsf{N}}(\mathsf{M}) \cap \mathsf{D}^{\star}(\mathsf{M})$ and $\mathsf{D}_{\mathsf{N}}(\mathsf{M})^{\star} := \mathsf{D}_{\mathsf{N}}(\mathsf{M}) \cap \mathsf{D}(\mathsf{M})^{\star}$.

Proposition 7.4.5 *If N is a thick abelian subcategory of M then $\mathsf{D}_{\mathsf{N}}(\mathsf{M})$ is a full triangulated subcategory of $\mathsf{D}(\mathsf{M})$.*

Proof Clearly $\mathsf{D}_{\mathsf{N}}(\mathsf{M})$ is closed under translations. Now suppose $M' \to M \to M'' \xrightarrow{\triangle}$ is a distinguished triangle in $\mathsf{D}(\mathsf{M})$ such that $M', M \in \mathsf{D}_{\mathsf{N}}(\mathsf{M})$; we have to show that M'' is also in $\mathsf{D}_{\mathsf{N}}(\mathsf{M})$. Consider the exact sequence $\mathrm{H}^i(M') \to \mathrm{H}^i(M) \to \mathrm{H}^i(M'') \to \mathrm{H}^{i+1}(M') \to \mathrm{H}^{i+1}(M)$. The four outer objects belong to N. Since N is a thick abelian subcategory of M, it follows (using Proposition 7.4.2) that $\mathrm{H}^i(M'') \in \mathsf{N}$. □

Example 7.4.6 Let A be a noetherian commutative ring. The category $\mathsf{Mod}_{\mathrm{f}}\, A$ of finitely generated modules is a thick abelian subcategory of $\mathsf{Mod}\, A$.

Example 7.4.7 Consider $\mathsf{Mod}\, \mathbb{Z} = \mathsf{Ab}$. As above we have the thick abelian subcategory $\mathsf{Ab}_{\mathrm{f}} = \mathsf{Mod}_{\mathrm{f}}\, \mathbb{Z}$ of finitely generated abelian groups. There is also the thick abelian subcategory $\mathsf{Ab}_{\mathrm{tors}}$ of torsion abelian groups (every element has a finite order). The intersection of $\mathsf{Ab}_{\mathrm{tors}}$ and Ab_{f} is the category $\mathsf{Ab}_{\mathrm{fin}}$ of finite abelian groups. This is also thick.

Example 7.4.8 Let X be a noetherian scheme (e.g. an algebraic variety over an algebraically closed field). Consider the abelian category $\mathsf{Mod}\, \mathcal{O}_X$ of $\mathcal{O}_X$-modules. In it there is the thick abelian subcategory $\mathsf{QCoh}\, \mathcal{O}_X$ of quasi-coherent sheaves, and in that there is the thick abelian subcategory $\mathsf{Coh}\,\mathcal{O}_X$ of coherent sheaves.

Passing to complexes, let us write $\mathbf{C}(X) := \mathbf{C}(\mathsf{Mod}\,\mathcal{O}_X)$, the DG category of unbounded complexes. Its strict subcategory is $\mathbf{C}_{\mathrm{str}}(X)$, and for each boundedness condition $\star$ we have the full subcategory $\mathbf{C}^{\star}_{\mathrm{str}}(X) \subseteq \mathbf{C}_{\mathrm{str}}(X)$. For the derived category we write $\mathbf{D}(X) := \mathbf{D}(\mathsf{Mod}\,\mathcal{O}_X)$. Inside it there are the full triangulated subcategories

$$\mathbf{D}_{\mathrm{c}}(X) := \mathbf{D}_{\mathsf{Coh}\,\mathcal{O}_X}(\mathsf{Mod}\,\mathcal{O}_X) \subseteq \mathbf{D}_{\mathrm{qc}}(X) := \mathbf{D}_{\mathsf{QCoh}\,\mathcal{O}_X}(\mathsf{Mod}\,\mathcal{O}_X).$$

One can show (see [62, Corollary II.7.19]) that the canonical functor $\mathbf{D}^+(\mathsf{QCoh}\,\mathcal{O}_X) \to \mathbf{D}^+_{\mathrm{qc}}(X)$ is an equivalence. The proof is based on the deep fact (see [62, Theorem II.7.18]) that $\mathsf{QCoh}\,\mathcal{O}_X$ has enough injectives relative to $\mathsf{Mod}\,\mathcal{O}_X$, in the sense of Definition 11.5.7. Then Theorem 11.5.8 says that every complex $\mathcal{M} \in \mathbf{D}^+_{\mathrm{qc}}(X)$ admits a quasi-isomorphism $\mathcal{M} \to \mathcal{I}$ in $\mathbf{C}_{\mathrm{str}}(X)$, where $\mathcal{I}$ is a bounded below complex of injective quasi-coherent $\mathcal{O}_X$-modules.

Then, using smart truncations and the fact that every quasi-coherent sheaf on X is the direct limit of its coherent subsheaves, one can show that the canonical functor $\mathbf{D}^{\mathrm{b}}(\mathsf{Coh}\,\mathcal{O}_X) \to \mathbf{D}^{\mathrm{b}}_{\mathrm{c}}(X)$ is an equivalence. See [70, Proposition 3.5]; when $X = \mathrm{Spec}(A)$ is affine see also Proposition 7.4.9 below. The category $\mathbf{D}^{\mathrm{b}}_{\mathrm{c}}(X)$ is the focus of a significant part of the research in modern algebraic geometry (e.g. birational geometry, see [70]).

A totally different approach yields this result: if X is separated and quasi-compact, then the canonical functor $\mathbf{D}(\mathsf{QCoh}\,\mathcal{O}_X) \to \mathbf{D}_{\mathrm{qc}}(X)$ is an equivalence. See [23, Corollary 5.5].

For a left noetherian ring A we write $\mathbf{D}_{\mathrm{f}}(\mathsf{Mod}\,A) := \mathbf{D}_{\mathsf{Mod}_{\mathrm{f}}\,A}(\mathsf{Mod}\,A)$.

Proposition 7.4.9 *Let A be a left noetherian ring and $\star \in \{-, \mathrm{b}\}$. Then the canonical functor $\mathbf{D}^{\star}(\mathsf{Mod}_{\mathrm{f}}\,A) \to \mathbf{D}_{\mathrm{f}}(\mathsf{Mod}\,A)^{\star}$ is an equivalence of triangulated categories.*

Proof Consider the functor $F : \mathbf{D}^-(\mathsf{Mod}_{\mathrm{f}}\,A) \to \mathbf{D}(\mathsf{Mod}\,A)$. Suppose $s : M \to L$ is a quasi-isomorphism in $\mathbf{K}(\mathsf{Mod}\,A)$, such that $L \in \mathbf{K}^-(\mathsf{Mod}_{\mathrm{f}}\,A)$. Then $M \in \mathbf{D}_{\mathrm{f}}(\mathsf{Mod}\,A)^-$. A bit later (in Theorem 11.4.40) we will prove that M admits a free resolution $P \to M$, where P is a bounded above complex of finitely generated free modules. Thus we get a quasi-isomorphism $t : P \to M$ with $P \in \mathbf{K}^-(\mathsf{Mod}_{\mathrm{f}}\,A)$. By Proposition 7.1.13 with condition (r) we conclude that F is fully faithful. This also shows that the essential image of F is $\mathbf{D}_{\mathrm{f}}(\mathsf{Mod}\,A)^-$.

Next consider the functor $G : \mathbf{D}^{\mathrm{b}}(\mathsf{Mod}_{\mathrm{f}}\,A) \to \mathbf{D}^-(\mathsf{Mod}_{\mathrm{f}}\,A)$. Suppose $s : L \to M$ is a quasi-isomorphism in $\mathbf{K}^-(\mathsf{Mod}_{\mathrm{f}}\,A)$ with $L \in \mathbf{K}^{\mathrm{b}}(\mathsf{Mod}_{\mathrm{f}}\,A)$. Say

$\mathrm{H}(L)$ is concentrated in the integer interval $[d_0, d_1]$. Then $t : M \to \mathrm{smt}^{\geq d_0}(M)$ is a quasi-isomorphism, and $\mathrm{smt}^{\geq d_0}(M) \in \mathsf{K}^{\mathrm{b}}(\mathsf{Mod}_{\mathrm{f}} A)$. By Proposition 7.1.13 with condition (l) we conclude that G is fully faithful. Therefore the composition $F \circ G : \mathsf{D}^{\mathrm{b}}(\mathsf{Mod}_{\mathrm{f}} A) \to \mathsf{D}(\mathsf{Mod}\, A)$ is fully faithful. Suitable truncations ($\mathrm{smt}^{\geq d_0}$ and $\mathrm{smt}^{\leq d_1}$) show that the essential image of $F \circ G$ is $\mathsf{D}_{\mathrm{f}}(\mathsf{Mod}\, A)^{\mathrm{b}}$. □

7.5 The Embedding of M in D(M)

Here again we only consider an abelian category M.

For $M, N \in \mathsf{M}$ there is no difference between the $\mathbb{K}$-modules $\mathrm{Hom}_{\mathsf{M}}(M, N)$, $\mathrm{Hom}_{\mathsf{C}(\mathsf{M})}(M, N)$ and $\mathrm{Hom}_{\mathsf{K}(\mathsf{M})}(M, N)$. Thus the canonical functors $\mathsf{M} \to \mathsf{C}(\mathsf{M})$ and $\mathsf{M} \to \mathsf{K}(\mathsf{M})$ are fully faithful. The same is true for $\mathsf{D}(\mathsf{M})$, but this requires a proof.

Let $\mathsf{D}(\mathsf{M})^0$ be the full subcategory of $\mathsf{D}(\mathsf{M})$ consisting of complexes whose cohomology is concentrated in degree 0. This is an additive subcategory of $\mathsf{D}(\mathsf{M})$.

Proposition 7.5.1 *The canonical functor* $\mathsf{M} \to \mathsf{D}(\mathsf{M})^0$ *is an equivalence.*

Proof Let's denote the canonical functor $\mathsf{M} \to \mathsf{D}(\mathsf{M})^0$ by F. Under the fully faithful embedding $\mathsf{M} \subseteq \mathsf{C}_{\mathrm{str}}(\mathsf{M})$, F is just the restriction of the functor $\tilde{\mathrm{Q}}$ from Definition 7.2.8.

The functor $\mathrm{H}^0 : \mathsf{D}(\mathsf{M}) \to \mathsf{M}$ satisfies $\mathrm{H}^0 \circ F = \mathrm{Id}_{\mathsf{M}}$. This implies that F is faithful.

Next we prove that F is full. Take any objects $M, N \in \mathsf{M}$ and a morphism $q : M \to N$ in $\mathsf{D}(\mathsf{M})$. By Proposition 7.2.10 we know that $q = \tilde{\mathrm{Q}}(a) \circ \tilde{\mathrm{Q}}(s)^{-1}$ for some morphisms $a : L \to N$ and $s : L \to M$ in $\mathsf{C}_{\mathrm{str}}(\mathsf{M})$, with s a quasi-isomorphism. Let $L' := \mathrm{smt}^{\leq 0}(L)$, the smart truncation of L. Since $L \in \mathsf{D}(\mathsf{M})^0$, the inclusion $u : L' \to L$ is a quasi-isomorphism in $\mathsf{C}_{\mathrm{str}}(\mathsf{M})$. Writing $a' := a \circ u$ and $s' := s \circ u$, we see that s' is a quasi-isomorphism, and $q = \tilde{\mathrm{Q}}(a') \circ \tilde{\mathrm{Q}}(s')^{-1}$.

Next let $L'' := \mathrm{smt}^{\geq 0}(L')$, the other smart truncation. The projection $v : L' \to L''$ is a surjective quasi-isomorphism in $\mathsf{C}_{\mathrm{str}}(\mathsf{M})$. Because L'' is a complex concentrated in degree 0, we can view it as an object of M. The morphisms a' and s' factor as $a' = a'' \circ v$ and $s' = s'' \circ v$, where $a'' : L'' \to N$ and $s'' : L'' \to M$ are morphisms in M. But s'' is a quasi-isomorphism in $\mathsf{C}_{\mathrm{str}}(\mathsf{M})$, and so it is actually an isomorphism in M. Therefore we have a morphism $a'' \circ (s'')^{-1} : M \to N$ in M, and

$$\tilde{\mathrm{Q}}(a'' \circ (s'')^{-1}) = \tilde{\mathrm{Q}}(a'') \circ \tilde{\mathrm{Q}}(s'')^{-1} = \tilde{\mathrm{Q}}(a') \circ \tilde{\mathrm{Q}}(s')^{-1} = q.$$

Finally we have to prove that every $L \in \mathbf{D}(\mathsf{M})^0$ is isomorphic, in $\mathbf{D}(\mathsf{M})$, to a complex L'' that's concentrated in degree 0. But we already showed it in the previous paragraphs. □

Proposition 7.5.2 *Let* M *be an abelian category and let* $0 \to L \xrightarrow{\phi} M \xrightarrow{\psi} N \to 0$ *be a diagram in* M. *The following conditions are equivalent*:

(i) *The diagram is an exact sequence.*

(ii) *There is a distinguished triangle* $L \xrightarrow{\tilde{\mathrm{Q}}(\phi)} M \xrightarrow{\tilde{\mathrm{Q}}(\psi)} N \xrightarrow{\theta} \mathrm{T}(L)$ *in* $\mathbf{D}(\mathsf{M})$.

Exercise 7.5.3 Prove Proposition 7.5.2.

The last two propositions say that the abelian category M, with its kernels and cokernels, can be recovered from the triangulated category $\mathbf{D}(\mathsf{M})$.

Remark 7.5.4 Assume that the diagram in Proposition 7.5.2 is an exact sequence. By Proposition 5.3.10, this sequence is split if and only if the morphism θ is zero. Furthermore, if M has enough injectives, then there is a canonical bijection between the set of isomorphism classes of extensions of N by L, and the set $\mathrm{Hom}_{\mathbf{D}(\mathsf{M})}(N, \mathrm{T}(L))$. See [62, Section I.6], or use Exercise 12.2.7, which says that $\mathrm{Hom}_{\mathbf{D}(\mathsf{M})}(N, \mathrm{T}(L)) \cong \mathrm{Ext}^1_{\mathsf{M}}(N, L)$.

7.6 The Opposite Derived Category Is Triangulated

Here we deal with a DG $\mathbb{K}$-ring A and a $\mathbb{K}$-linear abelian category M. In Section 5.6 we put a canonical triangulated structure on the opposite homotopy category $\mathbf{K}(A, \mathsf{M})^{\mathrm{op}}$. This structure is such that the flip functor

$$\mathrm{Flip} : \mathbf{K}(A, \mathsf{M})^{\mathrm{op}} \to \mathbf{K}(A^{\mathrm{op}}, \mathsf{M}^{\mathrm{op}}) = \mathbf{K}(A, \mathsf{M})^{\mathrm{flip}}$$

is an isomorphism of triangulated categories. Here, unlike in Section 5.6, we are omitting the overline decoration from Flip. In the current section we push this triangulated structure to the opposite derived category $\mathbf{D}(A, \mathsf{M})^{\mathrm{op}}$.

But first let us present two types of full subcategories $\mathsf{K} \subseteq \mathbf{K}(A, \mathsf{M})$ that are triangulated, and also $\mathsf{K}^{\mathrm{op}} \subseteq \mathbf{K}(A, \mathsf{M})^{\mathrm{op}}$ is triangulated.

For any boundedness condition $\star$ (see Definition 7.3.1), we know that $\mathbf{K}^\star(A, \mathsf{M})$ is a full triangulated subcategory of $\mathbf{K}(A, \mathsf{M})$. The notation $\mathbf{K}^\star(A, \mathsf{M})^{\mathrm{op}}$ refers to the opposite category of $\mathbf{K}^\star(A, \mathsf{M})$; thus, $\mathbf{K}^\star(A, \mathsf{M})^{\mathrm{op}}$ is the full subcategory of $\mathbf{K}(A, \mathsf{M})^{\mathrm{op}}$ on the DG modules satisfying the boundedness condition $\star$.

Proposition 7.6.1 *For every boundedness condition* $\star$, *the subcategory* $\mathbf{K}^\star(A, \mathsf{M})^{\mathrm{op}}$ *is triangulated in* $\mathbf{K}(A, \mathsf{M})^{\mathrm{op}}$.

Proof Let's write $\mathsf{K} := \mathbf{K}^{\star}(A, \mathsf{M})$ and $\mathsf{K}^{\mathrm{op}} := \mathbf{K}^{\star}(A, \mathsf{M})^{\mathrm{op}}$. Define $\mathsf{K}^{\mathrm{flip}} := \mathrm{Flip}(\mathsf{K}^{\mathrm{op}}) \subseteq \mathbf{K}(A^{\mathrm{op}}, \mathsf{M}^{\mathrm{op}})$. According to Theorem 3.9.16(2), we know that $\mathsf{K}^{\mathrm{flip}} = \mathbf{K}^{-\star}(A^{\mathrm{op}}, \mathsf{M}^{\mathrm{op}})$, where $-\star$ is the reversed boundedness condition. Thus $\mathsf{K}^{\mathrm{flip}}$ is a full triangulated subcategory of $\mathbf{K}(A^{\mathrm{op}}, \mathsf{M}^{\mathrm{op}})$. By definition of the triangulated structure on $\mathbf{K}(A, \mathsf{M})^{\mathrm{op}}$ we conclude that K^{op} is triangulated. □

Let $\mathsf{N} \subseteq \mathsf{M}$ be a thick abelian subcategory. Then $\mathbf{K}_{\mathsf{N}}(\mathsf{M})$, see Definition 7.4.4, is a full triangulated subcategory of $\mathbf{K}(\mathsf{M})$. This is just like Proposition 7.4.5. The opposite category $\mathbf{K}_{\mathsf{N}}(\mathsf{M})^{\mathrm{op}}$, on the same set of objects, is a full additive subcategory of $\mathbf{K}(\mathsf{M})^{\mathrm{op}}$.

Proposition 7.6.2 *Let* $\mathsf{N} \subseteq \mathsf{M}$ *be a thick abelian subcategory. Then* $\mathbf{K}_{\mathsf{N}}(\mathsf{M})^{\mathrm{op}}$ *is a full triangulated subcategory of* $\mathbf{K}(\mathsf{M})^{\mathrm{op}}$.

Proof Let's write $\mathsf{K} := \mathbf{K}_{\mathsf{N}}(\mathsf{M})$. Then $\mathsf{K}^{\mathrm{flip}} := \mathrm{Flip}(\mathsf{K}^{\mathrm{op}}) \subseteq \mathbf{K}(\mathsf{M}^{\mathrm{op}})$ is, by Theorem 3.9.16(4), the category $\mathbf{K}_{\mathsf{N}^{\mathrm{op}}}(\mathsf{M}^{\mathrm{op}})$. Thus $\mathsf{K}^{\mathrm{flip}}$ is a triangulated subcategory. By definition of the triangulated structure on $\mathbf{K}(\mathsf{M})^{\mathrm{op}}$ we conclude that K^{op} is triangulated. □

Later, in Chapter 10, we will see that several *resolving subcategories* are also full triangulated subcategories of $\mathbf{K}(A, \mathsf{M})^{\mathrm{op}}$. These are relevant to Theorems 8.4.9 and 8.4.23 below.

Taking intersections, we get more full triangulated subcategories of $\mathbf{K}(A, \mathsf{M})^{\mathrm{op}}$.

Recall that $\mathbf{S}(A, \mathsf{M})$ is the set of quasi-isomorphisms in $\mathbf{K}(A, \mathsf{M})$. Let $\mathbf{S}(A, \mathsf{M})^{\mathrm{op}}$ be the set of quasi-isomorphisms in $\mathbf{K}(A, \mathsf{M})^{\mathrm{op}}$. Note that being a quasi-isomorphism has nothing to do with the triangulated structure. Since a morphism ψ in $\mathbf{K}(A, \mathsf{M})$ is a quasi-isomorphism iff $\mathrm{Op}(\psi)$ is a quasi-isomorphism in $\mathbf{K}(A, \mathsf{M})^{\mathrm{op}}$, we have a bijection $\mathrm{Op} : \mathbf{S}(A, \mathsf{M}) \xrightarrow{\simeq} \mathbf{S}(A, \mathsf{M})^{\mathrm{op}}$. We know that $\mathbf{S}(A, \mathsf{M})$ is a left and right denominator set in $\mathbf{K}(A, \mathsf{M})$, by Proposition 7.1.1. Therefore, by Proposition 6.2.4, $\mathbf{S}(A, \mathsf{M})^{\mathrm{op}}$ is left and right denominator set in $\mathbf{K}(A, \mathsf{M})^{\mathrm{op}}$. We need to know more:

Proposition 7.6.3 *Let* K^{op} *be a full triangulated subcategory of* $\mathbf{K}(A, \mathsf{M})^{\mathrm{op}}$*, and define* $\mathsf{S}^{\mathrm{op}} := \mathsf{K}^{\mathrm{op}} \cap \mathbf{S}(A, \mathsf{M})^{\mathrm{op}}$*, the set of quasi-isomorphisms in* K^{op}*. Then:*

(1) S^{op} *is a denominator set of cohomological origin in* K^{op}.
(2) *The localized category* $\mathsf{D}^{\mathrm{op}} := (\mathsf{K}^{\mathrm{op}})_{\mathsf{S}^{\mathrm{op}}}$ *has a unique triangulated structure s.t. the localization functor* $\mathrm{Q}^{\mathrm{op}} : \mathsf{K}^{\mathrm{op}} \to \mathsf{D}^{\mathrm{op}}$ *is a triangulated functor.*

Proof On the flip side, i.e. in $\mathbf{K}(A^{\mathrm{op}}, \mathsf{M}^{\mathrm{op}})$, we know that the set of quasi-isomorphisms $\mathbf{S}(A^{\mathrm{op}}, \mathsf{M}^{\mathrm{op}})$ is a denominator set of cohomological origin. But by

Theorem 3.9.16(4) we have $\mathbf{S}(A, \mathsf{M})^{\mathrm{op}} = \mathrm{Flip}^{-1}(\mathbf{S}(A^{\mathrm{op}}, \mathsf{M}^{\mathrm{op}}))$, so this too is a denominator set of cohomological origin. Now we can use Theorem 7.1.3. □

Observe that $\mathsf{D}^{\mathrm{op}} = (\mathsf{K}_{\mathsf{S}})^{\mathrm{op}}$, by Proposition 6.1.5. The situation is summarized by the following commutative diagram of functors:

$$\begin{array}{ccccc} \mathsf{C}_{\mathrm{str}}^{\mathrm{op}} & \xrightarrow{\mathrm{P}^{\mathrm{op}}} \!\!\!\!\twoheadrightarrow & \mathsf{K}^{\mathrm{op}} & \xrightarrow{\mathrm{Q}^{\mathrm{op}}} & \mathsf{D}^{\mathrm{op}} \\ {\scriptstyle \mathrm{Flip}}\downarrow{\scriptstyle \cong} & & {\scriptstyle \mathrm{Flip}}\downarrow{\scriptstyle \cong} & & {\scriptstyle \mathrm{Flip}}\downarrow{\scriptstyle \cong} \\ \mathsf{C}_{\mathrm{str}}^{\mathrm{flip}} & \xrightarrow{\mathrm{P}^{\mathrm{flip}}} \!\!\!\!\twoheadrightarrow & \mathsf{K}^{\mathrm{flip}} & \xrightarrow{\mathrm{Q}^{\mathrm{flip}}} & \mathsf{D}^{\mathrm{flip}} \end{array} \tag{7.6.4}$$

Here C is the full subcategory of $\mathbf{C}(A, \mathsf{M})$ on the objects of K, and $\mathsf{C}^{\mathrm{flip}} := \mathrm{Flip}(\mathsf{C}) \subseteq \mathbf{C}(A, \mathsf{M})^{\mathrm{flip}} = \mathbf{C}(A^{\mathrm{op}}, \mathsf{M}^{\mathrm{op}})$. All these functors are bijective on objects. The vertical ones are bijective on morphisms too. The functors marked $\twoheadrightarrow$ are surjective on morphisms (i.e. they are full). The first vertical arrow is an isomorphism of abelian categories, and the other two vertical arrows are isomorphisms of triangulated categories.

Warning: Given K^{op} like in Proposition 7.6.3, there is no reason for the canonical triangulated functor $\mathsf{D}^{\mathrm{op}} \to \mathbf{D}(A, \mathsf{M})^{\mathrm{op}}$ to be fully faithful; see Proposition 7.1.13 for sufficient conditions.

Definition 7.6.5 Let $\mathsf{K} \subseteq \mathbf{K}(A, \mathsf{M})$ be a full additive subcategory s.t. K^{op} is a triangulated subcategory of $\mathbf{K}(A, \mathsf{M})^{\mathrm{op}}$, and let $\mathsf{S} := \mathsf{K} \cap \mathbf{S}(A, \mathsf{M})$. The category $\mathsf{D}^{\mathrm{op}} := (\mathsf{K}^{\mathrm{op}})_{\mathsf{S}^{\mathrm{op}}}$ is given the triangulated structure from Proposition 7.6.3.

For $\mathsf{K} = \mathbf{K}(A, \mathsf{M})$ we get a triangulated structure on $\mathsf{D}^{\mathrm{op}} = \mathbf{D}(A, \mathsf{M})^{\mathrm{op}}$.

Definition 7.6.6 Let $\mathsf{K} \subseteq \mathbf{K}(A, \mathsf{M})$ be a full additive subcategory s.t. K^{op} is a triangulated subcategory of $\mathbf{K}(A, \mathsf{M})^{\mathrm{op}}$. Define $\mathsf{D} := \mathsf{K}_{\mathsf{S}}$ and $\mathsf{D}^{\mathrm{op}} := (\mathsf{K}^{\mathrm{op}})_{\mathsf{S}^{\mathrm{op}}}$, where $\mathsf{S} := \mathsf{K} \cap \mathbf{S}(A, \mathsf{M})$. Let E be some triangulated category. A *contravariant triangulated functor* from D to E is, by definition, a triangulate functor $F : \mathsf{D}^{\mathrm{op}} \to \mathsf{E}$, where D^{op} has the canonical triangulated structure from Definition 7.6.5.

8

Derived Functors

Suppose $F : \mathsf{K}(A, \mathsf{M}) \to \mathsf{E}$ is a triangulated functor. Here A is a DG ring, M is an abelian category, and E is a triangulated category. In this chapter we define the right and left derived functors $\mathrm{R}F, \mathrm{L}F : \mathsf{D}(A, \mathsf{M}) \to \mathsf{E}$ of F. These are also triangulated functors, satisfying certain universal properties. We shall prove the uniqueness of the derived functors and their existence under suitable assumptions.

The universal properties of the derived functors are best stated in 2-categorical language. This will be explained in the first two sections. In Section 8.3 we define derived functors in the abstract setting (as opposed to the triangulated setup) and prove the main results for them. These results will then be specialized to various settings: triangulated functors (Section 8.4), contravariant triangulated functors (Section 8.5) and triangulated bifunctors (Section 9.2).

8.1 2-Categorical Notation

In this chapter we are going to do a lot of work with morphisms of functors (i.e. natural transformations). The language and notation of ordinary category theory that we used so far is not adequate for this purpose. Therefore we will now introduce notation from the theory of *2-categories*. (We will not give a definition of a 2-category here, but it is basically the structure of Cat that is mentioned below.) For more details on 2-categories the reader can look at [102] or [171, Section 1].

Consider the set Cat of all categories. The set theoretical aspects are neglected, as explained in Section 1.1. (Briefly, the precise solution is this: Cat is the set of all U-categories; so Cat is a subset of a bigger Grothendieck universe, say V, and it is a V-category.)

The set Cat is the set of objects of a 2-category. This means that in Cat there are two kinds of morphisms: 1-*morphisms* between objects, and 2-*morphisms* between 1-morphisms. There are several kinds of compositions, and these have several properties. All this will be explained below.

Suppose $\mathsf{C}_0, \mathsf{C}_1, \ldots$ are categories, namely objects of Cat. The 1-morphisms between them are the functors. The notation is as usual: $F : \mathsf{C}_0 \to \mathsf{C}_1$ denotes a functor.

Suppose $F, G : \mathsf{C}_0 \to \mathsf{C}_1$ are functors. The 2-morphisms from F to G are the morphisms of functors (i.e. the natural transformations), and the notation is $\eta : F \Rightarrow G$. The double arrow is the distinguishing notation for 2-morphisms. When specializing to an object $M \in \mathsf{C}_0$ we revert to the single arrow notation, namely $\eta_M : F(M) \to G(M)$ is the corresponding morphism in C_1. The diagram depicting this is

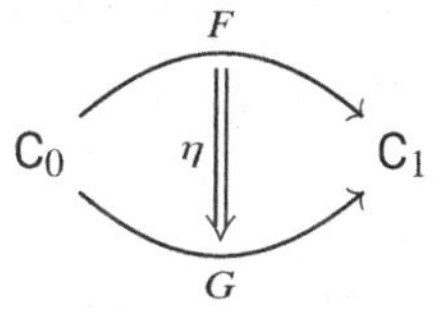

We shall refer to such a diagram as a 2-*diagram*.

Each object (category) C has its identity 1-morphism (functor) $\mathrm{Id}_{\mathsf{C}} : \mathsf{C} \to \mathsf{C}$. Each 1-morphism F has its identity 2-morphism (natural transformation) $\mathrm{id}_F : F \Rightarrow F$.

Now we consider compositions. For functors there is nothing new: given functors $F_1 : \mathsf{C}_0 \to \mathsf{C}_1$ and $F_2 : \mathsf{C}_1 \to \mathsf{C}_2$, their composition, which we now call *horizontal composition*, is the functor $F_2 \circ F_1 : \mathsf{C}_0 \to \mathsf{C}_2$. The diagram is

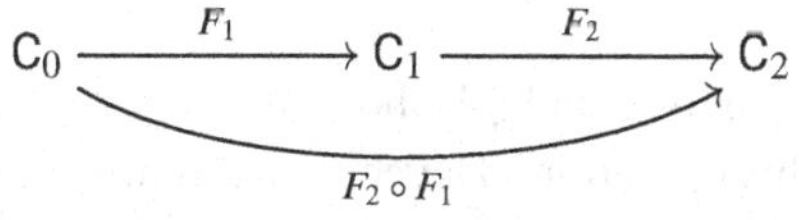

This can be viewed as a commutative 1-diagram, or as a shorthand for the 2-diagram

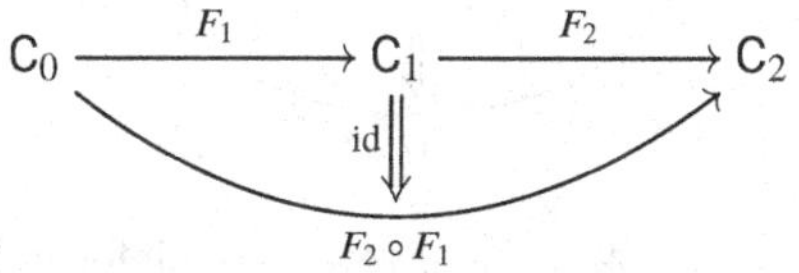

in which id is the identity 2-morphism of $F_2 \circ F_1$.

The complication begins with compositions of 2-morphisms. Suppose we are given 1-morphisms $F_i, G_i : \mathsf{C}_{i-1} \to \mathsf{C}_i$ and 2-morphisms $\eta_i : F_i \Rightarrow G_i$. In a diagram:

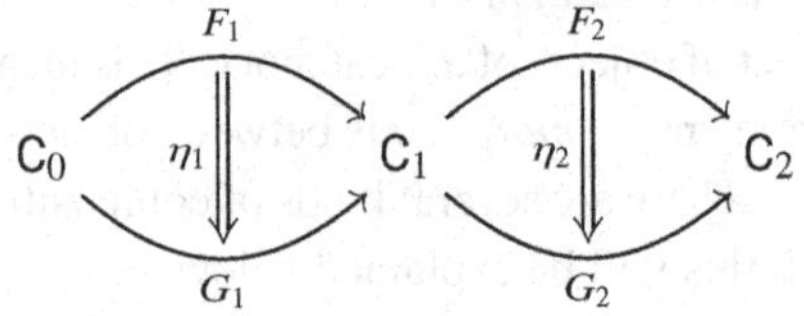

The *horizontal composition* is the morphism of functors $\eta_2 \circ \eta_1 : F_2 \circ F_1 \Rightarrow G_2 \circ G_1$. The diagram is

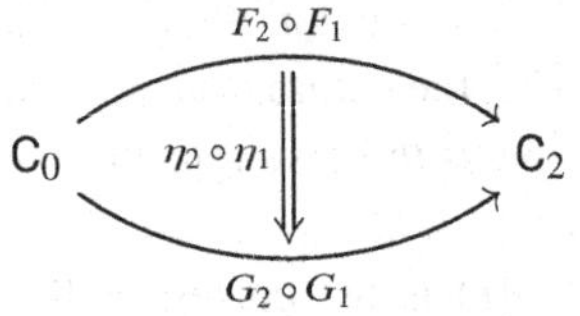

Exercise 8.1.1 For an object $M \in \mathsf{C}_0$, give an explicit formula for the morphism

$$(\eta_2 \circ \eta_1)_M : (F_2 \circ F_1)(M) \to (G_2 \circ G_1)(M)$$

in the category C_2.

Suppose we are given 1-morphisms $E, F, G : \mathsf{C}_0 \to \mathsf{C}_1$, and 2-morphisms $\zeta : E \Rightarrow F$ and $\eta : F \Rightarrow G$. The diagram depicting this is

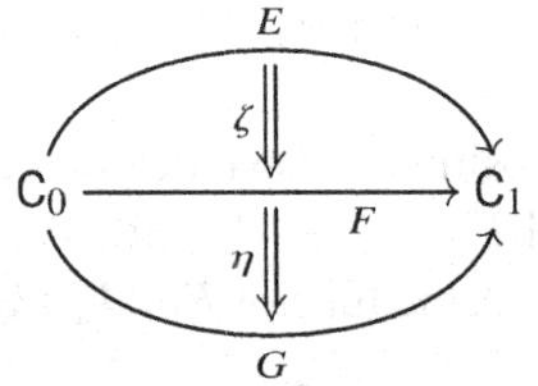

The *vertical composition* of ζ and η is the 2-morphism $\eta * \zeta : E \to G$. Notice the new symbol for this operation. The corresponding diagram is

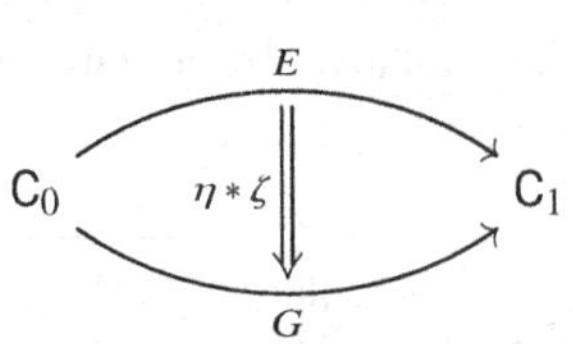

Exercise 8.1.2 For an object $M \in \mathsf{C}_0$, give an explicit formula for the morphism $(\eta * \zeta)_M : E(M) \to G(M)$ in the category C_1.

Something intricate occurs in the situation shown in the next diagram.

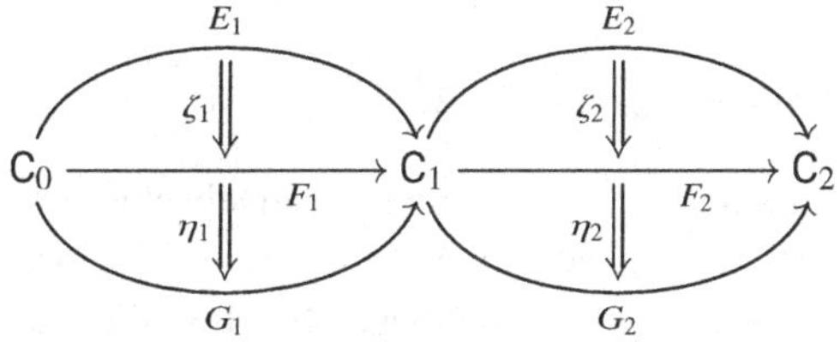

It turns out that

$$(\eta_2 * \zeta_2) \circ (\eta_1 * \zeta_1) = (\eta_2 \circ \eta_1) * (\zeta_2 \circ \zeta_1)$$

as morphisms $E_2 \circ E_1 \Rightarrow G_2 \circ G_1$. This is called the *exchange property*.

Exercise 8.1.3 Prove the exchange property.

Just like abstract categories, we can talk about triangulated categories. There is the 2-category TrCat of all ($\mathbb{K}$-linear) triangulated categories. The objects here are the triangulated categories (K, T); the 1-morphisms are the triangulated functors (F, τ); and the 2-morphisms are the morphisms of triangulated functors η. This is what we are going to use later in this chapter.

8.2 Functor Categories

In this section we isolate a part of 2-category theory. This simplifies the discussion greatly. All set theoretical issues (sizes of sets) are neglected. As before, this can be treated by introducing a bigger universe.

Definition 8.2.1 Given categories C and D, let $\mathsf{Fun}(\mathsf{C}, \mathsf{D})$ be the category whose objects are the functors $F : \mathsf{C} \to \mathsf{D}$. Given objects $F, G \in \mathsf{Fun}(\mathsf{C}, \mathsf{D})$, the morphisms $\eta : F \Rightarrow G$ in $\mathsf{Fun}(\mathsf{C}, \mathsf{D})$ are the morphisms of functors, i.e. the natural transformations.

For the case $\mathsf{D} = \mathsf{Set}$ this concept was already mentioned in Section 1.7.

In terms of the 2-category Cat from the previous section, C and D are objects of Cat; the objects of $\mathsf{Fun}(\mathsf{C}, \mathsf{D})$ are 1-morphisms in Cat; and the morphisms of $\mathsf{Fun}(\mathsf{C}, \mathsf{D})$ are 2-morphisms in Cat. Thus we made a "reduction of order," from 2 to 1, by passing from Cat to $\mathsf{Fun}(\mathsf{C}, \mathsf{D})$.

Suppose $G : \mathsf{C}' \to \mathsf{C}$ and $H : \mathsf{D} \to \mathsf{D}'$ are functors. There is an induced functor

$$\mathsf{Fun}(G, H) : \mathsf{Fun}(\mathsf{C}, \mathsf{D}) \to \mathsf{Fun}(\mathsf{C}', \mathsf{D}') \tag{8.2.2}$$

defined by $\mathsf{F}(G, H)(F) := H \circ F \circ G$.

Proposition 8.2.3 *If G and H are equivalences, then the functor* $\mathsf{Fun}(G, H)$ *in* (8.2.2) *is an equivalence.*

Exercise 8.2.4 Prove Proposition 8.2.3.

Recall that for a category C and a multiplicatively closed set of morphisms $\mathsf{S} \subseteq \mathsf{C}$ we denote by $\mathsf{C_S}$ the localization. It comes with the localization functor $\mathsf{Q} : \mathsf{C} \to \mathsf{C_S}$, which is the identity on objects. See Definition 6.1.2.

For a category E let $\mathsf{E}^{\times} \subseteq \mathsf{E}$ be the category of isomorphisms; it has all the objects, but its morphisms are just the isomorphisms in E.

Definition 8.2.5 Given categories C and E, a multiplicatively closed set of morphisms $\mathsf{S} \subseteq \mathsf{C}$, and a functor $F : \mathsf{C} \to \mathsf{E}$, we say that F *is localizable to* S if $F(\mathsf{S}) \subseteq \mathsf{E}^{\times}$. We denote by $\mathsf{Fun_S}(\mathsf{C}, \mathsf{E})$ the full subcategory of $\mathsf{Fun}(\mathsf{C}, \mathsf{E})$ on the functors that are localizable to S.

Here is a useful formulation of the universal property of localization of categories (see Definition 6.1.2). Recall that a functor is an isomorphism of categories iff it is an equivalence that is bijective on sets of objects.

Proposition 8.2.6 *Let* C *and* E *be categories, and let* $\mathsf{S} \subseteq \mathsf{C}$ *be a multiplicatively closed set of morphisms. Then the functor*

$$\mathsf{Fun}(\mathsf{Q}, \mathrm{Id}_{\mathsf{E}}) : \mathsf{Fun}(\mathsf{C_S}, \mathsf{E}) \to \mathsf{Fun_S}(\mathsf{C}, \mathsf{E})$$

is an isomorphism of categories.

Exercise 8.2.7 Prove Proposition 8.2.6.

By definition a bifunctor $F : \mathsf{C} \times \mathsf{D} \to \mathsf{E}$ is a functor from the product category $\mathsf{C} \times \mathsf{D}$. See Section 1.6. It will be useful to retain both meanings; so we shall write

$$\mathsf{BiFun}(\mathsf{C} \times \mathsf{D}, \mathsf{E}) := \mathsf{Fun}(\mathsf{C} \times \mathsf{D}, \mathsf{E}), \tag{8.2.8}$$

where in the first expression we recall that $\mathsf{C} \times \mathsf{D}$ is a product.

The next proposition describes bifunctors in a nonsymmetric fashion.

Proposition 8.2.9 *Let* C, D *and* E *be categories. There is an isomorphism of categories*

$$\Xi : \mathsf{Fun}(\mathsf{C} \times \mathsf{D}, \mathsf{E}) \to \mathsf{Fun}(\mathsf{C}, \mathsf{Fun}(\mathsf{D}, \mathsf{E}))$$

with the following formula: for a functor $F : \mathsf{C} \times \mathsf{D} \to \mathsf{E}$, *the functor*

$$\Xi(F) : \mathsf{C} \to \mathsf{Fun}(\mathsf{D}, \mathsf{E})$$

is $\Xi(F)(C) := F(C, -)$.

Exercise 8.2.10 Prove Proposition 8.2.9.

Proposition 8.2.11 *Let* C *and* D *be categories, and let* $\mathsf{S} \subseteq \mathsf{C}$ *and* $\mathsf{T} \subseteq \mathsf{D}$ *be multiplicatively closed sets of morphisms. Then* $\mathsf{S} \times \mathsf{T}$ *is a multiplicatively closed set of morphisms in* $\mathsf{C} \times \mathsf{D}$*, and the canonical functor*

$$\Theta : (\mathsf{C} \times \mathsf{D})_{\mathsf{S} \times \mathsf{T}} \to \mathsf{C}_{\mathsf{S}} \times \mathsf{D}_{\mathsf{T}}$$

is an isomorphism of categories.

Proof It is clear that $\mathsf{S} \times \mathsf{T} \subseteq \mathsf{C} \times \mathsf{D}$ is closed under compositions. Let's spell out how the functor Θ arises. The functor

$$\mathrm{Q}_{\mathsf{S} \times \mathsf{T}} : \mathsf{C} \times \mathsf{D} \to (\mathsf{C} \times \mathsf{D})_{\mathsf{S} \times \mathsf{T}}$$

is universal for functors $F : \mathsf{C} \times \mathsf{D} \to \mathsf{E}$ that invert $\mathsf{S} \times \mathsf{T}$; namely functors F such that $F(s, t) \in \mathsf{E}^{\times}$ for all $s \in \mathsf{S}$ and $t \in \mathsf{T}$. Since the functor

$$\mathrm{Q}_{\mathsf{S}} \times \mathrm{Q}_{\mathsf{T}} : \mathsf{C} \times \mathsf{D} \to \mathsf{C}_{\mathsf{S}} \times \mathsf{D}_{\mathsf{T}}$$

inverts $\mathsf{S} \times \mathsf{T}$, we get the functor Θ. We will prove that the functor $\mathrm{Q}_{\mathsf{S}} \times \mathrm{Q}_{\mathsf{T}}$ has the same universal property as $\mathrm{Q}_{\mathsf{S} \times \mathsf{T}}$; this will imply that Θ is an isomorphism.

Consider an arbitrary category E. Invoking Propositions 8.2.9 and 8.2.6 we get a commutative diagram

$$\begin{array}{ccccc}
\mathrm{Fun}(\mathsf{C}_{\mathsf{S}} \times \mathsf{D}_{\mathsf{T}}, \mathsf{E}) & \xrightarrow{\cong} & \mathrm{Fun}(\mathsf{C}_{\mathsf{S}}, \mathrm{Fun}(\mathsf{D}_{\mathsf{T}}, \mathsf{E})) & \xrightarrow{\cong} & \mathrm{Fun}_{\mathsf{S}}(\mathsf{C}, \mathrm{Fun}_{\mathsf{T}}(\mathsf{D}, \mathsf{E})) \\
\Big\downarrow{\scriptstyle \mathrm{Fun}(\mathrm{Q}_{\mathsf{S}} \times \mathrm{Q}_{\mathsf{T}}, \mathrm{Id}_{\mathsf{E}})} & & & & \Big\downarrow{\scriptstyle \text{f.f. emb.}} \\
\mathrm{Fun}(\mathsf{C} \times \mathsf{D}, \mathsf{E}) & & \xrightarrow[\cong]{\ \Xi\ } & & \mathrm{Fun}(\mathsf{C}, \mathrm{Fun}(\mathsf{D}, \mathsf{E}))
\end{array} \tag{8.2.12}$$

in which the right vertical arrow is a fully faithful embedding. Take any functor F in the bottom left corner of (8.2.12), and let $F' := \Xi(F)$, which lives in the bottom right corner. Now F' belongs to the top right corner of (8.2.12) iff $F'(s)(t) \in \mathsf{E}^{\times}$ for all $s \in \mathsf{S}$ and $t \in \mathsf{T}$. But $F'(s)(t) = F(s, t)$, so this happens iff F inverts $\mathsf{S} \times \mathsf{T}$. □

Denominator sets were introduced in Definition 6.2.14.

Proposition 8.2.13 *In the situation of Proposition* 8.2.11*, the following two conditions are equivalent*:

(i) *The multiplicatively closed sets* $\mathsf{S} \subseteq \mathsf{C}$ *and* $\mathsf{T} \subseteq \mathsf{D}$ *are left (resp. right) denominator sets.*
(ii) *The multiplicatively closed set* $\mathsf{S} \times \mathsf{T} \subseteq \mathsf{C} \times \mathsf{D}$ *is a left (resp. right) denominator set.*

Exercise 8.2.14 Prove Proposition 8.2.13.

8.3 Abstract Derived Functors

Here we deal with right and left derived functors in an abstract setup (as opposed to the triangulated setup).

Definition 8.3.1 (Right Derived Functor) Consider a category K and a multiplicatively closed set of morphisms $\mathsf{S} \subseteq \mathsf{K}$, with localization functor $\mathrm{Q} : \mathsf{K} \to \mathsf{K}_{\mathsf{S}}$. Let $F : \mathsf{K} \to \mathsf{E}$ be a functor. A *right derived functor of F with respect to* S is a pair $(\mathrm{R}F, \eta^{\mathrm{R}})$, where

$$\mathrm{R}F : \mathsf{K}_{\mathsf{S}} \to \mathsf{E}$$

is a functor, and

$$\eta^{\mathrm{R}} : F \Rightarrow \mathrm{R}F \circ \mathrm{Q}$$

is a morphism of functors, such that the following universal property holds:

(R) Given any pair (G, θ), consisting of a functor $G : \mathsf{K}_{\mathsf{S}} \to \mathsf{E}$ and a morphism of functors $\theta : F \Rightarrow G \circ \mathrm{Q}$, there is a unique morphism of functors $\mu : \mathrm{R}F \Rightarrow G$ such that $\theta = (\mu \circ \mathrm{id}_{\mathrm{Q}}) * \eta^{\mathrm{R}}$.

Pictorially: there is a 2-diagram

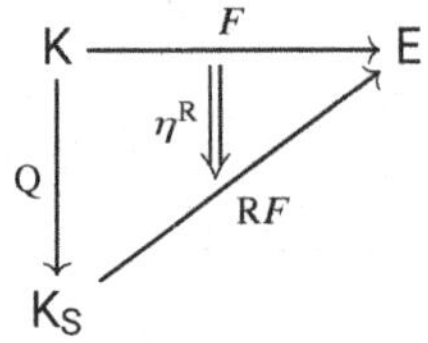

For any other pair (G, θ) there is a unique morphism μ that sits in this 2-diagram

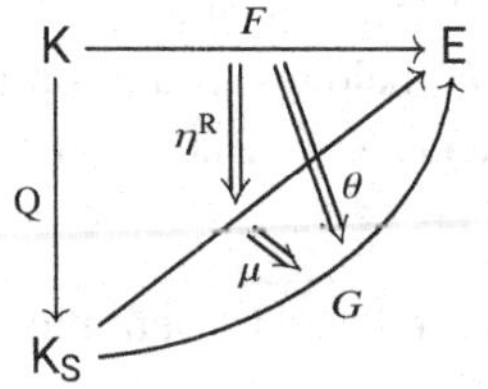

such that the diagram of 2-morphisms (with $*$ composition)

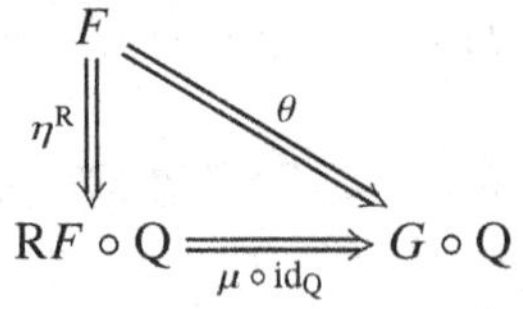

is commutative.

Proposition 8.3.2 *If a right derived functor* $(\mathrm{R}F, \eta^{\mathrm{R}})$ *exists, then it is unique, up to a unique isomorphism. Namely, if* (G, θ) *is another right derived functor of* F*, then there is a unique isomorphism of functors* $\mu : \mathrm{R}F \stackrel{\simeq}{\Rightarrow} G$ *such that* $\theta = (\mu \circ \mathrm{id}_{\mathrm{Q}}) * \eta^{\mathrm{R}}$.

Proof Despite the apparent complication of the situation, the usual argument for uniqueness of universals applies. To be explicit, given another right derived functor (G, θ), let $\mu : \mathrm{R}F \Rightarrow G$ be the unique morphism that's guaranteed by property (R) of the pair $(\mathrm{R}F, \eta^{\mathrm{R}})$. Then let $\nu : G \Rightarrow \mathrm{R}F$ be the unique morphism that's guaranteed by property (R) of the pair (G, θ). Then the morphisms $\mathrm{id}_{\mathrm{R}F}, \nu * \mu : \mathrm{R}F \Rightarrow \mathrm{R}F$ satisfy

$$(\mathrm{id}_{\mathrm{R}F} \circ \mathrm{id}_{\mathrm{Q}}) * \mu = \mu = ((\nu * \mu) \circ \mathrm{id}_{\mathrm{Q}}) * \mu.$$

The uniqueness in property (R) of the pair $(\mathrm{R}F, \eta^{\mathrm{R}})$ implies that $\mathrm{id}_{\mathrm{R}F} = \nu * \mu$. Likewise, $\mathrm{id}_G = \mu * \nu$. Hence μ is an isomorphism of functors. □

Here is a rather general existence result. Left denominator sets were introduced in Definition 6.2.16.

Theorem 8.3.3 (Existence of Right Derived Functor) *In the situation of Definition* 8.3.1, *assume there is a full subcategory* $\mathsf{J} \subseteq \mathsf{K}$ *such the following three conditions hold*:

(a) *The multiplicatively closed set* S *is a left denominator set in* K.
(b) *For every object* $M \in \mathsf{K}$ *there is a morphism* $\rho : M \to I$ *in* S, *with target* $I \in \mathsf{J}$.
(c) *If* ψ *is a morphism in* $\mathsf{S} \cap \mathsf{J}$, *then* $F(\psi)$ *is an isomorphism in* E.

Then the right derived functor

$$(\mathrm{R}F, \eta^{\mathrm{R}}) : \mathsf{K}_{\mathsf{S}} \to \mathsf{E}$$

exists. Moreover, for every object $I \in \mathsf{J}$ *the morphism*

$$\eta^{\mathrm{R}}_I : F(I) \to \mathrm{R}F(I)$$

in E *is an isomorphism.*

Condition (b) says that K *has enough right* J*-resolutions.* Condition (c) says that J *is an* F*-acyclic category.*

Theorem 8.3.3 is [76, Proposition 7.3.2]. However their notation is different: what we call "left denominator set," they call "right multiplicative system."

We need a definition and a few lemmas before giving the proof of the theorem. In them we assume the situation of the theorem.

Definition 8.3.4 In the situation of Theorem 8.3.3, by a *system of right* J-*resolutions* we mean a pair (I, ρ), where $I : \mathrm{Ob}(\mathsf{K}) \to \mathrm{Ob}(\mathsf{J})$ is a function, and $\rho = \{\rho_M\}_{M \in \mathrm{Ob}(\mathsf{K})}$ is a collection of morphisms $\rho_M : M \to I(M)$ in S. Moreover, if $M \in \mathrm{Ob}(\mathsf{J})$, then $I(M) = M$ and $\rho_M = \mathrm{id}_M$.

Since here K has enough right J-resolutions, it follows that systems of right J-resolutions (I, ρ) exist.

Let us introduce some new notation that will make the proofs more readable:

$$\mathsf{K}' := \mathsf{J}, \quad \mathsf{S}' := \mathsf{J} \cap \mathsf{S}, \quad \mathsf{D} := \mathsf{K}_{\mathsf{S}} \quad \text{and} \quad \mathsf{D}' := \mathsf{K}'_{\mathsf{S}'}. \tag{8.3.5}$$

The inclusion functor is $U : \mathsf{K}' \to \mathsf{K}$, and its localization is $V : \mathsf{D}' \to \mathsf{D}$. These sit in a commutative diagram

$$\begin{array}{ccc} \mathsf{K}' & \xrightarrow{U} & \mathsf{K} \\ {\scriptstyle Q'}\downarrow & & \downarrow{\scriptstyle Q} \\ \mathsf{D}' & \xrightarrow{V} & \mathsf{D} \end{array} \tag{8.3.6}$$

Lemma 8.3.7 *The multiplicatively closed set* S' *is a left denominator set in* K'.

Proof We need to verify conditions (LD1) and (LD2) in Definition 6.2.16.

(LD1): Given morphisms $a' : L' \to N'$ in K' and $s' : L' \to M'$ in S', we must find morphisms $b' : M' \to K'$ in K' and $t' : N' \to K'$ in S', such that $t' \circ a' = b' \circ s'$. Because $\mathsf{S} \subseteq \mathsf{K}$ satisfies this condition, we can find morphisms $b : M' \to K$ in K and $t : N' \to K$ in S such that $t \circ a' = b \circ s'$. There is a morphism $\rho : K \to K'$ in S with target $K' \in \mathsf{K}'$. Then the morphisms $t' := \rho \circ t$ and $b' := \rho \circ b$ satisfy $t' \circ a' = b' \circ s'$, and $t' \in \mathsf{S}'$.

(LD2): Given morphisms $a', b' : M' \to N'$ in K' and $s' : L' \to M'$ in S', which satisfy $a' \circ s' = b' \circ s'$, we must find a morphism $t' : N' \to K'$ in S' such that $t' \circ a' = t' \circ b'$. Because $\mathsf{S} \subseteq \mathsf{K}$ satisfies this condition, we can find a morphism $t : N' \to K$ in S such that $t \circ a' = t \circ b'$. There is a morphism $\rho : K \to K'$ in S with target $K' \in \mathsf{K}'$. Then the morphism $t' := \rho \circ t$ has the required property. □

Lemma 8.3.8 *The functor* $V : \mathsf{D}' \to \mathsf{D}$ *is an equivalence.*

Proof Condition (b) of the theorem implies that V is essentially surjective on objects. We need to prove that V is fully faithful. We shall use the left version of Proposition 6.2.34, namely S and S' are left denominator sets, and in condition (ii) the morphisms are $s : L' \to M$ and $t : M \to K'$. (See Remark 6.2.3 and Propositions 6.1.5 and 6.2.17 regarding side changes.) By Lemma 8.3.7 the left version of condition (i) of Proposition 6.2.34 holds. Condition (b) of the theorem implies the left version of condition (ii) of Proposition 6.2.34. Then the left version of Proposition 6.2.34 says that V is fully faithful. □

Lemma 8.3.9 *Suppose a system of right* K'-*resolutions* (I, ρ) *has been chosen. Then the function* $I : \mathrm{Ob}(K) \to \mathrm{Ob}(K')$ *extends uniquely to a functor* $I : D \to D'$, *such that* $I \circ V = \mathrm{Id}_{D'}$, *and* $Q(\rho) : \mathrm{Id}_D \Rightarrow V \circ I$ *is an isomorphism of functors. Therefore the functor* I *is a a quasi-inverse of* V.

The relevant 2-diagram is this:

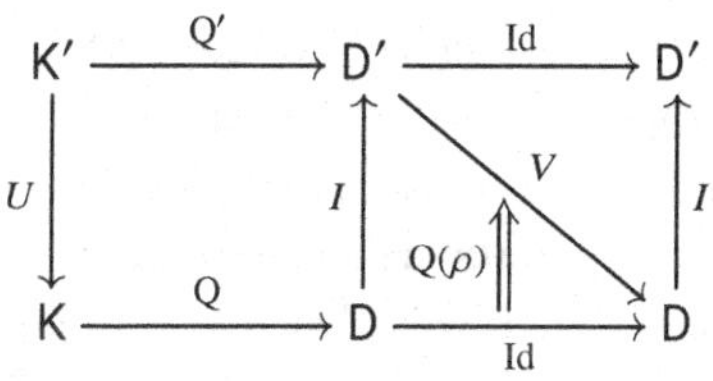

Recall that in a 2-diagram, an empty polygon means it is commutative, namely it can be filled with $\overset{\mathrm{id}}{\Longrightarrow}$.

Proof Consider a morphism $\psi : M \to N$ in D. Since $V : D' \to D$ is an equivalence, and since $V(I(M)) = I(M)$ and $V(I(N)) = I(N)$, there is a unique morphism $I(\psi) : I(M) \to I(N)$ in D' satisfying

$$V(I(\psi)) := Q(\rho_N) \circ \psi \circ Q(\rho_M)^{-1}. \tag{8.3.10}$$

in D.

Let us check that $I : D \to D'$ is really a functor. Suppose $\phi : L \to M$ and $\psi : M \to N$ are morphisms in D. Then

$$V(I(\psi) \circ I(\phi)) = V(I(\psi)) \circ V(I(\phi))$$
$$= (Q(\rho_N) \circ \psi \circ Q(\rho_M)^{-1}) \circ (Q(\rho_M) \circ \phi \circ Q(\rho_L)^{-1})$$
$$- Q(\rho_N) \circ (\psi \circ \phi) \circ Q(\rho_L)^{-1} = V(I(\psi \circ \phi)).$$

Since V is an equivalence, it follows that $I(\psi) \circ I(\phi) = I(\psi \circ \phi)$.

Because $\rho_{M'} : M' \to I(M')$ is the identity for any object $M' \in K'$, we see that there is equality $I \circ V = \mathrm{Id}_{D'}$. By the defining formula (8.3.10) of $I(\psi)$ we have a commutative diagram

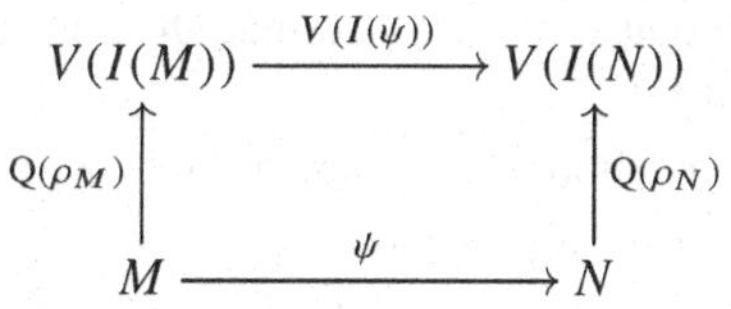

$$
\begin{array}{ccc}
V(I(M)) & \xrightarrow{\ \ V(I(\psi))\ \ } & V(I(N)) \\
{\scriptstyle Q(\rho_M)}\Big\uparrow & & \Big\uparrow{\scriptstyle Q(\rho_N)} \\
M & \xrightarrow{\ \ \psi\ \ } & N
\end{array}
$$

in D. Hence $Q(\rho) : \mathrm{Id}_D \Rightarrow V \circ I$ is an isomorphism of functors. □

Diagram (8.3.6) induces a commutative diagram of categories:

$$\begin{array}{ccc} \mathsf{Fun}(\mathsf{K}', \mathsf{E}) & \xleftarrow{\mathsf{Fun}(U,\mathrm{Id})} & \mathsf{Fun}(\mathsf{K}, \mathsf{E}) \\ \uparrow{\scriptstyle \text{f.f. emb}} & & {\scriptstyle \text{f.f. emb}}\uparrow \\ \mathsf{Fun}_{\mathsf{S}'}(\mathsf{K}', \mathsf{E}) & \xleftarrow[\text{equiv}]{\mathsf{Fun}(U,\mathrm{Id})} & \mathsf{Fun}_{\mathsf{S}}(\mathsf{K}, \mathsf{E}) \\ {\scriptstyle \mathsf{Fun}(\mathsf{Q}',\mathrm{Id})}\uparrow{\scriptstyle \text{isom}} & & {\scriptstyle \text{isom}}\uparrow{\scriptstyle \mathsf{Fun}(\mathsf{Q},\mathrm{Id})} \\ \mathsf{Fun}(\mathsf{D}', \mathsf{E}) & \xleftarrow[\text{equiv}]{\mathsf{Fun}(V,\mathrm{Id})} & \mathsf{Fun}(\mathsf{D}, \mathsf{E}) \end{array} \tag{8.3.11}$$

The vertical arrows marked "f.f. emb" are fully faithful embeddings by definition. According to Proposition 8.2.6, the vertical arrows marked "isom" are isomorphisms of categories. And by Lemma 8.3.8 and Proposition 8.2.3 the arrow $\mathsf{Fun}(V, \mathrm{Id})$ in the bottom row is an equivalence. As a consequence, the arrow $\mathsf{Fun}(U, \mathrm{Id})$ in the middle row is also an equivalence.

Let's introduce the notation $F' := F \circ U : \mathsf{K}' \to \mathsf{E}$. This functor is an object of the category in the middle left of diagram (8.3.11), since, by condition (c) of Theorem 8.3.3, it inverts S'.

Lemma 8.3.12 *Let $G : \mathsf{D} \to \mathsf{E}$ be a functor. Given a morphism $\theta' : F' \Rightarrow G \circ \mathsf{Q} \circ U$ of functors $\mathsf{K}' \to \mathsf{E}$, there is a unique morphism $\theta : F \Rightarrow G \circ \mathsf{Q}$ of functors $\mathsf{K} \to \mathsf{E}$ s.t. $\theta' = \theta \circ \mathrm{id}_U$.*

Note that G is an object in the category in the bottom right of diagram (8.3.11). The morphisms θ' and θ are in the middle left and top right, respectively, of this diagram.

Proof For every object $M \in \mathsf{K}$ there is a morphism $\rho : M \to I$ in S with target $I \in \mathsf{K}'$. A morphism of functors θ satisfying $\theta' = \theta \circ \mathrm{id}_U$ will make this diagram commutative:

$$\begin{array}{ccc} F(M) & \xrightarrow{\theta_M} & (G \circ \mathsf{Q})(M) \\ {\scriptstyle F(\rho)}\downarrow & & \cong\downarrow{\scriptstyle (G\circ\mathsf{Q})(\rho)} \\ F(I) & \xrightarrow{\theta'_I} & (G \circ \mathsf{Q})(I) \end{array} \tag{8.3.13}$$

We are using the facts that $I = U(I)$ and that $\mathsf{Q}(\rho)$ is an isomorphism. This proves the uniqueness of θ.

For existence, let us choose a system of right K'-resolutions (I, ρ), and define $\theta = \{\theta_M\}$ using (8.3.13), namely

$$\theta_M := (G \circ \mathsf{Q})(\rho_M)^{-1} \circ \theta'_{I(M)} \circ F(\rho_M). \tag{8.3.14}$$

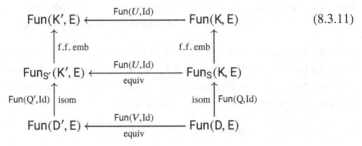

We must prove that this is indeed a morphism of functors $\mathsf{K} \to \mathsf{E}$. Namely, for a given morphism $\phi : M \to N$ in K, we have to prove that the diagram

$$\begin{array}{ccc} F(M) & \xrightarrow{\theta_M} & (G \circ \mathrm{Q})(M) \\ {\scriptstyle F(\phi)}\downarrow & & \downarrow{\scriptstyle (G\circ\mathrm{Q})(\phi)} \\ F(N) & \xrightarrow{\theta_N} & (G \circ \mathrm{Q})(N) \end{array} \tag{8.3.15}$$

in E is commutative.

Lemma 8.3.7 tells us that the morphism $I(\mathrm{Q}(\phi))$ in D' can be written as a left fraction

$$I(\mathrm{Q}(\phi)) = \mathrm{Q}'(\psi_1)^{-1} \circ \mathrm{Q}'(\psi_0) \tag{8.3.16}$$

of morphisms $\psi_0 \in \mathsf{K}'$ and $\psi_1 \in \mathsf{S}'$. We get these diagrams:

$$\begin{array}{ccc} M & \xrightarrow{\phi} & N \\ {\scriptstyle \rho_M}\downarrow & & \downarrow{\scriptstyle \rho_N} \\ I(M) & & I(N) \\ & {\scriptstyle \psi_0}\searrow \quad \swarrow{\scriptstyle \psi_1} & \\ & J & \end{array} \qquad \begin{array}{ccc} M & \xrightarrow{\mathrm{Q}(\phi)} & N \\ {\scriptstyle \mathrm{Q}(\rho_M)}\downarrow & & \downarrow{\scriptstyle \mathrm{Q}(\rho_N)} \\ I(M) & \xrightarrow{V(I(\mathrm{Q}(\phi)))} & I(N) \\ & {\scriptstyle \mathrm{Q}(\psi_0)}\searrow \quad \swarrow{\scriptstyle \mathrm{Q}(\psi_1)} & \\ & J & \end{array} \tag{8.3.17}$$

The first diagram is in the category K, and it might fail to be commutative. The second diagram is in the category D, and it is commutative: the bottom triangle commutes by formula (8.3.16) and the equality $V \circ \mathrm{Q}' = \mathrm{Q}$; and the top square commutes by formula (8.3.10). By condition (LO4) of the left Ore localization $\mathrm{Q} : \mathsf{K} \to \mathsf{D}$, there is a morphism $\psi : J \to L$ in S such that $\psi \circ \psi_0 \circ \rho_M = \psi \circ \psi_1 \circ \rho_N \circ \phi$ in K. There is the morphism $\rho_L : L \to I(L)$ in S, whose target $I(L)$ belongs to K'. Thus, after replacing the object J with $I(L)$, the morphism ψ_0 by $\rho_L \circ \psi \circ \psi_0$, and the morphism ψ_1 by $\rho_L \circ \psi \circ \psi_1$, and noting that the latter is a morphism in S', we can now assume that the first diagram in (8.3.17) commutative too.

Now we embed (8.3.15) in the bigger diagram (8.3.18) in E. Since $\rho_N \in \mathsf{S}$ and $\psi_1 \in \mathsf{S}'$, the morphisms $(G \circ \mathrm{Q})(\rho_N)$, $(G \circ \mathrm{Q})(\psi_1)$ and $F(\psi_1)$ are isomorphisms. The top and bottom squares in (8.3.18) are commutative by the definition of θ_M and θ_N, see formula (8.3.14). The left and right rounded shapes (those involving J) are commutative because the first diagram in (8.3.17) is commutative.

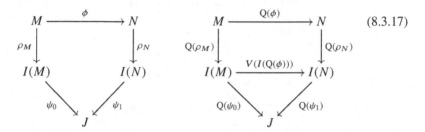

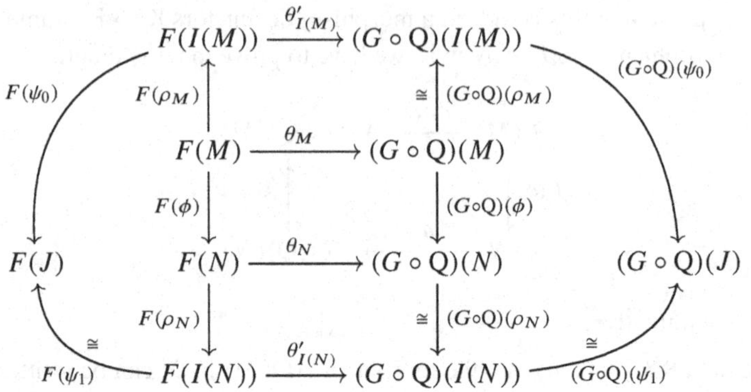

(8.3.18)

Since $\theta' : F' \Rightarrow G{\circ}Q \circ U$ is a morphism of functors, we have a commutative diagram

$$\begin{array}{ccc} F(I(M)) & \xrightarrow{\theta'_{I(M)}} & (G \circ Q)(I(M)) \\ {\scriptstyle F(\psi_0)}\downarrow & & \downarrow{\scriptstyle (G\circ Q)(\psi_0)} \\ F(J) & \xrightarrow{\theta'_J} & (G \circ Q)(J) \\ {\scriptstyle F(\psi_1)}\uparrow & & \uparrow{\scriptstyle (G\circ Q)(\psi_1)} \\ F(I(N)) & \xrightarrow{\theta'_{I(N)}} & (G \circ Q)(I(N)) \end{array} \tag{8.3.19}$$

When we erase the arrow θ'_J from diagram (8.3.19), we obtain the outer boundary of diagram (8.3.18). Therefore diagram (8.3.18) is commutative. In particular, the middle square of diagram (8.3.18) is commutative, and it is precisely diagram (8.3.15). □

Proof of Theorem 8.3.3 The proof is divided into four steps.

Step 1. Recall that the functor $F' = F \circ U$ lives in the middle left term in diagram (8.3.11). Because the arrow $\mathsf{Fun}(Q', \mathrm{Id})$ is an isomorphism, there is a unique functor $\mathrm{R}F'$ living in the bottom left term of diagram (8.3.11) that satisfies $\mathrm{R}F' \circ Q' = F'$. See next commutative diagram.

$$\begin{array}{ccc} \mathsf{K}' & \xrightarrow{F'} & \mathsf{E} \\ {\scriptstyle Q'}\downarrow & \nearrow{\scriptstyle \mathrm{R}F'} & \\ \mathsf{D}' & & \end{array} \tag{8.3.20}$$

Let $\eta' := \mathrm{id}_{F'}$. We claim that the pair $(\mathrm{R}F', \eta')$ is a right derived functor of F'. Indeed, suppose we are given a pair (G', θ'), where G' is a functor in the bottom left corner of diagram (8.3.11), and $\theta' : F' \Rightarrow G' \circ \mathrm{Q}'$ is a morphism in the top left corner of that diagram. See the 2-diagram (8.3.22). Because the function

$$\mathrm{Hom}_{\mathsf{Fun}(\mathsf{D}',\mathsf{E})}(\mathrm{R}F', G') \to \mathrm{Hom}_{\mathsf{Fun}(\mathsf{K}',\mathsf{E})}(F', G' \circ \mathrm{Q}') \tag{8.3.21}$$

is bijective – this is the left edge of diagram (8.3.11) – there is a unique morphism $\mu' : \mathrm{R}F' \Rightarrow G'$ that goes to θ' under (8.3.21).

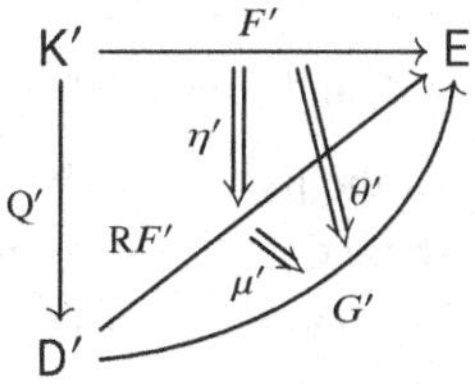

(8.3.22)

Step 2. Now we choose a system of right K'-resolutions (I, ρ), in the sense of Definition 8.3.4. By Lemma 8.3.9 we get an equivalence of categories $I : \mathsf{D} \to \mathsf{D}'$ that is a quasi-inverse to V, and an isomorphism of functors $\mathrm{Q}(\rho) : \mathrm{Id}_{\mathsf{D}} \xrightarrow{\simeq} V \circ I$. See the following 2-diagram (the solid arrows).

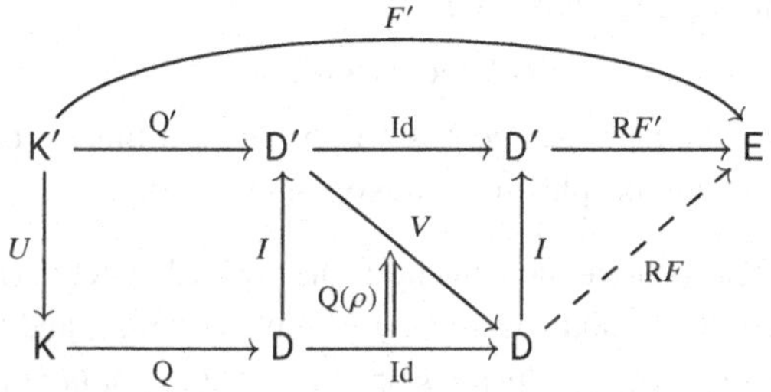

(8.3.23)

Define the functor

$$\mathrm{R}F := \mathrm{R}F' \circ I : \mathsf{D} \to \mathsf{E}. \tag{8.3.24}$$

It is the dashed arrow in diagram (8.3.23). So the functor $\mathrm{R}F$ lives in the bottom right corner of (8.3.11), and $\mathrm{R}F' = \mathrm{R}F \circ V$.

Step 3. Consider Lemma 8.3.12, with the functor $G := \mathrm{R}F$, and the morphism of functors

$$\eta' = \mathrm{id}_{F'} : F' \Rightarrow \mathrm{R}F' \circ \mathrm{Q}' = \mathrm{R}F \circ \mathrm{Q} \circ U.$$

The lemma says that there is a unique morphism of functors $\eta^{\mathrm{R}} : F \Rightarrow \mathrm{R}F \circ \mathrm{Q}$ s.t. $\eta^{\mathrm{R}} \circ \mathrm{id}_U = \eta'$.

We can give an explicit formula for the morphism of functors η^{R}. Take an object $M \in \mathsf{K}$. Then the morphism $\eta^{\mathrm{R}}_M : F(M) \to \mathrm{R}F(M) = F(I(M))$ in E is nothing but

$$\eta^{\mathrm{R}}_M := F(\rho_M). \tag{8.3.25}$$

Here is the calculation. By formula (8.3.14) we have

$$\eta^{\mathrm{R}}_M = (\mathrm{R}F \circ \mathrm{Q})(\rho_M)^{-1} \circ \eta'_{I(M)} \circ F(\rho_M).$$

But $\eta'_{I(M)} = \mathrm{id}_{F(I(M))}$,

$$I(\mathrm{Q}(\rho_M)) = \mathrm{Q}(\rho_{I(M)}) \circ \mathrm{Q}(\rho_M) \circ \mathrm{Q}(\rho_M)^{-1} = \mathrm{id}_{I(M)}$$

by (8.3.10), and

$$(\mathrm{R}F \circ \mathrm{Q})(\rho_M) = (\mathrm{R}F' \circ I \circ \mathrm{Q})(\rho_M) = \mathrm{R}F'(\mathrm{id}_{I(M)}) = \mathrm{id}_{F(I(M))}\,.$$

So everything else gets canceled, and we are left with (8.3.25).

Step 4. It remains to prove that the pair $(\mathrm{R}F, \eta^{\mathrm{R}})$ is a right derived functor of F. Suppose (G, θ) is a pair, where G is a functor in the category in the bottom right corner of diagram (8.3.11), and $\theta : F \Rightarrow G \circ \mathrm{Q}$ is a morphism in the top right corner of the diagram. We are looking for a morphism $\mu : \mathrm{R}F \Rightarrow G$ in the bottom right category in diagram (8.3.11) for which $\theta = (\mu \circ \mathrm{id}_{\mathrm{Q}}) * \eta^{\mathrm{R}}$. Let $G' := G \circ V$, and let $\theta' : F' \Rightarrow G' \circ \mathrm{Q}'$ be the morphism in the top left corner of (8.3.11) corresponding to θ. Because of the equivalence $\mathsf{Fun}(V, \mathrm{Id})$, finding such μ is the same as finding a morphism $\mu' : \mathrm{R}F' \Rightarrow G'$ in the bottom left category in diagram (8.3.11), satisfying

$$\theta' = (\mu' \circ \mathrm{id}_{\mathrm{Q}'}) * \eta'. \tag{8.3.26}$$

Finally, by step 1 the pair $(\mathrm{R}F', \eta')$ is a right derived functor of F'. This says that there is a unique morphism μ' satisfying (8.3.26). □

Definition 8.3.27 The construction of the right derived functor $(\mathrm{R}F, \eta^{\mathrm{R}})$ in the proof of Theorem 8.3.3, and specifically formulas (8.3.24) and (8.3.25), is called a *presentation of* $(\mathrm{R}F, \eta^{\mathrm{R}})$ *by the system of right* J*-resolutions* (I, ρ).

Of course, any other right derived functor of F (perhaps presented by another system of right J-resolutions) is uniquely isomorphic to $(\mathrm{R}F, \eta^{\mathrm{R}})$. This is according to Proposition 8.4.8. In Chapters 10 and 11 we shall give several existence results for right resolutions by suitable acyclic objects.

Now to left derived functors.

Definition 8.3.28 (Left Derived Functor) Consider a category K and a multiplicatively closed set of morphisms $\mathsf{S} \subseteq \mathsf{K}$, with localization functor $\mathrm{Q} : \mathsf{K} \to \mathsf{K}_{\mathsf{S}}$. Let $F : \mathsf{K} \to \mathsf{E}$ be a functor. A *left derived functor of F with respect to* S is a pair $(\mathrm{L}F, \eta^{\mathrm{L}})$, where

$$\mathrm{L}F : \mathsf{K}_{\mathsf{S}} \to \mathsf{E}$$

is a functor, and

$$\eta^{\mathrm{L}} : \mathrm{L}F \circ \mathrm{Q} \Rightarrow F$$

is a morphism of functors, such that the following universal property holds:

(L) Given any pair (G, θ), consisting of a functor $G : \mathsf{K}_\mathsf{S} \to \mathsf{E}$ and a morphism of functors $\theta : G \circ \mathrm{Q} \Rightarrow F$, there is a unique morphism of functors $\mu : G \Rightarrow \mathrm{L}F$ such that $\theta = \eta^{\mathrm{L}} * (\mu \circ \mathrm{id}_\mathrm{Q})$.

Pictorially: there is a 2-diagram

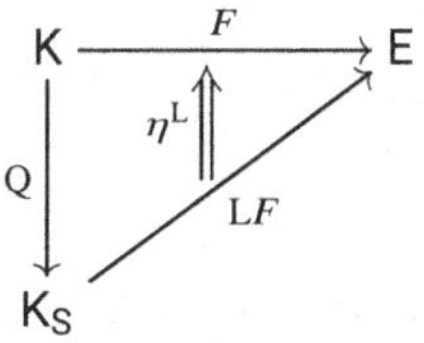

For any other pair (G, θ) there is a unique morphism μ that sits in this 2-diagram

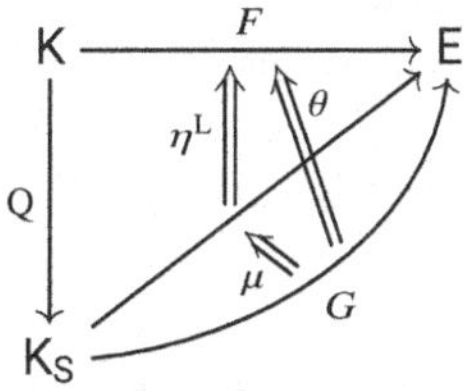

such that the diagram of 2-morphisms (with $*$ composition)

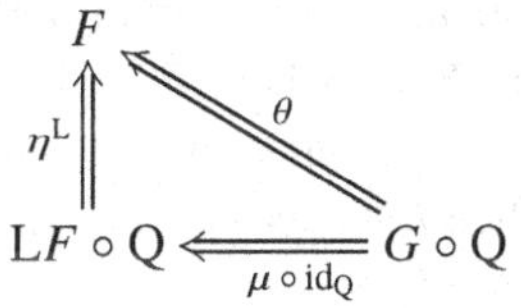

is commutative.

We see that a left derived functor with target E amounts to a right derived functor with target E^{op}. It means that new proofs are not needed.

Proposition 8.3.29 *If a left derived functor* $(\mathrm{L}F, \eta^{\mathrm{L}})$ *exists, then it is unique, up to a unique isomorphism. Namely, if* (G, θ) *is another left derived functor of* F*, then there is a unique isomorphism of functors* $\mu : G \xRightarrow{\simeq} \mathrm{L}F$ *such that* $\theta = \eta^{\mathrm{L}} * (\mu \circ \mathrm{id}_\mathrm{Q})$.

The proof is the same as that of Proposition 8.3.2, only some arrows have to be reversed.

Theorem 8.3.30 (Existence of Left Derived Functor) *In the situation of Definition* 8.3.28*, assume there is a full subcategory* $\mathsf{P} \subseteq \mathsf{K}$ *such the following three conditions hold*:

(a) *The multiplicatively closed set* S *is a right denominator set in* K.

(b) *For every object* $M \in \mathsf{K}$ *there is a morphism* $\rho : P \to M$ *in* S*, with source* $P \in \mathsf{P}$.

(c) *If ψ is a morphism in* $\mathsf{P} \cap \mathsf{S}$, *then* $F(\psi)$ *is an isomorphism in* E.
Then the left derived functor

$$(\mathrm{L}F, \eta^{\mathrm{L}}) : \mathsf{K}_{\mathsf{S}} \to \mathsf{E}$$

exists. Moreover, for every object $P \in \mathsf{P}$ *the morphism*

$$\eta^{\mathrm{L}}_P : \mathrm{L}F(P) \to F(P)$$

in E *is an isomorphism.*

Condition (b) says that K *has enough left* P*-resolutions*. Condition (c) says that P *is an* F*-acyclic category*.

The proof is the same as that of Theorem 8.3.3, only some arrows have to be reversed. We leave this as an exercise.

Exercise 8.3.31 Prove Theorem 8.3.30, including the lemmas leading to it. (Hint: replace E with E^{op}.)

For reference we give the next definition.

Definition 8.3.32 In the situation of Theorem 8.3.30, by a *system of left* P*-resolutions* we mean a pair (P, ρ), where $P : \mathrm{Ob}(\mathsf{K}) \to \mathrm{Ob}(\mathsf{P})$ is a function, and $\rho = \{\rho_M\}_{M \in \mathrm{Ob}(\mathsf{K})}$ is a collection of morphisms $\rho_M : P(M) \to M$ in S. Moreover, if $M \in \mathrm{Ob}(\mathsf{P})$, then $P(M) = M$ and $\rho_M = \mathrm{id}_M$.

When K has enough left P-resolutions, there exists a system of left P-resolutions (P, ρ) .

Definition 8.3.33 The construction of the left derived functor $(\mathrm{L}F, \eta^{\mathrm{L}})$ in the proof of Theorem 8.3.30 – i.e. the left variant of the proof of Theorem 8.3.3, and specifically the left versions of formulas (8.3.24) and (8.3.25) – is called a *presentation of* $(\mathrm{L}F, \eta^{\mathrm{L}})$ *by the system of left* P*-resolutions* (P, ρ).

Of course, any other left derived functor of F (perhaps presented by another system of left P-resolutions) is uniquely isomorphic to $(\mathrm{L}F, \eta^{\mathrm{L}})$. This is according to Proposition 8.3.29. In Chapters 10 and 11 we shall give several existence results for left resolutions by suitable acyclic objects.

Remark 8.3.34 The right derived functor $\mathrm{R}F$ from Definition 8.3.1 is a *left Kan extension* of F along Q. Likewise, the left derived functor $\mathrm{L}F$ from Definition 8.3.28 is a *right Kan extension* of F along Q. See [102, Chapter X].

8.4 Triangulated Derived Functors

In this section we specialize the definitions and results of the previous section to the case of triangulated functors between triangulated categories. There is a fixed nonzero commutative base ring $\mathbb{K}$, and all categories and functors here are $\mathbb{K}$-linear.

Triangulated functors and morphisms between them were introduced in Definition 5.2.1. Recall that a triangulated functor is a pair (F, τ), consisting of a linear functor F and a translation isomorphism τ that respect the distinguished triangles.

Here is the triangulated version of Definition 8.2.1.

Definition 8.4.1 Let K and L be triangulated categories. We define $\mathsf{TrFun}(\mathsf{K}, \mathsf{L})$ to be the category whose objects are the triangulated functors $(F, \tau) : \mathsf{K} \to \mathsf{L}$. Given objects (F, τ) and (G, ν) of $\mathsf{TrFun}(\mathsf{K}, \mathsf{L})$, the morphisms $\alpha : (F, \tau) \Rightarrow (G, \nu)$ in $\mathsf{TrFun}(\mathsf{K}, \mathsf{L})$ are the morphisms of triangulated functors.

Lemma 8.4.2 *Let $(F, \tau) : (\mathsf{K}, \mathrm{T}_{\mathsf{K}}) \to (\mathsf{L}, \mathrm{T}_{\mathsf{L}})$ be a triangulated functor between triangulated categories. Assume F is an equivalence (of abstract categories), with quasi-inverse $G : \mathsf{L} \to \mathsf{K}$, and with adjunction isomorphisms $\alpha : G \circ F \xRightarrow{\simeq} \mathrm{Id}_{\mathsf{K}}$ and $\beta : F \circ G \xRightarrow{\simeq} \mathrm{Id}_{\mathsf{L}}$. Then there is an isomorphism of functors $\nu : G \circ \mathrm{T}_{\mathsf{L}} \xRightarrow{\simeq} \mathrm{T}_{\mathsf{K}} \circ G$ such that $(G, \nu) : (\mathsf{L}, \mathrm{T}_{\mathsf{L}}) \to (\mathsf{K}, \mathrm{T}_{\mathsf{K}})$ is a triangulated functor, and α and β are isomorphisms of triangulated functors.*

Proof It is well-known that G is additive (or in our case, $\mathbb{K}$-linear); but since the proof is so easy, we shall reproduce it. Take any pair of objects $M, N \in \mathsf{L}$. We have to prove that the bijection

$$G_{M,N} : \mathrm{Hom}_{\mathsf{L}}(M, N) \to \mathrm{Hom}_{\mathsf{K}}(G(M), G(N))$$

is linear. But

$$G_{M,N} = (F_{G(M),G(N)})^{-1} \circ \mathrm{Hom}_{\mathsf{L}}(\beta_M, \beta_N^{-1})$$

as bijections (of sets) between these modules. Since $\mathrm{Hom}_{\mathsf{L}}(\beta_M, \beta_N^{-1})$ and $F^{-1}_{G(M),G(N)}$ are $\mathbb{K}$-linear, so is $G_{M,N}$.

We define the isomorphism of functors ν by the formula

$$\nu := (\alpha \circ \mathrm{id}_{\mathrm{T}_{\mathsf{K}} \circ G}) * (\mathrm{id}_G \circ \tau \circ \mathrm{id}_G)^{-1} * (\mathrm{id}_{G \circ \mathrm{T}_{\mathsf{L}}} \circ \beta)^{-1},$$

in terms of the 2-categorical notation. This gives rise to a commutative diagram of isomorphisms

$$\begin{array}{ccc} G \circ \mathrm{T_L} \circ F \circ G & \xLeftarrow{\mathrm{id} \circ \tau \circ \mathrm{id}} & G \circ F \circ \mathrm{T_K} \circ G \\ \Big\Downarrow{\scriptstyle \mathrm{id} \circ \beta} & & \Big\Downarrow{\scriptstyle \alpha \circ \mathrm{id}} \\ G \circ \mathrm{T_L} & \xRightarrow{\nu} & \mathrm{T_K} \circ G \end{array}$$

of additive functors $\mathsf{L} \to \mathsf{K}$. So the pair (G, ν) is a T-additive functor.

The verification that (G, ν) preserves triangles (in the sense of Definition 5.2.1(1)) is done like the proof of the additivity of G, but now using axiom (TR1)(a) from Definition 5.1.10. Also α and β are morphisms of triangulated functors (Definitions 5.2.1(2) and 5.1.5). We leave these verifications as an exercise. □

Exercise 8.4.3 Finish the proof above (the last assertions).

From here on in this section we shall usually keep the translation isomorphisms (such as τ and ν above) implicit.

Suppose we are given triangulated functors $U : \mathsf{K}' \to \mathsf{K}$ and $V : \mathsf{L} \to \mathsf{L}'$. There is an induced functor

$$\mathsf{TrFun}(U, V) : \mathsf{TrFun}(\mathsf{K}, \mathsf{L}) \to \mathsf{TrFun}(\mathsf{K}', \mathsf{L}');$$

the formula is the same as in (8.2.2).

Lemma 8.4.4 *If U and V are equivalences, then the functor* $\mathsf{TrFun}(U, V)$ *is an equivalence.*

Proof Use Proposition 8.2.3 and Lemma 8.4.2. □

Definition 8.4.5 Let K and L be triangulated categories, and let $\mathsf{S} \subseteq \mathsf{K}$ be a denominator set of cohomological origin. We define $\mathsf{TrFun}_{\mathsf{S}}(\mathsf{K}, \mathsf{L})$ to be the full subcategory of $\mathsf{TrFun}(\mathsf{K}, \mathsf{L})$ whose objects are the functors that are localizable to S, in the sense of Definition 8.2.5.

We know that the localization functor $\mathsf{Q} : \mathsf{K} \to \mathsf{K_S}$ is a left and right Ore localization.

Lemma 8.4.6 *Let* K *and* E *be triangulated categories, and let* $\mathsf{S} \subseteq \mathsf{K}$ *be a denominator set of cohomological origin. Then the functor*

$$\mathsf{TrFun}(\mathsf{Q}, \mathrm{Id}_{\mathsf{E}}) : \mathsf{TrFun}(\mathsf{K_S}, \mathsf{E}) \to \mathsf{TrFun}_{\mathsf{S}}(\mathsf{K}, \mathsf{E})$$

is an isomorphism of categories.

Proof Use Proposition 8.2.6 and Theorem 7.1.3. □

Here is the triangulated version of Definition 8.3.1.

Definition 8.4.7 Let $F : \mathsf{K} \to \mathsf{E}$ be a triangulated functor between triangulated categories, and let $\mathsf{S} \subseteq \mathsf{K}$ be a denominator set of cohomological origin. A *triangulated right derived functor of F with respect to* S is a triangulated functor $\mathrm{R}F : \mathsf{K}_{\mathsf{S}} \to \mathsf{E}$, together with a morphism $\eta^{\mathrm{R}} : F \Rightarrow \mathrm{R}F \circ \mathrm{Q}$ of triangulated functors $\mathsf{K} \to \mathsf{E}$. The pair $(\mathrm{R}F, \eta^{\mathrm{R}})$ must have this universal property:

(R) Given any pair (G, θ), consisting of a triangulated functor $G : \mathsf{K}_{\mathsf{S}} \to \mathsf{E}$ and a morphism of triangulated functors $\theta : F \Rightarrow G \circ \mathrm{Q}$, there is a unique morphism of triangulated functors $\mu : \mathrm{R}F \Rightarrow G$ such that $\theta = (\mu \circ \mathrm{id}_{\mathrm{Q}}) * \eta^{\mathrm{R}}$.

Proposition 8.4.8 *If a triangulated right derived functor* $(\mathrm{R}F, \eta^{\mathrm{R}})$ *exists, then it is unique, up to a unique isomorphism. Namely, if* (G, θ) *is another triangulated right derived functor of F, then there is a unique isomorphism of triangulated functors* $\mu : \mathrm{R}F \xRightarrow{\simeq} G$ *such that* $\theta = (\mu \circ \mathrm{id}_{\mathrm{Q}}) * \eta^{\mathrm{R}}$.

The proof is the same as that of Proposition 8.3.2. Now for the triangulated version of Theorem 8.3.3.

Theorem 8.4.9 *In the situation of Definition* 8.4.7, *assume there is a full triangulated subcategory* $\mathsf{J} \subseteq \mathsf{K}$ *with these two properties*:

(a) *Every object* $M \in \mathsf{K}$ *admits a morphism* $\rho : M \to I$ *in* S *with target* $I \in \mathsf{J}$.

(b) *If* ψ *is a morphism in* $\mathsf{J} \cap \mathsf{S}$, *then* $F(\psi)$ *is an isomorphism in* E.

Then the triangulated right derived functor $(\mathrm{R}F, \eta^{\mathrm{R}}) : \mathsf{K}_{\mathsf{S}} \to \mathsf{E}$ *of F with respect to* S *exists. Moreover, for every object* $I \in \mathsf{J}$ *the morphism* $\eta^{\mathrm{R}}_I : F(I) \to \mathrm{R}F(I)$ *in* E *is an isomorphism.*

Proof It will be convenient to change notation to that used in the proof of Theorem 8.3.3. Let's define $\mathsf{K}' := \mathsf{J}$, $\mathsf{S}' := \mathsf{K}' \cap \mathsf{S}$ and $\mathsf{D}' := \mathsf{K}'_{\mathsf{S}'}$. The localization functor of K' is $\mathrm{Q}' : \mathsf{K}' \to \mathsf{D}'$. The inclusion functor is $U : \mathsf{K}' \to \mathsf{K}$, and its localization is the functor $V : \mathsf{D}' \to \mathsf{D}$. We have this commutative diagram of triangulated functors between triangulated categories:

$$\begin{array}{ccc} \mathsf{K}' & \xrightarrow{U} & \mathsf{K} \\ {\scriptstyle \mathrm{Q}'}\downarrow & & \downarrow{\scriptstyle \mathrm{Q}} \\ \mathsf{D}' & \xrightarrow{V} & \mathsf{D} \end{array} \tag{8.4.10}$$

The functor U is fully faithful. By Lemma 8.3.8 the functor V is an equivalence.

Diagram (8.4.10) induces a commutative diagram of linear categories:

$$
\begin{array}{ccc}
\mathsf{TrFun}(\mathsf{K}',\mathsf{E}) & \xleftarrow{\mathsf{Fun}(U,\mathrm{Id})} & \mathsf{TrFun}(\mathsf{K},\mathsf{E}) \\
\uparrow \text{f.f. emb} & & \text{f.f. emb} \uparrow \\
\mathsf{TrFun}_{\mathsf{S}'}(\mathsf{K}',\mathsf{E}) & \xleftarrow[\text{equiv}]{\mathsf{Fun}(U,\mathrm{Id})} & \mathsf{TrFun}_{\mathsf{S}}(\mathsf{K},\mathsf{E}) \\
\mathsf{Fun}(\mathsf{Q}',\mathrm{Id}) \uparrow \text{isom} & & \text{isom} \uparrow \mathsf{Fun}(\mathsf{Q},\mathrm{Id}) \\
\mathsf{TrFun}(\mathsf{D}',\mathsf{E}) & \xleftarrow[\text{equiv}]{\mathsf{Fun}(V,\mathrm{Id})} & \mathsf{TrFun}(\mathsf{D},\mathsf{E})
\end{array} \qquad (8.4.11)
$$

By definition the arrows marked "f.f. emb" are fully faithful embeddings. According to Lemma 8.4.6, the arrows $\mathsf{Fun}(\mathsf{Q},\mathrm{Id})$ and $\mathsf{Fun}(\mathsf{Q}',\mathrm{Id})$ are isomorphisms of categories. By Lemma 8.4.4 the arrow $\mathsf{Fun}(V,\mathrm{Id})$ is an equivalence. It follows that the arrow $\mathsf{Fun}(U,\mathrm{Id})$ in the middle row is an equivalence too.

We know that $\mathsf{S} \subseteq \mathsf{K}$ is a left denominator set. Condition (b) of the theorem says that F sends morphisms in S' to isomorphisms in E. Condition (a) there says that there are enough right K'-resolutions in K. Thus we are in a position to use the abstract Theorem 8.3.3. It says that there is an abstract right derived functor $(\mathrm{R}F, \eta^{\mathrm{R}}) : \mathsf{D} = \mathsf{K}_{\mathsf{S}} \to \mathsf{E}$ of F with respect to S. However, going over the proof of Theorem 8.3.3, we see that all constructions there can be made within the triangulated setting, namely in diagram (8.4.11) instead of in diagram (8.3.11). Therefore $\mathrm{R}F$ is an object of the category in the bottom right corner of (8.4.11), and the morphism $\eta^{\mathrm{R}} : F \Rightarrow \mathrm{R}F \circ \mathrm{Q}$ is in the category in the top right corner of (8.4.11). The triangulated variant of step 4 in the proof of Theorem 8.3.3 shows that $(\mathrm{R}F, \eta^{\mathrm{R}})$ satisfies condition (R) of Definition 8.4.7. □

By slight abuse of notation, in the situation of Definition 8.4.7 we sometimes refer to the triangulated right derived functor $\mathrm{R}F$ just as "the right derived functor of F." As the next corollary shows, this does really cause a problem.

Corollary 8.4.12 *In the situation of Theorem* 8.4.9, *the triangulated right derived functor* $(\mathrm{R}F, \eta^{\mathrm{R}})$ *is also an abstract right derived functor of* F *with respect to* S, *in the sense of Definition* 8.3.1.

Proof This is seen in the proof of Theorem 8.4.9. □

In the situation of Theorem 8.4.9, let $\mathsf{K}^\star$ be a full triangulated subcategory of K. Define $\mathsf{S}^\star := \mathsf{K}^\star \cap \mathsf{S}$ and $\mathsf{J}^\star := \mathsf{K}^\star \cap \mathsf{J}$. Denote by $W : \mathsf{K}^\star \to \mathsf{K}$ the inclusion functor, and by $W_{\mathsf{S}^\star} : \mathsf{K}^\star_{\mathsf{S}^\star} \to \mathsf{K}_{\mathsf{S}}$ its localization. Warning: the functor $W_{\mathsf{S}^\star}$ is not necessarily fully faithful; cf. Proposition 7.1.13.

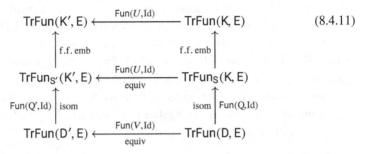

Proposition 8.4.13 *Assume that every object $M \in \mathsf{K}^\star$ admits a morphism $\rho : M \to I$ in $\mathsf{S}^\star$ with target $I \in \mathsf{J}^\star$. Then the pair* $(\mathrm{R}^\star F, \eta^\star) := (\mathrm{R}F \circ W_{\mathsf{S}^\star}, \eta^{\mathrm{R}} \circ \mathrm{id}_W)$ *is a right derived functor of* $F^\star := F \circ W : \mathsf{K}^\star \to \mathsf{E}$.

Exercise 8.4.14 Prove the last proposition. (Hint: Start by choosing a system of right $\mathsf{J}^\star$-resolutions in $\mathsf{K}^\star$. Then extend it to a system of right J-resolutions in K. Now follow the proofs of Theorems 8.4.9 and 8.3.3.)

Here is the specialization of Definition 8.4.7 to categories of DG modules. First:

Setup 8.4.15 We are given a DG ring A, an abelian category M, and a full triangulated subcategory $\mathsf{K} \subseteq \mathbf{K}(A, \mathsf{M})$. We write $\mathsf{S} := \mathsf{K} \cap \mathbf{S}(A, \mathsf{M})$, the set of quasi-isomorphisms in K, and $\mathsf{D} := \mathsf{K}_{\mathsf{S}}$, the derived category of K.

Definition 8.4.16 Under Setup 8.4.15, let E be a triangulated category, and let $F : \mathsf{K} \to \mathsf{E}$ be a triangulated functor. A *triangulated right derived functor* of F is a triangulated right derived functor $(\mathrm{R}F, \eta^{\mathrm{R}}) : \mathsf{D} \to \mathsf{E}$ of F with respect to S, in the sense of Definition 8.4.7.

Example 8.4.17 Suppose we start from an additive functor $F^0 : \mathsf{M} \to \mathsf{N}$ between abelian categories. We know how to extend it to a DG functor

$$\mathbf{C}^+(F^0) : \mathbf{C}^+(\mathsf{M}) \to \mathbf{C}^+(\mathsf{N}),$$

and this induces a triangulated functor

$$\mathbf{K}^+(F^0) : \mathbf{K}^+(\mathsf{M}) \to \mathbf{K}^+(\mathsf{N}).$$

By composing with the localization functor $\mathrm{Q}^+_{\mathsf{N}} : \mathbf{K}^+(\mathsf{N}) \to \mathbf{D}^+(\mathsf{N})$ we get a triangulated functor

$$F := \mathrm{Q}^+_{\mathsf{N}} \circ \mathbf{K}^+(F^0) : \mathbf{K}^+(\mathsf{M}) \to \mathbf{D}^+(\mathsf{N}). \tag{8.4.18}$$

Define $\mathsf{K} := \mathbf{K}^+(\mathsf{M})$, $\mathsf{S} := \mathbf{S}^+(\mathsf{M})$ and $\mathsf{E} := \mathbf{D}^+(\mathsf{N})$, so in this notation we have a triangulated functor $F : \mathsf{K} \to \mathsf{E}$, and we are in the situation of Definition 8.4.16. Note that the restriction of F to the full subcategory $\mathsf{M} \subseteq \mathbf{K}^+(\mathsf{M})$ coincides with the original functor F^0.

Assume that the abelian category M has enough injectives; namely that every object $M \in \mathsf{M}$ admits a monomorphism to an injective object. Let J be the full subcategory of K on the bounded below complexes of injective objects. We will prove later (see Corollary 10.1.19 and Theorem 11.5.8) that properties (a) and (b) of Theorem 8.4.9 hold for this J, regardless of what F is. Therefore the triangulated right derived functor

$$(\mathrm{R}^+F, \eta^+) : \mathbf{D}^+(\mathsf{M}) \to \mathbf{D}^+(\mathsf{N}) \tag{8.4.19}$$

exists, and for every complex $I \in \mathsf{J}$ the morphism

$$\eta^+_I : F(I) \to \mathrm{R}^+F(I) \tag{8.4.20}$$

in $\mathbf{D}^+(\mathsf{N})$ is an isomorphism.

Now assume the original functor $F^0 : \mathsf{M} \to \mathsf{N}$ is left exact. For each $q \geq 0$ we have the classical right derived functor $\mathrm{R}^q F^0 : \mathsf{M} \to \mathsf{N}$, and $\mathrm{R}^0 F^0 \cong F^0$. Formula (8.4.20) shows that for every $M \in \mathsf{M}$ and $q \geq 0$ there is an isomorphism $\mathrm{R}^q F^0(M) \cong \mathrm{H}^q(\mathrm{R}^+F(M))$ in N. As the object $M \in \mathsf{M}$ moves, this becomes an isomorphism $\mathrm{R}^q F^0 \cong \mathrm{H}^q \circ \mathrm{R}^+F$ of additive functors $\mathsf{M} \to \mathsf{N}$.

In this example we were careful to use distinct notations for the functor F^0 between abelian categories, and the induced triangulated functor $F = \mathrm{Q}^+_{\mathsf{N}} \circ \mathbf{K}^+(F^0)$. Later we will use the same notation for F^0 and F.

Our treatment of triangulated left derived functors will be brief: we will state the definitions and the main results, but won't give proofs, beyond a hint here and there on the passage from right to left derived functors.

Definition 8.4.21 Let $F : \mathsf{K} \to \mathsf{E}$ be triangulated functor between triangulated categories, and let $\mathsf{S} \subseteq \mathsf{K}$ be a denominator set of cohomological origin. A *triangulated left derived functor of F with respect to* S is a triangulated functor $\mathrm{L}F : \mathsf{K}_{\mathsf{S}} \to \mathsf{E}$, together with a morphism $\eta^{\mathrm{L}} : \mathrm{L}F \circ \mathrm{Q} \Rightarrow F$ of triangulated functors $\mathsf{K} \to \mathsf{E}$. The pair $(\mathrm{L}F, \eta^{\mathrm{L}})$ must have this universal property:

(L) Given any pair (G, θ), consisting of a triangulated functor $G : \mathsf{K}_{\mathsf{S}} \to \mathsf{E}$ and a morphism of triangulated functors $\theta : G \circ \mathrm{Q} \Rightarrow F$, there is a unique morphism of triangulated functors $\mu : G \Rightarrow \mathrm{L}F$ such that $\theta = \eta^{\mathrm{L}} * (\mu \circ \mathrm{id}_{\mathrm{Q}})$.

Proposition 8.4.22 *If a triangulated left derived functor* $(\mathrm{L}F, \eta^{\mathrm{L}})$ *exists, then it is unique, up to a unique isomorphism. Namely, if* (G, θ) *is another triangulated left derived functor of F, then there is a unique isomorphism of triangulated functors* $\mu : G \xRightarrow{\simeq} \mathrm{L}F$ *such that* $\theta = \eta^{\mathrm{L}} * (\mu \circ \mathrm{id}_{\mathrm{Q}})$.

The proof is the same as that of Proposition 8.4.8, with direction of arrows in E reversed.

Theorem 8.4.23 *In the situation of Definition* 8.4.21, *assume there is a full triangulated subcategory* $\mathsf{P} \subseteq \mathsf{K}$ *with these two properties*:

(a) *Every object $M \in \mathsf{K}$ admits a morphism $\rho : P \to M$ in S with source $P \in \mathsf{P}$.*

(b) *If ψ is a morphism in $\mathsf{P} \cap \mathsf{S}$, then $F(\psi)$ is an isomorphism in E.*

Then the triangulated left derived functor $(\mathrm{L}F, \eta^{\mathrm{L}}) : \mathsf{K}_{\mathsf{S}} \to \mathsf{E}$ *of F with respect to S exists. Moreover, for every object $P \in \mathsf{P}$ the morphism $\eta^{\mathrm{L}}_P : \mathrm{L}F(P) \to F(P)$ in E is an isomorphism.*

The proof is the same as that of Theorem 8.4.9, with direction of arrows in E reversed.

Corollary 8.4.24 *In the situation of Theorem* 8.4.23, *the triangulated left derived functor* $(\mathrm{L}F, \eta^{\mathrm{L}})$ *is also an abstract left derived functor of* F *with respect to* S, *in the sense of Definition* 8.3.28.

The proof is the same as that of Corollary 8.4.12, with direction of arrows in E reversed.

In the situation of Theorem 8.4.23, let $\mathsf{K}^\star$ be a full triangulated subcategory of K. Define $\mathsf{S}^\star := \mathsf{K}^\star \cap \mathsf{S}$ and $\mathsf{P}^\star := \mathsf{K}^\star \cap \mathsf{P}$. Denote by $W : \mathsf{K}^\star \to \mathsf{K}$ the inclusion functor, and by $W_{\mathsf{S}^\star} : \mathsf{K}^\star_{\mathsf{S}^\star} \to \mathsf{K}_{\mathsf{S}}$ its localization.

Proposition 8.4.25 *Assume that every* $M \in \mathsf{K}^\star$ *admits a morphism* $\rho : P \to M$ *in* $\mathsf{S}^\star$ *with source* $P \in \mathsf{P}^\star$. *Then the pair* $(\mathrm{L}^\star, \eta^\star) := (\mathrm{L}F \circ W_{\mathsf{S}^\star}, \eta^{\mathrm{L}} \circ \mathrm{id}_W)$ *is a triangulated left derived functor of* $F^\star := F \circ W : \mathsf{K}^\star \to \mathsf{E}$.

The proof is just like that of Proposition 8.4.13 (which was an exercise).

Here is the specialization of Definition 8.4.21 to categories of DG modules.

Definition 8.4.26 Under Setup 8.4.15, let E be a triangulated category, and let $F : \mathsf{K} \to \mathsf{E}$ be a triangulated functor. A *triangulated left derived functor* of F is a triangulated left derived functor $(\mathrm{L}F, \eta^{\mathrm{L}}) : \mathsf{D} \to \mathsf{E}$ of F with respect to S, in the sense of Definition 8.4.21.

Example 8.4.27 This is analogous to Example 8.4.17. We start with an additive functor $F^0 : \mathsf{M} \to \mathsf{N}$ between abelian categories. We know how to extend it to a DG functor

$$\mathbf{C}^-(F^0) : \mathbf{C}^-(\mathsf{M}) \to \mathbf{C}^-(\mathsf{N}),$$

and this induces a triangulated functor

$$\mathbf{K}^-(F^0) : \mathbf{K}^-(\mathsf{M}) \to \mathbf{K}^-(\mathsf{N}).$$

By composing with $\mathrm{Q}^-_{\mathsf{N}}$ we get a triangulated functor

$$F := \mathrm{Q}^-_{\mathsf{N}} \circ \mathbf{K}^-(F^0) : \mathbf{K}^-(\mathsf{M}) \to \mathbf{D}^-(\mathsf{N}). \tag{8.4.28}$$

Defining $\mathsf{K} := \mathbf{K}^-(\mathsf{M})$, $\mathsf{S} := \mathbf{S}^+(\mathsf{M})$ and $\mathsf{E} := \mathbf{D}^-(\mathsf{N})$, we have a triangulated functor $F : \mathsf{K} \to \mathsf{E}$, and the situation is that of Definition 8.4.26.

Assume that the abelian category M has enough projectives. Define P to be the full subcategory of K on the bounded above complexes of projective objects. We will prove later (in Corollary 10.2.14 and Theorem 11.3.6) that properties (a) and (b) of Theorem 8.4.23 hold in this situation. Therefore we have a left derived functor

$$(\mathrm{L}^-F, \eta^-) : \mathbf{D}^-(\mathsf{M}) \to \mathbf{D}^-(\mathsf{N}). \tag{8.4.29}$$

For any $P \in \mathsf{P}$ the morphism

$$\eta^-_P : \mathrm{L}^-F(P) \to F(P) \tag{8.4.30}$$

in E is an isomorphism.

Now assume the functor F^0 is right exact. Then F^0 has the classical left derived functors $\mathrm{L}_q F^0 : \mathsf{M} \to \mathsf{N}$ for all $q \geq 0$, and $\mathrm{L}_0 F^0 \cong F^0$. Formula (8.4.30) shows that for every $M \in \mathsf{M}$ there is an isomorphism $\mathrm{L}_q F^0(M) \cong \mathrm{H}^{-q}(\mathrm{L}^-F(M))$ in N. As M moves there is an isomorphism $\mathrm{L}_q F^0 \cong \mathrm{H}^{-q} \circ \mathrm{L}^-F$ of additive functors $\mathsf{M} \to \mathsf{N}$.

In this example, as in Example 8.4.14, we were careful to use distinct notations for the functor F^0 between abelian categories, and the induced triangulated functor $F = \mathrm{Q}^-_{\mathsf{N}} \circ \mathbf{K}^-(F^0)$. Later we will use the same notation for F^0 and F.

Remark 8.4.31 The reader might ask the following question at this point: Suppose we are given additive functors

$$\mathsf{L} \xrightarrow{F} \mathsf{M} \xrightarrow{G} \mathsf{N} \tag{8.4.32}$$

between abelian categories, and the categories M and N have enough injectives, so that the right derived functors

$$\mathbf{D}^+(\mathsf{L}) \xrightarrow{\mathrm{R}^+F} \mathbf{D}^+(\mathsf{M}) \xrightarrow{\mathrm{R}^+G} \mathbf{D}^+(\mathsf{N}), \qquad \mathrm{R}^+(G \circ F) : \mathbf{D}^+(\mathsf{L}) \to \mathbf{D}^+(\mathsf{N}) \tag{8.4.33}$$

exist (see Example 8.4.17 above). Does this diagram commute (up to an isomorphism of triangulated functors)? Similar questions can be asked about left derived functors, etc.

Here is an answer. Assume that F sends injectives in L to *right G-acyclic objects* of M. See Definition 16.1.3 (which talks about categories of graded modules, but the ideas are the same). By Lemma 16.1.5 we know that for each bounded below complex of right G-acyclic objects J, the canonical morphism $G(J) \to \mathrm{R}^+G(J)$ in $\mathbf{D}^+(\mathsf{M})$ is an isomorphism. Therefore, for every bounded below complex of injectives $I \in \mathbf{C}(\mathsf{L})$, letting $J := F(I)$, we have $J \cong \mathrm{R}^+F(I)$, and hence

$$\mathrm{R}^+(G \circ F)(I) \cong (G \circ F)(I) = G(F(I)) = G(J) \cong \mathrm{R}^+G(J) \cong \mathrm{R}^+G(\mathrm{R}^+F(I)).$$

Thus there is an isomorphism of triangulated functors

$$\mathrm{R}^+(G \circ F) \cong \mathrm{R}^+G \circ \mathrm{R}^+F : \mathbf{D}^+(\mathsf{L}) \to \mathbf{D}^+(\mathsf{N}). \tag{8.4.34}$$

The original application of the isomorphism (8.4.34), in [62, Proposition II.5.1], was this: given morphisms of schemes $X \xrightarrow{f} Y \xrightarrow{g} Z$, the derived pushforward functors satisfy

$$\mathrm{R}^+(g_* \circ f_*) \cong \mathrm{R}^+ g_* \circ \mathrm{R}^+ f_* \tag{8.4.35}$$

as functors $\mathbf{D}^+(X) \to \mathbf{D}^+(Z)$. This is true because the functor f_* sends injective $\mathcal{O}_X$-modules to flasque $\mathcal{O}_Y$-modules.

This pleasant phenomenon seems to fail in some important situations that we study in our book. For instance, take a noncommutative ring A and A-bimodules S, T that are tilting complexes but not invertible bimodules (see Section 14.5). Consider the abelian categories $\mathsf{L} = \mathsf{M} = \mathsf{N} := \mathbf{M}(A)$ and the functors $F := \mathrm{Hom}_A(T, -)$ and $G := \mathrm{Hom}_A(S, -)$. It is quite likely that the functors $\mathrm{R}^+ G \circ \mathrm{R}^+ F$ and $\mathrm{R}^+(G \circ F)$ are not isomorphic. A concrete instance in which this can be tested is the ring A from Example 18.6.24, and the bimodules $S = T := A^*$ there.

Remark 8.4.36 The composition of derived functors explained in the previous remark is also related to *spectral sequences*. Given the input (8.4.32), in classical homological algebra it was impossible to ask for something like the commutative diagram (8.4.33). The best approximation was to have a *convergent spectral sequence*

$$E_k^{i,j} = \mathrm{R}^j G \circ \mathrm{R}^i F \Rightarrow \mathrm{R}^n(G \circ F). \tag{8.4.37}$$

Indeed, under the assumption in the previous remark (namely that F sends injective objects to right G-acyclic objects), such a convergent spectral sequence does exist; see [101, Chapters XI–XII], [129, Section 11] or [158, Section 5.7]. In the geometric example above, for which the isomorphism (8.4.35) holds, this is the *Leray spectral sequence*.

There are several kinds of spectral sequences, but basically they are all of the same nature, i.e. approximations of the composition of derived functors.

Most traditional applications of spectral sequences are performed more effectively by filtrations and truncations of complexes, as can be seen in Chapter 12 of this book. Still, there are a few instances in which spectral sequences cannot be left out, such as the *niveau spectral sequence* in algebraic geometry (see [62, Chapter IV]), or its noncommutative relative, the *double Ext spectral sequence* for Auslander dualizing complexes (see Remark 18.5.13 and the paper [183]).

8.5 Contravariant Triangulated Derived Functors

In this section there is a fixed nonzero commutative base ring $\mathbb{K}$, and all categories and functors here are $\mathbb{K}$-linear. The next setup is assumed.

Setup 8.5.1 We are given a DG ring A, an abelian category M, and a full additive subcategory $\mathsf{K} \subseteq \mathbf{K}(A, \mathsf{M})$ s.t. $\mathsf{K}^{\mathrm{op}} \subseteq \mathbf{K}(A, \mathsf{M})^{\mathrm{op}}$ is triangulated. We write $\mathsf{S} := \mathsf{K} \cap \mathbf{S}(A, \mathsf{M})$, the set of quasi-isomorphisms in K, and $\mathsf{D}^{\mathrm{op}} := (\mathsf{K}^{\mathrm{op}})_{\mathsf{S}^{\mathrm{op}}}$, the derived category of K^{op}.

As explained in Sections 5.6 and 7.6, the categories K^{op} and D^{op} have canonical triangulated structures, and the localization functor $\mathrm{Q}^{\mathrm{op}} : \mathsf{K}^{\mathrm{op}} \to \mathsf{D}^{\mathrm{op}}$ is triangulated.

Definition 8.5.2 Let E be a triangulated category, and let $F : \mathsf{K}^{\mathrm{op}} \to \mathsf{E}$ be a triangulated functor. A *triangulated right derived functor* of F is a triangulated right derived functor $(\mathrm{R}F, \eta^{\mathrm{R}}) : \mathsf{D}^{\mathrm{op}} \to \mathsf{E}$ of F with respect to S^{op}, in the sense of Definition 8.4.7.

Likewise consider the following:

Definition 8.5.3 Let E be a triangulated category, and let $F : \mathsf{K}^{\mathrm{op}} \to \mathsf{E}$ be a triangulated functor. A *triangulated left derived functor* of F is a triangulated left derived functor $(\mathrm{L}F, \eta^{\mathrm{L}}) : \mathsf{D}^{\mathrm{op}} \to \mathsf{E}$ of F with respect to S^{op}, in the sense of Definition 8.4.21.

The uniqueness results (Propositions 8.4.8 and 8.4.22) and existence results (Theorems 8.4.9 and 8.4.23) apply in the contravariant case as well. We shall return to the existence question the end of Chapter 10, when we talk about *resolving subcategories*; this will make matters more concrete. For now we give one example.

Example 8.5.4 Let M and N be abelian categories, and let $F^0 : \mathsf{M} \to \mathsf{N}$ be a contravariant additive functor. This is the same as an additive functor $F^0 \circ \mathrm{Op} : \mathsf{M}^{\mathrm{op}} \to \mathsf{N}$. As in Theorem 3.9.13, F^0 gives rise to a contravariant DG functor

$$\mathbf{C}^-(F^0) : \mathbf{C}^-(\mathsf{M}) \to \mathbf{C}^+(\mathsf{N}).$$

This means, according to Proposition 3.9.3, that we have a DG functor

$$\mathbf{C}^-(F^0) \circ \mathrm{Op} : \mathbf{C}^-(\mathsf{M})^{\mathrm{op}} \to \mathbf{C}^+(\mathsf{N}).$$

Passing to homotopy categories, and using Theorem 5.6.10, we obtain a triangulated functor

$$\mathrm{Ho}(\mathbf{C}^-(F^0) \circ \mathrm{Op}) : \mathbf{K}^-(\mathsf{M})^{\mathrm{op}} \to \mathbf{K}^+(\mathsf{N})$$

(with a translation isomorphism $\bar{\tau}$ that we ignore now). Composing this with the localization functor Q_{N} we have a triangulated functor

$$F := \mathrm{Q}_{\mathsf{N}} \circ \mathrm{Ho}(\mathbf{C}^-(F^0) \circ \mathrm{Op}) : \mathbf{K}^-(\mathsf{M})^{\mathrm{op}} \to \mathbf{D}^+(\mathsf{N}). \tag{8.5.5}$$

Similarly there is a triangulated functor

$$F^{\mathrm{flip}} := \mathrm{Q}_{\mathsf{N}} \circ \mathrm{Ho}(\mathbf{C}^{+}(F^{0} \circ \mathrm{Op})) : \mathbf{K}^{+}(\mathsf{M}^{\mathrm{op}}) \to \mathbf{D}^{+}(\mathsf{N}). \tag{8.5.6}$$

Observe that this triangulated functor is of the same sort as the one we had in (8.4.18). The functors (8.5.5) and (8.5.6) fit into the following commutative diagram of triangulated functors

$$\begin{array}{ccc} \mathbf{K}^{-}(\mathsf{M})^{\mathrm{op}} & \xrightarrow{\;F\;} & \mathbf{D}^{+}(\mathsf{N}) \\ {\scriptstyle \overline{\mathrm{Flip}}}\big\downarrow{\scriptstyle \cong} & \nearrow{\scriptstyle F^{\mathrm{flip}}} & \\ \mathbf{K}^{+}(\mathsf{M}^{\mathrm{op}}) & & \end{array}$$

where $\overline{\mathrm{Flip}}$ is the isomorphism of triangulated categories from (5.6.3).

The triangulated right derived functor

$$\mathrm{R}^{-}F : \mathbf{D}^{-}(\mathsf{M})^{\mathrm{op}} \to \mathbf{D}^{+}(\mathsf{N}) \tag{8.5.7}$$

of the functor F from (8.5.5) would sit inside this diagram of triangulated functors, which is commutative up to a unique isomorphism:

$$\begin{array}{ccc} \mathbf{D}^{-}(\mathsf{M})^{\mathrm{op}} & \xrightarrow{\;\mathrm{R}^{-}F\;} & \mathbf{D}^{+}(\mathsf{N}) \\ {\scriptstyle \overline{\mathrm{Flip}}}\big\downarrow{\scriptstyle \cong} & \nearrow{\scriptstyle \mathrm{R}^{+}F^{\mathrm{flip}}} & \\ \mathbf{D}^{+}(\mathsf{M}^{\mathrm{op}}) & & \end{array}$$

Like in Example 8.4.17, to show existence of $\mathrm{R}^{+}F^{\mathrm{flip}}$ using Theorem 8.4.9, and to calculate it, we would need a full triangulated subcategory $\mathsf{J} \subseteq \mathbf{K}^{+}(\mathsf{M}^{\mathrm{op}})$, such that every $M \in \mathbf{K}^{+}(\mathsf{M}^{\mathrm{op}})$ admits a quasi-isomorphism $M \to I$ with $I \in \mathsf{J}$; and every quasi-isomorphism ψ in J goes to an isomorphism $F^{\mathrm{flip}}(\psi)$ in $\mathbf{D}^{+}(\mathsf{N})$. This translates, by the flip, to a full triangulated subcategory $\mathsf{P} \subseteq \mathbf{K}^{-}(\mathsf{M})$ such that $\mathsf{J} = \overline{\mathrm{Flip}}(\mathsf{P})$. The conditions on P are these: each $M \in \mathbf{K}^{-}(\mathsf{M})$ admits a quasi-isomorphism $P \to M$ with $P \in \mathsf{P}$; and each quasi-isomorphism ϕ in P goes to an isomorphism $F(\psi)$ in $\mathbf{D}^{+}(\mathsf{N})$. For instance, if M has enough projectives, then according to Corollary 11.3.19, the category P of bounded above complexes of projectives satisfies these conditions.

To calculate the derived functor $\mathrm{R}^{-}F$ in equation (8.5.7), we choose a system of left P-resolutions in $\mathbf{K}^{-}(\mathsf{M})$, as in Definition 8.3.32. Thus for every M we have a chosen quasi-isomorphism $\rho_M : P(M) \to M$ in $\mathbf{K}^{-}(\mathsf{M})$. We take $\mathrm{R}^{-}F(M) := F(P(M)) \in \mathbf{D}^{+}(\mathsf{N})$. The morphism $\eta^{-}_{M} : F(M) \to \mathrm{R}^{-}F(M)$ is $\eta^{-}_{M} := F(\rho_M) : F(M) \to F(P(M))$.

The story is very similar for the triangulated left derived functor

$$\mathrm{L}^{+}F : \mathbf{D}^{+}(\mathsf{M})^{\mathrm{op}} \to \mathbf{D}^{-}(\mathsf{N})$$

of

$$F := \mathrm{Q_N} \circ \mathrm{Ho}(\mathbf{C}^+(F^0) \circ \mathrm{Op}) : \mathbf{K}^+(\mathsf{M})^{\mathrm{op}} \to \mathbf{D}^-(\mathsf{N}),$$

so we won't tell it.

In later parts of the book we shall use the same symbol F to denote also the functors F^0, $F^0 \circ \mathrm{Op}$ and $\mathbf{C}^\star(F^0) \circ \mathrm{Op}$ from the example above. This will simplify our notation.

Though indirectly defined (in Definition 5.6.4), the translation functor $\mathrm{T^{op}}$ of $\mathbf{K}(A, \mathsf{M})^{\mathrm{op}}$ is known to have some useful properties. Note that the objects of $\mathbf{K}(A, \mathsf{M})^{\mathrm{op}}$ are the same as those of $\mathbf{K}(A, \mathsf{M})$.

Proposition 8.5.8 *For every object $M \in \mathbf{K}(A, \mathsf{M})^{\mathrm{op}}$ there are canonical isomorphisms $\mathrm{T^{op}}(M)^i \cong M^{i-1}$ and $\mathrm{H}^i(\mathrm{T^{op}}(M)) \cong \mathrm{H}^{i-1}(M)$ in M.*

Proof This is immediate from Theorem 3.9.16 and Definition 5.6.4. □

Here are several useful properties of the triangulated structure on $\mathbf{D}(A, \mathsf{M})^{\mathrm{op}}$.

Proposition 8.5.9 *Suppose*

$$0 \to L \xrightarrow{\phi} M \xrightarrow{\psi} N \to 0$$

is an exact sequence in $\mathbf{C}_{\mathrm{str}}(A, \mathsf{M})^{\mathrm{op}}$. Then there is a distinguished triangle

$$L \xrightarrow{\mathrm{Q^{op}}(\phi)} M \xrightarrow{\mathrm{Q^{op}}(\psi)} N \xrightarrow{\theta} \mathrm{T^{op}}(L)$$

in $\mathbf{D}(A, \mathsf{M})^{\mathrm{op}}$.

Proof Apply the functor Flip to the given exact sequence. By Theorem 3.9.16(3) we obtain a short exact sequence

$$0 \to \mathrm{Flip}(L) \xrightarrow{\mathrm{Flip}(\phi)} \mathrm{Flip}(M) \xrightarrow{\mathrm{Flip}(\psi)} \mathrm{Flip}(N) \to 0$$

in $\mathbf{C}_{\mathrm{str}}(A^{\mathrm{op}}, \mathsf{M}^{\mathrm{op}})$. According to Proposition 7.3.5, there is a distinguished triangle

$$\mathrm{Flip}(L) \xrightarrow{\mathrm{Q^{flip}}(\mathrm{Flip}(\phi))} \mathrm{Flip}(M) \xrightarrow{\mathrm{Q^{flip}}(\mathrm{Flip}(\psi))} \mathrm{Flip}(N) \xrightarrow{\theta'} \mathrm{T^{flip}}(\mathrm{Flip}(L))$$

in $\mathbf{D}(A^{\mathrm{op}}, \mathsf{M}^{\mathrm{op}})$. Take the morphism

$$\theta := \mathrm{Flip}^{-1}(\theta') : N \to \mathrm{Flip}^{-1}(\mathrm{T^{flip}}(\mathrm{Flip}(L))) = \mathrm{T^{op}}(L)$$

in $\mathbf{D}(A, \mathsf{M})^{\mathrm{op}}$. Since $\mathrm{Flip}^{-1}(\mathrm{Q^{flip}}(\mathrm{Flip}(\phi))) = \mathrm{Q^{op}}(\phi)$, and likewise for ψ, we have the desired distinguished triangle. □

Recall from Proposition 7.3.8 that if A is a nonpositive DG ring, then for any $M \in \mathbf{C}(A, \mathsf{M})$ and integer i the smart truncations $\mathrm{smt}^{\geq i}(M), \mathrm{smt}^{\leq i}(M) \in \mathbf{C}(A, \mathsf{M})$ exist.

Proposition 8.5.10 *Assume A is a nonpositive DG ring. For every object $M \in \mathsf{C}(A, \mathrm{M})$ there is a distinguished triangle*

$$\mathrm{smt}^{\geq i+1}(M) \to M \to \mathrm{smt}^{\leq i}(M) \to \mathrm{T}^{\mathrm{op}}(\mathrm{smt}^{\geq i+1}(M))$$

in $\mathsf{D}(A, \mathrm{M})^{\mathrm{op}}$. The truncations are performed in $\mathsf{C}(A, \mathrm{M})$ as above.

Proof Let $N := \mathrm{Flip}(M) \in \mathsf{D}(A^{\mathrm{op}}, \mathrm{M}^{\mathrm{op}})$. By Proposition 7.3.10 there is a distinguished triangle

$$\mathrm{smt}^{\leq i}(N) \to N \to \mathrm{smt}^{\geq i+1}(N) \to \mathrm{T}^{\mathrm{flip}}(\mathrm{smt}^{\leq i}(N))$$

in $\mathsf{D}(A^{\mathrm{op}}, \mathrm{M}^{\mathrm{op}})$, where now the truncations are done in $\mathsf{C}(A^{\mathrm{op}}, \mathrm{M}^{\mathrm{op}})$. The formulas defining the flip functor in the proof of Theorem 3.9.16 show that $\mathrm{Flip}^{-1}(\mathrm{smt}^{\leq i}(N)) = \mathrm{smt}^{\geq i+1}(M)$ and $\mathrm{Flip}^{-1}(\mathrm{smt}^{\geq i+1}(N)) = \mathrm{smt}^{\leq i}(M)$. □

9

DG and Triangulated Bifunctors

In this chapter we extend the theory of triangulated derived functors, which was studied in Chapter 8, to *triangulated bifunctors*. As stated in Convention 1.2.4, all linear structures and operations (such as categories and functors) are $\mathbb{K}$-linear, where $\mathbb{K}$ is a fixed commutative base ring. The symbol $\otimes$ means $\otimes_{\mathbb{K}}$.

9.1 DG Bifunctors

We already talked about bifunctors in Section 1.6. That was for categories without further structure. Here we will consider DG categories, and matters become a bit more complicated.

As a warmup, let us begin with linear bifunctors.

Definition 9.1.1 Let C_1 and C_2 be linear categories. We define the linear category $\mathsf{C}_1 \otimes \mathsf{C}_2$ as follows. The set of objects is

$$\operatorname{Ob}(\mathsf{C}_1 \otimes \mathsf{C}_2) := \operatorname{Ob}(\mathsf{C}_1) \times \operatorname{Ob}(\mathsf{C}_2).$$

For each pair of objects $(M_1, M_2), (N_1, N_2)$ in $\mathsf{C}_1 \otimes \mathsf{C}_2$, i.e. $M_i, N_i \in \operatorname{Ob}(\mathsf{C}_i)$, we let

$$\operatorname{Hom}_{\mathsf{C}_1 \otimes \mathsf{C}_2}((M_1, M_2), (N_1, N_2)) := \operatorname{Hom}_{\mathsf{C}_1}(M_1, N_1) \otimes \operatorname{Hom}_{\mathsf{C}_2}(M_2, N_2).$$

Given morphisms $\phi_i \in \operatorname{Hom}_{\mathsf{C}_i}(L_i, M_i)$ and $\psi_i \in \operatorname{Hom}_{\mathsf{C}_i}(M_i, N_i)$, for $i = 1, 2$, their tensors are morphisms

$$\phi_1 \otimes \phi_2 \in \operatorname{Hom}_{\mathsf{C}_1 \otimes \mathsf{C}_2}((L_1, L_2), (M_1, M_2))$$

and

$$\psi_1 \otimes \psi_2 \in \mathrm{Hom}_{\mathsf{C}_1 \otimes \mathsf{C}_2}\big((M_1, M_2), (N_1, N_2)\big).$$

Every morphism in $\mathsf{C}_1 \otimes \mathsf{C}_2$ is a sum of such pure tensors. We define the composition to be

$$\begin{aligned}(\psi_1 \otimes \psi_2) \circ (\phi_1 \otimes \phi_2) &:= (\psi_1 \circ \phi_1) \otimes (\psi_2 \circ \phi_2) \\ &\in \mathrm{Hom}_{\mathsf{C}_1 \otimes \mathsf{C}_2}\big((L_1, L_2), (N_1, N_2)\big).\end{aligned}$$

Definition 9.1.2 Let C_1, C_2 and D be linear categories. A *linear bifunctor*

$$F : \mathsf{C}_1 \times \mathsf{C}_2 \to \mathsf{D}$$

is, by definition, a linear functor $F : \mathsf{C}_1 \otimes \mathsf{C}_2 \to \mathsf{D}$.

Now to the DG situation.

Definition 9.1.3 Let C_1 and C_2 be DG categories. We define the DG category $\mathsf{C}_1 \otimes \mathsf{C}_2$ as follows: the set of objects is

$$\mathrm{Ob}(\mathsf{C}_1 \otimes \mathsf{C}_2) := \mathrm{Ob}(\mathsf{C}_1) \times \mathrm{Ob}(\mathsf{C}_2).$$

For each pair of objects $(M_1, M_2), (N_1, N_2)$ in $\mathsf{C}_1 \otimes \mathsf{C}_2$ we let

$$\mathrm{Hom}_{\mathsf{C}_1 \otimes \mathsf{C}_2}\big((M_1, M_2), (N_1, N_2)\big) := \mathrm{Hom}_{\mathsf{C}_1}(M_1, N_1) \otimes \mathrm{Hom}_{\mathsf{C}_2}(M_2, N_2).$$

The formula for the composition involves the Koszul sign rule. Given morphisms $\phi_i \in \mathrm{Hom}_{\mathsf{C}_i}(L_i, M_i)^{d_i}$ and $\psi_i \in \mathrm{Hom}_{\mathsf{C}_i}(M_i, N_i)^{e_i}$, for $i = 1, 2$, their tensors are morphisms

$$\phi_1 \otimes \phi_2 \in \mathrm{Hom}_{\mathsf{C}_1 \otimes \mathsf{C}_2}\big((L_1, L_2), (M_1, M_2)\big)^{d_1 + d_2}$$

and

$$\psi_1 \otimes \psi_2 \in \mathrm{Hom}_{\mathsf{C}_1 \otimes \mathsf{C}_2}\big((M_1, M_2), (N_1, N_2)\big)^{e_1 + e_2}.$$

We define the composition to be

$$\begin{aligned}(\psi_1 \otimes \psi_2) \circ (\phi_1 \otimes \phi_2) &:= (-1)^{d_1 \cdot e_2} \cdot (\psi_1 \circ \phi_1) \otimes (\psi_2 \circ \phi_2) \\ &\in \mathrm{Hom}_{\mathsf{C}_1 \otimes \mathsf{C}_2}\big((L_1, L_2), (N_1, N_2)\big)^{d_1 + d_2 + e_1 + e_2}.\end{aligned}$$

Example 9.1.4 Suppose C_1 and C_2 are single-object DG categories, say $\mathrm{Ob}(\mathsf{C}_i) = \{x_i\}$. Then $\mathsf{C}_1 \otimes \mathsf{C}_2$ also has a single-object, say $y := (x_1, x_2)$. The endomorphism DG rings satisfy

$$\mathrm{End}_{\mathsf{C}_1 \otimes \mathsf{C}_2}(y) = \mathrm{End}_{\mathsf{C}_1}(x_1) \otimes \mathrm{End}_{\mathsf{C}_2}(x_2).$$

See formula 3.1.20 and Example 3.3.10.

DG functors between DG categories were introduced in Definition 3.5.1.

Definition 9.1.5 Let C_1, C_2 and D be DG categories. A *DG bifunctor*

$$F : \mathsf{C}_1 \times \mathsf{C}_2 \to \mathsf{D}$$

is, by definition, a DG functor $F : \mathsf{C}_1 \otimes \mathsf{C}_2 \to \mathsf{D}$, where $\mathsf{C}_1 \otimes \mathsf{C}_2$ is the DG category from Definition 9.1.3.

Warning: due to the signs that the composition of odd morphisms acquires, a DG bifunctor F does not become, after we forget the DG structure, a linear bifunctor in the sense of Definition 9.1.2.

Proposition 9.1.6 *Given a DG bifunctor* $F : \mathsf{C}_1 \times \mathsf{C}_2 \to \mathsf{D}$, *there are induced linear bifunctors on the strict subcategories*

$$\operatorname{Str}(F) : \operatorname{Str}(\mathsf{C}_1) \times \operatorname{Str}(\mathsf{C}_2) \to \operatorname{Str}(\mathsf{D})$$

and on the homotopy categories

$$\operatorname{Ho}(F) : \operatorname{Ho}(\mathsf{C}_1) \times \operatorname{Ho}(\mathsf{C}_2) \to \operatorname{Ho}(\mathsf{D}).$$

Exercise 9.1.7 Prove this proposition.

Let F be a DG bifunctor as in Definition 9.1.5. If we fix an object $M_1 \in \mathsf{C}_1$, then $F(M_1, -) : \mathsf{C}_2 \to \mathsf{D}$ is a DG functor. Similarly if we fix the second argument of F.

For DG bifunctors there are several options for contravariance. In Section 3.9 we talked about contravariant DG functors, and the opposite DG category C^{op} of a given DG category C.

Definition 9.1.8 Let C_1, C_2 and D be DG categories. A *DG bifunctor that is contravariant in the first or the second argument* is, by definition, a DG bifunctor

$$F : \mathsf{C}_1^{\diamond_1} \times \mathsf{C}_2^{\diamond_2} \to \mathsf{D}$$

as in Definition 9.1.5, where the symbols $\diamond_1$ and $\diamond_2$ are either $\langle\text{empty}\rangle$ or op, as the case may be.

Here are the two main examples of DG bifunctors. We give each of them in the commutative version and the noncommutative version.

Example 9.1.9 Consider a commutative ring A. The category of complexes of A-modules is the DG category $\mathbf{C}(A)$, and we take $\mathsf{C}_1 = \mathsf{C}_2 = \mathsf{D} := \mathbf{C}(A)$. For each pair of objects $M_1, M_2 \in \mathbf{C}(A)$ there is an object $F(M_1, M_2) := M_1 \otimes_A M_2 \in \mathbf{C}(A)$. This is the usual tensor product of complexes. We define the action of F on morphisms as follows: given $\phi_i \in \operatorname{Hom}_{\mathbf{C}(A)}(M_i, N_i)^{k_i}$, we let

$$F(\phi_1, \phi_2) := \phi_1 \otimes \phi_2 \in \operatorname{Hom}_{\mathbf{C}(A)}(F(M_1, M_2), F(N_1, N_2))^{k_1+k_2}.$$

The result is a DG bifunctor

$$F = (- \otimes_A -) : \mathbf{C}(A) \times \mathbf{C}(A) \to \mathbf{C}(A).$$

Example 9.1.10 Consider DG rings A_0, A_1, A_2 (possibly noncommutative, but $\mathbb{K}$-central). Define the DG categories $\mathsf{C}_i := \mathbf{C}(A_{i-1} \otimes A_i^{\mathrm{op}})$ and $\mathsf{D} := \mathbf{C}(A_0 \otimes A_2^{\mathrm{op}})$. For each pair of objects $M_1 \in \mathsf{C}_1$ and $M_2 \in \mathsf{C}_2$ there is a DG $\mathbb{K}$-module $F(M_1, M_2) := M_1 \otimes_{A_1} M_2$; see Definition 3.3.23. This has a canonical DG $(A_0 \otimes A_2^{\mathrm{op}})$-module structure:

$$(a_0 \otimes a_2)\cdot(m_1 \otimes m_2) := (-1)^{j_2\cdot(k_1+k_2)}\cdot(a_0\cdot m_1) \otimes (m_2\cdot a_2)$$

for elements $a_i \in A_i^{j_i}$ and $m_i \in M_i^{k_i}$. In this way $F(M_1, M_2)$ becomes an object of D. We define the action of F on morphisms as follows: given $\phi_i \in \mathrm{Hom}_{\mathsf{C}_i}(M_i, N_i)^{k_i}$, we let

$$F(\phi_1, \phi_2) := \phi_1 \otimes \phi_2 \in \mathrm{Hom}_{\mathsf{D}}(F(M_1, M_2), F(N_1, N_2))^{k_1+k_2}.$$

The result is a DG bifunctor

$$F = (- \otimes_{A_1} -) : \mathbf{C}(A_0 \otimes A_1^{\mathrm{op}}) \times \mathbf{C}(A_1 \otimes A_2^{\mathrm{op}}) \to \mathbf{C}(A_0 \otimes A_2^{\mathrm{op}}).$$

Compare this example to the one-sided construction in Example 4.6.2.

Example 9.1.11 Again we take a commutative ring A, but now our bifunctor F arises from Hom, and so there is contravariance in the first argument. We define the DG categories $\mathsf{C}_1 = \mathsf{C}_2 = \mathsf{D} := \mathbf{C}(A)$, the same as in Example 9.1.9. For every pair of objects $M_1, M_2 \in \mathbf{C}(A)$ there is an object $F(M_1, M_2) := \mathrm{Hom}_A(M_1, M_2) \in \mathbf{C}(A)$. This is the usual Hom complex. We define the action of F on morphisms as follows: given

$$\phi_1 \in \mathrm{Hom}_{\mathsf{C}_1^{\mathrm{op}}}(M_1, N_1)^{k_1} = \mathrm{Hom}_{\mathbf{C}(A)^{\mathrm{op}}}(M_1, N_1)^{k_1} = \mathrm{Hom}_A(N_1, M_1)^{k_1}$$

and

$$\phi_2 \in \mathrm{Hom}_{\mathsf{C}_2}(M_2, N_2)^{k_2} = \mathrm{Hom}_{\mathbf{C}(A)}(M_2, N_2)^{k_2} = \mathrm{Hom}_A(M_2, N_2)^{k_2}$$

we let

$$\begin{aligned} F(\phi_1, \phi_2) &:= \mathrm{Hom}(\phi_1, \phi_2) \in \mathrm{Hom}_A(\mathrm{Hom}_A(M_1, M_2), \mathrm{Hom}_A(N_1, N_2))^{k_1+k_2} \\ &= \mathrm{Hom}_{\mathsf{D}}(F(M_1, M_2), F(N_1, N_2))^{k_1+k_2}. \end{aligned}$$

The result is a DG bifunctor

$$F = \mathrm{Hom}_A(-, -) : \mathbf{C}(A)^{\mathrm{op}} \times \mathbf{C}(A) \to \mathbf{C}(A).$$

Example 9.1.12 Consider DG rings A_0, A_1, A_2 (possibly noncommutative, but $\mathbb{K}$-central). There is a DG bifunctor

$$F := \mathrm{Hom}_{A_1}(-, -) : \mathbf{C}(A_1 \otimes A_0^{\mathrm{op}})^{\mathrm{op}} \times \mathbf{C}(A_1 \otimes A_2^{\mathrm{op}}) \to \mathbf{C}(A_0 \otimes A_2^{\mathrm{op}}).$$

The details here are so confusing that we just leave them out. (We shall come back to this bifunctor in Chapters 14–18, and then we shall have to deal with the details.)

9.2 Triangulated Bifunctors

T-additive categories and triangulated categories were defined in Chapter 5.

Suppose $(\mathsf{K}_1, \mathrm{T}_1)$ and $(\mathsf{K}_2, \mathrm{T}_2)$ are T-additive categories (linear over the base ring $\mathbb{K}$). There are two induced translation automorphisms on the category $\mathsf{K}_1 \times \mathsf{K}_2$:

$$\mathrm{T}_1(M_1, M_2) := (\mathrm{T}_1(M_1), M_2) \tag{9.2.1}$$

and

$$\mathrm{T}_2(M_1, M_2) := (M_1, \mathrm{T}_2(M_2)). \tag{9.2.2}$$

These two functors commute: $\mathrm{T}_2 \circ \mathrm{T}_1 = \mathrm{T}_1 \circ \mathrm{T}_2$.

Definition 9.2.3 Let $(\mathsf{K}_1, \mathrm{T}_1)$, $(\mathsf{K}_2, \mathrm{T}_2)$ and (L, T) be T-additive categories. A *T-additive bifunctor*

$$(F, \tau_1, \tau_2) : (\mathsf{K}_1, \mathrm{T}_1) \times (\mathsf{K}_2, \mathrm{T}_2) \to (\mathsf{L}, \mathrm{T})$$

is made up of an additive bifunctor $F : \mathsf{K}_1 \times \mathsf{K}_2 \to \mathsf{L}$, as in Definition 9.1.2, together with isomorphisms $\tau_i : F \circ \mathrm{T}_i \xrightarrow{\simeq} \mathrm{T} \circ F$ of bifunctors $\mathsf{K}_1 \times \mathsf{K}_2 \to \mathsf{L}$. The condition is that

$$\tau_1 \circ \tau_2 = -\tau_2 \circ \tau_1 \tag{$\dagger$}$$

as isomorphisms of bifunctors

$$F \circ \mathrm{T}_2 \circ \mathrm{T}_1 = F \circ \mathrm{T}_1 \circ \mathrm{T}_2 \xRightarrow{\simeq} \mathrm{T} \circ \mathrm{T} \circ F.$$

The reason for the sign appearing in condition (†) will become clear in the proof of Lemma 9.2.14.

Let (F, τ_1, τ_2) be a T-additive bifunctor as in Definition 9.2.3. If we fix an object $M_1 \in \mathsf{K}_1$, then

$$(F(M_1, -), \tau_2) : (\mathsf{K}_2, \mathrm{T}_2) \to (\mathsf{L}, \mathrm{T})$$

is a T-additive functor. Similarly if we fix the second argument of F.

In the next exercises we let the reader establish several operations on T-additive bifunctors.

Exercise 9.2.4 In the situation of Definition 9.2.3, suppose $(G, \tau) : (\mathsf{L}, \mathrm{T}) \to (\mathsf{L}', \mathrm{T}')$ is a T-additive functor into a fourth T-additive category $(\mathsf{L}', \mathrm{T}')$. Write the explicit formula for the T-additive bifunctor

$$(G, \tau) \circ (F, \tau_1, \tau_2) : (\mathsf{K}_1, \mathrm{T}_1) \times (\mathsf{K}_2, \mathrm{T}_2) \to (\mathsf{L}', \mathrm{T}').$$

This should be compared to Definition 5.1.4.

Exercise 9.2.5 In the situation of Definition 9.2.3, suppose

$$(F', \tau'_1, \tau'_2) : (\mathsf{K}_1, \mathrm{T}_1) \times (\mathsf{K}_2, \mathrm{T}_2) \to (\mathsf{L}, \mathrm{T})$$

is another T-additive bifunctor. Write the definition of a morphism of T-additive bifunctors $\zeta : (F, \tau_1, \tau_2) \to (F', \tau'_1, \tau'_2)$. Use Definition 5.1.4 as a template.

Exercise 9.2.6 Give a definition of a T-additive trifunctor. Show that if F and G are T-additive bifunctors, then $G(-, F(-,-))$ and $G(F(-,-),-)$ are T-additive trifunctors (whenever these compositions makes sense).

We now move to triangulated categories.

Definition 9.2.7 Let $(\mathsf{K}_1, \mathrm{T}_1)$, $(\mathsf{K}_2, \mathrm{T}_2)$ and (L, T) be triangulated categories. A *triangulated bifunctor*

$$(F, \tau_1, \tau_2) : (\mathsf{K}_1, \mathrm{T}_1) \times (\mathsf{K}_2, \mathrm{T}_2) \to (\mathsf{L}, \mathrm{T})$$

is a T-additive bifunctor that respects the triangulated structure in each argument. Namely, for every distinguished triangle

$$L_1 \xrightarrow{\alpha_1} M_1 \xrightarrow{\beta_1} N_1 \xrightarrow{\gamma_1} \mathrm{T}_1(L_1)$$

in K_1, and every object $L_2 \in \mathsf{K}_2$, the triangle

$$F(L_1, L_2) \xrightarrow{F(\alpha_1, \mathrm{id})} F(M_1, L_2) \xrightarrow{F(\beta_1, \mathrm{id})} F(N_1, L_2) \xrightarrow{\tau_1 \circ F(\gamma_1, \mathrm{id})} \mathrm{T}(F(L_1, L_2))$$

in L is distinguished; and the same for distinguished triangles in the second argument.

The operations on triangulated bifunctors are the same as those on T-additive bifunctors (see exercises above).

We now discuss triangulated bifunctors in our favorite setup: DG modules in abelian categories. This is done both in the covariant and in the contravariant direction, in either argument. Recall that the opposite homotopy category $\mathbf{K}(A, \mathsf{M})^{\mathrm{op}}$ was given a canonical triangulated structure in Section 5.6.

Setup 9.2.8 We are given this data:

- DG rings A_1 and A_2, and abelian categories M_1 and M_2.
- Direction indicators $\diamond_1$ and $\diamond_2$, which are either ⟨empty⟩ or op.
- Full subcategories $\mathsf{C}_i \subseteq \mathbf{C}(A_i, \mathsf{M}_i)$, whose homotopy categories $\mathsf{K}_i := \mathrm{Ho}(\mathsf{C}_i) \subseteq \mathbf{K}(A_i, \mathsf{M}_i)$ are such that $\mathsf{K}_i^{\diamond_i}$ is a triangulated subcategory of $\mathbf{K}(A_i, \mathsf{M}_i)^{\diamond_i}$.

The translation functor of the category $\mathbf{K}(A_i, \mathsf{M}_i)$ is denoted by T_i, and the translation functor of the category $\mathbf{K}(A_i, \mathsf{M}_i)^{\diamond_i}$ is denoted by $\mathrm{T}_i^{\diamond_i}$.

Definition 9.2.9 Under Setup 9.2.8, let L be another triangulated category. A *triangulated bifunctor from* $(\mathsf{K}_1, \mathsf{K}_2)$ *to* L *that is contravariant in the first or the second argument* is, by definition, a triangulated bifunctor

$$(F, \tau_1, \tau_2) : (\mathsf{K}_1^{\diamond_1}, \mathrm{T}_1^{\diamond_1}) \times (\mathsf{K}_2^{\diamond_2}, \mathrm{T}_2^{\diamond_2}) \to (\mathsf{L}, \mathrm{T})$$

as in Definition 9.2.7, where the contravariance is according to the direction indicators $\diamond_1$ and $\diamond_2$.

Next we explain how to obtain triangulated bifunctors from DG bifunctors. This is done in the following setup:

Setup 9.2.10 In addition to the data from Setup 9.2.8, we are given:

- A DG ring B and an abelian category N.
- A DG bifunctor $F : \mathsf{C}_1^{\diamond_1} \times \mathsf{C}_2^{\diamond_2} \to \mathbf{C}(B, \mathsf{N})$.

The translation functor of the category $\mathbf{K}(B, \mathsf{N})$ is denoted by T. For every object $(M_1, M_2) \in \mathsf{C}_1 \times \mathsf{C}_2$ there are isomorphisms

$$\tau_{i, M_1, M_2} : F(\mathrm{T}_i(M_1, M_2)) \xrightarrow{\simeq} \mathrm{T}(F(M_1, M_2)) \tag{9.2.11}$$

in $\mathbf{C}(B, \mathsf{N})$, arising from Definition 4.4.1. Let us make this explicit (only for $i = 2$, since the case $i = 1$ is similar). Fixing the object M_1 we obtain a DG functor $G : \mathsf{C}_2 \to \mathbf{C}(B, \mathsf{N})$, $G(M_2) := F(M_1, M_2)$. The isomorphism

$$\tau_{2, M_1, M_2} : G(\mathrm{T}_2(M_2)) \xrightarrow{\simeq} \mathrm{T}(G(M_2))$$

is then

$$\tau_{2, M_1, M_2} = \mathrm{t}_{G(M_2)} \circ G(\mathrm{t}_{M_2})^{-1}. \tag{9.2.12}$$

We are using the little t operator here.

Lemma 9.2.13 *Under Setups* 9.2.8 *and* 9.2.10, *assume that the direction indicators* $\diamond_1$ *and* $\diamond_2$ *are both* ⟨empty⟩. *Fix* $i \in \{1, 2\}$. *Letting the object* (M_1, M_2) *in* (9.2.11) *vary, we get an isomorphism* $\tau_i : F \circ \mathrm{T}_i \xrightarrow{\simeq} \mathrm{T} \circ F$ *of additive bifunctors* $\mathrm{Str}(\mathsf{C}_1) \times \mathrm{Str}(\mathsf{C}_2) \to \mathbf{C}_{\mathrm{str}}(B, \mathsf{N})$.

Proof This is an almost immediate consequence of the fact that the little t operators are morphisms of functors (see Theorem 4.1.8(2)), □

According to Proposition 9.1.6, the DG bifunctor F induces an additive bifunctor $F : \mathsf{K}_1^{\diamond_1} \times \mathsf{K}_2^{\diamond_2} \to \mathbf{K}(B, \mathsf{N})$ on the homotopy categories.

Lemma 9.2.14 *Under Setups* 9.2.8 *and* 9.2.10, *assume that the direction indicators* $\diamond_1$ *and* $\diamond_2$ *are both* ⟨empty⟩. *Then*

$$(F, \tau_1, \tau_2) : \mathsf{K}_1 \times \mathsf{K}_2 \to \mathbf{K}(B, \mathsf{N})$$

is a triangulated bifunctor.

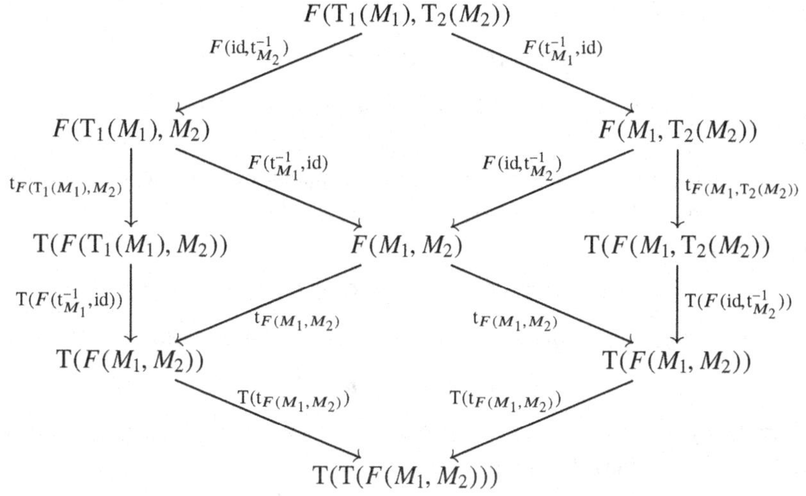

(9.2.15)

Proof The only challenge is to prove that (F, τ_1, τ_2) is a T-additive bifunctor; and in that, all we have to prove is that condition (†) in Definition 9.2.2 holds, namely that $\tau_1 \circ \tau_2 = -\tau_2 \circ \tau_1$. The rest hinges on single-argument considerations, which were already handled in Theorems 4.4.3 and 5.5.1.

So let us prove the equality (†). Choose a pair of objects (M_1, M_2). We have diagram (9.2.15) in the category $\mathsf{C}(B, \mathsf{N})$. Going from top to bottom on the left edge is the morphism $\tau_1 \circ \tau_2$, and going on the right edge is the morphism $\tau_2 \circ \tau_1$. The bottom diamond is trivially commutative. The two triangles, with common vertex at $F(M_1, M_2)$, are (-1)-commutative, because $\mathrm{t} : \mathrm{Id} \to \mathrm{T}$ is a degree -1 morphism of DG functors. Since they occur on both sides, these signs cancel each other. Finally, the top diamond is (-1)-commutative, because

$$(\mathrm{t}_{M_1}^{-1}, \mathrm{id}) \circ (\mathrm{id}, \mathrm{t}_{M_2}^{-1}) = (\mathrm{t}_{M_1}^{-1}, \mathrm{t}_{M_2}^{-1}) = -(\mathrm{id}, \mathrm{t}_{M_2}^{-1}) \circ (\mathrm{t}_{M_1}^{-1}, \mathrm{id}). \qquad \square$$

Theorem 9.2.16 *Under Setups 9.2.8 and 9.2.10, there are canonical translation isomorphisms τ_1 and τ_2 such that*

$$(F, \tau_1, \tau_2) : (\mathsf{K}_1^{\diamond_1}, \mathrm{T}_1^{\diamond_1}) \times (\mathsf{K}_2^{\diamond_2}, \mathrm{T}_2^{\diamond_2}) \to (\mathbf{K}(B, \mathsf{N}), \mathrm{T})$$

is a triangulated bifunctor.

Proof For $i = 1, 2$ let us define the indicator

$$\heartsuit_i := \begin{cases} \text{flip} & \text{if } \diamond_i = \text{op}, \\ \langle\text{empty}\rangle & \text{if } \diamond_i = \langle\text{empty}\rangle. \end{cases}$$

We get DG categories $\mathsf{C}_i^{\diamondsuit_i}$: if $\diamondsuit_i = \langle\text{empty}\rangle$ then $\mathsf{C}_i^{\diamondsuit_i} = \mathsf{C}_i \subseteq \mathbf{C}(A_i, \mathsf{M}_i)$, and if $\diamondsuit_i = \text{op}$ then

$$\mathsf{C}_i^{\diamondsuit_i} = \mathsf{C}_i^{\text{flip}} = \text{Flip}(\mathsf{C}_i^{\text{op}}) \subseteq \mathbf{C}(A_i, \mathsf{M}_i)^{\text{flip}} = \mathbf{C}(A_i^{\text{op}}, \mathsf{M}_i^{\text{op}}),$$

as in the proof of Theorem 5.6.10. Likewise, there are triangulated categories $\mathsf{K}_i^{\diamondsuit_i}$.

We start by composing F with the isomorphism of DG categories Flip^{-1} from Theorem 3.9.16, in each coordinate i for which $\diamondsuit_i = \text{op}$. This gives a new DG bifunctor

$$F' : \mathsf{C}_1^{\diamondsuit_1} \times \mathsf{C}_2^{\diamondsuit_2} \to \mathbf{C}(B, \mathsf{N}).$$

Now we can apply Lemma 9.2.14, to get a triangulated bifunctor

$$F' : \mathsf{K}_1^{\diamondsuit_1} \times \mathsf{K}_2^{\diamondsuit_2} \to \mathbf{K}(B, \mathsf{N}).$$

Specifically, this means that we have a pair of translations (τ'_1, τ'_2) such that (F', τ'_1, τ'_2) is a triangulated bifunctor.

Finally we compose F' with the isomorphism of triangulated categories Flip from formula (5.6.6), in each coordinate i for which $\diamondsuit_i = \text{op}$. This recovers the original bifunctor F on $\mathsf{K}_1^{\diamondsuit_1} \times \mathsf{K}_2^{\diamondsuit_2}$. There are translations τ_i, which are gotten from the translations τ'_i such as in the proof of Theorem 5.6.10. □

9.3 Derived Bifunctors

Here we explain what are right and left derived bifunctors of triangulated bifunctors. The definitions and results are very similar to the single-argument case. For the sake of simplicity, we shall mostly ignore the translation functors and the translation isomorphisms; enough was said about them in the previous section.

The next setup will be used throughout this section.

Setup 9.3.1 The following are given:

- Triangulated categories K_1, K_2 and E.
- A triangulated bifunctor $F : \mathsf{K}_1 \times \mathsf{K}_2 \to \mathsf{E}$.
- Denominator sets of cohomological origin $\mathsf{S}_1 \subseteq \mathsf{K}_1$ and $\mathsf{S}_2 \subseteq \mathsf{K}_2$.

We write $\mathsf{D}_i := (\mathsf{K}_i)_{\mathsf{S}_i}$, and $\mathsf{Q}_i : \mathsf{K}_i \to \mathsf{D}_i$ are the localization functors.

The localized category $\mathsf{D}_i := (\mathsf{K}_i)_{\mathsf{S}_i}$ is triangulated, and the localization functor $\mathsf{Q}_i : \mathsf{K}_i \to \mathsf{D}_i$ is triangulated. On the product categories we get a functor

$$\mathsf{Q}_1 \times \mathsf{Q}_2 : \mathsf{K}_1 \times \mathsf{K}_2 \to \mathsf{D}_1 \times \mathsf{D}_2 .$$

Before embarking on the definitions of the derived bifunctors, we want to introduce the relevant categories of functors. These are the bifunctor variants of what we had in Section 8.2.

Definition 9.3.2 We denote by $\mathsf{LinBiFun}(\mathsf{K}_1 \times \mathsf{K}_2, \mathsf{E})$ the category whose objects are the linear bifunctors $F : \mathsf{K}_1 \times \mathsf{K}_2 \to \mathsf{E}$. The morphisms are the obvious ones.

This is a linear category. Actually, this category is the same as the category $\mathsf{LinFun}(\mathsf{K}_1 \otimes \mathsf{K}_2, \mathsf{E})$ of linear functors $F : \mathsf{K}_1 \otimes \mathsf{K}_2 \to \mathsf{E}$, only with the extra information that the source is a product category.

Definition 9.3.3 We denote by $\mathsf{TrBiFun}(\mathsf{K}_1 \times \mathsf{K}_2, \mathsf{E})$ the category whose objects are the $\mathbb{K}$-linear triangulated bifunctors $(F, \tau_1, \tau_2) : \mathsf{K}_1 \times \mathsf{K}_2 \to \mathsf{E}$. The morphisms are those of T-additive bifunctors.

There is a functor

$$\mathsf{TrBiFun}(\mathsf{K}_1 \times \mathsf{K}_2, \mathsf{E}) \to \mathsf{LinBiFun}(\mathsf{K}_1 \times \mathsf{K}_2, \mathsf{E}) \tag{9.3.4}$$

that forgets (τ_1, τ_2). It is a faithful additive functor.

Suppose $U_i : \mathsf{K}'_i \to \mathsf{K}_i$ are triangulated functors between triangulated categories. We get an induced additive functor

$$\mathsf{Fun}(U_1 \times U_2, \mathrm{Id}_{\mathsf{E}}) : \mathsf{TrBiFun}(\mathsf{K}_1 \times \mathsf{K}_2, \mathsf{E}) \to \mathsf{TrBiFun}(\mathsf{K}'_1 \times \mathsf{K}'_2, \mathsf{E}) \tag{9.3.5}$$

The formula is $F \mapsto F \circ (U_1 \times U_2)$.

Lemma 9.3.6 *If the functors U_1 and U_2 are equivalences, then the functor in* (9.3.5) *is an equivalence.*

Proof This is basically the same as the proof of Lemma 8.4.4. □

As in Definition 8.2.5 we denote by

$$\mathsf{TrBiFun}_{\mathsf{S}_1 \times \mathsf{S}_2}(\mathsf{K}_1 \times \mathsf{K}_2, \mathsf{E}) \subseteq \mathsf{TrBiFun}(\mathsf{K}_1 \times \mathsf{K}_2, \mathsf{E})$$

the full subcategory on the triangulated bifunctors F such that $F(\mathsf{S}_1 \times \mathsf{S}_2) \subseteq \mathsf{E}^{\times}$.

Lemma 9.3.7 *The functor*

$$\mathsf{Fun}(\mathsf{Q}_1 \times \mathsf{Q}_2, \mathrm{Id}_{\mathsf{E}}) : \mathsf{TrBiFun}(\mathsf{D}_1 \times \mathsf{D}_2, \mathsf{E}) \to \mathsf{TrBiFun}_{\mathsf{S}_1 \times \mathsf{S}_2}(\mathsf{K}_1 \times \mathsf{K}_2, \mathsf{E})$$

is an isomorphism of categories.

Proof This is basically the same as the proof of Lemma 8.4.6, when combined with the isomorphism of triangulated categories $(\mathsf{K}_1 \times \mathsf{K}_2)_{\mathsf{S}_1 \times \mathsf{S}_2} \to \mathsf{D}_1 \times \mathsf{D}_2$ from Proposition 8.2.11. □

We are ready to talk about derived bifunctors.

Definition 9.3.8 Under Setup 9.3.1, a *triangulated right derived bifunctor of F with respect to* $\mathsf{S}_1 \times \mathsf{S}_2$ is a pair $(\mathrm{R}F, \eta^{\mathrm{R}})$, where

$$\mathrm{R}F : \mathsf{D}_1 \times \mathsf{D}_2 \to \mathsf{E}$$

is a triangulated bifunctor, and

$$\eta^{\mathrm{R}} : F \Rightarrow \mathrm{R}F \circ (\mathrm{Q}_1 \times \mathrm{Q}_2)$$

is a morphism of triangulated bifunctors, such that the following universal property holds:

(R) Given a pair (G, θ), consisting of a triangulated bifunctor $G : \mathsf{D}_1 \times \mathsf{D}_2 \to \mathsf{E}$ and a morphism of triangulated bifunctors $\theta : F \Rightarrow G \circ (\mathrm{Q}_1 \times \mathrm{Q}_2)$, there is a unique morphism of triangulated bifunctors $\mu : \mathrm{R}F \Rightarrow G$ such that $\theta = (\mu \circ \mathrm{id}_{\mathrm{Q}_1 \times \mathrm{Q}_2}) * \eta^{\mathrm{R}}$.

Here is a diagram showing property (R):

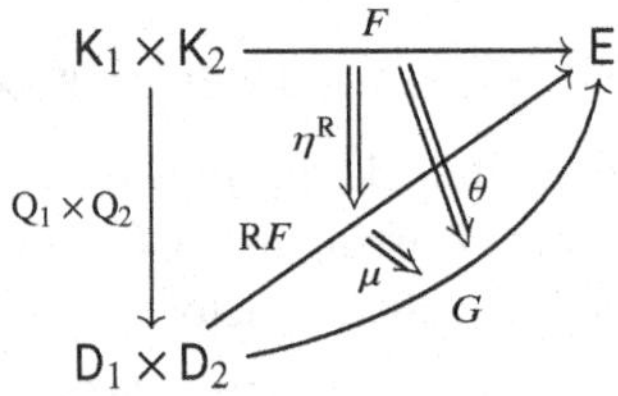

(9.3.9)

Proposition 9.3.10 *If a triangulated right derived bifunctor exists, then it is unique up to a unique isomorphism.*

Proof This is just like the proof of Proposition 8.4.8. We leave the small changes to the reader. □

Existence in general is like Theorem 8.4.9, but a bit more complicated.

Theorem 9.3.11 *Under Setup* 9.3.1, *assume there are full triangulated subcategories* $\mathsf{J}_1 \subseteq \mathsf{K}_1$ *and* $\mathsf{J}_2 \subset \mathsf{K}_2$ *with these two properties:*

(a) *If* $\phi_1 : I_1 \to J_1$ *is a morphism in* $\mathsf{J}_1 \cap \mathsf{S}_1$ *and* $\phi_2 : I_2 \to J_2$ *is a morphism in* $\mathsf{J}_2 \cap \mathsf{S}_2$, *then*

$$F(\phi_1, \phi_2) : F(I_1, I_2) \to F(J_1, J_2)$$

is an isomorphism in E.

(b) *For every* $p \in \{1, 2\}$ *and every object* $M_p \in \mathsf{K}_p$, *there exists a morphism* $\rho_p : M_p \to I_p$ *in* S_p *with target* $I_p \in \mathsf{J}_p$.

Then the triangulated right derived bifunctor

$$(\mathrm{R}F, \eta^{\mathrm{R}}) : \mathsf{D}_1 \times \mathsf{D}_2 \to \mathsf{E}$$

of F with respect to $\mathsf{S}_1 \times \mathsf{S}_2$ exists. Moreover, for every pair of objects $I_1 \in \mathsf{J}_1$ and $I_2 \in \mathsf{J}_2$ the morphism

$$\eta^{\mathrm{R}}_{I_1,I_2} : F(I_1, I_2) \to \mathrm{R}F(I_1, I_2)$$

in E is an isomorphism.

In applications we will see that either $\mathsf{J}_1 = \mathsf{K}_1$ or $\mathsf{J}_2 = \mathsf{K}_2$; namely we will only need to resolve in the second or in the first argument, respectively.

Proof It will be convenient to change notation. For $p = 1, 2$ let's define $\mathsf{K}'_p := \mathsf{J}_p$, $\mathsf{S}'_p := \mathsf{K}'_p \cap \mathsf{S}_p$ and $\mathsf{D}'_p := (\mathsf{K}'_p)_{\mathsf{S}'_p}$. The localization functors are $\mathrm{Q}'_p : \mathsf{K}'_p \to \mathsf{D}'_p$. The inclusions are $U_p : \mathsf{K}'_p \to \mathsf{K}_p$, and their localizations are the functors $V_p : \mathsf{D}'_p \to \mathsf{D}_p$. By Lemma 8.3.8 the functors V_p are equivalences.

The situation is depicted in these diagrams. We have this commutative diagram of products of triangulated functors between products of triangulated categories:

$$\begin{array}{ccc} \mathsf{K}'_1 \times \mathsf{K}'_2 & \xrightarrow{U_1 \times U_2} & \mathsf{K}_1 \times \mathsf{K}_2 \\ {\scriptstyle \mathrm{Q}'_1 \times \mathrm{Q}'_2}\big\downarrow & & \big\downarrow{\scriptstyle \mathrm{Q}_1 \times \mathrm{Q}_2} \\ \mathsf{D}'_1 \times \mathsf{D}'_2 & \xrightarrow{V_1 \times V_2} & \mathsf{D}_1 \times \mathsf{D}_2 \end{array} \tag{9.3.12}$$

The arrow $V_1 \times V_2$ is an equivalence. Diagram (9.3.12) induces a commutative diagram of linear categories:

$$\begin{array}{ccc} \mathsf{TrBiFun}(\mathsf{K}'_1 \times \mathsf{K}'_2, \mathsf{E}) & \xleftarrow{\mathsf{Fun}(U_1 \times U_2, \mathrm{Id})} & \mathsf{TrBiFun}(\mathsf{K}_1 \times \mathsf{K}_2, \mathsf{E}) \\ \big\uparrow{\scriptstyle \text{f.f. emb}} & & {\scriptstyle \text{f.f. emb}}\big\uparrow \\ \mathsf{TrBiFun}_{\mathsf{S}'_1 \times \mathsf{S}'_2}(\mathsf{K}'_1 \times \mathsf{K}'_2, \mathsf{E}) & \xleftarrow[\text{equiv}]{\mathsf{Fun}(U_1 \times U_2, \mathrm{Id})} & \mathsf{TrBiFun}_{\mathsf{S}_1 \times \mathsf{S}_2}(\mathsf{K}_1 \times \mathsf{K}_2, \mathsf{E}) \\ {\scriptstyle \mathsf{Fun}(\mathrm{Q}'_1 \times \mathrm{Q}'_2, \mathrm{Id})}\big\uparrow{\scriptstyle \text{isom}} & & {\scriptstyle \text{isom}}\big\uparrow{\scriptstyle \mathsf{Fun}(\mathrm{Q}_1 \times \mathrm{Q}_2, \mathrm{Id})} \\ \mathsf{TrBiFun}(\mathsf{D}'_1 \times \mathsf{D}'_2, \mathsf{E}) & \xleftarrow[\text{equiv}]{\mathsf{Fun}(V_1 \times V_2, \mathrm{Id})} & \mathsf{TrBiFun}(\mathsf{D}_1 \times \mathsf{D}_2, \mathsf{E}) \end{array} \tag{9.3.13}$$

According to Lemmas 9.3.6 and 9.3.7, the arrows in the diagram above that are marked "isom" or "equiv" are isomorphisms or equivalences, respectively. By definition the arrows marked "f.f. emb" are fully faithful embeddings.

We know that $\mathsf{S}_p \subseteq \mathsf{K}_p$ are left denominator sets. Therefore (see Proposition 8.2.13) $\mathsf{S}_1 \times \mathsf{S}_2 \subseteq \mathsf{K}_1 \times \mathsf{K}_2$ is a left denominator set. Condition (a) of Theorem 9.3.11 says that F sends morphisms in $\mathsf{S}'_1 \times \mathsf{S}'_2$ to isomorphisms in E. Condition (b) there says that there are enough right $(\mathsf{K}'_1 \times \mathsf{K}'_2)$-resolutions in $\mathsf{K}_1 \times \mathsf{K}_2$.

Thus we are in a position to use the abstract Theorem 8.3.3. It says that there is an abstract right derived functor $(\mathrm{R}F, \eta^{\mathrm{R}}) : \mathsf{D}_1 \times \mathsf{D}_2 \to \mathsf{E}$. However, going over the proof of Theorem 8.3.3, we see that all constructions there can be made within the triangulated setting, namely in diagram (9.3.13) instead of in diagram (8.3.11). Therefore $\mathrm{R}F$ is an object of the category in the bottom right corner of (9.3.13), and the morphism $\eta^{\mathrm{R}} : F \Rightarrow \mathrm{R}F \circ (\mathrm{Q}_1 \times \mathrm{Q}_2)$ is in the category in the top right corner of (9.3.13). □

We now treat left derived bifunctors. Due to the similarity to the right derived functors, we give only a few details here.

Definition 9.3.14 Under Setup 9.3.1, a *triangulated left derived bifunctor of F with respect to* $\mathsf{S}_1 \times \mathsf{S}_2$ is a pair $(\mathrm{L}F, \eta^{\mathrm{L}})$, where

$$\mathrm{L}F : \mathsf{D}_1 \times \mathsf{D}_2 \to \mathsf{E}$$

is a triangulated bifunctor, and

$$\eta^{\mathrm{L}} : \mathrm{L}F \circ (\mathrm{Q}_1 \times \mathrm{Q}_2) \Rightarrow F$$

is a morphism of triangulated bifunctors, such that the following universal property holds:

(L) Given a pair (G, θ), consisting of a triangulated bifunctor $G : \mathsf{D}_1 \times \mathsf{D}_2 \to \mathsf{E}$ and a morphism of triangulated bifunctors $\theta : G \circ (\mathrm{Q}_1 \times \mathrm{Q}_2) \Rightarrow F$, there is a unique morphism of triangulated bifunctors $\mu : G \Rightarrow \mathrm{L}F$ such that $\theta = \eta^{\mathrm{L}} * (\mu \circ \mathrm{id}_{\mathrm{Q}_1 \times \mathrm{Q}_2})$.

Proposition 9.3.15 *If a triangulated left derived bifunctor exists, then it is unique up to a unique isomorphism.*

Proof This is the opposite of Proposition 9.3.10, and we leave it to the reader to make the adjustments. □

Theorem 9.3.16 *In the situation of Definition* 9.3.14, *assume there are full triangulated subcategories* $\mathsf{P}_1 \subseteq \mathsf{K}_1$ *and* $\mathsf{P}_2 \subseteq \mathsf{K}_2$ *with these two properties:*

(a) *If* $\phi_1 : P_1 \to Q_1$ *is a morphism in* $\mathsf{P}_1 \cap \mathsf{S}_1$ *and* $\phi_2 : P_2 \to Q_2$ *is a morphism in* $\mathsf{P}_2 \cap \mathsf{S}_2$*, then*

$$F(\phi_1, \phi_2) : F(P_1, P_2) \to F(Q_1, Q_2)$$

is an isomorphism in E.

(b) *For every* $i \in \{1, 2\}$ *and every object* $M_i \in \mathsf{K}_i$*, there exists a morphism* $\rho_i : P_i \to M_i$ *in* S_i *with target* $P_i \in \mathsf{P}_i$.

Then the triangulated left derived bifunctor

$$(\mathrm{L}F, \eta^{\mathrm{L}}) : \mathsf{D}_1 \times \mathsf{D}_2 \to \mathsf{E}$$

of F with respect to $\mathsf{S}_1 \times \mathsf{S}_2$ exists. Moreover, for every pair of objects $P_1 \in \mathsf{P}_1$ and $P_2 \in \mathsf{P}_2$ the morphism

$$\eta^{\mathrm{L}}_{P_1,P_2} : \mathrm{L}F(P_1, P_2) \to F(P_1, P_2)$$

in E is an isomorphism.

Exercise 9.3.17 Prove Theorem 9.3.16. (Hint: This is the opposite of Theorem 9.3.11. The proof is analogous.)

In applications we will see that either $\mathsf{P}_1 = \mathsf{K}_1$ or $\mathsf{P}_2 = \mathsf{K}_2$; namely we will only need to resolve in the second or in the first argument, respectively.

There are more specialized definitions of triangulated derived bifunctors whose sources are categories of DG modules, like Definitions 8.4.16 and 8.4.26. We leave them to the reader.

Remark 9.3.18 Deriving triangulated bifunctors that are contravariant in one or two arguments involves passage to opposite categories; see Definition 9.2.9. Proposition 7.6.3 tells us that when $\mathsf{K}_i^{\mathrm{op}} \subseteq \mathbf{K}(A_i, \mathsf{M}_i)^{\mathrm{op}}$ are full triangulated subcategories, the sets $\mathsf{S}_i^{\mathrm{op}}$ of quasi-isomorphisms in them are of cohomological origin. So we are still within the conditions of Setup 9.3.1, and Theorems 9.3.11 and 9.3.16 can be applied. We leave it to the reader to work out the details in the various cases. However, the important case of the bifunctor $\mathrm{Hom}(-,-)$ will be given a full treatment in Section 12.2.

10

Resolving Subcategories of $\mathbf{K}(A, \mathsf{M})$

In this chapter we are back in the more concrete setting: A is a DG ring, and M is an abelian category, both over the commutative base ring $\mathbb{K}$; see Convention 1.2.4. We will define *K-projective* and *K-injective* DG modules in $\mathbf{K}(A, \mathsf{M})$. These DG modules form full triangulated subcategories of $\mathbf{K}(A, \mathsf{M})$ and are concrete versions of the abstract categories P and J that played important roles in Section 8.4. For $\mathbf{K}(A)$ we also define *K-flat DG modules*.

10.1 K-Injective DG Modules

For every integer p we have the pth cohomology functor $\mathrm{H}^p : \mathbf{C}_{\mathrm{str}}(A, \mathsf{M}) \to \mathsf{M}$. There is equality $\mathrm{H}^p = \mathrm{H}^0 \circ \mathrm{T}^p$, where T is the translation functor of $\mathbf{C}(A, \mathsf{M})$. The functors H^p pass to the homotopy category, and $\mathrm{H}^0 : \mathbf{K}(A, \mathsf{M}) \to \mathsf{M}$ is a cohomological functor, in the sense of Definition 5.2.5.

Definition 10.1.1 A DG module $N \in \mathbf{C}(A, \mathsf{M})$ is called *acyclic* if $\mathrm{H}^p(N) = 0$ for all p.

Definition 10.1.2 A DG module $I \in \mathbf{C}(A, \mathsf{M})$ is called *K-injective* if for every acyclic DG module $N \in \mathbf{C}(A, \mathsf{M})$, the DG $\mathbb{K}$-module $\mathrm{Hom}_{A,\mathsf{M}}(N, I)$ is acyclic.

The definition above characterizes K-injectives as objects of $\mathbf{C}(A, \mathsf{M})$. The next proposition shows that being K-injective is intrinsic to the triangulated category $\mathbf{K}(A, \mathsf{M})$, with the cohomological functor H^0 (which tells us which are the acyclic objects).

Proposition 10.1.3 *A DG module $I \in \mathbf{K}(A, \mathsf{M})$ is K-injective if and only if* $\operatorname{Hom}_{\mathbf{K}(A,\mathsf{M})}(N, I) = 0$ *for every acyclic DG module* $N \in \mathbf{K}(A, \mathsf{M})$.

Proof This is because for every integer p we have

$$\mathrm{H}^p(\operatorname{Hom}_{A,\mathsf{M}}(N, I)) \cong \mathrm{H}^0(\operatorname{Hom}_{A,\mathsf{M}}(\mathrm{T}^{-p}(N), I)) = \operatorname{Hom}_{\mathbf{K}(A,\mathsf{M})}(\mathrm{T}^{-p}(N), I),$$

and N is acyclic iff $\mathrm{T}^{-p}(N)$ is acyclic. □

Definition 10.1.4 Let $M \in \mathbf{C}(A, \mathsf{M})$. A *K-injective resolution* of M is a quasi-isomorphism $\rho : M \to I$ in $\mathbf{C}_{\mathrm{str}}(A, \mathsf{M})$, where I is a K-injective DG module.

Remark 10.1.5 Let $\rho : M \to I$ be a K-injective resolution in $\mathbf{C}_{\mathrm{str}}(A, \mathsf{M})$. By a slight abuse of notation, we shall often refer to the induced quasi-isomorphism $\mathrm{P}(\rho) : M \to I$ in $\mathbf{K}(A, \mathsf{M})$ as a K-injective resolution of M. Cf. Definitions 10.1.11 and 10.1.21 below.

In the next chapter we will prove existence of K-injective resolutions in several contexts. Here is an easy example of a K-injective complex.

Exercise 10.1.6 Let $I \in \mathbf{K}(\mathsf{M})$ be a complex of injective objects of M, with zero differential. Prove that I is K-injective.

Definition 10.1.7 Let K be a full subcategory of $\mathbf{K}(A, \mathsf{M})$. The full subcategory of K on the K-injective DG modules in it is denoted by $\mathsf{K}_{\mathrm{inj}}$. In other words, $\mathsf{K}_{\mathrm{inj}} = \mathbf{K}(A, \mathsf{M})_{\mathrm{inj}} \cap \mathsf{K}$.

Remark 10.1.8 Warning: the property of a DG module $I \in \mathsf{K}$ being K-injective is in general not intrinsic to the subcategory $\mathsf{K} \subseteq \mathbf{K}(A, \mathsf{M})$. This is because the condition in Proposition 10.1.3 (and in Definition 10.1.2) has to be tested against all acyclic DG modules $N \in \mathbf{K}(A, \mathsf{M})$.

Proposition 10.1.9 *If* K *is a full triangulated subcategory of* $\mathbf{K}(A, \mathsf{M})$, *then* $\mathsf{K}_{\mathrm{inj}}$ *is a full triangulated subcategory of* K.

Proof It suffices to prove that $\mathbf{K}(A, \mathsf{M})_{\mathrm{inj}}$ is a triangulated subcategory of $\mathbf{K}(A, \mathsf{M})$. It is easy to see that $\mathbf{K}(A, \mathsf{M})_{\mathrm{inj}}$ is closed under translations. Suppose $I \to J \to K \xrightarrow{\triangle}$ is a distinguished triangle in $\mathbf{K}(A, \mathsf{M})$ such that I, J are K-injective DG modules. We have to show that K is also K-injective. Take any acyclic DG module $N \in \mathbf{K}(A, \mathsf{M})$. There is an exact sequence

$$\operatorname{Hom}_{\mathbf{K}(A,\mathsf{M})}(N, J) \to \operatorname{Hom}_{\mathbf{K}(A,\mathsf{M})}(N, K) \to \operatorname{Hom}_{\mathbf{K}(A,\mathsf{M})}(N, \mathrm{T}(I))$$

in $\mathsf{Mod}\,\mathbb{K}$. Because J and $\mathrm{T}(I)$ are K-injectives, Proposition 10.1.3 says that

$$\operatorname{Hom}_{\mathbf{K}(A,\mathsf{M})}(N, J) = 0 = \operatorname{Hom}_{\mathbf{K}(A,\mathsf{M})}(N, \mathrm{T}(I)).$$

Therefore $\operatorname{Hom}_{\mathbf{K}(A,\mathsf{M})}(N, K) = 0$. But N is an arbitrary acyclic DG module, so K is K-injective. □

Example 10.1.10 Let $\star$ be some boundedness condition (namely b, + or −). We know that $\mathbf{K}^{\star}(A, \mathsf{M})$ is a full triangulated subcategory of $\mathbf{K}(A, \mathsf{M})$. Hence $\mathbf{K}^{\star}(A, \mathsf{M})_{\mathrm{inj}}$ is a triangulated subcategory too.

Definition 10.1.11 Let K be a full triangulated subcategory of $\mathbf{K}(A, \mathsf{M})$. We say that K *has enough K-injectives* if every DG module $M \in \mathsf{K}$ admits a K-injective resolution inside K. In other words, there is a quasi-isomorphism $\rho : M \to I$ where $I \in \mathsf{K}_{\mathrm{inj}}$.

Here is the crucial fact regarding K-injectives.

Lemma 10.1.12 *Let* $s : I \to M$ *be a quasi-isomorphism in* $\mathbf{K}(A, \mathsf{M})$*, and assume* I *is K-injective. Then* s *has a left inverse, namely there is a morphism* $t : M \to I$ *in* $\mathbf{K}(A, \mathsf{M})$ *such that* $t \circ s = \mathrm{id}_I$*. Moreover, this morphism* t *is unique.*

Proof Consider a distinguished triangle $I \xrightarrow{s} M \to N \xrightarrow{\triangle}$ in $\mathbf{K}(A, \mathsf{M})$ that's built on s. The long exact cohomology sequence tells us that N is an acyclic DG module. Therefore $\mathrm{Hom}_{\mathbf{K}(A,\mathsf{M})}(\mathrm{T}^p(N), I) = 0$ for all p. The exact sequence

$$\begin{gathered}\mathrm{Hom}_{\mathbf{K}(A,\mathsf{M})}(N, I) \to \mathrm{Hom}_{\mathbf{K}(A,\mathsf{M})}(M, I)\\ \to \mathrm{Hom}_{\mathbf{K}(A,\mathsf{M})}(I, I) \to \mathrm{Hom}_{\mathbf{K}(A,\mathsf{M})}(\mathrm{T}^{-1}(N), I)\end{gathered}$$

shows that the homomorphism

$$(-) \circ s : \mathrm{Hom}_{\mathbf{K}(A,\mathsf{M})}(M, I) \xrightarrow{\simeq} \mathrm{Hom}_{\mathbf{K}(A,\mathsf{M})}(I, I)$$

is a bijection. We take $t : M \to I$ to be the unique morphism in $\mathbf{K}(A, \mathsf{M})$ such that $t \circ s = \mathrm{id}_I$. □

Theorem 10.1.13 *Let* A *be a DG ring, let* M *be an abelian category, and let* K *be a full triangulated subcategory of* $\mathbf{K}(A, \mathsf{M})$*. Write* $\mathsf{S} := \mathsf{K} \cap \mathbf{S}(A, \mathsf{M})$*, and let* $\mathsf{D} := \mathsf{K}_{\mathsf{S}}$ *be the localization, with localization functor* $\mathrm{Q} : \mathsf{K} \to \mathsf{D}$*. Then for every* $M \in \mathsf{K}$ *and* $I \in \mathsf{K}_{\mathrm{inj}}$ *the homomorphism*

$$\mathrm{Q}_{M,I} : \mathrm{Hom}_{\mathsf{K}}(M, I) \to \mathrm{Hom}_{\mathsf{D}}(M, I)$$

is bijective.

Proof We know that $\mathrm{Q} : \mathsf{K} \to \mathsf{D}$ is a left Ore localization with respect to S. Suppose $q : M \to I$ is a morphism in $\mathsf{D} = \mathsf{K}_{\mathsf{S}}$. Let us present q as a left fraction: $q = \mathrm{Q}(s)^{-1} \circ \mathrm{Q}(a)$, where $a : M \to N$ and $s : I \to N$ are morphisms in K, and s is a quasi-isomorphism. By Lemma 10.1.12, s has a left inverse $t : M \to I$ in K. We get a morphism $t \circ a : M \to I$ in K, and an easy calculation shows that $\mathrm{Q}(t \circ a) = q$ in D. This proves surjectivity of the function $\mathrm{Q}_{M,I}$.

Now let's prove injectivity of $\mathrm{Q}_{M,I}$. If $a : M \to I$ is a morphism in K such that $\mathrm{Q}_{M,I}(a) = 0$, then by axiom (LO4) of left Ore localization (see

Definition 6.2.2) there is a quasi-isomorphism $s : I \to L$ in K such that $s \circ a = 0$ in K. Let t be the left inverse of s. Then $a = t \circ s \circ a = 0$ in K. □

Corollary 10.1.14 *In the situation of Theorem* 10.1.13, *the localization functor* $\mathrm{Q} : \mathsf{K}_{\mathrm{inj}} \to \mathsf{D}$ *is fully faithful.*

Proof Since $\mathsf{K}_{\mathrm{inj}}$ is a full subcategory of K, for every $J, I \in \mathsf{K}_{\mathrm{inj}}$ we have bijections

$$\mathrm{Hom}_{\mathsf{K}_{\mathrm{inj}}}(J, I) \xrightarrow{\simeq} \mathrm{Hom}_{\mathsf{K}}(J, I) \xrightarrow{\mathrm{Q}_{J,I}} \mathrm{Hom}_{\mathsf{D}}(J, I),$$

where the bijection $\mathrm{Q}_{J,I}$ is by the theorem above. □

Corollary 10.1.15 *In the situation of Theorem* 10.1.13, *if* K *has enough K-injectives, then the localization functor* $\mathrm{Q} : \mathsf{K}_{\mathrm{inj}} \to \mathsf{D}$ *is an equivalence.*

Proof By Corollary 10.1.14 the functor Q is fully faithful. The extra condition guarantees that Q is essentially surjective on objects. □

Corollary 10.1.16 *Let* $\star$ *be any boundedness condition. If* $\mathbf{K}^{\star}(A, \mathsf{M})$ *has enough K-injectives, then the triangulated functor*

$$\mathrm{Q} : \mathbf{K}^{\star}(A, \mathsf{M})_{\mathrm{inj}} \to \mathbf{D}^{\star}(A, \mathsf{M})$$

is an equivalence.

Proof Since $\mathbf{K}^{\star}(A, \mathsf{M})$ is a full triangulated subcategory of $\mathbf{K}(A, \mathsf{M})$, this is a special case of the Corollary 10.1.15. □

Remark 10.1.17 These last results are of tremendous importance, both theoretically and practically. In the theory, Corollary 10.1.16 shows that the localized category $\mathbf{D}^{\star}(A, \mathsf{M})$, which is too big to lie inside the original universe U (see Remark 6.2.20), is equivalent to a U-category. On the practical side, Corollary 10.1.14 means that among K-injective objects we do not need fractions to represent morphisms.

Corollary 10.1.18 *Let* $\mathsf{K} \subseteq \mathsf{K}'$ *be full triangulated subcategories of* $\mathbf{K}(A, \mathsf{M})$. *Define* $\mathsf{S} := \mathsf{K} \cap \mathbf{S}(A, \mathsf{M})$, $\mathsf{S}' := \mathsf{K}' \cap \mathbf{S}(A, \mathsf{M})$, $\mathsf{D} := \mathsf{K}_{\mathsf{S}}$ *and* $\mathsf{D}' := \mathsf{K}'_{\mathsf{S}'}$. *If* K *and* K′ *have enough K-injectives, then the canonical functor* $\mathsf{D} \to \mathsf{D}'$ *is fully faithful.*

Proof Combine Corollary 10.1.15 with the fact that the canonical functor $\mathsf{K} \to \mathsf{K}'$ is fully faithful. □

Corollary 10.1.19 *Let* $\phi : I \to J$ *be a morphism in* $\mathbf{C}_{\mathrm{str}}(A, \mathsf{M})$ *between K-injective objects. Then* ϕ *is a homotopy equivalence if and only if it is a quasi-isomorphism.*

Proof One implication is trivial. For the reverse implication, if ϕ is a quasi-isomorphism then it is an isomorphism in $\mathbf{D}(A, \mathsf{M})$, and by Corollary 10.1.14 for $\mathsf{K} = \mathbf{K}(A, \mathsf{M})$ we see that ϕ is an isomorphism in $\mathbf{K}(A, \mathsf{M})$. □

Theorem 10.1.20 *In the situation of Theorem* 10.1.13, *assume that* K *has enough K-injectives. Let* E *be a triangulated category, and let* $F : \mathsf{K} \to \mathsf{E}$ *be a triangulated functor. Then* F *has a right derived functor* $(\mathrm{R}F, \eta^{\mathrm{R}}) : \mathsf{D} \to \mathsf{E}$. *Furthermore, for every* $I \in \mathsf{K}_{\mathrm{inj}}$ *the morphism* $\eta^{\mathrm{R}}_I : F(I) \to \mathrm{R}F(I)$ *in* E *is an isomorphism.*

Proof We will use Theorem 8.4.9. In the notation of that theorem, let $\mathsf{J} := \mathsf{K}_{\mathrm{inj}}$. Condition (a) of that theorem holds (this is the "enough K-injectives" assumption). Next, Corollary 10.1.19 implies that every quasi-isomorphism $\phi : I \to J$ in $\mathsf{K}_{\mathrm{inj}}$ is actually an isomorphism. Therefore $F(\phi)$ is an isomorphism in E, and this is condition (b) of Theorem 8.4.9. □

Here is another useful definition. It is a variant of Definition 8.3.4.

Definition 10.1.21 Let K be a full triangulated subcategory of $\mathbf{K}(A, \mathsf{M})$, and assume K has enough K-injectives. A *system of K-injective resolutions* in K is a pair (I, ρ), where $I : \mathrm{Ob}(\mathsf{K}) \to \mathrm{Ob}(\mathsf{K}_{\mathrm{inj}})$ is a function, and $\rho = \{\rho_M\}_{M \in \mathrm{Ob}(\mathsf{K})}$ is a collection of quasi-isomorphisms $\rho_M : M \to I(M)$ in K. Moreover, if $M \in \mathrm{Ob}(\mathsf{K}_{\mathrm{inj}})$, then $I(M) = M$ and $\rho_M = \mathrm{id}_M$.

Example 10.1.22 Let A be any DG ring. We will prove later that $\mathbf{K}(A)$ has enough K-injectives. Therefore, given a triangulated functor $F : \mathbf{K}(A) \to \mathsf{E}$ into a triangulated category E, the right derived functor $(\mathrm{R}F, \eta^{\mathrm{R}}) : \mathbf{D}(A) \to \mathsf{E}$ exists. Suppose we choose a system of K-injective resolutions (I, ρ) in $\mathbf{K}(A)$. Then we get a presentation of $(\mathrm{R}F, \eta^{\mathrm{R}})$ as follows: $\mathrm{R}F(M) := F(I(M))$ and $\eta^{\mathrm{R}}_M := F(\rho_M)$.

The proposition below is a variant of Lemma 8.3.9.

Proposition 10.1.23 *In the situation of Theorem* 10.1.13, *suppose* K *has enough K-injectives, and let* (I, ρ) *be a system of K-injective resolutions in* K. *Then the function* I *extends uniquely to a triangulated functor* $I : \mathsf{D} \to \mathsf{K}_{\mathrm{inj}}$, *such that* $\mathrm{Id}_{\mathsf{K}_{\mathrm{inj}}} = I \circ \mathrm{Q}|_{\mathsf{K}_{\mathrm{inj}}}$, *and* $\rho : \mathrm{Id}_{\mathsf{D}} \Rightarrow \mathrm{Q} \circ I$ *is an isomorphism of triangulated functors.*

Proof The proof is the same as that of Lemma 8.3.9, except that here we use Corollary 10.1.15. □

The next corollary is a categorical interpretation of the last proposition.

Corollary 10.1.24 (Functorial K-Injective Resolutions) *Let K be a full triangulated subcategory of $\mathbf{K}(A, \mathsf{M})$, and assume K has enough K-injectives.*

(1) *There is a triangulated functor $I : \mathsf{K} \to \mathsf{K}$ and a morphism of triangulated functors $\rho : \mathrm{Id}_{\mathsf{K}} \Rightarrow I$, such that for every object $M \in \mathsf{K}$ the object $I(M)$ is K-injective, and the morphism $\rho_M : M \to I(M)$ is a quasi-isomorphism.*

(2) *If (I', ρ') is another such pair, then there is a unique isomorphism of triangulated functors $\zeta : I \xRightarrow{\simeq} I'$ such that $\rho' = \zeta \circ \rho$.*

Exercise 10.1.25 Prove Corollary 10.1.24.

Infinite direct sums in $\mathbf{C}_{\mathrm{str}}(A, \mathsf{M})$ were discussed in Proposition 3.8.15. Their existence depends only on the existence of infinite direct sums in M.

Theorem 10.1.26 *Let A be a DG ring, let M be an abelian category, and let $\{M_x\}_{x \in X}$ be a collection of objects of $\mathbf{C}(A, \mathsf{M})$. Assume that $\mathbf{C}(A, \mathsf{M})$ has enough K-injectives, and that the direct sum $M := \bigoplus_{x \in X} M_x$ exists in $\mathbf{C}_{\mathrm{str}}(A, \mathsf{M})$. Then M is the direct sum of the collection $\{M_x\}_{x \in X}$ in $\mathbf{D}(A, \mathsf{M})$.*

Proof Let's use these shorthands: $\mathsf{C}_{\mathrm{str}} := \mathbf{C}_{\mathrm{str}}(A, \mathsf{M})$, $\mathsf{K} := \mathbf{K}(A, \mathsf{M})$ and $\mathsf{D} := \mathbf{D}(A, \mathsf{M})$. For each index x there is a morphism $e_x : M_x \to M$ in $\mathsf{C}_{\mathrm{str}}$ that passes to a morphism $\mathrm{Q}(e_x) : M_x \to M$ in D. We need to verify that the collection of morphisms $\{\mathrm{Q}(e_x)\}_{x \in X}$ has the universal property of a coproduct.

Take any object $N \in \mathsf{D}$, and choose a K-injective resolution $N \to J$ in $\mathsf{C}_{\mathrm{str}}$. There is an isomorphism of DG $\mathbb{K}$-modules

$$\begin{aligned} \mathrm{Hom}_{A,\mathsf{M}}(M, J) &= \mathrm{Hom}_{A,\mathsf{M}}\Big(\bigoplus\nolimits_{x \in X} M_x,\, J\Big) \\ &\cong \prod\nolimits_{x \in X} \mathrm{Hom}_{A,\mathsf{M}}(M_x, J). \end{aligned} \tag{10.1.27}$$

From it we obtain isomorphisms of $\mathbb{K}$-modules

$$\begin{aligned} \mathrm{Hom}_{\mathsf{D}}(M, N) &\cong^{\dagger} \mathrm{Hom}_{\mathsf{K}}(M, J) = \mathrm{H}^0(\mathrm{Hom}_{A,\mathsf{M}}(M, J)) \\ &\cong^{\ddagger} \mathrm{H}^0\Big(\prod\nolimits_{x \in X} \mathrm{Hom}_{A,\mathsf{M}}(M_x, J)\Big) \\ &\cong^{\#} \prod\nolimits_{x \in X} \mathrm{H}^0(\mathrm{Hom}_{A,\mathsf{M}}(M_x, J)) \\ &= \prod\nolimits_{x \in X} \mathrm{Hom}_{\mathsf{K}}(M_x, J) \cong^{\dagger} \prod\nolimits_{x \in X} \mathrm{Hom}_{\mathsf{D}}(M_x, N). \end{aligned} \tag{10.1.28}$$

The isomorphisms $\cong^{\dagger}$ are due to Theorem 10.1.13; the isomorphism $\cong^{\ddagger}$ is from formula (10.1.27); and the isomorphism $\cong^{\#}$ is because the functor H^0 commutes with products. Since the composed isomorphism in (10.1.28) is induced by the collection of morphisms $\{\mathrm{Q}(e_x)\}_{x \in X}$, we see that the universal property of a coproduct holds. □

Remark 10.1.29 The concept of a K-injective complex (i.e. a K-injective object of $\mathbf{C}(\mathsf{M})$), as well as the similar concepts of K-projective and K-flat complexes that will be discussed below, were introduced by N. Spaltenstein [143] in 1988. At about the same time other authors (B. Keller [79], M. Bockstedt and A. Neeman [23], J. Bernstein and V. Lunts [20]) discovered this concept independently, with other names (such as *homotopically injective complex*). The texts [20] and [79] already talk about DG modules over DG rings.

10.2 K-Projective DG Modules

This section is dual to the previous one, and so we will be brief.

Definition 10.2.1 A DG module $P \in \mathbf{C}(A, \mathsf{M})$ is called *K-projective* if for every acyclic DG module $N \in \mathbf{C}(A, \mathsf{M})$, the DG $\mathbb{K}$-module $\mathrm{Hom}_{A,\mathsf{M}}(P, N)$ is acyclic.

Definition 10.2.2 Let $M \in \mathbf{C}(A, \mathsf{M})$. A *K-projective resolution* of M is a quasi-isomorphism $\rho : P \to M$ in $\mathbf{C}_{\mathrm{str}}(A, \mathsf{M})$, where P is a K-projective DG module.

Remark 10.1.5 on terminology applies here too.

Proposition 10.2.3 *A DG module* $P \in \mathbf{K}(A, \mathsf{M})$ *is K-projective if and only if* $\mathrm{Hom}_{\mathbf{K}(A,\mathsf{M})}(P, N) = 0$ *for every acyclic DG module* $N \in \mathbf{K}(A, \mathsf{M})$.

The proof is like that of Proposition 10.1.3.

Definition 10.2.4 Let K be a full subcategory of $\mathbf{K}(A, \mathsf{M})$. The full subcategory of K on the K-projective DG modules in it is denoted by $\mathsf{K}_{\mathrm{prj}}$. In other words, $\mathsf{K}_{\mathrm{prj}} = \mathbf{K}(A, \mathsf{M})_{\mathrm{prj}} \cap \mathsf{K}$.

The warning in Remark 10.1.8 applies here too.

Proposition 10.2.5 *If* K *is a full triangulated subcategory of* $\mathbf{K}(A, \mathsf{M})$, *then* $\mathsf{K}_{\mathrm{prj}}$ *is a full triangulated subcategory of* K.

The proof is like that of Proposition 10.1.9.

Example 10.2.6 Let $\star$ be some boundedness condition (namely b, + or −). Since $\mathbf{K}^{\star}(A, \mathsf{M})$ is a full triangulated subcategory of $\mathbf{K}(A, \mathsf{M})$, we see that $\mathbf{K}^{\star}(A, \mathsf{M})_{\mathrm{prj}}$ is a full triangulated subcategory too.

Definition 10.2.7 Let K be a full triangulated subcategory of $\mathbf{K}(A, \mathsf{M})$. We say that K *has enough K-projectives* if every DG module $M \in \mathsf{K}$ admits a K-projective resolution inside K. In other words, there is a quasi-isomorphism $\rho : P \to M$ where $P \in \mathsf{K}_{\mathrm{prj}}$.

Lemma 10.2.8 *Let $s : M \to P$ be a quasi-isomorphism in $\mathbf{K}(A, \mathsf{M})$, and assume P is K-projective. Then s has a right inverse; namely there is a morphism $t : P \to M$ in $\mathbf{K}(A, \mathsf{M})$ such that $s \circ t = \mathrm{id}_P$. Moreover, this morphism t is unique.*

The proof is almost the same as that of Lemma 10.1.12.

Theorem 10.2.9 *Let A be a DG ring, let M be an abelian category, and let K be a full triangulated subcategory of $\mathbf{K}(A, \mathsf{M})$. Write $\mathsf{S} := \mathsf{K} \cap \mathbf{S}(A, \mathsf{M})$, and let $\mathsf{D} := \mathsf{K}_{\mathsf{S}}$ be the localization, with localization functor $\mathrm{Q} : \mathsf{K} \to \mathsf{D}$. Then for every $M \in \mathsf{K}$ and $P \in \mathsf{K}_{\mathrm{prj}}$ the homomorphism*

$$\mathrm{Q}_{P,M} : \mathrm{Hom}_{\mathsf{K}}(P, M) \to \mathrm{Hom}_{\mathsf{D}}(P, M)$$

is bijective.

The proof is like that of Theorem 10.1.13, only now we use the fact that the functor Q is a right Ore localization.

The results from here to the end of the section are also proved like their K-injective counterparts in Section 10.1.

Corollary 10.2.10 *In the situation of Theorem 10.2.9, the localization functor $\mathrm{Q} : \mathsf{K}_{\mathrm{prj}} \to \mathsf{D}$ is fully faithful.*

Corollary 10.2.11 *In the situation of Theorem 10.2.9, if K has enough K-projectives, then the localization functor $\mathrm{Q} : \mathsf{K}_{\mathrm{prj}} \to \mathsf{D}$ is an equivalence.*

Corollary 10.2.12 *Let $\star$ be any boundedness condition. If $\mathbf{K}^{\star}(A, \mathsf{M})$ has enough K-projectives, then the triangulated functor*

$$\mathrm{Q} : \mathbf{K}^{\star}(A, \mathsf{M})_{\mathrm{prj}} \to \mathbf{D}^{\star}(A, \mathsf{M})$$

is an equivalence.

Corollary 10.2.13 *Let $\mathsf{K} \subseteq \mathsf{K}'$ be full triangulated subcategories of $\mathbf{K}(A, \mathsf{M})$. Define $\mathsf{S} := \mathsf{K} \cap \mathbf{S}(A, \mathsf{M})$, $\mathsf{S}' := \mathsf{K}' \cap \mathbf{S}(A, \mathsf{M})$, $\mathsf{D} := \mathsf{K}_{\mathsf{S}}$ and $\mathsf{D}' := \mathsf{K}'_{\mathsf{S}'}$. If K and K' have enough K-projectives, then the canonical functor $\mathsf{D} \to \mathsf{D}'$ is fully faithful.*

Corollary 10.2.14 *Let $\phi : P \to Q$ be a morphism in $\mathbf{C}_{\mathrm{str}}(A, \mathsf{M})$ between K-projective objects. Then ϕ is a homotopy equivalence if and only if it is a quasi-isomorphism.*

Theorem 10.2.15 *In the situation of Theorem* 10.2.9, *assume that* K *has enough K-projectives. Let* E *be a triangulated category, and let* $F : \mathsf{K} \to \mathsf{E}$ *be a triangulated functor. Then* F *has a left derived functor* $(\mathrm{L}F, \eta^{\mathrm{L}}) : \mathsf{D} \to \mathsf{E}$. *Furthermore, for every* $P \in \mathsf{K}_{\mathrm{prj}}$ *the morphism* $\eta^{\mathrm{L}}_P : \mathrm{L}F(P) \to F(P)$ *in* E *is an isomorphism.*

Definition 10.2.16 Let K be a full triangulated subcategory of $\mathbf{K}(A, \mathsf{M})$, and assume K has enough K-projectives. A *system of K-projective resolutions* in K is a pair (P, ρ), where $P : \mathrm{Ob}(\mathsf{K}) \to \mathrm{Ob}(\mathsf{K}_{\mathrm{prj}})$ is a function, and $\rho = \{\rho_M\}_{M \in \mathrm{Ob}(\mathsf{K})}$ is a collection of quasi-isomorphisms $\rho_M : P(M) \to M$ in K. Moreover, if $M \in \mathrm{Ob}(\mathsf{K}_{\mathrm{prj}})$, then $P(M) = M$ and $\rho_M = \mathrm{id}_M$.

Example 10.2.17 Let A be any DG ring. We will prove later that $\mathbf{K}(A)$ has enough K-projectives. Therefore, given a triangulated functor $F : \mathbf{K}(A) \to \mathsf{E}$ into a triangulated category E, the left derived functor $(\mathrm{L}F, \eta^{\mathrm{L}}) : \mathbf{D}(A) \to \mathsf{E}$ exists. Suppose we choose a system of K-projective resolutions (P, ρ) in $\mathbf{K}(A)$. Then we get a presentation of $(\mathrm{L}F, \eta^{\mathrm{R}})$ as follows: $\mathrm{L}F(M) := F(P(M))$ and $\eta^{\mathrm{L}}_M := F(\rho_M)$.

Proposition 10.2.18 *In the situation of Theorem* 10.2.9, *suppose* K *has enough K-projectives, and let* (P, ρ) *be a system of K-projective resolutions. Then the function* P *extends uniquely to a triangulated functor* $P : \mathsf{D} \to \mathsf{K}_{\mathrm{prj}}$, *such that* $\mathrm{Id}_{\mathsf{K}_{\mathrm{prj}}} = P \circ \mathrm{Q}|_{\mathsf{K}_{\mathrm{prj}}}$, *and* $\rho : \mathrm{Q} \circ P \Rightarrow \mathrm{Id}_{\mathsf{D}}$ *is an isomorphism of triangulated functors.*

Corollary 10.2.19 (Functorial K-Projective Resolutions) *Let* K *be a full triangulated subcategory of* $\mathbf{K}(A, \mathsf{M})$, *and assume* K *has enough K-projectives.*

(1) *There is a triangulated functor* $P : \mathsf{K} \to \mathsf{K}$ *and a morphism of triangulated functors* $\rho : P \Rightarrow \mathrm{Id}_{\mathsf{K}}$, *such that for every object* $M \in \mathsf{K}$ *the object* $P(M)$ *is K-projective, and the morphism* $\rho_M : P(M) \to M$ *is a quasi-isomorphism.*
(2) *If* (P', ρ') *is another such pair, then there is a unique isomorphism of triangulated functors* $\zeta : P' \xrightarrow{\simeq} P$ *such that* $\rho' = \rho \circ \zeta$.

10.3 K-Flat DG Modules

Recall that for a DG ring A, its opposite DG ring is A^{op}. The objects of $\mathbf{C}(A^{\mathrm{op}})$ are the right DG A-modules.

Definition 10.3.1 A DG module $P \in \mathbf{C}(A)$ is called *K-flat* if for every acyclic DG module $N \in \mathbf{C}(A^{\mathrm{op}})$, the DG $\mathbb{K}$-module $N \otimes_A P$ is acyclic.

Definition 10.3.2 Let $M \in \mathbf{C}(A)$. A *K-flat resolution* of M is a quasi-isomorphism $\rho : P \to M$ in $\mathbf{C}_{\mathrm{str}}(A)$, where P is a K-flat DG module.

Remark 10.1.5 on terminology applies here as well.

Definition 10.3.3 Let K be a full triangulated subcategory of $\mathbf{K}(A)$. We denote by $\mathsf{K}_{\mathrm{flat}}$ the full subcategory of $\mathbf{K}(A)$ on the K-flat complexes in it. Thus $\mathsf{K}_{\mathrm{flat}} = \mathsf{K} \cap \mathbf{K}(A)_{\mathrm{flat}}$.

The warning in Remark 10.1.8 applies here too.

Proposition 10.3.4 *If $P \in \mathbf{C}(A)$ is K-projective then it is K-flat.*

Proof Let $\mathbb{K}^*$ be an injective cogenerator of $\mathbf{M}(\mathbb{K}) = \mathsf{Mod}\,\mathbb{K}$. This means that $\mathbb{K}^*$ is an injective $\mathbb{K}$-module, such that every nonzero $\mathbb{K}$-module W admits a nonzero homomorphism $W \to \mathbb{K}^*$. See Examples 11.6.1 and 11.6.1. It is not hard to see that a DG $\mathbb{K}$-module W is acyclic if and only if $\mathrm{Hom}_{\mathbb{K}}(W, \mathbb{K}^*)$ is acyclic.

Take an acyclic complex $N \in \mathbf{C}(A^{\mathrm{op}})$. Then by Hom-tensor adjunction there is an isomorphism of DG $\mathbb{K}$-modules

$$\mathrm{Hom}_{\mathbb{K}}(N \otimes_A P, \mathbb{K}^*) \cong \mathrm{Hom}_A(P, \mathrm{Hom}_{\mathbb{K}}(N, \mathbb{K}^*)).$$

The right side is acyclic by our assumptions. Hence so is the left side. It follows that $N \otimes_A P$ is acyclic. □

Proposition 10.3.5 *A DG module $P \in \mathbf{K}(A)$ is K-flat iff*

$$\mathrm{Hom}_{\mathbf{K}(A)}(P, \mathrm{Hom}_{\mathbb{K}}(N, J)) = 0$$

for every acyclic $N \in \mathbf{C}(A^{\mathrm{op}})$ and every injective $J \in \mathsf{Mod}\,\mathbb{K}$.

Exercise 10.3.6 Prove Proposition 10.3.5. (Hint: look at the proof of Proposition 10.3.4.)

Proposition 10.3.7 *Let K be a full triangulated subcategory of $\mathbf{K}(A)$. Then $\mathsf{K}_{\mathrm{flat}}$ is a full triangulated subcategory of K.*

Exercise 10.3.8 Prove Proposition 10.3.7. (Hint: The challenge is to show that $\mathsf{K}_{\mathrm{flat}}$ is closed under distinguished triangles. Use Proposition 10.3.5.)

In Theorem 12.3.1 on the left derived bifunctor $(- \otimes^{\mathrm{L}}_A -)$ of the bifunctor $(- \otimes_A -)$ we will use K-flat resolutions.

Remark 10.3.9 In view of Proposition 10.3.4, the reader might wonder why we bother with K-flat DG modules. The reason is that there are situations in which there are enough K-flat objects but not enough K-projective objects.

One such situation is that of the category $\mathsf{C}(\mathcal{A}) = \mathsf{C}(\mathsf{Mod}\,\mathcal{A})$, where $(X, \mathcal{A})$ is a ringed space and $\mathsf{Mod}\,\mathcal{A}$ is the category of sheaves of $\mathcal{A}$-modules; see Example 11.3.16. Another situation is when K-flatness is needed is for DG bimodules over noncommutative DG rings, as in Chapters 14 and later.

10.4 Opposite Resolving Subcategories

In this section we explain how K-injective and K-projective DG modules in $\mathsf{K}(A, \mathsf{M})$ behave in the opposite category $\mathsf{K}(A, \mathsf{M})^{\mathrm{op}}$, and then use this to prove existence of contravariant triangulated derived functors.

Recall the flipping operation from Sections 3.9, 5.6 and 7.6. There is an isomorphism of DG categories

$$\mathrm{Flip} : \mathsf{C}(A, \mathsf{M})^{\mathrm{op}} \xrightarrow{\simeq} \mathsf{C}(A, \mathsf{M})^{\mathrm{flip}} = \mathsf{C}(A^{\mathrm{op}}, \mathsf{M}^{\mathrm{op}}), \tag{10.4.1}$$

and the corresponding isomorphisms of the homotopy categories

$$\mathrm{Flip} : \mathsf{K}(A, \mathsf{M})^{\mathrm{op}} \xrightarrow{\simeq} \mathsf{K}(A, \mathsf{M})^{\mathrm{flip}} = \mathsf{K}(A^{\mathrm{op}}, \mathsf{M}^{\mathrm{op}}) \tag{10.4.2}$$

and on the derived categories

$$\mathrm{Flip} : \mathsf{D}(A, \mathsf{M})^{\mathrm{op}} \xrightarrow{\simeq} \mathsf{D}(A, \mathsf{M})^{\mathrm{flip}} = \mathsf{D}(A^{\mathrm{op}}, \mathsf{M}^{\mathrm{op}}). \tag{10.4.3}$$

By definition of the triangulated structure on $\mathsf{K}(A, \mathsf{M})^{\mathrm{op}}$, the isomorphisms (10.4.2) and (10.4.3) are of triangulated categories.

The objects of $\mathsf{C}(A, \mathsf{M})$ and $\mathsf{C}(A, \mathsf{M})^{\mathrm{op}}$ are the same, and the property of a DG module M being acyclic is the same in both categories.

Lemma 10.4.4 *A DG module $M \in \mathsf{C}(A, \mathsf{M})$ is acyclic iff the DG module $\mathrm{Flip}(M) \in \mathsf{C}(A^{\mathrm{op}}, \mathsf{M}^{\mathrm{op}})$ is acyclic.*

Proof This follows from Theorem 3.9.16(4). □

Lemma 10.4.5 *A DG module $P \in \mathsf{K}(A, \mathsf{M})$ is K-projective iff the DG module $I := \mathrm{Flip}(P) \in \mathsf{K}(A^{\mathrm{op}}, \mathsf{M}^{\mathrm{op}})$ is K-injective.*

Proof According to Lemma 10.4.4, the functor Flip gives a bijection between the set of acyclic DG modules $M \in \mathsf{K}(A, \mathsf{M})$ and the set of acyclic DG modules $N := \mathrm{Flip}(M) \in \mathsf{K}(A^{\mathrm{op}}, \mathsf{M}^{\mathrm{op}})$. Since the functor (10.4.2) is an equivalence, we have an equality and an isomorphism

$$\mathrm{Hom}_{\mathsf{K}(A,\mathsf{M})}(P, M) = \mathrm{Hom}_{\mathsf{K}(A,\mathsf{M})^{\mathrm{op}}}(M, P) \cong \mathrm{Hom}_{\mathsf{K}(A^{\mathrm{op}},\mathsf{M}^{\mathrm{op}})}(N, I) \tag{10.4.6}$$

in $\mathsf{M}(\mathbb{K})$.

By Proposition 10.1.3, the DG module I is K-injective in $\mathsf{K}(A^{\mathrm{op}}, \mathsf{M}^{\mathrm{op}})$ iff for every acyclic DG module $N \in \mathsf{K}(A^{\mathrm{op}}, \mathsf{M}^{\mathrm{op}})$ we have $\mathrm{Hom}_{\mathsf{K}(A^{\mathrm{op}},\mathsf{M}^{\mathrm{op}})}(N, I) = 0$.

Similarly, by Proposition 10.2.3, the DG module P is K-projective in $\mathbf{K}(A, \mathsf{M})$ iff for every acyclic DG module $M \in \mathbf{K}(A, \mathsf{M})$ we have $\operatorname{Hom}_{\mathbf{K}(A,\mathsf{M})}(P, M) = 0$. In view of (10.4.6) we conclude that I is K-injective iff P is K-projective. □

Here are the contravariant modifications of Theorems 10.1.20 and 10.2.15.

Setup 10.4.7 We are given:

- A full additive subcategory $\mathsf{K} \subseteq \mathbf{K}(A, \mathsf{M})$, s.t. $\mathsf{K}^{\mathrm{op}} \subseteq \mathbf{K}(A, \mathsf{M})^{\mathrm{op}}$ is triangulated. The localization of K^{op} at the quasi-isomorphisms in it is denoted by D^{op}.
- A triangulated category E and a triangulated functor $F : \mathsf{K}^{\mathrm{op}} \to \mathsf{E}$.

Theorem 10.4.8 *Under Setup* 10.4.7, *assume that* K *has enough K-projectives. Then the triangulated right derived functor* $(\mathrm{R}F, \eta^{\mathrm{R}}) : \mathsf{D}^{\mathrm{op}} \to \mathsf{E}$ *of F exists. Furthermore, for every K-projective DG module $P \in \mathsf{K}$, the morphism* $\eta^{\mathrm{R}}_P : F(P) \to \mathrm{R}F(P)$ *in* E *is an isomorphism.*

Proof Let $\mathsf{K}^{\mathrm{flip}} := \mathrm{Flip}(\mathsf{K}^{\mathrm{op}}) \subseteq \mathbf{K}(A^{\mathrm{op}}, \mathsf{M}^{\mathrm{op}})$. This is a triangulated category. Its localization w.r.t. the quasi-isomorphisms is denoted by $\mathsf{D}^{\mathrm{flip}}$. We have isomorphisms of triangulated categories $\mathrm{Flip} : \mathsf{K}^{\mathrm{op}} \xrightarrow{\simeq} \mathsf{K}^{\mathrm{flip}}$ and $\mathrm{Flip} : \mathsf{D}^{\mathrm{op}} \xrightarrow{\simeq} \mathsf{D}^{\mathrm{flip}}$. Let $F^{\mathrm{flip}} := F \circ \mathrm{Flip}^{-1} : \mathsf{K}^{\mathrm{flip}} \to \mathsf{E}$. This is a triangulated functor.

Since Flip sends quasi-isomorphisms to quasi-isomorphisms, and by Lemma 10.4.5, each K-projective resolution $\rho : P \to M$ of M in K goes to a K-injective resolution $\mathrm{Flip}(\rho) : \mathrm{Flip}(M) \to \mathrm{Flip}(P)$ of $N := \mathrm{Flip}(M)$ in $\mathsf{K}^{\mathrm{flip}}$. We are given that K has enough K-projectives. Therefore $\mathsf{K}^{\mathrm{flip}}$ has enough K-injectives.

Now we can apply Theorem 10.1.20 to deduce that the right derived functor $(\mathrm{R}F^{\mathrm{flip}}, \eta^{\mathrm{flip}}) : \mathsf{D}^{\mathrm{flip}} \to \mathsf{E}$ of F^{flip} exists. Then

$$(\mathrm{R}F, \eta^{\mathrm{R}}) := (\mathrm{R}F^{\mathrm{flip}}, \eta^{\mathrm{flip}}) \circ \mathrm{Flip} : \mathsf{D}^{\mathrm{op}} \to \mathsf{E}$$

is a right derived functor of F.

The assertion about the morphism η^{R}_P is clear from the discussion above. □

Theorem 10.4.9 *Under Setup* 10.4.7, *assume that* K *has enough K-injectives. Then the left derived functor* $(\mathrm{L}F, \eta^{\mathrm{L}}) : \mathsf{D}^{\mathrm{op}} \to \mathsf{E}$ *of F exists. Furthermore, for every K-injective DG module $I \in \mathsf{K}$, the morphism* $\eta^{\mathrm{L}}_I : \mathrm{L}F(I) \to F(I)$ *in* E *is an isomorphism.*

Proof The same as that of the previous theorem, except that now we use Lemma 10.4.5 and Theorem 10.2.15. □

11

Existence of Resolutions

In this chapter we continue in the more concrete setting: $\mathbb{K}$ is a nonzero commutative base ring, A is a central DG $\mathbb{K}$-ring and M is a $\mathbb{K}$-linear abelian category; see Convention 1.2.4. The DG category $\mathbf{C}^{\star}(A, \mathsf{M})$ was introduced in Section 3.8. We will prove existence of K-projective and K-injective resolutions in $\mathbf{C}^{\star}(A, \mathsf{M})$, under suitable conditions.

11.1 Direct and Inverse Limits of Complexes

We shall have to work with limits in this chapter. Limits in abstract abelian and DG categories (not to mention triangulated categories) are a very delicate issue. We will try to be as concrete as possible, in order to avoid pitfalls and confusion. The notation we use for direct and inverse limits of systems indexed by $\mathbb{N}$ was presented in Section 1.8.

Proposition 11.1.1

(1) *Let* $\{M_k\}_{k \in \mathbb{N}}$ *be a direct system in* $\mathbf{C}_{\mathrm{str}}(A, \mathsf{M})$. *Assume that for every i the direct limit* $\lim_{k \to} M_k^i$ *exists in* M. *Then the direct limit* $M = \lim_{k \to} M_k$ *exists in* $\mathbf{C}_{\mathrm{str}}(A, \mathsf{M})$, *and in degree i it is* $M^i = \lim_{k \to} M_k^i$.

(2) *Let* $\{M_k\}_{k \in \mathbb{N}}$ *be an inverse system in* $\mathbf{C}_{\mathrm{str}}(A, \mathsf{M})$. *Assume that for every i the inverse limit* $\lim_{\leftarrow k} M_k^i$ *exists in* M. *Then the inverse limit* $M = \lim_{\leftarrow k} M_k$ *exists in* $\mathbf{C}_{\mathrm{str}}(A, \mathsf{M})$, *and in degree i it is* $M^i = \lim_{\leftarrow k} M_k^i$.

Proof We will only prove item (1); the proof of item (2) is similar. For each integer i define $M^i := \lim_{k \to} M_k^i \in \mathsf{M}$. By the universal property of the direct limit, the differentials $\mathrm{d} : M_k^i \to M_k^{i+1}$ induce differentials $\mathrm{d} : M^i \to M^{i+1}$, and in this way we obtain a complex $M := \{M^i\}_{i \in \mathbb{Z}} \in \mathbf{C}(\mathsf{M})$. Similarly, every element $a \in A^j$ induces morphisms $a : M^i \to M^{i+j}$ in M, and thus M becomes an object of $\mathbf{C}(A, \mathsf{M})$. There are morphisms $M_k \to M$ in $\mathbf{C}_{\mathrm{str}}(A, \mathsf{M})$, and it is easy to see that these make M into a direct limit of the system $\{M_k\}_{k \in \mathbb{N}}$. □

Since limits exist in $\mathsf{M} = \mathsf{Mod}\, \mathbb{K}$, the proposition above says that they exist in $\mathbf{C}_{\mathrm{str}}(A)$. Likewise, they exist in the category $\mathbf{G}_{\mathrm{str}}(\mathbb{K})$ of graded $\mathbb{K}$-modules.

We say that a direct system $\{M_k\}_{k \in \mathbb{N}}$ in M is *eventually stationary* if $\mu_k : M_k \to M_{k+1}$ are isomorphisms for large k. Similarly we can talk about an eventually stationary inverse system. The limit of an eventually stationary system (direct or inverse) always exists: it is M_k for large enough k.

Proposition 11.1.2

(1) *Let* $\{M_k\}_{k \in \mathbb{N}}$ *be a direct system in* $\mathbf{C}_{\mathrm{str}}(A, \mathsf{M})$. *Assume that for each* i *the direct system* $\{M_k^i\}_{k \in \mathbb{N}}$ *in* M *is eventually stationary. Then the direct limit* $M = \lim_{k \to} M_k$ *exists in* $\mathbf{C}_{\mathrm{str}}(A, \mathsf{M})$, *the direct limit* $\lim_{k \to} \mathrm{H}(M_k)$ *exists in* $\mathbf{G}_{\mathrm{str}}(\mathsf{M})$, *and the canonical morphism* $\lim_{k \to} \mathrm{H}(M_k) \to \mathrm{H}(M)$ *in* $\mathbf{G}_{\mathrm{str}}(\mathsf{M})$ *is an isomorphism.*

(2) *Let* $\{M_k\}_{k \in \mathbb{N}}$ *be an inverse system in* $\mathbf{C}_{\mathrm{str}}(A, \mathsf{M})$. *Assume that for each* i *the inverse system* $\{M_k^i\}_{k \in \mathbb{N}}$ *in* M *is eventually stationary. Then the inverse limit* $M = \lim_{\leftarrow k} M_k$ *exists in* $\mathbf{C}_{\mathrm{str}}(A, \mathsf{M})$, *the inverse limit* $\lim_{\leftarrow k} \mathrm{H}(M_k)$ *exists in* $\mathbf{G}_{\mathrm{str}}(\mathsf{M})$, *and the canonical morphism* $\mathrm{H}(M) \to \lim_{\leftarrow k} \mathrm{H}(M_k)$ *in* $\mathbf{G}_{\mathrm{str}}(\mathsf{M})$ *is an isomorphism.*

Proof Again, we only prove item (1). As mentioned above, for each i the limit $M^i = \lim_{k \to} M_k^i$ exists in M. By Proposition 11.1.1 the limit $M = \lim_{k \to} M_k$ exists in $\mathbf{C}_{\mathrm{str}}(A, \mathsf{M})$.

Regarding the cohomology: fix an integer i. Take k large enough such that $M_k^{i'} \to M_{k'}^{i'}$ are isomorphisms for all $k \le k'$ and $i - 1 \le i' \le i + 1$. Then $M_{k'}^{i'} \to M^{i'}$ are isomorphisms in this range, and therefore $\mathrm{H}^i(M_{k'}) \to \mathrm{H}^i(M)$ are isomorphisms for all $k \le k'$. We see that the direct system $\{\mathrm{H}^i(M_k)\}_{k \in \mathbb{N}}$ is eventually stationary, and its direct limit is $\mathrm{H}^i(M)$. □

When we drop the abstract abelian category M, i.e. when we work with $\mathsf{M} = \mathsf{Mod}\, \mathbb{K} = \mathbf{M}(\mathbb{K})$ and $\mathbf{C}_{\mathrm{str}}(A, \mathsf{M}) = \mathbf{C}_{\mathrm{str}}(A)$, there is no problem of existence of limits. The next proposition says that furthermore "direct limits are exact" in $\mathbf{C}_{\mathrm{str}}(A)$.

Proposition 11.1.3 *Let* $\{M_k\}_{k \in \mathbb{N}}$ *be a direct system in* $\mathbf{C}_{\mathrm{str}}(A)$, *with limit* $M = \lim_{k \to} M_k$. *Then the canonical homomorphism* $\lim_{k \to} \mathrm{H}(M_k) \to \mathrm{H}(M)$ *in* $\mathbf{G}_{\mathrm{str}}(\mathbb{K})$ *is bijective.*

Exercise 11.1.4 Prove Proposition 11.1.3. (Hint: forget the action of A, and work with complexes of $\mathbb{K}$-modules.)

Exactness of inverse limits tends to be much more complicated than that of direct limits, even for $\mathbb{K}$-modules. We always have to impose some condition on the inverse system to have exactness in the limit.

Let $(\{M_k\}_{k \in \mathbb{N}}, \{\mu_k\}_{k \in \mathbb{N}})$ be an inverse system in $\mathbf{M}(\mathbb{K})$. For every $k \leq l$ let $M_{k,l} \subseteq M_k$ be the image of the homomorphism $\mu_{k,l} : M_l \to M_k$. Note that there are inclusions $M_{k,l+1} \subseteq M_{k,l}$, so for fixed k we have an inverse system $\{M_{k,l}\}_{l \geq k}$ of submodules of M_k.

Definition 11.1.5 We say that an inverse system $\{M_k\}_{k \in \mathbb{N}}$ in $\mathbf{M}(\mathbb{K})$ has the *Mittag–Leffler property* if for every index k, the inverse system $\{M_{k,l}\}_{l \geq k}$ is eventually stationary.

Example 11.1.6 If the inverse system $(\{M_k\}_{k \in \mathbb{N}}, \{\mu_k\}_{k \in \mathbb{N}})$ in $\mathbf{M}(\mathbb{K})$ satisfies at least one of the following conditions, then it has the Mittag–Leffler property:

(a) The system has surjective transitions.

(b) The system is eventually stationary.

(c) For every $k \in \mathbb{N}$ there exists some $l \geq k$ such that $M_{k,l} = 0$. This is called the *trivial Mittag–Leffler property*, and one says that the system is *pro-zero*.

Theorem 11.1.7 (Mittag–Leffler Argument) *Let $\{M_k\}_{k \in \mathbb{N}}$ be an inverse system in $\mathbf{C}_{\mathrm{str}}(A)$, with inverse limit $M = \lim_{\leftarrow k} M_k$. Assume the system satisfies these two conditions:*

(a) *For every $i \in \mathbb{Z}$ the inverse system $\{M_k^i\}_{k \in \mathbb{N}}$ in $\mathbf{M}(\mathbb{K})$ has the Mittag–Leffler property.*

(b) *For every $i \in \mathbb{Z}$ the inverse system $\{\mathrm{H}^i(M_k)\}_{k \in \mathbb{N}}$ in $\mathbf{M}(\mathbb{K})$ has the Mittag–Leffler property.*

Then the canonical homomorphisms $\mathrm{H}^i(M) \to \lim_{\leftarrow k} \mathrm{H}^i(M_k)$ are bijective for all i.

Proof We can forget all about the graded A-module structure, and just view this as an inverse system in $\mathbf{C}_{\mathrm{str}}(\mathbb{Z})$, i.e. an inverse system of complexes of abelian groups. Now this is a special case of [75, Proposition 1.12.4] or [56, Chapitre 0_{III}, Proposition 13.2.3]. □

The most useful instance of the ML argument is this:

Corollary 11.1.8 *Let $\{M_k\}_{k \in \mathbb{N}}$ be an inverse system in $\mathbf{C}_{\mathrm{str}}(A)$, with inverse limit $M = \lim_{\leftarrow k} M_k$. Assume the system satisfies these two conditions:*

(a) *For every $i \in \mathbb{Z}$ the inverse system $\{M_k^i\}_{k \in \mathbb{N}}$ has surjective transitions.*

(b) *For every k the DG module M_k is acyclic.*

Then M is acyclic.

Proof Conditions (a) and (b) here imply conditions (a) and (b) of Theorem 11.1.7, respectively. □

Exercise 11.1.9 Prove Corollary 11.1.8 directly, without resorting to Theorem 11.1.7.

By *generalized integers* we mean elements of the ordered set $\mathbb{Z} \cup \{\pm\infty\}$. Recall that for a subset $S \subseteq \mathbb{Z}$, its infimum is $\inf(S) \in \mathbb{Z} \cup \{\pm\infty\}$, where $\inf(S) = +\infty$ iff $S = \varnothing$. Likewise, the supremum is $\sup(S) \in \mathbb{Z} \cup \{\pm\infty\}$, where $\sup(S) = -\infty$ iff $S = \varnothing$. For $i \in \mathbb{Z} \cup \{\pm\infty\}$ the expression $-i$ has a value in $\mathbb{Z} \cup \{\pm\infty\}$. If $i_0, i_1 \in \mathbb{Z} \cup \{\pm\infty\}$ satisfy $i_0 \leq i_1$, then $-i_1 \leq -i_0$. For $i_0, i_1 \in \mathbb{Z} \cup \{\infty\}$, the expressions $i_0 + i_1$ and $-i_0 - i_1$ have values in $\mathbb{Z} \cup \{\pm\infty\}$. A generalized integer i is called *finite* if $i < \infty$, i.e. if $i \in \mathbb{Z} \cup \{-\infty\}$.

Definition 11.1.10 Given $i_0 \in \mathbb{Z} \cup \{-\infty\}$ and $i_1 \in \mathbb{Z} \cup \{\infty\}$, with $i_0 \leq i_1$, the *integer interval* with these endpoints is the set

$$[i_0, i_1] := \{i \in \mathbb{Z} \mid i_0 \leq i \leq i_1\} \subseteq \mathbb{Z}.$$

There is also the empty integer interval $\varnothing$. When the elements of an integer interval represent degrees, we sometimes call it a *degree interval.*

We exclude the possibilities $i_0 = i_1 = \infty$ and $i_0 = i_1 = -\infty$ for endpoints of intervals. In calculus notation we have $[i_0, \infty] = [i_0, \infty) \cap \mathbb{Z}$, etc.; but we avoid the open interval notation on purpose. Whenever we refer to an integer interval $[i_0, i_1]$, it is always assumed that the generalized integers i_0 and i_1 satisfy the conditions in Definition 11.1.10. Further properties of integer intervals will be studied in Section 12.1.

Definition 11.1.11 Let $M = \{M^i\}_{i \in \mathbb{Z}}$ be a graded object in M, and let $S := \{i \in \mathbb{Z} \mid M^i \neq 0\} \subseteq \mathbb{Z}$.

(1) The *supremum* of M is $\sup(M) := \sup(S) \in \mathbb{Z} \cup \{\pm\infty\}$.
(2) The *infimum* of M is $\inf(M) := \inf(S) \in \mathbb{Z} \cup \{\pm\infty\}$.
(3) The *amplitude* of M is $\operatorname{amp}(M) := \sup(M) - \inf(M) \in \mathbb{N} \cup \{\pm\infty\}$.

Note that for $M = 0$ this reads $S = \varnothing$, $\inf(M) = \infty$, $\sup(M) = -\infty$ and $\operatorname{amp}(M) = -\infty$. Thus M is bounded (resp. bounded above, resp. bounded below) iff $\operatorname{amp}(M) < \infty$ (resp. $\sup(M) < \infty$, resp. $\inf(M) > -\infty$).

11.2 Totalizations

Consider a complex

$$(M, \partial) = (\cdots \to M^{-1} \xrightarrow{\partial^{-1}} M^0 \xrightarrow{\partial^0} M^1 \to \cdots) \tag{11.2.1}$$

with entries in the abelian category $\mathbf{C}_{\mathrm{str}}(A, \mathsf{M})$. This means that for each $i \in \mathbb{Z}$ there is a DG module $(M^i, \mathrm{d}_{M^i}) \in \mathbf{C}(A, \mathsf{M})$, and a morphism $\partial^i : M^i \to M^{i+1}$ in $\mathbf{C}_{\mathrm{str}}(A, \mathsf{M})$, such that $\partial^{i+1} \circ \partial^i = 0$. Recall that the strictness of ∂^i says that it has degree 0, and that

$$\partial^i \circ \mathrm{d}_{M^i} = \mathrm{d}_{M^{i+1}} \circ \partial^i. \tag{11.2.2}$$

In terms of our symbolic notation, (M, ∂) is an object of the DG category $\mathbf{C}(\mathbf{C}_{\mathrm{str}}(A, \mathsf{M}))$.

Each DG module (M^i, d_{M^i}) has its own internal structure:

$$(M^i, \mathrm{d}_{M^i}) = (\{M^{i,j}\}_{j \in \mathbb{Z}}, \{\mathrm{d}^j_{M^i}\}_{j \in \mathbb{Z}}). \tag{11.2.3}$$

Here $M^{i,j} \in \mathsf{M}$, $\mathrm{d}^j_{M^i} : M^{i,j} \to M^{i,j+1}$ is a morphism in M, and the relation $\mathrm{d}^{j+1}_{M^i} \circ \mathrm{d}^j_{M^i} = 0$ holds. There is an action of the DG ring A on M^i, prescribed by a DG $\mathbb{K}$-ring homomorphism $f_{M^i} : A \to \mathrm{End}_{\mathsf{M}}(M^i)$.

Sometimes we view M as a double complex, and then we refer to its rows and columns: M^i is the ith column of M, and $\{M^{i,j}\}_{i \in \mathbb{Z}}$ is its jth row. See diagram (11.2.4).

$$\begin{array}{ccccccccc}
 & & \vdots & & \vdots & & \vdots & & \\
 & & \uparrow & & \uparrow & & \uparrow & & \\
\cdots & \to & M^{-1,1} & \xrightarrow{\partial^{-1}} & M^{0,1} & \xrightarrow{\partial^{0}} & M^{1,1} & \to & \cdots \\
 & & \uparrow{\scriptstyle \mathrm{d}^0_{M^{-1}}} & & \uparrow{\scriptstyle \mathrm{d}^0_{M^{0}}} & & \uparrow{\scriptstyle \mathrm{d}^0_{M^{1}}} & & \\
\cdots & \to & M^{-1,0} & \xrightarrow{\partial^{-1}} & M^{0,0} & \xrightarrow{\partial^{0}} & M^{1,0} & \to & \cdots \\
 & & \uparrow & & \uparrow & & \uparrow & & \\
 & & \vdots & & \vdots & & \vdots & &
\end{array} \tag{11.2.4}$$

As explained in Proposition 11.1.1, if the abelian category M has countable direct sums, then so does $\mathbf{C}_{\mathrm{str}}(A, \mathsf{M})$. Recall that $A^\natural$ is the underlying graded ring of the DG ring A.

Definition 11.2.5 Assume that M has countable direct sums. Given a complex (M, ∂) with entries in $\mathbf{C}_{\mathrm{str}}(A, \mathsf{M})$, with notation as in (11.2.1), its *direct sum totalization* is the object

$$\mathrm{Tot}^{\oplus}(M, \partial) = (\mathrm{Tot}^{\oplus}(M), \mathrm{d}_{\mathrm{Tot}}) \in \mathbf{C}(A, \mathsf{M})$$

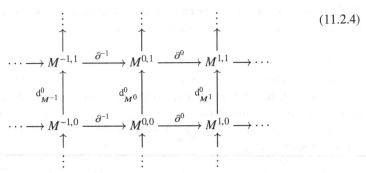

defined as follows, in four stages.

(1) There is a graded object

$$\mathrm{Tot}^{\oplus}(M) := \bigoplus_{i \in \mathbb{Z}} \mathrm{T}^{-i}(M^i) \in \mathbf{G}(A^{\natural}, \mathsf{M}).$$

(2) On each summand $\mathrm{T}^{-i}(M^i)$ of $\mathrm{Tot}^{\oplus}(M)$ there is a differential $\mathrm{d}_{\mathrm{T}^{-i}(M^i)}$, and we define the degree 1 operator $\mathrm{d}_M := \bigoplus_{i \in \mathbb{Z}} \mathrm{d}_{\mathrm{T}^{-i}(M^i)}$ on $\mathrm{Tot}^{\oplus}(M)$.

(3) For each i let

$$\mathrm{tot}(\partial)^i := \mathrm{t}^{-1}_{\mathrm{T}^{-(i+1)}(M^{i+1})} \circ \mathrm{T}^{-i}(\partial^i) : \mathrm{T}^{-i}(M^i) \to \mathrm{T}^{-(i+1)}(M^{i+1}).$$

We define the degree 1 operator $\mathrm{tot}(\partial) := \bigoplus_{i \in \mathbb{Z}} \mathrm{tot}(\partial)^i$ on the graded object $\mathrm{Tot}^{\oplus}(M)$.

(4) The differential $\mathrm{d}_{\mathrm{Tot}}$ on the graded object $\mathrm{Tot}^{\oplus}(M)$ is $\mathrm{d}_{\mathrm{Tot}} := \mathrm{d}_M + \mathrm{tot}(\partial)$.

For the definition to be valid we need to verify something:

Lemma 11.2.6 *The degree* 1 *operator* $\mathrm{d}_{\mathrm{Tot}}$ *makes the pair* $(\mathrm{Tot}^{\oplus}(M), \mathrm{d}_{\mathrm{Tot}})$ *into an object of* $\mathbf{C}(A, \mathsf{M})$.

Proof We have to check that $\mathrm{d}_{\mathrm{Tot}} \circ \mathrm{d}_{\mathrm{Tot}} = 0$ and that $\mathrm{tot}(\partial)$ is A-linear. Below we shall verify the first condition; the second verification is similar, and is left to the reader.

For a given $i \in \mathbb{Z}$ consider the composed morphism starting at the summand $\mathrm{T}^{-i}(M^i)$ of the total DG module:

$$\begin{aligned} \mathrm{T}^{-i}(M^i) &\xrightarrow{\mathrm{d}_{\mathrm{Tot}}} \mathrm{T}^{-i}(M^i) \oplus \mathrm{T}^{-(i+1)}(M^{i+1}) \\ &\xrightarrow{\mathrm{d}_{\mathrm{Tot}}} \mathrm{T}^{-i}(M^i) \oplus \mathrm{T}^{-(i+1)}(M^{i+1}) \oplus \mathrm{T}^{-(i+2)}(M^{i+2}). \end{aligned}$$

Let us view these direct sums as columns, with matrices of operators acting on them from the left. So we have to calculate the matrix product

$$\begin{bmatrix} \mathrm{d}_{\mathrm{T}^{-i}(M^i)} & 0 \\ \mathrm{t}^{-1} \circ \mathrm{T}^{-i}(\partial^i) & \mathrm{d}_{\mathrm{T}^{-(i+1)}(M^{i+1})} \\ 0 & \mathrm{t}^{-1} \circ \mathrm{T}^{-(i+1)}(\partial^{i+1}) \end{bmatrix} \circ \begin{bmatrix} \mathrm{d}_{\mathrm{T}^{-i}(M^i)} \\ \mathrm{t}^{-1} \circ \mathrm{T}^{-i}(\partial^i) \end{bmatrix}$$

The resulting column of operators has $\mathrm{d}_{\mathrm{T}^{-i}(M^i)} \circ \mathrm{d}_{\mathrm{T}^{-i}(M^i)} = 0$ in the first position. In the second position we have

$$\mathrm{t}^{-1} \circ \mathrm{T}^{-i}(\partial^i) \circ \mathrm{d}_{\mathrm{T}^{-i}(M^i)} + \mathrm{d}_{\mathrm{T}^{-(i+1)}(M^{i+1})} \circ \mathrm{t}^{-1} \circ \mathrm{T}^{-i}(\partial^i). \tag{11.2.7}$$

Using Proposition 4.1.12 we have

$$\mathrm{d}_{\mathrm{T}^{-(i+1)}(M^{i+1})} \circ \mathrm{t}^{-1} = -\mathrm{t}^{-1} \circ \mathrm{d}_{\mathrm{T}^{-i}(M^{i+1})} = -\mathrm{t}^{-1} \circ \mathrm{T}^{-i}(\mathrm{d}_{M^{i+1}}).$$

Therefore (11.2.7) equals

$$\mathrm{t}^{-1} \circ \mathrm{T}^{-i}(\partial^i \circ \mathrm{d}_{M^i} - \mathrm{d}_{M^{i+1}} \circ \partial^i),$$

which is zero by (11.2.2). In the bottom of the column we have

$$\mathrm{t}^{-1} \circ \mathrm{T}^{-(i+1)}(\partial^{i+1}) \circ \mathrm{t}^{-1} \circ \mathrm{T}^{-i}(\partial^{i}). \tag{11.2.8}$$

According to Definition 4.1.6, we know that

$$\mathrm{T}^{-(i+1)}(\partial^{i+1}) \circ \mathrm{t}^{-1} = \mathrm{t}^{-1} \circ \mathrm{T}(\mathrm{T}^{-(i+1)}(\partial^{i+1})) = \mathrm{t}^{-1} \circ \mathrm{T}^{-i}(\partial^{i+1})).$$

Therefore (11.2.8) equals $\mathrm{t}^{-1} \circ \mathrm{T}^{-i}(\partial^{i+1} \circ \partial^{i}) = 0$. □

Exercise 11.2.9 Finish the proof of the lemma (the A-linearity of $\mathrm{tot}(\partial)$).

Example 11.2.10 If $M^i = 0$ for $i \neq -1, 0$, then

$$\mathrm{Tot}^{\oplus}(M, \partial) = \mathrm{Cone}(\partial^{-1} : M^{-1} \to M^0),$$

the standard cone from Definition 4.2.1. In Proposition 11.2.13 there is a more general statement.

In Definition 7.3.6 we talked about smart truncations of DG modules. Here is another sort of truncations.

Definition 11.2.11 Let N be an abelian category. For a complex $N \in \mathbf{C}(\mathsf{N})$ its *stupid truncations* at an integer q are

$$\mathrm{stt}^{\leq q}(N) := (\cdots \to N^{q-1} \to N^q \to 0 \to 0 \to \cdots)$$

and

$$\mathrm{stt}^{\geq q}(N) := (\cdots \to 0 \to 0 \to N^q \to N^{q+1} \to \cdots).$$

These truncations fit into a short exact sequence

$$0 \to \mathrm{stt}^{\geq q}(N) \to N \to \mathrm{stt}^{\leq q-1}(N) \to 0 \tag{11.2.12}$$

in $\mathbf{C}_{\mathrm{str}}(\mathsf{N})$.

Warning: The stupid truncations do not apply to DG modules $N \in \mathbf{C}(B, \mathsf{N})$, unless the DG ring B is a ring. However, in what follows we will use stupid truncations in $\mathbf{C}(\mathsf{N})$, where N is the abelian category $\mathbf{C}_{\mathrm{str}}(A, \mathsf{M})$. This could be a bit confusing.

Proposition 11.2.13 *Let*

$$(M, \partial) = (\cdots \to M^{-1} \xrightarrow{\partial^{-1}} M^0 \xrightarrow{\partial^0} M^1 \to 0 \to \cdots)$$

be a complex with entries in the abelian category $\mathbf{C}_{\mathrm{str}}(A, \mathsf{M})$, *let*

$$(M', \partial) := (\cdots \to M^{-1} \xrightarrow{\partial^{-1}} M^0 \to 0 \to \cdots),$$

namely the stupid truncation of M below 0, *and let* $\rho : \mathrm{Tot}^{\oplus}(M', \partial) \to M^1$ *be the morphism in* $\mathbf{C}_{\mathrm{str}}(A, \mathsf{M})$ *induced by* $\partial^0 : M^0 \to M^1$. *Then there is an isomorphism*

$$\mathrm{T}(\mathrm{Tot}^{\oplus}(M, \partial)) \cong \mathrm{Cone}(-\rho : \mathrm{Tot}^{\oplus}(M', \partial) \to M^1) \tag{†}$$

in $\mathbf{C}_{\mathrm{str}}(A, \mathsf{M})$.

Proof By comparing translations we see that the objects on both sides of (†) match in each degree, and likewise for all the differentials that do not involve ∂^0. A calculation similar to that in the proof of Lemma 11.2.6 shows that there is a sign change there, and this is why we need $-\rho$ on the right side of (†). □

Corollary 11.2.14 *If, in the situation of Proposition* 11.2.13, *the DG module* $\mathrm{Tot}^{\oplus}(M, \partial)$ *is acyclic, then* $\rho : \mathrm{Tot}^{\oplus}(M', \partial) \to M^1$ *is a quasi-isomorphism.*

Proof If $\mathrm{Tot}^{\oplus}(M, \partial)$ is acyclic then so is $\mathrm{T}(\mathrm{Tot}^{\oplus}(M, \partial))$. The proposition implies that $-\rho$ is a quasi-isomorphism. Therefore ρ is a quasi-isomorphism. □

Definition 11.2.15 Let $M \in \mathbf{C}(A, \mathsf{M})$.

(1) A *filtration* on M is a direct system $\{F_j(M)\}_{j \geq -1}$ in $\mathbf{C}_{\mathrm{str}}(A, \mathsf{M})$ consisting of subobjects $F_j(M) \subseteq M$, indexed by the interval $[-1, \infty] \subseteq \mathbb{Z}$.
(2) We say that $M = \lim_{j \to} F_j(M)$ if the direct limit exists, and the canonical morphism $\lim_{j \to} F_j(M) \to M$ in $\mathbf{C}_{\mathrm{str}}(A, \mathsf{M})$ is an isomorphism.
(3) The filtration F induces, for each $j \geq 0$, the subquotient

$$\mathrm{Gr}^F_j(M) := F_j(M)/F_{j-1}(M) \in \mathbf{C}(A, \mathsf{M}).$$

Our filtrations will always be ascending. But sometimes they will be indexed by a subinterval $[j_0, j_1]$ of $[-1, \infty]$; and in this case $\mathrm{Gr}^F_j(M)$ will be defined only for $j \in [j_0 + 1, j_1]$.

In the next proposition we take $\mathsf{M} = \mathbf{M}(\mathbb{K})$; but really all we need is that M has exact countable direct limits.

Recall that for a DG A-module M we denote by $\mathrm{Z}(M)$ the object of cocycles of M. It is an object of $\mathbf{G}(\mathbb{K})$.

Proposition 11.2.16 *Let* (M, ∂) *be a complex with entries in the abelian category* $\mathbf{C}_{\mathrm{str}}(A)$, *with internal structure* (11.2.3). *Assume that the two conditions below hold*:

(i) *For every* $j \in \mathbb{Z}$ *the complex*

$$(\cdots \to M^{-1,j} \xrightarrow{\partial^{-1}} M^{0,j} \xrightarrow{\partial^0} M^{1,j} \to \cdots)$$

with entries in $\mathbf{M}(\mathbb{K})$ *is acyclic.*

(ii) *There is some* $j_1 \in \mathbb{Z}$ *such that for all* $j \geq j_1$ *the complex*

$$(\cdots \to \mathrm{Z}^j(M^{-1}) \xrightarrow{\partial^{-1}} \mathrm{Z}^j(M^0) \xrightarrow{\partial^0} \mathrm{Z}^j(M^1) \to \cdots)$$

with entries in $\mathbf{M}(\mathbb{K})$ *is acyclic.*

Then the DG A-module $\mathrm{Tot}^{\oplus}(M, \partial)$ *is acyclic.*

Proof The proof is done in two steps.

Step 1. Let us replace condition (ii) by the stronger condition:

(ii′) There is some $j_1' \in \mathbb{Z}$ such that $M^{i,j} = 0$ for all $j > j_1'$.

For every i we introduce a filtration $\{F_q(M^i)\}_{q \geq -1}$ on the DG $\mathbb{K}$-module M^i as follows:

$$F_q(M^i) := \operatorname{stt}^{\geq j_1'-q}(M^i) = (\cdots \to 0 \to M^{i,j_1'-q} \xrightarrow{\mathrm{d}_{M^i}} M^{i,j_1'-q+1} \to \cdots),$$

the stupid truncation above $j_1' - q$. These filtrations induces a filtration on the DG $\mathbb{K}$-module $\operatorname{Tot}^{\oplus}(M, \partial)$:

$$F_q(\operatorname{Tot}^{\oplus}(M)) := \operatorname{Tot}^{\oplus}(\cdots \to F_q(M^{-1}) \xrightarrow{\partial^{-1}} F_q(M^0) \xrightarrow{\partial^0} F_q(M^1) \to \cdots).$$

We note that $F_{-1}(\operatorname{Tot}^{\oplus}(M)) = 0$, and

$$\bigcup_{q \geq -1} F_q(\operatorname{Tot}^{\oplus}(M)) = \operatorname{Tot}^{\oplus}(M). \tag{11.2.17}$$

For every $q \geq 0$ the DG $\mathbb{K}$-module $\operatorname{Gr}_q^F(\operatorname{Tot}^{\oplus}(M))$ is isomorphic in $\mathbf{C}_{\mathrm{str}}(\mathbb{K})$, up to signs of differentials, to the complex in condition (i) with index $j = j_1' - q$, so it is acyclic. For $q = -1$ we trivially have an acyclic DG $\mathbb{K}$-module $F_{-1}(\operatorname{Tot}^{\oplus}(M))$. For every $q \geq 0$ there is an exact sequence

$$0 \to F_{q-1}(\operatorname{Tot}^{\oplus}(M)) \to F_q(\operatorname{Tot}^{\oplus}(M)) \to \operatorname{Gr}_q^F(\operatorname{Tot}^{\oplus}(M)) \to 0$$

in $\mathbf{C}_{\mathrm{str}}(\mathbb{K})$. By induction on q we conclude that $F_q(\operatorname{Tot}^{\oplus}(M))$ is acyclic. Finally, by (11.2.17), and the fact that cohomology commutes with direct limits in $\mathbf{C}_{\mathrm{str}}(\mathbb{K})$, we see that the DG module $\operatorname{Tot}^{\oplus}(M, \partial)$ is acyclic.

Step 2. Now we assume conditions (i) and (ii). For each i we introduce a filtration $\{G_q(M^i)\}_{q \geq 0}$ on the DG $\mathbb{K}$-module M^i as follows:

$$G_q(M^i) := \operatorname{smt}^{\leq j_1+q}(M^i) = (\cdots \to M^{i,j_1+q-1} \xrightarrow{\mathrm{d}_{M^i}} \mathrm{Z}^{j_1+q}(M^i) \to 0 \to \cdots),$$

the smart truncation below $j_1 + q$.

For every value of q the complex

$$G_q(M) := (\cdots \to G_q(M^{-1}) \xrightarrow{\partial^{-1}} G_q(M^0) \xrightarrow{\partial^0} G_q(M^1) \to \cdots)$$

satisfies conditions (i) and (ii′), with $j_1' := j_1 + q$. Indeed, for $j < j_1 + q$ the jth row of $G_q(M)$ is

$$(\cdots \to M^{-1,j} \xrightarrow{\partial^{-1}} M^{0,j} \xrightarrow{\partial^0} M^{1,j} \to \cdots),$$

and it is exact by condition (i). For $j = j_1 + q$ its jth row is

$$(\cdots \to \mathrm{Z}^j(M^{-1}) \xrightarrow{\partial^{-1}} \mathrm{Z}(M^0) \xrightarrow{\partial^0} \mathrm{Z}^j(M^1) \to \cdots),$$

and it is exact by condition (ii). And for $j > j_1 + q$ the jth row of $G_q(M)$ is zero. Therefore, by step 1, the DG module $G_q(\mathrm{Tot}^{\oplus}(M)) := \mathrm{Tot}^{\oplus}(G_q(M))$ is acyclic.

Finally, since $M = \lim_{q\to} G_q(M)$, it follows that

$$\mathrm{Tot}^{\oplus}(M) = \lim_{q\to} G_q(\mathrm{Tot}^{\oplus}(M)).$$

So $\mathrm{Tot}^{\oplus}(M)$ is acyclic. □

The second half of this section deals with products instead of direct sums. As explained in Proposition 11.1.1, if the abelian category M has countable products, then so does $\mathbf{C}_{\mathrm{str}}(A, \mathsf{M})$.

Definition 11.2.18 Assume that M has countable products. Given a complex (M, ∂) with entries in $\mathbf{C}_{\mathrm{str}}(A, \mathsf{M})$, with notation as in (11.2.1), its *product totalization* is the object

$$\mathrm{Tot}^{\Pi}(M, \partial) = (\mathrm{Tot}^{\Pi}(M), \mathrm{d}_{\mathrm{Tot}}) \in \mathbf{C}(A, \mathsf{M})$$

defined as follows, in four stages.

(1) There is a graded object

$$\mathrm{Tot}^{\Pi}(M) := \prod_{i\in\mathbb{Z}} \mathrm{T}^{-i}(M^i) \in \mathbf{G}(A^{\natural}, \mathsf{M}).$$

(2) On each factor $\mathrm{T}^{-i}(M^i)$ of $\mathrm{Tot}^{\Pi}(M)$ there is a differential $\mathrm{d}_{\mathrm{T}^{-i}(M^i)}$, and we let $\mathrm{d}_M := \prod_{i\in\mathbb{Z}} \mathrm{d}_{\mathrm{T}^{-i}(M^i)}$, which is a differential on $\mathrm{Tot}^{\Pi}(M)$.

(3) For each i let

$$\mathrm{tot}(\partial)^i := \mathrm{t}^{-1}_{\mathrm{T}^{-(i+1)}(M^{i+1})} \circ \mathrm{T}^{-i}(\partial^i) : \mathrm{T}^{-i}(M^i) \to \mathrm{T}^{-(i+1)}(M^{i+1}).$$

We define the degree 1 operator $\mathrm{tot}(\partial) := \prod_{i\in\mathbb{Z}} \mathrm{tot}(\partial)^i$ on the graded object $\mathrm{Tot}^{\Pi}(M)$.

(4) The differential $\mathrm{d}_{\mathrm{Tot}}$ on the graded object $\mathrm{Tot}^{\Pi}(M)$ is $\mathrm{d}_{\mathrm{Tot}} := \mathrm{d}_M + \mathrm{tot}(\partial)$.

Note that (if M also has countable direct sums) there is a canonical embedding

$$\mathrm{Tot}^{\oplus}(M) \subseteq \mathrm{Tot}^{\Pi}(M) \tag{11.2.19}$$

in $\mathbf{C}_{\mathrm{str}}(A, \mathsf{M})$. In case $M^i = 0$ for $|i| \gg 0$, then this is an equality.

For Definition 11.2.18 to be valid we need to verify something:

Lemma 11.2.20 *The degree* 1 *operator* $\mathrm{d}_{\mathrm{Tot}}$ *makes the pair* $(\mathrm{Tot}^{\Pi}(M), \mathrm{d}_{\mathrm{Tot}})$ *into an object of* $\mathbf{C}(A, \mathsf{M})$.

Proof This is the same calculation as in the proof of Lemma 11.2.6, heuristically using the inclusion (11.2.19). □

Proposition 11.2.21 *Let*

$$(M, \partial) = (\cdots \to 0 \to M^{-1} \xrightarrow{\partial^{-1}} M^0 \xrightarrow{\partial^0} M^1 \to \cdots)$$

be a complex with entries in the abelian category $\mathbf{C}_{\mathrm{str}}(A, \mathsf{M})$, *let*

$$(M', \partial) := (\cdots \to 0 \to M^0 \xrightarrow{\partial^0} M^1 \to \cdots),$$

namely the stupid truncation of M *above* 0, *and let* $\rho : M^{-1} \to \mathrm{Tot}^{\Pi}(M', \partial)$ *be the morphism in* $\mathbf{C}_{\mathrm{str}}(A, \mathsf{M})$ *induced by* $\partial^{-1} : M^{-1} \to M^0$. *Then there is an isomorphism*

$$\mathrm{Tot}^{\Pi}(M, \partial) \cong \mathrm{Cone}(\rho : M^{-1} \to \mathrm{Tot}^{\Pi}(M', \partial))$$

in $\mathbf{C}_{\mathrm{str}}(A, \mathsf{M})$.

Proof Like the proof of Proposition 11.2.13, only now there are no sign issues. □

This directly implies:

Corollary 11.2.22 *If, in the situation of Proposition* 11.2.21, *the DG module* $\mathrm{Tot}^{\Pi}(M, \partial)$ *is acyclic, then* $\rho : M^{-1} \to \mathrm{Tot}^{\Pi}(M', \partial)$ *is a quasi-isomorphism.*

In Section 1.3 we discussed quotients in categories. They are dual to subobjects. Similarly, we now introduce a notion dual to a filtration of an object.

Definition 11.2.23 Let $M \in \mathbf{C}(A, \mathsf{M})$.

(1) A *cofiltration* on M is an inverse system $\{F_j(M)\}_{j \geq -1}$ in $\mathbf{C}_{\mathrm{str}}(A, \mathsf{M})$ consisting of quotients $M \twoheadrightarrow F_j(M)$, indexed by the interval $[-1, \infty] \subseteq \mathbb{Z}$.
(2) We say that $M = \lim_{\leftarrow j} F_j(M)$ if the inverse limit exists, and the canonical morphism $M \to \lim_{\leftarrow j} F_j(M)$ in $\mathbf{C}_{\mathrm{str}}(A, \mathsf{M})$ is an isomorphism.
(3) The cofiltration F gives rise to the subquotients

$$\mathrm{Gr}_j^F(M) := \mathrm{Ker}(F_j(M) \to F_{j-1}(M)) \in \mathbf{C}(A, \mathsf{M}).$$

Recall that for a DG A-module M we denote by $\mathrm{Y}(M) := M / \mathrm{B}(M)$; this is the object of decocycles of M, and it belongs to $\mathbf{G}(\mathbb{K})$. See Definition 3.2.4 and equation 3.8.8.

Proposition 11.2.24 *Let* (M, ∂) *be a complex with entries in the abelian category* $\mathbf{C}_{\mathrm{str}}(A)$, *with internal structure* (11.2.3). *Assume that the two conditions below hold*:

(i) *For every* $j \in \mathbb{Z}$ *the complex*

$$(\cdots \to M^{-1,j} \xrightarrow{\partial^{-1}} M^{0,j} \xrightarrow{\partial^0} M^{1,j} \to \cdots)$$

with entries in $\mathbf{M}(\mathbb{K})$ *is acyclic.*

(ii) *There is some* $j_0 \in \mathbb{Z}$ *such that for all* $j \leq j_0$ *the complex*

$$(\cdots \to \mathrm{Y}^j(M^{-1}) \xrightarrow{\mathrm{Y}^j(\partial^{-1})} \mathrm{Y}^j(M^0) \xrightarrow{\mathrm{Y}^j(\partial^0)} \mathrm{Y}^j(M^1) \to \cdots)$$

with entries in $\mathbf{M}(\mathbb{K})$ *is acyclic.*

Then the DG A-module $\mathrm{Tot}^{\Pi}(M, \partial)$ *is acyclic.*

Proof The proof is divided into two steps.

Step 1. We replace condition (ii) by the stronger condition:

(ii′) There is some $j_0' \in \mathbb{Z}$ such that $M^{i,j} = 0$ for all $j < j_0'$.

For every i we introduce a cofiltration $\{F_q(M^i)\}_{q \geq -1}$ on the DG $\mathbb{K}$-module M^i as follows: $F_q(M^i) := \mathrm{stt}^{\leq j_0'+q}(M^i)$, the stupid truncation below $j_0' + q$. These cofiltrations induce a cofiltration on the DG $\mathbb{K}$-module $\mathrm{Tot}^{\Pi}(M, \partial)$:

$$F_q(\mathrm{Tot}^{\Pi}(M)) := \mathrm{Tot}^{\Pi}(\cdots \to F_q(M^{-1}) \xrightarrow{\partial^{-1}} F_q(M^0) \xrightarrow{\partial^0} F_q(M^1) \to \cdots).$$

We note that $F_{-1}(\mathrm{Tot}^{\Pi}(M)) = 0$, and

$$\lim_{\leftarrow q} F_q(\mathrm{Tot}^{\Pi}(M)) = \mathrm{Tot}^{\Pi}(M). \tag{11.2.25}$$

For every $q \geq 0$ the DG $\mathbb{K}$-module $\mathrm{Gr}_q^F(\mathrm{Tot}^{\Pi}(M))$ is isomorphic in $\mathbf{C}_{\mathrm{str}}(\mathbb{K})$ to the complex in condition (i) with index $j = j_0' + q$ (up to a sign change of the differentials), so it is acyclic. For $q = -1$ we trivially have an acyclic DG $\mathbb{K}$-module $F_{-1}(\mathrm{Tot}^{\Pi}(M))$. For every $q \geq 0$ there is an exact sequence

$$0 \to \mathrm{Gr}_q^F(\mathrm{Tot}^{\Pi}(M)) \to F_q(\mathrm{Tot}^{\Pi}(M)) \to F_{q-1}(\mathrm{Tot}^{\Pi}(M)) \to 0$$

in $\mathbf{C}_{\mathrm{str}}(\mathbb{K})$. By induction on q we conclude that $F_q(\mathrm{Tot}^{\Pi}(M))$ is acyclic. Finally, because the inverse system of complexes of $\mathbb{K}$-modules (11.2.25) has surjective transitions, the Mittag–Leffler argument (see Corollary 11.1.8) says that the DG module $\mathrm{Tot}^{\Pi}(M, \partial)$ is acyclic.

Step 2. Here we assume conditions (i) and (ii). For each i we introduce a cofiltration $\{G_q(M^i)\}_{q \geq 0}$ on the DG $\mathbb{K}$-module M^i as follows:

$$G_q(M^i) := \mathrm{smt}^{\geq j_0-q}(M^i) = (\cdots \to 0 \to \mathrm{Y}^{j_0-q}(M^i) \xrightarrow{\mathrm{d}_{M^i}} M^{i,j_0-q+1} \to \cdots),$$

the smart truncation above $j_0 - q$.

For each q the complex

$$G_q(M) := (\cdots \to G_q(M^{-1}) \xrightarrow{G_q(\partial^{-1})} G_q(M^0) \xrightarrow{G_q(\partial^0)} G_q(M^1) \to \cdots)$$

satisfies conditions (i) and (ii′), with $j_0' := j_0 - q$. Indeed, its jth row is

$$(\cdots \to M^{-1,j} \xrightarrow{\partial^{-1}} M^{0,j} \xrightarrow{\partial^{0}} M^{1,j} \to \cdots)$$

for $j > j_0 - q$; it is

$$(\cdots \to \mathrm{Y}^j(M^{-1}) \xrightarrow{\mathrm{Y}^j(\partial^{-1})} \mathrm{Y}(M^0) \xrightarrow{\mathrm{Y}^j(\partial^{0})} \mathrm{Y}^j(M^1) \to \cdots)$$

for $j = j_0 - q$; and it is zero for $j < j_0 - q$. All these rows are exact. Therefore, by step 1, the DG module $G_q(\mathrm{Tot}^{\Pi}(M)) := \mathrm{Tot}^{\Pi}(G_q(M))$ is acyclic.

Finally, since $M = \lim_{\leftarrow q} G_q(M)$, it follows that

$$\mathrm{Tot}^{\Pi}(M) = \lim_{\leftarrow q} G_q(\mathrm{Tot}^{\Pi}(M)).$$

This is an inverse system of complexes of $\mathbb{K}$-modules with surjective transitions. According to the Mittag–Leffler argument, the limit $\mathrm{Tot}^{\Pi}(M)$ is acyclic. □

11.3 K-Projective Resolutions in $\mathbf{C}^-(\mathsf{M})$

Recall that M is a $\mathbb{K}$-linear abelian category, and $\mathbf{C}(\mathsf{M})$ is the DG category of complexes in M. The strict category $\mathbf{C}_{\mathrm{str}}(\mathsf{M})$ is abelian.

The next definition is inspired by the work of B. Keller [79, Section 3.1].

Definition 11.3.1 Let P be an object of $\mathbf{C}(\mathsf{M})$.

(1) A *semi-projective filtration* on P is a filtration $F = \{F_j(P)\}_{j \geq -1}$ on P as an object of $\mathbf{C}_{\mathrm{str}}(\mathsf{M})$, such that:
- $F_{-1}(P) = 0$.
- Each $\mathrm{Gr}_j^F(P)$ is a complex of projective objects of M with zero differential.
- $P = \lim_{j\to} F_j(P)$ in $\mathbf{C}_{\mathrm{str}}(\mathsf{M})$.

(2) The complex P is called a *semi-projective complex* if it admits some semi-projective filtration.

K-projective objects in $\mathbf{C}(\mathsf{M})$ were introduced in Definition 10.2.1.

Theorem 11.3.2 *Let* M *be an abelian category, and let* P *be a semi-projective complex in* $\mathbf{C}(\mathsf{M})$. *Then* P *is K-projective.*

Proof The proof is in four steps.

Step 1. We start by proving that if $P = \mathrm{T}^k(Q)$, the translation of a projective object $Q \in \mathsf{M}$, then P is K-projective. This is easy: given an acyclic complex $N \in \mathbf{C}(\mathsf{M})$, we have

$$\mathrm{Hom}_{\mathsf{M}}(P, N) = \mathrm{Hom}_{\mathsf{M}}(\mathrm{T}^k(Q), N) \cong \mathrm{T}^{-k}(\mathrm{Hom}_{\mathsf{M}}(Q, N))$$

in $\mathbf{C}_{\mathrm{str}}(\mathbb{K})$. But $\mathrm{Hom}_{\mathsf{M}}(Q, -)$ is an exact functor $\mathsf{M} \to \mathbf{M}(\mathbb{K})$, so $\mathrm{Hom}_{\mathsf{M}}(Q, N)$ is an acyclic complex.

Step 2. Now P is a complex of projective objects of M with zero differential. This means that $P \cong \bigoplus_{k \in \mathbb{Z}} \mathrm{T}^k(Q_k)$ in $\mathbf{C}_{\mathrm{str}}(\mathsf{M})$, where each Q_k is a projective object in M. But then

$$\mathrm{Hom}_{\mathsf{M}}(P, N) \cong \prod_{k \in \mathbb{Z}} \mathrm{Hom}_{\mathsf{M}}(\mathrm{T}^k(Q_k), N)$$

in $\mathbf{C}_{\mathrm{str}}(\mathbb{K})$. This is an easy case of Proposition 1.8.6. By step 1 and the fact that a product of acyclic complexes in $\mathbf{C}_{\mathrm{str}}(\mathbb{K})$ is acyclic (itself an easy case of the Mittag–Leffler argument), we conclude that $\mathrm{Hom}_{\mathsf{M}}(P, N)$ is acyclic.

Step 3. Fix a semi-projective filtration $F = \{F_j(P)\}_{j \geq -1}$ on P. Here we prove that for every j the complex $F_j(P)$ is K-projective. This is done by induction on $j \geq -1$. For $j = -1$ it is trivial. For $j \geq 0$ there is an exact sequence

$$0 \to F_{j-1}(P) \to F_j(P) \to \mathrm{Gr}_j^F(P) \to 0 \tag{11.3.3}$$

in $\mathbf{C}_{\mathrm{str}}(\mathsf{M})$. In each degree $i \in \mathbb{Z}$ the exact sequence

$$0 \to F_{j-1}(P)^i \to F_j(P)^i \to \mathrm{Gr}_j^F(P)^i \to 0$$

in M splits, because $\mathrm{Gr}_j^F(P)^i$ is a projective object. Thus the exact sequence (11.3.3) is split exact in the abelian category $\mathbf{G}_{\mathrm{str}}(\mathsf{M})$ of graded objects in M.

Let $N \in \mathbf{C}(\mathsf{M})$ be an acyclic complex. Applying the functor $\mathrm{Hom}_{\mathsf{M}}(-, N)$ to the sequence of complexes (11.3.3) we obtain a sequence

$$0 \to \mathrm{Hom}_{\mathsf{M}}(\mathrm{Gr}_j^F(P), N) \to \mathrm{Hom}_{\mathsf{M}}(F_j(P), N) \to \mathrm{Hom}_{\mathsf{M}}(F_{j-1}(P), N) \to 0 \tag{11.3.4}$$

in $\mathbf{C}_{\mathrm{str}}(\mathbb{K})$. Because (11.3.3) is split exact in $\mathbf{G}_{\mathrm{str}}(\mathsf{M})$, the sequence (11.3.4) is split exact in $\mathbf{G}_{\mathrm{str}}(\mathbb{K})$. Therefore (11.3.4) is exact in $\mathbf{C}_{\mathrm{str}}(\mathbb{K})$.

By the induction hypothesis the complex $\mathrm{Hom}_{\mathsf{M}}(F_{j-1}(P), N)$ is acyclic. By step 2 the complex $\mathrm{Hom}_{\mathsf{M}}(\mathrm{Gr}_j^F(P), N)$ is acyclic. The long exact cohomology sequence associated to (11.3.4) shows that the complex $\mathrm{Hom}_{\mathsf{M}}(F_j(P), N)$ is acyclic too.

Step 4. We keep the semi-projective filtration $F = \{F_j(P)\}_{j \geq -1}$ from step 3. Take any acyclic complex $N \in \mathbf{C}(\mathsf{M})$. By Proposition 1.8.6 we know that

$$\mathrm{Hom}_{\mathsf{M}}(P, N) \cong \lim_{\leftarrow j} \mathrm{Hom}_{\mathsf{M}}(F_j(P), N)$$

in $\mathbf{C}_{\mathrm{str}}(\mathbb{K})$. According to step 3, the complexes $\mathrm{Hom}_{\mathsf{M}}(F_j(P), N)$ are all acyclic. The exactness of the sequences (11.3.4) implies that the inverse system $\{\mathrm{Hom}_{\mathsf{M}}(F_j(P), N)\}_{j \geq -1}$ in $\mathbf{C}_{\mathrm{str}}(\mathbb{K})$ has surjective transitions. Now the Mittag–Leffler argument (Corollary 11.1.8) says that the inverse limit complex $\mathrm{Hom}_{\mathsf{M}}(P, N)$ is acyclic. □

Proposition 11.3.5 *Let* M *be an abelian category. If $P \in \mathbf{C}(\mathsf{M})$ is a bounded above complex of projectives, then P is a semi-projective complex.*

Proof Say P is nonzero and $\sup(P) = i_1 \in \mathbb{Z}$. For $j \geq -1$ define $F_j(P) := \operatorname{stt}^{\geq i_1 - j}(P)$, the stupid truncation above $i_1 - j$ from Definition 11.2.11. Then $\{F_j(P)\}_{j \geq -1}$ is a semi-projective filtration on P. □

The next theorem is dual to [62, Lemma I.4.6(1)], in the sense of changing direction of arrows. We give a detailed proof.

Theorem 11.3.6 *Let* M *be an abelian category, and let* $\mathsf{P} \subseteq \mathsf{M}$ *be a full subcategory such that every object* $M \in \mathsf{M}$ *admits an epimorphism* $P \twoheadrightarrow M$ *from some object* $P \in \mathsf{P}$*. Then every complex* $M \in \mathbf{C}^-(\mathsf{M})$ *admits a quasi-isomorphism* $\rho : P \to M$ *in* $\mathbf{C}^-_{\mathrm{str}}(\mathsf{M})$, *such that* $\sup(P) = \sup(M)$, *and each* P^i *is an object of* P.

Proof This is divided into steps.

Step 1. We can assume that $M \neq 0$. After translating M, we can assume that $\sup(M) = 0$.

Step 2. Say the differential of the complex M is $\mathrm{d}^i_M : M^i \to M^{i+1}$. Let us choose an epimorphism $\rho^0 : P^0 \twoheadrightarrow M^0$ in M from some object $P^0 \in \mathsf{P}$. We get a morphism $\delta^0 : M^{-1} \oplus P^0 \to M^0$ whose components are d^{-1}_M and $-\rho^0$. Thus $\operatorname{Ker}(\delta^0) = M^{-1} \times_{M^0} P^0$, the fibered product. Next we choose an epimorphism $\psi^{-1} : P^{-1} \twoheadrightarrow \operatorname{Ker}(\delta^0)$ from some object $P^{-1} \in \mathsf{P}$. So there is an exact sequence

$$P^{-1} \xrightarrow{\psi^{-1}} M^{-1} \oplus P^0 \xrightarrow{\delta^0} M^0 \to 0. \tag{11.3.7}$$

The components of ψ^{-1} are denoted by $\rho^{-1} : P^{-1} \to M^{-1}$ and $\mathrm{d}^{-1}_P : P^{-1} \to P^0$. We have this commutative diagram

$$\begin{array}{ccccccc} & & P^{-1} & \xrightarrow{\mathrm{d}^{-1}_P} & P^0 & \longrightarrow & 0 \\ & & \downarrow{\scriptstyle \rho^{-1}} & & \downarrow{\scriptstyle \rho^0} & & \\ M^{-2} & \xrightarrow{\mathrm{d}^{-2}_M} & M^{-1} & \xrightarrow{\mathrm{d}^{-1}_M} & M^0 & \longrightarrow & 0 \end{array} \tag{11.3.8}$$

in M.

Step 3. This is the inductive step. Here $i \leq -1$, and we already have objects $P^i, \ldots, P^0$ in P, and morphisms $\rho^i, \ldots, \rho^0$ and $\mathrm{d}^i_P, \ldots, \mathrm{d}^{-1}_P$, which fit into this commutative diagram

$$\begin{array}{ccccccccc} & & P^i & \xrightarrow{\mathrm{d}^i_P} & P^{i+1} & \longrightarrow \cdots \xrightarrow{\mathrm{d}^{-1}_P} & P^0 & \longrightarrow & 0 \\ & & \downarrow{\scriptstyle \rho^i} & & \downarrow{\scriptstyle \rho^{i+1}} & & \downarrow{\scriptstyle \rho^0} & & \\ M^{i-1} & \xrightarrow{\mathrm{d}^{i-1}_M} & M^i & \xrightarrow{\mathrm{d}^i_M} & M^{i+1} & \longrightarrow \cdots \xrightarrow{\mathrm{d}^{-1}_M} & M^0 & \longrightarrow & 0 \end{array} \tag{11.3.9}$$

in M. Also $\mathrm{d}_P^{j+1} \circ \mathrm{d}_P^j = 0$ for all j in the interval $[i, -1]$

Define the morphism $\delta^i : M^{i-1} \oplus P^i \to M^i \oplus P^{i+1}$ to be the one with components d_M^{i-1}, $-\rho^i$ and $-\mathrm{d}_P^i$. Expressing direct sums of objects as columns, and letting matrices of morphisms act on them from the left, we have this matrix representation of δ^i :

$$\delta^i = \begin{bmatrix} \mathrm{d}_M^{i-1} & -\rho^i \\ 0 & -\mathrm{d}_P^i \end{bmatrix}. \tag{11.3.10}$$

So $\mathrm{Ker}(\delta^i) = M^{i-1} \times_{(M^i \oplus P^{i+1})} P^i$, the fibered product.

Let us choose an epimorphism $\psi^{i-1} : P^{i-1} \twoheadrightarrow \mathrm{Ker}(\delta^i)$ from an object $P^{i-1} \in \mathsf{P}$. We get an exact sequence

$$P^{i-1} \xrightarrow{\psi^{i-1}} M^{i-1} \oplus P^i \xrightarrow{\delta^i} M^i \oplus P^{i+1}. \tag{11.3.11}$$

The components of the morphism ψ^{i-1} are denoted by $\rho^{i-1} : P^{i-1} \to M^{i-1}$ and $\mathrm{d}_P^{i-1} : P^{i-1} \to P^i$. In matrix representation is

$$\psi^{i-1} = \begin{bmatrix} \rho^{i-1} \\ \mathrm{d}_P^{i-1} \end{bmatrix}. \tag{11.3.12}$$

In this way we obtain the slightly bigger diagram than (11.3.9):

$$\begin{array}{ccccccccccc} P^{i-1} & \xrightarrow{\mathrm{d}_P^{i-1}} & P^i & \xrightarrow{\mathrm{d}_P^i} & P^{i+1} & \longrightarrow \cdots \longrightarrow & P^0 & \longrightarrow & 0 \\ \downarrow{\scriptstyle \rho^{i-1}} & & \downarrow{\scriptstyle \rho^i} & & \downarrow{\scriptstyle \rho^{i+1}} & & \downarrow{\scriptstyle \rho^0} & & \\ M^{i-1} & \xrightarrow{\mathrm{d}_M^{i-1}} & M^i & \xrightarrow{\mathrm{d}_M^i} & M^{i+1} & \longrightarrow \cdots \longrightarrow & M^0 & \longrightarrow & 0 \end{array} \tag{11.3.13}$$

Because $\delta^i \circ \psi^{i-1} = 0$ in (11.3.11), it follows that $\mathrm{d}_P^i \circ \mathrm{d}_P^{i-1} = 0$, and also

$$\rho^i \circ \mathrm{d}_P^{i-1} = \mathrm{d}_M^{i-1} \circ \rho^{i-1}, \tag{11.3.14}$$

showing that diagram (11.3.13) is commutative.

Step 4. We carry out the construction in step 3 inductively for all $i \leq -1$, thus obtaining a diagram like (11.3.13) that goes infinitely to the left.

Letting $P^i := 0$ for positive i, the collection $P := \{P^i\}_{i \in \mathbb{Z}}$ becomes a complex, with differential $\mathrm{d}_P := \{\mathrm{d}_P^i\}_{i \in \mathbb{Z}}$. By equation (11.3.14), the collection $\rho := \{\rho^i\}_{i \in \mathbb{Z}}$ is a strict morphism of complexes $\rho : P \to M$.

Step 5. It remains to prove that ρ is a quasi-isomorphism.

Take any $i \leq 0$. Let us examine this diagram:

$$\begin{array}{ccccc} P^{i-1} & \xrightarrow{-\psi^{i-1}} & M^{i-1} \oplus P^i & \xrightarrow{\delta^i} & M^i \oplus P^{i+1} \\ \downarrow{\scriptstyle (0,\mathrm{id})} & & \downarrow{\scriptstyle \mathrm{id}} & & \downarrow{\scriptstyle \mathrm{id}} \\ M^{i-2} \oplus P^{i-1} & \xrightarrow{\delta^{i-1}} & M^{i-1} \oplus P^i & \xrightarrow{\delta^i} & M^i \oplus P^{i+1} \end{array} \tag{11.3.15}$$

Comparing the formula for δ^{i-1} in (11.3.10) to the formula for $-\psi^{i-1}$ in (11.3.12), we see that this diagram is commutative. An easy calculation using (11.3.14) shows that $\delta^i \circ \delta^{i-1} = 0$. The top row is exact, because (up to signs) for $i = 0$ it is part of the exact sequence (11.3.7), and for $i \leq -1$ it is the exact sequence (11.3.11). It follows that the bottom row is also exact.

Let $N = \{N^i\}_{i \in \mathbb{Z}}$ be the complex with components $N^i := M^{i-1} \oplus P^i$ for all i. The differential $\mathrm{d}_N = \{\mathrm{d}_N^i\}_{i \in \mathbb{Z}}$ is $\mathrm{d}_N^i := -\delta^i : N^i \to N^{i+1}$. We have $N^i = 0$ for $i \geq 2$. The exactness of the sequence (11.3.7) says that $\mathrm{H}^1(N) = 0$, and the exactness of the second row in (11.3.15) says that $\mathrm{H}^i(N) = 0$ for $i \leq 0$. Hence the complex N is acyclic. On the other hand, by the definition of the morphisms δ^i in (11.3.10), we see that N is just the standard cone on the strict morphism of complexes $\mathrm{T}^{-1}(\rho) : \mathrm{T}^{-1}(P) \to \mathrm{T}^{-1}(M)$. See Definition 4.2.1. Therefore $\mathrm{T}^{-1}(\rho)$ is a quasi-isomorphism, and so is ρ. □

Example 11.3.16 Let $(X, \mathcal{A})$ be a ringed space. Usually there aren't enough projective objects in the abelian category $\mathbf{M}(\mathcal{A}) = \mathsf{Mod}\,\mathcal{A}$ of sheaves of $\mathcal{A}$-modules on X; but there are always enough flat objects. A general method to produce flat sheaves is as follows. Given an open set $U \subseteq X$ with inclusion map $g : U \to X$, the extension by zero sheaf $g_!(\mathcal{A}|_U)$ is a flat $\mathcal{A}$-module. Let us call $g_!(\mathcal{A}|_U)$ a pseudo-free $\mathcal{A}$-module of pseudo-rank 1 and pseudo-support U . A *pseudo-free $\mathcal{A}$-module* is, by definition, a sheaf $\mathcal{P}$ which is a direct sum of rank 1 pseudo-free $\mathcal{A}$-modules (with varying pseudo-supports). A pseudo-free $\mathcal{A}$-module is flat, and every $\mathcal{A}$-module $\mathcal{M}$ admits an epimorphism $\mathcal{P} \twoheadrightarrow \mathcal{M}$ from some pseudo-free $\mathcal{A}$-module $\mathcal{P}$. More on this in [62, Proposition II.1.2] or [75, Proposition 2.4.12]. (Warning: in case $(X, \mathcal{O}_X)$ is a scheme, the pseudo-free $\mathcal{O}_X$-modules are usually not quasi-coherent.)

Now let $\mathbf{C}(\mathcal{A}) = \mathbf{C}(\mathsf{Mod}\,\mathcal{A})$ be the category of complexes of $\mathcal{A}$-modules. Because $\mathbf{M}(\mathcal{A})$ has enough flat objects, Theorem 11.3.6 tells us that every complex $\mathcal{M} \in \mathbf{C}^-(\mathcal{A})$ admits a quasi-isomorphism $\mathcal{P} \to \mathcal{M}$ from a bounded above complex of flat $\mathcal{A}$-modules $\mathcal{P}$. Similarly to Proposition 11.3.5 and Theorem 11.3.2, one can show that such a complex $\mathcal{P}$ is K-flat. The conclusion is that the subcategory $\mathbf{C}^-(\mathcal{A})$ has enough K-flat objects.

But in fact more is true: the category of unbounded complexes $\mathbf{C}(\mathcal{A})$ also has enough K-flat objects; see [143]. For an even more general statement see Remark 11.4.45 below.

Definition 11.3.17 Let M be an abelian category, and let $\mathsf{M}' \subseteq \mathsf{M}$ be a full abelian subcategory. We say that M' *has enough projectives relative to* M if every object $M \in \mathsf{M}'$ admits an epimorphism $P \twoheadrightarrow M$, where P is an object of M' that is projective in the bigger category M.

Of course, if M' has enough projectives relative to M then M' itself has enough projectives.

Thick abelian categories were defined in Definition 7.4.1. The next theorem is the opposite of [62, Lemma I.4.6(3)].

Theorem 11.3.18 *Let* M *be an abelian category, and let* $\mathsf{M}' \subseteq \mathsf{M}$ *be a thick abelian subcategory that has enough projectives relative to* M. *Let* $M \in \mathsf{C}(\mathsf{M})$ *be a complex with bounded above cohomology, such that* $\mathrm{H}^i(M) \in \mathsf{M}'$ *for all* i. *Then there is a quasi-isomorphism* $\rho : P \to M$ *in* $\mathsf{C}_{\mathrm{str}}(\mathsf{M})$, *where* $P \in \mathsf{C}^-(\mathsf{M}')$, *each* P^i *is projective in* M, *and* $\sup(P) = \sup(\mathrm{H}(M))$.

Proof The proof of Theorem 11.5.8, reversed, works here. (Despite being located later in the book, Theorem 11.5.8 is not logically dependent on the current theorem.) To be explicit, let us take $\mathsf{N} := \mathsf{M}^{\mathrm{op}}$ and $\mathsf{N}' := \mathsf{M}'^{\mathrm{op}}$. Since monomorphisms in M become epimorphisms in N, and projective objects in M become injective objects in N, the full abelian subcategory $\mathsf{N}' \subseteq \mathsf{N}$ satisfies assumptions of Theorem 11.5.8. By Theorem 3.9.16 we have a canonical isomorphism of categories $\mathsf{C}^+_{\mathrm{str}}(\mathsf{N}) \xrightarrow{\simeq} \mathsf{C}^-_{\mathrm{str}}(\mathsf{M})^{\mathrm{op}}$ that respects the cohomology functor. Thus a quasi-isomorphism $N \to J$ in $\mathsf{C}^+_{\mathrm{str}}(\mathsf{N})$ gives rise to a quasi-isomorphism $P \to M$ in $\mathsf{C}^-_{\mathrm{str}}(\mathsf{M})$. □

Corollary 11.3.19 *If* M *is an abelian category with enough projectives, then* $\mathsf{C}^-(\mathsf{M})$ *has enough K-projectives.*

Proof According to either Theorem 11.3.6 or Theorem 11.3.18, every $M \in \mathsf{C}^-(\mathsf{M})$ admits a quasi-isomorphism $P \to M$ from a bounded above complex of projectives P. Now use Proposition 11.3.5 and Theorem 11.3.2. □

Corollary 11.3.20 *Let* M *be an abelian category with enough projectives, and let* $M \in \mathsf{C}(\mathsf{M})$ *be a complex with bounded above cohomology. Then* M *has a K-projective resolution* $P \to M$, *such that* $\sup(P) = \sup(\mathrm{H}(M))$, *and every* P^i *is an projective object of* M.

Proof We may assume that $\mathrm{H}(M)$ is not zero. Let $i := \sup(\mathrm{H}(M)) \in \mathbb{Z}$, and take $N := \mathrm{smt}^{\leq i}(M)$, the smart truncation from Definition 7.3.6. Then $N \to M$ is a quasi-isomorphism and $\sup(N) = i$. According to either Theorem 11.3.6 or Theorem 11.3.18, there is a quasi-isomorphism $P \to N$, where P is a complex of projectives and $\sup(P) = i$. By Proposition 11.3.5 and Theorem 11.3.2 the complex P is K-projective. The composed quasi-isomorphism $P \to M$ is what we are looking for. □

Corollary 11.3.21 *Under the assumptions of Theorem* 11.3.18, *the canonical functor* $\mathsf{D}^-(\mathsf{M}') \to \mathsf{D}^-_{\mathsf{M}'}(\mathsf{M})$ *is an equivalence.*

Proof Consider the commutative diagram

$$\begin{array}{ccccc} \mathbf{K}^-(\mathsf{M}')_{\mathsf{M}\text{-prj}} & & \xrightarrow{\quad F \quad} & & \\ \downarrow{\scriptstyle \mathrm{Q}} & & & & \searrow \\ \mathbf{D}^-(\mathsf{M}') & \xrightarrow{G} & \mathbf{D}^-_{\mathsf{M}'}(\mathsf{M}) & \xrightarrow{\text{f.f.}} & \mathbf{D}(\mathsf{M}) \end{array}$$

where $\mathbf{K}^-(\mathsf{M}')_{\mathsf{M}\text{-prj}}$ is the homotopy category of bounded above complexes of objects of M' that are projective in M. Since these are K-projective complexes in $\mathbf{K}(\mathsf{M})$, the functors F and Q are fully faithful, by Corollary 10.2.10. The functor marked "f.f." is the fully faithful inclusion of this full subcategory. We conclude that the functor G is also fully faithful. The theorem tells us that the essential image of G is $\mathbf{D}^-_{\mathsf{M}'}(\mathsf{M})$. □

Here is an important instance where Theorem 11.3.18 and Corollary 11.3.21 apply.

Example 11.3.22 Let A be a left noetherian ring. Consider the abelian category $\mathsf{M} := \mathbf{M}(A) = \mathsf{Mod}\, A$, and the thick abelian subcategory $\mathsf{M}' := \mathbf{M}_{\mathrm{f}}(A) = \mathsf{Mod}_{\mathrm{f}}\, A$ of finitely generated modules. Then M' has enough projectives relative to M. Theorem 11.3.18 tells us that if $M \in \mathbf{C}(\mathsf{M})$ is a complex such that the modules $\mathrm{H}^i(M)$ are all finitely generated, and $\mathrm{H}^i(M) = 0$ for $i \gg 0$, then there is a resolution $P \to M$, where P is a bounded above complex of finitely generated projective modules. Corollary 11.3.21 says that the canonical functor

$$\mathbf{D}^-(\mathbf{M}_{\mathrm{f}}(A)) = \mathbf{D}^-(\mathsf{M}') \to \mathbf{D}^-_{\mathsf{M}'}(\mathsf{M}) = \mathbf{D}^-_{\mathrm{f}}(\mathbf{M}(A)) = \mathbf{D}^-_{\mathrm{f}}(A)$$

is an equivalence. See Example 11.4.44 for another approach to this problem.

We end this section with a result on K-flat complexes over rings.

Proposition 11.3.23 *Let A be a ring and P a bounded above complex of flat A-modules. Then P is a K-flat complex.*

Exercise 11.3.24 Prove this proposition. (Hint: see the proofs of Proposition 11.3.5 and Theorem 11.3.2.)

11.4 K-Projective Resolutions in $\mathbf{C}(A)$

In this section A is a DG $\mathbb{K}$-ring (without any vanishing assumption).

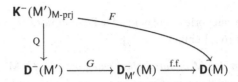

Let $A^\natural$ be the graded ring that we have after forgetting the differential of A. In Section 3.8 we introduced the functor $\mathrm{Und} : \mathbf{C}(A) \to \mathbf{G}(A^\natural)$ that forgets the differentials of DG modules. We shall now use the abbreviated form $M^\natural := \mathrm{Und}(M)$.

Recall that the translation $\mathrm{T}^{-i}(A)$ is a DG A-module in which the element $\mathrm{t}^{-i}(1_A)$ is in degree i. This element is a cocycle, and when we forget the differentials, the graded module $\mathrm{T}^{-i}(A)^\natural$ is free over the graded ring $A^\natural$, with basis $\mathrm{t}^{-i}(1_A)$. Therefore, for every DG A-module M there is a canonical isomorphism

$$\mathrm{Hom}_A(\mathrm{T}^{-i}(A), M) \cong \mathrm{T}^i(M) \tag{11.4.1}$$

in $\mathbf{C}_{\mathrm{str}}(\mathbb{K})$, and canonical isomorphisms

$$\mathrm{Hom}_{\mathbf{C}_{\mathrm{str}}(A)}(\mathrm{T}^{-i}(A), M) \cong \mathrm{Z}^0(\mathrm{Hom}_A(\mathrm{T}^{-i}(A), M)) \cong \mathrm{Z}^i(M) \tag{11.4.2}$$

in $\mathbf{M}(\mathbb{K})$. (Actually, (11.4.1) is an isomorphism in $\mathbf{C}_{\mathrm{str}}(A)$, but this uses the DG A-bimodule structure of $\mathrm{T}^{-i}(A)$.)

We begin with a definition that is very similar to Definition 11.3.1.

Definition 11.4.3 Let P be an object of $\mathbf{C}(\mathrm{A})$.

(1) We say that P is a *free DG A-module* if there is an isomorphism $P \cong \bigoplus_{s\in S} \mathrm{T}^{-i_s}(A)$ in $\mathbf{C}_{\mathrm{str}}(A)$, for some indexing set S and some collection of integers $\{i_s\}_{s\in S}$.

(2) A *semi-free filtration* on P is a filtration $F = \{F_j(P)\}_{j\geq -1}$ of P in $\mathbf{C}_{\mathrm{str}}(A)$, such that:
- $F_{-1}(P) = 0$.
- Each $\mathrm{Gr}_j^F(P)$ is a free DG A-module.
- $P = \lim_{j\to} F_j(P)$.

(3) The DG module P is called *semi-free* if it admits some semi-free filtration.

Note that the direct limit in item (2) is just the union of DG submodules.

There is an alternative description of semi-free filtrations, which is useful for understanding their structure. We will give it in Definition 11.4.7 below, after some combinatorial preliminaries.

The next definition has already appeared within Example 3.1.23.

Definition 11.4.4 A *graded set* is a set S that is partitioned into subsets: $S = \coprod_{i\in\mathbb{Z}} S^i$. The elements of S^i are said to have degree i, so that $S^i = \{s \in S \mid \deg(s) = i\}$.

A free DG A-module P as in Definition 11.4.3(1) can be described as follows. The indexing set S can be made into a graded set by defining $S^k := \{s \in S \mid i_s = k\}$. In other words, $\deg(s) = i_s$. We denote by $A\cdot s$ the free

DG A-module with basis s, so that $\mathrm{d}(s) = 0$, and there is an isomorphism $A \cdot s \xrightarrow{\simeq} \mathrm{T}^{-i_s}(A)$ in $\mathbf{C}_{\mathrm{str}}(A)$ sending $s \mapsto \mathrm{t}^{-i_s}(1_A)$. Writing

$$A \cdot S := \bigoplus_{s \in S} A \cdot s \tag{11.4.5}$$

we obtain $A \cdot S \cong P$ as DG A-modules. Note that $A^\natural \cdot S \cong P^\natural$ as graded $A^\natural$-modules.

Definition 11.4.6 Let S be a graded set. A *filtration* on S is a direct system $F = \{F_j(S)\}_{j \geq -1}$ of subsets of S, such that $F_{-1}(S) = \varnothing$ and $\bigcup_j F_j(S) = S$. The pair (S, F) is called a *filtered graded set.*

Note that each $F_j(S)$ is itself a graded set, with degree i component $F_j(S)^i := F_j(S) \cap S^i$.

Definition 11.4.7 Let P be a DG A-module. A *semi-basis* of P is a filtered graded set (S, F), together with an isomorphism $P^\natural \cong A^\natural \cdot S$ of graded $A^\natural$-modules, such that under this isomorphism we have $\mathrm{d}_P(F_j(S)) \subseteq A^\natural \cdot F_{j-1}(S)$ for every $j \geq 0$.

Proposition 11.4.8 *Let P be a DG A-module.*

(1) *If P has a semi-basis (S, F), then P has an induced semi-free filtration $\{F_j(P)\}_{j \geq -1}$ such that $F_j(P)^\natural = A^\natural \cdot F_j(S)$ as graded $A^\natural$-modules for every $j \geq -1$.*
(2) *Every semi-free filtration of P is induced by some semi-basis.*

Exercise 11.4.9 Prove Proposition 11.4.8.

Example 11.4.10 If A is a ring, then a free DG A-module P is a complex of free A-modules with zero differential. A semi-free DG A-module P is also a complex of free A-modules, but there is a differential on it, and there is a subtle condition on P imposed by the existence of a semi-free filtration. If the complex P happens to be bounded above, then it is automatically semi-free, with a filtration like the one in the proof of Proposition 11.3.5.

Exercise 11.4.11 Find a ring A, and a complex P of free A-modules, that is not semi-free. (Hint: Take the ring $A = \mathbb{K}[\epsilon]$ of dual numbers. Find a complex of free A-modules P that is acyclic but not null-homotopic. Now use Theorem 11.4.14 and Corollary 10.2.14 to deduce that P is not semi-free.)

Proposition 11.4.12 *Let A and B be DG rings, and let $P \in \mathbf{C}(A)$ and $Q \in \mathbf{C}(B)$ be semi-free DG modules. Then $P \otimes Q \in \mathbf{C}(A \otimes B)$ is a semi-free DG module.*

Exercise 11.4.13 Prove this proposition. (Hint: let $\{F_j(P)\}_{j\geq -1}$ and $\{G_k(Q)\}_{k\geq -1}$ be semi-free filtrations. Define

$$E_j(P \otimes Q) := \sum_{k=0}^{j} (F_k(P) \otimes G_{j-k}(Q)) \subseteq P \otimes Q.$$

Show that this is a semi-free filtration.)

Theorem 11.4.14 *Let P be an object of $\mathbf{C}(A)$. If P is semi-free, then it is K-projective.*

Proof This is similar to the proof of Theorem 11.3.2.

Step 1. We start by proving that if $P = \mathrm{T}^{-i}(A)$, a translation of A, then P is K-projective. This is easy: given an acyclic $N \in \mathbf{C}(A)$, we have

$$\mathrm{Hom}_A(P, N) = \mathrm{Hom}_A(\mathrm{T}^{-i}(A), N) \cong \mathrm{T}^i(\mathrm{Hom}_A(A, N)) \cong \mathrm{T}^i(N)$$

in $\mathbf{C}_{\mathrm{str}}(\mathbb{K})$, and this is acyclic.

Step 2. Now P is free, say $P \cong \bigoplus_{s\in S} \mathrm{T}^{-i_s}(A)$. Then

$$\mathrm{Hom}_A(P, N) \cong \prod_{s\in S} \mathrm{Hom}_A(\mathrm{T}^{-i_s}(A), N).$$

By step 1 and the fact that a product of acyclic complexes in $\mathbf{C}_{\mathrm{str}}(\mathbb{K})$ is acyclic, we conclude that $\mathrm{Hom}_{\mathsf{M}}(P, N)$ is acyclic.

Step 3. Fix a semi-free filtration $F = \{F_j(P)\}_{j\geq -1}$ of P. Here we prove that for every $j \geq -1$ the DG module $F_j(P)$ is K-projective. This is done by induction on $j \geq -1$. For $j = -1$ it is trivial. For $j \geq 0$ there is an exact sequence

$$0 \to F_{j-1}(P) \to F_j(P) \to \mathrm{Gr}_j^F(P) \to 0 \tag{11.4.15}$$

in the abelian category $\mathbf{C}_{\mathrm{str}}(A)$. Because $\mathrm{Gr}_j^F(P)$ is a free DG module, it is a projective object in the abelian category $\mathbf{G}_{\mathrm{str}}(A^\natural)$ of graded modules over the graded ring $A^\natural$. Therefore the sequence (11.4.15) is split exact in $\mathbf{G}_{\mathrm{str}}(A^\natural)$.

Let $N \in \mathbf{C}(A)$ be an acyclic DG module. Applying the functor $\mathrm{Hom}_A(-, N)$ to the sequence (11.4.15) we obtain a sequence

$$0 \to \mathrm{Hom}_A(\mathrm{Gr}_j^F(P), N) \to \mathrm{Hom}_A(F_j(P), N) \to \mathrm{Hom}_A(F_{j-1}(P), N) \to 0 \tag{11.4.16}$$

in $\mathbf{C}_{\mathrm{str}}(\mathbb{K})$. If we forget differentials this is a sequence in $\mathbf{G}_{\mathrm{str}}(\mathbb{K})$. Because (11.4.15) is split exact in $\mathbf{G}_{\mathrm{str}}(A^\natural)$, it follows that (11.4.16) is split exact in $\mathbf{G}_{\mathrm{str}}(\mathbb{K})$. Therefore (11.4.16) is exact in $\mathbf{C}_{\mathrm{str}}(\mathbb{K})$.

By the induction hypothesis the DG $\mathbb{K}$-module $\mathrm{Hom}_A(F_{j-1}(P), N)$ is acyclic. By step 2 the DG module $\mathrm{Hom}_A(\mathrm{Gr}_j^F(P), N)$ is acyclic. The long exact cohomology sequence associated to (11.4.16) shows that the DG module $\mathrm{Hom}_A(F_j(P), N)$ is acyclic too.

Step 4. We keep the semi-free filtration $F = \{F_j(P)\}_{j \geq -1}$ from step 3. Take any acyclic $N \in \mathbf{C}(\mathrm{M})$. By Proposition 1.8.6 we know that

$$\mathrm{Hom}_A(P, N) \cong \lim_{\leftarrow j} \mathrm{Hom}_A(F_j(P), N)$$

in $\mathbf{C}_{\mathrm{str}}(\mathbb{K})$. According to step 3, the complexes $\mathrm{Hom}_A(F_j(P), N)$ are all acyclic. The exactness of the sequences (11.4.16) implies that the inverse system $\{\mathrm{Hom}_A(F_j(P), N)\}_{j \geq -1}$ in $\mathbf{C}_{\mathrm{str}}(\mathbb{K})$ has surjective transitions. Now the Mittag–Leffler argument (Corollary 11.1.8) says that the inverse limit complex $\mathrm{Hom}_A(P, N)$ is acyclic. □

Here is a result similar to Theorem 11.3.6.

Theorem 11.4.17 *Let A be a DG ring. Every $M \in \mathbf{C}(A)$ admits a quasi-isomorphism $\rho : P \to M$ in $\mathbf{C}_{\mathrm{str}}(A)$ from a semi-free DG A-module P.*

The proof of this theorem was communicated to us by B. Keller. First we need a construction and a lemma.

Let us define the DG A-module

$$C := \mathrm{Cone}(\mathrm{id} : \mathrm{T}^{-1}(A) \to \mathrm{T}^{-1}(A)), \tag{11.4.18}$$

the standard cone of the identity automorphism of $\mathrm{T}^{-1}(A)$ in $\mathbf{C}_{\mathrm{str}}(A)$. Since

$$C = \mathrm{T}^{-1}(A) \oplus \mathrm{T}(\mathrm{T}^{-1}(A)) = \mathrm{T}^{-1}(A) \oplus A$$

as graded objects, we have the elements $e_0 := 1_A \in A \subseteq C^0$ and $e_1 := \mathrm{t}^{-1}(1_A) \in \mathrm{T}^{-1}(A)^1 \subseteq C^1$. They satisfy $\mathrm{d}_C(e_0) = e_1$. Note that the DG module C is semi-free, with semi-free filtration

$$F_j(C) := \begin{cases} 0 & \text{if } j = -1 \\ \mathrm{T}^{-1}(A) & \text{if } j = 0 \\ C & \text{if } j \geq 1. \end{cases} \tag{11.4.19}$$

As a graded $A^\natural$-module, $C^\natural$ is free, with basis (e_0, e_1). Of course, the DG module C is contractible.

Lemma 11.4.20 *Let $M \in \mathbf{C}(A)$.*

(1) *There is a homomorphism $\psi : Q \to M$ in $\mathbf{C}_{\mathrm{str}}(A)$, such that $\mathrm{Z}(\psi) : \mathrm{Z}(Q) \to \mathrm{Z}(M)$ is surjective, and $Q = \bigoplus_{s \in S} \mathrm{T}^{-i_s}(A)$ for a collection of integers $\{i_s\}_{s \in S}$.*

(2) *There is a surjective homomorphism $\psi' : Q' \to M$ in $\mathbf{C}_{\mathrm{str}}(A)$, such that $Q' := \bigoplus_{s \in S'} \mathrm{T}^{-i_s}(C)$ for some collection of integers $\{i_s\}_{s \in S'}$, where C is the DG A-module from formula* (11.4.18).

To clarify the notation in the lemma, the collections of integers $\{i_s\}_{s \in S}$ and $\{i_s\}_{s \in S'}$ are distinct; namely the function $i : s \mapsto i_s$ is $i : S \sqcup S' \to \mathbb{Z}$.

Proof

(1) For every cocycle $m \in \mathrm{Z}^i(M)$ there a homomorphism $\psi_m : \mathrm{T}^{-i}(A) \to M$ that sends the element $\mathrm{t}^{-i}(1_A) \in \mathrm{Z}^i(\mathrm{T}^{-i}(A))$ to m. Thus, if $\{m_s\}_{s \in S}$ is a collection of homogeneous cocycles that generates $\mathrm{Z}(M)$ as a $\mathbb{K}$-module, we get a homomorphism ψ as claimed.

(2) For every element $m \in M^i$ there is a homomorphism $\psi'_m : \mathrm{T}^{-i}(C) \to M$ in $\mathbf{C}_{\mathrm{str}}(A)$ that sends $\mathrm{t}^{-i}(e_0) \mapsto m$ and $\mathrm{t}^{-i}(e_1) \mapsto (-1)^i \cdot \mathrm{d}_M(m)$. Hence, by taking a collection $\{m_s\}_{s \in S'}$ of homogeneous elements that generates M as a $\mathbb{K}$-module, we get a homomorphism ψ' as claimed. □

The DG A-module Q in item (1) of the lemma is free. The DG A-module Q' in item (2) is semi-free, with semi-free filtration induced by that of C, namely

$$F_j(Q') := \bigoplus_{s \in S'} \mathrm{T}^{-i_s}(F_j(C)) \tag{11.4.21}$$

for $j \geq -1$, where $F_j(C)$ is from (11.4.19).

Proof of Theorem 11.4.17 The proof is in several steps.

Step 1. We are going to produce an exact sequence

$$\cdots \to Q^{-2} \xrightarrow{\partial^{-2}} Q^{-1} \xrightarrow{\partial^{-1}} Q^0 \xrightarrow{\eta} M \to 0 \tag{11.4.22}$$

in $\mathbf{C}_{\mathrm{str}}(A)$ with these properties:

(a) The sequence

$$\cdots \to \mathrm{Z}(Q^{-2}) \xrightarrow{\mathrm{Z}(\partial^{-2})} \mathrm{Z}(Q^{-1}) \xrightarrow{\mathrm{Z}(\partial^{-1})} \mathrm{Z}(Q^0) \xrightarrow{\mathrm{Z}(\eta)} \mathrm{Z}(M) \to 0$$

in $\mathbf{G}_{\mathrm{str}}(\mathbb{K})$ is exact.

(b) For every $i \leq 0$ the DG A-module Q^i has the following decomposition in $\mathbf{C}_{\mathrm{str}}(A)$: $Q^i = Q_i \oplus Q'_i$, where $Q_i = \bigoplus_{s \in S_i} \mathrm{T}^{-j_s}(A)$ and $Q'_i = \bigoplus_{s \in S'_i} \mathrm{T}^{-j_s}(C)$ for some collection of integers $\{j_s\}_{s \in S_i \sqcup S'_i}$.

This will be done recursively on $i \leq 0$.

By Lemma 11.4.20 we can find DG A-modules Q_0, Q'_0 and homomorphisms $\psi_0 : Q_0 \to M$, $\psi'_0 : Q'_0 \to M$, such that both $\mathrm{Z}(\psi_0)$ and ψ'_0 are surjective. The DG modules Q_0, Q'_0 are of the required form (property (b) with $i = 0$). We let $Q^0 := Q_0 \oplus Q'_0$ and $\eta := \psi_0 \oplus \psi'_0$.

Let $N^0 := \mathrm{Ker}(\eta)$. Then the sequence

$$0 \to N^0 \to Q^0 \xrightarrow{\eta} M \to 0 \tag{11.4.23}$$

in $\mathbf{C}_{\mathrm{str}}(A)$ is exact. The sequence

$$0 \to \mathrm{Z}(N^0) \to \mathrm{Z}(Q^0) \xrightarrow{\mathrm{Z}(\eta)} \mathrm{Z}(M) \to 0$$

in $\mathbf{G}_{\mathrm{str}}(\mathbb{K})$ is also exact. Indeed, the exactness at $\mathrm{Z}(N^0)$ and $\mathrm{Z}(Q^0)$ is because the functor $\mathrm{Z} : \mathbf{C}_{\mathrm{str}}(\mathbb{K}) \to \mathbf{G}_{\mathrm{str}}(\mathbb{K})$ is left exact (see Proposition 3.8.14), and the sequence (11.4.23) is exact. The exactness at $\mathrm{Z}(M)$ is by the surjectivity of $\mathrm{Z}(\psi_0) : \mathrm{Z}(Q_0) \to \mathrm{Z}(M)$.

Next we repeat this procedure with N^0 instead of M, to obtain an exact sequence

$$0 \to N^{-1} \to Q^{-1} \xrightarrow{\partial^{-1}} N^0 \to 0, \tag{11.4.24}$$

with Q^{-1} of the required form (property (b) with $i = -1$), and such that

$$0 \to \mathrm{Z}(N^{-1}) \to \mathrm{Z}(Q^{-1}) \xrightarrow{\mathrm{Z}(\partial^{-1})} \mathrm{Z}(N^0) \to 0$$

is also exact. We then splice (11.4.23) and (11.4.24) to get the sequence

$$0 \to N^{-1} \to Q^{-1} \xrightarrow{\partial^{-1}} Q^0 \xrightarrow{\eta} M.$$

Continuing recursively we obtain an exact sequence (11.4.22) that has properties (a) and (b).

Step 2. We view the exact sequence (11.4.22) as an acyclic complex with entries in $\mathbf{C}_{\mathrm{str}}(A)$ that has Q^0 in degree 0. Define the total DG A-modules

$$P := \mathrm{Tot}^{\oplus}(\cdots \to Q^{-2} \xrightarrow{\partial^{-2}} Q^{-1} \xrightarrow{\partial^{-1}} Q^0 \to 0 \to \cdots)$$

and

$$P^{\mathrm{aug}} := \mathrm{Tot}^{\oplus}(\cdots \to Q^{-2} \xrightarrow{\partial^{-2}} Q^{-1} \xrightarrow{\partial^{-1}} Q^0 \xrightarrow{\eta} M \to 0 \to \cdots).$$

Because the sequence (11.4.22) and the sequence appearing in property (a) are exact, we can use Proposition 11.2.16 (with any $j_1 \in \mathbb{Z}$). The conclusion is that the DG A-module P^{aug} is acyclic. Then, by Corollary 11.2.14, the homomorphism $\rho : P \to M$ in $\mathbf{C}_{\mathrm{str}}(A)$ that is induced by η is a quasi-isomorphism.

Step 3. It remains to produce a semi-free filtration $\{F_j(P)\}_{j \geq -1}$ on the DG A-module P. Our formula is this: $F_{-1}(P) := 0$ of course. For $k \geq 0$ we let

$$F_{2\cdot k}(P) := \Big(\bigoplus\nolimits_{l=0}^{-(k-1)} \mathrm{T}^{-l}(Q^l) \Big) \oplus \mathrm{T}^k(Q_{-k} \oplus F_0(Q'_{-k})),$$

where $F_0(Q'_{-k})$ comes from (11.4.21), and

$$F_{2\cdot k+1}(P) := \bigoplus\nolimits_{l=0}^{-k} \mathrm{T}^{-l}(Q^l).$$

We leave it to the reader to verify that this is a semi-free filtration. □

Exercise 11.4.25 Finish step 3 of the proof.

Corollary 11.4.26 *Let A be a DG ring. The category $\mathbf{C}(A)$ has enough K-projectives.*

Proof Combine Theorems 11.4.14 and 11.4.17. □

Recall that a DG ring A is nonpositive if $A^i = 0$ for all $i > 0$. According to Proposition 7.3.8, when A is nonpositive, the smart truncation functors exist.

Corollary 11.4.27 *If the DG ring A is nonpositive, then every $M \in \mathbf{C}(A)$ admits a quasi-isomorphism $\rho : P \to M$ in $\mathbf{C}_{\mathrm{str}}(A)$ from a semi-free DG A-module P, such that* $\sup(P) = \sup(\mathrm{H}(M))$.

Proof Let $i_1 := \sup(\mathrm{H}(M))$. If $i_1 = \pm\infty$ then there is nothing to prove beyond Theorem 11.4.17; so let us assume that $i_1 \in \mathbb{Z}$. Define $M' := \mathrm{smt}^{\leq i_1}(M)$. Then $M' \to M$ is a quasi-isomorphism, and $\sup(M') = i_1$. By replacing M with M' we can now assume that $\sup(M) = i_1$.

Since $\sup(M) \leq i_1$, in step 1 of the proof of the theorem we can take the DG module Q_0 to be concentrated in degrees $\leq i_1$. Another consequence of the fact that $\sup(M) \leq i_1$ is that $M^{i_1} = \mathrm{Z}^{i_1}(M)$. This means that the homomorphism $\psi_0 : Q_0^{i_1} \to M^{i_1}$ is already surjective; so we don't need to cover M^{i_1} by Q_0', and thus we can take Q_0' to be concentrated in degrees $\leq i_1$. In this way we can arrange to have $\sup(Q^0) \leq i_1$. But then we also have $\sup(N^0) \leq i_1$. Thus recursively we can arrange to have $\sup(Q^l) \leq i_1$ for all l. Therefore in step 2 of the proof we get $\sup(P) \leq i_1$.

Finally, because $\sup(\mathrm{H}(M)) = i_1$ we must have $\sup(P) = i_1$. □

Definition 11.4.28 Let A be a DG ring. A DG A-module P is called a *pseudo-finite free DG A-module* if there are numbers $i_1 \in \mathbb{Z}$ and $r_i \in \mathbb{N}$ such that $P \cong \bigoplus_{i \leq i_1} \mathrm{T}^{-i}(A)^{\oplus r_i}$ in $\mathbf{C}_{\mathrm{str}}(A)$.

Definition 11.4.29 Let A be a DG ring and let P be a DG A-module.

(1) A *pseudo-finite semi-free filtration* on P is a semi-free filtration $F = \{F_j(P)\}_{j \geq -1}$ satisfying this finiteness condition: there are numbers $i_1 \in \mathbb{Z}$ and $r_j \in \mathbb{N}$ such that $\mathrm{Gr}_j^F(P) \cong \mathrm{T}^{-i_1+j}(A)^{\oplus r_j}$ in $\mathbf{C}_{\mathrm{str}}(A)$ for all $j \in \mathbb{N}$.

(2) We call P a *pseudo-finite semi-free DG A-module* if it admits some pseudo-finite semi-free filtration.

Note that in item (1) of the definition above, for every $j \geq -1$ we have

$$F_j(P)^\natural \cong \bigoplus_{0 \leq k \leq j} \mathrm{Gr}_j^F(P)^\natural \cong \bigoplus_{0 \leq k \leq j} \mathrm{T}^{-i_1+k}(A^\natural)^{\oplus r_k} \tag{11.4.30}$$

as graded $A^\natural$-modules.

Proposition 11.4.31 *Let P be a DG A-module. The two conditions below are equivalent.*

(i) *P is a pseudo-finite semi-free DG A-module.*

(ii) *P admits a semi-basis (S, F) such that S is bounded above, say by $i_1 \in \mathbb{Z}$; each of the sets S^i is finite; and $F_j(S) = \bigcup_{i \in [i_1 - j, i_1]}$ for every $j \geq -1$.*

Exercise 11.4.32 Prove Proposition 11.4.31. (Hint: the numbers i_1 in Definition 11.4.29(1) and in condition (ii) are the same.)

A graded $A^\natural$-module $P^\natural$ is called *pseudo-finite free* if there is an isomorphism of graded $A^\natural$-modules $P^\natural \cong \bigoplus_{j \geq 0} \mathrm{T}^{-i_1+j}(A^\natural)^{\oplus r_j}$ for some numbers $i_1 \in \mathbb{Z}$ and $r_j \in \mathbb{N}$.

Proposition 11.4.33 *Let A be a DG ring and let P be a DG A-module. The following two conditions are equivalent:*

(i) *The DG A-module P is pseudo-finite semi-free.*

(ii) *The DG A-module P admits a semi-free filtration $F = \{F_j(P)\}_{j \geq -1}$ such that each $\mathrm{Gr}_j^F(P)$ is a pseudo-finite free DG A-module, and this asymptotic formula holds: $\lim_{j \to \infty} \sup(\mathrm{Gr}_j^F(P)) = -\infty$.*

If A is nonpositive, the conditions above are equivalent to:

(iii) *The graded $A^\natural$-module $P^\natural$ is pseudo-finite free.*

If A is a ring (i.e. $A^i = 0$ for all $i \neq 0$), the conditions above are equivalent to:

(iv) *P is a bounded above complex of finitely generated free A-modules.*

Exercise 11.4.34 Prove the proposition above. (Hint: rephrase these conditions in terms of semi-bases.)

Remark 11.4.35 Condition (iv) of Proposition 11.4.33 is a hint for the reason we chose the name "pseudo-finite" in Definition 11.4.29, and also the name "pseudo-noetherian" in Definitions 11.4.36 and 11.4.37 below (in conjunction with Theorem 11.4.40). Indeed, in [21] and [144], a complex of modules over a ring A is called *pseudo-coherent* if it is isomorphic, in $\mathbf{D}(A)$, to a bounded above complex of finite free A-modules.

For a generalization of the notion of pseudo-coherent complex to DG modules, see Definition 12.9.1.

Definition 11.4.36 A graded ring A is called *left pseudo-noetherian* if it is nonpositive, the ring A^0 is left noetherian, and each A^i is a finitely generated (left) A^0-module.

Definition 11.4.37 Let A be a DG ring.

(1) We call A a *cohomologically left pseudo-noetherian DG ring* if its cohomology $\mathrm{H}(A)$ is a left pseudo-noetherian graded ring.

(2) Suppose A is cohomologically left pseudo-noetherian. We denote by $\mathsf{D}_{\mathrm{f}}(A)$ the full subcategory of $\mathsf{D}(A)$ on the DG modules M such that $\mathrm{H}^i(M)$ is a finitely generated $\mathrm{H}^0(A)$-module for every i.

Of course, $\mathsf{D}_{\mathrm{f}}(A)$ is a full triangulated subcategory of $\mathsf{D}(A)$. As usual, we combine indicators: $\mathsf{D}^{\star}_{\mathrm{f}}(A) := \mathsf{D}_{\mathrm{f}}(A) \cap \mathsf{D}^{\star}(A)$.

Example 11.4.38 Suppose A^0 is a nonzero noetherian commutative ring. Let $X = \coprod_{i \leq 0} X^i$ be a nonpositive graded set, such that each graded piece X^i is finite. Consider the strongly commutative graded polynomial ring $A^0[X] := A^0 \otimes \mathbb{K}[X]$ from Example 3.1.23. The graded ring $A^0[X]$ is (left and right) pseudo-noetherian, but if X is infinite then $A^0[X]$ is not noetherian (on either side).

Likewise, let $Y = \coprod_{i<0} Y^i$ be a negative graded set, such that each graded piece Y^i is finite. Take the noncommutative graded polynomial ring $A^0\langle Y\rangle := A^0 \otimes \mathbb{K}\langle Y\rangle$ from Example 3.1.23 and Definition 12.8.1. The graded ring $A^0\langle Y\rangle$ is (left and right) pseudo-noetherian, but if Y has at least two elements, then $A^0\langle Y\rangle$ is not noetherian (on either side).

Remark 11.4.39 Let A be a DG ring whose cohomology $\mathrm{H}(A)$ is a nonpositive graded ring. Define $A' := \mathrm{smt}^{\leq 0}(A)$, the smart truncation of A below 0. It is easy to see that A' is a DG subring of A, and the inclusion $A' \to A$ is a DG ring quasi-isomorphism. By Theorem 12.7.2 there is an equivalence of triangulated categories $\mathsf{D}(A) \to \mathsf{D}(A')$. This says that for many purposes we can assume that A itself is a nonpositive DG ring.

Theorem 11.4.40 *Let A be a nonpositive cohomologically left pseudo-noetherian DG ring, and let $M \in \mathsf{D}^{-}_{\mathrm{f}}(A)$. Then there is a quasi-isomorphism $P \to M$ in $\mathsf{C}_{\mathrm{str}}(A)$, where P is a pseudo-finite semi-free DG A-module, and* $\sup(P) = \sup(\mathrm{H}(M))$.

Proof The proof is divided into five parts.

Part 1. We may assume that $\mathrm{H}(M) \neq 0$, so that $i_1 := \sup(\mathrm{H}(M)) \in \mathbb{Z}$. We are going to construct a direct system

$$0 = F_{-1}(P) \subseteq F_0(P) \subseteq F_1(P) \subseteq \cdots$$

of pseudo-finite semi-free DG A-modules, together with a direct system of DG A-module homomorphisms $F_j(\rho) : F_j(P) \to M$ for $j \geq 0$, having these properties:

(a) Let $j \geq 0$. The homomorphism $\mathrm{H}^i(F_j(\rho)) : \mathrm{H}^i(F_j(P)) \to \mathrm{H}^i(M)$ is surjective for all i, and bijective for all $i > i_1 - j$.

(b) $F_0(P)$ is a pseudo-finite free DG A-module with $\sup(F_0(P)) = i_1$.

(c) Let $j \geq 1$. There is an isomorphism $F_j(P)/F_{j-1}(P) \cong \mathrm{T}^{-i_1+j}(A)^{\oplus r_j}$ in $\mathbf{C}_{\mathrm{str}}(A)$, for some $r_j \in \mathbb{N}$.

This will be done recursively on $j \geq 0$. .

Part 2. Here we deal with the cases $j = -1, 0$. For $j = -1$ we take $F_{-1}(P) := 0$ of course.

Now we deal with $F_0(P)$. For every $i \leq i_1$ we choose finitely many generators of the $\mathrm{H}^0(A)$-module $\mathrm{H}^i(M)$, and then we lift them to elements of $\mathrm{Z}^i(M)$. These choices give rise to a pseudo-finite free DG A-module $F_0(P)$ and a homomorphism $F_0(\rho) : F_0(P) \to M$ in $\mathbf{C}_{\mathrm{str}}(A)$, such that $\sup(F_0(P)) = i_1$, and the homomorphism $\mathrm{H}(F_0(\rho)) : \mathrm{H}(F_0(P)) \to \mathrm{H}(M)$ in $\mathbf{G}_{\mathrm{str}}(\mathbb{K})$ is surjective. Condition (a) holds for $j = 0$, and so does condition (b). Condition (c) is not applicable here.

Part 3. Suppose that $j \geq 0$, and we have a direct system

$$0 = F_{-1}(P) \subseteq F_0(P) \subseteq F_1(P) \subseteq \cdots \subseteq F_j(P)$$

of semi-free DG A-modules, together with a direct system of DG A-module homomorphisms $F_{j'}(\rho) : F_{j'}(P) \to M$ for $j' \in [0, j]$, satisfying conditions (a)-(c) with index j' instead of j. In this part we will construct the DG A-module $F_{j+1}(P)$.

Let

$$\bar{N} := \mathrm{Ker}(\mathrm{H}^{i_1-j}(F_j(\rho)) : \mathrm{H}^{i_1-j}(F_j(P)) \to \mathrm{H}^{i_1-j}(M)). \tag{11.4.41}$$

Because $F_j(P)$ is a pseudo-finite semi-free DG A-module and A is a cohomologically left pseudo-noetherian DG ring, it follows that $\bar{N}$ is a finitely generated $\mathrm{H}^0(A)$-module. We choose elements $\bar{n}_1, \ldots, \bar{n}_{r_{j+1}} \in \bar{N}$ that generate $\bar{N}$ as an $\mathrm{H}^0(A)$-module. Then we lift them to elements $n_1, \ldots, n_{r_{j+1}} \in \mathrm{Z}^{i_1-j}(F_j(P))$.

Consider the free DG A-module $Q' := \mathrm{T}^{-i_1+j}(A)^{\oplus r_{j+1}}$ with basis $(e'_1, \ldots, e'_{r_{j+1}})$ concentrated in degree $i_1 - j$. There is a homomorphism $\phi : Q' \to F_j(P)$ in $\mathbf{C}_{\mathrm{str}}(A)$ with formula $\phi(e'_k) := n_k$. Define

$$F_{j+1}(P) := \mathrm{Cone}(\phi : Q' \to F_j(P)), \tag{11.4.42}$$

the standard cone of ϕ. The DG A-module $F_{j+1}(P)$ is pseudo-finite semi-free. Condition (b) still holds, because $F_0(P)$ did not change. And $F_{j+1}(P)/F_j(P) \cong Q$, where

$$Q := \mathrm{T}(Q') \cong \mathrm{T}^{-i_1+j+1}(A)^{\oplus r_{j+1}}, \tag{11.4.43}$$

so condition (c) holds also for $j + 1$.

Part 4. We continue part 3. Here we construct the homomorphism $F_{j+1}(\rho) : F_{j+1}(P) \to M$ in $\mathbf{C}_{\mathrm{str}}(A)$. For every $k \in [1, r_{j+1}]$ let $e_k := \mathrm{t}_{Q'}(e'_k) \in Q^{i_1-j-1}$. The sequence $(e_1, \ldots, e_{r_{j+1}})$ is a basis of the free DG A-module Q. We have

$\mathrm{d}_Q(e_k) = 0$ and $\mathrm{t}_Q^{-1}(e_k) = e'_k \in Q'$, so by the definition of the differential of the cone we have

$$\mathrm{d}_{F_{j+1}(P)}(e_k) = \mathrm{d}_Q(e_k) + \phi(\mathrm{t}_Q^{-1}(e_k)) = \phi(e'_k) = n_k.$$

Now the cohomology class $\bar{n}_k$ of n_k belongs to $\bar{N}$, so $\mathrm{H}^{i_1-j}(F_j(\rho))(\bar{n}_k) = 0$. This means that the element $F_j(\rho)(n_k)$ is a coboundary in M, and we can lift it to an element $m_k \in M^{i_1-j-1}$.

By definition of the cone we have $F_{j+1}(P)^{\natural} = F_j(P)^{\natural} \oplus Q^{\natural}$ as graded $A^{\natural}$-modules. We define the homomorphism $F_{j+1}(\rho) : F_{j+1}(P)^{\natural} \to M^{\natural}$ in $\mathbf{G}_{\mathrm{str}}(A^{\natural})$ to be the extension of $F_j(\rho)$ such that $F_{j+1}(\rho)(e_k) := m_k$. Then $F_{j+1}(\rho) : F_{j+1}(P) \to M$ is actually a homomorphism in $\mathbf{C}_{\mathrm{str}}(A)$.

There is equality $F_{j+1}(P)^i = F_j(P)^i$ in all degrees $i \geq i_1 - j$. Therefore $\mathrm{H}^i(F_{j+1}(\rho)) = \mathrm{H}^i(F_j(\rho))$ for all $i > i_1 - j$, and these homomorphisms are bijective by assumption. The homomorphism $\mathrm{Z}^{i_1-j}(F_j(P)) \to \mathrm{Z}^{i_1-j}(F_{j+1}(P))$ is surjective, and hence $\mathrm{H}^{i_1-j}(F_j(P)) \to \mathrm{H}^{i_1-j}(F_{j+1}(P))$ is also surjective. Because the submodule $\bar{N} \subseteq \mathrm{H}^{i_1-j}(F_j(P))$ goes to zero in $\mathrm{H}^{i_1-j}(F_{j+1}(P))$, we have the following commutative diagram:

$$\begin{array}{ccccccccc}
0 & \longrightarrow & \bar{N} & \longrightarrow & \mathrm{H}^{i_1-j}(F_j(P)) & \xrightarrow{\mathrm{H}^{i_1-j}(F_j(\rho))} & \mathrm{H}^{i_1-j}(M) & \longrightarrow & 0 \\
 & & \downarrow & & \downarrow{\scriptstyle \mathrm{surj}} & & \downarrow{\scriptstyle \mathrm{id}} & & \\
0 & \longrightarrow & 0 & \longrightarrow & \mathrm{H}^{i_1-j}(F_{j+1}(P)) & \xrightarrow{\mathrm{H}^{i_1-j}(F_{j+1}(\rho))} & \mathrm{H}^{i_1-j}(M) & \longrightarrow & 0
\end{array}$$

The top row is exact, by (11.4.41). Therefore the bottom row is exact, i.e. $\mathrm{H}^{i_1-j}(F_{j+1}(\rho))$ is bijective. We conclude that condition (a) holds with index $j + 1$.

Part 5. Define $P := \lim_{j\to} F_j(P)$ and $\rho := \lim_{j\to} F_j(\rho)$ in $\mathbf{C}_{\mathrm{str}}(A)$. Condition (c) implies that for every i the direct system $\{F_j(P^i)\}_{j\geq -1}$ is eventually stationary. Hence, by condition (a), the homomorphism $\mathrm{H}^i(\rho)$ is bijective. So ρ is a quasi-isomorphism.

By condition (b) and (c) the filtration $\{F_j(P)\}_{j\geq -1}$ on P satisfies condition (ii) of Proposition 11.4.33. Therefore P is a pseudo-finite semi-free DG A-module. □

Example 11.4.44 A special yet very important case of Theorem 11.4.40 is this: A is a left noetherian ring, and M is a complex of A-modules with bounded above cohomology, such that each $\mathrm{H}^i(M)$ is a finitely generated A-module. Then M has a resolution $P \to M$, where P is a complex of finitely generated free A-modules, and $\sup(P) = \sup(\mathrm{H}(M))$. Compare this to Example 11.3.22.

Remark 11.4.45 In Example 11.3.16 we discussed the existence of K-flat resolutions of unbounded complexes of sheaves of modules. Here is a more general assertion.

Consider a *DG ringed space* $(X, \mathcal{A})$, namely $\mathcal{A}$ is a sheaf of DG rings on a topological space X. There are no commutativity or boundedness conditions on $\mathcal{A}$. By combining the concept of a semi-free DG module from Definition 11.4.3 with the concept of a pseudo-free module from Example 11.3.16, we get something new: a *semi-pseudo-free DG $\mathcal{A}$-module*. This is a K-flat DG $\mathcal{A}$-module. The arguments in [179] imply that every DG $\mathcal{A}$-module $\mathcal{M}$ admits a quasi-isomorphism $\mathcal{P} \to \mathcal{M}$ from a semi-pseudo-free DG $\mathcal{A}$-module $\mathcal{P}$. Therefore $\mathbf{C}(\mathcal{A})$ has enough K-flat objects. A detailed proof of this assertion will be published in the future.

11.5 K-Injective Resolutions in $\mathbf{C}^+(\mathsf{M})$

In this section M is an abelian category, and $\mathbf{C}(\mathsf{M})$ is the category of complexes in M. Cofiltrations and their limits were introduced in Definition 11.2.23.

Definition 11.5.1 Let I be a complex in $\mathbf{C}(\mathsf{M})$.

(1) A *semi-injective cofiltration* on I is a cofiltration $G = \{G_q(I)\}_{q \geq -1}$ in $\mathbf{C}_{\mathrm{str}}(\mathsf{M})$ such that:
   - $G_{-1}(I) = 0$.
   - Each $\mathrm{Gr}^G_q(I)$ is a complex of injective objects of M with zero differential.
   - $I = \lim_{\leftarrow q} G_q(I)$.

(2) The complex I is called a *semi-injective complex* if it admits some semi-injective cofiltration.

Theorem 11.5.2 *Let* M *be an abelian category, and let* I *be a semi-injective complex in* $\mathbf{C}(\mathsf{M})$. *Then* I *is K-injective.*

Proof The proof is very similar to that of Theorem 11.3.2.

Step 1. We start by proving that if $I = \mathrm{T}^p(J)$, the translation of an injective object $J \in \mathsf{M}$, then I is K-injective. This is easy: given an acyclic complex $N \in \mathbf{C}(\mathsf{M})$, we have

$$\mathrm{Hom}_{\mathsf{M}}(N, I) = \mathrm{Hom}_{\mathsf{M}}(N, \mathrm{T}^p(J)) \cong \mathrm{T}^p(\mathrm{Hom}_{\mathsf{M}}(N, J))$$

in $\mathbf{C}_{\mathrm{str}}(\mathbb{K})$. But $\mathrm{Hom}_{\mathsf{M}}(-, J)$ is an exact functor $\mathsf{M} \to \mathbf{M}(\mathbb{K})$, so $\mathrm{Hom}_{\mathsf{M}}(N, J)$ is an acyclic complex.

Step 2. Now I is a complex of injective objects of M with zero differential. This means that $I \cong \prod_{p\in\mathbb{Z}} \mathrm{T}^p(J_p)$ in $\mathbf{C}_{\mathrm{str}}(\mathsf{M})$, where each J_p is an injective object in M. But then

$$\mathrm{Hom}_{\mathsf{M}}(N, I) \cong \prod_{p\in\mathbb{Z}} \mathrm{Hom}_{\mathsf{M}}(N, \mathrm{T}^p(J_p)).$$

This is an easy case of Proposition 1.8.6(2). By step 1 and the fact that a product of acyclic complexes in $\mathbf{C}_{\mathrm{str}}(\mathbb{K})$ is acyclic (itself an easy case of the Mittag–Leffler argument), we conclude that $\mathrm{Hom}_{\mathsf{M}}(N, I)$ is acyclic.

Step 3. Fix a semi-injective cofiltration $G = \{G_q(I)\}_{q\geq -1}$ of I. Here we prove that for every q the complex $G_q(I)$ is K-injective. This is done by induction on q. For $q = -1$ it is trivial. For $q \geq 0$ there is an exact sequence of complexes

$$0 \to \mathrm{Gr}_q^G(I) \to G_q(I) \to G_{q-1}(I) \to 0 \tag{11.5.3}$$

in $\mathbf{C}_{\mathrm{str}}(\mathsf{M})$. In each degree $p \in \mathbb{Z}$ the exact sequence

$$0 \to \mathrm{Gr}_q^G(I)^p \to G_q(I)^p \to G_{q-1}(I)^p \to 0$$

in M splits, because $\mathrm{Gr}_q^G(I)^p$ is an injective object. Thus the exact sequence (11.5.3) is split in the category $\mathbf{G}_{\mathrm{str}}(\mathsf{M})$ of graded objects in M.

Let $N \in \mathbf{C}(\mathsf{M})$ be an acyclic complex. Applying the functor $\mathrm{Hom}_{\mathsf{M}}(N, -)$ to the sequence of complexes (11.5.3) we obtain a sequence

$$0 \to \mathrm{Hom}_{\mathsf{M}}(N, \mathrm{Gr}_q^G(I)) \to \mathrm{Hom}_{\mathsf{M}}(N, G_q(I)) \to \mathrm{Hom}_{\mathsf{M}}(N, G_{q-1}(I)) \to 0 \tag{11.5.4}$$

in $\mathbf{C}_{\mathrm{str}}(\mathbb{K})$. Because (11.5.3) is split exact in $\mathbf{G}_{\mathrm{str}}(\mathsf{M})$, the sequence (11.5.4) is split exact in $\mathbf{G}_{\mathrm{str}}(\mathbb{K})$. Therefore (11.5.4) is exact in $\mathbf{C}_{\mathrm{str}}(\mathbb{K})$.

By the induction hypothesis the complex $\mathrm{Hom}_{\mathsf{M}}(N, G_{q-1}(I))$ is acyclic. By step 2 the complex $\mathrm{Hom}_{\mathsf{M}}(N, \mathrm{Gr}_q^G(I))$ is acyclic. The long exact cohomology sequence associated to (11.5.4) shows that the complex $\mathrm{Hom}_{\mathsf{M}}(N, G_q(I))$ is acyclic too.

Step 4. We keep the semi-injective cofiltration $G = \{G_q(I)\}_{q\geq -1}$ from step 3. Take any acyclic complex $N \in \mathbf{C}(\mathsf{M})$. By Proposition 1.8.6 we know that

$$\mathrm{Hom}_{\mathsf{M}}(N, I) \cong \lim_{\leftarrow q} \mathrm{Hom}_{\mathsf{M}}(N, G_q(I))$$

in $\mathbf{C}_{\mathrm{str}}(\mathbb{K})$. According to step 3, the complexes $\mathrm{Hom}_{\mathsf{M}}(N, G_q(I))$ are all acyclic. The exactness of the sequences (11.5.4) implies that the inverse system $\{\mathrm{Hom}_{\mathsf{M}}(N, G_q(I))\}_{q\geq -1}$ in $\mathbf{C}_{\mathrm{str}}(\mathbb{K})$ has surjective transitions. Now the Mittag–Leffler argument (Corollary 11.1.8) says that the inverse limit complex $\mathrm{Hom}_{\mathsf{M}}(N, I)$ is acyclic. □

Proposition 11.5.5 *Let* M *be an abelian category. If* I *is a bounded below complex of injectives, then* I *is a semi-injective complex.*

Proof We can assume that $I \neq 0$. Let $p_0 := \inf(I) \in \mathbb{Z}$. For $q \geq -1$ let $G_q(I) := \mathrm{stt}^{\leq p_0+q}(I)$, the stupid truncation from Definition 11.2.11. The cofiltration $G = \{G_q(I)\}_{q \geq -1}$ is semi-injective. □

The next theorem is [62, Lemma I.4.6(1)]. See also [75, Proposition 1.7.7(i)].

Theorem 11.5.6 *Let* M *be an abelian category, and let* $\mathsf{J} \subseteq \mathsf{M}$ *be a full subcategory such that every object* $M \in \mathsf{M}$ *admits a monomorphism* $M \rightarrowtail I$ *to some object* $I \in \mathsf{J}$. *Then every complex* $M \in \mathbf{C}^+(\mathsf{M})$ *admits a quasi-isomorphism* $\rho : M \to I$ *in* $\mathbf{C}^+_{\mathrm{str}}(\mathsf{M})$, *such that* $\inf(I) = \inf(M)$, *and each* I^p *is an object of* J.

Proof The proof is the same as that of Theorem 11.3.6, except for a mechanical reversal of arrows. To be more explicit, let us take $\mathsf{N} := \mathsf{M}^{\mathrm{op}}$ and $\mathsf{Q} := \mathsf{J}$. Since monomorphisms in M become epimorphisms in N, the full subcategory $\mathsf{Q} \subseteq \mathsf{N}$ satisfies the assumptions of Theorem 11.3.6. By Theorem 3.9.16 we have a canonical isomorphism of categories $\mathbf{C}^-_{\mathrm{str}}(\mathsf{N}) \xrightarrow{\simeq} \mathbf{C}^+_{\mathrm{str}}(\mathsf{M})^{\mathrm{op}}$. Thus a quasi-isomorphism $Q \to N$ in $\mathbf{C}^-_{\mathrm{str}}(\mathsf{N})$ gives rise to a quasi-isomorphism $M \to I$ in $\mathbf{C}^+_{\mathrm{str}}(\mathsf{M})$. □

Definition 11.5.7 Let M be an abelian category, and let $\mathsf{M}' \subseteq \mathsf{M}$ be a full abelian subcategory. We say that M' *has enough injectives relative to* M if every object $M \in \mathsf{M}'$ admits a monomorphism $M \rightarrowtail I$, where I is an object of M' that is injective in the bigger category M.

Of course, in this situation the category M' itself has enough injectives.

Thick abelian categories were defined in Definition 7.4.1. The next theorem is [62, Lemma I.4.6(3)]. See also [75, Proposition 1.7.11].

Theorem 11.5.8 *Let* M *be an abelian category, and let* $\mathsf{M}' \subseteq \mathsf{M}$ *be a thick abelian subcategory that has enough injectives relative to* M. *Let* $M \in \mathbf{C}(\mathsf{M})$ *be a complex with bounded below cohomology, such that* $\mathrm{H}^i(M) \in \mathsf{M}'$ *for all* i. *Then there is a quasi-isomorphism* $\rho : M \to I$ *in* $\mathbf{C}_{\mathrm{str}}(\mathsf{M})$, *such that* $I \in \mathbf{C}^+(\mathsf{M}')$, *each* I^p *is an injective object in* M, *and* $\inf(I) = \inf(\mathrm{H}(M))$.

Before the proof we need some auxiliary material.

Suppose we are given morphisms $\psi_1 : K \to L_1$ and $\psi_2 : K \to L_2$ in M. The *fibered coproduct* is the object

$$L_1 \oplus_K L_2 := \mathrm{Coker}((\psi_1, -\psi_2) : K \to L_1 \oplus L_2) \tag{11.5.9}$$

in M. It has an obvious universal property. The commutative diagram

$$\begin{array}{ccc} K & \xrightarrow{\psi_1} & L_1 \\ {\scriptstyle \psi_2}\downarrow & & \downarrow{\scriptstyle \epsilon_1} \\ L_2 & \xrightarrow{\epsilon_2} & L_1 \oplus_K L_2 \end{array} \tag{11.5.10}$$

in which ϵ_i are the morphisms induced by the embeddings $L_i \to L_1 \oplus L_2$, is sometimes called a *pushout diagram*.

Lemma 11.5.11 *In the situation above*:

(1) *The sequence*

$$\mathrm{Ker}(\psi_1) \xrightarrow{\psi_2} L_2 \xrightarrow{\epsilon_2} L_1 \oplus_K L_2 \xrightarrow{\pi_1} \mathrm{Coker}(\psi_1) \to 0,$$

in which π_1 is the morphism induced by the projection $L_1 \oplus L_2 \to L_1$, is exact.

(2) *Suppose $L_1 \to L'_1$ be a monomorphism. Then the induced morphism $L_1 \oplus_K L_2 \to L'_1 \oplus_K L_2$ is a monomorphism.*

Exercise 11.5.12 Prove Lemma 11.5.11. (Hint: use the first sheaf trick, Proposition 2.4.9.)

Proof of Theorem 11.5.8 The proof is an adaptation of the proof of [75, Proposition 1.7.11]. In the proof we use the objects of cocycles $\mathrm{Z}^p(L)$, coboundaries $\mathrm{B}^p(L)$ and decocycles $\mathrm{Y}^p(L)$ that are associated to a complex L and an integer p; see Definition 3.2.4.

Step 1. We can assume that $\mathrm{H}(M) \neq 0$. By translating M, we may further assume that $\inf(\mathrm{H}(M)) = 0$. Then, after replacing M with its smart truncation $\mathrm{smt}^{\geq 0}(M)$, we can assume that $\inf(M) = 0$.

We are going to construct an inverse system

$$F_p(I) = (\cdots \to 0 \to I^0 \xrightarrow{\mathrm{d}_I^0} I^1 \xrightarrow{\mathrm{d}_I^1} \cdots \xrightarrow{\mathrm{d}_I^{p-1}} I^p \to 0 \to \cdots) \tag{11.5.13}$$

in $\mathbf{C}_{\mathrm{str}}(\mathsf{M}')$, indexed by $p \geq -1$, such that the objects $I^p \in \mathsf{M}'$ are injective in M. The morphism $F_{p+1}(I) \to F_p(I)$ will send the object $I^{p+1} = F_{p+1}(I)^{p+1}$ to 0, and will be the identity in all other degrees. Simultaneously we will construct an inverse system of morphisms

$$F_p(\phi) : M \to F_p(I) \tag{11.5.14}$$

in $\mathbf{C}_{\mathrm{str}}(\mathsf{M})$. The construction will be inductive. The morphisms $F_p(\phi)$ will satisfy this condition:

(♡) The morphism $\mathrm{H}^q(F_p(\phi)) : \mathrm{H}^q(M) \to \mathrm{H}^q(F_p(I))$ is a monomorphism for all $q \leq p$ and an isomorphism for all $q \leq p - 1$.

Then the complex $I := \lim_{\leftarrow p} F_p(I)$ in $\mathbf{C}(\mathsf{M}')$ is going to be of the required kind, and the morphism $\phi := \lim_{\leftarrow p} F_p(\phi)$ in $\mathbf{C}_{\mathrm{str}}(\mathsf{M})$ is going to be a quasi-isomorphism. Note that these inverse limits are innocent, since in each degree q the inverse systems $\{F_p(I)^q\}_{p\geq -1}$ and $\{F_p(\phi)^q\}_{p\geq -1}$ are eventually stationary.

Step 2. In this step we begin the construction. Here $p \geq -1$, and we have a complex $F_p(I)$ as in (11.5.13) of the kind described above, and a morphism $F_p(\phi) : M \to F_p(I)$ in $\mathbf{C}_{\mathrm{str}}(\mathsf{M})$ satisfying condition (♡). If $p = -1$ then $F_{-1}(I) = 0$ and $F_{-1}(\phi) = 0$ of course.

We claim that the objects $F_p(I)^q$, $\mathrm{H}^q(F_p(I))$, $\mathrm{B}^q(F_p(I))$, $\mathrm{Y}^q(F_p(I))$ and $\mathrm{Z}^q(F_p(I))$ belong to M' for all $q \in [0, p+1]$. For $q = p+1$ all these objects are zero. For $F_p(I)^q = I^q$ it is given. For $\mathrm{H}^p(F_p(I)) = \mathrm{Coker}(\mathrm{d}_I^{p-1})$ it is because M' is a full abelian subcategory of M. For $\mathrm{H}^q(F_p(I))$ when $q < p$ we use the isomorphisms $\mathrm{H}^q(F_p(\phi))$ and the fact that $\mathrm{H}^q(M) \in \mathsf{M}'$. As for the rest of the objects listed, this is shown by descending induction on q, starting from $q = p$, using these short exact sequences

$$0 \to \mathrm{Z}^q(F_p(I)) \to F_p(I)^q \to \mathrm{B}^{q+1}(F_p(I)) \to 0,$$

$$0 \to \mathrm{B}^q(F_p(I)) \to F_p(I)^q \to \mathrm{Y}^q(F_p(I)) \to 0,$$

$$0 \to \mathrm{B}^q(F_p(I)) \to \mathrm{Z}^q(F_p(I)) \to \mathrm{H}^q(F_p(I)) \to 0,$$

and the fact that M' is a full abelian subcategory of M.

Step 3. The construction continues. Let us denote the components of the morphism $F_p(\phi)$ by $\phi^q := F_p(\phi)^q : M^q \to F_p(I)^q = I^q$ for $q \in [0, p]$; outside this degree range we have $F_p(\phi)^q = 0$.

The differential $\mathrm{d}_M^p : M^p \to M^{p+1}$ induces a morphism $\psi_1 : \mathrm{Y}^p(M) \to \mathrm{Z}^{p+1}(M)$. There is also the morphism $\psi_2 := \mathrm{Y}^p(\phi) : \mathrm{Y}^p(M) \to \mathrm{Y}^p(F_p(I))$. We define N to be the fibered coproduct

$$N := \mathrm{Z}^{p+1}(M) \oplus_{\mathrm{Y}^p(M)} \mathrm{Y}^p(F_p(I)) \tag{11.5.15}$$

relative to the morphisms ψ_1 and ψ_2. In the notation of Lemma 11.5.11, with the objects $K := \mathrm{Y}^p(M)$, $L_1 := \mathrm{Z}^{p+1}(M)$ and $L_2 := \mathrm{Y}^p(F_p(I))$, and with the morphism ψ_1, ψ_2 that we have just defined, there are morphisms $\epsilon_1 : \mathrm{Z}^{p+1}(M) \to N$, $\epsilon_2 : \mathrm{Y}^p(F_p(I)) \to N$ and $\pi_1 : N \to \mathrm{Coker}(\psi_1) = \mathrm{H}^{p+1}(M)$. According to Lemma 11.5.11(1), there is an exact sequence

$$\mathrm{H}^p(M) \xrightarrow{\psi_2} \mathrm{Y}^p(F_p(I)) \xrightarrow{\epsilon_2} N \xrightarrow{\pi_1} \mathrm{H}^{p+1}(M) \to 0 \tag{11.5.16}$$

in M. However, letting $\beta : \mathrm{H}^p(F_p(I)) \to \mathrm{Y}^p(F_p(I))$ be the canonical monomorphism, there is equality $\psi_2 = \beta \circ \mathrm{H}^p(F_p(\phi))$. By condition (♡) with indices $q = p$ we know that $\mathrm{H}^p(F_p(\phi))$ is a monomorphism. Therefore ψ_2 is a monomorphism, and (11.5.16) extends to a slightly longer exact sequence

$$0 \to \mathrm{H}^p(M) \xrightarrow{\psi_2} \mathrm{Y}^p(F_p(I)) \xrightarrow{\epsilon_2} N \xrightarrow{\pi_1} \mathrm{H}^{p+1}(M) \to 0 \tag{11.5.17}$$

in M.

Let $\gamma : \mathrm{Z}^{p+1}(M) \rightarrowtail M^{p+1}$ be the canonical monomorphism, and define $\psi_1^+ := \gamma \circ \psi_1 : \mathrm{Y}^p(M) \to M^{p+1}$. Let $\psi_2^+ := \psi_2 : \mathrm{Y}^p(M) \to \mathrm{Y}^p(F_p(I))$. We have the corresponding fibered coproduct

$$N^+ := M^{p+1} \oplus_{\mathrm{Y}^p(M)} \mathrm{Y}^p(F_p(I)). \tag{11.5.18}$$

The monomorphism $\gamma : \mathrm{Z}^{p+1}(M) \rightarrowtail M^{p+1}$ induces a morphism $\omega : N \to N^+$ between these fibered coproducts, and Lemma 11.5.11(2) says that ω is also a monomorphism.

Step 4. We now produce the object $I^{p+1} \in \mathsf{M}'$. In step 2 we showed that $\mathrm{Y}^p(F_p(I)) \in \mathsf{M}'$. It is given that $\mathrm{H}^q(M) \in \mathsf{M}'$ for all q. The exact sequence (11.5.16) and Proposition 7.4.2 say that $N \in \mathsf{M}'$ too. By assumption there is a monomorphism $\chi : N \rightarrowtail I^{p+1}$ into some object $I^{p+1} \in \mathsf{M}'$ that's injective in M. Because $\omega : N \to N^+$ is a monomorphism, there is a morphism $\chi^+ : N^+ \to I^{p+1}$ such that $\chi^+ \circ \omega = \chi$.

We have a canonical epimorphism $\alpha_I : I^p \twoheadrightarrow \mathrm{Y}^p(F_p(I))$. Define the morphism $\mathrm{d}_I^p : I^p \to I^{p+1}$ by the formula $\mathrm{d}_I^p := \chi \circ \epsilon_2 \circ \alpha_I$. Since $\alpha_I \cdot \mathrm{d}_I^{p-1} = 0$, we see that $\mathrm{d}_I^p \circ \mathrm{d}_I^{p-1} = 0$. This is depicted in the commutative diagram below.

$$\begin{tikzcd} I^{p-1} \arrow[r, "\mathrm{d}_I^{p-1}"'] \arrow[rr, bend left, "0"] & I^p \arrow[r, "\alpha_I"] \arrow[rrr, bend right, "\mathrm{d}_I^p"'] & \mathrm{Y}^p(F_p(I)) \arrow[r, "\epsilon_2"] & N \arrow[r, "\chi"] & I^{p+1} \end{tikzcd}$$

Therefore we obtain a new complex

$$F_{p+1}(I) := (\cdots \to 0 \to I^0 \xrightarrow{\mathrm{d}_I^0} I^1 \xrightarrow{\mathrm{d}_I^1} \cdots \xrightarrow{\mathrm{d}_I^{p-1}} I^p \xrightarrow{\mathrm{d}_I^p} I^{p+1} \to 0 \to \cdots) \tag{11.5.19}$$

in $\mathbf{C}(\mathsf{M}')$, and it projects onto $F_p(I)$.

Step 5. In this step we construct the morphism $\phi^{p+1} : M^{p+1} \to I^{p+1}$. Let $\epsilon_1^+ : M^{p+1} \to N^+$ be the morphism that's part of the fibered coproduct (11.5.18). We define the morphism $\phi^{p+1} : M^{p+1} \to I^{p+1}$ by the formula $\phi^{p+1} := \chi^+ \circ \epsilon_1^+$.

Let's examine the next diagram, where $\alpha_M : M^p \twoheadrightarrow \mathrm{Y}^p(M)$ is the canonical epimorphism, and $\psi_1^+, \psi_2^+ = \psi_2, \epsilon_1^+, \epsilon_2^+$ and ϵ_2 are the morphisms involved in the two fibered coproducts.

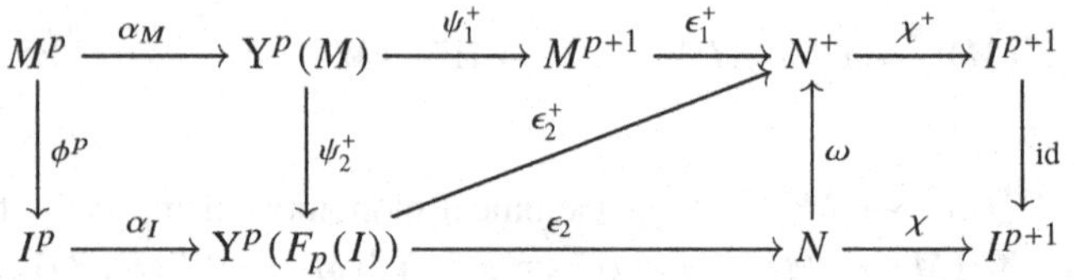

$$
\begin{array}{ccccccccc}
M^p & \xrightarrow{\alpha_M} & \mathrm{Y}^p(M) & \xrightarrow{\psi_1^+} & M^{p+1} & \xrightarrow{\epsilon_1^+} & N^+ & \xrightarrow{\chi^+} & I^{p+1} \\
\downarrow{\scriptstyle \phi^p} & & \downarrow{\scriptstyle \psi_2^+} & & \nearrow{\scriptstyle \epsilon_2^+} & & \uparrow{\scriptstyle \omega} & & \downarrow{\scriptstyle \mathrm{id}} \\
I^p & \xrightarrow{\alpha_I} & \mathrm{Y}^p(F_p(I)) & & \xrightarrow{\quad\epsilon_2\quad} & & N & \xrightarrow{\chi} & I^{p+1}
\end{array}
\tag{11.5.20}
$$

The left square is commutative because $\alpha : \mathrm{Id} \to \mathrm{Y}$ is a morphism of functors. The triangle that has $\mathrm{Y}^p(M)$ as a vertex is commutative because N^+ is the fibered coproduct. The other triangle, the one that has N as a vertex, is commutative because ω is a morphism between fibered coproducts. The right square is commutative by the defining property of χ^+. So the whole diagram is commutative.

We now take part of the commutative diagram (11.5.20), and add to it three curved arrows.

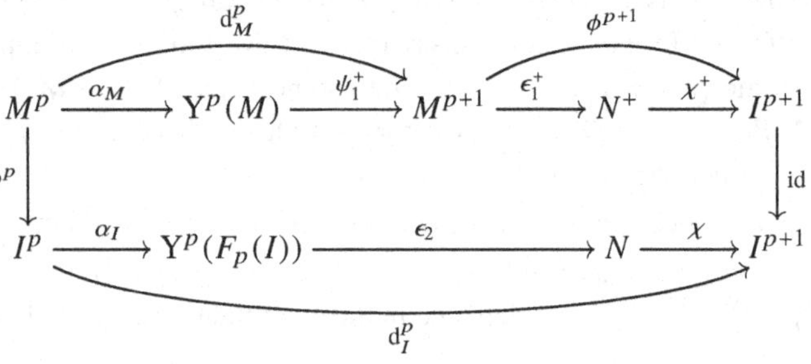

$$
\begin{array}{ccccccccc}
 & & \overset{\mathrm{d}_M^p}{\frown} & & & & \overset{\phi^{p+1}}{\frown} & & \\
M^p & \xrightarrow{\alpha_M} & \mathrm{Y}^p(M) & \xrightarrow{\psi_1^+} & M^{p+1} & \xrightarrow{\epsilon_1^+} & N^+ & \xrightarrow{\chi^+} & I^{p+1} \\
\downarrow{\scriptstyle \phi^p} & & & & & & & & \downarrow{\scriptstyle \mathrm{id}} \\
I^p & \xrightarrow{\alpha_I} & \mathrm{Y}^p(F_p(I)) & & \xrightarrow{\quad\epsilon_2\quad} & & N & \xrightarrow{\chi} & I^{p+1} \\
 & & & & \underset{\mathrm{d}_I^p}{\smile} & & & &
\end{array}
\tag{11.5.21}
$$

The curved polygons are commutative, as can be seen from the definitions of the morphisms ψ_1^+, ϕ^{p+1} and d_I^p. We conclude that $\mathrm{d}_I^p \circ \phi^p = \phi^{p+1} \circ \mathrm{d}_M^p$, so there is a new morphism of complexes $F_{p+1}(\phi) : M \to F_{p+1}(I)$ whose degree $p+1$ component is ϕ^{p+1}.

Step 6. Here we prove that $\mathrm{H}^{p+1}(F_{p+1}(\phi))$ is a monomorphism and $\mathrm{H}^p(F_{p+1}(\phi))$ is an isomorphism. This will verify condition (♡) with index $p+1$.

Consider the fibered coproduct

$$N^\diamond := I^{p+1} \oplus_{\mathrm{Y}^p(F_p(I))} \mathrm{Y}^p(F_p(I)) \tag{11.5.22}$$

of the morphisms $\psi_1^\diamond$ and $\psi_2^\diamond$, where $\psi_1^\diamond : \mathrm{Y}^p(F_p(I)) \to I^{p+1}$ is the morphism induced by d_I^p, and $\psi_2^\diamond$ is the identity of $\mathrm{Y}^p(F_p(I))$. Note that $\epsilon_1^\diamond : I^{p+1} \to N^\diamond$ is an isomorphism, and $\mathrm{Coker}(\psi_1^\diamond) = \mathrm{Y}^{p+1}(F_{p+1}(I)) = \mathrm{H}^{p+1}(F_{p+1}(I))$. The morphisms $\phi^{p+1} : \mathrm{Z}^{p+1}(M) \to I^{p+1}$ and $\mathrm{Y}^p(F_p(\phi)) : \mathrm{Y}^p(M) \to \mathrm{Y}^p(F_p(I))$ induce a morphism $\omega^\diamond : N \to N^\diamond$ between the fibered coproducts (11.5.15) and (11.5.22). According to Lemma 11.5.11(1), there is an induced commutative diagram

$$\begin{array}{ccccccccccc} 0 & \longrightarrow & \mathrm{H}^p(M) & \xrightarrow{\psi_2} & \mathrm{Y}^p(F_p(I)) & \xrightarrow{\epsilon_2} & N & \xrightarrow{\pi_2} & \mathrm{H}^{p+1}(M) & \longrightarrow & 0 \\ & & \downarrow{\scriptstyle \mathrm{H}^p(F_{p+1}(\phi))} & & \downarrow{\scriptstyle \mathrm{id}} & & \downarrow{\scriptstyle \omega^\diamond} & & \downarrow{\scriptstyle \mathrm{H}^{p+1}(F_{p+1}(\phi))} & & \\ 0 & \longrightarrow & \mathrm{H}^p(F_{p+1}(I)) & \xrightarrow{\psi_2^\diamond} & \mathrm{Y}^p(F_p(I)) & \xrightarrow{\epsilon_2^\diamond} & N^\diamond & \xrightarrow{\pi_2^\diamond} & \mathrm{H}^{p+1}(F_{p+1}(I)) & \longrightarrow & 0 \end{array} \tag{11.5.23}$$

The first row is an exact sequence, since it is (11.5.17). The second row is also exact: exactness at the last three objects is by Lemma 11.5.11(1). As for exactness at $\mathrm{H}^p(F_{p+1}(I))$, we use the fact that the morphism $\mathrm{Y}^p(F_{p+1}(I)) \to \mathrm{Y}^p(F_p(I))$ is an isomorphism, and its composition with the canonical embedding $\mathrm{H}^p(F_{p+1}(I)) \rightarrowtail \mathrm{Y}^p(F_{p+1}(I))$ is $\psi_2^\diamond$, so the latter is a monomorphism. A little calculation shows that $\omega^\diamond = \epsilon_1^\diamond \circ \chi$, so it is a monomorphism.

Finally, a diagram chase in (11.5.23), using the first sheaf trick, shows that $\mathrm{H}^{p+1}(F_{p+1}(\phi))$ is a monomorphism, and that $\mathrm{H}^p(F_{p+1}(\phi))$ is an isomorphism. □

Corollary 11.5.24 *Under the assumptions of Theorem* 11.5.8, *the canonical functor* $\mathbf{D}^+(\mathsf{M}') \to \mathbf{D}^+_{\mathsf{M}'}(\mathsf{M})$ *is an equivalence.*

Proof This is very similar to the proof of Corollary 11.3.21, and there is no need to repeat it. □

Here is an important instance in which Theorem 11.5.8 and Corollary 11.5.24 apply.

Example 11.5.25 Let $(X, \mathcal{O}_X)$ be a noetherian scheme. Associated to it are these abelian categories: the category $\mathsf{M} := \mathsf{Mod}\, \mathcal{O}_X$ of $\mathcal{O}_X$-modules, and the thick abelian subcategory $\mathsf{M}' := \mathsf{QCoh}\, \mathcal{O}_X$ of quasi-coherent $\mathcal{O}_X$-modules. According to [62, Proposition II.7.6], the category M' has enough injectives relative to M. By Corollary 11.5.24 the canonical functor

$$\mathbf{D}(\mathsf{QCoh}\, \mathcal{O}_X) \to \mathbf{D}^+_{\mathrm{qc}}(\mathsf{Mod}\, \mathcal{O}_X)$$

is an equivalence.

For a more general statement, using other methods, see [23, Corollary 5.5].

Corollary 11.5.26 *If* M *is an abelian category with enough injectives, then* $\mathbf{C}^+(\mathsf{M})$ *has enough K-injectives.*

Proof According to either Theorem 11.5.6 or Theorem 11.5.8, every $M \in \mathbf{C}^+(\mathsf{M})$ admits a quasi-isomorphism $M \to I$ to bounded below complex of injectives I. Now use Proposition 11.5.5 and Theorem 11.5.2. □

Corollary 11.5.27 *Let* M *be an abelian category with enough injectives, and let* $M \in \mathbf{C}(\mathsf{M})$ *be a complex with bounded below cohomology. Then* M *has a K-injective resolution* $M \to I$*, such that* $\inf(I) = \inf(\mathrm{H}(M))$*, and every* I^p *is an injective object of* M.

Proof We may assume that $\mathrm{H}(M)$ is nonzero. Let $p := \inf(\mathrm{H}(M)) \in \mathbb{Z}$, and let $N := \mathrm{smt}^{\geq p}(M)$, the smart truncation from Definition 7.3.6. So $M \to N$ is a quasi-isomorphism, and $\inf(N) = p$. According to either Theorem 11.5.6 or Theorem 11.5.8, there is a quasi-isomorphism $N \to I$, where I is a complex of injectives and $\inf(I) = p$. By Proposition 11.5.5 and Theorem 11.5.2 the complex I is K-injective. The composed quasi-isomorphism $M \to I$ is what we are looking for. □

11.6 K-Injective Resolutions in $\mathbf{C}(A)$

Recall that we are working over a nonzero commutative base ring $\mathbb{K}$, and A is a central DG $\mathbb{K}$-ring.

An *injective cogenerator* of the abelian category $\mathbf{M}(\mathbb{K}) = \mathsf{Mod}\,\mathbb{K}$ is an injective $\mathbb{K}$-module $\mathbb{K}^*$ with this property: if M is a nonzero $\mathbb{K}$-module, then $\mathrm{Hom}_{\mathbb{K}}(M, \mathbb{K}^*)$ is nonzero. These always exist. Here are a few examples.

Example 11.6.1 For every $\mathbb{K}$ there is a canonical choice for an injective cogenerator: $\mathbb{K}^* := \mathrm{Hom}_{\mathbb{Z}}(\mathbb{K}, \mathbb{Q}/\mathbb{Z})$. See proof of Theorem 2.7.13. Usually this is a very big module!

Example 11.6.2 Assume $\mathbb{K}$ is a complete noetherian local ring, with maximal ideal $\mathfrak{m}$ and residue field $\Bbbk = \mathbb{K}/\mathfrak{m}$. In this case we would prefer to take the smallest possible injective cogenerator $\mathbb{K}^*$, and this is the injective hull of $\Bbbk$ as a $\mathbb{K}$-module.

Here are some special cases. If $\mathbb{K}$ is a field, then $\mathbb{K}^* = \mathbb{K} = \Bbbk$. If $\mathbb{K} = \widehat{\mathbb{Z}}_p$, the ring of p-adic integers, then $\Bbbk = \mathbb{F}_p$, and $\mathbb{K}^* \cong \widehat{\mathbb{Q}}_p/\widehat{\mathbb{Z}}_p$, which is the p-primary part of $\mathbb{Q}/\mathbb{Z}$. If $\mathbb{K}$ contains some field, then there exists a ring homomorphism $\Bbbk \to \mathbb{K}$ that lifts the canonical surjection $\mathbb{K} \to \Bbbk$. (This is by the Cohen Structure Theorem, see [46, Theorem 7.7].) After choosing such a lifting, there is an isomorphism of $\mathbb{K}$-modules $\mathbb{K}^* \cong \mathrm{Hom}^{\mathrm{cont}}_{\Bbbk}(\mathbb{K}, \Bbbk)$, where continuity is for the $\mathfrak{m}$-adic topology on $\mathbb{K}$ and the discrete topology on $\Bbbk$.

In this section we have the following standing convention.

Convention 11.6.3 There is a fixed injective cogenerator $\mathbb{K}^*$ of $\mathbf{M}(\mathbb{K})$.

We view $\mathbb{K}^*$ as a DG $\mathbb{K}$-module concentrated in degree 0 (with zero differential). For every $p \in \mathbb{Z}$ there is the DG $\mathbb{K}$-module $\mathrm{T}^{-p}(\mathbb{K}^*)$, which is concentrated in degree p.

It will be convenient to blur the distinction between DG $\mathbb{K}$-modules with zero differentials and graded $\mathbb{K}$-modules. Namely, if N is a DG $\mathbb{K}$-module such that $\mathrm{d}_N = 0$, we will identify N with the graded modules $N^\natural$, $\mathrm{H}(N)$, etc.

Definition 11.6.4 For a DG $\mathbb{K}$-module V we define the DG $\mathbb{K}$-module $V^* := \mathrm{Hom}_{\mathbb{K}}(V, \mathbb{K}^*)$.

The operation $(-)^*$ is an exact contravariant functor from $\mathbf{C}_{\mathrm{str}}(\mathbb{K})$ to itself, and also from $\mathbf{G}_{\mathrm{str}}(\mathbb{K})$ to itself. Note that given $M \in \mathbf{C}(A)$, its dual M^* is an object of $\mathbf{C}(A^{\mathrm{op}})$, so that we have an exact functor

$$(-)^* : \mathbf{C}_{\mathrm{str}}(A)^{\mathrm{op}} \to \mathbf{C}_{\mathrm{str}}(A^{\mathrm{op}}). \tag{11.6.5}$$

Definition 11.6.6 A DG $\mathbb{K}$-module W is called *cofree* if there is an isomorphism $W \cong V^*$ in $\mathbf{C}_{\mathrm{str}}(\mathbb{K})$, for some free DG $\mathbb{K}$-module V.

Since the differential of a free DG $\mathbb{K}$-module V is zero, we see that a cofree DG module W also has zero differential.

Lemma 11.6.7 *Let V be a free DG $\mathbb{K}$-module, and let $W := V^*$.*

(1) *If $V \cong \bigoplus_{s \in S} \mathrm{T}^{p_s}(\mathbb{K})$ for some indexing set S and some collection of integers $\{p_s\}_{s \in S}$, then $W \cong \prod_{s \in S} \mathrm{T}^{-p_s}(\mathbb{K}^*)$.*
(2) *As an object of the abelian category $\mathbf{G}_{\mathrm{str}}(\mathbb{K})$, W is injective.*

In item (1) the isomorphisms can be considered either in $\mathbf{C}_{\mathrm{str}}(\mathbb{K})$ or in $\mathbf{G}_{\mathrm{str}}(\mathbb{K})$; it doesn't matter, because V and W have zero differentials. However, item (2) is true only in the category $\mathbf{G}_{\mathrm{str}}(\mathbb{K})$.

Lemma 11.6.8 *Let $\phi : U \to V$ be a homomorphism in $\mathbf{G}_{\mathrm{str}}(\mathbb{K})$.*

(1) *ϕ is injective iff $\phi^* : V^* \to U^*$ is surjective.*
(2) *ϕ is surjective iff $\phi^* : V^* \to U^*$ is injective.*
(3) *The canonical homomorphism $U \to U^{**} = (U^*)^*$ in $\mathbf{G}_{\mathrm{str}}(\mathbb{K})$ is injective.*
(4) *There exists an injective homomorphism $U \rightarrowtail W$ in $\mathbf{G}_{\mathrm{str}}(\mathbb{K})$ into some cofree DG $\mathbb{K}$-module W.*

Exercise 11.6.9 Prove Lemmas 11.6.7 and 11.6.8.

Definition 11.6.10 Let W be a cofree DG $\mathbb{K}$-module. The *cofree DG A-module coinduced from W* is the DG A-module $I_W := \mathrm{Hom}_{\mathbb{K}}(A, W)$. There is a homomorphism $\theta_W : I_W \to W$ in $\mathbf{C}_{\mathrm{str}}(\mathbb{K})$, with formula $\theta_W(\chi) := \chi(1_A) \in W$.

Definition 11.6.11 A DG A-module J is called *cofree* if there is an isomorphism $J \cong I_W$ in $\mathbf{C}_{\mathrm{str}}(A)$ for some cofree DG $\mathbb{K}$-module W.

Lemma 11.6.12 *Consider the free DG $\mathbb{K}$-module $V := \bigoplus_{s \in S} \mathrm{T}^{p_s}(\mathbb{K})$ and the cofree DG $\mathbb{K}$-module $W := V^*$. There are canonical isomorphisms*

$$I_W \cong (A \otimes V)^* \cong \Big(\bigoplus\nolimits_{s \in S} \mathrm{T}^{p_s}(A)\Big)^* \cong \prod\nolimits_{s \in S} \mathrm{T}^{-p_s}(A^*)$$

in $\mathbf{C}_{\mathrm{str}}(A)$.

Exercise 11.6.13 Prove the lemma above.

Lemma 11.6.14 *Let W be a cofree DG $\mathbb{K}$-module, and let M be a DG A-module. The homomorphism*

$$\mathrm{Hom}(\mathrm{id}_M, \theta_W) : \mathrm{Hom}_A(M, I_W) \to \mathrm{Hom}_{\mathbb{K}}(M, W)$$

in $\mathbf{C}_{\mathrm{str}}(\mathbb{K})$ is an isomorphism.

Proof The homomorphism

$$\mathrm{Hom}(\mathrm{id}_M, \theta_W) : \mathrm{Hom}_A(M, \mathrm{Hom}_{\mathbb{K}}(A, W)) \xrightarrow{\simeq} \mathrm{Hom}_{\mathbb{K}}(M, W)$$

is just adjunction for the DG ring homomorphism $\mathbb{K} \to A$, so it is bijective. □

Recall that $\mathbf{G}_{\mathrm{str}}(A^{\natural})$ is the abelian category whose objects are the graded $A^{\natural}$-modules, and the morphisms are the A-linear homomorphisms of degree 0. The forgetful functor $\mathrm{Und} : \mathbf{C}_{\mathrm{str}}(A) \to \mathbf{G}_{\mathrm{str}}(A^{\natural})$, $M \mapsto M^{\natural}$, is faithful.

Lemma 11.6.15 *Let I be a cofree DG A-module. Then $I^{\natural}$ is an injective object of $\mathbf{G}_{\mathrm{str}}(A^{\natural})$.*

Proof We can assume that $I = I_W$ for some cofree DG $\mathbb{K}$-module W. For every $M \in \mathbf{G}_{\mathrm{str}}(A^{\natural})$ there are isomorphisms

$$\begin{aligned} \mathrm{Hom}_{\mathbf{G}_{\mathrm{str}}(A^{\natural})}(M, I_W^{\natural}) &= \mathrm{Hom}_A(M, I_W)^0 \\ &\cong^{\heartsuit} \mathrm{Hom}_{\mathbb{K}}(M, W)^0 = \prod_{p \in \mathbb{Z}} \mathrm{Hom}_{\mathbb{K}}(M^p, W^p) \end{aligned}$$

in $\mathbf{M}(\mathbb{K})$. The isomorphism $\cong^{\heartsuit}$ is by Lemma 11.6.14. For every p the functor $\mathbf{G}_{\mathrm{str}}(A^{\natural}) \to \mathbf{M}(\mathbb{K})$, $M \mapsto M^p$, is exact. Because each W^p is an injective object of $\mathbf{M}(\mathbb{K})$, the contravariant functor $\mathrm{Hom}_{\mathbb{K}}(-, W^p)$ from $\mathbf{M}(\mathbb{K})$ to itself is exact. And the product of exact functors into $\mathbf{M}(\mathbb{K})$ is exact. We conclude that the functor $\mathrm{Hom}_{\mathbf{G}_{\mathrm{str}}(A^{\natural})}(-, I_W^{\natural})$ is exact. □

Recall that for a DG $\mathbb{K}$-module M we have the object of decocycles

$$\mathrm{Y}(M) := \mathrm{Coker}(\mathrm{d}_M : \mathrm{T}^{-1}(M) \to M) = M / \mathrm{B}(M) \in \mathbf{G}(\mathbb{K}).$$

Lemma 11.6.16 *Let W be a cofree DG $\mathbb{K}$-module, let M be a DG A-module, and let $\chi : \mathrm{Y}(M) \to W$ be a homomorphism in $\mathbf{G}_{\mathrm{str}}(\mathbb{K})$. Then there is a unique homomorphism $\psi : M \to I_W$ in $\mathbf{C}_{\mathrm{str}}(A)$, such that the diagram*

$$\mathrm{Y}(M) \xrightarrow{\mathrm{Y}(\psi)} \mathrm{Y}(I_W) \xrightarrow{\mathrm{Y}(\theta_W)} \mathrm{Y}(W) = W, \qquad \chi : \mathrm{Y}(M) \to \mathrm{Y}(W)$$

in $\mathbf{G}_{\mathrm{str}}(\mathbb{K})$ is commutative.

Proof Since the differentials of W and $\mathrm{Y}(M)$ are zero, the canonical homomorphism

$$\alpha : \mathrm{Hom}_{\mathbb{K}}(\mathrm{Y}(M), W)^0 = \mathrm{Z}^0(\mathrm{Hom}_{\mathbb{K}}(\mathrm{Y}(M), W)) \to \mathrm{Z}^0(\mathrm{Hom}_{\mathbb{K}}(M, W)),$$

that's induced by the canonical surjection $M \twoheadrightarrow \mathrm{Y}(M)$, is bijective. This gives us a unique homomorphism $\alpha(\chi) : M \to W$ in $\mathbf{C}_{\mathrm{str}}(\mathbb{K})$. Next we use Lemma 11.6.14 to obtain a unique $\psi : M \to I_W$ in $\mathbf{C}_{\mathrm{str}}(A)$ s.t. $\theta_W \circ \psi = \alpha(\chi)$. This ψ is what we are looking for. □

The next definition is dual to Definition 11.4.3.

Definition 11.6.17 Let I be an object of $\mathbf{C}(A)$.

(1) A *semi-cofree cofiltration* on I is a cofiltration $G = \{G_q(I)\}_{q \geq -1}$ on I in $\mathbf{C}_{\mathrm{str}}(A)$ such that:
- $G_{-1}(I) = 0$.
- Each $\mathrm{Gr}^G_q(I)$ is a cofree DG A-module.
- $I = \lim_{\leftarrow q} G_q(I)$.

(2) The DG A-module I is called a *semi-cofree* if it admits a semi-cofree cofiltration.

Proposition 11.6.18 *If I is a semi-cofree DG A-module, then $I^\natural$ is an injective object in the abelian category $\mathbf{G}_{\mathrm{str}}(A^\natural)$.*

Proposition 11.6.19 *Assume A is a ring. If I is a semi-cofree DG A-module, then each I^p is an injective A-module.*

Exercise 11.6.20 Prove the two propositions above.

Theorem 11.6.21 *Let I be an object of $\mathbf{C}(A)$. If I is semi-cofree, then it is K-injective.*

Proof The proof is very similar to those of Theorems 11.3.2 and 11.4.14. But because the arguments involve limits, we shall give the full proof.

Step 1. Suppose I is cofree; say $I \cong \prod_{s \in S} \mathrm{T}^{-p_s}(A^*)$. Lemma 11.6.14 implies that for every DG A-module N there is an isomorphism

$$\mathrm{Hom}_A(N, I) \cong \prod_{s \in S} \mathrm{Hom}_{\mathbb{K}}(\mathrm{T}^{p_s}(N), \mathbb{K}^*)$$

of graded $\mathbb{K}$-modules. It follows that if N is acyclic, then so is $\mathrm{Hom}_A(N, I)$.

Step 2. Fix a semi-cofree cofiltration $G = \{G_q(I)\}_{q \geq -1}$ of I. Here we prove that for every $q \geq -1$ the DG module $G_q(I)$ is K-injective. This is done by induction on $q \geq -1$. For $q = -1$ it is trivial. For $q \geq 0$ there is an exact sequence

$$0 \to \mathrm{Gr}_q^G(I) \to G_q(I) \to G_{q-1}(I) \to 0 \tag{11.6.22}$$

in the category $\mathbf{C}_{\mathrm{str}}(A)$. Because $\mathrm{Gr}_q^G(I)$ is a cofree DG A-module, it is an injective object in the abelian category $\mathbf{G}_{\mathrm{str}}(A^\natural)$; see Lemma 11.6.15. Therefore the sequence (11.6.22) is split exact in $\mathbf{G}_{\mathrm{str}}(A^\natural)$.

Let $N \in \mathbf{C}(A)$ be an acyclic DG module. Applying the functor $\mathrm{Hom}_A(N, -)$ to the sequence (11.6.22) we obtain a sequence

$$0 \to \mathrm{Hom}_A(N, \mathrm{Gr}_q^G(I)) \to \mathrm{Hom}_A(N, G_q(I)) \to \mathrm{Hom}_A(N, G_{q-1}(I)) \to 0 \tag{11.6.23}$$

in $\mathbf{C}_{\mathrm{str}}(\mathbb{K})$. If we forget differentials this is a sequence in $\mathbf{G}_{\mathrm{str}}(\mathbb{K})$. Because (11.6.22) is split exact in $\mathbf{G}_{\mathrm{str}}(A^\natural)$, it follows that (11.6.23) is split exact in $\mathbf{G}_{\mathrm{str}}(\mathbb{K})$. Therefore (11.6.23) is exact in $\mathbf{C}_{\mathrm{str}}(\mathbb{K})$.

By the induction hypothesis the DG $\mathbb{K}$-module $\mathrm{Hom}_A(N, G_{q-1}(I))$ is acyclic. By step 1 the DG $\mathbb{K}$-module $\mathrm{Hom}_A(N, \mathrm{Gr}_q^G(I))$ is acyclic. The long exact cohomology sequence associated to (11.6.23) shows that the DG $\mathbb{K}$-module $\mathrm{Hom}_A(N, G_q(I))$ is acyclic too.

Step 3. We keep the semi-cofree cofiltration $G = \{G_q(I)\}_{q \geq -1}$ from step 2. Take any acyclic $N \in \mathbf{C}(A)$. By Proposition 1.8.6 we know that

$$\mathrm{Hom}_A(N, I) \cong \lim_{\leftarrow j} \mathrm{Hom}_A(N, G_q(I))$$

in $\mathbf{C}_{\mathrm{str}}(\mathbb{K})$. According to step 2, the complexes $\mathrm{Hom}_A(N, G_q(I))$ are all acyclic. The exactness of the sequences (11.6.23) implies that the inverse system $\{\mathrm{Hom}_A(N, G_q(I))\}_{q \geq -1}$ in $\mathbf{C}_{\mathrm{str}}(\mathbb{K})$ has surjective transitions. Now the Mittag–Leffler argument (Corollary 11.1.8) says that the inverse limit complex $\mathrm{Hom}_A(N, I)$ is acyclic. □

Theorem 11.6.24 *Let A be a DG ring. Every DG A-module M admits a quasi-isomorphism $\rho : M \to I$ in $\mathbf{C}_{\mathrm{str}}(A)$ to a semi-cofree DG A-module I.*

The proof of this theorem was communicated to us by B. Keller. We need a lemma first.

The semi-free DG A-module C was defined in formula (11.4.18). Its dual DG A-module $C^* = \mathrm{Hom}_{\mathbb{K}}(C, \mathbb{K}^*)$ is a semi-cofree DG A-module, with semi-cofree cofiltration $G_q(C^*) := F_q(C)^*$, where $F_q(C)$ is from (11.4.19).

Lemma 11.6.25 *Let $M \in \mathbf{C}(A)$.*

(1) *There is a homomorphism $\psi : M \to J$ in $\mathbf{C}_{\mathrm{str}}(A)$, such that $\mathrm{Z}(\psi) : \mathrm{Z}(M) \to \mathrm{Z}(J)$ is injective, and $J = \prod_{s \in S} \mathrm{T}^{-p_s}(A^*)$ for a collection of integers $\{p_s\}_{s \in S}$.*

(2) *There is an injective homomorphism $\psi' : M \to J'$ in $\mathbf{C}_{\mathrm{str}}(A)$, such that $J' = \prod_{s \in S'} \mathrm{T}^{-p_s}(C^*)$ for some collection of integers $\{p_s\}_{s \in S'}$.*

To clarify the notation in the lemma, we have two distinct collections of integers, $\{p_s\}_{s \in S}$ and $\{p_s\}_{s \in S'}$.

Proof

(1) By Lemma 11.6.8 there is an injective homomorphism $\chi : \mathrm{Y}(M) \rightarrowtail W$ for some cofree graded $\mathbb{K}$-module $W = \prod_{s \in S} \mathrm{T}^{-p_s}(\mathbb{K}^*)$. Let $J := I_W$. According to Lemma 11.6.16, there is a homomorphism $\psi : M \to J$ in $\mathbf{C}_{\mathrm{str}}(A)$ s.t. $\mathrm{Y}(\theta_W) \circ \mathrm{Y}(\psi) = \chi$. Because χ is injective, so is $\mathrm{Y}(\psi)$.

(2) Consider the DG A^{op}-module M^*. By Lemma 11.4.20 there is a collection of integers $\{p_s\}_{s \in S'}$, and a surjective homomorphism $\phi : Q \to M^*$ in $\mathbf{C}_{\mathrm{str}}(A^{\mathrm{op}})$, where $Q := \bigoplus_{s \in S'} \mathrm{T}^{p_s}(C)$. Dualizing we get a DG A-module $J' := Q^* \cong \prod_{s \in S'} \mathrm{T}^{-p_s}(C^*)$, and an injective homomorphism $\phi^* : M^{**} \to Q^* = J'$ in $\mathbf{C}_{\mathrm{str}}(A)$. Composing ϕ^* with the canonical embedding $M \rightarrowtail M^{**}$ gives us $\psi' : M \rightarrowtail J'$. □

The DG A-module J in item (1) of the lemma is cofree. The DG A-module J' in item (2) is semi-cofree, with semi-cofree cofiltration induced by that of C^*, namely

$$G_q(J') := \prod_{s \in S'} \mathrm{T}^{-p_s}(G_q(C^*)). \tag{11.6.26}$$

Proof of Theorem 11.6.24

Step 1. We are going to produce an exact sequence

$$0 \to M \xrightarrow{\eta} J^0 \xrightarrow{\partial^0} J^1 \xrightarrow{\partial^1} J^2 \to \cdots \tag{11.6.27}$$

in $\mathbf{C}_{\mathrm{str}}(A)$ with these properties:

(a) The sequence

$$0 \to \mathrm{Y}(M) \xrightarrow{\mathrm{Y}(\eta)} \mathrm{Y}(J^0) \xrightarrow{\mathrm{Y}(\partial^0)} \mathrm{Y}(J^1) \xrightarrow{\mathrm{Y}(\partial^1)} \mathrm{Y}(J^2) \to \cdots$$

is exact in $\mathbf{G}_{\mathrm{str}}(\mathbb{K})$.

(b) For every $p \geq 0$ the DG A-module J^p has the following decomposition in $\mathbf{C}_{\mathrm{str}}(A)$: $J^p = J_p \oplus J'_p$, where $J_p = \prod_{s \in S_p} \mathrm{T}^{-q_s}(A^*)$ and $J'_p = \prod_{s \in S'_p} \mathrm{T}^{-q_s}(C^*)$ for some collection of integers $\{q_s\}_{s \in S_p \sqcup S'_p}$.

This will be done recursively on p.

By Lemma 11.6.25 we can find DG A-modules J_0, J'_0 and homomorphisms $\psi_0 : M \to J_0$, $\psi'_0 : M \to J'_0$, such that both $\mathrm{Y}(\psi_0)$ and ψ'_0 are injective. The DG modules J_0, J'_0 are of the required form (property (b) with $p = 0$). We let $J^0 := J_0 \oplus J'_0$ and $\eta := \psi_0 \oplus \psi'_0$.

Let $N^0 := \mathrm{Coker}(\eta)$. Then the sequence

$$0 \to M \xrightarrow{\eta} J^0 \to N^0 \to 0 \tag{11.6.28}$$

in $\mathbf{C}_{\mathrm{str}}(A)$ is exact. The sequence

$$0 \to \mathrm{Y}(M) \xrightarrow{\mathrm{Y}(\eta)} \mathrm{Y}(J^0) \to \mathrm{Y}(N^0) \to 0$$

in $\mathbf{G}_{\mathrm{str}}(\mathbb{K})$ is also exact. Indeed, the exactness at $\mathrm{Y}(J^0)$ and $\mathrm{Y}(N^0)$ is because the functor $\mathrm{Y} : \mathbf{C}_{\mathrm{str}}(\mathbb{K}) \to \mathbf{G}_{\mathrm{str}}(\mathbb{K})$ is right exact (see Proposition 3.8.14), and the sequence (11.6.28) is exact. The exactness at $\mathrm{Y}(M)$ is by the injectivity of $\mathrm{Y}(\psi_0) : \mathrm{Y}(M) \to \mathrm{Y}(J_0)$.

Next we repeat this procedure with N^0 instead of M, to obtain an exact sequence

$$0 \to N^0 \xrightarrow{\partial^0} J^1 \to N^1 \to 0, \tag{11.6.29}$$

with J^1 of the required form (property (b) with $p = 1$), and such that

$$0 \to \mathrm{Y}(N^0) \xrightarrow{\mathrm{Y}(\partial^0)} \mathrm{Y}(J^1) \to \mathrm{Y}(N^1) \to 0$$

is also exact. We then splice (11.6.28) and (11.6.29) to get the exact sequence

$$0 \to M \xrightarrow{\eta} J^0 \xrightarrow{\partial^0} J^1 \to N^1 \to 0.$$

Continuing recursively we obtain an exact sequence (11.6.27) that has properties (a) and (b).

Step 2. We view the exact sequence (11.6.27) as an acyclic complex with entries in $\mathbf{C}_{\mathrm{str}}(A)$ that has J^0 in degree 0. Define the total DG A-modules

$$I := \mathrm{Tot}^{\Pi}(\cdots \to 0 \to J^0 \xrightarrow{\partial^0} J^1 \xrightarrow{\partial^1} J^2 \to \cdots)$$

and

$$I^{\mathrm{aug}} := \mathrm{Tot}^{\Pi}(\cdots \to 0 \to M \xrightarrow{\eta} J^0 \xrightarrow{\partial^0} J^1 \xrightarrow{\partial^1} J^2 \to \cdots).$$

Because the sequence (11.6.27) and the sequence appearing in property (a) are exact, we can use Proposition 11.2.24 (with any $j_0 \in \mathbb{Z}$). The conclusion is that the DG A-module I^{aug} is acyclic. Then, by Corollary 11.2.22, the homomorphism $\rho : M \to I$ in $\mathbf{C}_{\mathrm{str}}(A)$ is a quasi-isomorphism.

Step 3. It remains to produce a semi-cofree cofiltration $\{G_q(I)\}_{q \geq -1}$ on the DG A-module I. Our formula is this: $G_{-1}(I) := 0$ of course. For $r \geq 0$ we let

$$G_{2 \cdot r}(I) := \Big(\bigoplus\nolimits_{p=0}^{r-1} \mathrm{T}^{-p}(J^p) \Big) \oplus \mathrm{T}^{-r}(J_r \oplus G_0(J'_r)),$$

where $G_0(J'_r)$ comes from (11.6.26), and

$$G_{2 \cdot r+1}(I) := \bigoplus\nolimits_{p=0}^{r} \mathrm{T}^{-p}(J^p).$$

We leave it to the reader to verify that this is a semi-cofree cofiltration. □

Corollary 11.6.30 *Let A be a DG ring. The category $\mathbf{C}(A)$ has enough K-injectives.*

Proof Combine Theorems 11.6.21 and 11.6.24. □

Recall that a DG ring A is nonpositive if $A^i = 0$ for all $i > 0$. When A is nonpositive, the DG module A^* is concentrated in degrees ≥ 0.

Corollary 11.6.31 *If the DG ring A is nonpositive, then every $M \in \mathbf{C}(A)$ admits a quasi-isomorphism $\rho : M \to I$ in $\mathbf{C}_{\mathrm{str}}(A)$ to a semi-cofree DG A-module I, such that* $\inf(I) = \inf(\mathrm{H}(M))$.

Proof Let $p_0 := \inf(\mathrm{H}(M))$. If $p_0 = \pm\infty$ then there is nothing to prove beyond Theorem 11.6.24; so let us assume that $p_0 \in \mathbb{Z}$. Define $M' := \mathrm{smt}^{\geq p_0}(M)$. Then $M \to M'$ is a quasi-isomorphism, and $\inf(M') = p_0$. By replacing M with M' we can now assume that $\inf(M) = p_0$.

Since $\inf(M) = p_0$, it follows that $M^{p_0} = \mathrm{Y}^{p_0}(M)$. This means that in step 1 of the proof of the theorem, the homomorphism $\psi_0 : M^{p_0} \to J_0^{p_0}$ is already injective. So, like in the proof of Corollary 11.4.27, we can arrange to have $\inf(J^0) \geq p_0$. Then, recursively, we can arrange to have $\inf(J^q) \geq p_0$ for all q. Therefore in step 2 of the proof we get $\inf(I) \geq p_0$.

Finally, because $\inf(\mathrm{H}(M)) = p_0$ we must have $\inf(I) = p_0$. □

We end with the K-injective analogue of Remark 11.4.45.

Remark 11.6.32 Suppose $(X, \mathcal{A})$ is a ringed space. We know that the abelian category $\mathbf{M}(\mathcal{A}) = \mathsf{Mod}\,\mathcal{A}$ of sheaves of $\mathcal{A}$-modules on X has enough injective objects. Theorem 11.5.6 tells us that the subcategory $\mathbf{C}^+(\mathcal{A})$ has enough K-injective objects. But in fact more is true: the category of unbounded complexes $\mathbf{C}(\mathcal{A})$ also has enough K-injective objects; see [143]. (There is another, more abstract proof of this statement: because $\mathbf{M}(\mathcal{A})$ is a Grothendieck abelian category, it follows that $\mathbf{C}(\mathcal{A})$ has enough K-injectives. See [76, Corollary 14.1.8].)

Now let's consider a DG ringed space $(X, \mathcal{A})$. Fix an injective cogenerator $\mathbb{K}^*$ of $\mathbf{M}(\mathbb{K})$. For a point $x \in X$, consider the cofree DG $\mathcal{A}_x$-module $I_x := \mathrm{Hom}_{\mathbb{K}}(\mathcal{A}_x, \mathbb{K}^*)$. Pushing this module forward along the inclusion $\{x\} \subseteq X$ we get a constant sheaf $\mathcal{I}_x$ with support the closed set $\overline{\{x\}}$, which we refer to as the cofree DG $\mathcal{A}$-module associated to x. By definition, a *cofree DG $\mathcal{A}$-module* is a product of sheaves $\mathcal{I}_x$ (for varying points $x \in X$). Combining this notion with that of a semi-cofree DG module, we can talk about *semi-cofree DG $\mathcal{A}$-modules*. Like Theorem 11.6.21, one can show that a semi-cofree DG $\mathcal{A}$-module $\mathcal{I}$ is K-injective.

Assume that the topological space X is finite dimensional, in the sense that there is a natural number d such that for every open set $U \subseteq X$, every sheaf of $\mathbb{K}$-module $\mathcal{N}$ on U, and every $p > d$, the sheaf cohomology $\mathrm{H}^p(U, \mathcal{N})$ vanishes. This condition is quite mild – it is satisfied when X is a finite dimensional noetherian scheme, and also when X is a subspace of a finite dimensional topological manifold. Then a geometric variant of Theorem 11.6.24 is true: every DG $\mathcal{A}$-module $\mathcal{M}$ admits a quasi-isomorphism $\mathcal{M} \to \mathcal{I}$ to a semi-cofree DG $\mathcal{A}$-module $\mathcal{I}$. The finite dimensionality of X is needed to invoke the ML argument. A detailed proof of this assertion will be published in the future.

12

Adjunctions, Equivalences and Cohomological Dimension

In this chapter we discuss the derived Hom and tensor bifunctors in several situations and show how these bifunctors are related in adjunction formulas. Cohomological dimensions of functors and DG modules are introduced, and they are used to prove several theorems. Given a homomorphism between two DG rings, we study the derived restriction, induction and coinduction functors between the derived categories of these DG rings. We prove that for a quasi-isomorphism of DG rings the restriction functor is an equivalence. We also prove the existence of NC semi-free DG ring resolutions.

As before, we work over a nonzero commutative base ring $\mathbb{K}$, and Convention 1.2.4 is in place.

12.1 Boundedness Conditions Revisited

This section is about notation. We start by introducing some combinatorial definitions that will enable a precise treatment of cohomological dimensions of objects and functors (in Section 12.4).

We also change our notation for boundedness conditions in derived categories (Definition 12.1.6) and for the translation functor (Definition 12.1.7). These changes will simplify and improve our discussion (while hopefully not creating confusion). See Remark 12.1.8 for an explanation.

Integer intervals were introduced in Definition 11.1.10. *Boundedness conditions* were introduced in Sections 3.8 and 7.3. We need to refine these notions.

Definition 12.1.1 Let S be an integer interval.

(1) We call S a *bounded above integer interval* if either $S = [i_0, i_1]$ with $i_1 \in \mathbb{Z}$, or if $S = \varnothing$.
(2) We call S a *bounded below integer interval* if either $S = [i_0, i_1]$ with $i_0 \in \mathbb{Z}$, or if $S = \varnothing$.
(3) We call S a *bounded integer interval* if it is both bounded above and bounded below; i.e. either $S = [i_0, i_1]$ with $i_0, i_1 \in \mathbb{Z}$, or $S = \varnothing$.

The next definition was already mentioned in passing in Section 3.8.

Definition 12.1.2 The boundedness conditions in Definition 12.1.1 have abbreviations, which we call *boundedness indicators*.

(1) The indicator "$-$" stands for "bounded above."
(2) The indicator "$+$" stands for "bounded below."
(3) The indicator "b" stands for "bounded."
(4) The indicator "⟨empty⟩" (the empty symbol) stands for the empty condition (no condition at all).

The symbol "$\star$" is used to denote an unspecified boundedness indicator, i.e. it is a placeholder for one of the boundedness indicators in Definition 12.1.2.

Definition 12.1.3 Let S be an integer interval.

(1) If $S = [i_0, i_1]$, then its *length* is $i_1 - i_0 \in \mathbb{N} \cup \{\infty\}$. If $S = \varnothing$, then its length is $-\infty$.
(2) If $S = [i_0, i_1]$, then the *reversed integer interval* is $-S := [-i_1, -i_0]$. If $S = \varnothing$, then its reverse is also $\varnothing$.
(3) If $S = [i_0, i_1]$, and if $T = [j_0, j_1]$ is a second nonempty integer interval, then their sum is $S + T := [i_0 + j_0, i_1 + j_1]$. If $S = \varnothing$, and it T is a second integer interval, then the sum is $S + T := \varnothing$.

Thus an integer interval S (whether empty or not) has finite length if and only if it is bounded.

Definition 12.1.4 Let $M = \{M^i\}_{i \in \mathbb{Z}}$ be a graded object in an abelian category M, i.e. $M \in \mathbf{G}(\mathsf{M})$.

(1) We say that M is *concentrated* in an integer interval S if the set $\{i \in \mathbb{Z} \mid M^i \neq 0\}$ is contained in S.
(2) The *concentration* of M is the smallest integer interval $\operatorname{con}(M)$ in which M is concentrated.
(3) We say that M is of *boundedness type* $\star$, for some boundedness indicator $\star$ if the integer interval $\operatorname{con}(M)$ satisfies the boundedness condition $\star$.

Observe that $\mathrm{con}(M) = \varnothing$ if and only if $M = 0$. Definition 12.1.4(3) agrees with the boundedness conditions on a graded module M from Sections 3.8 and 7.3.

In Definition 11.1.11 we introduced the notions $\inf(M)$ and $\sup(M)$ for a graded object M. Using them we can describe the concentration $\mathrm{con}(M) = [i_0, i_1]$ when M is nonzero: $i_0 = \inf(M)$ and $i_1 = \sup(M)$. The amplitude $\mathrm{amp}(M)$ is the length of the interval $\mathrm{con}(M)$.

Recall that to a DG module $M \in \mathbf{C}(A, \mathsf{M})$ we associate the graded object $M^\natural \in \mathbf{G}(\mathsf{M})$ that is gotten by forgetting the differential. The cohomology $\mathrm{H}(M)$ is another object of $\mathbf{G}(\mathsf{M})$.

Definition 12.1.5 Let A be a DG ring, M an abelian category, and $M \in \mathbf{C}(A, \mathsf{M})$.

(1) The *concentration* of M and the *boundedness type* of M are the same as in Definition 12.1.4, when M is seen as an object of $\mathbf{G}(\mathsf{M})$.
(2) The *cohomological concentration* and the *cohomological boundedness type* of M are the concentration and boundedness type, respectively, of the graded object $\mathrm{H}(M) \in \mathbf{G}(\mathsf{M})$, in the sense of Definition 12.1.4.

The next definition is in conflict with Definitions 7.3.3 and 7.3.4; but we already warned (in Remark 7.3.13) that this change will take place. See Remark 12.1.8 for further discussion.

Definition 12.1.6 Let A be a DG ring, M an abelian category, and $\star$ a boundedness indicator. We denote by $\mathbf{D}^\star(A, \mathsf{M})$ the full subcategory of $\mathbf{D}(A, \mathsf{M})$ on the DG modules M of cohomological boundedness type $\star$, as in Definition 12.1.5 above.

Thus, for example, a DG module M belongs to $\mathbf{D}^{\mathrm{b}}(A, \mathsf{M})$ if and only if $\mathrm{con}(\mathrm{H}(M))$ is a bounded integer interval. As explained in Section 7.3, $\mathbf{D}^\star(A, \mathsf{M})$ is a full triangulated subcategory of $\mathbf{D}(A, \mathsf{M})$.

The notations $\mathbf{C}^\star(A, \mathsf{M})$ and $\mathbf{K}^\star(A, \mathsf{M})$ remain the same, as in Definitions 3.8.20 and 7.3.1. Namely, these are the full subcategories of $\mathbf{C}(A, \mathsf{M})$ and $\mathbf{K}(A, \mathsf{M})$, respectively, on the DG modules M of boundedness type $\star$.

Definition 12.1.7 Let A be a DG ring and M an abelian category. For a DG module $M \in \mathbf{C}(A, \mathsf{M})$ and an integer i, we write $M[i] := \mathrm{T}^i(M)$, the ith translation of M, as in Definition 4.1.4.

This notation applies also to the homotopy category $\mathbf{K}(A, \mathsf{M})$, the derived category $\mathbf{D}(A, \mathsf{M})$, and every other T-additive category.

The notation $M[i]$ makes it difficult to use the little t operator, and to talk about translation isomorphisms, but hopefully we won't require them anymore.

Remark 12.1.8 We should say a few explanatory words on the change in Definition 12.1.6, which conflicts with earlier definitions.

Recall that in Definitions 7.3.3 and 7.3.4 we made a distinction between

$$\mathbf{D}^{\star}(A, \mathsf{M}) = \mathbf{K}^{\star}(A, \mathsf{M})_{\mathbf{S}^{\star}(A,\mathsf{M})}, \tag{12.1.9}$$

the localization of $\mathbf{K}^{\star}(A, \mathsf{M})$ with respect to the quasi-isomorphisms in it, and $\mathbf{D}(A, \mathsf{M})^{\star}$, the full subcategory of $\mathbf{D}(A, \mathsf{M})$ on the complexes of cohomological boundedness type $\star$. This distinction was necessary in the early part of the book, when we were studying foundations and resolutions. Now that the foundations are behind us, we no longer have a need for the subcategory $\mathbf{D}^{\star}(-)$ in the sense of formula (12.1.9), i.e. the definition based on boundedness type.

Still, one might ask: Why make the notational switch from $\mathbf{D}(-)^{\star}$ to $\mathbf{D}^{\star}(-)$, for the definition based on cohomological boundedness type? There are two reasons for that. The first is typographic: the notation $\mathbf{D}(-)^{\star}$ becomes cumbersome when coupled with other indicators, such as $\mathbf{D}_{\mathrm{f}}(-)^{\star}$ or $(\mathbf{D}(-)^{\star})^{\mathrm{op}}$. After Definition 12.1.6 we can refer to these categories by the more elegant expressions $\mathbf{D}^{\star}_{\mathrm{f}}(-)$ or $\mathbf{D}^{\star}(-)^{\mathrm{op}}$, respectively.

Second, the expression $\mathbf{D}(-)^{\star}$ is a novelty, and nobody else uses it, whereas the notation $\mathbf{D}^{\star}(-)$ is common. According to Proposition 7.3.12, in most situations the two distinct meanings of $\mathbf{D}^{\star}(-)$, the old one and the new one, refer to nested subcategories of $\mathbf{D}(-)$ whose inclusion is an equivalence. Furthermore, in a few recent texts on derived categories (such as [144, Definition tag=05RU] and [172, Definition 1.11]) we find the same usage of the expression $\mathbf{D}^{\star}(-)$ as in Definition 12.1.6.

12.2 The Bifunctor RHom

Consider a DG ring A and an abelian category M. Like in Example 9.1.11 we get a DG bifunctor

$$F := \mathrm{Hom}_{A,\mathsf{M}}(-,-) : \mathbf{C}(A, \mathsf{M})^{\mathrm{op}} \times \mathbf{C}(A, \mathsf{M}) \to \mathbf{C}(\mathbb{K}).$$

In Definition 5.6.4 we put a triangulated structure on the opposite homotopy category $\mathbf{K}(A, \mathsf{M})^{\mathrm{op}}$. According to Theorem 9.2.16, there is an induced triangulated bifunctor

$$F = \mathrm{Hom}_{A,\mathsf{M}}(-,-) : \mathbf{K}(A, \mathsf{M})^{\mathrm{op}} \times \mathbf{K}(A, \mathsf{M}) \to \mathbf{K}(\mathbb{K}).$$

Postcomposing with the localization functor $\mathrm{Q} : \mathbf{K}(\mathbb{K}) \to \mathbf{D}(\mathbb{K})$, we obtain a triangulated bifunctor

$$F = \mathrm{Hom}_{A,\mathsf{M}}(-,-) : \mathbf{K}(A, \mathsf{M})^{\mathrm{op}} \times \mathbf{K}(A, \mathsf{M}) \to \mathbf{D}(\mathbb{K}).$$

Next we pick full additive subcategories $\mathsf{K}_1, \mathsf{K}_2 \subseteq \mathbf{K}(A, \mathsf{M})$ s.t. $\mathsf{K}_1^{\mathrm{op}} \subseteq \mathbf{K}(A, \mathsf{M})^{\mathrm{op}}$ and $\mathsf{K}_2 \subseteq \mathbf{K}(A, \mathsf{M})$ are triangulated. In practice this choice would be by some boundedness conditions; for instance $\mathsf{K}_1 := \mathbf{K}^-(\mathsf{M})$ or $\mathsf{K}_2 := \mathbf{K}^+(\mathsf{M})$, cf. Corollaries 11.3.19 and 11.5.26, respectively. We want to construct the right derived bifunctor of the triangulated bifunctor

$$F = \mathrm{Hom}_{A,\mathsf{M}}(-,-) : \mathsf{K}_1^{\mathrm{op}} \times \mathsf{K}_2 \to \mathbf{D}(\mathbb{K}).$$

This is done in the next theorem.

Theorem 12.2.1 *Let A be a DG $\mathbb{K}$-ring, let M be a $\mathbb{K}$-linear abelian category, and let $\mathsf{K}_1, \mathsf{K}_2 \subseteq \mathbf{K}(A, \mathsf{M})$ be full additive subcategories, such that $\mathsf{K}_1^{\mathrm{op}}$ is a full triangulated subcategory of $\mathbf{K}(A, \mathsf{M})^{\mathrm{op}}$, and K_2 is a full triangulated subcategory of $\mathbf{K}(A, \mathsf{M})$. Assume either that K_1 has enough K-projectives, or that K_2 has enough K-injectives. Let D_i denote the localization of K_i with respect to the quasi-isomorphisms in it. Then the triangulated bifunctor*

$$\mathrm{Hom}_{A,\mathsf{M}}(-,-) : \mathsf{K}_1^{\mathrm{op}} \times \mathsf{K}_2 \to \mathbf{D}(\mathbb{K})$$

has a right derived bifunctor

$$\mathrm{RHom}_{A,\mathsf{M}}(-,-) : \mathsf{D}_1^{\mathrm{op}} \times \mathsf{D}_2 \to \mathbf{D}(\mathbb{K}).$$

Moreover, if $P_1 \in \mathsf{K}_1$ is K-projective, or if $I_2 \in \mathsf{K}_2$ is K-injective, then the morphism

$$\eta^{\mathrm{R}}_{P_1,I_2} : \mathrm{Hom}_{A,\mathsf{M}}(P_1, I_2) \to \mathrm{RHom}_{A,\mathsf{M}}(P_1, I_2)$$

in $\mathbf{D}(\mathbb{K})$ is an isomorphism.

Proof If K_2 has enough K-injectives, then we can take $\mathsf{J}_2 := \mathsf{K}_{2,\mathrm{inj}}$, the full subcategory on the K-injectives inside K_2. And we take $\mathsf{J}_1 := \mathsf{K}_1$. We claim that the conditions of Theorem 9.3.11 are satisfied. Condition (b) is simply the assumption that K_2 has enough K-injectives. As for condition (a): this is Lemma 12.2.2 below, with its condition (ii).

On the other hand, if K_1 has enough K-projectives, then we can take $\mathsf{J}_1 := \mathsf{K}_{1,\mathrm{prj}}$, the full subcategory on the K-projectives inside K_1. And we take $\mathsf{J}_2 := \mathsf{K}_2$. We claim that the conditions of Theorem 9.3.11 are satisfied for $\mathsf{J}_1^{\mathrm{op}} \subseteq \mathsf{K}_1^{\mathrm{op}}$. Condition (b) is simply the assumption that K_1 has enough K-projectives: a quasi-isomorphism $\rho : P \to M$ in K_1 becomes a quasi-isomorphism $\mathrm{Op}(\rho) : \mathrm{Op}(M) \to \mathrm{Op}(P)$ in $\mathsf{K}_1^{\mathrm{op}}$, and $\mathrm{Op}(P) \in \mathsf{J}_1^{\mathrm{op}}$. As for condition (a): this is Lemma 12.2.2 below, with its condition (i).

The last assertion also follows from 12.2.2. □

Lemma 12.2.2 *Suppose $\phi_1 : Q_1 \to P_1$ and $\phi_2 : I_2 \to J_2$ are quasi-isomorphisms in $\mathbf{C}(A, \mathsf{M})$, and either of the conditions below holds:*

(i) *Q_1 and P_1 are both K-projective.*

(ii) *I_2 and J_2 are both K-injective.*

Then the homomorphism

$$\mathrm{Hom}_{A,\mathsf{M}}(\phi_1, \phi_2) : \mathrm{Hom}_{A,\mathsf{M}}(P_1, I_2) \to \mathrm{Hom}_{A,\mathsf{M}}(Q_1, J_2)$$

in $\mathbf{C}(\mathbb{K})$ is a quasi-isomorphism.

Proof We will only prove the case where Q_1, P_1 are both K-projective; the other case is very similar.

The homomorphism in question factors as follows:

$$\mathrm{Hom}_{A,\mathsf{M}}(\phi_1, \phi_2) = \mathrm{Hom}_{A,\mathsf{M}}(\phi_1, \mathrm{id}_{J_2}) \circ \mathrm{Hom}_{A,\mathsf{M}}(\mathrm{id}_{P_1}, \phi_2).$$

It suffices to prove that each of the factors is a quasi-isomorphism. This can be done by a messy direct calculation, but we will provide an indirect proof that relies on properties of the homotopy category $\mathsf{K} := \mathbf{K}(A, \mathsf{M})$ that were already established.

Let K_2 be the standard cone on the homomorphism $\phi_2 : I_2 \to J_2$. So K_2 is acyclic. Because P_1 is K-projective, it follows that $\mathrm{Hom}_{A,\mathsf{M}}(P_1, K_2)$ is acyclic. Thus for every integer l we have

$$\mathrm{Hom}_{\mathsf{K}}(P_1[-l], K_2) \cong \mathrm{H}^l(\mathrm{Hom}_{A,\mathsf{M}}(P_1, K_2)) = 0. \tag{12.2.3}$$

Next, there is a distinguished triangle

$$I_2 \xrightarrow{\phi_2} J_2 \xrightarrow{\beta_2} K_2 \xrightarrow{\gamma_2} I_2[1] \tag{12.2.4}$$

in K. Applying the cohomological functor $\mathrm{Hom}_{\mathsf{K}}(P_1[-l], -)$ to the distinguished triangle (12.2.4) yields a long exact sequence, as explained in Section 5.2. From it we deduce that the homomorphisms

$$\mathrm{Hom}_{\mathsf{K}}(P_1[-l], I_2) \to \mathrm{Hom}_{\mathsf{K}}(P_1[-l], J_2)$$

are bijective for all l. Using the isomorphisms like (12.2.3) for I_2 and J_2 we see that

$$\mathrm{Hom}_{A,\mathsf{M}}(\mathrm{id}_{P_1}, \phi_2) : \mathrm{Hom}_{A,\mathsf{M}}(P_1, I_2) \to \mathrm{Hom}_{A,\mathsf{M}}(P_1, J_2)$$

is a quasi-isomorphism.

According to Corollary 10.2.14, the homomorphism $\phi_1 : Q_1 \to P_1$ is a homotopy equivalence; so it is an isomorphism in K. Therefore for every integer l the homomorphism

$$\mathrm{Hom}_{\mathsf{K}}(Q_1, J_2[l]) \to \mathrm{Hom}_{\mathsf{K}}(P_1, J_2[l])$$

is bijective. As above we conclude that

$$\mathrm{Hom}_{A,\mathsf{M}}(\phi_1, \mathrm{id}_{J_2}) : \mathrm{Hom}_{A,\mathsf{M}}(Q_1, J_2) \to \mathrm{Hom}_{A,\mathsf{M}}(P_1, J_2)$$

is a quasi-isomorphism. □

Remark 12.2.5 Theorem 12.2.1 should be viewed as a template. It has commutative and noncommutative variants, which we shall study later. And there are geometric variants in which the source and target are categories of sheaves.

We end this chapter with the connection between RHom and morphisms in the derived category.

Definition 12.2.6 Under the assumptions of Theorem 12.2.1, for DG modules $M_1 \in \mathsf{K}_1$ and $M_2 \in \mathsf{K}_2$, and for an integer i, we write

$$\mathrm{Ext}^i_{A,\mathrm{M}}(M_1, M_2) := \mathrm{H}^i(\mathrm{RHom}_{A,\mathrm{M}}(M_1, M_2)) \in \mathbf{M}(\mathbb{K}).$$

Exercise 12.2.7 Let A be a ring. Prove that for modules $M_1, M_2 \in \mathbf{M}(A)$ the $\mathbb{K}$-module $\mathrm{Ext}^i_A(M_1, M_2)$ defined above is canonically isomorphic to the module in the classical definition. Moreover, this is true regardless of the choices of the subcategories K_1 and K_2, as long as $M_i \in \mathsf{K}_i$.

Corollary 12.2.8 *Under the assumptions of Theorem* 12.2.1, *there is an isomorphism*

$$\mathrm{Ext}^0_{A,\mathrm{M}}(-,-) \xrightarrow{\simeq} \mathrm{Hom}_{\mathbf{D}(A,\mathrm{M})}(-,-)$$

of additive bifunctors $\mathsf{D}_1^{\mathrm{op}} \times \mathsf{D}_2 \to \mathbf{M}(\mathbb{K})$.

Exercise 12.2.9 Prove Corollary 12.2.8. (Hint: use Theorems 10.1.13 and 10.2.9.)

12.3 The Bifunctor $\otimes^{\mathrm{L}}$

Consider a DG ring A. Like in Example 9.1.10 we get a DG bifunctor

$$F := (- \otimes_A -) : \mathbf{C}(A^{\mathrm{op}}) \times \mathbf{C}(A) \to \mathbf{C}(\mathbb{K}).$$

Passing to homotopy categories, and postcomposing with $\mathrm{Q} : \mathbf{K}(\mathbb{K}) \to \mathbf{D}(\mathbb{K})$, we obtain a triangulated bifunctor

$$F = (- \otimes_A -) : \mathbf{K}(A^{\mathrm{op}}) \times \mathbf{K}(A) \to \mathbf{D}(\mathbb{K}).$$

Next we pick full triangulated subcategories $\mathsf{K}_1 \subseteq \mathbf{K}(A^{\mathrm{op}})$ and $\mathsf{K}_2 \subseteq \mathbf{K}(A)$. In practice this choice would be by some boundedness conditions; for instance $\mathsf{K}_1 := \mathbf{C}^-(A^{\mathrm{op}})$ or $\mathsf{K}_2 := \mathbf{C}^-(A)$, cf. Corollary 11.3.19. We want to construct the left derived bifunctor of the triangulated bifunctor

$$F = (- \otimes_A -) : \mathsf{K}_1 \times \mathsf{K}_2 \to \mathbf{D}(\mathbb{K}).$$

This is done in the next theorem.

Theorem 12.3.1 *Let A be a DG $\mathbb{K}$-ring, and let $\mathsf{K}_1 \subseteq \mathbf{K}(A^{\mathrm{op}})$ and $\mathsf{K}_2 \subseteq \mathbf{K}(A)$ be full triangulated subcategories. Assume that either K_1 or K_2 has enough K-flat objects. Let D_i denote the localization of K_i with respect to the quasi-isomorphisms in it. Then the triangulated bifunctor*

$$(- \otimes_A -) : \mathsf{K}_1 \times \mathsf{K}_2 \to \mathbf{D}(\mathbb{K})$$

has a left derived bifunctor

$$(- \otimes^{\mathrm{L}}_A -) : \mathsf{D}_1 \times \mathsf{D}_2 \to \mathbf{D}(\mathbb{K}).$$

Moreover, if either $P_1 \in \mathsf{K}_1$ or $P_2 \in \mathsf{K}_2$ is K-flat, then the morphism

$$\eta^{\mathrm{L}}_{P_1,P_2} : P_1 \otimes^{\mathrm{L}}_A P_2 \to P_1 \otimes_A P_2$$

in $\mathbf{D}(\mathbb{K})$ is an isomorphism.

Proof If K_2 has enough K-flats, then we can take $\mathsf{P}_2 := \mathsf{K}_{2,\mathrm{flat}}$, the full subcategory on the K-flats inside K_2. And we take $\mathsf{P}_1 := \mathsf{K}_1$. We claim that the conditions of Theorem 9.3.16 are satisfied. Condition (b) is simply the assumption that K_2 has enough K-flats. As for condition (a): this is Lemma 12.3.2 below, with its condition (ii).

The other case is proved the same way (but replacing sides). The last assertion also follows from 12.3.2. □

Lemma 12.3.2 *Suppose $\phi_1 : P_1 \to Q_1$ and $\phi_2 : P_2 \to Q_2$ are quasi-isomorphisms in $\mathbf{C}(A^{\mathrm{op}})$ and $\mathbf{C}(A)$, respectively, and either of the conditions below holds:*

(i) *Q_1 and P_1 are both K-flat.*
(ii) *P_2 and Q_2 are both K-flat.*

Then the homomorphism

$$\phi_1 \otimes \phi_2 : P_1 \otimes_A P_2 \to Q_1 \otimes_A Q_2$$

in $\mathbf{C}(\mathbb{K})$ is a quasi-isomorphism.

Proof We will only prove the lemma under condition (i); the other case is very similar. The homomorphism in question factors as follows:

$$\phi_1 \otimes \phi_2 = (\phi_1 \otimes \mathrm{id}_{P_2}) \circ (\mathrm{id}_{P_1} \otimes \phi_2).$$

It suffices to prove that each of the factors is a quasi-isomorphism. This can be done by a messy direct calculation, but we will provide an indirect proof that relies on properties of the DG categories $\mathbf{C}(A^{\mathrm{op}})$ and $\mathbf{C}(A)$ that were already established.

First we shall prove that $\mathrm{id}_{P_1} \otimes \phi_2$ is a quasi-isomorphism. Let R_2 be the standard cone on the strict homomorphism $\phi_2 : P_2 \to Q_2$. So there is a standard triangle

$$P_2 \xrightarrow{\phi_2} Q_2 \to R_2 \to P_2[1] \tag{12.3.3}$$

in $\mathbf{C}_{\mathrm{str}}(A)$, and R_2 is acyclic. Applying the DG functor $P_1 \otimes_A (-)$ to the triangle (12.3.3), and using Theorem 4.5.5, we see that there is a standard triangle

$$P_1 \otimes_A P_2 \xrightarrow{\mathrm{id}_{P_1} \otimes \phi_2} P_1 \otimes_A Q_2 \to P_1 \otimes_A R_2 \to (P_1 \otimes_A P_2)[1] \qquad (12.3.4)$$

in $\mathbf{C}(\mathbb{K})$. This becomes a distinguished triangle in the triangulated category $\mathbf{K}(\mathbb{K})$. Thus there is a long exact sequence in cohomology associated to (12.3.4). Because P_1 is K-flat, it follows that $P_1 \otimes_A R_2$ is acyclic. We conclude that $\mathrm{H}^i(\mathrm{id}_{P_1} \otimes \phi_2)$ is bijective for all i.

Now we shall prove that $\phi_1 \otimes \mathrm{id}_{P_2}$ is a quasi-isomorphism. Let $R_1 \in \mathbf{C}(A^{\mathrm{op}})$ be the standard cone on the homomorphism $\phi_1 : P_1 \to Q_1$. It is both acyclic and K-flat. Using standard triangles like (12.3.3) and (12.3.4) we reduce the problem to showing that $R_1 \otimes_A P_2$ is acyclic. According to Corollary 11.4.26 and Proposition 10.3.4, there is a quasi-isomorphism $\tilde{P}_2 \to P_2$ in $\mathbf{C}(A)$ from some K-flat DG module $\tilde{P}_2$. As already proved in the previous paragraph, since R_1 is K-flat, the homomorphism $R_1 \otimes_A \tilde{P}_2 \to R_1 \otimes_A P_2$ is a quasi-isomorphism. But R_1 is acyclic and $\tilde{P}_2$ is K-flat, and therefore $R_1 \otimes_A \tilde{P}_2$ is acyclic. We conclude that $R_1 \otimes_A P_2$ is acyclic, as required. □

Remark 12.3.5 Theorem 12.3.1 should be viewed as a template. It has commutative and noncommutative variants, which we will talk about later. And there are geometric variants where the source and target are categories of sheaves.

Definition 12.3.6 Under the assumptions of Theorem 12.3.1, for DG modules $M_1 \in \mathsf{K}_1$ and $M_2 \in \mathsf{K}_2$, and for an integer i, we write

$$\mathrm{Tor}_i^A(M_1, M_2) := \mathrm{H}^{-i}(M_1 \otimes_A^{\mathrm{L}} M_2) \in \mathbf{M}(\mathbb{K}).$$

Exercise 12.3.7 Let A be a ring. Prove that for modules $M_1 \in \mathbf{M}(A^{\mathrm{op}})$ and $M_2 \in \mathbf{M}(A)$ the $\mathbb{K}$-module $\mathrm{Tor}_i^A(M_1, M_2)$ defined above is canonically isomorphic to the module in classical definition. Moreover, this is true regardless of the choices of the subcategories K_1 and K_2, as long as $M_i \in \mathsf{K}_i$.

Recall that the category $\mathbf{D}(A)$ admits infinite direct sums.

Proposition 12.3.8 *For every $M \in \mathbf{D}(A^{\mathrm{op}})$ the functor*

$$M \otimes_A^{\mathrm{L}} (-) : \mathbf{D}(A) \to \mathbf{D}(\mathbb{K})$$

commutes with infinite direct sums.

Proof Fix a K-flat resolution $P \to M$ in $\mathbf{C}_{\mathrm{str}}(A^{\mathrm{op}})$, so we have an isomorphism of functors $M \otimes_A^{\mathrm{L}} (-) \cong P \otimes_A (-)$. The functor

$$P \otimes_A (-) : \mathbf{C}_{\mathrm{str}}(A) \to \mathbf{C}_{\mathrm{str}}(\mathbb{K})$$

commutes with infinite direct sums. But by Theorem 10.1.26 the direct sums in $\mathbf{C}_{\mathrm{str}}(-)$ and in $\mathbf{D}(-)$ are the same. □

12.4 Cohomological Dimensions of Functors and Objects

The material here is a refinement of the notion of "way-out functors" from [62, Section I.7]. Most of it is taken from [62] and [172]. As always, there is a fixed base ring $\mathbb{K}$, and Convention 1.2.4 is in force.

Integer intervals and their properties were introduced in Section 12.1.

Definition 12.4.1 Let A, B be DG rings, let M, N be abelian categories, and let $\mathsf{C}_0 \subseteq \mathsf{C} \subseteq \mathbf{D}(A, \mathsf{M})$ be full subcategories.

(1) Let $F : \mathsf{C} \to \mathbf{D}(B, \mathsf{N})$ be an additive functor, and let S be an integer interval. We say that F *has cohomological displacement at most* S *relative to* C_0 if

$$\operatorname{con}(\mathrm{H}(F(M))) \subseteq \operatorname{con}(\mathrm{H}(M)) + S$$

for every $M \in \mathsf{C}_0$.

(2) Let $F : \mathsf{C}^{\mathrm{op}} \to \mathbf{D}(B, \mathsf{N})$ be an additive functor, and let S be an integer interval. We say that F *has cohomological displacement at most* S *relative to* C_0 if

$$\operatorname{con}(\mathrm{H}(F(M))) \subseteq -\operatorname{con}(\mathrm{H}(M)) + S$$

for every $M \in \mathsf{C}_0$.

(3) Let F be as in item (1) or (2). The *cohomological displacement of* F *relative to* C_0 is the smallest integer interval S for which F has cohomological displacement at most S relative to C_0.

(4) Let S be the cohomological displacement of F relative to C_0. The *cohomological dimension of* F *relative to* C_0 is defined to be the length of the integer interval S.

(5) In case $\mathsf{C}_0 = \mathsf{C}$, we omit the clause "relative to C_0" in all items above.

To emphasize the most important case of item (4) of the definition, *the functor* F *has finite cohomological dimension if its cohomological displacement is bounded.*

Example 12.4.2 The functor F is the zero functor iff it has cohomological displacement $\varnothing$ and cohomological dimension $-\infty$.

Example 12.4.3 Let $F : \mathbf{D}(A, \mathsf{M}) \to \mathbf{D}(B, \mathsf{N})$ be an additive functor of finite cohomological dimension. Then for each boundedness indicator $\star$ we have $F(\mathbf{D}^{\star}(A, \mathsf{M})) \subseteq \mathbf{D}^{\star}(B, \mathsf{N})$.

Example 12.4.4 Consider a nonzero commutative ring $A = B$, and the abelian category $\mathsf{M} = \mathsf{N} := \mathbf{M}(\mathbb{K})$. So $\mathbf{D}(A, \mathsf{M}) = \mathbf{D}(B, \mathsf{N}) = \mathbf{D}(A)$. Take $\mathsf{C} := \mathbf{D}(A)$.

For the covariant case (item (1) in Definition 12.4.1) take a nonzero projective module P, and let

$$F := \mathrm{RHom}_A(P \oplus P[1], -) : \mathsf{D}(A) \to \mathsf{D}(A).$$

Then F has cohomological displacement $[0, 1]$. For the contravariant case (item (2)) take a nonzero injective module I, and let

$$F := \mathrm{RHom}_A(-, I \oplus I[1]) : \mathsf{D}(A)^{\mathrm{op}} \to \mathsf{D}(A).$$

Then F has cohomological displacement $[-1, 0]$. In both cases the cohomological dimension of F is 1.

Example 12.4.5 Suppose A and B are rings and $F : \mathsf{M}(A) \to \mathsf{M}(B)$ is a left exact additive functor. We get a triangulated functor $\mathrm{R}F : \mathsf{D}(A) \to \mathsf{D}(B)$, and $\mathrm{H}^i(\mathrm{R}F(M)) = \mathrm{R}^i F(M)$ for all $M \in \mathsf{M}(A)$. Taking $\mathsf{C} := \mathsf{M}(A)$, with its canonical embedding into $\mathsf{D}(A)$, the cohomological dimension of $\mathrm{R}F$ relative to $\mathsf{M}(A)$ equals the usual right cohomological dimension of the functor F.

Next is an example about triangulated functors of finite cohomological dimensions. Another example, of a contravariant triangulated functor of finite cohomological dimension, is the duality functor D associated to a dualizing complex; see Sections 13.1, 17.1 and 18.1.

Example 12.4.6 Let A be a commutative ring, let $\boldsymbol{a} = (a_1, \ldots, a_n)$ be a finite sequence of elements of A, and let $\mathfrak{a} \subseteq A$ be the ideal generated by $\boldsymbol{a}$. The $\mathfrak{a}$-*adic completion functor* $\Lambda_{\mathfrak{a}}$ and the $\mathfrak{a}$-*torsion functor* $\Gamma_{\mathfrak{a}}$ have derived functors $\mathrm{L}\Lambda_{\mathfrak{a}}, \mathrm{R}\Gamma_{\mathfrak{a}} : \mathsf{D}(A) \to \mathsf{D}(A)$.

We call the sequence $\boldsymbol{a}$ *weakly proregular* if the Koszul complexes associated to powers of $\boldsymbol{a}$ satisfy a rather complicated asymptotic formula – see [118, Definition 4.21]. If A is noetherian then weak proregularity is automatic; but this condition is also satisfied in many important non-noetherian cases.

When $\boldsymbol{a}$ is weakly proregular, the functors $\mathrm{L}\Lambda_{\mathfrak{a}}$ and $\mathrm{R}\Gamma_{\mathfrak{a}}$ have finite cohomological dimensions, bounded by the length of $\boldsymbol{a}$. Furthermore, the *commutative MGM Equivalence* holds. To explain it, we need to introduce two full triangulated subcategories of $\mathsf{D}(A)$:

- The category $\mathsf{D}(A)_{\mathrm{com}}$ of *derived* $\mathfrak{a}$-*adically complete complexes*. These are the complexes M such that the canonical morphism $M \to \mathrm{L}\Lambda_{\mathfrak{a}}(M)$ is an isomorphism.
- The category $\mathsf{D}(A)_{\mathrm{tor}}$ of *derived* $\mathfrak{a}$-*torsion complexes*. These are the complexes M such that the canonical morphism $\mathrm{R}\Gamma_{\mathfrak{a}}(M) \to M$ is an isomorphism.

The MGM Equivalence says that the functor $\mathrm{R}\Gamma_{\mathfrak{a}} : \mathsf{D}(A)_{\mathrm{com}} \to \mathsf{D}(A)_{\mathrm{tor}}$ is an equivalence of triangulated categories, with quasi-inverse $\mathrm{L}\Lambda_{\mathfrak{a}}$. The name

"MGM" stands for "Matlis–Greenlees–May." This duality was worked out in the paper [118] by M. Porta, L. Shaul and A. Yekutieli; expanding earlier work by E. Matlis; A. Grothendieck; J. Greenlees and P. May; L. Alonso, A. Jeremias and J. Lipman; and P. Schenzel.

In Section 16.6 we give a noncommutative variant of the MGM Equivalence, adapted from the paper [157] by R. Vyas and Yekutieli.

Remark 12.4.7 Assume that in Definition 12.4.1(1) we take $A = B = \mathbb{K}$, $\mathsf{C} = \mathbf{D}(\mathsf{M})$, and $F : \mathbf{D}(\mathsf{M}) \to \mathbf{D}(\mathsf{N})$ is a triangulated functor. The functor F has bounded below (resp. above) cohomological displacement if and only if it is a *way-out right* (resp. *left*) functor, in the sense of [62, Section I.7, Definition]. For instance, if F is a way-out right functor, with bounding integers n_1 and n_2 (as defined in [62]), then the cohomological displacement of F is contained in the integer interval $[n_1 - n_2, \infty]$. Conversely, if F has cohomological displacement at most $[i_0, \infty]$ for some integer i_0, then F is way-out right, and for every $n_1 \in \mathbb{Z}$ the integer $n_2 := n_1 - i_0$ satisfies the condition in [62]. We find likewise for way-out left functors.

Definition 12.4.8 Let $\star_1, \star_2$ be boundedness indicators, and assume the right derived bifunctor

$$\mathrm{RHom}_{A,\mathsf{M}} : \mathbf{D}^{\star_1}(A, \mathsf{M})^{\mathrm{op}} \times \mathbf{D}^{\star_2}(A, \mathsf{M}) \to \mathbf{D}(\mathbb{K})$$

exists. Let S be an integer interval of length $i \in \mathbb{N} \cup \{\pm\infty\}$.

(1) Let $M \in \mathbf{D}^{\star_1}(A, \mathsf{M})$, and let $\mathsf{C} \subseteq \mathbf{D}^{\star_2}(A, \mathsf{M})$ be a full subcategory. We say that M has *projective concentration* S and *projective dimension* i relative to C if the functor

$$\mathrm{RHom}_{A,\mathsf{M}}(M, -)|_{\mathsf{C}} : \mathsf{C} \to \mathbf{D}(\mathbb{K})$$

has cohomological displacement $-S$.

(2) Let $M \in \mathbf{D}^{\star_2}(A, \mathsf{M})$, and let $\mathsf{C} \subseteq \mathbf{D}^{\star_1}(A, \mathsf{M})$ be a full subcategory. We say that M has *injective concentration* S and *injective dimension* i relative to C if the functor

$$\mathrm{RHom}_{A,\mathsf{M}}(-, M)|_{\mathsf{C}^{\mathrm{op}}} : \mathsf{C}^{\mathrm{op}} \to \mathbf{D}(\mathbb{K})$$

has cohomological displacement S.

(3) If $\mathsf{C} = \mathbf{D}(A, \mathsf{M})$, then we omit the clause "relative to C" in items (1) and (2).

Example 12.4.9 Continuing with the setup of Example 12.4.4, the DG module $P \oplus P[1]$ (resp. $I \oplus I[1]$) has projective (resp. injective) concentration $[-1, 0]$.

Example 12.4.10 Let A be a DG ring such that $\mathrm{H}(A)$ is nonzero, and consider the free DG module $P := A \in \mathbf{D}(A)$. The functor $F := \mathrm{RHom}_A(P, -)$:

$\mathbf{D}(A) \to \mathbf{D}(\mathbb{K})$ is isomorphic to the forgetful functor, so it has cohomological displacement $[0, 0]$ and cohomological dimension 0. Thus the DG module P has projective concentration $[0, 0]$ and projective dimension 0. Note however that the cohomology $\mathrm{H}(P)$ could be unbounded!

Proposition 12.4.11 *Let* M *be an abelian category with enough injectives. The following are equivalent for* $M \in \mathsf{M}$:

(i) *M is an injective object of* M.

(ii) $\mathrm{Ext}^p_{\mathsf{M}}(N, M) = 0$ *for every* $N \in \mathsf{M}$ *and every* $p \geq 1$.

(iii) $\mathrm{Ext}^1_{\mathsf{M}}(N, M) = 0$ *for every* $N \in \mathsf{M}$.

Note that by Corollary 11.5.27 and Definition 12.2.6, the $\mathbb{K}$-modules $\mathrm{Ext}^p_{\mathsf{M}}(N, M)$ exist.

Exercise 12.4.12 Prove Proposition 12.4.11. (Hint: the proof is just like in the case $\mathsf{M} = \mathbf{M}(A)$. It can be found in standard textbooks on homological algebra.)

Proposition 12.4.13 *Let* M *be an abelian category with enough injectives. The following are equivalent for* $M \in \mathbf{D}^+(\mathsf{M})$:

(i) *M has finite injective dimension.*

(ii) *M has finite injective dimension relative to* M.

(iii) *There is a quasi-isomorphism* $M \to I$ *in* $\mathbf{C}_{\mathrm{str}}(\mathsf{M})$ *to a bounded complex of injective A-modules I.*

Note that by Corollary 11.5.27 we can apply Definition 12.4.8 with boundedness type $\star_2 = +$, so we can talk about the injective dimension of M.

Proof

(i) $\Rightarrow$ (ii): This is trivial, since $\mathsf{M} \subseteq \mathbf{D}(\mathsf{M})$.

(ii) $\Rightarrow$ (iii): We may assume that $\mathrm{H}(M)$ is nonzero. Let S be the injective concentration of the complex M relative to M, as in Definition 12.4.8; this is a bounded integer interval (possibly empty, a priory).

Let q be an integer such that $\mathrm{H}^q(M) \neq 0$. Consider the object $N := \mathrm{Z}^q(M) \in \mathsf{M}$ and the canonical monomorphism $N \to M^q$. This can be viewed as a morphism $\phi : N[-q] \to M$ in $\mathbf{C}_{\mathrm{str}}(\mathsf{M})$, and then $\mathrm{H}^q(\phi) : N = \mathrm{H}^q(N[-q]) \to \mathrm{H}^q(M)$ is a nonzero morphism in M. Therefore $\mathrm{Q}(\phi) : N[-q] \to M$ is a nonzero morphism in $\mathbf{D}(\mathsf{M})$, so this is a nonzero element of $\mathrm{H}^q(\mathrm{RHom}_{\mathsf{M}}(N, M))$. Looking at Definition 12.4.8 we see that $q \in S$. Since the interval S is nonempty and bounded, it has to be $S = [q_0, q_1]$ for some integers $q_0 \leq \inf(\mathrm{H}(M))$ and $q_1 \geq \sup(\mathrm{H}(M))$.

According to Corollary 11.5.27, there is quasi-isomorphism $M \to J$, where J is a complex of injective objects of M, and $\inf(J) = \inf(\mathrm{H}(M)) \geq q_0$. Take

$I := \operatorname{smt}^{\leq q_1}(J)$, the smart truncation from Definition 7.3.6. Then the canonical homomorphism $I \to J$ is a quasi-isomorphism. The complex I is concentrated in the integer interval $[q_0, q_1]$, and $I^q = J^q$ is an injective object for all $q < q_1$.

Let us prove that $I^{q_1} = \mathrm{Z}^{q_1}(J)$ is also an injective object of M. Classically we would use a cosyzygy argument. Here we use another trick. Define $I' := \operatorname{stt}^{\leq q_1 - 1}(I)$, the stupid truncation of I from Definition 11.2.11. so

$$I' = (\cdots 0 \to I^{q_0} \to \cdots \to I^{q_1 - 1} \to 0 \to \cdots).$$

This is a bounded complex of injective objects. Consider the short exact sequence

$$0 \to I^{q_1}[-q_1] \to I \to I' \to 0$$

in $\mathbf{C}^+_{\mathrm{str}}(\mathsf{M})$. According to Proposition 7.3.5, this gives a distinguished triangle

$$I^{q_1}[-q_1] \to I \to I' \xrightarrow{\triangle} \tag{12.4.14}$$

in $\mathbf{D}^+(\mathsf{M})$. Take any object $L \in \mathsf{M}$. Applying $\operatorname{RHom}_{\mathsf{M}}(L, -)$ to the distinguished triangle (12.4.14) and then taking cohomologies, we get a long exact sequence

$$\cdots \to \operatorname{Ext}^{q+q_1-1}_{\mathsf{M}}(L, I') \to \operatorname{Ext}^{q}_{\mathsf{M}}(L, I^{q_1}) \to \operatorname{Ext}^{q+q_1}_{\mathsf{M}}(L, I) \to \cdots$$

in $\mathbf{M}(\mathbb{K})$. For every $q > 0$ the $\mathbb{K}$-module $\operatorname{Ext}^{q+q_1-1}_{\mathsf{M}}(L, I')$ vanishes trivially. By the definition of the interval $[q_0, q_1]$, and the existence of an isomorphism $M \cong I$ in $\mathbf{D}(\mathsf{M})$, for every $q > 0$ the $\mathbb{K}$-module $\operatorname{Ext}^{q+q_1}_{\mathsf{M}}(L, I)$ is zero. Hence $\operatorname{Ext}^{q}_{\mathsf{M}}(L, I^{q_1}) = 0$ for all $q > 0$. This proves that the object I^{q_1} is injective (see Proposition 12.4.11).

We have quasi-isomorphisms $M \to J$ and $I \to J$. Since I is K-injective, there is a quasi-isomorphism $M \to I$.

(iii) ⇒ (i): This is also trivial. □

Proposition 12.4.15 *Let* M *be an abelian category with enough projectives. The following are equivalent for* $M \in \mathsf{M}$:

(i) *M is a projective object of* M.

(ii) $\operatorname{Ext}^i_{\mathsf{M}}(M, N) = 0$ *for every* $N \in \mathsf{M}$ *and every* $i \geq 1$.

(iii) $\operatorname{Ext}^1_{\mathsf{M}}(M, N) = 0$ *for every* $N \in \mathsf{M}$.

Note that by Corollary 11.3.20 and Definition 12.2.6, the $\mathbb{K}$-modules $\operatorname{Ext}^i_{\mathsf{M}}(M, N)$ exist.

Proposition 12.4.16 *Let* M *be an abelian category with enough projectives. The following are equivalent for* $M \in \mathbf{D}^-(\mathsf{M})$:

(i) *M has finite projective dimension.*

(ii) *M has finite projective dimension relative to* M.

(iii) *There is a quasi-isomorphism $P \to M$ in $\mathbf{C}_{\mathrm{str}}(\mathsf{M})$ from a bounded complex of projective objects P.*

Note that by Corollary 11.3.20 we can apply Definition 12.4.8 with boundedness type $\star_1 = -$, so we can talk about the projective dimension of M.

Exercise 12.4.17 Prove Propositions 12.4.15 and 12.4.16. (Hint: compare to Propositions 12.4.11 and 12.4.13. Or look in any standard textbook on homological algebra.)

Definition 12.4.18 Let $\star_1, \star_2$ be boundedness indicators, and assume the left derived bifunctor

$$(- \otimes^{\mathrm{L}}_A -) : \mathbf{D}^{\star_1}(A^{\mathrm{op}}) \times \mathbf{D}^{\star_2}(A) \to \mathbf{D}(\mathbb{K})$$

exists. Let S be an integer interval of length $i \in \mathbb{N} \cup \{\pm\infty\}$.

(1) Let $M \in \mathbf{D}^{\star_2}(A)$, and let $\mathsf{C} \subseteq \mathbf{D}^{\star_1}(A^{\mathrm{op}})$ be a full subcategory. We say that M has *flat concentration* S and *flat dimension* i relative to C if the functor $(- \otimes^{\mathrm{L}}_A M)|_{\mathsf{C}} : \mathsf{C} \to \mathbf{D}(\mathbb{K})$ has cohomological displacement S.
(2) If $\mathsf{C} = \mathbf{D}(A^{\mathrm{op}})$, then we omit the clause "relative to C."

Proposition 12.4.19 *Let A be a ring. The following are equivalent for $M \in \mathbf{D}(A)$:*

(i) *M has finite flat dimension.*
(ii) *M has finite flat dimension relative to $\mathbf{M}(A^{\mathrm{op}})$.*
(iii) *There is an isomorphism $P \cong M$ in $\mathbf{D}(A)$, where P is a bounded complex of flat A-modules.*

Exercise 12.4.20 Prove Proposition 12.4.19. (The proof is similar to that of Proposition 12.4.16. It can be found in many standard books on homological algebra.)

Remark 12.4.21 Propositions 12.4.13, 12.4.16 and 12.4.19 do not seem to have analogues for $M \in \mathbf{D}(A)$, when A is a DG ring that is not concentrated in degree 0.

12.5 Theorems on Functors Satisfying Finiteness Conditions

In this section we apply the definitions from the previous section (on cohomological dimensions of functors), combined with noetherian conditions, to prove Theorems 12.5.2, 12.5.7 and 12.5.8. These results (or their analogues) were originally proved in [62, Section I.7].

As always, there is a fixed base ring $\mathbb{K}$, and Convention 1.2.4 is in force.

Definition 12.5.1 Suppose A is a left noetherian ring.

(1) We denote by $\mathbf{M}_{\mathrm{f}}(A)$ the full subcategory of $\mathbf{M}(A) = \operatorname{Mod} A$ on the finitely generated modules.

(2) We denote by $\mathbf{D}_{\mathrm{f}}(A)$ the full subcategory of $\mathbf{D}(A) = \mathbf{D}(\operatorname{Mod} A)$ on the complexes with cohomology modules in $\mathbf{M}_{\mathrm{f}}(A)$.

Because A is left noetherian, the category $\mathbf{M}_{\mathrm{f}}(A)$ is a thick abelian subcategory of $\mathbf{M}(A)$, and the category $\mathbf{D}_{\mathrm{f}}(A)$ is a full triangulated subcategory of $\mathbf{D}(A)$. When viewed as a left module, $A \in \mathbf{M}_{\mathrm{f}}(A) \subseteq \mathbf{D}^{\mathrm{b}}_{\mathrm{f}}(A)$. Note the similarity to Definition 11.4.37.

Theorem 12.5.2 *Let A be a left noetherian ring, let N be an abelian category, let $\star$ be a boundedness indicator, let $F, G : \mathbf{D}^{\star}_{\mathrm{f}}(A) \to \mathbf{D}(\mathsf{N})$ be triangulated functors, and let $\zeta : F \to G$ be a morphism of triangulated functors. Assume that the morphism $\zeta_A : F(A) \to G(A)$ in $\mathbf{D}(\mathsf{N})$ is an isomorphism.*

(1) *If $\star = -$, and if F and G have bounded above cohomological displacements, then $\zeta_M : F(M) \to G(M)$ is an isomorphism for every $M \in \mathbf{D}^{-}_{\mathrm{f}}(A)$.*

(2) *If $\star = \langle\text{empty}\rangle$, and if F and G have bounded cohomological displacements, then ζ_M is an isomorphism for every $M \in \mathbf{D}_{\mathrm{f}}(A)$.*

We shall require the next lemmas for the proof of the theorem.

Lemma 12.5.3 *Let D be a triangulated category, let $F, G : \mathsf{D} \to \mathbf{D}(\mathsf{N})$ be triangulated functors, let $\zeta : F \to G$ be a morphism of triangulated functors, and let $L \xrightarrow{\phi} M \to N \xrightarrow{\triangle}$ be a distinguished triangle in D.*

(1) *If the morphisms ζ_L and ζ_M are both isomorphisms, then ζ_N is an isomorphism.*

(2) *Let j be an integer. If $\mathrm{H}^{j-1}(F(N))$, $\mathrm{H}^{j-1}(G(N))$, $\mathrm{H}^{j}(F(N))$ and $\mathrm{H}^{j}(G(N))$ are all zero, and if $\mathrm{H}^{j}(\zeta_L)$ is an isomorphism, then $\mathrm{H}^{j}(\zeta_M)$ is an isomorphism.*

Proof (1) In $\mathbf{D}(\mathsf{N})$ we get the commutative diagram

$$\begin{array}{ccccccc} F(L) & \longrightarrow & F(M) & \longrightarrow & F(N) & \longrightarrow & F(L)[1] \\ \downarrow{\scriptstyle \zeta_L} & & \downarrow{\scriptstyle \zeta_M} & & \downarrow{\scriptstyle \zeta_N} & & \downarrow{\scriptstyle \zeta_L[1]} \\ G(L) & \longrightarrow & G(M) & \longrightarrow & G(N) & \longrightarrow & G(L)[1] \end{array} \tag{12.5.4}$$

with horizontal distinguished triangles. According to Proposition 5.3.2, ζ_N is an isomorphism.

(2) Passing to cohomologies in (12.5.4) we have a commutative diagram with exact rows

$$\begin{array}{ccccccc}
\mathrm{H}^{j-1}(F(N)) & \longrightarrow & \mathrm{H}^{j}(F(L)) & \xrightarrow{\mathrm{H}^{j}(F(\phi))} & \mathrm{H}^{j}(F(M)) & \longrightarrow & \mathrm{H}^{j}(F(N)) \\
\downarrow{\scriptstyle \mathrm{H}^{j-1}(\zeta_N)} & & \downarrow{\scriptstyle \mathrm{H}^{j}(\zeta_L)} & & \downarrow{\scriptstyle \mathrm{H}^{j}(\zeta_M)} & & \downarrow{\scriptstyle \mathrm{H}^{j}(\zeta_N)} \\
\mathrm{H}^{j-1}(G(N)) & \longrightarrow & \mathrm{H}^{j}(G(L)) & \xrightarrow{\mathrm{H}^{j}(G(\phi))} & \mathrm{H}^{j}(G(M)) & \longrightarrow & \mathrm{H}^{j}(G(N))
\end{array}$$

in N. The vanishing assumption implies that $\mathrm{H}^j(F(\phi))$ and $\mathrm{H}^j(G(\phi))$ are isomorphisms. Hence $\mathrm{H}^j(\zeta_M)$ is an isomorphism. □

Lemma 12.5.5 *Let D be a triangulated category, let $F, G : \mathsf{D} \to \mathbf{D}(\mathsf{N})$ be triangulated functors, and let $\zeta : F \to G$ be a a morphism of triangulated functors. The following conditions are equivalent for $M \in \mathsf{D}$:*

(i) *ζ_M is an isomorphism.*

(ii) *$\zeta_{M[i]}$ is an isomorphism for every integer i.*

(iii) *The morphism $\mathrm{H}^j(\zeta_M) : \mathrm{H}^j(F(M)) \to \mathrm{H}^j(G(M))$ is an isomorphism for every integer j.*

Proof The equivalence (i) ⇔ (ii) is because both F and G are triangulated functors. The equivalence (i) ⇔ (iii) is because the functor $\mathrm{H} : \mathbf{D}(\mathsf{N}) \to \mathbf{G}_{\mathrm{str}}(\mathsf{N})$ is conservative; see Corollary 7.2.13. □

Proof of Theorem 12.5.2

(1) First assume P is a bounded complex of finitely generated free A-modules. Then P is obtained from A by finitely many standard cones and translations. By Lemmas 12.5.3(1) and 12.5.5 it follows that ζ_P is an isomorphism.

Next let P be a bounded above complex of finitely generated free A-modules. Choose some integer j. Let i_1 be an integer such that the integer interval $[-\infty, i_1]$ contains the cohomological displacements of F and G. Define $P' := \mathrm{stt}^{\leq j-i_1-2}(P)$, the stupid truncation of P below $j - i_1 - 2$; and let $P'' := \mathrm{stt}^{\geq j-i_1-1}(P)$, the complementary stupid truncation. See Definition 11.2.11. According to Proposition 7.3.5, the short exact sequence (11.2.12) gives a distinguished triangle $P'' \to P \to P' \xrightarrow{\triangle}$ in $\mathbf{D}^-_{\mathrm{f}}(A)$. The complex P'' is a bounded complex of finitely generated free A-modules, so we already know that $\zeta_{P''}$ is an isomorphism. Hence $\mathrm{H}^j(\zeta_{P''})$ is an isomorphism. On the other hand $\mathrm{H}(P')$ is concentrated in the degree interval $[-\infty, j - i_1 - 2]$. Therefore $\mathrm{H}^k(F(P')) = \mathrm{H}^k(G(P')) = 0$ for all $k \geq j - 1$. By Lemma 12.5.3(2), $\mathrm{H}^j(\zeta_P)$ is an isomorphism. Because j is arbitrary, Lemma 12.5.5 says that ζ_P is an isomorphism.

Now take an arbitrary $M \in \mathbf{D}^-_{\mathrm{f}}(A)$. By Theorem 11.4.40 and Example 11.4.44 there is a resolution $P \to M$, where P is a bounded above complex of finitely generated free A-modules. Since ζ_P is an isomorphism, so is ζ_M.

(2) Now we assume that the functors F and G have finite cohomological dimensions. Take any complex $M \in \mathbf{D}_{\mathrm{f}}(A)$. By Lemma 12.5.5 it suffices to prove that $\mathrm{H}^j(\zeta_M)$ is an isomorphism for every integer j.

Let $[i_0, i_1]$ be a bounded integer interval that contains the cohomological displacements of the functors F and G. Define $M'' := \mathrm{smt}^{\leq j-i_0}(M)$, the smart truncation of M below $j-i_0$; and let $M' := \mathrm{smt}^{\geq j-i_0+1}(M)$, the complementary smart truncation. See Definition 7.3.6. According to Proposition 7.3.10, there is a distinguished triangle $M'' \to M \to M' \xrightarrow{\triangle}$ in $\mathbf{D}_{\mathrm{f}}(A)$. The cohomologies of these complexes satisfy $\mathrm{H}^i(M'') = \mathrm{H}^i(M)$ and $\mathrm{H}^i(M') = 0$ for $i \leq j - i_0$; and $\mathrm{H}^i(M'') = 0$ and $\mathrm{H}^i(M') = \mathrm{H}^i(M)$ for $i \geq j - i_0 + 1$. Note that $M'' \in \mathbf{D}^-_{\mathrm{f}}(A)$.

By part (1) we know that $\zeta_{M''}$ is an isomorphism, and therefore also $\mathrm{H}^j(\zeta_{M''})$ is an isomorphism. The cohomology $\mathrm{H}(M')$ is concentrated in the degree interval $[j - i_0 + 1, \infty]$, and therefore the cohomologies $\mathrm{H}(F(M'))$ and $\mathrm{H}(G(M'))$ are concentrated in the interval $[j + 1, \infty]$. In particular the objects $\mathrm{H}^{j-1}(F(M'))$, $\mathrm{H}^{j-1}(G(M'))$, $\mathrm{H}^j(F(M'))$ and $\mathrm{H}^j(G(M'))$ are zero. By Lemma 12.5.3(2), $\mathrm{H}^j(\zeta_M)$ is an isomorphism. □

The triangulated structure of $\mathbf{D}(A)^{\mathrm{op}}$ was introduced in Section 7.6. By our notational conventions, $\mathbf{D}_{\mathrm{f}}(A)^{\mathrm{op}}$ is the full subcategory of $\mathbf{D}(A)^{\mathrm{op}}$ on the complexes with finitely generated cohomology modules, and it is triangulated, by Propositions 7.6.2 and 7.6.3.

Exercise 12.5.6 State and prove the contravariant modification of Theorem 12.5.2. (Hint: study the proofs of Theorems 12.5.2 and 12.5.7.)

Theorem 12.5.7 *Let A be a left noetherian ring, let N be an abelian category, let $\mathsf{N}_0 \subseteq \mathsf{N}$ be a thick abelian subcategory, let $\star$ be a boundedness condition, and let $F : \mathbf{D}^{\star}_{\mathrm{f}}(A)^{\mathrm{op}} \to \mathbf{D}(\mathsf{N})$ be a triangulated functor. Assume that $F(A)$ belongs to $\mathbf{D}_{\mathsf{N}_0}(\mathsf{N})$.*

(1) *If $\star = -$, and if F has bounded below cohomological displacement, then $F(M)$ belongs to $\mathbf{D}_{\mathsf{N}_0}(\mathsf{N})$ for every $M \in \mathbf{D}^-_{\mathrm{f}}(A)$.*
(2) *If $\star = \langle\text{empty}\rangle$, and if F has bounded cohomological displacement, then $F(M)$ belongs to $\mathbf{D}_{\mathsf{N}_0}(\mathsf{N})$ for every $M \in \mathbf{D}_{\mathrm{f}}(A)$.*

Proof

(1) First assume P is a bounded complex of finitely generated free A-modules. The complex P is obtained from the free module A by finitely many extensions and translations in $\mathbf{C}_{\mathrm{str}}(A)^{\mathrm{op}}$. These come either from stupid truncations

as in (11.2.12) or from breaking up finite direct sums. According to Proposition 8.5.9, P is obtained from A by finitely many standard cones and translations in $\mathbf{D}^{\star}_{\mathrm{f}}(A)^{\mathrm{op}}$. Since $\mathbf{D}_{\mathsf{N}_0}(\mathsf{N})$ is a full triangulated subcategory and F is a triangulated functor, it follows that $F(P) \in \mathbf{D}_{\mathsf{N}_0}(\mathsf{N})$.

Next let P be a bounded above complex of finitely generated free A-modules. Choose some integer j. We want to prove that $\mathrm{H}^j(F(P)) \in \mathsf{N}_0$. Let i_0 be an integer such that the integer interval $[i_0, \infty]$ contains the cohomological displacement of F. Define $P' := \mathrm{stt}^{\leq -j-1+i_0}(P)$, the stupid truncation of P below $-j-1+i_0$; and let $P'' := \mathrm{stt}^{\geq j+i_0}(P)$, the complementary stupid truncation. These truncations are done in the category $\mathbf{C}_{\mathrm{str}}(A)$. The short exact sequence (11.2.12) gives, upon applying Op, a short exact sequence in $\mathbf{C}_{\mathrm{str}}(A)^{\mathrm{op}}$. According to Proposition 8.5.9, there is a distinguished triangle $P' \to P \to P'' \xrightarrow{\triangle}$ in $\mathbf{D}^{-}_{\mathrm{f}}(A)^{\mathrm{op}}$. Since F is a triangulated functor, there is a distinguished triangle $F(P') \to F(P) \to F(P'') \xrightarrow{\triangle}$ in $\mathbf{D}(\mathsf{N})$. The complex P'' is a bounded complex of finitely generated free A-modules, so we already know that $F(P'') \in \mathbf{D}_{\mathsf{N}_0}(\mathsf{N})$, and in particular $\mathrm{H}^j(F(P'')) \in \mathsf{N}_0$. On the other hand $\mathrm{H}(P')$ is concentrated in the interval $[-\infty, -j-1+i_0]$. Therefore $\mathrm{H}(F(P'))$ is concentrated in the interval $[j+1, \infty]$, and in particular $\mathrm{H}^{j-1}(F(P')) = \mathrm{H}^j(F(P')) = 0$. As we saw in the proof of Lemma 12.5.3(2), $\mathrm{H}^j(F(P'')) \to \mathrm{H}^j(F(P))$ is an isomorphism. The conclusion is that $\mathrm{H}^j(F(P)) \in \mathsf{N}_0$.

Now take an arbitrary $M \in \mathbf{D}^{-}_{\mathrm{f}}(A)$. There is a quasi-isomorphism $P \to M$, where P is a bounded above complex of finitely generated free A-modules. So $F(M) \cong F(P)$, and thus $F(M) \in \mathbf{D}_{\mathsf{N}_0}(\mathsf{N})$.

(2) Now we assume that the functor F has finite cohomological dimension. Take any complex $M \in \mathbf{D}_{\mathrm{f}}(A)$. We want to prove that for every $j \in \mathbb{Z}$ the object $\mathrm{H}^j(F(M))$ lies in N_0.

Let $[i_0, i_1]$ be a bounded integer interval that contains the cohomological displacement of the functor F. Define $M'' := \mathrm{smt}^{\leq -j+1+i_1}(M)$, the smart truncation of M below $-j+1+i_1$; and let $M' := \mathrm{smt}^{\geq -j+2+i_1}(M)$, the complementary smart truncation. These truncations are preformed in the category $\mathbf{C}_{\mathrm{str}}(A)$. By Proposition 8.5.10 there is a distinguished triangle $M' \to M \to M'' \xrightarrow{\triangle}$ in $\mathbf{D}_{\mathrm{f}}(A)^{\mathrm{op}}$. Since F is a triangulated functor, there is a distinguished triangle $F(M') \to F(M) \to F(M'') \xrightarrow{\triangle}$ in $\mathbf{D}(\mathsf{N})$. The cohomology of M' is concentrated in the integer interval $[-j+2+i_1, \infty]$, and therefore the cohomology of $F(M')$ is concentrated in the interval $[-\infty, j-2]$. In particular the objects $\mathrm{H}^{j-1}(F(M'))$ and $\mathrm{H}^j(F(M'))$ are zero. By the proof of Lemma 12.5.3(2), the morphism $\mathrm{H}^j(F(M'')) \to \mathrm{H}^j(F(M))$ is an isomorphism. But $M'' \in \mathbf{D}^{-}_{\mathrm{f}}(A)$, so as we proved in part (1), its cohomologies are inside N_0. □

Here is the covariant version of Theorem 12.5.7.

Theorem 12.5.8 *Let A be a left noetherian ring, let N be an abelian category, let $\mathsf{N}_0 \subseteq \mathsf{N}$ be a thick abelian subcategory, let $\star$ be a boundedness indicator, and let $F : \mathbf{D}^{\star}_{\mathrm{f}}(A) \to \mathbf{D}(\mathsf{N})$ be a triangulated functor. Assume that $F(A)$ belongs to $\mathbf{D}_{\mathsf{N}_0}(\mathsf{N})$.*

(1) *If $\star = -$, and if F has bounded above cohomological displacement, then $F(M)$ belongs to $\mathbf{D}_{\mathsf{N}_0}(\mathsf{N})$ for every $M \in \mathbf{D}^{-}_{\mathrm{f}}(A)$.*
(2) *If $\star = \langle \text{empty} \rangle$, and if F has bounded cohomological displacement, then $F(M)$ belongs to $\mathbf{D}_{\mathsf{N}_0}(\mathsf{N})$ for every $M \in \mathbf{D}_{\mathrm{f}}(A)$.*

Exercise 12.5.9 Prove Theorem 12.5.8. (Hint: study the proofs of Theorems 12.5.2 and 12.5.7.)

Remark 12.5.10 In Theorems 12.5.2, 12.5.7 and 12.5.8 the source category is $\mathbf{D}^{\star}_{\mathrm{f}}(A)$, for a left noetherian ring A. There are variants of these theorems in which the source category is $\mathbf{D}^{\star}_{\mathrm{f}}(A)$, for a left pseudo-noetherian DG ring A; and also with source category $\mathbf{D}^{\star}_{\mathsf{M}_0}(\mathsf{M})$ for suitable abelian categories $\mathsf{M}_0 \subseteq \mathsf{M}$. We leave it to the reader to investigate these variations.

12.6 Derived Restriction, Induction and Coinduction Functors

As before, there is a nonzero commutative base ring $\mathbb{K}$. All DG rings are $\mathbb{K}$-central, and all homomorphisms between them are over $\mathbb{K}$.

Definition 12.6.1 Let $u : A \to B$ be a DG ring homomorphism.

(1) The *restriction functor* is the $\mathbb{K}$-linear DG functor $\mathrm{Rest}_u = \mathrm{Rest}_{B/A} : \mathbf{C}(B) \to \mathbf{C}(A)$, that is the identity on the underlying DG $\mathbb{K}$-modules, and the A-action is via u.
(2) Since the functor Rest_u is exact, it extends to a triangulated functor $\mathrm{Rest}_u : \mathbf{D}(B) \to \mathbf{D}(A)$.

Definition 12.6.2 Let $u : A \to B$ be a DG ring homomorphism.

(1) The *induction functor* is the DG functor $\mathrm{Ind}_u = \mathrm{Ind}_{B/A} : \mathbf{C}(A) \to \mathbf{C}(B)$ with formula $\mathrm{Ind}_u := B \otimes_A (-)$.
(2) The *derived induction functor* $\mathrm{LInd}_u = \mathrm{LInd}_{B/A} : \mathbf{D}(A) \to \mathbf{D}(B)$ is the triangulated left derived functor of Ind_u, namely $\mathrm{LInd}_u := B \otimes^{\mathrm{L}}_A (-)$.

For every $M \in \mathbf{D}(A)$ there is the canonical morphism $\eta^{\mathrm{L}}_M : \mathrm{LInd}_u(M) \to \mathrm{Ind}_u(M)$ in $\mathbf{D}(B)$, which is part of the left derived functor.

Definition 12.6.3 Let $u : A \to B$ be a DG ring homomorphism, and let $M \in \mathsf{D}(A)$ and $N \in \mathsf{D}(B)$ be DG modules. A morphism $\lambda : M \to \operatorname{Rest}_u(N)$ in $\mathsf{D}(A)$ is called a *forward morphism in* $\mathsf{D}(A)$ *over* u.

Similarly there are forward morphisms over u in the categories $\mathsf{C}(A)$, $\mathsf{C}_{\mathrm{str}}(A)$ and $\mathsf{K}(A)$.

We often omit the functor Rest_u, and just talk about a forward morphism $\lambda : M \to N$ in $\mathsf{D}(A)$ over u, etc.

For any DG module $M \in \mathsf{C}(A)$ there is a canonical forward homomorphism

$$\mathrm{q}_{u,M} : M \to B \otimes_A M = \operatorname{Ind}_u(M), \quad \mathrm{q}_{u,M}(m) := 1_B \otimes m \tag{12.6.4}$$

in $\mathsf{C}_{\mathrm{str}}(A)$ over u. Now let $N \in \mathsf{C}(B)$. The usual change of ring adjunction formula gives a canonical isomorphism

$$\operatorname{fadj}_{u,M,N} : \operatorname{Hom}_{\mathsf{C}(A)}(M, N) \xrightarrow{\simeq} \operatorname{Hom}_{\mathsf{C}(B)}(B \otimes_A M, N) \tag{12.6.5}$$

in $\mathsf{C}_{\mathrm{str}}(\mathbb{K})$. It is characterized by the property that for every forward morphism $\lambda : M \to N$ there is equality $\operatorname{fadj}_{u,M,N}(\lambda) \circ \mathrm{q}_{u,M} = \lambda$. We refer to the isomorphism $\operatorname{fadj}_{u,M,N}$ as *forward adjunction*. Since the isomorphism $\operatorname{fadj}_{u,M,N}$ is functorial in M and N, we see that the functor Ind_u is a left adjoint of Rest_u.

The next theorem shows that this is also true on the derived level.

Theorem 12.6.6 *Let $u : A \to B$ be a homomorphism of DG $\mathbb{K}$-rings.*

(1) *For each $M \in \mathsf{D}(A)$ there is a unique forward morphism*

$$\mathrm{q}^{\mathrm{L}}_{u,M} : M \to B \otimes^{\mathrm{L}}_A M = \operatorname{LInd}_u(M)$$

in $\mathsf{D}(A)$ over u, called the canonical forward morphism, *which is functorial in M, and satisfies $\eta^{\mathrm{L}}_M \circ \mathrm{q}^{\mathrm{L}}_{u,M} = \mathrm{q}_{u,M}$.*

(2) *For each $M \in \mathsf{D}(A)$ and $N \in \mathsf{D}(B)$ there is a unique $\mathbb{K}$-linear isomorphism*

$$\operatorname{fadj}^{\mathrm{L}}_{u,M,N} : \operatorname{Hom}_{\mathsf{D}(A)}(M, \operatorname{Rest}_u(N)) \xrightarrow{\simeq} \operatorname{Hom}_{\mathsf{D}(B)}(\operatorname{LInd}_u(M), N),$$

called derived forward adjunction, *such that $\operatorname{fadj}^{\mathrm{L}}_{u,M,N}(\lambda) \circ \mathrm{q}^{\mathrm{L}}_{u,M} = \lambda$ for all forward morphisms $\lambda : M \to N$ in $\mathsf{D}(A)$ over u. The isomorphism $\operatorname{fadj}^{\mathrm{L}}_{u,M,N}$ is functorial in M and N.*

(3) *The functor* $\operatorname{LInd}_u : \mathsf{D}(A) \to \mathsf{D}(B)$ *is left adjoint to* Rest_u.

Here are the commutative diagrams in $\mathsf{D}(A)$ illustrating the theorem.

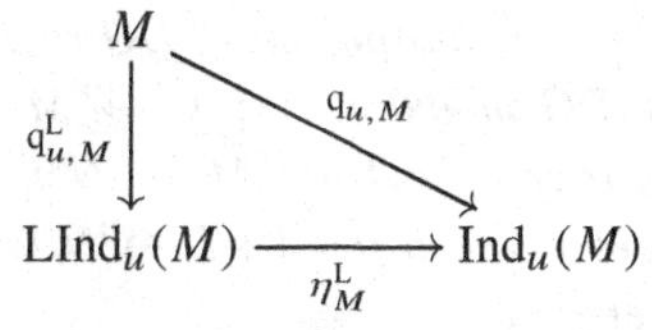

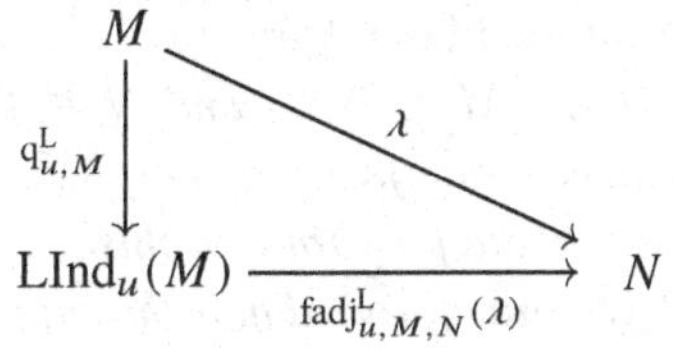

Proof

(1) Let us choose a system of K-projective resolutions in $\mathbf{C}(A)$, as in Definition 10.2.16, and use it to present the derived functor LInd_u. To construct an isomorphism $\mathrm{q}^{\mathrm{L}}_{u,M}$ which is functorial in M, it suffices to consider a K-projective DG A-module $M = P$. But then $\mathrm{LInd}_u(P) = \mathrm{Ind}_u(P)$, $\eta^{\mathrm{L}}_P = \mathrm{id}_P$, and we have no choice but to define $\mathrm{q}^{\mathrm{L}}_{u,P} := \mathrm{q}_{u,P}$.

(2) We keep the system of K-projective resolutions. As in item (1), it is enough to say what is $\mathrm{fadj}^{\mathrm{L}}_{u,P,N}$ for a K-projective DG A-module P. Then there is a canonical isomorphism

$$\mathrm{Q}_A : \mathrm{Hom}_{\mathbf{K}(A)}(P, N) \xrightarrow{\simeq} \mathrm{Hom}_{\mathbf{D}(A)}(P, N).$$

Since $B \otimes_A P = \mathrm{LInd}_u(P)$ is a K-projective DG B-module, we also have a canonical isomorphism

$$\mathrm{Q}_B : \mathrm{Hom}_{\mathbf{K}(B)}(B \otimes_A P, N) \xrightarrow{\simeq} \mathrm{Hom}_{\mathbf{D}(B)}(B \otimes_A P, N).$$

But the usual adjunction formula (12.6.5) induces, by passing to 0th cohomology on both sides, an isomorphism

$$\mathrm{H}^0(\mathrm{fadj}_{u,P,N}) : \mathrm{Hom}_{\mathbf{K}(A)}(P, N) \xrightarrow{\simeq} \mathrm{Hom}_{\mathbf{K}(B)}(B \otimes_A P, N),$$

and it satisfies $\mathrm{H}^0(\mathrm{fadj}_{u,P,N})(\lambda) \circ \mathrm{q}_{u,P} = \lambda$ for any forward morphism $\lambda : P \to N$ in $\mathbf{K}(A)$. We define

$$\mathrm{fadj}^{\mathrm{L}}_{u,P,N} := \mathrm{Q}_B \circ \mathrm{H}^0(\mathrm{fadj}_{u,P,N}) \circ \mathrm{Q}_A^{-1}.$$

(3) Since $\mathrm{fadj}^{\mathrm{L}}_{u,M,N}$ is a bifunctorial isomorphism, it determines an adjunction. □

Definition 12.6.7 Let $u : A \to B$ be a DG ring homomorphism, and let $M \in \mathbf{D}(A)$ and $N \in \mathbf{D}(B)$ be DG modules. A forward morphism $\lambda : M \to N$ in $\mathbf{D}(A)$ over u is called a *nondegenerate forward morphism* if the corresponding morphism

$$\mathrm{fadj}^{\mathrm{L}}_{u,M,N}(\lambda) : B \otimes^{\mathrm{L}}_A M = \mathrm{LInd}_u(M) \to N$$

in $\mathbf{D}(B)$ is an isomorphism.

Example 12.6.8 Given $M \in \mathbf{D}(A)$, let $N := B \otimes^{\mathrm{L}}_A M \in \mathbf{D}(B)$. The canonical derived forward morphism $\mathrm{q}^{\mathrm{L}}_{u,M} : M \to N$ is a nondegenerate forward morphism in $\mathbf{D}(A)$ over u, because $\mathrm{fadj}^{\mathrm{L}}_{u,M,N}(\mathrm{q}^{\mathrm{L}}_{u,M}) = \mathrm{id}_N$.

Proposition 12.6.9 *Suppose $A \xrightarrow{u} B \xrightarrow{v} C$ are homomorphisms of DG rings, $L \in \mathbf{D}(A)$, $M \in \mathbf{D}(B)$ and $N \in \mathbf{D}(C)$ are DG modules, $\lambda : L \to M$ is a nondegenerate forward morphism in $\mathbf{D}(A)$ over u, and $\mu : M \to N$ is a nondegenerate forward morphism in $\mathbf{D}(B)$ over v. Then $\mu \circ \lambda : L \to N$ is a nondegenerate forward morphism in $\mathbf{D}(A)$ over $v \circ u$.*

Proof Since λ is a nondegenerate forward morphism, it induces an isomorphism $N \cong B \otimes^{\mathrm{L}}_A M$. We can thus assume that $N = B \otimes^{\mathrm{L}}_A M$ and that $\lambda = \mathrm{q}^{\mathrm{L}}_{u,M} : M \to B \otimes^{\mathrm{L}}_A M$. Similarly, we can assume that $\mu = \mathrm{q}^{\mathrm{L}}_{v,N} : N \to C \otimes^{\mathrm{L}}_B N$. But now $\mu \circ \lambda = \mathrm{q}^{\mathrm{L}}_{v,N} \circ \mathrm{q}^{\mathrm{L}}_{u,M} = \mathrm{q}^{\mathrm{L}}_{v \circ u,L} : L \to N$, and this is known to be a nondegenerate forward morphism (see Example 12.6.8). □

Definition 12.6.10 Let $u : A \to B$ be a homomorphism of DG $\mathbb{K}$-rings.

(1) The *coinduction functor* is the $\mathbb{K}$-linear DG functor $\mathrm{CInd}_u = \mathrm{CInd}_{B/A} : \mathbf{C}(A) \to \mathbf{C}(B)$ with formula $\mathrm{CInd}_u := \mathrm{Hom}_A(B, -)$.

(2) The *derived coinduction functor* $\mathrm{RCInd}_u = \mathrm{RCInd}_{B/A} : \mathbf{D}(A) \to \mathbf{D}(B)$ is the triangulated right derived functor of CInd_u.

Thus for each $M \in \mathbf{D}(A)$ we have $\mathrm{RCInd}_u(M) = \mathrm{RHom}_A(B, M)$. There is the canonical morphism $\eta^{\mathrm{R}}_M : \mathrm{CInd}_u(M) \to \mathrm{RCInd}_u(M)$ in $\mathbf{D}(B)$, which is part of the right derived functor,

Definition 12.6.11 Let $u : A \to B$ be a DG ring homomorphism, and let $M \in \mathbf{D}(A)$ and $N \in \mathbf{D}(B)$ be DG modules. A morphism $\theta : \mathrm{Rest}_u(N) \to M$ in $\mathbf{D}(A)$ is called a *backward* (or *trace*) *morphism in* $\mathbf{D}(A)$ *over* u.

Similarly there are backward morphisms over u in the categories $\mathbf{C}(A)$, $\mathbf{C}_{\mathrm{str}}(A)$ and $\mathbf{K}(A)$.

We shall use the terms *backward morphism* and *trace morphism* synonymously. We often omit the functor Rest_u, and just talk about a backward morphism $\theta : N \to M$ in $\mathbf{D}(A)$ over u.

For any DG module $M \in \mathbf{C}(A)$ there is a canonical backward homomorphism $\mathrm{tr}_{u,M} : \mathrm{CInd}_u(M) = \mathrm{Hom}_A(B, M) \to M$, $\mathrm{tr}_{u,M}(\phi) := \phi(1_B)$ in $\mathbf{C}_{\mathrm{str}}(A)$ over u. Now let $N \in \mathbf{C}(B)$. The usual change of ring adjunction formula gives a canonical isomorphism

$$\mathrm{badj}_{u,M,N} : \mathrm{Hom}_{\mathbf{C}(A)}(N, M) \xrightarrow{\simeq} \mathrm{Hom}_{\mathbf{C}(B)}(N, \mathrm{Hom}_A(B, M))$$

in $\mathbf{C}_{\mathrm{str}}(\mathbb{K})$. It is characterized by the property that for every backward morphism $\theta : N \to M$ there is equality $\mathrm{tr}_{u,M} \circ \mathrm{badj}_{u,M,N}(\theta) = \theta$. We refer to the isomorphism $\mathrm{badj}_{u,M,N}$ as *backward adjunction*. Since the isomorphism $\mathrm{badj}_{u,M,N}$ is functorial in M and N, we see that the functor CInd_u is a right adjoint of Rest_u.

The next theorem shows that this is also true on the derived level.

Theorem 12.6.12 *Let $u : A \to B$ be a homomorphism of DG $\mathbb{K}$-rings.*

(1) *For each $M \in \mathbf{D}(A)$ there is a unique backward morphism*

$$\mathrm{tr}^{\mathrm{R}}_{u,M} : \mathrm{RCInd}_u(M) = \mathrm{RHom}_A(B, M) \to M$$

in $\mathbf{D}(A)$ over u, called the canonical backward morphism, *which is functorial in M, and satisfies $\mathrm{tr}^{\mathrm{R}}_{u,M} \circ \eta^{\mathrm{R}}_M = \mathrm{tr}_{u,M}$.*

(2) *For each* $M \in \mathsf{D}(A)$ *and* $N \in \mathsf{D}(B)$ *there is a unique* $\mathbb{K}$*-linear isomorphism*

$$\operatorname{badj}^{\mathrm{R}}_{u,M,N} : \operatorname{Hom}_{\mathsf{D}(A)}(\operatorname{Rest}_u(N), M) \xrightarrow{\simeq} \operatorname{Hom}_{\mathsf{D}(B)}(N, \operatorname{RCInd}_u(M)),$$

called derived backward adjunction, *such that* $\operatorname{tr}^{\mathrm{R}}_{u,M} \circ \operatorname{badj}^{\mathrm{R}}_{u,M,N}(\theta) = \theta$ *for all backward morphisms* $\theta : N \to M$ *in* $\mathsf{D}(A)$ *over* u*. The isomorphism* $\operatorname{badj}^{\mathrm{R}}_{u,M,N}$ *is functorial in* M *and* N*.*

(3) *The functor* $\operatorname{RCInd}_u : \mathsf{D}(A) \to \mathsf{D}(B)$ *is right adjoint to* Rest_u*.*

Here are the commutative diagrams in $\mathsf{D}(A)$ illustrating the theorem.

$$\begin{array}{ccc} \operatorname{CInd}_u(M) & \xrightarrow{\eta^{\mathrm{R}}_M} & \operatorname{RCInd}_u(M) \\ & {\scriptstyle \operatorname{tr}_{u,M}} \searrow & \downarrow {\scriptstyle \operatorname{tr}^{\mathrm{R}}_{u,M}} \\ & & M \end{array} \qquad \begin{array}{ccc} N & \xrightarrow{\operatorname{badj}^{\mathrm{R}}_{u,M,N}(\theta)} & \operatorname{RCInd}_u(M) \\ & {\scriptstyle \theta} \searrow & \downarrow {\scriptstyle \operatorname{tr}^{\mathrm{R}}_{u,M}} \\ & & M \end{array}$$

Proof The proof is very similar to that of Theorem 12.6.6. The only change is that now we choose a system of K-injective resolutions in $\mathsf{C}(A)$. For a K-injective DG A-module I, the DG B-module $\operatorname{Hom}_A(B, I)$ is K-injective, and it represents $\operatorname{RCInd}_u(M)$. □

Definition 12.6.13 Let $u : A \to B$ be a DG ring homomorphism, and let $M \in \mathsf{D}(A)$ and $N \in \mathsf{D}(B)$ be DG modules. A backward morphism $\theta : N \to M$ in $\mathsf{D}(A)$ over u is called *a nondegenerate backward morphism* if the corresponding morphism

$$\operatorname{badj}^{\mathrm{R}}_{u,M,N}(\theta) : N \to \operatorname{RHom}_A(B, M) = \operatorname{RCInd}_u(M)$$

in $\mathsf{D}(B)$ is an isomorphism.

Example 12.6.14 Given $M \in \mathsf{D}(A)$, let $N := \operatorname{RHom}_A(B, M) \in \mathsf{D}(B)$. The canonical backward morphism $\operatorname{tr}^{\mathrm{R}}_{u,M} : N \to M$ is a nondegenerate backward morphism in $\mathsf{D}(A)$ over u, because $\operatorname{badj}^{\mathrm{R}}_{u,M,N}(\operatorname{tr}^{\mathrm{R}}_{u,M}) = \operatorname{id}_N$.

Proposition 12.6.15 *Suppose* $A \xrightarrow{u} B \xrightarrow{v} C$ *are homomorphisms of DG rings,* $L \in \mathsf{D}(A)$*,* $M \in \mathsf{D}(B)$ *and* $N \in \mathsf{D}(C)$ *are DG modules,* $\theta : M \to L$ *is a nondegenerate backward morphism in* $\mathsf{D}(A)$ *over* u*, and* $\zeta : N \to M$ *is a nondegenerate backward morphism in* $\mathsf{D}(B)$ *over* v*. Then* $\theta \circ \zeta : N \to L$ *is a nondegenerate backward morphism in* $\mathsf{D}(A)$ *over* $v \circ u$*.*

Proof Since θ is a nondegenerate backward morphism, it induces an isomorphism $M \cong \operatorname{RHom}_A(B, L)$. We can thus assume that $M = \operatorname{RHom}_A(B, L)$ and that $\theta = \operatorname{tr}^{\mathrm{R}}_{u,L}$. Similarly, we can assume that $N = \operatorname{RHom}_B(C, M)$ and that $\zeta = \operatorname{tr}^{\mathrm{R}}_{v,M}$. But now $\theta \circ \zeta = \operatorname{tr}^{\mathrm{R}}_{u,L} \circ \operatorname{tr}^{\mathrm{R}}_{v,M} = \operatorname{tr}^{\mathrm{R}}_{v \circ u,L} : N \to L$, and this is known to be a nondegenerate backward morphism (see Example 12.6.14). □

Example 12.6.16 If $A = B$ and $u = \mathrm{id}_A$, then backward and forward morphisms over u are just morphisms in $\mathbf{D}(A)$. Nondegenerate (backward or forward) morphisms are just isomorphisms in $\mathbf{D}(A)$.

12.7 DG Ring Quasi-Isomorphisms

Suppose A and B are DG $\mathbb{K}$-rings. A homomorphism of DG $\mathbb{K}$-rings $u : A \to B$ induces a homomorphism of graded $\mathbb{K}$-rings $\mathrm{H}(u) : \mathrm{H}(A) \to \mathrm{H}(B)$; cf. Example 3.3.20.

Definition 12.7.1 A DG ring homomorphism $u : A \to B$ is called a *quasi-isomorphism of DG rings* if $\mathrm{H}(u)$ is an isomorphism.

Here is a fundamental result, due to B. Keller in 1994 [79], and independently to V. Hinich in 1997 [66]. It is the justification behind the use of DG ring resolutions (see Remarks 13.4.24, 14.3.24 and 18.1.30).

Theorem 12.7.2 *Let $u : A \to B$ be a quasi-isomorphism of DG $\mathbb{K}$-rings. Then:*

(1) *The restriction functor* $\mathrm{Rest}_u : \mathbf{D}(B) \to \mathbf{D}(A)$ *is an equivalence of $\mathbb{K}$-linear triangulated categories, with quasi-inverse* LInd_u.

(2) *For every $M \in \mathbf{D}(B^{\mathrm{op}})$ and $N \in \mathbf{D}(B)$ there is an isomorphism $M \otimes^{\mathrm{L}}_A N \xrightarrow{\simeq} M \otimes^{\mathrm{L}}_B N$ in $\mathbf{D}(\mathbb{K})$. This isomorphism is functorial in M and N.*

(3) *For every $M, N \in \mathbf{D}(B)$ there is an isomorphism* $\mathrm{RHom}_A(M, N) \xrightarrow{\simeq} \mathrm{RHom}_B(M, N)$ *in $\mathbf{D}(\mathbb{K})$. This isomorphism is functorial in M and N.*

Notice that the restriction functor Rest_u is suppressed in parts (2) and (3) of the theorem. The proof hinges on the next lemma.

Lemma 12.7.3 *Let $u : A \to B$ be a quasi-isomorphism of DG $\mathbb{K}$-rings, let $N \in \mathbf{D}(B)$, and let $\rho : P \to N$ be a K-projective resolution in $\mathbf{C}_{\mathrm{str}}(A)$.*

(1) *The homomorphism $\mathrm{q}_{u,P} : P \to B \otimes_A P$ in $\mathbf{C}_{\mathrm{str}}(A)$ from equation (12.6.4), and the homomorphism $\psi : B \otimes_A P \to N$, $\psi(b \otimes p) := b \cdot \rho(p)$, in $\mathbf{C}_{\mathrm{str}}(B)$, are both quasi-isomorphisms.*

(2) *$B \otimes_A P$ is a K-projective DG B-module.*

Proof

(1) Look at the commutative diagram

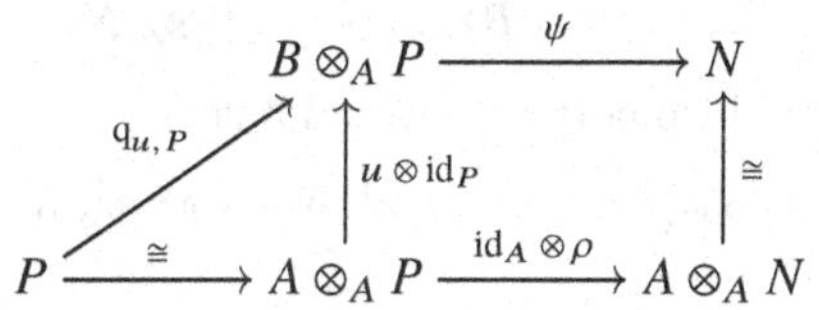

in $\mathbf{C}_{\mathrm{str}}(A)$, in which the arrows marked $\cong$ are the canonical isomorphisms. The homomorphism $u \otimes \mathrm{id}_P$ is a quasi-isomorphism, because u is a quasi-isomorphism and P is K-flat over A. The homomorphism $\mathrm{id}_A \otimes \rho$ is a quasi-isomorphism, because ρ is a quasi-isomorphism and A is K-flat over itself. Therefore $\mathrm{q}_{u,P}$ and ψ are quasi-isomorphisms.

(2) Take an acyclic DG B-module N. There is an isomorphism

$$\mathrm{Hom}_B(B \otimes_A P, N) \cong \mathrm{Hom}_A(P, N)$$

in $\mathbf{C}_{\mathrm{str}}(\mathbb{K})$, coming from adjunction for the DG ring homomorphism u. By assumption the DG module $\mathrm{Hom}_A(P, N)$ is acyclic. This shows that $B \otimes_A P$ is K-projective over B. □

Proof of Theorem 12.7.2

(1) Take any $N \in \mathbf{D}(B)$. Choose a K-projective resolution $\rho : P \to N$ in $\mathbf{C}_{\mathrm{str}}(A)$, so that $(\mathrm{LInd}_u \circ \mathrm{Rest}_u)(N) \cong B \otimes_A P$ in $\mathbf{D}(B)$. By the lemma we have a quasi-isomorphism $\psi : B \otimes_A P \to N$ in $\mathbf{C}_{\mathrm{str}}(B)$. This means that we have an isomorphism

$$\mathrm{Q}(\psi) : (\mathrm{LInd}_u \circ \mathrm{Rest}_u)(N) \xrightarrow{\simeq} N$$

in $\mathbf{D}(B)$, and it is functorial in N.

On the other hand, starting from a DG module $M \in \mathbf{D}(A)$, and choosing a K-projective resolution $\rho : P \to M$ in $\mathbf{C}_{\mathrm{str}}(A)$, we have an isomorphism $\mathrm{LInd}_u(M) \cong B \otimes_A P$ in $\mathbf{D}(B)$. So there is an isomorphism $(\mathrm{Rest}_u \circ \mathrm{LInd}_u)(M) \cong B \otimes_A P$ in $\mathbf{D}(A)$. According to the lemma, there is a quasi-isomorphism $\mathrm{q}_{u,P} : P \to B \otimes_A P$ in $\mathbf{C}_{\mathrm{str}}(A)$. We obtain an isomorphism

$$\mathrm{Q}(\rho) \circ \mathrm{Q}(\mathrm{q}_{u,P})^{-1} : (\mathrm{Rest}_u \circ \mathrm{LInd}_u)(M) \xrightarrow{\simeq} M$$

in $\mathbf{D}(A)$, and it is functorial in M.

(2) Choose a K-projective resolution $\rho : P \to M$ in $\mathbf{C}(A^{\mathrm{op}})$. This produces an isomorphism $\phi_1 : P \otimes_A N \xrightarrow{\simeq} M \otimes^{\mathrm{L}}_A N$ in $\mathbf{D}(\mathbb{K})$. The DG module $P \otimes_A B \in \mathbf{C}(B^{\mathrm{op}})$ is K-projective, and there is a quasi-isomorphism $\psi : P \otimes_A B \to M$ in $\mathbf{C}_{\mathrm{str}}(B^{\mathrm{op}})$, by the lemma (with B^{op} instead of B). We get an isomorphism

$$\phi_2 := \mathrm{Q}(\psi) \otimes^{\mathrm{L}}_B \mathrm{id}_N : (P \otimes_A B) \otimes_B N \xrightarrow{\simeq} M \otimes^{\mathrm{L}}_B N$$

in $\mathbf{D}(\mathbb{K})$. There is also the canonical isomorphism

$$\phi_3 : (P \otimes_A B) \otimes_B N \xrightarrow{\simeq} P \otimes_A N$$

in $\mathbf{D}(\mathbb{K})$. The functorial isomorphism we want is

$$\phi_2 \circ \phi_3^{-1} \circ \phi_1^{-1} : M \otimes^{\mathrm{L}}_A N \xrightarrow{\simeq} M \otimes^{\mathrm{L}}_B N$$

in $\mathbf{D}(\mathbb{K})$.

(3) Let us choose a K-projective resolution $\rho : P \to M$ in $\mathbf{C}(A)$. This produces an isomorphism

$$\phi_1 : \mathrm{RHom}_A(M, N) \xrightarrow{\simeq} \mathrm{Hom}_A(P, N)$$

in $\mathbf{D}(\mathbb{K})$. The DG module $B \otimes_A P \in \mathbf{C}(B)$ is K-projective, and the homomorphism $\psi : B \otimes_A P \to M$ in $\mathbf{C}_{\mathrm{str}}(B)$ is a quasi-isomorphism (again using the lemma). In this way we have an isomorphism

$$\phi_2 : \mathrm{Hom}_B(B \otimes_A P, N) \xrightarrow{\simeq} \mathrm{RHom}_B(M, N)$$

in $\mathbf{D}(\mathbb{K})$. And there is the canonical isomorphism of adjunction

$$\phi_3 : \mathrm{Hom}_B(B \otimes_A P, N) \xrightarrow{\simeq} \mathrm{Hom}_A(P, N)$$

in $\mathbf{D}(\mathbb{K})$. The functorial isomorphism we want is

$$\phi_2 \circ \phi_3^{-1} \circ \phi_1 : \mathrm{RHom}_A(M, N) \xrightarrow{\simeq} \mathrm{RHom}_B(M, N). \qquad \square$$

12.8 Existence of DG Ring Resolutions

In Theorem 12.7.2 we saw that a DG ring quasi-isomorphism $A \to B$ does not change the derived categories: the restriction functor $\mathbf{D}(B) \to \mathbf{D}(A)$ is an equivalence of triangulated categories.

Sometimes it is advantageous to replace a given DG ring by a quasi-isomorphic one (see Remarks 12.8.21, 12.8.22, 13.4.24, 14.3.24 and 18.1.30). In this section we study *semi-free DG rings*, and *semi-free DG ring resolutions*.

As before, we fix a nonzero commutative base ring $\mathbb{K}$. All DG rings are $\mathbb{K}$-central, namely we work in the category $\mathsf{DGRng}/_{\mathrm{c}}\,\mathbb{K}$. We use the shorthand "NC" for "noncommutative."

Graded sets and filtered graded sets were introduced in Definitions 11.4.4 and 11.4.6, respectively. The next definition was mentioned briefly in Example 3.1.23.

Definition 12.8.1 Let X be a graded set. The *NC polynomial ring on X* is the graded $\mathbb{K}$-ring $\mathbb{K}\langle X \rangle$.

Perhaps we should say a few words on the structure of the ring $\mathbb{K}\langle X \rangle$. Given $n \geq 0$ and elements $x_1, \ldots, x_n \in X$, their product $x_1 \cdots x_n \in \mathbb{K}\langle X \rangle$ is called a monomial (or a word), and its degree is

$$\deg(x_1 \cdots x_n) := \deg(x_1) + \cdots + \deg(x_n) \in \mathbb{Z}. \qquad (12.8.2)$$

For $n = 0$ the unique monomial is denoted by 1. As a graded $\mathbb{K}$-module, $\mathbb{K}\langle X \rangle$ is free, with basis the set of monomials. Multiplication in $\mathbb{K}\langle X \rangle$ is by concatenation of monomials, extended $\mathbb{K}$-bilinearly.

Definition 12.8.3 Let (X, F) be a filtered graded set. There is an induced filtration $F = \{F_j(\mathbb{K}\langle X\rangle)\}_{j\geq -1}$ on $\mathbb{K}\langle X\rangle$ by $\mathbb{K}$-submodules, as follows:

$$F_j(\mathbb{K}\langle X\rangle) := \begin{cases} 0 & \text{if } j = -1 \\ \mathbb{K}\langle F_j(X)\rangle & \text{if } j \geq 0. \end{cases}$$

For every $j \geq 0$ the filtered piece $F_j(\mathbb{K}\langle X\rangle)$ is a graded subring of $\mathbb{K}\langle X\rangle$. And of course $\mathbb{K}\langle X\rangle = \bigcup_j F_j(\mathbb{K}\langle X\rangle)$.

Recall that given a DG ring A, we denote by $A^\natural$ the graded ring gotten by forgetting the differential d_A. Here is the main definition of this section.

Definition 12.8.4 A *NC semi-free DG $\mathbb{K}$-ring* is a DG $\mathbb{K}$-ring A that admits an isomorphism of graded $\mathbb{K}$-rings $A^\natural \cong \mathbb{K}\langle X\rangle$, for some filtered graded set (X, F), such that under this isomorphism there is an inclusion $\mathrm{d}_A(F_j(X)) \subseteq F_{j-1}(\mathbb{K}\langle X\rangle)$ for every $j \geq 0$. Such a filtered graded set (X, F) is called a *multiplicative semi-basis* of A.

In this book we are mostly interested in semi-free DG rings because of the proposition below.

Proposition 12.8.5 *If A is a NC semi-free DG $\mathbb{K}$-ring, then as a DG $\mathbb{K}$-module A is semi-free, and hence K-flat.*

Proof Suppose we are given an isomorphism $A^\natural \cong \mathbb{K}\langle X\rangle$ for some filtered graded set (X, F), as in Definition 12.8.4. Let Y be the set of monomials in the elements of X. The set Y is graded, see (12.8.2), and it forms a basis of $A^\natural$ as a graded $\mathbb{K}$-module. However, since X could have elements of positive degree, this is not enough to make A into a semi-free DG $\mathbb{K}$-module; we need to put a suitable filtration on Y. Cf. Proposition 11.4.8.

For an element $x \in X$ let $\mathrm{ord}^F(x) := \min\{j \mid x \in F_j(X)\} \in \mathbb{N}$. Thus $F_j(X) = \{x \in X \mid \mathrm{ord}^F(x) \leq j\}$. We extend this order function to monomials by $\mathrm{ord}^F(x_1 \cdots x_n) := \sum_i \mathrm{ord}^F(x_i)$. The element 1 (the monomial of length 0) gets order 0. Then we define $G_j(Y) := \{y \in Y \mid \mathrm{ord}^F(y) \leq j\}$. This is a filtration $G = \{G_j(Y)\}_{j\geq -1}$ of the graded set Y, and it induces a filtration $G = \{G_j(A)\}_{j\geq -1}$ on A by $\mathbb{K}$-submodules, where $G_j(A) \subseteq A$ is defined to be the $\mathbb{K}$-linear span of $G_j(Y)$.

The graded Leibniz rule, together with the inclusion $\mathrm{d}_A(F_j(X)) \subseteq F_{j-1}(\mathbb{K}\langle X\rangle)$ from Definition 12.8.4, imply that $\mathrm{d}_A(G_j(A)) \subseteq G_{j-1}(A)$ for every $j \geq 0$. And $\mathrm{Gr}^G_j(A)$ is a free DG $\mathbb{K}$-module, with basis $\mathrm{Gr}^G_j(Y) := \{y \in Y \mid \mathrm{ord}^F(y) = j\}$. Thus G is a semi-free filtration on the DG $\mathbb{K}$-module A, and A is a semi-free DG $\mathbb{K}$-module; see Definition 11.4.3.

Finally, according to Theorem 11.4.14 and Proposition 10.3.4, A is a K-flat DG $\mathbb{K}$-module. □

Definition 12.8.6 Let A be a DG $\mathbb{K}$-ring. A *NC semi-free DG ring resolution* of A relative to $\mathbb{K}$ is a quasi-isomorphism $u : \tilde{A} \to A$ of DG $\mathbb{K}$-rings, where $\tilde{A}$ is a NC semi-free DG $\mathbb{K}$-ring. We also call $u : \tilde{A} \to A$ a *NC semi-free resolution of A in* $\mathsf{DGRng}/_{\mathrm{c}}\, \mathbb{K}$.

The important result of this section is the next existence theorem.

Theorem 12.8.7 *Let A be a DG $\mathbb{K}$-ring. There exists a NC semi-free DG ring resolution $u : \tilde{A} \to A$ of A relative to $\mathbb{K}$.*

The proof of the theorem comes after two lemmas. If X and Y are graded sets, a degree i function $f : X \to Y$ is a function such that $f(X^j) \subseteq Y^{j+i}$ for all j. Likewise, a degree i function $f : X \to B$ to a graded ring B is a function such that $f(X^j) \subseteq B^{j+i}$ for all j.

Lemma 12.8.8 *Let X be a graded set, and consider the graded $\mathbb{K}$-ring $B := \mathbb{K}\langle X \rangle$. Suppose $\mathrm{d}_X : X \to B$ is a degree 1 function.*

(1) *The function d_X extends uniquely to a $\mathbb{K}$-linear degree 1 derivation $\mathrm{d}_B : B \to B$.*

(2) *The derivation d_B is a differential on B, i.e. $\mathrm{d}_B \circ \mathrm{d}_B = 0$, if and only if $\mathrm{d}_B(\mathrm{d}_X((x))) = 0$ for every $x \in X$.*

Proof

(1) For elements $x_1, \ldots, x_n \in X$, with $k_j := \deg(x_j)$, define

$$\mathrm{d}_B(x_1 \cdots x_n) := \sum_{j=1}^{n} (-1)^{k_1+\cdots+k_{j-1}} \cdot x_1 \cdots x_{j-1} \cdot \mathrm{d}_X(x_j) \cdot x_{j+1} \cdots x_n. \quad (12.8.9)$$

This extends uniquely to a $\mathbb{K}$-linear homomorphism $\mathrm{d}_B : B \to B$ of degree 1. A calculation shows that the graded Leibniz rule holds, so this is degree 1 derivation.

(2) It is enough to verify that $\mathrm{d}_B \circ \mathrm{d}_B = 0$ on monomials. This is an elementary calculation, using formula (12.8.9). □

Lemma 12.8.10 *Let $B, C \in \mathsf{DGRng}/_{\mathrm{c}}\, \mathbb{K}$. Assume that B is NC semi-free, with a multiplicative semi-basis (X, F). Let $w : X \to C$ be a degree 0 function.*

(1) *The function w extends uniquely to a graded $\mathbb{K}$-ring homomorphism $w : B^{\natural} \to C^{\natural}$.*

(2) *The graded ring homomorphism w becomes a DG ring homomorphism $w : B \to C$ if and only if $\mathrm{d}_C(w(x)) = w(\mathrm{d}_B(x))$ for every $x \in X$.*

Proof Item (1) is clear. As for item (2), it is enough to verify that $\mathrm{d}_C \circ w = w \circ \mathrm{d}_B$ on monomials $x_1 \cdots x_n \in B$. This is done by induction on n, using the graded Leibniz rule. □

Proof of Theorem 12.8.7 The strategy is this: we are going to construct an increasing sequence of semi-free DG rings

$$F_0(\tilde{A}) \subseteq F_1(\tilde{A}) \subseteq F_2(\tilde{A}) \subseteq \cdots, \tag{12.8.11}$$

together with a compatible system of DG ring homomorphisms $F_j(u) : F_j(\tilde{A}) \to A$. At the same time we will construct an increasing sequence of graded sets

$$\varnothing = F_{-1}(X) \subseteq F_0(X) \subseteq F_1(X) \subseteq F_2(X) \subseteq \cdots, \tag{12.8.12}$$

such that there is equality of graded rings $F_j(\tilde{A})^{\natural} = \mathbb{K}\langle F_j(X)\rangle$ for every $j \geq 0$.

Then we will define the DG ring

$$\tilde{A} := \lim_{j\to} F_j(\tilde{A}), \tag{12.8.13}$$

the DG ring homomorphism

$$u := \lim_{j\to} F_j(u) : \tilde{A} \to A, \tag{12.8.14}$$

and the graded set

$$X := \lim_{j\to} F_j(X). \tag{12.8.15}$$

The graded set X is filtered by $F := \{F_j(X)\}_{j\geq -1}$. By the construction (X, F) will be a multiplicative semi-basis of $\tilde{A}$, so that the latter is a semi-free DG ring. The construction will also ensure that u is a quasi-isomorphism. All this will be done in several steps.

Step 1. In this step we deal with $j = 0$. Let's choose a collection $\{\bar{a}_x\}_{x\in F_0(X)}$ of nonzero homogeneous elements of $\mathrm{H}(A)$ that generates it as a $\mathbb{K}$-ring. We make $F_0(X)$ into a graded set by declaring $\deg(x) := \deg(\bar{a}_x)$ for $x \in F_0(X)$. Next, for each x we choose a homogeneous cocycle $a_x \in \mathrm{Z}(A)$ that represents the cohomology class $\bar{a}_x$. In this way we obtain a degree 0 function $F_0(u) : F_0(X) \to A$, $F_0(u)(x) := a_x$.

Define the DG ring $F_0(\tilde{A}) := \mathbb{K}\langle F_0(X)\rangle$, with zero differential. According to Lemma 12.8.10, the function $F_0(u) : F_0(X) \to A$ extends uniquely to a DG ring homomorphism $F_0(u) : F_0(\tilde{A}) \to A$. The choice of the collection $\{\bar{a}_x\}_{x\in F_0(X)}$ guarantees that the graded $\mathbb{K}$-ring homomorphism $\mathrm{H}(F_0(u)) : \mathrm{H}(F_0(\tilde{A})) \to \mathrm{H}(A)$ is surjective.

Step 2. In this step and the next one we deal with the induction. Let $j \geq 0$. The assumption is that we have a semi-free DG ring $F_j(\tilde{A})$, with a multiplicative semi-basis

$$\varnothing = F_{-1}(X) \subseteq F_0(X) \subseteq \cdots \subseteq F_{j-1}(X) \subseteq F_j(X) \subseteq F_j(X) \subseteq \cdots \tag{12.8.16}$$

(constant from level j onward), and a DG ring homomorphism $F_j(u) : F_j(\tilde{A}) \to A$ such that the graded ring homomorphism $\mathrm{H}(F_j(u)) : \mathrm{H}(F_j(\tilde{A})) \to \mathrm{H}(A)$ is surjective.

Define the graded two-sided ideal $K_j := \mathrm{Ker}(\mathrm{H}(F_j(u))) \subseteq \mathrm{H}(F_j(\tilde{A}))$. So there is an exact sequence of graded $\mathrm{H}(F_j(\tilde{A}))$-bimodules

$$0 \to K_j \to \mathrm{H}(F_j(\tilde{A})) \xrightarrow{\mathrm{H}(F_j(u))} \mathrm{H}(A) \to 0. \tag{12.8.17}$$

Choose a collection $\{\bar{a}_y\}_{y \in Y_{j+1}}$ of nonzero homogeneous elements of K_j that generates it as an $\mathrm{H}(F_j(\tilde{A}))$-bimodule (i.e. as a two-sided ideal). We make Y_{j+1} into a graded set by declaring $\deg(y) := \deg(\bar{a}_y) - 1$ for $y \in Y_{j+1}$. Next, for each element $y \in Y_{j+1}$ we choose a homogeneous cocycle $a_y \in \mathrm{Z}(F_j(\tilde{A}))$ that represents the cohomology class $\bar{a}_y$. In this way we obtain a collection $\{a_y\}_{y \in Y_{j+1}}$ of homogeneous cocycles in $F_j(\tilde{A})$.

Define the graded set $F_{j+1}(X) := F_j(X) \sqcup Y_{j+1}$ and the graded ring $F_{j+1}(\tilde{A})^\natural := \mathbb{K}\langle F_{j+1}(X)\rangle$. By definition $F_j(\tilde{A})^\natural$ is a graded subring of $F_{j+1}(\tilde{A})^\natural$. Next, define the degree 1 function $\mathrm{d}_{F_{j+1}(X)} : F_{j+1}(X) \to F_{j+1}(\tilde{A})$ by the formula

$$\mathrm{d}_{F_{j+1}(X)}(x) := \begin{cases} \mathrm{d}_{F_j(\tilde{A})}(x) & \text{if } x \in F_j(X) \\ a_x & \text{if } x \in Y_{j+1}. \end{cases}$$

In this formula we are using the collection of homogeneous cocycles $\{a_y\}_{y \in Y_{j+1}}$ that has been chosen above. According to Lemma 12.8.8, the function $\mathrm{d}_{F_{j+1}(X)}$ extends uniquely to a differential on $F_{j+1}(\tilde{A})$, making it into a DG ring. And of course $F_j(\tilde{A}) \subseteq F_{j+1}(\tilde{A})$ is a DG subring.

Let us denote by $v_j : F_j(\tilde{A}) \to F_{j+1}(\tilde{A})$ the inclusion. Because the cocycles $a_y \in F_j(\tilde{A})$, for $y \in Y_{j+1}$, become coboundaries in $F_{j+1}(\tilde{A})$, we see that

$$K_j \subseteq \mathrm{Ker}(\mathrm{H}(v_j)) \subseteq \mathrm{H}(F_j(\tilde{A})). \tag{12.8.18}$$

Observe that the DG ring $F_{j+1}(\tilde{A})$ is semi-free, with multiplicative semi-basis

$$\varnothing = F_{-1}(X) \subseteq F_0(X) \subseteq \cdots \subseteq F_j(X) \subseteq F_{j+1}(X) \subseteq F_{j+1}(X) \subseteq \cdots,$$

constant from level $j + 1$ onward.

Step 3. In this step we continue step 2, to construct the DG ring homomorphism $F_{j+1}(u) : F_{j+1}(\tilde{A}) \to A$. For each element $y \in Y_{j+1}$ the cohomology class $\bar{a}_y \in \mathrm{H}(F_j(\tilde{A}))$ belongs to the kernel of $\mathrm{H}(F_j(u))$. This means that the cocycle $F_j(u)(a_y) \in A$ is a coboundary. So we can find a homogeneous element $b_y \in A$ such that $\mathrm{d}_A(b_y) = F_j(u)(a_y)$. Define the degree 0 function $F_{j+1}(u) : F_{j+1}(X) \to A$ by the formula

$$F_{j+1}(u)(x) := \begin{cases} F_j(u)(x) & \text{if } x \in F_j(X) \\ b_x & \text{if } x \in Y_{j+1}. \end{cases}$$

By Lemma 12.8.10 this function extends uniquely to a DG ring homomorphism $F_{j+1}(u) : F_{j+1}(\tilde{A}) \to A$. Note that the restriction of $F_{j+1}(u)$ to $F_j(\tilde{A})$ coincides with $F_j(u)$.

Step 4. Applying steps 2-3 recursively, we obtain direct systems $\{F_j(\tilde{A})\}_{j\geq 0}$ and $\{F_j(u)\}_{j\geq 0}$ of DG rings and homomorphisms, and we can define the DG ring $\tilde{A}$ by formula (12.8.13), and the DG ring homomorphism $\tilde{u}$ by formula (12.8.14). The graded set X from formula (12.8.15) is filtered, and by the construction in steps 1 and 2 the filtered graded set (X, F) is a multiplicative semi-basis of $\tilde{A}$.

It remains to prove that the graded ring homomorphism $\mathrm{H}(u) : \mathrm{H}(\tilde{A}) \to \mathrm{H}(A)$ is bijective. Recall that for each $j \geq 0$ we denoted by v_j the inclusion $F_j(\tilde{A}) \subseteq F_{j+1}(\tilde{A})$. Define the graded ring $L^j := \mathrm{Im}(\mathrm{H}(v_j)) \subseteq \mathrm{H}(F_{j+1}(\tilde{A}))$. We get a commutative diagram

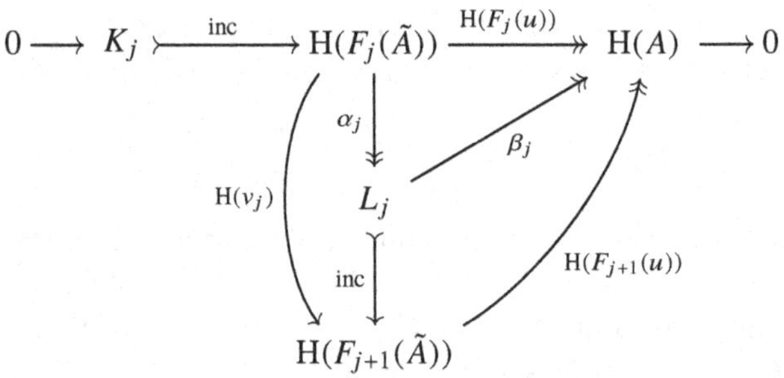

in $\mathbf{G}_{\mathrm{str}}(\mathbb{K})$. The top row is an exact sequence (it is (12.8.17)). Because α_j is surjective, there is equality $\mathrm{Ker}(\beta_j) = \alpha_j(\mathrm{Ker}(\mathrm{H}(F_j(u))) = \alpha_j(K_j)$. But by formula (12.8.18) we know that $\alpha_j(K_j) = 0$. The conclusion is that $\beta_j : L_j \to \mathrm{H}(A)$ is bijective. Hence $\{L_j\}_{j\geq 0}$ is a constant direct system, and $\lim_{j\to} \beta_j : \lim_{j\to} L_j \to \mathrm{H}(A)$ is bijective.

Now the direct systems $\{L_j\}_{j\geq 0}$ and $\{\mathrm{H}(F_j(\tilde{A}))\}_{j\geq 0}$ are sandwiched. Therefore the second direct system has a limit too, and it is the same limit; i.e.

$$\lim_{j\to} \mathrm{H}(F_j(u)) : \lim_{j\to} \mathrm{H}(F_j(\tilde{A})) \to \mathrm{H}(A)$$

is bijective. Because cohomology commutes with direct limits, and by formula (12.8.13), this implies that $\mathrm{H}(u)$ is bijective. □

Exercise 12.8.19 Modify the proof of Theorem 12.8.7 to make the quasi-isomorphism $u : \tilde{A} \to A$ surjective.

Corollary 12.8.20 *If* $\mathrm{H}(A)$ *is nonpositive, then* A *has a semi-free DG ring resolution* $\tilde{A} \to A$ *such that* $\tilde{A}$ *is nonpositive.*

Proof Under this assumption, the graded set X in the proof of Theorem 12.8.7 is nonpositive. □

Remark 12.8.21 NC semi-free DG rings have important lifting properties within $\mathsf{DGRng}/_{\mathrm{c}}\,\mathbb{K}$. V. Hinich, in [66], calls them *standard cofibrant objects* of the Quillen model structure on $\mathsf{DGRng}/_{\mathrm{c}}\,\mathbb{K}$.

One can also consider the full subcategory $\mathsf{DGRng}^{\leq 0}/_{\mathrm{c}}\,\mathbb{K}$ of nonpositive NC DG rings. NC nonpositive semi-free DG rings are studied in detail in [173]. This is done in a slightly more general context: the base ring $\mathbb{K}$ is allowed to be a commutative DG ring (see next remark), and $\mathbb{K} \to A$ is a central homomorphism of DG rings (i.e. the image of $\mathbb{K}$ is in $\mathrm{Cent}(A)$). The lifting properties are the same.

See Remark 6.2.38 for the related A_∞ theory.

Remark 12.8.22 There are also *commutative semi-free DG rings*. Recall that a commutative DG ring in this book means nonpositive and strongly commutative; see Definitions 3.1.22 and 3.3.4. The commutative DG rings form the category $\mathsf{DGRng}^{\leq 0}_{\mathrm{sc}}/\mathbb{K}$, which is a full subcategory of $\mathsf{DGRng}/_{\mathrm{c}}\,\mathbb{K}$.

A nonpositive graded set S gives rise to the commutative polynomial ring $\mathbb{K}[S]$, see Example 3.1.23. A commutative DG ring $\tilde{A}$ is called commutative semi-free if it admits a graded ring isomorphism $\tilde{A}^{\natural} \cong \mathbb{K}[S]$ for some nonpositive graded set S. The graded set S is automatically filtered by degree, namely $F_j(S) := \bigcup_{i \geq -j} S^i$. This implies (cf. Proposition 12.8.5) that $\tilde{A}$ is semi-free as a DG $\mathbb{K}$-module .

In [173] it is proved that every $A \in \mathsf{DGRng}^{\leq 0}_{\mathrm{sc}}/\mathbb{K}$ admits a surjective quasi-isomorphism $\tilde{A} \to A$ from a commutative semi-free DG ring $\tilde{A}$. It is also proved that the commutative semi-free DG rings have good lifting properties in $\mathsf{DGRng}^{\leq 0}_{\mathrm{sc}}/\mathbb{K}$. The situation studied in [173] is a bit more general: the base ring $\mathbb{K}$ is allowed to be itself a commutative DG ring; so the theory becomes relative.

If in Example 6.2.35 we take the topological space X to be a single point, then the category $\mathsf{DGRng}^{\leq 0}_{\mathrm{sc}}/\mathbb{K}_X$ discussed there coincides with $\mathsf{DGRng}^{\leq 0}_{\mathrm{sc}}/\mathbb{K}$, and the semi-pseudo-free DG $\mathbb{K}_X$-rings are just the semi-free commutative DG rings mentioned here.

Unlike the geometric setting, the category $\mathsf{DGRng}^{\leq 0}_{\mathrm{sc}}/\mathbb{K}$ admits a Quillen model structure (at least when $\mathbb{K}$ contains $\mathbb{Q}$), and the calculus of fractions for the derived category $\mathbf{D}(\mathsf{DGRng}^{\leq 0}_{\mathrm{sc}}/\mathbb{K})$ can be deduced from it.

The fully faithful embedding

$$\mathsf{DGRng}^{\leq 0}_{\mathrm{sc}}/\mathbb{K} \to \mathsf{DGRng}^{\leq 0}/_{\mathrm{c}}\,\mathbb{K}$$

induces a functor

$$\mathbf{D}(\mathsf{DGRng}^{\leq 0}_{\mathrm{sc}}/\mathbb{K}) \to \mathbf{D}(\mathsf{DGRng}^{\leq 0}/_{\mathrm{c}}\,\mathbb{K})$$

between the derived categories, which presumably is neither full nor faithful (but we did not explore this question seriously).

12.9 The Derived Tensor-Evaluation Morphism

In this section we present an important theorem (Theorem 12.9.10) that will reappear – with modifications – several times in the book. Before presenting it, we need a few finiteness and boundedness conditions, some old and some new.

Pseudo-finite semi-free DG modules were introduced in Definition 11.4.29. The saturated full triangulated subcategory generated by a set of objects was defined in Definition 5.3.20.

Definition 12.9.1 Let A be a DG ring. A DG A-module L is called *derived pseudo-finite* if it belongs to the saturated full triangulated subcategory of $\mathbf{D}(A)$ generated by the pseudo-finite semi-free DG A-modules.

Example 12.9.2 Suppose A is a nonpositive cohomologically left pseudo-noetherian DG ring. Then the derived pseudo-finite DG A-modules are precisely the objects of $\mathbf{D}^{-}_{\mathrm{f}}(A)$. One direction is by Theorem 11.4.40, and the other direction is trivial.

Example 12.9.3 If L is an algebraically perfect DG A-module (to be defined later, see Definition 14.1.3) then it is derived pseudo-finite.

Remark 12.9.4 Let L be a DG A-module. Following [21], let us say that L is *pseudo-coherent* if L is isomorphic in $\mathbf{D}(A)$ to a pseudo-finite semi-free DG A-module P. Clearly pseudo-coherent implies derived pseudo-finite. According to [144, Lemma tag=064X] the converse is true when A is a ring; but we do not know if the converse is true in general.

Anyhow, the attribute "derived pseudo-finite" is very useful, and it can replace "pseudo-coherent" for most applications. For instance, see condition (i) in Theorem 12.9.10 below.

In Definition 12.4.18 we saw the notion of bounded below flat concentration of a DG module. Here is a refinement of it.

Definition 12.9.5 Let A be a DG ring.

(1) We say that a DG A-module P has *tensor displacement* inside an integer interval $[i_0, i_1]$ if for every $N \in \mathbf{C}(A^{\mathrm{op}})$ this inclusion of integer intervals holds:

$$\operatorname{con}(N \otimes_A P) \subseteq \operatorname{con}(N) + [i_0, i_1].$$

(2) We say that a DG A-module P has *bounded below tensor displacement* if it has tensor displacement inside the integer interval $[i_0, \infty]$ for some $i_0 \in \mathbb{Z}$.

(3) A DG A-module M is said to have *derived bounded below tensor displacement* if M belongs to the saturated full triangulated subcategory of $\mathbf{D}(A)$ generated by the K-flat DG A-modules P that have bounded below tensor displacement.

Example 12.9.6 The DG A-module $M := A$ has derived bounded below tensor displacement, regardless of the boundedness of A or $\mathrm{H}(A)$.

More generally, if M is an algebraically perfect DG A-module (see Definition 14.1.3) then it has derived bounded below tensor displacement.

Example 12.9.7 If A is a ring, and M is a complex of A-modules of finite flat dimension, then M has derived bounded below tensor displacement. See Proposition 12.4.19.

Remark 12.9.8 Clearly, if M has derived bounded below tensor displacement, then it has bounded below flat concentration. We don't know if the converse is true in general.

Remark 12.9.9 Definition 12.9.5 can be adapted to other kinds of boundedness conditions. It can also be adapted to Hom displacement, in the first or second argument, to give refined notions of projective and injective displacements.

Theorem 12.9.10 (Tensor-Evaluation) *Let A and B be DG rings, and let $L \in \mathbf{D}(A)$, $M \in \mathbf{D}(A \otimes B^{\mathrm{op}})$ and $N \in \mathbf{D}(B)$ be DG modules. There is unique morphism*

$$\mathrm{ev}^{\mathrm{R,L}}_{L,M,N} : \mathrm{RHom}_A(L, M) \otimes^{\mathrm{L}}_B N \to \mathrm{RHom}_A(L, M \otimes^{\mathrm{L}}_B N)$$

in $\mathbf{D}(\mathbb{K})$, called derived tensor-evaluation, *which enjoys properties* (a)-(c) *below.*

(a) *The morphism $\mathrm{ev}^{\mathrm{R,L}}_{L,M,N}$ is functorial in the objects L, M, N.*

(b) *If L is K-projective over A and N is K-flat over B, then $\mathrm{ev}^{\mathrm{R,L}}_{L,M,N} = \mathrm{Q}(\mathrm{ev}_{L,M,N})$, where*

$$\mathrm{ev}_{L,M,N} : \mathrm{Hom}_A(L, M) \otimes_B N \to \mathrm{Hom}_A(L, M \otimes_B N)$$

is the obvious homomorphism in $\mathbf{C}_{\mathrm{str}}(\mathbb{K})$

(c) *If all conditions* (i)-(iii) *below hold, then* $\mathrm{ev}^{\mathrm{R,L}}_{L,M,N}$ *is an isomorphism.*
  (i) *The DG A-module L is derived pseudo-finite (Definition* 12.9.1).
  (ii) *The DG A-B-bimodule M is cohomologically bounded below (Definition* 12.1.5(2)).
  (iii) *The DG B-module N has derived bounded below tensor displacement (Definition* 12.9.5(3)).

We need a lemma first.

Lemma 12.9.11 *Suppose we are given:*

- *A pseudo-finite semi-free DG A-module P, with pseudo-finite semi-free filtration* $\{F_j(P)\}_{j\geq -1}$.
- *A bounded below DG A-B-bimodule M.*
- *A DG B-module Q with bounded below tensor displacement.*

Then for every degree l there is an index j, depending on l, such that the canonical homomorphisms

$$(\mathrm{Hom}_A(P, M) \otimes_B Q)^l \to (\mathrm{Hom}_A(F_j(P), M) \otimes_B Q)^l$$

and

$$(\mathrm{Hom}_A(P, M \otimes_B Q))^l \to (\mathrm{Hom}_A(F_j(P), M \otimes_B Q))^l$$

in $\mathbf{M}(\mathbb{K})$ *are bijective.*

Proof In the notation of Definition 11.4.29, there are numbers $i_1 \in \mathbb{Z}$ and $r_k \in \mathbb{N}$ such that

$$(P/F_j(P))^\natural \cong \bigoplus_{k\geq j+1} \mathrm{T}^{-i_1+k}(A^\natural)^{\oplus r_k} \tag{12.9.12}$$

in $\mathbf{G}_{\mathrm{str}}(A^\natural)$ for all $j \geq 0$. In terms of bases, the graded-free $A^\natural$-module $(P/F_j(P))^\natural$ has a basis concentrated in degrees $\leq i_1 - j - 1$.

Say M is concentrated in the degree interval $[s_0, \infty]$, and Q has tensor displacement inside the degree interval $[t_0, \infty]$, for some $s_0, t_0 \in \mathbb{Z}$. From equation (12.9.12) we see that for every j, both graded $\mathbb{K}$-modules

$$\mathrm{Hom}_A(P/F_j(P), M) \otimes_B Q \quad \text{and} \quad \mathrm{Hom}_A(P/F_j(P), M \otimes_B Q)$$

are concentrated in the degree interval

$$[-i_1 + j + 1 + s_0 + t_0, \infty] \subseteq \mathbb{Z}.$$

The split short exact sequence

$$0 \to F_j(P)^\natural \to P^\natural \to (P/F_j(P))^\natural \to 0$$

in $\mathbf{G}_{\mathrm{str}}(A)$ tells us that for every $j \geq l + i_1 - s_0 - t_0$ the homomorphisms in question are bijective. □

Proof of Theorem 12.9.10 To construct the morphism $\mathrm{ev}^{\mathrm{R,L}}_{L,M,N}$ we choose a K-projective resolution $P \to L$ in $\mathbf{C}_{\mathrm{str}}(A)$ and a K-projective resolution $Q \to N$ in

$\mathbf{C}_{\mathrm{str}}(B)$. These choices are unique up to homotopy. Then $\mathrm{ev}^{\mathrm{R,L}}_{L,M,N}$ is represented by the obvious homomorphism

$$\mathrm{ev}_{P,M,Q} : \mathrm{Hom}_A(P, M) \otimes_B Q \to \mathrm{Hom}_A(P, M \otimes_B Q)$$

in $\mathbf{C}_{\mathrm{str}}(\mathbb{K})$. The morphism $\mathrm{ev}^{\mathrm{R,L}}_{L,M,N}$ in $\mathbf{D}(\mathbb{K})$ does not depend on the choices, and hence it is functorial; this is property (a). Property (b) is clearly satisfied. Properties (a) and (b) imply that the morphism $\mathrm{ev}^{\mathrm{R,L}}_{L,M,N}$ is unique.

Property (c) is proved in several steps.

Step 1. Let P be a finite semi-free DG A-module; namely P admits a pseudo-finite semi-free filtration $F = \{F_j(P)\}_{j\geq -1}$ (see Definition 11.4.29) such that $P = F_{j_1}(P)$ for some $j_1 \in \mathbb{N}$. Then the graded $A^{\natural}$-module $P^{\natural}$ is finite free. This implies that

$$\mathrm{ev}_{P,M,Q} : \mathrm{Hom}_A(P, M) \otimes_B Q \to \mathrm{Hom}_A(P, M \otimes_B Q)$$

is an isomorphism in $\mathbf{C}_{\mathrm{str}}(\mathbb{K})$ for all $M \in \mathbf{C}(A \otimes B^{\mathrm{op}})$ and $Q \in \mathbf{C}(B)$.

Step 2. Assume that M is bounded below. Let Q be a K-flat DG B-module of bounded below tensor displacement, and let P be a pseudo-finite semi-free DG A-module, with pseudo-finite semi-free filtration $F = \{F_j(P)\}_{j\geq -1}$. By Lemma 12.9.11, for every $l \in \mathbb{Z}$ there is an index j such that

$$(\mathrm{Hom}_A(P, M) \otimes_B Q)^l = (\mathrm{Hom}_A(F_j(P), M) \otimes_B Q)^l$$

and

$$(\mathrm{Hom}_A(P, M \otimes_B Q))^l = (\mathrm{Hom}_A(F_j(P), M \otimes_B Q))^l.$$

By step 1 the homomorphism

$$\mathrm{ev}_{F_j(P),M,Q} : \mathrm{Hom}_A(F_j(P), M) \otimes_B Q \to \mathrm{Hom}_A(F_j(P), M \otimes_B Q)$$

is bijective. We conclude that

$$\mathrm{ev}_{P,M,Q} : (\mathrm{Hom}_A(P, M) \otimes_B Q)^l \to \mathrm{Hom}_A(P, M \otimes_B Q)^l$$

is bijective. Since l is arbitrary, it follows that $\mathrm{ev}_{P,M,Q}$ is an isomorphism in $\mathbf{C}_{\mathrm{str}}(\mathbb{K})$.

Step 3. In step 2 we proved that the morphism

$$\mathrm{ev}^{\mathrm{R,L}}_{L,M,N} : \mathrm{RHom}_A(L, M) \otimes^{\mathrm{L}}_B N \to \mathrm{RHom}_A(L, M \otimes^{\mathrm{L}}_B N)$$

in $\mathbf{D}(\mathbb{K})$ is an isomorphism if M is bounded below, $L = P$ is a pseudo-finite semi-free DG A-module, and $N = Q$ is K-flat DG B-module of bounded below tensor displacement. If M is only assumed to satisfy condition (ii), then it is quasi-isomorphic to a suitable smart truncation of it which is bounded below; therefore $\mathrm{ev}^{\mathrm{R,L}}_{L,M,N}$ is an isomorphism under this assumption too.

Fixing M and N, the full subcategory of $\mathbf{D}(A)$ on the objects L for which $\mathrm{ev}^{\mathrm{R,L}}_{L,M,N}$ is an isomorphism is a saturated full triangulated subcategory (by Proposition 5.3.22). Therefore $\mathrm{ev}^{\mathrm{R,L}}_{L,M,N}$ is an isomorphism for every L, M, N

such that L satisfies condition (i), M satisfies condition (ii), and $N = Q$ is K-flat of bounded below tensor displacement.

Now we fix L and M that satisfy condition (i) and (ii), respectively. The full subcategory of $\mathbf{D}(B)$ on the objects N for which $\mathrm{ev}^{\mathrm{R,L}}_{L,M,N}$ is an isomorphism is a saturated full triangulated subcategory. By the previous paragraph, this subcategory contains the K-flat complexes Q of bounded below tensor displacement. Therefore it contains all the complexes N satisfying condition (iii). □

Remark 12.9.13 Let us explain the syntax of the expression $\mathrm{ev}^{\mathrm{R,L}}_{L,M,N}$ in Theorem 12.9.10. The letters "ev" stand for "evaluation." The letters "R" and "L" in the superscript refer to the left and right derived functors, whereas the letters "L," "M" and "N" in the subscript refer to the DG modules involved. Note the different fonts. This sort of syntax is used frequently in the book; cf. the expressions $\mathrm{hm}^{\mathrm{R}}_{M}$ in Definition 13.1.5, $\mathrm{ev}^{\mathrm{R,R}}$ in Lemma 13.1.16, $\mathrm{ev}^{\mathrm{R,L}}_{\mathfrak{m},(-)}$ in Theorem 16.4.3 and $\mathrm{fadj}^{\mathrm{L}}_{u,M,N}$ in Theorem 12.6.6.

12.10 Hom-Tensor Formulas for Weakly Commutative DG Rings

In this section we assume that all DG rings are weakly commutative (see Definition 3.3.4). For such a DG ring A, its 0th cohomology $\mathrm{H}^0(A)$ is a commutative ring.

Proposition 12.10.1 *For a weakly commutative DG ring A, the category $\mathbf{D}(A)$ is $\mathrm{H}^0(A)$-linear.*

Proof This is true already for the homotopy category. Indeed, for any $M, N \in \mathbf{C}(A)$ the $\mathbb{K}$-module $\mathrm{H}^0(\mathrm{Hom}_A(M, N))$ is an $\mathrm{H}^0(A)$-module; and composition in $\mathbf{K}(A)$ is $\mathrm{H}^0(A)$-bilinear. □

Let A be a DG ring. There is a canonical isomorphism $u : A \xrightarrow{\simeq} A^{\mathrm{op}}$ to its opposite. The formula is $u(a) := (-1)^i \cdot a$ for $a \in A^i$. This implies that every left DG A-module can be made into a right DG A-module, and vice-versa. The formula relating the left and right actions is $m \cdot a = (-1)^{i \cdot j} \cdot a \cdot m$ for $a \in A^i$ and $m \in M^j$. On the level of categories we obtain an isomorphism of DG categories $\mathbf{C}(A) \cong \mathbf{C}(A^{\mathrm{op}})$, which is the identity functor on the underlying DG $\mathbb{K}$-modules.

Weak commutativity makes the tensor and Hom functors A-bilinear (in the graded sense), and therefore their derived functors have more structure: they are $\mathrm{H}^0(A)$-bilinear triangulated bifunctors

$$(- \otimes^{\mathrm{L}}_A -) : \mathbf{D}(A) \times \mathbf{D}(A) \to \mathbf{D}(A) \tag{12.10.2}$$

and

$$\mathrm{RHom}_A(-,-) : \mathbf{D}(A)^{\mathrm{op}} \times \mathbf{D}(A) \to \mathbf{D}(A). \qquad (12.10.3)$$

When the DG rings in Theorem 12.7.2 are weakly commutative, this result can be amplified:

Proposition 12.10.4 *In the situation of Theorem* 12.7.2, *assume that A and B are weakly commutative.*

(1) *The restriction functor* $\mathrm{Rest}_u : \mathbf{D}(B) \to \mathbf{D}(A)$ *is an equivalence of* $\mathrm{H}^0(A)$*-linear triangulated categories, with quasi-inverse* LInd_u.

(2) *For every* $M, N \in \mathbf{D}(B)$ *there is an isomorphism* $M \otimes^{\mathrm{L}}_A N \xrightarrow{\simeq} M \otimes^{\mathrm{L}}_B N$ *in* $\mathbf{D}(A)$. *This isomorphism is functorial in M and N.*

(3) *For every* $M, N \in \mathbf{D}(B)$ *there is an isomorphism* $\mathrm{RHom}_A(M,N) \xrightarrow{\simeq} \mathrm{RHom}_B(M,N)$ *in* $\mathbf{D}(A)$. *This isomorphism is functorial in M and N.*

Proof

(1) The ring homomorphism $\mathrm{H}^0(u) : \mathrm{H}^0(A) \to \mathrm{H}^0(B)$ makes $\mathbf{D}(B)$ into an $\mathrm{H}^0(A)$-linear category, and then $\mathrm{Rest}_u : \mathbf{D}(B) \to \mathbf{D}(A)$ becomes an $\mathrm{H}^0(A)$-linear functor.

(2,3) Going over the steps in the proof of Theorem 12.7.2, we see that all the moves are A-linear (in the graded sense). □

Suppose $A \to B$ is a DG ring homomorphism. For every $M \in \mathbf{C}(A)$ and $N \in \mathbf{C}(B)$, their tensor product $M \otimes_A N$ is a DG B-module, via the action on N. Thus we get a DG bifunctor

$$(- \otimes_A -) : \mathbf{C}(A) \times \mathbf{C}(B) \to \mathbf{C}(B). \qquad (12.10.5)$$

There is an isomorphism

$$M \otimes_A N \xrightarrow{\simeq} N \otimes_A M \qquad (12.10.6)$$

in $\mathbf{C}_{\mathrm{str}}(B)$, which uses the Koszul sign rule $m \otimes n \mapsto (-1)^{i \cdot j} \cdot n \otimes m$ for homogeneous elements $m \in M^i$ and $n \in N^j$.

Proposition 12.10.7 *Let $A \to B$ be a homomorphism of weakly commutative DG rings. The DG bifunctor* (12.10.5) *has a left derived bifunctor*

$$(- \otimes^{\mathrm{L}}_A -) : \mathbf{D}(A) \times \mathbf{D}(B) \to \mathbf{D}(B).$$

This is an $\mathrm{H}^0(A)$*-bilinear triangulated bifunctor.*

Proof The left derived bifunctor exists by Theorem 9.3.16, using K-flat resolutions in the first argument (i.e. in $\mathbf{C}_{\mathrm{str}}(A)$). As for the $\mathrm{H}^0(A)$-bilinearity: this is already comes in the homotopy category level, namely for the bifunctor

$$(- \otimes_A -) : \mathbf{K}(A) \times \mathbf{K}(B) \to \mathbf{K}(B). \qquad □$$

Similarly we have an $\mathrm{H}^0(A)$-bilinear triangulated bifunctor

$$(- \otimes^{\mathrm{L}}_A -) : \mathbf{D}(B) \times \mathbf{D}(A) \to \mathbf{D}(B).$$

When $A = B$ these operations coincide with (12.10.2).

Proposition 12.10.8 (Symmetry) *Let $A \to B$ be a homomorphism of weakly commutative DG rings. For every $M \in \mathbf{D}(A)$ and $N \in \mathbf{D}(B)$ there is an isomorphism $M \otimes^{\mathrm{L}}_A N \cong N \otimes^{\mathrm{L}}_A M$ in $\mathbf{D}(B)$. This isomorphisms is functorial in M and N.*

Proof Use the isomorphism (12.10.6). □

Proposition 12.10.9 (Associativity) *Let $A_1 \to A_2 \to A_3$ be homomorphisms between weakly commutative DG rings. For each i let $M_i \in \mathbf{D}(A_i)$. There is an isomorphism*

$$(M_1 \otimes^{\mathrm{L}}_{A_1} M_2) \otimes^{\mathrm{L}}_{A_2} M_3 \cong M_1 \otimes^{\mathrm{L}}_{A_1} (M_2 \otimes^{\mathrm{L}}_{A_2} M_3)$$

in $\mathbf{D}(A_3)$. This isomorphisms is functorial in the M_i.

Proof Choose K-flat resolutions $P_i \to M_i$ in $\mathbf{C}(A_i)$. Then we have isomorphisms

$$\begin{aligned}(M_1 \otimes^{\mathrm{L}}_{A_1} M_2) \otimes^{\mathrm{L}}_{A_2} M_3 &\cong (P_1 \otimes_{A_1} P_2) \otimes_{A_2} P_3 \\ &\cong^{\dagger} P_1 \otimes_{A_1} (P_2 \otimes_{A_2} P_3) \cong M_1 \otimes^{\mathrm{L}}_{A_1} (M_2 \otimes^{\mathrm{L}}_{A_2} M_3)\end{aligned}$$

in $\mathbf{D}(A_3)$. The isomorphism $\cong^{\dagger}$ is due to the usual associativity of $\otimes$. □

Suppose $A \to B$ is a DG ring homomorphism. For every $M \in \mathbf{C}(A)$ and $N \in \mathbf{C}(B)$, the DG $\mathbb{K}$-module $\mathrm{Hom}_A(N, M)$ has a DG B-module structure, via the action on N. Thus we get a DG bifunctor

$$\mathrm{Hom}_A(-,-) : \mathbf{C}(B)^{\mathrm{op}} \times \mathbf{C}(A) \to \mathbf{C}(B). \tag{12.10.10}$$

Proposition 12.10.11 *Let $A \to B$ be a homomorphism of weakly commutative DG rings. The DG bifunctor* (12.10.10) *has a right derived bifunctor*

$$\mathrm{RHom}_A(-,-) : \mathbf{D}(B)^{\mathrm{op}} \times \mathbf{D}(A) \to \mathbf{D}(B).$$

This is an $\mathrm{H}^0(A)$-bilinear triangulated bifunctor.

Proof The right derived bifunctor exists by Theorem 9.3.11, using K-injective resolutions in the second argument (i.e. in $\mathbf{C}_{\mathrm{str}}(A)$). □

Proposition 12.10.12 (Hom-Tensor Adjunction) *Let $A_1 \to A_2 \to A_3$ be homomorphisms between weakly commutative DG rings. For each i let $M_i \in \mathbf{D}(A_i)$. There is an isomorphism*

$$\mathrm{RHom}_{A_2}(M_3, \mathrm{RHom}_{A_1}(M_2, M_1)) \cong \mathrm{RHom}_{A_1}(M_2 \otimes^{\mathrm{L}}_{A_2} M_3, M_1)$$

in $\mathbf{D}(A_3)$. This isomorphism is functorial in the M_i.

Proof Choose a K-flat resolution $P_2 \to M_2$ in $\mathbf{C}(A_2)$, and a K-injective resolution $M_1 \to I_1$ in $\mathbf{C}(A_1)$. An easy adjunction calculation show that $\operatorname{Hom}_{A_1}(P_2, I_1)$ is K-injective in $\mathbf{C}(A_2)$. Then we have isomorphisms

$$\begin{aligned} \operatorname{RHom}_{A_2}(M_3, \operatorname{RHom}_{A_1}(M_2, M_1)) &\cong \operatorname{Hom}_{A_2}(M_3, \operatorname{Hom}_{A_1}(P_2, I_1)) \\ &\cong^{\dagger} \operatorname{Hom}_{A_1}(P_2 \otimes_{A_2} M_3, I_1) \cong \operatorname{RHom}_{A_1}(M_2 \otimes^{\mathrm{L}}_{A_2} M_3, M_1) \end{aligned}$$

in $\mathbf{D}(A_3)$. The isomorphism $\cong^{\dagger}$ is due to the usual Hom-tensor adjunction. □

Remark 12.10.13 It is not hard to show that $\mathbf{D}(A)$ is a closed symmetric monoidal category, with monoidal operation $(- \otimes^{\mathrm{L}}_{A} -)$ and monoidal unit object A. It needs just a bit more than Propositions 12.10.9, 12.10.8 and 12.10.12.

For a noncommutative version of this monoidal structure see Remark 14.3.23 and Definition 15.3.13.

We end this section with a weakly commutative variant of Theorem 12.9.10. Cohomologically pseudo-noetherian nonpositive DG rings were introduced in Definition 11.4.37. DG modules with derived bounded below tensor displacement were defined in Definition 12.9.5.

Theorem 12.10.14 *Let $A \to B$ be a homomorphism between weakly commutative DG rings, and let $L \in \mathbf{D}(A)$ and $M, N \in \mathbf{D}(B)$ be DG modules. There is a morphism*

$$\operatorname{ev}^{\mathrm{R,L}}_{L,M,N} : \operatorname{RHom}_A(L, M) \otimes^{\mathrm{L}}_{B} N \to \operatorname{RHom}_A(L, M \otimes^{\mathrm{L}}_{B} N)$$

in $\mathbf{D}(B)$, called derived tensor-evaluation, *which is functorial in these DG modules. Moreover, if conditions* (i)-(iii) *below hold, then $\operatorname{ev}^{\mathrm{R,L}}_{L,M,N}$ is an isomorphism.*

(i) *The DG ring A is nonpositive and cohomologically pseudo-noetherian, and $L \in \mathbf{D}^{-}_{\mathrm{f}}(A)$.*
(ii) *The DG B-module M is in $\mathbf{D}^{+}(B)$.*
(iii) *The DG B-module N has derived bounded below tensor displacement.*

Proof Choose a K-projective resolution $P \to L$ in $\mathbf{C}_{\mathrm{str}}(A)$ and a K-flat resolution $Q \to N$ in $\mathbf{C}_{\mathrm{str}}(B)$. Consider the obvious homomorphism

$$\operatorname{ev}_{P,M,Q} : \operatorname{Hom}_A(P, M) \otimes_B Q \to \operatorname{Hom}_A(P, M \otimes_B Q)$$

in $\mathbf{C}_{\mathrm{str}}(B)$. We take $\operatorname{ev}^{\mathrm{R,L}}_{L,M,N} := \mathrm{Q}(\operatorname{ev}_{P,M,Q})$ in $\mathbf{D}(B)$. The morphism $\operatorname{ev}^{\mathrm{R,L}}_{L,M,N}$ does not depend on the choices, and hence it is functorial.

Now assume that conditions (i)-(iii) hold. Because the restriction functor $\mathbf{D}(B) \to \mathbf{D}(\mathbb{K})$ is conservative, it is enough to show that $\operatorname{ev}^{\mathrm{R,L}}_{L,M,N}$ is an isomorphism in $\mathbf{D}(\mathbb{K})$. According to Theorem 11.4.40, L is a derived pseudo-finite DG A-module (cf. Example 12.9.2). Now we can use Theorem 12.9.10. □

13

Dualizing Complexes over Commutative Rings

In this chapter we finally explain what was outlined, as a motivating discussion, in Section 0.2 of the Introduction. Dualizing complexes are perhaps the most compelling reason to study derived categories. In the commutative ring setting of the current chapter the technicalities are milder than in the geometric setting (see Remark 13.5.18) and the noncommutative ring setting (which is treated in Chapters 17 and 18).

In the first and third sections of this chapter we talk about *dualizing complexes* and *residue complexes*, respectively, over commutative rings. This material is based on the original treatment by A. Grothendieck in [62], but with a much more detailed discussion. Sandwiched between them is a reminder of the classification of injective modules, due to E. Matlis.

The last two sections are on *rigid dualizing complexes* in the sense of M. Van den Bergh.

In this chapter all rings are commutative $\mathbb{K}$-rings by default (see Convention 1.2.4). However, the base ring $\mathbb{K}$ will remain implicit most of the time.

13.1 Dualizing Complexes

Let A be a commutative ring. The category of A-modules is $\mathbf{M}(A) = \operatorname{Mod} A$. Because A is commutative, the Hom bifunctor has its target in $\mathbf{M}(A)$:

$$\operatorname{Hom}_A(-,-) : \mathbf{M}(A)^{\mathrm{op}} \times \mathbf{M}(A) \to \mathbf{M}(A),$$

and likewise for the right derived bifunctor:

$$\operatorname{RHom}_A(-,-) : \mathbf{D}(A)^{\mathrm{op}} \times \mathbf{D}(A) \to \mathbf{D}(A).$$

See Proposition 12.10.11.

Let $M \in \mathbf{C}(A)$. The DG A-module $\operatorname{End}_A(M) = \operatorname{Hom}_A(M, M)$ is a noncommutative DG central A-ring; namely there is a central DG ring homomorphism

$$\operatorname{hm}_M : A \to \operatorname{Hom}_A(M, M) \tag{13.1.1}$$

called *homothety*. When we forget the ring structure, hm_M becomes a homomorphism in $\mathbf{C}_{\mathrm{str}}(A)$.

Definition 13.1.2 Given a complex $M \in \mathbf{D}(A)$, the *derived homothety morphism*

$$\operatorname{hm}^{\mathrm{R}}_M : A \to \operatorname{RHom}_A(M, M)$$

is the morphism in $\mathbf{D}(A)$ with this formula: $\operatorname{hm}^{\mathrm{R}}_M := \eta^{\mathrm{R}}_{M,M} \circ \mathrm{Q}(\operatorname{hm}_M)$.

In other words, the diagram

$$A \xrightarrow{\mathrm{Q}(\operatorname{hm}_M)} \operatorname{Hom}_A(M, M) \xrightarrow{\eta^{\mathrm{R}}_{M,M}} \operatorname{RHom}_A(M, M), \qquad A \xrightarrow{\operatorname{hm}^{\mathrm{R}}_M} \operatorname{RHom}_A(M, M)$$

in $\mathbf{D}(A)$ is commutative. The letter "R" in the superscript of the expression $\operatorname{hm}^{\mathrm{R}}_M$ refers to "right derived functor."

Exercise 13.1.3 Prove that if $\rho : M \to I$ is a K-injective resolution, then the diagram

$$\begin{array}{ccccc} A & \xrightarrow{\mathrm{Q}(\operatorname{hm}_I)} & \operatorname{Hom}_A(I, I) & \xrightarrow[\cong]{\eta^{\mathrm{R}}_{I,I}} & \operatorname{RHom}_A(I, I) \\ & \searrow^{\operatorname{hm}^{\mathrm{R}}_M} & & & \downarrow \cong \ \operatorname{RHom}(\mathrm{Q}(\rho), \mathrm{Q}(\rho)^{-1}) \\ & & & & \operatorname{RHom}_A(M, M) \end{array}$$

in $\mathbf{D}(A)$ is commutative.

Exercise 13.1.4 Formulate and prove a version of the previous exercise with a K-projective resolution of M.

Definition 13.1.5 A complex $M \in \mathbf{D}(A)$ is said to have the *derived Morita property* if the derived homothety morphism $\operatorname{hm}^{\mathrm{R}}_M : A \to \operatorname{RHom}_A(M, M)$ in $\mathbf{D}(A)$ is an isomorphism.

Proposition 13.1.6 *The following conditions are equivalent for a complex $M \in \mathbf{D}(A)$:*

(i) *M has the derived Morita property.*
(ii) *The homothety ring homomorphism $A \to \operatorname{End}_{\mathbf{D}(A)}(M)$ is bijective, and $\operatorname{Hom}_{\mathbf{D}(A)}(M, M[p]) = 0$ for all $p \neq 0$.*

(iii) *The* A*-module* $\mathrm{H}^0(\mathrm{RHom}_A(M, M))$ *is free of rank* 1*, with basis the element* id_M*, and* $\mathrm{H}^p(\mathrm{RHom}_A(M, M)) = 0$ *for every* $p \neq 0$.

Exercise 13.1.7 Prove Proposition 13.1.6. (Hint: see Corollary 12.2.8 and the preceding material.)

Remark 13.1.8 In some texts (e.g. in [13]), a complex M with the derived Morita property is called a *semi-dualizing complex*. This name is only partly justified, because this property occurs in the definition of a dualizing complex – see Definition 13.1.9 below. However, there is a whole other class of complexes with the derived Morita property – these are the *tilting complexes*. Often these two classes of complexes are disjoint. More on these notions, and their noncommutative variants, can be found in Chapters 14 and 18 of the book.

From here on the commutative ring A is assumed to be noetherian. Recall that $\mathbf{D}_{\mathrm{f}}(A)$ is the full subcategory of $\mathbf{D}(A)$ on the complexes with finitely generated cohomology modules. The subcategory $\mathbf{D}_{\mathrm{f}}(A)$ is triangulated.

The next definition first appeared in [62, Section V.2]. The injective dimension of a complex was defined in Definition 12.4.8.

Definition 13.1.9 Let A be a noetherian commutative ring. A complex of A-modules R is called a *dualizing complex* if it has the following three properties:

(i) $R \in \mathbf{D}^{\mathrm{b}}_{\mathrm{f}}(A)$.
(ii) R has finite injective dimension.
(iii) R has the derived Morita property.

Recall that in the traditional literature (e.g. [105]), a noetherian ring A is called *regular* if all its local rings $A_{\mathfrak{p}}$ are regular local rings. The *Krull dimension* of A is the dimension of the scheme $\mathrm{Spec}(A)$; namely the supremum of the lengths of strictly ascending chains of prime ideals in A. In practice we never see regular rings that are not finite dimensional (there are only pretty exotic examples of them). The following definition will simplify matters for us:

Definition 13.1.10 We shall say that a noetherian commutative ring A is *regular* if it has finite Krull dimension, and all its local rings $A_{\mathfrak{p}}$ are regular local rings, in the sense of [105].

Every field $\mathbb{K}$, and the ring of integers $\mathbb{Z}$, are regular rings. If A is regular, then so is the polynomial ring $A[t_1, \ldots, t_n]$ in $n < \infty$ variables, and also the localization of A at every multiplicatively closed set. See [105, Chapter 7].

As proved by Serre (see [105, Theorem 19.2]) a regular ring A has *finite global cohomological dimension*. This means that there is a number $d \in \mathbb{N}$

such that for all modules $M, N \in \mathbf{M}(A)$ and all $q > d$, the modules $\operatorname{Ext}^q_A(M, N)$ vanish. This implies that every A-module M has injective, projective and flat dimensions $\leq d$. It follows that every complex $M \in \mathbf{D}^{\mathrm{b}}(A)$ has finite injective, projective and flat dimensions (see Definitions 12.4.8 and 12.4.18).

Example 13.1.11 Let A be a regular ring. Taking $R := A$ we see that R satisfies condition (ii) of Definition 13.1.9. The other two conditions hold regardless of the regularity of A. Thus the complex $R = A$ is a dualizing complex over the ring A.

In Section 0.2 of the Introduction we used this fact for the ring $A = \mathbb{Z}$.

Definition 13.1.12 Given a dualizing complex $R \in \mathbf{D}(A)$, the *duality functor* associated to it is the triangulated functor

$$D : \mathbf{D}(A)^{\mathrm{op}} \to \mathbf{D}(A), \quad D := \operatorname{RHom}_A(-, R).$$

The notation "D" deliberately keeps the dualizing complex R implicit. Note that upon applying the functor D to the object $A \in \mathbf{D}(A)$ we get $D(A) = R$.

Let us choose a K-injective resolution $\rho : R \to I$. There is an isomorphism of triangulated functors

$$\operatorname{pres}_I : D \xrightarrow{\simeq} \operatorname{Hom}_A(-, I) \tag{13.1.13}$$

from $\mathbf{D}(A)^{\mathrm{op}}$ to $\mathbf{D}(A)$, which we call a presentation of D. For every $M \in \mathbf{D}(A)$ the diagram

$$\begin{array}{ccc} D(M) & \xrightarrow{\mathrm{id}} & \operatorname{RHom}_A(M, R) \\ {\scriptstyle \operatorname{pres}_{I,M}}\downarrow{\scriptstyle \cong} & & {\scriptstyle \cong}\downarrow{\scriptstyle \operatorname{RHom}(\mathrm{id}_M, \mathrm{Q}(\rho))} \\ \operatorname{Hom}_A(M, I) & \xrightarrow[\cong]{\eta^{\mathrm{R}}_{M,I}} & \operatorname{RHom}_A(M, I) \end{array} \tag{13.1.14}$$

is commutative. (Note that we can choose I to be a bounded complex of injectives, by Proposition 12.4.13.)

Let $M, I \in \mathbf{C}(A)$. There is a homomorphism

$$\operatorname{ev}_{M,I} : M \to \operatorname{Hom}_A(\operatorname{Hom}_A(M, I), I) \tag{13.1.15}$$

in $\mathbf{C}_{\mathrm{str}}(A)$, called *Hom-evaluation*, with formula

$$\operatorname{ev}_{M,N}(m)(\phi) := (-1)^{p \cdot q} \cdot \phi(m)$$

for $m \in M^p$ and $\phi \in \operatorname{Hom}_A(M, I)^q$.

Lemma 13.1.16 *Let R be a dualizing complex over A, with associated duality functor D.*

(1) *The functor $D \circ D : \mathbf{D}(A) \to \mathbf{D}(A)$ is triangulated.*

(2) *There is a unique morphism* $\mathrm{ev}^{\mathrm{R,R}} : \mathrm{Id} \to D \circ D$ *of triangulated functors from* $\mathbf{D}(A)$ *to itself, called* derived Hom-evaluation, *such that for every K-injective resolution* $\rho : R \to I$, *and every complex* $M \in \mathbf{D}(A)$, *the diagram*

$$\begin{array}{ccc} M & \xrightarrow{\ \mathrm{ev}^{\mathrm{R,R}}_M\ } & D(D(M)) \\ & \searrow^{\mathrm{Q}(\mathrm{ev}_{M,I})} & \downarrow \cong \\ & & \mathrm{Hom}_A(\mathrm{Hom}_A(M, I), I) \end{array}$$

is commutative. Here the vertical isomorphism is a double application of the presentation pres_I.

(3) *For* $M = A$ *there is equality* $\mathrm{ev}^{\mathrm{R,R}}_A = \mathrm{hm}^{\mathrm{R}}_R$ *of morphisms* $A \to (D \circ D)(A)$ *in* $\mathbf{D}(A)$.

Proof (1) Choose a K-injective resolution $\rho : R \to I$. Let's write $F := \mathrm{Hom}_A(\mathrm{Hom}_A(-, I), I)$. The functor F can be seen as a DG functor from $\mathbf{C}(A)$ to itself, that is then made into a functor from $\mathbf{D}(A)$ to itself. So by Theorem 5.5.1(1), F is a triangulated functor (i.e. there is a translation isomorphism τ_F, etc.). Using the presentation (13.1.13) twice we get an isomorphism $D \circ D \cong F$ of functors from $\mathbf{D}(A)$ to itself. This makes $D \circ D$ triangulated.

(2) With the K-injective resolution $\rho : R \to I$ above, we define the morphism ev so as to make the diagram commutative. According to Theorem 5.5.1(2), ev is a morphism of triangulated functors. Because the resolution $\rho : R \to I$ is unique up to homotopy, we get the same morphism ev regardless of choice of resolution.

(3) This is clear from Exercise 13.1.3. □

Exercise 13.1.17 Suppose that A is a regular ring, and we take the dualizing complex $R := A$. Let P be a bounded complex of finitely generated projective A-modules. Show that:

(1) There is a canonical isomorphism $D(P) \cong \mathrm{Hom}_A(P, A)$ in $\mathbf{D}(A)$.
(2) There is a canonical isomorphism

$$(D \circ D)(P) \cong \mathrm{Hom}_A(\mathrm{Hom}_A(P, A), A)$$

in $\mathbf{D}(A)$.
(3) Under the isomorphism in item (2), the morphism $\mathrm{ev}^{\mathrm{R,R}}_P$ goes to $\mathrm{Q}(\mathrm{ev}_{P,A})$.
(4) The homomorphism ev_P is a quasi-isomorphism.

Here is the first important result regarding dualizing complexes.

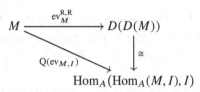

Theorem 13.1.18 *Suppose R is a dualizing complex over the noetherian commutative ring A, with associated duality functor D. Then for every complex $M \in \mathsf{D}_{\mathrm{f}}(A)$ the following hold:*

(1) *The complex $D(M)$ belongs to $\mathsf{D}_{\mathrm{f}}(A)$.*

(2) *The morphism*

$$\mathrm{ev}_M^{R,R} : M \to D(D(M))$$

in $\mathsf{D}(A)$ is an isomorphism.

Proof (1) Condition (ii) of Definition 13.1.9 says that the functor D has finite cohomological dimension. Condition (i) says that $D(A) \in \mathsf{D}_{\mathrm{f}}(A)$. The assertion follows from Theorem 12.5.7, with $\mathsf{N}_0 := \mathsf{M}_{\mathrm{f}}(A)$.

(2) The composition $D \circ D$ is a triangulated functor with finite cohomological dimension (at most twice the injective dimension of R). The cohomological dimension of the identity functor Id is 0 (if $A \neq 0$). By condition (iii) of Definition 13.1.9 we know that ev_A is an isomorphism. Now we can use Theorem 12.5.2. □

Corollary 13.1.19 *Under the assumptions of Theorem 13.1.18, let $\star$ be one of the boundedness conditions* b, +, − *or* ⟨empty⟩, *and let $-\star$ be the reversed boundedness condition. Then the functor*

$$D : \mathsf{D}_{\mathrm{f}}^{\star}(A)^{\mathrm{op}} \to \mathsf{D}_{\mathrm{f}}^{-\star}(A)$$

is an equivalence of triangulated categories.

Proof The previous theorem tells us that D is its own quasi-inverse. The claim about the boundedness holds because D has finite cohomological dimension. □

We saw that dualizing complexes exist over regular rings. This fact is used for the very general existence result Theorem 13.1.34. But first we need some preparation.

Here are some adjunction properties related to homomorphisms between commutative rings. These are enhancements of the material from Section 12.10.

Recall that a ring homomorphism $u : A \to B$ gives rise to a forgetful functor $\mathrm{Rest}_u : \mathsf{D}(B) \to \mathsf{D}(A)$. This functor is going to be implicit in the discussion below.

Lemma 13.1.20 *Let $A \to B$ be a ring homomorphism.*

(1) *If $I \in \mathsf{C}(A)$ is K-injective, then $J := \mathrm{Hom}_A(B, I) \in \mathsf{C}(B)$ is K-injective.*

(2) *Given* $M \in \mathsf{D}(A)$, *let us define* $N := \mathrm{RHom}_A(B, M) \in \mathsf{D}(B)$. *Then there is an isomorphism* $\mathrm{RHom}_B(-, N) \cong \mathrm{RHom}_A(-, M)$ *of triangulated functors* $\mathsf{D}(B)^{\mathrm{op}} \to \mathsf{D}(B)$.

Proof (1) This is an adjunction calculation. Suppose $L \in \mathsf{C}(B)$ is acyclic. There are isomorphisms

$$\mathrm{Hom}_B(L, J) \cong \mathrm{Hom}_B(L, \mathrm{Hom}_A(B, I)) \cong \mathrm{Hom}_A(L, I) \tag{13.1.21}$$

in $\mathsf{C}(B)$. Since I is K-injective over A, the complex $\mathrm{Hom}_A(L, I)$ is acyclic.

(2) Choose a K-injective resolution $M \to I$ in $\mathsf{C}(A)$. Let $J := \mathrm{Hom}_A(B, I)$. Then $N \to J$ is a K-injective resolution in $\mathsf{C}(B)$. There are isomorphisms of triangulated functors

$$\mathrm{RHom}_A(-, M) \cong \mathrm{Hom}_A(-, I) \tag{13.1.22}$$

and

$$\mathrm{RHom}_B(-, N) \cong \mathrm{Hom}_B(-, J), \tag{13.1.23}$$

where the functors (13.1.22) are contravariant functors from $\mathsf{D}(A)$ to itself, and the functors (13.1.23) are contravariant functors from $\mathsf{D}(B)$ to itself. But given $L \in \mathsf{C}(B)$ we can view $\mathrm{Hom}_A(L, I)$ as a complex of B-modules, and in this way the functors (13.1.22) become contravariant triangulated functors from $\mathsf{D}(B)$ to itself. Formula (13.1.21) shows that the functors (13.1.22) and (13.1.23) are isomorphic. □

Lemma 13.1.24 *Let* $A \to B$ *be a flat ring homomorphism, let* $M \in \mathsf{D}^-_{\mathrm{f}}(A)$, *and let* $N \in \mathsf{D}^+(A)$. *Then there is an isomorphism*

$$\mathrm{RHom}_A(M, N) \otimes_A B \cong \mathrm{RHom}_B(B \otimes_A M, B \otimes_A N)$$

in $\mathsf{D}(B)$. *This isomorphism is functorial in* M *and* N.

Proof Theorem 12.10.14 tells us that

$$\mathrm{RHom}_A(M, N) \otimes_A B \cong \mathrm{RHom}_A(M, B \otimes_A N)$$

in $\mathsf{D}(B)$. Then, by forward adjunction (Theorem 12.6.6) we get

$$\mathrm{RHom}_A(M, B \otimes_A N) \cong \mathrm{RHom}_B(B \otimes_A M, B \otimes_A N).$$

The functoriality is clear. □

Lemma 13.1.25 *Let* I *be an* A-*module. The following conditions are equivalent:*

(i) I *is injective.*
(ii) *For every finitely generated* A-*module* M *the module* $\mathrm{Ext}^1_A(M, I)$ *is zero.*

Exercise 13.1.26 Prove Lemma 13.1.25. (Hint: use the Baer criterion, Theorem 2.7.10.)

Lemma 13.1.27 *The injective dimension of a complex $N \in \mathbf{D}(A)$ equals the cohomological dimension of the functor*

$$\mathrm{RHom}_A(-, N)|_{\mathbf{M}_{\mathrm{f}}(A)^{\mathrm{op}}} : \mathbf{M}_{\mathrm{f}}(A)^{\mathrm{op}} \to \mathbf{D}(A).$$

Proof By definition the injective dimension of N, say d, is the cohomological dimension of the functor $\mathrm{RHom}_A(-, N) : \mathbf{D}(A)^{\mathrm{op}} \to \mathbf{D}(A)$. Let d' be the cohomological dimension of the functor $\mathrm{RHom}_A(-, N)|_{\mathbf{M}_{\mathrm{f}}(A)^{\mathrm{op}}}$. Obviously the inequality $d \geq d'$ holds. For the reverse inequality we may assume that $\mathrm{H}(N)$ is nonzero and $d' < \infty$. This implies that there are integers $q_1 = q_0 + d'$ such that for every $M \in \mathbf{M}_{\mathrm{f}}(A)$ there is an inclusion

$$\mathrm{con}(\mathrm{H}(\mathrm{RHom}_A(M, N))) \subseteq [q_0, q_1].$$

In particular, for $M = A$, we get $\mathrm{con}(\mathrm{H}(N)) \subseteq [q_0, q_1]$. Let $N \to J$ be an injective resolution in $\mathbf{C}(A)$ with $\inf(J) \geq q_0$. Take $I := \mathrm{smt}^{\leq q_1}(J)$, the smart truncation from Definition 7.3.6. The proof of Proposition 12.4.13, plus Lemma 13.1.25, show that $N \to I$ is an injective resolution. But then $\mathrm{RHom}_A(-, N) \cong \mathrm{Hom}_A(-, I)$, so this functor has cohomological displacement in the interval $[q_0, q_1]$ that has length d'. □

Recall that a ring homomorphism $A \to B$ is called *finite* if it makes B into a finitely generated A-module.

Proposition 13.1.28 *Let $A \to B$ be a finite ring homomorphism, and let R_A be a dualizing complex over A. Then the complex*

$$R_B := \mathrm{RHom}_A(B, R_A) \in \mathbf{D}(B)$$

is a dualizing complex over B.

Proof Consider the duality functors $D_A := \mathrm{RHom}_A(-, R_A)$ and $D_B := \mathrm{RHom}_B(-, R_B)$. As explained in the proof of Lemma 13.1.20(2), they are isomorphic as contravariant triangulated functors from $\mathbf{D}(B)$ to itself. Since $R_B = D_A(B)$ and $B \in \mathbf{D}^{\mathrm{b}}_{\mathrm{f}}(A)$, by Corollary 13.1.19, we have $R_B \in \mathbf{D}^{\mathrm{b}}_{\mathrm{f}}(A)$. But then also $R_B \in \mathbf{D}^{\mathrm{b}}_{\mathrm{f}}(B)$. Next, because $D_B(L) \cong D_A(L)$ for all $L \in \mathbf{D}(B)$, this implies that the cohomological dimension of D_B is at most that of D_A, which is finite. We see that the injective dimension of the complex R_B is finite. Lastly, there is an isomorphism $D_B \circ D_B \cong D_A \circ D_A$ as functors from $\mathbf{D}^{\mathrm{b}}_{\mathrm{f}}(B)$ to itself, and hence $\mathrm{ev}^{\mathrm{R,R}} : \mathrm{Id} \to D_B \circ D_B$ is an isomorphism. Applying this to the object $B \in \mathbf{D}^{\mathrm{b}}_{\mathrm{f}}(B)$ we see that $\mathrm{hm}^{\mathrm{R}}_{R_B} = \mathrm{ev}^{\mathrm{R,R}}_B : B \to (D_B \circ D_B)(B)$ is an isomorphism. So R_B has the derived Morita property. The conclusion is that R_B is a dualizing complex over B. □

Recall that a ring homomorphism $A \to B$ is called a *localization* if B is isomorphic, as an A-ring, to the localization A_S of A with respect to some multiplicatively closed subset S.

Proposition 13.1.29 *Let $A \to B$ be a localization ring homomorphism, and let R_A be a dualizing complex over A. Then the complex*

$$R_B := B \otimes_A R_A \in \mathbf{D}(B)$$

is a dualizing complex over B.

Proof It is clear that $R_B \in \mathbf{D}^{\mathrm{b}}_{\mathrm{f}}(B)$. By Lemma 13.1.27, to compute the injective dimension of R_B it is enough to look at $\mathrm{RHom}_B(M, R_B)$ for $M \in \mathbf{M}_{\mathrm{f}}(B)$. We can find a finitely generated A-submodule $M' \subseteq M$ such that $B \cdot M' = M$; and then $M \cong B \otimes_A M'$. Lemma 13.1.24 tells us that

$$\mathrm{RHom}_B(M, R_B) \cong \mathrm{RHom}_A(M', R_A) \otimes_A B.$$

We conclude that the injective dimension of R_B is at most that of R_A, which is finite. Lastly, by the same lemma we get an isomorphism

$$\mathrm{RHom}_B(R_B, R_B) \cong \mathrm{RHom}_A(R_A, R_A) \otimes_A B,$$

and it is compatible with the morphisms from B. Thus R_B has the derived Morita property. □

Recall that a ring homomorphism $A \to B$ is called *finite type* if B is finitely generated as an A ring (i.e. B is a quotient of the polynomial ring $A[t_1, \ldots, t_n]$ for some $n < \infty$).

Definition 13.1.30 A ring homomorphism $u : A \to B$ is called *essentially finite type* (EFT) if it can be factored into $u = u_{\mathrm{loc}} \circ u_{\mathrm{ft}}$, where $u_{\mathrm{ft}} : A \to B_{\mathrm{ft}}$ is a finite type ring homomorphism, and $u_{\mathrm{loc}} : B_{\mathrm{ft}} \to B$ is a localization homomorphism.

This is illustrated in the following commutative diagram.

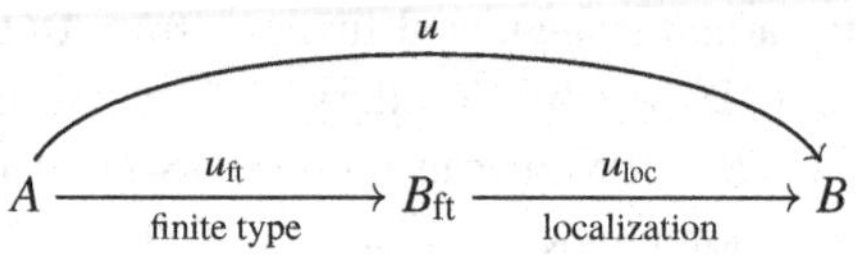

Proposition 13.1.31 *Let $u : A \to B$ be an EFT ring homomorphism.*

(1) *Let $v : B \to C$ be another EFT ring homomorphism. Then $v \circ u : A \to C$ is an EFT ring homomorphism.*
(2) *Let $A \to A'$ be a ring homomorphism, and define $B' := A' \otimes_A B$. Then the induced ring homomorphism $u' : A' \to B'$ is EFT.*

(3) *If the ring A is noetherian, then B is also noetherian.*

Exercise 13.1.32 Prove Proposition 13.1.31.

Example 13.1.33 Let $\mathbb{K}$ be a noetherian ring, let X be a finite type $\mathbb{K}$-scheme, and let $x \in X$ be a point. Then the local ring $\mathcal{O}_{X,x}$ is an EFT $\mathbb{K}$-ring.

Theorem 13.1.34 *Let $\mathbb{K}$ be a regular noetherian commutative ring, and let A be an essentially finite type commutative $\mathbb{K}$-ring. Then A has a dualizing complex.*

See Definition 13.1.10 regarding regular rings.

Proof The ring homomorphism $\mathbb{K} \to A$ can be factored as $\mathbb{K} \to A_{\mathrm{pl}} \to A_{\mathrm{ft}} \to A$, where $A_{\mathrm{pl}} = \mathbb{K}[t_1, \ldots, t_n]$ is a polynomial ring, $A_{\mathrm{pl}} \to A_{\mathrm{ft}}$ is surjective, and $A_{\mathrm{ft}} \to A$ is a localization. (The subscripts stand for "polynomial" and "finite type," respectively.) According to [105, Theorem 19.5], the ring A_{pl} is regular; so, as shown in Example 13.1.11, the complex $R_{\mathrm{pl}} := A_{\mathrm{pl}}$ is a dualizing complex over A_{pl}.

Define $R_{\mathrm{ft}} := \mathrm{RHom}_{A_{\mathrm{pl}}}(A_{\mathrm{ft}}, R_{\mathrm{pl}}) \in \mathbf{D}(A_{\mathrm{ft}})$. By Proposition 13.1.28 this is a dualizing complex over A_{ft}. Finally define $R := A \otimes_{A_{\mathrm{ft}}} R_{\mathrm{ft}} \in \mathbf{D}^{\mathrm{b}}_{\mathrm{f}}(A)$. By Proposition 13.1.29 this is dualizing complex over A. □

The proof of Theorem 13.1.34 might give the impression that A could have a lot of nonisomorphic dualizing complexes. This is not quite true, as the next theorem demonstrates.

Theorem 13.1.35 *Let A be a noetherian commutative ring with connected spectrum, and let R and R' be dualizing complexes over A. Then there is a rank 1 projective A-module L and an integer d, such that $R' \cong R \otimes_A L[d]$ in $\mathbf{D}(A)$.*

We need some lemmas before proving the theorem.

Lemma 13.1.36 (Künneth Trick) *Let $M, M' \in \mathbf{D}^-(A)$, and let $i, i' \in \mathbb{Z}$ be such that* $\sup(\mathrm{H}(M)) \leq i$ *and* $\sup(\mathrm{H}(M')) \leq i'$. *Then*

$$\mathrm{H}^{i+i'}(M \otimes^{\mathrm{L}}_A M') \cong \mathrm{H}^i(M) \otimes_A \mathrm{H}^{i'}(M')$$

as A-modules.

Exercise 13.1.37 Prove Lemma 13.1.36.

Lemma 13.1.38 (Projective Truncation Trick) *Let $M \in \mathbf{D}^-(A)$, with $i_1 :=$* $\sup(\mathrm{H}(M)) \in \mathbb{Z}$. *Assume the A-module $P := \mathrm{H}^{i_1}(M)$ is projective. Then there is an isomorphism $M \cong \mathrm{smt}^{\leq i_1 - 1}(M) \oplus P[-i_1]$ in $\mathbf{D}(A)$.*

Exercise 13.1.39 Prove Lemma 13.1.38. (Hint: first replace M with its truncation $\operatorname{smt}^{\leq i_1}(M)$. Then prove that P is a direct summand of M^{i_1}.)

By a *principal open set* in the affine scheme in $\operatorname{Spec}(A)$ we mean a set of the form $\operatorname{Spec}(A_s)$, where A_s is the localization of A at an element $s \in A$. Thus

$$\operatorname{Spec}(A_s) = \{\mathfrak{p} \in \operatorname{Spec}(A) \mid s \notin \mathfrak{p}\}.$$

Lemma 13.1.40 *Let $M, M' \in \mathbf{M}_{\mathrm{f}}(A)$, and let $\mathfrak{p} \subseteq A$ be a prime ideal.*

(1) *If $M_{\mathfrak{p}} \neq 0$ and $M'_{\mathfrak{p}} \neq 0$ then $M_{\mathfrak{p}} \otimes_{A_{\mathfrak{p}}} M'_{\mathfrak{p}} \neq 0$.*
(2) *If $M_{\mathfrak{p}} \otimes_{A_{\mathfrak{p}}} M'_{\mathfrak{p}} \cong A_{\mathfrak{p}}$ then $M_{\mathfrak{p}} \cong M'_{\mathfrak{p}} \cong A_{\mathfrak{p}}$.*
(3) *If $M_{\mathfrak{p}} \cong A_{\mathfrak{p}}$, then there is a principal open neighborhood* $\operatorname{Spec}(A_s)$ *of* $\mathfrak{p}$ *in* $\operatorname{Spec}(A)$ *such that $M_s \cong A_s$ as A_s-modules.*

Exercise 13.1.41 Prove Lemma 13.1.40. (Hint: use the Nakayama Lemma.)

Here is a pretty difficult technical lemma.

Lemma 13.1.42 *In the situation of the theorem, let $M, M' \in \mathbf{D}^-_{\mathrm{f}}(A)$ satisfy $M \otimes^{\mathrm{L}}_A M' \cong A$ in $\mathbf{D}(A)$. Then $M \cong L[d]$ in $\mathbf{D}(A)$ for some rank 1 projective A-module L and an integer d.*

Proof For each prime $\mathfrak{p} \subseteq A$ let $M_{\mathfrak{p}} := A_{\mathfrak{p}} \otimes_A M$, and define $e_{\mathfrak{p}} := \sup(\mathrm{H}(M_{\mathfrak{p}}))$ in $\mathbb{Z} \cup \{-\infty\}$. Define the generalized integer $e'_{\mathfrak{p}}$ similarly.

Fix one prime $\mathfrak{p}$. The associativity and symmetry of the left derived tensor product imply the existence of these isomorphisms

$$\begin{aligned} M_{\mathfrak{p}} \otimes^{\mathrm{L}}_{A_{\mathfrak{p}}} M'_{\mathfrak{p}} &= (A_{\mathfrak{p}} \otimes^{\mathrm{L}}_A M) \otimes^{\mathrm{L}}_{A_{\mathfrak{p}}} (A_{\mathfrak{p}} \otimes^{\mathrm{L}}_A M') \\ &\cong A_{\mathfrak{p}} \otimes^{\mathrm{L}}_A (M \otimes^{\mathrm{L}}_A M') \cong A_{\mathfrak{p}} \otimes^{\mathrm{L}}_A A \cong A_{\mathfrak{p}} \end{aligned} \tag{13.1.43}$$

in $\mathbf{D}(A_{\mathfrak{p}})$. Since $A_{\mathfrak{p}} \neq 0$, it follows that $\mathrm{H}(M_{\mathfrak{p}}) \neq 0$ and $\mathrm{H}(M'_{\mathfrak{p}}) \neq 0$. So $e_{\mathfrak{p}}, e'_{\mathfrak{p}} \in \mathbb{Z}$, and $\mathrm{H}^{e_{\mathfrak{p}}}(M_{\mathfrak{p}}), \mathrm{H}^{e'_{\mathfrak{p}}}(M'_{\mathfrak{p}})$ are nonzero finitely generated $A_{\mathfrak{p}}$-modules. By Lemma 13.1.40(1) we know that

$$\mathrm{H}^{e_{\mathfrak{p}}}(M_{\mathfrak{p}}) \otimes_{A_{\mathfrak{p}}} \mathrm{H}^{e'_{\mathfrak{p}}}(M'_{\mathfrak{p}}) \neq 0.$$

According to Lemma 13.1.36, we have

$$\mathrm{H}^{e_{\mathfrak{p}}}(M_{\mathfrak{p}}) \otimes_{A_{\mathfrak{p}}} \mathrm{H}^{e'_{\mathfrak{p}}}(M'_{\mathfrak{p}}) \cong \mathrm{H}^{(e_{\mathfrak{p}}+e'_{\mathfrak{p}})}(M_{\mathfrak{p}} \otimes^{\mathrm{L}}_{A_{\mathfrak{p}}} M'_{\mathfrak{p}}) \cong \mathrm{H}^{(e_{\mathfrak{p}}+e'_{\mathfrak{p}})}(A_{\mathfrak{p}}).$$

But $A_{\mathfrak{p}}$ is concentrated in degree 0; this forces $e_{\mathfrak{p}} + e'_{\mathfrak{p}} = 0$ and

$$\mathrm{H}^{e_{\mathfrak{p}}}(M_{\mathfrak{p}}) \otimes_{A_{\mathfrak{p}}} \mathrm{H}^{e'_{\mathfrak{p}}}(M'_{\mathfrak{p}}) \cong A_{\mathfrak{p}}$$

in $\mathbf{D}(A_{\mathfrak{p}})$. By Lemma 13.1.40(2), we now see that $\mathrm{H}^{e_{\mathfrak{p}}}(M_{\mathfrak{p}}) \cong \mathrm{H}^{e'_{\mathfrak{p}}}(M'_{\mathfrak{p}}) \cong A_{\mathfrak{p}}$. According to Lemma 13.1.38, there are isomorphisms

$$M_{\mathfrak{p}} \cong A_{\mathfrak{p}}[-e_{\mathfrak{p}}] \oplus \operatorname{smt}^{\leq e_{\mathfrak{p}}-1}(M_{\mathfrak{p}}) \tag{13.1.44}$$

and

$$M'_{\mathfrak{p}} \cong A_{\mathfrak{p}}[-e'_{\mathfrak{p}}] \oplus \operatorname{smt}^{\leq e'_{\mathfrak{p}}-1}(M'_{\mathfrak{p}})$$

in $\mathbf{D}(A_{\mathfrak{p}})$. These, with (13.1.43), give an isomorphism

$$(A_{\mathfrak{p}}[-e_{\mathfrak{p}}] \oplus \operatorname{smt}^{\leq e_{\mathfrak{p}}-1}(M_{\mathfrak{p}})) \otimes^{\mathrm{L}}_{A_{\mathfrak{p}}} (A_{\mathfrak{p}}[-e'_{\mathfrak{p}}] \oplus \operatorname{smt}^{\leq e'_{\mathfrak{p}}-1}(M'_{\mathfrak{p}})) \cong A_{\mathfrak{p}}. \tag{13.1.45}$$

The left side of (13.1.45) is the direct sum of four objects. Passing to the cohomology of (13.1.45) we see that $N := \mathrm{H}(\operatorname{smt}^{\leq e_{\mathfrak{p}}-1}(M_{\mathfrak{p}})[-e'_{\mathfrak{p}}])$ is a direct summand of $A_{\mathfrak{p}}$. But, since $e'_{\mathfrak{p}} + e_{\mathfrak{p}} = 0$, the graded module N is concentrated in the degree interval $[\infty, -1]$. It follows that $N = 0$. Therefore, by (13.1.44), we deduce that

$$M_{\mathfrak{p}} \cong A_{\mathfrak{p}}[-e_{\mathfrak{p}}]. \tag{13.1.46}$$

The calculation above works for every prime $\mathfrak{p}$. From (13.1.46) we get

$$A_{\mathfrak{p}} \otimes_A \mathrm{H}^i(M) \cong \mathrm{H}^i(M_{\mathfrak{p}}) \cong \begin{cases} A_{\mathfrak{p}} & \text{if } i = e_{\mathfrak{p}}, \\ 0 & \text{otherwise.} \end{cases} \tag{13.1.47}$$

We now use Lemma 13.1.40(3) to deduce that for every prime $\mathfrak{p}$ there is an open neighborhood $U_{\mathfrak{p}}$ of $\mathfrak{p}$ in $\operatorname{Spec}(A)$ such that $\mathrm{H}^{e_{\mathfrak{p}}}(M_{\mathfrak{q}}) \cong A_{\mathfrak{q}}$ for all $\mathfrak{q} \in U_{\mathfrak{p}}$. This implies, by equation (13.1.47), that $e_{\mathfrak{q}} = e_{\mathfrak{p}}$. Therefore $\mathfrak{p} \mapsto e_{\mathfrak{p}}$ is a locally constant function $\operatorname{Spec}(A) \to \mathbb{Z}$. We assumed that $\operatorname{Spec}(A)$ is connected, and this implies that this is a constant function, say $e_{\mathfrak{p}} = -d$ for some integer d.

Define $L := \mathrm{H}^{-d}(M) \in \mathbf{M}_{\mathrm{f}}(A)$. Using truncation we see that $M \cong L[d]$ in $\mathbf{D}(A)$. We know that $L_{\mathfrak{p}} \cong A_{\mathfrak{p}}$ for all primes $\mathfrak{p}$. Finally, Lemma 13.1.24 says that the A-module L is projective. □

Remark 13.1.48 Lemma 13.1.42 is actually true in much greater generality: the ring A does not have to be noetherian, and we do not have to assume that the complexes M and M' have bounded above or finite cohomology. The proof is harder (see the proof of Theorem 14.5.17).

Proof of Theorem 13.1.35 Define the duality functors $D := \operatorname{RHom}_A(-, R)$ and $D' := \operatorname{RHom}_A(-, R')$; these are finite dimensional contravariant triangulated functors from $\mathbf{D}_{\mathrm{f}}(A)$ to itself. And define $F := D' \circ D$ and $F' := D \circ D'$, which are finite dimensional (covariant) triangulated functors from $\mathbf{D}_{\mathrm{f}}(A)$ to itself. Let

$$M := F(A) = D'(D(A)) = \operatorname{RHom}_A(R, R') \tag{13.1.49}$$

and

$$M' := F'(A) = D(D'(A)) = \operatorname{RHom}_A(R', R).$$

These are objects of $\mathbf{D}^{\mathrm{b}}_{\mathrm{f}}(A)$.

For every object $N \in \mathbf{D}(A)$ there is a morphism

$$\psi_N : N \otimes^{\mathrm{L}}_A \operatorname{RHom}_A(R, R') \to \operatorname{RHom}_A(\operatorname{RHom}_A(N, R), R')$$

defined as follows: we choose a K-projective resolution $P \to N$ and a K-injective resolution $R' \to I'$. Then ψ_N is represented by the obvious homomorphism of complexes

$$P \otimes_A \operatorname{Hom}_A(R, I') \to \operatorname{Hom}_A(\operatorname{Hom}_A(P, R), I').$$

As N changes, ψ_N is a morphism of triangulated functors $\psi : (-) \otimes^{\mathrm{L}}_A M \to D' \circ D = F$. For $N = A$ the morphism ψ_A is an isomorphism, by equation (13.1.49). The functor F has finite cohomological dimension, and the functor $(-) \otimes^{\mathrm{L}}_A M$ has bounded above cohomological displacement. According to Theorem 12.5.2, the morphism ψ_N is an isomorphism for every $N \in \mathbf{D}^-_{\mathrm{f}}(A)$. In particular this is true for $N := M'$. So, using Theorem 13.1.18, we obtain

$$M' \otimes^{\mathrm{L}}_A M \cong (D' \circ D)(M') \cong (D' \circ D \circ D \circ D')(A) \cong A.$$

According to Lemma 13.1.42, there is an isomorphism $M \cong L[d]$. Finally, using the isomorphism ψ_R, we get

$$R \otimes_A L[d] \cong F(R) = D'(D(R)) \cong D'(A) = R'.$$

□

What if $\operatorname{Spec}(A)$ has more than one connected component?

Definition 13.1.50 Let A be a ring.

(1) We say that A has a *finite connected component decomposition* if

$$\operatorname{Spec}(A) = \coprod_{i=1}^{m} \operatorname{Spec}(A_i),$$

a finite disjoint union of nonempty connected closed subschemes.

(2) If A has a finite connected component decomposition as above, then the *connected component decomposition* of A is the ring isomorphism $A \cong \prod_{i=1}^{m} A_i$.

Of course, if A has a finite connected component decomposition, then this decomposition is unique, up to renumbering.

Proposition 13.1.51 *If A is noetherian, then it has a finite connected component decomposition.*

Exercise 13.1.52

(1) Prove the proposition above.

(2) Find a ring A that does not have a finite connected component decomposition.

Corollary 13.1.53 *Let R and R' be dualizing complexes over A, and let $A = \prod_{i=1}^{m} A_i$ be the connected component decomposition of A. Then there is an isomorphism*

$$R' \cong R \otimes_A (L_1[d_1] \oplus \cdots \oplus L_m[d_m])$$

in $\mathsf{D}(A)$*, where each* L_i *is a rank* 1 *projective* A_i*-module, and each* d_i *is an integer. Furthermore, the modules* L_i *are unique up to isomorphism, and the integers* d_i *are unique.*

Exercise 13.1.54 Prove Corollary 13.1.53.

Remark 13.1.55 A rank 1 projective A-module L is also called an *invertible* A*-module*. This is because L is invertible for the tensor product. Recall that the group of isomorphism classes of invertible A-modules is the *commutative Picard group* $\mathrm{Pic}_A(A)$.

The *commutative derived Picard group* $\mathrm{DPic}_A(A)$ is the abelian group $\mathrm{Pic}_A(A) \times \mathbb{Z}^m$ that classifies dualizing complexes over A, as in Corollary 13.1.53.

Now assume that A is *noncommutative*, and flat central over a commutative ring $\mathbb{K}$. There are noncommutative versions of dualizing complexes and of "invertible" complexes, which are called *tilting complexes*. The latter form the nonabelian group $\mathrm{DPic}_{\mathbb{K}}(A)$, and it classifies noncommutative dualizing complexes. See [126], [127], [79], [165] and [132]. We shall study this material in Chapters 14 and 18 of the book.

Remark 13.1.56 The lack of uniqueness of dualizing complexes has always been a source of difficulty. A certain uniqueness or functoriality is needed already for proving the existence of dualizing complexes on schemes.

In [62] Grothendieck utilized local and global duality in order to formulate a suitable uniqueness of dualizing complexes. This approach was very cumbersome (even without providing details!).

Since then, there have been a few approaches in the literature to attack this difficulty. Generally speaking, these approaches come in two flavors:

- *Representability.* This started with P. Deligne's Appendix to [62] and continued most notably in the work of A. Neeman, J. Lipman and their coauthors. See [114], [98] and their references.
- *Explicit Constructions.* Mostly in the early work of Lipman et al., including [97] and [96], and in the work of A. Yekutieli [162], [164] and [168].

In Section 13.5 of the book we will present *rigid residue complexes*, for which there is a built-in uniqueness, and even functoriality (see Remark 13.5.17).

13.2 Interlude: The Matlis Classification of Injective Modules

We start with a few facts about injective modules over rings that are neither commutative nor noetherian. Sources for this material are [129] and [93].

Definition 13.2.1

(1) Let M be an A-module. A submodule $N \subseteq M$ is called an *essential submodule* if for every nonzero submodule $L \subseteq M$, the intersection $N \cap L$ is nonzero. In this case we also say that M is an *essential extension* of N.

(2) An *essential monomorphism* is a monomorphism $\phi : N \rightarrowtail M$ whose image is an essential submodule of M.

(3) Let M be an A-module. An *injective hull* (or *injective envelope*) of M is an injective module I, together with an essential monomorphism $M \rightarrowtail I$.

Proposition 13.2.2 *Every A-module M admits an injective hull.*

Proof See [129, Theorem 3.30] or [93, Section 3.D]. □

There is a weak uniqueness result for injective hulls.

Proposition 13.2.3 *Let M be an A-module, and suppose $\phi : M \rightarrowtail I$ and $\phi' : M \rightarrowtail I'$ are monomorphisms into injective modules.*

(1) *If ϕ is essential, then there is a monomorphism $\psi : I \rightarrowtail I'$ such that $\psi \circ \phi = \phi'$.*

(2) *If ϕ' is also essential, then ψ above is an isomorphism.*

Exercise 13.2.4 Prove Proposition 13.2.3.

In classical homological algebra we talk about the minimal injective resolution of a module M. Let us recall it. We start by taking the injective hull $\phi : M \rightarrowtail I^0$. This gives an exact sequence

$$0 \to M \xrightarrow{\phi} I^0 \to M^1 \to 0,$$

where M^1 is the cokernel of ϕ. Then we take the injective hull $M^1 \rightarrowtail I^1$, and this gives a longer exact sequence

$$0 \to M \to I^0 \to I^1 \to M^2 \to 0,$$

and so on. We want to generalize this idea to complexes.

Definition 13.2.5

(1) A *minimal complex of injective A-modules* is a bounded below complex of injective modules I, such that for every integer q the submodule of cocycles $\mathrm{Z}^q(I) \subseteq I^q$ is essential.

(2) Let $M \in \mathbf{D}^+(A)$. A *minimal injective resolution* of M is a quasi-isomorphism $M \to I$ into a minimal complex of injectives I.

Proposition 13.2.6 *Let* $M \in \mathbf{D}^+(A)$.

(1) *There exists a minimal injective resolution* $\phi : M \to I$.

(2) *If* $\phi' : M \to I'$ *is another minimal injective resolution, then there is an isomorphism* $\psi : I \to I'$ *in* $\mathbf{C}_{\mathrm{str}}(A)$ *such that* $\phi' = \psi \circ \phi$.

(3) *If M has finite injective dimension, then it has a bounded minimal injective resolution I.*

Proof (1) We know that there is a quasi-isomorphism $M \to J$ where J is a bounded below complex of injective modules. For every q let E^q be an injective hull of $\mathrm{Z}^q(J)$. By Proposition 13.2.3(1) we can assume that E^q sits inside J^q like this: $\mathrm{Z}^q(J) \subseteq E^q \subseteq J^q$. Since E^q is injective, we can decompose J^q into a direct sum: $J^q \cong E^q \oplus K^q$. The homomorphism $\mathrm{d}_J^q : K^q \to J^{q+1}$ is a monomorphism since $K^q \cap \mathrm{Z}^q(J) = 0$. And the image $\mathrm{d}_J^q(K^q)$ is contained in E^{q+1}. Thus $\mathrm{d}_J^q(K^q)$ is a direct summand of E^{q+1}, and this shows that the quotient

$$I^{q+1} := E^{q+1}/\mathrm{d}_J^q(K^q) \cong J^{q+1}/(K^{q+1} \oplus \mathrm{d}_J^q(K^q))$$

is an injective module. The canonical surjection of graded modules $\pi : J \to I$ is a homomorphism of complexes, with kernel the acyclic complex

$$\bigoplus_{q \in \mathbb{Z}} \big(K^q[-q] \xrightarrow{\mathrm{d}_J^q} \mathrm{d}_J^q(K^q)[-q-1] \big).$$

Therefore π is a quasi-isomorphism. A short calculation shows that I is a minimal complex of injectives, i.e. $\mathrm{Z}^q(I) \subseteq I^q$ is essential.

(2) See next exercise. (We will not need this fact.)

(3) According to Proposition 12.4.13, the complex J that appears in item (1) can be chosen to be bounded. □

Exercise 13.2.7 Prove Proposition 13.2.6(2).

Remark 13.2.8 Important: the isomorphisms ψ in Propositions 13.2.3 and 13.2.6 are not unique (see next exercise). We will see below (in Section 13.5) that a rigid residue complex is a minimal complex of injectives that has no nontrivial rigid automorphisms.

Exercise 13.2.9 Take $A := \mathbb{K}[[t]]$, the power series ring over a field $\mathbb{K}$. Let $M := A/(t)$, the trivial module (the residue field viewed as an A-module).

(1) Find a minimal injective resolution $M \to I$.

(2) Find a nontrivial automorphism of the complex I in $\mathbf{C}_{\mathrm{str}}(A)$ that fixes the submodule $M \subseteq I^0$.

Now we add the noetherian condition.

Proposition 13.2.10 *Assume A is a left noetherian ring. Let $\{I_z\}_{z \in Z}$ be a collection of injective A-modules. Then $I := \bigoplus_{z \in Z} I_z$ is an injective A-module.*

Exercise 13.2.11 Prove Proposition 13.2.10. (Hint: use the Baer criterion.)

From here on in this section all rings are noetherian commutative. For them much more can be said.

Recall that a module M is called *indecomposable* if it is not the direct sum of two nonzero modules.

Definition 13.2.12 Let $\mathfrak{a} \subseteq A$ be an ideal.

(1) Let M be an A-module. The $\mathfrak{a}$*-torsion submodule* of M is the submodule $\Gamma_{\mathfrak{a}}(M)$ consisting of the elements that are annihilated by powers of $\mathfrak{a}$. Thus

$$\Gamma_{\mathfrak{a}}(M) = \lim_{i \to} \mathrm{Hom}_A(A/\mathfrak{a}^i, M) \subseteq M.$$

(2) If $\Gamma_{\mathfrak{a}}(M) = M$ then M is called an $\mathfrak{a}$*-torsion module.*

(3) The functor $\Gamma_{\mathfrak{a}} : \mathbf{M}(A) \to \mathbf{M}(A)$ is called the $\mathfrak{a}$*-torsion functor.*

Here are some important properties of the torsion functor.

Proposition 13.2.13 *Let $\mathfrak{a}$ be an ideal in A.*

(1) *The functor $\Gamma_{\mathfrak{a}}$ is left exact.*

(2) *The functor $\Gamma_{\mathfrak{a}}$ commutes with infinite direct sums.*

Exercise 13.2.14 Prove the proposition above.

Perhaps the most important theorem about injective modules over noetherian commutative rings is the following structural result due to E. Matlis [104] from 1958. See also [148, Section V.4], [93, Sections 3.F and 3.I], [105, Section 18] and [30].

For a prime ideal $\mathfrak{p} \subseteq A$ we write $\boldsymbol{k}(\mathfrak{p}) := A_{\mathfrak{p}}/\mathfrak{p}_{\mathfrak{p}}$, the residue field of $\mathfrak{p}$.

Theorem 13.2.15 (Matlis) *Let A be a noetherian commutative ring.*

(1) *Let $\mathfrak{p} \subseteq A$ be a prime ideal, and let $J(\mathfrak{p})$ be the injective hull of the $A_{\mathfrak{p}}$-module $\boldsymbol{k}(\mathfrak{p})$. Then, as an A-module, $J(\mathfrak{p})$ is injective, indecomposable and $\mathfrak{p}$-torsion.*

(2) *Suppose I is an indecomposable injective A-module. Then $I \cong J(\mathfrak{p})$ for a unique prime ideal $\mathfrak{p} \subseteq A$.*
(3) *Every injective A module I is a direct sum of indecomposable injective A-modules.*

Theorem 13.2.15 tells us that every injective A-module I can be written as a direct sum

$$I \cong \bigoplus_{\mathfrak{p} \in \operatorname{Spec}(A)} J(\mathfrak{p})^{\oplus \mu_{\mathfrak{p}}} \tag{13.2.16}$$

for a collection of cardinal numbers $\{\mu_{\mathfrak{p}}\}_{\mathfrak{p} \in \operatorname{Spec}(A)}$, called the *Bass numbers*. General counting tricks can show that the multiplicity $\mu_{\mathfrak{p}}$ is an invariant of I. But we can be more precise:

Proposition 13.2.17 *Let I be an injective A-module, with direct sum decomposition* (13.2.16). *Then for every $\mathfrak{p}$ there is equality*

$$\mu_{\mathfrak{p}} = \operatorname{rank}_{\boldsymbol{k}(\mathfrak{p})}\big(\operatorname{Hom}_{A_{\mathfrak{p}}}(\boldsymbol{k}(\mathfrak{p}), A_{\mathfrak{p}} \otimes_A I)\big).$$

Proof Consider another prime $\mathfrak{q}$. If $\mathfrak{q} \not\subseteq \mathfrak{p}$ then there is an element $a \in \mathfrak{q} - \mathfrak{p}$, and then a is both invertible and locally nilpotent on $A_{\mathfrak{p}} \otimes_A J(\mathfrak{q})$. This implies that $A_{\mathfrak{p}} \otimes_A J(\mathfrak{q}) = 0$. On the other hand, if $\mathfrak{q} \subseteq \mathfrak{p}$, then $A_{\mathfrak{p}} \otimes_A J(\mathfrak{q}) \cong J(\mathfrak{q})$. Therefore $A_{\mathfrak{p}} \otimes_A I \cong \bigoplus_{\mathfrak{q} \subseteq \mathfrak{p}} J(\mathfrak{p})^{\oplus \mu_{\mathfrak{p}}}$.

Next, if $\mathfrak{q} \subsetneq \mathfrak{p}$, then there is an element $b \in \mathfrak{p} - \mathfrak{q}$, and it is both invertible and zero on the module $\operatorname{Hom}_{A_{\mathfrak{p}}}(\boldsymbol{k}(\mathfrak{p}), J(\mathfrak{q}))$. The implication is that this module is zero.

Finally, if $\mathfrak{q} = \mathfrak{p}$ then we have

$$\operatorname{Hom}_{A_{\mathfrak{p}}}(\boldsymbol{k}(\mathfrak{p}), J(\mathfrak{p})) \cong \operatorname{Hom}_{A_{\mathfrak{p}}}(\boldsymbol{k}(\mathfrak{p}), \boldsymbol{k}(\mathfrak{p})) \cong \boldsymbol{k}(\mathfrak{p}),$$

because the inclusion $\boldsymbol{k}(\mathfrak{p}) \subseteq J(\mathfrak{p})$ is essential.

Since $\boldsymbol{k}(\mathfrak{p})$ is a finitely generated $A_{\mathfrak{p}}$-module, the functor $\operatorname{Hom}_{A_{\mathfrak{p}}}(\boldsymbol{k}(\mathfrak{p}), -)$ commutes with infinite direct sums. Therefore, putting all these cases together, we see that

$$\operatorname{Hom}_{A_{\mathfrak{p}}}(\boldsymbol{k}(\mathfrak{p}), A_{\mathfrak{p}} \otimes_A I)) \cong \boldsymbol{k}(\mathfrak{p})^{\oplus \mu_{\mathfrak{p}}}$$

as $\boldsymbol{k}(\mathfrak{p})$-modules. □

13.3 Residue Complexes

In this section A is a noetherian commutative ring. Here we introduce residue complexes (called residual complexes in [62]). Most of the material is taken from the original [62]. In Example 13.3.12 we will see the relation between the geometry of $\operatorname{Spec}(A)$ and the structure of dualizing complexes over A (continuing Example 0.2.19 from the Introduction).

Lemma 13.3.1 *Let R be a dualizing complex over A and let $\mathfrak{p} \subseteq A$ be a prime ideal. We write $R_{\mathfrak{p}} := A_{\mathfrak{p}} \otimes_A R \in \mathbf{D}(A_{\mathfrak{p}})$. There is an integer d such that*

$$\operatorname{Ext}^i_{A_{\mathfrak{p}}}(\boldsymbol{k}(\mathfrak{p}), R_{\mathfrak{p}}) \cong \begin{cases} \boldsymbol{k}(\mathfrak{p}) & \text{if } i = -d, \\ 0 & \text{otherwise.} \end{cases}$$

Proof By Proposition 13.1.29, $R_{\mathfrak{p}}$ is a dualizing complex over the local ring $A_{\mathfrak{p}}$. And by Proposition 13.1.28, $S := \operatorname{RHom}_{A_{\mathfrak{p}}}(\boldsymbol{k}(\mathfrak{p}), R_{\mathfrak{p}})$ is a dualizing complex over the residue field $\boldsymbol{k}(\mathfrak{p})$. Since $\boldsymbol{k}(\mathfrak{p})$ is a field, it is a regular ring, and so it is a dualizing complex over itself. Theorem 13.1.35 tells us that $S \cong \boldsymbol{k}(\mathfrak{p})[d]$ in $\mathbf{D}(\boldsymbol{k}(\mathfrak{p}))$ for some integer d. □

Definition 13.3.2 The number d in Lemma 13.3.1 is called the *dimension of* $\mathfrak{p}$ *relative to* R, and is denoted by $\dim_R(\mathfrak{p})$. In this way we obtain a function

$$\dim_R : \operatorname{Spec}(A) \to \mathbb{Z},$$

called the *dimension function* associated to R.

Let us recall a few notions regarding the combinatorics of prime ideals in a ring A. A prime ideal $\mathfrak{q}$ is an *immediate specialization* of another prime $\mathfrak{p}$ if $\mathfrak{p} \subsetneqq \mathfrak{q}$, and there is no other prime $\mathfrak{p}'$ such that $\mathfrak{p} \subsetneqq \mathfrak{p}' \subsetneqq \mathfrak{q}$. In other words, if the dimension of the local ring $A_{\mathfrak{q}}/\mathfrak{p}_{\mathfrak{q}}$ is 1.

A *chain of prime ideals* in A is a sequence $(\mathfrak{p}_0, \ldots, \mathfrak{p}_n)$ of primes such that $\mathfrak{p}_i \subsetneqq \mathfrak{p}_{i+1}$ for all i. The number n is the *length* of the chain. The chain is called *saturated* if for each i the prime $\mathfrak{p}_{i+1}$ is an immediate specialization of $\mathfrak{p}_i$.

Theorem 13.3.3 *Let R be a dualizing complex over A and let $\mathfrak{p}, \mathfrak{q} \subseteq A$ be prime ideals. Assume that $\mathfrak{q}$ is an immediate specialization of $\mathfrak{p}$. Then*

$$\dim_R(\mathfrak{q}) = \dim_R(\mathfrak{p}) - 1.$$

To prove this theorem, we need a baby version of local cohomology: codimension 1 only.

Let $\mathfrak{a}$ be an ideal in A. The torsion functor $\Gamma_{\mathfrak{a}}$ has a right derived functor $\mathrm{R}\Gamma_{\mathfrak{a}}$. For every complex $M \in \mathbf{D}(A)$, the module $\mathrm{H}^p_{\mathfrak{a}}(M) := \mathrm{H}^p(\mathrm{R}\Gamma_{\mathfrak{a}}(M))$ is called the *pth cohomology of M with support in* $\mathfrak{a}$. In case A is a local ring and $\mathfrak{m}$ is its maximal ideal, then $\mathrm{H}^p_{\mathfrak{m}}(M)$ is also called the *local cohomology of M*.

Now suppose $\mathfrak{a}$ is a principal ideal in A, generated by an element a. Let $A_a = A[a^{-1}]$ be the localized ring. For any A-module M we write $M_a = A_a \otimes_A M$. There is a canonical exact sequence

$$0 \to \Gamma_{\mathfrak{a}}(M) \to M \to M_a. \tag{13.3.4}$$

Lemma 13.3.5 *Let $\mathfrak{a} = (a)$ be a principal ideal in A.*

(1) *For every injective module I the sequence $0 \to \Gamma_{\mathfrak{a}}(I) \to I \to I_a \to 0$ is exact.*

(2) *For every $M \in \mathsf{D}^+(A)$ there is a long exact sequence of A-modules*

$$\cdots \to \mathrm{H}^p_{\mathfrak{a}}(M) \to \mathrm{H}^p(M) \to \mathrm{H}^p(M_a) \to \mathrm{H}^{p+1}_{\mathfrak{a}}(M) \to \cdots .$$

Proof (1) Let $J(\mathfrak{q})$ be an indecomposable injective A-module. According to Theorem 13.2.15(1), if $a \in \mathfrak{q}$ then $\Gamma_{\mathfrak{a}}(J(\mathfrak{q})) = J(\mathfrak{q})$ and $J(\mathfrak{q})_a = 0$. But if $a \notin \mathfrak{q}$ then $J(\mathfrak{q}) = J(\mathfrak{q})_a$ and $\Gamma_{\mathfrak{a}}(J(\mathfrak{q})) = 0$. By Theorem 13.2.15 and Proposition 13.2.13(2) we see that each injective module I breaks up into a direct sum $I = \Gamma_{\mathfrak{a}}(I) \oplus I_a$, and this proves that the sequence is split exact.

(2) Choose a resolution $M \to I$ by a bounded below complex of injectives. We obtain an exact sequence of complexes as in item (1). The long exact sequence in cohomology

$$\cdots \to \mathrm{H}^p(\Gamma_{\mathfrak{a}}(I)) \to \mathrm{H}^p(I) \to \mathrm{H}^p(I_a) \to \mathrm{H}^{p+1}(\Gamma_{\mathfrak{a}}(I)) \to \cdots$$

is what we want. □

Lemma 13.3.6 *Suppose A is an integral domain, with fraction field K, such that $A \neq K$. Then K is not a finitely generated A-module.*

Proof Let $a \in A$ be a nonzero element that is not invertible. Then

$$A \subsetneq a^{-1} \cdot A \subsetneq a^{-2} \cdot A \subsetneq \cdots \subseteq K$$

is an infinite ascending sequence of A-submodules of K. □

Lemma 13.3.7 *For every ideal $\mathfrak{a}$ and every $M \in \mathsf{D}(A)$ there is an isomorphism of A-modules $\mathrm{H}^p_{\mathfrak{a}}(M) \cong \lim_{k\to} \mathrm{Ext}^p_A(A/\mathfrak{a}^k, M)$.*

Proof Choose a K-injective resolution $M \to I$. Then, using the fact that cohomology commutes with direct limits, we have

$$\begin{aligned} \mathrm{H}^p_{\mathfrak{a}}(M) &\cong \mathrm{H}^p(\Gamma_{\mathfrak{a}}(I)) \cong \mathrm{H}^p(\lim_{k\to} \mathrm{Hom}_A(A/\mathfrak{a}^k, I)) \\ &\cong \lim_{k\to} \mathrm{H}^p(\mathrm{Hom}_A(A/\mathfrak{a}^k, I)) \cong \lim_{k\to} \mathrm{Ext}^p_A(A/\mathfrak{a}^k, M). \end{aligned}$$

□

Lemma 13.3.8 *Assume A is local, with maximal ideal $\mathfrak{m}$. Let R be a dualizing complex over A, and let $d := \dim_R(\mathfrak{m})$. Then for every $i \neq -d$ the local cohomology $\mathrm{H}^i_{\mathfrak{m}}(R)$ vanishes.*

See Remark 13.3.24 for more about $\mathrm{H}^{-d}_{\mathfrak{m}}(R)$.

Proof We know that

$$\mathrm{Ext}^i_A(\boldsymbol{k}(\mathfrak{m}), R) \cong \begin{cases} \boldsymbol{k}(\mathfrak{m}) & \text{if } i = -d, \\ 0 & \text{otherwise.} \end{cases}$$

Let N be a finite length A-module. Since N is gotten from the residue field $\boldsymbol{k}(\mathfrak{m})$ by finitely many extensions, induction on the length of N shows that $\mathrm{Ext}^i_A(N, R) = 0$ for all $i \neq -d$. This holds in particular for $N := A/\mathfrak{m}^k$. Now use Lemma 13.3.7. □

Proof of Theorem 13.3.3 Define $d := \dim_R(\mathfrak{q})$ and $e := \dim_R(\mathfrak{p})$. We need to prove that $e = d + 1$.

By definition, d is the unique integer s.t. $\mathrm{Ext}^{-d}_{A_\mathfrak{q}}(\boldsymbol{k}(\mathfrak{q}), R_\mathfrak{q}) \neq 0$. Let's define $\bar{A} := A/\mathfrak{p}$. We know that $\bar{R} := \mathrm{RHom}_A(\bar{A}, R)$ is a dualizing complex over $\bar{A}$. There are isomorphisms

$$\begin{aligned} \mathrm{RHom}_{A_\mathfrak{q}}(\boldsymbol{k}(\mathfrak{q}), R_\mathfrak{q}) &\cong \mathrm{RHom}_{\bar{A}_\mathfrak{q}}(\boldsymbol{k}(\mathfrak{q}), \mathrm{RHom}_{A_\mathfrak{q}}(\bar{A}_\mathfrak{q}, R_\mathfrak{q})) \\ &\cong \mathrm{RHom}_{\bar{A}_\mathfrak{q}}(\boldsymbol{k}(\mathfrak{q}), \bar{R}_\mathfrak{q}) \end{aligned}$$

in $\mathsf{D}(\bar{A}_\mathfrak{q})$, coming from adjunction for the homomorphism $A_\mathfrak{q} \to \bar{A}_\mathfrak{q}$. There is also an isomorphism

$$\mathrm{RHom}_{A_\mathfrak{p}}(\boldsymbol{k}(\mathfrak{p}), R_\mathfrak{p}) \cong \mathrm{RHom}_{\bar{A}_\mathfrak{p}}(\boldsymbol{k}(\mathfrak{p}), \bar{R}_\mathfrak{p})$$

in $\mathsf{D}(\bar{A}_\mathfrak{p})$. Hence we can replace A and R with $\bar{A}_\mathfrak{q}$ and $\bar{R}_\mathfrak{q}$, respectively.

Now we have $\mathfrak{p} = 0$ and $A = A_\mathfrak{q}$. Thus A is a 1-dimensional local integral domain, with only two primes ideals: $0 = \mathfrak{p}$ and the maximal ideal $\mathfrak{q}$. Take any nonzero element $a \in \mathfrak{q}$. Then the localization A_a is the field of fractions of A, i.e. $A_a = \boldsymbol{k}(\mathfrak{p})$. On the other hand, letting $\mathfrak{a} := (a) \subsetneq A$, the quotient $A/\mathfrak{a}$ is an artinian local ring. So $A/\mathfrak{a}$ is a finite length A-module, the ideal $\mathfrak{a}$ is $\mathfrak{q}$-primary, and $\Gamma_\mathfrak{a} = \Gamma_\mathfrak{q}$.

By Lemma 13.3.5 there is an exact sequence of A-modules

$$\cdots \to \mathrm{H}^{-e}_\mathfrak{a}(R) \to \mathrm{H}^{-e}(R) \xrightarrow{\phi} \mathrm{H}^{-e}(R_a) \to \mathrm{H}^{-e+1}_\mathfrak{a}(R) \to \cdots.$$

Since $a \neq 0$, there are equalities $A_a = A_\mathfrak{p} = \mathrm{Frac}(A) = \boldsymbol{k}(\mathfrak{p})$. Then $\mathrm{H}^{-e}(R_a) \cong \boldsymbol{k}(\mathfrak{p})$, and this is not a finitely generated A-module by Lemma 13.3.6. On the other hand the A-module $\mathrm{H}^{-e}(R)$ is finitely generated. We conclude that homomorphism ϕ is not surjective, and thus $\mathrm{H}^{-e+1}_\mathfrak{a}(R) \neq 0$. But $\mathrm{H}^{-e+1}_\mathfrak{a}(R) = \mathrm{H}^{-e+1}_\mathfrak{q}(R)$, so according to Lemma 13.3.8, we must have $-e + 1 = -d$. Thus $e = d + 1$ as claimed. □

Corollary 13.3.9 *If A has a dualizing complex, then the Krull dimension of A is finite. More precisely, if R is a dualizing complex over A, then* $\dim(A)$ *is at most the injective dimension of R.*

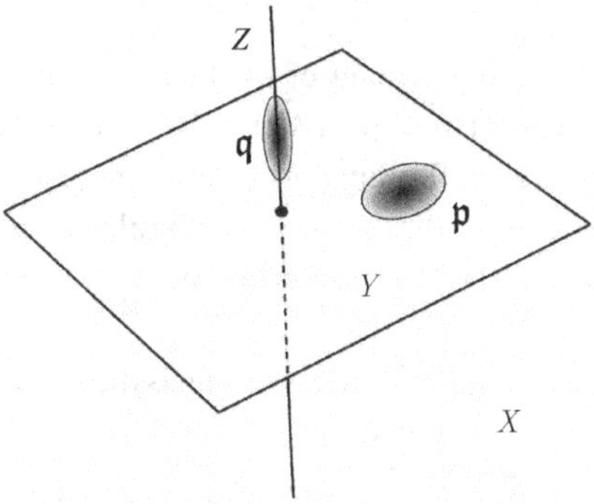

Figure 13.1. An algebraic variety X that is connected but not equidimensional: it has irreducible components Y and Z of dimensions 2 and 1, respectively. The generic points $\mathfrak{p} \in Y$ and $\mathfrak{q} \in Z$ are shown.

Proof Let $[i_0, i_1]$ be the injective concentration of the complex R. See Definition 12.4.8. This is a bounded interval. Since

$$\operatorname{Ext}^i_{A_\mathfrak{p}}(\boldsymbol{k}(\mathfrak{p}), R_\mathfrak{p}) \cong \operatorname{Ext}^i_A(A/\mathfrak{p}, R)_\mathfrak{p},$$

we see that $\dim_R(\mathfrak{p}) \in -[i_0, i_1] = [-i_1, -i_0]$.

Let $(\mathfrak{p}_0, \ldots, \mathfrak{p}_n)$ be a chain of prime ideals in A. Because A is noetherian, we can squeeze more primes into this chain, until after finitely many steps it becomes saturated. According to Theorem 13.3.3, for a saturated chain we have $n = \dim_R(\mathfrak{p}_0) - \dim_R(\mathfrak{p}_n)$. Therefore $n \leq i_1 - i_0$. □

Definition 13.3.10 The ring A is called *catenary* if for every pair of primes $\mathfrak{p} \subseteq \mathfrak{q}$ there is a number $n_{\mathfrak{p},\mathfrak{q}}$ such that for every saturated chain $(\mathfrak{p}_0, \ldots, \mathfrak{p}_n)$ with $\mathfrak{p}_0 = \mathfrak{p}$ and $\mathfrak{p}_n = \mathfrak{q}$, there is equality $n = n_{\mathfrak{p},\mathfrak{q}}$.

Corollary 13.3.11 *If A has a dualizing complex, then it is catenary.*

Proof Let R be a dualizing complex over A. The proof of the previous corollary shows that the number $n_{\mathfrak{p},\mathfrak{q}} = \dim_R(\mathfrak{p}) - \dim_R(\mathfrak{q})$ has the desired property. □

Example 13.3.12 This is a continuation of Example 0.2.19 from the Introduction. Consider the ring $A = \mathbb{R}[t_1, t_2, t_3]/(t_3 \cdot t_1, t_3 \cdot t_2)$. The affine algebraic variety $X = \operatorname{Spec}(A) \subseteq \mathbf{A}^3_\mathbb{R}$ is shown in Figure 13.1. It is the union of a plane Y and a line Z, meeting at the origin.

Since the ring A is finite type over the field $\mathbb{R}$, it has a dualizing complex R. We will now prove that there is some integer i s.t. $\mathrm{H}^i(R)$ and $\mathrm{H}^{i+1}(R)$ are nonzero.

Define the prime ideals $\mathfrak{m} := (t_1, t_2, t_3)$, $\mathfrak{q} := (t_1, t_2)$ and $\mathfrak{p} := (t_3)$. Thus $\mathfrak{m}$ is the origin, $\mathfrak{q}$ is the generic point of the line $Z = \operatorname{Spec}(A/\mathfrak{q})$, and $\mathfrak{p}$ is the generic point of the plane $Y = \operatorname{Spec}(A/\mathfrak{p})$. By translating R as needed, we can assume that $\dim_R(\mathfrak{m}) = 0$. Since $\mathfrak{m}$ is an immediate specialization of $\mathfrak{q}$, Theorem 13.3.3 tells us that $\dim_R(\mathfrak{q}) = 1$. Similarly, since every line in Y passing through the origin gives rise to a saturated chain $(\mathfrak{p}, \mathfrak{q}', \mathfrak{m})$, we see that $\dim_R(\mathfrak{p}) = 2$.

Since $\mathfrak{q}$ is the generic point of Z, its local ring is the residue field: $A_\mathfrak{q} = \boldsymbol{k}(\mathfrak{q})$. We know that $\dim_R(\mathfrak{q}) = 1$. Hence

$$\boldsymbol{k}(\mathfrak{q}) \cong \operatorname{Ext}^{-1}_{A_\mathfrak{q}}(\boldsymbol{k}(\mathfrak{q}), R_\mathfrak{q}) = \operatorname{Ext}^{-1}_{A_\mathfrak{q}}(A_\mathfrak{q}, R_\mathfrak{q}) \cong \operatorname{Ext}^{-1}_{A}(A, R)_\mathfrak{q} \cong \mathrm{H}^{-1}(R)_\mathfrak{q}.$$

Therefore $\mathrm{H}^{-1}(R) \neq 0$. A similar calculation involving $\mathfrak{p}$ shows that $\mathrm{H}^{-2}(R) \neq 0$.

Example 13.3.13 Let A be a local ring, with maximal ideal $\mathfrak{m}$ and residue field $\boldsymbol{k}(\mathfrak{m})$. Recall that A is called *Gorenstein* if the free module A has finite injective dimension. The ring A is called *Cohen–Macaulay* if its depth is equal to its dimension, where the depth of A is the minimal integer i such that $\operatorname{Ext}^i_A(\boldsymbol{k}(\mathfrak{m}), A) \neq 0$. It is known that Gorenstein implies Cohen–Macaulay. See [105] for details.

As is our usual practice (cf. Definition 13.1.10) we shall say that a noetherian commutative ring A is Cohen–Macaulay (resp. Gorenstein) if it has finite Krull dimension, and all its local rings $A_\mathfrak{p}$ are Cohen–Macaulay (resp. Gorenstein) local rings, as defined above.

Assume A has a connected spectrum, and let R be a dualizing complex over A. Grothendieck showed in [62, Section V.9] that A is a Cohen–Macaulay ring iff $R \cong L[d]$ for some finitely generated module L and some integer d; the proof is not easy. It is however pretty easy to prove (using Theorem 13.1.35) that A is a Gorenstein ring iff $R \cong L[d]$ for some invertible module L and some integer d.

There is a lot more to say about the relation between the CM (Cohen–Macaulay) property and duality. See Remark 13.3.26

Recall that for each $\mathfrak{p} \in \operatorname{Spec}(A)$ we denote by $J(\mathfrak{p})$ the corresponding indecomposable injective A-module.

Definition 13.3.14 A *residue complex* over A is a complex of A-modules $\mathcal{K}$ having these properties:

(i) $\mathcal{K}$ is a dualizing complex.

(ii) For every integer d there is an isomorphism of A-modules

$$\mathcal{K}^{-d} \cong \bigoplus_{\substack{\mathfrak{p} \in \operatorname{Spec}(A) \\ \dim_{\mathcal{K}}(\mathfrak{p}) = d}} J(\mathfrak{p})\,.$$

The reason we like residue complexes is this:

Theorem 13.3.15 *Suppose $\mathcal{K}$ and $\mathcal{K}'$ are residue complexes over A that have the same dimension function. Then the homomorphism*

$$\mathrm{Q} : \mathrm{Hom}_{\mathbf{C}_{\mathrm{str}}(A)}(\mathcal{K}, \mathcal{K}') \to \mathrm{Hom}_{\mathbf{D}(A)}(\mathcal{K}, \mathcal{K}')$$

is bijective.

In other words, for each morphism $\psi : \mathcal{K} \to \mathcal{K}'$ in $\mathbf{D}(A)$ there is a unique strict homomorphism of complexes $\phi : \mathcal{K} \to \mathcal{K}'$ such that $\psi = \mathrm{Q}(\phi)$.

Proof Since the complex $\mathcal{K}'$ is K-injective, by Theorem 10.1.13, we know that the homomorphism

$$\mathrm{Q} : \mathrm{Hom}_{\mathbf{K}(A)}(\mathcal{K}, \mathcal{K}') \to \mathrm{Hom}_{\mathbf{D}(A)}(\mathcal{K}, \mathcal{K}')$$

is bijective. And by definition the homomorphism

$$\mathrm{P} : \mathrm{Hom}_{\mathbf{C}_{\mathrm{str}}(A)}(\mathcal{K}, \mathcal{K}') \to \mathrm{Hom}_{\mathbf{K}(A)}(\mathcal{K}, \mathcal{K}')$$

is surjective. It remains to prove that $\mathrm{Hom}_A(\mathcal{K}, \mathcal{K}')^{-1} = 0$, i.e. here are no nonzero degree -1 homomorphisms $\gamma : \mathcal{K} \to \mathcal{K}'$.

The residue complexes $\mathcal{K}$ and $\mathcal{K}'$ decompose into indecomposable summands by the formula in property (ii) of Definition 13.3.14. A homomorphism $\gamma : \mathcal{K} \to \mathcal{K}'$ of degree -1 is nonzero iff at least one of its components $\gamma_{\mathfrak{p},\mathfrak{q}} : J(\mathfrak{p}) \to J(\mathfrak{q})$ is nonzero, for some $J(\mathfrak{p}) \subseteq \mathcal{K}^{-i}$ and $J(\mathfrak{q}) \subseteq \mathcal{K}'^{-i-1}$. Denoting by dim the dimension function of both these dualizing complexes, we have $\dim(\mathfrak{p}) = i$ and $\dim(\mathfrak{q}) = i + 1$. But the lemma below says that $\mathfrak{q}$ has to be a specialization of $\mathfrak{p}$. Therefore, as in the proof of Corollary 13.3.9, there is an inequality in the oppose direction: $\dim(\mathfrak{p}) \geq \dim(\mathfrak{q})$. We see that it is impossible to have a nonzero degree -1 homomorphism $\gamma : \mathcal{K} \to \mathcal{K}'$. □

Lemma 13.3.16 *Let $\mathfrak{p}, \mathfrak{q}$ be prime ideals. If there is a nonzero homomorphism $\gamma : J(\mathfrak{p}) \to J(\mathfrak{q})$, then $\mathfrak{q}$ is a specialization of $\mathfrak{p}$.*

Proof Assume $\mathfrak{q}$ is not a specialization of $\mathfrak{p}$; i.e. $\mathfrak{p} \not\subseteq \mathfrak{q}$. So there is an element $a \in \mathfrak{p} - \mathfrak{q}$. Let $\gamma : J(\mathfrak{p}) \to J(\mathfrak{q})$ be a homomorphism, and consider the module $N := \gamma(J(\mathfrak{p})) \subseteq J(\mathfrak{q})$. Since $J(\mathfrak{p})$ is $\mathfrak{p}$-torsion, the element a acts on N locally-nilpotently. On the other hand, $J(\mathfrak{q})$ is a module over $A_{\mathfrak{q}}$, so a acts invertibly on $J(\mathfrak{q})$, and hence it has zero annihilator in N. The conclusion is that $N = 0$. □

Here is a general existence theorem. Minimal injective resolutions were defined in Definition 13.2.5. Their existence was proved in Proposition 13.2.6.

Theorem 13.3.17 *Suppose the ring A has a dualizing complex R. Let $R \to \mathcal{K}$ be a minimal injective resolution of R. Then $\mathcal{K}$ is a residue complex over A.*

The proof is after two lemmas.

Lemma 13.3.18 *Let $S \subseteq A$ be a multiplicatively closed set, with localization A_S. For each complex of A-modules M we write $M_S := A_S \otimes_A M$.*

(1) *If I is an injective A-module, then I_S is an injective A_S-module.*
(2) *If I is an injective A-module and $M \subseteq I$ is an essential A-submodule, then $M_S \subseteq I_S$ is an essential A_S-submodule.*
(3) *If I is a minimal complex of injective A-modules, then I_S is a minimal complex of injective A_S-modules.*

Proof (1) By Theorem 13.2.15 there is a direct sum decomposition $I \cong I' \oplus I''$, where $I' \cong \bigoplus_{\mathfrak{p} \cap S = \varnothing} J(\mathfrak{p})^{\oplus \mu_{\mathfrak{p}}}$ and $I'' \cong \bigoplus_{\mathfrak{p} \cap S \neq \varnothing} J(\mathfrak{p})^{\oplus \mu_{\mathfrak{p}}}$. If $\mathfrak{p} \cap S = \varnothing$ then $J(\mathfrak{p}) \cong J(\mathfrak{p})_S$ is an injective A_S-module; and if $\mathfrak{p} \cap S \neq \varnothing$ then $J(\mathfrak{p})_S = 0$. We see that $I_S \cong I'$ is an injective A_S-module.

(2) Denote by $\lambda : I \to I_S$ the canonical homomorphism. Under the decomposition $I \cong I' \oplus I''$ above, $\lambda|_{I'} : I' \to I_S$ is an isomorphism.

Let L be a nonzero A_S-submodule of I_S. Consider the A-submodule $L' := (\lambda|_{I'})^{-1}(L) \subseteq I'$; so $\lambda|_{L'} : L' \to L$ is an isomorphism and $L' \neq 0$. Because $M \subseteq I$ is essential, the intersection $M \cap L'$ is nonzero. But then $\lambda(M \cap L')$ is a nonzero submodule of $M_S \cap L$, so $M_S \cap L$ is nonzero.

(3) By part (1) the complex I_S is a bounded below complex of injective A_S-modules. Exactness of localization shows that $\mathrm{Z}^n(I_S) = \mathrm{Z}^n(I)_S$ inside I^n_S; so by part (2) the inclusion $\mathrm{Z}^n(I_S) \rightarrowtail I^n_S$ is essential. □

Lemma 13.3.19 *Let $\mathfrak{a} \subseteq A$ be an ideal, and define $B := A/\mathfrak{a}$.*

(1) *If I is an injective A-module, then $J :=$ $\mathrm{Hom}_A(B, I)$ *is an injective B-module.*
(2) *Let I and J be as above. If $M \subseteq I$ is an essential A-submodule, then $N :=$ $\mathrm{Hom}_A(B, M)$ *is an essential B-submodule of J.*
(3) *If I is a minimal complex of injective A-modules, then $J :=$ $\mathrm{Hom}_A(B, I)$ *is a minimal complex of injective B-modules,*

Proof (1) This is immediate from adjunction.

(2) We identify J and N with the submodules of I and M, respectively, that are annihilated by $\mathfrak{a}$. Let $L \subseteq J$ be a nonzero B-submodule. Then L is a nonzero A-submodule of I. Because M is essential, the intersection $L \cap M$ is nonzero. But $L \cap M$ is annihilated by $\mathfrak{a}$, so it sits inside N, and in fact $L \cap M = L \cap N$.

(3) By part (1) the complex J is a bounded below complex of injective B-modules. Left exactness of $\mathrm{Hom}_A(B, -)$ shows that $\mathrm{Z}^n(J) = \mathrm{Hom}_A(B, \mathrm{Z}^n(I))$ inside J^n; so by part (2) the inclusion $\mathrm{Z}^n(J) \rightarrowtail J^n$ is essential. □

Proof of Theorem 13.3.17 Since $\mathcal{K} \cong R$ in $\mathbf{D}(A)$, it follows that $\mathcal{K}$ is a dualizing complex. To show that $\mathcal{K}$ has property (ii) of Definition 13.3.14, we have to count multiplicities. For every $\mathfrak{p}$ and d let $\mu_{\mathfrak{p},d}$ be the multiplicity of $J(\mathfrak{p})$ in $\mathcal{K}^{-d}$, so that

$$\mathcal{K}^{-d} \cong \bigoplus_{\mathfrak{p} \in \operatorname{Spec}(A)} J(\mathfrak{p})^{\oplus \mu_{\mathfrak{p},d}}.$$

We have to prove that

$$\mu_{\mathfrak{p},d} = \begin{cases} 1 & \text{if } \dim_{\mathcal{K}}(\mathfrak{p}) = d, \\ 0 & \text{otherwise.} \end{cases} \tag{13.3.20}$$

Now by Lemma 13.3.18(3) the complex $\mathcal{K}_{\mathfrak{p}} = A_{\mathfrak{p}} \otimes_A \mathcal{K}$ is a minimal complex of injective $A_{\mathfrak{p}}$-modules. Because $\mathcal{K}_{\mathfrak{p}}$ is K-injective over $A_{\mathfrak{p}}$, we get

$$\operatorname{Ext}^{-d}_{A_{\mathfrak{p}}}(\boldsymbol{k}(\mathfrak{p}), R_{\mathfrak{p}}) \cong \mathrm{H}^{-d}(\operatorname{Hom}_{A_{\mathfrak{p}}}(\boldsymbol{k}(\mathfrak{p}), \mathcal{K}_{\mathfrak{p}}))$$

as $\boldsymbol{k}(\mathfrak{p})$-modules. By Lemma 13.3.19(3) the complex $\operatorname{Hom}_{A_{\mathfrak{p}}}(\boldsymbol{k}(\mathfrak{p}), \mathcal{K}_{\mathfrak{p}})$ is a minimal complex of injective $\boldsymbol{k}(\mathfrak{p})$-modules. It is easy to see (and we leave this verification to the reader) that a minimal complex of injectives over a field must have zero differential. Therefore

$$\mathrm{H}^{-d}(\operatorname{Hom}_{A_{\mathfrak{p}}}(\boldsymbol{k}(\mathfrak{p}), \mathcal{K}_{\mathfrak{p}})) \cong \operatorname{Hom}_{A_{\mathfrak{p}}}(\boldsymbol{k}(\mathfrak{p}), \mathcal{K}_{\mathfrak{p}}^{-d}).$$

Now by arguments like in the proofs of Lemmas 13.3.18(1) and 13.3.16 we know that

$$\operatorname{Hom}_{A_{\mathfrak{p}}}(\boldsymbol{k}(\mathfrak{p}), J(\mathfrak{q})_{\mathfrak{p}}) \cong \begin{cases} \boldsymbol{k}(\mathfrak{p}) & \text{if } \mathfrak{q} = \mathfrak{p}, \\ 0 & \text{otherwise.} \end{cases}$$

It follows that

$$\operatorname{Hom}_{A_{\mathfrak{p}}}(\boldsymbol{k}(\mathfrak{p}), \mathcal{K}_{\mathfrak{p}}^{-d}) \cong \boldsymbol{k}(\mathfrak{p})^{\oplus \mu_{\mathfrak{p},d}}.$$

We see that

$$\operatorname{rank}_{\boldsymbol{k}(\mathfrak{p})}(\operatorname{Ext}^{-d}_{A_{\mathfrak{p}}}(\boldsymbol{k}(\mathfrak{p}), R_{\mathfrak{p}})) = \mu_{\mathfrak{p},d}.$$

But by Definition 13.3.2 this number satisfies (13.3.20). □

Corollary 13.3.21 *If $\mathcal{K}$ is a residue complex over A, then it is a minimal complex of injective A-modules.*

Proof Let $\phi : \mathcal{K} \to \mathcal{K}'$ be a minimal injective resolution of $\mathcal{K}$. According to Theorem 13.3.17, $\mathcal{K}'$ is also a residue complex. Now $\mathrm{Q}(\phi) : \mathcal{K} \to \mathcal{K}'$ is an isomorphism in $\mathbf{D}(A)$, so by Theorem 13.3.15 we know that $\phi : \mathcal{K} \to \mathcal{K}'$ is an isomorphism in $\mathbf{C}_{\mathrm{str}}(A)$. □

Exercise 13.3.22 Find a direct proof of Corollary 13.3.21, without resorting to Theorems 13.3.17 and 13.3.15. (Hint: look at the proof of Proposition 13.2.6.)

We end this section with three remarks.

Remark 13.3.23 Here is a brief explanation of *Matlis Duality*. For more details see [62, Section V.5], [105, Theorem 18.6] or [30, Section 10.2]. Assume A is a complete local ring with maximal ideal $\mathfrak{m}$. As usual, the category of finitely generated A-modules is $\mathbf{M}_{\mathrm{f}}(A)$. There is also the category $\mathbf{M}_{\mathrm{a}}(A)$ of artinian A-modules. These are full abelian subcategories of $\mathbf{M}(A)$. Note that these subcategories are characterized by dual properties: the objects of $\mathbf{M}_{\mathrm{f}}(A)$ are noetherian, i.e. they satisfy the ascending chain condition; and the objects of $\mathbf{M}_{\mathrm{a}}(A)$ satisfy the descending chain condition.

Consider the indecomposable injective module $J(\mathfrak{m})$. The functor $D := \mathrm{Hom}_A(-, J(\mathfrak{m}))$ is exact of course. Matlis Duality asserts that $D : \mathbf{M}_{\mathrm{f}}(A)^{\mathrm{op}} \to \mathbf{M}_{\mathrm{a}}(A)$ is an equivalence, with quasi-inverse $D : \mathbf{M}_{\mathrm{a}}(A)^{\mathrm{op}} \to \mathbf{M}_{\mathrm{f}}(A)$.

Later in this book we present a noncommutative graded version of Matlis Duality – this is Theorem 15.2.34.

Remark 13.3.24 We now provide a brief discussion of *Local Duality*, based on [62, Section V.6]. (There is a weaker variant of this result, for modules instead of complexes, which can be found in [30, Theorem 11.2.6].) Again A is local, with maximal ideal $\mathfrak{m}$. Let R be a dualizing complex over A. By translating R we can assume that $\dim_R(\mathfrak{m}) = 0$. Lemma 13.3.8 tells us that $\mathrm{H}^i_{\mathfrak{m}}(R) = 0$ for all $i \neq 0$. A calculation, which relies on Matlis duality, shows that $\mathrm{H}^0_{\mathfrak{m}}(R) \cong J(\mathfrak{m})$, the indecomposable injective corresponding to $\mathfrak{m}$.

Let us fix an isomorphism $\beta : \mathrm{H}^0_{\mathfrak{m}}(R) \xrightarrow{\simeq} J(\mathfrak{m})$. This induces a morphism

$$\theta_M : \mathrm{R}\Gamma_{\mathfrak{m}}(M) \to \mathrm{Hom}_A(\mathrm{RHom}_A(M, R), J(\mathfrak{m})), \tag{13.3.25}$$

functorial in $M \in \mathbf{D}^+(A)$. The Local Duality Theorem [62, Theorem V.6.2] says that θ_M is an isomorphism if $M \in \mathbf{D}^+_{\mathrm{f}}(A)$.

Here is a modern take on this theorem: we can construct the morphism θ_M for all $M \in \mathbf{D}(A)$. Let's replace R by the residue complex $\mathcal{K}$ (the minimal injective resolution of R). Then β is just an A-module isomorphism $\beta : \mathcal{K}^0 \xrightarrow{\simeq} J(\mathfrak{m})$. For each complex M we choose a K-injective resolution $M \to I(M)$. Then θ_M is represented by the homomorphism

$$\tilde{\theta}_M : \Gamma_{\mathfrak{m}}(I(M)) \to \mathrm{Hom}_A(\mathrm{Hom}_A(I(M), \mathcal{K}), \mathcal{K}^0)$$

in $\mathbf{C}_{\mathrm{str}}(A)$ that sends an element $u \in \Gamma_{\mathfrak{m}}(I(M))^p$ and a homomorphism $\phi \in \mathrm{Hom}_A(I(M), \mathcal{K})^{-p}$ to $(-1)^p \cdot \phi(u) \in \mathcal{K}^0$.

We know that the functor $\mathrm{R}\Gamma_{\mathfrak{m}}$ has finite cohomological dimension, bounded by the number of generators of the ideal $\mathfrak{m}$; see [54] or [118]. The functor $\mathrm{RHom}_A(-, R)$ has finite cohomological dimension, which is the injective dimension of R. And the functor $\mathrm{Hom}_A(-, J(\mathfrak{m}))$ has cohomological dimension 0. Since $A \in \mathsf{D}^+_{\mathrm{f}}(A)$, the Local Duality Theorem from [62] tells us that θ_A is an isomorphism. Now we can apply Theorem 12.5.2 to conclude that θ_M is an isomorphism for every $M \in \mathsf{D}_{\mathrm{f}}(A)$.

Finally, let us mention that in Section 17.2 there is a noncommutative graded version of Local Duality.

Remark 13.3.26 Here is more on the CM (Cohen–Macaulay) property and duality. Let A be a noetherian ring with connected spectrum. Assume A has a dualizing complex R, and corresponding dimension function $\dim_R$.

Consider a complex $M \in \mathsf{D}^{\mathrm{b}}_{\mathrm{f}}(A)$. In [62] Grothendieck defines M to be a *CM complex with respect to* R if for every prime ideal $\mathfrak{p} \subseteq A$ and every $i \neq -\dim_R(\mathfrak{p})$ the local cohomology satisfies $\mathrm{H}^i_{\mathfrak{p}}(M_{\mathfrak{p}}) = 0$.

It is proved in [62] that when A is a regular ring, $R = A$, and M is a finitely generated A-module, then M is a CM module (in the conventional sense, see [105]) iff it is a CM complex.

Let $\mathsf{D}^0_{\mathrm{f}}(A)$ be the full subcategory of $\mathsf{D}^{\mathrm{b}}_{\mathrm{f}}(A)$ on the complexes M such that $\mathrm{H}^i(M) = 0$ for all $i \neq 0$. We know that $\mathsf{D}^0_{\mathrm{f}}(A)$ is equivalent to $\mathsf{M}_{\mathrm{f}}(A) = \mathsf{Mod}_{\mathrm{f}}\, A$. In [187] it was proved that the following are equivalent for a complex $M \in \mathsf{D}^{\mathrm{b}}_{\mathrm{f}}(A)$:

(i) The complex M is CM w.r.t. R.
(ii) The complex $\mathrm{RHom}_A(M, R)$ belongs to $\mathsf{D}^0_{\mathrm{f}}(A)$.

It follows that the CM complexes form an abelian subcategory of $\mathsf{D}^{\mathrm{b}}_{\mathrm{f}}(A)$, dual to $\mathsf{M}_{\mathrm{f}}(A)$. In fact, they are the heart of a perverse t-structure on $\mathsf{D}^{\mathrm{b}}_{\mathrm{f}}(A)$, and hence they deserve to be called *perverse finitely generated A-modules*. Geometrically, on the scheme $X := \mathrm{Spec}(A)$, the CM complexes inside $\mathsf{D}^{\mathrm{b}}_{\mathrm{c}}(X)$ form a stack of abelian categories, and so they are *perverse coherent sheaves*. All this is explained in [187, section 6].

13.4 Van den Bergh Rigidity

As we saw in Theorem 13.1.35, a dualizing complex R over a noetherian commutative ring A is not unique. This lack of uniqueness (not to mention any sort of functoriality!) was the source of major difficulties in [62], first for gluing dualizing complexes on schemes, and then for producing the trace morphisms associated to maps of schemes.

In 1997, M. Van den Bergh [153] discovered the idea of *rigidity* for dualizing complexes. This was done in the context of noncommutative ring theory: A is a noncommutative noetherian ring, central over a base field $\mathbb{K}$. The theory of noncommutative rigid dualizing complexes was developed further in several papers of J.J. Zhang and A. Yekutieli, among them [183] and [187]. We shall talk about this noncommutative theory in Chapter 18 of the book.

Here we will deal with the commutative side only, which turns out to be extremely powerful. Before explaining it, let us first observe that this is one of the rare cases in which an idea originating from noncommutative algebra had significant impact in commutative algebra and algebraic geometry.

In this section we study Van den Bergh rigidity in the following context:

Setup 13.4.1 A is a commutative ring, and B is a flat commutative A-ring.

We introduce the notion of *rigid complex over B relative to A*, and describe some of its properties. In Section 13.5 we will discuss *rigid dualizing and residue complexes*. This material is adapted from the papers [188], [189] and [175].

The theory of rigid complexes does not really require the assumption that B is flat over A, but flatness makes the theory much easier. See Remark 13.4.24.

Consider the enveloping ring $B \otimes_A B$. It comes equipped with a few ring homomorphisms:

$$B \xrightarrow{\eta_i} B \otimes_A B \xrightarrow{\epsilon} B, \tag{13.4.2}$$

where $\eta_0(b) := b \otimes 1$, $\eta_1(b) := 1 \otimes b$, and $\epsilon(b_0 \otimes b_1) := b_0 \cdot b_1$. We view B as a module over $B \otimes_A B$ via ϵ. Of course, $\epsilon \circ \eta_i = \mathrm{id}_B$.

Suppose we are given B-modules M_0 and M_1. Then the tensor product $M_0 \otimes_A M_1$ is a $(B \otimes_A B)$-module. In this way we get an additive bifunctor

$$(- \otimes_A -) : \mathbf{M}(B) \times \mathbf{M}(B) \to \mathbf{M}(B \otimes_A B).$$

Passing to complexes, and then to homotopy categories, we obtain a triangulated bifunctor

$$(- \otimes_A -) : \mathbf{K}(B) \times \mathbf{K}(B) \to \mathbf{K}(B \otimes_A B). \tag{13.4.3}$$

Lemma 13.4.4 *The bifunctor* (13.4.3) *has a left derived bifunctor*

$$(- \otimes^{\mathrm{L}}_A -) : \mathbf{D}(B) \times \mathbf{D}(B) \to \mathbf{D}(B \otimes_A B).$$

If either M_0 or M_1 is a complex of B-modules that is K-flat over A, then the morphism $\eta^{\mathrm{L}}_{M_0,M_1} : M_0 \otimes^{\mathrm{L}}_A M_1 \to M_0 \otimes_A M_1$ in $\mathbf{D}(B \otimes_A B)$ is an isomorphism.

Proof This is a variant of Theorem 12.3.1. We know by Corollary 11.4.26 and Proposition 10.3.4 that every complex $M \in \mathbf{C}(B)$ admits a K-flat resolution $P \to M$. Because B is flat over A, the complex P is also K-flat over A. By Theorem 9.3.16 the left derived bifunctor $(- \otimes^{\mathrm{L}}_A -)$ exists, and the condition on $\eta^{\mathrm{L}}_{M_0,M_1}$ holds. □

Remark 13.4.5 The innocuous looking Lemma 13.4.4 is actually of tremendous importance. Without the flatness of $A \to B$ we could do very little homological algebra of bimodules. Getting around the lack of flatness requires the use of flat DG ring resolutions, as explained in Remark 13.4.24.

Every module $L \in \mathbf{M}(B)$ has an action by $B \otimes_A B$ coming from the homomorphism ϵ in (13.4.2). Consider now a module $N \in \mathbf{M}(B \otimes_A B)$. The $\mathbb{K}$-module N has two possible B-module structures, coming from the homomorphisms η_i. Thus the $\mathbb{K}$-module $\operatorname{Hom}_{B\otimes_A B}(L, N)$ has three possible B-module structures: there is one action from the B-module structure on L, and there are two from the B-module structures on N.

Lemma 13.4.6 *The three B-module structures on* $\operatorname{Hom}_{B\otimes_A B}(L, N)$ *coincide.*

Exercise 13.4.7 Prove the lemma.

We are mostly interested in the B-module $L = B$. As the module N changes, we get an additive functor

$$\operatorname{Hom}_{B\otimes_A B}(B, -) : \mathbf{M}(B \otimes_A B) \to \mathbf{M}(B).$$

Passing to complexes, and then to homotopy categories, we get a triangulated functor

$$\operatorname{Hom}_{B\otimes_A B}(B, -) : \mathbf{K}(B \otimes_A B) \to \mathbf{K}(B).$$

This has a right derived functor

$$\operatorname{RHom}_{B\otimes_A B}(B, -) : \mathbf{D}(B \otimes_A B) \to \mathbf{D}(B), \tag{13.4.8}$$

that is calculated by K-injective resolutions. Namely if $I \in \mathbf{C}(B \otimes_A B)$ is a K-injective complex, then the morphism

$$\eta^{\mathrm{R}}_{B,I} : \operatorname{Hom}_{B\otimes_A B}(B, I) \to \operatorname{RHom}_{B\otimes_A B}(B, I)$$

in $\mathbf{D}(B)$ is an isomorphism.

By composing the bifunctor $(- \otimes^{\mathrm{L}}_A -)$ from Lemma 13.4.4 and the functor $\operatorname{RHom}_{B\otimes_A B}(B, -)$ from (13.4.8) we obtain a triangulated bifunctor

$$\operatorname{RHom}_{B\otimes_A B}(B, - \otimes^{\mathrm{L}}_A -) : \mathbf{D}(B) \times \mathbf{D}(B) \to \mathbf{D}(B). \tag{13.4.9}$$

Definition 13.4.10 Under Setup 13.4.1, the *squaring operation* is the functor

$$\operatorname{Sq}_{B/A} : \mathbf{D}(B) \to \mathbf{D}(B)$$

defined as follows:

(1) For a complex $M \in \mathbf{D}(B)$, its square is the complex

$$\operatorname{Sq}_{B/A}(M) := \operatorname{RHom}_{B\otimes_A B}(B, M \otimes^{\mathrm{L}}_A M) \in \mathbf{D}(B).$$

(2) For a morphism $\phi : M \to N$ in $\mathsf{D}(B)$, its square is the morphism

$$\mathrm{Sq}_{B/A}(\phi) := \mathrm{RHom}_{B \otimes_A B}(B, \phi \otimes^{\mathrm{L}}_A \phi) : \mathrm{Sq}_{B/A}(M) \to \mathrm{Sq}_{B/A}(N)$$

in $\mathsf{D}(B)$.

It will be good to have an explicit formulation of the squaring operation. Let us first choose a K-projective resolution $\sigma : P \to M$ in $\mathsf{C}(B)$. Note that P is unique up to homotopy equivalence, and σ is unique up to homotopy. Since B is flat over A, the complex P is K-flat over A. We get an isomorphism $\mathrm{pres}_P : P \otimes_A P \xrightarrow{\simeq} M \otimes^{\mathrm{L}}_A M$ in $\mathsf{D}(B \otimes_A B)$, which we call a *presentation*. It is uniquely characterized by the commutativity of the diagram

$$\begin{array}{ccc}
M \otimes^{\mathrm{L}}_A M & \xrightarrow{\eta^{\mathrm{L}}_{M,M}} & M \otimes_A M \\
{\scriptstyle \mathrm{Q}(\sigma) \otimes^{\mathrm{L}}_A \mathrm{Q}(\sigma)}\ \uparrow \cong & & \uparrow\ {\scriptstyle \mathrm{Q}(\sigma \otimes_A \sigma)} \\
P \otimes^{\mathrm{L}}_A P & \xrightarrow[\eta^{\mathrm{L}}_{P,P}]{\cong} & P \otimes_A P
\end{array}$$

with the diagonal isomorphism $\mathrm{pres}_P : P \otimes_A P \xrightarrow{\cong} M \otimes^{\mathrm{L}}_A M$, in $\mathsf{D}(B \otimes_A B)$.

Next we choose a K-injective resolution $\rho : P \otimes_A P \to I$ in $\mathsf{C}(B \otimes_A B)$. The complex I is unique up to homotopy equivalence, and the homomorphism ρ is unique up to homotopy. The resolution ρ gives rise to an isomorphism

$$\mathrm{pres}_I : \mathrm{Hom}_{B \otimes_A B}(B, I) \xrightarrow{\simeq} \mathrm{RHom}_{B \otimes_A B}(B, P \otimes_A P)$$

in $\mathsf{D}(B)$ such that the diagram

$$\begin{array}{ccc}
\mathrm{Hom}_{B \otimes_A B}(B, P \otimes_A P) & \xrightarrow{\eta^{\mathrm{R}}_{B, P \otimes_A P}} & \mathrm{RHom}_{B \otimes_A B}(B, P \otimes_A P) \\
{\scriptstyle \mathrm{Q}(\mathrm{Hom}(\mathrm{id}, \rho))}\ \downarrow & & \downarrow \cong\ {\scriptstyle \mathrm{RHom}(\mathrm{id}, \mathrm{Q}(\rho))} \\
\mathrm{Hom}_{B \otimes_A B}(B, I) & \xrightarrow[\eta^{\mathrm{R}}_{B,I}]{\cong} & \mathrm{RHom}_{B \otimes_A B}(B, I)
\end{array}$$

with the diagonal isomorphism $\mathrm{pres}_I : \mathrm{Hom}_{B \otimes_A B}(B, I) \xrightarrow{\cong} \mathrm{RHom}_{B \otimes_A B}(B, P \otimes_A P)$, is commutative.

The combination of the presentations pres_P and pres_I gives an isomorphism

$$\mathrm{pres}_{P,I} : \mathrm{Hom}_{B \otimes_A B}(B, I) \xrightarrow{\simeq} \mathrm{Sq}_{B/A}(M)$$

in $\mathsf{D}(B)$, which we also call a presentation.

Let $\phi : M \to N$ be a morphism in $\mathsf{D}(B)$. The morphism $\mathrm{Sq}_{B/A}(\phi)$ can also be made explicit using presentations. For that we need to choose a K-projective resolution $\sigma_N : Q \to N$ in $\mathsf{C}(B)$, and a K-injective resolution $\rho_N : Q \otimes_A Q \to J$ in $\mathsf{C}(B \otimes_A B)$. These provide us with a presentation

$$\mathrm{pres}_{Q,J} : \mathrm{Hom}_{B \otimes_A B}(M, J) \xrightarrow{\simeq} \mathrm{Sq}_{B/A}(N).$$

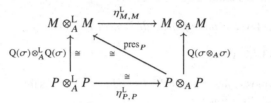

There are homomorphisms $\tilde{\phi} : P \to Q$ in $\mathbf{C}_{\mathrm{str}}(B)$, and $\chi : I \to J$ in $\mathbf{C}_{\mathrm{str}}(B \otimes_A B)$, both unique up to homotopy, such that the diagrams

$$\begin{CD} P @>{\mathrm{Q}(\sigma)}>{\cong}> M \\ @V{\mathrm{Q}(\tilde{\phi})}VV @VV{\phi}V \\ Q @>{\mathrm{Q}(\sigma_N)}>{\cong}> N \end{CD} \qquad \begin{CD} M \otimes^{\mathrm{L}}_A M @<{\mathrm{pres}_P}<{\cong}< P \otimes_A P @>{\mathrm{Q}(\rho)}>{\cong}> I \\ @V{\phi \otimes^{\mathrm{L}}_A \phi}VV @V{\mathrm{Q}(\tilde{\phi} \otimes_A \tilde{\phi})}VV @VV{\mathrm{Q}(\chi)}V \\ N \otimes^{\mathrm{L}}_A N @<{\mathrm{pres}_Q}<{\cong}< Q \otimes_A Q @>{\mathrm{Q}(\rho_N)}>{\cong}> J \end{CD}$$

in $\mathbf{D}(B)$ and $\mathbf{D}(B \otimes_A B)$, respectively, are commutative. See Sections 10.1 and 10.2. Then the diagram

$$\begin{CD} \mathrm{Hom}_{B \otimes_A B}(B, I) @>{\mathrm{pres}_{P,I}}>{\cong}> \mathrm{Sq}_{B/A}(M) \\ @V{\mathrm{Q}(\mathrm{Hom}(\mathrm{id}, \chi))}VV @VV{\mathrm{Sq}_{B/A}(\phi)}V \\ \mathrm{Hom}_{B \otimes_A B}(B, J) @>{\mathrm{pres}_{Q,J}}>{\cong}> \mathrm{Sq}_{B/A}(N) \end{CD} \tag{13.4.11}$$

in $\mathbf{D}(B)$ is commutative.

The squaring operation is not an additive functor. In fact, it is a *quadratic functor*:

Theorem 13.4.12 *Let $\phi : M \to N$ be a morphism in $\mathbf{D}(B)$ and let $b \in B$. Then*

$$\mathrm{Sq}_{B/A}(b \cdot \phi) = b^2 \cdot \mathrm{Sq}_{B/A}(\phi),$$

as morphisms $\mathrm{Sq}_{B/A}(M) \to \mathrm{Sq}_{B/A}(N)$ in $\mathbf{D}(B)$.

Proof We shall use presentations. Let $\tilde{\phi} : P \to Q$ be a homomorphism in $\mathbf{C}_{\mathrm{str}}(B)$ that represents ϕ, as above. Then the homomorphism $b \cdot \tilde{\phi} : P \to Q$ $\mathbf{C}_{\mathrm{str}}(B)$ represents $b \cdot \phi$. Tensoring we get a homomorphism

$$(b \cdot \tilde{\phi}) \otimes_A (b \cdot \tilde{\phi}) : P \otimes_A P \to Q \otimes_A Q$$

$\mathbf{C}_{\mathrm{str}}(B \otimes_A B)$. But

$$(b \cdot \tilde{\phi}) \otimes_A (b \cdot \tilde{\phi}) = (b \otimes b) \cdot (\tilde{\phi} \otimes_A \tilde{\phi}).$$

Hence on the K-injectives we get the homomorphism $(b \otimes b) \cdot \chi : I \to J$ $\mathbf{C}_{\mathrm{str}}(B \otimes_A B)$. We conclude that

$$\mathrm{Hom}_{B \otimes_A B}(\mathrm{id}_B, (b \otimes b) \cdot \chi) : \mathrm{Hom}_{B \otimes_A B}(B, I) \to \mathrm{Hom}_{B \otimes_A B}(B, J)$$

represents $\mathrm{Sq}_{B/A}(b \cdot \phi)$. Finally, by Lemma 13.4.6 we know that

$$\begin{aligned} \mathrm{Hom}_{B \otimes_A B}(\mathrm{id}_B, (b \otimes b) \cdot \chi) &= \mathrm{Hom}_{B \otimes_A B}(b^2 \cdot \mathrm{id}_B, \chi) \\ &= b^2 \cdot \mathrm{Hom}_{B \otimes_A B}(\mathrm{id}_B, \chi). \end{aligned}$$

□

Definition 13.4.13 Let $M \in \mathsf{D}(B)$. A *rigidifying isomorphism* for M over B relative to A is an isomorphism $\rho : M \xrightarrow{\simeq} \mathrm{Sq}_{B/A}(M)$ in $\mathsf{D}(B)$.

Definition 13.4.14 A *rigid complex* over B relative to A is a pair (M, ρ), consisting of a complex $M \in \mathsf{D}(B)$ and a rigidifying isomorphism $\rho : M \xrightarrow{\simeq} \mathrm{Sq}_{B/A}(M)$ in $\mathsf{D}(B)$.

Definition 13.4.15 Suppose (M, ρ) and (N, σ) are rigid complexes over B relative to A. A *morphism of rigid complexes* $\phi : (M, \rho) \to (N, \sigma)$ is a morphism $\phi : M \to N$ in $\mathsf{D}(B)$, such that the diagram

$$\begin{array}{ccc} M & \xrightarrow[\cong]{\rho} & \mathrm{Sq}_{B/A}(M) \\ \phi \downarrow & & \downarrow \mathrm{Sq}_{B/A}(\phi) \\ N & \xrightarrow[\cong]{\sigma} & \mathrm{Sq}_{B/A}(N) \end{array}$$

in $\mathsf{D}(B)$ is commutative.

The category of rigid complexes over B relative to A is denoted by $\mathsf{D}(B)_{\mathrm{rig}/A}$.

Recall (Definition 13.1.5) that a complex $M \in \mathsf{D}(B)$ has the derived Morita property if the derived homothety morphism $\mathrm{hm}^{\mathrm{R}}_M : B \to \mathrm{RHom}_B(M, M)$ in $\mathsf{D}(B)$ is an isomorphism.

Theorem 13.4.16 *Let (M, ρ) be a rigid complex over B relative to A. If M has the derived Morita property, then the only automorphism of (M, ρ) in $\mathsf{D}(B)_{\mathrm{rig}/A}$ is the identity.*

Proof Let $\phi : (M, \rho) \xrightarrow{\simeq} (M, \rho)$ be an automorphism in $\mathsf{D}(B)_{\mathrm{rig}/A}$. By Proposition 13.1.6, there is a unique invertible element $b \in B$ such that $\phi = b \cdot \mathrm{id}_M$, as morphisms $M \to M$ in $\mathsf{D}(B)$.

Next, according to Theorem 13.4.12, we have

$$\mathrm{Sq}_{B/A}(\phi) = \mathrm{Sq}_{B/A}(b \cdot \mathrm{id}_M) = b^2 \cdot \mathrm{Sq}_{B/A}(\mathrm{id}_M).$$

Plugging this into the diagram in Definition 13.4.15 we get a commutative diagram

$$\begin{array}{ccc} M & \xrightarrow[\cong]{\rho} & \mathrm{Sq}_{B/A}(M) \\ b \cdot \mathrm{id}_M \downarrow & & \downarrow b^2 \cdot \mathrm{id}_M \\ M & \xrightarrow[\cong]{\rho} & \mathrm{Sq}_{B/A}(M) \end{array}$$

in $\mathsf{D}(B)$. Once more using Proposition 13.1.6 we see that $b^2 = b$. Because b is an invertible element, it follows that $b = 1$. Thus $\phi = \mathrm{id}_M$. □

Example 13.4.17 Assume $B = A$, and take $M := B$. Then $B \otimes_A B \cong B$, $M \otimes^{\mathrm{L}}_A M \cong M$, and there are canonical isomorphisms

$$\mathrm{Sq}_{B/A}(M) = \mathrm{RHom}_{B \otimes_A B}(B, M \otimes^{\mathrm{L}}_A M) \cong \mathrm{Hom}_B(B, M) \cong M.$$

Thus the pair (M, id_M) belongs to $\mathbf{D}(B)_{\mathrm{rig}/A}$. Furthermore, the complex M has the derived Morita property, so Theorem 13.4.16 applies.

Remark 13.4.18 To the reader who might object to Example 13.4.17 for being silly, we have two things to say. First, the existence theorem for rigid dualizing complexes 13.5.7, whose proof is beyond the scope of this book, starts with the *tautological rigid structure* of $\mathbb{K} \in \mathbf{D}(\mathbb{K})$, where $\mathbb{K}$ is the regular base ring. The rigid structure is then propagated to all flat essentially finite type (FEFT) $\mathbb{K}$-rings using *induction* and *coinduction* of rigid structures. See the exercise and examples below for an indication of these procedures.

The second fact we want to point out is that when A is an FEFT $\mathbb{K}$-ring, then *A has exactly one rigid complex $(M, \rho) \in \mathbf{D}(A)_{\mathrm{rig}/\mathbb{K}}$* (up to a unique isomorphism) that is nonzero on each connected component of $\mathrm{Spec}(A)$; and it is the rigid dualizing complex of A (which will be defined in the next section). This is proved in [188], [13] and [139].

Exercise 13.4.19 Assume $A \neq 0$, and take $B := A[t_1, \ldots, t_n]$, the polynomial ring in n variables.

(1) Let $C := B \otimes_A B$, and let I be the kernel of the multiplication homomorphism $\epsilon : C \to B$. Show that I is generated by the sequence $\boldsymbol{c} = (c_1, \ldots, c_n)$, where $c_j := t_j \otimes 1 - 1 \otimes t_j$.

(2) Show that the Koszul complex $\mathrm{K}(C; \boldsymbol{c})$ is a free resolution of B over C. (Hint: for $1 \leq m \leq n$ define $B_m := A[t_1, \ldots, t_m]$ and $C_m := B_m \otimes_A B_m$. By direct calculation show that $\mathrm{K}(C_1; \boldsymbol{c}_1) \to B_1$ is a quasi-isomorphism. Next, there is a ring isomorphism $B_m \otimes_A B_1 \xrightarrow{\simeq} B_{m+1}$, $1 \otimes t_1 \mapsto t_{m+1}$. This induces a ring isomorphism $C_m \otimes_A C_1 \cong C_{m+1}$ and an isomorphism of complexes

$$\mathrm{K}(C_m; \boldsymbol{c}_m) \otimes_A \mathrm{K}(C_1; \boldsymbol{c}_1) \cong \mathrm{K}(C_{m+1}; \boldsymbol{c}_{m+1}).$$

Use this last formula and induction on m to prove that $\mathrm{K}(C_m; \boldsymbol{c}_m) \to B_m$ is a quasi-isomorphism.)

(3) Prove that

$$\mathrm{Ext}^i_C(B, C) \cong \begin{cases} B & \text{if } i = n \\ 0 & \text{if } i \neq n. \end{cases}$$

(Hint: use the Koszul resolution from above.)

(4) Conclude from item (3) that the complex $B[n] \in \mathbf{D}(B)$ is rigid relative to A; namely that there is a rigidifying isomorphism $\rho : B[n] \xrightarrow{\simeq} \mathrm{Sq}_{B/A}(B[n])$ in $\mathbf{D}(B)$.

Example 13.4.20 Let A be a nonzero noetherian ring, and let $B := A[t_1, \ldots, t_n]$ as in the exercise above. Let $\Delta_{B/A} := \Omega^n_{B/A}$, the module of degree n differential forms of B relative to A. It is the nth exterior power of $\Omega^1_{B/A}$, so it is a free B module of rank 1, with basis $\mathrm{d}(t_1) \wedge \cdots \wedge \mathrm{d}(t_n)$. By the previous example, there exists some rigidifying isomorphism ρ for the complex $\Delta_{B/A}[n]$.

However, it can be shown (see [188] or [175]) that the complex $\Delta_{B/A}[n]$ has a *canonical rigidifying isomorphism relative to A*. In other words, there is a rigidifying isomorphism $\rho : \Delta_{B/A}[n] \xrightarrow{\simeq} \mathrm{Sq}_{B/A}(\Delta_{B/A}[n])$ in $\mathbf{D}(B)$ which is invariant under A-ring automorphisms of B.

This rigidifying isomorphism ρ is an incarnation of Grothendieck's Fundamental Local Isomorphism [62, Proposition II.7.2] and the Residue Isomorphism [62, Theorem II.8.2].

It is well-known that a finitely generated module M over a noetherian ring A is flat iff it is projective. See [105, Corollary to Theorem 7.12].

Example 13.4.21 Let A be a noetherian ring, and let $A \to B$ be a finite flat ring homomorphism. So B is a finitely generated projective A-module. Define $\Delta_{B/A} := \mathrm{Hom}_A(B, A) \in \mathbf{M}(B)$. It can be shown (see [175]) that the complex $\Delta_{B/A}$ has a canonical rigidifying isomorphism relative to A. In other words, there is a rigidifying isomorphism $\rho : \Delta_{B/A} \xrightarrow{\simeq} \mathrm{Sq}_{B/A}(\Delta_{B/A})$ in $\mathbf{D}(B)$ which is invariant under A-ring automorphisms of B.

Remark 13.4.22 The squaring operation $\mathrm{Sq}_{B/A}$ is functorial also in the ring B, in two directions. For this remark we assume that the ring A is noetherian, and the rings B and C are flat essentially finite type A-rings. (See Remark 13.4.24 regarding the flatness.)

Suppose $u : B \to C$ is a finite A-ring homomorphism, and $\theta : N \to M$ is a backward (or trace) morphism in $\mathbf{D}(B)$ over u, in the sense of Definition 12.6.11. Then there is a backward morphism

$$\mathrm{Sq}_{u/A}(\theta) : \mathrm{Sq}_{C/A}(N) \to \mathrm{Sq}_{B/A}(M)$$

in $\mathbf{D}(B)$ over u. The morphism $\mathrm{Sq}_{u/A}(\theta)$ is functorial in u and θ in an obvious sense. Furthermore, under suitable finiteness conditions, if θ is a nondegenerate backward morphism (see Definition 12.6.13), then so is $\mathrm{Sq}_{u/A}(\theta)$. We call this property the *trace functoriality* of the squaring operation.

The trace functoriality of the squaring operation allows us to talk about *rigid trace morphisms*. With $u : B \to C$ as above, suppose that $(M, \rho) \in \mathbf{D}(B)_{\mathrm{rig}/A}$ and $(N, \sigma) \in \mathbf{D}(C)_{\mathrm{rig}/A}$. A *rigid trace morphism* $\theta : (N, \sigma) \to (M, \rho)$ over u

relative to A is a trace morphism $\theta : N \to M$ over u such that the diagram

$$\begin{array}{ccc} N & \xrightarrow[\cong]{\sigma} & \mathrm{Sq}_{C/A}(N) \\ {\scriptstyle \theta}\downarrow & & \downarrow{\scriptstyle \mathrm{Sq}_{u/A}(\theta)} \\ M & \xrightarrow[\cong]{\rho} & \mathrm{Sq}_{B/A}(M) \end{array}$$

in $\mathsf{D}(B)$ is commutative.

Next, suppose $v : B \to C$ is an essentially étale homomorphism (i.e. formally étale, plus the EFT condition that was already assumed) of A-rings, and $\lambda : M \to N$ is a forward morphism over v, in the sense of Definition 12.6.3. Then there is a forward morphism

$$\mathrm{Sq}_{v/A}(\lambda) : \mathrm{Sq}_{B/A}(M) \to \mathrm{Sq}_{C/A}(N)$$

in $\mathsf{D}(B)$ over u. The morphism $\mathrm{Sq}_{v/A}(\lambda)$ is functorial in v and λ in an obvious sense. Furthermore, under suitable finiteness conditions, if λ is a nondegenerate forward morphism (see Definition 12.6.7), then so is $\mathrm{Sq}_{v/A}(\lambda)$. We call this property the *étale functoriality* of the squaring operation.

The étale functoriality of the squaring operation allows us to talk about *rigid localization morphisms*. With $v : B \to C$ as above, suppose that $(M, \rho) \in \mathsf{D}(B)_{\mathrm{rig}/A}$, and $(N, \sigma) \in \mathsf{D}(C)_{\mathrm{rig}/A}$. A *rigid localization morphism* $\lambda : (M, \rho) \to (N, \sigma)$ over v relative to A is a forward morphism $\lambda : N \to M$ over v such that the diagram

$$\begin{array}{ccc} M & \xrightarrow[\cong]{\rho} & \mathrm{Sq}_{B/A}(M) \\ {\scriptstyle \lambda}\downarrow & & \downarrow{\scriptstyle \mathrm{Sq}_{v/A}(\lambda)} \\ N & \xrightarrow[\cong]{\sigma} & \mathrm{Sq}_{C/A}(N) \end{array}$$

in $\mathsf{D}(B)$ is commutative.

These functorialities of the squaring operation are hard to prove. An imprecise study of them was made in the papers [188] and [189]. A correct treatment of the trace functoriality can be found in the paper [173]; and a correct treatment of the étale functoriality will appear in [175].

Remark 13.4.23 The squaring operation is related to *Hochschild cohomology*. Assume for simplicity that A is a field and M is a B-module. Then for each i the cohomology

$$\mathrm{H}^i(\mathrm{Sq}_{B/A}(M)) = \mathrm{Ext}^i_{B \otimes_A B}(B, M \otimes_A M)$$

is the ith Hochschild cohomology with values in the B-bimodule $M \otimes_A M$. For more on this material see [14], [137] and [138].

Remark 13.4.24 It is possible to avoid the assumption that B is flat over A. This is done by choosing a commutative DG ring $\tilde{B}$ that is K-flat as a DG A-module, with a DG ring quasi-isomorphism $\tilde{B} \to B$ over A. Such resolutions always exist (see [188] or [173]). Then we take

$$\mathrm{Sq}_{B/A}(M) := \mathrm{RHom}_{\tilde{B} \otimes_A \tilde{B}}(B, M \otimes^{\mathrm{L}}_A M). \tag{13.4.25}$$

This was the construction used by Zhang and Yekutieli in the paper [188].

Unfortunately there was a serious error in [188]: we did not prove that formula (13.4.25) is independent of the resolution $\tilde{B}$. This error was discovered, and corrected, by L. Avramov, S. Iyengar, J. Lipman and S. Nayak in their paper [14].

There were ensuing errors in [188] regarding the functoriality of the squaring operation in the ring B (as described in Remark 13.4.22). The paper [14] did not treat such functoriality at all, and the constructions and proofs of the trace functoriality were corrected only in our recent paper [173]. It is worthwhile to mention that the correct proofs (both in [14] and [173]) rely on *noncommutative DG rings* and *DG bimodules* over them; and the squaring operation is replaced by the *rectangle operation*.

Because the nonflat case is so much more complicated, we have decided not to reproduce it in the book. The interested reader can look up the research papers [173], [175], [176] and [177], the survey article [168], or the lecture notes [174].

A general treatment of derived categories of bimodules, based on K-flat DG ring resolutions, will be in the paper [178].

13.5 Rigid Dualizing and Residue Complexes

In this section we make these assumptions:

Setup 13.5.1 $\mathbb{K}$ is a nonzero regular noetherian commutative ring (Definition 13.1.10). All other rings are in the category $\mathsf{Rng}_{\mathrm{c}}/_{\mathrm{feft}}\,\mathbb{K}$ of FEFT commutative $\mathbb{K}$-rings, where "FEFT" is short for "flat essentially finite type."

Definition 13.5.2 A *rigid dualizing complex* over A relative to $\mathbb{K}$ is a rigid complex (R, ρ) over A relative to $\mathbb{K}$, as in Definition 13.4.14, such that R is a dualizing complex over A, in the sense of Definition 13.1.9.

See Remark 13.4.24 regarding the flatness condition.

Definition 13.5.3 For $A \in \mathsf{Rng}_{\mathrm{c}}/_{\mathrm{feft}}\,\mathbb{K}$ we write $A^{\mathrm{en}} := A \otimes_{\mathbb{K}} A$ for the *enveloping ring* of A relative to $\mathbb{K}$.

With this notation the square of a complex $M \in \mathsf{D}(A)$ relative to $\mathbb{K}$ is

$$\mathrm{Sq}_{A/\mathbb{K}}(M) = \mathrm{RHom}_{A^{\mathrm{en}}}(A, M \otimes^{\mathrm{L}}_{\mathbb{K}} M) \in \mathsf{D}(A).$$

Recall the category $\mathsf{D}(A)_{\mathrm{rig}/\mathbb{K}}$ of rigid complexes over A relative to $\mathbb{K}$, from Definition 13.4.15.

Theorem 13.5.4 (Uniqueness) *Let A be a flat essentially finite type ring over the regular noetherian ring $\mathbb{K}$. If A has a rigid dualizing complex (R, ρ), then it is unique up to a unique isomorphism in $\mathsf{D}(A)_{\mathrm{rig}/\mathbb{K}}$.*

Proof Suppose (R', ρ') is another rigid dualizing complex over A relative to $\mathbb{K}$. Let $A = \prod_{i=1}^{r} A_i$ be the connected component decomposition of A. Corollary 13.1.53 says that $R' \cong R \otimes^{\mathrm{L}}_{A} P$, where $P \cong \bigoplus_{i=1}^{r} L_i[n_i]$ for integers n_i and rank 1 projective A_i-modules L_i. There is an isomorphism

$$R' \otimes^{\mathrm{L}}_{\mathbb{K}} R' = (R \otimes^{\mathrm{L}}_{A} P) \otimes^{\mathrm{L}}_{\mathbb{K}} (R \otimes^{\mathrm{L}}_{A} P) \cong (R \otimes^{\mathrm{L}}_{\mathbb{K}} R) \otimes^{\mathrm{L}}_{A^{\mathrm{en}}} (P \otimes^{\mathrm{L}}_{\mathbb{K}} P) \quad (13.5.5)$$

in $\mathsf{D}(A^{\mathrm{en}})$, and $P \otimes^{\mathrm{L}}_{\mathbb{K}} P$ has finite flat dimension over A^{en}. So we have this sequence of isomorphisms in $\mathsf{D}(A)$:

$$\begin{aligned}
R \otimes^{\mathrm{L}}_{A} P &\cong R' \cong \mathrm{Sq}_{A/\mathbb{K}}(R') = \mathrm{RHom}_{A^{\mathrm{en}}}(A, R' \otimes^{\mathrm{L}}_{\mathbb{K}} R') \\
&\cong^{\diamond} \mathrm{RHom}_{A^{\mathrm{en}}}(A, (R \otimes^{\mathrm{L}}_{\mathbb{K}} R) \otimes^{\mathrm{L}}_{A^{\mathrm{en}}} (P \otimes^{\mathrm{L}}_{\mathbb{K}} P)) \\
&\cong^{\dagger} \mathrm{RHom}_{A^{\mathrm{en}}}(A, R \otimes^{\mathrm{L}}_{\mathbb{K}} R) \otimes^{\mathrm{L}}_{A^{\mathrm{en}}} (P \otimes^{\mathrm{L}}_{\mathbb{K}} P) \qquad (13.5.6) \\
&= \mathrm{Sq}_{A/\mathbb{K}}(R) \otimes^{\mathrm{L}}_{A^{\mathrm{en}}} (P \otimes^{\mathrm{L}}_{\mathbb{K}} P) \\
&\cong R \otimes^{\mathrm{L}}_{A^{\mathrm{en}}} (P \otimes^{\mathrm{L}}_{\mathbb{K}} P) \cong R \otimes^{\mathrm{L}}_{A} P \otimes^{\mathrm{L}}_{A} P.
\end{aligned}$$

The isomorphism $\cong^{\diamond}$ is by (13.5.5), and the isomorphism $\cong^{\dagger}$ is by Theorem 12.10.14. We also used the rigidifying isomorphisms of R and R'. Now

$$\mathrm{RHom}_{A}(R, R \otimes^{\mathrm{L}}_{A} P) \cong \mathrm{RHom}_{A}(R, R) \otimes^{\mathrm{L}}_{A} P \cong P,$$

again using Theorem 12.10.14, and by the derived Morita property of R. Likewise, there are isomorphisms

$$\begin{aligned}
\mathrm{RHom}_{A}(R, R \otimes^{\mathrm{L}}_{A} P \otimes^{\mathrm{L}}_{A} P) &\cong \mathrm{RHom}_{A}(R, R \otimes^{\mathrm{L}}_{A} P) \otimes^{\mathrm{L}}_{A} P \\
&\cong \mathrm{RHom}_{A}(R, R) \otimes^{\mathrm{L}}_{A} P \otimes^{\mathrm{L}}_{A} P \cong P \otimes^{\mathrm{L}}_{A} P
\end{aligned}$$

in $\mathsf{D}(A)$. Thus, together with (13.5.6), we deduce that $P \otimes^{\mathrm{L}}_{A} P \cong P$. But then on each connected component A_i we have

$$L_i[n_i] \cong L_i[n_i] \otimes_A L_i[n_i] = (L_i \otimes_A L_i)[2 \cdot n_i].$$

This implies that $L_i \cong A_i$ and $n_i = 0$. We see that actually $P \cong A$, so there is an isomorphism $\phi_0 : R \xrightarrow{\simeq} R'$ in $\mathsf{D}(A)$.

The isomorphism ϕ_0 might not be rigid; but due to the derived Morita property of the complex R', there is an invertible element $a \in A$ such that

$\mathrm{Sq}_{A/\mathbb{K}}(\phi_0) \circ \rho = a \cdot \rho' \circ \phi_0$ as isomorphisms $R \xrightarrow{\simeq} \mathrm{Sq}_{A/\mathbb{K}}(R')$. Define $\phi := a^{-1} \cdot \phi_0$. Then, according to Theorem 13.4.12, we have

$$\mathrm{Sq}_{A/\mathbb{K}}(\phi) \circ \rho = a^{-2} \cdot \mathrm{Sq}_{A/\mathbb{K}}(\phi_0) \circ \rho = a^{-2} \cdot a \cdot \rho' \circ \phi_0 = \rho' \circ \phi.$$

We see that $\phi : (R, \rho) \xrightarrow{\simeq} (R', \rho')$ is a rigid isomorphism. Its uniqueness is already known by Theorem 13.4.16, since R' has the derived Morita property. □

Theorem 13.5.7 (Existence) *Let A be a flat essentially finite type ring over the regular noetherian ring $\mathbb{K}$. Then A has a rigid dualizing complex (R_A, ρ_A).*

The proof of this theorem is beyond the scope of the book, as it requires the study of *induced and coinduced rigid structures*. For a proof see [175]. (There is an incomplete proof in [189, Theorem 1.1].) The next two examples provide proofs in some cases.

Example 13.5.8 Suppose $A = \mathbb{K}[t_1, \ldots, t_n]$, the polynomial ring in n variables. (More generally, we can take A to be any essentially smooth $\mathbb{K}$-ring of relative dimension n). Define the complex $R_A := \Omega^n_{A/\mathbb{K}}[n]$. According to Example 13.4.20, there is a rigidifying isomorphism $\rho_A : R_A \xrightarrow{\simeq} \mathrm{Sq}_{A/\mathbb{K}}(R_A)$. Because A is a regular ring, it follows that R_A is a dualizing complex. We see that (R_A, ρ_A) is a rigid dualizing complex.

Example 13.5.9 Suppose A is a finite type $\mathbb{K}$-ring. As explained in Example 18.5.6, A has a noncommutative rigid dualizing complex that is actually A-central. Thus it is a rigid dualizing complex over A relative to $\mathbb{K}$ in the sense of Definition 13.5.2.

The dimension function $\dim_R$ relative to a dualizing complex R was introduced in Definition 13.3.2. If $R' \cong R$, then of course the dimension functions satisfy $\dim_{R'} = \dim_R$. In view of the previous theorem, the next definition is valid.

Definition 13.5.10 Let $A \in \mathsf{Rng_c}/_{\mathrm{feft}}\,\mathbb{K}$. The *rigid dimension function relative to* $\mathbb{K}$ is the function

$$\mathrm{rig.dim}_{\mathbb{K}} : \mathrm{Spec}(A) \to \mathbb{Z}$$

given by the formula $\mathrm{rig.dim}_{\mathbb{K}}(\mathfrak{p}) := \dim_R(\mathfrak{p})$, where R is any rigid dualizing complex over A relative to $\mathbb{K}$.

The next examples (taken from [175]) give an idea what the rigid dimension function looks like.

Example 13.5.11 Let $\mathbb{K}$ be a field and A a finite type $\mathbb{K}$-ring. Then for every $\mathfrak{p} \in \operatorname{Spec}(A)$ there is equality $\operatorname{rig.dim}_{\mathbb{K}}(\mathfrak{p}) = \dim(A/\mathfrak{p})$, where the latter is the Krull dimension of the ring $A/\mathfrak{p}$.

Example 13.5.12 Again $\mathbb{K}$ is a field, but now L is a finitely generated field extension of $\mathbb{K}$; or in other words, L is a field in $\mathsf{Rng}_{\mathrm{c}}/_{\mathrm{feft}}\,\mathbb{K}$. Then, writing $\mathfrak{p} := (0) \subseteq L$, we have $\operatorname{rig.dim}_{\mathbb{K}}(\mathfrak{p}) = \operatorname{tr.deg}_{\mathbb{K}}(L)$, the transcendence degree of the field extension $\mathbb{K} \to L$.

Example 13.5.13 Take $\mathbb{K} = A = \mathbb{Z}$. For a maximal ideal $\mathfrak{m} = (p) \subseteq \mathbb{Z}$ we have $\operatorname{rig.dim}_{\mathbb{K}}(\mathfrak{m}) = -1$; and for the generic ideal $\mathfrak{p} = (0) \subseteq \mathbb{Z}$ we have $\operatorname{rig.dim}_{\mathbb{K}}(\mathfrak{p}) = 0$.

Residue complexes were introduced in Section 13.3.

Definition 13.5.14 A *rigid residue complex* over A relative to $\mathbb{K}$ is a rigid complex $(\mathcal{K}_A, \rho_A)$ over A relative to $\mathbb{K}$, such that $\mathcal{K}_A$ is a residue complex over A.

Using the rigid dimension function relative to $\mathbb{K}$, we have this decomposition of the A-module $\mathcal{K}_A^{-i}$ for each i : $\mathcal{K}_A^{-i} \cong \bigoplus_{\operatorname{rig.dim}_{\mathbb{K}}(\mathfrak{p})=i} J(\mathfrak{p})$, where $J(\mathfrak{p})$ is the indecomposable injective module corresponding to the prime ideal $\mathfrak{p}$.

In Definition 13.4.15 we introduced the category $\mathbf{D}(A)_{\mathrm{rig}/\mathbb{K}}$. Recall that the objects of $\mathbf{D}(A)_{\mathrm{rig}/\mathbb{K}}$ are rigid complexes (M, ρ) over A relative to $\mathbb{K}$; and the morphisms $\phi : (M, \rho) \to (N, \sigma)$ are the morphisms $\phi : M \to N$ in $\mathbf{D}(A)$ for which there is equality $\sigma \circ \phi = \operatorname{Sq}_{A/\mathbb{K}}(\phi) \circ \rho$. Rigid residue complexes live, or rather move, in another category.

Definition 13.5.15 The category $\mathbf{C}(A)_{\mathrm{rig}/\mathbb{K}}$ is defined as follows. Its objects are the rigid complexes (M, ρ) over A relative to $\mathbb{K}$. Given two objects (M, ρ) and (N, σ), a morphism $\phi : (M, \rho) \to (N, \sigma)$ in $\mathbf{C}(A)_{\mathrm{rig}/\mathbb{K}}$ is a morphism $\phi : M \to N$ in $\mathbf{C}_{\mathrm{str}}(A)$, such that the diagram

$$\begin{array}{ccc} M & \xrightarrow[\cong]{\rho} & \operatorname{Sq}_{A/\mathbb{K}}(M) \\ {\scriptstyle \mathrm{Q}(\phi)}\downarrow & & \downarrow{\scriptstyle \operatorname{Sq}_{A/\mathbb{K}}(\mathrm{Q}(\phi))} \\ N & \xrightarrow[\cong]{\sigma} & \operatorname{Sq}_{A/\mathbb{K}}(N) \end{array}$$

in $\mathbf{D}(A)$ is commutative.

Let us emphasize the hybrid nature of the category $\mathbf{C}(A)_{\mathrm{rig}/\mathbb{K}}$: the morphisms are homomorphisms of complexes (literally degree 0 homomorphisms of graded A-modules $\phi : M \to N$ that commute with the differentials); but they must satisfy a compatibility condition (rigidity) in the derived category.

Theorem 13.5.16 *Let A be an FEFT ring over the regular noetherian ring $\mathbb{K}$. The ring A has a rigid residue complex $(\mathcal{K}_A, \rho_A)$ relative to $\mathbb{K}$, and it is unique, up to a unique isomorphism in $\mathsf{C}(A)_{\mathrm{rig}/\mathbb{K}}$.*

Proof Existence: by Theorem 13.5.7 there is a rigid dualizing complex (R_A, ρ_A) over A relative to $\mathbb{K}$. Let $\mathcal{K}_A$ be the minimal injective resolution of the complex R_A. According to Theorem 13.3.17, $\mathcal{K}_A$ is a residue complex. It inherits the rigidifying isomorphism ρ_A from R_A. So the pair $(\mathcal{K}_A, \rho_A)$ is a rigid residue complex over A relative to $\mathbb{K}$.

Uniqueness: suppose $(\mathcal{K}', \rho')$ is another rigid residue complex over A relative to $\mathbb{K}$. Theorem 13.5.4 tells us that there is a unique isomorphism $\phi : (\mathcal{K}_A, \rho_A) \xrightarrow{\simeq} (\mathcal{K}', \rho')$ in $\mathsf{D}(A)_{\mathrm{rig}/\mathbb{K}}$. Now the dimension functions of these two residue complexes are equal (both are $\mathrm{rig.dim}_{\mathbb{K}}$). So by Theorem 13.3.15 the function

$$\mathrm{Q} : \mathrm{Hom}_{\mathsf{C}(A)_{\mathrm{rig}/\mathbb{K}}}((\mathcal{K}_A, \rho_A), (\mathcal{K}', \rho')) \to \mathrm{Hom}_{\mathsf{D}(A)_{\mathrm{rig}/\mathbb{K}}}((\mathcal{K}_A, \rho_A), (\mathcal{K}', \rho'))$$

is bijective. Thus ϕ lifts uniquely to an isomorphism in $\mathsf{C}(A)_{\mathrm{rig}/\mathbb{K}}$. □

Remark 13.5.17 The rigid residue complex $\mathcal{K}_A$ is functorial in the ring A in two different ways, which we briefly explain here. As in Setup 13.5.1, we work in the category $\mathsf{Rng}_{\mathrm{c}}/_{\mathrm{feft}}\,\mathbb{K}$. The flatness condition can be removed (at a price, see Remark 13.4.24), but it seems that the assumption that the base ring $\mathbb{K}$ is regular, and the other rings are EFT over it, are necessary.

Suppose $u : A \to B$ is a finite homomorphism in $\mathsf{Rng}_{\mathrm{c}}/_{\mathrm{feft}}\,\mathbb{K}$. Then there is a homomorphism $\mathrm{tr}_{B/A} : \mathcal{K}_B \to \mathcal{K}_A$ in $\mathsf{C}_{\mathrm{str}}(A)$, called the *rigid trace homomorphism*. It is a nondegenerate rigid trace morphism over u (see Remark 13.4.22), namely it respects the rigidifying isomorphisms, and the induced homomorphism $\mathcal{K}_B \to \mathrm{Hom}_A(B, \mathcal{K}_A)$ in $\mathsf{C}_{\mathrm{str}}(B)$ is an isomorphism. Furthermore, generalizing Theorem 13.5.16, $\mathrm{tr}_{B/A}$ is the unique nondegenerate rigid trace morphism over u. If $B \to C$ is another finite homomorphism in $\mathsf{Rng}_{\mathrm{c}}/_{\mathrm{feft}}\,\mathbb{K}$, then $\mathrm{tr}_{B/A} \circ \mathrm{tr}_{C/B} = \mathrm{tr}_{C/A}$.

Suppose $v : A \to A'$ is an essentially étale homomorphism in $\mathsf{Rng}_{\mathrm{c}}/_{\mathrm{feft}}\,\mathbb{K}$. Then there is a homomorphism $\mathrm{q}_{A'/A} : \mathcal{K}_A \to \mathcal{K}_{A'}$ in $\mathsf{C}_{\mathrm{str}}(A)$, called the *rigid localization homomorphism*. It is a nondegenerate rigid localization morphism over v (see Remark 13.4.22), namely it respects the rigidifying isomorphisms, and the induced homomorphism $A' \otimes_A \mathcal{K}_A \to \mathcal{K}_{A'}$ in $\mathsf{C}_{\mathrm{str}}(A')$ is an isomorphism. Furthermore, $\mathrm{q}_{A'/A}$ is the unique nondegenerate rigid localization morphism over v. If $A' \to A''$ is another essentially étale homomorphism in $\mathsf{Rng}_{\mathrm{c}}/_{\mathrm{feft}}\,\mathbb{K}$, then $\mathrm{q}_{A''/A'} \circ \mathrm{q}_{A'/A} = \mathrm{q}_{A''/A}$.

The two functorialities of the rigid residue complex $\mathcal{K}_A$ commute with each other, in the following sense. Suppose we are given the first commutative diagram below in $\mathsf{Rng}_{\mathrm{c}}/_{\mathrm{feft}}\,\mathbb{K}$, which is cocartesian (i.e. $B' \cong A' \otimes_A B$), $A \to B$ is finite, and $A \to A'$ is essentially étale. Then the second diagram in $\mathbf{C}_{\mathrm{str}}(A)$ is commutative.

$$
\begin{array}{ccc}
A & \longrightarrow & B \\
\downarrow & & \downarrow \\
A' & \longrightarrow & B'
\end{array}
\qquad\qquad
\begin{array}{ccc}
\mathcal{K}_A & \xleftarrow{\;\mathrm{tr}_{B/A}\;} & \mathcal{K}_B \\
{\scriptstyle \mathrm{q}_{A'/A}}\downarrow & & \downarrow{\scriptstyle \mathrm{q}_{B'/B}} \\
\mathcal{K}_{A'} & \xleftarrow{\;\mathrm{tr}_{B'/A'}\;} & \mathcal{K}_{B'}
\end{array}
$$

Imprecise proofs of these assertions can be found in the papers [188] and [189]. A correct treatment will appear in [175].

Remark 13.5.18 The functorialities of the rigid residue complexes that were outlined in Remark 13.4.22 enable similar construction for schemes. In this remark we shall dispense with the flatness condition (invoking Remark 13.4.24); so $\mathbb{K}$ is a nonzero regular noetherian base ring (e.g. a field or the ring of integers $\mathbb{Z}$), and we consider the category $\mathsf{Sch}/_{\mathrm{eft}}\,\mathbb{K}$ of EFT $\mathbb{K}$-schemes.

Let X be an EFT $\mathbb{K}$-scheme. A *rigid residue complex* on X is a pair $(\mathcal{K}_X, \boldsymbol{\rho}_X)$, consisting of a bounded complex of injective quasi-coherent $\mathcal{O}_X$-modules $\mathcal{K}_X$, and a *rigid structure* $\boldsymbol{\rho}_X$ on $\mathcal{K}_X$, to be explained below. For every affine open set $U \subseteq X$ which is *strictly EFT*, namely $A := \Gamma(U, \mathcal{O}_X)$ is an EFT $\mathbb{K}$-ring, the complex $\mathcal{K}_A := \Gamma(U, \mathcal{K}_X)$ is equipped with a rigidifying isomorphism ρ_U, making it into a rigid residue complex over A relative to $\mathbb{K}$. If $U' \subseteq U$ is a smaller strictly EFT open set of X, then the restriction homomorphism $\mathcal{K}_A \to \mathcal{K}_{A'}$ must be the unique rigid localization homomorphism, for the given rigidifying isomorphisms ρ_U and $\rho_{U'}$. The collection of these rigidifying isomorphisms $\boldsymbol{\rho}_X := \{\rho_U\}$ is the rigid structure on $\mathcal{K}_X$. The scheme X admits a rigid residue complex $(\mathcal{K}_X, \boldsymbol{\rho}_X)$, and it is unique up to a unique rigid isomorphism.

If $f : Y \to X$ is a finite map in $\mathsf{Sch}/_{\mathrm{eft}}\,\mathbb{K}$, then there is the rigid trace homomorphism $\mathrm{tr}_f : f_*(\mathcal{K}_Y) \to \mathcal{K}_X$ in $\mathbf{C}_{\mathrm{str}}(X) := \mathbf{C}_{\mathrm{str}}(\mathsf{Mod}\,\mathcal{O}_X)$; and it is nondegenerate, in the sense that the induced homomorphism $f_*(\mathcal{K}_Y) \to \mathcal{H}om_X(f_*(\mathcal{O}_Y), \mathcal{K}_X)$ in $\mathbf{C}_{\mathrm{str}}(X)$ is an isomorphism. If $g : X' \to X$ is an essentially étale map in $\mathsf{Sch}/_{\mathrm{eft}}\,\mathbb{K}$, then there is the rigid localization homomorphism $\mathrm{q}_g : \mathcal{K}_X \to g_*(\mathcal{K}_{X'})$ in $\mathbf{C}_{\mathrm{str}}(X)$; and it is nondegenerate, in the sense that the induced homomorphism $g^*(\mathcal{K}_X) \to \mathcal{K}_{X'}$ is an isomorphism in $\mathbf{C}_{\mathrm{str}}(X')$. In this way the rigid residue complexes become a sheaf on the small étale site of X.

Actually, an infinitesimal local construction gives, for every morphism $f : Y \to X$ $\mathsf{Sch}/_{\mathrm{eft}}\,\mathbb{K}$, the *ind-rigid trace homomorphism* $\mathrm{tr}_f : f_*(\mathcal{K}_Y) \to \mathcal{K}_X$ in $\mathbf{G}_{\mathrm{str}}(X)$; namely this is a homomorphism of graded quasi-coherent sheaves on X. The *Residue Theorem* says that when f is proper, the homomorphism

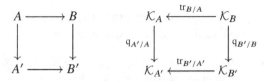

tr_f commutes with the differentials. And the *Duality Theorem* says that in the proper case, the ind-rigid trace tr_f induces an isomorphism

$$\mathrm{R}f_*(\mathrm{R}\mathcal{H}om_Y(\mathcal{N}, \mathcal{K}_Y)) \to \mathrm{R}\mathcal{H}om_X(\mathrm{R}f_*(\mathcal{N}), \mathcal{K}_X)$$

in $\mathsf{D}(X)$ for every $\mathcal{N} \in \mathsf{D}^{\mathrm{b}}_{\mathrm{c}}(X)$. This explicit rigid version of the original theorems of Grothendieck from [62] will appear in [176].

Remark 13.5.19 In this remark we consider *Deligne–Mumford stacks*, of finite type over the regular noetherian base ring $\mathbb{K}$. Due to the étale functoriality of the rigid residue complexes, as explained in Remark 13.5.17, every such stack $\mathfrak{X}$ admits a rigid residue complex $(\mathcal{K}_{\mathfrak{X}}, \rho_{\mathfrak{X}})$. Here the rigid structure $\rho_{\mathfrak{X}} = \{\rho_{(U,g)}\}$ is indexed by the étale maps $g : U \to \mathfrak{X}$ from FT affine $\mathbb{K}$-schemes U.

To a map $f : \mathfrak{Y} \to \mathfrak{X}$ between DM stacks, we associate the ind-rigid trace $\mathrm{tr}_f : f_*(\mathcal{K}_{\mathfrak{Y}}) \to \mathcal{K}_{\mathfrak{X}}$, which is a homomorphism of graded quasi-coherent sheaves on $\mathfrak{X}$. In order to construct the ind-rigid trace for stacks we need another property of rigid residue complexes over rings, which we call *étale codescent*. Suppose $v : A \to A'$ is a faithfully étale ring homomorphism of EFT $\mathbb{K}$-rings. Let $w_1, w_2 : A' \to A' \otimes_A A'$ be the two inclusions. Étale codescent says that in every degree i the sequence of A-module homomorphisms

$$\mathcal{K}^i_{A' \otimes_A A'} \xrightarrow{\mathrm{tr}_{w_1} - \mathrm{tr}_{w_2}} \mathcal{K}^i_{A'} \xrightarrow{\mathrm{tr}_v} \mathcal{K}^i_A \to 0$$

is exact.

As for the Residue Theorem: we can only prove it for a proper map of DM stacks $f : \mathfrak{Y} \to \mathfrak{X}$ which is *coarsely schematic*. Likewise, we can only prove the Duality Theorem for a proper map of DM stacks f which is coarsely schematic and *tame*. These conditions, and a sketch of the proofs, can be found in the lecture notes [174]. These results shall be written in detail in the future paper [177]. Observe that duality is not expected to hold without the tameness condition; but the coarsely schematic condition seems to be a temporary technical hitch.

14

Perfect and Tilting DG Modules over NC DG Rings

Perfect and tilting complexes over noncommutative rings are among the important applications of derived categories to ring theory. In this chapter of the book we study the more general concepts of perfect DG modules and tilting DG bimodules over noncommutative DG rings.

Throughout the chapter we adhere to the following convention, which extends Convention 1.2.4.

Convention 14.0.1 There is a fixed a nonzero commutative base ring $\mathbb{K}$ (e.g. a field, or the ring of integers $\mathbb{Z}$). By default, all linear objects and operations are $\mathbb{K}$-linear. This means that the DG rings are DG central $\mathbb{K}$-rings (Definition 3.3.1), the DG modules are $\mathbb{K}$-linear, the DG bimodules are central over $\mathbb{K}$ and all homomorphisms between them are $\mathbb{K}$-linear. Likewise, all linear categories and functors are $\mathbb{K}$-linear. The symbol $\otimes$ stands for $\otimes_{\mathbb{K}}$.

All the results in this chapter specialize to rings: the category $\mathsf{Rng}/_{\mathrm{c}}\,\mathbb{K}$ of *central* $\mathbb{K}$*-rings* (also known as associative unital $\mathbb{K}$-algebras) is a full subcategory of the category $\mathsf{DGRng}/_{\mathrm{c}}\,\mathbb{K}$ of DG central $\mathbb{K}$-rings.

From section 14.3 onward we shall impose a flatness condition on our DG rings.

14.1 Algebraically Perfect DG Modules

Here we define *algebraically perfect DG modules*, and prove several of their characterizations, in Theorems 14.1.22, 14.1.26 and 14.1.33. See Remark 14.1.4 regarding nomenclature.

Semi-free filtrations of DG modules were introduced in Definition 11.4.3.

Definition 14.1.1 Let A be a DG ring and P a DG A-module.

(1) Let $F = \{F_j(P)\}_{j \geq -1}$ be a semi-free filtration of P. We say that F has *finite extension length* if there is a number $j_1 \in \mathbb{N}$ such that $F_{j_1}(P) = P$.

(2) The DG module P is called *semi-free of finite extension length* if it admits some semi-free filtration F that has finite extension length.

Recall that for a DG module M and an integer k we write $M[k] := \mathrm{T}^k(M)$, the kth translation of M. See Definition 12.1.7.

Definition 14.1.2 Let A be a DG ring and P a DG A-module.

(1) We call P a *finite free DG A-module* if there is an isomorphism $P \cong \bigoplus_{i=1}^{r} A[-k_i]$ in $\mathbf{C}_{\mathrm{str}}(A)$ for some $r \in \mathbb{N}$ and $k_i \in \mathbb{Z}$.

(2) A *finite semi-free filtration* on P is a semi-free filtration $F = \{F_j(P)\}_{j \geq -1}$ that has these properties:
   - Each $\mathrm{Gr}_j^F(P)$ is a finite free DG A-module.
   - The filtration F has finite extension length (Definition 14.1.1(1)).

(3) The DG module P is called *finite semi-free* if it admits some finite semi-free filtration.

Saturated full triangulated subcategories, and those generated by a set of objects, were introduced in Definitions 5.3.19 and 5.3.20.

Definition 14.1.3 A DG A-module L is called *algebraically perfect* if L belongs to the saturated full triangulated subcategory of $\mathbf{D}(A)$ generated by the DG module A.

Remark 14.1.4 The definition above is new. Earlier texts used the term *perfect*, and this referred, somewhat ambiguously, to several distinct properties of DG modules, or complexes, that are sometimes equivalent to each other. See Theorems 14.1.22, 14.1.26 and 14.1.33. For commutative DG rings there is another notion – that of a *geometrically perfect* DG module; see Remark 14.1.25.

The projective dimension of a DG module was introduced in Definition 12.4.8.

Proposition 14.1.5 *If $L \in \mathbf{D}(A)$ is algebraically perfect, then it has finite projective dimension.*

Proof The projective dimension of the DG module A is zero. If L has finite projective dimension then so do all its translates $L[k]$. If there's a distinguished triangle $L' \to L'' \to L \xrightarrow{\triangle}$ s.t. both L' and L'' have finite projective dimension, then so does L. If L is a direct summand of L', and L' has finite projective dimension, then so does L. Now use Proposition 5.3.21. □

Recall that the category $\mathbf{D}(A)$ has arbitrary direct sums, and they are the direct sums in $\mathbf{C}_{\mathrm{str}}(A)$; see Theorem 10.1.26 and Corollary 11.6.30.

Definition 14.1.6 A DG A-module L is called a *compact object of* $\mathbf{D}(A)$ if the functor

$$\mathrm{Hom}_{\mathbf{D}(A)}(L, -) : \mathbf{D}(A) \to \mathbf{M}(\mathbb{K})$$

commutes with infinite direct sums.

Explanation: Let $\{M_z\}_{z \in Z}$ be a collection of objects of $\mathbf{D}(A)$, with direct sum $M := \bigoplus_{z \in Z} M_z$ and embeddings $e_z : M_z \to M$. The canonical homomorphism of $\mathbb{K}$-modules

$$\begin{gathered} \Phi_L : \bigoplus_{z \in Z} \mathrm{Hom}_{\mathbf{D}(A)}(L, M_z) \to \mathrm{Hom}_{\mathbf{D}(A)}(L, M)\,, \\ \Phi_L := \{\mathrm{Hom}(\mathrm{id}_L, e_z)\}_{z \in Z} \end{gathered} \tag{14.1.7}$$

is always injective. Compactness of L says that Φ_L is bijective. Some texts use the name "small" instead of "compact"; see Remark 14.1.36.

Proposition 14.1.8 *Let A and B be DG rings, $F : \mathbf{D}(A) \to \mathbf{D}(B)$ an equivalence of triangulated categories, and $L \in \mathbf{D}(A)$. Then L is a compact object of $\mathbf{D}(A)$ if and only if $F(L)$ is a compact object of $\mathbf{D}(B)$.*

Proof Suppose $\{N_z\}_{z \in Z}$ is a collection of objects of $\mathbf{D}(B)$. Because F is an equivalence of categories, we can assume that $N_z = F(M_z)$ for some $M_z \in \mathbf{D}(A)$. The equivalence F respects coproducts. Namely, if $M = \bigoplus_{z \in Z} M_z$ in $\mathbf{D}(A)$, with embeddings $e_z : M_z \to M$, then the object $N := F(M) \in \mathbf{D}(B)$, with the morphisms $F(e_z) : N_z \to N$, is a coproduct of the collection of objects $\{N_z)\}_{z \in Z}$. There is a commutative diagram of $\mathbb{K}$-modules

$$\begin{array}{ccc} \bigoplus_{z \in Z} \mathrm{Hom}_{\mathbf{D}(A)}(L, M_z) & \xrightarrow{\ \Phi_L\ } & \mathrm{Hom}_{\mathbf{D}(A)}(L, M) \\ {\scriptstyle \bigoplus F}\downarrow{\scriptstyle \cong} & & {\scriptstyle \cong}\downarrow{\scriptstyle F} \\ \bigoplus_{z \in Z} \mathrm{Hom}_{\mathbf{D}(B)}(F(L), N_z) & \xrightarrow{\ \Phi_{F(L)}\ } & \mathrm{Hom}_{\mathbf{D}(B)}(F(L), N) \end{array}$$

We see that if Φ_L is an isomorphism, then so is $\Phi_{F(L)}$.

For the reverse direction we do the same, but now for a quasi-inverse $G : \mathbf{D}(B) \to \mathbf{D}(A)$ of F, the object $F(L) \in \mathbf{D}(B)$ and the object $L \cong G(F(L)) \in \mathbf{D}(A)$. □

Here are a few lemmas about compact objects that will be needed for the proof of Theorem 14.1.22.

Lemma 14.1.9 *Let L be a compact object of $\mathbf{D}(A)$, let $\{M_z\}_{z \in Z}$ be a collection of objects of $\mathbf{D}(A)$, let $M := \bigoplus_{z \in Z} M_z$, and let $\phi : L \to M$ be a morphism in $\mathbf{D}(A)$. Then there is a finite subset $Z_0 \subseteq Z$, such that writing $M_0 := \bigoplus_{z \in Z_0} M_z$, and denoting by $\gamma_0 : M_0 \to M$ the inclusion, there exists a morphism $\phi_0 : L \to M_0$ in $\mathbf{D}(A)$ satisfying $\phi = \mathrm{Q}(\gamma_0) \circ \phi_0$.*

Proof We are given $\phi \in \mathrm{Hom}_{\mathbf{D}(A)}(L, M)$. Using the canonical isomorphism (14.1.7) that we get by the compactness of L, we can express ϕ as a finite sum, say $\phi = \Phi_L(\sum_{z \in Z_0} \phi_z)$, where $Z_0 \subseteq Z$ is a finite subset, and $\phi_z \in \mathrm{Hom}_{\mathbf{D}(A)}(L, M_z)$. The morphism

$$\phi_0 := \sum\nolimits_{z \in Z_0} \phi_z : L \to \bigoplus\nolimits_{z \in Z_0} M_z = M_0$$

in $\mathbf{D}(A)$ has the required property. □

Remark 14.1.10 It might be helpful to say some words about the lemma. The inclusion $\gamma_0 : M_0 \to M$ is a (split) monomorphism in the abelian category $\mathbf{C}_{\mathrm{str}}(A)$. Suppose we choose a K-projective resolution $P \to L$. The morphisms $\phi : L \to M$ and $\phi_0 : L \to M_0$ in $\mathbf{D}(A)$ are represented by homomorphisms $\tilde{\phi} : P \to M$ and $\tilde{\phi}_0 : P \to M_0$ in $\mathbf{C}_{\mathrm{str}}(A)$. There is no reason for the homomorphism $\tilde{\phi}$ to factor through M_0 in $\mathbf{C}_{\mathrm{str}}(A)$. All we can say is that *there is a homotopy $\tilde{\phi} \Rightarrow \gamma_0 \circ \tilde{\phi}_0$*.

Later, after proving Theorem 14.1.22, we will know that there is a finite semi-free DG A-module Q, with homomorphisms $P \xrightarrow{\alpha} Q \xrightarrow{\beta} P$ in $\mathbf{C}_{\mathrm{str}}(A)$, and with a homotopy $\beta \circ \alpha \Rightarrow \mathrm{id}_P$. The homomorphism $\tilde{\phi} \circ \beta : Q \to M$ does factor through some finite direct sum M_0, and hence so does $\tilde{\phi} \circ \beta \circ \alpha : P \to M$. The homotopy $\beta \circ \alpha \Rightarrow \mathrm{id}_P$ induces the homotopy $\tilde{\phi} \Rightarrow \gamma_0 \circ \tilde{\phi}_0$ that was mentioned above.

The next definition is taken from [23].

Definition 14.1.11 Suppose $(\{M_i\}_{i \in \mathbb{N}}, \{\mu_i\}_{i \in \mathbb{N}})$ is a direct system in $\mathbf{D}(A)$. Let

$$\phi : \bigoplus\nolimits_{i \in \mathbb{N}} M_i \to \bigoplus\nolimits_{i \in \mathbb{N}} M_i$$

be the morphism in $\mathbf{D}(A)$ defined by

$$\phi|_{M_i} := (\mathrm{id}, -\mu_i) : M_i \to M_i \oplus M_{i+1}.$$

A *homotopy colimit* of the direct system $\{M_i\}_{i \in \mathbb{N}}$ is an object $M \in \mathbf{D}(A)$ that is a cone on ϕ, in the sense of Definition 5.3.7. Namely M sits in some distinguished triangle

$$\bigoplus_{i\in\mathbb{N}} M_i \xrightarrow{\phi} \bigoplus_{i\in\mathbb{N}} M_i \xrightarrow{\psi} M \xrightarrow{\triangle}$$

in $\mathbf{D}(A)$.

Observe that a homotopy colimit exists, and it is unique up to a nonunique isomorphism in $\mathbf{D}(A)$. See Corollary 5.3.8.

Lemma 14.1.12 *If* $(\{M_i\}_{i\in\mathbb{N}}, \{\mu_i\}_{i\in\mathbb{N}})$ *is a direct system in* $\mathbf{C}_{\mathrm{str}}(A)$*, and we take* $M := \lim_{i\to} M_i$*, the direct limit in* $\mathbf{C}_{\mathrm{str}}(A)$*, then* M *is a homotopy colimit of the direct system* $(\{M_i\}_{i\in\mathbb{N}}, \{\mathrm{Q}(\mu_i)\}_{i\in\mathbb{N}})$ *in* $\mathbf{D}(A)$

Proof Let

$$\tilde{\phi} : \bigoplus_{i\in\mathbb{N}} M_i \to \bigoplus_{i\in\mathbb{N}} M_i$$

be the morphism in $\mathbf{C}_{\mathrm{str}}(A)$ defined like in Definition 14.1.11 above; so the morphism ϕ in that definition equals $\mathrm{Q}(\tilde{\phi})$.

An easy calculation shows that

$$0 \to \bigoplus_{i\in\mathbb{N}} M_i \xrightarrow{\tilde{\phi}} \bigoplus_{i\in\mathbb{N}} M_i \xrightarrow{\tilde{\psi}} M \to 0,$$

where $\tilde{\psi}$ is the canonical homomorphism, is an exact sequence in $\mathbf{C}_{\mathrm{str}}(A)$. By Proposition 7.3.5 there is a distinguished triangle

$$\bigoplus_{i\in\mathbb{N}} M_i \xrightarrow{\mathrm{Q}(\tilde{\phi})} \bigoplus_{i\in\mathbb{N}} M_i \xrightarrow{\mathrm{Q}(\tilde{\psi})} M \xrightarrow{\triangle}$$

in $\mathbf{D}(A)$. □

Lemma 14.1.13 *Let* L *be a compact object of* $\mathbf{D}(A)$*, and let* $\{M_i\}_{i\in\mathbb{N}}$ *be a direct system in* $\mathbf{D}(A)$*, with homotopy colimit* M*. Then the morphism* ψ *in Definition* 14.1.11 *induces an isomorphism*

$$\lim_{i\to} \mathrm{Hom}_{\mathbf{D}(A)}(L, M_i) \xrightarrow{\simeq} \mathrm{Hom}_{\mathbf{D}(A)}(L, M)$$

in $\mathbf{M}(\mathbb{K})$.

Proof Applying the cohomological functor $\mathrm{Hom}_{\mathbf{D}(A)}(L, -)$ to the distinguished triangle in Definition 14.1.11, we get long exact sequence of $\mathbb{K}$-modules

$$\begin{aligned} &\cdots \xrightarrow{\delta^{k-1}} \mathrm{Hom}_{\mathbf{D}(A)}\Big(L, \Big(\bigoplus_{i\in\mathbb{N}} M_i\Big)[k]\Big) \\ &\xrightarrow{\mathrm{Hom}(L,\phi[k])} \mathrm{Hom}_{\mathbf{D}(A)}\Big(L, \Big(\bigoplus_{i\in\mathbb{N}} M_i\Big)[k]\Big) \\ &\xrightarrow{\mathrm{Hom}(L,\psi[k])} \mathrm{Hom}_{\mathbf{D}(A)}(L, M[k]) \xrightarrow{\delta^k} \cdots . \end{aligned} \tag{14.1.14}$$

Using the compactness of L and the resulting isomorphism Φ_L, we obtain from (14.1.14) the long exact sequence of $\mathbb{K}$-modules

$$\begin{aligned} \cdots \xrightarrow{\delta^{k-1}} & \bigoplus\nolimits_{i \in \mathbb{N}} \operatorname{Hom}_{\mathsf{D}(A)}(L, M_i[k]) \\ \xrightarrow{\phi^k} & \bigoplus\nolimits_{i \in \mathbb{N}} \operatorname{Hom}_{\mathsf{D}(A)}(L, M_i[k]) \\ \xrightarrow{\psi^k} & \operatorname{Hom}_{\mathsf{D}(A)}(L, M[k]) \xrightarrow{\delta^k} \cdots . \end{aligned} \tag{14.1.15}$$

We know that for every k the homomorphism ϕ^k above is injective. This implies that the connecting homomorphisms δ^k in (14.1.15) are all zero. Hence, when we take $k = 0$, and we replace the two occurrences of "$\cdots$" in (14.1.15) by "0," we get a short exact sequence. Finally, a calculation like in the proof of Lemma 14.1.12 shows that

$$\operatorname{Coker}(\phi^0) \cong \lim_{i \to} \operatorname{Hom}_{\mathsf{D}(A)}(L, M_i). \qquad \square$$

Lemma 14.1.16 *Let L be a compact object of $\mathsf{D}(A)$, let M be a DG A-module, and let $\{F_j(M)\}_{j \geq -1}$ be a filtration of M in $\mathsf{C}_{\mathrm{str}}(A)$ such that $\bigcup_j F_j(M) = M$. Given a morphism $\phi : L \to M$ in $\mathsf{D}(A)$, there is an index $j_1 \in \mathbb{N}$, and a morphism $\phi_{j_1} : L \to F_{j_1}(M)$ in $\mathsf{D}(A)$, such that $\phi = \mathrm{Q}(\gamma_{j_1}) \circ \phi_{j_1}$, where $\gamma_{j_1} : F_{j_1}(M) \to M$ is the inclusion.*

Proof Here $M \cong \lim_{j \to} F_j(M)$ in $\mathsf{C}_{\mathrm{str}}(A)$. By Lemma 14.1.12 we know that M is a homotopy colimit of the direct system $\{F_j(M)\}_{j \geq -1}$. According to Lemma 14.1.13, we get a bijection

$$\lim_{j \to} \operatorname{Hom}_{\mathsf{D}(A)}(L, F_j(M)) \cong \operatorname{Hom}_{\mathsf{D}(A)}(L, M). \tag{14.1.17}$$

So there is some $j_1 \in \mathbb{N}$ and some $\phi_{j_1} \in \operatorname{Hom}_{\mathsf{D}(A)}(L, F_{j_1}(M))$ that goes to ϕ under the bijection (14.1.17). $\square$

The next lemma is adapted from [144, Lemma tag=09R2].

Lemma 14.1.18 *Let L be a compact object of $\mathsf{D}(A)$, let P be a semi-free DG A-module of finite extension length, and let $\phi : L \to P$ be a morphism in $\mathsf{D}(A)$. Then there is a finite semi-free DG A-module P', with a morphism $\phi' : L \to P'$ in $\mathsf{D}(A)$ and a monomorphism $\gamma' : P' \to P$ in $\mathsf{C}_{\mathrm{str}}(A)$, such that $\phi = \mathrm{Q}(\gamma') \circ \phi'$.*

Proof Let $\{F_j(P)\}_{j \geq -1}$ be a semi-free filtration of P of finite extension length; say $F_{j_1}(P) = P$. We will construct DG submodules

$$P_0 \subseteq P_1 \subseteq \cdots \subseteq P_{j_1} \subseteq P_{j_1+1}$$

of P, such that properties (a)-(c) below hold for every $k \in [0, j_1 + 1]$. For each such k and each $j \geq -1$ we define $F_j(P_k) := F_j(P) \cap P_k$.

(a) The filtration $\{F_j(P_k)\}_{j\geq -1}$ is a semi-free filtration on P_k.
(b) The free DG A-module $\mathrm{Gr}^F_j(P_k)$ is finite free for all $j \geq k$.
(c) The morphism ϕ factors through P_k; namely there is a morphism $\phi_k : L \to P_k$ in $\mathbf{D}(A)$ s.t. $\phi = \mathrm{Q}(\gamma_k) \circ \phi_k$, where $\gamma_k : P_k \to P$ is the inclusion.

For $k = 0$, the DG module $P' := P_0$ is finite semi-free. The morphism $\phi' := \phi_0 : L \to P'$ in $\mathbf{D}(A)$ and the inclusion $\gamma' := \gamma_0 : P' \to P$ in $\mathbf{C}_{\mathrm{str}}(A)$ will satisfy $\phi = \mathrm{Q}(\gamma') \circ \phi'$.

The construction of the DG modules P_k and the morphisms ϕ_k is by descending induction on k, starting from $k = j + 1$. We start with $P_{j_1+1} := P$ and $\phi_{j_1+1} := \phi$. Properties (a)-(c) hold trivially here.

Now to the inductive step. Suppose that $k \in [1, \ldots, j+1]$, and we already found a DG submodule $P_k \subseteq P$ and a morphism $\phi_k : L \to P_k$ as required. We are going to produce a DG submodule $P_{k-1} \subseteq P_k$ such that $\mathrm{Gr}^F_j(P_{k-1}) = \mathrm{Gr}^F_j(P_k)$ for all $j \neq k-1$, and $\mathrm{Gr}^F_{k-1}(P_{k-1})$ is a finite free DG submodule of $\mathrm{Gr}^F_{k-1}(P_k)$. This means that we have to shrink $F_{k-1}(P_k)$ suitably.

For every $j \geq 0$ the DG A-module $\mathrm{Gr}^F_j(P_k)$ is free, and we choose a basis for it, namely a collection $\{\bar{p}_z\}_{z \in Z_j}$ of elements such that $\mathrm{Gr}^F_j(P_k) = \bigoplus_{z \in Z_j} A \cdot \bar{p}_z$ in $\mathbf{C}_{\mathrm{str}}(A)$. Moreover, we choose these bases such that the indexing set Z_j is finite for every $j \geq k$ (this is possible by property (b)), and Z_j is empty for every $j \geq j_1 + 1$ (due to the fact that $F_{j_1}(P) = P$).

By definition $\mathrm{Gr}^F_j(P_k) = F_j(P_k)/F_{j-1}(P_k)$, so we can lift the basis $\{\bar{p}_z\}_{z \in Z_j}$ of $\mathrm{Gr}^F_j(P_k)$ to a collection $\{p_z\}_{z \in Z_j}$ of homogeneous elements $p_z \in F_j(P_k)$. Let's define the indexing set $Z := \coprod_{j \geq k} Z_j$. Then there are decompositions

$$P_k^\natural = F_{k-2}(P_k)^\natural \oplus \Big(\bigoplus\nolimits_{z \in Z_{k-1}} A^\natural \cdot p_z\Big) \oplus \Big(\bigoplus\nolimits_{z \in Z} A^\natural \cdot p_z\Big) \tag{14.1.19}$$

and

$$F_{k-1}(P_k)^\natural = F_{k-2}(P_k)^\natural \oplus \Big(\bigoplus\nolimits_{z \in Z_{k-1}} A^\natural \cdot p_z\Big)$$

in $\mathbf{G}_{\mathrm{str}}(A^\natural)$, and each $A^\natural \cdot p_z$ is a free graded $A^\natural$-module of rank 1.

Because Z is a finite set, there is a finite subset $Y \subseteq Z_{k-1}$ such that

$$\mathrm{d}(p_z) \in F_{k-2}(P_k) \oplus \Big(\bigoplus\nolimits_{y \in Y} A \cdot p_y\Big) \oplus \Big(\bigoplus\nolimits_{z' \in Z} A \cdot p_{z'}\Big)$$

for every $z \in Z$. For every $z \in Z_{k-1}$ the element $\bar{p}_z \in \mathrm{Gr}^F_{k-1}(P_k)$ satisfies $\mathrm{d}(\bar{p}_z) = 0$, and therefore the element $p_z \in F_{k-1}(P_k)$ satisfies $\mathrm{d}(p_z) \in F_{k-2}(P_k)$. In particular this is true for p_y with $y \in Y$. And $F_{k-2}(P_k)$ is a DG submodule of P_k. Combining all these observations we see that the graded submodule

$$Q^\natural := F_{k-2}(P_k)^\natural \oplus \Big(\bigoplus\nolimits_{y \in Y} A^\natural \cdot p_y\Big) \oplus \Big(\bigoplus\nolimits_{z \in Z} A^\natural \cdot p_z\Big) \tag{14.1.20}$$

of $P_k^\natural$ actually underlies a DG submodule $Q \subseteq P_k$. (Compare this to the decomposition (14.1.19) of $P_k^\natural$.)

Consider the quotient DG A-module $N := P_k/Q$, with the canonical projection $\alpha : P_k \to N$, that's an epimorphism in $\mathbf{C}_{\mathrm{str}}(A)$. Define the set $X := Z_{k-1} - Y$ and the elements $n_x := \alpha(p_x) \in N$. Then $N^\natural$ is a free graded $A^\natural$-module, with basis indexed by X, i.e. $N^\natural = \bigoplus_{x \in X} A^\natural \cdot n_x$. Furthermore, because $\mathrm{d}(p_x) \in F_{k-2}(P_k) \subseteq Q$ for all $x \in X$, we have $\mathrm{d}(n_x) = 0$; so actually N is a free DG A-module with basis $\{n_x\}_{x \in X}$.

Next consider the morphism $\theta : L \to N$ in $\mathbf{D}(A)$ with formula $\theta := \mathrm{Q}(\alpha) \circ \phi_k$. By Lemma 14.1.9 there is a finite subset $X_0 \subseteq X$ such that θ factors through the DG submodule $N_0 := \bigoplus_{x \in X_0} A \cdot n_x$. In other words, there's a morphism $\theta_0 : L \to N_0$ in $\mathbf{D}(A)$ s.t. $\theta = \mathrm{Q}(\beta_0) \circ \theta_0$, where $\beta_0 : N_0 \to N$ is the inclusion (a monomorphism in $\mathbf{C}_{\mathrm{str}}(A)$). Define $\bar{N} := N/N_0 = \mathrm{Coker}(\beta_0)$, and let $\delta : N \to \bar{N}$ be the projection (an epimorphism in $\mathbf{C}_{\mathrm{str}}(A)$). Since $\delta \circ \beta_0 = 0$ in $\mathbf{C}_{\mathrm{str}}(A)$, it follows that

$$\mathrm{Q}(\delta) \circ \theta = \mathrm{Q}(\delta) \circ \mathrm{Q}(\beta_0) \circ \theta_0 = \mathrm{Q}(\delta \circ \beta_0) \circ \theta_0 = 0$$

in $\mathbf{D}(A)$.

The subset $Y \cup X_0$ of Z_{k-1} is finite. We define

$$P^\natural_{k-1} := F_{k-2}(P_k)^\natural \oplus \Big(\bigoplus\nolimits_{z \in Y \cup X_0} A^\natural \cdot p_z\Big) \oplus \Big(\bigoplus\nolimits_{z \in Z} A^\natural \cdot p_z\Big). \qquad (14.1.21)$$

Then $P^\natural_{k-1}$ is the underlying graded module of a DG submodule $P_{k-1} \subseteq P_k$, slightly bigger than Q. (Compare this formula to (14.1.19) and (14.1.20).) By construction the DG module P_{k-1} has properties (a)-(b) with index $k-1$. Indeed,

$$\mathrm{Gr}^F_j(P_{k-1}) \cong \begin{cases} \mathrm{Gr}^F_j(P_k) & \text{if } j \geq k \text{ or } j \leq k-2 \\ \bigoplus_{z \in Y \cup X_0} A \cdot \bar{p}_z & \text{if } j = k-1 \end{cases}$$

Let us denote by $\epsilon : P_{k-1} \to P_k$ the inclusion. There is a short exact sequence $0 \to P_{k-1} \xrightarrow{\epsilon} P_k \xrightarrow{\delta \circ \alpha} \bar{N} \to 0$ in $\mathbf{C}_{\mathrm{str}}(A)$, and this becomes a distinguished triangle $P_{k-1} \xrightarrow{\epsilon} P_k \xrightarrow{\delta \circ \alpha} \bar{N} \xrightarrow{\triangle}$ in $\mathbf{D}(A)$. Applying the cohomological functor $\mathrm{Hom}_{\mathbf{D}(A)}(L, -)$ to this distinguished triangle gives rise to a long exact sequence

$$\cdots \to \mathrm{Hom}_{\mathbf{D}(A)}(L, P_{k-1}) \to \mathrm{Hom}_{\mathbf{D}(A)}(L, P_k) \to \mathrm{Hom}_{\mathbf{D}(A)}(L, \bar{N}) \to \cdots.$$

The morphism ϕ_k belongs to the middle term here. We know that

$$\mathrm{Q}(\delta \circ \alpha) \circ \phi_k = \mathrm{Q}(\delta) \circ \mathrm{Q}(\alpha) \circ \phi_k = \mathrm{Q}(\delta) \circ \theta = 0.$$

Thus there is a morphism $\phi_{k-1} : L \to P_{k-1}$ in $\mathbf{D}(A)$ s.t. $\phi_k = \mathrm{Q}(\epsilon) \circ \phi_{k-1}$. The morphism ϕ_{k-1} has property (c) with index $k-1$. □

The derived tensor-evaluation morphism $\mathrm{ev}^{\mathrm{R,L}}_{L,M,N}$ was introduced in Theorem 12.9.10. Finite semi-free, algebraically perfect and compact DG modules were defined in Definitions 14.1.2(3), 14.1.3 and 14.1.6, respectively.

Theorem 14.1.22 *Let A be a DG ring and let L be a DG A-module. The following four conditions are equivalent.*

(i) *L is algebraically perfect.*

(ii) *L is a direct summand in $\mathsf{D}(A)$ of a finite semi-free DG A-module.*

(iii) *For every DG ring B, every DG module $M \in \mathsf{D}(A \otimes B^{\mathrm{op}})$ and every DG module $N \in \mathsf{D}(B)$, the tensor-evaluation morphism*

$$\mathrm{ev}^{\mathrm{R,L}}_{L,M,N} : \mathrm{RHom}_A(L, M) \otimes^{\mathrm{L}}_B N \to \mathrm{RHom}_A(L, M \otimes^{\mathrm{L}}_B N)$$

in $\mathsf{D}(\mathbb{K})$ is an isomorphism.

(iv) *L is a compact object of $\mathsf{D}(A)$.*

Proof

(ii) $\Rightarrow$ (i): We use Proposition 5.3.21. A finite free DG A-module is algebraically perfect. A finite semi-free DG module is obtained by finitely many cones from finite free DG modules, so it is also algebraically perfect. Finally, our DG module L is a direct summand of a finite semi-free DG module, so it is algebraically perfect.

(i) $\Rightarrow$ (iii): Fixing M and N, we have two triangulated functors $F, G : \mathsf{D}(A) \to \mathsf{D}(\mathbb{K})$, namely $F := \mathrm{RHom}_A(-, M) \otimes^{\mathrm{L}}_B N$ and $G := \mathrm{RHom}_A(-, M \otimes^{\mathrm{L}}_B N)$. There is a morphism of triangulated functors $\epsilon := \mathrm{ev}^{\mathrm{R,L}}_{(-),M,N} : F \to G$, and we want to prove that $\epsilon_L : F(L) \to G(L)$ is an isomorphism. Trivially for $L := A$ the morphism ϵ_L is an isomorphism. According to Proposition 5.3.22, the morphism ϵ_L is an isomorphism for every algebraically perfect DG module L.

(iii) $\Rightarrow$ (iv): Let $\{N_z\}_{z \in Z}$ be a collection of objects of $\mathsf{D}(A)$. Define $N := \bigoplus_{z \in Z} N_z$. We have to prove that the canonical homomorphism of $\mathbb{K}$-modules

$$\Phi_L : \bigoplus_{z \in Z} \mathrm{Hom}_{\mathsf{D}(A)}(L, N_z) \to \mathrm{Hom}_{\mathsf{D}(A)}(L, N)$$

is bijective. Take $B := A$ and $M := A \in \mathsf{D}(A \otimes A^{\mathrm{op}})$. Consider the commutative diagram of $\mathbb{K}$-modules (14.1.23). The homomorphism θ is bijective because both $(- \otimes^{\mathrm{L}}_A -)$ and H^0 respect all direct sums (see Theorem 10.1.26 and Propositions 11.1.3 and 12.3.8). The homomorphisms $\mathrm{H}^0(\mathrm{ev}^{\mathrm{R,L}}_{L,M,N_z})$ and $\mathrm{H}^0(\mathrm{ev}^{\mathrm{R,L}}_{L,M,N})$ are bijective by condition (iii). The homomorphisms ζ_{N_z} and ζ_N are bijective because $A \otimes^{\mathrm{L}}_A (-) \cong \mathrm{Id}$ and by Corollary 12.2.8. We conclude that Φ_L is bijective.

(iv) $\Rightarrow$ (ii): Choose a semi-free resolution $\rho : P \to L$ in $\mathsf{C}_{\mathrm{str}}(A)$. We have an isomorphism $\phi := \mathrm{Q}(\rho)^{-1} : L \to P$ in $\mathsf{D}(A)$. Let $\{F_j(P)\}_{j \geq -1}$ be a semi-free filtration on P. According to Lemma 14.1.16, there is an

$$\begin{array}{ccc}
\bigoplus_{z\in Z} \mathrm{H}^0(\mathrm{RHom}_A(L, A) \otimes^{\mathrm{L}}_A N_z) & \xrightarrow[\cong]{\ \theta\ } & \mathrm{H}^0(\mathrm{RHom}_A(L, A) \otimes^{\mathrm{L}}_A N) \\
{\scriptstyle \bigoplus \mathrm{H}^0(\mathrm{ev}^{\mathrm{R,L}}_{L,M,N_z})}\downarrow{\scriptstyle \cong} & & {\scriptstyle \cong}\downarrow{\scriptstyle \mathrm{H}^0(\mathrm{ev}^{\mathrm{R,L}}_{L,M,N})} \\
\bigoplus_{z\in Z} \mathrm{H}^0(\mathrm{RHom}_A(L, A \otimes^{\mathrm{L}}_A N_z)) & \longrightarrow & \mathrm{H}^0(\mathrm{RHom}_A(L, A \otimes^{\mathrm{L}}_A N)) \\
{\scriptstyle \bigoplus \zeta_{N_z}}\downarrow{\scriptstyle \cong} & & {\scriptstyle \cong}\downarrow{\scriptstyle \zeta_N} \\
\bigoplus_{z\in Z} \mathrm{Hom}_{\mathbf{D}(A)}(L, N_z) & \xrightarrow{\ \Phi_L\ } & \mathrm{Hom}_{\mathbf{D}(A)}(L, N)
\end{array} \tag{14.1.23}$$

index j_1 such that, letting $P' := F_{j_1}(P)$, and letting $\gamma' : P' \to P$ be the inclusion, there exists a morphism $\phi' : L \to P'$ in $\mathbf{D}(A)$ satisfying $\phi = \mathrm{Q}(\gamma') \circ \phi'$.

Now the DG module P' is semi-free of finite extension length. Lemma 14.1.18 says that there is a finite semi-free DG A-module P'', with a morphism $\phi'' : L \to P''$ in $\mathbf{D}(A)$, and a monomorphism $\gamma'' : P'' \to P'$ in $\mathbf{C}_{\mathrm{str}}(A)$, such that $\phi' = \mathrm{Q}(\gamma'') \circ \phi''$.

Consider the morphism $\psi'' : P'' \to L$ in $\mathbf{D}(A)$ defined by $\psi'' := \mathrm{Q}(\rho) \circ \mathrm{Q}(\gamma') \circ \mathrm{Q}(\gamma'')$. We have equality

$$\psi'' \circ \phi'' = \mathrm{Q}(\rho) \circ \mathrm{Q}(\gamma') \circ \mathrm{Q}(\gamma'') \circ \phi'' = \mathrm{Q}(\rho) \circ \phi = \mathrm{id}_L .$$

Thus L is a retract in $\mathbf{D}(A)$ of the finite semi-free DG module P''. But by Theorem 5.3.18, L is then a direct summand of P'' in $\mathbf{D}(A)$. □

Corollary 14.1.24 *Let A and B be DG rings, let $F : \mathbf{D}(A) \to \mathbf{D}(B)$ be an equivalence of triangulated categories, and let $L \in \mathbf{D}(A)$. Then L is an algebraically perfect DG A-module if and only if $F(L)$ is an algebraically perfect DG B-module.*

Proof Combine Theorem 14.1.22 and Proposition 14.1.8. □

Remark 14.1.25 Assume A is a *commutative DG ring* (i.e. nonpositive and strongly commutative, see Definitions 3.1.22 and 3.3.4) . Then $\bar{A} := \mathrm{H}^0(A)$ is a commutative ring, and there is a DG ring homomorphism $A \to \bar{A}$. As explained in [172], given a multiplicatively closed set $\bar{S} \subseteq \bar{A}$, letting $S \subseteq A^0$ be the preimage of $\bar{S}$, the localized commutative DG ring $A_S := A \otimes_{A^0} A^0_S$ satisfies $\mathrm{H}^0(A_S) = \bar{A}_{\bar{S}}$. We are going to use the convenient notation $A_{\bar{S}} := A_S$. In case $\bar{S} = \{\bar{s}^k\}_{k\in\mathbb{N}}$ for some element $\bar{s} \in \bar{A}$, we shall write $A_{\bar{s}} := A_{\bar{S}}$.

A sequence of elements $(\bar{s}_1, \ldots, \bar{s}_n)$ in $\bar{A}$ is called a *covering sequence* if $\mathrm{Spec}(\bar{A}) = \bigcup_i \mathrm{Spec}(\bar{A}_{\bar{s}_i})$. This just means that $\bar{A} = \sum_i \bar{A} \cdot \bar{s}_i$. For each i there is a localized DG ring $A_{\bar{s}_i}$.

A DG module $L \in \mathsf{D}(A)$ is called *geometrically perfect* if for some covering sequence $(\bar{s}_1, \ldots, \bar{s}_n)$ of $\bar{A}$ there are finite semi-free DG $A_{\bar{s}_i}$-modules P_i, and isomorphisms $P_i \cong A_{\bar{s}_i} \otimes_A L$ in $\mathsf{D}(A_{\bar{s}_i})$.

In case A is a commutative ring, for which $A = A^0 = \bar{A}$, the definition becomes simpler: a complex $L \in \mathsf{D}(A)$ is geometrically perfect if there is a covering sequence $(s_1, \ldots, s_n)$ of A, bounded complexes of finite free A_{s_i}-modules P_i, and isomorphisms $P_i \cong A_{s_i} \otimes_A L$ in $\mathsf{D}(A_{s_i})$. This is the classical definition of a perfect complex of A-modules, see [21].

According to [172, Corollary 5.21], a DG A-module L is geometrically perfect iff it is algebraically perfect. And according to [172, Theorem 5.11], L is geometrically perfect iff $L \in \mathsf{D}^-(A)$, and its derived reduction $\bar{A} \otimes^{\mathrm{L}}_A L \in \mathsf{D}(\bar{A})$ is geometrically perfect.

In the special case of a ring we can say more.

Theorem 14.1.26 *Let A be a ring and let L be a DG A-module. The following two conditions are equivalent.*

(i) *L is an algebraically perfect DG A-module.*
(ii) *L is isomorphic in $\mathsf{D}(A)$ to a bounded complex of finitely generated projective A-modules.*

The proof is after this lemma. The concentration $\mathrm{con}(M)$ of a graded module was defined in Definition 12.1.4.

Lemma 14.1.27 *In the situation of the theorem, if L is algebraically perfect and $\mathrm{H}(L) \neq 0$, then $\mathrm{con}(\mathrm{H}(L)) = [i_0, i_1]$ for some integers $i_0 \le i_1$, and $\mathrm{H}^{i_1}(L)$ is a finitely presented A-module.*

Proof According to Theorem 14.1.22, L is a direct summand in $\mathsf{D}(A)$ of a finite semi-free DG A-module P. But A is a ring, so P is a bounded complex of (finitely generated free) A-modules. This implies that $\mathrm{H}(P)$ is bounded. But $\mathrm{H}(L)$ is a direct summand, in $\mathsf{G}_{\mathrm{str}}(A)$, of $\mathrm{H}(P)$; so it is also bounded. Because $\mathrm{H}(L)$ is nonzero, its concentration is a bounded nonempty integer interval $[i_0, i_1]$. Note that $i_1 = \sup(\mathrm{H}(L))$.

Corollary 11.4.27 says that there is a semi-free resolution $Q \to L$ with $\sup(Q) = i_1$. As in the proof of the implication (iv) $\Rightarrow$ (ii) in the proof of Theorem 14.1.22, there is a finite semi-free DG A-module Q'', with a monomorphism $Q'' \to Q$ in $\mathsf{C}_{\mathrm{str}}(A)$, such that L is a direct summand of Q'' in $\mathsf{D}(A)$. Now Q'' is a bounded complex of finitely generated free A-modules, with $\sup(Q'') = i_1$, and hence $\mathrm{H}^{i_1}(Q'')$ is a finitely presented A-module. But $\mathrm{H}^{i_1}(L)$ is a direct summand of $\mathrm{H}^{i_1}(Q'')$, so it too is a finitely presented A-module. □

Proof of the Theorem The implication (ii) $\Rightarrow$ (i) is easy; compare to the proof of the implication (ii) $\Rightarrow$ (i) of Theorem 14.1.22.

The implication (i) $\Rightarrow$ (ii) will be proved in two steps.

Step 1. Here L is an A-module, which is algebraically perfect as a DG A-module. According to Proposition 14.1.5, the DG module L has finite projective dimension, say $d \in \mathbb{N}$. (The case $L \cong 0$ can be excluded.) This is also the projective dimension of the module L in the classical sense (see Proposition 12.4.16). By Lemma 14.1.27 the module L is finitely generated. We will proceed by induction on d.

If $d = 0$ then $P := L$ is already a finitely generated projective A-module. If $d \geq 1$, then we can find a finite free A-module P^0, with a surjection $\eta : P^0 \to L$ in $\mathbf{M}(A)$. Let $L' := \mathrm{Ker}(\eta)$, so we have a short exact sequence $0 \to L' \to P^0 \xrightarrow{\eta} L \to 0$ in $\mathbf{M}(A)$. According to Proposition 7.3.5, this becomes a distinguished triangle

$$L' \to P^0 \xrightarrow{\mathrm{Q}(\eta)} L \xrightarrow{\triangle} \tag{14.1.28}$$

in $\mathbf{D}(A)$. Because both L and P^0 are perfect, we see that the DG A-module L' is also perfect. Thus L' is a finitely generated A-module of finite projective dimension. However, by the usual syzygy argument (cf. [130, Section 5.1.1]) – or by examination of the long exact sequence gotten by applying $\mathrm{Hom}_{\mathbf{D}(A)}(-, N)$, for an A-module N, to the distinguished triangle (14.1.28) – we see that the projective dimension of L' is $\leq d - 1$. Induction says that there is an isomorphism $L' \cong P'$ in $\mathbf{D}(A)$ for some bounded complex of finitely generated projective A-modules P'. Plugging this isomorphism into the distinguished triangle (14.1.28) we obtain a distinguished triangle

$$P' \xrightarrow{\psi} P^0 \xrightarrow{\mathrm{Q}(\eta)} L \xrightarrow{\triangle} \tag{14.1.29}$$

in $\mathbf{D}(A)$. Because P' is K-projective, there is a homomorphism $\tilde{\psi} : P' \to P^0$ in $\mathbf{C}_{\mathrm{str}}(A)$ such that $\psi = \mathrm{Q}(\tilde{\psi})$. Let $P := \mathrm{Cone}(\tilde{\psi})$, the standard cone on $\tilde{\psi}$. Then P is a bounded complex of finitely generated projective A-modules, and the distinguished triangle (14.1.29) gives rise to an isomorphism $P \cong L$ in $\mathbf{D}(A)$.

Step 2. Here L is an arbitrary algebraically perfect DG A-module, nonzero in $\mathbf{D}(A)$. By Lemma 14.1.27 we have $\mathrm{con}(\mathrm{H}(L)) = [i_0, i_1]$ for integers $i_0 \leq i_1$, and $\mathrm{H}^{i_1}(L)$ is a finitely generated A-module. Let $n := \mathrm{amp}(\mathrm{H}(L)) = i_1 - i_0 \in \mathbb{N}$. We proceed by induction on n.

If $n = 0$ then L is isomorphic to the translation of an A-module, and we are done by step 1. On the other hand, if $n \geq 1$, then choose a finite free A-module P^{i_1}, with a surjection $\bar{\eta} : P^{i_1} \to \mathrm{H}^{i_1}(L)$ in $\mathbf{M}(A)$. This can be lifted to a homomorphism $\eta : P^{i_1}[-i_1] \to L$ in $\mathbf{C}_{\mathrm{str}}(A)$. Consider the DG A-module

$L' := \operatorname{Cone}(\eta)$. There is a distinguished triangle

$$P^{i_1}[-i_1] \xrightarrow{\mathrm{Q}(\eta)} L \to L' \xrightarrow{\triangle} \tag{14.1.30}$$

in $\mathbf{D}(A)$. The DG module L' is also algebraically perfect. There is a long exact cohomology sequence

$$\cdots \to 0 \to \mathrm{H}^{i_0}(L) \to \mathrm{H}^{i_0}(L') \to \cdots \to 0 \to \mathrm{H}^{i_1-2}(L) \to \mathrm{H}^{i_1-2}(L')$$
$$\to 0 \to \mathrm{H}^{i_1-1}(L) \to \mathrm{H}^{i_1-1}(L') \to P^{i_1} \xrightarrow{\bar{\eta}} \mathrm{H}^{i_1}(L) \to \mathrm{H}^{i_1}(L') \to 0 \to \cdots$$

Because $\bar{\eta}$ is surjective, we know that $\mathrm{H}^{i_1}(L') = 0$. Hence $\operatorname{con}(\mathrm{H}(L')) \subseteq [i_0, i_1 - 1]$ and $\operatorname{amp}(\mathrm{H}(L')) \leq n - 1$.

The induction assumption says that there is an isomorphism $L' \cong P'$ in $\mathbf{D}(A)$ for some bounded complex of finitely generated projective A-modules P'. Plugging this isomorphism into the triangle (14.1.30), and then turning it, we obtain a distinguished triangle

$$P'[-1] \xrightarrow{\psi} P^{i_1}[-i_1] \xrightarrow{\mathrm{Q}(\eta)} L \xrightarrow{\triangle} \tag{14.1.31}$$

in $\mathbf{D}(A)$. Because $P'[-1]$ is K-projective, there is a homomorphism $\tilde{\psi} : P'[-1] \to P^{i_1}[-i_1]$ in $\mathbf{C}_{\mathrm{str}}(A)$ such that $\psi = \mathrm{Q}(\tilde{\psi})$. Let $P := \operatorname{Cone}(\tilde{\psi})$, the standard cone on $\tilde{\psi}$. Then P is a bounded complex of finitely generated projective A-modules, and the distinguished triangle (14.1.31) gives rise to an isomorphism $P \cong L$ in $\mathbf{D}(A)$. □

Corollary 14.1.32 *Let A be a ring and let L be a complex of A-modules. The following two conditions are equivalent.*

(i) *L is a compact object of $\mathbf{D}(A)$.*
(ii) *L is isomorphic in $\mathbf{D}(A)$ to a bounded complex of finitely generated projective A-modules.*

Proof Combine Theorems 14.1.26 and 14.1.22. □

Another special case is when the DG ring A is nonpositive and cohomologically left pseudo-noetherian; see Definition 11.4.37.

Theorem 14.1.33 *Let A be a cohomologically left pseudo-noetherian nonpositive DG ring, and let L be a DG A-module. The following two conditions are equivalent.*

(i) *L is an algebraically perfect DG A-module.*
(ii) *L has finite projective dimension and belongs to $\mathbf{D}^-_{\mathrm{f}}(A)$.*

Proof

(i) ⇒ (ii): Since $A \in \mathbf{D}^-_{\mathrm{f}}(A)$ and $\mathbf{D}^-_{\mathrm{f}}(A)$ is a saturated full triangulated subcategory of $\mathbf{D}(A)$, we see that $L \in \mathbf{D}^-_{\mathrm{f}}(A)$. By Proposition 14.1.5, L has finite projective dimension.

(ii) ⇒ (i): We may assume that $\mathrm{H}(L) \neq 0$. Let $i_1 := \sup(\mathrm{H}(L)) \in \mathbb{Z}$, and let $n \in \mathbb{N}$ be the projective dimension of L. By Theorem 11.4.40 there is a quasi-isomorphism $\rho : P \to L$ in $\mathbf{C}_{\mathrm{str}}(A)$ from a pseudo-finite semi-free DG A-module P such that $\sup(P) = i_1$. Let $\{F_j(P)\}_{j \geq -1}$ be a pseudo-finite semi-free filtration of P as in Definition 11.4.29(1), with the same upper bound i_1. Define $P' := F_{i_1+n}(P)$ and $P'' := P/P'$. The DG A-module P'' is concentrated in the degree interval $[-\infty, i_1 - n - 1]$, and therefore

$$\mathrm{con}(\mathrm{H}(\mathrm{RHom}_A(L, P''))) \subseteq [-\infty, -1].$$

This implies that

$$\mathrm{Hom}_{\mathbf{D}(A)}(L, P'') \cong \mathrm{H}^0(\mathrm{RHom}_A(L, P'')) = 0. \tag{14.1.34}$$

The DG A-module P' is finite semi-free. Let $\gamma' : P' \to P$ be the inclusion. From the distinguished triangle $P' \xrightarrow{\mathrm{Q}(\gamma')} P \to P'' \xrightarrow{\triangle}$ in $\mathbf{D}(A)$ we get an exact sequence of $\mathbb{K}$-modules

$$\mathrm{Hom}_{\mathbf{D}(A)}(L, P') \to \mathrm{Hom}_{\mathbf{D}(A)}(L, P) \to \mathrm{Hom}_{\mathbf{D}(A)}(L, P'').$$

The isomorphism $\phi := \mathrm{Q}(\rho)^{-1} : L \to P$ belongs to the middle term above. But by (14.1.34) we see that ϕ comes from some morphism $\phi' : L \to P'$ in $\mathbf{D}(A)$. Because $\mathrm{Q}(\rho \circ \gamma') \circ \phi' = \mathrm{id}_L$, and using Theorem 5.3.18, we see that L is a direct summand of P' in $\mathbf{D}(A)$. Theorem 14.1.22 says that L is algebraically perfect. □

Corollary 14.1.35 *Let A be a left noetherian ring, and let L be a DG A-module. The following two conditions are equivalent.*

(i) *L is an algebraically perfect DG A-module.*
(ii) *L has finite projective dimension and belongs to $\mathbf{D}^{\mathrm{b}}_{\mathrm{f}}(A)$.*

Proof Combine Theorems 14.1.26 and 14.1.33. □

Remark 14.1.36 Here are some historical notes on perfect and compact objects.

Perfect complexes in algebraic geometry were introduced by A. Grothendieck et al. in [21]. For a scheme X, let $\mathbf{D}_{\mathrm{qc}}(X)$ be the derived category of complexes of $\mathcal{O}_X$-modules with quasi-coherent cohomology. The definition of a perfect complex given in [21] was this: $\mathcal{L}$ is perfect if locally it is isomorphic to a bounded complex of finite rank free $\mathcal{O}_X$-modules. This coincides with our notion of geometrically perfect complex, as in Remark 14.1.25, when A is a commutative ring and $X = \mathrm{Spec}(A)$.

The idea of defining finiteness properties of an object L in a category D via the commutation of the functor $\mathrm{Hom}_{\mathsf{D}}(L, -)$ with suitable coproducts goes back to the early days of category theory. Objects L such that $\mathrm{Hom}_{\mathsf{D}}(L, -)$

commutes with coproducts were called *small.* Grothendieck [53] used small objects in his construction of injective resolutions in what was eventually called a Grothendieck abelian category. P. Freyd [47] used small objects in the proof of the Freyd–Mitchell Theorem. In algebraic topology, small objects were used by E. Brown [31] in his celebrated Representability Theorem.

J. Rickard [126] proved a slightly weaker version of our Corollary 14.1.32. Shortly afterward, B. Keller [79] deduced from A. Neeman's work [112] a result that is essentially our Theorem 14.1.22 – he did not have condition (iii), but on the other hand he considered the more general derived category $\mathbf{D}(\mathsf{A})$ of DG modules over a DG category A, whereas we only consider $\mathbf{D}(A)$ for a DG ring A, which is a single-object DG category.

R.W. Thomason [150] discovered that the perfect complexes in $\mathbf{D}_{\mathrm{qc}}(X)$, when X is a quasi-projective scheme over a ring $\mathbb{K}$, are precisely the compact objects in $\mathbf{D}_{\mathrm{qc}}(X)$. Neeman [112] realized the connection between the work of Thomason and that of the topologists A.K. Bousfield and D. Ravenel. The terminology switch from *small object* to *compact object* seems to have taken place in [112]. Neeman's book [113] is essentially devoted to the study of α-compactly generated triangulated categories, for a regular cardinal number α, and to generalizations of the Brown Representability Theorem. Note that what we called "compact object" in Definition 14.1.6 is, in the framework of [112], an "$\aleph_0$-compact object."

The paper [26] by A. Bondal and M. Van den Bergh was very influential in promoting the role of compact objects in algebraic geometry. In the last 20 years there has been a proliferation in the presence of compact (or perfect) objects in research in the areas of derived algebraic geometry, noncommutative algebraic geometry and mathematical physics – much of this due to the influence of M. Kontsevich. For more details we recommend looking in the online reference [115].

14.2 Derived Morita Theory

Recall that Convention 14.0.1 is in effect. In particular all DG rings are $\mathbb{K}$-central, and $\otimes = \otimes_{\mathbb{K}}$.

Definition 14.2.1 Let A and B be DG rings. The objects of the category $\mathbf{C}(A \otimes B^{\mathrm{op}})$ are called *DG A-B-bimodules*.

There is a commutative diagram (14.2.2) in $\mathsf{DGRng}/_{\mathrm{c}}\,\mathbb{K}$.

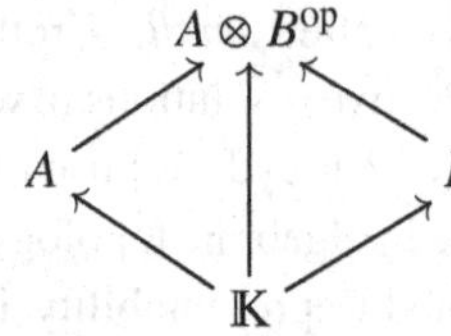

(14.2.2)

By restriction there is a commutative diagram of DG functors, and a commutative diagram of triangulated functors, see (14.2.3).

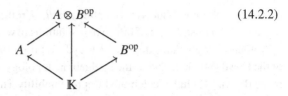

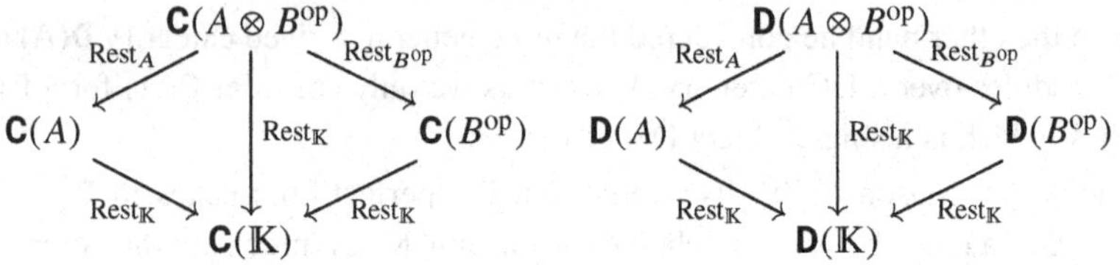

(14.2.3)

There are several manipulations of DG bimodules that we are going to use. First we note that $(A^{\mathrm{op}})^{\mathrm{op}} = A$. Next, given DG $\mathbb{K}$-modules M and N, let

$$\mathrm{br}_{M,N} : M \otimes N \xrightarrow{\simeq} N \otimes M \tag{14.2.4}$$

be the isomorphism in $\mathbf{C}(\mathbb{K})$ defined by

$$\mathrm{br}_{M,N}(m \otimes n) := (-1)^{i \cdot j} \cdot n \otimes m \tag{14.2.5}$$

for homogeneous elements $m \in M^i$ and $n \in N^j$. In Section 3.1 the isomorphism $\mathrm{br}_{M,N}$ was called the braiding of the symmetric monoidal category $\mathbf{C}(\mathbb{K})$. If $M \in \mathbf{C}(A^{\mathrm{op}})$, then we can view M either as a left DG A^{op}-module, or as a right DG A-module. The opposite holds for $N \in \mathbf{C}(A)$. It is not hard to see that there is an isomorphism

$$M \otimes_A N \xrightarrow{\simeq} N \otimes_{A^{\mathrm{op}}} M \tag{14.2.6}$$

in $\mathbf{C}(\mathbb{K})$ with the same formula (14.2.5). This works also for DG rings: there are isomorphisms

$$A \otimes B^{\mathrm{op}} \xrightarrow{\simeq} B^{\mathrm{op}} \otimes A \tag{14.2.7}$$

and

$$A \otimes B^{\mathrm{op}} \xrightarrow{\simeq} (B \otimes A^{\mathrm{op}})^{\mathrm{op}} \tag{14.2.8}$$

in $\mathsf{DGRng}/_{\mathrm{c}}\, \mathbb{K}$, with a formulas like (14.2.5).

There is a DG bifunctor

$$(- \otimes_B -) : \mathbf{C}(A \otimes B^{\mathrm{op}}) \times \mathbf{C}(B) \to \mathbf{C}(A). \tag{14.2.9}$$

Here is a variation of Theorem 12.3.1.

Proposition 14.2.10 *The bifunctor* (14.2.9) *has a triangulated left derived bifunctor*

$$(- \otimes^{\mathrm{L}}_{B} -) : \mathbf{D}(A \otimes B^{\mathrm{op}}) \times \mathbf{D}(B) \to \mathbf{D}(A).$$

If $P \in \mathbf{D}(B)$ is a K-flat DG module, then for every $M \in \mathbf{D}(A \otimes B^{\mathrm{op}})$ the morphism

$$\eta^{\mathrm{L}}_{M,P} : M \otimes^{\mathrm{L}}_{B} P \to M \otimes_{B} P$$

in $\mathbf{D}(A)$ is an isomorphism.

Proof This is because $\mathbf{C}(B)$ has enough K-flat objects. See Theorem 9.3.16 and Lemma 12.3.2. □

Remark 14.2.11 There is a delicate issue here. Even though the category $\mathbf{C}(A \otimes B^{\mathrm{op}})$ has enough K-flat objects, they can not be used to calculate $(- \otimes^{\mathrm{L}}_{B} -)$. The reason is this: given $N \in \mathbf{C}(B)$, the DG $(A \otimes B^{\mathrm{op}})$-modules that are acyclic for the DG functor $(-) \otimes_{B} N$ are those that are *K-flat over* B^{op}. In general, a K-flat DG $(A \otimes B^{\mathrm{op}})$-module M is not K-flat as a B^{op}-module.

A sufficient condition for a K-flat DG $(A \otimes B^{\mathrm{op}})$-module M to be K-flat over B^{op} is that A is K-flat as a DG $\mathbb{K}$-module. In Section 14.3 we will make heavy use of this fact.

There is another DG bifunctor we shall want to use:

$$\mathrm{Hom}_{A}(-,-) : \mathbf{C}(A \otimes B^{\mathrm{op}})^{\mathrm{op}} \times \mathbf{C}(A) \to \mathbf{C}(B). \tag{14.2.12}$$

Here is a variation of Theorem 12.2.1.

Proposition 14.2.13 *The bifunctor* (14.2.12) *has a triangulated right derived bifunctor*

$$\mathrm{RHom}_{A}(-,-) : \mathbf{D}(A \otimes B^{\mathrm{op}})^{\mathrm{op}} \times \mathbf{D}(A) \to \mathbf{D}(B).$$

If $I \in \mathbf{D}(A)$ is a K-injective DG module, then for every $M \in \mathbf{D}(A \otimes B^{\mathrm{op}})$ the morphism

$$\eta^{\mathrm{R}}_{M,I} : \mathrm{Hom}_{A}(M, I) \to \mathrm{RHom}_{A}(M, I)$$

in $\mathbf{D}(B)$ is an isomorphism.

Proof This is because $\mathbf{C}(A)$ has enough K-injective objects. See Theorem 9.2.16, Theorem 9.3.11 and Lemma 12.2.2. □

Remark 14.2.14 Like in Remark 14.2.11, even though the category $\mathbf{C}(A \otimes B^{\mathrm{op}})$ has enough K-projective DG modules, they can not be used to calculate $\mathrm{RHom}_{A}(-, N)$. This is because a K-projective DG $(A \otimes B^{\mathrm{op}})$-module P is not, in general, K-projective over A.

A sufficient condition for a K-projective DG $(A \otimes B^{\mathrm{op}})$-module P to be K-projective over A is that B is K-projective over $\mathbb{K}$. This is a bit stronger than the condition that B is K-flat over $\mathbb{K}$.

Proposition 14.2.15 *Let $L \in \mathsf{D}(A \otimes B^{\mathrm{op}})$, and consider the $\mathbb{K}$-linear triangulated functors*

$$F := \mathrm{RHom}_A(L, -) : \mathsf{D}(A) \to \mathsf{D}(B)$$

and

$$G := L \otimes^{\mathrm{L}}_B (-) : \mathsf{D}(B) \to \mathsf{D}(A).$$

Then:

(1) *The functor F is a right adjoint of G.*
(2) *The functor F is an equivalence if and only if the functor G is an equivalence.*

Proof

(1) Given $M \in \mathsf{D}(B)$ and $N \in \mathsf{D}(A)$ we choose a K-projective resolution $P \to M$ in $\mathsf{C}_{\mathrm{str}}(B)$ and a K-injective resolution $N \to I$ in $\mathsf{C}_{\mathrm{str}}(A)$. We get isomorphisms $F(N) \cong F(I) \cong \mathrm{Hom}_A(L, I)$ and $G(M) \cong G(P) \cong L \otimes_B P$ in $\mathsf{D}(B)$ and $\mathsf{D}(A)$, respectively. These give rise to isomorphisms of $\mathbb{K}$-modules

$$\begin{aligned} \mathrm{Hom}_{\mathsf{D}(B)}(M, F(N)) &\cong^{\diamond} \mathrm{H}^0(\mathrm{RHom}_B(M, F(N))) \\ &\cong \mathrm{H}^0(\mathrm{Hom}_B(P, \mathrm{Hom}_A(L, I))) \cong^{\dagger} \mathrm{H}^0(\mathrm{Hom}_A(L \otimes_B P, I)). \\ &\cong \mathrm{H}^0(\mathrm{RHom}_A(G(M), N)) \cong^{\diamond} \mathrm{Hom}_{\mathsf{D}(A)}(G(M), N) \end{aligned}$$

The isomorphism $\cong^{\dagger}$ is due to the noncommutative Hom-tensor adjunction (see [129, Theorem 2.11]), and the isomorphisms $\cong^{\diamond}$ are by Corollary 12.2.8. The composed isomorphism

$$\mathrm{Hom}_{\mathsf{D}(B)}(M, F(N)) \cong \mathrm{Hom}_{\mathsf{D}(A)}(G(M), N)$$

is functorial in M and N.

(2) Clear from (1). □

Definition 14.2.16 Let A and B be DG rings. A DG module $L \in \mathsf{D}(A \otimes B^{\mathrm{op}})$ is called a *pretilting DG A-B-bimodule* if it satisfies the equivalent conditions in Proposition 14.2.15(2).

Remark 14.2.17 The literature has several different meanings for the word "tilting." See Remark 14.2.39 for a historical survey. The commonly perceived meaning of a tilting object, say as in the book [3], is very close to what we call a pretilting DG bimodule (when A and B are rings).

Notice that there is a lack of symmetry in the definition of a pretilting DG A-B-bimodule L. This lack of symmetry will disappear when we talk about *tilting DG bimodules* in Section 14.3.

Definition 14.2.18 Let A and B be DG rings, and let $M \in \mathbf{C}(A \otimes B^{\mathrm{op}})$. We say that M is *K-flat* (resp. *K-injective*, resp. *K-projective*, resp. *semi-free*, resp. *algebraically perfect*) *over* A, or *on the* A *side*, if $\operatorname{Rest}_A(M) \in \mathbf{C}(A)$ is K-flat (resp. K-injective, resp. K-projective, resp. semi-free, resp. algebraically perfect). Likewise, we define the properties *on the* B^{op} *side*.

Proposition 14.2.19 *Suppose* $L \in \mathbf{D}(A \otimes B^{\mathrm{op}})$ *is a pretilting DG A-B-bimodule. Then L is an algebraically perfect DG A-module.*

Proof Under the equivalence $G : \mathbf{D}(B) \to \mathbf{D}(A)$ from Proposition 14.2.15 we have $G(B) \cong L$. Of course, $B \in \mathbf{D}(B)$ is an algebraically perfect DG B-module. Now use Corollary 14.1.24. □

Observe that an object $N \in \mathbf{D}(A)$ is *nonzero* if it is not a zero object of the category $\mathbf{D}(A)$, i.e. if $N \not\cong 0$ in $\mathbf{D}(A)$. This is equivalent to the condition that $\mathrm{H}(N) \neq 0$. The clarification is important for the next definition.

Definition 14.2.20 Let $\mathsf{E} \subseteq \mathbf{D}(A)$ be a full triangulated subcategory that's closed under infinite direct sums.

(1) An object $L \in \mathbf{D}(A)$ is called a *compact object relative to* E if the functor

$$\operatorname{Hom}_{\mathbf{D}(A)}(L, -) : \mathsf{E} \to \mathbf{M}(\mathbb{K})$$

commutes with infinite direct sums.

(2) An object $L \in \mathbf{D}(A)$ is called a *generator relative to* E if for every nonzero object $N \in \mathsf{E}$ there exists some $i \in \mathbb{Z}$ such that $\operatorname{Hom}_{\mathbf{D}(A)}(L, N[i]) \neq 0$.

(3) An object $L \in \mathsf{E}$ is called a *compact generator of* E if it is both a compact object relative to E and a generator relative to E.

Example 14.2.21 For every positive integer r the free DG module $P := A^r$ is a compact generator of $\mathsf{E} := \mathbf{D}(A)$.

Example 14.2.22 Assume A is a commutative noetherian ring and $\boldsymbol{a} = (a_1, \ldots, a_n)$ is a finite sequence of elements in it. (Or, more generally, A is commutative and $\boldsymbol{a}$ is a *weakly proregular sequence*, as in [118].) Let $\mathfrak{a} \subseteq A$ be the ideal generated by $\boldsymbol{a}$. A complex $M \in \mathbf{D}(A)$ is called *cohomologically* $\mathfrak{a}$*-torsion* if all its cohomology modules $\mathrm{H}^i(M)$ are $\mathfrak{a}$-torsion. Consider the full subcategory $\mathsf{E} := \mathbf{D}_{\mathfrak{a}\text{-tor}}(A)$ of $\mathbf{D}(A)$ on the cohomologically $\mathfrak{a}$-torsion complexes. This is full triangulated subcategory that's closed under infinite direct sums. The Koszul complex $\mathrm{K}(A; \boldsymbol{a})$ is a compact generator of $\mathbf{D}_{\mathfrak{a}\text{-tor}}(A)$. See [119, Proposition 5.1].

Proposition 14.2.23 *Let A and B be DG rings,* $F : \mathbf{D}(A) \to \mathbf{D}(B)$ *an equivalence of triangulated categories,* $\mathsf{E} \subseteq \mathbf{D}(A)$ *a full triangulated subcategory that's closed under infinite direct sums, and* $L \in \mathbf{D}(A)$.

(1) *L is a compact object relative to E if and only if $F(L)$ is a compact object relative to $F(\mathsf{E})$.*
(2) *L is a generator relative to E if and only if $F(L)$ is a generator relative to $F(\mathsf{E})$.*

Proof For item (1) use the proof of Proposition 14.1.8. Item (2) is trivial. □

Definition 14.2.24 Let A and B be DG rings, let $\mathsf{E} \subseteq \mathbf{D}(A)$ be a full triangulated subcategory that is closed under infinite direct sums, and let $L \in \mathbf{D}(A \otimes B^{\mathrm{op}})$.

(1) We say that L is a *compact object relative to* E *on the A side* if $\mathrm{Rest}_A(L) \in \mathbf{D}(A)$ is a compact object relative to E, in the sense of Definition 14.2.20(1).
(2) We say that L is a *generator relative to* E *on the A side* if $\mathrm{Rest}_A(L) \in \mathbf{D}(A)$ is a generator relative to E, in the sense of Definition 14.2.20(2).
(3) We say that L is a *compact generator of* E *on the A side* if $\mathrm{Rest}_A(L) \in \mathbf{D}(A)$ is a compact generator of E, in the sense of Definition 14.2.20(3).

Lemma 14.2.25 *Let $\mathsf{E} \subseteq \mathbf{D}(A)$ be a full triangulated subcategory that is closed under infinite direct sums, and let $L \in \mathbf{D}(A \otimes B^{\mathrm{op}})$. The following two conditions are equivalent.*

(i) *The functor* $\mathrm{RHom}_A(L, -)|_{\mathsf{E}} : \mathsf{E} \to \mathbf{D}(B)$ *commutes with infinite direct sums.*
(ii) *L is a compact object relative to E on the A side.*

Proof Let's write $F := \mathrm{RHom}_A(L, -)$. We know that for every $M \in \mathbf{D}(A)$ and $j \in \mathbb{Z}$ there are isomorphisms

$$\mathrm{Hom}_{\mathbf{D}(A)}(L, M[j]) \cong \mathrm{H}^0(F(M[j])) \cong \mathrm{H}^j(F(M))$$

in $\mathbf{M}(\mathbb{K})$, and they are functorial in M. We see that L is a compact object relative to E on the A side iff the functor $\mathrm{H} \circ F|_{\mathsf{E}} : \mathsf{E} \to \mathbf{G}_{\mathrm{str}}(\mathbb{K})$ commutes with infinite direct sums. But the functor $\mathrm{H} : \mathbf{D}(B) \to \mathbf{G}_{\mathrm{str}}(\mathbb{K})$ commutes with infinite direct sums and is conservative (see Corollary 7.2.13). So $\mathrm{H} \circ F|_{\mathsf{E}}$ commutes with infinite direct sums iff $F|_{\mathsf{E}}$ does. □

Lemma 14.2.26 *Let A and B be DG rings, let $F, G : \mathbf{D}(A) \to \mathbf{D}(B)$ be triangulated functors that commute with infinite direct sums, and let $\eta : F \to G$ be a morphism of triangulated functors. Assume that $\eta_A : F(A) \to G(A)$ is an isomorphism. Then η is an isomorphism.*

Proof Let us denote by E the full subcategory of $\mathbf{D}(A)$ on the objects M such that $\eta_M : F(M) \to G(M)$ is an isomorphism. We have to prove that $\mathsf{E} = \mathbf{D}(A)$.

By Proposition 5.3.22, the category E is a full triangulated subcategory of $\mathbf{D}(A)$. Since both functors F, G commute with infinite direct sums, we know that E is closed under infinite direct sums.

We are given that the free DG module A belongs to E. Hence every free DG A-module P is in E. Because E is triangulated, it follows that every semi-free DG A-module P of finite extension length (Definition 14.1.1) belongs to it.

Next consider an arbitrary semi-free DG A-module P. Let $\{F_j(P)\}_{j \geq -1}$ be a semi-free filtration of P (see Definition 11.4.3). Then P is a homotopy colimit of the direct system $\{F_j(P)\}_{j \geq -1}$, and we have a distinguished triangle

$$\bigoplus\nolimits_{j \in \mathbb{N}} F_j(P) \xrightarrow{\phi} \bigoplus\nolimits_{j \in \mathbb{N}} F_j(P) \to P \xrightarrow{\triangle} \tag{14.2.27}$$

in $\mathbf{D}(A)$; see Definition 14.1.11. Each $F_j(P)$ is a semi-free DG A-module of finite extension length. It follows that $P \in \mathsf{E}$.

Finally, every DG A-module M admits a quasi-isomorphism $P \to M$ with P semi-free. Therefore $M \in \mathsf{E}$. □

Lemma 14.2.28 *Let A and B be DG rings, let E be a full triangulated subcategory of $\mathbf{D}(A)$ which is closed under infinite direct sums and isomorphisms, and let $G : \mathbf{D}(B) \to \mathbf{D}(A)$ be a triangulated functor that commutes with infinite direct sums. Assume that $G(B) \in \mathsf{E}$. Then the essential image of G is contained in E.*

Proof This is like the proof of Lemma 14.2.26. □

Let A be a DG ring. Given a DG A-module M, the DG ring $\mathrm{End}_A(M)$ acts on M from the left; and thus $B := \mathrm{End}_A(M)^{\mathrm{op}}$ acts on M from the right. Because the right action of B on M commutes with the left action of A on it, we see that $M \in \mathbf{C}(A \otimes B^{\mathrm{op}})$.

Recall that the category $\mathbf{C}(A)$ has enough K-projective and K-injective objects.

Theorem 14.2.29 (Derived Morita) *Let A be a DG ring, let $\mathsf{E} \subseteq \mathbf{D}(A)$ be a full triangulated subcategory which is closed under infinite direct sums, and let L be a compact generator of E. Choose an isomorphism $L \cong P$ in $\mathbf{D}(A)$, where the DG A-module P is either K-projective or K-injective, and define the DG ring $B := \mathrm{End}_A(P)^{\mathrm{op}}$. Consider the triangulated functors*

$$F := \mathrm{RHom}_A(P, -) : \mathbf{D}(A) \to \mathbf{D}(B)$$

and

$$G := P \otimes^{\mathrm{L}}_B (-) : \mathbf{D}(B) \to \mathbf{D}(A).$$

Then the functor

$$F|_{\mathsf{E}} : \mathsf{E} \to \mathbf{D}(B)$$

is an equivalence, with quasi-inverse G.

This theorem is a variant of [79, Theorem 8.2] by B. Keller. In fact, Keller worked in the more general setting of a DG category A, whereas we only treat a DG ring A. For the history of these ideas see Remark 14.2.39. A comparison to classical Morita Theory can be found in Example 14.2.36.

Note that if we choose P to be a K-projective DG A-module, then $F \cong \mathrm{Hom}_A(P, -)$. However, even then, P will rarely be K-flat as a DG B^{op}-module, so G is only the left derived functor of $P \otimes_B (-)$.

Proof The proof is in four steps.

Step 1. We can assume that the category E is closed under isomorphisms in $\mathbf{D}(A)$. The functor G commutes with infinite direct sums (see Proposition 12.3.8), and $G(B) \cong L \in \mathsf{E}$. According to Lemma 14.2.28, the image of G is contained in E.

Step 2. We already know that the functor F is right adjoint to G, by Proposition 14.2.15. The corresponding morphisms of triangulated functors are denoted by

$$\theta : \mathrm{Id}_{\mathbf{D}(B)} \to F \circ G \quad \text{and} \quad \zeta : G \circ F \to \mathrm{Id}_{\mathbf{D}(A)} . \tag{14.2.30}$$

By step 1 we see that $F|_{\mathsf{E}} : \mathsf{E} \to \mathbf{D}(B)$ is the right adjoint of $G : \mathbf{D}(B) \to \mathsf{E}$. Hence (14.2.30) restricts to morphisms of triangulated functors

$$\theta : \mathrm{Id}_{\mathbf{D}(B)} \to F|_{\mathsf{E}} \circ G \quad \text{and} \quad \zeta : G \circ F|_{\mathsf{E}} \to \mathrm{Id}_{\mathsf{E}} . \tag{14.2.31}$$

We will prove that θ and ζ in (14.2.31) are isomorphisms. By Lemma 14.2.25 the functor $F|_{\mathsf{E}}$ commutes with infinite direct sums. Therefore both $F|_{\mathsf{E}} \circ G$ and $G \circ F|_{\mathsf{E}}$ commute with infinite direct sums.

Step 3. Now we will prove that θ is an isomorphism of functors; i.e. for every $N \in \mathbf{D}(B)$ the morphism $\theta_N : N \to (F|_{\mathsf{E}} \circ G)(N)$ is an isomorphism. The functors $\mathrm{Id}_{\mathbf{D}(B)}$ and $F|_{\mathsf{E}} \circ G$ both commute with infinite direct sums. Therefore, by Lemma 14.2.26, it suffices to check that θ_B is an isomorphism in $\mathbf{D}(B)$. But θ_B is represented (both when P is K-projective and when it is K-injective) by the canonical homomorphism $\tilde{\theta}_B : B \to \mathrm{Hom}_A(P, P \otimes_B B)$, in $\mathbf{C}_{\mathrm{str}}(B)$, which is clearly bijective.

Step 4. Finally we will prove that ζ is an isomorphism of functors. Take any $M \in \mathsf{E}$, and consider the distinguished triangle

$$(G \circ F|_{\mathsf{E}})(M) \xrightarrow{\zeta_M} M \to M' \xrightarrow{\triangle} \tag{14.2.32}$$

in E, in which $M' \in \mathsf{E}$ is the cone of ζ_M (see Definition 5.3.7).

Applying F and using the functorial isomorphism θ we get a distinguished triangle

$$F|_{\mathsf{E}}(M) \xrightarrow{\mathrm{id}_{F|_{\mathsf{E}}(M)}} F|_{\mathsf{E}}(M) \to F|_{\mathsf{E}}(M') \xrightarrow{\triangle} \tag{14.2.33}$$

in $\mathbf{D}(B)$. This implies that $F|_{\mathsf{E}}(M') \cong 0$ in $\mathbf{D}(B)$, and thus, after applying the forgetful functor, we get $F|_{\mathsf{E}}(M') \cong 0$ in $\mathbf{D}(\mathbb{K})$. But $F|_{\mathsf{E}}(M') \cong \mathrm{RHom}_A(L, M')$, so $\mathrm{Hom}_{\mathbf{D}(A)}(L, M'[i]) = 0$ for all i. Because L is generator of E, it follows that $M' \cong 0$ in E. Going back to the distinguished triangle (14.2.32) we conclude that ζ_M is an isomorphism. □

Remark 14.2.34 Suppose that in the theorem we were to choose some other isomorphism $L \cong P'$ in $\mathbf{D}(A)$ to a K-projective or K-injective DG A-module P. Then the DG ring $B' := \mathrm{End}_A(P')^{\mathrm{op}}$ would be related to B as follows. Without loss of generality we can assume that either P is K-projective or P' is K-injective. Then there is a quasi-isomorphism $\phi : P \to P'$ in $\mathbf{C}_{\mathrm{str}}(A)$, unique up to homotopy, that respects the given isomorphisms $L \cong P$ and $L \cong P'$ in $\mathbf{D}(A)$. Define the DG bimodule $N := \mathrm{Hom}_A(P, P') \in \mathbf{C}(B \otimes B'^{\mathrm{op}})$, and the matrix DG ring $C := \begin{bmatrix} B & N[-1] \\ 0 & B' \end{bmatrix}$, which has a differential depending on ϕ. Then the obvious DG ring homomorphisms $C \to B$ and $C \to B'$ are quasi-isomorphisms. See [119, Proposition 3.3].

Example 14.2.35 In case $\mathsf{E} = \mathbf{D}(A)$, the theorem shows that the DG bimodule $P \in \mathbf{D}(A \otimes B^{\mathrm{op}})$ is a pretilting DG A-B-bimodule.

Example 14.2.36 Assume A is a ring, and $P \in \mathbf{M}(A) = \mathsf{Mod}\, A$ is a *progenerator*, i.e. P is a finitely generated projective A-module, such that every nonzero A-module N admits a nonzero homomorphism $P \to N$. Let $B := \mathrm{End}_A(P)^{\mathrm{op}}$. Because P is a compact generator of $\mathbf{D}(A)$, the theorem says that

$$F := \mathrm{Hom}_A(P, -) : \mathbf{D}(A) \to \mathbf{D}(B)$$

is an equivalence of triangulated categories.

Classical Morita Theory (see e.g. [130, Section 4.1]) says that F restricts to an equivalence of abelian categories

$$F = \mathrm{Hom}_A(P, -) : \mathbf{M}(A) \to \mathbf{M}(B).$$

Furthermore, classical Morita Theory tells us that there is a bijection between the set of isomorphism classes of such A-B-bimodules P, and the set of isomorphism classes of linear equivalences $F : \mathbf{M}(A) \to \mathbf{M}(B)$. This last assertion is an open problem in derived Morita theory; see Remark 14.4.36.

Remark 14.2.37 Here is a geometric variant of Theorem 14.2.29. Let $(X, \mathcal{O}_X)$ be a scheme. We denote by $\mathbf{D}_{\mathrm{qc}}(X) = \mathbf{D}_{\mathrm{qc}}(\mathsf{Mod}\, \mathcal{O}_X)$ the derived category of (unbounded) complexes of $\mathcal{O}_X$-modules with quasi-coherent cohomology sheaves. This is a triangulated category with infinite direct sums, and with enough K-injective resolutions (by [143, Theorem 4.5]).

Suppose $\mathcal{L} \in \mathbf{D}_{\mathrm{qc}}(X)$ is a compact generator. Choose a K-injective resolution $\mathcal{L} \to \mathcal{I}$ in $\mathbf{C}_{\mathrm{str}}(X)$, and define the DG ring $B := \mathrm{End}_X(\mathcal{I})^{\mathrm{op}}$. Then the functor

$$\mathrm{RHom}_X(\mathcal{I}, -) : \mathbf{D}_{\mathrm{qc}}(X) \to \mathbf{D}(B) \tag{14.2.38}$$

is an equivalence, with quasi-inverse

$$\mathcal{I} \otimes^{\mathrm{L}}_B (-) : \mathbf{D}(B) \to \mathbf{D}_{\mathrm{qc}}(X).$$

This result is stated as [26, Corollary 3.1.8], without such a precise formulation, and the proof is attributed to B. Keller. Presumably the proof of Theorem 14.2.29 above, with slight changes, would work also in this geometric context.

Having a compact generator of $\mathbf{D}_{\mathrm{qc}}(X)$ is quite a general feature – by [26, Theorem 3.1.1] this is true if X is quasi-compact and quasi-separated.

The first such result is perhaps due to A. Beilinson [15], who showed that for $X = \mathbf{P}^n_{\mathbb{K}}$, the n-dimensional projective space over a field $\mathbb{K}$, the sheaf

$$\mathcal{L} := \bigoplus\nolimits_{i=0}^{n} \mathcal{O}_X(i)$$

is a compact generator of $\mathbf{D}_{\mathrm{qc}}(X)$. The DG ring B appearing in the equivalence (14.2.38) is actually a finite $\mathbb{K}$-ring (the path ring of a Kronecker quiver modulo commutation relations).

Remark 14.2.39 Tilting theory and derived Morita Theory have their origins in the *representation theory of finite dimensional algebras* (in our terminology these are "finite central $\mathbb{K}$-rings," where $\mathbb{K}$ is a base field). Among the first examples of tilting are the *reflection functors* in the paper [19] by I.N. Bernstein, I.M. Gelfand and V.A. Ponomarev. Later these functors were understood to be of the form $\mathrm{Hom}_A(T, -)$ for a *tilting A-module T*.

These ideas were generalized by M. Auslander, M. Platzeck and I. Reiten [11], and later by S. Brenner and M.C.R. Butler [29], who coined the term *tilting functor*. These concepts were further clarified by D. Happel, C.M. Ringel and K. Bongartz (see [58], [59] and [61]). In [58] Happel showed that for a tilting A-module T, with opposite endomorphism ring $B := \mathrm{End}_A(T)^{\mathrm{op}}$, the functor $\mathrm{RHom}_A(T, -)$ is an equivalence of triangulated categories $\mathbf{D}^{\mathrm{b}}_{\mathrm{f}}(A) \to \mathbf{D}^{\mathrm{b}}_{\mathrm{f}}(B)$. This result was slightly generalized by E. Cline, B. Parshall and L. Scott [41].

J. Rickard [126], [127] was the first to talk about *tilting complexes* (as opposed to tilting modules). He introduced two-sided tilting complexes (which we will discuss in Section 14.4), and proved the celebrated Theorem 14.4.30 (under some boundedness conditions). The generalization of derived Morita Theory to unbounded derived categories, and from rings to DG categories, was done by B. Keller [79]. In this seminal paper Keller also gave new construction of two-sided tilting complexes, and characterized algebraic triangulated categories with a compact generator as those that are equivalent to derived categories of DG rings.

For a thorough survey of tilting theory see the book [3].

14.3 DG Bimodules over K-Flat DG Rings

Recall that Convention 14.0.1 is in effect. In particular all DG rings are $\mathbb{K}$-central, and $\otimes = \otimes_{\mathbb{K}}$.

Definition 14.3.1 Let A be a DG $\mathbb{K}$-ring. We call A a *K-flat DG $\mathbb{K}$-ring* if A is K-flat as a DG $\mathbb{K}$-module (see Definition 10.3.1). The category of K-flat central DG $\mathbb{K}$-rings is denoted by $\mathsf{DGRng}/_{\mathrm{fc}}\, \mathbb{K}$.

Example 14.3.2 If $\mathbb{K}$ is a *field*, then every DG $\mathbb{K}$-ring A is K-flat over $\mathbb{K}$.

Example 14.3.3 If A is a *semi-free* DG $\mathbb{K}$-ring (either commutative or noncommutative), then A is a K-flat DG $\mathbb{K}$-ring. See Definition 12.8.4, Proposition 12.8.5 and Remark 12.8.22.

Example 14.3.4 If A is a *nonpositive* DG $\mathbb{K}$-ring, and each A^i is a *flat* $\mathbb{K}$-module, then A is a K-flat DG $\mathbb{K}$-ring. This includes the case of a flat $\mathbb{K}$-ring A. See Proposition 11.3.23.

Example 14.3.5 A very special case of Examples 14.3.2 and 14.3.4 is when $\mathbb{K}$ is a field and A is a $\mathbb{K}$-ring (better known as a unital associative $\mathbb{K}$-algebra). This setting is the one commonly used in ring theory, see [8], [10], [161], [153], [145], [183].

Definition 14.3.6 Let A and B be K-flat DG $\mathbb{K}$-rings. The *derived category of DG B-A-bimodules* is the $\mathbb{K}$-linear triangulated category $\mathsf{D}(B \otimes A^{\mathrm{op}})$.

See Remark 14.3.24 regarding the derived category of DG B-A-bimodules in the nonflat case.

Definition 14.3.7 For a K-flat DG $\mathbb{K}$-ring A we write $A^{\mathrm{en}} := A \otimes A^{\mathrm{op}}$, and call it the *enveloping DG ring* of A (relative to $\mathbb{K}$).

A explained in formula (14.2.8), the enveloping DG ring has a canonical isomorphism $A^{\mathrm{en}} \xrightarrow{\simeq} (A^{\mathrm{en}})^{\mathrm{op}}$.

From here to the end of this chapter we assume the next convention, which strengthens Convention 14.0.1.

Convention 14.3.8 In addition to the stipulations of Convention 14.0.1, we also assume by default that all DG rings are *K-flat central DG $\mathbb{K}$-rings*.

Recall that a homomorphism of DG rings $f : A \to B$ induces a forgetful functor $\mathrm{Rest}_f : \mathbf{C}(B) \to \mathbf{C}(A)$ that we call *restriction*. This is an exact DG functor, and thus it induces a triangulated functor

$$\mathrm{Rest}_f : \mathbf{D}(B) \to \mathbf{D}(A). \tag{14.3.9}$$

Proposition 14.3.10 *Let $f : A \to B$ be a DG ring homomorphism. The triangulated functor* Rest_f *is conservative. Namely a morphism $\phi : M \to N$ in $\mathbf{D}(B)$ is an isomorphism if and only if the morphism*

$$\mathrm{Rest}_f(\phi) : \mathrm{Rest}_f(M) \to \mathrm{Rest}_f(N)$$

in $\mathbf{D}(A)$ is an isomorphism.

Proof We know by Corollary 7.2.13 that the functor $\mathrm{H} : \mathbf{D}(B) \to \mathbf{G}(\mathbb{K})$ is conservative. But $\mathrm{H} = \mathrm{H} \circ \mathrm{Rest}_f$. □

Lemma 14.3.11 *Let A and B be DG rings.*

1. *If $P \in \mathbf{C}(A \otimes B^{\mathrm{op}})$ is K-flat, then P is K-flat over A.*
2. *If $I \in \mathbf{C}(A \otimes B^{\mathrm{op}})$ is K-injective, then I is K-injective over A.*
3. *If B is K-projective as a DG $\mathbb{K}$-module, and if $P \in \mathbf{C}(A \otimes_{\mathbb{K}} B^{\mathrm{op}})$ is K-projective, then P is K-projective over A.*
4. *If B is a semi-free DG $\mathbb{K}$-ring, and if $P \in \mathbf{C}(A \otimes_{\mathbb{K}} B^{\mathrm{op}})$ is semi-free, then P is semi-free over A.*

Proof (1) and (2) are direct consequences of the canonical isomorphisms

$$M \otimes_A P \cong (M \otimes B) \otimes_{A \otimes B} P$$

and

$$\mathrm{Hom}_A(N, I) \cong \mathrm{Hom}_{A \otimes B}(N \otimes B, I)$$

in $\mathbf{C}_{\mathrm{str}}(\mathbb{K})$, for $M \in \mathbf{C}(A^{\mathrm{op}})$ and $N \in \mathbf{C}(A)$, together with the fact that B is K-flat over $\mathbb{K}$. Items (3) and (4) are left as an exercise. □

Exercise 14.3.12 Prove items (3) and (4) in this lemma.

Proposition 14.3.13 *Let A, B and C be DG rings.*

(1) *The DG bifunctor*

$$(- \otimes_B -) : \mathbf{C}(A \otimes B^{\mathrm{op}}) \times \mathbf{C}(B \otimes C^{\mathrm{op}}) \to \mathbf{C}(A \otimes C^{\mathrm{op}})$$

has a triangulated left derived bifunctor

$$((- \otimes^{\mathrm{L}}_B -), \eta^{\mathrm{L}}) : \mathbf{D}(A \otimes B^{\mathrm{op}}) \times \mathbf{D}(B \otimes C^{\mathrm{op}}) \to \mathbf{D}(A \otimes C^{\mathrm{op}}).$$

(2) *Given $M \in \mathbf{C}(A \otimes B^{\mathrm{op}})$ and $N \in \mathbf{C}(B \otimes C^{\mathrm{op}})$, such that M is K-flat over B^{op} or N is K-flat over B, the morphism*

$$\eta^{\mathrm{L}}_{M,N} : M \otimes^{\mathrm{L}}_B N \to M \otimes_B N$$

in $\mathbf{D}(A \otimes C^{\mathrm{op}})$ is an isomorphism.

(3) *Suppose we are given DG ring homomorphisms $A' \to A$, $f : B' \to B$ and $C' \to C$, such that f is a quasi-isomorphism. Then the diagram*

$$\begin{array}{ccc} \mathbf{D}(A \otimes B^{\mathrm{op}}) \times \mathbf{D}(B \otimes C^{\mathrm{op}}) & \xrightarrow{(- \otimes^{\mathrm{L}}_B -)} & \mathbf{D}(A \otimes C^{\mathrm{op}}) \\ \downarrow{\scriptstyle \mathrm{Rest} \times \mathrm{Rest}} & & \downarrow{\scriptstyle \mathrm{Rest}} \\ \mathbf{D}(A' \otimes B'^{\mathrm{op}}) \times \mathbf{D}(B' \otimes C'^{\mathrm{op}}) & \xrightarrow{(- \otimes^{\mathrm{L}}_{B'} -)} & \mathbf{D}(A' \otimes C'^{\mathrm{op}}) \end{array}$$

is commutative up to an isomorphism of triangulated bifunctors.

(4) *Suppose D is another DG ring. Then there is an isomorphism*

$$((- \otimes^{\mathrm{L}}_B -) \otimes^{\mathrm{L}}_C -) \cong (- \otimes^{\mathrm{L}}_B (- \otimes^{\mathrm{L}}_C -))$$

of triangulated trifunctors

$$\mathbf{D}(A \otimes B^{\mathrm{op}}) \times \mathbf{D}(B \otimes C^{\mathrm{op}}) \times \mathbf{D}(C \otimes D^{\mathrm{op}}) \to \mathbf{D}(A \otimes D^{\mathrm{op}}).$$

Proof

(1) The DG bifunctor $(- \otimes_B -)$ induces a triangulated bifunctor

$$(- \otimes_B -) : \mathbf{K}(A \otimes B^{\mathrm{op}}) \times \mathbf{K}(B \otimes C^{\mathrm{op}}) \to \mathbf{K}(A \otimes C^{\mathrm{op}})$$

on the homotopy categories, in the obvious way. By Corollary 11.4.26, Proposition 10.3.4 and Lemma 14.3.11(1), every DG bimodule $N \in \mathbf{C}(B \otimes C^{\mathrm{op}})$ admits a quasi-isomorphism $\rho_N : P \to N$, where $P \in \mathbf{C}(B \otimes C^{\mathrm{op}})$ is K-flat over B. Given $M \in \mathbf{D}(A \otimes B^{\mathrm{op}})$, let us define the object

$$M \otimes^{\mathrm{L}}_B N := M \otimes_B P \in \mathbf{D}(A \otimes C^{\mathrm{op}}),$$

with the morphism

$$\eta^{\mathrm{L}}_{M,N} := \mathrm{id}_M \otimes_B \rho_N : M \otimes^{\mathrm{L}}_B N \to M \otimes_B N.$$

The pair $((- \otimes^{\mathrm{L}}_B -), \eta^{\mathrm{L}})$ is a left derived bifunctor of $(- \otimes_B -)$.

(2) Under either assumption the homomorphism $\mathrm{id}_M \otimes_B \rho_N$ is a quasi-isomorphism; cf. Lemma 12.3.2.

(3) This is similar to the proof of item (2) of Theorem 12.7.2. Take $M \in \mathsf{C}(A \otimes B^{\mathrm{op}})$ and $N \in \mathsf{C}(B \otimes C^{\mathrm{op}})$. Let $\rho_N : P \to N$ be the resolution from item (1). Choose a resolution $\rho'_P : P' \to P$ in $\mathsf{C}_{\mathrm{str}}(B' \otimes C'^{\mathrm{op}})$ where P' is K-flat over B'. Then we have a canonical isomorphism

$$\operatorname{Rest}(M) \otimes^{\mathrm{L}}_{B'} \operatorname{Rest}(N) \cong M \otimes_{B'} P'$$

in $\mathsf{D}(A' \otimes C'^{\mathrm{op}})$. Now in the commutative diagram

$$\begin{array}{ccc} B' \otimes_{B'} P' & \xrightarrow{\ \cong\ } & P' \\ {\scriptstyle f \otimes \mathrm{id}_{P'}} \big\downarrow & & \big\downarrow {\scriptstyle \rho'_P} \\ B \otimes_{B'} P' & \xrightarrow{\ \mathrm{id}_B \otimes \rho'_P\ } & P \end{array}$$

in $\mathsf{C}_{\mathrm{str}}(B')$ the vertical arrows are quasi-isomorphisms, because P' is K-flat over B'. Therefore the bottom arrow $\mathrm{id}_B \otimes \rho'_P$ is a quasi-isomorphism.

We now look at this commutative diagram

$$M \otimes_{B'} P' \xrightarrow{\ \cong\ } M \otimes_B B \otimes_{B'} P' \xrightarrow{\ \phi\ } M \otimes_B P \qquad \left(\text{curved arrow } \mathrm{id}_M \otimes \rho'_P : M \otimes_{B'} P' \to M \otimes_B P\right)$$

in $\mathsf{C}_{\mathrm{str}}(A' \otimes C'^{\mathrm{op}})$, where $\phi := \mathrm{id}_M \otimes \mathrm{id}_B \otimes \rho'_P$. Because both P and $B \otimes_{B'} P'$ are K-flat over B and $\mathrm{id}_B \otimes \rho'_P$ is a quasi-isomorphism, Lemma 12.3.2 tells us that ϕ is a quasi-isomorphism. Hence $\mathrm{id}_M \otimes \rho'_P$ is a quasi-isomorphism. We get the desired canonical isomorphism

$$\mathrm{Q}(\mathrm{id}_M \otimes \rho'_P) : \operatorname{Rest}(M) \otimes^{\mathrm{L}}_{B'} \operatorname{Rest}(N) \xrightarrow{\simeq} \operatorname{Rest}(M \otimes^{\mathrm{L}}_B N)$$

in $\mathsf{D}(A' \otimes C'^{\mathrm{op}})$.

(4) Given M, N, P as above and $L \in \mathsf{C}(C \otimes D^{\mathrm{op}})$, we choose a quasi-isomorphism $\rho_L : Q \to L$, where $Q \in \mathsf{C}(C \otimes D^{\mathrm{op}})$ is K-flat over C. A small calculation shows that $P \otimes_C Q$ is K-flat over B. The desired isomorphism

$$(M \otimes^{\mathrm{L}}_B N) \otimes^{\mathrm{L}}_C L \cong M \otimes^{\mathrm{L}}_B (N \otimes^{\mathrm{L}}_C L)$$

in $\mathsf{D}(A \otimes D^{\mathrm{op}})$ comes from the obvious isomorphism

$$(M \otimes_B P) \otimes_C Q \cong M \otimes_B (P \otimes_C Q)$$

in $\mathsf{C}_{\mathrm{str}}(A \otimes D^{\mathrm{op}})$. □

Here is some notation: suppose we are given morphisms $\phi : M' \to M$ in $\mathsf{D}(A \otimes B^{\mathrm{op}})$ and $\psi : N' \to N$ in $\mathsf{D}(B \otimes C^{\mathrm{op}})$. The result of applying the bifunctor $(- \otimes^{\mathrm{L}}_B -)$ is the morphism

$$\phi \otimes^{\mathrm{L}}_B \psi : M' \otimes^{\mathrm{L}}_B N' \to M \otimes^{\mathrm{L}}_B N \tag{14.3.14}$$

in $\mathsf{D}(A \otimes C^{\mathrm{op}})$.

Proposition 14.3.15 *Let A, B and C be DG rings.*

(1) *The DG bifunctor*

$$\mathrm{Hom}_B(-,-) : \mathbf{C}(B \otimes A^{\mathrm{op}})^{\mathrm{op}} \times \mathbf{C}(B \otimes C^{\mathrm{op}}) \to \mathbf{C}(A \otimes C^{\mathrm{op}})$$

has a triangulated right derived bifunctor

$$(\mathrm{RHom}_B(-,-), \eta^{\mathrm{R}}) : \mathbf{D}(B \otimes A^{\mathrm{op}})^{\mathrm{op}} \times \mathbf{D}(B \otimes C^{\mathrm{op}}) \to \mathbf{D}(A \otimes C^{\mathrm{op}}).$$

(2) *Given $M \in \mathbf{C}(B \otimes A^{\mathrm{op}})$ and $N \in \mathbf{C}(B \otimes C^{\mathrm{op}})$, such that either M is K-projective over B or N is K-injective over B, the morphism*

$$\eta^{\mathrm{R}}_{M,N} : \mathrm{Hom}_B(M, N) \to \mathrm{RHom}_B(M, N)$$

in $\mathbf{D}(A \otimes C^{\mathrm{op}})$ is an isomorphism.

(3) *Suppose we are given DG ring homomorphisms $A' \to A$, $f : B' \to B$ and $C' \to C$, such that f is a quasi-isomorphism. Then the diagram*

$$\begin{array}{ccc} \mathbf{D}(B \otimes A^{\mathrm{op}})^{\mathrm{op}} \times \mathbf{D}(B \otimes C^{\mathrm{op}}) & \xrightarrow{\mathrm{RHom}_B(-,-)} & \mathbf{D}(A \otimes C^{\mathrm{op}}) \\ \Big\downarrow{\scriptstyle \mathrm{Rest} \times \mathrm{Rest}} & & \Big\downarrow{\scriptstyle \mathrm{Rest}} \\ \mathbf{D}(B' \otimes A'^{\mathrm{op}})^{\mathrm{op}} \times \mathbf{D}(B' \otimes C'^{\mathrm{op}}) & \xrightarrow{\mathrm{RHom}_{B'}(-,-)} & \mathbf{D}(A' \otimes C'^{\mathrm{op}}) \end{array}$$

is commutative up to an isomorphism of triangulated bifunctors.

Exercise 14.3.16 Prove Proposition 14.3.15. Hints: this is similar to the proof of items (1)-(3) of Proposition 14.3.13, but now we rely on Corollary 11.6.30 and Lemma 14.3.11(2) for the existence of resolutions $\rho_N : N \to I$ in $\mathbf{C}_{\mathrm{str}}(B \otimes C^{\mathrm{op}})$ by DG bimodules I that are K-injective over B. For item (3) use Lemma 12.2.2.

Again there is related notation: suppose we are given morphisms $\phi : M \to M'$ in $\mathbf{D}(B \otimes A^{\mathrm{op}})$ and $\psi : N' \to N$ in $\mathbf{D}(B \otimes C^{\mathrm{op}})$. The result of applying the bifunctor $\mathrm{RHom}_B(-,-)$ is the morphism

$$\mathrm{RHom}_B(\phi, \psi) : \mathrm{RHom}_B(M', N') \to \mathrm{RHom}_B(M, N) \tag{14.3.17}$$

in $\mathbf{D}(A \otimes C^{\mathrm{op}})$.

Proposition 14.3.18 *Let A, B, C and D be DG rings. For $M \in \mathbf{D}(B \otimes A^{\mathrm{op}})$, $N \in \mathbf{D}(C \otimes B^{\mathrm{op}})$ and $L \in \mathbf{D}(C \otimes D^{\mathrm{op}})$ there is an isomorphism*

$$\mathrm{RHom}_B(M, \mathrm{RHom}_C(N, L)) \cong \mathrm{RHom}_C(N \otimes^{\mathrm{L}}_B M, L)$$

in $\mathbf{D}(A \otimes D^{\mathrm{op}})$ called derived Hom-tensor adjunction. *This isomorphism is functorial in the objects M, N, L.*

Proof We choose a quasi-isomorphism $L \to J$ in $\mathbf{C}_{\mathrm{str}}(C \otimes D^{\mathrm{op}})$ into a DG module J that is K-injective over C, and a quasi-isomorphism $Q \to N$ in

$\mathbf{C}_{\mathrm{str}}(C \otimes B^{\mathrm{op}})$ from a DG module Q that is K-flat over B^{op}. A calculation shows that the DG bimodule $\mathrm{Hom}_C(Q, J)$ is K-injective over B. The desired isomorphism comes from the obvious adjunction isomorphism

$$\mathrm{Hom}_B(M, \mathrm{Hom}_C(Q, J)) \cong \mathrm{Hom}_C(Q \otimes_B M, J)$$

in $\mathbf{C}_{\mathrm{str}}(A \otimes D^{\mathrm{op}})$. □

Proposition 14.3.19 *Let A, B, C and D be DG rings. For $L \in \mathbf{D}(A \otimes C^{\mathrm{op}})$, $M \in \mathbf{D}(A \otimes B^{\mathrm{op}})$ and $N \in \mathbf{D}(B \otimes D^{\mathrm{op}})$ there is a morphism*

$$\mathrm{ev}^{\mathrm{R,L}}_{L,M,N} : \mathrm{RHom}_A(L, M) \otimes^{\mathrm{L}}_B N \to \mathrm{RHom}_A(L, M \otimes^{\mathrm{L}}_B N)$$

in $\mathbf{D}(C \otimes D^{\mathrm{op}})$, called derived tensor-evaluation. *This morphism is functorial in the objects L, M, N. Moreover, after applying the restriction functor $\mathbf{D}(C \otimes D^{\mathrm{op}}) \to \mathbf{D}(\mathbb{K})$, the morphism $\mathrm{ev}^{\mathrm{R,L}}_{L,M,N}$ coincides with the morphism from Theorem* 12.9.10.

The reason we can give stronger statement here, as compared to Theorem 12.9.10, is because here our DG rings are K-flat over $\mathbb{K}$.

Proof Choose a quasi-isomorphism $Q \to N$ in $\mathbf{C}_{\mathrm{str}}(B \otimes D^{\mathrm{op}})$ from a DG module Q that is K-flat over B, a quasi-isomorphism $M \to I$ in $\mathbf{C}_{\mathrm{str}}(A \otimes B^{\mathrm{op}})$ into a DG module I that is K-injective over A, and a quasi-isomorphism $\rho : I \otimes_B Q \to J$ in $\mathbf{C}_{\mathrm{str}}(A \otimes D^{\mathrm{op}})$ into a DG module J that is K-injective over A. Then $\mathrm{ev}^{\mathrm{R,L}}_{L,M,N}$ is represented by the composed homomorphism

$$\mathrm{Hom}_A(L, I) \otimes_B Q \xrightarrow{\mathrm{ev}_{L,I,Q}} \mathrm{Hom}_A(L, I \otimes_B Q) \xrightarrow{\tilde{\rho}} \mathrm{Hom}_A(L, J)$$

in $\mathbf{C}_{\mathrm{str}}(C \otimes D^{\mathrm{op}})$, where $\tilde{\rho} := \mathrm{Hom}(\mathrm{id}_L, \rho)$.

When we forget C and D, then we can choose a K-projective resolution $P \to L$ in $\mathbf{C}_{\mathrm{str}}(A)$. Then we have a commutative diagram

$$\begin{array}{ccccc} \mathrm{Hom}_A(P, I) \otimes_B Q & \xrightarrow{\mathrm{ev}_{P,I,Q}} & \mathrm{Hom}_A(P, I \otimes_B Q) & \xrightarrow{\text{q.i.}} & \mathrm{Hom}_A(P, J) \\ \uparrow{\scriptstyle \text{q.i.}} & & \uparrow & & \uparrow{\scriptstyle \text{q.i.}} \\ \mathrm{Hom}_A(L, I) \otimes_B Q & \xrightarrow{\mathrm{ev}_{L,I,Q}} & \mathrm{Hom}_A(L, I \otimes_B Q) & \xrightarrow{\tilde{\rho}} & \mathrm{Hom}_A(L, J) \end{array}$$

in $\mathbf{C}_{\mathrm{str}}(\mathbb{K})$, in which the arrows marked "q.i." are quasi-isomorphisms. By comparing to the proof of Theorem 12.9.10 this proves that last assertion. □

According to Propositions 14.3.13 and 14.3.15, there are triangulated bifunctors

$$(- \otimes^{\mathrm{L}}_A -) : \mathbf{D}(A^{\mathrm{en}}) \times \mathbf{D}(A^{\mathrm{en}}) \to \mathbf{D}(A^{\mathrm{en}}), \tag{14.3.20}$$

$$(- \otimes^{\mathrm{L}}_A -) : \mathbf{D}(A^{\mathrm{en}}) \times \mathbf{D}(A \otimes B^{\mathrm{op}}) \to \mathbf{D}(A \otimes B^{\mathrm{op}}) \tag{14.3.21}$$

and

$$\mathrm{RHom}_A(-,-) : \mathsf{D}(A^{\mathrm{en}})^{\mathrm{op}} \times \mathsf{D}(A \otimes B^{\mathrm{op}}) \to \mathsf{D}(A \otimes B^{\mathrm{op}}). \tag{14.3.22}$$

Remark 14.3.23 The category $\mathsf{D}(A^{\mathrm{en}})$, with the tensor operation $(- \otimes^{\mathrm{L}}_A -)$, is a *monoidal category*. Moreover, $\mathsf{D}(A^{\mathrm{en}})$ is a *biclosed monoidal category*: the two internal Hom operations are $\mathrm{RHom}_A(-,-)$ and $\mathrm{RHom}_{A^{\mathrm{op}}}(-,-)$. If A is weakly commutative, then $\mathsf{D}(A^{\mathrm{en}})$ is a symmetric monoidal category. See [157, Section 5], [102] and [115].

Remark 14.3.24 Here is an outline of our way to handle derived categories of DG bimodules in the absence of flatness. The idea is to choose K-flat resolutions $\tilde{A} \to A$ and $\tilde{B} \to B$ in the category $\mathsf{DGRng}/_{\mathrm{c}}\,\mathbb{K}$ of DG central $\mathbb{K}$-rings. This can be done by Theorem 12.8.7 and Proposition 12.8.5. Then the *derived category of DG A-B-bimodules* is the triangulated category $\mathsf{D}(\tilde{A} \otimes \tilde{B}^{\mathrm{op}})$. Note that the restriction functors $\mathsf{D}(\tilde{A}) \to \mathsf{D}(A)$ and $\mathsf{D}(\tilde{B}^{\mathrm{op}}) \to \mathsf{D}(B^{\mathrm{op}})$ are equivalences. This says that we have triangulated bifunctors

$$(- \otimes^{\mathrm{L}}_{\tilde{B}} -) : \mathsf{D}(\tilde{A} \otimes \tilde{B}^{\mathrm{op}}) \times \mathsf{D}(B) \to \mathsf{D}(A)$$

and

$$\mathrm{RHom}_{\tilde{A}}(-,-) : \mathsf{D}(\tilde{A} \otimes \tilde{B}^{\mathrm{op}})^{\mathrm{op}} \times \mathsf{D}(A) \to \mathsf{D}(B).$$

We also have the "tilde" versions of the functors (14.3.20), (14.3.21) and (14.3.22). These are collectively called the *package of standard derived functors*.

The triangulated category $\mathsf{D}(\tilde{A} \otimes \tilde{B}^{\mathrm{op}})$ is independent of the resolutions, up to a canonical equivalence. The argument is this. Suppose we are given K-flat resolutions $\tilde{A}_i \to A$ and $\tilde{B}_i \to B$ in the category $\mathsf{DGRng}/_{\mathrm{c}}\,\mathbb{K}$, indexed by $i \in I$ for some set I. For each $i, j \in I$, the DG module $A \otimes^{\mathrm{L}}_{\mathbb{K}} B$ has an outside action by the DG ring $\tilde{A}_j \otimes \tilde{B}_j^{\mathrm{op}}$ and an inside action the DG ring $\tilde{A}_i \otimes \tilde{B}_i^{\mathrm{op}}$. (See Section 18.2 regarding outside and inside actions.) These give rise to a DG bimodule

$$T_{i,j} := A \otimes^{\mathrm{L}}_{\mathbb{K}} B \in \mathsf{D}((\tilde{A}_j \otimes \tilde{B}_j^{\mathrm{op}}) \otimes (\tilde{A}_i \otimes \tilde{B}_i^{\mathrm{op}})^{\mathrm{op}}).$$

It turns out that $T_{i,j}$ is a *tilting DG bimodule*, as defined in the next section; and hence the triangulated functor

$$G_{i,j} := T_{i,j} \otimes^{\mathrm{L}}_{\tilde{A}_i \otimes \tilde{B}_i^{\mathrm{op}}} (-) : \mathsf{D}(\tilde{A}_i \otimes \tilde{B}_i^{\mathrm{op}}) \to \mathsf{D}(\tilde{A}_j \otimes \tilde{B}_j^{\mathrm{op}}) \tag{14.3.25}$$

is an equivalence. For a third index $k \in I$ there is a canonical isomorphism

$$T_{j,k} \otimes^{\mathrm{L}}_{\tilde{A}_j \otimes \tilde{B}_j^{\mathrm{op}}} T_{i,j} \cong T_{i,k} \quad \text{in} \quad \mathsf{D}((\tilde{A}_k \otimes \tilde{B}_k^{\mathrm{op}}) \otimes (\tilde{A}_i \otimes \tilde{B}_i^{\mathrm{op}})^{\mathrm{op}}).$$

Thus there are isomorphisms of triangulated functors

$$G_{j,k} \circ G_{i,j} \xrightarrow{\simeq} G_{i,k}, \tag{14.3.26}$$

and they satisfy the pentagon axiom. The equivalences $G_{i,j}$ respect the packages of standard derived functors.

We see that there is a $\mathbb{K}$-linear triangulated category that we will symbolically denote by $\mathbf{D}(A \otimes^{\mathrm{L}}_{\mathbb{K}} B^{\mathrm{op}})$, and it is canonically equivalent to all the triangulated categories $\mathbf{D}(\tilde{A}_i \otimes \tilde{B}_i^{\mathrm{op}})$. For more details see the lecture notes [180] or the paper [178].

Remark 14.3.27 Actually, we can say more. Recall the 2-category $\mathsf{TrCat}/\mathbb{K}$ of $\mathbb{K}$-linear triangulated categories that was briefly mentioned at the end of Section 8.1. There is a pseudofunctor

$$\mathrm{DerCat} : \mathsf{DGRng}/_{\mathrm{c}}\,\mathbb{K} \to \mathsf{TrCat}/\mathbb{K} \tag{14.3.28}$$

that sends a DG ring A to the triangulated category $\mathrm{DerCat}(A) := \mathbf{D}(A)$, and a homomorphism $f : A \to B$ in $\mathsf{DGRng}/_{\mathrm{c}}\,\mathbb{K}$ is sent to the triangulated functor

$$\mathrm{DerCat}(f) := \mathrm{LInd}_f = B \otimes^{\mathrm{L}}_{A} (-) : \mathbf{D}(A) \to \mathbf{D}(B).$$

Note that the category $\mathsf{DGRng}/_{\mathrm{c}}\,\mathbb{K}$ is not linear (the morphisms sets have no linear structure), whereas the 2-category $\mathsf{TrCat}/\mathbb{K}$ is linear (in the sense that the sets of 2-morphisms are $\mathbb{K}$-modules).

Define $\mathbf{D}(\mathsf{DGRng}/_{\mathrm{c}}\,\mathbb{K})$ to be the categorical localization of the category $\mathsf{DGRng}/_{\mathrm{c}}\,\mathbb{K}$ with respect to the quasi-isomorphisms in it. Then the pseudofunctor DerCat from (14.3.28) localizes, in the categorical sense, to a pseudofunctor

$$\mathrm{DerCat} : \mathbf{D}(\mathsf{DGRng}/_{\mathrm{c}}\,\mathbb{K}) \to \mathsf{TrCat}/\mathbb{K}.$$

The (nonlinear) bifunctor

$$(- \otimes_{\mathbb{K}} -) : (\mathsf{DGRng}/_{\mathrm{c}}\,\mathbb{K}) \times (\mathsf{DGRng}/_{\mathrm{c}}\,\mathbb{K}) \to \mathsf{DGRng}/_{\mathrm{c}}\,\mathbb{K}$$

has a left derived bifunctor

$$(- \otimes^{\mathrm{L}}_{\mathbb{K}} -) : \mathbf{D}(\mathsf{DGRng}/_{\mathrm{c}}\,\mathbb{K}) \times \mathbf{D}(\mathsf{DGRng}/_{\mathrm{c}}\,\mathbb{K}) \to \mathbf{D}(\mathsf{DGRng}/_{\mathrm{c}}\,\mathbb{K}).$$

Given a pair of DG rings A and B, we thus have DG ring $A \otimes^{\mathrm{L}}_{\mathbb{K}} B^{\mathrm{op}}$, well-defined up to a unique isomorphism in $\mathbf{D}(\mathsf{DGRng}/_{\mathrm{c}}\,\mathbb{K})$. The derived category of DG A-B-bimodules described at the end of Remark 14.3.24 is canonically equivalent to $\mathrm{DerCat}(A \otimes^{\mathrm{L}}_{\mathbb{K}} B^{\mathrm{op}})$ in $\mathsf{TrCat}/\mathbb{K}$.

14.4 Tilting DG Bimodules

We continue with Convention 14.3.8. In particular all DG rings are $\mathbb{K}$-flat central over the base ring $\mathbb{K}$, $\otimes = \otimes_{\mathbb{K}}$, and $A^{\mathrm{en}} = A \otimes A^{\mathrm{op}}$ for a DG ring A.

Definition 14.4.1 Let A and B be DG rings. An object $T \in \mathsf{D}(B \otimes A^{\mathrm{op}})$ is called a *tilting DG B-A-bimodule* if there exists some object $S \in \mathsf{D}(A \otimes B^{\mathrm{op}})$, and isomorphisms $S \otimes^{\mathrm{L}}_{B} T \cong A$ in $\mathsf{D}(A^{\mathrm{en}})$ and $T \otimes^{\mathrm{L}}_{A} S \cong B$ in $\mathsf{D}(B^{\mathrm{en}})$.

When $B = A$, so that $B \otimes A^{\mathrm{op}} = A^{\mathrm{en}}$, we use the term *tilting DG A-bimodule* as an abbreviation for the term *tilting DG A-A-bimodule.*

It is clear from the symmetry of the definition that the object $S \in \mathsf{D}(A \otimes B^{\mathrm{op}})$ is a tilting DG A-B-bimodule.

Lemma 14.4.2 *Let $T \in \mathsf{D}(B \otimes A^{\mathrm{op}})$ and $S, S' \in \mathsf{D}(A \otimes B^{\mathrm{op}})$ satisfy $S \otimes^{\mathrm{L}}_{B} T \cong A$ in $\mathsf{D}(A^{\mathrm{en}})$ and $T \otimes^{\mathrm{L}}_{A} S' \cong B$ in $\mathsf{D}(B^{\mathrm{en}})$. Then $S' \cong S$ in $\mathsf{D}(A \otimes B^{\mathrm{op}})$. Therefore T is a tilting DG B-A-bimodule, and S is a tilting DG A-B-bimodule.*

Proof Using the associativity of derived tensor products from Proposition 14.3.13(4) we have isomorphisms

$$S \cong S \otimes^{\mathrm{L}}_{B} B \cong S \otimes^{\mathrm{L}}_{B} (T \otimes^{\mathrm{L}}_{A} S') \cong (S \otimes^{\mathrm{L}}_{B} T) \otimes^{\mathrm{L}}_{A} S' \cong A \otimes^{\mathrm{L}}_{A} S' \cong S'$$

in $\mathsf{D}(A \otimes B^{\mathrm{op}})$. □

The lemma implies that the tilting DG A-B-bimodule S in Definition 14.4.1 is unique up to isomorphism.

Definition 14.4.3 The tilting DG A-B-bimodule S in Definition 14.4.1 is called the *quasi-inverse of T.*

Here are some general properties of tilting DG bimodules.

Proposition 14.4.4 *Let A, B, C be DG rings, let T be a tilting DG B-A-bimodule, and let S be a tilting DG C-B-bimodule. Then $S \otimes^{\mathrm{L}}_{B} T$ is a tilting C-A-bimodule.*

Proof Let $T^{\vee}$ be the quasi-inverse of T, and let $S^{\vee}$ be the quasi-inverse of S. Then

$$\begin{aligned}(T^{\vee} \otimes^{\mathrm{L}}_{B} S^{\vee}) \otimes^{\mathrm{L}}_{C} (S \otimes^{\mathrm{L}}_{B} T) &\cong T^{\vee} \otimes^{\mathrm{L}}_{B} (S^{\vee} \otimes^{\mathrm{L}}_{C} S) \otimes^{\mathrm{L}}_{B} T \\ &\cong T^{\vee} \otimes^{\mathrm{L}}_{B} B \otimes^{\mathrm{L}}_{B} T \cong A\end{aligned}$$

in $\mathsf{D}(A^{\mathrm{en}})$. Likewise,

$$(S \otimes^{\mathrm{L}}_{B} T) \otimes^{\mathrm{L}}_{A} (T^{\vee} \otimes^{\mathrm{L}}_{B} S^{\vee}) \cong C$$

in $\mathsf{D}(C^{\mathrm{en}})$. □

Proposition 14.4.5 *Suppose T is a tilting DG B-A-bimodule, and S is a tilting DG A-B-bimodule. Then the triangulated functor*

$$G_{T,S} : \mathsf{D}(A^{\mathrm{en}}) \to \mathsf{D}(B^{\mathrm{en}})$$

with formula $G_{T,S}(M) := T \otimes^{\mathrm{L}}_{A} M \otimes^{\mathrm{L}}_{A} S$ is an equivalence.

Proposition 14.4.6 *Suppose T is a tilting DG B-A-bimodule, with quasi-inverse $T^{\vee}$. Fix an isomorphism $T^{\vee} \otimes^{\mathrm{L}}_{B} T \xrightarrow{\simeq} A$ in $\mathbf{D}(A^{\mathrm{en}})$. Then for every $M_1, M_2 \in \mathbf{D}(A^{\mathrm{en}})$ there is an isomorphism*

$$G_{T,T^{\vee}}(M_1 \otimes^{\mathrm{L}}_{A} M_2) \xrightarrow{\simeq} G_{T,T^{\vee}}(M_1) \otimes^{\mathrm{L}}_{B} G_{T,T^{\vee}}(M_2)$$

in $\mathbf{D}(B^{\mathrm{en}})$. This isomorphism is bifunctorial in (M_1, M_2).

Exercise 14.4.7 Prove Propositions 14.4.5 and 14.4.6.

Definition 14.4.8 Let A be a K-flat central DG $\mathbb{K}$-ring. The *noncommutative derived Picard group of A relative to* $\mathbb{K}$ is the group $\mathrm{DPic}_{\mathbb{K}}(A)$, whose elements are the isomorphism classes in $\mathbf{D}(A^{\mathrm{en}})$ of the tilting DG A-bimodules. The multiplication in this group is induced from $(- \otimes^{\mathrm{L}}_{A} -)$, and the unit element is the class of the DG bimodule A.

This definition is valid by Proposition 14.4.4. Propositions 14.4.5 and 14.4.6 say that:

Corollary 14.4.9 *A tilting DG B-A-bimodule T, with quasi-inverse $T^{\vee}$, and an isomorphism $T^{\vee} \otimes^{\mathrm{L}}_{B} T \cong A$ in $\mathbf{D}(A^{\mathrm{en}})$, induce a group isomorphism*

$$\mathrm{DPic}_{\mathbb{K}}(A) \xrightarrow{\simeq} \mathrm{DPic}_{\mathbb{K}}(B).$$

We don't know much about the structure of the derived Picard group $\mathrm{DPic}_{\mathbb{K}}(A)$ in general. However, when A is either a ring or a commutative DG ring, there are a few strong structural results – see next section.

Remark 14.4.10 If A is not K-flat over the base ring $\mathbb{K}$, then the derived Picard group should be defined using a K-flat resolution $\tilde{A} \to A$, as explained in Remark 14.3.24. Namely $\mathrm{DPic}_{\mathbb{K}}(A)$ is the group whose elements are the isomorphism classes in $\mathbf{D}(\tilde{A}^{\mathrm{en}})$ of the tilting DG $\tilde{A}$-bimodules, etc. This group is independent, up to a canonical group isomorphism, of the resolution $\tilde{A} \to A$.

The derived homothety morphism in the commutative setting, and the related derived Morita property, were defined in Section 13.1. In the current noncommutative setting these notions become more involved, as we shall now see.

Given a DG bimodule $M \in \mathbf{C}(B \otimes_{\mathbb{K}} A^{\mathrm{op}})$, there are DG ring homomorphisms

$$\mathrm{hm}_{M,A^{\mathrm{op}}} : A^{\mathrm{op}} \to \mathrm{End}_{B}(M) = \mathrm{Hom}_{B}(M, M)$$

and

$$\mathrm{hm}_{M,B} : B \to \mathrm{End}_{A^{\mathrm{op}}}(M) = \mathrm{Hom}_{A^{\mathrm{op}}}(M, M)$$

that we call the *noncommutative homothety homomorphisms through A^{op} and through B*, respectively. When we forget the ring structures, these become homomorphisms

$$\mathrm{hm}_{M,A^{\mathrm{op}}} : A \to \mathrm{Hom}_B(M, M) \tag{14.4.11}$$

and

$$\mathrm{hm}_{M,B} : B \to \mathrm{Hom}_{A^{\mathrm{op}}}(M, M) \tag{14.4.12}$$

in $\mathbf{C}_{\mathrm{str}}(A^{\mathrm{en}})$ and $\mathbf{C}_{\mathrm{str}}(B^{\mathrm{en}})$, respectively.

Definition 14.4.13 Let $M \in \mathbf{D}(B \otimes A^{\mathrm{op}})$.

(1) The *noncommutative derived homothety morphism of M through A^{op}* is the morphism

$$\mathrm{hm}^{\mathrm{R}}_{M,A^{\mathrm{op}}} := \eta^{\mathrm{R}}_{M,M} \circ \mathrm{Q}(\mathrm{hm}_{M,A^{\mathrm{op}}}) : A \to \mathrm{RHom}_B(M, M)$$

in $\mathbf{D}(A^{\mathrm{en}})$.

(2) The *noncommutative derived homothety morphism of M through B* is the morphism

$$\mathrm{hm}^{\mathrm{R}}_{M,B} := \eta^{\mathrm{R}}_{M,M} \circ \mathrm{Q}(\mathrm{hm}_{M,B}) : B \to \mathrm{RHom}_{A^{\mathrm{op}}}(M, M)$$

in $\mathbf{D}(B^{\mathrm{en}})$.

Here is a commutative diagram in $\mathbf{D}(A^{\mathrm{en}})$ depicting the noncommutative derived homothety morphism through A^{op}.

$$\begin{array}{ccc} A & & \\ {\scriptstyle \mathrm{Q}(\mathrm{hm}_{M,A^{\mathrm{op}}})}\downarrow & \searrow^{\mathrm{hm}^{\mathrm{R}}_{M,A^{\mathrm{op}}}} & \\ \mathrm{Hom}_B(M, M) & \xrightarrow{\eta^{\mathrm{R}}_{M,M}} & \mathrm{RHom}_B(M, M) \end{array}$$

The morphism $\eta^{\mathrm{R}}_{M,M}$ is part of the right derived bifunctor $\mathrm{RHom}_B(-,-)$; see Proposition 14.3.15.

Definition 14.4.14 Let A and B be DG rings, and let M be an object of $\mathbf{D}(B \otimes A^{\mathrm{op}})$.

(1) We say that M has the *noncommutative derived Morita property on the B side* if the derived homothety morphism

$$\mathrm{hm}^{\mathrm{R}}_{M,A^{\mathrm{op}}} : A \to \mathrm{RHom}_B(M, M)$$

in $\mathbf{D}(A^{\mathrm{en}})$ is an isomorphism.

(2) We say that M has the *noncommutative derived Morita property on the A^{op} side* if the derived homothety morphism

$$\mathrm{hm}^{\mathrm{R}}_{M,B} : B \to \mathrm{RHom}_{A^{\mathrm{op}}}(M, M)$$

in $\mathbf{D}(B^{\mathrm{en}})$ is an isomorphism.

(3) We say that M has the *noncommutative derived Morita property on both sides* if it has the noncommutative derived Morita property on the B side and on the A^{op} side.

Recall the restriction functors from diagram (14.2.3). Compact generators were introduced in Definition 14.2.20(3). The next definition resembles Definition 14.2.24.

Definition 14.4.15 Let A and B be DG rings, and let T be an object of $\mathsf{D}(B \otimes A^{\mathrm{op}})$.

(1) We say that T is a *compact generator of* $\mathsf{D}(B)$, or a *compact generator on the B side*, if $\mathrm{Rest}_B(T) \in \mathsf{D}(B)$ is a compact generator.
(2) We say that T is a *compact generator of* $\mathsf{D}(A^{\mathrm{op}})$, or a *compact generator on the* A^{op} *side*, if $\mathrm{Rest}_{A^{\mathrm{op}}}(T) \in \mathsf{D}(A^{\mathrm{op}})$ is a compact generator.
(3) We say that T is a *compact generator on both sides* if it is a compact generator on the B side and on the A^{op} side.

The next theorem is similar to results appearing in [127] and [84]. It is a derived version of the classical result for invertible bimodules over a ring (see Example 14.2.36 or [130, Section 4.1]).

In Definition 14.2.16 we introduced the notion of pretilting DG bimodules, and in Definition 14.4.1 we introduced tilting DG bimodules.

Theorem 14.4.16 *Let A and B be K-flat DG central $\mathbb{K}$-rings. The following three conditions are equivalent for an object $T \in \mathsf{D}(B \otimes A^{\mathrm{op}})$.*

(i) *The DG B-A-bimodule T is tilting.*
(ii) *The DG B-A-bimodule T is pretilting.*
(iii) *The DG B-A-bimodule T is a compact generator on the B side, and it has the noncommutative derived Morita property on the B side.*

Proof

(i) ⇒ (ii): Let $S \in \mathsf{D}(A \otimes B^{\mathrm{op}})$ be the quasi-inverse of T. The functor

$$G_T := T \otimes^{\mathrm{L}}_A (-) : \mathsf{D}(A) \to \mathsf{D}(B) \tag{14.4.17}$$

has a quasi-inverse

$$G_S := S \otimes^{\mathrm{L}}_B (-) : \mathsf{D}(B) \to \mathsf{D}(A). \tag{14.4.18}$$

So G_T is an equivalence, and the DG bimodule T is pretilting.

(ii) ⇒ (iii): By definition, the functor G_T from formula (14.4.17) is an equivalence of triangulated categories. Also $G_T(A) \cong T$ in $\mathsf{D}(B)$. Because A is a compact generator of $\mathsf{D}(A)$, Proposition 14.1.8 says that T is a compact generator of $\mathsf{D}(B)$.

Next we shall prove that T has the noncommutative derived Morita property on the B side, namely that the derived homothety morphism

$$\mathrm{hm}^{\mathrm{R}}_{M,A^{\mathrm{op}}} : A \to \mathrm{RHom}_B(T,T) \tag{14.4.19}$$

in $\mathbf{D}(A^{\mathrm{en}})$ is an isomorphism. According to Proposition 14.3.10, the restriction functor $\mathrm{Rest}_A : \mathbf{D}(A^{\mathrm{en}}) \to \mathbf{D}(A)$ is conservative. Therefore, to prove that the morphism $\mathrm{hm}^{\mathrm{R}}_{M,A^{\mathrm{op}}}$ is an isomorphism in $\mathbf{D}(A^{\mathrm{en}})$, we can forget the A^{op}-module structures on the DG bimodules A and $\mathrm{RHom}_B(T,T)$, and just prove that $\mathrm{Rest}_A(\mathrm{hm}^{\mathrm{R}}_{M,A^{\mathrm{op}}})$ is an isomorphism in $\mathbf{D}(A)$.

According to Proposition 14.2.15, the functor

$$F_T := \mathrm{RHom}_B(T,-) : \mathbf{D}(B) \to \mathbf{D}(A), \tag{14.4.20}$$

which is the right adjoint of G_T, is an equivalence. So there is an isomorphism of triangulated functors $\zeta : \mathrm{Id}_{\mathbf{D}(A)} \xrightarrow{\simeq} F_T \circ G_T$ from $\mathbf{D}(A)$ to itself. There is a diagram

$$\begin{array}{ccccc} A & & \xrightarrow{\quad \zeta_A,\ \cong \quad} & & (F_T \circ G_T)(A) \\ \Big\downarrow{\scriptstyle \mathrm{Rest}_A(\mathrm{hm}^{\mathrm{R}}_{M,A^{\mathrm{op}}})} & & & & \Big\| \\ \mathrm{RHom}_B(T,T) & \xrightarrow{\cong} & \mathrm{RHom}_B(T, T \otimes^{\mathrm{L}}_A A) & \xrightarrow{\cong} & (F_T \circ G_T)(A) \end{array}$$

in $\mathbf{D}(A)$, and a calculation (with elements, using a K-flat resolution $P \to T$ in $\mathbf{C}_{\mathrm{str}}(B \otimes A^{\mathrm{op}})$) shows that it is commutative. Because ζ_A is an isomorphism, we conclude that $\mathrm{Rest}_A(\mathrm{hm}^{\mathrm{R}}_{M,A^{\mathrm{op}}})$ is an isomorphism.

(iii) $\Rightarrow$ (ii): Choose a resolution $T \to I$ in $\mathbf{C}(B \otimes A^{\mathrm{op}})$ such that I is K-injective over B; this is possible by Lemma 14.3.11(2). (If A is K-projective over $\mathbb{K}$ then we can also choose a resolution $P \to T$ in $\mathbf{C}(B \otimes A^{\mathrm{op}})$ such that P is K-projective over B; and then the proof can proceed with P instead of I.) Then the noncommutative derived homothety morphism $\mathrm{hm}^{\mathrm{R}}_{T,A^{\mathrm{op}}}$ in $\mathbf{D}(A^{\mathrm{en}})$ is represented by the canonical DG ring homomorphism $g : A \to A_I$, where $A_I := \mathrm{End}_B(I)^{\mathrm{op}}$. The noncommutative derived Morita property on the B side says that g is a quasi-isomorphism. Hence, by Theorem 12.7.2, the restriction functor $\mathrm{Rest}_g : \mathbf{D}(A_I) \to \mathbf{D}(A)$ is an equivalence.

Because $I \cong T$ is a compact generator of $\mathbf{D}(B)$, according to Theorem 14.2.29, we know that the functor $F_I := \mathrm{RHom}_B(I,-) : \mathbf{D}(B) \to \mathbf{D}(A_I)$ is an equivalence. Taking F_T to be the functor from (14.4.20), we have a diagram of functors

$$\begin{array}{ccc} \mathbf{D}(B) & & \\ \Big\downarrow{\scriptstyle F_I} & \searrow{\scriptstyle F_T} & \\ \mathbf{D}(A_I) & \xrightarrow{\mathrm{Rest}_g} & \mathbf{D}(A) \end{array}$$

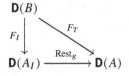

that is commutative up to isomorphism (cf. Theorem 12.7.2(3)). Therefore F_T is an equivalence, and thus T is pretilting.

(ii) $\Rightarrow$ (i): We assume that T is a pretilting DG B-A-bimodule, i.e. the functors G_T and F_T, from (14.4.17) and (14.4.20), respectively, are equivalences. Define the DG bimodule

$$S := \mathrm{RHom}_B(T, B) \in \mathbf{D}(A \otimes B^{\mathrm{op}}). \tag{14.4.21}$$

Consider the tensor-evaluation morphism

$$\mathrm{ev}^{\mathrm{R,L}}_{T,B,T} : \mathrm{RHom}_B(T, B) \otimes^{\mathrm{L}}_B T \to \mathrm{RHom}_B(T, B \otimes^{\mathrm{L}}_B T) \tag{14.4.22}$$

in $\mathbf{D}(A^{\mathrm{en}})$ from Proposition 14.3.19. Because the restriction functor $\mathrm{Rest}_{\mathbb{K}} : \mathbf{D}(A^{\mathrm{en}}) \to \mathbf{D}(\mathbb{K})$ is conservative, to prove that $\mathrm{ev}^{\mathrm{R,L}}_{T,B,T}$ is an isomorphism in $\mathbf{D}(A^{\mathrm{en}})$, it suffices to prove it for $\mathrm{Rest}_{\mathbb{K}}(\mathrm{ev}^{\mathrm{R,L}}_{T,B,T})$. But by Proposition 14.2.19 the DG module T is algebraically perfect on the B side; so by Theorem 14.1.22, the morphism $\mathrm{Rest}_{\mathbb{K}}(\mathrm{ev}^{\mathrm{R,L}}_{T,B,T})$ is an isomorphism. Thus we get these isomorphisms

$$\begin{aligned} S \otimes^{\mathrm{L}}_B T &= \mathrm{RHom}_B(T, B) \otimes^{\mathrm{L}}_B T \\ &\xrightarrow{\mathrm{ev}^{\mathrm{R,L}}_{T,B,T}} \mathrm{RHom}_B(T, B \otimes^{\mathrm{L}}_B T) \cong \mathrm{RHom}_B(T, T) \end{aligned} \tag{14.4.23}$$

in $\mathbf{D}(A^{\mathrm{en}})$. On the other hand, in the proof of "(ii) $\Rightarrow$ (iii)" above we already showed (formula (14.4.19)) that $A \cong \mathrm{RHom}_B(T, T)$ in $\mathbf{D}(A^{\mathrm{en}})$. Combining this with (14.4.23) we deduce that

$$S \otimes^{\mathrm{L}}_B T \cong A \ \text{ in } \ \mathbf{D}(A^{\mathrm{en}}). \tag{14.4.24}$$

Next, consider the functor G_S from formula (14.4.18), where now S is the DG bimodule from (14.4.21). The same arguments as above – those used for $\mathrm{ev}^{\mathrm{R,L}}_{T,B,T}$ – show that for every $N \in \mathbf{D}(B)$ the morphism

$$\mathrm{ev}^{\mathrm{R,L}}_{T,B,N} : \mathrm{RHom}_B(T, B) \otimes^{\mathrm{L}}_B N \to \mathrm{RHom}_B(T, B \otimes^{\mathrm{L}}_B N)$$

in $\mathbf{D}(B)$ is an isomorphism. So, like (14.4.23), we get isomorphisms

$$\begin{aligned} G_S(N) &= S \otimes^{\mathrm{L}}_B N = \mathrm{RHom}_B(T, B) \otimes^{\mathrm{L}}_B N \\ &\xrightarrow{\mathrm{ev}^{\mathrm{R,L}}_{T,B,N}} \mathrm{RHom}_B(T, B \otimes^{\mathrm{L}}_B N) \cong \mathrm{RHom}_B(T, N) = F_T(N) \end{aligned} \tag{14.4.25}$$

in $\mathbf{D}(A)$. Since these isomorphisms are functorial in N, we see that $G_S \cong F_T$ as functors. Therefore G_S is an equivalence, and S is a pretilting DG A-B-bimodule.

The same proof, but now for the pretilting DG A-B-bimodule S instead of for T, shows that the DG B-A-bimodule

$$T' := \mathrm{RHom}_A(S, A) \in \mathbf{D}(B \otimes A^{\mathrm{op}}) \tag{14.4.26}$$

satisfies the corresponding version of (14.4.24), i.e. $T' \otimes^{\mathrm{L}}_A S \cong B$ in $\mathbf{D}(B^{\mathrm{en}})$. By Lemma 14.4.2 the DG B-A-bimodule $T \cong T'$ is tilting. □

Corollary 14.4.27 *Let T be a tilting DG B-A-bimodule. Then its quasi-inverse is the DG A-B-bimodule $S := \mathrm{RHom}_B(T, B)$.*

Proof This was shown in the proof of the implication "(ii) ⇒ (i)" above. □

Corollary 14.4.28 *Let T be a tilting DG B-A-bimodule. Then T is a compact generator on both sides, and it has the noncommutative derived Morita property on both sides.*

Proof By Theorem 14.4.16, T is a compact generator on the B side, and it has the noncommutative derived Morita property on the B side.

Using the DG ring isomorphism $A^{\mathrm{op}} \otimes B \xrightarrow{\simeq} B \otimes A^{\mathrm{op}}$ of (14.2.7), we can also view T as an object of $\mathbf{D}(A^{\mathrm{op}} \otimes B)$. And as such, T is a tilting DG A^{op}-B^{op}-bimodule. Now Theorem 14.4.16 tells us that T is a compact generator on the A^{op} side, and it has the noncommutative derived Morita property on the A^{op} side. □

The example of classical Morita Theory was already given; see Example 14.2.36. Next is an example that goes in another direction altogether.

Example 14.4.29 Suppose $f : A \to B$ is a quasi-isomorphism in $\mathsf{DGRng}/_{\mathrm{fc}}\, \mathbb{K}$. Then $T := B \in \mathbf{D}(B \otimes A^{\mathrm{op}})$ is a tilting DG B-A-bimodule. The equivalence

$$F_T := \mathrm{RHom}_B(T, -) : \mathbf{D}(B) \to \mathbf{D}(A)$$

is just the restriction functor Rest_f; and the equivalence

$$G_T := T \otimes^{\mathrm{L}}_A (-) : \mathbf{D}(A) \to \mathbf{D}(B)$$

is just the derived induction functor LInd_f.

Theorem 14.4.30 (Rickard–Keller) *Let A and B be K-flat DG central $\mathbb{K}$-rings. Assume that there exists a $\mathbb{K}$-linear equivalence of triangulated categories $F : \mathbf{D}(A) \to \mathbf{D}(B)$, and that $\mathrm{H}^i(A) = 0$ for all $i \neq 0$. Then there exists a tilting DG B-A-bimodule T.*

This theorem is very similar to J. Rickard's [127, Theorem 6.4] and to B. Keller's [79, Corollary 9.2]. See Remark 14.4.35 for a brief discussion.

Proof We begin with a warning: in the proof we will construct some new DG rings, and they might fail to be K-flat over $\mathbb{K}$, thus violating Convention 14.3.8.

Consider the object $L := F(A) \in \mathbf{D}(B)$. It is a compact generator, by Propositions 14.1.8 and 14.2.23. Let us choose a K-projective resolution $P \to L$ in $\mathbf{C}_{\mathrm{str}}(B)$, and define the DG ring $\tilde{A} := \mathrm{End}_B(P)^{\mathrm{op}}$. So $P \in \mathbf{D}(B \otimes_{\mathbb{K}} \tilde{A}^{\mathrm{op}})$.

Because F is an equivalence, we get ring isomorphisms

$$\mathrm{H}^0(A) \cong \mathrm{End}_{\mathbf{D}(A)}(A)^{\mathrm{op}} \xrightarrow{F} \mathrm{End}_{\mathbf{D}(B)}(L)^{\mathrm{op}} \cong \mathrm{End}_{\mathbf{D}(B)}(P)^{\mathrm{op}} \cong \mathrm{H}^0(\tilde{A}). \tag{14.4.31}$$

For the same reason we have $\mathrm{H}^i(\tilde{A}) \cong \mathrm{H}^i(A) = 0$ for all $i \neq 0$. Define the DG rings $A_1 := \mathrm{smt}^{\leq 0}(A)$ and $\tilde{A}_1 := \mathrm{smt}^{\leq 0}(\tilde{A})$, and the rings $A_2 := \mathrm{H}^0(A)$ and $\tilde{A}_2 := \mathrm{H}^0(\tilde{A})$. There are canonical quasi-isomorphisms $A_1 \to A$, $A_1 \to A_2$, $\tilde{A}_1 \to \tilde{A}$ and $\tilde{A}_1 \to \tilde{A}_2$ in $\mathsf{DGRng}/_{\mathrm{c}}\,\mathbb{K}$. Equation (14.4.31) produces an isomorphism $A_2 \cong \tilde{A}_2$ in $\mathsf{Rng}/_{\mathrm{c}}\,\mathbb{K}$. Now the DG ring B is K-flat over $\mathbb{K}$, so we have induced quasi-isomorphisms

$$B \otimes A^{\mathrm{op}} \leftarrow B \otimes A_1^{\mathrm{op}} \to B \otimes A_2^{\mathrm{op}} \cong B \otimes \tilde{A}_2^{\mathrm{op}} \leftarrow B \otimes \tilde{A}_1^{\mathrm{op}} \to B \otimes \tilde{A}^{\mathrm{op}}$$

in $\mathsf{DGRng}/_{\mathrm{c}}\,\mathbb{K}$. According to Theorem 12.7.2, we get $\mathbb{K}$-linear equivalences of triangulated categories

$$\begin{aligned} &\mathbf{D}(B \otimes A^{\mathrm{op}}) \to \mathbf{D}(B \otimes A_1^{\mathrm{op}}) \to \mathbf{D}(B \otimes A_2^{\mathrm{op}}) \\ &\to \mathbf{D}(B \otimes \tilde{A}_2^{\mathrm{op}}) \to \mathbf{D}(B \otimes \tilde{A}_1^{\mathrm{op}}) \to \mathbf{D}(B \otimes \tilde{A}^{\mathrm{op}}). \end{aligned}$$

There is an object $T \in \mathbf{D}(B \otimes A^{\mathrm{op}})$ that corresponds to $P \in \mathbf{D}(B \otimes \tilde{A}^{\mathrm{op}})$ under this chain of equivalences. All these equivalence restrict to the identity functor of $\mathbf{D}(B)$, and hence there is an isomorphism

$$\mathrm{Rest}_B(T) \cong \mathrm{Rest}_B(P) \cong L \tag{14.4.32}$$

in $\mathbf{D}(B)$.

Since L is a compact generator on the B side, the same is true for T. Finally, by equation (14.4.32), and using the equivalence F, we know that

$$\mathrm{H}^i(\mathrm{RHom}_B(T,T)) \cong \mathrm{H}^i(\mathrm{RHom}_B(L,L)) \cong \mathrm{H}^i(\mathrm{RHom}_A(A,A)) = 0 \tag{14.4.33}$$

for all $i \neq 0$. By equation (14.4.31) we know that

$$\mathrm{H}^0(\mathrm{RHom}_B(T,T)) \cong \mathrm{End}_{\mathbf{D}(B)}(L) \cong \mathrm{End}_{\mathbf{D}(A)}(A) \cong \mathrm{H}^0(A) \tag{14.4.34}$$

as $\mathrm{H}^0(A)$-bimodules. Once we trace the various morphisms, formulas (14.4.33) and (14.4.34) imply that the derived homothety morphism $\mathrm{hm}^{\mathrm{R}}_{T,A^{\mathrm{op}}} : A \to \mathrm{RHom}_B(T,T)$ in $\mathsf{D}(A^{\mathrm{en}})$ is an isomorphism. So T has the derived Morita property on the B side. By Theorem 14.4.16, T is a tilting DG B-A-bimodule. □

Remark 14.4.35 J. Rickard, in [126, Theorem 6.4], considers rings A and B, and he does not assume that they are flat over a base ring $\mathbb{K}$. So instead of producing a tilting complex T, as we do in Theorem 14.4.30 above, he actually produces a pretilting complex. Another difference is that Rickard considers an equivalence $F : \mathbf{D}^{\mathrm{b}}(A) \to \mathbf{D}^{\mathrm{b}}(B)$, whereas we look at the unbounded derived categories.

B. Keller, in [79, Corollary 9.2], works in much greater generality than we do: instead of DG rings A and B, he looks at DG categories A and B. Also he does not require A to be K-flat over $\mathbb{K}$.

The apparent flatness limitation in our Theorem 14.4.30 can be effectively overcome using K-flat resolutions $\tilde{A} \to A$ and $\tilde{B} \to B$ in $\mathsf{DGRng}/_{\mathrm{c}}\mathbb{K}$, as explained in Remark 14.3.24. Since the restriction functors $\mathbf{D}(A) \to \mathbf{D}(\tilde{A})$ and $\mathbf{D}(B) \to \mathbf{D}(\tilde{B})$ are equivalences, we get an equivalence $\tilde{F} : \mathbf{D}(\tilde{A}) \to \mathbf{D}(\tilde{B})$. Theorem 14.4.30 produces a tilting DG bimodule $T \in \mathbf{D}(\tilde{B} \otimes \tilde{A}^{\mathrm{op}})$.

Remark 14.4.36 In the situation of Theorem 14.4.30, it is not known whether one can find a tilting DG B-A-bimodule T such that $F \cong T \otimes^{\mathrm{L}}_{A} (-)$ as triangulated functors. This question was already raised by Rickard in [126], and still remains open, except for a few special cases (see [109] for hereditary rings, and [40] for triangular rings).

If we drop the condition that the cohomology of A is concentrated in degree 0, then there is a counterexample to the assertion of Theorem 14.4.30, due to B. Shipley [140, Section 5], based on her joint work with D. Dugger [45]. Here is its translation to our terminology. She considers the DG ring $A = \mathbb{Z}[\bar{x}] := \mathbb{Z}\langle x \rangle / (x^4)$, where x is a variable of degree -1, and $\mathrm{d}(\bar{x}) := 2$. (Note that A is weakly commutative but not strongly commutative, since $\bar{x}^2 \neq 0$.) The second DG ring is $\mathrm{H}(A)$. A calculation shows that $\mathrm{H}(A) \cong \mathbb{F}_2[\bar{y}] := \mathbb{F}_2\langle y \rangle / (y^2)$, where y is a variable of degree -2, and $\mathrm{d}(\bar{y}) := 0$. For the base ring we take $\mathbb{K} := \mathbb{Z}$. Because $\mathrm{H}(A)$ is not K-flat over $\mathbb{K}$, we choose a K-flat resolution $B \to \mathrm{H}(A)$ over $\mathbb{K}$.

Shipley asserts that on the one hand there is an equivalence of triangulated categories $\mathbf{D}(A) \to \mathbf{D}(B)$, and on the other hand there does not exist a DG bimodule $T \in \mathbf{D}(B \otimes A^{\mathrm{op}})$ that is a compact generator on the B side, and has the noncommutative derived Morita property on the B side (condition (iii) of Theorem 14.4.16). Thus there does not exist a tilting DG B-A-bimodule.

Remark 14.4.37 In algebraic geometry the role of tensoring with tilting bimodule complexes is played by *Fourier-Mukai transforms*. This theory is at the core of contemporary birational geometry. See the survey [64] or the book [70].

14.5 Tilting Bimodule Complexes over Rings

We continue with Convention 14.3.8, but in this section we only look at rings. So our rings are flat central over the base ring $\mathbb{K}$, $\otimes = \otimes_{\mathbb{K}}$, and $A^{\mathrm{en}} = A \otimes A^{\mathrm{op}}$ for a ring A. All bimodules are $\mathbb{K}$-central, and all additive functors are $\mathbb{K}$-linear.

Rather than speaking about DG bimodules, here we speak about complexes of bimodules. By default our rings are nonzero.

In classical Morita Theory, an *invertible B-A-bimodule* is a bimodule P for which there exists an A-B-bimodule Q, such that $Q \otimes_B P \cong A$ in $\mathsf{M}(A^{\mathrm{en}})$ and $P \otimes_A Q \cong B$ in $\mathsf{M}(B^{\mathrm{en}})$. It is known that P is a finitely generated projective module over B and over A^{op}; and the reverse is true for Q. Furthermore, every $\mathbb{K}$-linear equivalence $F : \mathsf{M}(A) \to \mathsf{M}(B)$ is isomorphic, as a functor, to the functor $P \otimes_A (-)$ for an invertible bimodule P, and this P is unique up to a unique isomorphism. See [130, Section 4.1] for proofs.

We see that invertible bimodules are a very special kind of tilting bimodule complexes. Proposition 14.5.3 clarifies matters. But first a lemma, which is a noncommutative variant of Lemma 13.1.36

Lemma 14.5.1 (Künneth Trick) *Let $M \in \mathsf{D}^-(B \otimes A^{\mathrm{op}})$ and $N \in \mathsf{D}^-(A \otimes C^{\mathrm{op}})$, and let $i_1, j_1 \in \mathbb{Z}$ be such that* $\sup(\mathrm{H}(M)) \leq i_1$ *and* $\sup(\mathrm{H}(N)) \leq j_1$. *Then*

$$\mathrm{H}^{i_1+j_1}(M \otimes^{\mathrm{L}}_A N) \cong \mathrm{H}^{i_1}(M) \otimes_A \mathrm{H}^{j_1}(N)$$

in $\mathsf{M}(B \otimes C^{\mathrm{op}})$.

Exercise 14.5.2 Prove Lemma 14.5.1.

Proposition 14.5.3 *The following conditions are equivalent for a tilting bimodule complex $T \in \mathsf{D}(B \otimes A^{\mathrm{op}})$, with quasi-inverse $S = T^{\vee}$.*

(i) *There is an isomorphism $T \cong P$ in $\mathsf{D}(B \otimes A^{\mathrm{op}})$ for some invertible B-A-bimodule P.*
(ii) $\mathrm{H}^0(T)$ *is a projective B-module, and* $\mathrm{H}^i(T) = 0$ *for all $i \neq 0$.*
(iii) $\mathrm{H}^i(T) = 0$ *and* $\mathrm{H}^i(S) = 0$ *for all $i \neq 0$.*

Proof

(i) $\Rightarrow$ (ii): this is trivial.

(ii) $\Rightarrow$ (iii): Let $P := \mathrm{H}^0(T) \in \mathsf{M}(B \otimes A^{\mathrm{op}})$, so there is an isomorphism $T \cong P$ in $\mathsf{D}(B \otimes A^{\mathrm{op}})$. According to Corollary 14.4.27, we have $S \cong \mathrm{RHom}_B(T, B) \cong \mathrm{Hom}_B(P, B)$.

(iii) $\Rightarrow$ (i): Define $P := \mathrm{H}^0(T) \in \mathsf{M}(B \otimes A^{\mathrm{op}})$ and $Q := \mathrm{H}^0(S) \in \mathsf{M}(A \otimes B^{\mathrm{op}})$. The Künneth Trick (Lemma 14.5.1) says that $P \otimes_B Q \cong \mathrm{H}^0(T \otimes^{\mathrm{L}}_B S) \cong \mathrm{H}^0(A) = A$ and $Q \otimes_A P \cong \mathrm{H}^0(S \otimes^{\mathrm{L}}_A T) \cong \mathrm{H}^0(B) = B$ in $\mathsf{M}(A^{\mathrm{en}})$ and $\mathsf{M}(B^{\mathrm{en}})$, respectively. Thus P and Q are invertible bimodules. But $T \cong P$ in $\mathsf{D}(B \otimes A^{\mathrm{op}})$. □

The *Jacobson radical* of a ring A is the intersection of all maximal left (or right) ideals of A. See [71, Section 4.4.2] or [130, Section 2.5].

Definition 14.5.4 A ring A, with Jacobson radical $\mathfrak{r}$, is called *local* if $A/\mathfrak{r}$ is a simple artinian ring.

Note that this definition is wider than what is found in some books (e.g. [130]), where the condition is that $A/\mathfrak{r}$ is a division ring.

Example 14.5.5 If C is a commutative local ring with maximal idea $\mathfrak{m}$, and $n \geq 1$, then the matrix ring $A := \mathrm{Mat}_{n \times n}(C)$ is local, with Jacobson radical $\mathfrak{m} \cdot A = \mathrm{Mat}_{n \times n}(\mathfrak{m})$.

Lemma 14.5.6 *Let A be a local ring, let M be a nonzero finitely generated right A-module, and let N be a nonzero finitely generated left A-module. Then $M \otimes_A N$ is nonzero.*

Proof Define $K := A/\mathfrak{r}$, which is a simple artinian ring. By the noncommutative Nakayama Lemma (see [130, Proposition 2.5.24]) the right K-module $M \otimes_A K$ is nonzero, and also the left K-module $K \otimes_A N$ is nonzero. There is a canonical surjection

$$M \otimes_A N \to (M \otimes_A K) \otimes_K (K \otimes_A N).$$

It suffices to prove that the target is nonzero. Thus we can assume that $A = K$ is a simple artinian ring.

Every left module over K is a direct sum of simple ones; so we can assume that N is simple. Likewise, we can assume that M is a simple right K-module.

There is a ring isomorphism $K \cong \mathrm{Mat}_{n \times n}(D)$, where D is a division ring and n is a positive integer. The simple left K-module N is isomorphic to D^n, seen as a column module; thus it is in fact a K-D-bimodule. Similarly $M \cong D^n$, seen as a row module; thus it is a D-K-bimodule. By an easy calculation (this is an elementary case of Morita equivalence) there is an isomorphism $M \otimes_K N \cong D$ of D-D-bimodules. This is nonzero. □

Theorem 14.5.7 ([132], [165]) *Let A and B be flat central $\mathbb{K}$-rings, with A local, and let T be a tilting B-A-bimodule complex. Then there is an isomorphism $T \cong P[n]$ in $\mathbf{D}(B \otimes A^{\mathrm{op}})$ for some invertible B-A-bimodule P and some integer n.*

See Remark 14.5.32 regarding the history of this theorem.

Proof Let $S \in \mathbf{D}(A \otimes B^{\mathrm{op}})$ be the quasi-inverse of T. From Corollary 14.4.28 and Theorem 14.1.22 we know that T and S are algebraically perfect complexes on both sides. By Lemma 14.1.27 we know that $\mathrm{con}(\mathrm{H}(T)) = [i_0, i_1]$ and $\mathrm{con}(\mathrm{H}(S)) = [j_0, j_1]$ for some integers $i_0 \leq i_1$ and $j_0 \leq j_1$; and also that $P := \mathrm{H}^{i_1}(T)$ and $Q := \mathrm{H}^{j_1}(S)$ are finitely presented modules on both sides.

We now use the Künneth trick to obtain an isomorphism

$$P \otimes_A Q \cong \mathrm{H}^{i_1+j_1}(T \otimes^{\mathrm{L}}_A S) \cong \mathrm{H}^{i_1+j_1}(B)$$

in $\mathsf{M}(B^{\mathrm{en}})$. By Lemma 14.5.6 we know that $P \otimes_A Q$ is nonzero. Therefore $i_1 + j_1 = 0$, and $P \otimes_A Q \cong B$ in $\mathbf{M}(B^{\mathrm{en}})$. For similar reasons, on the reverse side we have $Q \otimes_B P \cong \mathrm{H}^0(S \otimes^{\mathrm{L}}_B T) \cong A$ in $\mathbf{M}(A^{\mathrm{en}})$. Thus P and Q are invertible bimodules.

We now restrict T to $\mathbf{D}(A^{\mathrm{op}})$. Because its top cohomology module $\mathrm{H}^{i_1}(T) = P$ is projective, we can split it off (see Lemma 13.1.38, which holds also for a noncommutative ring), to get an isomorphism $T \cong T' \oplus P[-i_1]$ in $\mathbf{D}(A^{\mathrm{op}})$, with $\sup(\mathrm{H}(T')) \leq i_1 - 1$. Similarly there's an isomorphism $S \cong S' \oplus Q[-j_1]$ in $\mathbf{D}(A)$ with $\sup(\mathrm{H}(S')) \leq j_1 - 1$. It follows that

$$\begin{aligned} B &\cong T \otimes^{\mathrm{L}}_A S \cong (T' \oplus P[-i_1]) \otimes^{\mathrm{L}}_A (S' \oplus Q[-j_1]) \\ &\cong (T' \otimes^{\mathrm{L}}_A S') \oplus (P[-i_1] \otimes_A S') \oplus (T' \otimes_A Q[-j_1]) \oplus B \end{aligned}$$

in $\mathbf{D}(\mathbb{K})$. The object $B \in \mathbf{D}(\mathbb{K})$ has cohomology concentrated in degree 0. The direct summand $T' \otimes_A Q[-j_1]$ has cohomology concentrated in degrees ≤ -1, so its cohomology must be zero, and therefore $T' \otimes_A Q[-j_1] = 0$ in $\mathbf{D}(\mathbb{K})$. But Q is an invertible bimodule, and this forces $T' = 0$ in $\mathbf{D}(A^{\mathrm{op}})$. The conclusion is that $\mathrm{H}^i(T) = 0$ for all $i \neq i_1$.

Returning to the category $\mathbf{D}(A \otimes B^{\mathrm{op}})$ we see that $T \cong P[-i_1]$. Finally we take $n := -i_1$. □

Corollary 14.5.8 *Let A and B be rings, with A local. If there is an equivalence of triangulated categories $\mathbf{D}(A) \to \mathbf{D}(B)$, then there is an equivalence of abelian categories $\mathbf{M}(A) \to \mathbf{M}(B)$. In other words, A and B are Morita equivalent.*

Proof According to Theorem 14.4.30, there exists a tilting B-A-bimodule complex T. Theorem 14.5.7 says that $T \cong P[n]$ for an invertible B-A-bimodule P. Then $P \otimes_A (-) : \mathbf{M}(A) \to \mathbf{M}(B)$ is an equivalence of abelian categories. □

Definition 14.5.9 For a central $\mathbb{K}$-ring A we define the *noncommutative Picard group of A relative to* $\mathbb{K}$ to be the group $\mathrm{Pic}_{\mathbb{K}}(A)$, whose elements are the isomorphism classes in $\mathbf{M}(A^{\mathrm{en}})$ of the invertible bimodules. The operation is induced by $(- \otimes_A -)$, and the unit element is the class of A.

The derived Picard group was introduced in Definition 14.4.8.

Corollary 14.5.10 *If A is a local ring, then*

$$\mathrm{DPic}_{\mathbb{K}}(A) = \mathrm{Pic}_{\mathbb{K}}(A) \times \mathbb{Z}.$$

Proof By Theorem 14.5.7 there is a surjective group homomorphism $\mathrm{Pic}_{\mathbb{K}}(A) \times \mathbb{Z} \to \mathrm{DPic}_{\mathbb{K}}(A)$, defined by $(P, n) \mapsto P[n]$. It is injective because the functor $\mathbf{M}(A^{\mathrm{en}}) \to \mathbf{D}(A^{\mathrm{en}})$ is fully faithful. □

The center of a ring A is denoted by $\operatorname{Cent}(A)$. Of course, $\operatorname{Cent}(A) = \operatorname{Cent}(A^{\mathrm{op}})$. Given a complex $M \in \mathbf{D}(B \otimes A^{\mathrm{op}})$, there are ring homomorphisms

$$\mathrm{chm}^{\mathbf{D}}_{M,A^{\mathrm{op}}} : \operatorname{Cent}(A) \to \operatorname{End}_{\mathbf{D}(B\otimes A^{\mathrm{op}})}(M) \tag{14.5.11}$$

and

$$\mathrm{chm}^{\mathbf{D}}_{M,B} : \operatorname{Cent}(B) \to \operatorname{End}_{\mathbf{D}(B\otimes A^{\mathrm{op}})}(M) \tag{14.5.12}$$

that we call the *central homotheties* through A^{op} and B, respectively. The formulas are the obvious ones: for an element $a \in \operatorname{Cent}(A)$ the action on M is $\mathrm{chm}^{\mathbf{D}}_{M,A^{\mathrm{op}}}(a)(m) := m \cdot a$ for $m \in M^i$, and likewise for $\mathrm{chm}^{\mathbf{D}}_{M,B}$.

Lemma 14.5.13 *If T is a tilting B-A-bimodule complex, then the ring homomorphisms $\mathrm{chm}^{\mathbf{D}}_{M,A^{\mathrm{op}}}$ and $\mathrm{chm}^{\mathbf{D}}_{M,B}$ are both isomorphisms.*

Proof Let S be the quasi-inverse of T. The functor

$$G := (-) \otimes^{\mathrm{L}}_A S : \mathbf{D}(B \otimes A^{\mathrm{op}}) \to \mathbf{D}(B \otimes B^{\mathrm{op}}) = \mathbf{D}(B^{\mathrm{en}})$$

is an equivalence. It induces a ring isomorphism

$$G : \operatorname{End}_{\mathbf{D}(B\otimes A^{\mathrm{op}})}(T) \xrightarrow{\simeq} \operatorname{End}_{\mathbf{D}(B^{\mathrm{en}})}(B),$$

and $G \circ \mathrm{chm}^{\mathbf{D}}_{T,B} = \mathrm{chm}^{\mathbf{D}}_{B,B}$. But $\operatorname{End}_{\mathbf{D}(B^{\mathrm{en}})}(B) \cong \operatorname{End}_{\mathbf{M}(B^{\mathrm{en}})}(B)$, and the ring homomorphism $\operatorname{Cent}(B) \to \operatorname{End}_{\mathbf{M}(B^{\mathrm{en}})}(B)$ is bijective.

The proof for $\mathrm{chm}^{\mathbf{D}}_{T,A^{\mathrm{op}}}$ is similar. □

Definition 14.5.14 Given a tilting B-A-bimodule complex T, we denote by $g_T : \operatorname{Cent}(A) \to \operatorname{Cent}(B)$ the ring isomorphism such that $\mathrm{chm}^{\mathbf{D}}_{T,A^{\mathrm{op}}} = \mathrm{chm}^{\mathbf{D}}_{T,B} \circ g_T$.

Even though the localization of a noncommutative ring B with respect to a multiplicatively closed subset $Z \subseteq B$ is problematic in general, there is no difficulty at all if $Z \subseteq \operatorname{Cent}(B)$. In this case, letting $C := \operatorname{Cent}(B)$, we get canonical A-ring isomorphisms $B_Z \cong C_Z \otimes_C B \cong B \otimes_C C_Z$.

Lemma 14.5.15 *Let T be a tilting B-A-bimodule complex, and assume that the ring A is commutative. Let $Z \subseteq A$ be a multiplicatively closed set, and define $A' := A_Z$ and $B' := B_{g_T(Z)}$. Then the complex $T' := B' \otimes_B T \otimes_A A'$ in $\mathbf{D}(B' \otimes A'^{\mathrm{op}})$ is a tilting B'-A'-bimodule complex.*

Proof The cohomology $\mathrm{H}(T)$ is a central graded A-bimodule, where the left action of A on $\mathrm{H}(T)$ is via the isomorphism g_T. (Warning: the complex of bimodules T need not be central over A.) The flatness of $A \to A'$ and $B \to B'$ gives

$$\mathrm{H}(T') = \mathrm{H}(B' \otimes_B T \otimes_A A') \cong B' \otimes_B \mathrm{H}(T) \otimes_A A' \cong A' \otimes_A \mathrm{H}(T) \otimes_A A'.$$

Therefore $\mathrm{H}(T) \otimes_A A' \to \mathrm{H}(T')$ is an isomorphism of graded modules, and this implies that

$$T \otimes_A A' \to T' \tag{14.5.16}$$

is an isomorphism in $\mathbf{D}(B \otimes A'^{\mathrm{op}})$.

Let $S := \mathrm{RHom}_B(T, B)$ in $\mathbf{D}(A \otimes B^{\mathrm{op}})$ be the quasi-inverse of T. In analogy to Definition 14.5.14, there is a ring isomorphism $h_S : A \to \mathrm{Cent}(B)$ such that $\mathrm{chm}^{\mathbf{D}}_{S,A} = \mathrm{chm}^{\mathbf{D}}_{S,B^{\mathrm{op}}} \circ h_S$. The cohomology $\mathrm{H}(S)$ is a central graded A-bimodule, where the right action of A on $\mathrm{H}(S)$ is via h_S.

Let us consider the derived tensor product $S \otimes^{\mathrm{L}}_B T \in \mathbf{D}(A^{\mathrm{en}})$, and the four ways that an element $a \in A$ acts on it as an object of $\mathbf{D}(A^{\mathrm{en}})$. The action of a from the right side of T the same as the action of $g_T(a) \in \mathrm{Cent}(B)$ from the left side of T. This is also the action of $g_T(a)$ from the right side of S, and also the action of $h_S^{-1}(g_T(a))$ from the left on S. But there are ring isomorphisms

$$\mathrm{End}_{\mathbf{D}(A^{\mathrm{en}})}(S \otimes^{\mathrm{L}}_B T) \cong \mathrm{End}_{\mathbf{D}(A^{\mathrm{en}})}(A) \cong A,$$

and a acts on A the same from both sides. The conclusion is that $h_S^{-1}(g_T(a)) = a$. Because a is an arbitrary element, we see that $g_T = h_S$ as isomorphisms $A \xrightarrow{\simeq} \mathrm{Cent}(B)$. So we have a single ring isomorphism $A \cong \mathrm{Cent}(B)$, for which both $\mathrm{H}(T)$ and $\mathrm{H}(S)$ are central graded A-bimodules.

Define $S' := A' \otimes_A S \otimes_B B' \in \mathbf{D}(A' \otimes B'^{\mathrm{op}})$. The same arguments as for T show that $A' \otimes_A S \to S'$ is an isomorphism in $\mathbf{D}(A' \otimes B^{\mathrm{op}})$.

We now calculate:

$$\begin{aligned} S' \otimes^{\mathrm{L}}_{B'} T' &= (A' \otimes_A S \otimes_B B') \otimes^{\mathrm{L}}_{B'} (B' \otimes_B T \otimes_A A') \\ &\cong (A' \otimes_A S) \otimes^{\mathrm{L}}_B (B' \otimes_B T \otimes_A A') \cong^{\dagger} (A' \otimes_A S) \otimes^{\mathrm{L}}_B (T \otimes_A A') \\ &\cong A' \otimes_A (S \otimes^{\mathrm{L}}_B T) \otimes_A A') \cong A' \otimes_A A \otimes_A A' \cong A' \end{aligned}$$

in $\mathbf{D}(A'^{\mathrm{en}})$. We used the associativity of $(- \otimes^{\mathrm{L}}_{(-)} -)$ several times, and also the ring isomorphism $A' \otimes_A A' \cong A'$. The isomorphism $\cong^{\dagger}$ is by formula (14.5.16). Similarly we show that $T' \otimes^{\mathrm{L}}_{A'} S' \cong B'$ in $\mathbf{D}(B'^{\mathrm{en}})$. □

Theorem 14.5.17 ([132], [165]) *Let A and B be flat central $\mathbb{K}$-rings, and assume A is commutative with connected spectrum. Let T be a tilting B-A-bimodule complex. Then there is an isomorphism $T \cong P[n]$ in $\mathbf{D}(B \otimes A^{\mathrm{op}})$, for some invertible B-A-bimodule P and some integer n.*

See Remark 14.5.32 regarding the history of this theorem.

Proof Because A is commutative, we have $A = A^{\mathrm{op}}$. The ring homomorphism g_T makes B into a central A-ring. We may assumed that $A \neq 0$, and hence $T \neq 0$. The complex T is algebraically perfect over A, so it has bounded cohomology, say with $\sup(\mathrm{H}(T)) = i_1 \in \mathbb{Z}$. By Lemma 14.1.27 the A-module $P := \mathrm{H}^{i_1}(T)$ is finitely presented.

For a prime $\mathfrak{p} \in \operatorname{Spec}(A)$, with corresponding local ring $A_{\mathfrak{p}}$, we write $P_{\mathfrak{p}} := P \otimes_A A_{\mathfrak{p}}$. Define $Y \subseteq \operatorname{Spec}(A)$ to be the support of P, i.e.

$$Y := \{\mathfrak{p} \in \operatorname{Spec}(A) \mid P_{\mathfrak{p}} \neq 0\}.$$

This is a nonempty set. Since P is finitely generated, it follows that Y is a closed subset of $\operatorname{Spec}(A)$; see [28, Proposition II.4.17].

Take any prime $\mathfrak{p} \in Y$, and let $B_{\mathfrak{p}} := B \otimes_A A_{\mathfrak{p}}$. Then, by Lemma 14.5.15, the complex

$$T_{\mathfrak{p}} := B_{\mathfrak{p}} \otimes_B T \otimes_A A_{\mathfrak{p}} \in \mathbf{D}(B_{\mathfrak{p}} \otimes A_{\mathfrak{p}}^{\mathrm{op}})$$

is a tilting $B_{\mathfrak{p}}$-$A_{\mathfrak{p}}$-bimodule complex. Since $\mathrm{H}^{i_1}(T_{\mathfrak{p}}) \cong P_{\mathfrak{p}} \neq 0$, Theorem 14.5.7 implies that

$$T_{\mathfrak{p}} \cong P_{\mathfrak{p}}[-i_1] \in \mathbf{D}(B_{\mathfrak{p}} \otimes A_{\mathfrak{p}}^{\mathrm{op}}), \tag{14.5.18}$$

and that $P_{\mathfrak{p}}$ is an invertible $B_{\mathfrak{p}}$-$A_{\mathfrak{p}}$-bimodule. In particular, $P_{\mathfrak{p}}$ is a nonzero finitely generated projective $A_{\mathfrak{p}}$-module. Thus $P_{\mathfrak{p}}$ is a free $A_{\mathfrak{p}}$-module, of rank $r_{\mathfrak{p}} > 0$. Recall that P is a finitely presented A-module. According to [28, Section II.5.1, Corollary], there is an open neighborhood U of $\mathfrak{p}$ in $\operatorname{Spec}(A)$ on which P is free of rank $r_{\mathfrak{p}}$. In particular $P_{\mathfrak{q}} \neq 0$ for all $\mathfrak{q} \in U$. Therefore $U \subseteq Y$.

The conclusion is that Y is also open in $\operatorname{Spec}(A)$. Since $\operatorname{Spec}(A)$ is connected, it follows that $Y = \operatorname{Spec}(A)$. Another conclusion is that P is projective as an A-module – see [28, Section II.5.2, Theorem 1].

Going back to equation (14.5.18) we see that $\mathrm{H}^i(T)_{\mathfrak{p}} \cong \mathrm{H}^i(T_{\mathfrak{p}}) = 0$ for all $i \neq i_1$. Therefore $\mathrm{H}^i(T) = 0$ for $i \neq i_1$. By truncation we get an isomorphism $T \cong P[n]$ in $\mathbf{D}(B \otimes A^{\mathrm{op}})$, where $n := -i_1$. Finally, by Proposition 14.5.3 the B-A-bimodule P is invertible. □

Let A be a commutative ring. An A-module P can be viewed as a central A-bimodule. If P is a rank 1 projective A-module, then as a bimodule it is invertible. The usual Picard group of A (see [63, Section II.6]) is then the subgroup $\operatorname{Pic}_A(A)$ of $\operatorname{Pic}_{\mathbb{K}}(A)$, whose elements are the isomorphism classes of the central A-bimodules; and we refer to it here as the *commutative Picard group of A*.

Exercise 14.5.19 Let A be a commutative ring. We denote by $\operatorname{Aut}_{\mathbb{K}}(A)$ the group of ring automorphisms of A (i.e. the Galois group). Show that

$$\operatorname{Pic}_{\mathbb{K}}(A) \cong \operatorname{Aut}_{\mathbb{K}}(A) \ltimes \operatorname{Pic}_A(A).$$

(Hints: (1) Every invertible A-bimodule P is isomorphic to $P' \otimes_A P''$, where P' is free of rank 1 as a left A-module, and P'' is a central invertible bimodule.

This is proved like Theorem 14.5.17. (2) If P' is an invertible A-bimodule that's free of rank 1 as a left A-module, then it is also free as a right A-module, with the same basis element $e \in P'$. The proof is like those of Lemma 14.5.6 and Theorem 14.5.17. Conclude that $P' \cong A(\gamma)$, where $\gamma \in \mathrm{Aut}_{\mathbb{K}}(A)$, and $A(\gamma)$ is the twisted bimodule that is the left A-module A, with right action $p \cdot a := \gamma(a) \cdot p$ for $p \in A(\gamma)$ and $a \in A$.)

Corollary 14.5.20 *Let A be a commutative ring with connected spectrum. Then*

$$\mathrm{DPic}_{\mathbb{K}}(A) = \mathrm{Pic}_{\mathbb{K}}(A) \times \mathbb{Z}.$$

Proof The same as that of Corollary 14.5.10. □

Corollary 14.5.21 *Let A and B be rings, and assume that A is commutative ring with connected spectrum. If there is an equivalence of triangulated categories $\mathsf{D}(A) \to \mathsf{D}(B)$, then there is an equivalence of abelian categories $\mathsf{M}(A) \to \mathsf{M}(B)$. In other words, A and B are Morita equivalent.*

Proof According to Theorem 14.4.30, there exists a tilting B-A-bimodule complex T. Theorem 14.5.17 says that $T \cong P[n]$ for an invertible B-A-bimodule P. Then $P \otimes_A (-) : \mathsf{M}(A) \to \mathsf{M}(B)$ is $\mathbb{K}$-linear equivalence of abelian categories. Moreover, P is a projective A-module of rank r for some positive integer r, and $B \cong \mathrm{End}_A(P)^{\mathrm{op}}$. □

Remark 14.5.22 In the situation of Corollary 14.5.21, we can view B as a central A-ring, using the homomorphism g_T from Definition 14.5.14. Then B is an Azumaya A-ring; see [130, Section 5.3]. Moreover, letting $X := \mathrm{Spec}(A)$, and letting $\mathcal{B}$ be the sheafification of B to X, then $\mathcal{B}$ is a trivial Azumaya $\mathcal{O}_X$-ring, in the sense of [108, Section IV.2].

Let A be a commutative ring. If $\mathrm{Spec}(A)$ has a finite connected component decomposition, in the sense of Definition 13.1.50(1), then the connected component decomposition of A is $A = \prod_{i=1}^{m} A_i$, see Definition 13.1.50(2).

Corollary 14.5.23 *Let A and B be rings. Assume A is commutative, and it has a finite connected component decomposition $A = \prod_{i=1}^{m} A_i$. Let T be a tilting B-A-bimodule complex. We consider B as a central A-ring via the isomorphism g_T from Definition 14.5.14. Then $T \cong \bigoplus_{i=1}^{m} P_i[n_i]$ in $\mathsf{D}(B \otimes A^{\mathrm{op}})$, where $B_i := A_i \otimes_A B$, P_i is an invertible B_i-A_i-bimodule, and $n_i \in \mathbb{Z}$.*

Corollary 14.5.24 *Let A be a commutative ring, and assume it has a finite connected component decomposition $A = \prod_{i=1}^{m} A_i$. Then*

$$\mathrm{DPic}_{\mathbb{K}}(A) = \mathrm{Pic}_{\mathbb{K}}(A) \times \mathbb{Z}^m.$$

Corollary 14.5.25 *Let A and B be rings. Assume that A is commutative and it has a finite connected component decomposition. If there is an equivalence of triangulated categories $\mathsf{D}(A) \to \mathsf{D}(B)$, then there is an equivalence of abelian categories $\mathsf{M}(A) \to \mathsf{M}(B)$.*

Exercise 14.5.26 Prove Corollaries 14.5.23, 14.5.24 and 14.5.25. Cf. Exercise 13.1.54.

The general case of these last three corollaries, namely when A does not have a finite connected component decomposition, is dealt with in [172, Theorem 6.13].

Remark 14.5.27 The results in this section have variants in which the flatness condition on the $\mathbb{K}$-rings A and B is dropped.

For instance, Theorem 14.5.7 can be modified as follows: If A and B are rings, with A local, and if T is a pretilting B-A-bimodule complex, then there is an isomorphism $T \cong P[n]$ in $\mathsf{D}(A \otimes B^{\mathrm{op}})$ for some invertible B-A-bimodule P and some integer n. Here is an outline of the proof. One chooses nonpositive K-flat resolutions $\tilde{A} \to A$ and $\tilde{B} \to B$ in $\mathsf{DGRng}/_{\mathrm{c}}\mathbb{K}$. Next a version of the Künneth trick (Lemma 14.5.1) for nonpositive DG rings has to be proved: For $M \in \mathsf{D}^{-}(\tilde{B} \otimes \tilde{A}^{\mathrm{op}})$ and $N \in \mathsf{D}^{-}(\tilde{A} \otimes \tilde{B}^{\mathrm{op}})$ satisfying $\sup(\mathrm{H}(M)) \leq i_1$ and $\sup(\mathrm{H}(N)) \leq j_1$, there is an isomorphism

$$\mathrm{H}^{i_1+j_1}(M \otimes^{\mathrm{L}}_{\tilde{A}} N) \cong \mathrm{H}^{i_1}(M) \otimes_A \mathrm{H}^{j_1}(N)$$

in $\mathsf{M}(B^{\mathrm{en}})$. Then the proof of Theorem 14.5.7 is repeated, for the tilting DG bimodule $T \in \mathsf{D}(\tilde{A} \otimes \tilde{B}^{\mathrm{op}})$.

The proofs of the other results are along the same lines.

Remark 14.5.28 Assume A is a commutative DG ring (Definitions 3.1.22 and 3.3.4), with reduction $\bar{A} = \mathrm{H}^0(A)$. Here we do not need to assume flatness over the base ring $\mathbb{K}$. A DG A-module can be viewed as a central DG A-bimodule, so we have a derived tensor product

$$(- \otimes^{\mathrm{L}}_{A} -) : \mathsf{D}(A) \times \mathsf{D}(A) \to \mathsf{D}(A).$$

This operation is symmetric, i.e. $S \otimes^{\mathrm{L}}_{A} T \cong T \otimes^{\mathrm{L}}_{A} S$.

A DG A-module T is called a *tilting DG A-module* if there is some $S \in \mathsf{D}(A)$ such that $S \otimes^{\mathrm{L}}_{A} T \cong A$. The *commutative derived Picard group* of A is the group $\mathrm{DPic}_A(A)$, whose elements are the isomorphism classes in $\mathsf{D}(A)$ of the tilting DG modules, the multiplication is induced by $(- \otimes^{\mathrm{L}}_{A} -)$, and the unit is the class of A. It is an abelian group.

In [172, Theorem 6.14] it was proved that the group homomorphism $\mathrm{DPic}_A(A) \to \mathrm{DPic}_{\bar{A}}(\bar{A})$, which induced by the derived reduction functor

$\bar{A} \otimes^{\mathrm{L}}_{A} (-)$, is bijective. If the commutative ring $\bar{A}$ has a finite connected component decomposition $A = \prod_{i=1}^{m} A_i$, then there is an isomorphism

$$\mathrm{DPic}_{\bar{A}}(\bar{A}) \cong \mathrm{Pic}_{\bar{A}}(\bar{A}) \times \mathbb{Z}^m.$$

This is according to Corollary 14.5.24 with $\mathbb{K} = \bar{A}$.

As Corollaries 14.5.10 and 14.5.20 show, if the ring A is either local, or commutative with connected spectrum, then the group $\mathrm{DPic}_{\mathbb{K}}(A)$ is not very interesting: it is

$$\mathrm{DPic}_{\mathbb{K}}(A) = \mathrm{Pic}_{\mathbb{K}}(A) \times \langle \sigma \rangle, \tag{14.5.29}$$

where $\mathrm{Pic}_{\mathbb{K}}(A)$ is the classical contribution, and $\langle \sigma \rangle \cong \mathbb{Z}$ is the subgroup generated by the element σ, which is the class of the tilting complex $A[1]$.

The next example shows that matters are very different when A is neither commutative nor local.

Example 14.5.30 Let $\mathbb{K}$ be an algebraically closed field and let n be an integer ≥ 2. Consider the ring A of upper-triangular $n \times n$ matrices. For $n = 2$ it is $A = \left[\begin{smallmatrix} \mathbb{K} & \mathbb{K} \\ 0 & \mathbb{K} \end{smallmatrix}\right]$. The group $\mathrm{DPic}_{\mathbb{K}}(A)$ contains a new element in this case. The classical NC Picard group $\mathrm{Pic}_{\mathbb{K}}(A)$ is trivial here. But the bimodule $A^* := \mathrm{Hom}_{\mathbb{K}}(A, \mathbb{K})$ is tilting, and its class $\mu \in \mathrm{DPic}_{\mathbb{K}}(A)$ satisfies

$$\mu^{n+1} = \sigma^{n-1}. \tag{14.5.31}$$

This was calculated by A. Yekutieli (for $n = 2$) and E. Kreines (for $n \geq 3$) in [165]. When $n = 2$ this says that $\sigma = \mu^3$, so the group $\mathrm{DPic}_{\mathbb{K}}(A)$ is larger than the classical part $\mathrm{Pic}_{\mathbb{K}}(A) \times \langle \sigma \rangle = \langle \sigma \rangle \cong \mathbb{Z}$.

Later, in [109], J.-I. Miyachi and Yekutieli showed that the group $\mathrm{DPic}_{\mathbb{K}}(A)$ is abelian, it is generated by σ and μ, and the only relation is (14.5.31). Note that the ring A is isomorphic to the path ring $\mathbb{K}[Q]$ of the Dynkin quiver Q of type A_n with all arrows going in the same direction. The paper [109] contains calculations of the groups $\mathrm{DPic}_{\mathbb{K}}(A)$ for several other types of path rings of quivers.

M. Kontsevich interprets the relation (14.5.31) as follows: the *fractional Calabi–Yau dimension* of the "noncommutative smooth proper space" $\mathbf{D}^{\mathrm{b}}_{\mathrm{f}}(A)$ is $\frac{n-1}{n+1}$. More on this in Remark 18.6.20 and Example 18.6.24.

Remark 14.5.32 Here are a few historical notes on Theorems 14.5.7 and 14.5.17. The concept of derived Picard group was discovered independently, around 1997, by R. Rouquier and A. Zimmermann [132] (who had used the notation $\mathrm{TrPic}(A)$ for this group) and Yekutieli [165]. The motivations of the two teams of authors were very different: Rouquier and Zimmermann were interested in invariants of finite dimensional algebras, whereas Yekutieli was trying to classify noncommutative dualizing complexes (See Chapter 18).

Remark 14.5.33 Suppose $\mathbb{K}$ is an algebraically closed field and A is a finite $\mathbb{K}$-ring. It turns out that in this case the group $\mathrm{DPic}_{\mathbb{K}}(A)$ is a *locally algebraic group*. Namely there is a connected algebraic group $\mathrm{DPic}^0_{\mathbb{K}}(A)$, and $\mathrm{DPic}_{\mathbb{K}}(A)$ is a (usually infinite) disjoint union of connected algebraic varieties, that are left and right cosets of $\mathrm{DPic}^0_{\mathbb{K}}(A)$. This same sort of geometric structure can be found in the *Picard scheme* of a smooth projective algebraic variety X, in which the identity component is the *Jacobian variety* of X. But whereas the Picard scheme is an abelian group scheme, the group $\mathrm{DPic}_{\mathbb{K}}(A)$ is not abelian; and this means that there is a nontrivial geometric action of $\mathrm{DPic}^0_{\mathbb{K}}(A)$, by conjugation, on the connected components.

The result above is in the paper [167], and it is based on the papers [69] by B. Huisgen-Zimmermann and M. Saorin, and [131] by Rouquier.

Later B. Keller [81] used an abstraction of this idea – essentially viewing the derived Picard group as a derived group stack (evaluating it on commutative artinian DG rings) – to prove that the *Hochschild cohomology* of A is the *Lie algebra*, in the derived sense, of $\mathrm{DPic}_{\mathbb{K}}(A)$. This allowed him to prove that the Hochschild cohomology of A, with its *Gerstenhaber Lie bracket*, is a derived invariant of A.

15

Algebraically Graded Noncommutative Rings

In this chapter we study *algebraically graded rings* and the categories of *algebraically graded modules* over them. These are the graded rings that appear in standard texts on commutative and noncommutative algebra and are distinct from the *cohomologically graded rings* that underlie *DG rings*.

In the subsequent chapters of the book (Chapters 16 and 17) we shall concentrate on *connected graded rings* (traditionally known as *connected graded algebras*), which behave like "complete local rings" within the algebraically graded context.

The interest in algebraically graded rings, and especially in connected graded rings, stems from the prominent role they played in *noncommutative algebraic geometry*, as it was developed by M. Artin and his collaborators since around 1990. See the papers [8], [10] and [145].

15.1 Categories of Algebraically Graded Modules

Let $\mathbb{K}$ be a nonzero commutative base ring. An *algebraically graded* $\mathbb{K}$-*module* is a $\mathbb{K}$-module M with a direct sum decomposition $M = \bigoplus_{i \in \mathbb{Z}} M_i$ into $\mathbb{K}$-submodules. The submodule M_i is called the homogeneous component of M of *algebraic degree* i. If $m \in M_i$ is a nonzero element, then we write $\deg(m) = i$.

In Section 3.1 we discussed *cohomologically graded rings and modules*. There are three features that distinguish between cohomologically graded $\mathbb{K}$-modules and algebraically graded $\mathbb{K}$-modules. The first distinguishing feature is the use of upper versus lower indices. The second is that in an algebraically graded module there is never a differential involved; i.e. it is not the underlying graded module of a DG module. The third and most important change is in *signs of permutations*, and thus in *commutativity*. In the cohomologically graded setting the Koszul sign rule dictates that for graded $\mathbb{K}$-modules M and N, the braiding isomorphism $\mathrm{br}_{M,N} : M \otimes N \to N \otimes M$, where $\otimes := \otimes_{\mathbb{K}}$, is $\mathrm{br}_{M,N}(m \otimes n) = (-1)^{i \cdot j} \cdot n \otimes m$ for homogeneous elements $m \in M^i$ and $n \in N^j$. But in the algebraically graded setting there are no signs: $\mathrm{br}_{M,N}(m \otimes n) := n \otimes m$ for $m \in M_i$ and $n \in N_j$. This is reflected in Definition 15.1.13 below.

Let $M = \bigoplus_{i \in \mathbb{Z}} M_i$ be an algebraically graded $\mathbb{K}$-module. A nonzero element $m \in M$ can be expressed uniquely as a sum

$$m = m_1 + \cdots + m_r, \tag{15.1.1}$$

such that $r \geq 1$; each m_i is a nonzero homogeneous element; and $\deg(m_i) < \deg(m_{i+1})$. This is called the *homogeneous component decomposition* of m. Of course, m is homogeneous if and only if $r = 1$.

Suppose M and N are algebraically graded $\mathbb{K}$-modules. A $\mathbb{K}$-linear homomorphism $\phi : M \to N$ is said to be of *algebraic degree* i if $\phi(M_j) \subseteq N_{j+i}$ for all $j \in \mathbb{Z}$. The $\mathbb{K}$-module of all $\mathbb{K}$-linear homomorphisms of degree i is denoted by $\mathrm{Hom}_{\mathbb{K}}(M, N)_i$. Taking the direct sum we get an algebraically graded $\mathbb{K}$-module

$$\mathrm{Hom}_{\mathbb{K}}(M, N) := \bigoplus_{i \in \mathbb{Z}} \mathrm{Hom}_{\mathbb{K}}(M, N)_i. \tag{15.1.2}$$

Definition 15.1.3 Let M and N be algebraically graded $\mathbb{K}$-modules. A *strict homomorphism of algebraically graded* $\mathbb{K}$*-modules* $\phi : M \to N$ is $\mathbb{K}$-linear homomorphism of algebraic degree 0.

The *category of algebraically graded* $\mathbb{K}$*-modules* is the category $\mathbf{M}(\mathbb{K}, \mathrm{gr})$, whose objects are the algebraically graded $\mathbb{K}$-modules, and whose morphisms are the strict homomorphisms.

Thus

$$\mathrm{Hom}_{\mathbf{M}(\mathbb{K},\mathrm{gr})}(M, N) = \mathrm{Hom}_{\mathbb{K}}(M, N)_0.$$

It is easy to see that $\mathbf{M}(\mathbb{K}, \mathrm{gr})$ is a $\mathbb{K}$-linear abelian category. The kernels and cokernels are degreewise.

Suppose $M, N \in \mathbf{M}(\mathbb{K}, \mathrm{gr})$. Their tensor product is also algebraically graded:

$$(M \otimes N)_i = \bigoplus_{j \in \mathbb{Z}} (M_j \otimes N_{i-j})$$

and

$$M \otimes N = \bigoplus_{i \in \mathbb{Z}} (M \otimes N)_i.$$

Definition 15.1.4 An *algebraically graded central* $\mathbb{K}$*-ring* is a central $\mathbb{K}$-ring A, equipped with a direct sum decomposition $A = \bigoplus_{i \in \mathbb{Z}} A_i$ into $\mathbb{K}$-submodules, such that $1_A \in A_0$, and $A_i \cdot A_j \subseteq A_{i+j}$ for all i, j.

A *homomorphism of algebraically graded central* $\mathbb{K}$*-rings* $f : A \to B$ is a $\mathbb{K}$-ring homomorphism such that $f(A_i) \subseteq B_i$ for all i.

The algebraically graded central $\mathbb{K}$-rings form a category, which we denote by $\mathsf{Rng}_{\mathrm{gr}}/_{\mathrm{c}}\,\mathbb{K}$.

The base ring $\mathbb{K}$ is an algebraically graded ring, concentrated in degree 0. It is the initial object of the category $\mathsf{Rng}_{\mathrm{gr}}/_{\mathrm{c}}\,\mathbb{K}$.

In the traditional ring theory literature (e.g. [106], [130], [8], [10], [161] and [182]) the algebraically graded central $\mathbb{K}$-rings were called "associative unital graded $\mathbb{K}$-algebras."

Example 15.1.5 Let M be an algebraically graded $\mathbb{K}$-module. Then $\operatorname{End}_{\mathbb{K}}(M) := \operatorname{Hom}_{\mathbb{K}}(M, M)$ is an algebraically graded central $\mathbb{K}$-ring. The grading is according to (15.1.2), and multiplication is composition.

Definition 15.1.6 Let A be an algebraically graded central $\mathbb{K}$-ring.

(1) An *algebraically graded left* A*-module* is a left A-module M, equipped with a direct sum decomposition $M = \bigoplus_{i \in \mathbb{Z}} M_i$ into $\mathbb{K}$-submodules, such that $A_i \cdot M_j \subseteq M_{i+j}$ for all i, j.

(2) Suppose M and N are algebraically graded left A-modules. A $\mathbb{K}$-linear homomorphism $\phi : M \to N$ of degree i is said to be a *homomorphism of algebraically graded* A*-modules* if $\phi(a \cdot m) = a \cdot \phi(m)$ for all $a \in A$ and $m \in M$.

(3) The $\mathbb{K}$-module of all A-linear homomorphisms of degree i is denoted by $\operatorname{Hom}_A(M, N)_i$. Taking the direct sum we get an algebraically graded $\mathbb{K}$-module

$$\operatorname{Hom}_A(M, N) := \bigoplus_{i \in \mathbb{Z}} \operatorname{Hom}_A(M, N)_i.$$

Note that there is an inclusion $\operatorname{Hom}_A(M, N) \subseteq \operatorname{Hom}_{\mathbb{K}}(M, N)$ in $\mathbf{M}(\mathbb{K}, \mathrm{gr})$.

Definition 15.1.7 Let A be an algebraically graded central $\mathbb{K}$-ring.

(1) Suppose M and N are algebraically graded left A-modules. The elements of $\operatorname{Hom}_A(M, N)_0$ are called *strict homomorphisms of algebraically graded* A*-modules*.

(2) The category of algebraically graded left A-modules, with strict homomorphisms, is denoted by $\mathbf{M}(A, \mathrm{gr})$.

Thus

$$\operatorname{Hom}_{\mathbf{M}(A,\mathrm{gr})}(M, N) = \operatorname{Hom}_A(M, N)_0. \tag{15.1.8}$$

Clearly, to put an algebraically graded left A-module structure on an algebraically graded $\mathbb{K}$-module M is the same as to give a homomorphism $A \to \operatorname{End}_{\mathbb{K}}(M)$ in $\mathsf{Rng}_{\mathrm{gr}}/_{\mathrm{c}}\,\mathbb{K}$.

Algebraically graded right A-modules, and algebraically graded A-B-bimodules, are defined similarly to Definition 15.1.6. There are no signs anywhere. All homomorphisms are $\mathbb{K}$-linear, and all bimodules are central over $\mathbb{K}$.

Definition 15.1.9 Let A be an algebraically graded ring. Given an algebraically graded A-module M and an integer i, the *ith degree twist of M* is the algebraically graded A-module $M(i)$ whose degree j homogeneous component is $M(i)_j := M_{i+j}$. The action of A is not changed.

The degree twist is an automorphism of the abelian category $\mathbf{M}(A, \mathrm{gr})$. And

$$\operatorname{Hom}_{\mathbf{M}(A,\mathrm{gr})}(M, N(i)) = \operatorname{Hom}_A(M, N)_i.$$

There is a similar twisting for graded right modules and for graded bimodules.

There is a functor $\mathrm{Ungr} : \mathsf{Rng}_{\mathrm{gr}}/_{\mathrm{c}}\,\mathbb{K} \to \mathsf{Rng}/_{\mathrm{c}}\,\mathbb{K}$ that forgets the grading. For a fixed ring $A \in \mathsf{Rng}_{\mathrm{gr}}/_{\mathrm{c}}\,\mathbb{K}$ there is a forgetful functor

$$\mathrm{Ungr} : \mathbf{M}(A, \mathrm{gr}) \to \mathbf{M}(\mathrm{Ungr}(A)). \tag{15.1.10}$$

This is an exact faithful $\mathbb{K}$-linear functor.

Remark 15.1.11 The functor Ungr in (15.1.10) is usually not full, for two reasons. First, there could be nonzero homomorphisms of nonzero degree, so that

$$\operatorname{Hom}_{\mathbf{M}(A,\mathrm{gr})}(M, N) = \operatorname{Hom}_A(M, N)_0 \subsetneq \operatorname{Hom}_A(M, N).$$

Second, it is easy find an example where

$$\mathrm{Ungr}(\operatorname{Hom}_A(M, N)) \subsetneq \operatorname{Hom}_{\mathrm{Ungr}(A)}(\mathrm{Ungr}(M), \mathrm{Ungr}(N)).$$

Just take $M := \bigoplus_{i\in\mathbb{Z}} A(i)$ and $N := A$. Then

$$\mathrm{Ungr}(\operatorname{Hom}_A(M, N)) \cong \bigoplus_{i\in\mathbb{Z}} \mathrm{Ungr}(A)$$

and

$$\operatorname{Hom}_{\mathrm{Ungr}(A)}(\mathrm{Ungr}(M), \mathrm{Ungr}(N)) \cong \prod_{i\in\mathbb{Z}} \mathrm{Ungr}(A)$$

in $\mathbf{M}(\mathbb{K})$.

Remark 15.1.12 Homomorphisms between algebraically graded modules and cohomologically graded modules are distinct, not only in the position of the degree indices. Compare Definition 15.1.6(2) to Definition 3.1.31. This distinction is due to the Koszul sign rule, which is present only in the cohomologically graded setting.

Definition 15.1.13 Let A and B be algebraically graded central $\mathbb{K}$-rings.

(1) The *opposite ring* of A is the central $\mathbb{K}$-ring A^{op}, which is the same algebraically graded $\mathbb{K}$-module as A, with an isomorphism $\mathrm{op} := \mathrm{id} : A \xrightarrow{\simeq} A^{\mathrm{op}}$ in $\mathbf{M}(\mathbb{K}, \mathrm{gr})$. The multiplication $\cdot^{\mathrm{op}}$ of A^{op} is

$$\mathrm{op}(a_1) \cdot^{\mathrm{op}} \mathrm{op}(a_2) := \mathrm{op}(a_2 \cdot a_1).$$

(2) The ring A is called a *commutative graded ring* if $a_2 \cdot a_1 = a_1 \cdot a_2$ for all $a_1, a_2 \in A$.

(3) The tensor product $A \otimes B$ is made into an algebraically graded ring with multiplication

$$(a_1 \otimes b_1) \cdot (a_2 \otimes b_2) := (a_1 \cdot a_2) \otimes (b_1 \otimes b_2).$$

(4) The *enveloping* ring of A is the algebraically graded ring $A^{\mathrm{en}} := A \otimes A^{\mathrm{op}}$.

In terms of opposite rings, an algebraically graded ring A is commutative if and only if $A = A^{\mathrm{op}}$. The category of algebraically graded right A-modules is identified with $\mathbf{M}(A^{\mathrm{op}}, \mathrm{gr})$; and the category of $\mathbb{K}$-central algebraically graded A-B-bimodules is identified with $\mathbf{M}(A \otimes B^{\mathrm{op}}, \mathrm{gr})$.

There are graded ring homomorphisms $A \to A \otimes B \leftarrow B$, and corresponding restriction functors

$$\mathbf{M}(A, \mathrm{gr}) \xleftarrow{\mathrm{Rest}_A} \mathbf{M}(A \otimes B, \mathrm{gr}) \xrightarrow{\mathrm{Rest}_B} \mathbf{M}(B, \mathrm{gr}). \tag{15.1.14}$$

We shall usually omit any explicit reference to these restriction functors.

Given $M \in \mathbf{M}(A, \mathrm{gr})$ and $N \in \mathbf{M}(B, \mathrm{gr})$, the tensor product $M \otimes N$ belongs to $\mathbf{M}(A \otimes B, \mathrm{gr})$.

Suppose $M \in \mathbf{M}(A^{\mathrm{op}}, \mathrm{gr})$ and $N \in \mathbf{M}(A, \mathrm{gr})$. Their tensor product over A is also algebraically graded:

$$M \otimes_A N = \bigoplus_{i \in \mathbb{Z}} (M \otimes_A N)_i \in \mathbf{M}(\mathbb{K}, \mathrm{gr}), \tag{15.1.15}$$

where $(M \otimes_A N)_i$ is the image of $(M \otimes N)_i$ under the canonical surjection $M \otimes N \twoheadrightarrow M \otimes_A N$ in $\mathbf{M}(\mathbb{K}, \mathrm{gr})$. See Lemma 3.1.30 for the corresponding statement in the cohomologically graded setting.

There is an isomorphism of algebraically graded $\mathbb{K}$-rings

$$(A^{\mathrm{en}})^{\mathrm{op}} \xrightarrow{\simeq} A^{\mathrm{en}}, \quad a_1 \otimes a_2 \mapsto a_2 \otimes a_1. \tag{15.1.16}$$

From here on in this chapter, and in Chapters 16 and 17, we assume the following convention.

Convention 15.1.17 We fix a nonzero commutative base ring $\mathbb{K}$. The symbol $\otimes$ means $\otimes_{\mathbb{K}}$. All graded rings are assumed to be algebraically graded central $\mathbb{K}$-rings, and all graded ring homomorphisms are over $\mathbb{K}$. In other words, we work inside the category $\mathsf{Rng}_{\mathrm{gr}}/_{\mathrm{c}}\, \mathbb{K}$.

Let A be a graded ring. Its enveloping ring is the graded ring $A^{\mathrm{en}} := A \otimes A^{\mathrm{op}}$. All graded A-modules are assumed by default to be algebraically graded left A-modules. For $M, N \in \mathbf{M}(A, \mathrm{gr})$ the expression $\mathrm{Hom}_A(M, N)$ refers to the graded $\mathbb{K}$-module in Definition 15.1.6(3). All graded bimodules are assumed to be $\mathbb{K}$-central; i.e. a graded A-B-bimodule is an object of $\mathbf{M}(A \otimes B^{\mathrm{op}}, \mathrm{gr})$.

Proposition 15.1.18 *Let A be a graded ring.*

(1) *The category* $\mathbf{M}(A, \mathrm{gr})$ *is abelian.*
(2) *The category* $\mathbf{M}(A, \mathrm{gr})$ *has inverse limits.*
(3) *The category* $\mathbf{M}(A, \mathrm{gr})$ *has direct limits. Moreover, direct limits in* $\mathbf{M}(A, \mathrm{gr})$ *commute with the ungrading functor* (15.1.10).

Proof (1) The kernels and cokernels in $\mathbf{M}(A, \mathrm{gr})$ are degreewise, so the abelian property of $\mathbf{M}(A, \mathrm{gr})$ is inherited from the abelian property of $\mathbf{M}(\mathbb{K})$.

(2) Inverse limits in $\mathbf{M}(A, \mathrm{gr})$ are taken degreewise. Namely, given an inverse system $\{M_x\}_{x \in X^{\mathrm{op}}}$ of objects of $\mathbf{M}(A, \mathrm{gr})$, indexed by a directed set X, let $M_i := \lim_{\leftarrow x} (M_x)_i \in \mathbf{M}(\mathbb{K})$. Then $M := \bigoplus_{i \in \mathbb{Z}} M_i \in \mathbf{M}(A, \mathrm{gr})$ is the inverse limit of $\{M_x\}_{x \in X^{\mathrm{op}}}$.

(3) Direct limits in $\mathbf{M}(A, \mathrm{gr})$ are taken degreewise. Namely, given a direct system $\{N_y\}_{y \in Y}$ of objects of $\mathbf{M}(A, \mathrm{gr})$, indexed by a directed set Y, let $N_i := \lim_{y \to} (N_y)_i \in \mathbf{M}(\mathbb{K})$. Then $N := \bigoplus_{i \in \mathbb{Z}} N_i \in \mathbf{M}(A, \mathrm{gr})$ is the direct limit of $\{N_y\}_{y \in Y}$.

Because direct sums and direct limits commute with each other, it follows that $\lim_{y \to}$ commutes with Ungr. □

Definition 15.1.19 A graded ring A is called *nonnegative* if $A_i = 0$ for all $i < 0$; i.e. if $A = \bigoplus_{i \geq 0} A_i$.

If A is a nonnegative graded ring A, then the two-sided graded ideal $\mathfrak{m} := \bigoplus_{i \geq 1} A_i$ is called the *augmentation ideal.* It is the kernel of the surjective graded ring homomorphism $A \to A_0$, which we call the *augmentation homomorphism.*

The augmentation ideal of the graded ring A^{op} is denoted by $\mathfrak{m}^{\mathrm{op}}$.

Definition 15.1.20 Let A be a graded ring A and let $M = \bigoplus_{i \in \mathbb{Z}} M_i$ be a graded A-module.

(1) We call M a *bounded above* graded module if $M_i = 0$ for $i \gg 0$.
(2) We call M a *bounded below* graded module if $M_i = 0$ for $i \ll 0$.
(3) We call M a *bounded* graded module if it is both bounded above and bounded below.

Example 15.1.21 If A is a nonnegative graded ring, then as a graded module A is bounded below. For such A, the bounded below graded modules behave like finitely generated modules over a noetherian local commutative ring – see Proposition 15.1.22 below.

The next proposition is the first version of the graded Nakayama Lemma that we present; the second is Proposition 15.2.30.

Proposition 15.1.22 *Let A be a nonnegative graded ring, with augmentation ideal $\mathfrak{m}$. Let M be a bounded below graded A-module, and let $M' \subseteq M$ be a graded A-submodule. If $M' + \mathfrak{m} \cdot M = M$ then $M' = M$.*

Proof Say $M = \bigoplus_{i \geq i_0} M_i$ for an integer i_0. We will prove that $M'_i = M_i$ as A_0-modules for all i, by induction on $i \geq i_0 - 1$. For $i = i_0 - 1$ this is clear because both these A_0-modules are zero.

Now take $i \geq i_0$, and assume that $M'_j = M_j$ for all $j < i$. Suppose $m \in M_i$ is a nonzero element. Because $M = M' + \mathfrak{m} \cdot M$, we can express m as follows: $m = m' + \sum_{k=1}^{r} a_k \cdot m_k$, where $m' \in M'_i$; $r \geq 0$; $a_k \in \mathfrak{m}$ and $m_k \in M$ are nonzero homogeneous elements; and $\deg(a_k) + \deg(m_k) = i$. But then $\deg(m_k) < i$, so by the induction hypothesis we have $m_k \in M'$. We conclude that $m \in M'_i$. □

Definition 15.1.23 Let A be a graded ring.

(1) A graded A-module M is called a *graded-finite A-module* if M can be generated by finitely many homogeneous elements.
(2) The full subcategory of $\mathbf{M}(A, \mathrm{gr})$ on the graded-finite A-modules is denoted by $\mathbf{M}_{\mathrm{f}}(A, \mathrm{gr})$.
(3) A is called a *left graded-noetherian ring* if every graded left ideal $\mathfrak{a} \subseteq A$ is a graded-finite A-module, in the sense of item (1) above.
(4) A is called a *graded-noetherian ring* if both A and A^{op} are left graded-noetherian rings.

An obvious modification of item (1) above gives the definition of a graded-finite right A-module. Of course, the ring A is called a right graded-noetherian ring if A^{op} is a left graded-noetherian ring.

Proposition 15.1.24 *If A is left graded-noetherian ring, then $\mathbf{M}_{\mathrm{f}}(A, \mathrm{gr})$ is a thick abelian subcategory of $\mathbf{M}(A, \mathrm{gr})$.*

Exercise 15.1.25 Prove this proposition.

Proposition 15.1.26 *Let A be a graded ring and M a graded A-module. The following two conditions are equivalent.*

(i) *M is a graded-finite A-module, in the sense of Definition* 15.1.23(1).

(ii) $\mathrm{Ungr}(M)$ *is a finitely generated* $\mathrm{Ungr}(A)$*-module.*

Exercise 15.1.27 Prove this proposition.

Theorem 15.1.28 *Let A be a graded ring. The following two conditions are equivalent.*

(i) *The graded ring A is left graded-noetherian, in the sense of Definition* 15.1.23(3).
(ii) *The ring* $\mathrm{Ungr}(A)$ *is left noetherian.*

Proof The case when A is nonnegative is proved in many textbooks, and it is a nice exercise for the reader.

The general case is much harder, and the few proofs in the literature are quite complicated. Here is a short proof, which is due to S. Snigerov.

The implication (ii) $\Rightarrow$ (i) is trivial. So let us assume that A is left graded-noetherian, and we shall prove that $\mathrm{Ungr}(A)$ is left noetherian. The proof proceeds in a few steps.

Step 1. Consider an A_0-submodule $V \subseteq A_j$ for some j, and let $\mathfrak{b} := A \cdot V \subseteq A$, which is a graded left ideal of A. By assumption, $\mathfrak{b}$ is generated by finitely many homogeneous elements. An easy calculation shows that $V = \mathfrak{b} \cap A_j$, and that V is finitely generated as an A_0-module. The consequence is that the ring A_0 is left noetherian, and that every A_j is a finitely generated A_0-module.

Step 2. Now we fix a nonzero left ideal $L \subseteq A$. We are going to find a finite number of generators for L as an A-module.

For a nonzero element $a \in L$ we consider its homogeneous component decomposition (15.1.1), and we define its bottom degree component $\mathrm{bot}(a) := a_1$, and its top degree component $\mathrm{top}(a) := a_r$. Let $\mathfrak{a}_{\mathrm{top}}$ be the left ideal in A generated by the elements $\mathrm{top}(a)$, where a runs over all the nonzero elements of L. By assumption the graded ideal $\mathfrak{a}_{\mathrm{top}}$ is graded-finite; so we can choose a finite collection $\{a_x\}_{x \in X_{\mathrm{top}}}$ of nonzero elements of L, such that their top degree components $\mathrm{top}(a_x)$ generate the left ideal $\mathfrak{a}_{\mathrm{top}}$. Now let

$$d_1 := \max \{\deg(\mathrm{top}(a_x)) \mid x \in X_{\mathrm{top}}\} \in \mathbb{Z}.$$

Step 3. Define $\mathfrak{a}_{\mathrm{bot}}$ to be the left ideal in A generated by the elements $\mathrm{bot}(a)$, where a runs over all the nonzero elements of $L \cap A_{\leq d_1}$. (It is possible that there are no such elements.) By assumption the graded left ideal $\mathfrak{a}_{\mathrm{bot}}$ is graded-finite; so we can choose a finite collection $\{a_x\}_{x \in X_{\mathrm{bot}}}$ of nonzero elements of $L \cap A_{\leq d_1}$, such that their bottom degree components $\mathrm{bot}(a_x)$ generate the left ideal $\mathfrak{a}_{\mathrm{bot}}$. Now we let

$$d_0 := \min (\{\deg(\mathrm{bot}(a_x)) \mid x \in X_{\mathrm{bot}}\} \cup \{0\}) \in \mathbb{Z}.$$

Step 4. By step 1 the A_0-module $V := L \cap (\bigoplus_{i=d_0}^{d_1} A_i)$ is finitely generated. We choose a finite collection $\{a_x\}_{x \in X_{\mathrm{mid}}}$ that generates V as an A_0-module.

Step 5. Define the finite indexing set $X := X_{\mathrm{top}} \sqcup X_{\mathrm{bot}} \sqcup X_{\mathrm{mid}}$. We claim that the collection $\{a_x\}_{x \in X}$ generates the left ideal L. Take any element $a \in L$. By subtracting from a a finite linear combination of elements from $\{a_x\}_{x \in X_{\mathrm{top}}}$, with coefficients taken from $A_{\geq 1}$, we can assume that $a \in L \cap A_{\leq d_1}$. Next, by subtracting from a a finite linear combination of elements from $\{a_x\}_{x \in X_{\mathrm{bot}}}$, with coefficients taken from $A_{\leq -1}$, we can assume that $a \in V$, the A_0-submodule from step 4. So a is finite linear combination of elements from $\{a_x\}_{x \in X_{\mathrm{mid}}}$, with coefficients taken from A_0. □

In view of Proposition 15.1.26 and Theorem 15.1.28, we can use this more relaxed terminology without danger of ambiguity:

Definition 15.1.29 Let A be a graded ring.

(1) We call A a *left noetherian graded ring* if it is left graded-noetherian, and likewise for the attributes "right noetherian graded ring" and "noetherian graded ring."

(2) Let M be a graded A-module. We call M a *finite graded A-module* if it is graded-finite, and likewise for right modules.

We do not assume the rings here are noetherian. Partly this is because of the facts in the remark below. Instead we resort to *derived graded-pseudo-finite* modules and complexes, see Definition 15.3.22.

Remark 15.1.30 Unlike the commutative theory, the Hilbert Basis Theorem does not hold in the noncommutative setting. Thus, a finitely generated $\mathbb{K}$-ring A need not be noetherian; see Example 15.2.19. Likewise, if A and B are noetherian finitely generated $\mathbb{K}$-rings, the tensor product $A \otimes B$ need not be noetherian; there is a counterexample in [128]. See [7] for a deep discussion of permanence properties of NC noetherian rings.

We now pass to complexes of algebraically graded modules, retaining Convention 15.1.17. In Section 3.6 we introduced the category $\mathbf{C}(\mathsf{M})$ of complexes with entries in an abelian category M. This is a DG category. Its strict subcategory $\mathbf{C}_{\mathrm{str}}(\mathsf{M})$ is abelian. Here we take the abelian category $\mathsf{M} := \mathbf{M}(A, \mathrm{gr})$ for a graded ring A.

Definition 15.1.31 Let A be a graded ring.

(1) The category of complexes with entries in the abelian category $\mathbf{M}(A, \mathrm{gr})$ is the DG category $\mathbf{C}(A, \mathrm{gr}) := \mathbf{C}(\mathbf{M}(A, \mathrm{gr}))$.

(2) The strict subcategory of $\mathbf{C}(A, \mathrm{gr})$ is the abelian category $\mathbf{C}_{\mathrm{str}}(A, \mathrm{gr}) := \mathbf{C}_{\mathrm{str}}(\mathbf{M}(A, \mathrm{gr}))$.
(3) Given a boundedness indicator $\star$, we denote by $\mathbf{C}^{\star}(A, \mathrm{gr})$ the full subcategory of $\mathbf{C}(A, \mathrm{gr})$ on the complexes M having boundedness condition $\star$.

We now try to make Definition 15.1.31 more explicit. An object $M \in \mathbf{C}(A, \mathrm{gr})$ has these direct sum decompositions:

$$M = \bigoplus_{i \in \mathbb{Z}} M^i = \bigoplus_{i,j \in \mathbb{Z}} M^i_j. \tag{15.1.32}$$

Here each $M^i \in \mathbf{M}(A, \mathrm{gr})$, and each $M^i_j \in \mathbf{M}(\mathbb{K})$. We call M^i_j the homogeneous component of M of bidegree $\left[\begin{smallmatrix} i \\ j \end{smallmatrix}\right] \in \left[\begin{smallmatrix} \mathbb{Z} \\ \mathbb{Z} \end{smallmatrix}\right] = \mathbb{Z}^2$. The upper index i is called the *cohomological degree*, and the lower index j is called the *algebraic degree*. The differential d_M of the complex M is an A-linear homomorphism $\mathrm{d}_M : M \to M$ of bidegree $\left[\begin{smallmatrix} 1 \\ 0 \end{smallmatrix}\right]$ that satisfies $\mathrm{d}_M \circ \mathrm{d}_M = 0$.

A morphism $\phi : M \to N$ in the strict category $\mathbf{C}_{\mathrm{str}}(A, \mathrm{gr})$ has bidegree $\left[\begin{smallmatrix} 0 \\ 0 \end{smallmatrix}\right]$ and satisfies $\phi \circ \mathrm{d}_M = \mathrm{d}_N \circ \phi$.

Lastly, to clarify item (3) in the definition, a complex M belongs to $\mathbf{C}^{+}(A, \mathrm{gr})$ if $M^i = 0$ for $i \ll 0$, etc. Indicators can be combined, so $\mathbf{C}^{\star}_{\mathrm{str}}(A, \mathrm{gr})$ is a thick abelian subcategory of $\mathbf{C}_{\mathrm{str}}(A, \mathrm{gr})$.

Like for any abelian category, we have a homotopy category and a derived category. Here is the specific notation:

Definition 15.1.33 Let A be a graded ring.
(1) The homotopy category of complexes with entries in the abelian category $\mathbf{M}(A, \mathrm{gr})$ is the triangulated category $\mathbf{K}(A, \mathrm{gr}) := \mathbf{K}(\mathbf{M}(A, \mathrm{gr}))$.
(2) The derived category of the abelian category $\mathbf{M}(A, \mathrm{gr})$ is the triangulated category $\mathbf{D}(A, \mathrm{gr}) := \mathbf{D}(\mathbf{M}(A, \mathrm{gr}))$.
(3) Given a boundedness indicator $\star$, we denote by $\mathbf{D}^{\star}(A, \mathrm{gr})$ the full triangulated subcategory of $\mathbf{D}(A, \mathrm{gr})$ on the complexes M whose cohomology $\mathrm{H}(M)$ has boundedness condition $\star$.

As always for derived categories, there are functors

$$\mathbf{C}_{\mathrm{str}}(A, \mathrm{gr}) \xrightarrow{\mathrm{P}} \mathbf{K}(A, \mathrm{gr}) \xrightarrow{\mathrm{Q}} \mathbf{D}(A, \mathrm{gr}), \qquad \tilde{\mathrm{Q}} : \mathbf{C}_{\mathrm{str}}(A, \mathrm{gr}) \to \mathbf{D}(A, \mathrm{gr}) \tag{15.1.34}$$

By abuse of notation we sometimes write Q rather than $\tilde{\mathrm{Q}}$; or, when there is no danger of confusion, we suppress the categorical localization functors Q and $\tilde{\mathrm{Q}}$ altogether.

There is yet another category associated to A : it is the category $\mathbf{G}(A, \mathrm{gr}) := \mathbf{G}(\mathbf{M}(A, \mathrm{gr}))$ of cohomologically graded objects in $\mathbf{M}(A, \mathrm{gr})$. It has its strict subcategory $\mathbf{G}_{\mathrm{str}}(A, \mathrm{gr})$. This is the target of the cohomology functor

$$\mathrm{H} : \mathbf{C}_{\mathrm{str}}(A, \mathrm{gr}) \to \mathbf{G}_{\mathrm{str}}(A, \mathrm{gr}).$$

If $A \to B$ is a graded ring homomorphism that makes B into a finite left A-module, and if A is left noetherian, then of course B is also left noetherian. Here is a partial converse:

Theorem 15.1.35 *Let A be a nonnegative graded ring, and let $a \in A$ be a nonzero homogeneous central element of positive degree. Define the nonnegative graded ring $B := A/(a)$. If B is left noetherian, then A is also left noetherian.*

This result is very similar to [8, Lemma 8.2]. There are three changes: we do not assume that the element a is regular (i.e. not a zero-divisor), and we do not assume that A is connected graded (only nonnegative). On the other hand, in [8] the element a is only required to be normalizing, not central. Our proof is essentially copied from [8].

Proof Assume, for the sake of contradiction, that the graded ring A is not left noetherian. Then (taking Theorem 15.1.28 into account) there is a graded left ideal $L \subseteq A$ that is not finitely generated. By Zorn's Lemma we can assume that L is maximal with this property. Thus every graded left ideal $L^{\dagger}$ such that $L \subsetneq L^{\dagger} \subseteq A$ must be finitely generated.

Consider the graded A-module $M := A/L$. Every graded A-module $M^{\dagger} \subseteq M$ is of the form $M^{\dagger} = L^{\dagger}/L$, and therefore $M^{\dagger}$ is a finitely generated A-module. In other words, M is a noetherian object of $\mathbf{M}(A, \mathrm{gr})$.

Let $i := \deg(a)$. We examine the commutative diagram (15.1.36) in $\mathbf{M}(A, \mathrm{gr})$. Viewing the columns of this diagram as complexes, whose nonzero terms are in cohomological degrees 0 and 1, this is a short exact sequence in the abelian category $\mathbf{C}_{\mathrm{str}}(A, \mathrm{gr})$.

$$\begin{array}{ccccccccc} 0 & \longrightarrow & L & \longrightarrow & A & \longrightarrow & M & \longrightarrow & 0 \\ & & \uparrow {\scriptstyle a\cdot(-)} & & \uparrow {\scriptstyle a\cdot(-)} & & \uparrow {\scriptstyle a\cdot(-)} & & \\ 0 & \longrightarrow & L(-i) & \longrightarrow & A(-i) & \longrightarrow & M(-i) & \longrightarrow & 0 \end{array} \tag{15.1.36}$$

Define the graded A-modules $\bar{M}$ and $\bar{L}$ by these exact sequences:

$$0 \to \bar{M} \to M(-i) \xrightarrow{a\cdot(-)} M$$

and

$$L(-i) \xrightarrow{a\cdot(-)} L \to \bar{L} \to 0. \tag{15.1.37}$$

The long exact cohomology sequence of (15.1.36) gives rise to this exact sequence in $\mathbf{M}(A, \mathrm{gr})$:

$$\bar{M} \xrightarrow{\phi} \bar{L} \xrightarrow{\psi} B. \tag{15.1.38}$$

Because M is a noetherian object of $\mathbf{M}(A, \mathrm{gr})$, so is its degree twist $M(-i)$, and also the subobject $\bar{M} \subseteq M(-i)$. On the other hand, we are given that the ring B is noetherian; so it is a noetherian object of $\mathbf{M}(A, \mathrm{gr})$. The exact sequence (15.1.38) shows that $\bar{L}$ is also a noetherian object of $\mathbf{M}(A, \mathrm{gr})$. We see that $\bar{L}$ is a finitely generated A-module.

Finally, let us choose finitely many homogeneous generators of the graded A-module $\bar{L}$, and then lift them to homogeneous elements of L. Let $L' \subseteq L$ be the A-submodule generated by this finite collection of elements. The exact sequence (15.1.37) shows that $L = (a \cdot L) + L'$. The graded A-module L is bounded below, because $L \subseteq A$ and A is nonnegative. The element a belongs to the augmentation ideal $\mathfrak{m}$ of A. Therefore, by Proposition 15.1.22 (the graded Nakayama Lemma), we have $L' = L$. This contradicts the assumption that L is not finitely generated. □

15.2 Properties of Algebraically Graded Modules

In this section we continue with Convention 15.1.17, to which we add:

Convention 15.2.1 The base ring $\mathbb{K}$ is a field.

See Remark 15.3.35 regarding the requirement that $\mathbb{K}$ be a field. In a sense the category $\mathbf{M}(\mathbb{K}, \mathrm{gr})$ is boring under Convention 15.2.1 :

Proposition 15.2.2 *For an object $M \in \mathbf{M}(\mathbb{K}, \mathrm{gr})$ there is an isomorphism $M \simeq \bigoplus_{x \in X} \mathbb{K}(-i_x)$, where X is an indexing set and $\{i_x\}_{x \in X}$ is a collection of integers.*

Proof This is a variation on the usual proof for $\mathbb{K}$-modules: by Zorn's Lemma there is a maximal linearly independent collection $\{m_x\}_{x \in X}$ of homogeneous elements of M. This is a basis of M. The number i_x is the degree of m_x. □

Definition 15.2.3 Let A be a graded ring. We define the graded bimodule

$$A^* := \mathrm{Hom}_{\mathbb{K}}(A, \mathbb{K}) \in \mathbf{M}(A^{\mathrm{en}}, \mathrm{gr}),$$

see formula (15.1.2).

Definition 15.2.4 Let A be a graded ring.

(1) A graded A-module P is called a *graded-free A-module* if $P \cong \bigoplus_{x \in X} A(-i_x)$ in $\mathbf{M}(A, \text{gr})$ for some collection $\{i_x\}_{x \in X}$ of integers.
(2) A graded A-module P is called a *graded-projective A-module* if it is a projective object in the abelian category $\mathbf{M}(A, \text{gr})$.
(3) A graded A-module I is called a *graded-cofree A-module* if $I \cong \prod_{x \in X} A^*(-p_x)$ in $\mathbf{M}(A, \text{gr})$ for some collection $\{p_x\}_{x \in X}$ of integers.
(4) A graded A-module I is called a *graded-injective A-module* if it is an injective object in the abelian category $\mathbf{M}(A, \text{gr})$.
(5) A graded A-module P is called a *graded-flat A-module* if the functor

$$(-) \otimes_A P : \mathbf{M}(A^{\text{op}}, \text{gr}) \to \mathbf{M}(\mathbb{K}, \text{gr})$$

is exact.

Note that a graded-free A-module P is of the form $P \cong A \otimes V$ for some $V \in \mathbf{M}(\mathbb{K}, \text{gr})$; and a graded-cofree A-module I is of the form $I \cong \text{Hom}_{\mathbb{K}}(A \otimes W, \mathbb{K})$ for some $W \in \mathbf{M}(\mathbb{K}, \text{gr})$.

Proposition 15.2.5 *Let A and B be graded rings.*
(1) *There is an isomorphism*

$$\text{Hom}_A(-, A^*) \cong \text{Hom}_{\mathbb{K}}(-, \mathbb{K})$$

of functors

$$\mathbf{M}(A \otimes B^{\text{op}}, \text{gr})^{\text{op}} \to \mathbf{M}(B \otimes A^{\text{op}}, \text{gr}).$$

(2) *The graded bimodule A^* is graded-injective over A, namely it is an injective object in the category* $\mathbf{M}(A, \text{gr})$.

Proof This is by the adjunction isomorphism

$$\text{Hom}_A(M, \text{Hom}_{\mathbb{K}}(A, \mathbb{K})) \cong \text{Hom}_{\mathbb{K}}(M, \mathbb{K})$$

in $\mathbf{M}(\mathbb{K}, \text{gr})$. □

The ungrading functor (15.1.10) preserves finiteness (i.e. being a finitely generated module). It also preserves freeness and projectivity, as the next proposition shows.

Proposition 15.2.6 *Let A be a graded ring, and let* $P \in \mathbf{M}(A, \text{gr})$.
(1) *If P is a graded-free A-module, then* $\text{Ungr}(P)$ *is a free* $\text{Ungr}(A)$*-module.*
(2) *P is a graded-projective A-module if and only if it is a direct summand, in* $\mathbf{M}(A, \text{gr})$*, of a graded-free A-module.*
(3) *If P is a graded-projective A-module, then* $\text{Ungr}(P)$ *is a projective* $\text{Ungr}(A)$*-module.*

Exercise 15.2.7 Prove Proposition 15.2.6.

Remark 15.2.8 In the situation of Proposition 15.2.6, one can show that P is a graded-flat A-module if and only if $\mathrm{Ungr}(P)$ is a flat $\mathrm{Ungr}(A)$-module. The proof is not easy. See [111, Proposition 3.4.5].

For injectives things are much more complicated, as the next theorem and example show.

Theorem 15.2.9 (Van den Bergh) *Let A be a left noetherian nonnegative graded ring, and let I be graded-injective A-module. Then* $\mathrm{Ungr}(I)$ *has injective dimension at most* 1 *as an* $\mathrm{Ungr}(A)$*-module.*

Proof See [181]. A more general statement can be found in [142]. □

Example 15.2.10 Consider the graded ring $A := \mathbb{K}[t]$, the ring of polynomials in a variable t of degree 1. The graded A-module $I := \mathbb{K}[t, t^{-1}]$ is graded-injective, but $\mathrm{Ungr}(I)$ has injective dimension 1 over $\mathrm{Ungr}(A)$.

Proposition 15.2.11

(1) *A graded-free A-module P is graded-projective.*
(2) *A graded-projective A-module P is graded-flat.*
(3) *A graded-cofree A-module I is graded-injective.*

Exercise 15.2.12 Prove this proposition.

Proposition 15.2.13 *Let A and B be graded rings.*

(1) *If $P \in \mathbf{M}(A \otimes B, \mathrm{gr})$ is projective, then it is projective in $\mathbf{M}(A, \mathrm{gr})$.*
(2) *If $P \in \mathbf{M}(A \otimes B, \mathrm{gr})$ is flat, then it is flat in $\mathbf{M}(A, \mathrm{gr})$.*
(3) *If $I \in \mathbf{M}(A \otimes B, \mathrm{gr})$ is injective, then it is injective in $\mathbf{M}(A, \mathrm{gr})$.*

Proof (1) Take $N \in \mathbf{C}(A, \mathrm{gr})$. We have isomorphisms

$$\begin{aligned} \mathrm{Hom}_A(P, N) &\cong \mathrm{Hom}_{A\otimes B}(P, \mathrm{Hom}_A(A \otimes B, N)) \\ &\cong \mathrm{Hom}_{A\otimes B}(P, \mathrm{Hom}_{\mathbb{K}}(B, N)) \end{aligned}$$

in $\mathbf{C}_{\mathrm{str}}(\mathbb{K}, \mathrm{gr})$. As an object of $\mathbf{M}(\mathbb{K}, \mathrm{gr})$, B is projective. Hence, if N is an acyclic complex, then so is $\mathrm{Hom}_A(P, N)$.

(2, 3) Like item (1), but using the fact that B is graded-flat over $\mathbb{K}$, and using suitable adjunction isomorphisms. □

Definition 15.2.14 Let A be a graded ring.

(1) Let M be a graded A-module, and let $N \subseteq M$ be a graded submodule. We say that N is a *graded-essential submodule* of M if for every nonzero graded submodule $M' \subseteq M$ the intersection $N \cap M'$ is nonzero.

(2) A homomorphism $\phi : N \to M$ in $\mathbf{M}(A, \mathrm{gr})$ is called a *graded-essential monomorphism* if ϕ is an injective homomorphism, and $\phi(N)$ is a graded-essential submodule of M.

Item (2) of the definition is a special case of the usual definition of an essential monomorphism in an abelian category.

Exercise 15.2.15 Let $N \subseteq M$ be graded A-modules. Show that N is a graded-essential submodule of M if and only if $\mathrm{Ungr}(N)$ is an essential $\mathrm{Ungr}(A)$-submodule of $\mathrm{Ungr}(M)$. (Hint: prove that for every nonzero element $m \in M$ there is a homogeneous element $a \in A$ such that $a \cdot m \in N$ and $a \cdot m \neq 0$. Do this by induction on the number r appearing in the decomposition (15.1.1). See the proof of [111, Lemma 3.3.13], which is not totally correct, but can be fixed using our hint.)

Proposition 15.2.16 *Let A be a graded ring.*

(1) *The category $\mathbf{M}(A, \mathrm{gr})$ has enough projectives.*
(2) *The category $\mathbf{M}(A, \mathrm{gr})$ has enough injectives. Moreover, every $M \in \mathbf{M}(A, \mathrm{gr})$ has an injective hull, namely there is a graded-essential monomorphism $M \rightarrowtail I$ to a graded-injective module I.*

Proof (1) Given a module $M \in \mathbf{M}(A, \mathrm{gr})$ and a homogeneous element $m \in M_i$, there is a homomorphism $A(-i) \to M$ that sends $1_A \mapsto m$. Therefore, by taking a sufficiently large direct sum of degree twists of A, there is a surjection $P \twoheadrightarrow M$ from a graded-free module P. And graded-free modules are graded-projective, by Proposition 15.2.11.

(2) For every nonzero homogeneous element $m \in M_q$ there is a homomorphism $M \to \mathbb{K}(-q)$ in $\mathbf{M}(\mathbb{K}, \mathrm{gr})$ that is nonzero on m. By Proposition 15.2.5(1) we get a homomorphism $M \to A^*(-q)$ in $\mathbf{M}(A, \mathrm{gr})$ that is nonzero on m. Thus, by taking a sufficiently large product $I := \prod_{x \in X} A^*(-q_x)$ in $\mathbf{M}(A, \mathrm{gr})$ we obtain a monomorphism $M \rightarrowtail I$. And according Proposition 15.2.11, the graded-cofree object I is injective in $\mathbf{M}(A, \mathrm{gr})$.

As for injective hulls: this is the same as in the ungraded case; see [129, Theorem 3.30] or [93, Section 3.D]. □

Now a structural result on injective objects in $\mathbf{M}(A, \mathrm{gr})$.

Theorem 15.2.17 *Let A be a left noetherian graded ring.*

(1) *If $\{I_x\}_{x \in X}$ is a collection of graded-injective A-modules, then $I := \bigoplus_{x \in X} I_x$ is a graded-injective A-module.*
(2) *Every graded-injective A-module I is a direct sum of indecomposable graded-injective A-modules.*

Proof This is because $\mathsf{M}(A, \mathrm{gr})$ is a locally noetherian Grothendieck abelian category. See [148, Section V.4]. Item (1) can be proved directly using the graded version of the Baer criterion (Theorem 2.7.10). □

Definition 15.2.18 A *connected graded* $\mathbb{K}$*-ring* is a nonnegative graded $\mathbb{K}$-ring A (see Conventions 15.1.17 and 15.2.1, and Definition 15.1.19), such that the ring homomorphism $\mathbb{K} \to A_0$ is bijective, and A_i is a finitely generated $\mathbb{K}$-module for all $i \geq 0$.

Here are a few examples of connected graded rings.

Example 15.2.19 Consider the noncommutative polynomial ring $A := \mathbb{K}\langle x_1, \ldots, x_n \rangle$ in $n \geq 1$ variables, all of algebraic degree 1. This is a connected graded $\mathbb{K}$-ring. If $n \geq 2$ then A is not commutative and not noetherian (on either side). If $n = 1$ then $\mathbb{K}\langle x_1 \rangle = \mathbb{K}[x_1]$, see next example.

Example 15.2.20 Let $A := \mathbb{K}[x_1, \ldots, x_n]$, the commutative polynomial ring in $n \geq 1$ variables, all of algebraic degree 1. This is a connected graded $\mathbb{K}$-ring, commutative and noetherian.

Example 15.2.21 Let $\mathbb{K}[t]$ be the commutative polynomial ring in a variable t of algebraic degree 1. Next let

$$A := \mathbb{K}[t]\langle x, y \rangle / (y \cdot x - x \cdot y - t^2),$$

where x and y are noncommuting variables of algebraic degree 1 (that commute with t). This is the *homogeneous first Weyl algebra*. It is a connected graded $\mathbb{K}$-ring, noetherian, but not commutative. Indeed, if $\mathrm{char}(\mathbb{K}) = 0$, then the center of A is $\mathbb{K}[t]$.

Example 15.2.22 Let $\mathfrak{g}$ be a finite Lie algebra over $\mathbb{K}$. Choose a basis $v_1, \ldots, v_n$ for $\mathfrak{g}$ as a $\mathbb{K}$-module, and let $\gamma_{i,j,k} \in \mathbb{K}$ be the constants describing the Lie bracket, i.e.

$$[v_i, v_j] = \sum_k \gamma_{i,j,k} \cdot v_k \in \mathfrak{g}.$$

Let $\mathbb{K}[t]$ be the commutative polynomial ring in a variable t of algebraic degree 1. Let $x_1, \ldots, x_n$ be variables of algebraic degree 1, and define the graded ring

$$A := \mathbb{K}[t]\langle x_1, \ldots, x_n \rangle \,/\, (I),$$

where I is the two-sided ideal generated by the quadratic elements

$$x_i \cdot x_j - x_j \cdot x_i - \sum_{k=1}^{n} \gamma_{i,j,k} \cdot x_k \cdot t.$$

This is the *homogeneous universal enveloping ring* of $\mathfrak{g}$. The quotient $A/(t-1)$ is the universal enveloping ring $\mathrm{U}(\mathfrak{g})$. The ring A is connected graded, noetherian, but often noncommutative (since the Lie algebra $\mathfrak{g}$ embeds into the quotient ring $\mathrm{U}(\mathfrak{g})$ with its commutator Lie bracket).

Example 15.2.23 Suppose A and B are both connected graded $\mathbb{K}$-rings. Then $A \otimes B$ is a connected graded $\mathbb{K}$-ring.

Definition 15.2.24 A graded $\mathbb{K}$-module $M = \bigoplus_{i \in \mathbb{Z}} M_i$ is called *degreewise finite* if the $\mathbb{K}$-module M_i is finitely generated for every $i \in \mathbb{Z}$.

The full subcategory of $\mathbf{M}(\mathbb{K}, \mathrm{gr})$ on the degreewise finite modules is denoted by $\mathbf{M}_{\mathrm{dwf}}(\mathbb{K}, \mathrm{gr})$.

It is obvious that the category $\mathbf{M}_{\mathrm{dwf}}(\mathbb{K}, \mathrm{gr})$ is a thick abelian subcategory of $\mathbf{M}(\mathbb{K}, \mathrm{gr})$, closed under subobjects and quotient objects.

Definition 15.2.25 For $M \in \mathbf{M}(\mathbb{K}, \mathrm{gr})$ we let

$$M^* := \mathrm{Hom}_{\mathbb{K}}(M, \mathbb{K}) \in \mathbf{M}(\mathbb{K}, \mathrm{gr}).$$

This is the *graded $\mathbb{K}$-linear dual of M*.

According to formula (15.1.2), we have $M^* = \bigoplus_{i \in \mathbb{Z}} (M^*)_i$, where

$$(M^*)_i = \mathrm{Hom}_{\mathbb{K}}(M, \mathbb{K})_i = \mathrm{Hom}_{\mathbb{K}}(M_{-i}, \mathbb{K}) = (M_{-i})^*.$$

Lemma 15.2.26 *If $M \in \mathbf{M}_{\mathrm{dwf}}(\mathbb{K}, \mathrm{gr})$ then $M^* \in \mathbf{M}_{\mathrm{dwf}}(\mathbb{K}, \mathrm{gr})$, and the Hom-evaluation homomorphism $\mathrm{ev}_M : M \to M^{**} = (M^*)^*$ in $\mathbf{M}_{\mathrm{dwf}}(\mathbb{K}, \mathrm{gr})$ is bijective. Thus $(-)^*$ is a duality of the category $\mathbf{M}_{\mathrm{dwf}}(\mathbb{K}, \mathrm{gr})$.*

The easy proof is omitted.

Definition 15.2.27 Let A be a graded ring. A module $M \in \mathbf{M}(A, \mathrm{gr})$ is called *degreewise $\mathbb{K}$-finite* if it is degreewise finite as a graded $\mathbb{K}$-module.

The full subcategory of $\mathbf{M}(A, \mathrm{gr})$ on the degreewise $\mathbb{K}$-finite modules is denoted by $\mathbf{M}_{\mathrm{dwf}}(A, \mathrm{gr})$.

Of course, $\mathbf{M}_{\mathrm{dwf}}(A, \mathrm{gr})$ is a thick abelian subcategory of $\mathbf{M}(A, \mathrm{gr})$, closed under subobjects and quotients.

Proposition 15.2.28 *Let A be a graded ring. If $M \in \mathbf{M}_{\mathrm{dwf}}(A, \mathrm{gr})$ then $M^* \in \mathbf{M}_{\mathrm{dwf}}(A^{\mathrm{op}}, \mathrm{gr})$, and the Hom-evaluation homomorphism $\mathrm{ev}_M : M \to M^{**}$ in $\mathbf{M}(A, \mathrm{gr})$ is bijective. Thus*

$$(-)^* : \mathbf{M}_{\mathrm{dwf}}(A, \mathrm{gr})^{\mathrm{op}} \to \mathbf{M}_{\mathrm{dwf}}(A^{\mathrm{op}}, \mathrm{gr})$$

is an equivalence of categories.

Exercise 15.2.29 Prove this proposition. (Hint: use Lemma 15.2.26.)

Noncommutative connected graded rings are very similar to complete commutative local rings. Here is the second variant of the graded Nakayama Lemma (the first was Proposition 15.1.22).

Proposition 15.2.30 *Let A be a connected graded ring, and let $M \in \mathbf{M}(A, \mathrm{gr})$ be a bounded below graded module. Define $V := \mathbb{K} \otimes_A M \in \mathbf{M}(\mathbb{K}, \mathrm{gr})$.*

(1) *There is a surjection $\pi : A \otimes V \twoheadrightarrow M$ in $\mathbf{M}(A, \mathrm{gr})$, such that the diagram*

$$\begin{array}{ccc} A \otimes V & \xrightarrow{\ \pi\ } & M \\ {\scriptstyle \beta}\downarrow & & \downarrow{\scriptstyle \alpha} \\ V & \xrightarrow{\ \mathrm{id}\ } & V \end{array}$$

in which the morphisms α and β are the canonical projections, is commutative.

(2) *If $V = 0$ then $M = 0$.*

(3) *If M is a graded-projective A-module, then the surjection π above is bijective. Hence M is a graded-free A-module.*

Note that the projection β is induced by the augmentation homomorphism $A \to \mathbb{K}$, so $\mathrm{Ker}(\beta) = \mathfrak{m} \otimes V$. Likewise, $\mathrm{Ker}(\alpha) = \mathfrak{m} \cdot M \subseteq M$.

Proof

(1) Say $M = \bigoplus_{i \geq i_0} M_i$. Choose a section $\pi_0 : V \rightarrowtail M$ in $\mathbf{M}(\mathbb{K}, \mathrm{gr})$ of the canonical projection $\alpha : M \twoheadrightarrow V$. Since $A \otimes V$ is a graded-free A-module, π_0 extends to a homomorphism $\pi : A \otimes V \to M$ in $\mathbf{M}(A, \mathrm{gr})$. See the commutative diagram below, in which $\gamma(v) := 1_A \otimes v$.

$$\begin{array}{ccc} V & & \\ {\scriptstyle \gamma}\downarrow & \searrow^{\pi_0} & \\ A \otimes V & \xrightarrow{\ \pi\ } & M \end{array}$$

By Proposition 15.1.22 the homomorphism π is surjective.

(2) Clear from the existence of the surjection π.

(3) Let $N := \mathrm{Ker}(\pi) \subseteq A \otimes V$, so there is a short exact sequence

$$0 \to N \to A \otimes V \xrightarrow{\pi} M \to 0$$

in $\mathbf{M}(A, \mathrm{gr})$, and N is a bounded below graded A-module. If M is graded-projective, then this sequence is split. It follows that $\mathbb{K} \otimes_A N = 0$. By part (2) we see that $N = 0$. □

Definition 15.2.31 Let A be a connected graded $\mathbb{K}$-ring, and let $M \in \mathbf{M}(A, \mathrm{gr})$. The *socle* of M is the graded A-submodule $\mathrm{Soc}(M) := \mathrm{Hom}_A(\mathbb{K}, M) \subseteq M$.

In other words, $\mathrm{Soc}(M) = \{m \in M \mid \mathfrak{m}\cdot m = 0\}$. The A-module structure of $\mathrm{Soc}(M)$ is through the augmentation homomorphism $A \to \mathbb{K}$; so Proposition 15.2.2 applies to it.

Here is a dual to the graded Nakayama Lemma.

Proposition 15.2.32 *Let A be a connected graded $\mathbb{K}$-ring, let $N \in \mathbf{M}(A, \mathrm{gr})$ be a bounded above graded module, and let $W := \mathrm{Soc}(N)$.*

(1) *There is an injection $\sigma : N \rightarrowtail \mathrm{Hom}_{\mathbb{K}}(A, W)$ in $\mathbf{M}(A, \mathrm{gr})$, such that the diagram*

$$\begin{array}{ccc} W & \xrightarrow{\ \mathrm{id}\ } & W \\ {\scriptstyle\alpha}\downarrow & & \downarrow{\scriptstyle\beta} \\ N & \xrightarrow{\ \sigma\ } & \mathrm{Hom}_{\mathbb{K}}(A, W) \end{array}$$

in which α and β are the canonical embeddings, is commutative.

(2) *If $W = 0$ then $N = 0$.*

(3) *If N is a graded-injective A-module, then the homomorphism σ above is bijective. Hence N is a graded-cofree A-module.*

Note that β is induced from the augmentation homomorphism $A \to \mathbb{K}$, and $\mathrm{Im}(\beta) = \mathrm{Soc}(\mathrm{Hom}_{\mathbb{K}}(A, W))$.

Proof

(1) Say $N = \bigoplus_{i \le i_1} N_i$ for some $i_1 \in \mathbb{Z}$. Choose a splitting $\sigma_0 : N \twoheadrightarrow W$ in $\mathbf{M}(\mathbb{K}, \mathrm{gr})$ of the inclusion $\alpha : W \rightarrowtail N$. By adjunction there is an isomorphism

$$\mathrm{Hom}_{\mathbb{K}}(N, W) \cong \mathrm{Hom}_A(N, \mathrm{Hom}_{\mathbb{K}}(A, W))$$

in $\mathbf{M}(\mathbb{K}, \mathrm{gr})$, so there is a homomorphism $\sigma : N \to \mathrm{Hom}_{\mathbb{K}}(A, W)$ in $\mathbf{M}(A, \mathrm{gr})$ that lifts σ_0. See the commutative diagram below, where $\gamma(\chi) := \chi(1_A)$.

$$\begin{array}{ccc} N & \xrightarrow{\ \sigma\ } & \mathrm{Hom}_{\mathbb{K}}(A, W) \\ & \searrow^{\sigma_0} & \downarrow{\scriptstyle\gamma} \\ & & W \end{array}$$

A descending induction on degree, starting from i_1, shows that σ is an injection.

(2) Clear from the existence of σ.

(3) Let $M := \mathrm{Coker}(\sigma)$. If N is graded-injective, then the exact sequence

$$0 \to N \xrightarrow{\sigma} \mathrm{Hom}_{\mathbb{K}}(A, W) \to M \to 0$$

in $\mathbf{M}(A, \mathrm{gr})$ is split. Applying the additive functor Soc to it, we see that $\mathrm{Soc}(M) = 0$. Since M is a bounded above graded module, item (2) says that $M = 0$. □

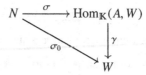

Definition 15.2.33 Let A be a graded ring. A module $N \in \mathbf{M}(A^{\mathrm{op}}, \mathrm{gr})$ is called a *cofinite graded A^{op}-module* if $N \cong M^*$ for some finite graded A-module M.

The full subcategory of $\mathbf{M}(A^{\mathrm{op}}, \mathrm{gr})$ on the cofinite graded A^{op}-modules is denoted by $\mathbf{M}_{\mathrm{cof}}(A^{\mathrm{op}}, \mathrm{gr})$.

Recall that an object M in an abelian category M is called *noetherian* (resp. *artinian*) if it satisfies the ascending (resp. descending) chain condition on subobjects. Let us denote the corresponding full subcategories of M by $\mathsf{M}_{\mathrm{not}}$ and $\mathsf{M}_{\mathrm{art}}$, respectively. Then $\mathsf{M}_{\mathrm{not}}, \mathsf{M}_{\mathrm{art}} \subseteq \mathsf{M}$ are thick abelian subcategories, closed under subobjects and quotient objects.

Here is a variant of the Matlis Duality Theorem (see Remark 13.3.23):

Theorem 15.2.34 (NC Graded Matlis Duality) *Assume A is a left noetherian connected graded ring.*

(1) *The category $\mathbf{M}_{\mathrm{f}}(A, \mathrm{gr})$ is a thick abelian subcategory of $\mathbf{M}_{\mathrm{dwf}}(A, \mathrm{gr})$, closed under subobjects and quotient objects.*
(2) *The category $\mathbf{M}_{\mathrm{cof}}(A^{\mathrm{op}}, \mathrm{gr})$ is a thick abelian subcategory of $\mathbf{M}_{\mathrm{dwf}}(A^{\mathrm{op}}, \mathrm{gr})$, closed under subobjects and quotient objects.*
(3) *The functor*

$$(-)^* : \mathbf{M}_{\mathrm{f}}(A, \mathrm{gr})^{\mathrm{op}} \to \mathbf{M}_{\mathrm{cof}}(A^{\mathrm{op}}, \mathrm{gr})$$

is an equivalence.
(4) *The modules in $\mathbf{M}_{\mathrm{f}}(A, \mathrm{gr})$ are the noetherian objects in the abelian category $\mathbf{M}(A, \mathrm{gr})$.*
(5) *The modules in $\mathbf{M}_{\mathrm{cof}}(A^{\mathrm{op}}, \mathrm{gr})$ are the artinian objects in the abelian category $\mathbf{M}(A^{\mathrm{op}}, \mathrm{gr})$.*

Proof

(1) Since A is connected, as a left module it belongs to $\mathbf{M}_{\mathrm{dwf}}(A, \mathrm{gr})$. Therefore every finite graded A-module belongs to $\mathbf{M}_{\mathrm{dwf}}(A, \mathrm{gr})$, i.e. $\mathbf{M}_{\mathrm{f}}(A, \mathrm{gr}) \subseteq \mathbf{M}_{\mathrm{dwf}}(A, \mathrm{gr})$.

Because A is left noetherian, it follows that $\mathbf{M}_{\mathrm{f}}(A, \mathrm{gr})$ is a thick abelian subcategory of $\mathbf{M}(A, \mathrm{gr})$, closed under subobjects and quotient objects.

(2, 3) By definition $\mathbf{M}_{\mathrm{cof}}(A^{\mathrm{op}}, \mathrm{gr})$ is the essential image under the functor $(-)^*$ of the category $\mathbf{M}_{\mathrm{f}}(A, \mathrm{gr})$. Now use the equivalence of Proposition 15.2.28 and part (1).

(4) This is proved the same way as for ungraded left noetherian rings.

(5) We know by (2, 3, 4) that the objects of $\mathbf{M}_{\mathrm{cof}}(A^{\mathrm{op}}, \mathrm{gr})$ are artinian objects in $\mathbf{M}(A^{\mathrm{op}}, \mathrm{gr})$.

For the opposite direction, let $N \in \mathbf{M}(A^{\mathrm{op}}, \mathrm{gr})$ be an artinian object. The descending chain condition forces N to be bounded above, namely $N_i = 0$ for $i \gg 0$. Consider the socle $W := \mathrm{Hom}_{A^{\mathrm{op}}}(\mathbb{K}, N)$. This is a graded A^{op}-submodule of N, isomorphic to $\bigoplus_{x \in X} \mathbb{K}(-i_x)$ for some indexing set X and a collection of integers $\{i_x\}_{x \in X}$. By the descending chain condition the set X has to be finite. Hence

$$\bigoplus_{x \in X} A^*(-i_x) \cong A^* \otimes W \cong \mathrm{Hom}_{\mathbb{K}}(A, W).$$

Proposition 15.2.32 tells us that there is an injective homomorphism $N \rightarrowtail \bigoplus_{x \in X} A^*(-i_x)$. We know that $A^*(-i_x) \in \mathbf{M}_{\mathrm{cof}}(A, \mathrm{gr})$; and hence so is N. □

Remark 15.2.35 If A is connected graded but not left noetherian, then there are cofinite graded A^{op}-modules that are not artinian. Take the ring $A := \mathbb{K}\langle x_1, x_2 \rangle$ from Example 15.2.19. Then $A^* \in \mathbf{M}(A^{\mathrm{op}}, \mathrm{gr})$ is cofinite but not artinian.

15.3 Resolutions and Derived Functors

In this section we continue with Conventions 15.1.17 and 15.2.1. Thus $\mathbb{K}$ is a base field and $\otimes = \otimes_{\mathbb{K}}$. In the next definitions we collect the algebraically graded versions of some definitions from Chapters 10 and 11.

Definition 15.3.1 Let A be a graded ring, and let $P \in \mathbf{C}(A, \mathrm{gr})$.

(1) The complex P is called a *graded-free complex* if all the graded A-modules P^i are graded-free, and the differential is zero.
(2) The complex P is called a *semi-graded-free complex* if it admits a filtration $\{F_j(P)\}_{j \geq -1}$ in $\mathbf{C}_{\mathrm{str}}(A, \mathrm{gr})$, such that $F_{-1}(P) = 0$, $P = \bigcup_j F_j(P)$, and $\mathrm{Gr}_j^F(P)$ is a graded-free complex for every j.
(3) The complex P is called a *K-graded-projective complex* if for every acyclic complex $N \in \mathbf{C}(A, \mathrm{gr})$, the complex $\mathrm{Hom}_A(P, N)$ is acyclic.

Definition 15.3.2 Let A be a graded ring, and let $I \in \mathbf{C}(A, \mathrm{gr})$.

(1) The complex I is called a *graded-cofree complex* if all the A-modules I^p are graded-cofree, and the differential is zero.
(2) The complex I is called a *semi-graded-cofree complex* if it admits a cofiltration $\{F_q(I)\}_{q \geq -1}$ in $\mathbf{C}_{\mathrm{str}}(A, \mathrm{gr})$, such that $F_{-1}(I) = 0$, each $\mathrm{Gr}_q^F(I)$ is graded-cofree complex, and $I = \lim_{\leftarrow q} F_q(I)$.
(3) The complex I is called a *K-graded-injective complex* if for every acyclic complex $N \in \mathbf{C}(A, \mathrm{gr})$, the complex $\mathrm{Hom}_A(N, I)$ is acyclic.

Definition 15.3.3 A complex $P \in \mathbf{C}(A, \mathrm{gr})$ is called a *K-graded-flat complex* if for every acyclic complex $N \in \mathbf{C}(A^{\mathrm{op}}, \mathrm{gr})$ the complex $N \otimes_A P$ is acyclic.

Theorem 15.3.4 *Let A be a graded ring, and let $M \in \mathbf{C}(A, \mathrm{gr})$.*

(1) *There exists a quasi-isomorphism $\rho : P \to M$ in $\mathbf{C}_{\mathrm{str}}(A, \mathrm{gr})$, where P is a semi-graded-free complex, and $\sup(P) = \sup(\mathrm{H}(M))$.*
(2) *There exists a quasi-isomorphism $\rho : M \to I$ in $\mathbf{C}_{\mathrm{str}}(A, \mathrm{gr})$, where I is a semi-graded-cofree complex, and $\inf(I) = \inf(\mathrm{H}(M))$.*

Proof

(1) This is a modification of Theorem 11.4.17 and Corollary 11.4.27. The proof is the same, after a few obvious modifications.

(2) This is a modification of Theorem 11.6.24 and Corollary 11.6.31. The proof is the same, after a few obvious modifications. The injective object $\mathbb{K}^*$ we use here is $\mathbb{K}^* := \mathbb{K}$ of course. □

The various kinds of complexes mentioned in the definitions above are related as follows.

Theorem 15.3.5 *Let A be a graded ring.*

(1) *A complex $P \in \mathbf{C}(A, \mathrm{gr})$ is K-graded-projective if and only if it is isomorphic in $\mathbf{K}(A, \mathrm{gr})$ to a semi-graded-free complex Q.*
(2) *If $P \in \mathbf{C}(A, \mathrm{gr})$ is K-graded-projective, then it is K-graded-flat.*
(3) *If $P \in \mathbf{C}(A, \mathrm{gr})$ is K-graded-projective, then $P^* \in \mathbf{C}(A^{\mathrm{op}}, \mathrm{gr})$ is K-graded-injective.*
(4) *A complex $I \in \mathbf{C}(A, \mathrm{gr})$ is K-graded-injective iff it is isomorphic in $\mathbf{K}(A, \mathrm{gr})$ to a semi-graded-cofree complex J.*

Proof

(1) By Theorem 11.4.14, which applies to this context too, a semi-graded-free complex Q is K-graded-projective. And by Proposition 10.2.3, a complex $P \in \mathbf{C}(A, \mathrm{gr})$ is K-graded-projective iff $\mathrm{Hom}_{\mathbf{K}(A,\mathrm{gr})}(P, N) = 0$ for every acyclic complex $N \in \mathbf{C}(A, \mathrm{gr})$.

First let's assume that there is an isomorphism $P \cong Q$ in $\mathbf{K}(A, \mathrm{gr})$ to a semi-graded-free complex Q. Then for every acyclic complex $N \in \mathbf{C}(A, \mathrm{gr})$ we get

$$\mathrm{Hom}_{\mathbf{K}(A,\mathrm{gr})}(P, N) \cong \mathrm{Hom}_{\mathbf{K}(A,\mathrm{gr})}(Q, N) = 0.$$

Therefore P is K-graded-projective.

Conversely, let's assume that P is K-graded-projective. By Theorem 15.3.4 (1) there is a quasi-isomorphism $\rho : Q \to P$ in $\mathbf{C}_{\mathrm{str}}(A, \mathrm{gr})$ from a semi-graded-free complex Q. By Corollary 10.2.14, $\mathrm{P}(\rho) : Q \to P$ is an isomorphism in $\mathbf{K}(A, \mathrm{gr})$.

(2) Take $N \in \mathbf{C}(A^{\mathrm{op}}, \mathrm{gr})$. We have the adjunction isomorphism

$$\begin{aligned}(N \otimes_A P)^* &= \mathrm{Hom}_{\mathbb{K}}(N \otimes_A P, \mathbb{K}) \\ &\cong \mathrm{Hom}_A(P, \mathrm{Hom}_{\mathbb{K}}(N, \mathbb{K})) = \mathrm{Hom}_A(P, N^*)\end{aligned}$$

in $\mathbf{C}_{\mathrm{str}}(\mathbb{K}, \mathrm{gr})$. If N is acyclic then so are N^*, $\mathrm{Hom}_A(P, N^*)$ and $(N \otimes_A P)^*$. Since the functor $(-)^*$ is faithfully exact, we see that $N \otimes_A P$ is acyclic too. Therefore P is K-graded-flat.

(3) Take $N \in \mathbf{C}(A^{\mathrm{op}}, \mathrm{gr})$. We have the adjunction isomorphism

$$\begin{aligned}\mathrm{Hom}_{A^{\mathrm{op}}}(N, P^*) &= \mathrm{Hom}_{A^{\mathrm{op}}}(N, \mathrm{Hom}_{\mathbb{K}}(P, \mathbb{K})) \\ &\cong \mathrm{Hom}_{\mathbb{K}}(N \otimes_A P, \mathbb{K}) = (N \otimes_A P)^*\end{aligned}$$

in $\mathbf{C}_{\mathrm{str}}(\mathbb{K}, \mathrm{gr})$. By item (2) P is K-graded-flat. If N is acyclic then so are $N \otimes_A P$, $(N \otimes_A P)^*$ and $\mathrm{Hom}_{A^{\mathrm{op}}}(N, P^*)$. Therefore P^* is K-graded-injective.

(4) By Theorem 11.6.21, which applies to this context too, a semi-graded-cofree complex J is K-graded-injective. And by Proposition 10.1.3, a complex $I \in \mathbf{C}(A, \mathrm{gr})$ is K-graded-injective iff $\mathrm{Hom}_{\mathbf{K}(A,\mathrm{gr})}(N, I) = 0$ for every acyclic complex $N \in \mathbf{C}(A, \mathrm{gr})$.

First let's assume that there is an isomorphism $I \cong J$ in $\mathbf{K}(A, \mathrm{gr})$ to a semi-graded-cofree complex J. Then for every acyclic complex $N \in \mathbf{C}(A, \mathrm{gr})$ we get

$$\mathrm{Hom}_{\mathbf{K}(A,\mathrm{gr})}(N, I) \cong \mathrm{Hom}_{\mathbf{K}(A,\mathrm{gr})}(N, J) = 0.$$

Therefore I is K-graded-injective.

Conversely, let's assume that I is K-graded-injective. By Theorem 15.3.4(2) there is a quasi-isomorphism $\rho : I \to J$ in $\mathbf{C}_{\mathrm{str}}(A, \mathrm{gr})$ to a semi-graded-cofree complex J. By Corollary 10.1.19, $\mathrm{P}(\rho) : I \to J$ is an isomorphism in $\mathbf{K}(A, \mathrm{gr})$. □

The next proposition tells us that some resolving properties of complexes are preserved by the restriction functor $\mathbf{C}(A \otimes B, \mathrm{gr}) \to \mathbf{C}(A, \mathrm{gr})$.

Proposition 15.3.6 *Let A and B be graded rings.*

(1) *If* $P \in \mathbf{C}(A \otimes B, \mathrm{gr})$ *is K-graded-projective, then it is K-graded-projective in* $\mathbf{C}(A, \mathrm{gr})$.
(2) *If* $P \in \mathbf{C}(A \otimes B, \mathrm{gr})$ *is K-graded-flat, then it is K-graded-flat in* $\mathbf{C}(A, \mathrm{gr})$.
(3) *If* $I \in \mathbf{C}(A \otimes B, \mathrm{gr})$ *is K-graded-injective, then it is K-graded-injective in* $\mathbf{C}(A, \mathrm{gr})$.

Proof

(1) Take $N \in \mathbf{C}(A, \mathrm{gr})$. We have isomorphisms

$$\begin{aligned}\mathrm{Hom}_A(P, N) &\cong \mathrm{Hom}_{A \otimes B}(P, \mathrm{Hom}_A(A \otimes B, N)) \\ &\cong \mathrm{Hom}_{A \otimes B}(P, \mathrm{Hom}_{\mathbb{K}}(B, N))\end{aligned}$$

in $\mathbf{C}_{\mathrm{str}}(\mathbb{K}, \mathrm{gr})$. As an object of $\mathbf{C}(\mathbb{K}, \mathrm{gr})$, B is K-graded-projective. Hence, if N is acyclic, then so is $\mathrm{Hom}_A(P, N)$.

(2, 3) Like item (1), but using the fact that B is K-graded-flat over $\mathbb{K}$, and using other adjunction isomorphisms. □

Corollary 15.3.7 *Let A and B be graded rings, and let $M \in \mathbf{C}(A \otimes B, \mathrm{gr})$.*

(1) *There exists a quasi-isomorphism $P \to M$ in $\mathbf{C}_{\mathrm{str}}(A \otimes B, \mathrm{gr})$, where P is a K-graded-projective complex over A, every P^i is a graded-projective module over A, and* $\sup(P) = \sup(\mathrm{H}(M))$.
(2) *There exists a quasi-isomorphism $P \to M$ in $\mathbf{C}_{\mathrm{str}}(A \otimes B, \mathrm{gr})$, where P is a K-graded-flat complex over A, every P^i is a graded-flat module over A, and* $\sup(P) = \sup(\mathrm{H}(M))$.
(3) *There exists a quasi-isomorphism $M \to I$ in $\mathbf{C}_{\mathrm{str}}(A \otimes B, \mathrm{gr})$, where I is a K-graded-injective complex over A, every I^p is a graded-injective module over A, and* $\inf(I) = \inf(\mathrm{H}(M))$.

Proof

(1) Take the resolution $P \to M$ from Theorem 15.3.4(1). In view of Theorem 15.3.5(1), Proposition 15.3.6(1) and Proposition 15.2.6(2), the complex P has the required properties.

(2) Use item (1), plus Theorem 15.3.5(2).

(3) Take the resolution $M \to I$ from Theorem 15.3.4(2). This will do, by Theorem 15.3.5(4), Proposition 15.3.6(3), Proposition 15.2.11(3) and Proposition 15.2.13(3). □

Corollary 15.3.8 *Let A, B and C be graded rings.*

(1) *The triangulated right derived bifunctor*

$$\mathrm{RHom}_A(-,-) : \mathbf{D}(A \otimes B^{\mathrm{op}}, \mathrm{gr})^{\mathrm{op}} \times \mathbf{D}(A \otimes C^{\mathrm{op}}, \mathrm{gr}) \to \mathbf{D}(B \otimes C^{\mathrm{op}}, \mathrm{gr})$$

exists.

(2) *Let $M \in \mathbf{D}(A \otimes B^{\mathrm{op}}, \mathrm{gr})$ and $N \in \mathbf{D}(A \otimes C^{\mathrm{op}}, \mathrm{gr})$. Assume that either M is K-graded-projective over A, or N is K-graded-injective over A. Then the morphism*

$$\eta^{\mathrm{R}}_{M,N} : \mathrm{Hom}_A(M, N) \to \mathrm{RHom}_A(M, N)$$

in $\mathbf{D}(B \otimes C^{\mathrm{op}}, \mathrm{gr})$ is an isomorphism.

(3) *Let $B' \to B$ and $C' \to C$ be graded $\mathbb{K}$-ring homomorphisms. Then the diagram*

$$\begin{array}{ccc} \mathsf{D}(A \otimes B^{\mathrm{op}}, \mathrm{gr})^{\mathrm{op}} \times \mathsf{D}(A \otimes C^{\mathrm{op}}, \mathrm{gr}) & \xrightarrow{\mathrm{RHom}_A(-,-)} & \mathsf{D}(B \otimes C^{\mathrm{op}}, \mathrm{gr}) \\ \downarrow{\scriptstyle \mathrm{Rest} \times \mathrm{Rest}} & & \downarrow{\scriptstyle \mathrm{Rest}} \\ \mathsf{D}(A \otimes B'^{\mathrm{op}}, \mathrm{gr})^{\mathrm{op}} \times \mathsf{D}(A \otimes C'^{\mathrm{op}}, \mathrm{gr}) & \xrightarrow{\mathrm{RHom}_A(-,-)} & \mathsf{D}(B' \otimes C'^{\mathrm{op}}, \mathrm{gr}) \end{array}$$

is commutative up to isomorphism.

Proof The same arguments used in the proof of Proposition 14.3.15 work here, because there exist enough acyclic resolutions – see Corollary 15.3.7. □

Corollary 15.3.9 *The category $\mathsf{D}(A, \mathrm{gr})$ has infinite direct sums. On objects, the direct sum is degreewise, for the $\mathbb{Z}^2$ grading.*

Proof The proof of Theorem 10.1.26 works here. □

Corollary 15.3.10 *Let A, B and C be graded rings.*

(1) *The triangulated left derived bifunctor*

$$(- \otimes^{\mathrm{L}}_A -) : \mathsf{D}(B \otimes A^{\mathrm{op}}, \mathrm{gr}) \times \mathsf{D}(A \otimes C^{\mathrm{op}}, \mathrm{gr}) \to \mathsf{D}(B \otimes C^{\mathrm{op}}, \mathrm{gr})$$

exists.

(2) *Let $M \in \mathsf{D}(B \otimes A^{\mathrm{op}}, \mathrm{gr})$ and $N \in \mathsf{D}(A \otimes C^{\mathrm{op}}, \mathrm{gr})$. Assume that either M is K-graded-flat over A^{op}, or N is K-graded-flat over A. Then the morphism*

$$\eta^{\mathrm{L}}_{M,N} : M \otimes^{\mathrm{L}}_A N \to M \otimes_A N$$

in $\mathsf{D}(B \otimes C^{\mathrm{op}}, \mathrm{gr})$ is an isomorphism.

(3) *Let $B' \to B$ and $C' \to C$ be graded $\mathbb{K}$-ring homomorphisms. Then the diagram*

$$\begin{array}{ccc} \mathsf{D}(B \otimes A^{\mathrm{op}}, \mathrm{gr}) \times \mathsf{D}(A \otimes C^{\mathrm{op}}, \mathrm{gr}) & \xrightarrow{(- \otimes^{\mathrm{L}}_A -)} & \mathsf{D}(B \otimes C^{\mathrm{op}}, \mathrm{gr}) \\ \downarrow{\scriptstyle \mathrm{Rest} \times \mathrm{Rest}} & & \downarrow{\scriptstyle \mathrm{Rest}} \\ \mathsf{D}(B' \otimes A^{\mathrm{op}}, \mathrm{gr}) \times \mathsf{D}(A \otimes C'^{\mathrm{op}}, \mathrm{gr}) & \xrightarrow{(- \otimes^{\mathrm{L}}_A -)} & \mathsf{D}(B' \otimes C'^{\mathrm{op}}, \mathrm{gr}) \end{array}$$

is commutative up to isomorphism.

Proof The same arguments used in the proof of Proposition 14.3.13 work here, because there exist enough acyclic resolutions – see Corollary 15.3.7. □

Corollary 15.3.11 *Let A, B and C be graded rings.*

(1) *Given* $M \in \mathbf{M}(A \otimes B^{\mathrm{op}}, \mathrm{gr})$ *and* $N \in \mathbf{M}(A \otimes C^{\mathrm{op}}, \mathrm{gr})$*, for every* $p \in \mathbb{N}$ *there is a canonical isomorphism*

$$\mathrm{Ext}^p_A(M, N) \cong \mathrm{H}^p(\mathrm{RHom}_A(M, N))$$

in $\mathbf{M}(B \otimes C^{\mathrm{op}}, \mathrm{gr})$*. Here* $\mathrm{Ext}^p_A(-, -)$ *is the classical pth right derived bifunctor of* $\mathrm{Hom}_A(-, -)$*.*

(2) *Given* $M \in \mathbf{M}(B \otimes A^{\mathrm{op}}, \mathrm{gr})$ *and* $N \in \mathbf{M}(A \otimes C^{\mathrm{op}}, \mathrm{gr})$*, for every* $p \in \mathbb{N}$ *there is a canonical isomorphism*

$$\mathrm{Tor}^A_p(M, N) \cong \mathrm{H}^{-p}(M \otimes^{\mathrm{L}}_A N)$$

in $\mathbf{M}(B \otimes C^{\mathrm{op}}, \mathrm{gr})$*. Here* $\mathrm{Tor}^A_p(-, -)$ *is the classical pth left derived bifunctor of* $(- \otimes_A -)$*.*

Proof This is clear from Corollary 15.3.8(2) and Corollary 15.3.10(2), respectively. □

Corollary 15.3.12 *Let A, B, C and D be graded rings.*

(1) *Let* $L \in \mathbf{D}(A \otimes B^{\mathrm{op}}, \mathrm{gr})$*,* $M \in \mathbf{D}(B \otimes C^{\mathrm{op}}, \mathrm{gr})$ *and* $N \in \mathbf{D}(C \otimes D^{\mathrm{op}}, \mathrm{gr})$*. There is an isomorphism*

$$L \otimes^{\mathrm{L}}_B (M \otimes^{\mathrm{L}}_C N) \cong (L \otimes^{\mathrm{L}}_B M) \otimes^{\mathrm{L}}_C N$$

in $\mathbf{D}(A \otimes D^{\mathrm{op}}, \mathrm{gr})$*. This isomorphism is functorial in* L, M, N*.*

(2) *Let* $L \in \mathbf{D}(B \otimes A^{\mathrm{op}}, \mathrm{gr})$*,* $M \in \mathbf{D}(C \otimes B^{\mathrm{op}}, \mathrm{gr})$ *and* $N \in \mathbf{D}(C \otimes D^{\mathrm{op}}, \mathrm{gr})$*. There is an isomorphism*

$$\mathrm{RHom}_B(L, \mathrm{RHom}_C(M, N)) \cong \mathrm{RHom}_C(M \otimes^{\mathrm{L}}_B L, N)$$

in $\mathbf{D}(A \otimes D^{\mathrm{op}}, \mathrm{gr})$*. This isomorphism is functorial in* L, M, N*.*

Proof Both assertions are immediate consequences of the associativity of the tensor product for modules, and the Hom-tensor adjunction for modules. □

The triangulated category $\mathbf{D}(A^{\mathrm{en}}, \mathrm{gr})$ has a biclosed monoidal structure on it, with monoidal operation $(- \otimes^{\mathrm{L}}_A -)$ and monoidal unit A. See Remark 14.3.23 for the nongraded variant. In the next definition we present the pertinent operations related to this structure; these will be needed for several constructions. We do not need to know anything about the concept of monoidal structure beyond this terminology. The interested reader can look it up in [102]; the previous corollaries (almost) guarantee that the monoidal axioms hold in this context.

Definition 15.3.13 (Monoidal Operations) Let A and B be graded rings, and let $M \in \mathbf{D}(A \otimes B^{\mathrm{op}}, \mathrm{gr})$ and $N \in \mathbf{D}(B \otimes A^{\mathrm{op}}, \mathrm{gr})$ be complexes.

(1) The *left unitor* isomorphism for M is the obvious isomorphism $\mathrm{lu} : A \otimes^{\mathrm{L}}_A M \xrightarrow{\simeq} M$ in $\mathbf{D}(A \otimes B^{\mathrm{op}}, \mathrm{gr})$.

(2) The *right unitor* isomorphism for N is the obvious isomorphism ru : $N \otimes^{\mathrm{L}}_{A} A \xrightarrow{\simeq} N$ in $\mathbf{D}(B \otimes A^{\mathrm{op}}, \mathrm{gr})$.
(3) The *left co-unitor* isomorphism for M is the obvious isomorphism lcu : $\mathrm{RHom}_{A}(A, M) \xrightarrow{\simeq} M$ in $\mathbf{D}(A \otimes B^{\mathrm{op}}, \mathrm{gr})$.

Here is a notion that is dual to a minimal complex of injectives (see Definition 13.2.5). Recall that for a complex P, the module of degree i coboundaries is $\mathrm{B}^{i}(P) = \mathrm{Im}(\mathrm{d}_{P}^{i-1} : P^{i-1} \to P^{i}) \subseteq P^{i}$.

Definition 15.3.14 Let A be a connected graded ring, with augmentation ideal $\mathfrak{m}$. A *minimal complex of graded-free A-modules* is a bounded above complex P, such that for every i the graded module P^{i} is graded-free, and there is an inclusion $\mathrm{B}^{i}(P) \subseteq \mathfrak{m} \cdot P^{i}$.

Proposition 15.3.15 *Let A be a connected graded ring, and let P be a bounded above complex of graded-free A-modules. The following conditions are equivalent:*
(i) *P is a minimal complex of graded-free A-modules.*
(ii) *The complex $\mathbb{K} \otimes_{A} P$ has zero differential.*

Exercise 15.3.16 Prove this proposition.

Definition 15.3.17 Let A be a connected graded ring, and let $M \in \mathbf{M}(A, \mathrm{gr})$. A *minimal graded-free resolution* of M is a quasi-isomorphism $\rho : P \to M$ in $\mathbf{C}_{\mathrm{str}}(A, \mathrm{gr})$ from a minimal complex of graded-free A-modules P.

Proposition 15.3.18 *Let A be a connected graded ring, and let $M \in \mathbf{M}(A, \mathrm{gr})$ be a bounded below graded module. Then:*
(1) *M has a minimal graded-free resolution $\eta : P \to M$.*
(2) *For every i there is an isomorphism $P^{i} \cong A \otimes \mathrm{Tor}^{A}_{-i}(\mathbb{K}, M)$ in $\mathbf{M}(A, \mathrm{gr})$.*
(3) *The minimal graded-free resolution $\eta : P \to M$ in item (1) above is unique, up to a non unique isomorphism in $\mathbf{C}_{\mathrm{str}}(A, \mathrm{gr})$.*

Proof

(1) Let $V_0 := \mathbb{K} \otimes_{A} M$ and $P^{0} := A \otimes V_0$. By Proposition 15.2.30 there is a surjection $\eta : P^{0} \to M$. Let $M^{-1} := \mathrm{Ker}(\eta) \subseteq P^{0}$. By construction we have $M^{-1} \subseteq \mathfrak{m} \otimes V_0 = \mathfrak{m} \cdot P^{0}$. And of course M^{-1} is a bounded below graded A-module.

Next let $V_1 := \mathbb{K} \otimes_{A} M^{-1}$ and $P^{-1} := A \otimes V_1$. By Proposition 15.2.30 there is a surjection $\mathrm{d}^{-1} : P^{-1} \to M^{-1}$. The graded module $M^{-2} := \mathrm{Ker}(\mathrm{d}^{-1}) \subseteq P^{-1}$ satisfies $M^{-2} \subseteq \mathfrak{m} \otimes V_1 = \mathfrak{m} \cdot P^{-1}$. And so on.

(2) Due to minimality, the differential of the complex $\mathbb{K} \otimes_A P$ is zero; see Proposition 15.3.15. Therefore

$$\mathrm{Tor}^A_{-i}(\mathbb{K}, M) \cong \mathrm{H}^i(\mathbb{K} \otimes^{\mathrm{L}}_A M) \cong \mathrm{H}^i(\mathbb{K} \otimes_A P) \cong \mathbb{K} \otimes_A P^i \cong V_{-i}.$$

(3) Suppose $\eta' : P' \to M$ is some other minimal graded-free resolution. Because P and P' are K-projective in $\mathbf{C}(A, \mathrm{gr})$, there is a homotopy equivalence $\phi : P \to P'$ in $\mathbf{C}_{\mathrm{str}}(A, \mathrm{gr})$. Therefore

$$\mathrm{id} \otimes \phi : \mathbb{K} \otimes_A P \to \mathbb{K} \otimes_A P' \tag{15.3.19}$$

is a quasi-isomorphism. But by Proposition 15.3.15, these complexes have zero differentials, so (15.3.19) is an isomorphism in $\mathbf{C}_{\mathrm{str}}(\mathbb{K}, \mathrm{gr})$. The graded Nakayama Lemma (Proposition 15.2.30(3)) implies that $\phi : P \to P'$ is an isomorphism in $\mathbf{C}_{\mathrm{str}}(A, \mathrm{gr})$. □

Exercise 15.3.20 Try to generalize Proposition 15.3.18 to a complex $M \in \mathbf{C}(A, \mathrm{gr})$. The trick is to find the correct boundedness conditions on M.

Corollary 15.3.21 *Assume A is a left noetherian connected graded ring, and M is a finite graded A-module. Then in the minimal graded-free resolution $\eta : P \to M$, the graded-free A-modules P^i are all finite.*

Proof In the proof of Proposition 15.3.18(1), the graded $\mathbb{K}$-modules V_i are all finite. □

Here is a graded version of Definitions 11.4.29 and 12.9.1.

Definition 15.3.22 Let A be a graded ring.

(1) A complex $P \in \mathbf{C}(A, \mathrm{gr})$ is called *pseudo-finite semi-graded-free* if it is a bounded above complex of finite graded-free A-modules.
(2) A complex $L \in \mathbf{D}(A, \mathrm{gr})$ is called *derived graded-pseudo-finite* if it belongs to the saturated full triangulated subcategory of $\mathbf{D}(A, \mathrm{gr})$ generated by the pseudo-finite semi-graded-free complexes.

Example 15.3.23 If the graded ring A is connected and left noetherian, then the derived graded-pseudo-finite complexes over A are precisely the objects of $\mathbf{D}^-_{\mathrm{f}}(A, \mathrm{gr})$. See Theorems 11.3.18 or 11.4.40.

Proposition 15.3.24 *Assume A is a left noetherian connected graded ring. Consider A as a graded A^{en}-module. Then in the minimal graded-free resolution $\rho : P \to A$ of A over A^{en}, the graded-free A^{en}-modules P^i are all finite.*

The subtle point is that the graded ring A^{en} might not be left noetherian – see Remark 15.1.30.

Proof Let us write $P^i \cong A^{\mathrm{en}} \otimes V_{-i}$, where $V_{-i} \in \mathbf{M}(\mathbb{K}, \mathrm{gr})$. We shall prove that all the V_{-i} are finite graded $\mathbb{K}$-modules.

When we restrict the quasi-isomorphism $\rho : P \to A$ to the category $\mathbf{C}_{\mathrm{str}}(A^{\mathrm{op}}, \mathrm{gr})$, it becomes a homotopy equivalence. Hence $\tau := \rho \otimes \mathrm{id} : P \otimes_A \mathbb{K} \to A \otimes_A \mathbb{K}$ is a quasi-isomorphism in $\mathbf{C}_{\mathrm{str}}(A, \mathrm{gr})$. Writing $Q := P \otimes_A \mathbb{K} \in \mathbf{C}(A, \mathrm{gr})$, we get a graded-free resolution $\tau : Q \to \mathbb{K}$ over A. Note that $Q^i \cong A \otimes V_{-i}$ in $\mathbf{C}_{\mathrm{str}}(A, \mathrm{gr})$.

Now

$$\mathbb{K} \otimes_A Q \cong \mathbb{K} \otimes_A P \otimes_A \mathbb{K} \cong \mathbb{K} \otimes_{A^{\mathrm{en}}} P \tag{15.3.25}$$

in $\mathbf{C}_{\mathrm{str}}(\mathbb{K}, \mathrm{gr})$. Because P is a minimal complex of graded-free A^{en}-modules, the complex $\mathbb{K} \otimes_{A^{\mathrm{en}}} P$ has zero differential; see Proposition 15.3.15. But then the complex $\mathbb{K} \otimes_A Q$ has zero differential. We conclude that Q is a minimal complex of graded-free A-modules, and hence $\tau : Q \to \mathbb{K}$ is a minimal graded-free resolution over A. Corollary 15.3.21 says that the graded A-modules Q^i are finite; and hence the graded $\mathbb{K}$-modules V_{-i} are finite. □

Corollary 15.3.26 *If A is a left noetherian connected graded ring, then A is a derived graded-pseudo-finite complex over A^{en}.*

Proof This is an immediate consequence of Proposition 15.3.24, which says that the complex P is pseudo-finite graded-semi-free over A^{en}. □

Theorem 15.3.27 *Let A, B, C and D be graded rings. For complexes $L \in \mathbf{D}(A \otimes C^{\mathrm{op}}, \mathrm{gr})$, $M \in \mathbf{D}(A \otimes B^{\mathrm{op}}, \mathrm{gr})$ and $N \in \mathbf{D}(B \otimes D^{\mathrm{op}}, \mathrm{gr})$ there is a morphism*

$$\mathrm{ev}^{\mathrm{R,L}}_{L,M,N} : \mathrm{RHom}_A(L, M) \otimes^{\mathrm{L}}_B N \to \mathrm{RHom}_A(L, M \otimes^{\mathrm{L}}_B N)$$

in $\mathbf{D}(C \otimes D^{\mathrm{op}}, \mathrm{gr})$, called derived graded tensor-evaluation. *The morphism $\mathrm{ev}^{\mathrm{R,L}}_{L,M,N}$ is functorial in the objects L, M, N. Moreover, if all three conditions below hold, then $\mathrm{ev}^{\mathrm{R,L}}_{L,M,N}$ is an isomorphism.*

(i) *The complex L is derived graded-pseudo-finite over A.*
(ii) *The complex M has bounded below cohomology.*
(iii) *The complex N has finite graded-flat dimension over B.*

Just to clarify, condition (ii) says that $\mathrm{H}^i(M) = 0$ for $i \ll 0$. We do not care about the boundedness (in algebraic degree) of the graded modules $\mathrm{H}^i(M)$.

Proof The proof of just like that of Theorem 12.9.10, but working in the graded derived categories. For condition (iii), notice that if N has finite graded-flat dimension over B, then there is an isomorphism $P \cong N$ in $\mathbf{D}(B, \mathrm{gr})$, where P is a bounded complex of graded-flat B-modules. □

Definition 15.3.28 Let A be a graded ring.

(1) The full subcategory of $\mathbf{D}(A, \mathrm{gr})$ on the complexes M whose cohomology modules $\mathrm{H}^p(M)$ belong to $\mathbf{M}_{\mathrm{dwf}}(A, \mathrm{gr})$ is denoted by $\mathbf{D}_{\mathrm{dwf}}(A, \mathrm{gr})$.
(2) The full subcategory of $\mathbf{D}(A, \mathrm{gr})$ on the complexes whose cohomology modules belong to $\mathbf{M}_{\mathrm{f}}(A, \mathrm{gr})$ is denoted by $\mathbf{D}_{\mathrm{f}}(A, \mathrm{gr})$.
(3) The full subcategory of $\mathbf{D}(A, \mathrm{gr})$ on the complexes whose cohomology modules belong to $\mathbf{M}_{\mathrm{cof}}(A, \mathrm{gr})$ is denoted by $\mathbf{D}_{\mathrm{cof}}(A, \mathrm{gr})$.

Because $\mathbf{M}_{\mathrm{dwf}}(A, \mathrm{gr})$ is a a thick abelian subcategory of $\mathbf{M}(A, \mathrm{gr})$, it follows that $\mathbf{D}_{\mathrm{dwf}}(A, \mathrm{gr})$ is a full triangulated subcategory of $\mathbf{D}(A, \mathrm{gr})$, and likewise for $\mathbf{D}_{\mathrm{f}}(A, \mathrm{gr})$ when A is left noetherian, and for $\mathbf{D}_{\mathrm{cof}}(A, \mathrm{gr})$ when A is right noetherian connected.

Because $(-)^*$ is exact, it induces a triangulated functor

$$(-)^* : \mathbf{D}(A, \mathrm{gr})^{\mathrm{op}} \to \mathbf{D}(A^{\mathrm{op}}, \mathrm{gr}). \tag{15.3.29}$$

Proposition 15.3.30 *Let A be a graded ring.*
(1) *If $M \in \mathbf{D}_{\mathrm{dwf}}(A, \mathrm{gr})$ then $M^* \in \mathbf{D}_{\mathrm{dwf}}(A^{\mathrm{op}}, \mathrm{gr})$.*
(2) *The functor*

$$(-)^* : \mathbf{D}_{\mathrm{dwf}}(A, \mathrm{gr})^{\mathrm{op}} \to \mathbf{D}_{\mathrm{dwf}}(A^{\mathrm{op}}, \mathrm{gr})$$

is an equivalence of $\mathbb{K}$-linear triangulated categories.
(3) *If A is left noetherian connected, then the functor*

$$(-)^* : \mathbf{D}_{\mathrm{f}}(A, \mathrm{gr})^{\mathrm{op}} \to \mathbf{D}_{\mathrm{cof}}(A^{\mathrm{op}}, \mathrm{gr})$$

is an equivalence of $\mathbb{K}$-linear triangulated categories.

Exercise 15.3.31 Prove this proposition. (Hint: use Proposition 15.2.28 and Theorem 15.2.34.)

The next result is from [153].

Theorem 15.3.32 *Let A, B and C be graded $\mathbb{K}$-rings.*
(1) *For $M \in \mathbf{D}(A \otimes B^{\mathrm{op}}, \mathrm{gr})$ and $N \in \mathbf{D}(A \otimes C^{\mathrm{op}}, \mathrm{gr})$, there is a morphism*

$$\theta_{M,N} : \mathrm{RHom}_A(M, N) \to \mathrm{RHom}_{A^{\mathrm{op}}}(N^*, M^*)$$

in $\mathbf{D}(B \otimes C^{\mathrm{op}}, \mathrm{gr})$, that is functorial in M and N.
(2) *If $M \in \mathbf{D}_{\mathrm{dwf}}(A \otimes B^{\mathrm{op}}, \mathrm{gr})$ and $N \in \mathbf{D}_{\mathrm{dwf}}(A \otimes C^{\mathrm{op}}, \mathrm{gr})$ then $\theta_{M,N}$ is an isomorphism.*

Proof

(1) Choose a K-graded-projective resolution $\rho : P \to M$ in $\mathbf{C}_{\mathrm{str}}(A \otimes B^{\mathrm{op}}, \mathrm{gr})$. By Proposition 15.3.6 the complex P is K-graded-projective over A; so there is a canonical isomorphism

$$\phi_\rho : \mathrm{RHom}_A(M, N) \xrightarrow{\simeq} \mathrm{Hom}_A(P, N)$$

in $\mathbf{D}(B \otimes C^{\mathrm{op}}, \mathrm{gr})$. $\mathbb{K}$-linear duality gives a homomorphism

$$\tilde{\theta}_{P,N} : \mathrm{Hom}_A(P, N) \to \mathrm{Hom}_{A^{\mathrm{op}}}(N^*, P^*)$$

in $\mathbf{C}_{\mathrm{str}}(B \otimes C^{\mathrm{op}}, \mathrm{gr})$.

Now $\rho^* : M^* \to P^*$ is a quasi-isomorphism in $\mathbf{C}_{\mathrm{str}}(B \otimes A^{\mathrm{op}}, \mathrm{gr})$, and according to Proposition 15.3.6(1) and Theorem 15.3.5(3), the complex P^* is K-graded-injective over A^{op}. Therefore there is a canonical isomorphism

$$\psi_{\rho^*} : \mathrm{Hom}_{A^{\mathrm{op}}}(N^*, P^*) \xrightarrow{\simeq} \mathrm{RHom}_{A^{\mathrm{op}}}(N^*, M^*)$$

in $\mathbf{D}(B \otimes C^{\mathrm{op}}, \mathrm{gr})$. The resulting morphism

$$\theta_{M,N} := \psi_{\rho^*} \circ \mathrm{Q}(\tilde{\theta}_{P,N}) \circ \phi_\rho : \mathrm{RHom}_A(M, N) \to \mathrm{RHom}_{A^{\mathrm{op}}}(M^*, N^*)$$

in $\mathbf{D}(B \otimes C^{\mathrm{op}}, \mathrm{gr})$ is functorial in M and N.

(2) Because the restriction functors between the derived categories are conservative (see Proposition 14.3.10), we can forget the rings B and C, and just consider $\theta_{M,N}$ as a morphism $\mathbf{D}(\mathbb{K}, \mathrm{gr})$.

For every integer i there is a canonical isomorphism

$$\mathrm{H}^i(\mathrm{RHom}_A(M, N)) \cong \mathrm{Hom}_{\mathbf{D}_{\mathrm{dwf}}(A,\mathrm{gr})}(M, N[i]). \tag{15.3.33}$$

Using the isomorphism $N^*[-i] \cong N[i]^*$ in $\mathbf{D}(A^{\mathrm{op}}, \mathrm{gr})$ we also have

$$\begin{aligned} \mathrm{H}^i(\mathrm{RHom}_{A^{\mathrm{op}}}(N^*, M^*)) &\cong \mathrm{Hom}_{\mathbf{D}_{\mathrm{dwf}}(A^{\mathrm{op}},\mathrm{gr})}(N^*, M^*[i]) \\ &\cong \mathrm{Hom}_{\mathbf{D}_{\mathrm{dwf}}(A^{\mathrm{op}},\mathrm{gr})}(N[i]^*, M^*). \end{aligned} \tag{15.3.34}$$

The construction of $\theta_{M,N}$ in the proof of item (1) above shows that the $\mathbb{K}$-linear homomorphism

$$\mathrm{H}^i(\theta_{M,N}) : \mathrm{H}^i(\mathrm{RHom}_A(M, N)) \to \mathrm{H}^i(\mathrm{RHom}_{A^{\mathrm{op}}}(N^*, M^*))$$

becomes, after using the isomorphisms (15.3.33) and (15.3.34), the duality morphism

$$\mathrm{Hom}_{\mathbf{D}_{\mathrm{dwf}}(A,\mathrm{gr})}(M, N[i]) \to \mathrm{Hom}_{\mathbf{D}_{\mathrm{dwf}}(A^{\mathrm{op}},\mathrm{gr})}(N[i]^*, M^*).$$

According to Proposition 15.3.30(2), this is an isomorphism. □

Remark 15.3.35 In this chapter, and the two chapters following it, we work over a base field $\mathbb{K}$. This simplifies a lot of the constructions. It is however possible to relax this condition, and to work over a noetherian commutative base ring $\mathbb{K}$ that is not a field. But then there are (at least) two complications. The first complication is that we will have to fix an injective cogenerator $\mathbb{K}^*$ of $\mathbf{M}(\mathbb{K})$ (cf. Section 11.6). Matlis Duality will take the form $M^* := \mathrm{Hom}_{\mathbb{K}}(M, \mathbb{K}^*)$, and $\mathrm{Hom}_{\mathbb{K}}(-, -)$ might also involve continuity in a delicate way.

The second complication is that for the homological algebra to be effective, the graded $\mathbb{K}$-rings would have to be *flat*. The absence of flatness poses serious difficulties. This can be tackled using *K-flat DG ring resolutions*; and these DG rings would have to be bigraded (with upper and lower indices, as explained in Section 15.1). This sort of resolution has not been studied yet, as far as we know; but see [180] for some preliminary ideas.

15.4 Artin–Schelter Regular Graded Rings

In this section we talk about an important class of graded rings: the *Artin–Schelter regular graded rings*. These are connected graded noncommutative rings that homologically are similar to commutative polynomial rings. A great deal of noncommutative ring theory and homological algebra grew out of the study of this class of rings. See Remark 15.4.23 for some historical notes.

We continue with Conventions 15.1.17 and 15.2.1. Thus all graded rings are central over the base field $\mathbb{K}$.

Suppose A is a connected graded ring. Recall that the augmentation ideal of A is denoted by $\mathfrak{m}$. The opposite ring of A is A^{op}, and the enveloping ring is A^{en}. We view $A/\mathfrak{m} \cong \mathbb{K}$ as a graded A^{en}-module.

Definition 15.4.1 Let A be a noetherian connected graded ring.

(1) The graded ring A is called a *regular graded ring* if it has finite graded global cohomological dimension; namely if there is a natural number d such that $\mathrm{Ext}^i_A(M,N) = 0$ and $\mathrm{Ext}^i_{A^{\mathrm{op}}}(M',N') = 0$ for all $i > d$, $M, N \in \mathbf{M}(A, \mathrm{gr})$ and $M', N' \in \mathbf{M}(A^{\mathrm{op}}, \mathrm{gr})$. The smallest such natural number d is called the *graded global dimension* of the ring A.

(2) The graded ring A is called a *Gorenstein graded ring* if the graded bimodule A has finite graded-injective dimension over A and over A^{op}; namely if there is a natural number d such that $\mathrm{Ext}^i_A(M,A) = 0$ and $\mathrm{Ext}^i_{A^{\mathrm{op}}}(M',A) = 0$ for all $i > d$, $M \in \mathbf{M}(A,\mathrm{gr})$ and $M' \in \mathbf{M}(A^{\mathrm{op}},\mathrm{gr})$. The smallest such natural number d is called the *graded-injective dimension* of the ring A.

Recall that the graded-projective dimension of a graded A-module M is the smallest generalized integer $d \in \mathbb{N} \cup \{\pm\infty\}$ such that $\mathrm{Ext}^i_A(M,N) = 0$ for all integers $i > d$ and all $N \in \mathbf{M}(A,\mathrm{gr})$.

The minimal graded-free resolution of a graded module was introduced in Definitions 15.3.14 and 15.3.17.

Theorem 15.4.2 *Let A be a noetherian connected graded ring, and let d be a natural number. The following five conditions are equivalent.*

(i) *The graded ring A is regular, of graded global dimension d.*

(ii) *The graded-projective dimension of the graded A-module* $\mathbb{K}$ *is* d.
(iii) *The graded* $\mathbb{K}$*-modules* $\operatorname{Tor}_i^A(\mathbb{K}, \mathbb{K})$ *vanish for all* $i > d$, *but not for* $i = d$.
(iv) *The graded* $\mathbb{K}$*-modules* $\operatorname{Ext}_A^i(\mathbb{K}, \mathbb{K})$ *vanish for all* $i > d$, *but not for* $i = d$.
(v) *Let* $P \to A$ *be the minimal graded-free resolution of the graded* A^{en}*-module* A. *Then the graded* A^{en}*-modules* P^{-i} *vanish for all* $i > d$, *but not for* $i = d$.

Observe that in this theorem, conditions (i), (iii) and (v) are op-symmetric, but not conditions (ii) and (iv).

Proof Consider the minimal graded-free resolution $\rho : P \to A$ of A over A^{en}. We express the graded-free A^{en}-module P^{-i}, for $i \geq 0$, as

$$P^{-i} = A^{\mathrm{en}} \otimes V_i \cong A \otimes V_i \otimes A \tag{15.4.3}$$

with $V_i \in \mathbf{M}_{\mathrm{f}}(\mathbb{K}, \mathrm{gr})$. The finiteness of the graded $\mathbb{K}$-modules V_i is by Proposition 15.3.24. In formula (15.4.3) the ring A^{en} acts on itself from the left, and it acts on $A \otimes V_i \otimes A$ by the outside action, i.e.

$$(a_1 \otimes \mathrm{op}(a_2)) \cdot (b_1 \otimes v \otimes b_2) = (a_1 \cdot b_1) \otimes v \otimes (b_2 \cdot a_2)$$

for $a_1 \otimes \mathrm{op}(a_2) \in A^{\mathrm{en}}$ and $b_1 \otimes v \otimes b_2 \in A \otimes V_i \otimes A$.

As explained in the proof of Proposition 15.3.24, in the category $\mathbf{C}_{\mathrm{str}}(A^{\mathrm{op}}, \mathrm{gr})$ the homomorphism $\rho : P \to A$ is a homotopy equivalence. Hence, taking a module $M \in \mathbf{M}(A, \mathrm{gr})$, and writing $Q_M := P \otimes_A M$, we get a quasi-isomorphism

$$\tau_M := \rho \otimes \mathrm{id}_M : Q_M = P \otimes_A M \to A \otimes_A M \cong M$$

in $\mathbf{C}_{\mathrm{str}}(A, \mathrm{gr})$. Because M is a graded-free $\mathbb{K}$-module, it follows that every

$$Q_M^{-i} = P^{-i} \otimes_A M \cong A \otimes V_i \otimes M$$

is also a graded-free A-module. In this way we obtain a (functorial) graded-free resolution $\tau_M : Q_M \to M$ in $\mathbf{C}_{\mathrm{str}}(A, \mathrm{gr})$.

Assume now that condition (v) holds. This means that $V_i = 0$ for all $i > d$, and $V_d \neq 0$. The graded-free resolution Q_M has length d for every nonzero $M \in \mathbf{M}(A, \mathrm{gr})$. We see that

$$\operatorname{Ext}_A^i(M, N) \cong \mathrm{H}^i(\operatorname{Hom}_A(Q_M, N)) = 0$$

for all $N \in \mathbf{M}(A, \mathrm{gr})$ and $i > d$. By the op-symmetry of condition (v), it also follows that $\operatorname{Ext}_{A^{\mathrm{op}}}^i(M', N') = 0$ for all $M', N' \in \mathbf{M}(A^{\mathrm{op}}, \mathrm{gr})$ and $i > d$. This says that the graded global dimension of A is at most d. To get equality, we test for $M = N = \mathbb{K}$. In this case we know that $Q_{\mathbb{K}}$ is a minimal graded-free resolution of $\mathbb{K}$ over A (see the proof of Proposition 15.3.24), so the complex

$\mathbb{K} \otimes_A Q_{\mathbb{K}}$ has trivial differential. Therefore

$$\begin{aligned}\operatorname{Ext}^i_A(\mathbb{K}, \mathbb{K}) &\cong \mathrm{H}^i(\operatorname{Hom}_A(Q_{\mathbb{K}}, \mathbb{K})) \\ &\cong \mathrm{H}^i(\operatorname{Hom}_{\mathbb{K}}(\mathbb{K} \otimes_A Q_{\mathbb{K}}, \mathbb{K})) \cong \operatorname{Hom}_{\mathbb{K}}(V_i, \mathbb{K}).\end{aligned} \tag{15.4.4}$$

This is nonzero for $i = d$. Likewise over A^{op}. We have thus verified conditions (i), (ii) and (iv). Condition (iii) is verified similarly: the resolution $Q_{\mathbb{K}}$ shows that $\operatorname{Tor}^A_i(\mathbb{K}, \mathbb{K}) \cong V_i$ as graded $\mathbb{K}$-modules.

Let us prove the implication (i) $\Rightarrow$ (v): since $\operatorname{Ext}^i_A(\mathbb{K}, \mathbb{K}) = 0$ for all $i > d$, equations (15.4.4) and (15.4.3) say that $P^{-i} = 0$ in this range. On the other hand, there is some pair $M, N \in \mathbf{M}(A, \mathrm{gr})$ such that $\operatorname{Ext}^d_A(M, N) \neq 0$, and this says that $P^{-d} \neq 0$.

The other implications are proved similarly, and we leave them as an exercise for the reader. □

Exercise 15.4.5 Finish the proof of Theorem 15.4.2.

The next condition first appeared in the paper [6] by M. Artin and W. Schelter. Throughout, "AS" is an abbreviation for "Artin–Schelter."

Definition 15.4.6 Let A be a noetherian connected graded ring. We say that A satisfies the *AS condition* if there are integers n and l such that

$$\operatorname{Ext}^n_A(\mathbb{K}, A) \cong \operatorname{Ext}^n_{A^{\mathrm{op}}}(\mathbb{K}, A) \cong \mathbb{K}(l)$$

as graded $\mathbb{K}$-modules, and

$$\operatorname{Ext}^i_A(\mathbb{K}, A) \cong \operatorname{Ext}^i_{A^{\mathrm{op}}}(\mathbb{K}, A) \cong 0$$

for all $i \neq n$. The number n is called the *AS dimension* of A, and the number l is called the *AS index* of A.

In derived category terms, the AS condition says that

$$\operatorname{RHom}_A(\mathbb{K}, A) \cong \operatorname{RHom}_{A^{\mathrm{op}}}(\mathbb{K}, A) \cong \mathbb{K}(l)[-n] \tag{15.4.7}$$

in $\mathbf{D}(\mathbb{K}, \mathrm{gr})$.

Definition 15.4.8 Let A be a noetherian connected graded ring.

(1) We say that A is an *AS regular graded ring* of dimension n if it is a regular graded ring (Definition 15.4.1(1)), and it satisfies the *AS condition* (Definition 15.4.6) with AS dimension n.

(2) We say that A is an *AS Gorenstein graded ring* of dimension n if it is a Gorenstein graded ring (Definition 15.4.1(2)), and it satisfies the *AS condition* (Definition 15.4.6) with AS dimension n.

As we shall see later, in Section 16.5, the AS condition implies that the local cohomology of graded A-modules behaves as though A is commutative. In the paper [10] by Artin and J. J. Zhang, subsequent to [6], this behavior was formalized as the χ *condition*; see Definition 16.5.14 below.

Some texts do not make the assumption that the connected graded ring A is noetherian, as we did in Definition 15.4.8. However, without this assumption the theory is not as rich.

It is natural to ask whether the AS dimension of the ring A coincides with its graded global dimension (in the AS regular case), or with its graded-injective dimension (in the AS Gorenstein case). The next propositions say that these numbers agree.

Proposition 15.4.9 *Let A be an AS regular graded ring. Then the AS dimension of A equals its graded global dimension.*

Proof Let n be the graded global dimension of A. Let $P \to A$ be the minimal graded-free resolution of the graded A^{en}-module A. By Theorem 15.4.2 we know that the graded A^{en}-modules P^{-i} vanish for all $i > n$, but not for $i = n$. Let $Q_{\mathbb{K}} \to \mathbb{K}$ be the induced graded-free resolution of the A-module $\mathbb{K}$ from the proof of Theorem 15.4.2. Then the complex $Q_{\mathbb{K}}$ is concentrated in cohomological degrees $[-n, 0]$, and $Q_{\mathbb{K}}^{-n} \neq 0$. Also, as shown there, $Q_{\mathbb{K}} \to \mathbb{K}$ is a minimal graded-free resolution of $\mathbb{K}$ over A.

Now let $T := \mathrm{Hom}_A(Q_{\mathbb{K}}, A) \in \mathbf{C}(A^{\mathrm{op}}, \mathrm{gr})$, so $\mathrm{RHom}_A(\mathbb{K}, A) \cong T$ in $\mathbf{D}(\mathbb{K}, \mathrm{gr})$. A quick calculation shows that T is a minimal complex of graded-free A^{op}-modules. This complex is concentrated in cohomological degrees $[0, n]$, and $T^n \neq 0$. Therefore $\mathrm{Ext}^n_A(\mathbb{K}, A) \cong \mathrm{H}^n(T) \neq 0$. A similar calculation shows that $\mathrm{Ext}^n_{A^{\mathrm{op}}}(\mathbb{K}, A) \neq 0$. We see that the AS dimension of A is n. □

Proposition 15.4.10 *Let A be an AS Gorenstein graded ring. Then the AS dimension of A equals its graded-injective dimension.*

Proof This is an immediate consequence of [72, Theorem 4.5]. □

In Examples 17.3.15 and 17.3.16 we will present some important kinds of AS regular rings.

Definition 15.4.11 Let A be a nonzero graded ring. A central element a of A is called a *regular central element* if it is not a zero-divisor; i.e. $a \cdot b = 0$ implies $b = 0$.

If the element $a \in A$ is not central, then we have to distinguish between left and right regularity of a.

The next two theorems discuss the relation between the rings A and B from the following setup.

Setup 15.4.12 Let A be a noetherian connected graded ring, and let $a \in A$ be a homogeneous regular central element of positive degree. Define the connected graded ring $B := A/(a)$.

Theorem 15.4.13 *Assume Setup* 15.4.12.

(1) *If B is a regular graded ring of graded global dimension $n - 1$, then A is a regular graded ring of graded global dimension n.*

(2) *Moreover, if B is an AS regular graded ring of dimension $n - 1$, then A is an AS regular graded of dimension n.*

Proof

(1) By Theorem 15.4.2, to prove that A is a regular graded ring of graded global dimension n, it is enough to show that the cohomology of the complex $\mathbb{K} \otimes^{\mathrm{L}}_A \mathbb{K} \in \mathbf{D}(\mathbb{K}, \mathrm{gr})$ has concentration $[-n, 0]$. Because a graded-free resolution of $\mathbb{K}$ over A is also a free resolution in the ungraded sense, and because we are only interested in the vanishing of cohomology, we can forget the algebraic grading, and just look at the complex $\mathbb{K} \otimes^{\mathrm{L}}_A \mathbb{K} \in \mathbf{D}(\mathbb{K})$. Similarly, because B is a regular graded ring of graded global dimension $n - 1$, we know that the complex $\mathbb{K} \otimes^{\mathrm{L}}_B \mathbb{K} \in \mathbf{D}(\mathbb{K})$ satisfies

$$\operatorname{con}(\mathrm{H}(\mathbb{K} \otimes^{\mathrm{L}}_B \mathbb{K})) = [-n + 1, 0]. \tag{15.4.14}$$

Let $\tilde{B} := \mathrm{K}(A, a)$ be the Koszul complex on the regular central element $a \in A$; namely

$$\tilde{B} = (\cdots \to 0 \to A \xrightarrow{a \cdot (-)} A \to 0 \to \cdots),$$

concentrated in cohomological degrees -1 and 0. This is a DG ring: as a cohomologically graded ring we have $\tilde{B} := A \otimes \mathbb{K}[x]$, where $\mathbb{K}[x]$ is the strongly commutative polynomial ring on the variable x that has cohomological degree -1 (see Examples 3.1.23 and 3.3.9). The differential is $\mathrm{d}(x) := a$. The canonical DG ring homomorphism $\tilde{B} \to B$ is a quasi-isomorphism. Also $\tilde{B}$ is semi-free DG module over A and over A^{op}. According to Theorem 12.7.2(2), there is an isomorphism

$$\mathbb{K} \otimes^{\mathrm{L}}_B \mathbb{K} \cong \mathbb{K} \otimes^{\mathrm{L}}_{\tilde{B}} \mathbb{K} \tag{15.4.15}$$

in $\mathbf{D}(\mathbb{K})$.

By the associativity of the derived tensor product, and because $\mathbb{K}$ is a DG A-module via the DG ring homomorphism $A \to \tilde{B}$, there are isomorphisms

$$\mathbb{K} \otimes^{\mathrm{L}}_A \mathbb{K} \cong \mathbb{K} \otimes^{\mathrm{L}}_A (\tilde{B} \otimes^{\mathrm{L}}_{\tilde{B}} \mathbb{K}) \cong (\mathbb{K} \otimes^{\mathrm{L}}_A \tilde{B}) \otimes^{\mathrm{L}}_{\tilde{B}} \mathbb{K} \tag{15.4.16}$$

in $\mathbf{D}(\mathbb{K})$. Now the image of the element a in $\mathbb{K}$ is zero, so we obtain these isomorphisms

$$\mathbb{K} \otimes^{\mathrm{L}}_A \tilde{B} \cong \mathbb{K} \otimes_A \tilde{B} \cong (\mathbb{K} \xrightarrow{0 \cdot (-)} \mathbb{K}) \cong \mathbb{K}[1] \oplus \mathbb{K}$$

in $\mathsf{D}(\tilde{B}^{\mathrm{op}})$. Substituting this into (15.4.16) we get

$$\mathbb{K} \otimes^{\mathrm{L}}_{A} \mathbb{K} \cong (\mathbb{K}[1] \oplus \mathbb{K}) \otimes^{\mathrm{L}}_{\tilde{B}} \mathbb{K} \cong (\mathbb{K} \otimes^{\mathrm{L}}_{\tilde{B}} \mathbb{K})[1] \oplus (\mathbb{K} \otimes^{\mathrm{L}}_{\tilde{B}} \mathbb{K})$$

in $\mathsf{D}(\mathbb{K})$. Hence

$$\mathrm{H}^{i}(\mathbb{K} \otimes^{\mathrm{L}}_{A} \mathbb{K}) \cong \mathrm{H}^{i+1}(\mathbb{K} \otimes^{\mathrm{L}}_{B} \mathbb{K}) \oplus \mathrm{H}^{i}(\mathbb{K} \otimes^{\mathrm{L}}_{B} \mathbb{K}) \tag{15.4.17}$$

as $\mathbb{K}$-modules. This, together with (15.4.14), imply that $\operatorname{con}(\mathrm{H}(\mathbb{K} \otimes^{\mathrm{L}}_{A} \mathbb{K})) = [-n, 0]$, as required.

(2) Now B is an AS regular graded ring of dimension $n - 1$. As explained in part (1) of the proof, we may neglect the algebraic grading, and we only need to prove that

$$\mathrm{RHom}_{A}(\mathbb{K}, A) \cong \mathbb{K}[-n] \tag{15.4.18}$$

in $\mathsf{D}(\mathbb{K})$.

Let $\tilde{B}$ be the DG ring from above. Because $\mathbb{K}$ is a DG A-module via the DG ring homomorphism $A \to \tilde{B}$, there is an adjunction isomorphism

$$\mathrm{RHom}_{A}(\mathbb{K}, A) \cong \mathrm{RHom}_{\tilde{B}}(\mathbb{K}, \mathrm{RHom}_{A}(\tilde{B}, A)) \tag{15.4.19}$$

in $\mathsf{D}(\mathbb{K})$. Now $\tilde{B}$ is a semi-free DG module over A, so

$$\mathrm{RHom}_{A}(\tilde{B}, A) \cong \mathrm{Hom}_{A}(\tilde{B}, A) \cong (A \xrightarrow{(-)\cdot a} A) \cong B[-1]$$

in $\mathsf{D}(\tilde{B})$. Substituting this into (15.4.19), and using Theorem 12.7.2(2), we get

$$\mathrm{RHom}_{A}(\mathbb{K}, A) \cong \mathrm{RHom}_{\tilde{B}}(\mathbb{K}, B[-1]) \cong \mathrm{RHom}_{B}(\mathbb{K}, B)[-1]$$

in $\mathsf{D}(\mathbb{K})$. The AS condition for B says that $\mathrm{RHom}_{B}(\mathbb{K}, B) \cong \mathbb{K}[-n+1]$ in $\mathsf{D}(\mathbb{K})$. Hence the isomorphism (15.4.18) holds. □

Remark 15.4.20 The proof of the theorem above can actually yield more, with some modifications. If we were to make the DG ring $\tilde{B}$ into a bigraded object, with both algebraic and cohomological grading, then we would obtain an isomorphism (15.4.15) in the derived category $\mathsf{D}(\tilde{B}, \mathrm{gr})$ of bigraded DG $\tilde{B}$-modules. The problem is that we did not develop this kind of theory in the book...

The outcome of this more refined construction would be that formula (15.4.17) is an isomorphism in $\mathsf{M}(\mathbb{K}, \mathrm{gr})$. This would give the ability to compare the AS indexes of A and B. As can be guessed (cf. the easy case of Example 17.3.16, taking a to be one of the variables), if the AS index of A is l, then the AS index of B is $l - \deg(a)$.

Theorem 15.4.21 *Assume Setup* 15.4.12.

(1) *If A is a Gorenstein graded ring of graded-injective dimension n, then B is a Gorenstein graded ring of graded-injective dimension at most $n - 1$.*

(2) *Moreover, if A is an AS Gorenstein graded ring of dimension n, then B is an AS Gorenstein graded ring of dimension $n - 1$.*

Exercise 15.4.22 Prove Theorem 15.4.21. (Hint: study the proof of Theorem 15.4.13.)

Remark 15.4.23 Here are some historical notes on the material in this section. Around 1985, M. Artin and W. Schelter [6] began a systematic study of connected graded noncommutative rings (better known as graded algebras), with the hope of obtaining a classification in low dimensions, in analogy to the commutative projective geometry of curves and surfaces. This is how the concept of *Artin–Schelter regular graded rings* was born. Closely related concepts are *Sklyanin algebras* and *Koszul algebras*.

The area of research initiated by Artin and Schelter is often called *NC algebraic geometry*. Indeed, in 1994 Artin and J. J. Zhang [10] introduced the concept of *NC projective scheme*. Given a noetherian connected graded ring A, let $\mathsf{M}_{\mathrm{tor}}(A, \mathrm{gr}) \subseteq \mathsf{M}(A, \mathrm{gr})$ be the full subcategory on the $\mathfrak{m}$-torsion modules. The quotient abelian category $\mathsf{M}(A, \mathrm{gr})/\mathsf{M}_{\mathrm{tor}}(A, \mathrm{gr})$ is viewed as the category of quasi-coherent sheaves on an imaginary scheme $\mathrm{Proj}(A)$. Note that if A happens to be commutative, then, by the famous theorem of J.P. Serre, this quotient category is canonically equivalent to $\mathsf{QCoh}\, X$, where $X = \mathrm{Proj}(A)$ is the genuine projective scheme. The geometric metaphor of Artin and Zhang turned out to be very useful in predicting "geometric properties" of connected graded rings, which were later proved algebraically, working in $\mathsf{M}(A, \mathrm{gr})$ or its derived category $\mathsf{D}(A, \mathrm{gr})$.

For AS regular graded rings of dimension ≤ 3 generated in degree 1, the classification was pretty much complete by 1991, in papers by Artin, jointly with J. Tate and M. Van den Bergh; see the papers [8] and [9], and the survey [145]. Research on AS graded rings of higher dimensions, and with generators in various degrees, was later conducted by J.T. Stafford, S.P. Smith, Zhang, and their students and collaborators.

In 2006, V. Ginzburg published the highly influential preprint [52] on CY (Calabi–Yau) algebras. Since then, the research on AS regular rings and their relatives has merged with the research on CY rings and categories. (It turns out that for connected graded rings, the twisted CY rings are precisely the AS regular rings.) See Remark 18.6.20 for more on CY theory. Two of the latest papers on graded CY rings are [123] and [124].

In the subsequent chapters of the book we shall deal with a few ideas that emerged as by-products of the Artin school of NC algebraic geometry: the χ condition, balanced dualizing complexes and rigid dualizing complexes.

16

Derived Torsion over NC Graded Rings

As already mentioned, connected graded noncommutative rings are similar in several aspects to complete local commutative rings. In this chapter we concentrate on the derived $\mathfrak{m}$-torsion functors associated to a connected graded ring A with augmentation ideal $\mathfrak{m}$.

Some of the definitions and results here are quite old, going back to papers from the 1990s. Other definitions and results are adaptations of the connected graded setting of the work in the recent paper [157].

The χ *condition* on a noncommutative noetherian connected graded ring A was introduced by M. Artin and J. J. Zhang [10] to guarantee that the *noncommutative projective scheme* $\operatorname{Proj}(A)$ will have good "geometric properties," resembling the classical commutative case. In this chapter and the subsequent one, we will focus on another property that the χ condition implies: the *symmetry of derived torsion*; namely that from the point of view of derived torsion, the ring A looks commutative. It is important to say that this phenomenon is *only seen in the derived category of graded bimodules* – on the elementary level the ring A is often terribly noncommutative (cf. Examples 16.5.25 and 16.5.26), and the bimodules are far from being central!

In this chapter we follow Conventions 15.1.17 and 15.2.1. So $\mathbb{K}$ is a base field, and all rings are algebraically graded central $\mathbb{K}$-rings. The definitions and results from Chapter 15 will be used here freely. See Remark 15.3.35 regarding the possibility of replacing the base field $\mathbb{K}$ with a base commutative ring.

16.1 Quasi-Compact Finite Dimensional Functors

The content of this section is adapted from [157, Section 1]. We state the results in the context of algebraically graded rings and modules, since here we need them for the graded $\mathfrak{m}$-torsion functor $F := \Gamma_{\mathfrak{m}}$; yet these results hold in greater generality (see [157]).

Recall from Section 15.1 that given a graded ring A, the category of graded A-modules is $\mathsf{M}(A, \mathrm{gr})$. This is a $\mathbb{K}$-linear abelian category. The category of complexes in $\mathsf{M}(A, \mathrm{gr})$ is $\mathsf{C}(A, \mathrm{gr})$, the homotopy category is $\mathsf{K}(A, \mathrm{gr})$, and the derived category is $\mathsf{D}(A, \mathrm{gr})$.

In this section we consider the following setup:

Setup 16.1.1 *A and B are graded rings, and $F : \mathsf{M}(A, \mathrm{gr}) \to \mathsf{M}(B, \mathrm{gr})$ is a left exact linear functor.*

The functor F extends to a triangulated functor $F : \mathsf{K}(A, \mathrm{gr}) \to \mathsf{K}(B, \mathrm{gr})$ between the homotopy categories. Because there are enough K-injective resolutions, F has a right derived functor $(\mathrm{R}F, \eta^{\mathrm{R}})$. Let us recall what this means: $\mathrm{R}F : \mathsf{D}(A, \mathrm{gr}) \to \mathsf{D}(B, \mathrm{gr})$ is a triangulated functor, and $\eta^{\mathrm{R}} : \mathrm{Q} \circ F \to \mathrm{R}F \circ \mathrm{Q}$ is a morphism of triangulated functors $\mathsf{K}(A, \mathrm{gr}) \to \mathsf{D}(B, \mathrm{gr})$ that has a certain universal property (see Definition 8.4.7). The right derived functor $(\mathrm{R}F, \eta^{\mathrm{R}})$ can be constructed using a K-injective presentation: for each complex M we choose a K-injective resolution $\rho_M : M \to I_M$ in $\mathsf{C}_{\mathrm{str}}(A, \mathrm{gr})$, and then we take $\mathrm{R}F(M) := F(I_M)$ and $\eta^{\mathrm{R}}_M := F(\rho_M)$.

Recall that the functor F has the classical right derived functors $\mathrm{R}^q F : \mathsf{M}(A, \mathrm{gr}) \to \mathsf{M}(B, \mathrm{gr})$. These can be expressed as follows: $\mathrm{R}^q F(M) = \mathrm{H}^q(\mathrm{R}F(M))$ for a module $M \in \mathsf{M}(A, \mathrm{gr})$. Since F is left exact, the canonical homomorphism $F \to \mathrm{R}^0 F$ is an isomorphism.

Here is a classical definition:

Definition 16.1.2 The *right cohomological dimension of F* is

$$d := \sup\{q \in \mathbb{N} \mid \mathrm{R}^q F \neq 0\} \in \mathbb{N} \cup \{\pm\infty\}.$$

If $F \neq 0$ then $d \in \mathbb{N} \cup \{\infty\}$, but for $F = 0$ the dimension is $d = -\infty$. The generalized integer d is said to be finite if $d < \infty$; this terminology is designed so that the zero functor will have finite right cohomological dimension.

Here is another standard definition.

Definition 16.1.3 A module $I \in \mathsf{M}(A, \mathrm{gr})$ is called a *right F-acyclic object* if $\mathrm{R}^q F(I) = 0$ for all $q > 0$.

Of course, every injective object of $\mathsf{M}(A, \mathrm{gr})$ is right F-acyclic. But often there are many more right F-acyclic objects.

Definition 16.1.4 A complex $I \in \mathsf{C}(A, \mathrm{gr})$ is called a *right F-acyclic complex* if the morphism $\eta^{\mathrm{R}}_I : F(I) \to \mathrm{R}F(I)$ in $\mathsf{D}(B, \mathrm{gr})$ is an isomorphism.

Similarly, a K-injective complex is right F-acyclic, but often there are many more. It is easy to see that a module $I \in \mathsf{M}(A, \mathrm{gr})$ is a right F-acyclic object in the sense of Definition 16.1.3 if and only if it is right F-acyclic as a complex, in the sense of Definition 16.1.4.

Lemma 16.1.5 *Let $I \in \mathsf{C}(A, \mathrm{gr})$ be a complex such that each of the modules I^q is right F-acyclic. If I is a bounded below complex, or if F has finite right cohomological dimension, then I is a right F-acyclic complex.*

Proof This is contained in the proof of [62, Corollary I.5.3], but for the sake of completeness we give a proof here.

We can assume that neither F nor I are zero. First let us assume that I is bounded below. We can find a quasi-isomorphism $I \to J$ in $\mathsf{C}_{\mathrm{str}}(A, \mathrm{gr})$, where J is a bounded below complex of injective objects of $\mathsf{M}(A, \mathrm{gr})$, and thus it is a K-injective complex. We must show that $F(I) \to F(J)$ is a quasi-isomorphism. This amounts to showing that the complex $F(K)$ is acyclic, where K is the standard cone on the quasi-isomorphism $I \to J$. Thus we reduce the problem to proving that for an acyclic bounded below complex I made up of right F-acyclic objects, the complex $F(I)$ is acyclic.

Say $\inf(I) = d_0 \in \mathbb{Z}$. For any q let $\mathrm{Z}^q(I) := \mathrm{Ker}(I^q \to I^{q+1})$, the object of degree q cocycles. The acyclicity of the complex I says that we have short exact sequences

$$0 \to \mathrm{Z}^q(I) \to I^q \to \mathrm{Z}^{q+1}(I) \to 0.$$

By induction on $q \geq d_0 - 1$, using the long exact cohomology sequence, we prove that each $\mathrm{Z}^q(I)$ is a right F-acyclic object. Therefore the sequences

$$0 \to F(\mathrm{Z}^q(I)) \to F(I^q) \to F(\mathrm{Z}^{q+1}(I)) \to 0$$

are exact for all q. Splicing together three sequences like this, for $q-1$, q and $q+1$, shows that the sequence $F(I^{q-1}) \to F(I^q) \to F(I^{q+1})$ is exact. Thus the complex $F(I)$ is acyclic.

Now I is no longer bounded below, but the right cohomological dimension of F is finite, say $d \in \mathbb{N}$. Let $I \to J$ be a quasi-isomorphism in $\mathsf{C}_{\mathrm{str}}(A, \mathrm{gr})$, where J is a K-injective complex made up of injective objects (see Corollary 15.3.7(3)). We must show that $F(I) \to F(J)$ is a quasi-isomorphism. As above, by replacing I with the standard cone on $I \to J$, we reduce the problem to proving that for an acyclic complex I made up of right F-acyclic objects, the complex $F(I)$ is acyclic.

Fix an integer q. Let $M := \mathrm{Z}^q(I)$, and let

$$J := (\cdots \to 0 \to I^q \to I^{q+1} \to \cdots),$$

the complex with I^q placed in degree 0 (so J is a shift of a stupid truncation of I). By the previous part of the proof we know that $\mathrm{R}F(J) \cong F(J)$ in $\mathsf{D}(B, \mathrm{gr})$. We have a quasi-isomorphism $M \to J$. Hence, for every $p > d$, by our assumption on F we have

$$0 = \mathrm{H}^p(\mathrm{R}F(M)) \cong \mathrm{H}^p(\mathrm{R}F(J)) \cong \mathrm{H}^p(F(J)) \cong \mathrm{H}^{p+q}(F(I)).$$

Since q was chosen arbitrarily, we conclude that $F(I)$ is acyclic. □

Recall the *cohomological dimension* of the triangulated functor $\mathrm{R}F : \mathsf{D}(A, \mathrm{gr}) \to \mathsf{D}(B, \mathrm{gr})$ from Definition 12.4.1. In what follows we use the various notions introduced in Section 12.4, adapted to the graded setting, such as the concentration interval $\mathrm{con}(M) \subseteq \mathbb{Z}$ of an object $M \in \mathsf{G}(A, \mathrm{gr})$.

Lemma 16.1.6 *The cohomological dimension of the right derived functor* $\mathrm{R}F$ *equals the right cohomological dimension of* F.

Proof As explained in Example 12.4.5, the cohomological dimension of $(\mathrm{R}F)|_{\mathsf{M}(A)}$ equals the right cohomological dimension of the functor F. We need to prove the reverse inequality.

We can assume that $F \neq 0$, and it has finite right cohomological dimension, say $d \in \mathbb{N}$. We shall prove that the cohomological displacement of the derived functor $\mathrm{R}F$ is contained in the interval $[d_0, d_1] := [0, d]$. To be explicit, we shall prove that for every complex $M \in \mathsf{D}(A, \mathrm{gr})$ the integer interval $\mathrm{con}(\mathrm{H}(\mathrm{R}F(M)))$ is contained in the interval $\mathrm{con}(\mathrm{H}(M)) + [0, d]$. This is done by cases, and we can assume that $M \neq 0$.

Case 1. If M is not in $\mathsf{D}^-(A, \mathrm{gr})$, i.e. $\mathrm{con}(\mathrm{H}(M)) = [d_0, \infty]$ for some $d_0 \in \mathbb{Z} \cup \{-\infty\}$, then we can take a K-injective resolution $M \to I$ such that $\inf(I) = d_0$. Then $\mathrm{H}(\mathrm{R}F(M)) = \mathrm{H}(F(I))$ is concentrated in $[d_0, \infty]$.

Case 2. Here $M \in \mathsf{D}^-(A, \mathrm{gr})$, so that $\mathrm{con}(\mathrm{H}(M)) = [d_0, d_1]$ for some $d_0 \in \mathbb{Z} \cup \{-\infty\}$ and $d_1 \in \mathbb{Z}$. We now take a K-injective resolution $M \to I$ such that $\inf(I) = d_0$ and I is made up of injective objects. We must prove that $\mathrm{H}^q(F(I)) = 0$ for all $q > d_1 + d$. This is done as in the proof of Lemma 16.1.5. Let $N := \mathrm{Z}^{d_1}(I)$, and let

$$J := (\cdots \to 0 \to I^{d_1} \to I^{d_1+1} \to \cdots),$$

the complex with I^{d_1} placed in degree 0. So there is a quasi-isomorphism $N \to J$, and it is an injective resolution of N. Hence for every $p > d$ we have

$$0 = \mathrm{H}^p(\mathrm{R}F(N)) \cong \mathrm{H}^p(F(J)) \cong \mathrm{H}^{d_1+p}(F(I)). \qquad \square$$

Definition 16.1.7 Let M and N be linear categories with arbitrarily direct sums. A linear functor $G : \mathsf{M} \to \mathsf{N}$ is called *quasi-compact* if it commutes with infinite direct sums. Namely, for every collection $\{M_x\}_{x \in X}$ of objects of M, the canonical morphism

$$\bigoplus\nolimits_{x \in X} G(M_x) \to G\Big(\bigoplus\nolimits_{x \in X} M_x\Big)$$

in N is an isomorphism.

The name "quasi-compact functor" is inspired by the property of pushforward of quasi-coherent sheaves along a quasi-compact map of schemes.

Lemma 16.1.8 *Assume that the functors $\mathrm{R}^q F$ are quasi-compact, for all $q > 0$. Let $\{I_x\}_{x \in X}$ be a collection of right F-acyclic objects of $\mathbf{M}(A, \mathrm{gr})$. Then the object $I := \bigoplus_{x \in X} I_x \in \mathbf{M}(A, \mathrm{gr})$ is right F-acyclic.*

Proof Take some $q > 0$. Because $\mathrm{R}^q F$ is quasi-compact, the canonical homomorphism $\bigoplus_{x \in X} \mathrm{R}^q F(I_x) \to \mathrm{R}^q F(I)$ in N is an isomorphism. But by assumption, $\mathrm{R}^q F(I_x) = 0$ for all x. $\square$

Lemma 16.1.9 *Assume the functors $\mathrm{R}^q F$ are quasi-compact for all $q \geq 0$. Let $\{I_x\}_{x \in X}$ be a collection of right F-acyclic complexes in $\mathbf{C}(A, \mathrm{gr})$, and define $I := \bigoplus_{x \in X} I_x \in \mathbf{C}(A, \mathrm{gr})$. If I is a bounded below complex, or if F has finite right cohomological dimension, then I is a right F-acyclic complex.*

Proof For every index x let us choose a quasi-isomorphism $\phi_x : I_x \to J_x$, where J_x is a K-injective complex consisting of injective objects, and $\inf(J_x) \geq \inf(I_x)$. Define $J := \bigoplus_{x \in X} J_x$. We get a quasi-isomorphism $\phi := \bigoplus_{x \in X} \phi_x : I \to J$ in $\mathbf{C}_{\mathrm{str}}(A, \mathrm{gr})$. By construction, if I is a bounded below complex, then so is J.

For every $x \in X$ there is a commutative diagram

$$\begin{array}{ccc} F(I_x) & \xrightarrow{F(\phi_x)} & F(J_x) \\ \Big\downarrow{\scriptstyle \eta^{\mathrm{R}}_{I_x}} & & \Big\downarrow{\scriptstyle \eta^{\mathrm{R}}_{J_x}} \\ \mathrm{R}F(I_x) & \xrightarrow{\mathrm{R}F(\phi_x)} & \mathrm{R}F(J_x) \end{array}$$

in $\mathbf{D}(B, \mathrm{gr})$. The vertical arrows are isomorphisms because both I_x and J_x are right F-acyclic complexes. The morphism $\mathrm{R}F(\phi_x)$ is also an isomorphism. It follows that $F(\phi_x)$ is an isomorphism in $\mathbf{D}(B, \mathrm{gr})$; and therefore it is a quasi-isomorphism in $\mathbf{C}_{\mathrm{str}}(B, \mathrm{gr})$.

Next consider this commutative diagram in $\mathbf{C}_{\mathrm{str}}(B, \mathrm{gr})$:

$$\begin{array}{ccc} \bigoplus_{x \in X} F(I_x) & \xrightarrow{\bigoplus_{x \in X} F(\phi_x)} & \bigoplus_{x \in X} F(J_x) \\ \downarrow & & \downarrow \\ F(I) & \xrightarrow{F(\phi)} & F(J) \end{array}$$

Because each $F(\phi_x)$ is a quasi-isomorphism, it follows that the top horizontal arrow is a quasi-isomorphism. Because the functor $F = \mathrm{R}^0 F$ is quasi-compact, the vertical arrows are isomorphisms. We conclude that $F(\phi)$ is a quasi-isomorphism in $\mathbf{C}_{\mathrm{str}}(B, \mathrm{gr})$.

Finally we look at this commutative diagram in $\mathbf{D}(B, \mathrm{gr})$:

$$\begin{array}{ccc} F(I) & \xrightarrow{F(\phi)} & F(J) \\ {\scriptstyle \eta^{\mathrm{R}}_I} \downarrow & & \downarrow {\scriptstyle \eta^{\mathrm{R}}_J} \\ \mathrm{R}F(I) & \xrightarrow{\mathrm{R}F(\phi)} & \mathrm{R}F(J) \end{array}$$

We know that the morphisms $F(\phi)$ and $\mathrm{R}F(\phi)$ are isomorphisms. For every p, the object $J^p = \bigoplus_{x \in X} J^p_x$ in $\mathbf{M}(A, \mathrm{gr})$ is a direct sum of injective objects. So according to Lemma 16.1.8, J^p is a right F-acyclic object. By Lemma 16.1.5, in either of the two cases, the complex J is a right F-acyclic complex. This says that the morphism η^{R}_J is an isomorphism. We conclude that the morphism η^{R}_I is an isomorphism too, and this says that I is a right F-acyclic complex. □

Theorem 16.1.10 *Under Setup* 16.1.1, *assume the functor F has finite right cohomological dimension, and the functors $\mathrm{R}^q F$, for all $q \geq 0$, are quasi-compact. Then the triangulated functor* $\mathrm{R}F : \mathbf{D}(A, \mathrm{gr}) \to \mathbf{D}(B, \mathrm{gr})$ *has finite cohomological dimensional and is quasi-compact.*

Proof By Lemma 16.1.6 the functor $\mathrm{R}F$ is finite dimensional. We need to prove that $\mathrm{R}F$ commutes with infinite direct sums. Namely, consider a collection $\{M_x\}_{x \in X}$ of complexes in $\mathbf{C}(A, \mathrm{gr})$, and let $M := \bigoplus_{x \in X} M_x$. We have to prove that the canonical morphism $\bigoplus_{x \in X} \mathrm{R}F(M_x) \to \mathrm{R}F(M)$ in $\mathbf{D}(B, \mathrm{gr})$ is an isomorphism.

For every index x we choose a quasi-isomorphism $\phi_x : M_x \to I_x$ in $\mathbf{C}_{\mathrm{str}}(A, \mathrm{gr})$, where I_x is a K-injective complex. Define $I := \bigoplus_{x \in X} I_x$, so there is a quasi-isomorphism $\phi : M \to I$ in $\mathbf{C}_{\mathrm{str}}(A, \mathrm{gr})$. We get a commutative diagram

$$\begin{array}{ccc} \bigoplus_{x \in X} \mathrm{R}F(M_x) & \xrightarrow{\bigoplus \mathrm{R}F(\phi_x)} & \bigoplus_{x \in X} \mathrm{R}F(I_x) \\ \downarrow & & \downarrow \\ \mathrm{R}F(M) & \xrightarrow{\mathrm{R}F(\phi)} & \mathrm{R}F(I) \end{array}$$

in $\mathbf{D}(B, \mathrm{gr})$, in which the horizontal arrows are isomorphisms. Therefore it suffices to prove that the canonical morphism

$$\bigoplus\nolimits_{x \in X} \mathrm{R}F(I_x) \to \mathrm{R}F(I) \tag{16.1.11}$$

in $\mathbf{D}(B, \mathrm{gr})$ is an isomorphism.

Consider this commutative diagram in $\mathbf{D}(B, \mathrm{gr})$:

$$\begin{array}{ccc} \bigoplus_{x \in X} F(I_x) & \xrightarrow{\bigoplus \eta^{\mathrm{R}}_{I_x}} & \bigoplus_{x \in X} \mathrm{R}F(I_x) \\ \downarrow & & \downarrow \\ F(I) & \xrightarrow{\eta^{\mathrm{R}}_{I}} & \mathrm{R}F(I) \end{array}$$

For each x the morphism $\eta^{\mathrm{R}}_{I_x}$ is an isomorphism; and hence the top horizontal arrow is an isomorphism. Because $F = \mathrm{R}^0 F$ is quasi-compact, the left vertical arrow is an isomorphism (in $\mathbf{C}_{\mathrm{str}}(B, \mathrm{gr})$, and so also in $\mathbf{D}(B, \mathrm{gr})$). By Lemma 16.1.9 the complex I is right F-acyclic, and therefore η^{R}_{I} is an isomorphism. Hence the remaining arrow is an isomorphism; but this is the morphism (16.1.11). □

16.2 Weakly Stable and Idempotent Copointed Functors

The content of this section is adapted from [157, Section 2]. Recall that we follow Conventions 15.1.17 and 15.2.1. As in Section 16.1, the results of this section apply in much greater generality. We state the results in the special context of algebraically graded rings and modules, since we need them for the graded $\mathfrak{m}$-torsion functor $F := \Gamma_{\mathfrak{m}}$.

Definition 16.2.1 Let A be a graded ring and $F : \mathbf{M}(A, \mathrm{gr}) \to \mathbf{M}(A, \mathrm{gr})$ a left exact linear functor.

(1) The functor F is called *stable* if for every injective object $I \in \mathbf{M}(A, \mathrm{gr})$, the object $F(I)$ is injective.
(2) The functor F is called *weakly stable* if for every injective object $I \in \mathbf{M}(A, \mathrm{gr})$, the object $F(I)$ is right F-acyclic, in the sense of Definition 16.1.3.

Clearly, stable implies weakly stable.

Remark 16.2.2 The name "stable" comes from the theory of *torsion classes*; see [148, Section VI.7]. Indeed, a torsion class is called stable if the corresponding torsion functor F is a stable functor, as defined above.

The name "weakly stable" was coined in [157], where it was shown that for a finitely generated ideal $\mathfrak{a}$ in a commutative ring A, the torsion functor $\Gamma_{\mathfrak{a}}$ is weakly stable iff the ideal $\mathfrak{a}$ is *weakly proregular*. We should mention the well-known fact that if the commutative ring A is noetherian, then for every ideal $\mathfrak{a} \subseteq A$ the torsion functor $\Gamma_{\mathfrak{a}}$ is stable (cf. [63, Lemma III.3.2]; or use the Matlis classification of injective A-modules in Section 13.2).

Definition 16.2.3 Let M be a linear category (e.g. $\mathbf{M}(A, \mathrm{gr})$ or $\mathbf{D}(A, \mathrm{gr})$ for a graded ring A), with identity functor Id_{M}.

(1) A *copointed linear functor* on M is a pair (F, σ), consisting of a linear functor $F : \mathsf{M} \to \mathsf{M}$ and a morphism of functors $\sigma : F \to \mathrm{Id}_{\mathsf{M}}$.

(2) The copointed linear functor (F, σ) is called *idempotent* if the morphisms

$$\sigma_{F(M)},\ F(\sigma_M) : F(F(M)) \to F(M)$$

are isomorphisms for all objects $M \in \mathsf{M}$.

(3) If M is a triangulated category, F is a triangulated functor, and σ is a morphism of triangulated functors, then we call (F, σ) a *copointed triangulated functor*.

The name "copointed" is explained in Remark 16.2.11.

Weak stability and idempotence together have the following effect.

Lemma 16.2.4 *Let (F, σ) be an idempotent copointed linear functor on $\mathbf{M}(A, \mathrm{gr})$, and assume that F is left exact and weakly stable. If I is a right F-acyclic object of $\mathbf{M}(A, \mathrm{gr})$, then $F(I)$ is also a right F-acyclic object of $\mathbf{M}(A, \mathrm{gr})$.*

Proof Choose an injective resolution $\rho : I \to J$; i.e. J is a complex of injectives concentrated in nonnegative degrees, and ρ is a quasi-isomorphism in $\mathbf{C}_{\mathrm{str}}(A, \mathrm{gr})$. Since $\mathrm{H}^q(F(J)) \cong \mathrm{R}^q F(I)$, and since I is a right F-acyclic object, we see that the homomorphism of complexes $F(\rho) : F(I) \to F(J)$ is a quasi-isomorphism. Therefore both $F(\rho)$ and $\mathrm{R}F(F(\rho))$ are isomorphisms in $\mathbf{D}(A, \mathrm{gr})$.

The weak stability of F implies that $F(J)$ is a bounded below complex of right F-acyclic objects. According to Lemma 16.1.5, the complex $F(J)$ is right F-acyclic. This means that the morphism $\eta^{\mathrm{R}}_{F(J)} : F(J) \to \mathrm{R}F(J)$ is an isomorphism.

The idempotence of the copointed functor F says that the morphisms $\sigma_{F(I)}$ and $\sigma_{F(J)}$ are isomorphisms in $\mathbf{C}_{\mathrm{str}}(A, \mathrm{gr})$. We get a commutative diagram in $\mathbf{D}(A, \mathrm{gr})$:

$$\begin{array}{ccccc} F(I) & \xleftarrow[\cong]{\sigma_{F(I)}} & F(F(I)) & \xrightarrow{\eta^{\mathrm{R}}_{F(I)}} & \mathrm{R}F(F(I)) \\ {\scriptstyle F(\rho)}\downarrow{\scriptstyle \cong} & & \downarrow{\scriptstyle F(F(\rho))} & & {\scriptstyle \mathrm{R}F(F(\rho))}\downarrow{\scriptstyle \cong} \\ F(J) & \xleftarrow[\cong]{\sigma_{F(J)}} & F(F(J)) & \xrightarrow[\cong]{\eta^{\mathrm{R}}_{F(J)}} & \mathrm{R}F(F(J)) \end{array}$$

We conclude that $\eta^{\mathrm{R}}_{F(I)}$ is an isomorphism; and this means that the object $F(I)$ is right F-acyclic. □

The next lemma is a generalization of [118, Proposition 3.10].

Lemma 16.2.5 *Suppose we are given a copointed linear functor (F, σ) on $\mathbf{M}(A, \mathrm{gr})$. Then there is a unique morphism $\sigma^{\mathrm{R}} : \mathrm{R}F \to \mathrm{Id}_{\mathbf{D}(A,\mathrm{gr})}$ of triangulated functors from $\mathbf{D}(A, \mathrm{gr})$ to itself, satisfying this condition: for every complex $M \in \mathbf{D}(A, \mathrm{gr})$ there is equality $\sigma^{\mathrm{R}}_M \circ \eta^{\mathrm{R}}_M = \sigma_M$ of morphisms $F(M) \to M$ in $\mathbf{D}(A, \mathrm{gr})$.*

In a commutative diagram:

$$\begin{array}{ccc} F(M) & \xrightarrow{\eta^{\mathrm{R}}_M} & \mathrm{R}F(M) \\ & {\scriptstyle \sigma_M}\searrow & \downarrow{\scriptstyle \sigma^{\mathrm{R}}_M} \\ & & M \end{array} \tag{16.2.6}$$

Proof The existence of the morphism σ^{R} comes for free from the universal property of the right derived functor. Still, for later reference, we give the construction.

For a K-injective complex I the morphism $\eta^{\mathrm{R}}_I : F(I) \to \mathrm{R}F(I)$ in $\mathbf{D}(A, \mathrm{gr})$ is an isomorphism, and we define $\sigma^{\mathrm{R}}_I : \mathrm{R}F(I) \to I$ to be $\sigma^{\mathrm{R}}_I := \sigma_I \circ (\eta^{\mathrm{R}}_I)^{-1}$.

For an arbitrary complex M we choose a quasi-isomorphism $\rho : M \to I$ into a K-injective complex, and then we let $\sigma^{\mathrm{R}}_M := \rho^{-1} \circ \sigma^{\mathrm{R}}_I \circ \mathrm{R}F(\rho)$ in $\mathbf{D}(A, \mathrm{gr})$. The corresponding commutative diagram in $\mathbf{D}(A, \mathrm{gr})$ is (16.2.7).

$$\begin{array}{ccccc} & & \overset{\sigma_M}{\frown} & & \\ F(M) & \xrightarrow{\eta^{\mathrm{R}}_M} & \mathrm{R}F(M) & \xrightarrow{\sigma^{\mathrm{R}}_M} & M \\ {\scriptstyle F(\rho)}\downarrow & & {\scriptstyle \mathrm{R}F(\rho)}\downarrow{\scriptstyle \cong} & & {\scriptstyle \rho}\downarrow{\scriptstyle \cong} \\ F(I) & \xrightarrow[\cong]{\eta^{\mathrm{R}}_I} & \mathrm{R}F(I) & \xrightarrow{\sigma^{\mathrm{R}}_I} & I \\ & & \underset{\sigma_I}{\smile} & & \end{array} \tag{16.2.7}$$

It is easy to see that the collection of morphisms $\{\sigma^{\mathrm{R}}_M\}_{M \in \mathsf{D}(A,\mathrm{gr})}$ has the desired properties. □

In this way we obtain a copointed triangulated functor $(\mathrm{R}F, \sigma^{\mathrm{R}})$ on $\mathsf{D}(A, \mathrm{gr})$.

Theorem 16.2.8 *Let (F, σ) be an idempotent copointed linear functor on $\mathsf{M}(A, \mathrm{gr})$, and assume that F is left exact and weakly stable.*

(1) *The copointed triangulated functor $(\mathrm{R}F, \sigma^{\mathrm{R}})$ on $\mathsf{D}^+(A, \mathrm{gr})$ is idempotent.*
(2) *If F has finite right cohomological dimension, then the copointed triangulated functor $(\mathrm{R}F, \sigma^{\mathrm{R}})$ on $\mathsf{D}(A, \mathrm{gr})$ is idempotent.*

Proof Let $M \in \mathsf{D}(A, \mathrm{gr})$. Choose a K-injective resolution $M \to I$ in $\mathsf{C}_{\mathrm{str}}(A, \mathrm{gr})$, such that I is a complex consisting of injective objects of $\mathsf{M}(A, \mathrm{gr})$, and $\inf(I) = \inf(\mathrm{H}(M))$. It suffices to prove that the morphisms

$$\sigma^{\mathrm{R}}_{\mathrm{R}F(I)},\ \mathrm{R}F(\sigma^{\mathrm{R}}_I) : \mathrm{R}F(\mathrm{R}F(I)) \to \mathrm{R}F(I)$$

in $\mathsf{D}(A, \mathrm{gr})$ are isomorphisms.

Note that by our choice, if $M \in \mathsf{D}^+(A, \mathrm{gr})$ then I is a bounded below complex. Since each I^q is an injective object, it is right F-acyclic. Because F is a weakly stable functor, each of the objects $F(I^q)$ is right F-acyclic too. If the functor F has finite right cohomological dimension, or if I is bounded below, Lemma 16.1.5 says that the complexes I and $F(I)$ are both right F-acyclic complexes.

Consider the diagram

$$\begin{array}{ccccc} F(F(I)) & \xrightarrow{\eta^{\mathrm{R}}_{F(I)}} & \mathrm{R}F(F(I)) & \xrightarrow{\mathrm{R}F(\eta^{\mathrm{R}}_I)} & \mathrm{R}F(\mathrm{R}F(I)) \\ \Big\downarrow{\scriptstyle F(\sigma_I)} & & \Big\downarrow{\scriptstyle \mathrm{R}F(\sigma_I)} & & \Big\downarrow{\scriptstyle \mathrm{R}F(\sigma^{\mathrm{R}}_I)} \\ F(I) & \xrightarrow{\eta^{\mathrm{R}}_I} & \mathrm{R}F(I) & \xrightarrow{\mathrm{id}} & \mathrm{R}F(I) \end{array} \tag{16.2.9}$$

in $\mathsf{D}(A, \mathrm{gr})$. The left square is commutative: it is gotten from the vertical morphism $\sigma_I : F(I) \to I$, to which we apply in the horizontal direction the morphism of functors $\eta^{\mathrm{R}} : F \to \mathrm{R}F$. The right square is also commutative: it comes from applying the functor $\mathrm{R}F$ to the commutative diagram

$$\begin{array}{ccc} F(I) & \xrightarrow{\eta^{\mathrm{R}}_I} & \mathrm{R}F(I) \\ \Big\downarrow{\scriptstyle \sigma_I} & & \Big\downarrow{\scriptstyle \sigma^{\mathrm{R}}_I} \\ I & \xrightarrow{\mathrm{id}} & I \end{array}$$

that characterizes σ^{R}_I. Because I and $F(I)$ are right F-acyclic complexes, the morphisms η^{R}_I and $\eta^{\mathrm{R}}_{F(I)}$ are isomorphisms. Hence $\mathrm{R}F(\eta^{\mathrm{R}}_I)$ is an isomorphism.

So the horizontal morphisms in the diagram 16.2.9 are all isomorphisms. We are given that F is idempotent, and thus $F(\sigma_I)$ is an isomorphism. The conclusion of this discussion is that $\mathrm{R}F(\sigma^{\mathrm{R}}_I)$ is an isomorphism.

Next, let $\phi : \mathrm{R}F(I) \to J$ be an isomorphism in $\mathbf{D}(A, \mathrm{gr})$ to a K-injective complex J such that $\inf(J) = \inf(\mathrm{H}(\mathrm{R}F(I)))$. We know that $\eta^{\mathrm{R}}_I : F(I) \to \mathrm{R}F(I)$ is an isomorphism, and therefore the composed morphism $\phi \circ \eta^{\mathrm{R}}_I : F(I) \to J$ is an isomorphism in $\mathbf{D}(A, \mathrm{gr})$. Let $\tilde{\psi} : F(I) \to J$ be a quasi-isomorphism in $\mathbf{C}_{\mathrm{str}}(A, \mathrm{gr})$ representing $\phi \circ \eta^{\mathrm{R}}_I$. Since $F(I)$ and J are both right F-acyclic complexes, it follows that $F(\tilde{\psi}) : F(F(I)) \to F(J)$ is a quasi-isomorphism. Consider the following commutative diagram:

$$\begin{array}{ccccccc} F(F(I)) & \xrightarrow{F(\tilde{\psi})} & F(J) & \xrightarrow{\eta^{\mathrm{R}}_J} & \mathrm{R}F(J) & \xleftarrow{\mathrm{R}F(\phi)} & \mathrm{R}F(\mathrm{R}F(I)) \\ \downarrow{\scriptstyle \sigma_{F(I)}} & & \downarrow{\scriptstyle \sigma_J} & & \downarrow{\scriptstyle \sigma^{\mathrm{R}}_J} & & \downarrow{\scriptstyle \sigma^{\mathrm{R}}_{\mathrm{R}F(I)}} \\ F(I) & \xrightarrow{\tilde{\psi}} & J & \xrightarrow{\mathrm{id}} & J & \xleftarrow{\phi} & \mathrm{R}F(I) \end{array}$$

in $\mathbf{D}(A, \mathrm{gr})$. All horizontal arrows here are isomorphisms. We are given that F is idempotent, and thus $\sigma_{F(I)}$ is an isomorphism. The conclusion is that $\sigma^{\mathrm{R}}_{\mathrm{R}F(I)}$ is an isomorphism. □

We end this section with a notion dual to "copointed functor."

Definition 16.2.10 Let M be a linear category (e.g. $\mathbf{M}(A, \mathrm{gr})$ or $\mathbf{D}(A, \mathrm{gr})$), with identity functor Id_{M}.

(1) A *pointed linear functor* on M is a pair (G, τ), consisting of a linear functor $G : \mathsf{M} \to \mathsf{M}$ and a morphism of functors $\tau : \mathrm{Id}_{\mathsf{M}} \to G$.

(2) The pointed linear functor (G, τ) is called *idempotent* if the morphisms

$$\tau_{G(M)}, G(\tau_M) : G(M) \to G(G(M))$$

are isomorphisms for all objects $M \in \mathsf{M}$.

(3) If M is a triangulated category, G is a triangulated functor, and τ is a morphism of triangulated functors, then we call (G, τ) a *pointed triangulated functor*.

Remark 16.2.11 Idempotent copointed functors already appeared in the literature under another name: *idempotent comonads*. Another name for (nearly) the same notion is a (Bousfield) colocalization functor, see e.g. [90]. Dually, idempotent pointed functors are the same thing as *idempotent monads*. See [115] for a discussion of these concepts.

In [76, Section 4.1], what we call an idempotent pointed functor is called a *projector*. It is proved there that for an idempotent pointed functor (G, τ), and an object $M \in \mathsf{M}$, the isomorphisms $\tau_{G(M)}$ and $G(\tau_M)$ are equal. The same proof (with arrows reversed) shows that for an idempotent copointed functor (F, σ), and an object M, the morphisms $\sigma_{F(M)}$ and $F(\sigma_M)$ are equal. We shall not require these facts.

16.3 Graded Torsion: Weak Stability and Idempotence

In this section we begin the study of derived torsion over a connected graded ring. We adhere to Conventions 15.1.17 and 15.2.1. Thus $\mathbb{K}$ is a field, and all graded $\mathbb{K}$-rings are algebraically graded central $\mathbb{K}$-rings. The symbol $\otimes$ means $\otimes_{\mathbb{K}}$. The enveloping ring of a graded $\mathbb{K}$-ring A is the graded $\mathbb{K}$-ring $A^{\mathrm{en}} = A \otimes A^{\mathrm{op}}$. The category of left algebraically graded A-modules is $\mathbf{M}(A, \mathrm{gr})$, and it is a $\mathbb{K}$-linear abelian category. Its derived category is $\mathbf{D}(A, \mathrm{gr})$.

Recall the notion of left noetherian connected graded $\mathbb{K}$-ring from Definitions 15.1.29 and 15.2.18. In this section we will assume the following setup:

Setup 16.3.1 We are given a left noetherian connected graded $\mathbb{K}$-ring A, with augmentation ideal $\mathfrak{m} = \bigoplus_{i \geq 1} A_i \subsetneq A$. We are also given graded $\mathbb{K}$-rings B and C.

The graded rings B and C are auxiliary – they are just placeholders, allowing us to write more general formulas. For instance, $\mathbf{M}(A \otimes B^{\mathrm{op}}, \mathrm{gr})$ is $\mathbf{M}(A^{\mathrm{en}}, \mathrm{gr})$ when we take $B := A$, but it is $\mathbf{M}(A, \mathrm{gr})$ when $B := \mathbb{K}$.

Since A is left noetherian, the augmentation ideal $\mathfrak{m}$ is a finite graded left A-module.

We shall often make implicit use of the canonical isomorphisms

$$A \otimes_A M \xrightarrow{\simeq} M \quad \text{and} \quad \mathrm{Hom}_A(A, M) \xrightarrow{\simeq} M \tag{16.3.2}$$

in $\mathbf{M}(A \otimes B^{\mathrm{op}}, \mathrm{gr})$. There are similar canonical isomorphisms

$$A \otimes^{\mathrm{L}}_A M \xrightarrow{\simeq} M \quad \text{and} \quad \mathrm{RHom}_A(A, M) \xrightarrow{\simeq} M \tag{16.3.3}$$

in $\mathbf{D}(A \otimes B^{\mathrm{op}}, \mathrm{gr})$. Also we view $\mathbb{K} \cong A/\mathfrak{m}$ as a graded A^{en}-module via the augmentation ring homomorphism $A \to \mathbb{K}$.

We are interested in graded $\mathfrak{m}$-torsion. This makes sense for bimodules, and hence the slightly complicated definition below.

Definition 16.3.4 Under Setup 16.3.1:

(1) Let $M \in \mathbf{M}(A \otimes B^{\mathrm{op}}, \mathrm{gr})$. An element $m \in M$ is called an $\mathfrak{m}$-*torsion element* if $\mathfrak{m}^j \cdot m = 0$ for $j \gg 0$.
(2) The set of $\mathfrak{m}$-torsion elements of M is denoted by $\Gamma_{\mathfrak{m}}(M)$, and it is called the $\mathfrak{m}$-*torsion submodule* of M.

(3) We call M an $\mathfrak{m}$-*torsion module* if $\Gamma_{\mathfrak{m}}(M) = M$.
(4) We denote by $\mathbf{M}_{\mathrm{tor}}(A \otimes B^{\mathrm{op}}, \mathrm{gr})$ the full subcategory of $\mathbf{M}(A \otimes B^{\mathrm{op}}, \mathrm{gr})$ on the $\mathfrak{m}$-torsion modules.

The fact that $\Gamma_{\mathfrak{m}}(M)$ is a graded $(A \otimes B^{\mathrm{op}})$-submodule of M is easy to see. Indeed, we can express the torsion submodule as follows:

$$\Gamma_{\mathfrak{m}}(M) = \lim_{j \to} \mathrm{Hom}_A(A/\mathfrak{m}^j, M) \subseteq \mathrm{Hom}_A(A, M), \tag{16.3.5}$$

using the identification (16.3.2), where $j \geq 1$, $\mathfrak{m}^j$ is the j-fold product $\mathfrak{m} \cdots \mathfrak{m} \subseteq A$, $A \to A/\mathfrak{m}^{j+1} \to A/\mathfrak{m}^j$ are the canonical graded ring surjections, and the operation $\lim_{j\to}$ is in the category $\mathbf{M}(A \otimes B^{\mathrm{op}}, \mathrm{gr})$.

It is clear that

$$\Gamma_{\mathfrak{m}} : \mathbf{M}(A \otimes B^{\mathrm{op}}, \mathrm{gr}) \to \mathbf{M}_{\mathrm{tor}}(A \otimes B^{\mathrm{op}}, \mathrm{gr})$$

is a linear functor, and the inclusions $\Gamma_{\mathfrak{m}}(M) \subseteq M$ assemble into a monomorphism $\sigma : \Gamma_{\mathfrak{m}} \rightarrowtail \mathrm{Id}$ of functors from $\mathbf{M}(A \otimes B^{\mathrm{op}}, \mathrm{gr})$ to itself.

The ring B plays an auxiliary role in Definition 16.3.4. Because there is no mention of torsion for B^{op}-modules, the notation $\mathbf{M}_{\mathrm{tor}}(A \otimes B^{\mathrm{op}}, \mathrm{gr})$ is not ambiguous. However, in Section 16.5, where torsion for B^{op} modules does become an option, we will switch to the richer notation $\mathbf{M}_{(\mathrm{tor},..)}(A \otimes B^{\mathrm{op}}, \mathrm{gr})$. See Definition 16.5.5.

Lemma 16.3.6 *Under Setup* 16.3.1 *the following hold.*

(1) *The functor* $\Gamma_{\mathfrak{m}}$ *commutes with graded ring homomorphisms* $C \to B$, *in the sense that the diagram*

$$\begin{array}{ccc} \mathbf{M}(A \otimes B^{\mathrm{op}}, \mathrm{gr}) & \xrightarrow{\Gamma_{\mathfrak{m}}} & \mathbf{M}_{\mathrm{tor}}(A \otimes B^{\mathrm{op}}, \mathrm{gr}) \\ {\scriptstyle \mathrm{Rest}} \downarrow & & \downarrow {\scriptstyle \mathrm{Rest}} \\ \mathbf{M}(A \otimes C^{\mathrm{op}}, \mathrm{gr}) & \xrightarrow{\Gamma_{\mathfrak{m}}} & \mathbf{M}_{\mathrm{tor}}(A \otimes C^{\mathrm{op}}, \mathrm{gr}) \end{array}$$

is commutative.

(2) *The functor* $\Gamma_{\mathfrak{m}}$ *is idempotent, in the sense that for every* $M \in \mathbf{M}(A \otimes B^{\mathrm{op}}, \mathrm{gr})$ *the homomorphisms*

$$\sigma_{\Gamma_{\mathfrak{m}}(M)},\ \Gamma_{\mathfrak{m}}(\sigma_M) : \Gamma_{\mathfrak{m}}(\Gamma_{\mathfrak{m}}(M)) \to \Gamma_{\mathfrak{m}}(M)$$

are bijective and are equal to each other.

(3) *Let* $\phi : M \to N$ *be a homomorphism in* $\mathbf{M}(A \otimes B^{\mathrm{op}}, \mathrm{gr})$, *such that* M *is* $\mathfrak{m}$-*torsion. Then there is a unique homomorphism* $\phi' : M \to \Gamma_{\mathfrak{m}}(N)$ *satisfying* $\phi = \sigma_N \circ \phi'$.

(4) *The functor* $\Gamma_{\mathfrak{m}}$ *is left exact.*

(5) *The functor* $\Gamma_{\mathfrak{m}}$ *commutes with direct limits.*

(6) *The category* $\mathbf{M}_{\mathrm{tor}}(A \otimes B^{\mathrm{op}}, \mathrm{gr})$ *is a thick abelian subcategory of* $\mathbf{M}(A \otimes B^{\mathrm{op}}, \mathrm{gr})$*, closed under taking subobjects and quotients.*

Exercise 16.3.7 Prove Lemma 16.3.6. (Hint: for item (6) you will need the fact that A is left noetherian.)

Recall that for a module $M \in \mathbf{M}(A, \mathrm{gr})$, its socle is $\mathrm{Soc}(M) = \mathrm{Hom}_A(\mathbb{K}, M)$. The socle is a functor, and there is a monomorphism $\mathrm{Soc} \rightarrowtail \Gamma_{\mathfrak{m}}$ of functors from $\mathbf{M}(A, \mathrm{gr})$ to itself.

Lemma 16.3.8 *Let* $M \in \mathbf{M}(A, \mathrm{gr})$*, and let* $W := \mathrm{Soc}(M)$*. If* M *is an* $\mathfrak{m}$*-torsion module, then* W *is a graded-essential submodule of* M*.*

Proof Take any nonzero graded A-submodule $M' \subseteq M$. Let $m \in M'$ be some nonzero homogeneous element. Because M is a torsion module, yet $m \neq 0$, there is a unique $j \in \mathbb{N}$ such that $\mathfrak{m}^{j+1} \cdot m = 0$ but $\mathfrak{m}^j \cdot m \neq 0$. So there is a homogeneous element $a \in \mathfrak{m}^j$ such that $a \cdot m \neq 0$. The element $a \cdot m$ is then a nonzero homogeneous element in $M' \cap W$. □

The $\mathbb{K}$-linear dual (see Definition 15.2.25) of the ring A is the graded bimodule $A^* = \mathrm{Hom}_{\mathbb{K}}(A, \mathbb{K}) \in \mathbf{M}(A^{\mathrm{en}}, \mathrm{gr})$. The augmentation ring homomorphism $A \to \mathbb{K}$ induces, by dualizing, a canonical monomorphism

$$\mathbb{K} \rightarrowtail A^* \tag{16.3.9}$$

in $\mathbf{M}(A^{\mathrm{en}}, \mathrm{gr})$. For every $W \in \mathbf{M}(\mathbb{K}, \mathrm{gr})$ there is an induced monomorphism

$$W \cong W \otimes \mathbb{K} \rightarrowtail W \otimes A^* \tag{16.3.10}$$

in $\mathbf{M}(A^{\mathrm{en}}, \mathrm{gr})$.

Lemma 16.3.11 *Under Setup* 16.3.1*, let* $W \in \mathbf{M}(\mathbb{K}, \mathrm{gr})$*, and define* $I := A^* \otimes W \in \mathbf{M}(A, \mathrm{gr})$*. Consider the canonical monomorphism* $W \rightarrowtail I$ *in* $\mathbf{M}(A, \mathrm{gr})$ *from* (16.3.10)*. Then*:

(1) I *is a graded-injective* A*-module.*
(2) I *is an* $\mathfrak{m}$*-torsion* A*-module.*
(3) $W = \mathrm{Soc}(I)$.
(4) W *is a graded-essential submodule of* I*.*

Proof

(1) Say $W = \bigoplus_{x \in X} \mathbb{K}(-i_x)$. Then $I \cong \bigoplus_{x \in X} A^*(-i_x)$ in $\mathbf{M}(A, \mathrm{gr})$. According to Proposition 15.2.5(2) and Theorem 15.2.17(1), the graded A-module I is graded-injective.

(2) The A-module A^* is $\mathfrak{m}$-torsion, and a direct sum of torsion modules is torsion (see Lemma 16.3.6(5)).

(3) This is because $\mathbb{K} = \mathrm{Soc}(A^*)$, via the canonical monomorphism (16.3.9).

(4) Combine items (2) and (3) with Lemma 16.3.8. □

Proposition 16.3.12 *Under Setup* 16.3.1, *let* $M \in \mathbf{M}(A, \mathrm{gr})$, *and let* $W := \mathrm{Soc}(M)$. *The three conditions below are equivalent.*

(i) *M is an $\mathfrak{m}$-torsion graded A-module.*
(ii) *W is a graded-essential A-submodule of M.*
(iii) *There is a graded-essential monomorphism $\epsilon : M \rightarrowtail A^* \otimes W$ in $\mathbf{M}(A, \mathrm{gr})$ that restricts to the identity on W.*

Proof

(i) ⇒ (ii): This is Lemma 16.3.8.

(ii) ⇒ (iii): By Lemma 16.3.11 the graded A-module $A^* \otimes W$ is graded-injective. Hence there is a homomorphism $\epsilon : M \to A^* \otimes W$ in $\mathbf{M}(A, \mathrm{gr})$ that makes the diagram

$$\begin{array}{ccc} W & \rightarrowtail & M \\ \big\downarrow & \swarrow_{\epsilon} & \\ A^* \otimes W & & \end{array}$$

in $\mathbf{M}(A, \mathrm{gr})$ commutative. Let $M' := \mathrm{Ker}(\epsilon)$. If M' were nonzero, then, because $W \subseteq M$ is a graded-essential submodule, we would have $W \cap M' \neq 0$. But $W \to A^* \otimes W$ is a monomorphism. We conclude that ϵ is a monomorphism too.

Finally, by Lemma 16.3.11(4) we know that W is a graded-essential submodule of $A^* \otimes W$. It follows that the monomorphism ϵ is graded-essential.

(iii) ⇒ (i): The graded A-module A^* is torsion. We know that the subcategory $\mathbf{M}_{\mathrm{tor}}(A, \mathrm{gr})$ of $\mathbf{M}(A, \mathrm{gr})$ is closed under taking arbitrary direct sums and subobjects. Hence M is $\mathfrak{m}$-torsion. □

Cofinite graded modules were defined in Definition 15.2.33.

Corollary 16.3.13 *Assume A is a noetherian connected graded ring. Let $M \in \mathbf{M}_{\mathrm{tor}}(A, \mathrm{gr})$. The two conditions below are equivalent.*

(i) *M is a cofinite graded A-module.*
(ii) *$W := \mathrm{Soc}(M)$ is a finite graded $\mathbb{K}$-module.*

Proof Graded Matlis Duality (Theorem 15.2.34) tells us that the cofinite graded A-modules are the artinian objects in the abelian category $\mathbf{M}(A, \mathrm{gr})$.

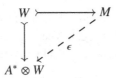

(i) ⇒ (ii): We are given that M is a cofinite graded A-module. Hence so is its submodule W. Since W is an A-module via $\mathbb{K}$, it must be an artinian graded $\mathbb{K}$-module. So W is finite over $\mathbb{K}$.

(ii) ⇒ (i): Since A^* is a cofinite graded A-module and $A^* \otimes W$ is a finite direct sum of degree twists of A^*, we see that $A^* \otimes W$ is a cofinite graded A-module. By Proposition 16.3.12 we know that M is isomorphic to a graded submodule of $A^* \otimes W$; so M is also a cofinite graded A-module. □

Remark 16.3.14 We do not know if the corollary remains true if A is just left noetherian.

Now we pass to derived categories and functors.

Definition 16.3.15 Under Setup 16.3.1, we denote by $\mathbf{D}_{\mathrm{tor}}(A \otimes B^{\mathrm{op}}, \mathrm{gr})$ the full subcategory of $\mathbf{D}(A \otimes B^{\mathrm{op}}, \mathrm{gr})$ on the complexes M whose cohomology modules $\mathrm{H}^p(M)$ belong to $\mathbf{M}_{\mathrm{tor}}(A \otimes B^{\mathrm{op}}, \mathrm{gr})$ for all p.

Proposition 16.3.16 *Under Setup 16.3.1 the following hold.*

(1) *The functor $\Gamma_{\mathfrak{m}}$ has a triangulated right derived functor*

$$\mathrm{R}\Gamma_{\mathfrak{m}} : \mathbf{D}(A \otimes B^{\mathrm{op}}, \mathrm{gr}) \to \mathbf{D}(A \otimes B^{\mathrm{op}}, \mathrm{gr}).$$

If $I \in \mathbf{D}(A \otimes B^{\mathrm{op}}, \mathrm{gr})$ is K-graded-injective over A, then the morphism $\eta^{\mathrm{R}}_I : \Gamma_{\mathfrak{m}}(I) \to \mathrm{R}\Gamma_{\mathfrak{m}}(I)$ is an isomorphism.

(2) *There is a unique morphism $\sigma^{\mathrm{R}} : \mathrm{R}\Gamma_{\mathfrak{m}} \to \mathrm{Id}$ of triangulated functors from $\mathbf{D}(A \otimes B^{\mathrm{op}}, \mathrm{gr})$ to itself, such that for every $M \in \mathbf{D}(A \otimes B^{\mathrm{op}}, \mathrm{gr})$ there is equality $\sigma^{\mathrm{R}}_M \circ \eta^{\mathrm{R}}_M = \mathrm{Q}(\sigma_M)$ of morphisms $\Gamma_{\mathfrak{m}}(M) \to M$ in $\mathbf{D}(A \otimes B^{\mathrm{op}}, \mathrm{gr})$.*

(3) *The category $\mathbf{D}_{\mathrm{tor}}(A \otimes B^{\mathrm{op}}, \mathrm{gr})$ is a full triangulated subcategory of $\mathbf{D}(A \otimes B^{\mathrm{op}}, \mathrm{gr})$, and it contains the image of the functor $\mathrm{R}\Gamma_{\mathfrak{m}}$.*

(4) *The functor $\mathrm{R}\Gamma_{\mathfrak{m}}$ commutes with graded $\mathbb{K}$-ring homomorphisms $C \to B$, in the sense that the diagram*

$$\begin{array}{ccc} \mathbf{D}(A \otimes B^{\mathrm{op}}, \mathrm{gr}) & \xrightarrow{\mathrm{R}\Gamma_{\mathfrak{m}}} & \mathbf{D}_{\mathrm{tor}}(A \otimes B^{\mathrm{op}}, \mathrm{gr}) \\ \downarrow{\scriptstyle \mathrm{Rest}} & & \downarrow{\scriptstyle \mathrm{Rest}} \\ \mathbf{D}(A \otimes C^{\mathrm{op}}, \mathrm{gr}) & \xrightarrow{\mathrm{R}\Gamma_{\mathfrak{m}}} & \mathbf{D}_{\mathrm{tor}}(A \otimes C^{\mathrm{op}}, \mathrm{gr}) \end{array}$$

is commutative up to isomorphism, and likewise for the morphism of triangulated functors σ^{R}.

(5) *For every $M \in \mathbf{D}(A \otimes B^{\mathrm{op}}, \mathrm{gr})$ and $p \in \mathbb{Z}$ there is an isomorphism*

$$\mathrm{H}^p(\mathrm{R}\Gamma_{\mathfrak{m}}(M)) \cong \lim_{j \to} \mathrm{Ext}^p_A(A/\mathfrak{m}^j, M)$$

in $\mathbf{M}(A \otimes B^{\mathrm{op}}, \mathrm{gr})$. This isomorphism is functorial in M.

Proof (1) By Corollary 15.3.8(3) every $M \in \mathbf{C}(A \otimes B^{\mathrm{op}}, \mathrm{gr})$ has a resolution $\rho : M \to I$ in $\mathbf{C}_{\mathrm{str}}(A \otimes B^{\mathrm{op}}, \mathrm{gr})$ by a complex I that is K-graded-injective over A. If $\psi : I \to I'$ is a quasi-isomorphism in $\mathbf{C}_{\mathrm{str}}(A \otimes B^{\mathrm{op}}, \mathrm{gr})$ between such complexes, then it is a homotopy equivalence in $\mathbf{C}_{\mathrm{str}}(A, \mathrm{gr})$, and hence $\Gamma_{\mathfrak{m}}(\psi)$ is a quasi-isomorphism. We see that the full subcategory $\mathsf{J} \subseteq \mathbf{K}(A \otimes B^{\mathrm{op}}, \mathrm{gr})$ on these complexes I satisfies the conditions of Theorem 8.4.9. Therefore the right derived functor $\mathrm{R}\Gamma_{\mathfrak{m}}$ exists, and η^{R}_I is an isomorphism for all $I \in \mathsf{J}$.

(2) This is a special case of Lemma 16.2.5.

(3) Clear from Lemma 16.3.6(6).

(4) This is immediate from items (1) and (2).

(5) Use formula (16.3.5), and the fact that cohomology commutes with direct limits. □

Definition 16.3.17 Under Setup 16.3.1, the *ith local cohomology functor* is the functor

$$\mathrm{H}^i_{\mathfrak{m}} := \mathrm{R}^i\Gamma_{\mathfrak{m}} : \mathbf{M}(A \otimes B^{\mathrm{op}}, \mathrm{gr}) \to \mathbf{M}(A \otimes B^{\mathrm{op}}, \mathrm{gr}),$$

the ith right derived functor of $\Gamma_{\mathfrak{m}}$.

Corollary 16.3.18 *For every $i \in \mathbb{N}$ there is an isomorphism $\mathrm{H}^i_{\mathfrak{m}} \cong \mathrm{H}^i \circ \mathrm{R}\Gamma_{\mathfrak{m}}$ of functors from $\mathbf{M}(A \otimes B^{\mathrm{op}}, \mathrm{gr})$ to itself.*

Proof This is immediate from Proposition 16.3.16(1). □

Proposition 16.3.19 *For every $i \in \mathbb{N}$ the functor*

$$\mathrm{H}^i_{\mathfrak{m}} : \mathbf{M}(A \otimes B^{\mathrm{op}}, \mathrm{gr}) \to \mathbf{M}(A \otimes B^{\mathrm{op}}, \mathrm{gr})$$

commutes with direct limits.

Proof In view of Proposition 16.3.16(4), with $C := \mathbb{K}$, we can assume that $B = \mathbb{K}$, so $A \otimes B^{\mathrm{op}} = A$. Since direct limits commute with each other, by Proposition 16.3.16(5), it suffices to prove that for every i, j the functor $\mathrm{Ext}^i_A(A/\mathfrak{m}^j, -)$ commutes with direct limits. Now A is left noetherian, so we can find a resolution $P \to A/\mathfrak{m}^j$ by a nonpositive complex, where each P^i is a finite graded-free A-module. And the functor

$$\mathrm{Ext}^i_A(A/\mathfrak{m}^j, -) \cong \mathrm{H}^i \circ \mathrm{Hom}_A(P, -) : \mathbf{M}(A, \mathrm{gr}) \to \mathbf{M}(A, \mathrm{gr})$$

commutes with infinite direct limits. □

Definition 16.1.2, when used in our context, says that the right cohomological dimension of the functor $\Gamma_{\mathfrak{m}}$ is

$$d := \sup \{p \in \mathbb{N} \mid \mathrm{H}^p_{\mathfrak{m}} \neq 0\} \in \mathbb{N} \cup \{\pm\infty\}. \tag{16.3.20}$$

The cohomological dimension of a triangulated functor was introduced in Definition 12.4.1. A triangulated functor is called *quasi-compact* if it commutes with infinite direct sums (Definition 16.1.7).

Theorem 16.3.21 *Under Setup* 16.3.1, *assume the torsion functor* $\Gamma_{\mathfrak{m}} : \mathbf{M}(A, \mathrm{gr}) \to \mathbf{M}(A, \mathrm{gr})$ *has finite right cohomological dimension. Then the derived torsion functor*

$$\mathrm{R}\Gamma_{\mathfrak{m}} : \mathbf{D}(A \otimes B^{\mathrm{op}}, \mathrm{gr}) \to \mathbf{D}(A \otimes B^{\mathrm{op}}, \mathrm{gr})$$

is quasi-compact and has finite cohomological dimension.

Proof According to Proposition 16.3.16(4), the functor

$$\Gamma_{\mathfrak{m}} : \mathbf{M}(A \otimes B^{\mathrm{op}}, \mathrm{gr}) \to \mathbf{M}(A \otimes B^{\mathrm{op}}, \mathrm{gr})$$

also has finite right cohomological dimension. The right derived functors

$$\mathrm{R}^q\Gamma_{\mathfrak{m}} = \mathrm{H}^q_{\mathfrak{m}} : \mathbf{M}(A \otimes B^{\mathrm{op}}, \mathrm{gr}) \to \mathbf{M}(A \otimes B^{\mathrm{op}}, \mathrm{gr})$$

are quasi-compact by Proposition 16.3.19. So Theorem 16.1.10 applies. □

Lemma 16.3.22 *Let* $I \in \mathbf{M}(A, \mathrm{gr})$, *and let* $W := \mathrm{Soc}(I)$. *The following three conditions are equivalent.*

(i) *I is graded-injective and $\mathfrak{m}$-torsion.*
(ii) *I is graded-injective and W is an essential submodule of I.*
(iii) *There is an isomorphism $I \cong A^* \otimes W$ in $\mathbf{C}(A, \mathrm{gr})$ which is the identity on W.*

In condition (iii) we view W as a submodule of $A^* \otimes W$ using the canonical embedding (16.3.10).

Proof

(i) ⇒ (ii): This is Lemma 16.3.8.

(ii) ⇒ (iii): By Proposition 16.3.12 there is an essential monomorphism $\epsilon : I \to A^* \otimes W$ in $\mathbf{M}(A, \mathrm{gr})$ that extends the identity of W. Because I is graded-injective, the monomorphism ϵ is split, by some epimorphism $\sigma : A^* \otimes W \to I$. But W is a graded-essential submodule of $A^* \otimes W$, and therefore σ is also a monomorphism. We conclude that both σ and ϵ are isomorphisms.

(iii) ⇒ (i): According to Lemma 16.3.11, the graded A-module $A^* \otimes W$ is graded-injective and $\mathfrak{m}$-torsion. □

Lemma 16.3.23 *Let $M \in \mathbf{M}(A, \mathrm{gr})$ be an $\mathfrak{m}$-torsion module, and let I be an injective hull of M in $\mathbf{M}(A, \mathrm{gr})$. Then $I \cong A^* \otimes W$, where W is the socle of M. In particular, I is an $\mathfrak{m}$-torsion module.*

Proof By Proposition 16.3.12 there is a graded-essential monomorphism $\epsilon : M \rightarrowtail A^* \otimes W$ in $\mathbf{M}(A, \mathrm{gr})$. From Lemma 16.3.11(1) we know that $A^* \otimes W$ is a graded-injective module over A. So by the uniqueness (up to isomorphism) of graded-injective hulls, we get an isomorphism $I \cong A^* \otimes W$. Finally, by Lemma 16.3.11(2), $A^* \otimes W$ is $\mathfrak{m}$-torsion. □

Lemma 16.3.24 *Let I be a graded-injective A-module. Then there is an isomorphism $\Gamma_{\mathfrak{m}}(I) \cong A^* \otimes W$ in $\mathbf{M}(A, \mathrm{gr})$, where W is the socle of I.*

Proof Consider the essential monomorphism $\epsilon : W \rightarrowtail A^* \otimes W$ from Lemma 16.3.11(4), and the monomorphism $\tau : W \rightarrowtail I$. Because I is graded-injective, there is a homomorphism $\psi : A^* \otimes W \to I$ in $\mathbf{M}(A, \mathrm{gr})$ such that $\psi \circ \epsilon = \tau$. Since ϵ is a graded-essential monomorphism, the usual calculation (see the proof of the implication (ii) $\Rightarrow$ (iii) in Proposition 16.3.12) shows that ψ is a monomorphism. Let $J \subseteq I$ be the image of ψ, so $\psi : A^* \otimes W \xrightarrow{\simeq} J$. We know (from Lemma 16.3.11) that J is graded-injective and $\mathfrak{m}$-torsion. Hence there is a direct sum decomposition $I = J \oplus J'$ in $\mathbf{M}(A, \mathrm{gr})$, and $\Gamma_{\mathfrak{m}}(J) = J$.

By Proposition 16.3.12 we know that $W \cap J' = \mathrm{Soc}(J')$ is a graded-essential submodule of $\Gamma_{\mathfrak{m}}(J')$; but $W \subseteq J$, so $W \cap J' = 0$, and therefore $\Gamma_{\mathfrak{m}}(J') = 0$. The conclusion is that $\Gamma_{\mathfrak{m}}(I) = J$. □

Theorem 16.3.25 *Under Setup* 16.3.1, *the torsion functor*

$$\Gamma_{\mathfrak{m}} : \mathbf{M}(A, \mathrm{gr}) \to \mathbf{M}(A, \mathrm{gr})$$

is stable, namely for every graded-injective A-module I the $\mathfrak{m}$-torsion submodule $\Gamma_{\mathfrak{m}}(I)$ is graded-injective.

Proof This is immediate from Lemma 16.3.24 and Lemma 16.3.11(1). □

Here is the graded NC version of Definition 13.2.5.

Definition 16.3.26

(1) A *minimal complex of graded-injective A-modules* is a bounded below complex I of graded-injective A-modules, such that for every p the submodule $\mathrm{Z}^p(I) \subseteq I^p$ is graded-essential.

(2) Let $M \in \mathbf{D}^+(A, \mathrm{gr})$. A *minimal graded-injective resolution* of M is a quasi-isomorphism $\rho : M \to I$ in $\mathbf{C}_{\mathrm{str}}(A, \mathrm{gr})$, where I is a minimal complex of graded-injective A-modules.

Lemma 16.3.27 *Let $M \in \mathbf{D}^+(A, \mathrm{gr})$. Then M has a minimal graded-injective resolution.*

Proof The proof of Proposition 13.2.6 works here too. □

There is uniqueness for minimal graded-injective resolutions, but we won't need it.

For the next proposition, and for later results, it will be convenient to use this notation:

Notation 16.3.28 If the torsion functor $\Gamma_{\mathfrak{m}}$ has finite right cohomological dimension, then we let $\star :=$ ⟨empty⟩, the empty boundedness indicator; otherwise we let $\star := +$. Thus $\mathbf{D}^\star$ is either $\mathbf{D}$ or $\mathbf{D}^+$.

Proposition 16.3.29 *With Notation* 16.3.28, *let $M \in \mathbf{D}^\star(A \otimes B^{\mathrm{op}}, \mathrm{gr})$. The following conditions are equivalent:*

(i) *$M \in \mathbf{D}^\star_{\mathrm{tor}}(A \otimes B^{\mathrm{op}}, \mathrm{gr})$.*

(ii) *The morphism $\sigma^{\mathrm{R}}_M : \mathrm{R}\Gamma_{\mathfrak{m}}(M) \to M$ in $\mathbf{D}(A \otimes B^{\mathrm{op}}, \mathrm{gr})$ is an isomorphism.*

Proof The implication (ii) ⇒ (i) is due to Proposition 16.3.16(3).

For the other implication, we can forget the ring B; it is not relevant. First let us assume that $M \in \mathbf{D}^+_{\mathrm{tor}}(A, \mathrm{gr})$. By Lemma 16.3.23 the injective hull in $\mathbf{M}(A, \mathrm{gr})$ of a torsion module is torsion. Then, according to Theorem 11.5.8, there is a quasi-isomorphism $M \to I$ in $\mathbf{C}^+_{\mathrm{str}}(A, \mathrm{gr})$, where I is a complex of torsion graded-injective A-modules. It follows that homomorphism $\sigma_I : \Gamma_{\mathfrak{m}}(I) \to I$ in $\mathbf{C}^+_{\mathrm{str}}(A, \mathrm{gr})$, which represents σ^{R}_M, is an isomorphism.

When $\mathrm{H}(M)$ is not bounded below, but $\Gamma_{\mathfrak{m}}$ has finite right cohomological dimension, we can use smart truncation, as in the proof of Theorem 12.5.2(2). □

Proposition 16.3.30 *With Notation* 16.3.28, *Let $M \in \mathbf{D}^\star_{\mathrm{tor}}(A \otimes B^{\mathrm{op}}, \mathrm{gr})$ and $N \in \mathbf{D}^+(A \otimes C^{\mathrm{op}}, \mathrm{gr})$. Then there is an isomorphism*

$$\mathrm{RHom}_A(M, N) \xrightarrow{\simeq} \mathrm{RHom}_A(M, \mathrm{R}\Gamma_{\mathfrak{m}}(N))$$

in $\mathbf{D}(B \otimes C^{\mathrm{op}}, \mathrm{gr})$, and it is functorial in M and N.

Proof Choose a resolution $M \to I_M$ in $\mathbf{C}^\star_{\mathrm{str}}(A \otimes B^{\mathrm{op}}, \mathrm{gr})$ where I_M is K-graded-injective over A, each I^p_M is graded-injective over A, and $\inf(I_M) = \inf(\mathrm{H}((M))$; this is possible by Corollary 15.3.7(3). Choose the same sort of resolution $N \to I_N$ in $\mathbf{C}^+_{\mathrm{str}}(A \otimes C^{\mathrm{op}}, \mathrm{gr})$. Then

$$\mathrm{RHom}_A(M, N) \cong \mathrm{Hom}_A(I_M, I_N) \tag{16.3.31}$$

in $\mathbf{D}(B \otimes C^{\mathrm{op}}, \mathrm{gr})$, and $\mathrm{R}\Gamma_{\mathfrak{m}}(N) \cong \Gamma_{\mathfrak{m}}(I_N)$ in $\mathbf{D}(A \otimes C^{\mathrm{op}}, \mathrm{gr})$. By Theorem 16.3.25 each $\Gamma_{\mathfrak{m}}(I^p_N)$ is graded-injective as an A-module, so $\Gamma_{\mathfrak{m}}(I_N)$ is a bounded below complex of graded-injective A-modules, and therefore it is a K-graded-injective complex over A. We see that

$$\mathrm{RHom}_A(M, \mathrm{R}\Gamma_{\mathfrak{m}}(N)) \cong \mathrm{Hom}_A(I_M, \Gamma_{\mathfrak{m}}(I_N)) \tag{16.3.32}$$

in $\mathbf{D}(B \otimes C^{\mathrm{op}}, \mathrm{gr})$.

By Proposition 16.3.29 (in both cases of Notation 16.3.28) the homomorphism $\sigma_{I_M} : \Gamma_{\mathfrak{m}}(I_M) \to I_M$ is a quasi-isomorphism. Hence in the commutative diagram

$$\begin{array}{ccc} \mathrm{Hom}_A(I_M, I_N) & \xrightarrow{\mathrm{Hom}(\sigma_{I_M}, \mathrm{id})} & \mathrm{Hom}_A(\Gamma_{\mathfrak{m}}(I_M), I_N) \\ {\scriptstyle \mathrm{Hom}(\mathrm{id}, \sigma_{I_N})} \uparrow & & \uparrow {\scriptstyle \mathrm{Hom}(\mathrm{id}, \sigma_{I_N})} \\ \mathrm{Hom}_A(I_M, \Gamma_{\mathfrak{m}}(I_N)) & \xrightarrow{\mathrm{Hom}(\sigma_{I_M}, \mathrm{id})} & \mathrm{Hom}_A(\Gamma_{\mathfrak{m}}(I_M), \Gamma_{\mathfrak{m}}(I_N)) \end{array}$$

in $\mathbf{C}_{\mathrm{str}}(B \otimes C^{\mathrm{op}}, \mathrm{gr})$ the two horizontal arrows are quasi-isomorphisms. The right vertical arrow is an isomorphism in $\mathbf{C}_{\mathrm{str}}(B \otimes C^{\mathrm{op}}, \mathrm{gr})$ by Lemma 16.3.6(3). It follows that the left vertical arrow is a quasi-isomorphism. Combining this with the isomorphisms (16.3.31) and (16.3.32) we obtain the desired isomorphism above. □

The next definition is a specialization of Definitions 16.1.3, 16.1.4 and 16.2.1.

Definition 16.3.33 Under Setup 16.3.1:

(1) A module $N \in \mathbf{M}(A \otimes B^{\mathrm{op}}, \mathrm{gr})$ is called *graded-$\mathfrak{m}$-flasque* if it is a right $\Gamma_{\mathfrak{m}}$-acyclic object, i.e. if $\mathrm{H}^q_{\mathfrak{m}}(N) = 0$ for all $q > 0$.

(2) The functor

$$\Gamma_{\mathfrak{m}} : \mathbf{M}(A \otimes B^{\mathrm{op}}, \mathrm{gr}) \to \mathbf{M}(A \otimes B^{\mathrm{op}}, \mathrm{gr})$$

is called *weakly stable* if for every graded-injective module $I \in \mathbf{M}(A \otimes B^{\mathrm{op}}, \mathrm{gr})$, the module $\Gamma_{\mathfrak{m}}(I) \in \mathbf{M}(A \otimes B^{\mathrm{op}}, \mathrm{gr})$ is graded-$\mathfrak{m}$-flasque.

(3) A complex $N \in \mathbf{D}(A \otimes B^{\mathrm{op}}, \mathrm{gr})$ is called a *K-graded-$\mathfrak{m}$-flasque complex* if it is a right $\Gamma_{\mathfrak{m}}$-acyclic complex, i.e. if the morphism $\eta^{\mathrm{R}}_N : \Gamma_{\mathfrak{m}}(N) \to \mathrm{R}\Gamma_{\mathfrak{m}}(N)$ in $\mathbf{D}(A \otimes B^{\mathrm{op}}, \mathrm{gr})$ is an isomorphism.

The term *flasque* in this meaning seems to have been first used in [190].

Example 16.3.34 Suppose $M \in \mathbf{M}(A, \mathrm{gr})$ is an $\mathfrak{m}$-torsion module. Consider its minimal graded-injective resolution $0 \to M \to I^0 \to I^1 \to \cdots$. By Lemma 16.3.23 and induction on q, we see that all the graded modules I^q are $\mathfrak{m}$-torsion; in fact $I^q \cong A^* \otimes W^q$ for some $W^q \in \mathbf{M}(\mathbb{K}, \mathrm{gr})$. Thus $\mathrm{R}\Gamma_{\mathfrak{m}}(M) \cong \Gamma_{\mathfrak{m}}(I) = I \cong M$, so $\mathrm{R}^q\Gamma_{\mathfrak{m}}(M) = 0$ for all $q > 0$, and M is a graded-$\mathfrak{m}$-flasque A-module.

Theorem 16.3.35 *Under Setup* 16.3.1, *the torsion functor*

$$\Gamma_{\mathfrak{m}} : \mathbf{M}(A \otimes B^{\mathrm{op}}, \mathrm{gr}) \to \mathbf{M}(A \otimes B^{\mathrm{op}}, \mathrm{gr})$$

is weakly stable.

Proof Take an injective object $I \in \mathbf{M}(A \otimes B^{\mathrm{op}}, \mathrm{gr})$. According to Proposition 15.2.13(3), the module I is graded-injective over A. By Lemma 16.3.24 we know that there is an isomorphism $\Gamma_{\mathfrak{m}}(I) \cong A^* \otimes W$ in $\mathbf{M}(A, \mathrm{gr})$ for some graded $\mathbb{K}$-module W. This is a graded-injective A-module, and therefore $\Gamma_{\mathfrak{m}}(I)$ is graded-$\mathfrak{m}$-flasque as an object of $\mathbf{M}(A, \mathrm{gr})$. Finally, by Proposition 16.3.16(4), the property of being graded-$\mathfrak{m}$-flasque is insensitive to the restriction functor $\mathbf{M}(A \otimes B^{\mathrm{op}}, \mathrm{gr}) \to \mathbf{M}(A, \mathrm{gr})$, so $\Gamma_{\mathfrak{m}}(I)$ is graded-$\mathfrak{m}$-flasque as an object of $\mathbf{M}(A \otimes B^{\mathrm{op}}, \mathrm{gr})$. □

In Proposition 16.3.16(2) we presented a morphism $\sigma^{\mathrm{R}} : \mathrm{R}\Gamma_{\mathfrak{m}} \to \mathrm{Id}$ of triangulated functors from $\mathbf{D}^{\star}(A \otimes B^{\mathrm{op}}, \mathrm{gr})$ to itself. Here we are using Notation 16.3.28. According to Definition 16.2.3, the pair $(\mathrm{R}\Gamma_{\mathfrak{m}}, \sigma^{\mathrm{R}})$ is a copointed triangulated functor. Recall that $(\mathrm{R}\Gamma_{\mathfrak{m}}, \sigma^{\mathrm{R}})$ is an idempotent copointed triangulated functor if for every complex $M \in \mathbf{D}^{\star}(A \otimes B^{\mathrm{op}}, \mathrm{gr})$ the morphisms

$$\sigma^{\mathrm{R}}_{\mathrm{R}\Gamma_{\mathfrak{m}}(M)}, \mathrm{R}\Gamma_{\mathfrak{m}}(\sigma^{\mathrm{R}}_M) : \mathrm{R}\Gamma_{\mathfrak{m}}(\mathrm{R}\Gamma_{\mathfrak{m}}(M)) \to \mathrm{R}\Gamma_{\mathfrak{m}}(M)$$

in $\mathbf{D}(A \otimes B^{\mathrm{op}}, \mathrm{gr})$ are isomorphisms.

Corollary 16.3.36 ([157]) *With Notation* 16.3.28, *the copointed triangulated functor* $(\mathrm{R}\Gamma_{\mathfrak{m}}, \sigma^{\mathrm{R}})$ *on* $\mathbf{D}^{\star}(A \otimes B^{\mathrm{op}}, \mathrm{gr})$ *is idempotent.*

Proof By Theorem 16.3.35 the functor $\Gamma_{\mathfrak{m}}$ is weakly stable, and by Lemma 16.3.6(2) the copointed functor $(\Gamma_{\mathfrak{m}}, \sigma)$ is idempotent. Thus Theorem 16.2.8 applies. □

We end the section with an example showing a torsion functor as in Theorem 16.3.35 which is weakly stable but not stable. The example is a slight modification of an example in [157].

Example 16.3.37 Let $A := \mathbb{K}[t]$, the polynomial ring in a variable t of degree 1. It is connected noetherian, and its augmentation ideal $\mathfrak{m}$ is generated by t. Let $B := \mathbb{K}[s_0, s_1, \ldots]$, the commutative polynomial ring in countably many variables $s_0, s_1, \ldots$ of degree 1. So B is a graded ring, and it is not noetherian. Theorem 16.3.35 tells us that the functor

$$\Gamma_{\mathfrak{m}} : \mathbf{M}(A \otimes B, \mathrm{gr}) \to \mathbf{M}(A \otimes B, \mathrm{gr})$$

is weakly stable. (Since B is commutative, it doesn't matter whether we write B or B^{op}.) We will prove that $\Gamma_{\mathfrak{m}}$ is not stable.

Consider a countable collection $\{I_p\}_{p \in \mathbb{N}}$ of graded-injective B-modules. Then $I := \prod_{p \in \mathbb{N}} I_p(-p)$ is a graded-injective B-module, and therefore $J := \operatorname{Hom}_B(A \otimes B, I)$ is a graded-injective $(A \otimes B)$-module. If the functor $\Gamma_{\mathfrak{m}}$ on $\mathbf{M}(A \otimes B, \mathrm{gr})$ were stable, this would imply that $\Gamma_{\mathfrak{m}}(J)$ is also a graded-injective $(A \otimes B)$-module. Since $B \to A \otimes B$ is flat, $\Gamma_{\mathfrak{m}}(J)$ is then a graded-injective B-module (see Proposition 15.2.13).

As graded B-modules there are isomorphisms

$$A \otimes B \cong \bigoplus\nolimits_{p \in \mathbb{N}} B \cdot t^p \cong \bigoplus\nolimits_{p \in \mathbb{N}} B(-p).$$

Therefore

$$J = \operatorname{Hom}_B(A \otimes B, I) \cong \prod\nolimits_{p \in \mathbb{N}} I(p)$$

as graded B-modules, and

$$\Gamma_{\mathfrak{m}}(J) \cong \lim_{p \to} \Big(\prod\nolimits_{q=0}^{p} I(q) \Big) \cong \bigoplus\nolimits_{p \in \mathbb{N}} I(p).$$

Now for each p the graded-module I_p is a direct summand of $I(p)$, and hence $\bigoplus_{p \in \mathbb{N}} I_p$ is a direct summand of $\Gamma_{\mathfrak{m}}(J)$, as graded B-modules. Therefore $\bigoplus_{p \in \mathbb{N}} I_p$ is a graded-injective B-module.

The conclusion is that a countable direct sum of graded-injective B-modules is graded-injective. According to the Bass–Papp Theorem [93, Theorem 3.46], which holds also in the graded sense (cf. Proposition 15.1.26), the graded ring B is noetherian. This is a contradiction.

16.4 Representability of Derived Torsion

The results of this section are an adaptation of results from [157, Section 7] to the connected graded setting.

In Chapter 17 we are going to use the representability of derived torsion to give an alternative proof of M. Van den Bergh's theorem on existence of balanced dualizing complexes (see Theorem 17.3.19).

Recall that Conventions 15.1.17 and 15.2.1 are in force. In this section we assume the following strengthening of Setup 16.3.1.

Setup 16.4.1 We are given a left noetherian connected graded $\mathbb{K}$-ring A, with augmentation ideal $\mathfrak{m} = \bigoplus_{i \geq 1} A_i \subseteq A$, such that the torsion functor $\Gamma_{\mathfrak{m}}$ has finite right cohomological dimension. We are also given a graded $\mathbb{K}$-ring B.

As before, the graded ring B plays an auxiliary role, a placeholder for A or $\mathbb{K}$.

Definition 16.4.2 Under Setup 16.4.1, the *dedualizing complex of* A is the complex $P_A := \mathrm{R}\Gamma_{\mathfrak{m}}(A) \in \mathbf{D}(A^{\mathrm{en}}, \mathrm{gr})$.

The name "dedualizing" is due to L. Positselsky [121] . Regarding the lack of left-right symmetry in this definition, see Corollary 16.5.11 and Proposition 16.5.19.

As a special case of Proposition 16.3.16, there is a morphism $\sigma^{\mathrm{R}}_A : P_A = \mathrm{R}\Gamma_{\mathfrak{m}}(A) \to A$ in $\mathbf{D}(A^{\mathrm{en}}, \mathrm{gr})$.

Theorem 16.4.3 (Representability of Derived Torsion, [157]) *Under Setup* 16.4.1 *there is a unique isomorphism*

$$\mathrm{ev}^{\mathrm{R,L}}_{\mathfrak{m},(-)} : P_A \otimes^{\mathrm{L}}_A (-) \xrightarrow{\simeq} \mathrm{R}\Gamma_{\mathfrak{m}}(-)$$

of triangulated functors from $\mathbf{D}(A \otimes B^{\mathrm{op}}, \mathrm{gr})$ *to itself, called* derived $\mathfrak{m}$-torsion tensor-evaluation, *such that for every complex* $M \in \mathbf{D}(A \otimes B^{\mathrm{op}}, \mathrm{gr})$ *the diagram*

$$\begin{array}{ccc} P_A \otimes^{\mathrm{L}}_A M & \xrightarrow[\cong]{\mathrm{ev}^{\mathrm{R,L}}_{\mathfrak{m},M}} & \mathrm{R}\Gamma_{\mathfrak{m}}(M) \\ {\scriptstyle \sigma^{\mathrm{R}}_A \otimes^{\mathrm{L}}_A \mathrm{id}_M} \downarrow & & \downarrow {\scriptstyle \sigma^{\mathrm{R}}_M} \\ A \otimes^{\mathrm{L}}_A M & \xrightarrow[\cong]{\mathrm{lu}} & M \end{array} \qquad (\heartsuit)$$

is commutative.

The isomorphism marked "lu" in diagram ($\heartsuit$) is the left unitor isomorphism of the closed monoidal structure, see equation (16.3.3).

The proof of the theorem requires the next lemma.

Lemma 16.4.4 *Let* $F, G : \mathbf{D}(A, \mathrm{gr}) \to \mathbf{D}(B, \mathrm{gr})$ *be quasi-compact triangulated functors, and let* $\eta : F \to G$ *be a morphism of triangulated functors. Assume that* $\eta_A : F(A) \to G(A)$ *is an isomorphism. Then* η *is an isomorphism.*

Proof We are given that η_A is an isomorphism. Because the algebraic degree twist $M \mapsto M(i)$ is an automorphism of $\mathbf{D}(A, \mathrm{gr})$, it follows that $\eta_{A(i)}$ is an isomorphism for every integer i, and likewise for the cohomological degree shift (i.e. the translation $M \mapsto M[i]$ on complexes), so $\eta_{A(i)[j]}$ is an isomorphism for every integer j. Both functors are quasi-compact, and therefore η_P is an isomorphism for every graded-free complex of A-modules $P \cong \bigoplus_{x \in X} A(i_x)[j_x]$.

Suppose we are given a distinguished triangle $M' \to M \to M'' \xrightarrow{\vartriangle}$ in $\mathbf{D}(A, \mathrm{gr})$, such that two of the three morphisms $\eta_{M'}$, η_M and $\eta_{M''}$ are isomorphisms. Because η is a morphism of triangulated functors, it follows that third morphism is also an isomorphism.

Next consider a semi-graded-free DG module P. Take a filtration $\{F_j(P)\}_{j\geq -1}$ of P as in Definition 15.3.1(2). For every j we have a distinguished triangle

$$F_{j-1}(P) \xrightarrow{\theta_j} F_j(P) \to \mathrm{Gr}^F_j(P) \xrightarrow{\triangle}$$

in $\mathbf{D}(A, \mathrm{gr})$, where $\theta_j : F_{j-1}(P) \to F_j(P)$ is the inclusion. Since $\mathrm{Gr}^F_j(P)$ is a graded-free complex, by induction on j, we conclude that $\eta_{F_j(P)}$ is an isomorphism for every $j \geq 0$. The homotopy colimit construction (see Definition 14.1.11) gives a distinguished triangle

$$\bigoplus_{j\in\mathbb{N}} F_j(P) \xrightarrow{\Theta} \bigoplus_{j\in\mathbb{N}} F_j(P) \to P \xrightarrow{\triangle}$$

in $\mathbf{D}(A, \mathrm{gr})$, where

$$\Theta|_{F_{j-1}(P)} := (\mathrm{id}, -\theta_j) : F_{j-1}(P) \to F_{j-1}(P) \oplus F_j(P).$$

By quasi-compactness we know that $\eta_{\bigoplus F_j(P)}$ is an isomorphism. Therefore η_P is an isomorphism.

Finally, every $M \in \mathbf{D}(A, \mathrm{gr})$ admits an isomorphism $M \cong P$ with P semi-graded-free. Therefore η_M is an isomorphism. □

Proof of Theorem 16.4.3

Step 1. We begin by constructing the morphism of triangulated functors $\mathrm{ev}^{\mathrm{R,L}}_{\mathfrak{m},(-)}$. For each complex $M \in \mathbf{C}(A\otimes B^{\mathrm{op}}, \mathrm{gr})$ we choose a K-graded-injective resolution $\rho_M : M \to I_M$ and a K-graded-flat resolution $\theta_M : Q_M \to M$, both in $\mathbf{C}_{\mathrm{str}}(A \otimes B^{\mathrm{op}}, \mathrm{gr})$. Note that I_M is K-graded-$\mathfrak{m}$-flasque over A, and Q_M is K-graded-flat over A. We use these choices for presentations of the right derived functor

$$(\mathrm{R}\Gamma_{\mathfrak{m}}, \eta^{\mathrm{R}}) : \mathbf{D}(A \otimes B^{\mathrm{op}}, \mathrm{gr}) \to \mathbf{D}(A \otimes B^{\mathrm{op}}, \mathrm{gr})$$

and the left derived bifunctor

$$(- \otimes^{\mathrm{L}}_A -, \eta^{\mathrm{L}}) : \mathbf{D}(A^{\mathrm{en}}, \mathrm{gr}) \times \mathbf{D}(A \otimes B^{\mathrm{op}}, \mathrm{gr}) \to \mathbf{D}(A \otimes B^{\mathrm{op}}, \mathrm{gr}).$$

Let us also choose a K-graded-injective resolution $\psi : A \to J$ in $\mathbf{C}_{\mathrm{str}}(A^{\mathrm{en}}, \mathrm{gr})$. With these choices we have the following presentations: $P_A = \Gamma_{\mathfrak{m}}(J)$,

$$\mathrm{R}\Gamma_{\mathfrak{m}}(M) = \Gamma_{\mathfrak{m}}(I_M) \tag{16.4.5}$$

and

$$P_A \otimes^{\mathrm{L}}_A M = \Gamma_{\mathfrak{m}}(J) \otimes_A Q_M. \tag{16.4.6}$$

Suppose $N \in \mathbf{C}(A \otimes B^{\mathrm{op}}, \mathrm{gr})$ is some complex. Given homogeneous elements $x \in \Gamma_{\mathfrak{m}}(J)$ and $n \in N$, the tensor $x \otimes n$ belongs to $\Gamma_{\mathfrak{m}}(J \otimes_A N)$. In this way we obtain a homomorphism

$$\mathrm{ev}_{\mathfrak{m},J,N} : \Gamma_{\mathfrak{m}}(J) \otimes_A N \to \Gamma_{\mathfrak{m}}(J \otimes_A N) \tag{16.4.7}$$

in $\mathbf{C}_{\mathrm{str}}(A \otimes B^{\mathrm{op}}, \mathrm{gr})$, which is functorial in N.

Consider a complex $M \in \mathbf{C}(A \otimes B^{\mathrm{op}}, \mathrm{gr})$. The choices we made give rise to this solid diagram

$$\begin{array}{ccccc} A \otimes_A Q_M & \xrightarrow{\mathrm{lu}} & Q_M & \xrightarrow{\theta_M} & M \\ \downarrow{\scriptstyle \psi \otimes_A \mathrm{id}} & & & & \downarrow{\scriptstyle \rho_M} \\ J \otimes_A Q_M & & \overset{\chi_M}{\dashrightarrow} & & I_M \end{array} \tag{16.4.8}$$

in $\mathbf{C}_{\mathrm{str}}(A \otimes B^{\mathrm{op}}, \mathrm{gr})$. The homomorphisms $\psi \otimes_A \mathrm{id}$, $\theta_M \circ \mathrm{lu}$ and ρ_M are quasi-isomorphisms. Because I_M is K-injective in $\mathbf{C}(A \otimes B^{\mathrm{op}}, \mathrm{gr})$, there is a quasi-isomorphism $\chi_M : J \otimes_A Q_M \to I_M$ that makes this diagram commutative up to homotopy.

We now form diagram (16.4.9) in $\mathbf{C}_{\mathrm{str}}(A \otimes B^{\mathrm{op}}, \mathrm{gr})$. Here $\mathrm{ev}_{\mathfrak{m},J,Q_M}$ is

$$\begin{array}{ccccc} \Gamma_{\mathfrak{m}}(J) \otimes_A Q_M & \xrightarrow{\mathrm{ev}_{\mathfrak{m},J,Q_M}} & \Gamma_{\mathfrak{m}}(J \otimes_A Q_M) & \xrightarrow{\Gamma_{\mathfrak{m}}(\chi_M)} & \Gamma_{\mathfrak{m}}(I_M) \\ \downarrow{\scriptstyle \sigma_J \otimes_A \mathrm{id}} & & \downarrow{\scriptstyle \sigma_{(J \otimes_A Q_M)}} & & \downarrow{\scriptstyle \sigma_{I_M}} \\ J \otimes_A Q_M & \xrightarrow{\mathrm{id}} & J \otimes_A Q_M & \xrightarrow{\chi_M} & I_M \\ \uparrow{\scriptstyle \psi \otimes_A \mathrm{id}} & & \uparrow{\scriptstyle \psi \otimes_A \mathrm{id}} & & \uparrow{\scriptstyle \rho_M} \\ A \otimes_A Q_M & \xrightarrow{\mathrm{id}} & A \otimes_A Q_M & \xrightarrow{\theta_M \circ \mathrm{lu}} & M \end{array} \tag{16.4.9}$$

the homomorphism from (16.4.7). The diagram (16.4.9) is commutative up to homotopy. (Actually all small squares, except the bottom right one, are commutative in the strict sense.) The vertical arrows $\psi \otimes_A \mathrm{id}$ and ρ_M are quasi-isomorphisms. Passing to $\mathbf{D}(A \otimes B^{\mathrm{op}}, \mathrm{gr})$ we get a commutative diagram, with vertical isomorphisms between the second and third rows. The diagram with the four extreme objects only is the one we are looking for. By construction it is a commutative diagram in $\mathbf{D}(A \otimes B^{\mathrm{op}}, \mathrm{gr})$, and it is functorial in M. The morphism

$$\mathrm{ev}^{\mathrm{R,L}}_{\mathfrak{m},M} : P_A \otimes^{\mathrm{L}}_A M \to \mathrm{R}\Gamma_{\mathfrak{m}}(M)$$

that we want is represented by

$$\Gamma_{\mathfrak{m}}(\chi_M) \circ \mathrm{ev}_{\mathfrak{m},J,Q_M} : \Gamma_{\mathfrak{m}}(J) \otimes_A Q_M \to \Gamma_{\mathfrak{m}}(I_M).$$

See (16.4.5) and (16.4.6).

Step 2. In this step we prove that $\mathrm{ev}^{\mathrm{R,L}}_{\mathfrak{m},M}$ is an isomorphism for every complex $M \in \mathbf{D}(A \otimes B^{\mathrm{op}}, \mathrm{gr})$. Because the functor Rest is conservative, it suffices to prove that the morphism

$$\operatorname{Rest}(\mathrm{ev}^{\mathrm{R,L}}_{\mathfrak{m},M}) : \operatorname{Rest}(P_A \otimes^{\mathrm{L}}_A M) \to \operatorname{Rest}(\mathrm{R}\Gamma_{\mathfrak{m}}(M))$$

in $\mathbf{D}(A, \mathrm{gr})$ is an isomorphism. Going over all the details of the construction above, and noting that $\rho_M : M \to I_M$ and $\theta_M : Q_M \to M$ are K-flat and K-injective resolutions, respectively, also in the category $\mathbf{C}_{\mathrm{str}}(A, \mathrm{gr})$, we might as well forget about the ring B.

So now we are in the case $B = \mathbb{K}$, $A \otimes B^{\mathrm{op}} = A$, and we want to prove that $\mathrm{ev}^{\mathrm{R,L}}_{\mathfrak{m},M}$ is an isomorphism for every $M \in \mathbf{D}(A, \mathrm{gr})$. By Theorem 16.3.21 the functor $\mathrm{R}\Gamma_{\mathfrak{m}}$ on $\mathbf{D}(A, \mathrm{gr})$ is quasi-compact. The functor $P \otimes^{\mathrm{L}}_A (-)$ is also quasi-compact. This means that we can use Lemma 16.4.4, and it tells us that it suffices to prove that $\mathrm{ev}^{\mathrm{R,L}}_{\mathfrak{m},M}$ is an isomorphism for $M = A$.

Let us examine the morphism $\mathrm{ev}^{\mathrm{R,L}}_{\mathfrak{m},A}$, i.e. $\mathrm{ev}^{\mathrm{R,L}}_{\mathfrak{m},M}$ for $M = A$. We can choose the K-flat resolution $\theta_A : Q_A \to A$ in $\mathbf{C}_{\mathrm{str}}(A, \mathrm{gr})$ to be the identity of A. Also, we can choose the K-injective resolution $\rho_A : A \to I_A$ in $\mathbf{C}_{\mathrm{str}}(A, \mathrm{gr})$ to be the restriction of $\psi : A \to J$. Then the homomorphism $\chi_A : J \otimes_A Q_A \to I_A$ in diagram (16.4.8) can be chosen to be $\chi_A = \mathrm{id} \otimes_A \mathrm{id}$. We get a commutative diagram

$$\begin{CD}
\Gamma_{\mathfrak{m}}(J) \otimes_A Q_A @>{\mathrm{ev}_{\mathfrak{m},J,Q_A}}>> \Gamma_{\mathfrak{m}}(J \otimes_A Q_A) @>{\Gamma_{\mathfrak{m}}(\chi_A)}>> \Gamma_{\mathfrak{m}}(I_A) \\
@V{\mathrm{id} \otimes_A \theta_A}VV @V{\Gamma_{\mathfrak{m}}(\mathrm{id} \otimes_A \theta_A)}VV @V{\mathrm{id}}VV \\
\Gamma_{\mathfrak{m}}(J) \otimes_A A @>{\mathrm{ev}_{\mathfrak{m},J,A}}>> \Gamma_{\mathfrak{m}}(J \otimes_A A) @>{\Gamma_{\mathfrak{m}}(\mathrm{ru})}>> \Gamma_{\mathfrak{m}}(J)
\end{CD}$$

in $\mathbf{C}_{\mathrm{str}}(A, \mathrm{gr})$. Here $\mathrm{ru} : J \otimes_A A \xrightarrow{\simeq} J$ is the canonical isomorphism (the right unitor). The horizontal arrows in the second row, and the vertical arrows, are all bijective. We conclude that $\Gamma_{\mathfrak{m}}(\chi_A) \circ \mathrm{ev}_{\mathfrak{m},J,Q_A}$ is bijective. Hence $\mathrm{ev}^{\mathrm{R,L}}_{\mathfrak{m},A}$ is an isomorphism in $\mathbf{D}(A, \mathrm{gr})$.

Step 3. It remains to prove the uniqueness of $\mathrm{ev}^{\mathrm{R,L}}_{\mathfrak{m},(-)}$. By applying the functor $\mathrm{R}\Gamma_{\mathfrak{m}}$ to the diagram ($\heartsuit$), and then making use of the morphism of functors $\sigma^{\mathrm{R}} : \mathrm{R}\Gamma_{\mathfrak{m}} \to \mathrm{Id}$, we obtain the commutative diagram (16.4.10) in $\mathbf{D}(A \otimes B^{\mathrm{op}}, \mathrm{gr})$. The vertical morphisms $\sigma^{\mathrm{R}}_{\mathrm{R}\Gamma_{\mathfrak{m}}(M)}$ and $\mathrm{R}\Gamma_{\mathfrak{m}}(\sigma^{\mathrm{R}}_M)$ are isomorphisms because of the idempotence (Corollary 16.3.36). Therefore the other two vertical arrows are isomorphisms. We see that the isomorphism $\mathrm{ev}^{\mathrm{R,L}}_{\mathfrak{m},M}$ can be expressed as the composition of other isomorphisms; so it is unique. □

The next result is due to Van den Bergh [153]. We give another proof, based on Theorem 16.4.3. Recall the $\mathbb{K}$-linear duality functor

$$(-)^* : \mathbf{D}(A \otimes B^{\mathrm{op}}, \mathrm{gr})^{\mathrm{op}} \to \mathbf{D}(B \otimes A^{\mathrm{op}}, \mathrm{gr}).$$

$$
\begin{array}{ccc}
P_A \otimes^{\mathrm{L}}_A M & \xrightarrow[\cong]{\ \mathrm{ev}^{\mathrm{R,L}}_{\mathfrak{m},M}\ } & \mathrm{R}\Gamma_{\mathfrak{m}}(M) \\
{\scriptstyle \sigma^{\mathrm{R}}_{(P_A \otimes^{\mathrm{L}}_A M)}} \Big\uparrow & & \Big\uparrow {\scriptstyle \sigma^{\mathrm{R}}_{\mathrm{R}\Gamma_{\mathfrak{m}}(M)}} \\
\mathrm{R}\Gamma_{\mathfrak{m}}(P_A \otimes^{\mathrm{L}}_A M) & \xrightarrow[\cong]{\ \mathrm{R}\Gamma_{\mathfrak{m}}(\mathrm{ev}^{\mathrm{R,L}}_{\mathfrak{m},M})\ } & \mathrm{R}\Gamma_{\mathfrak{m}}(\mathrm{R}\Gamma_{\mathfrak{m}}(M)) \\
{\scriptstyle \mathrm{R}\Gamma_{\mathfrak{m}}(\sigma^{\mathrm{R}}_A \otimes^{\mathrm{L}}_A \mathrm{id}_M)} \Big\downarrow & & \Big\downarrow {\scriptstyle \mathrm{R}\Gamma_{\mathfrak{m}}(\sigma^{\mathrm{R}}_M)} \\
\mathrm{R}\Gamma_{\mathfrak{m}}(A \otimes^{\mathrm{L}}_A M) & \xrightarrow[\cong]{\ \mathrm{R}\Gamma_{\mathfrak{m}}(\mathrm{lu})\ } & \mathrm{R}\Gamma_{\mathfrak{m}}(M)
\end{array}
\tag{16.4.10}
$$

Its formula is

$$(-)^* := \mathrm{Hom}_{\mathbb{K}}(-, \mathbb{K}) \cong \mathrm{Hom}_A(-, A^*).$$

Corollary 16.4.11 (Van den Bergh's Local Duality, [153]) *Under Setup* 16.4.1, *for every complex* $M \in \mathbf{D}(A \otimes B^{\mathrm{op}}, \mathrm{gr})$ *there is an isomorphism*

$$\mathrm{R}\Gamma_{\mathfrak{m}}(M)^* \cong \mathrm{RHom}_A(M, (P_A)^*)$$

in $\mathbf{D}(B \otimes A^{\mathrm{op}}, \mathrm{gr})$. *This isomorphism is functorial in* M.

Proof We have the following functorial isomorphisms

$$
\begin{aligned}
\mathrm{R}\Gamma_{\mathfrak{m}}(M)^* &\cong^{\dagger} (P_A \otimes^{\mathrm{L}}_A M)^* = \mathrm{Hom}_{\mathbb{K}}(P_A \otimes^{\mathrm{L}}_A M, \mathbb{K}) \\
&\cong \mathrm{RHom}_A(P_A \otimes^{\mathrm{L}}_A M, A^*) \\
&\cong^{\heartsuit} \mathrm{RHom}_A(M, \mathrm{RHom}_A(P_A, A^*)) \\
&\cong \mathrm{RHom}_A(M, (P_A)^*)
\end{aligned}
$$

in $\mathbf{D}(B \otimes A^{\mathrm{op}}, \mathrm{gr})$. The isomorphism $\cong^{\dagger}$ is according to Theorem 16.4.3. The isomorphism $\cong^{\heartsuit}$ comes from derived Hom-tensor adjunction. □

16.5 Symmetry of Derived Torsion

The results of this section are an adaptation of results from [157, Section 8] to the connected graded setting. Among other things, we prove that the χ condition of Artin–Zhang implies symmetry of derived torsion.

In Chapter 17 below we are going to use the symmetry of derived torsion to give an alternative proof of Van den Bergh's theorem on existence of balanced dualizing complexes (see Theorem 17.3.19).

We continue with Conventions 15.1.17 and 15.2.1. Thus $\mathbb{K}$ is a field, and all graded rings are algebraically graded central $\mathbb{K}$-rings. The following setup is assumed in this section:

Setup 16.5.1 We are given a noetherian connected graded $\mathbb{K}$-ring A, with augmentation ideal $\mathfrak{m}$. The augmentation ideal of the opposite ring A^{op} is $\mathfrak{m}^{\mathrm{op}}$.

The assumption that A is noetherian means that both A and A^{op} are left noetherian. The enveloping ring of A is $A^{\mathrm{en}} = A \otimes A^{\mathrm{op}}$, and it is often not noetherian.

As explained in the previous section, there is a torsion functor $\Gamma_{\mathfrak{m}} : \mathbf{M}(A, \mathrm{gr}) \to \mathbf{M}(A, \mathrm{gr})$. But by the same token, there is also a torsion functor $\Gamma_{\mathfrak{m}^{\mathrm{op}}} : \mathbf{M}(A^{\mathrm{op}}, \mathrm{gr}) \to \mathbf{M}(A^{\mathrm{op}}, \mathrm{gr})$. On the category $\mathbf{M}(A^{\mathrm{en}}, \mathrm{gr})$ we have two torsion functors:

$$\Gamma_{\mathfrak{m}}, \Gamma_{\mathfrak{m}^{\mathrm{op}}} : \mathbf{M}(A^{\mathrm{en}}, \mathrm{gr}) \to \mathbf{M}(A^{\mathrm{en}}, \mathrm{gr}).$$

These functors commute with each other. Indeed, for any $M \in \mathbf{M}(A^{\mathrm{en}}, \mathrm{gr})$ there is equality $\Gamma_{\mathfrak{m}}(\Gamma_{\mathfrak{m}^{\mathrm{op}}}(M)) = \Gamma_{\mathfrak{m}^{\mathrm{op}}}(\Gamma_{\mathfrak{m}}(M))$ of graded A^{en}-submodules of M.

Definition 16.5.2 Under Setup 16.5.1, we say that A has *finite local cohomological dimension* if the torsion functors $\Gamma_{\mathfrak{m}}$ and $\Gamma_{\mathfrak{m}^{\mathrm{op}}}$, on $\mathbf{M}(A, \mathrm{gr})$ and $\mathbf{M}(A^{\mathrm{op}}, \mathrm{gr})$, respectively, have finite right cohomological dimensions, in the sense of Definition 16.1.2. Namely if there is a natural number d such that $\mathrm{H}^p_{\mathfrak{m}} = 0$ and $\mathrm{H}^p_{\mathfrak{m}^{\mathrm{op}}} = 0$ for all $p > d$. See also formula (16.3.20).

Note that the vanishing of the derived functors $\mathrm{H}^p_{\mathfrak{m}} = \mathrm{R}^p\Gamma_{\mathfrak{m}}$, and thus the cohomological dimension of the functor $\Gamma_{\mathfrak{m}}$, is the same on bimodules and on modules, by Proposition 16.3.16(4), and likewise for $\mathrm{H}^p_{\mathfrak{m}^{\mathrm{op}}}$.

As explained in Proposition 16.3.16(1), there are derived torsion functors

$$\mathrm{R}\Gamma_{\mathfrak{m}}, \mathrm{R}\Gamma_{\mathfrak{m}^{\mathrm{op}}} : \mathbf{D}(A^{\mathrm{en}}, \mathrm{gr}) \to \mathbf{D}(A^{\mathrm{en}}, \mathrm{gr}). \tag{16.5.3}$$

There is no reason for these derived functors to commute with each other in general.

Definition 16.5.4 Under Setup 16.5.1, the morphisms from Proposition 16.3.16(2) associated to the derived functors in formula (16.5.3) are denoted by $\sigma^{\mathrm{R}}_{\mathfrak{m}} : \mathrm{R}\Gamma_{\mathfrak{m}} \to \mathrm{Id}$ and $\sigma^{\mathrm{R}}_{\mathfrak{m}^{\mathrm{op}}} : \mathrm{R}\Gamma_{\mathfrak{m}^{\mathrm{op}}} \to \mathrm{Id}$.

Definition 16.5.5 Under Setup 16.5.1:

(1) We denote by $\mathbf{M}_{(\mathrm{tor},..)}(A^{\mathrm{en}})$ the full subcategory of $\mathbf{M}(A^{\mathrm{en}})$ on the bimodules M that are $\mathfrak{m}$-torsion as A-modules; i.e. such that $\mathrm{Rest}_A(M) \in \mathbf{M}_{\mathrm{tor}}(A)$.

(2) We denote by $\mathbf{M}_{(..,\mathrm{tor})}(A^{\mathrm{en}})$ the full subcategory of $\mathbf{M}(A^{\mathrm{en}})$ on the bimodules M that are $\mathfrak{m}^{\mathrm{op}}$-torsion as A^{op}-modules.

(3) We let

$$\mathbf{M}_{(\mathrm{tor},\mathrm{tor})}(A^{\mathrm{en}}) := \mathbf{M}_{(\mathrm{tor},..)}(A^{\mathrm{en}}) \cap \mathbf{M}_{(..,\mathrm{tor})}(A^{\mathrm{en}}).$$

(4) Let $\star, \diamond$ be torsion indicators, i.e. "tor" or "..". We denote by $\mathbf{D}_{(\star,\diamond)}(A^{\mathrm{en}})$ the full subcategory of $\mathbf{D}(A^{\mathrm{en}})$ on the complexes M such that $\mathrm{H}^p(M) \in \mathbf{M}_{(\star,\diamond)}(A^{\mathrm{en}})$ for every integer p.

Trivially, for $M \in \mathbf{D}(A^{\mathrm{en}}, \mathrm{gr})$ the A-modules $\mathrm{H}^p(\mathrm{R}\Gamma_{\mathfrak{m}}(M))$ are $\mathfrak{m}$-torsion, and the A^{op}-modules $\mathrm{H}^p(\mathrm{R}\Gamma_{\mathfrak{m}^{\mathrm{op}}}(M))$ are $\mathfrak{m}^{\mathrm{op}}$-torsion. Sometimes more happens:

Definition 16.5.6 Under Setup 16.5.1, a complex $M \in \mathbf{D}(A^{\mathrm{en}}, \mathrm{gr})$ is said to have *weakly symmetric derived* $\mathfrak{m}$*-torsion* if these two conditions hold:

- For every $p \in \mathbb{Z}$ the bimodule $\mathrm{H}^p(\mathrm{R}\Gamma_{\mathfrak{m}}(M))$ is $\mathfrak{m}^{\mathrm{op}}$-torsion.
- For every $p \in \mathbb{Z}$ the bimodule $\mathrm{H}^p(\mathrm{R}\Gamma_{\mathfrak{m}^{\mathrm{op}}}(M))$ is $\mathfrak{m}$-torsion.

In terms of Definition 16.5.5, a complex $M \in \mathbf{D}(A^{\mathrm{en}}, \mathrm{gr})$ has weakly symmetric derived $\mathfrak{m}$-torsion if

$$\mathrm{R}\Gamma_{\mathfrak{m}}(M),\ \mathrm{R}\Gamma_{\mathfrak{m}^{\mathrm{op}}}(M) \in \mathbf{D}_{(\mathrm{tor},\mathrm{tor})}(A^{\mathrm{en}}, \mathrm{gr}).$$

Definition 16.5.7 Under Setup 16.5.1, a complex $M \in \mathbf{D}(A^{\mathrm{en}}, \mathrm{gr})$ is said to have *symmetric derived* $\mathfrak{m}$*-torsion* if there is an isomorphism

$$\epsilon_M : \mathrm{R}\Gamma_{\mathfrak{m}}(M) \xrightarrow{\simeq} \mathrm{R}\Gamma_{\mathfrak{m}^{\mathrm{op}}}(M)$$

in $\mathbf{D}(A^{\mathrm{en}}, \mathrm{gr})$, such that the diagram

$$\begin{array}{ccc} \mathrm{R}\Gamma_{\mathfrak{m}}(M) & \xrightarrow[\cong]{\epsilon_M} & \mathrm{R}\Gamma_{\mathfrak{m}^{\mathrm{op}}}(M) \\ & \searrow^{\sigma^{\mathrm{R}}_{\mathfrak{m},M}} & \downarrow{\scriptstyle \sigma^{\mathrm{R}}_{\mathfrak{m}^{\mathrm{op}},M}} \\ & & M \end{array}$$

in $\mathbf{D}(A^{\mathrm{en}}, \mathrm{gr})$ is commutative. Such an isomorphism ϵ_M is called a *symmetry isomorphism*.

Of course, if M has symmetric derived $\mathfrak{m}$-torsion, then it has weakly symmetric derived $\mathfrak{m}$-torsion.

Theorem 16.5.8 (Symmetric Derived Torsion, [157]) *Let A be a noetherian connected graded $\mathbb{K}$-ring of finite local cohomological dimension. If $M \in \mathbf{D}(A^{\mathrm{en}}, \mathrm{gr})$ has weakly symmetric derived $\mathfrak{m}$-torsion, then M has symmetric derived $\mathfrak{m}$-torsion.*

Moreover, the symmetry isomorphism ϵ_M is unique, and it is functorial in such complexes M.

Proof In Theorem 16.4.3, with $B := A$, we have an isomorphism

$$\mathrm{ev}^{\mathrm{R,L}}_{\mathfrak{m},(-)} : P_A \otimes^{\mathrm{L}}_A (-) \xrightarrow{\simeq} \mathrm{R}\Gamma_{\mathfrak{m}}(-)$$

of triangulated functors from $\mathbf{D}(A^{\mathrm{en}}, \mathrm{gr})$ to itself. The same theorem, but with the roles of A and A^{op} exchanged, gives an isomorphism

$$\mathrm{ev}^{\mathrm{R,L}}_{\mathfrak{m}^{\mathrm{op}},(-)} : (-) \otimes^{\mathrm{L}}_A P_{A^{\mathrm{op}}} \xrightarrow{\simeq} \mathrm{R}\Gamma_{\mathfrak{m}^{\mathrm{op}}}(-)$$

of triangulated functors from $\mathbf{D}(A^{\mathrm{en}}, \mathrm{gr})$ to itself. Here $P_{A^{\mathrm{op}}} := \mathrm{R}\Gamma_{\mathfrak{m}^{\mathrm{op}}}(A) \in \mathbf{D}(A^{\mathrm{en}}, \mathrm{gr})$.

Consider diagram (16.5.9) in $\mathbf{D}(A^{\mathrm{en}}, \mathrm{gr})$. In this diagram, α is the associativity isomorphism of the derived tensor product, and ru, lu are the monoidal unitor isomorphisms – all belonging to the monoidal structure of $\mathbf{D}(A^{\mathrm{en}}, \mathrm{gr})$. A quick check, using K-flat resolutions and elements in them, shows that this is a commutative diagram.

$$\begin{array}{ccc}
(P_A \otimes^{\mathrm{L}}_A M) \otimes^{\mathrm{L}}_A P_{A^{\mathrm{op}}} & \xrightarrow[\simeq]{\alpha} & P_A \otimes^{\mathrm{L}}_A (M \otimes^{\mathrm{L}}_A P_{A^{\mathrm{op}}}) \\
\Big\downarrow{\scriptstyle \mathrm{id} \otimes^{\mathrm{L}}_A \sigma^{\mathrm{R}}_{\mathfrak{m}^{\mathrm{op}},A}} & & \Big\downarrow{\scriptstyle \sigma^{\mathrm{R}}_{\mathfrak{m},A} \otimes^{\mathrm{L}}_A \mathrm{id}} \\
(P_A \otimes^{\mathrm{L}}_A M) \otimes^{\mathrm{L}}_A A & & A \otimes^{\mathrm{L}}_A (M \otimes^{\mathrm{L}}_A P_{A^{\mathrm{op}}}) \\
\Big\downarrow{\scriptstyle \mathrm{ru}} & & \Big\downarrow{\scriptstyle \mathrm{lu}} \\
P_A \otimes^{\mathrm{L}}_A M & & M \otimes^{\mathrm{L}}_A P_{A^{\mathrm{op}}} \\
& \searrow{\scriptstyle \sigma^{\mathrm{R}}_{\mathfrak{m},A} \otimes^{\mathrm{L}}_A \mathrm{id}} & \Big\downarrow{\scriptstyle \mathrm{id} \otimes^{\mathrm{L}}_A \sigma^{\mathrm{R}}_{\mathfrak{m}^{\mathrm{op}},A}} \\
& & M
\end{array} \tag{16.5.9}$$

By Theorem 16.4.3, diagram (16.5.9) gives rise to diagram (16.5.10), with its solid arrows only. This new diagram in $\mathbf{D}(A^{\mathrm{en}}, \mathrm{gr})$ (solid arrows only) is isomorphic to the previous one (without its second row), via the various isomorphisms $\mathrm{ev}^{\mathrm{R,L}}_{\mathfrak{m},(-)}$ and $\mathrm{ev}^{\mathrm{R,L}}_{\mathfrak{m}^{\mathrm{op}},(-)}$.

Because $\mathrm{R}\Gamma_{\mathfrak{m}}(M) \in \mathbf{D}_{(\mathrm{tor},\mathrm{tor})}(A^{\mathrm{en}}, \mathrm{gr})$, Proposition 16.3.29, applied with A^{op} instead of with A, says that $\sigma^{\mathrm{R}}_{\mathfrak{m}^{\mathrm{op}},\mathrm{R}\Gamma_{\mathfrak{m}}(M)}$ is an isomorphism. Likewise, because $\mathrm{R}\Gamma_{\mathfrak{m}^{\mathrm{op}}}(M) \in \mathbf{D}_{(\mathrm{tor},\mathrm{tor})}(A^{\mathrm{en}}, \mathrm{gr})$, Proposition 16.3.29 says that $\sigma^{\mathrm{R}}_{\mathfrak{m},\mathrm{R}\Gamma_{\mathfrak{m}^{\mathrm{op}}}(M)}$ is an isomorphism. We define ϵ_M to be the unique isomorphism (the dashed arrow) that makes diagram (16.5.10) commutative.

Diagram (16.5.10) shows that ϵ_M is unique: it has to be

$$\epsilon_M = \sigma^{\mathrm{R}}_{\mathfrak{m},\mathrm{R}\Gamma_{\mathfrak{m}^{\mathrm{op}}}(M)} \circ \alpha' \circ (\sigma^{\mathrm{R}}_{\mathfrak{m}^{\mathrm{op}},\mathrm{R}\Gamma_{\mathfrak{m}}(M)})^{-1}.$$

The functoriality of ϵ_M is a consequence of the functoriality of diagram (16.5.10). □

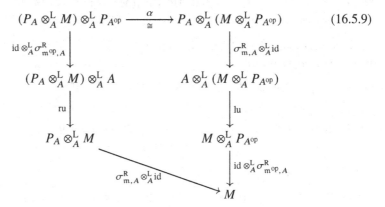

$$\begin{array}{ccc}
\mathrm{R}\Gamma_{\mathfrak{m}^{\mathrm{op}}}(\mathrm{R}\Gamma_{\mathfrak{m}}(M)) & \xrightarrow[\cong]{\alpha'} & \mathrm{R}\Gamma_{\mathfrak{m}}(\mathrm{R}\Gamma_{\mathfrak{m}^{\mathrm{op}}}(M)) \\
\Big\downarrow {\scriptstyle \sigma^{\mathrm{R}}_{\mathfrak{m}^{\mathrm{op}}, \mathrm{R}\Gamma_{\mathfrak{m}}(M)}} & & \Big\downarrow {\scriptstyle \sigma^{\mathrm{R}}_{\mathfrak{m}, \mathrm{R}\Gamma_{\mathfrak{m}^{\mathrm{op}}}(M)}} \\
\mathrm{R}\Gamma_{\mathfrak{m}}(M) & \overset{\epsilon_M}{\dashrightarrow} & \mathrm{R}\Gamma_{\mathfrak{m}^{\mathrm{op}}}(M) \\
 & {\scriptstyle \sigma^{\mathrm{R}}_{\mathfrak{m},M}} \searrow & \Big\downarrow {\scriptstyle \sigma^{\mathrm{R}}_{\mathfrak{m}^{\mathrm{op}},M}} \\
 & & M
\end{array} \tag{16.5.10}$$

Corollary 16.5.11 *Under the assumptions of Theorem* 16.5.8, *if the bimodule A has has weakly symmetric derived $\mathfrak{m}$-torsion, then there is a unique isomorphism $\epsilon_A : P_A \xrightarrow{\simeq} P_{A^{\mathrm{op}}}$ in $\mathbf{D}(A^{\mathrm{en}}, \mathrm{gr})$ such that $\sigma^{\mathrm{R}}_{\mathfrak{m},A} = \sigma^{\mathrm{R}}_{\mathfrak{m}^{\mathrm{op}},A} \circ \epsilon_A$ as morphisms $P_A \to A$.*

Proof Take $M = A$ in the theorem. □

The socle functor from Definition 15.2.31 extends to complexes.

Lemma 16.5.12 *Let I be a minimal complex of graded-injective A-modules, and let $W := \mathrm{Soc}(I)$. Then the differential of the complex W is zero.*

Proof Take some nonzero homogeneous element $w \in W^p$, say $w \in W^p_q$. Let $U := \mathbb{K} \cdot w \subseteq W^p_q$, so U is a nonzero graded A-submodule of I^p. Because $\mathrm{Z}^p(I)$ is an essential graded A-submodule of I^p, we must have $\mathrm{Z}^p(I) \cap U \neq 0$. Thus $w \in \mathrm{Z}^p(I)$, so $\mathrm{d}(w) = 0$. □

Cofinite graded A-modules were defined in Definition 15.2.33. By graded Matlis duality (Theorem 15.2.34) the cofinite graded A-modules are the artinian objects of $\mathbf{M}(A, \mathrm{gr})$. The full subcategory $\mathbf{M}_{\mathrm{cof}}(A, \mathrm{gr})$ of cofinite modules is a thick abelian subcategory of $\mathbf{M}(A, \mathrm{gr})$, closed under subobjects and quotients.

Lemma 16.5.13 *Consider a complex $M \in \mathbf{D}^+(A, \mathrm{gr})$ and an integer p.*

1. *If $\mathrm{Ext}^p_A(\mathbb{K}, M)$ is a finite graded $\mathbb{K}$-module, then $\mathrm{H}^p_{\mathfrak{m}}(M)$ is a cofinite graded A-module.*
2. *If $\mathrm{H}^q_{\mathfrak{m}}(M)$ is a cofinite graded A-module for every $q \leq p$, then $\mathrm{Ext}^q_A(\mathbb{K}, M)$ is a finite graded $\mathbb{K}$-module for every $q \leq p$.*

Proof

(1) Let $M \to I$ be a minimal graded-injective resolution, and let $W := \mathrm{Soc}(I)$. Then $\mathrm{Ext}^p_A(\mathbb{K}, M) \cong \mathrm{H}^p(W)$. But by Lemma 16.5.12 we know that the differential of W is zero, so in fact $\mathrm{Ext}^p_A(\mathbb{K}, M) \cong W^p$. Therefore W is a finite graded $\mathbb{K}$-module.

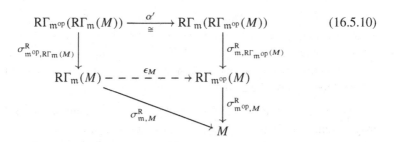

Next, let $J := \Gamma_{\mathfrak{m}}(I)$. This complex has the property that $\mathrm{H}^p_{\mathfrak{m}}(M) \cong \mathrm{H}^p(J)$. According to Lemma 16.3.24, we have an isomorphism $J^p \cong A^* \otimes W^p$. Since A^* is a cofinite graded A-module and J^p is a finite direct sum of degree twists of A^*, it is also a cofinite graded A-module. But $\mathrm{H}^p_{\mathfrak{m}}(M)$ is a subquotient of J^p, so it too is a cofinite graded A-module.

(2) We continue with the minimal graded-injective resolution $M \to I$. We may assume that $\mathrm{H}(M) \neq 0$. Let $p_0 := \inf(\mathrm{H}(M))$; so $p_0 = \inf(I)$. Take the subcomplexes $W \subseteq J \subseteq I$ as above. We know that $J^q \cong A^* \otimes W^q$ for every q. We will prove that W^q is a finite graded $\mathbb{K}$-module for every $q \leq p$, by induction on q, starting with $q = p_0$.

So take an integer q in the range $p_0 \leq q \leq p$, and assume that W^{q-1} is a finite graded $\mathbb{K}$-module. There is an exact sequence

$$J^{q-1} \xrightarrow{\mathrm{d}} \mathrm{Z}^q(J) \to \mathrm{H}^q_{\mathfrak{m}}(M) \to 0$$

in $\mathbf{M}(A, \mathrm{gr})$. Since $J^{q-1} \cong A^* \otimes W^{q-1}$, this is a cofinite graded A-module, as we already noticed. We also know that $\mathrm{H}^q_{\mathfrak{m}}(M)$ is cofinite. It follows that $\mathrm{Z}^q(J)$ is a cofinite graded A-module. But by Lemma 16.5.12 we have $W^q \subseteq \mathrm{Z}^q(J)$, so W^q is a cofinite graded A-module annihilated by $\mathfrak{m}$. Therefore W^q is a finite graded $\mathbb{K}$-module. □

Definition 16.5.14 (Artin–Zhang, [10]) Assume Setup 16.5.1.

1. We say that the ring A satisfies the *left χ condition* if for every $M \in \mathbf{M}_{\mathrm{f}}(A, \mathrm{gr})$ and every integer p, the graded $\mathbb{K}$-module $\mathrm{Ext}^p_A(\mathbb{K}, M)$ is finite.
2. The ring A is said to satisfy the *χ condition* if both graded rings A and A^{op} satisfy the left χ condition.

Another way to state the left χ condition is this: $M \in \mathbf{M}_{\mathrm{f}}(A, \mathrm{gr})$ implies $\mathrm{RHom}_A(\mathbb{K}, M) \in \mathbf{D}_{\mathrm{f}}(\mathbb{K}, \mathrm{gr})$.

Remark 16.5.15 What we call the "left χ condition" was actually called the "left χ° condition" in [10, Definition 3.2]. Their "left χ condition" (in [10, Definition 3.7]) is much more complicated to state. However, for a left noetherian connected graded ring A (as we have here) these two conditions are equivalent, by [10, Proposition 3.11(2)].

Definition 16.5.16 Assume Setup 16.5.1 .

1. We say that the ring A satisfies the *left special χ condition* if for every integer p, the graded $\mathbb{K}$-module $\mathrm{Ext}^p_A(\mathbb{K}, A)$ is finite.
2. The ring A is said to satisfy the *special χ condition* if both graded rings A and A^{op} satisfy the left special χ condition.

Proposition 16.5.17 *The following two conditions are equivalent.*

(i) *A satisfies the left χ condition.*
(ii) *For every $M \in \mathbf{M}_{\mathrm{f}}(A, \mathrm{gr})$ and every p the graded A-module $\mathrm{H}^p_{\mathfrak{m}}(M)$ is cofinite.*

Proof This is immediate from Lemma 16.5.13. □

Definition 16.5.18 We denote by $\mathbf{D}^{\mathrm{b}}_{(\mathrm{f},\mathrm{f})}(A^{\mathrm{en}}, \mathrm{gr})$ the full subcategory of $\mathbf{D}^{\mathrm{b}}(A^{\mathrm{en}}, \mathrm{gr})$ on the complexes of bimodules M whose cohomology bimodules $\mathrm{H}^p(M)$ are finite graded modules over A and over A^{op}.

Proposition 16.5.19

(1) *If A satisfies the χ condition, then it satisfies the special χ condition.*
(2) *If A satisfies the χ condition, then every $M \in \mathbf{D}^{\mathrm{b}}_{(\mathrm{f},\mathrm{f})}(A^{\mathrm{en}}, \mathrm{gr})$ has weakly symmetric derived $\mathfrak{m}$-torsion.*
(3) *If A satisfies the special χ condition, then the bimodule A has weakly symmetric derived $\mathfrak{m}$-torsion.*

Proof

(1) This is trivial.

(2) Take some $M \in \mathbf{D}^{\mathrm{b}}_{(\mathrm{f},\mathrm{f})}(A^{\mathrm{en}}, \mathrm{gr})$. Using smart truncation and induction on the amplitude of cohomology, we can assume that $M \in \mathbf{M}_{(\mathrm{f},\mathrm{f})}(A^{\mathrm{en}}, \mathrm{gr})$.

The χ condition tells us that $\mathrm{Ext}^p_A(\mathbb{K}, M)$ and $\mathrm{Ext}^p_{A^{\mathrm{op}}}(\mathbb{K}, M)$ are finite as graded $\mathbb{K}$-modules, for all p. For every $q \geq 1$ the graded bimodule $A/\mathfrak{m}^q$ is gotten from $\mathbb{K}$ by finitely many degree twists and extensions in $\mathbf{M}(A^{\mathrm{en}}, \mathrm{gr})$. Therefore $\mathrm{Ext}^p_A(A/\mathfrak{m}^q, M)$ and $\mathrm{Ext}^p_{A^{\mathrm{op}}}(A/\mathfrak{m}^q, M)$ are finite as graded $\mathbb{K}$-modules, for all p and all $q \geq 1$. Thus as left and as right graded A-modules, $\mathrm{Ext}^p_A(A/\mathfrak{m}^q, M)$ and $\mathrm{Ext}^p_{A^{\mathrm{op}}}(A/\mathfrak{m}^q, M)$ are of finite length, and thus torsion on both sides. Passing to the direct limit using Proposition 16.3.16(5), we see that $\mathrm{H}^p_{\mathfrak{m}}(M)$ and $\mathrm{H}^p_{\mathfrak{m}^{\mathrm{op}}}(M)$ are torsion A-modules on both sides.

(3) This is just a special case of the proof of item (2). □

Lemma 16.5.20 *Assume the functor $\Gamma_{\mathfrak{m}}$ has finite cohomological dimension, and A has the left special χ condition. Then for every $M \in \mathbf{D}_{\mathrm{f}}(A, \mathrm{gr})$ and every integer i, the graded A-module $\mathrm{H}^i_{\mathfrak{m}}(M)$ is cofinite.*

Proof Fix an integer i. We will prove that the graded A-module $\mathrm{H}^i_{\mathfrak{m}}(M)$ is cofinite. Say $\Gamma_{\mathfrak{m}}$ has cohomological dimension $\leq d$ for some natural number d.

Step 1. Consider $M' := \mathrm{smt}^{\leq i+1}(M)$, the smart truncation of the complex M below $i + 1$. There is a distinguished triangle $M' \to M \to M'' \xrightarrow{\triangle}$ in $\mathbf{D}(A, \mathrm{gr})$, and hence there is a distinguished triangle

$$\mathrm{R}\Gamma_{\mathfrak{m}}(M') \to \mathrm{R}\Gamma_{\mathfrak{m}}(M) \to \mathrm{R}\Gamma_{\mathfrak{m}}(M'') \xrightarrow{\triangle} .$$

Because $\mathrm{H}^j_{\mathfrak{m}}(M'') = 0$ for $j \leq i+1$, we see that $\mathrm{H}^i_{\mathfrak{m}}(M) \cong \mathrm{H}^i_{\mathfrak{m}}(M')$. Therefore, by replacing M with M', we can assume that $M \in \mathbf{D}^-_{\mathrm{f}}(A, \mathrm{gr})$.

Step 2. Now $M \in \mathbf{D}^-_{\mathrm{f}}(A, \mathrm{gr})$. Choose a resolution $P \to M$, where P is a bounded above complex of finite graded-free A-modules. Let $P' := \mathrm{stt}^{\geq i-d-2}(P)$, the stupid truncation of P above $i-d-2$. Then $\mathrm{H}^i_{\mathfrak{m}}(M) \cong \mathrm{H}^i(P')$. But P' is gotten from A by finitely many cones, finite direct sums, translations and degree twists. Since A has the left special χ condition, and using Lemma 16.5.13(1), the graded A-module $\mathrm{H}^j_{\mathfrak{m}}(A)$ is cofinite for every j. Therefore $\mathrm{H}^j_{\mathfrak{m}}(P')$ is cofinite for every j; including $j = i$. □

Here is a converse to the trivial implication in Proposition 16.5.19(1), under an extra finiteness condition.

Proposition 16.5.21 *If A satisfies the special χ condition and has finite local cohomological dimension, then A satisfies the χ condition.*

Proof This is an immediate consequence of Lemma 16.5.20 and Proposition 16.5.17, applied to both A and A^{op}. □

AS regular graded rings were introduced in Definition 15.4.8.

Corollary 16.5.22 *Let A be an AS regular graded ring. Then A satisfies the χ condition and has finite local cohomological dimension.*

Proof The special χ condition is an immediate consequence of the AS condition (Definition 15.4.6). The graded global cohomological dimension of A bounds its local cohomological dimension, so the latter is finite. By Proposition 16.5.21, A satisfies the χ condition. □

Actually this is true also for AS Gorenstein rings, but the proof is harder. See Corollary 17.3.14 and Corollary 17.2.13.

The χ condition, and the finiteness of local cohomological dimension, occur in many important examples of noncommutative graded rings. The main results that ensure it involve a lifting argument – Theorems 15.4.13 and 17.4.35. Here are several examples.

Example 16.5.23 Let $A := \mathbb{K}[x_1, \ldots, x_n]$, the commutative polynomial ring in $n \geq 1$ variables from Example 15.2.20. This is an AS regular graded ring of dimension n. By Corollary 16.5.22 the graded ring A has the χ condition and finite local cohomological dimension.

Example 16.5.24 If B is a commutative noetherian connected graded $\mathbb{K}$-ring, then there is a finite homomorphism $A \to B$ from a polynomial ring A as in the previous example. By Corollaries 17.4.16 and 17.4.17 the ring B satisfies the χ condition and has finite local cohomological dimension.

Example 16.5.25 Let A be the homogeneous Weyl algebra from Example 15.2.21. The element $t \in A$ is central regular of degree 1, and $A/(t) \cong \mathbb{K}[x, y]$, the commutative polynomial ring in two variables of degree 1. The ring $\mathbb{K}[x, y]$ is AS regular of dimension 2, so by Theorem 15.4.13 the ring A is AS regular of dimension 3. By Corollary 16.5.22 the graded ring A has the χ condition and finite local cohomological dimension.

As noted before, in characteristic 0 the ring A is very noncommutative.

Example 16.5.26 Let A be the homogeneous universal enveloping algebra from Example 15.2.22. The element $t \in A$ is central regular of degree 1, and $A/(t) \cong \mathbb{K}[x_1, \ldots, x_n]$, the commutative polynomial ring in n variables. The polynomial ring is AS regular of dimension n, so by Theorem 15.4.13 the ring A is AS regular of dimension $n + 1$. By Corollary 16.5.22 the graded ring A has the χ condition and finite local cohomological dimension.

As noted in Example 15.2.22, the ring A could be very noncommutative, if $\mathfrak{g}$ is very nonabelian (e.g. semisimple).

We end this section with a result that ties the χ condition to symmetry of derived torsion. Recall that $\mathbf{D}_{(\mathrm{f,f})}(A^{\mathrm{en}}, \mathrm{gr})$ is the full subcategory of $\mathbf{D}(A^{\mathrm{en}}, \mathrm{gr})$ on the complexes M such that for every p the graded bimodule $\mathrm{H}^p(M)$ is finite over A and A^{op}.

Theorem 16.5.27 *Under Setup* 16.5.1, *assume A satisfies the χ condition and has finite local cohomological dimension. Then there is a unique isomorphism*

$$\epsilon : \mathrm{R}\Gamma_{\mathfrak{m}} \xrightarrow{\simeq} \mathrm{R}\Gamma_{\mathfrak{m}^{\mathrm{op}}}$$

of triangulated functors $\mathbf{D}^{\mathrm{b}}_{(\mathrm{f,f})}(A^{\mathrm{en}}, \mathrm{gr}) \to \mathbf{D}(A^{\mathrm{en}}, \mathrm{gr})$, *such that the diagram*

$$\begin{array}{ccc} \mathrm{R}\Gamma_{\mathfrak{m}} & \xrightarrow[\cong]{\ \epsilon\ } & \mathrm{R}\Gamma_{\mathfrak{m}^{\mathrm{op}}} \\ & \searrow{\scriptstyle \sigma^{\mathrm{R}}_{\mathfrak{m}}} & \downarrow{\scriptstyle \sigma^{\mathrm{R}}_{\mathfrak{m}^{\mathrm{op}}}} \\ & & \mathrm{Id} \end{array}$$

is commutative.

Proof According to Proposition 16.5.19(2), every complex $M \in \mathbf{D}^{\mathrm{b}}_{(\mathrm{f,f})}(A^{\mathrm{en}}, \mathrm{gr})$ has weakly symmetric derived $\mathfrak{m}$-torsion. By Theorem 16.5.8 such a complex M has symmetric derive torsion, and the symmetry isomorphism ϵ_M is unique and functorial. □

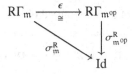

16.6 NC MGM Equivalence

The *commutative MGM Equivalence* was explained in Example 12.4.6. A noncommutative variant of this theory was recently developed by R. Vyas and A. Yekutieli [157], and in this section we give an adaptation of it to the case of a NC connected graded ring A. In Chapter 17 below we shall use the NC MGM Equivalence to prove the existence of balanced trace morphisms (Theorem 17.4.22), and in Chapter 18 the NC MGM Equivalence will help us prove that a balanced dualizing complex is graded rigid (Theorem 18.4.5).

We continue with Conventions 15.1.17 and 15.2.1, so $\mathbb{K}$ is a base field, and A and B are graded $\mathbb{K}$-rings.

Recall (from Remark 14.3.23) that the derived category $\mathbf{D}(A^{\mathrm{en}}, \mathrm{gr})$ has a biclosed monoidal structure on it, with monoidal operation $(- \otimes^{\mathrm{L}}_{A} -)$ and monoidal unit A. For complexes $M \in \mathbf{D}(A \otimes B^{\mathrm{op}}, \mathrm{gr})$ and $N \in \mathbf{D}(B \otimes A^{\mathrm{op}}, \mathrm{gr})$ the left and right unitor isomorphisms are $\mathrm{lu} : A \otimes^{\mathrm{L}}_{A} M \xrightarrow{\simeq} M$ and $\mathrm{ru} : M \otimes^{\mathrm{L}}_{A} A \xrightarrow{\simeq} M$, respectively. The left co-unitor isomorphism is $\mathrm{lcu} : \mathrm{RHom}_{A}(A, M) \xrightarrow{\simeq} M$.

Definition 16.6.1 ([157])

(1) A *copointed object* in the monoidal category $\mathbf{D}(A^{\mathrm{en}}, \mathrm{gr})$ is a pair (P, σ), consisting of an object $P \in \mathbf{D}(A^{\mathrm{en}}, \mathrm{gr})$ and a morphism $\sigma : P \to A$ in $\mathbf{D}(A^{\mathrm{en}}, \mathrm{gr})$.

(2) The copointed object (P, σ) is called *idempotent* if the morphisms

$$\mathrm{lu} \circ (\sigma \otimes^{\mathrm{L}}_{A} \mathrm{id}),\ \mathrm{ru} \circ (\mathrm{id} \otimes^{\mathrm{L}}_{A} \sigma) : \ P \otimes^{\mathrm{L}}_{A} P \to P$$

in $\mathbf{D}(A^{\mathrm{en}}, \mathrm{gr})$ are both isomorphisms.

Remark 16.6.2 Here is an explanation of the name "copointed." In a monoidal category, with unit object A, a *point* of an object P is a morphism $A \to P$. It thus makes sense to refer to a morphism in the dual direction, say $\sigma : P \to A$, as a *copoint* of P.

Definition 16.6.3 Let (P, σ) be a copointed object in $\mathbf{D}(A^{\mathrm{en}}, \mathrm{gr})$.

(1) Define the triangulated functors

$$F, G : \mathbf{D}(A \otimes B^{\mathrm{op}}, \mathrm{gr}) \to \mathbf{D}(A \otimes B^{\mathrm{op}}, \mathrm{gr})$$

to be $F := P \otimes^{\mathrm{L}}_{A} (-)$ and $G := \mathrm{RHom}_{A}(P, -)$.

(2) Let $\sigma : F \to \mathrm{Id}$ and $\tau : \mathrm{Id} \to G$ be the morphisms of triangulated functors from $\mathbf{D}(A \otimes B^{\mathrm{op}}, \mathrm{gr})$ to itself that are induced by the morphism $\sigma : P \to A$. Namely

$$\sigma_{M} := \mathrm{lu} \circ (\sigma \otimes^{\mathrm{L}}_{A} \mathrm{id}_{M}) : F(M) = P \otimes^{\mathrm{L}}_{A} M \to M$$

and

$$\tau_M := \mathrm{RHom}_A(\sigma, \mathrm{id}_M) \circ \mathrm{lcu}^{-1} : M \to \mathrm{RHom}_A(P, M) = G(M).$$

We refer to (F, σ) and (G, τ) as the *copointed and pointed triangulated functors induced by the copointed object* (P, σ).

Item (2) of the definition is shown in the commutative diagrams below in the category $\mathbf{D}(A \otimes B^{\mathrm{op}}, \mathrm{gr})$.

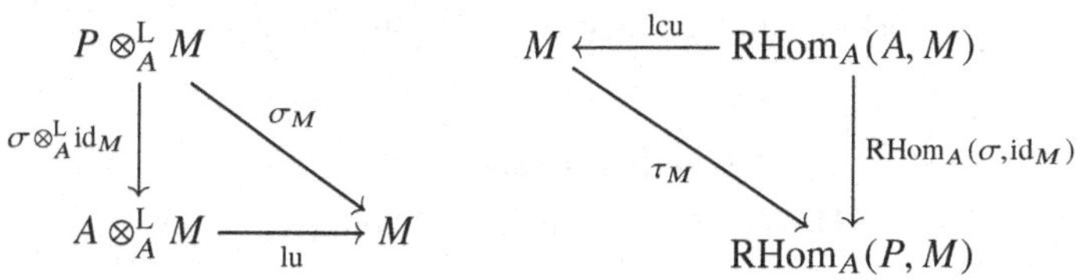

Definition 16.6.4 Let (F, σ) and (G, τ) be the copointed and pointed triangulated functors on $\mathbf{D}(A \otimes B^{\mathrm{op}}, \mathrm{gr})$ from Definition 16.6.3.

(1) Let $\mathbf{D}(A \otimes B^{\mathrm{op}}, \mathrm{gr})_F$ be the full triangulated subcategory of $\mathbf{D}(A \otimes B^{\mathrm{op}}, \mathrm{gr})$ on the set of objects
$$\{M \mid \sigma_M : F(M) \to M \text{ is an isomorphism}\}.$$

(2) Let $\mathbf{D}(A \otimes B^{\mathrm{op}}, \mathrm{gr})_G$ be the full triangulated subcategory of $\mathbf{D}(A \otimes B^{\mathrm{op}}, \mathrm{gr})$ on the set of objects
$$\{M \mid \tau_M : M \to G(M) \text{ is an isomorphism}\}.$$

Idempotent (co)pointed triangulated functors were introduced in Definitions 16.2.3 and 16.2.10.

Lemma 16.6.5 *If the copointed object (P, σ) is idempotent, then the copointed triangulated functor (F, σ) and the pointed triangulated functor (G, τ) on $\mathbf{D}(A \otimes B^{\mathrm{op}}, \mathrm{gr})$ are idempotent.*

Proof For $M \in \mathbf{D}(A \otimes B^{\mathrm{op}}, \mathrm{gr})$ there are equalities (up to the associativity isomorphism of $(- \otimes^{\mathrm{L}}_A -)$ that should be inserted in the locations marked by "†"):

$$F(\sigma_M) = \mathrm{id}_P \otimes^{\mathrm{L}}_A (\mathrm{lu} \circ (\sigma \otimes^{\mathrm{L}}_A \mathrm{id}_M)) =^{\dagger} (\mathrm{ru} \circ (\mathrm{id}_P \otimes^{\mathrm{L}}_A \sigma)) \otimes^{\mathrm{L}}_A \mathrm{id}_M \quad (16.6.6)$$

and

$$\sigma_{F(M)} = \mathrm{lu} \circ (\sigma \otimes^{\mathrm{L}}_A (\mathrm{id}_P \otimes^{\mathrm{L}}_A \mathrm{id}_M)) =^{\dagger} (\mathrm{lu} \circ (\sigma \otimes^{\mathrm{L}}_A \mathrm{id}_P)) \otimes^{\mathrm{L}}_A \mathrm{id}_M \quad (16.6.7)$$

of morphisms

$$F(F(M)) = P \otimes^{\mathrm{L}}_A P \otimes^{\mathrm{L}}_A M \to F(M) = P \otimes^{\mathrm{L}}_A M$$

in $\mathbf{D}(A \otimes B^{\mathrm{op}}, \mathrm{gr})$. Because both $\mathrm{lu} \circ (\sigma \otimes^{\mathrm{L}}_A \mathrm{id}_P)$ and $\mathrm{ru} \circ (\mathrm{id}_P \otimes^{\mathrm{L}}_A \sigma)$ are isomorphisms in $\mathrm{D}(A^{\mathrm{en}})$, it follows that the morphisms $F(\sigma_M)$ and $\sigma_{F(M)}$ are isomorphisms in $\mathrm{D}(A \otimes B^{\mathrm{op}})$.

There are also equalities (up to the associativity and adjunction isomorphisms of $(- \otimes^{\mathrm{L}}_A -)$ and $\mathrm{RHom}_A(-,-)$ that should be inserted in the locations marked by "‡"):

$$\begin{aligned} G(\tau_M) &= \mathrm{RHom}_A(\mathrm{id}_P, \mathrm{RHom}_A(\sigma, \mathrm{id}_M)) \circ \mathrm{RHom}_A(\mathrm{id}_P, \mathrm{lcu}) \\ &=^{\ddagger} \mathrm{RHom}_A(\sigma \otimes^{\mathrm{L}}_A \mathrm{id}_P, \mathrm{id}_M) \circ \mathrm{RHom}_A(\mathrm{lu}, \mathrm{id}_M) \end{aligned} \tag{16.6.8}$$

and

$$\begin{aligned} \tau_{G(M)} &= \mathrm{RHom}_A(\sigma, \mathrm{id}_{\mathrm{RHom}_A(P,M)}) \circ \mathrm{lcu} \\ &=^{\ddagger} \mathrm{RHom}_A(\mathrm{id}_P \otimes^{\mathrm{L}}_A \sigma, \mathrm{id}_M) \circ \mathrm{RHom}_A(\mathrm{ru}, \mathrm{id}_M) \end{aligned} \tag{16.6.9}$$

of morphisms

$$\begin{aligned} G(M) = \mathrm{RHom}_A(P, M) &\to G(G(M)) \\ &= \mathrm{RHom}_A(P, \mathrm{RHom}_A(P, M)) \cong \mathrm{RHom}_A(P \otimes^{\mathrm{L}}_A P, M) \end{aligned}$$

in $\mathbf{D}(A \otimes B^{\mathrm{op}}, \mathrm{gr})$. Because both $\mathrm{lu} \circ (\sigma \otimes^{\mathrm{L}}_A \mathrm{id}_P)$ and $\mathrm{ru} \circ (\mathrm{id}_P \otimes^{\mathrm{L}}_A \sigma)$ are isomorphisms in $\mathbf{D}(A^{\mathrm{en}}, \mathrm{gr})$, it follows that the morphisms $G(\tau_M)$ and $\tau_{G(M)}$ are isomorphisms in $\mathbf{D}(A \otimes B^{\mathrm{op}}, \mathrm{gr})$. □

Lemma 16.6.10 *Consider the functors F and G on $\mathbf{D}(A \otimes B^{\mathrm{op}}, \mathrm{gr})$ from Definition* 16.6.3. *For every $M, N \in \mathbf{D}(A \otimes B^{\mathrm{op}}, \mathrm{gr})$ there is a bijection*

$$\mathrm{Hom}_{\mathbf{D}(A \otimes B^{\mathrm{op}}, \mathrm{gr})}(F(M), N) \cong \mathrm{Hom}_{\mathbf{D}(A \otimes B^{\mathrm{op}}, \mathrm{gr})}(M, G(N)),$$

and it is functorial in M and N.

Proof Choose a K-injective resolution $N \to J$ in $\mathbf{C}(A \otimes B^{\mathrm{op}}, \mathrm{gr})$, and a K-flat resolution $\tilde{P} \to P$ in $\mathbf{C}(A^{\mathrm{en}}, \mathrm{gr})$. The usual Hom-tensor adjunction gives rise to an isomorphism

$$\mathrm{Hom}_{A \otimes B^{\mathrm{op}}}(\tilde{P} \otimes_A M, J) \cong \mathrm{Hom}_{A \otimes B^{\mathrm{op}}}(M, \mathrm{Hom}_A(\tilde{P}, J)) \tag{16.6.11}$$

in $\mathbf{C}_{\mathrm{str}}(\mathbb{K}, \mathrm{gr})$. From this we deduce that $\mathrm{Hom}_A(\tilde{P}, J)$ is K-injective in $\mathbf{C}(A \otimes B^{\mathrm{op}}, \mathrm{gr})$. We see that the isomorphism (16.6.11) represents an isomorphism

$$\mathrm{RHom}_{A \otimes B^{\mathrm{op}}}(P \otimes^{\mathrm{L}}_A M, N) \cong \mathrm{RHom}_{A \otimes B^{\mathrm{op}}}(M, \mathrm{RHom}_A(P, N)) \tag{16.6.12}$$

in $\mathbf{D}(\mathbb{K}, \mathrm{gr})$. Taking H^0 in (16.6.12) gives us the isomorphism

$$\mathrm{Hom}_{\mathbf{D}(A \otimes B^{\mathrm{op}}, \mathrm{gr})}(P \otimes^{\mathrm{L}}_A M, N) \cong \mathrm{Hom}_{\mathbf{D}(A \otimes B^{\mathrm{op}}, \mathrm{gr})}(M, \mathrm{RHom}_A(P, N))$$

in $\mathbf{M}(\mathbb{K}, \mathrm{gr})$. This is what we want. □

Lemma 16.6.13 *Consider the functors F and G on $\mathsf{D}(A \otimes B^{\mathrm{op}}, \mathrm{gr})$ from Definition* 16.6.3. *Assume that the copointed object (P, ρ) is idempotent. Then the kernel of F equals the kernel of G. Namely for every $M \in \mathsf{D}(A \otimes B^{\mathrm{op}}, \mathrm{gr})$ we have $F(M) = 0$ if and only if $G(M) = 0$.*

Proof We shall use the adjunction formula from Lemma 16.6.10, with $N = M$.

First assume $F(M) = 0$. Then $\mathrm{Hom}_{\mathsf{D}(A \otimes B^{\mathrm{op}}, \mathrm{gr})}(F(M), M)$ is zero, and by Lemma 16.6.10 we see that $\mathrm{Hom}_{\mathsf{D}(A \otimes B^{\mathrm{op}}, \mathrm{gr})}(M, G(M))$ is zero too. This implies that the morphism $\tau_M : M \to G(M)$ is zero. Applying G to it we deduce that the morphism $G(\tau_M) : G(M) \to G(G(M))$ is zero. But by Lemma 16.6.5 the pointed functor (G, τ) is idempotent, and this means that $G(\tau_M)$ is an isomorphism. Therefore $G(M) = 0$.

Now assume that $G(M) = 0$. Again using Lemma 16.6.10, but now in the reverse direction, we see that the morphism $\sigma_M : F(M) \to M$ is zero. Therefore the morphism $F(\sigma_M) : F(F(M)) \to F(M)$ is zero. But by Lemma 16.6.5 the copointed functor (F, σ) is idempotent, and this means that $F(\sigma_M)$ is an isomorphism. Therefore $F(M) = 0$. □

Recall that Convention 15.1.17 is in effect.

Theorem 16.6.14 (Abstract Equivalence, [157]) *Let A and B be graded $\mathbb{K}$-rings, and let (P, σ) be an idempotent copointed object in $\mathsf{D}(A^{\mathrm{en}}, \mathrm{gr})$. Consider the triangulated functors*

$$F, G : \mathsf{D}(A \otimes B^{\mathrm{op}}, \mathrm{gr}) \to \mathsf{D}(A \otimes B^{\mathrm{op}}, \mathrm{gr})$$

and the categories $\mathsf{D}(A \otimes B^{\mathrm{op}}, \mathrm{gr})_F$ and $\mathsf{D}(A \otimes B^{\mathrm{op}}, \mathrm{gr})_G$ from Definitions 16.6.3 *and* 16.6.4. *The following hold:*

(1) *The functor G is a right adjoint to F.*
(2) *The copointed triangulated functor (F, σ) and the pointed triangulated functor (G, τ) are idempotent.*
(3) *The categories $\mathsf{D}(A \otimes B^{\mathrm{op}}, \mathrm{gr})_F$ and $\mathsf{D}(A \otimes B^{\mathrm{op}}, \mathrm{gr})_G$ are the essential images of the functors F and G, respectively.*
(4) *The functor*

$$F : \mathsf{D}(A \otimes B^{\mathrm{op}}, \mathrm{gr})_G \to \mathsf{D}(A \otimes B^{\mathrm{op}}, \mathrm{gr})_F$$

is an equivalence of triangulated categories, with quasi-inverse G.

Proof (1) This is Lemma 16.6.10.

(2) This is Lemma 16.6.5.

(3) Take an object $M \in \mathsf{D}(A \otimes B^{\mathrm{op}}, \mathrm{gr})_G$. Then $M \cong G(M)$, so that M is in the essential image of G. Conversely, suppose there is an isomorphism

$\phi : M \xrightarrow{\simeq} G(N)$ for some $N \in \mathsf{D}(A \otimes B^{\mathrm{op}}, \mathrm{gr})$. We have to prove that τ_M is an isomorphism. There is a commutative diagram

$$\begin{array}{ccc} M & \xrightarrow{\phi} & G(N) \\ {\scriptstyle \tau_M}\downarrow & & \downarrow{\scriptstyle \tau_{G(N)}} \\ G(M) & \xrightarrow{G(\phi)} & G(G(N)) \end{array}$$

in $\mathsf{D}(A \otimes B^{\mathrm{op}}, \mathrm{gr})$ with horizontal isomorphisms. The idempotence of (G, τ) says that the morphism $\tau_{G(N)}$ is an isomorphism. Therefore τ_M is an isomorphism.

A similar argument (with reversed arrows, and using the idempotence of (F, σ)) tells us that the essential image of F is $\mathsf{D}(A \otimes B^{\mathrm{op}}, \mathrm{gr})_F$.

(4) The morphism $\sigma : P \to A$ sits inside a distinguished triangle

$$P \xrightarrow{\sigma} A \to N \xrightarrow{\triangle} \tag{16.6.15}$$

in $\mathsf{D}(A^{\mathrm{en}}, \mathrm{gr})$. Let us apply the functor $P \otimes^{\mathrm{L}}_A (-)$ to (16.6.15). We get a distinguished triangle

$$P \otimes^{\mathrm{L}}_A P \xrightarrow{\mathrm{ru} \circ (\mathrm{id} \otimes^{\mathrm{L}}_A \sigma)} P \to P \otimes^{\mathrm{L}}_A N \xrightarrow{\triangle}$$

in $\mathsf{D}(A^{\mathrm{en}}, \mathrm{gr})$. By the idempotence of (P, σ), the first morphism above is an isomorphism; and hence $P \otimes^{\mathrm{L}}_A N = 0$. Therefore for every $M \in \mathsf{D}(A \otimes B^{\mathrm{op}}, \mathrm{gr})$ the complex $F(N \otimes^{\mathrm{L}}_A M) = P \otimes^{\mathrm{L}}_A N \otimes^{\mathrm{L}}_A M$ is zero. Lemma 16.6.13 tells us that

$$G(N \otimes^{\mathrm{L}}_A M) = \mathrm{RHom}_A(P, N \otimes^{\mathrm{L}}_A M) \tag{16.6.16}$$

is zero.

Now we go back to the distinguished triangle (16.6.15) and we apply to it the functor $(-) \otimes^{\mathrm{L}}_A M$, and then the functor $\mathrm{RHom}_A(P, -)$. The result is the distinguished triangle

$$\mathrm{RHom}_A(P, P \otimes^{\mathrm{L}}_A M) \xrightarrow{\alpha_M} \mathrm{RHom}_A(P, M) \to \mathrm{RHom}_A(P, N \otimes^{\mathrm{L}}_A M) \xrightarrow{\triangle}$$

in $\mathsf{D}(A \otimes B^{\mathrm{op}}, \mathrm{gr})$. By (16.6.16) the third object in this triangle is zero, and it follows that $\alpha_M : G(F(M)) \to G(M)$ is an isomorphism. If moreover $M \in \mathsf{D}(A \otimes B^{\mathrm{op}}, \mathrm{gr})_G$, then τ_M is an isomorphism too, and thus we have an isomorphism

$$\tau_M^{-1} \circ \alpha_M : G(F(M)) \to M \tag{16.6.17}$$

in $\mathsf{D}(A \otimes B^{\mathrm{op}}, \mathrm{gr})_G$ that's functorial in M.

Similarly, if we apply the functor $(-) \otimes^{\mathrm{L}}_A P$ to (16.6.15), we get a distinguished triangle

$$P \otimes^{\mathrm{L}}_A P \xrightarrow{\mathrm{lu} \circ (\sigma \otimes^{\mathrm{L}}_A \mathrm{id})} P \to N \otimes^{\mathrm{L}}_A P \xrightarrow{\triangle}$$

in $\mathsf{D}(A^{\mathrm{en}}, \mathrm{gr})$. By the idempotence condition, the first morphism above is an isomorphism; and hence $N \otimes^{\mathrm{L}}_A P = 0$. Therefore for every $M \in \mathsf{D}(A \otimes B^{\mathrm{op}}, \mathrm{gr})$ the complex

$$G(\mathrm{RHom}_A(N, M)) = \mathrm{RHom}_A(P, \mathrm{RHom}_A(N, M)) \cong \mathrm{RHom}_A(N \otimes^{\mathrm{L}}_A P, M)$$

is zero. Lemma 16.6.13 tells us that

$$F(\mathrm{RHom}_A(N, M)) = P \otimes^{\mathrm{L}}_A \mathrm{RHom}_A(N, M) \tag{16.6.18}$$

is zero.

Next we apply the functor $\mathrm{RHom}_A(-, M)$, and then the functor $P \otimes^{\mathrm{L}}_A (-)$, to the distinguished triangle (16.6.15). We obtain distinguished triangle

$$P \otimes^{\mathrm{L}}_A \mathrm{RHom}_A(N, M) \to P \otimes^{\mathrm{L}}_A M \xrightarrow{\beta_M} P \otimes^{\mathrm{L}}_A \mathrm{RHom}_A(P, M) \xrightarrow{\triangle}$$

in $\mathsf{D}(A \otimes B^{\mathrm{op}}, \mathrm{gr})$. By (16.6.18) the first object in this triangle is zero, and so $\beta_M : F(M) \to F(G(M))$ is an isomorphism. If moreover $M \in \mathsf{D}(A \otimes B^{\mathrm{op}}, \mathrm{gr})_F$, then σ_M is an isomorphism too, and thus we have an isomorphism

$$\beta_M \circ \sigma_M^{-1} : M \to F(G(M)) \tag{16.6.19}$$

in $\mathsf{D}(A \otimes B^{\mathrm{op}}, \mathrm{gr})_F$ that's functorial in M.

The isomorphisms (16.6.17) and (16.6.19) tell us that F and G are equivalences between $\mathsf{D}(A \otimes B^{\mathrm{op}}, \mathrm{gr})_G$ and $\mathsf{D}(A \otimes B^{\mathrm{op}}, \mathrm{gr})_F$, quasi-inverse to each other. □

We now leave the abstract setting, and return to $\mathfrak{m}$-torsion. So we are in the situation of Setup 16.4.1. The dedualizing complex $P_A := \Gamma_{\mathfrak{m}}(A) \in \mathsf{D}(A^{\mathrm{en}}, \mathrm{gr})$ from Definition 16.4.2 is equipped with the morphism $\sigma^{\mathrm{R}}_A : P_A = \mathrm{R}\Gamma_{\mathfrak{m}}(A) \to A$ from Proposition 16.3.16.

Definition 16.6.20 The pair $(P_A, \sigma^{\mathrm{R}}_A)$ is called the *dedualizing copointed object* of the monoidal category $\mathsf{D}(A^{\mathrm{en}}, \mathrm{gr})$.

Theorem 16.6.21 ([157]) *Under Setup* 16.4.1, *the dedualizing copointed object* $(P_A, \sigma^{\mathrm{R}}_A)$ *in the monoidal category* $\mathsf{D}(A^{\mathrm{en}}, \mathrm{gr})$ *is idempotent.*

Proof Let's write $\sigma := \sigma^{\mathrm{R}}_A$ and $P := P_A$. We shall start by proving that

$$\mathrm{lu} \circ (\sigma \otimes^{\mathrm{L}}_A \mathrm{id}) : P \otimes^{\mathrm{L}}_A P \to P \tag{16.6.22}$$

is isomorphism in $\mathsf{D}(A^{\mathrm{en}}, \mathrm{gr})$. Because the forgetful functor $\mathrm{Rest} : \mathsf{D}(A^{\mathrm{en}}, \mathrm{gr}) \to \mathsf{D}(A, \mathrm{gr})$ is conservative, it is enough if we prove that $\mathrm{Rest}(\mathrm{lu} \circ (\sigma \otimes^{\mathrm{L}}_A \mathrm{id}))$ is an isomorphism. Let us introduce the temporary notation $P' := \mathrm{Rest}(P) \in \mathsf{D}(A, \mathrm{gr})$. With this notation, what we have to show is that

$$\mathrm{lu} \circ (\sigma \otimes^{\mathrm{L}}_A \mathrm{id}) : P \otimes^{\mathrm{L}}_A P' \to P' \tag{16.6.23}$$

is an isomorphism in $\mathsf{D}(A, \mathrm{gr})$.

Consider Theorem 16.4.3 with $B = \mathbb{K}$ and $M = P' \in \mathsf{D}(A, \mathrm{gr})$. There is a commutative diagram

$$\begin{array}{ccc} P \otimes^{\mathrm{L}}_A P' & \xrightarrow{\mathrm{ev}^{\mathrm{R,L}}_{\mathfrak{m},P'}} & \mathrm{R}\Gamma_{\mathfrak{m}}(P') \\ {\scriptstyle \sigma \otimes^{\mathrm{L}}_A \mathrm{id}} \downarrow & & \downarrow {\scriptstyle \sigma^{\mathrm{R}}_{P'}} \\ A \otimes^{\mathrm{L}}_A P' & \xrightarrow{\mathrm{lu}} & P' \end{array}$$

in $\mathsf{D}(A, \mathrm{gr})$, and the horizontal arrows are isomorphisms. It suffices to prove that $\sigma^{\mathrm{R}}_{P'} : \mathrm{R}\Gamma_{\mathfrak{m}}(P') \to P'$ is an isomorphism in $\mathsf{D}(A, \mathrm{gr})$. But there is an isomorphism $P' \cong \mathrm{R}\Gamma_{\mathfrak{m}}(A')$, where $A' := \mathrm{Rest}(A) \in \mathsf{D}(A, \mathrm{gr})$. So what we need to prove is that

$$\sigma^{\mathrm{R}}_{\mathrm{R}\Gamma_{\mathfrak{m}}(A')} : \mathrm{R}\Gamma_{\mathfrak{m}}(\mathrm{R}\Gamma_{\mathfrak{m}}(A')) \to \mathrm{R}\Gamma_{\mathfrak{m}}(A')$$

is an isomorphism in $\mathsf{D}(A, \mathrm{gr})$. This is true because the copointed triangulated functor $(\mathrm{R}\Gamma_{\mathfrak{m}}, \sigma^{\mathrm{R}})$ on $\mathsf{D}(A, \mathrm{gr})$ is idempotent; see Corollary 16.3.36 with $B := \mathbb{K}$.

Now we are going to prove that

$$\mathrm{ru} \circ (\mathrm{id} \otimes^{\mathrm{L}}_A \sigma) : P \otimes^{\mathrm{L}}_A P \to P \tag{16.6.24}$$

is an isomorphism in $\mathsf{D}(A^{\mathrm{en}}, \mathrm{gr})$.

We have this commutative (up to a canonical isomorphism) diagram in $\mathsf{D}(A^{\mathrm{en}}, \mathrm{gr})$:

$$\begin{array}{ccc} P \otimes^{\mathrm{L}}_A P & \xleftarrow{\mathrm{id} \otimes^{\mathrm{L}}_A \mathrm{ru}} & P \otimes^{\mathrm{L}}_A P \otimes^{\mathrm{L}}_A A \\ {\scriptstyle \mathrm{id} \otimes^{\mathrm{L}}_A \sigma} \downarrow & & \downarrow {\scriptstyle \mathrm{id} \otimes^{\mathrm{L}}_A \sigma \otimes^{\mathrm{L}}_A \mathrm{id}} \\ P \otimes^{\mathrm{L}}_A A & \xleftarrow{\mathrm{id} \otimes^{\mathrm{L}}_A \mathrm{ru}} & P \otimes^{\mathrm{L}}_A A \otimes^{\mathrm{L}}_A A \end{array} \tag{16.6.25}$$

Using Theorem 16.4.3 with $B = A$ and $M = A \in \mathsf{D}(A^{\mathrm{en}}, \mathrm{gr})$, we have this commutative diagram in $\mathsf{D}(A^{\mathrm{en}}, \mathrm{gr})$:

$$\begin{array}{ccc} P_A \otimes^{\mathrm{L}}_A A & \xrightarrow{\mathrm{ev}^{\mathrm{R,L}}_{\mathfrak{m},A}} & \mathrm{R}\Gamma_{\mathfrak{m}}(A) \\ {\scriptstyle \sigma \otimes^{\mathrm{L}}_A \mathrm{id}} \downarrow & & \downarrow {\scriptstyle \sigma^{\mathrm{R}}_A} \\ A \otimes^{\mathrm{L}}_A A & \xrightarrow{\mathrm{lu}} & A \end{array} \tag{16.6.26}$$

Applying the functor $P \otimes^{\mathrm{L}}_A (-)$ to this diagram, we obtain this commutative diagram

$$\begin{CD} P \otimes^{\mathrm{L}}_{A} P \otimes^{\mathrm{L}}_{A} A @>{\mathrm{id} \otimes^{\mathrm{L}}_{A} \mathrm{ev}^{\mathrm{R,L}}_{\mathfrak{m},A}}>> P \otimes^{\mathrm{L}}_{A} \mathrm{R}\Gamma_{\mathfrak{m}}(A) \\ @V{\mathrm{id} \otimes^{\mathrm{L}}_{A} \sigma \otimes^{\mathrm{L}}_{A} \mathrm{id}}VV @VV{\mathrm{id} \otimes^{\mathrm{L}}_{A} \sigma^{\mathrm{R}}_{A}}V \\ P \otimes^{\mathrm{L}}_{A} A \otimes^{\mathrm{L}}_{A} A @>{\mathrm{id} \otimes^{\mathrm{L}}_{A} \mathrm{lu}}>> P \otimes^{\mathrm{L}}_{A} A \end{CD} \tag{16.6.27}$$

in $\mathbf{D}(A^{\mathrm{en}}, \mathrm{gr})$. The last move is using the fact that $\mathrm{ev}^{\mathrm{R,L}}_{\mathfrak{m},(-)}$ is an isomorphism of functors; this yields the next commutative diagram:

$$\begin{CD} P \otimes^{\mathrm{L}}_{A} \mathrm{R}\Gamma_{\mathfrak{m}}(A) @>{\mathrm{ev}^{\mathrm{R,L}}_{\mathfrak{m},\mathrm{R}\Gamma_{\mathfrak{m}}(A)}}>> \mathrm{R}\Gamma_{\mathfrak{m}}(\mathrm{R}\Gamma_{\mathfrak{m}}(A)) \\ @V{\mathrm{id} \otimes^{\mathrm{L}}_{A} \sigma^{\mathrm{R}}_{A}}VV @VV{\mathrm{R}\Gamma_{\mathfrak{m}}(\sigma^{\mathrm{R}}_{A})}V \\ P \otimes^{\mathrm{L}}_{A} A @>{\mathrm{ev}^{\mathrm{R,L}}_{\mathfrak{m},A}}>> \mathrm{R}\Gamma_{\mathfrak{m}}(A) \end{CD} \tag{16.6.28}$$

All horizontal arrows in diagrams (16.6.25), (16.6.26), (16.6.27) and (16.6.28) are isomorphisms. Since ru is an isomorphism, to prove that (16.6.24) is an isomorphism, it is enough to prove that the morphism $\mathrm{id} \otimes^{\mathrm{L}}_{A} \sigma$, which is the left vertical arrow in diagram (16.6.25), is an isomorphism. Now this last morphism coincides with the left vertical arrow in diagram (16.6.28). Therefore it is enough to prove that the right vertical arrow in diagram (16.6.28) is an isomorphism. This is

$$\mathrm{R}\Gamma_{\mathfrak{m}}(\sigma^{\mathrm{R}}_{A}) : \mathrm{R}\Gamma_{\mathfrak{m}}(\mathrm{R}\Gamma_{\mathfrak{m}}(A)) \to \mathrm{R}\Gamma_{\mathfrak{m}}(A). \tag{16.6.29}$$

Because the functor Rest is conservative, it suffices to prove that

$$\mathrm{R}\Gamma_{\mathfrak{m}}(\sigma^{\mathrm{R}}_{A'}) : \mathrm{R}\Gamma_{\mathfrak{m}}(\mathrm{R}\Gamma_{\mathfrak{m}}(A')) \to \mathrm{R}\Gamma_{\mathfrak{m}}(A')$$

is an isomorphism in $\mathbf{D}(A, \mathrm{gr})$, where, as before, we write $A' := \mathrm{Rest}(A) \in \mathbf{D}(A, \mathrm{gr})$. This is true because the copointed triangulated functor $(\mathrm{R}\Gamma_{\mathfrak{m}}, \sigma^{\mathrm{R}})$ on $\mathbf{D}(A, \mathrm{gr})$ is idempotent. □

Definition 16.6.30 Under Setup 16.4.1, consider the dedualizing copointed object $(P_A, \sigma^{\mathrm{R}}_{A})$ in the monoidal category $\mathbf{D}(A^{\mathrm{en}}, \mathrm{gr})$; see Definition 16.6.20.

(1) The *abstract derived* $\mathfrak{m}$*-adic completion functor* is the triangulated pointed functor $(\mathrm{ADC}_{\mathfrak{m}}, \tau)$ on $\mathbf{D}(A \otimes B^{\mathrm{op}}, \mathrm{gr})$ that's induced by the copointed object $(P_A, \sigma^{\mathrm{R}}_{A})$, in the sense of Definition 16.6.3

(2) The full subcategory of $\mathbf{D}(A \otimes B^{\mathrm{op}}, \mathrm{gr})$ on the complexes M such that $\tau_M : M \to \mathrm{ADC}_{\mathfrak{m}}(M)$ is an isomorphism is denoted by $\mathbf{D}(A \otimes B^{\mathrm{op}}, \mathrm{gr})_{\mathrm{com}}$.

To put item (1) in explicit terms, $\mathrm{ADC}_{\mathfrak{m}}(M) = \mathrm{RHom}_A(P_A, M)$, and the morphism $\tau_M : M \to \mathrm{ADC}_{\mathfrak{m}}(M)$ is $\tau_M = \mathrm{RHom}_A(\sigma^{\mathrm{R}}_{A}, \mathrm{id}_M)$.

We already have the copointed triangulated functor $(\mathrm{R}\Gamma_{\mathfrak{m}}, \sigma^{\mathrm{R}})$ on the category $\mathbf{D}(A \otimes B^{\mathrm{op}}, \mathrm{gr})$. In analogy with Definition 16.6.30(2) we make the next definition.

Definition 16.6.31 Under Setup 16.4.1, we denote by $\mathbf{D}(A \otimes B^{\mathrm{op}}, \mathrm{gr})_{\mathrm{tor}}$ the full subcategory of $\mathbf{D}(A \otimes B^{\mathrm{op}}, \mathrm{gr})$ on the complexes M such that $\sigma^{\mathrm{R}}_M : \mathrm{R}\Gamma_{\mathfrak{m}}(M) \to M$ is an isomorphism.

Proposition 16.3.29 says that

$$\mathbf{D}(A \otimes B^{\mathrm{op}}, \mathrm{gr})_{\mathrm{tor}} = \mathbf{D}_{\mathrm{tor}}(A \otimes B^{\mathrm{op}}, \mathrm{gr}),$$

where the latter is the full subcategory of $\mathbf{D}(A \otimes B^{\mathrm{op}}, \mathrm{gr})$ on the complexes M whose cohomology modules are $\mathfrak{m}$-torsion.

Theorem 16.6.32 (NC MGM Equivalence, [157]) *Under Setup* 16.4.1 *the following hold.*

(1) *The categories* $\mathbf{D}(A \otimes B^{\mathrm{op}}, \mathrm{gr})_{\mathrm{com}}$ *and* $\mathbf{D}(A \otimes B^{\mathrm{op}}, \mathrm{gr})_{\mathrm{tor}}$ *are the essential images of the functors* $\mathrm{ADC}_{\mathfrak{m}}$ *and* $\mathrm{R}\Gamma_{\mathfrak{m}}$, *respectively.*
(2) *The pointed triangulated functor* $(\mathrm{ADC}_{\mathfrak{m}}, \tau)$ *and the copointed triangulated functor* $(\mathrm{R}\Gamma_{\mathfrak{m}}, \sigma^{\mathrm{R}})$ *are idempotent.*
(3) *The functor*

$$\mathrm{R}\Gamma_{\mathfrak{m}} : \mathbf{D}(A \otimes B^{\mathrm{op}}, \mathrm{gr})_{\mathrm{com}} \to \mathbf{D}(A \otimes B^{\mathrm{op}}, \mathrm{gr})_{\mathrm{tor}}$$

is an equivalence of triangulated categories, with quasi-inverse $\mathrm{ADC}_{\mathfrak{m}}$.

Proof By Definition 16.6.30, the pointed triangulated functor $(\mathrm{ADC}_{\mathfrak{m}}, \tau)$ is the one induced by the copointed object $(P_A, \sigma^{\mathrm{R}}_A)$. And by Theorem 16.4.3, the copointed triangulated functor $(\mathrm{R}\Gamma_{\mathfrak{m}}, \sigma^{\mathrm{R}})$ is the one induced by the copointed object $(P_A, \sigma^{\mathrm{R}}_A)$. According to Theorem 16.6.21, the copointed object $(P_A, \sigma^{\mathrm{R}}_A)$ is idempotent. This means that we can use Theorem 16.6.14 on abstract equivalence. Item (1) here is then a special case of item (3) of Theorem 16.6.14; item (2) here is a special case of item (2) of Theorem 16.6.14; and item (3) here is a special case of item (4) of Theorem 16.6.14. □

Example 16.6.33 In Example 12.4.6 we presented the commutative MGM Equivalence. Recall the setting: a commutative ring A, and an ideal $\mathfrak{a} \subseteq A$ generated by a weakly proregular sequence $\boldsymbol{a}$. The module category there was $\mathbf{M}(A)$. If we want to transfer this to the connected graded setting of the current section, where the module category is $\mathbf{M}(A, \mathrm{gr})$, then we should take A to be a commutative connected graded $\mathbb{K}$-ring, and $\boldsymbol{a}$ is a sequence of homogeneous elements that generates the augmentation ideal $\mathfrak{m}$. Then $A = \mathbb{K}[\boldsymbol{a}]$ is a finitely generated commutative $\mathbb{K}$-ring, and thus noetherian.

Because A is commutative, left A-modules are the same as central A-bimodules. This is the reason we work with the category $\mathbf{M}(A, \mathrm{gr})$ and not with $\mathbf{M}(A^{\mathrm{en}}, \mathrm{gr})$.

Regardless of which setting we choose (graded or ungraded), the dedualizing complex P_A over the ring A (as an object of $\mathbf{D}(A, \mathrm{gr})$ or $\mathbf{D}(A)$, respectively) can be made explicit: $P_A \cong \mathrm{K}^{\vee}_{\infty}(A; \boldsymbol{a}) \cong \mathrm{Tel}(A; \boldsymbol{a})$, where $\mathrm{K}^{\vee}_{\infty}(A; \boldsymbol{a})$ is the *infinite dual Koszul complex* and $\mathrm{Tel}(A; \boldsymbol{a})$ is the *telescope complex.* Since the telescope complex is a bounded complex of free A-modules (of countable rank), the derived torsion and completion functors have particularly nice presentations: $\mathrm{R}\Gamma_{\mathfrak{m}} = \mathrm{Tel}(A; \boldsymbol{a}) \otimes_A (-)$ and $\mathrm{ADC}_{\mathfrak{m}} = \mathrm{Hom}_A(\mathrm{Tel}(A; \boldsymbol{a}), -)$. Also, the abstract derived completion functor $\mathrm{ADC}_{\mathfrak{m}}$ is canonically isomorphic to $\mathrm{L}\Lambda_{\mathfrak{m}}$, the left derived functor of the $\mathfrak{m}$-adic completion functor $\Lambda_{\mathfrak{m}}$. See [118] for details.

17

Balanced Dualizing Complexes over NC Graded Rings

Let $A = \bigoplus_{i \geq 0} A_i$ be a noncommutative noetherian connected graded $\mathbb{K}$-ring, with augmentation ideal $\mathfrak{m} = \bigoplus_{i \geq 1} A_i$. A *balanced dualizing complex* over A is a dualizing complex R over A, in the noncommutative graded sense, that satisfies the *Noncommutative Graded Local Duality Theorem* (Theorem 17.2.7 below). Balanced dualizing complexes were introduced in 1992 by A. Yekutieli [161].

In this chapter we define balanced dualizing complexes, and then we prove their uniqueness and existence, the Local Duality Theorem, and the trace functoriality. One of the main results of this chapter is Corollary 17.3.24, which says that the following two properties are equivalent:

(i) The ring A satisfies the χ condition, and it has finite local cohomological dimension.

(ii) The ring A has a balanced dualizing complex.

This result is the product of the combined efforts of Yekutieli, J. J. Zhang and M. Van den Bergh. The χ condition was already discussed in Section 16.5. M. Artin and Zhang had introduced the χ condition and the finiteness of local cohomological dimension in their 1994 paper [10], to ensure that the *noncommutative projective scheme* $\operatorname{Proj}(A)$ has "good geometric properties." On the other hand, as mentioned above, balanced dualizing complexes were designed to satisfy the Noncommutative Local Duality Theorem. It is remarkable that in the end, these two properties turned out to be equivalent.

Throughout this chapter we adhere to Conventions 15.1.17 and 15.2.1. Thus $\mathbb{K}$ is a base field, and all graded rings are algebraically graded central $\mathbb{K}$-rings, as defined in Chapter 15. See Remark 17.3.40 regarding the complete case (A is adically complete instead of graded) and the arithmetic case (the base ring $\mathbb{K}$ is not a field).

17.1 Graded NC Dualizing Complexes

In this section we introduce *graded noncommutative dualizing complexes*. This is a generalization of Grothendieck's original commutative definition from [62] (see Chapter 13).

Consider a graded $\mathbb{K}$-ring A. Complexes of graded A-modules were studied in Sections 15.1 and 15.3, and here we use these constructions and results. Let us recall that $\mathbf{M}(A, \mathrm{gr})$ is the abelian category of graded A-modules, and $\mathbf{D}(A, \mathrm{gr})$ is its derived category.

Suppose B is a second graded ring. Given complexes $M \in \mathbf{D}(A \otimes B^{\mathrm{op}}, \mathrm{gr})$ and $N \in \mathbf{D}(A^{\mathrm{en}}, \mathrm{gr})$, there is the *noncommutative derived Hom-evaluation morphism*

$$\mathrm{ev}^{\mathrm{R,R}}_{M,N} : M \to \mathrm{RHom}_{A^{\mathrm{op}}}(\mathrm{RHom}_A(M, N), N) \tag{17.1.1}$$

in $\mathbf{D}(A \otimes B^{\mathrm{op}}, \mathrm{gr})$. For a choice of a K-injective resolution $N \to I$ in $\mathbf{C}_{\mathrm{str}}(A^{\mathrm{en}}, \mathrm{gr})$, the morphism $\mathrm{ev}^{\mathrm{R,R}}_{M,N}$ is represented by the homomorphism

$$\mathrm{ev}_{M,I} : M \to \mathrm{Hom}_{A^{\mathrm{op}}}(\mathrm{Hom}_A(M, I), I)$$

in $\mathbf{C}_{\mathrm{str}}(A \otimes B^{\mathrm{op}}, \mathrm{gr})$, whose formula is $\mathrm{ev}_{M,I}(m)(\phi) := (-1)^{p \cdot q} \cdot \phi(m)$ for $m \in M^p$ and $\phi \in \mathrm{Hom}_A(M, I)^q$.

In the special case $B = A$ and $M = A \in \mathbf{D}(A^{\mathrm{en}}, \mathrm{gr})$ we have the canonical isomorphisms $N \cong \mathrm{RHom}_A(A, N) \cong \mathrm{RHom}_{A^{\mathrm{op}}}(A, N)$, i.e. the left and right co-unitor isomorphisms of the biclosed monoidal structure. Then the derived NC Hom-evaluation morphism $\mathrm{ev}^{\mathrm{R,R}}_{M,N}$ from (17.1.1) specializes to the *NC derived homothety morphism through A*:

$$\mathrm{hm}^{\mathrm{R}}_{N,A} = \mathrm{ev}^{\mathrm{R,R}}_{A,N} : A \to \mathrm{RHom}_{A^{\mathrm{op}}}(N, N) \tag{17.1.2}$$

in $\mathbf{D}(A^{\mathrm{en}}, \mathrm{gr})$. Similarly, after exchanging A and A^{op}, we recover the *NC derived homothety morphism through A^{op}*:

$$\mathrm{hm}^{\mathrm{R}}_{N,A^{\mathrm{op}}} = \mathrm{ev}^{\mathrm{R,R}}_{A,N} : A \to \mathrm{RHom}_A(N, N), \tag{17.1.3}$$

also in $\mathbf{D}(A^{\mathrm{en}}, \mathrm{gr})$. This should be compared to Definition 14.4.13, which refers to the ungraded NC setting.

Definition 17.1.4 ([161]) Let A be a noetherian graded ring. A *graded NC dualizing complex* over A is a complex $R \in \mathbf{D}^{\mathrm{b}}(A^{\mathrm{en}}, \mathrm{gr})$ with the following three properties:

(i) *Finiteness of cohomology*: for every integer p the A-bimodule $\mathrm{H}^p(R)$ is a finite graded module over A and over A^{op}.
(ii) *Finite injective dimension*: the complex R has finite graded-injective dimension over A and over A^{op}.
(iii) *NC Derived Morita property*: the noncommutative derived homothety morphisms

$$\mathrm{hm}^{\mathrm{R}}_{R,A} : A \to \mathrm{RHom}_{A^{\mathrm{op}}}(R, R)$$

and

$$\mathrm{hm}^{\mathrm{R}}_{R,A^{\mathrm{op}}} : A \to \mathrm{RHom}_{A}(R, R)$$

in $\mathbf{D}(A^{\mathrm{en}}, \mathrm{gr})$ are both isomorphisms.

Condition (i) can be restated as $R \in \mathbf{D}_{(\mathrm{f},\mathrm{f})}(A^{\mathrm{en}}, \mathrm{gr})$. It can be shown (using Corollary 15.3.7(3) and smart truncation, as in the proof of Proposition 12.4.13) that condition (ii) is equivalent to the existence of an isomorphism $R \cong I$ in $\mathbf{D}(A^{\mathrm{en}}, \mathrm{gr})$, where I is a bounded complex, and every bimodule I^p is graded-injective over A and over A^{op}.

Definition 17.1.5 In the situation of Definition 17.1.4, given another graded ring B, the *duality functors* associated to the dualizing complex R are the triangulated functors

$$D_A : \mathbf{D}(A \otimes B^{\mathrm{op}}, \mathrm{gr})^{\mathrm{op}} \to \mathbf{D}(B \otimes A^{\mathrm{op}}, \mathrm{gr}), \quad D_A := \mathrm{RHom}_A(-, R)$$

and

$$D_{A^{\mathrm{op}}} : \mathbf{D}(B \otimes A^{\mathrm{op}}, \mathrm{gr})^{\mathrm{op}} \to \mathbf{D}(A \otimes B^{\mathrm{op}}, \mathrm{gr}), \quad D_{A^{\mathrm{op}}} := \mathrm{RHom}_{A^{\mathrm{op}}}(-, R).$$

Since the dualizing complex R is usually clear from the context, we can omit it from the notation of the duality functors D_A and $D_{A^{\mathrm{op}}}$. We can also omit R from the notation of the related derived Hom-evaluation morphisms, and just write $\mathrm{ev}^{\mathrm{R,R}}_M : M \to D_{A^{\mathrm{op}}}(D_A(M))$ and $\mathrm{ev}^{\mathrm{R,R}}_{M'} : M' \to D_A(D_{A^{\mathrm{op}}}(M'))$ for complexes $M \in \mathbf{D}(A \otimes B^{\mathrm{op}}, \mathrm{gr})$ and $M' \in \mathbf{D}(B \otimes A^{\mathrm{op}}, \mathrm{gr})$.

Theorem 17.1.6 *Let A and B be graded rings, with A noetherian. Let $R \in \mathbf{D}(A^{\mathrm{en}}, \mathrm{gr})$ be a graded NC dualizing complex over A, with associated duality functors D_A and $D_{A^{\mathrm{op}}}$. Let $\star$ be a boundedness indicator, and let $M \in \mathbf{D}^{\star}_{(\mathrm{f},..)}(A \otimes B^{\mathrm{op}}, \mathrm{gr})$. Then the following hold:*
(1) *The complex $D_A(M)$ belongs to $\mathbf{D}^{-\star}_{(..,\mathrm{f})}(B \otimes A^{\mathrm{op}}, \mathrm{gr})$, where $-\star$ is the reversed boundedness indicator.*
(2) *The derived Hom-evaluation morphism $\mathrm{ev}^{\mathrm{R,R}}_M : M \to D_{A^{\mathrm{op}}}(D_A(M))$ in $\mathbf{D}(A \otimes B^{\mathrm{op}}, \mathrm{gr})$ is an isomorphism.*

(3) *The functor*

$$D_A : \mathsf{D}^{\star}_{(\mathrm{f},..)}(A \otimes B^{\mathrm{op}}, \mathrm{gr})^{\mathrm{op}} \to \mathsf{D}^{-\star}_{(..,\mathrm{f})}(B \otimes A^{\mathrm{op}}, \mathrm{gr})$$

is an equivalence of triangulated categories, with quasi-inverse $D_{A^{\mathrm{op}}}$.

Proof

(1) We can forget the ring B. Since the functor $D_A : \mathsf{D}(A, \mathrm{gr})^{\mathrm{op}} \to \mathsf{D}(A^{\mathrm{op}}, \mathrm{gr})$ has finite cohomological dimension, and since $D_A(A) = R \in \mathsf{M}_{\mathrm{f}}(A^{\mathrm{op}}, \mathrm{gr})$, this is a consequence of Theorem 12.5.7(2), slightly modified to handle the algebraically graded situation.

(2) Because the restriction functor $\mathsf{D}(A \otimes B^{\mathrm{op}}, \mathrm{gr}) \to \mathsf{D}(A, \mathrm{gr})$ is conservative, we can forget about the ring B. The derived Morita property says that $\mathrm{ev}^{\mathrm{R,R}}_M : M \to D_{A^{\mathrm{op}}}(D_A(M))$ is an isomorphism for $M = A$. The functors Id and $D_{A^{\mathrm{op}}} \circ D_A$ have finite cohomological dimensions. The assertion is then a consequence of Theorem 12.5.2(2), slightly modified to handle the algebraically graded situation.

(3) This is clear from items (2) and (3). □

We shall require a notion of twisting in the NC graded setting.

Definition 17.1.7 Let ϕ be an automorphism of A in the category $\mathsf{Rng}_{\mathrm{gr}}/_{\mathrm{c}}\,\mathbb{K}$, and let i be an integer. The *(ϕ, i)-twist of the bimodule A* is the object $A(\phi, i) \in \mathsf{M}(A^{\mathrm{en}}, \mathrm{gr})$ defined as follows: as a left graded A-module it is the graded-free A-module $A(i)$ with basis element e in degree $-i$. The right A-module action is given by the formula $e \cdot^{\phi} a := \phi(a) \cdot e$ for $a \in A$.

To be more explicit, an element $m \in A(\phi, i)$ can be expressed uniquely as $m = b \cdot e$ with $b \in A$. Then $m \cdot^{\phi} a = (b \cdot e) \cdot^{\phi} a = (b \cdot \phi(a)) \cdot e \in A(\phi, i)$.

Definition 17.1.8 An *invertible graded A-bimodule* is a bimodule $L \in \mathsf{M}(A^{\mathrm{en}}, \mathrm{gr})$ such that there exists some $L^{\vee} \in \mathsf{M}(A^{\mathrm{en}}, \mathrm{gr})$ satisfying $L \otimes_A L^{\vee} \cong L^{\vee} \otimes_A L \cong A$ in $\mathsf{M}(A^{\mathrm{en}}, \mathrm{gr})$. The bimodule $L^{\vee}$ is called a quasi-inverse of L.

Proposition 17.1.9 *Assume A is connected graded. Let L be an invertible graded A-bimodule. Then $L \cong A(\phi, i)$, for a unique automorphism ϕ and a unique integer i.*

Proof Take a quasi-inverse $L^{\vee}$ of L. Since the tensor product commutes with the ungrading functor, we see that $L \otimes_A L^{\vee} \cong L^{\vee} \otimes_A L \cong A$ in $\mathsf{M}(\mathrm{Ungr}(A)^{\mathrm{en}})$. Classical Morita Theory (see e.g. [130, Chapter 4]) says that L and $L^{\vee}$ are finite projective $\mathrm{Ungr}(A)$-modules on both sides. Hence they are flat over A and A^{op}.

Consider the obvious surjection

$$\mathbb{K} \cong \mathbb{K} \otimes_A L^\vee \otimes_A L \twoheadrightarrow (\mathbb{K} \otimes_A L^\vee \otimes_A \mathbb{K}) \otimes (\mathbb{K} \otimes_A L)$$

in $\mathbf{M}(A^{\mathrm{en}}, \mathrm{gr})$. Since L is a nonzero finite graded A-module and A^{op}-module, by the graded Nakayama Lemma (Proposition 15.2.30), the $\mathbb{K}$-modules $\mathbb{K} \otimes_A L^\vee \otimes_A \mathbb{K}$ and $\mathbb{K} \otimes_A L$ are nonzero. Therefore $\mathbb{K} \otimes_A L$ is a graded-free $\mathbb{K}$-module of rank 1, so $\mathbb{K} \otimes_A L \cong \mathbb{K}(i)$ in $\mathbf{M}(\mathbb{K}, \mathrm{gr})$ for some integer i. But L is a flat A-module, and therefore, as in the classical commutative proof, we get an isomorphism $L \cong A(i)$ in $\mathbf{M}(A, \mathrm{gr})$.

By symmetry we also have an isomorphism $L \cong A(i)$ in $\mathbf{M}(A^{\mathrm{op}}, \mathrm{gr})$. Thus there is an element $e \in L_{-i}$ which is a basis of L on both sides. For $a \in A$ let $\phi(a) \in A$ be the unique element such that $e \cdot a = \phi(a) \cdot e$. Then ϕ is a $\mathbb{K}$-ring automorphism of A, and $L \cong A(\phi, i)$ in $\mathbf{M}(A^{\mathrm{en}}, \mathrm{gr})$.

Because the only invertible elements of A are the nonzero elements of $A_0 \cong \mathbb{K}$ and these are central elements, A does not have nontrivial inner automorphisms. Thus the automorphism ϕ is unique. □

The next theorem is a variant of Theorem 13.1.35.

Theorem 17.1.10 (Uniqueness of NC Graded DC, [161]) *Let A be a noetherian connected graded ring, and let R and R' be graded NC dualizing complexes over A. Then there is an isomorphism*

$$R' \cong R \otimes_A A(\phi, i)[j]$$

in $\mathbf{D}(A^{\mathrm{en}}, \mathrm{gr})$, for a unique automorphism ϕ and unique integers i, j.

For the proof we shall need a few lemmas. Recall that a module $N \in \mathbf{M}(A, \mathrm{gr})$ is called bounded below if $N_i = 0$ for $i \ll 0$.

Lemma 17.1.11 *Let $T, T' \in \mathbf{D}^-(A^{\mathrm{en}}, \mathrm{gr})$ satisfy these conditions:*

- *For every k the graded bimodules $\mathrm{H}^k(T)$ and $\mathrm{H}^k(T')$ are bounded below.*
- *There are isomorphisms $T \otimes^{\mathrm{L}}_A T' \cong T' \otimes^{\mathrm{L}}_A T \cong A$ in $\mathbf{D}(A^{\mathrm{en}}, \mathrm{gr})$.*

Then $T \cong A(\phi, i)[j]$, for a unique automorphism ϕ and unique integers i, j.

Proof This is similar to the proof of Lemma 13.1.42. Define $j_1 := \sup(\mathrm{H}(T))$ and $j_1' := \sup(\mathrm{H}(T'))$. By smart truncation we can assume that $j_1 = \sup(T)$ and $j_1' := \sup(T')$. The Künneth trick (see Lemma 13.1.36; it works also in the graded setting) we have

$$\mathrm{H}^{j_1}(T) \otimes_A \mathrm{H}^{j_1'}(T') \cong \mathrm{H}^{j_1+j_1'}(T \otimes^{\mathrm{L}}_A T') \cong \mathrm{H}^{j_1+j_1'}(A)$$

in $\mathbf{M}(A^{\mathrm{en}}, \mathrm{gr})$. The graded Nakayama Lemma (Proposition 15.2.30) implies that $\mathrm{H}^{j_1}(T) \otimes_A \mathrm{H}^{j_1'}(T') \neq 0$. Hence we must have $j_1 + j_1' = 0$ and $\mathrm{H}^{j_1}(T) \otimes_A \mathrm{H}^{j_1'}(T') \cong A$. By symmetry, we also have $\mathrm{H}^{j_1'}(T') \otimes_A \mathrm{H}^{j_1}(T) \cong A$. According to Proposition 17.1.9, we know that $\mathrm{H}^{j_1}(T) \cong L$, where $L := A(\phi, i)$ for some ϕ and i. Therefore $\mathrm{H}^{j'_1}(T') \cong L'$, where $L' := A(\phi^{-1}, -i)$.

We now forget the left A-module structure of T, and the right A-module structure of T'. By the projective truncation trick (Lemma 13.1.38) we get an isomorphism $T \cong L[-j_1] \oplus N$ in $\mathbf{D}(A^{\mathrm{op}}, \mathrm{gr})$, where $\sup(N) \leq j_1 - 1$. Similarly there is an isomorphism $T' \cong L'[-j'_1] \oplus N'$ in $\mathbf{D}(A, \mathrm{gr})$, where $\sup(N') \leq j'_1 - 1$. Then there is an isomorphism

$$\begin{aligned} A &\cong \mathrm{H}(T \otimes^{\mathrm{L}}_A T') \cong (L \otimes_A L') \\ &\oplus (L[-j_1] \otimes_A \mathrm{H}(N')) \oplus (\mathrm{H}(N) \otimes_A L'[-j'_1]) \oplus \mathrm{H}(N \otimes^{\mathrm{L}}_A N') \end{aligned} \tag{17.1.12}$$

in $\mathbf{G}_{\mathrm{str}}(\mathbb{K}, \mathrm{gr})$. But A is concentrated in cohomological degree 0, and this forces the three summands in the second line of (17.1.12) to be zero. Since L' is graded-free of rank 1 over A, it follows that $\mathrm{H}(N) = 0$. Therefore $\mathrm{H}^k(T) = 0$ for all $k \neq j_1$. Letting $j := -j_1$ we get $T \cong \mathrm{H}^{-j}(T)[j] \cong L[j] = A(\phi, i)[j]$ in $\mathbf{D}(A^{\mathrm{en}}, \mathrm{gr})$. □

Lemma 17.1.13 *Let $R, R' \in \mathbf{D}(A^{\mathrm{en}}, \mathrm{gr})$ be graded NC dualizing complexes, and let $M \in \mathbf{D}(A^{\mathrm{en}}, \mathrm{gr})$. Then there is a morphism*

$$\psi_M : M \otimes^{\mathrm{L}}_A \mathrm{RHom}_A(R, R') \xrightarrow{\simeq} \mathrm{RHom}_A(\mathrm{RHom}_{A^{\mathrm{op}}}(M, R), R')$$

in $\mathbf{D}(A^{\mathrm{en}}, \mathrm{gr})$ that is functorial in M. If $M \in \mathbf{D}^{-}_{(..,\mathrm{f})}(A^{\mathrm{en}}, \mathrm{gr})$ then ψ_M is an isomorphism.

Proof This is the NC graded version of the isomorphism that was used in the proof of Theorem 13.1.35. To define ψ_M, we choose a K-projective resolution $P \to M$, and K-injective resolutions $R \to I$ and $R' \to I'$, all in $\mathbf{C}_{\mathrm{str}}(A^{\mathrm{en}}, \mathrm{gr})$. Then ψ_M is represented by the obvious homomorphism

$$\tilde{\psi}_{P,I,I'} : P \otimes_A \mathrm{Hom}_A(I, I') \to \mathrm{Hom}_A(\mathrm{Hom}_{A^{\mathrm{op}}}(P, I), I')$$

in $\mathbf{C}_{\mathrm{str}}(A^{\mathrm{en}}, \mathrm{gr})$.

To show that ψ_M is an isomorphism when $M \in \mathbf{D}^{-}_{(..,\mathrm{f})}(A^{\mathrm{en}}, \mathrm{gr})$, we can forget the A-module structure on M, and view ψ_M as a morphism in $\mathbf{D}(\mathbb{K}, \mathrm{gr})$. So we have a morphism ψ_M between triangulated functors $\mathbf{D}^{-}_{\mathrm{f}}(A^{\mathrm{op}}, \mathrm{gr}) \to \mathbf{D}(\mathbb{K}, \mathrm{gr})$. It is clear that ψ_A is an isomorphism. By Theorem 12.5.2(1), which is valid also in the algebraically graded context, ψ_M is an isomorphism for every $M \in \mathbf{D}^{-}_{\mathrm{f}}(A^{\mathrm{op}}, \mathrm{gr})$. □

Proof of Theorem 17.1.10 Define the duality functors

$$D_A := \mathrm{RHom}_A(-, R), \quad D_{A^{\mathrm{op}}} := \mathrm{RHom}_{A^{\mathrm{op}}}(-, R),$$

$$D'_A := \mathrm{RHom}_A(-, R'), \quad D'_{A^{\mathrm{op}}} := \mathrm{RHom}_{A^{\mathrm{op}}}(-, R').$$

These are contravariant triangulated functors from $\mathsf{D}(A^{\mathrm{en}}, \mathrm{gr})$ to itself. Then define the complexes $T := (D'_A \circ D_{A^{\mathrm{op}}})(A) = \mathrm{RHom}_A(R, R')$ and $T' := (D_A \circ D'_{A^{\mathrm{op}}})(A) = \mathrm{RHom}_A(R', R)$ in $\mathsf{D}(A^{\mathrm{en}}, \mathrm{gr})$. Note that by Theorem 17.1.6 we have $T, T' \in \mathsf{D}^{\mathrm{b}}_{(\mathrm{f},..)}(A^{\mathrm{en}}, \mathrm{gr})$. Hence the graded bimodules $\mathrm{H}^k(T)$ and $\mathrm{H}^k(T')$ are all bounded below.

According to Lemma 17.1.13 and Theorem 17.1.6 (applied twice), there are isomorphisms

$$\begin{aligned} T' \otimes^{\mathrm{L}}_A T &\cong (D'_A \circ D_{A^{\mathrm{op}}})(T') = (D'_A \circ D_{A^{\mathrm{op}}} \circ D_A \circ D'_{A^{\mathrm{op}}})(A) \\ &\cong (D'_A \circ D'_{A^{\mathrm{op}}})(A) \cong A \end{aligned}$$

in $\mathsf{D}(A^{\mathrm{en}}, \mathrm{gr})$. By symmetry, there is also an isomorphism $T \otimes^{\mathrm{L}}_A T' \cong A$ in $\mathsf{D}(A^{\mathrm{en}}, \mathrm{gr})$. By Lemma 17.1.11 there is an isomorphism $T \cong A(\phi, i)[j]$ in $\mathsf{D}(A^{\mathrm{en}}, \mathrm{gr})$, for a unique automorphism ϕ and unique integers i, j.

Finally we obtain these isomorphisms

$$\begin{aligned} R \otimes_A A(\phi, i)[j] &\cong R \otimes^{\mathrm{L}}_A T \cong^{\dagger} (D'_A \circ D_{A^{\mathrm{op}}})(R) \\ &\cong (D'_A \circ D_{A^{\mathrm{op}}} \circ D_A)(A) \cong D'_A(A) = R' \end{aligned}$$

in $\mathsf{D}(A^{\mathrm{en}}, \mathrm{gr})$. The isomorphism $\cong^{\dagger}$ is from Lemma 17.1.13. □

17.2 Balanced DC: Definition, Uniqueness and Local Duality

Our goal in this section is to relate graded NC dualizing complexes with derived torsion. We continue with Conventions 15.1.17 and 15.2.1, and we assume the following setup:

Setup 17.2.1 $\mathbb{K}$ is a base field, and A is a noetherian connected graded $\mathbb{K}$-ring, with augmentation ideal $\mathfrak{m}$. The augmentation ideal of the opposite ring A^{op} is $\mathfrak{m}^{\mathrm{op}}$.

Recall the graded bimodule $A^* = \mathrm{Hom}_{\mathbb{K}}(A, \mathbb{K})$.

Definition 17.2.2 ([161]) Under Setup 17.2.1, a *balanced dualizing complex* over A is a pair (R, β), where:

(B1) $R \in \mathsf{D}(A^{\mathrm{en}}, \mathrm{gr})$ is a graded NC dualizing complex over A (Definition 17.1.4), with symmetric derived $\mathfrak{m}$-torsion (Definition 16.5.7).

(B2) $\beta : \mathrm{R}\Gamma_{\mathfrak{m}}(R) \xrightarrow{\simeq} A^*$ is an isomorphism in $\mathsf{D}(A^{\mathrm{en}}, \mathrm{gr})$, called a *balancing isomorphism.*

The lack of left-right symmetry in item (B2) of this definition will be removed in Corollary 17.2.16 below.

Remark 17.2.3 Definition 17.2.2 is a bit more sophisticated than the original definition in [161]. There the condition was that $\mathrm{R}\Gamma_{\mathfrak{m}}(R) \cong \mathrm{R}\Gamma_{\mathfrak{m}^{\mathrm{op}}}(R) \cong A^*$ in $\mathbf{D}(A^{\mathrm{en}}, \mathrm{gr})$, but a balancing isomorphism β was not specified. The current improved definition is influenced by later research, especially [183] and [157].

The next theorem is implicit in [161] and [183].

Theorem 17.2.4 (Uniqueness of BDC) *Under Setup* 17.2.1, *suppose that* (R, β) *and* (R', β') *are balanced dualizing complexes over A. Then there is a unique isomorphism* $\psi : R' \xrightarrow{\simeq} R$ *in* $\mathbf{D}(A^{\mathrm{en}}, \mathrm{gr})$, *such that* $\beta \circ \mathrm{R}\Gamma_{\mathfrak{m}}(\psi) = \beta'$ *as isomorphisms* $\mathrm{R}\Gamma_{\mathfrak{m}}(R') \xrightarrow{\simeq} A^*$ *in* $\mathbf{D}(A^{\mathrm{en}}, \mathrm{gr})$.

Proof By Theorem 17.1.10 there is an isomorphism $\psi^\dagger : R' \xrightarrow{\simeq} R \otimes_A L$ in $\mathbf{D}(A^{\mathrm{en}}, \mathrm{gr})$, where $L = A(\phi, i)[j]$ is a translate of a twisted bimodule. Applying the functor $\mathrm{R}\Gamma_{\mathfrak{m}}$ to $\psi^\dagger$ we get a diagram of isomorphisms

$$\begin{array}{ccccc} \mathrm{R}\Gamma_{\mathfrak{m}}(R') & \xrightarrow[\cong]{\mathrm{R}\Gamma_{\mathfrak{m}}(\psi^\dagger)} & \mathrm{R}\Gamma_{\mathfrak{m}}(R \otimes_A L) & \xrightarrow[\cong]{\gamma} & \mathrm{R}\Gamma_{\mathfrak{m}}(R) \otimes_A L \\ {\scriptstyle \beta'}\big\downarrow{\scriptstyle \cong} & & & & {\scriptstyle \beta \otimes \mathrm{id}}\big\downarrow{\scriptstyle \cong} \\ A^* & & & & A^* \otimes_A L \end{array} \tag{17.2.5}$$

in $\mathbf{D}(A^{\mathrm{en}}, \mathrm{gr})$. The isomorphism γ is the obvious one. Therefore $A^* \cong A^* \otimes_A L$ in $\mathbf{M}(A^{\mathrm{en}}, \mathrm{gr})$. Due to the NC Graded Matlis Duality (Theorem 15.2.34) we have $L \cong A$ in $\mathbf{M}(A^{\mathrm{en}}, \mathrm{gr})$.

Let us now rewrite diagram (17.2.5) with $L = A$. This is the solid diagram

$$\begin{array}{ccc} \mathrm{R}\Gamma_{\mathfrak{m}}(R') & \xrightarrow[\cong]{\mathrm{R}\Gamma_{\mathfrak{m}}(\psi^\dagger)} & \mathrm{R}\Gamma_{\mathfrak{m}}(R) \\ {\scriptstyle \beta'}\big\downarrow{\scriptstyle \cong} & & {\scriptstyle \beta}\big\downarrow{\scriptstyle \cong} \\ A^* & \overset{c\cdot(-)}{\underset{\cong}{\dashrightarrow}} & A^* \end{array} \tag{17.2.6}$$

in $\mathbf{D}(A^{\mathrm{en}}, \mathrm{gr})$. Because the automorphisms of A^* in $\mathbf{D}(A^{\mathrm{en}}, \mathrm{gr})$ are multiplication by nonzero elements of $\mathbb{K}$, there is a unique $c \in \mathbb{K}^\times$ for which the diagram (17.2.6) is commutative. Hence $\psi := c^{-1} \cdot \psi^\dagger$ is the unique isomorphism $R' \xrightarrow{\simeq} R$ satisfying $\beta \circ \mathrm{R}\Gamma_{\mathfrak{m}}(\psi) = \beta'$. □

The next theorem is a noncommutative version of Grothendieck's Local Duality Theorem [62, Theorem V.6.2]. Balanced dualizing complexes were invented in order to make this theorem hold. Theorem 17.2.7 first appeared in [161], in a slightly different formulation, and with a complicated proof. The current formulation of the theorem, as well as the proof, are taken from [39, Proposition 3.4]. We thank R. Vyas for pointing out this proof to us.

Theorem 17.2.7 (Local Duality) *Under Setup* 17.2.1, *let* (R, β) *be a balanced dualizing complex over A. There is a morphism*

$$\xi : \mathrm{RHom}_A(-, R) \to (\mathrm{R}\Gamma_{\mathfrak{m}}(-))^*$$

of triangulated functors $\mathbf{D}(A, \mathrm{gr})^{\mathrm{op}} \to \mathbf{D}(A^{\mathrm{op}}, \mathrm{gr})$, *such that for every* $M \in \mathbf{D}^+_{\mathrm{f}}(A, \mathrm{gr})$ *the morphism*

$$\xi_M : \mathrm{RHom}_A(M, R) \to (\mathrm{R}\Gamma_{\mathfrak{m}}(M))^*$$

in $\mathbf{D}(A^{\mathrm{op}}, \mathrm{gr})$ *is an isomorphism.*

Note that this result resembles Van den Bergh's Local Duality, Corollary 16.4.11 – but this is misleading: the assumptions are not the same.

Proof The proof is divided into three steps.

Step 1. Choose K-injective resolutions $R \to J$ and $\rho : \Gamma_{\mathfrak{m}}(J) \to K$ in $\mathbf{C}_{\mathrm{str}}(A^{\mathrm{en}}, \mathrm{gr})$. For each $M \in \mathbf{D}(A, \mathrm{gr})$ choose a K-injective resolution $M \to I_M$ in $\mathbf{C}_{\mathrm{str}}(A, \mathrm{gr})$. There is an obvious homomorphism

$$\phi_M : \mathrm{Hom}_A(I_M, J) \to \mathrm{Hom}_A(\Gamma_{\mathfrak{m}}(I_M), \Gamma_{\mathfrak{m}}(J))$$

in $\mathbf{C}_{\mathrm{str}}(A^{\mathrm{op}}, \mathrm{gr})$. Next we have the homomorphism

$$\mathrm{Hom}_A(\mathrm{id}, \rho) : \mathrm{Hom}_A(\Gamma_{\mathfrak{m}}(I_M), \Gamma_{\mathfrak{m}}(J)) \to \mathrm{Hom}_A(\Gamma_{\mathfrak{m}}(I_M), K)$$

in $\mathbf{C}_{\mathrm{str}}(A^{\mathrm{op}}, \mathrm{gr})$. Composing these homomorphisms, and then going to the derived category, we obtain a morphism

$$\begin{aligned} \theta_M := \mathrm{Q}(\mathrm{Hom}_A(\mathrm{id}, \rho) \circ \phi_M) &: \mathrm{RHom}_A(M, R) \\ &\to \mathrm{RHom}_A(\mathrm{R}\Gamma_{\mathfrak{m}}(M), \mathrm{R}\Gamma_{\mathfrak{m}}(R)) \end{aligned} \tag{17.2.8}$$

in $\mathbf{D}(A^{\mathrm{op}}, \mathrm{gr})$. We have the balancing isomorphism

$$\beta : \mathrm{R}\Gamma_{\mathfrak{m}}(R) \xrightarrow{\simeq} A^* \tag{17.2.9}$$

in $\mathbf{D}(A^{\mathrm{en}}, \mathrm{gr})$. According to Proposition 15.2.5, there is an isomorphism

$$\alpha_M : \mathrm{RHom}_A(\mathrm{R}\Gamma_{\mathfrak{m}}(R), A^*) \xrightarrow{\simeq} (\mathrm{R}\Gamma_{\mathfrak{m}}(R))^*$$

in $\mathbf{D}(A^{\mathrm{op}}, \mathrm{gr})$, and it is functorial in M. Let us define the morphism

$$\xi_M := \alpha_M \circ \mathrm{RHom}(\mathrm{id}, \beta) \circ \theta_M : \mathrm{RHom}_A(M, R) \to (\mathrm{R}\Gamma_{\mathfrak{m}}(M))^* \tag{17.2.10}$$

in $\mathbf{D}(A^{\mathrm{op}}, \mathrm{gr})$. By construction the morphism ξ_M is functorial in M.

Step 2. Take the complex $M := R$. By definition R has the NC derived Morita property on the A side. This implies that the element $\mathrm{id}_R \in \mathrm{H}^0(\mathrm{RHom}_A(R, R))$ is a basis of this rank 1 graded-free A^{op}-module. Because of the isomorphism (17.2.9) and by graded Matlis Duality, we know that $\mathrm{R}\Gamma_{\mathfrak{m}}(R)$ has the NC derived Morita property on the A side. Therefore the element

$$\mathrm{id}_{\mathrm{R}\Gamma_{\mathfrak{m}}(R)} \in \mathrm{H}^0(\mathrm{RHom}_A(\mathrm{R}\Gamma_{\mathfrak{m}}(R), \mathrm{R}\Gamma_{\mathfrak{m}}(R)))$$

is a basis of this rank 1 graded-free A^{op}-module. The construction of the morphism θ_R in (17.2.8) shows that $\mathrm{H}^0(\theta_R)(\mathrm{id}_R) = \mathrm{id}_{\mathrm{R}\Gamma_{\mathfrak{m}}(R)}$. We conclude that $\mathrm{H}^0(\theta_R)$ is an isomorphism. Since all other cohomologies of the complexes below vanish, it follows that

$$\theta_R : \mathrm{RHom}_A(R, R) \to \mathrm{RHom}_A(\mathrm{R}\Gamma_{\mathfrak{m}}(R), \mathrm{R}\Gamma_{\mathfrak{m}}(R)) \tag{17.2.11}$$

is an isomorphism in $\mathbf{D}(A^{\mathrm{op}}, \mathrm{gr})$. Therefore, in view of (17.2.10), ξ_R is also an isomorphism.

Step 3. Define the functor

$$F := (\mathrm{R}\Gamma_{\mathfrak{m}}(-))^* : \mathbf{D}(A, \mathrm{gr})^{\mathrm{op}} \to \mathbf{D}(A^{\mathrm{op}}, \mathrm{gr}).$$

Thus $\xi : D_A \to F$ is a morphism of triangulated functors, and we must prove that $\xi_M : D_A(M) \to F(M)$ is an isomorphism for every $M \in \mathbf{D}^+_{\mathrm{f}}(A, \mathrm{gr})$. We know that $D_{A^{\mathrm{op}}} : \mathbf{D}^-_{\mathrm{f}}(A^{\mathrm{op}}, \mathrm{gr})^{\mathrm{op}} \to \mathbf{D}^+_{\mathrm{f}}(A, \mathrm{gr})$ is an equivalence. Thus it suffices to prove that the morphism

$$\zeta := \xi \circ \mathrm{id}_{D_{A^{\mathrm{op}}}} : D_A \circ D_{A^{\mathrm{op}}} \to F \circ D_{A^{\mathrm{op}}} \tag{17.2.12}$$

of functors $\mathbf{D}^-_{\mathrm{f}}(A^{\mathrm{op}}, \mathrm{gr}) \to \mathbf{D}(A^{\mathrm{op}}, \mathrm{gr})$ is an isomorphism.

Consider the object $A \in \mathbf{D}^-_{\mathrm{f}}(A^{\mathrm{op}}, \mathrm{gr})$. By formula (17.2.12) the diagram

$$\begin{array}{ccc}
(D_A \circ D_{A^{\mathrm{op}}})(A) & \xrightarrow{\ \mathrm{id}\ } & \mathrm{RHom}_A(R, R) \\
\Big\downarrow{\scriptstyle \zeta_A} & & \Big\downarrow{\scriptstyle \xi_R} \\
(F \circ D_{A^{\mathrm{op}}})(A) & \xrightarrow{\ \mathrm{id}\ } & (\mathrm{R}\Gamma_{\mathfrak{m}}(R))^*
\end{array}$$

in $\mathbf{D}(A^{\mathrm{op}}, \mathrm{gr})$ is commutative. In step 2 we proved that ξ_R is an isomorphism. Hence ζ_A is an isomorphism.

The functors D_A and $D_{A^{\mathrm{op}}}$ have bounded cohomological displacements, and hence so does their composition $D_A \circ D_{A^{\mathrm{op}}}$. (In fact the composition is isomorphic to the identity functor, so it has cohomological displacement $[0, 0]$.) The functor $\mathrm{R}\Gamma_{\mathfrak{m}}$ has bounded below cohomological displacement, and the contravariant functor $(-)^*$ has cohomological displacement $[0, 0]$, so their composition F has bounded above cohomological displacement. We conclude that the functors $D_A \circ D_{A^{\mathrm{op}}}$ and $F \circ D_{A^{\mathrm{op}}}$ have bounded above cohomological displacements. We have already established that ζ_A is an isomorphism. Theorem 12.5.2(1), which is valid also in the algebraically graded setting, says that ζ_N is an isomorphism for every $N \in \mathbf{D}^-_{\mathrm{f}}(A^{\mathrm{op}}, \mathrm{gr})$. □

Corollary 17.2.13 ([182]) *Under Setup* 17.2.1, *assume A has a balanced dualizing complex. Then A satisfies the χ condition, and it has finite local cohomological dimension.*

Proof Given $M \in \mathsf{M}_{\mathrm{f}}(A, \mathrm{gr})$, Theorem 17.2.7 provides an isomorphism

$$\xi_M : \mathrm{RHom}_A(M, R) \xrightarrow{\simeq} (\mathrm{R}\Gamma_{\mathfrak{m}}(M))^*$$

in $\mathsf{D}(A^{\mathrm{op}}, \mathrm{gr})$. This gives an isomorphism

$$\mathrm{H}^{-p}(\xi_M) : \mathrm{Ext}_A^{-p}(M, R) \xrightarrow{\simeq} (\mathrm{H}^p_{\mathfrak{m}}(M))^* \tag{17.2.14}$$

in $\mathsf{M}(A^{\mathrm{op}}, \mathrm{gr})$ for every p. By Theorem 17.1.6(1) we know that $\mathrm{Ext}_A^{-p}(M, R) \in \mathsf{M}_{\mathrm{f}}(A^{\mathrm{op}}, \mathrm{gr})$. Using NC graded Matlis Duality (Theorem 15.2.34) we get

$$\mathrm{H}^p_{\mathfrak{m}}(M) \cong (\mathrm{Ext}_A^{-p}(M, R))^* \in \mathsf{M}_{\mathrm{cof}}(A, \mathrm{gr}). \tag{17.2.15}$$

By Proposition 16.5.17 we conclude that A satisfies the left χ condition. The op-symmetry of the situation implies that A^{op} also satisfies the left χ condition.

Because $R \in \mathsf{D}^{\mathrm{b}}(A^{\mathrm{en}}, \mathrm{gr})$ and is nonzero, we have $q_0 := \inf(\mathrm{H}(R)) \in \mathbb{Z}$. So $\mathrm{Ext}^q_A(M, R) = 0$ for all $M \in \mathsf{M}(A, \mathrm{gr})$ and $q < q_0$. Taking $p := -q$ and using (17.2.15) we get $\mathrm{H}^p(\mathrm{R}\Gamma_{\mathfrak{m}}(M)) = \mathrm{H}^p_{\mathfrak{m}}(M) = 0$ for all $p > -q_0$. So the functor $\mathrm{R}\Gamma_{\mathfrak{m}}$ has finite cohomological dimension. The op-symmetry of the situation implies that the functor $\mathrm{R}\Gamma_{\mathfrak{m}^{\mathrm{op}}}$ also has finite cohomological dimension. □

Corollary 17.2.16 *Let (R, β) be a balanced dualizing complex over A. Then there is a unique symmetry isomorphism $\epsilon_R : \mathrm{R}\Gamma_{\mathfrak{m}}(R) \xrightarrow{\simeq} \mathrm{R}\Gamma_{\mathfrak{m}^{\mathrm{op}}}(R)$ in $\mathsf{D}(A^{\mathrm{en}}, \mathrm{gr})$, in the sense of Definition* 16.5.7. *Hence there is a unique isomorphism $\beta^{\mathrm{op}} : \mathrm{R}\Gamma_{\mathfrak{m}^{\mathrm{op}}}(R) \xrightarrow{\simeq} A^*$ $\mathsf{D}(A^{\mathrm{en}}, \mathrm{gr})$ such that $\beta^{\mathrm{op}} \circ \epsilon_R = \beta$.*

Proof By Corollary 17.2.13 the ring A has finite local cohomological dimension. So Theorem 16.5.8 applies. □

17.3 Balanced DC: Existence

In this section we are going to prove the existence of balanced dualizing complexes in two important situations. We shall adhere to Conventions 15.1.17 and 15.2.1, and to Setup 17.2.1.

The next lemma is [161, Proposition 4.4].

Lemma 17.3.1 *Let R be a graded NC dualizing complex over A, with duality functors $D_A, D_{A^{\mathrm{op}}} : \mathsf{D}(A^{\mathrm{en}}, \mathrm{gr}) \to \mathsf{D}(A^{\mathrm{en}}, \mathrm{gr})$. The following four conditions are equivalent.*

(i) *There is an isomorphism $D_A(\mathbb{K}) \cong \mathbb{K}$ in $\mathsf{D}(A^{\mathrm{en}}, \mathrm{gr})$.*

(i′) *There is an isomorphism $D_{A^{\mathrm{op}}}(\mathbb{K}) \cong \mathbb{K}$ in $\mathsf{D}(A^{\mathrm{en}}, \mathrm{gr})$.*

(ii) *There is an isomorphism $\mathrm{R}\Gamma_{\mathfrak{m}}(R) \cong A^* \otimes_A L$ in $\mathsf{D}(A^{\mathrm{en}}, \mathrm{gr})$, for some graded invertible A-bimodule L generated in algebraic degree 0.*

(ii′) *There is an isomorphism* $\mathrm{R}\Gamma_{\mathfrak{m}^{\mathrm{op}}}(R) \cong L' \otimes_A A^*$ *in* $\mathsf{D}(A^{\mathrm{en}}, \mathrm{gr})$, *for some graded invertible A-bimodule* L' *generated in algebraic degree* 0.

Note that by Proposition 17.1.9, a graded invertible A-bimodule L generated in algebraic degree 0 is isomorphic to $A(\phi, 0)$ for some automorphism ϕ.

Proof

(i) ⇒ (i′): It is given that $D_A(\mathbb{K}) \cong \mathbb{K}$. By Theorem 17.1.6 we know that $\mathbb{K} \cong D_{A^{\mathrm{op}}}(D_A(\mathbb{K}))$. Together we get $D_{A^{\mathrm{op}}}(\mathbb{K}) \cong \mathbb{K}$.

(i′) ⇒ (i): The same, by op-symmetry (replacing A with A^{op}).

((i) and (i′)) ⇒ (ii): Recall that a graded invertible bimodule L is free of rank 1 on both sides. So it suffices to prove separately that

$$\mathrm{R}\Gamma_{\mathfrak{m}}(R) \cong A^* \ \text{ in } \mathsf{D}(A, \mathrm{gr}) \tag{17.3.2}$$

and

$$\mathrm{R}\Gamma_{\mathfrak{m}}(R) \cong A^* \ \text{ in } \mathsf{D}(A^{\mathrm{op}}, \mathrm{gr}). \tag{17.3.3}$$

Let $R \to I$ be a minimal graded-injective resolution of R over A, so that $\mathrm{R}\Gamma_{\mathfrak{m}}(R) \cong \Gamma_{\mathfrak{m}}(I)$ in $\mathsf{D}(A, \mathrm{gr})$. By Lemma 16.5.12 the subcomplex $W := \mathrm{Soc}(I) = \mathrm{Hom}_A(\mathbb{K}, I) \subseteq I$ has zero differential. But $D_A(\mathbb{K}) = \mathrm{RHom}_A(\mathbb{K}, R) \cong \mathrm{Hom}_A(\mathbb{K}, I) = W$ in $\mathsf{D}(A, \mathrm{gr})$. From condition (i) we see that $W \cong \mathbb{K}$ in $\mathsf{D}(A, \mathrm{gr})$; and therefore also $W \cong \mathbb{K}$ in $\mathsf{C}_{\mathrm{str}}(A, \mathrm{gr})$. According to Lemma 16.3.24 and Proposition 16.3.12, $W^p \subseteq \Gamma_{\mathfrak{m}}(I^p)$ is an essential submodule. Again using Lemma 16.3.24 we conclude that $\Gamma_{\mathfrak{m}}(I) \cong A^*$ in $\mathsf{C}_{\mathrm{str}}(A^{\mathrm{en}}, \mathrm{gr})$. Hence the isomorphism (17.3.2) holds.

For the other isomorphism, let $\mathsf{M}_{\mathrm{f}/\mathbb{K}}(A, \mathrm{gr}) \subseteq \mathsf{M}(A, \mathrm{gr})$ be the full subcategory on the modules that are finite over $\mathbb{K}$, or in other words, the finite length A-modules. Conditions (i) and (i′) imply, using induction on length of modules, that the functor $\mathrm{H}^0 \circ D_A : \mathsf{M}_{\mathrm{f}/\mathbb{K}}(A, \mathrm{gr})^{\mathrm{op}} \to \mathsf{M}_{\mathrm{f}/\mathbb{K}}(A^{\mathrm{op}}, \mathrm{gr})$ is an equivalence, with quasi-inverse $\mathrm{H}^0 \circ D_{A^{\mathrm{op}}}$. For every $j \geq 1$ we have $A/\mathfrak{m}^j \in \mathsf{M}_{\mathrm{f}/\mathbb{K}}(A, \mathrm{gr})$. The duality provides an isomorphism

$$\mathrm{Hom}_{A^{\mathrm{op}}}(\mathbb{K}, \mathrm{H}^0(D_A(A/\mathfrak{m}^j))) \cong \mathrm{Hom}_A(A/\mathfrak{m}^j, \mathbb{K}) \cong \mathbb{K}.$$

This we rewrite as

$$\mathrm{Hom}_{A^{\mathrm{op}}}(\mathbb{K}, \mathrm{Ext}^0_A(A/\mathfrak{m}^j, R)) \cong \mathbb{K}.$$

By functoriality we can pass to the limit. Using Proposition 16.3.16(5) we obtain an isomorphism $\mathrm{Hom}_{A^{\mathrm{op}}}(\mathbb{K}, \mathrm{H}^0_{\mathfrak{m}}(R)) \cong \mathbb{K}$. I.e. the socle of the graded A^{op}-module $\mathrm{H}^0_{\mathfrak{m}}(R)$ is $\mathbb{K}$. By Proposition 16.3.12 there is an essential monomorphism $\psi : \mathrm{H}^0_{\mathfrak{m}}(R) \rightarrowtail A^*$ in $\mathsf{M}(A^{\mathrm{op}}, \mathrm{gr})$. But from the previous calculation we already know that $\mathrm{H}^0_{\mathfrak{m}}(R) \cong A^*$ in $\mathsf{M}(A, \mathrm{gr})$, and hence also in $\mathsf{M}(\mathbb{K}, \mathrm{gr})$.

As these are degreewise finite graded $\mathbb{K}$-modules, ψ must be bijective. We conclude that the isomorphism (17.3.3) holds.

((i) and (i′)) ⇒ (ii′): The same, by op-symmetry.

(ii) ⇒ (i): By Proposition 16.3.30 there is an isomorphism

$$\mathrm{RHom}_A(\mathbb{K}, R) \cong \mathrm{RHom}_A(\mathbb{K}, \mathrm{R}\Gamma_{\mathfrak{m}}(R))$$

in $\mathsf{D}(A^{\mathrm{en}}, \mathrm{gr})$. Now condition (ii) implies that $\mathrm{R}\Gamma_{\mathfrak{m}}(R) \cong A^*$ in $\mathsf{D}(A, \mathrm{gr})$. So we get

$$\mathrm{RHom}_A(\mathbb{K}, R) \cong \mathrm{RHom}_A(\mathbb{K}, A^*) \cong \mathrm{Hom}_A(\mathbb{K}, A^*) \cong \mathbb{K}$$

in $\mathsf{D}(\mathbb{K}, \mathrm{gr})$. This implies that $\mathrm{RHom}_A(\mathbb{K}, R) \cong \mathbb{K}$ in $\mathsf{D}(A^{\mathrm{en}}, \mathrm{gr})$.

(ii′) ⇒ (i′): The same, by op-symmetry. □

Definition 17.3.4 ([161]) Under Setup 17.2.1, a *prebalanced dualizing complex* over A is a graded NC dualizing complex R over A that satisfies the equivalent conditions of Lemma 17.3.1.

Here is the first of two existence theorems for balanced dualizing complexes. The second is Theorem 17.3.19.

Theorem 17.3.5 (Existence of BDC, [161]) *Under Setup* 17.2.1, *assume that A has a prebalanced dualizing complex. Then A has a balanced dualizing complex.*

The proof comes after the next lemma. Let $\mathsf{M}_{\mathrm{f}/\mathbb{K}}(A^{\mathrm{en}}, \mathrm{gr}) \subseteq \mathsf{M}(A^{\mathrm{en}}, \mathrm{gr})$ be the full subcategory on the bimodules that are finite over $\mathbb{K}$.

Lemma 17.3.6 *Suppose R is a prebalanced dualizing complex over A such that $\mathrm{R}\Gamma_{\mathfrak{m}}(R) \cong A^*$ in $\mathsf{D}(A^{\mathrm{en}}, \mathrm{gr})$. Then there is an isomorphism $D_A \cong \mathrm{Hom}_{\mathbb{K}}(-, \mathbb{K})$ of functors $\mathsf{M}_{\mathrm{f}/\mathbb{K}}(A^{\mathrm{en}}, \mathrm{gr})^{\mathrm{op}} \to \mathsf{M}_{\mathrm{f}/\mathbb{K}}(A^{\mathrm{en}}, \mathrm{gr})$.*

Proof Let us choose an isomorphism $\beta : \mathrm{R}\Gamma_{\mathfrak{m}}(R) \xrightarrow{\simeq} A^*$ in $\mathsf{D}(A^{\mathrm{en}}, \mathrm{gr})$. Take $M \in \mathsf{M}_{\mathrm{f}/\mathbb{K}}(A^{\mathrm{en}}, \mathrm{gr})$. There is a sequence of functorial isomorphisms

$$\begin{aligned} D_A(M) &= \mathrm{RHom}_A(M, R) \cong^{\dagger} \mathrm{RHom}_A(M, \mathrm{R}\Gamma_{\mathfrak{m}}(R)) \\ &\cong^{\ddagger} \mathrm{RHom}_A(M, A^*) \cong \mathrm{Hom}_{\mathbb{K}}(M, \mathbb{K}) \end{aligned}$$

in $\mathsf{D}(A^{\mathrm{en}}, \mathrm{gr})$. The isomorphism $\cong^{\dagger}$ is by Proposition 16.3.30, and the isomorphism $\cong^{\ddagger}$ comes from β. □

Proof of Theorem 17.3.5 The proof is in two steps.

Step 1. Let R' be a prebalanced dualizing complex over A. This means that there is an isomorphism

$$\mathrm{R}\Gamma_{\mathfrak{m}}(R') \cong A^* \otimes_A L \tag{17.3.7}$$

in $\mathsf{D}(A^{\mathrm{en}}, \mathrm{gr})$, for some graded invertible A-bimodule L generated in algebraic degree 0. Define $L^{\vee} := \mathrm{Hom}_A(L, A)$. Thus, if $L \cong A(\phi, i)$, then $L^{\vee} \cong A(\phi^{-1}, -i)$ in $\mathsf{M}(A^{\mathrm{en}}, \mathrm{gr})$.

Next let

$$R := R' \otimes_A L^{\vee} \in \mathsf{D}(A^{\mathrm{en}}, \mathrm{gr}), \tag{17.3.8}$$

which is a graded dualizing complex. It is clear from (17.3.7) that there is an isomorphism

$$\beta : \mathrm{R}\Gamma_{\mathfrak{m}}(R) \xrightarrow{\simeq} A^* \text{ in } \mathsf{D}(A^{\mathrm{en}}, \mathrm{gr}), \tag{17.3.9}$$

and this is the balancing isomorphism that we take. To complete the proof that (R, β) is a balanced dualizing complex, it remains to prove that

$$\mathrm{R}\Gamma_{\mathfrak{m}^{\mathrm{op}}}(R) \cong A^* \text{ in } \mathsf{D}(A^{\mathrm{en}}, \mathrm{gr}). \tag{17.3.10}$$

This will be done in the second step. (We cannot use Theorem 16.5.8 on symmetric derived torsion, since that would involve circular reasoning.)

Step 2. The graded dualizing complex R is prebalanced, by (17.3.9). So according to Lemma 17.3.1, there is an isomorphism

$$\mathrm{R}\Gamma_{\mathfrak{m}^{\mathrm{op}}}(R) \cong L^{\diamond} \otimes_A A^* \tag{17.3.11}$$

in $\mathsf{D}(A^{\mathrm{en}}, \mathrm{gr})$, for some invertible graded A-bimodule $L^{\diamond}$ generated in algebraic degree 0.

For every $j \geq 1$ the bimodule $A/\mathfrak{m}^j$ belongs to $\mathsf{M}_{\mathrm{f}/\mathbb{K}}(A^{\mathrm{en}}, \mathrm{gr})$, and it is $\mathfrak{m}$-torsion and $\mathfrak{m}^{\mathrm{op}}$-torsion. Therefore we get isomorphisms

$$\begin{aligned} D_{A^{\mathrm{op}}}(A/\mathfrak{m}^j) &= \mathrm{RHom}_{A^{\mathrm{op}}}(A/\mathfrak{m}^j, R) \\ &\cong^{(\mathrm{a})} \mathrm{RHom}_{A^{\mathrm{op}}}(A/\mathfrak{m}^j, \mathrm{R}\Gamma_{\mathfrak{m}^{\mathrm{op}}}(R)) \\ &\cong^{(\mathrm{b})} \mathrm{RHom}_{A^{\mathrm{op}}}(A/\mathfrak{m}^j, L^{\diamond} \otimes_A A^*) \\ &\cong L^{\diamond} \otimes_A \mathrm{RHom}_{A^{\mathrm{op}}}(A/\mathfrak{m}^j, A^*) \\ &\cong L^{\diamond} \otimes_A \mathrm{Hom}_{\mathbb{K}}(A/\mathfrak{m}^j, \mathbb{K}) = L^{\diamond} \otimes_A (A/\mathfrak{m}^j)^* \end{aligned} \tag{17.3.12}$$

in $\mathsf{D}(A^{\mathrm{en}}, \mathrm{gr})$. Explanation: the isomorphism $\cong^{(\mathrm{a})}$ is by Proposition 16.3.30 (transcribed to the ring A^{op}); and the isomorphism $\cong^{(\mathrm{b})}$ is from equation (17.3.11).

There are also isomorphisms

$$\begin{aligned} A/\mathfrak{m}^j &\cong^1 D_A(D_{A^{\mathrm{op}}}(A/\mathfrak{m}^j)) \cong^2 D_A(L^{\diamond} \otimes_A (A/\mathfrak{m}^j)^*) \\ &\cong^3 (L^{\diamond} \otimes_A (A/\mathfrak{m}^j)^*)^* \cong^4 (A/\mathfrak{m}^j) \otimes_A (L^{\diamond})^{\vee} \end{aligned} \tag{17.3.13}$$

in $\mathsf{M}(A^{\mathrm{en}}, \mathrm{gr})$, where $(L^{\diamond})^{\vee} := \mathrm{Hom}_A(L^{\diamond}, A)$. Here is how they arise: the isomorphism $\cong^1$ is by Theorem 17.1.6; the isomorphism $\cong^2$ is from equation (17.3.12); and the isomorphism $\cong^3$ is due to Lemma 17.3.6. For the isomorphism $\cong^4$, say $L^\diamond \cong A(\nu, k)$; then $(L^\diamond)^\vee \cong A(\nu^{-1}, -k)$ and $(L^\diamond)^* \cong A^*(\nu^{-1}, -k) = A^* \otimes_A A(\nu^{-1}, -k)$.

Because the isomorphisms (17.3.13) hold for every $j \geq 1$, it follows that $(L^\diamond)^\vee \cong A$ in $\mathbf{M}(A^{\mathrm{en}}, \mathrm{gr})$, and hence $L^\diamond \cong A$ in $\mathbf{M}(A^{\mathrm{en}}, \mathrm{gr})$. Then the isomorphism (17.3.11) becomes (17.3.10). □

Recall the AS Gorenstein graded rings from Definition 15.4.8. For an AS Gorenstein graded ring A there are isomorphisms

$$\mathrm{RHom}_A(\mathbb{K}, A) \cong \mathrm{RHom}_{A^{\mathrm{op}}}(\mathbb{K}, A) \cong \mathbb{K}(l)[-n]$$

in $\mathbf{D}(\mathbb{K}, \mathrm{gr})$. The numbers n and l are called the dimension and AS index of A, respectively.

Corollary 17.3.14 ([161]) *Let A be an AS Gorenstein graded ring, of dimension n and with AS index l. Then there is a unique automorphism ϕ of A in $\mathsf{Rng}_{\mathrm{gr}}/_{\mathrm{c}}\,\mathbb{K}$ such that the complex $R_A := A(\phi, -l)[n] \in \mathbf{D}(A^{\mathrm{en}}, \mathrm{gr})$ is a balanced dualizing complex over A.*

Proof The complex $R' := A(-l)[n] \in \mathbf{D}(A^{\mathrm{en}}, \mathrm{gr})$ is a prebalanced dualizing complex. Theorem 17.3.5 asserts that A has a balanced dualizing complex R_A. The construction of R_A in formula (17.3.8) in the proof of Theorem 17.3.5 shows that $R_A = R' \otimes_A A(\phi, 0) \cong A(\phi, -l)[n]$ for some automorphism ϕ of A.

The uniqueness of n, l and ϕ is a consequence of Theorems 17.2.4 and 17.1.10. □

Here are two important examples of AS regular graded rings and their balanced dualizing complexes.

Example 17.3.15 Suppose A is an *elliptic 3-dimensional Artin–Schelter regular graded ring*; see [8]. There is a normalizing regular element $g \in A_3$, and it defines an automorphism ϕ_g of A by the formula $\phi_g(a) \cdot g = g \cdot a$ for $a \in A$. There is a constant $\lambda \in \mathbb{K}^\times$ arising from the elliptic curve associated to A, and it defines an automorphism ϕ_λ of A by the formula $\phi_\lambda(a) := \lambda^i \cdot a$ for $a \in A_i$. According to [161, Theorem 7.18], the balanced dualizing complex of A is $R_A = A(\phi_g \circ \phi_\lambda, -3)[3]$.

Example 17.3.16 Let $A := \mathbb{K}[t_1, \ldots, t_n]$, the commutative polynomial ring in n variables, of algebraic degrees $\deg(t_i) = l_i \geq 1$. This is an Artin–Schelter regular graded ring, and a calculation with a Koszul complex shows that $\mathrm{RHom}_A(\mathbb{K}, A) \cong \mathbb{K}(l)[-n]$ in $\mathbf{D}(A^{\mathrm{en}}, \mathrm{gr})$, where $l := \sum_{i=1}^{n} l_i$. Because $\mathrm{R}\Gamma_{\mathfrak{m}}(A) \cong A^*(l)[-n]$ is a central A-bimodule, it follows that the balanced dualizing dualizing complex of A is $R_A := A(-l)[n]$.

If we want to tie this with the commutative theory from Chapter 13, and to algebraic geometry, then we should take $R_A := \Omega^n_{A/\mathbb{K}}[n]$, where $\Omega^n_{A/\mathbb{K}}$ is the graded-free A-module generated by the differential form $\mathrm{d}(t_1) \wedge \cdots \wedge \mathrm{d}(t_n)$ that has algebraic degree l.

Now an example that relies on Example 17.3.16.

Example 17.3.17 Suppose B is a commutative noetherian connected graded $\mathbb{K}$-ring. We can find a finite homomorphism $A \to B$ in $\mathsf{Rng}_{\mathrm{gr}}/_{\mathrm{c}}\,\mathbb{K}$ from a commutative polynomial ring A. Let $R_A := A(-l)[n]$ be the balanced dualizing complex of A, as in the previous example. Define $R_B := \mathrm{RHom}_A(B, R_A) \in \mathbf{D}(B^{\mathrm{en}}, \mathrm{gr})$, where we use a commutative graded-injective resolution $R_A \to I$, so that R_B is a complex of central graded B-modules. By the arguments of Proposition 13.1.28, R_B is graded NC dualizing complex over B. A calculation that we won't perform (but which is an easy variant of the proof of Theorem 17.4.22 on the balanced trace morphism) shows that R_B is balanced.

The moral is that the commutative algebraically graded duality theory embeds within the noncommutative algebraically graded duality theory.

The next theorem is an important converse to Corollary 17.2.13.

Recall that for a graded NC dualizing complex R, the derived homothety morphism through A^{op}, namely the morphism $\mathrm{hm}^{\mathrm{R}}_{R,A^{\mathrm{op}}} : A \to \mathrm{RHom}_A(R, R)$ in $\mathbf{D}(A^{\mathrm{en}}, \mathrm{gr})$, is an isomorphism. See condition (iii) of Definition 17.1.4. By Van den Bergh's Local Duality (Corollary 16.4.11), for every complex $M \in \mathbf{D}(A^{\mathrm{en}}, \mathrm{gr})$ there is an isomorphism

$$\delta_M : \mathrm{RHom}_A(M, (P_A)^*) \xrightarrow{\simeq} \mathrm{R}\Gamma_{\mathfrak{m}}(M)^* \tag{17.3.18}$$

in $\mathbf{D}(A^{\mathrm{en}}, \mathrm{gr})$. Here $P_A = \mathrm{R}\Gamma_{\mathfrak{m}}(A) \in \mathbf{M}(A^{\mathrm{en}}, \mathrm{gr})$ is the dedualizing complex of A.

Theorem 17.3.19 (Existence of BDC, Van den Bergh [153]) *Under Setup* 17.2.1, *assume the ring A satisfies the special χ condition, and it has finite local cohomological dimension. Then A has a balanced dualizing complex (R_A, β_A).*

More explicitly, the complex $R_A := (P_A)^* \in \mathbf{D}(A^{\mathrm{en}}, \mathrm{gr})$ is a graded dualizing complex over A with symmetric derived $\mathfrak{m}$-torsion. It has a balancing isomorphism $\beta_A : \mathrm{R}\Gamma_{\mathfrak{m}}(R_A) \xrightarrow{\simeq} A^*$, which is the unique isomorphism in $\mathbf{D}(A^{\mathrm{en}}, \mathrm{gr})$ *such that the diagram*

$$A \xrightarrow[\mathrm{hm}^{\mathrm{R}}_{R,A^{\mathrm{op}}}]{\cong} \mathrm{RHom}_A(R_A, R_A) \xrightarrow[\delta_{R_A}]{\cong} \mathrm{R}\Gamma_{\mathfrak{m}}(R_A)^*, \qquad A \xrightarrow{(\beta_A)^*} \mathrm{R}\Gamma_{\mathfrak{m}}(R_A)^*$$

in $\mathsf{D}(A^{\mathrm{en}}, \mathrm{gr})$ *is commutative.*

Proof The proof proceeds in five steps.

Step 1. We are given that A has finite local cohomological dimension and it satisfies the special χ condition. By Proposition 16.5.19, the bimodule A has weakly symmetric derived $\mathfrak{m}$-torsion. Theorem 16.5.8 says that the bimodule A has symmetric derived $\mathfrak{m}$-torsion; namely that there is an isomorphism

$$\epsilon_A : P_A = \mathrm{R}\Gamma_{\mathfrak{m}}(A) \xrightarrow{\simeq} \mathrm{R}\Gamma_{\mathfrak{m}^{\mathrm{op}}}(A) \tag{17.3.20}$$

in $\mathsf{D}^{\mathrm{b}}(A^{\mathrm{en}}, \mathrm{gr})$.

The left χ condition for A, with Proposition 16.5.17, tell us that the complex $\mathrm{R}\Gamma_{\mathfrak{m}}(A)$ has cofinite cohomology modules over A. By the same token, the left χ condition for A^{op} tells us that the complex $\mathrm{R}\Gamma_{\mathfrak{m}^{\mathrm{op}}}(A)$ has cofinite cohomology modules over A^{op}. Using the isomorphism (17.3.20) we conclude that $P_A \in \mathsf{D}^{\mathrm{b}}_{(\mathrm{cof},\mathrm{cof})}(A^{\mathrm{en}}, \mathrm{gr})$. Now Graded Matlis Duality (Theorem 15.2.34) says that the complex R_A satisfies

$$R_A = (P_A)^* \in \mathsf{D}^{\mathrm{b}}_{(\mathrm{f},\mathrm{f})}(A^{\mathrm{en}}, \mathrm{gr}). \tag{17.3.21}$$

This is condition (i) of Definition 17.1.4.

Step 2. Because the functor $\mathrm{R}\Gamma_{\mathfrak{m}}$ has finite cohomological dimension, Van den Bergh's Local Duality (Corollary 16.4.11) says that the functor $\mathrm{RHom}_A(-, R) \cong \mathrm{R}\Gamma_{\mathfrak{m}}(-)^*$ also has finite cohomological dimension. This means that the complex R_A has finite graded-injective dimension over A. Similarly, because $\mathrm{R}\Gamma_{\mathfrak{m}^{\mathrm{op}}}$ has finite cohomological dimension, the complex R_A has finite graded-injective dimension over A^{op}. This is condition (ii) of Definition 17.1.4.

Step 3. Now we shall prove that R_A has the derived Morita property on the A side. We have these isomorphisms in the category $\mathsf{D}(A^{\mathrm{en}}, \mathrm{gr})$:

$$\begin{aligned} \mathrm{RHom}_A(R_A, R_A) &= \mathrm{RHom}_A(\mathrm{R}\Gamma_{\mathfrak{m}}(A)^*, \mathrm{R}\Gamma_{\mathfrak{m}}(A)^*) \\ &\cong^1 \mathrm{RHom}_{A^{\mathrm{op}}}(\mathrm{R}\Gamma_{\mathfrak{m}}(A), \mathrm{R}\Gamma_{\mathfrak{m}}(A)) \\ &\cong^2 \mathrm{RHom}_{A^{\mathrm{op}}}(\mathrm{R}\Gamma_{\mathfrak{m}^{\mathrm{op}}}(A), \mathrm{R}\Gamma_{\mathfrak{m}^{\mathrm{op}}}(A)) \\ &\cong^3 \mathrm{RHom}_{A^{\mathrm{op}}}(\mathrm{R}\Gamma_{\mathfrak{m}^{\mathrm{op}}}(A), A) \\ &\cong^4 \mathrm{RHom}_A(A^*, \mathrm{R}\Gamma_{\mathfrak{m}^{\mathrm{op}}}(A)^*) \\ &\cong^5 \mathrm{RHom}_A(A^*, \mathrm{R}\Gamma_{\mathfrak{m}}(A)^*) \\ &\cong^6 (A^*)^* \cong^7 A\,. \end{aligned} \tag{17.3.22}$$

The isomorphisms $\cong^1$, $\cong^4$ and $\cong^7$ are from graded Matlis duality (Theorem 15.2.34); the isomorphisms $\cong^2$ and $\cong^5$ are by the symmetry isomorphism ϵ_A; the isomorphism $\cong^3$ is due to Proposition 16.3.30; and the isomorphism $\cong^6$ is due to Van den Bergh's Local Duality (Corollary 16.4.11). These isomorphisms respect the derived homothety morphisms through A^{op} (in lines 1, 5 and 6), the derived homothety morphisms through A (in lines 2, 3 and 4); and the canonical isomorphisms $A \to (A^*)^*$ and $A \to A$ in line 7. (Compare to the proof of Theorem 18.1.25, and the use of Lemma 18.1.21 in it.) The conclusion is that the derived homothety morphism $\mathrm{hm}^{\mathrm{R}}_{R,A^{\mathrm{op}}} : A \to \mathrm{RHom}_A(R_A, R_A)$ is an isomorphism. This establishes the derived Morita property on the A side for R_A.

Step 4. Replacing A with A^{op}, using the symmetry isomorphism ϵ_A, the same calculation as in step 3 shows that R_A has the derived Morita property on the A^{op} side. Thus condition (iii) of Definition 17.1.4 is verified. We conclude that R_A is a graded NC dualizing complex.

Step 5. In this step we prove that R_A is balanced. By formula (17.3.21), the derived Morita property on the A side, and Van den Bergh's Local Duality (Corollary 16.4.11), there are isomorphisms

$$A \xrightarrow[\cong]{\mathrm{hm}^{\mathrm{R}}_{R_A,A^{\mathrm{op}}}} \mathrm{RHom}_A(R_A, R_A) \xrightarrow[\cong]{\delta_{R_A}} (\mathrm{R}\Gamma_{\mathfrak{m}}(R_A))^* \tag{17.3.23}$$

in $\mathbf{D}(A^{\mathrm{en}}, \mathrm{gr})$. See the isomorphism (17.3.18) with $M := R_A = (P_A)^*$. After we apply the functor $(-)^*$ to (17.3.23), Graded Matlis Duality (Theorem 15.2.34) says that the morphism

$$\beta := (\delta_{R_A} \circ \mathrm{hm}^{\mathrm{R}}_{R_A,A^{\mathrm{op}}})^* : \Gamma_{\mathfrak{m}}(R_A) \to A^*$$

in $\mathbf{D}(A^{\mathrm{en}}, \mathrm{gr})$ is an isomorphism. This is a balancing isomorphism for R_A, so item (B2) of Definition 17.2.2 is satisfied.

Similarly, the Derived Morita property on the A^{op} side, Van den Bergh's Local Duality for A^{op}, and the symmetry isomorphism ϵ_A give us these isomorphisms

$$A \xrightarrow[\cong]{\mathrm{hm}^{\mathrm{R}}_{R_A,A}} \mathrm{RHom}_{A^{\mathrm{op}}}(R_A, R_A) \xrightarrow[\cong]{\delta_{R_A}} (\mathrm{R}\Gamma_{\mathfrak{m}^{\mathrm{op}}}(R_A))^*$$

in $\mathbf{D}(A^{\mathrm{en}}, \mathrm{gr})$. By Graded Matlis Duality, this yields an isomorphism $\mathrm{R}\Gamma_{\mathfrak{m}^{\mathrm{op}}}(R_A) \cong A^*$. We see that R_A has weakly symmetric derived torsion; and hence, by Theorem 16.5.8, R_A has symmetric derived torsion. This is condition (B1) of Definition 17.2.2. □

Corollary 17.3.24 (Two Equivalent Properties) *Under Setup* 17.2.1, *the two properties below are equivalent*:

(i) *The graded ring A satisfies the χ condition, and it has finite local cohomological dimension.*
(ii) *The graded ring A has a balanced dualizing complex.*

Proof The implication (i) ⇒ (ii) is the Van den Bergh Existence Theorem 17.3.19. The reverse implication is Corollary 17.2.13. □

Here are several other consequences of Theorem 17.3.19. Recall that a dualizing complex R_A over A gives rise to the associated duality functors

$$D_A, D_{A^{\mathrm{op}}} : \mathbf{D}(A^{\mathrm{en}}, \mathrm{gr})^{\mathrm{op}} \to \mathbf{D}(A^{\mathrm{en}}, \mathrm{gr}), \tag{17.3.25}$$

namely $D_A = \mathrm{RHom}_A(-, R_A)$ and $D_{A^{\mathrm{op}}} = \mathrm{RHom}_{A^{\mathrm{op}}}(-, R_A)$.

Theorem 17.3.26 *Under Setup* 17.2.1, *let R_A be a balanced dualizing complex over A, with associated duality functors D_A and $D_{A^{\mathrm{op}}}$.*

(1) *There is an isomorphism $D_A \cong D_{A^{\mathrm{op}}}$ between the triangulated functors in* (17.3.25).
(2) *If $M \in \mathbf{D}^{\mathrm{b}}_{(\mathrm{f,f})}(A^{\mathrm{en}}, \mathrm{gr})$ then $D_A(M)$ belongs to $\mathbf{D}^{\mathrm{b}}_{(\mathrm{f,f})}(A^{\mathrm{en}}, \mathrm{gr})$.*
(3) *The functor*

$$D_A : \mathbf{D}^{\mathrm{b}}_{(\mathrm{f,f})}(A^{\mathrm{en}}, \mathrm{gr})^{\mathrm{op}} \to \mathbf{D}^{\mathrm{b}}_{(\mathrm{f,f})}(A^{\mathrm{en}}, \mathrm{gr})$$

is an equivalence of triangulated categories, and it is its own quasi-inverse.

Proof

(1) By Theorems 17.3.19 and 17.2.4 there is a canonical isomorphism $R_A \cong (P_A)^*$ in $\mathbf{D}(A^{\mathrm{en}}, \mathrm{gr})$.

Take a complex $M \in \mathbf{D}^{\mathrm{b}}_{(\mathrm{f,f})}(A^{\mathrm{en}}, \mathrm{gr})$. We know that A has the χ condition and finite local cohomological dimension. By Proposition 16.5.19 the complex M has symmetric derived torsion. This means that there's an isomorphism $\epsilon_M : \mathrm{R}\Gamma_{\mathfrak{m}}(M) \xrightarrow{\simeq} \mathrm{R}\Gamma_{\mathfrak{m}^{\mathrm{op}}}(M)$ in $\mathbf{D}(A^{\mathrm{en}}, \mathrm{gr})$. By Theorem 16.4.3 and its opposite version (namely the theorem applied to the ring A^{op} instead of A), this translates to an isomorphism

$$P_A \otimes^{\mathrm{L}}_A M \cong M \otimes^{\mathrm{L}}_A P_A \tag{17.3.27}$$

in $\mathbf{D}(A^{\mathrm{en}}, \mathrm{gr})$. Applying $(-)^*$ to (17.3.27) we obtain an isomorphism

$$(P_A \otimes^{\mathrm{L}}_A M)^* \cong (M \otimes^{\mathrm{L}}_A P_A)^* \tag{17.3.28}$$

in $\mathbf{D}(A^{\mathrm{en}}, \mathrm{gr})$. Now by Hom-tensor adjunction we get isomorphisms

$$\begin{aligned} (P_A \otimes^{\mathrm{L}}_A M)^* &= \mathrm{Hom}_{\mathbb{K}}(P_A \otimes^{\mathrm{L}}_A M, \mathbb{K}) \\ &\cong \mathrm{RHom}_A(M, \mathrm{RHom}_{\mathbb{K}}(P_A, \mathbb{K})) \\ &\cong \mathrm{RHom}_A(M, (P_A)^*) \cong \mathrm{RHom}_A(M, R_A) = D_A(M). \end{aligned} \tag{17.3.29}$$

Similarly there are isomorphisms

$$\begin{aligned} (M \otimes^{\mathrm{L}}_A P_A)^* &= \mathrm{Hom}_{\mathbb{K}}(M \otimes^{\mathrm{L}}_A P_A, \mathbb{K}) \\ &\cong \mathrm{RHom}_{A^{\mathrm{op}}}(M, \mathrm{RHom}_{\mathbb{K}}(P_A, \mathbb{K})) \\ &\cong \mathrm{RHom}_{A^{\mathrm{op}}}(M, (P_A)^*) \cong \mathrm{RHom}_{A^{\mathrm{op}}}(M, R_A) = D_{A^{\mathrm{op}}}(M). \end{aligned} \tag{17.3.30}$$

All these isomorphisms are functorial in the argument M. By combining the isomorphisms (17.3.28), (17.3.29) and (17.3.30) we get an isomorphism of functors $D_A \cong D_{A^{\mathrm{op}}}$.

(2) By Theorem 17.1.6 we know that $D_A(M)$ belongs to $\mathbf{D}^{\mathrm{b}}_{(..,\mathrm{f})}(A^{\mathrm{en}}, \mathrm{gr})$. The same theorem, applied to the ring A^{op}, tells us that $D_{A^{\mathrm{op}}}(M)$ belongs to $\mathbf{D}^{\mathrm{b}}_{(\mathrm{f},..)}(A^{\mathrm{en}}, \mathrm{gr})$. Thus $D_A(M) \cong D_{A^{\mathrm{op}}}(M) \in \mathbf{D}^{\mathrm{b}}_{(\mathrm{f},\mathrm{f})}(A^{\mathrm{en}}, \mathrm{gr})$.

(3) We already know, by Theorem 17.1.6, that $D_{A^{\mathrm{op}}} \circ D_A \cong \mathrm{Id}$ as triangulated functors from $\mathbf{D}^{\mathrm{b}}_{(\mathrm{f},\mathrm{f})}(A^{\mathrm{en}}, \mathrm{gr})$ to itself. By item (2) the essential image of D_A is $\mathbf{D}^{\mathrm{b}}_{(\mathrm{f},\mathrm{f})}(A^{\mathrm{en}}, \mathrm{gr})$. And by item (1) the functors D_A and $D_{A^{\mathrm{op}}}$ are isomorphic on this triangulated category. □

Lemma 17.3.31 *Under Setup* 17.2.1, *assume A satisfies the* χ *condition and has finite local cohomological dimension. Then there are inclusions* $\mathbf{D}_{\mathrm{f}}(A, \mathrm{gr}) \subseteq \mathbf{D}(A, \mathrm{gr})_{\mathrm{com}}$ *and* $\mathbf{D}_{\mathrm{cof}}(A, \mathrm{gr}) \subseteq \mathbf{D}(A, \mathrm{gr})_{\mathrm{tor}}$.

Proof The inclusion $\mathbf{D}_{\mathrm{cof}}(A, \mathrm{gr}) \subseteq \mathbf{D}_{\mathrm{tor}}(A, \mathrm{gr})$ is trivial. And Proposition 16.3.29 says that $\mathbf{D}_{\mathrm{tor}}(A, \mathrm{gr}) = \mathbf{D}(A, \mathrm{gr})_{\mathrm{tor}}$.

Now to the complete complexes. Take a complex $M \in \mathbf{D}_{\mathrm{f}}(A, \mathrm{gr})$. Then there are isomorphisms

$$\begin{aligned}
\mathrm{ADC}_{\mathfrak{m}}(M) &= \mathrm{RHom}_A(P_A, M) \\
&\cong^1 \mathrm{RHom}_{A^{\mathrm{op}}}(M^*, (P_A)^*) = \mathrm{RHom}_{A^{\mathrm{op}}}(M^*, R_A) \\
&\cong^2 \mathrm{RHom}_{A^{\mathrm{op}}}(M^*, \mathrm{R}\Gamma_{\mathfrak{m}^{\mathrm{op}}}(R_A)) \\
&\cong^3 \mathrm{RHom}_{A^{\mathrm{op}}}(M^*, A^*) \cong (M^*)^* \cong^4 M
\end{aligned} \tag{17.3.32}$$

in $\mathbf{D}(A, \mathrm{gr})$. The isomorphism $\cong^1$ is due to Theorem 15.3.32, which applies because $M, P_A \in \mathbf{C}_{\mathrm{dwf}}(A, \mathrm{gr})$. The isomorphism $\cong^2$ comes from Proposition 16.3.30, and it applies since $M^* \in \mathbf{D}_{\mathrm{cof}}(A^{\mathrm{op}}, \mathrm{gr}) \subseteq \mathbf{D}_{\mathrm{tor}}(A^{\mathrm{op}}, \mathrm{gr})$ and $R_A \in \mathbf{D}^{\mathrm{b}}(A^{\mathrm{op}}, \mathrm{gr})$. The isomorphism $\cong^3$ is because R_A is a balanced dualizing complex. The isomorphism $\cong^4$ is by Proposition 15.3.30, which applies because $M \in \mathbf{C}_{\mathrm{dwf}}(A, \mathrm{gr})$. We see that M is in the essential image of the functor $\mathrm{ADC}_{\mathfrak{m}}$. According to Theorem 16.6.32(1), it follows that $M \in \mathbf{D}^{\mathrm{b}}(A, \mathrm{gr})_{\mathrm{com}}$. □

The next theorem is a refinement of the NC graded MGM Equivalence (Theorem 16.6.32). Given another graded ring B, there are triangulated functors

$$\mathrm{R}\Gamma_{\mathfrak{m}}, \mathrm{ADC}_{\mathfrak{m}} : \mathbf{D}(A \otimes B^{\mathrm{op}}, \mathrm{gr}) \to \mathbf{D}(A \otimes B^{\mathrm{op}}, \mathrm{gr}).$$

They are the derived $\mathfrak{m}$-torsion and the abstract derived $\mathfrak{m}$-adic completion, respectively.

Theorem 17.3.33 *Under Setup* 17.2.1, *assume A satisfies the χ condition and has finite local cohomological dimension. Let B be some graded ring.*

(1) *If* $M \in \mathbf{D}^{\mathrm{b}}_{(\mathrm{f},..)}(A \otimes B^{\mathrm{op}}, \mathrm{gr})$ *then* $\mathrm{R}\Gamma_{\mathfrak{m}}(M) \in \mathbf{D}^{\mathrm{b}}_{(\mathrm{cof},..)}(A \otimes B^{\mathrm{op}}, \mathrm{gr})$.

(2) *For* $N \in \mathbf{D}^{\mathrm{b}}_{(\mathrm{cof},..)}(A \otimes B^{\mathrm{op}}, \mathrm{gr})$ *there is an isomorphism* $\mathrm{ADC}_{\mathfrak{m}}(N) \cong D_{A^{\mathrm{op}}}(N^*)$ *in* $\mathbf{D}(A \otimes B^{\mathrm{op}}, \mathrm{gr})$, *and it is functorial in N.*

(3) *If* $N \in \mathbf{D}^{\mathrm{b}}_{(\mathrm{cof},..)}(A \otimes B^{\mathrm{op}}, \mathrm{gr})$ *then* $\mathrm{ADC}_{\mathfrak{m}}(N) \in \mathbf{D}^{\mathrm{b}}_{(\mathrm{f},..)}(A \otimes B^{\mathrm{op}}, \mathrm{gr})$.

(4) *The functor*

$$\mathrm{R}\Gamma_{\mathfrak{m}} : \mathbf{D}^{\mathrm{b}}_{(\mathrm{f},..)}(A \otimes B^{\mathrm{op}}, \mathrm{gr}) \to \mathbf{D}^{\mathrm{b}}_{(\mathrm{cof},..)}(A \otimes B^{\mathrm{op}}, \mathrm{gr})$$

is an equivalence of triangulated categories, with quasi-inverse $\mathrm{ADC}_{\mathfrak{m}}$.

Proof

(1) The ring B is irrelevant, so we may omit it. The χ condition and finiteness of local cohomology imply that $\mathrm{R}\Gamma_{\mathfrak{m}}(M) \in \mathbf{D}^{\mathrm{b}}_{\mathrm{cof}}(A, \mathrm{gr})$.

(2) Take a complex $N \in \mathbf{D}^{\mathrm{b}}_{\mathrm{cof}}(A \otimes B^{\mathrm{op}}, \mathrm{gr})$. We have these isomorphisms

$$\begin{aligned}\mathrm{ADC}_{\mathfrak{m}}(N) &= \mathrm{RHom}_{A}(P_A, N) \cong^{\dagger} \mathrm{RHom}_{A^{\mathrm{op}}}(N^*, (P_A)^*) \\ &= \mathrm{RHom}_{A^{\mathrm{op}}}(N^*, R_A) = D_{A^{\mathrm{op}}}(N^*)\end{aligned}$$

in $\mathbf{D}(A \otimes B^{\mathrm{op}}, \mathrm{gr})$. The isomorphism $\cong^{\dagger}$ is due to Theorem 15.3.32, which applies because $N \in \mathbf{C}_{\mathrm{dwf}}(A \otimes B^{\mathrm{op}}, \mathrm{gr})$ and $P_A \in \mathbf{C}_{\mathrm{dwf}}(A^{\mathrm{en}}, \mathrm{gr})$.

(3) Again we can forget about B. Take $N \in \mathbf{D}^{\mathrm{b}}_{\mathrm{cof}}(A, \mathrm{gr})$. Then $N^* \in \mathbf{D}_{\mathrm{f}}(A^{\mathrm{op}}, \mathrm{gr})$, so by Theorem 17.1.6 we have $D_{A^{\mathrm{op}}}(N^*) \in \mathbf{D}^{\mathrm{b}}_{\mathrm{f}}(A, \mathrm{gr})$. But by item (2) we know that $\mathrm{ADC}_{\mathfrak{m}}(N) \cong D_{A^{\mathrm{op}}}(N^*)$.

(4) Items (1) and (3) tell us that the functors $\mathrm{R}\Gamma_{\mathfrak{m}}$ and $\mathrm{ADC}_{\mathfrak{m}}$ send the categories $\mathbf{D}^{\mathrm{b}}_{(\mathrm{f},..)}(A \otimes B^{\mathrm{op}}, \mathrm{gr})$ and $\mathbf{D}^{\mathrm{b}}_{(\mathrm{cof},..)}(A \otimes B^{\mathrm{op}}, \mathrm{gr})$ to each other, respectively. Lemma 17.3.31 says that the NC MGM Equivalence (Theorem 16.6.32) applies here, namely that there is an isomorphism of triangulated functors $\mathrm{ADC}_{\mathfrak{m}} \circ \mathrm{R}\Gamma_{\mathfrak{m}} \cong \mathrm{Id}$ from $\mathbf{D}^{\mathrm{b}}_{(\mathrm{f},..)}(A \otimes B^{\mathrm{op}}, \mathrm{gr})$ to itself, and an isomorphism of triangulated functors $\mathrm{R}\Gamma_{\mathfrak{m}} \circ \mathrm{ADC}_{\mathfrak{m}} \cong \mathrm{Id}$ from $\mathbf{D}^{\mathrm{b}}_{(\mathrm{cof},..)}(A \otimes B^{\mathrm{op}}, \mathrm{gr})$ to itself. □

In Theorem 16.5.27 we saw that there is an isomorphism $\epsilon : \mathrm{R}\Gamma_{\mathfrak{m}} \xrightarrow{\simeq} \mathrm{R}\Gamma_{\mathfrak{m}^{\mathrm{op}}}$ of triangulated functors $\mathbf{D}^{\mathrm{b}}_{(\mathrm{f},\mathrm{f})}(A^{\mathrm{en}}) \to \mathbf{D}(A^{\mathrm{en}}, \mathrm{gr})$. The next corollary produces a similar isomorphism for the abstract derived completion functors

$$\mathrm{ADC}_{\mathfrak{m}}, \mathrm{ADC}_{\mathfrak{m}^{\mathrm{op}}} : \mathbf{D}(A^{\mathrm{en}}, \mathrm{gr}) \to \mathbf{D}(A^{\mathrm{en}}, \mathrm{gr}), \tag{17.3.34}$$

where we define $\mathrm{ADC}_{\mathfrak{m}^{\mathrm{op}}} := \mathrm{RHom}_{A^{\mathrm{op}}}(P_A, -)$.

Corollary 17.3.35 *Under Setup* 17.2.1, *assume A satisfies the χ condition and has finite local cohomological dimension. Then there is an isomorphism γ : $\mathrm{ADC}_{\mathfrak{m}} \xrightarrow{\simeq} \mathrm{ADC}_{\mathfrak{m}^{\mathrm{op}}}$ of triangulated functors $\mathbf{D}^{\mathrm{b}}_{(\mathrm{cof,cof})}(A^{\mathrm{en}}, \mathrm{gr}) \to \mathbf{D}(A^{\mathrm{en}}, \mathrm{gr})$.*

Proof Take a complex $N \in \mathbf{D}^{\mathrm{b}}_{(\mathrm{cof,cof})}(A^{\mathrm{en}}, \mathrm{gr})$. By Theorem 17.3.33(2), there is an isomorphism $\mathrm{ADC}_{\mathfrak{m}}(N) \cong D_{A^{\mathrm{op}}}(N^*)$ in $\mathbf{D}^{\mathrm{b}}(A^{\mathrm{en}}, \mathrm{gr})$. Similarly (replacing A with A^{op}) there is an isomorphism $\mathrm{ADC}_{\mathfrak{m}^{\mathrm{op}}}(N) \cong D_A(N^*)$ in $\mathbf{D}^{\mathrm{b}}(A^{\mathrm{en}}, \mathrm{gr})$. Now $N^* \in \mathbf{D}^{\mathrm{b}}_{(\mathrm{f,f})}(A^{\mathrm{en}}, \mathrm{gr})$, so by Theorem 17.3.26 we have $D_{A^{\mathrm{op}}}(N^*) \cong D_A(N^*)$. Since we relied on functorial isomorphisms, this is also a functorial isomorphism. □

Corollary 17.3.36 *Assume A satisfies the χ condition and has finite local cohomological dimension. Then the functor*

$$\mathrm{R}\Gamma_{\mathfrak{m}} : \mathbf{D}^{\mathrm{b}}_{(\mathrm{f,f})}(A^{\mathrm{en}}, \mathrm{gr}) \to \mathbf{D}^{\mathrm{b}}_{(\mathrm{cof,cof})}(A^{\mathrm{en}}, \mathrm{gr})$$

is an equivalence of triangulated categories, with quasi-inverse $\mathrm{ADC}_{\mathfrak{m}}$.

Proof Taking $B := A$ in Theorem 17.3.33(4) we get an equivalence

$$\mathrm{R}\Gamma_{\mathfrak{m}} : \mathbf{D}^{\mathrm{b}}_{(\mathrm{f},..)}(A^{\mathrm{en}}, \mathrm{gr}) \to \mathbf{D}^{\mathrm{b}}_{(\mathrm{cof},..)}(A^{\mathrm{en}}, \mathrm{gr}), \tag{17.3.37}$$

with quasi-inverse $\mathrm{ADC}_{\mathfrak{m}}$. From the same theorem, but switching A and A^{op}, we get an equivalence

$$\mathrm{R}\Gamma_{\mathfrak{m}^{\mathrm{op}}} : \mathbf{D}^{\mathrm{b}}_{(..,\mathrm{f})}(A^{\mathrm{en}}, \mathrm{gr}) \to \mathbf{D}^{\mathrm{b}}_{(..,\mathrm{cof})}(A^{\mathrm{en}}, \mathrm{gr}),$$

with quasi-inverse $\mathrm{ADC}_{\mathfrak{m}^{\mathrm{op}}}$. According to Theorem 16.5.27, there is an isomorphism $\mathrm{R}\Gamma_{\mathfrak{m}}(M) \cong \mathrm{R}\Gamma_{\mathfrak{m}^{\mathrm{op}}}(M)$ for $M \in \mathbf{D}^{\mathrm{b}}_{(\mathrm{f,f})}(A^{\mathrm{en}}, \mathrm{gr})$. Therefore, for such a complex M we have $\mathrm{R}\Gamma_{\mathfrak{m}}(M) \in \mathbf{D}^{\mathrm{b}}_{(\mathrm{cof,cof})}(A^{\mathrm{en}}, \mathrm{gr})$. Corollary 17.3.35 says that $\mathrm{ADC}_{\mathfrak{m}}(N) \cong \mathrm{ADC}_{\mathfrak{m}^{\mathrm{op}}}(N)$ for a complex $N \in \mathbf{D}^{\mathrm{b}}_{(\mathrm{cof,cof})}(A^{\mathrm{en}}, \mathrm{gr})$; so for such a complex N we have $\mathrm{ADC}_{\mathfrak{m}}(N) \in \mathbf{D}^{\mathrm{b}}_{(\mathrm{f,f})}(A^{\mathrm{en}}, \mathrm{gr})$. The conclusion is that the equivalence (17.3.37) restricts to the equivalence stated in the corollary. □

Question 17.3.38 A consequence of Theorem 16.4.3 is that the complex P_A has finite flat dimension over A. Is it true that the complex P_A has finite projective dimension over A? In other words, does the functor

$$\mathrm{ADC}_{\mathfrak{m}} = \mathrm{RHom}_A(P_A, -) : \mathbf{D}(A, \mathrm{gr}) \to \mathbf{D}(A, \mathrm{gr})$$

have finite cohomological dimension?

In the commutative weakly proregular case this is true – see Examples 12.4.6 and 16.6.33. Indeed, if $\mathfrak{a}$ is a WPR ideal in a commutative ring A, then the cohomological dimension of the functor $\mathrm{ADC}_{\mathfrak{a}} \cong \mathrm{L}\Lambda_{\mathfrak{a}}$ is at most the length of a finite generating sequence of $\mathfrak{a}$.

Remark 17.3.39 The proof of Theorem 17.3.19 provided here is not the original proof from [153]. Our proof, which relies on the NC MGM Equivalence, seems to allow a generalization to the case of a base ring $\mathbb{K}$ that is not a field (as long as the ring A is flat over $\mathbb{K}$). See [157] for some results in this direction.

Remark 17.3.40 Q.S. Wu and J.J. Zhang [159], [160] studied prebalanced dualizing complexes over complete semilocal noncommutative rings that contain a field. Some of their technical results are included in Section 17.4 below.

An ongoing project of R. Vyas and A. Yekutieli (a continuation of [157]) aims to study balanced dualizing complexes over complete semilocal noncommutative rings in the arithmetic setting, namely without the presence of a base field. A prototypical example is the ring $A = \mathbb{K}[[G]]/I$, where $\mathbb{K} = \widehat{\mathbb{Z}}_p$, G is a compact p-adic Lie group, $\mathbb{K}[[G]]$ is the noncommutative Iwasawa algebra, and $I \subseteq \mathbb{K}[[G]]$ is some two-sided ideal. Because the ring A could fail to be flat over $\mathbb{K}$, the balanced dualizing complex of A will be an object of the *derived category of bimodules* $\mathsf{D}(\tilde{A}^{\mathrm{en}})$, where $\tilde{A} \to A$ is a K-flat DG ring resolution of A and $\tilde{A}^{\mathrm{en}} := \tilde{A} \otimes_{\mathbb{K}} \tilde{A}^{\mathrm{op}}$; see [180].

Remark 17.3.41 Suppose A is a commutative noetherian complete local ring, with maximal ideal $\mathfrak{m}$. In the paper [1] the authors discuss two kinds of dualizing complexes over A: the usual dualizing complexes (see Section 13.1), which they call *c-dualizing complexes*, which have finite (and thus complete) cohomology modules; and the *t-dualizing complexes*, which have cofinite (and thus torsion) cohomology modules. The torsion injective module $I = A^*$ (the injective hull of the residue field) is a t-dualizing complex in their sense. One of the results of [1] is that the MGM Equivalence (which they, in that early paper, called *GM Duality*) exchanges t-dualizing complexes and c-dualizing complexes. In particular, by applying the derived $\mathfrak{m}$-adic completion functor $\mathrm{L}\Lambda_{\mathfrak{m}}$ one obtains a c-dualizing complex $R := \mathrm{L}\Lambda_{\mathfrak{m}}(A^*)$.

It was observed by Vyas that the Van den Bergh Existence Theorem (Theorem 17.3.19) can be understood as a noncommutative variant of that very same result. In our noncommutative graded setting we must replace the true derived $\mathfrak{m}$-adic completion functor $\mathrm{L}\Lambda_{\mathfrak{m}}$ with its abstract avatar $\mathrm{ADC}_{\mathfrak{m}}$. The graded A-bimodule A^* is a t-dualizing complex (if we adjust the definition from [1] to the NC connected graded setting), and indeed, as Theorem 17.3.19 says, the complex

$$R_A := \mathrm{ADC}_{\mathfrak{m}}(A^*) = \mathrm{RHom}_A(P_A, A^*) = (P_A)^*$$

is a graded NC dualizing complex (as defined in Definition 17.1.4).

17.4 Balanced Trace Morphisms

In this section we adhere to Conventions 15.1.17 and 15.2.1.

Definition 17.4.1 A graded ring homomorphism $f : A \to B$ is called *finite* if it makes B into a finite graded A-module on both sides.

Throughout this section we assume the next setup:

Setup 17.4.2 We are given noetherian connected graded $\mathbb{K}$-rings A and B, with augmentation ideals $\mathfrak{m}$ and $\mathfrak{n}$, respectively, and a finite graded $\mathbb{K}$-ring homomorphism $f : A \to B$. We are also given a homomorphism of graded $\mathbb{K}$-rings $g : C \to D$.

The rings C and D will play auxiliary roles, to allow us to handle bimodules. In practice C will be either A or $\mathbb{K}$, and D will be either B or $\mathbb{K}$. The graded ring homomorphism $f \otimes g^{\mathrm{op}} : A \otimes C^{\mathrm{op}} \to B \otimes D^{\mathrm{op}}$ induces a restriction functor

$$\mathrm{Rest} = \mathrm{Rest}_{f \otimes g^{\mathrm{op}}} : \mathbf{D}(B \otimes D^{\mathrm{op}}, \mathrm{gr}) \to \mathbf{D}(A \otimes C^{\mathrm{op}}, \mathrm{gr}).$$

This restriction functor will often remain implicit.

Theorem 17.4.3 *Assume Setup* 17.4.2.

(1) *There is a canonical morphism*

$$\mathrm{Rest} \circ \mathrm{R}\Gamma_{\mathfrak{n}} \to \mathrm{R}\Gamma_{\mathfrak{m}} \circ \mathrm{Rest}$$

of triangulated functors $\mathbf{D}(B \otimes D^{\mathrm{op}}, \mathrm{gr}) \to \mathbf{D}(A \otimes C^{\mathrm{op}}, \mathrm{gr})$.

(2) *Let* $N \in \mathbf{D}(B \otimes D^{\mathrm{op}}, \mathrm{gr})$, *and assume either that* $N \in \mathbf{D}^{+}(B \otimes D^{\mathrm{op}}, \mathrm{gr})$ *or that the functor* $\Gamma_{\mathfrak{m}}$ *has finite right cohomological dimension. Then the canonical morphism* $\mathrm{R}\Gamma_{\mathfrak{n}}(N) \to \mathrm{R}\Gamma_{\mathfrak{m}}(N)$ *in* $\mathbf{D}(A \otimes C^{\mathrm{op}}, \mathrm{gr})$ *is an isomorphism.*

For the proof we need a few lemmas.

Lemma 17.4.4 *For every* $N \in \mathbf{M}(B \otimes D^{\mathrm{op}}, \mathrm{gr})$ *there is equality* $\Gamma_{\mathfrak{m}}(N) = \Gamma_{\mathfrak{n}}(N)$ *of* $(B \otimes D^{\mathrm{op}})$*-submodules of* N.

Proof Take any integer $q \geq 1$, and let $\mathfrak{m}^q := \mathfrak{m} \cdots \mathfrak{m}$, the q-fold power of the ideal $\mathfrak{m}$. Because B is a finite A-module, it follows that $B/(\mathfrak{m}^q \cdot B) \cong (A/\mathfrak{m}^q) \otimes_A B$ is a finite module over $A/\mathfrak{m}^q$, and hence it is a finite $\mathbb{K}$-module. This implies that the graded module $B/(\mathfrak{m}^q \cdot B)$ is concentrated in a finite algebraic degree interval, say $[0, i_1]$. So for every $i > i_1$ there is equality $(\mathfrak{m}^q \cdot B)_i = B_i$. Taking $q' := \max(q, i_1 + 1)$ we get $\mathfrak{n}^{q'} \subseteq \mathfrak{m}^q \cdot B$.

The argument above shows that for every $q \geq 1$ there exists some $q' \geq q$ for which $\mathfrak{n}^{q'} \subseteq \mathfrak{m}^q \cdot B$. Trivially there is an inclusion $\mathfrak{m}^q \cdot B \subseteq \mathfrak{n}^q$. These inclusions yield

$$\mathrm{Hom}_B(B/\mathfrak{n}^{q'}, N) \subseteq \mathrm{Hom}_A(A/\mathfrak{m}^q, N) \subseteq \mathrm{Hom}_B(B/\mathfrak{n}^q, N) \subseteq N.$$

Combined with formula (16.3.5) the assertion is proved. □

Recall the ML condition from Definition 11.1.5. An inverse system $\{N_q\}_{q \in \mathbb{N}}$ in $\mathbf{M}(\mathbb{K}, \mathrm{gr})$ is said to have the *trivial ML property*, or is called *prozero*, if for every q there is some $q' \geq q$ such that the homomorphism $N_{q'} \to N_q$ is zero. See [158, Definition 3.5.6].

Lemma 17.4.5 *Let $\{N_q\}_{q \in \mathbb{N}}$ be an inverse system in $\mathbf{M}(\mathbb{K}, \mathrm{gr})$ that has the ML property. The following are equivalent:*

(i) *The inverse system has the trivial ML property.*

(ii) $\lim_{\leftarrow q} N_q = 0$.

Lemma 17.4.6 *Let $\{N_q\}_{q \in \mathbb{N}}$ be an inverse system in $\mathbf{M}(A, \mathrm{gr})$ that has the trivial ML property, and let $M \in \mathbf{M}(A, \mathrm{gr})$. Then $\lim_{q \to} \mathrm{Hom}_A(N_q, M) = 0$.*

Exercise 17.4.7 Prove Lemmas 17.4.5 and 17.4.6.

For a finite graded A-module M, and for numbers $p, q \in \mathbb{N}$, we define

$$F_{p,q}(M) := \mathrm{Tor}_p^A(A/\mathfrak{m}^{q+1}, M) = \mathrm{H}^{-p}((A/\mathfrak{m}^{q+1}) \otimes_A^{\mathrm{L}} M) \in \mathbf{M}(A, \mathrm{gr}). \tag{17.4.8}$$

Then, fixing p, we define

$$F_p(M) := \lim_{\leftarrow q} F_{p,q}(M) \in \mathbf{M}(A, \mathrm{gr}). \tag{17.4.9}$$

In this way we obtain functors

$$F_{p,q}, F_p : \mathbf{M}_{\mathrm{f}}(A, \mathrm{gr}) \to \mathbf{M}(A, \mathrm{gr}). \tag{17.4.10}$$

Lemmas 17.4.11 and 17.4.15 below are essentially [159, Lemma 6.4 and Proposition 6.5].

Lemma 17.4.11 *Let M be a finite graded A-module and let $p \geq 1$. Then the inverse system $\{F_{p,q}(M)\}_{q \in \mathbb{N}}$ has the trivial ML property.*

Proof

Step 1. Here we show that for fixed p, the inverse system $\{F_{p,q}(M)\}_{q \in \mathbb{N}}$ satisfies the ML condition.

Because A is left noetherian and M is a finite graded A-module, we can find a resolution $\cdots \to L^{-1} \to L^0 \to M \to 0$ in $\mathbf{M}(A, \mathrm{gr})$, such that each L^{-p} is a finite graded-free A-module. Then

$$F_{p,q}(M) = \mathrm{H}^{-p}((A/\mathfrak{m}^{q+1}) \otimes_A^{\mathrm{L}} M) \cong \mathrm{H}^{-p}((A/\mathfrak{m}^{q+1}) \otimes_A L)$$

is a finite $(A/\mathfrak{m}^{q+1})$-module. So as a graded A-module it is artinian. For fixed q, and for all $q' \geq q$, the descending chain condition on the submodules $\operatorname{Im}(F_{p,q'}(M) \to F_{p,q}(M)) \subseteq F_{p,q}(M)$ tells us that this eventually stabilizes. This is the ML property.

Step 2. Let $0 \to M' \to M \to M'' \to 0$ be a short exact sequence in $\mathbf{M}_{\mathrm{f}}(A, \mathrm{gr})$. For each q we have a long exact Tor sequence

$$\begin{aligned} \cdots &\to F_{1,q}(M') \to F_{1,q}(M) \to F_{1,q}(M'') \\ &\to F_{0,q}(M') \to F_{0,q}(M) \to F_{0,q}(M'') \to 0. \end{aligned} \tag{17.4.12}$$

As q varies, these long exact sequences form an inverse system. By step 1 we know that in every position the ML property is satisfied. Therefore by Theorem 11.1.7, applied in each degree separately, i.e. to the inverse systems in $\mathbf{M}(\mathbb{K})$, the inverse limit

$$\begin{aligned} \cdots &\to F_1(M') \to F_1(M) \to F_1(M'') \\ &\to F_0(M') \to F_0(M) \to F_0(M'') \to 0 \end{aligned} \tag{17.4.13}$$

is an exact sequence.

Step 3. For a fixed degree k, we have $(\mathfrak{m}^{q+1} \cdot M)_k = 0$ when $q \gg 0$. This is because M is a finite graded A-module, and thus it is bounded below in algebraic degree. It follows that the canonical homomorphism $M_k \to (M/\mathfrak{m}^{q+1} \cdot M)_k$ is bijective for $q \gg 0$. Going to the limit in q we see that the canonical homomorphism

$$M \to F_0(M) = \lim_{\leftarrow q} F_{0,q}(M) = \lim_{\leftarrow q} (M/\mathfrak{m}^{q+1} \cdot M)$$

is bijective. Therefore there is an isomorphism of functors $F_0 \cong \mathrm{Id}$. We deduce that F_0 is an exact functor on $\mathbf{M}_{\mathrm{f}}(A, \mathrm{gr})$.

Step 4. We now prove, by induction on p, that the functor F_p is zero for every $p \geq 1$.

Given a module $M \in \mathbf{M}_{\mathrm{f}}(A, \mathrm{gr})$, we choose a surjection $L \twoheadrightarrow M$ from a finite graded-free module L. We get a short exact sequence $0 \to N \to L \to M \to 0$ in $\mathbf{M}_{\mathrm{f}}(A, \mathrm{gr})$. Hence, by step 2, there is a long exact sequence

$$\begin{aligned} \cdots &\to F_2(N) \to F_2(L) \to F_2(M) \\ &\xrightarrow{\delta^1} F_1(N) \to F_1(L) \to F_1(M) \\ &\xrightarrow{\delta^0} F_0(N) \to F_0(L) \to F_0(M) \to 0 \end{aligned} \tag{17.4.14}$$

in $\mathbf{M}(A, \mathrm{gr})$; cf. the exact sequence (17.4.13). Now $F_{p,q}(L) = 0$ for all $q \geq 0$ and all $p > 0$, because L is flat; so in the limit we get $F_p(L) = 0$ for all $p > 0$.

Since the functor F_0 is exact (by step 3), we see that $\delta^0 = 0$. Also we have $F_1(L) = 0$. Therefore $F_1(M) = 0$.

But M was an arbitrary object of $\mathsf{M}_{\mathrm{f}}(A, \mathrm{gr})$, so in fact the functor F_1 is zero. This implies that in the exact sequence (17.4.14) we have $F_1(N) = 0$. We also have $F_2(L) = 0$. This implies that $F_2(M) = 0$. And so on. This finishes the proof that $F_p = 0$ for all $p \geq 1$.

Step 5. Take some $M \in \mathsf{M}_{\mathrm{f}}(A, \mathrm{gr})$. We know that $F_p(M) = 0$ for every $p \geq 1$. By step 1 we also know that for every p inverse system $\{F_{p,q}(M)\}_{q \in \mathbb{N}}$ has the ML property. Lemma 17.4.5 says that for every $p \geq 1$ the inverse system $\{F_{p,q}(M)\}_{q \in \mathbb{N}}$ has the trivial ML property. □

Lemma 17.4.15 *If J is a graded-injective B-module, then as an A-module, J is graded-$\mathfrak{m}$-flasque.*

Proof Take some $p > 0$. We have to show that $\mathrm{R}^p\Gamma_{\mathfrak{m}}(J) = 0$. Now

$$\mathrm{R}^p\Gamma_{\mathfrak{m}}(J) \cong \lim_{q \to} \mathrm{Ext}^p_A(A/\mathfrak{m}^{q+1}, J).$$

But

$$\mathrm{Ext}^p_A(A/\mathfrak{m}^{q+1}, J) = \mathrm{H}^p(\mathrm{RHom}_A(A/\mathfrak{m}^{q+1}, J)).$$

By derived Hom-tensor adjunction we have a canonical isomorphism

$$\mathrm{RHom}_A(A/\mathfrak{m}^{q+1}, J) \cong \mathrm{RHom}_B(B \otimes^{\mathrm{L}}_A (A/\mathfrak{m}^{q+1}), J)$$

in $\mathsf{D}(A, \mathrm{gr})$. Because J is graded-injective, we have

$$\mathrm{RHom}_B(B \otimes^{\mathrm{L}}_A (A/\mathfrak{m}^{q+1}), J) \cong \mathrm{Hom}_B(B \otimes^{\mathrm{L}}_A (A/\mathfrak{m}^{q+1}), J),$$

and also

$$\begin{aligned} \mathrm{H}^p(\mathrm{Hom}_B(B \otimes^{\mathrm{L}}_A (A/\mathfrak{m}^{q+1}), J) &\cong \mathrm{Hom}_B(\mathrm{H}^{-p}(B \otimes^{\mathrm{L}}_A (A/\mathfrak{m}^{q+1})), J) \\ &\cong \mathrm{Hom}_B(\mathrm{Tor}^A_p(B, A/\mathfrak{m}^{q+1}), J). \end{aligned}$$

Putting it all together we get

$$\mathrm{R}^p\Gamma_{\mathfrak{m}}(J) \cong \lim_{q \to} \mathrm{Hom}_B(F_{p,q}(B), J),$$

where $F_{p,q}(B) := \mathrm{Tor}^A_p(B, A/\mathfrak{m}^{q+1})$, like in formula (17.4.8). Now B is a finite A^{op}-module. According to Lemma 17.4.11, applied to the ring A^{op} instead of A, the inverse system $\{F_{p,q}(M)\}_{q \in \mathbb{N}}$ has the trivial ML property. By Lemma 17.4.6 we see that $\lim_{q \to} \mathrm{Hom}_B(F_{p,q}(B), J) = 0$. □

Proof of Theorem 17.4.3

(1) For a complex $N \in \mathsf{D}(B \otimes D^{\mathrm{op}}, \mathrm{gr})$ we choose a K-injective resolution $\rho : N \to J$ in $\mathsf{C}_{\mathrm{str}}(B \otimes D^{\mathrm{op}}, \mathrm{gr})$. Then we choose a K-injective resolution $\sigma : J \to I$ in $\mathsf{C}_{\mathrm{str}}(A \otimes C^{\mathrm{op}}, \mathrm{gr})$. (We are hiding the restriction functor Rest here.) So $\sigma \circ \rho : N \to I$ is a K-injective resolution in $\mathsf{C}_{\mathrm{str}}(A \otimes C^{\mathrm{op}}, \mathrm{gr})$. We get presentations $\mathrm{R}\Gamma_{\mathfrak{n}}(N) = \Gamma_{\mathfrak{n}}(J)$ and $\mathrm{R}\Gamma_{\mathfrak{m}}(N) = \Gamma_{\mathfrak{m}}(I)$. By Lemma 17.4.4

there is equality $\Gamma_{\mathfrak{m}}(J) = \Gamma_{\mathfrak{n}}(J)$. The morphism $\mathrm{R}\Gamma_{\mathfrak{n}}(N) \to \mathrm{R}\Gamma_{\mathfrak{m}}(N)$ in $\mathbf{D}(A \otimes C^{\mathrm{op}}, \mathrm{gr})$ we are looking for is the one represented by the homomorphism $\Gamma_{\mathfrak{m}}(\sigma) : \Gamma_{\mathfrak{n}}(J) = \Gamma_{\mathfrak{m}}(J) \to \Gamma_{\mathfrak{m}}(I)$ in $\mathbf{C}_{\mathrm{str}}(A \otimes C^{\mathrm{op}}, \mathrm{gr})$. It does not depend on the choices made.

(2) Now we choose the K-injective resolution $\rho : N \to J$ in $\mathbf{C}_{\mathrm{str}}(B \otimes D^{\mathrm{op}}, \mathrm{gr})$ more carefully: we take the semi-graded-cofree resolution from Theorem 15.3.4 (2). Then the complex J is K-injective in $\mathbf{C}(B \otimes D^{\mathrm{op}}, \mathrm{gr})$, each J^p is injective in $\mathbf{M}(B \otimes D^{\mathrm{op}}, \mathrm{gr})$, and $\inf(J) = \inf(\mathrm{H}(N))$. According to Lemma 17.4.15, J is a complex of graded-$\mathfrak{m}$-flasque modules. By Lemma 16.1.5, under either of the two conditions, the complex J is graded-$\mathfrak{m}$-flasque. Therefore the homomorphism $\Gamma_{\mathfrak{m}}(\sigma) : \Gamma_{\mathfrak{m}}(J) \to \Gamma_{\mathfrak{m}}(I)$ in $\mathbf{C}_{\mathrm{str}}(A \otimes C^{\mathrm{op}}, \mathrm{gr})$ is a quasi-isomorphism. □

Corollary 17.4.16 *If the functor $\Gamma_{\mathfrak{m}}$ has finite right cohomological dimension, then the functor $\Gamma_{\mathfrak{n}}$ has finite right cohomological dimension.*

Proof Part 2 of Theorem 17.4.3 implies that the right cohomological dimension of $\Gamma_{\mathfrak{n}}$ is at most the right cohomological dimension of $\Gamma_{\mathfrak{m}}$. □

Corollary 17.4.17 *If the ring A satisfies the left χ condition, then so does the ring B.*

Proof A graded B-module N is finite iff it is finite as a graded A-module, and the same for cofiniteness. Let $N \in \mathbf{M}_{\mathrm{f}}(B, \mathrm{gr})$. By Proposition 16.5.17 the graded A-modules $\mathrm{H}^p_{\mathfrak{m}}(N)$ are all cofinite. But by part 2 of Theorem 17.4.3 we know that $\mathrm{H}^p_{\mathfrak{n}}(N) \cong \mathrm{H}^p_{\mathfrak{m}}(N)$ as graded A-modules. So the graded B-modules $\mathrm{H}^p_{\mathfrak{n}}(N)$ are all cofinite. Again using Proposition 16.5.17 we see that B satisfies the left χ condition. □

Corollary 17.4.18 *If the ring A has a balanced dualizing complex (R_A, β_A), then the ring B has a balanced dualizing complex (R_B, β_B).*

Proof According to Corollary 17.3.24, A satisfies the χ condition, and it has finite local cohomological dimension. Corollaries 17.4.16 and 17.4.17, applied to the finite graded ring homomorphisms $A \to B$ and $A^{\mathrm{op}} \to B^{\mathrm{op}}$, tell us that B satisfies the χ condition, and it has finite local cohomological dimension. Again using Corollary 17.3.24, now in the reverse direction, we conclude that B has a balanced dualizing complex. □

The graded ring homomorphism $f : A \to B$ induces, by applying the functor $(-)^* = \mathrm{Hom}_{\mathbb{K}}(-, \mathbb{K})$, a homomorphism $f^* : B^* \to A^*$ in $\mathbf{M}(A^{\mathrm{en}}, \mathrm{gr})$.

Definition 17.4.19 Under Setup 17.4.2, assume A and B have balanced dualizing complexes (R_A, β_A) and (R_B, β_B), respectively. A *balanced trace morphism* is a morphism

$$\mathrm{tr}_f = \mathrm{tr}_{B/A} : R_B \to R_A$$

in $\mathsf{D}(A^{\mathrm{en}}, \mathrm{gr})$, such that the diagram

$$\begin{array}{ccc} \mathrm{R}\Gamma_{\mathfrak{m}}(R_B) & \xrightarrow{\mathrm{R}\Gamma_{\mathfrak{m}}(\mathrm{tr}_f)} & \mathrm{R}\Gamma_{\mathfrak{m}}(R_A) \\ {\scriptstyle \beta_B}\downarrow{\scriptstyle \cong} & & {\scriptstyle \cong}\downarrow{\scriptstyle \beta_A} \\ B^* & \xrightarrow{f^*} & A^* \end{array}$$

in $\mathsf{D}(A^{\mathrm{en}}, \mathrm{gr})$ is commutative. Here we use the canonical isomorphism $\mathrm{R}\Gamma_{\mathfrak{n}}(R_B) \cong \mathrm{R}\Gamma_{\mathfrak{m}}(R_B)$ from Theorem 17.4.3(2).

In Section 12.6 we studied backward morphisms for objects in derived categories. This makes sense also for complexes of graded bimodules. Namely, given complexes $M \in \mathsf{D}(A \otimes C^{\mathrm{op}}, \mathrm{gr})$ and $N \in \mathsf{D}(B \otimes D^{\mathrm{op}}, \mathrm{gr})$, we can talk about a *backward morphism* $\theta : N \to M$ in $\mathsf{D}(A \otimes C^{\mathrm{op}}, \mathrm{gr})$. The backward morphism θ induces two morphisms by adjunction. First there is the morphism $\mathrm{badj}^{\mathrm{R}}_A(\theta) : N \to \mathrm{RHom}_A(B, M)$ in $\mathsf{D}(B \otimes C^{\mathrm{op}}, \mathrm{gr})$; this is the *derived backward adjunction morphism on the A side*. The second is $\mathrm{badj}^{\mathrm{R}}_{C^{\mathrm{op}}}(\theta) : N \to \mathrm{RHom}_{C^{\mathrm{op}}}(D, M)$ in $\mathsf{D}(A \otimes D^{\mathrm{op}}, \mathrm{gr})$; this is the *derived backward adjunction morphism on the C^{op} side*. The constructions are just like those in Theorem 12.6.12(2), but now using the graded bimodule resolutions from Section 15.3.

Definition 17.4.20 Consider a backward morphism $\theta : N \to M$ in $\mathsf{D}(A \otimes C^{\mathrm{op}}, \mathrm{gr})$.

(1) θ is called a *nondegenerate backward morphism on the A side* if the morphism $\mathrm{badj}^{\mathrm{R}}_A(\theta)$ is an isomorphism.
(2) θ is called a *nondegenerate backward morphism on the C^{op} side* if the morphism $\mathrm{badj}^{\mathrm{R}}_{C^{\mathrm{op}}}(\theta)$ is an isomorphism.
(3) θ is said to be a *nondegenerate backward morphism on both sides* if it is nondegenerate both on the A side and on the C^{op} side.

Example 17.4.21 The backward morphism $f^* : B^* \to A^*$ in $\mathsf{D}(A^{\mathrm{en}}, \mathrm{gr})$ is a nondegenerate backward morphism on both sides.

Theorem 17.4.22 *The following hold in the situation of Definition* 17.4.19.

(1) *There exists a unique balanced trace morphism* $\mathrm{tr}_f : R_B \to R_A$.
(2) *The balanced trace morphism* tr_f *is a nondegenerate backward morphism on both sides.*

We shall need several lemmas for the proof. In analogy to Definition 16.4.2, the dedualizing complex of B is $P_B := \mathrm{R}\Gamma_{\mathfrak{n}}(B) \cong \mathrm{R}\Gamma_{\mathfrak{n}^{\mathrm{op}}}(B)$ in $\mathbf{D}(B^{\mathrm{en}}, \mathrm{gr})$.

Lemma 17.4.23 *There are canonical isomorphisms* $P_B \cong B \otimes^{\mathrm{L}}_A P_A$ *and* $P_B \cong P_A \otimes^{\mathrm{L}}_A B$ *in* $\mathbf{D}(B \otimes A^{\mathrm{op}}, \mathrm{gr})$ *and* $\mathbf{D}(A \otimes B^{\mathrm{op}}, \mathrm{gr})$, *respectively.*

Proof According to Theorems 17.4.3(2) and 16.4.3, there are isomorphisms $P_B = \mathrm{R}\Gamma_{\mathfrak{n}}(B) \cong \mathrm{R}\Gamma_{\mathfrak{m}}(B) \cong P_A \otimes^{\mathrm{L}}_A B$ in $\mathbf{D}(A \otimes B^{\mathrm{op}}, \mathrm{gr})$. The same theorems, but applied to the opposite rings, give us isomorphisms $P_B \cong \mathrm{R}\Gamma_{\mathfrak{n}^{\mathrm{op}}}(B) \cong \mathrm{R}\Gamma_{\mathfrak{m}^{\mathrm{op}}}(B) \cong B \otimes^{\mathrm{L}}_A P_A$ in $\mathbf{D}(B \otimes A^{\mathrm{op}}, \mathrm{gr})$. □

Like Definition 16.6.30, the abstract derived $\mathfrak{n}$-adic completion functor is

$$\mathrm{ADC}_{\mathfrak{n}} := \mathrm{RHom}_B(P_B, -) : \mathbf{D}(B \otimes D^{\mathrm{op}}, \mathrm{gr}) \to \mathbf{D}(B \otimes D^{\mathrm{op}}, \mathrm{gr}).$$

Lemma 17.4.24 *For* $N \in \mathbf{D}(B \otimes D, \mathrm{gr})$ *there is an isomorphism* $\mathrm{ADC}_{\mathfrak{n}}(N) \cong \mathrm{ADC}_{\mathfrak{m}}(N)$ *in* $\mathbf{D}(A \otimes D^{\mathrm{op}}, \mathrm{gr})$. *It is functorial in* N.

Proof We have isomorphisms

$$\begin{aligned} \mathrm{ADC}_{\mathfrak{n}}(N) &= \mathrm{RHom}_B(P_B, N) \cong^1 \mathrm{RHom}_B(B \otimes^{\mathrm{L}}_A P_A, N) \\ &\cong^2 \mathrm{RHom}_A(P_A, N) = \mathrm{ADC}_{\mathfrak{m}}(N) \end{aligned}$$

in $\mathbf{D}(A \otimes D^{\mathrm{op}}, \mathrm{gr})$. The isomorphism $\cong^1$ comes from Lemma 17.4.23, and the isomorphism $\cong^2$ is by adjunction. □

In the next lemma we take $C = D := \mathbb{K}$.

Lemma 17.4.25 *Let* $N \in \mathbf{D}^+_{\mathrm{f}}(B, \mathrm{gr})$, *let* $M \in \mathbf{D}^+_{\mathrm{f}}(A, \mathrm{gr})$, *and let* $\theta : N \to M$ *be a backward morphism in* $\mathbf{D}(A, \mathrm{gr})$. *The following conditions are equivalent:*

(i) θ *is a nondegenerate backward morphism on the* A *side.*

(ii) *The morphism* $\mathrm{R}\Gamma_{\mathfrak{m}}(\theta) : \mathrm{R}\Gamma_{\mathfrak{m}}(N) \to \mathrm{R}\Gamma_{\mathfrak{m}}(M)$ *in* $\mathbf{D}(A, \mathrm{gr})$ *is a nondegenerate backward morphism on the* A *side.*

Proof Let us choose bounded below injective resolutions $N \to J$ and $M \to I$ in $\mathbf{C}_{\mathrm{str}}(B, \mathrm{gr})$ and $\mathbf{C}_{\mathrm{str}}(A, \mathrm{gr})$, respectively. Then the backward morphism θ is represented by a homomorphism $\tilde{\theta} : J \to I$ in $\mathbf{C}_{\mathrm{str}}(A, \mathrm{gr})$.

The morphism $\mathrm{badj}^{\mathrm{R}}_A(\theta) : N \to \mathrm{RHom}_A(B, M)$ in $\mathbf{D}(B, \mathrm{gr})$ is represented by the homomorphism $\mathrm{badj}_A(\tilde{\theta}) : J \to \mathrm{Hom}_A(B, I)$ in $\mathbf{C}_{\mathrm{str}}(B, \mathrm{gr})$. The formula for it it this: $\mathrm{badj}_A(\tilde{\theta})(j)(b) = \tilde{\theta}(b \cdot j)$ for $j \in J$ and $b \in B$. Now $\mathrm{Hom}_A(B, I)$ is a K-graded-injective complex over B, by adjunction. By Lemma 17.4.15 the complexes J and $\mathrm{Hom}_A(B, I)$ are graded-$\mathfrak{m}$-flasque. So the morphism

$$\mathrm{R}\Gamma_{\mathfrak{m}}(\mathrm{badj}^{\mathrm{R}}_A(\theta)) : \mathrm{R}\Gamma_{\mathfrak{m}}(N) \to \mathrm{R}\Gamma_{\mathfrak{m}}(\mathrm{RHom}_A(B, M))$$

is represented by

$$\Gamma_{\mathfrak{m}}(\operatorname{badj}_A(\tilde{\theta})) : \Gamma_{\mathfrak{m}}(J) \to \Gamma_{\mathfrak{m}}(\operatorname{Hom}_A(B, I)). \tag{17.4.26}$$

The morphism $\mathrm{R}\Gamma_{\mathfrak{m}}(\theta)$ is represented by the homomorphism $\Gamma_{\mathfrak{m}}(\tilde{\theta}) : \Gamma_{\mathfrak{m}}(J) \to \Gamma_{\mathfrak{m}}(I)$. Since the complex $\Gamma_{\mathfrak{m}}(I)$ is K-graded-injective over A (by stability, see Theorem 16.3.25), it follows that $\operatorname{badj}^{\mathrm{R}}_A(\mathrm{R}\Gamma_{\mathfrak{m}}(\theta))$ is represented by

$$\operatorname{badj}_A(\Gamma_{\mathfrak{m}}(\tilde{\theta})) : \Gamma_{\mathfrak{m}}(J) \to \operatorname{Hom}_A(B, \Gamma_{\mathfrak{m}}(I)). \tag{17.4.27}$$

A calculation, using the fact that B is a finite A-module, shows that

$$\Gamma_{\mathfrak{m}}(\operatorname{Hom}_A(B, I)) = \operatorname{Hom}_A(B, \Gamma_{\mathfrak{m}}(I)) \tag{17.4.28}$$

as subcomplexes of $\operatorname{Hom}_A(B, I)$. Let us consider the diagram

$$\begin{array}{ccccc} \Gamma_{\mathfrak{m}}(J) & \xrightarrow{\operatorname{badj}_A(\Gamma_{\mathfrak{m}}(\tilde{\theta}))} & \operatorname{Hom}_A(B, \Gamma_{\mathfrak{m}}(I)) & & \\ & \searrow^{\Gamma_{\mathfrak{m}}(\operatorname{badj}_A(\tilde{\theta}))} & \downarrow {=} & \searrow^{\subseteq} & \\ & & \Gamma_{\mathfrak{m}}(\operatorname{Hom}_A(B, I)) & \xrightarrow{\subseteq} & \operatorname{Hom}_A(B, I) \end{array} \tag{17.4.29}$$

in $\mathbf{C}_{\mathrm{str}}(A, \mathrm{gr})$. It is commutative: the triangle on the right is just the equality (17.4.28). And an element $j \in \Gamma_{\mathfrak{m}}(J)$ goes by all paths to the homomorphism $b \mapsto \tilde{\theta}(b \cdot j)$ in $\operatorname{Hom}_A(B, I)$.

Passing to the derived category, (17.4.29) gives the commutative diagram

$$\begin{array}{ccc} \mathrm{R}\Gamma_{\mathfrak{m}}(N) & \xrightarrow{\operatorname{badj}^{\mathrm{R}}_A(\mathrm{R}\Gamma_{\mathfrak{m}}(\theta))} & \mathrm{RHom}_A(B, \mathrm{R}\Gamma_{\mathfrak{m}}(M)) \\ & \searrow^{\mathrm{R}\Gamma_{\mathfrak{m}}(\operatorname{badj}^{\mathrm{R}}_A(\theta))} & \downarrow \cong \\ & & \mathrm{R}\Gamma_{\mathfrak{m}}(\mathrm{RHom}_A(B, M)) \end{array} \tag{17.4.30}$$

in $\mathbf{D}(A, \mathrm{gr})$.

Since both N and $\mathrm{RHom}_A(B, M)$ belong to $\mathbf{D}^{+}_{\mathrm{f}}(A, \mathrm{gr})$, by Lemma 17.3.31, they belong to $\mathbf{D}^{+}(A, \mathrm{gr})_{\mathrm{com}}$. By Theorem 16.6.32 (the MGM Equivalence) we know that the morphism $\operatorname{badj}^{\mathrm{R}}_A(\theta)$ is an isomorphism if and only if the morphism $\mathrm{R}\Gamma_{\mathfrak{m}}(\operatorname{badj}^{\mathrm{R}}_A(\theta))$ is an isomorphism. Diagram (17.4.30) says that this happens if and only if $\operatorname{badj}^{\mathrm{R}}_A(\mathrm{R}\Gamma_{\mathfrak{m}}(\theta))$ is an isomorphism. Going back to the definition, we conclude that θ is nondegenerate if and only if $\mathrm{R}\Gamma_{\mathfrak{m}}(\theta)$ is nondegenerate. □

Proof of Theorem 17.4.22

(1) According to Theorems 17.3.19 and 17.2.4, there is an isomorphism $R_A \cong (P_A)^* \cong \mathrm{ADC}_{\mathfrak{m}}(A^*)$ in $\mathbf{D}(A^{\mathrm{en}}, \mathrm{gr})$. The same theorems, together with Lemma 17.4.24 (taking $C := A$ and $D := B$), say that $R_B \cong (P_B)^* \cong \mathrm{ADC}_{\mathfrak{n}}(B^*) \cong \mathrm{ADC}_{\mathfrak{m}}(B^*)$ in $\mathbf{D}(A^{\mathrm{en}}, \mathrm{gr})$.

The equivalence in Corollary 17.3.36 says that there are isomorphisms

$$\mathrm{R}\Gamma_{\mathfrak{m}}(R_B) \cong (\mathrm{R}\Gamma_{\mathfrak{m}} \circ \mathrm{ADC}_{\mathfrak{m}})(B^*) \cong B^* \tag{17.4.31}$$

and

$$\mathrm{R}\Gamma_{\mathfrak{m}}(R_A) \cong (\mathrm{R}\Gamma_{\mathfrak{m}} \circ \mathrm{ADC}_{\mathfrak{m}})(A^*) \cong A^* \tag{17.4.32}$$

in $\mathbf{D}(A^{\mathrm{en}}, \mathrm{gr})$. And that there is a unique morphism $\theta : R_B \to R_A$ in $\mathbf{D}(A^{\mathrm{en}}, \mathrm{gr})$ such that the diagram

$$\begin{array}{ccc} \mathrm{R}\Gamma_{\mathfrak{m}}(R_B) & \xrightarrow{\mathrm{R}\Gamma_{\mathfrak{m}}(\theta)} & \mathrm{R}\Gamma_{\mathfrak{m}}(R_A) \\ \cong\downarrow & & \downarrow\cong \\ B^* & \xrightarrow{f^*} & A^* \end{array} \tag{17.4.33}$$

in $\mathbf{D}(A^{\mathrm{en}}, \mathrm{gr})$, in which the vertical isomorphisms are (17.4.31) and (17.4.32), is commutative.

But to get a balanced trace morphism tr_f we need a commutative diagram like (17.4.33) in which the vertical isomorphisms are β_B and β_A, respectively. Because the automorphisms of A^* and B^* in $\mathbf{D}(A^{\mathrm{en}}, \mathrm{gr})$ are multiplication by nonzero elements of $\mathbb{K}$, there is a unique $c \in \mathbb{K}^{\times}$ such that $\mathrm{tr}_f := c \cdot \theta : R_B \to R_A$ is a balanced trace morphism.

(2) To see that tr_f is a nondegenerate backward morphism on the A side, we can forget the A^{op}-structure. So we can view tr_f as a backward morphism $\mathrm{tr}_f : R_B \to R_A$ in $\mathbf{D}(A, \mathrm{gr})$. Now $\mathrm{R}\Gamma_{\mathfrak{m}}(\mathrm{tr}_f) = \beta_A^{-1} \circ f^* \circ \beta_B$. In Example 17.4.21 we saw that $f^* : B^* \to A^*$ is a nondegenerate backward morphism in $\mathbf{D}(A, \mathrm{gr})$. The morphisms β_A and β_B are isomorphisms. Therefore

$$\mathrm{R}\Gamma_{\mathfrak{m}}(\mathrm{tr}_f) : \mathrm{R}\Gamma_{\mathfrak{m}}(R_B) \to \mathrm{R}\Gamma_{\mathfrak{m}}(R_A)$$

is a nondegenerate backward morphism in $\mathbf{D}(A, \mathrm{gr})$. According to Lemma 17.4.25, the morphism tr_f is a nondegenerate backward morphism in $\mathbf{D}(A, \mathrm{gr})$.

As for the A^{op} side: the symmetry from Corollary 17.3.35 gives isomorphisms $R_A \cong \mathrm{ADC}_{\mathfrak{m}^{\mathrm{op}}}(A^*)$ and $R_B \cong \mathrm{ADC}_{\mathfrak{m}^{\mathrm{op}}}(B^*)$ in $\mathbf{D}(A^{\mathrm{en}}, \mathrm{gr})$. Going over the constructions in the proof of item (1) above, we see that there is equality $\mathrm{R}\Gamma_{\mathfrak{m}^{\mathrm{op}}}(\mathrm{tr}_f) = d \cdot \beta_A^{-1} \circ f^* \circ \beta_B$ for some $d \in \mathbb{K}^{\times}$. Therefore $\mathrm{R}\Gamma_{\mathfrak{m}^{\mathrm{op}}}(\mathrm{tr}_f)$ is a nondegenerate backward morphism $\mathbf{D}(A^{\mathrm{op}}, \mathrm{gr})$. Now Lemma 17.4.25, transcribed to the ring A^{op}, says that tr_f is a nondegenerate backward morphism in $\mathbf{D}(A^{\mathrm{op}}, \mathrm{gr})$. □

Corollary 17.4.34 *Let $A \xrightarrow{f} B \xrightarrow{g} C$ be finite homomorphisms in $\mathsf{Rng}_{\mathrm{gr}}/_{\mathrm{c}}\mathbb{K}$, and assume these rings have balanced dualizing complexes (R_A, β_A), (R_B, β_B) and (R_C, β_C), respectively. Then $\mathrm{tr}_{g \circ f} = \mathrm{tr}_f \circ \mathrm{tr}_g$ as morphisms $R_C \to R_A$ in $\mathbf{D}(A^{\mathrm{en}}, \mathrm{gr})$.*

Proof The follows from the uniqueness in Theorem 17.4.22 and the fact that $(g \circ f)^* = f^* \circ g^*$ as homomorphisms $C^* \to A^*$ in $\mathbf{M}(A, \mathrm{gr})$. □

We end this section with a partial converse to Corollaries 17.4.16 and 17.4.17. Recall that a central element $a \in A$ is called *regular* if it is not a zero divisor, i.e. the only element $b \in A$ such that $a \cdot b = 0$ is $b = 0$.

Theorem 17.4.35 *Let A be a noetherian connected graded ring, and let $a \in A$ be a regular homogeneous central element of positive degree. Define the connected graded ring $B := A/(a)$. If B satisfies the χ condition and it has finite local cohomological dimension, then A also satisfies the χ condition and it has finite local cohomological dimension.*

This is [10, Theorem 8.8]. The original proof does not use derived categories, and is quite different from the proof below.

Proof By op-symmetry it suffices to prove that A satisfies the left χ condition and that the functor $\mathrm{R}\Gamma_{\mathfrak{m}}$ has finite cohomological dimension.

Let $d \in \mathbb{N}$ be the local cohomological dimension of B, and let $i \geq 1$ be the degree of the element a. In view of Propositions 16.5.17 and 16.3.19, all we need to prove is that the following two conditions hold for every $M \in \mathbf{M}_{\mathrm{f}}(A, \mathrm{gr})$:

(i) $\mathrm{H}^p_{\mathfrak{m}}(M)$ is a cofinite A-module for every p.

(ii) $\mathrm{H}^p_{\mathfrak{m}}(M) = 0$ for every $p > d + 1$.

Let us denote the augmentation ideal of B by $\mathfrak{n}$. The short exact sequence $0 \to A \xrightarrow{a \cdot (-)} A(i) \to B(i) \to 0$ in $\mathbf{M}(A^{\mathrm{en}}, \mathrm{gr})$ is viewed as a distinguished triangle $A \xrightarrow{a \cdot (-)} A(i) \to B(i) \xrightarrow{\triangle}$ in $\mathbf{D}(A^{\mathrm{en}}, \mathrm{gr})$. Applying the functor $(-) \otimes^{\mathrm{L}}_A M$ to it, we obtain a distinguished triangle

$$M \xrightarrow{a \cdot (-)} M(i) \to N \xrightarrow{\triangle} \tag{17.4.36}$$

in $\mathbf{D}(A, \mathrm{gr})$, where by definition $N := B(i) \otimes^{\mathrm{L}}_A M \in \mathbf{D}(B, \mathrm{gr})$. Note that $\mathrm{H}^p(N) \in \mathbf{M}_{\mathrm{f}}(B, \mathrm{gr})$ for all p, and $\mathrm{H}^p(N) = 0$ unless $-1 \leq p \leq 0$.

Next we apply the functor $\mathrm{R}\Gamma_{\mathfrak{m}}$ to (17.4.36), and use Theorem 17.4.3, to obtain this distinguished triangle

$$\mathrm{R}\Gamma_{\mathfrak{m}}(M) \xrightarrow{a \cdot (-)} \mathrm{R}\Gamma_{\mathfrak{m}}(M(i)) \to \mathrm{R}\Gamma_{\mathfrak{n}}(N) \xrightarrow{\triangle} \tag{17.4.37}$$

in $\mathbf{D}(A, \mathrm{gr})$. Taking cohomologies in (17.4.37) we obtain this exact sequence

$$\mathrm{H}^{p-1}_{\mathfrak{n}}(N) \to \mathrm{H}^p_{\mathfrak{m}}(M) \xrightarrow{a \cdot (-)} \mathrm{H}^p_{\mathfrak{m}}(M(i)) \to \mathrm{H}^p_{\mathfrak{n}}(N) \tag{17.4.38}$$

in $\mathbf{M}(A, \mathrm{gr})$. Let use write

$$K^p := \mathrm{Ker}(\mathrm{H}^p_{\mathfrak{m}}(M) \xrightarrow{a \cdot (-)} \mathrm{H}^p_{\mathfrak{m}}(M(i))) \subseteq \mathrm{H}^p_{\mathfrak{m}}(M).$$

Since $a \in \mathfrak{m}$, there is an inclusion

$$\mathrm{Soc}(\mathrm{H}^p_{\mathfrak{m}}(M)) = \mathrm{Hom}_A(\mathbb{K}, \mathrm{H}^p_{\mathfrak{m}}(M)) \subseteq K^p. \tag{17.4.39}$$

For $p > d + 1$ we have $\mathrm{H}^{p-1}_{\mathfrak{n}}(N) = 0$ and $\mathrm{H}^p_{\mathfrak{n}}(N) = 0$. The exact sequence (17.4.38) says that the homomorphism $a \cdot (-) : \mathrm{H}^p_{\mathfrak{m}}(M) \to \mathrm{H}^p_{\mathfrak{m}}(M(i))$ is bijective. So $K^p = 0$, and by (17.4.39) we see that $\mathrm{Soc}(\mathrm{H}^p_{\mathfrak{m}}(M)) = 0$. But $\mathrm{H}^p_{\mathfrak{m}}(M)$ is an $\mathfrak{m}$-torsion module, so according to Proposition 16.3.12, we get $\mathrm{H}^p_{\mathfrak{m}}(M) = 0$. This establishes condition (ii).

Finally take any p. By the χ condition for B and by Proposition 16.5.17 we know that $\mathrm{H}^{p-1}_{\mathfrak{n}}(N)$ is a cofinite graded B-module; and thus it is a cofinite graded A-module. According to the NC Graded Matlis Duality (Theorem 15.2.34), the cofinite graded A-modules are the artinian objects in the abelian category $\mathbf{M}(A, \mathrm{gr})$. From the exact sequence (17.4.38) we obtain a surjection $\mathrm{H}^{p-1}_{\mathfrak{n}}(N) \to K^p$. Therefore K^p is a cofinite graded A-module. The inclusion (17.4.39) says that $\mathrm{Soc}(\mathrm{H}^p_{\mathfrak{m}}(M))$ is a cofinite graded A-module; and thus it is a finite graded $\mathbb{K}$-module. Hence, by Corollary 16.3.13, $\mathrm{H}^p_{\mathfrak{m}}(M)$ is a cofinite graded A-module. This is condition (i). □

18

Rigid Noncommutative Dualizing Complexes

In this chapter of the book we are going to work with noncommutative (namely not necessarily commutative) rings. We shall often use the abbreviation "NC" for "noncommutative." The goal is to introduce *rigid NC dualizing complexes* in the sense of M. Van den Bergh [153] and to prove their existence and uniqueness under certain conditions.

18.1 Noncommutative Dualizing Complexes

Here we will define NC dualizing complexes over a NC ring A. The definition is almost identical to that in the NC graded setting (see Section 17.1). We will prove that the NC dualizing complexes over A are parameterized by the derived Picard group of A. The content of this section is adapted from the paper [165].

Let $\mathbb{K}$ be a commutative ring. Recall (from Definition 1.2.1) that a *central* $\mathbb{K}$-*ring* is a ring A equipped with a ring homomorphism $\mathbb{K} \to \operatorname{Cent}(A)$, where $\operatorname{Cent}(A)$ is the center of A. In more traditional texts, a central $\mathbb{K}$-ring is called an *associative unital* $\mathbb{K}$-*algebra*. The category of central $\mathbb{K}$-rings, with $\mathbb{K}$-ring homomorphisms, is denoted by $\mathsf{Rng}/_{\mathrm{c}}\, \mathbb{K}$.

Here is the convention that is in force throughout this chapter:

Convention 18.1.1 There is a base field $\mathbb{K}$. All rings are by default central $\mathbb{K}$-rings, and all homomorphisms between $\mathbb{K}$-rings are over $\mathbb{K}$; namely we work within the category $\mathsf{Rng}/_{\mathrm{c}}\, \mathbb{K}$. All bimodules are $\mathbb{K}$-central, and all additive functors are $\mathbb{K}$-linear. We use the notation $\otimes$ for $\otimes_{\mathbb{K}}$. For a ring A we write $A^{\mathrm{en}} := A \otimes A^{\mathrm{op}}$, the enveloping ring of A.

See Remark 18.1.30 regarding the assumption that the base ring $\mathbb{K}$ is a field.

Recall that a NC ring A is called noetherian if it is both left noetherian and right noetherian. See Remark 15.1.30 regarding the failure of the noetherian property to be preserved under finitely generated ring extensions.

Let A be a ring. As before, $\mathbf{M}(A)$ is the category of left A-modules. The category of right A-modules is $\mathbf{M}(A^{\mathrm{op}})$, and the category of A-bimodules is $\mathbf{M}(A^{\mathrm{en}})$. Given another ring B, the category of A-B-bimodules is $\mathbf{M}(A \otimes B^{\mathrm{op}})$. These are $\mathbb{K}$-linear abelian categories.

The derived category of $\mathbf{M}(A)$ is $\mathbf{D}(A) := \mathbf{D}(\mathbf{M}(A))$; its objects are the complexes of (left) A-modules. The other derived categories are $\mathbf{D}(A^{\mathrm{op}}) := \mathbf{D}(\mathbf{M}(A^{\mathrm{op}}))$, $\mathbf{D}(A^{\mathrm{en}}) := \mathbf{D}(\mathbf{M}(A^{\mathrm{en}}))$ and $\mathbf{D}(A \otimes B^{\mathrm{op}}) := \mathbf{D}(\mathbf{M}(A \otimes B^{\mathrm{op}}))$.

Consider the canonical ring antiautomorphism $\mathrm{op} : B \to B^{\mathrm{op}}$, that is the identity on the underlying $\mathbb{K}$-module. There is a canonical ring isomorphism

$$A \otimes B^{\mathrm{op}} \xrightarrow{\simeq} B^{\mathrm{op}} \otimes A, \quad a \otimes \mathrm{op}(b) \mapsto \mathrm{op}(b) \otimes a. \tag{18.1.2}$$

The induced isomorphism on the bimodule categories $\mathbf{M}(A \otimes B^{\mathrm{op}}) \xrightarrow{\simeq} \mathbf{M}(B^{\mathrm{op}} \otimes A)$ is the identity on the underlying $\mathbb{K}$-modules. Indeed, for $M \in \mathbf{M}(A \otimes B^{\mathrm{op}})$, $m \in M$ and $a \otimes \mathrm{op}(b) \in A \otimes B^{\mathrm{op}}$, we have $(\mathrm{op}(b) \otimes a) \cdot m = a \cdot m \cdot b = (a \otimes \mathrm{op}(b)) \cdot m$. Now take $B = A$ in (18.1.2). Since $(A^{\mathrm{op}})^{\mathrm{op}} = A$, there is a canonical ring isomorphism

$$A^{\mathrm{en}} = A \otimes A^{\mathrm{op}} \xrightarrow{\simeq} A^{\mathrm{op}} \otimes A = (A^{\mathrm{op}})^{\mathrm{en}} \tag{18.1.3}$$

and an induced isomorphism on the bimodule categories

$$\mathbf{M}(A^{\mathrm{en}}) \xrightarrow{\simeq} \mathbf{M}((A^{\mathrm{op}})^{\mathrm{en}}). \tag{18.1.4}$$

All these categorical relations pass to derived categories.

The ring A^{en} is isomorphic to its opposite ring. There is a ring isomorphism

$$\mathrm{trop} : A^{\mathrm{en}} \xrightarrow{\simeq} (A^{\mathrm{en}})^{\mathrm{op}} \tag{18.1.5}$$

with this formula: for $a_1, a_2 \in A$ it sends the element $a_1 \otimes \mathrm{op}(a_2) \in A^{\mathrm{en}}$ to

$$\mathrm{trop}(a_1 \otimes \mathrm{op}(a_2)) := \mathrm{op}(a_2 \otimes \mathrm{op}(a_1)) \in (A^{\mathrm{en}})^{\mathrm{op}}. \tag{18.1.6}$$

On the underlying $\mathbb{K}$-module $A \otimes A$ this is just the transposition $a_1 \otimes a_2 \mapsto a_2 \otimes a_1$, and hence the name "transpose opposite" and the acronym "trop."

Usually an A-A-bimodule M is seen as a left A^{en}-module, with this action of an element $a_1 \otimes \mathrm{op}(a_2) \in A^{\mathrm{en}}$ on an element $m \in M$:

$$(a_1 \otimes \mathrm{op}(a_2)) \cdot m = a_1 \cdot m \cdot a_2. \tag{18.1.7}$$

But using the ring isomorphism trop from formula (18.1.5) we can also consider M as a left $(A^{\mathrm{en}})^{\mathrm{op}}$-module, and thus as a right A^{en}-module. Explicitly, the right action of an element $a_2 \otimes \mathrm{op}(a_1) \in A^{\mathrm{en}}$ on an element $m \in M$ is:

$$\begin{aligned} m \cdot (a_2 \otimes \mathrm{op}(a_1)) &=^{\heartsuit} \mathrm{op}(a_2 \otimes \mathrm{op}(a_1)) \cdot m \\ &=^{\dagger} (a_1 \otimes \mathrm{op}(a_2)) \cdot m =^{\diamond} a_1 \cdot m \cdot a_2. \end{aligned} \tag{18.1.8}$$

In equality $=^{\heartsuit}$ we move from a right action by A^{en} to the left action by $(A^{\mathrm{en}})^{\mathrm{op}}$, and in equality $=^{\dagger}$ we made use of formula (18.1.6) to move from $(A^{\mathrm{en}})^{\mathrm{op}}$ to A^{en}. Finally, the equality $=^{\diamond}$ is formula (18.1.7). This game of musical chairs will be played in Sections 18.5 and 18.2.

Let A and B be rings. The canonical ring homomorphisms

$$A \to A \otimes B^{\mathrm{op}} \leftarrow B^{\mathrm{op}}$$

induce restriction functors

$$\mathbf{D}(A) \xleftarrow{\mathrm{Rest}_A} \mathbf{D}(A \otimes B^{\mathrm{op}}) \xrightarrow{\mathrm{Rest}_{B^{\mathrm{op}}}} \mathbf{D}(B^{\mathrm{op}}). \tag{18.1.9}$$

See the commutative diagram (14.2.3). We shall usually keep these restriction functors implicit, and instead use terminology like that in Definition 14.2.18.

In Definition 12.5.1 we introduced this notation: for a ring A we denote by $\mathbf{M}_{\mathrm{f}}(A)$ the full subcategory of $\mathbf{M}(A)$ on the finite (i.e. finitely generated) A-modules. And $\mathbf{D}_{\mathrm{f}}(A)$ is the full subcategory of $\mathbf{D}(A)$ on the complexes of A-modules M such that $\mathrm{H}^i(M) \in \mathbf{M}_{\mathrm{f}}(A)$ for every i.

If A is left noetherian, then $\mathbf{M}_{\mathrm{f}}(A)$ is a thick abelian subcategory of $\mathbf{M}(A)$; and hence $\mathbf{D}_{\mathrm{f}}(A)$ is a full triangulated subcategory of $\mathbf{D}(A)$. As usual we can combine indicators: $\mathbf{D}^{\star}_{\mathrm{f}}(A) := \mathbf{D}_{\mathrm{f}}(A) \cap \mathbf{D}^{\star}(A)$, where $\star$ is some boundedness indicator ($+$, $-$, b or ⟨empty⟩).

Now to finiteness of A-B-bimodules. The expression $\mathbf{M}_{(\mathrm{f},..)}(A \otimes B^{\mathrm{op}})$ denotes the full subcategory of $\mathbf{M}(A \otimes B^{\mathrm{op}})$ on the bimodules that are finite over A; and the expression $\mathbf{M}_{(..,\mathrm{f})}(A \otimes B^{\mathrm{op}})$ denotes the full subcategory on the bimodules that are finite over B^{op}. Of course, we define

$$\mathbf{M}_{(\mathrm{f},\mathrm{f})}(A \otimes B^{\mathrm{op}}) := \mathbf{M}_{(\mathrm{f},..)}(A \otimes B^{\mathrm{op}}) \cap \mathbf{M}_{(..,\mathrm{f})}(A \otimes B^{\mathrm{op}}).$$

The same notation shall apply to derived categories.

To a complex of bimodules $M \in \mathbf{D}(A \otimes B^{\mathrm{op}})$, we associated two derived homothety morphisms in Section 14.4. Let us recall them. The derived homothety morphism through B^{op} is the morphism

$$\mathrm{hm}^{\mathrm{R}}_{M,B^{\mathrm{op}}} : B \to \mathrm{RHom}_A(M, M) \tag{18.1.10}$$

in $\mathbf{D}(B^{\mathrm{en}})$; and the derived homothety morphism through A is the morphism

$$\mathrm{hm}^{\mathrm{R}}_{M,A} : A \to \mathrm{RHom}_{B^{\mathrm{op}}}(M, M) \tag{18.1.11}$$

in $\mathbf{D}(A^{\mathrm{en}})$. According to Definition 14.4.14, the complex M has the noncommutative derived Morita property on the A side (resp. on the B^{op} side) if the morphism $\mathrm{hm}^{\mathrm{R}}_{M,B^{\mathrm{op}}}$ (resp. $\mathrm{hm}^{\mathrm{R}}_{M,A}$) is an isomorphism.

Definition 18.1.12 ([165]) Let A be a noetherian ring. A *noncommutative dualizing complex over* A is an object $R \in \mathbf{D}^{\mathrm{b}}(A^{\mathrm{en}})$ satisfying the following three conditions:

(i) For every p the A-bimodule $\mathrm{H}^p(R)$ is finite over A and over A^{op}.
(ii) The complex R has finite injective dimension over A and over A^{op}.
(iii) The complex of bimodules R has the noncommutative derived Morita property on both sides.

Note that conditions (i) and (ii) are one-sided, namely they refer separately to $\mathrm{Rest}_A(R) \in \mathbf{D}(A)$ and to $\mathrm{Rest}_{A^{\mathrm{op}}}(R) \in \mathbf{D}(A^{\mathrm{op}})$; whereas condition (iii) is more complicated, and we will study it further in Lemma 18.1.21 below. Also note that condition (i) can be stated as "R belongs to $\mathbf{D}_{(\mathrm{f,f})}(A^{\mathrm{en}})$."

Before continuing, we need to say something about the op-symmetry of Definition 18.1.12. Given a complex $R \in \mathbf{D}(A^{\mathrm{en}})$, by formula (18.1.4) the complex R is also an object of $\mathbf{D}((A^{\mathrm{op}})^{\mathrm{en}})$, and we could ask whether R is a dualizing complex over the ring A^{op}. The answer is positive of course, because all three conditions are op-symmetric. We record this fact as the next proposition.

Proposition 18.1.13 *If R is a NC dualizing complex over A, then R is also a NC dualizing complex over A^{op}.*

Example 18.1.14 A NC ring A is called *regular* if it has finite global cohomological dimension on both sides. This is very similar to the graded definition (see Definition 15.4.1). Let us recall what it means: there is some natural number d such that $\mathrm{Ext}^i_A(M, N) = 0$ for all $i > d$ and all $M, N \in \mathbf{M}(A)$; and also $\mathrm{Ext}^i_{A^{\mathrm{op}}}(M', N') = 0$ for all $i > d$ and all $M', N' \in \mathbf{M}(A^{\mathrm{op}})$. The smallest such number d is the global cohomological dimension of A.

Such rings are easy to find; for instance, if $\mathbb{K}$ is a field, and $\mathfrak{g}$ is finite Lie algebra over $\mathbb{K}$, then the *universal enveloping ring* $A := \mathrm{U}(\mathfrak{g})$ is regular; the global cohomological dimension of A is $d := \mathrm{rank}_{\mathbb{K}}(\mathfrak{g})$.

A weaker condition is this: A is called *Gorenstein* if A has finite injective dimension both as a left module and as a right module over itself. Namely there is some natural number d such that $\mathrm{Ext}^i_A(M, A) = 0$ for all $i > d$ and all $M \in \mathbf{M}(A)$; and also $\mathrm{Ext}^i_{A^{\mathrm{op}}}(M', A) = 0$ for all $i > d$ and all $M' \in \mathbf{M}(A^{\mathrm{op}})$.

Assume A is Gorenstein. Then the complex of bimodules $R := A$ satisfies condition (ii) of Definition 18.1.12. Conditions (i) and (iii) of this definition are automatically true. We conclude that $R := A$ is a NC dualizing complex over itself.

Definition 18.1.15 Let A be a noetherian ring, let R be a NC dualizing complex over A, and let B be a second ring. The *duality functors* associated to R are the triangulated functors

$$D_A : \mathbf{D}(A \otimes B^{\mathrm{op}})^{\mathrm{op}} \to \mathbf{D}(B \otimes A^{\mathrm{op}}), \quad D_A := \mathrm{RHom}_A(-, R)$$

and

$$D_{A^{\mathrm{op}}} : \mathbf{D}(B \otimes A^{\mathrm{op}})^{\mathrm{op}} \to \mathbf{D}(A \otimes B^{\mathrm{op}}), \quad D_{A^{\mathrm{op}}} := \mathrm{RHom}_{A^{\mathrm{op}}}(-, R).$$

Notice that the expressions D_A and $D_{A^{\mathrm{op}}}$ leave R and B implicit; this is because the dualizing complex R is taken to be fixed, and the second ring B is less important.

The NC derived Hom-evaluation morphisms were already discussed in Section 17.1, in the algebraically graded context. The nongraded definition is very similar. Given a complex $M \in \mathbf{D}(A \otimes B^{\mathrm{op}})$, there is a morphism

$$\mathrm{ev}^{\mathrm{R,R}}_M : M \to D_{A^{\mathrm{op}}}(D_A(M)) = \mathrm{RHom}_{A^{\mathrm{op}}}(\mathrm{RHom}_A(M, R), R) \qquad (18.1.16)$$

in $\mathbf{D}(A \otimes B^{\mathrm{op}})$, which is functorial in M. If we choose a K-injective resolution $R \to I$ in $\mathbf{D}(A^{\mathrm{en}})$, then we have an isomorphism

$$D_{A^{\mathrm{op}}}(D_A(M)) \cong \mathrm{Hom}_{A^{\mathrm{op}}}(\mathrm{Hom}_A(M, I), I)$$

in $\mathbf{D}(A \otimes B^{\mathrm{op}})$, and the morphism $\mathrm{ev}^{\mathrm{R,R}}_M$ is represented by the Hom-evaluation homomorphism

$$\mathrm{ev}_{M,I} : M \to \mathrm{Hom}_{A^{\mathrm{op}}}(\mathrm{Hom}_A(M, I), I)$$

in $\mathbf{C}_{\mathrm{str}}(A \otimes B^{\mathrm{op}})$. For this we use the results of Section 14.3, which apply because A and B are flat over $\mathbb{K}$.

In the special case when $B = A$ and $M = A$, the morphism $\mathrm{ev}^{\mathrm{R,R}}_A$ recovers the derived homothety morphism through A, see (18.1.11); i.e.

$$\mathrm{hm}^{\mathrm{R}}_{R,A} = \mathrm{ev}^{\mathrm{R,R}}_A : A \to \mathrm{RHom}_{A^{\mathrm{op}}}(R, R) \qquad (18.1.17)$$

in $\mathbf{D}(A^{\mathrm{en}})$. This formula makes implicit use of the left co-unitor isomorphism $\mathrm{lcu} : \mathrm{RHom}_A(A, R) \xrightarrow{\simeq} R$ in $\mathbf{D}(A^{\mathrm{en}})$. The same holds for the homothety morphism $\mathrm{hm}^{\mathrm{R}}_{R,A^{\mathrm{op}}}$.

Theorem 18.1.18 *Let A be a noetherian ring, let R be a NC dualizing complex over A, let B be a second ring, and let D_A and $D_{A^{\mathrm{op}}}$ be the duality functors associated to R. Let $\star$ be a boundedness indicator, and let $M \in \mathbf{D}^{\star}_{(\mathrm{f},..)}(A \otimes B^{\mathrm{op}})$. Then the following hold:*

(1) *The complex $D_A(M)$ belongs to $\mathbf{D}^{-\star}_{(..,\mathrm{f})}(B \otimes A^{\mathrm{op}})$, where $-\star$ is the reversed boundedness indicator.*

(2) *The derived Hom-evaluation morphism* $\mathrm{ev}_M^{\mathrm{R,R}} : M \to D_{A^{\mathrm{op}}}(D_A(M))$ *in* $\mathbf{D}(A \otimes B^{\mathrm{op}})$ *is an isomorphism.*

Proof The proof is the same as that of Theorem 17.1.6, with the obvious modifications (i.e. neglecting the algebraic grading). It all boils down to Theorems 12.5.7 and 12.5.2. □

Corollary 18.1.19 *Under the assumptions of Theorem* 18.1.18, *the functor*

$$D_A : \mathbf{D}^{\star}_{(\mathrm{f},..)}(A \otimes B^{\mathrm{op}})^{\mathrm{op}} \to \mathbf{D}^{-\star}_{(..,\mathrm{f})}(B \otimes A^{\mathrm{op}})$$

is an equivalence of triangulated categories, with quasi-inverse $D_{A^{\mathrm{op}}}$.

Proof First we note that the op-symmetry (Proposition 18.1.13) implies that Theorem 18.1.18 is true after exchanging A with A^{op}. Thus $\mathrm{ev}_N^{\mathrm{R,R}} : N \to D_A(D_{A^{\mathrm{op}}}(N))$ is an isomorphism for every $N \in \mathbf{D}^{-\star}_{(..,\mathrm{f})}(B \otimes A^{\mathrm{op}})$. Now the assertion is clear. □

We need a better understanding of the derived homothety morphisms. Suppose $M \in \mathbf{C}(A \otimes B^{\mathrm{op}})$. The right action of B on M is the homothety ring homomorphism through B^{op}, $\mathrm{hm}_{M,B^{\mathrm{op}}} : B^{\mathrm{op}} \to \mathrm{End}_{\mathbf{C}_{\mathrm{str}}(A)}(M)$. Namely for elements $b \in B$ and $m \in M^i$ we have $\mathrm{hm}_{M,B^{\mathrm{op}}}(\mathrm{op}(b))(m) = m \cdot b \in M^i$. By composing $\mathrm{hm}_{M,B^{\mathrm{op}}}$ with the localization functor $\mathrm{Q} : \mathbf{C}_{\mathrm{str}}(A) \to \mathbf{D}(A)$ we obtain the ring homomorphism

$$\mathrm{hm}^{\mathbf{D}}_{M,B^{\mathrm{op}}} := \mathrm{Q} \circ \mathrm{hm}_{M,B^{\mathrm{op}}} : B^{\mathrm{op}} \to \mathrm{End}_{\mathbf{D}(A)}(M). \tag{18.1.20}$$

Lemma 18.1.21 *Let* $M \in \mathbf{D}(A \otimes B^{\mathrm{op}})$. *The three conditions below are equivalent.*

(i) *M has the derived Morita property on the A side, i.e.* (18.1.10) *is an isomorphism.*
(ii) *For every $p \neq 0$ the module* $\mathrm{Hom}_{\mathbf{D}(A)}(M, M[p])$ *is zero, and the ring homomorphism* $\mathrm{hm}^{\mathbf{D}}_{M,B^{\mathrm{op}}} : B^{\mathrm{op}} \to \mathrm{End}_{\mathbf{D}(A)}(M)$ *is bijective.*
(iii) *For every $p \neq 0$ the module* $\mathrm{H}^p(\mathrm{RHom}_A(M, M))$ *is zero, and the B^{op}-module* $\mathrm{H}^0(\mathrm{RHom}_A(M, M))$ *is free, with basis the element* id_M.

Proof This is clear from the canonical isomorphisms

$$\mathrm{Hom}_{\mathbf{D}(A)}(M, M[p]) \cong \mathrm{H}^p(\mathrm{RHom}_A(M, M))$$

of Corollary 12.2.8. □

Lemma 18.1.22 *Let* $T \in \mathbf{D}(A^{\mathrm{en}})$, *and consider the triangulated functor*

$$G := T \otimes^{\mathrm{L}}_A (-) : \mathbf{D}(A \otimes B^{\mathrm{op}}) \to \mathbf{D}(A \otimes B^{\mathrm{op}}).$$

For every $M \in \mathsf{D}(A \otimes B^{\mathrm{op}})$, *the diagram*

$$\begin{array}{ccc} B^{\mathrm{op}} & \xrightarrow{\ \mathrm{hm}^{\mathsf{D}}_{M,B^{\mathrm{op}}}\ } & \mathrm{End}_{\mathsf{D}(A)}(M) \\ & \searrow^{\mathrm{hm}^{\mathsf{D}}_{G(M),B^{\mathrm{op}}}} & \downarrow G \\ & & \mathrm{End}_{\mathsf{D}(A)}(G(M)) \end{array}$$

of ring homomorphisms is commutative.

Proof This is because the action of B^{op} on M is categorical, i.e. by endomorphisms in $\mathsf{D}(A)$. □

Lemma 18.1.23 *Let* $R \in \mathsf{D}(A^{\mathrm{en}})$, *and consider the triangulated functor*

$$D := \mathrm{RHom}_A(-, R) : \mathsf{D}(A \otimes B^{\mathrm{op}})^{\mathrm{op}} \to \mathsf{D}(B \otimes A^{\mathrm{op}}).$$

For every $M \in \mathsf{D}(A \otimes B^{\mathrm{op}})$, *the diagram*

$$\begin{array}{ccc} B^{\mathrm{op}} & \xrightarrow{\ \mathrm{hm}^{\mathsf{D}}_{M,B^{\mathrm{op}}}\ } & \mathrm{End}_{\mathsf{D}(A)}(M) \\ & \searrow^{\mathrm{Op}(\mathrm{hm}^{\mathsf{D}}_{D(M),B})} & \downarrow D \\ & & \mathrm{End}_{\mathsf{D}(A^{\mathrm{op}})}(D(M))^{\mathrm{op}} \end{array}$$

of ring homomorphisms is commutative.

Proof Same as the previous lemma, only here D is contravariant. □

Tilting DG bimodules were defined in Definition 14.4.1. Here we call them *tilting complexes*. Given a tilting complex $T \in \mathsf{D}(A^{\mathrm{en}})$, we know that its quasi-inverse $T^{\vee}$ satisfies $T^{\vee} \cong \mathrm{RHom}_A(T, A) \cong \mathrm{RHom}_{A^{\mathrm{op}}}(T, A)$ in $\mathsf{D}(A^{\mathrm{en}})$. See Corollary 14.4.27 with $B = A$, and use the ring isomorphism (18.1.3).

Lemma 18.1.24 *Let T be a tilting complex complex over A, with quasi-inverse* $T^{\vee}$. *Then there is an isomorphism* $T \otimes^{\mathrm{L}}_A (-) \cong \mathrm{RHom}_A(T^{\vee}, -)$ *of triangulated functors from* $\mathsf{D}(A \otimes B^{\mathrm{op}})$ *to itself.*

Proof Let's write $S := T^{\vee}$. Then $T \cong S^{\vee} \cong \mathrm{RHom}_A(S, A)$. We know that S is perfect on the A side (see Corollary 14.4.28), so by Theorem 14.1.22 we have

$$\begin{aligned} T \otimes^{\mathrm{L}}_A M &\cong \mathrm{RHom}_A(S, A) \otimes^{\mathrm{L}}_A M \\ &\cong \mathrm{RHom}_A(S, A \otimes^{\mathrm{L}}_A M) \cong \mathrm{RHom}_A(T^{\vee}, M). \end{aligned}$$ □

Theorem 18.1.25 ([165]) *Let A be a noetherian ring, and let R be a NC dualizing complex over A.*

(1) *Given a tilting complex T over A, the complex $R' := T \otimes^{\mathrm{L}}_{A} R$ is also a NC dualizing complex over A.*
(2) *Given a second NC dualizing complex R' over A, the complex $T := \mathrm{RHom}_{A}(R, R')$ is a tilting complex over A, and $R' \cong T \otimes^{\mathrm{L}}_{A} R$ in $\mathsf{D}(A^{\mathrm{en}})$.*
(3) *If T is a tilting complex over A, and if $T \otimes^{\mathrm{L}}_{A} R \cong R$ in $\mathsf{D}(A^{\mathrm{en}})$, then $T \cong A$ in $\mathsf{D}(A^{\mathrm{en}})$.*

Proof

(1) We need to verify that R' satisfies conditions (i)-(iii) of Definition 18.1.12.

Consider the equivalence of triangulated categories $G := T \otimes^{\mathrm{L}}_{A} (-) : \mathsf{D}(A) \to \mathsf{D}(A)$. The functor G has finite cohomological dimension (see Theorem 14.1.22, Corollary 14.4.28 and Proposition 14.1.5). Now $G(A) = T$, and $T \in \mathsf{D}^{\mathrm{b}}_{\mathrm{f}}(A)$ by Theorem 14.1.26, so Theorem 12.5.8 tells us that $G(M) \in \mathsf{D}^{\mathrm{b}}_{\mathrm{f}}(A)$ for every $M \in \mathsf{D}^{\mathrm{b}}_{\mathrm{f}}(A)$. Taking $M := R$ we see that $R' = G(R) \in \mathsf{D}^{\mathrm{b}}_{\mathrm{f}}(A)$.

Let $T^{\vee}$ be the quasi-inverse of T. By Lemma 18.1.24, with $B = A$, we see that

$$R' = T \otimes^{\mathrm{L}}_{A} R \cong \mathrm{RHom}_{A}(T^{\vee}, R) = D_{A}(T^{\vee}) \tag{18.1.26}$$

in $\mathsf{D}(A^{\mathrm{en}})$. The tilting complex $T^{\vee}$ belongs to $\mathsf{D}^{\mathrm{b}}_{\mathrm{f}}(A)$, so according to Theorem 18.1.18(1) with $B = \mathbb{K}$, we see that $D_{A}(T^{\vee}) \in \mathsf{D}^{\mathrm{b}}_{\mathrm{f}}(A^{\mathrm{op}})$. Thus $R' \in \mathsf{D}^{\mathrm{b}}_{\mathrm{f}}(A^{\mathrm{op}})$. We have now verified condition (i) of Definition 18.1.12 for R'.

To prove that R' has finite injective dimension over A, we need to find a uniform bound on the cohomology of $\mathrm{RHom}_{A}(M, R')$, for all $M \in \mathsf{M}(A)$. This is the same as finding a natural number d such that $\mathrm{Hom}_{\mathsf{D}(A)}(M, R'[i]) = 0$ for all $M \in \mathsf{M}(A)$ and $|i| > d$. We shall use the equivalence of triangulated categories

$$G^{\vee} := T^{\vee} \otimes^{\mathrm{L}}_{A} (-) \cong \mathrm{RHom}_{A}(T, -) : \mathsf{D}(A) \to \mathsf{D}(A).$$

Since $R \cong G^{\vee}(R')$, we obtain a $\mathbb{K}$-module isomorphism

$$G^{\vee} : \mathrm{Hom}_{\mathsf{D}(A)}(M, R'[i]) \xrightarrow{\simeq} \mathrm{Hom}_{\mathsf{D}(A)}(G^{\vee}(M), R[i])$$

for every i. The functor $G^{\vee}$ has finite cohomological dimension, and R has finite injective dimension; their sum d serves as the required bound.

Now let us prove that R' has finite injective dimension over A^{op}. Take a module $M \in \mathsf{M}(A^{\mathrm{op}})$. We have these isomorphisms for every integer i :

$$\begin{aligned}
\mathrm{H}^{i}(\mathrm{RHom}_{A^{\mathrm{op}}}(M, R')) &\cong^{1} \mathrm{Hom}_{\mathsf{D}(A^{\mathrm{op}})}(M, R'[i]) \\
&\cong^{2} \mathrm{Hom}_{\mathsf{D}(A^{\mathrm{op}})}(M, D_{A}(T^{\vee})[i]) \\
&\cong^{3} \mathrm{Hom}_{\mathsf{D}(A)}(T^{\vee}, D_{A^{\mathrm{op}}}(M)[i]) \\
&\cong^{4} \mathrm{H}^{i}(\mathrm{RHom}_{A}(T^{\vee}, \mathrm{RHom}_{A^{\mathrm{op}}}(M, R))) \\
&\cong^{5} \mathrm{H}^{i}(T \otimes^{\mathrm{L}}_{A} \mathrm{RHom}_{A^{\mathrm{op}}}(M, R)) \\
&= \mathrm{H}^{i}((G \circ D_{A^{\mathrm{op}}})(M)).
\end{aligned}$$

The justifications are as follows:

$\cong^1$: This is by Corollary 12.2.8.
$\cong^2$: This is by formula (18.1.26).
$\cong^3$: This relies on Corollary 18.1.19.
$\cong^4$: Again we use Corollary 12.2.8.
$\cong^5$: It is due to Lemma 18.1.24.

The functors G and $D_{A^{\mathrm{op}}}$ have finite cohomological dimensions, so we have a uniform bound on the cohomology of $(G \circ D_{A^{\mathrm{op}}})(M)$. This completes the proof that R' satisfies condition (ii) of Definition 18.1.12.

To prove that R' has the derived Morita property on the A side, we use the fact that R has the derived Morita property on the A side, and we invoke Lemma 18.1.21 with $B := A$ and $M := R$. Since $R' = G(R)$, for every $i \in \mathbb{Z}$ we have an isomorphism of $\mathbb{K}$-modules

$$G : \operatorname{Hom}_{\mathbf{D}(A)}(R, R[i]) \xrightarrow{\simeq} \operatorname{Hom}_{\mathbf{D}(A)}(R', R'[i]),$$

and this is zero when $i \neq 0$. For $i = 0$ we use Lemma 18.1.22: there is a commutative diagram of rings

$$\begin{array}{ccc} A^{\mathrm{op}} & \xrightarrow[\cong]{\mathrm{hm}^{\mathbf{D}}_{A^{\mathrm{op}},R}} & \operatorname{End}_{\mathbf{D}(A)}(R) \\ & \searrow^{\mathrm{hm}^{\mathbf{D}}_{A^{\mathrm{op}},R'}} & \cong \downarrow G \\ & & \operatorname{End}_{\mathbf{D}(A)}(R') \end{array}$$

and the vertical arrow G is an isomorphism. Therefore $\mathrm{hm}^{\mathbf{D}}_{A^{\mathrm{op}},R'}$ is a ring isomorphism.

Finally, to prove that R' has the derived Morita property on the A^{op} side we use the fact that the tilting complex $T^{\vee}$ has the derived Morita property on the A side (by Corollary 14.4.28), together with Lemma 18.1.21, applied to the complexes $T^{\vee} \in \mathbf{M}(A \otimes A^{\mathrm{op}})$ and $R' \in \mathbf{M}(A^{\mathrm{op}} \otimes A)$. Recall from (18.1.26) that $R' \cong D_A(T^{\vee})$ in $\mathbf{D}(A^{\mathrm{en}})$. By Corollary 18.1.19, for every $i \in \mathbb{Z}$ we have an isomorphism of $\mathbb{K}$-modules

$$D_A : \operatorname{Hom}_{\mathbf{D}(A)}(T^{\vee}, T^{\vee}[i]) \xrightarrow{\simeq} \operatorname{Hom}_{\mathbf{D}(A^{\mathrm{op}})}(R', R'[i]),$$

and this is zero when $i \neq 0$. For $i = 0$ we use Lemma 18.1.23: there is a commutative diagram of rings

$$\begin{array}{ccc} A^{\mathrm{op}} & \xrightarrow[\cong]{\mathrm{hm}^{\mathbf{D}}_{A^{\mathrm{op}},T^{\vee}}} & \operatorname{End}_{\mathbf{D}(A)}(T^{\vee}) \\ & \searrow^{\mathrm{Op}(\mathrm{hm}^{\mathbf{D}}_{A,R'})} & \cong \downarrow D_A \\ & & \operatorname{End}_{\mathbf{D}(A^{\mathrm{op}})}(R')^{\mathrm{op}} \end{array}$$

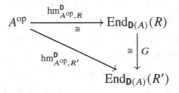

It follows that $\mathrm{hm}^{\mathbf{D}}_{A,R'} : A \to \mathrm{End}_{\mathbf{D}(A^{\mathrm{op}})}(R')$ is a ring isomorphism.

(2) The proof of this item is a minor modification of the proof of Theorem 17.1.10. Indeed, an ungraded version of Lemma 17.1.13 holds here, and we use Theorem 18.1.18 instead of Theorem 17.1.6.

(3) Assume that $T \otimes^{\mathrm{L}}_A R \cong R$ in $\mathbf{D}(A^{\mathrm{en}})$. Of course, $R = D_A(A)$. Formula (18.1.26) says that $D_A(A) \cong D_A(T^{\vee})$ in $\mathbf{D}(A^{\mathrm{en}})$. By Corollary 18.1.19 we conclude that $A \cong T^{\vee}$ in $\mathbf{D}(A^{\mathrm{en}})$. But then $A \cong T$ in $\mathbf{D}(A^{\mathrm{en}})$. □

The op-symmetry that gave rise to Proposition 18.1.13 implies that several variations of Theorem 18.1.25 are true. We wish to write down only one of them, which will be needed later.

Corollary 18.1.27 *Let A be a noetherian ring, and let R and R' be NC dualizing complexes over A. Then there is a tilting complex T' such that $R' \cong R \otimes^{\mathrm{L}}_A T'$ in $\mathbf{D}(A^{\mathrm{en}})$.*

Proof This is a trivial (yet confusing) consequence of item (2) of Theorem 18.1.25, when viewing R and R' as NC dualizing complexes over the ring A^{op}. □

Recall that a left action of a group G on a nonempty set X is called *simply transitive* if there is one G-orbit in X, and the stabilizer of every point $x \in X$ is trivial. In other words, for every (or equivalently, for some) point $x \in X$ the action function $G \times \{x\} \to X$, $(g, x) \mapsto g(x)$, is bijective.

Corollary 18.1.28 *Let A be a ring, and assume it has at least one NC dualizing complex. Then the left action of the group $\mathrm{DPic}_{\mathbb{K}}(A)$ on the set of isomorphism classes of NC dualizing complexes, induced by the bifunctor $(T, R) \mapsto T \otimes^{\mathrm{L}}_A R$, is simply transitive.*

Proof By Theorem 18.1.25(1) this is a well-defined action: $T \otimes^{\mathrm{L}}_A R$ is a NC dualizing complex. The action is transitive by Theorem 18.1.25(2), and it has trivial stabilizers by Theorem 18.1.25(3). □

Remark 18.1.29 Suppose A is a noetherian commutative $\mathbb{K}$-ring. A noncommutative dualizing complex over A is not the same as a commutative dualizing complex over A, in the sense of Definition 13.1.9. Indeed, there is a $\mathbb{K}$-ring homomorphism $A^{\mathrm{en}} \to A$, and this induces a restriction functor $\mathbf{D}(A) \to \mathbf{D}(A^{\mathrm{en}})$. If $R \in \mathbf{D}(A)$ is a commutative dualizing complex over A, then its image in $\mathbf{D}(A^{\mathrm{en}})$ is a noncommutative dualizing complex over A. However, according to Theorem 18.1.25(1), given any tilting complex T over A, the complex $R' := T \otimes^{\mathrm{L}}_A R$ is also a NC dualizing complex over A. And R' can very easily fail to be in the essential image of $\mathsf{D}(A)$; e.g. by taking $T := A(\psi)$, the twist by an automorphism ψ of the trivial bimodule A.

Remark 18.1.30 In case the base ring $\mathbb{K}$ is not a field, but A is a *flat noetherian* $\mathbb{K}$*-ring*, then the definitions and results in this section are still valid.

In case the ring A is noetherian, but is not flat over the base ring $\mathbb{K}$, we can still define NC dualizing complexes over it. For that we choose a K-flat DG ring resolution $\tilde{A} \to A$ over $\mathbb{K}$; see Theorem 12.8.7 and Proposition 12.8.5. A NC dualizing complex over A is then a complex $R \in \mathsf{D}(\tilde{A}^{\mathrm{en}})$ that satisfies the three conditions of Definition 18.1.12. As explained in Remark 14.3.24, the triangulated category $\mathsf{D}(\tilde{A}^{\mathrm{en}})$ is independent of the resolution $\tilde{A}$, up to a canonical equivalence. This means that the dualizing complex R is also independent of the resolution.

It is expected that Theorems 18.1.18 and 18.1.25 will still hold in this general situation; but we did not verify this assertion.

However, flatness alone is not sufficient for many of the results in the following sections (mainly in Section 18.5, in which we rely on results from Chapter 17, which assumed Convention 15.1.17). Therefore we decided to assume $\mathbb{K}$ is a field in the current section as well.

18.2 Rigid NC DC: Definition and Uniqueness

In this section we introduce rigid NC dualizing complexes, and prove their uniqueness. The material is mostly from Van den Bergh's paper [153], with some improvements coming from [165]. We continue with Convention 18.1.1; in particular, $\mathbb{K}$ is a base field, and all rings are central over $\mathbb{K}$.

Let A be a ring, and let $M, N \in \mathsf{M}(A^{\mathrm{en}})$. The $\mathbb{K}$-module $M \otimes N$ has two A-module structures and two A^{op}-module structures, all commuting with each other. We organize them into an *outside action* and an *inside action* of the ring A^{en}. Here are the formulas: given elements $m \in M$, $n \in N$ and $a_1, a_2 \in A$, we let

$$(a_1 \otimes \mathrm{op}(a_2)) \cdot^{\mathrm{out}} (m \otimes n) := (a_1 \cdot m) \otimes (n \cdot a_2) \tag{18.2.1}$$

and

$$(a_1 \otimes \mathrm{op}(a_2)) \cdot^{\mathrm{in}} (m \otimes n) := (m \cdot a_2) \otimes (a_1 \cdot n). \tag{18.2.2}$$

In these formulas we are using the canonical antiautomorphism $\mathrm{op} : A \to A^{\mathrm{op}}$. We shall denote the two copies of A^{en} that act on $M \otimes N$ by

$$A^{\mathrm{en,out}} = A^{\mathrm{out}} \otimes A^{\mathrm{op,out}} \tag{18.2.3}$$

and

$$A^{\mathrm{en,in}} = A^{\mathrm{in}} \otimes A^{\mathrm{op,in}}. \tag{18.2.4}$$

Namely the ring $A^{\mathrm{en,out}}$ acts on $M \otimes N$ by the outside action (18.2.1), and the ring $A^{\mathrm{en,in}}$ acts by the inside action (18.2.2).

Let us define the ring

$$A^{\mathrm{four}} := A^{\mathrm{en,out}} \otimes A^{\mathrm{en,in}}. \tag{18.2.5}$$

There are ring homomorphisms

$$A^{\mathrm{en,out}} \to A^{\mathrm{four}} \leftarrow A^{\mathrm{en,in}}. \tag{18.2.6}$$

With this notation we have $M \otimes N \in \mathbf{M}(A^{\mathrm{four}})$. The action of the ring A^{four} on the module $M \otimes N$ is this, explicitly: an element

$$(a_1 \otimes \mathrm{op}(a_2)) \otimes (a_3 \otimes \mathrm{op}(a_4)) \in A^{\mathrm{four}}$$

acts on an element $m \otimes n \in M \otimes N$ by

$$((a_1 \otimes \mathrm{op}(a_2)) \otimes (a_3 \otimes \mathrm{op}(a_4))) \cdot (m \otimes n) = (a_1 \cdot m \cdot a_4) \otimes (a_3 \cdot n \cdot a_2).$$

In some situations we prefer to view the inside action of A^{en} on $M \otimes N$ as a right action. This is possible by the isomorphism trop from formula (18.1.5). To be explicit, for elements $m \otimes n \in M \otimes N$ and $a_3 \otimes \mathrm{op}(a_4) \in A^{\mathrm{en,in}}$ the action is as follows, using formula (18.1.8):

$$\begin{aligned}(m \otimes n) \cdot^{\mathrm{in}} (a_3 \otimes \mathrm{op}(a_4)) &= (a_4 \otimes \mathrm{op}(a_3)) \cdot^{\mathrm{in}} (m \otimes n) \\ &= (m \cdot a_3) \otimes (a_4 \cdot n).\end{aligned} \tag{18.2.7}$$

Example 18.2.8 Suppose we take $M = N := A \in \mathbf{M}(A^{\mathrm{en}})$. Then $P := M \otimes N = A \otimes A$ can also be viewed as $P \cong A^{\mathrm{en}}$, and then in it an A^{en}-bimodule. The left A^{en}-module structure on $P \cong A^{\mathrm{en}}$ is this: for a module element $b_1 \otimes b_2 \in P$ and a ring element $a_1 \otimes \mathrm{op}(a_2) \in A^{\mathrm{en}}$ we have

$$(a_1 \otimes \mathrm{op}(a_2)) \cdot (b_1 \otimes b_2) = (a_1 \cdot b_1) \otimes (b_2 \cdot a_2);$$

so this is the outside action on $P = A \otimes A$. The right A^{en}-module structure on $P \cong A^{\mathrm{en}}$ is this: for a ring element $a_3 \otimes \mathrm{op}(a_4) \in A^{\mathrm{en}}$ we have

$$(b_1 \otimes b_2) \cdot (a_3 \otimes \mathrm{op}(a_4)) = (b_1 \cdot a_3) \otimes (a_4 \cdot b_2);$$

so this is the inside action on $P = A \otimes A$, cf. (18.2.7).

Given bimodules $L, M, N \in \mathbf{M}(A^{\mathrm{en}})$, we denote by $\mathrm{Hom}_{A^{\mathrm{en,out}}}(L, M \otimes N)$ the $\mathbb{K}$-module of homomorphisms $\phi : L \to M \otimes N$ that are A^{en}-linear for the outside action of A^{en} on $M \otimes N$. This is an A^{en}-module by the inside action on $M \otimes N$. Namely

$$\mathrm{Hom}_{A^{\mathrm{en,out}}}(L, M \otimes N) \in \mathbf{M}(A^{\mathrm{en,in}}) = \mathbf{M}(A^{\mathrm{en}}).$$

All these four-module operations extend in an obvious way to complexes, and can be derived.

Definition 18.2.9 Let A be a ring and $M \in \mathsf{D}(A^{\mathrm{en}})$. The *NC square of M* is the complex

$$\mathrm{Sq}_A(M) := \mathrm{RHom}_{A^{\mathrm{en,out}}}(A, M \otimes M) \in \mathsf{D}(A^{\mathrm{en,in}}) = \mathsf{D}(A^{\mathrm{en}}).$$

Of course, the square is relative to the base field $\mathbb{K}$, which is implicit in the formulas.

The square of the complex M is calculated as follows. First note that M is K-flat over $\mathbb{K}$, since $\mathbb{K}$ is a field; so $M \otimes M = M \otimes^{\mathrm{L}}_{\mathbb{K}} M \in \mathsf{D}(A^{\mathrm{four}})$. We choose a K-injective resolution $M \otimes M \to I$ in $\mathsf{C}_{\mathrm{str}}(A^{\mathrm{four}})$. The complex I is unique up to a homotopy equivalence that itself is unique up to homotopy (in other words, I is unique up to a unique isomorphism in $\mathsf{K}(A^{\mathrm{four}})$). Because of the flatness of the ring homomorphisms (18.2.6), the complex I is K-injective over $A^{\mathrm{en,out}}$, and

$$\mathrm{Sq}_A(M) = \mathrm{Hom}_{A^{\mathrm{en,out}}}(A, I) \in \mathsf{D}(A^{\mathrm{en,in}}) = \mathsf{D}(A^{\mathrm{en}}). \tag{18.2.10}$$

Definition 18.2.11 Let $\phi : M \to N$ be a morphism in $\mathsf{D}(A^{\mathrm{en}})$. The *NC square of ϕ* is the morphism

$$\mathrm{Sq}_A(\phi) := \mathrm{RHom}_{A^{\mathrm{en,out}}}(\mathrm{id}_A, \phi \otimes \phi) : \mathrm{Sq}_A(M) \to \mathrm{Sq}_A(N)$$

in $\mathsf{D}(A^{\mathrm{en}})$.

The square of the morphism ϕ is calculated as follows. Choose a K-projective resolution $P \to M$ in $\mathsf{C}_{\mathrm{str}}(A^{\mathrm{en}})$. Then ϕ is represented by a homomorphism $\tilde{\phi} : P \to N$ in $\mathsf{C}_{\mathrm{str}}(A^{\mathrm{en}})$. The morphism $\phi \otimes \phi : M \otimes M \to N \otimes N$ in $\mathsf{D}(A^{\mathrm{four}})$ is represented by the homomorphism $\tilde{\phi} \otimes \tilde{\phi} : P \otimes P \to N \otimes N$ in $\mathsf{C}_{\mathrm{str}}(A^{\mathrm{four}})$. Next we choose K-injective resolutions $P \otimes P \to I$ and $N \otimes N \to J$ in $\mathsf{C}_{\mathrm{str}}(A^{\mathrm{four}})$. The homomorphism $\tilde{\phi} \otimes \tilde{\phi}$ induces a homomorphism $\tilde{\psi} : I \to J$ in $\mathsf{C}_{\mathrm{str}}(A^{\mathrm{four}})$. Note that $\tilde{\psi}$ is unique up to homotopy. Finally

$$\mathrm{Sq}_A(\phi) = \mathrm{Q}(\mathrm{Hom}_{A^{\mathrm{en,out}}}(\mathrm{id}_A, \tilde{\psi})) :$$
$$\mathrm{Sq}_A(M) = \mathrm{Hom}_{A^{\mathrm{en,out}}}(A, I) \to \mathrm{Sq}_A(N) = \mathrm{Hom}_{A^{\mathrm{en,out}}}(A, J)$$

in $\mathsf{D}(A^{\mathrm{en,in}}) = \mathsf{D}(A^{\mathrm{en}})$.

Definition 18.2.12 Let A be a ring. A *NC rigid complex over A* is a pair (M, ρ), where $M \in \mathsf{D}(A^{\mathrm{en}})$, and $\rho : M \xrightarrow{\simeq} \mathrm{Sq}_A(M)$ is an isomorphism in $\mathsf{D}(A^{\mathrm{en}})$, called a *NC rigidifying isomorphism*.

Definition 18.2.13 Let (M, ρ) and (N, σ) be NC rigid complexes over A. A *NC rigid morphism over A*

$$\phi : (M, \rho) \to (N, \sigma)$$

is a morphism $\phi : M \to N$ in $\mathbf{D}(A^{\mathrm{en}})$, such that the diagram

$$\begin{array}{ccc} M & \xrightarrow{\rho} & \mathrm{Sq}_A(M) \\ {\scriptstyle \phi}\downarrow & & \downarrow{\scriptstyle \mathrm{Sq}_A(\phi)} \\ N & \xrightarrow{\sigma} & \mathrm{Sq}_A(N) \end{array}$$

in $\mathbf{D}(A^{\mathrm{en}})$ is commutative.

Definition 18.2.14 Let A be a ring. The category of *NC rigid complexes over A* is the category whose objects are the NC rigid complexes from Definition 18.2.12, and whose morphisms are the NC rigid morphisms Definition 18.2.13.

Definition 18.2.15 (Van den Bergh [153]) Let A be a noetherian ring. A *rigid NC dualizing complex over A* is a NC rigid complex (R, ρ), such that R is a NC dualizing complex over A, in the sense of Definition 18.1.12.

Again, we recall that the notion of rigidity is relative to the base field $\mathbb{K}$.

Recall that the center of the ring A is denoted by $\mathrm{Cent}(A)$. Of course, $\mathrm{Cent}(A) = \mathrm{Cent}(A^{\mathrm{op}})$. Given a bimodule $M \in \mathbf{M}(A^{\mathrm{en}})$, there are two ring homomorphisms

$$\mathrm{chm}_{M,A}, \mathrm{chm}_{M,A^{\mathrm{op}}} : \mathrm{Cent}(A) \to \mathrm{End}_{\mathbf{M}(A^{\mathrm{en}})}(M),$$

which we call the *central homotheties* through A and through A^{op}, respectively. In terms of elements they are $\mathrm{chm}_{M,A}(a)(m) := a \cdot m$ and $\mathrm{chm}_{M,A^{\mathrm{op}}}(a)(m) := m \cdot a$ for $a \in \mathrm{Cent}(A)$ and $m \in M$.

Example 18.2.16 For the bimodule $M = A$, both ring homomorphisms

$$\mathrm{chm}_{A,A}, \mathrm{chm}_{A,A^{\mathrm{op}}} : \mathrm{Cent}(A) \to \mathrm{End}_{\mathbf{M}(A^{\mathrm{en}})}(A)$$

are isomorphisms, and they are equal.

The ring homomorphisms $\mathrm{chm}_{M,A}$ and $\mathrm{chm}_{M,A^{\mathrm{op}}}$ extend to complexes, namely for $M \in \mathbf{C}(A^{\mathrm{en}})$ there are ring homomorphisms

$$\mathrm{chm}_{M,A}, \mathrm{chm}_{M,A^{\mathrm{op}}} : \mathrm{Cent}(A) \to \mathrm{End}_{\mathbf{C}_{\mathrm{str}}(A^{\mathrm{en}})}(M).$$

By postcomposing them with the categorical localization functor Q we obtain ring homomorphisms

$$\mathrm{chm}^{\mathbf{D}}_{M,A}, \mathrm{chm}^{\mathbf{D}}_{M,A^{\mathrm{op}}} : \mathrm{Cent}(A) \to \mathrm{End}_{\mathbf{D}(A^{\mathrm{en}})}(M).$$

These central homotheties were already used in Section 14.5.

Example 18.2.17 For the complex of bimodules $M = A$, both ring homomorphisms

$$\mathrm{chm}^{\mathbf{D}}_{M,A}, \mathrm{chm}^{\mathbf{D}}_{M,A^{\mathrm{op}}} : \mathrm{Cent}(A) \to \mathrm{End}_{\mathbf{D}(A^{\mathrm{en}})}(A)$$

are isomorphisms, and they are equal. This is because of the fully faithful embedding $\mathbf{M}(A^{\mathrm{en}}) \to \mathbf{D}(A^{\mathrm{en}})$ and the previous example.

Lemma 18.2.18 *Given $R \in \mathbf{D}(A^{\mathrm{en}})$, consider the functor*

$$D := \mathrm{Hom}_A(-, R) : \mathbf{D}(A^{\mathrm{en}})^{\mathrm{op}} \to \mathbf{D}(A^{\mathrm{en}}).$$

Then for every $M \in \mathbf{D}(A^{\mathrm{en}})$ the diagram of rings

$$\begin{array}{ccc} \mathrm{Cent}(A) & \xrightarrow{\mathrm{chm}^{\mathbf{D}}_{M,A^{\mathrm{op}}}} & \mathrm{End}_{\mathbf{D}(A^{\mathrm{en}})}(M) \\ & \searrow^{\mathrm{chm}^{\mathbf{D}}_{D(M),A}} & \downarrow D \\ & & \mathrm{End}_{\mathbf{D}(A^{\mathrm{en}})}(D(M))^{\mathrm{op}} \end{array}$$

is commutative.

Proof Choose a K-injective resolution $R \to I$ in $\mathbf{C}_{\mathrm{str}}(A^{\mathrm{en}})$, and consider the functor $\tilde{D} := \mathrm{Hom}_A(-, I)$. So $D \cong \tilde{D}$ as functors $\mathbf{D}(A^{\mathrm{en}})^{\mathrm{op}} \to \mathbf{D}(A^{\mathrm{en}})$. Because of the contravariance of $\tilde{D}$, the diagram of rings

$$\begin{array}{ccc} \mathrm{Cent}(A) & \xrightarrow{\mathrm{chm}_{M,A^{\mathrm{op}}}} & \mathrm{End}_{\mathbf{C}_{\mathrm{str}}(A^{\mathrm{en}})}(M) \\ & \searrow^{\mathrm{chm}_{\tilde{D}(M),A}} & \downarrow \tilde{D} \\ & & \mathrm{End}_{\mathbf{C}_{\mathrm{str}}(A^{\mathrm{en}})}(\tilde{D}(M))^{\mathrm{op}} \end{array}$$

is commutative. After applying the localization functor Q to the vertical arrow $\tilde{D}$ in the diagram above, we obtain the commutative diagram on the derived level. □

Lemma 18.2.19 *Given $M, N \in \mathbf{D}(A^{\mathrm{en}})$, consider the complex*

$$L := \mathrm{RHom}_{A^{\mathrm{en,out}}}(A, M \otimes N) \in \mathbf{D}(A^{\mathrm{en,in}}) = \mathbf{D}(A^{\mathrm{en}}).$$

Let $a, a', a'' \in \mathrm{Cent}(A)$ be elements such that $\mathrm{chm}^{\mathbf{D}}_{N,A}(a) = \mathrm{chm}^{\mathbf{D}}_{N,A^{\mathrm{op}}}(a')$ in $\mathrm{End}_{\mathbf{D}(A^{\mathrm{en}})}(N)$ and $\mathrm{chm}^{\mathbf{D}}_{M,A}(a') = \mathrm{chm}^{\mathbf{D}}_{M,A^{\mathrm{op}}}(a'')$ in $\mathrm{End}_{\mathbf{D}(A^{\mathrm{en}})}(M)$. Then $\mathrm{chm}^{\mathbf{D}}_{L,A}(a) = \mathrm{chm}^{\mathbf{D}}_{L,A^{\mathrm{op}}}(a'')$ in $\mathrm{End}_{\mathbf{D}(A^{\mathrm{en}})}(L)$.

Proof The proof resembles that of Lemma 14.5.15. Consider the object $M \otimes N \in \mathbf{D}(A^{\mathrm{four}})$. There are four ring homomorphisms

$$\mathrm{chm}^{\mathbf{D}}_{M\otimes N,B} : \mathrm{Cent}(A) \to \mathrm{End}_{\mathbf{D}(A^{\mathrm{four}})}(M \otimes N),$$

corresponding to these four options for the ring B, in terms of formulas (18.2.3) and (18.2.4): $B := A^{\mathrm{out}}, A^{\mathrm{op,out}}, A^{\mathrm{in}}, A^{\mathrm{op,in}}$. We know that

$$\begin{aligned} \mathrm{chm}^{\mathbf{D}}_{M\otimes N,A^{\mathrm{in}}}(a) &= \mathrm{chm}^{\mathbf{D}}_{M\otimes N,A^{\mathrm{op,out}}}(a'), \\ \mathrm{chm}^{\mathbf{D}}_{M\otimes N,A^{\mathrm{out}}}(a') &= \mathrm{chm}^{\mathbf{D}}_{M\otimes N,A^{\mathrm{op,in}}}(a''). \end{aligned} \tag{18.2.20}$$

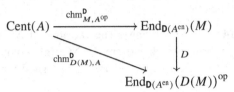

Let $M \otimes N \to J$ be a K-injective resolution in $\mathbf{C}_{\mathrm{str}}(A^{\mathrm{four}})$, so

$$\operatorname{End}_{\mathbf{D}(A^{\mathrm{four}})}(M \otimes N) \cong \mathrm{H}^0(\operatorname{End}_{\mathbf{C}_{\mathrm{str}}(A^{\mathrm{four}})}(J))$$

as rings. Formulas (18.2.20) imply that

$$\begin{aligned} &\mathrm{chm}_{J,A^{\mathrm{in}}}(a) \sim \mathrm{chm}_{J,A^{\mathrm{op,out}}}(a'), \\ &\mathrm{chm}_{J,A^{\mathrm{out}}}(a') \sim \mathrm{chm}_{J,A^{\mathrm{op,in}}}(a''), \end{aligned} \tag{18.2.21}$$

where "$\sim$" means the homotopy relation on degree 0 cocycles in the DG ring $\operatorname{End}_{A^{\mathrm{four}}}(J)$.

Now let us write $\tilde{L} = \operatorname{Hom}_{A^{\mathrm{en,out}}}(A, J) \in \mathbf{C}(A^{\mathrm{en,in}})$. The complex $\tilde{L}$ is K-injective over $A^{\mathrm{en,in}}$, again by the flatness of the ring homomorphism (18.2.6), and $\tilde{L} \cong L$ in $\mathbf{D}(A^{\mathrm{en,in}})$. Hence there is a ring isomorphism

$$\mathrm{H}^0(\operatorname{End}_{A^{\mathrm{en,in}}}(\tilde{L})) \cong \operatorname{End}_{\mathbf{D}(A^{\mathrm{en,in}})}(L). \tag{18.2.22}$$

Now A is a central bimodule over its center. This implies that the two actions of $\mathrm{Cent}(A)$ on $\tilde{L}$ that come from $\mathrm{chm}_{J,A^{\mathrm{out}}}$ and $\mathrm{chm}_{J,A^{\mathrm{op,out}}}$ are equal. The two actions of $\mathrm{Cent}(A)$ on $\tilde{L}$ that come from $\mathrm{chm}_{J,A^{\mathrm{in}}}$ and $\mathrm{chm}_{J,A^{\mathrm{op,in}}}$ are the actions $\mathrm{chm}_{\tilde{L},A^{\mathrm{in}}}$ and $\mathrm{chm}_{\tilde{L},A^{\mathrm{op,in}}}$, respectively. From formulas (18.2.21) we conclude that $\mathrm{chm}_{\tilde{L},A^{\mathrm{in}}}(a) \sim \mathrm{chm}_{\tilde{L},A^{\mathrm{op,in}}}(a'')$. By formula (18.2.22) we see that $\mathrm{chm}^{\mathbf{D}}_{L,A^{\mathrm{in}}}(a) = \mathrm{chm}^{\mathbf{D}}_{L,A^{\mathrm{op,in}}}(a'')$ as claimed. □

Theorem 18.2.23 ([165]) *Let A be a noetherian ring, and let R be a NC dualizing complex over A.*

(1) *The two ring homomorphisms*

$$\mathrm{chm}^{\mathbf{D}}_{R,A},\ \mathrm{chm}^{\mathbf{D}}_{R,A^{\mathrm{op}}} : \mathrm{Cent}(A) \to \operatorname{End}_{\mathbf{D}(A^{\mathrm{en}})}(R)$$

are bijective.

(2) *If R is a rigid NC dualizing complex, then* $\mathrm{chm}^{\mathbf{D}}_{R,A} = \mathrm{chm}^{\mathbf{D}}_{R,A^{\mathrm{op}}}$.

Proof

(1) Recall the duality functor D_A from Definition 18.1.15, with $B = A$. By Corollary 18.1.19 it gives rise to an equivalence $D_A : \mathbf{D}^{\mathrm{b}}_{(\mathrm{f},..)}(A^{\mathrm{en}})^{\mathrm{op}} \to \mathbf{D}^{\mathrm{b}}_{(..,\mathrm{f})}(A^{\mathrm{en}})$. By Lemma 18.2.18, for every $M \in \mathbf{D}^{\mathrm{b}}_{(\mathrm{f},..)}(A^{\mathrm{en}})$ we get a commutative diagram of rings

$$\begin{array}{ccc} \mathrm{Cent}(A) & \xrightarrow{\ \mathrm{chm}^{\mathbf{D}}_{M,A^{\mathrm{op}}}\ } & \operatorname{End}_{\mathbf{D}(A^{\mathrm{en}})}(M) \\ & \searrow^{\mathrm{chm}^{\mathbf{D}}_{D_A(M),A}} & \downarrow{\cong}\ D_A \\ & & \operatorname{End}_{\mathbf{D}(A^{\mathrm{en}})}(D_A(M))^{\mathrm{op}} \end{array}$$

Note that the vertical arrow D_A is an isomorphism. Taking $M = R$ we have $D_A(R) \cong A$ in $\mathbf{D}(A^{\mathrm{en}})$. As we saw in Example 18.2.17, the homomorphism $\mathrm{chm}^{\mathbf{D}}_{A,A}$ is bijective. This proves that $\mathrm{chm}^{\mathbf{D}}_{R,A^{\mathrm{op}}}$ is bijective.

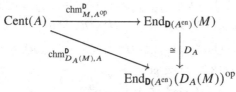

We find likewise for $\operatorname{chm}^{\mathbf{D}}_{R,A}$, but this time we use the duality functor $D_{A^{\mathrm{op}}}$.

(2) Now there is a rigidifying isomorphism $\rho : R \xrightarrow{\simeq} \mathrm{Sq}_A(R)$ in $\mathbf{D}(A^{\mathrm{en}})$. Let $f : \mathrm{Cent}(A) \to \mathrm{Cent}(A)$ be the ring automorphism $f := (\operatorname{chm}^{\mathbf{D}}_{R,A^{\mathrm{op}}})^{-1} \circ \operatorname{chm}^{\mathbf{D}}_{R,A}$. We need to prove that $f = \mathrm{id}$.

For this we shall use Lemma 18.2.19, with $M = N := R$, so $L \cong \mathrm{Sq}_A(R)$ in $\mathbf{D}(A^{\mathrm{en}})$. Take an element $a \in \mathrm{Cent}(A)$, and let $a' := f(a)$ and $a'' := f(a')$ in $\mathrm{Cent}(A)$. The lemma says that $\operatorname{chm}^{\mathbf{D}}_{L,A}(a) = \operatorname{chm}^{\mathbf{D}}_{L,A^{\mathrm{op}}}(a'')$. But by the definition of f and a' we have $\operatorname{chm}^{\mathbf{D}}_{R,A}(a) = \operatorname{chm}^{\mathbf{D}}_{R,A^{\mathrm{op}}}(a')$. Since $R \cong L$ in $\mathbf{D}(A^{\mathrm{en}})$, and since $\operatorname{chm}^{\mathbf{D}}_{R,A^{\mathrm{op}}}$ is bijective, we conclude that $a' = a''$. This means that $f(a) = f(f(a))$, so $a = f(a)$. Finally, the element a was arbitrary, and hence $f = \mathrm{id}$. □

Lemma 18.2.24 *Let (R, ρ) and (R', ρ') be rigid dualizing complexes, let $\phi : R \to R'$ be a morphism in $\mathbf{D}(A^{\mathrm{en}})$, and let $a \in \mathrm{Cent}(A)$. Then*

$$\mathrm{Sq}_A(\phi \circ \operatorname{chm}^{\mathbf{D}}_{R,A}(a)) = \mathrm{Sq}_A(\phi) \circ \operatorname{chm}^{\mathbf{D}}_{\mathrm{Sq}_A(R),A}(a^2)$$

as morphisms $\mathrm{Sq}_A(R) \to \mathrm{Sq}_A(R')$ in $\mathbf{D}(A^{\mathrm{en}})$.

Proof Because Sq_A is a functor, there is equality

$$\mathrm{Sq}_A(\phi \circ \operatorname{chm}^{\mathbf{D}}_{R,A}(a)) = \mathrm{Sq}_A(\phi) \circ \mathrm{Sq}_A(\operatorname{chm}^{\mathbf{D}}_{R,A}(a)).$$

So we can assume that $R' = R$ and $\phi = \mathrm{id}_R$. We need to prove that the equality

$$\mathrm{Sq}_A(\operatorname{chm}^{\mathbf{D}}_{R,A}(a)) = \operatorname{chm}^{\mathbf{D}}_{\mathrm{Sq}_A(R),A}(a^2) \tag{18.2.25}$$

holds for these endomorphisms of $\mathrm{Sq}_A(R)$ in $\mathbf{D}(A^{\mathrm{en}})$.

From Definition 18.2.11, applied to the morphism $\operatorname{chm}^{\mathbf{D}}_{R,A}(a) : R \to R$ in $\mathbf{D}(A^{\mathrm{en}})$, we see that

$$\mathrm{Sq}_A(\operatorname{chm}^{\mathbf{D}}_{R,A}(a)) = \operatorname{chm}^{\mathbf{D}}_{\mathrm{Sq}_A(R),A^{\mathrm{out}}}(a) \circ \operatorname{chm}^{\mathbf{D}}_{\mathrm{Sq}_A(R),A^{\mathrm{in}}}(a) \tag{18.2.26}$$

as endomorphisms of $\mathrm{Sq}_A(R)$ in $\mathbf{D}(A^{\mathrm{en}}) = \mathbf{D}(A^{\mathrm{en,in}})$. But because R is a rigid dualizing complex, Theorem 18.2.11(2) says that $\operatorname{chm}^{\mathbf{D}}_{R,A}(a) = \operatorname{chm}^{\mathbf{D}}_{R,A^{\mathrm{op}}}(a)$. Therefore, in formula (18.2.26) we can replace the action of a through A^{out} with its action through $A^{\mathrm{op,in}}$. This gives

$$\mathrm{Sq}_A(\operatorname{chm}^{\mathbf{D}}_{R,A}(a)) = \operatorname{chm}^{\mathbf{D}}_{\mathrm{Sq}_A(R),A^{\mathrm{op,in}}}(a) \circ \operatorname{chm}^{\mathbf{D}}_{\mathrm{Sq}_A(R),A^{\mathrm{in}}}(a). \tag{18.2.27}$$

Next, using Lemma 18.2.19, with $M = N := R$ and $a' = a'' := a$, we obtain

$$\operatorname{chm}^{\mathbf{D}}_{\mathrm{Sq}_A(R),A^{\mathrm{op,in}}}(a) = \operatorname{chm}^{\mathbf{D}}_{\mathrm{Sq}_A(R),A^{\mathrm{in}}}(a).$$

Plugging this into (18.2.27) we get

$$\mathrm{Sq}_A(\operatorname{chm}^{\mathbf{D}}_{R,A}(a)) = \operatorname{chm}^{\mathbf{D}}_{\mathrm{Sq}_A(R),A^{\mathrm{in}}}(a)^2.$$

Finally, since $\operatorname{chm}^{\mathbf{D}}_{\mathrm{Sq}_A(R),A^{\mathrm{in}}}$ is a ring homomorphism, we end up with (18.2.25). □

In Definition 12.9.1 we introduced derived pseudo-finite DG modules over a DG ring A. Recall that a DG module $M \in \mathbf{D}(A)$ is called derived pseudo-finite if it belongs to the saturated full triangulated subcategory of $\mathbf{D}(A)$ generated by the pseudo-finite semi-free DG modules. When A is a ring, the pseudo-finite semi-free DG A-modules are precisely the bounded above complexes of finite free A-modules.

Theorem 18.2.28 (Uniqueness, [153], [165]) *Let A be a noetherian ring, and assume A is a derived pseudo-finite complex over A^{en}. Suppose (R, ρ) is a NC rigid dualizing complex over A. Then (R, ρ) is unique, up to a unique NC rigid isomorphism.*

Proof Suppose (R', ρ') is another NC rigid dualizing complex over A. According to Theorem 18.1.25 and Corollary 18.1.27, there are tilting complexes T and T' over A such that

$$R' \cong T \otimes^{\mathrm{L}}_A R \cong R \otimes^{\mathrm{L}}_A T' \tag{18.2.29}$$

in $\mathbf{D}(A^{\mathrm{en}})$. We have the following sequence of isomorphisms in $\mathbf{D}(A^{\mathrm{en}})$:

$$\begin{aligned} T \otimes^{\mathrm{L}}_A R \cong^1 R' &\cong^2 \mathrm{RHom}_{A^{\mathrm{en}},\mathrm{out}}(A, R' \otimes R') \\ &\cong^1 \mathrm{RHom}_{A^{\mathrm{en}},\mathrm{out}}(A, (R \otimes^{\mathrm{L}}_A T') \otimes (T \otimes^{\mathrm{L}}_A R)) \\ &\cong^3 \mathrm{RHom}_{A^{\mathrm{en}},\mathrm{out}}(A, (R \otimes R) \otimes^{\mathrm{L}}_{A^{\mathrm{en}}} (T' \otimes T)) \\ &\cong^4 \mathrm{RHom}_{A^{\mathrm{en}},\mathrm{out}}(A, R \otimes R) \otimes^{\mathrm{L}}_{A^{\mathrm{en}}} (T' \otimes T) \\ &\cong^5 R \otimes^{\mathrm{L}}_{A^{\mathrm{en}}} (T' \otimes T) \cong^6 T \otimes^{\mathrm{L}}_A R \otimes^{\mathrm{L}}_A T'. \end{aligned} \tag{18.2.30}$$

Here are the explanations for the various isomorphisms:

$\cong^1$: These are due to (18.2.29).

$\cong^2$: This is the rigidifying isomorphism ρ'.

$\cong^3$: This isomorphism is by rearranging the tensor factors, and in it the inside action of A^{en} on $R \otimes R$ is viewed as a right action, and this matches the left outside action of A^{en} on $T' \otimes T$. Cf. formulas (18.2.7) and (18.2.1).

$\cong^4$: It is an application of the tensor-evaluation isomorphism (see Theorem 12.9.10). This is where we need A to be a derived pseudo-finite complex over A^{en} – so that condition (i) of Theorem 12.9.10 will hold. Condition (ii) that theorem holds since the complex $R \otimes R$ has bounded cohomology. As for condition (iii) of the theorem: the complex $T' \otimes T$ is a tilting complex over the ring A^{en}, so it has finite flat dimension, and thus it certainly has derived bounded below tensor displacement.

$\cong^5$: This uses the rigidifying isomorphism ρ.

$\cong^6$: It is a rearranging of the tensor factors. Recall that the left action of A^{en} on $T' \otimes T$ is the outside action, cf. formula(18.2.1).

Let $T^\vee$ be the quasi-inverse of T. After applying $T^\vee \otimes^{\mathrm{L}}_A (-)$ to the isomorphisms (18.2.30) we get $R \cong R \otimes^{\mathrm{L}}_A T'$ in $\mathbf{D}(A^{\mathrm{en}})$. Equation (18.2.29) tells us that there is an isomorphism $\phi^\dagger : R \xrightarrow{\simeq} R'$ in $\mathbf{D}(A^{\mathrm{en}})$.

The isomorphism $\phi^\dagger$ above need not be rigid. What we do know is that both morphisms $\rho' \circ \phi^\dagger, \mathrm{Sq}_A(\phi^\dagger) \circ \rho : R \to \mathrm{Sq}_A(R')$ are isomorphisms in $\mathbf{D}(A^{\mathrm{en}})$. By Theorem 18.2.23 the automorphisms of R in $\mathbf{D}(A^{\mathrm{en}})$ are all of the form $\mathrm{chm}^{\mathbf{D}}_{R,A}(a)$, for elements $a \in \mathrm{Cent}(A)^\times$. Thus there is a unique invertible central element $a \in A$ such that

$$\mathrm{Sq}_A(\phi^\dagger) \circ \rho = \rho' \circ \phi^\dagger \circ \mathrm{chm}^{\mathbf{D}}_{R,A}(a). \tag{18.2.31}$$

Define the isomorphism

$$\phi := \phi^\dagger \circ \mathrm{chm}^{\mathbf{D}}_{R,A}(a^{-1}) : R \xrightarrow{\simeq} R'$$

in $\mathbf{D}(A^{\mathrm{en}})$. Then we have the following equalities:

$$\begin{aligned}
\mathrm{Sq}_A(\phi) \circ \rho &= \mathrm{Sq}_A(\phi^\dagger \circ \mathrm{chm}^{\mathbf{D}}_{R,A}(a^{-1})) \circ \rho \\
&=^{\text{(i)}} \mathrm{Sq}_A(\phi^\dagger) \circ \mathrm{chm}^{\mathbf{D}}_{\mathrm{Sq}_A(R),A}(a^{-2}) \circ \rho \\
&=^{\text{(ii)}} \mathrm{Sq}_A(\phi^\dagger) \circ \rho \circ \mathrm{chm}^{\mathbf{D}}_{R,A}(a^{-2}) \\
&=^{\text{(iii)}} \rho' \circ \phi^\dagger \circ \mathrm{chm}^{\mathbf{D}}_{R,A}(a) \circ \mathrm{chm}^{\mathbf{D}}_{R,A}(a^{-2}) \\
&=^{\text{(iv)}} \rho' \circ \phi^\dagger \circ \mathrm{chm}^{\mathbf{D}}_{R,A}(a^{-1}) = \rho' \circ \phi.
\end{aligned}$$

The equality $=^{\text{(i)}}$ is due to Lemma 18.2.24. In the equality $=^{\text{(ii)}}$ we have used the fact that ρ is a $\mathrm{Cent}(A)$-linear morphism for the action through A, so ρ commutes with $\mathrm{chm}^{\mathbf{D}}_{R,A}(a^{-2})$. Equality $=^{\text{(iii)}}$ is from (18.2.31). And equality $=^{\text{(iv)}}$ is because $\mathrm{chm}^{\mathbf{D}}_{R,A}$ is a ring homomorphism. We see that $\phi : R \xrightarrow{\simeq} R'$ is a rigid isomorphism. The uniqueness of the element a implies the uniqueness of the rigid isomorphism ϕ (by the same sort of calculation). □

Remark 18.2.32 As mentioned in Remark 18.1.30, for the purposes of Section 18.1, the only requirement on the base ring $\mathbb{K}$ is that the central $\mathbb{K}$-ring A is flat over it.

In this section that is not enough. Suppose that A is flat over $\mathbb{K}$. In order to prove the uniqueness of a rigid dualizing complex $R \in \mathbf{D}^{\mathrm{b}}(A^{\mathrm{en}})$, utilizing Theorem 12.9.10, we need that the complex $R \otimes^{\mathrm{L}}_{\mathbb{K}} R \in \mathbf{D}(A^{\mathrm{en}})$ will have bounded below cohomology. The only way we know to guarantee this is if the base ring $\mathbb{K}$ is *regular*. Notice that the regularity requirement is also needed in the commutative theory (see Setup 13.5.1 and Remark 13.5.18).

Presumably the regularity of the base ring $\mathbb{K}$ is the only obstruction. The lack of flatness of A can be handled using a suitable DG ring resolution $\tilde{A} \to A$, as explained in Remark 18.1.30. Thus we predict that the definitions and results of this section will remain valid if $\mathbb{K}$ is a regular nonzero commutative base ring, and A is a noetherian central $\mathbb{K}$-ring.

18.3 Interlude: Graded Rings of Laurent Type

Algebraically graded rings were introduced in Chapter 15, and we now return to them, and refer to them simply as graded rings. In this chapter we specialize to a particular kind of graded ring that is needed for the next two sections. The two main results here are Theorems 18.3.10 and 18.3.20.

We continue with Convention 18.1.1; in particular, $\mathbb{K}$ is a base field, and all rings are central over $\mathbb{K}$.

Definition 18.3.1 A graded ring $\tilde{B}$ is called a *graded ring of Laurent type* if there is an invertible central element $\tilde{c}$ in $\tilde{B}$ of degree 1. Such an element $\tilde{c}$ is called a *uniformizer* of $\tilde{B}$.

Of course, if $\tilde{B}$ is a graded ring of Laurent type, then $\tilde{B}_i \cdot \tilde{B}_j = \tilde{B}_{i+j}$ for all $i, j \in \mathbb{Z}$. A graded ring with this property is called a *strongly graded ring*.

The next lemma describes the structure of graded rings of Laurent type. Let $\mathbb{K}[t, t^{-1}]$ be the ring of Laurent polynomials in the degree 1 variable t. Given a ring B (not graded), we define the graded ring of Laurent type $B[t, t^{-1}] := B \otimes_{\mathbb{K}} \mathbb{K}[t, t^{-1}]$. The uniformizer is t, and the degree 0 component is B.

Lemma 18.3.2 *Let $\tilde{B}$ be a graded ring of Laurent type, with uniformizer $\tilde{c}$, and define $B := \tilde{B}_0$, the degree 0 subring. Then there is a unique isomorphism of graded rings $B[t, t^{-1}] \xrightarrow{\simeq} \tilde{B}$ that is the identity on B, and sends $t \mapsto \tilde{c}$.*

Exercise 18.3.3 Prove Lemma 18.3.2.

Note that for $\tilde{B}$ as in the lemma, there is also a surjective ring homomorphism $\tilde{B} \to B$ that sends $\tilde{c} \mapsto 1$. This is not a graded ring homomorphism.

Recall that given a graded ring $\tilde{B}$, the category of graded $\tilde{B}$-modules is $\mathbf{M}(\tilde{B}, \mathrm{gr})$. For a graded module $\tilde{M} \in \mathbf{M}(\tilde{B}, \mathrm{gr})$, let us write $\mathrm{Deg}_0(\tilde{M}) := \tilde{M}_0$, the homogeneous component of degree 0 of $\tilde{M}$.

Lemma 18.3.4 *Assume $\tilde{B}$ is a graded ring of Laurent type, with $B := \tilde{B}_0$.*

(1) *The functor $\mathrm{Deg}_0 : \mathbf{M}(\tilde{B}, \mathrm{gr}) \to \mathbf{M}(B)$ is an equivalence of abelian categories, with quasi-inverse $N \mapsto \tilde{B} \otimes_B N \cong \mathbb{K}[t, t^{-1}] \otimes N$.*

(2) *There is a bifunctorial isomorphism*

$$\mathrm{Deg}_0(\mathrm{Hom}_{\tilde{B}}(\tilde{M}, \tilde{N})) \xrightarrow{\simeq} \mathrm{Hom}_B(\mathrm{Deg}_0(\tilde{M}), \mathrm{Deg}_0(\tilde{N}))$$

for $\tilde{M}, \tilde{N} \in \mathbf{M}(\tilde{B}, \mathrm{gr})$. It sends $\tilde{\phi} \mapsto \tilde{\phi}|_{\mathrm{Deg}_0(\tilde{M})}$.

(3) *If the graded ring $\tilde{B}$ is left noetherian, then the ring B is left noetherian, and the functor* $\mathrm{Deg}_0 : \mathbf{M}_\mathrm{f}(\tilde{B}, \mathrm{gr}) \to \mathbf{M}_\mathrm{f}(B)$ *is an equivalence of abelian categories.*

Since Deg_0 is exact, it extends to derived categories.

Lemma 18.3.5 *Assume $\tilde{B}$ is a graded ring of Laurent type, with $B := \tilde{B}_0$.*

(1) *For every boundedness condition $\star$, the functor* $\mathrm{Deg}_0 : \mathbf{D}^\star(\tilde{B}, \mathrm{gr}) \to \mathbf{D}^\star(B)$ *is an equivalence of triangulated categories.*
(2) *If the graded ring $\tilde{B}$ is left noetherian, then* $\mathrm{Deg}_0 : \mathbf{D}^\star_\mathrm{f}(\tilde{B}, \mathrm{gr}) \to \mathbf{D}^\star_\mathrm{f}(B)$ *is an equivalence.*

Exercise 18.3.6 Prove Lemmas 18.3.4 and 18.3.5.

Lemma 18.3.7 *Let $\tilde{B}$ and $\tilde{C}$ be graded rings, with $B := \tilde{B}_0$ and $C := \tilde{C}_0$. Assume $\tilde{B}$ is a graded ring of Laurent type. For $\tilde{M}, \tilde{N} \in \mathbf{D}(\tilde{B} \otimes \tilde{C}^{\mathrm{op}}, \mathrm{gr})$ there is an isomorphism*

$$\mathrm{Deg}_0(\mathrm{RHom}_{\tilde{B}}(\tilde{M}, \tilde{N})) \xrightarrow{\simeq} \mathrm{RHom}_B(\mathrm{Deg}_0(\tilde{M}), \mathrm{Deg}_0(\tilde{N}))$$

in $\mathbf{D}(C^{\mathrm{en}})$, which is functorial in these complexes.

Proof Choose a semi-graded-free resolution $\tilde{P} \to \tilde{M}$ over $\tilde{B} \otimes \tilde{C}^{\mathrm{op}}$. Because $\tilde{C}$ is a graded-free $\mathbb{K}$-module, it follows that the complex $\tilde{P}$ is semi-graded-free over $\tilde{B}$. Since $B = \mathrm{Deg}_0(\tilde{B})$, the complex $\mathrm{Deg}_0(\tilde{P})$ is semi-free over the ring B; so $\mathrm{Deg}_0(\tilde{P}) \to \mathrm{Deg}_0(\tilde{M})$ is a semi-free resolution over B. We get these isomorphisms:

$$\begin{aligned} &\mathrm{Deg}_0(\mathrm{RHom}_{\tilde{B}}(\tilde{M}, \tilde{N})) \cong \mathrm{Deg}_0(\mathrm{Hom}_{\tilde{B}}(\tilde{P}, \tilde{N})) \\ &\quad \cong^{\dagger} \mathrm{Hom}_B(\mathrm{Deg}_0(\tilde{P}), \mathrm{Deg}_0(\tilde{N})) \cong \mathrm{RHom}_B(\mathrm{Deg}_0(\tilde{M}), \mathrm{Deg}_0(\tilde{N})) \end{aligned}$$

in $\mathbf{D}(C^{\mathrm{en}})$. The isomorphism $\cong^{\dagger}$ comes from Lemma 18.3.4(2). □

Lemma 18.3.8 *Let $\tilde{B}$ be a graded ring, and let $\tilde{c}$ be a central homogeneous element of $\tilde{B}$ of positive degree. Then the central element $\tilde{c} - 1$ is regular in $\tilde{B}$.*

Proof Take a nonzero element $\tilde{b} \in \tilde{B}$, with homogeneous component decomposition $\tilde{b} = \tilde{b}_1 + \cdots + \tilde{b}_r$, see (15.1.1). So either $r = 1$ and $\tilde{b} = \tilde{b}_1$ is homogeneous, or $r \geq 2$, and then $\deg(\tilde{b}_1) < \deg(\tilde{b}_i)$ for all $2 \leq i \leq r$. Then the homogeneous component decomposition of $\tilde{b} \cdot (\tilde{c} - 1)$ is

$$\tilde{b} \cdot (\tilde{c} - 1) = -\tilde{b}_1 + (\text{higher degree terms}).$$

The element $-\tilde{b}_1$ is nonzero, and therefore $\tilde{b} \cdot (\tilde{c} - 1) \neq 0$. □

Derived pseudo-finite complexes were introduced in Definition 12.9.1.

Lemma 18.3.9 *Let A, B, C be rings; let $M_1, M_2 \in \mathbf{D}(A)$; let $N_1 \in \mathbf{D}(B)$; and let $N_2 \in \mathbf{D}(B \otimes C)$. There is a morphism*

$$\theta : \mathrm{RHom}_A(M_1, M_2) \otimes \mathrm{RHom}_B(N_1, N_2) \to \mathrm{RHom}_{A \otimes B}(M_1 \otimes N_1, M_2 \otimes N_2)$$

in $\mathbf{D}(C)$, which is functorial in these complexes.

If M_1 is derived pseudo-finite over A, N_1 is derived pseudo-finite over B, and the complexes M_2 and N_2 have bounded below cohomologies, then θ is an isomorphism.

Proof Choose semi-free resolutions $P_1 \to M_1$ and $Q_1 \to N_1$, over A and B, respectively. Then $P_1 \otimes Q_1 \to M_1 \otimes N_1$ is a semi-free resolution over $A \otimes B$; see Proposition 11.4.12. The morphism θ is represented by the obvious homomorphism

$$\tilde{\theta} : \mathrm{Hom}_A(P_1, M_2) \otimes \mathrm{Hom}_B(Q_1, N_2) \to \mathrm{Hom}_{A \otimes B}(P_1 \otimes Q_1, M_2 \otimes N_2)$$

in $\mathbf{C}_{\mathrm{str}}(C)$.

Now assume that $\mathrm{H}(M_2)$ and $\mathrm{H}(N_2)$ are bounded below. By replacing M_2 and N_2 with suitable smart truncations, we can assume these are bounded below complexes. We fix them.

If P_1 and Q_1 are pseudo-finite semi-free, then $\tilde{\theta}$ is an isomorphism in $\mathbf{C}_{\mathrm{str}}(C)$, by the usual calculation; see Lemma 12.9.11.

If we fix N_1, then the complexes M_1 for which θ is an isomorphism form a saturated full triangulated subcategory of $\mathbf{D}(A)$. Likewise, if we fix M_1, then the complexes N_1 for which θ is an isomorphism form a saturated full triangulated subcategory of $\mathbf{D}(B)$. See Proposition 5.3.22. Therefore, as in the proof of Theorem 12.9.10, θ is an isomorphism whenever M_1 is derived pseudo-finite over A and N_1 is derived pseudo-finite over B. □

The inside and outside actions from Section 18.2 make sense also in the graded setting. Here is the first main result of this section.

Theorem 18.3.10 *Let $\tilde{B}$ be a graded ring of Laurent type, with $B := \tilde{B}_0$. Assume that B is a derived pseudo-finite complex over B^{en}. Given complexes $\tilde{M}, \tilde{N} \in \mathbf{D}^+(\tilde{B}^{\mathrm{en}}, \mathrm{gr})$, there is an isomorphism*

$$\begin{aligned} &\mathrm{Deg}_0(\mathrm{RHom}_{\tilde{B}^{\mathrm{en,out}}}(\tilde{B}, \tilde{M} \otimes \tilde{N})) \\ &\quad \xrightarrow{\simeq} \mathrm{RHom}_{B^{\mathrm{en,out}}}(B, \mathrm{Deg}_0(\tilde{M}) \otimes \mathrm{Deg}_0(\tilde{N}))[-1] \end{aligned}$$

in $\mathbf{D}(B^{\mathrm{en,in}}) = \mathbf{D}(B^{\mathrm{en}})$, which is functorial in these complexes.

Proof Say $\tilde{c}$ is a uniformizer of $\tilde{B}$. The ring $\tilde{B}^{\mathrm{en}} = \tilde{B} \otimes \tilde{B}^{\mathrm{op}}$ is actually $\mathbb{Z}^2$-graded. Taking the degree 0 component is for the total degree, so the ring

$\operatorname{Deg}_0(\tilde{B}^{\mathrm{en}})$ retains a $\mathbb{Z}$-grading, which we shall call the *hidden grading*. For the hidden grading there is a graded ring isomorphism

$$B^{\mathrm{en}}[s, s^{-1}] = B^{\mathrm{en}} \otimes \mathbb{K}[s, s^{-1}] \xrightarrow{\simeq} \operatorname{Deg}_0(\tilde{B}^{\mathrm{en}}), \tag{18.3.11}$$

where s is a variable of hidden degree 1, and it goes to the element $\tilde{c} \otimes \tilde{c}^{-1} \in \operatorname{Deg}_0(\tilde{B}^{\mathrm{en}})$.

Similarly for complexes: given $\tilde{M}, \tilde{N} \in \mathbf{D}(\tilde{B}^{\mathrm{en}}, \mathrm{gr})$, there is a canonical isomorphism

$$\operatorname{Deg}_0(\tilde{M}) \otimes \operatorname{Deg}_0(\tilde{N}) \otimes \mathbb{K}[s, s^{-1}] \xrightarrow{\simeq} \operatorname{Deg}_0(\tilde{M} \otimes \tilde{N}) \tag{18.3.12}$$

in $\mathbf{D}(B^{\mathrm{four}}[s, s^{-1}])$. We are neglecting the hidden grading from this stage onward in the proof.

According to Lemma 18.3.7 – which applies because $\tilde{B}^{\mathrm{en}}$ is a graded ring of Laurent type – there is an isomorphism

$$\operatorname{Deg}_0(\operatorname{RHom}_{\tilde{B}^{\mathrm{en,out}}}(\tilde{B}, \tilde{M} \otimes \tilde{N})) \cong \operatorname{RHom}_{\operatorname{Deg}_0(\tilde{B}^{\mathrm{en,out}})}(B, \operatorname{Deg}_0(\tilde{M} \otimes \tilde{N})) \tag{18.3.13}$$

in $\mathbf{D}(\operatorname{Deg}_0(\tilde{B}^{\mathrm{en,in}}))$. After applying the restriction functor $\mathbf{D}(\operatorname{Deg}_0(\tilde{B}^{\mathrm{en,in}})) \to \mathbf{D}(B^{\mathrm{en,in}}) = \mathbf{D}(B^{\mathrm{en}})$, (18.3.13) becomes an isomorphism in $\mathbf{D}(B^{\mathrm{en}})$. We then have the following isomorphisms in $\mathbf{D}(B^{\mathrm{en}})$:

$$\begin{aligned} &\operatorname{RHom}_{\operatorname{Deg}_0(\tilde{B}^{\mathrm{en,out}})}(B, \operatorname{Deg}_0(\tilde{M} \otimes \tilde{N})) \\ &\quad \cong^{\dagger} \operatorname{RHom}_{B^{\mathrm{en,out}} \otimes \mathbb{K}[s,s^{-1}]}\big(B \otimes \mathbb{K}, (\operatorname{Deg}_0(\tilde{M}) \otimes \operatorname{Deg}_0(\tilde{N})) \otimes \mathbb{K}[s, s^{-1}]\big) \\ &\quad \cong^{\ddagger} \operatorname{RHom}_{B^{\mathrm{en,out}}}(B, \operatorname{Deg}_0(\tilde{M}) \otimes \operatorname{Deg}_0(\tilde{N})) \otimes \operatorname{RHom}_{\mathbb{K}[s,s^{-1}]}(\mathbb{K}, \mathbb{K}[s, s^{-1}]). \end{aligned} \tag{18.3.14}$$

The isomorphism $\cong^{\dagger}$ comes from formulas (18.3.11) and (18.3.12). The isomorphism $\cong^{\ddagger}$ is by Lemma 18.3.9; it justified because B is a derived pseudo-finite complex over B^{en}, and of course $\mathbb{K}$ is a derived pseudo-finite complex over the noetherian ring $\mathbb{K}[s, s^{-1}]$.

The element $s - 1 \in \mathbb{K}[s, s^{-1}]$ is regular, so we have the Koszul resolution

$$0 \to \mathbb{K}[s, s^{-1}] \xrightarrow{(s-1)\cdot(-)} \mathbb{K}[s, s^{-1}] \to \mathbb{K} \to 0.$$

We use it to calculate

$$\operatorname{RHom}_{\mathbb{K}[s,s^{-1}]}(\mathbb{K}, \mathbb{K}[s, s^{-1}]) \cong \mathbb{K}[-1]$$

in $\mathbf{D}(\mathbb{K})$. Plugging this into (18.3.14) finishes the proof. □

From here until the end of this section we consider a connected graded ring $\tilde{A}$, with a degree 1 central element $\tilde{c} \in \tilde{A}$. Define the ring $A := \tilde{A}/(\tilde{c} - 1)$, and the canonical ring surjection $f : \tilde{A} \to A$. Of course, the ring A is not graded.

Let $\tilde{A}_{\tilde{c}}$ be the localization of $\tilde{A}$ with respect to $\tilde{c}$, i.e. inverting the homogeneous multiplicatively closed set $\{\tilde{c}^i\}_{i\in\mathbb{N}}$. Because the image of $\tilde{c}$ in $\tilde{A}_{\tilde{c}}$ is an invertible central element of degree 1, the graded ring $\tilde{A}_{\tilde{c}}$ is a graded ring of Laurent type, with uniformizer $\tilde{c}$. We let $\operatorname{inc} : \operatorname{Deg}_0(\tilde{A}_{\tilde{c}}) = (\tilde{A}_{\tilde{c}})_0 \to \tilde{A}_{\tilde{c}}$ be the inclusion of the degree 0 component of $\tilde{A}_{\tilde{c}}$. There is also the ring homomorphism $f_{\tilde{c}} : \tilde{A}_{\tilde{c}} \to A$, $f_{\tilde{c}}(\tilde{c}) = 1$.

Lemma 18.3.15

(1) *The ring homomorphism*

$$g := f_{\tilde{c}} \circ \operatorname{inc} : \operatorname{Deg}_0(\tilde{A}_{\tilde{c}}) = (\tilde{A}_{\tilde{c}})_0 \to A$$

is an isomorphism.

(2) *The ring isomorphism g extends to a graded ring isomorphism $\tilde{g} : \tilde{A}_{\tilde{c}} \xrightarrow{\simeq} A[t, t^{-1}]$ that sends $\tilde{c} \mapsto t$.*

The situation is shown in this commutative diagram of rings:

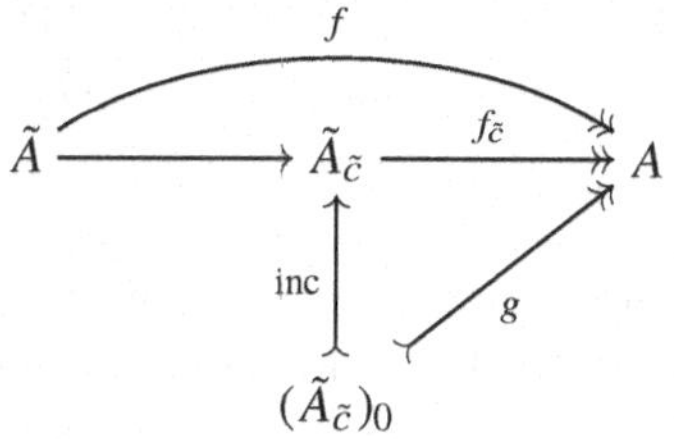

Proof In this proof we work with the graded ring of Laurent type $\tilde{B} := \tilde{A}_{\tilde{c}}$ and its uniformizer $\tilde{c}$.

(1) The homomorphism $f : \tilde{A} \to A$ is surjective, so every nonzero $a \in A$ can be written as $a = f(\tilde{b})$ for some nonzero $\tilde{b} \in \tilde{A}$. Consider the homogeneous component decomposition

$$\tilde{b} = \tilde{b}_1 + \cdots + \tilde{b}_r \tag{18.3.16}$$

of $\tilde{b}$, as in (15.1.1), with $r \geq 1$. Define $a' := \sum_{i=1}^{r} \tilde{b}_i \cdot \tilde{c}^{-\deg(\tilde{b}_i)} \in (\tilde{A}_{\tilde{c}})_0$. Then $a = g(a')$. We see that g is surjective.

Next we look at

$$\operatorname{Ker}(g) = \operatorname{Ker}(f_{\tilde{c}}) \cap (\tilde{A}_{\tilde{c}})_0 \subseteq (\tilde{A}_{\tilde{c}})_0. \tag{18.3.17}$$

The ideal $\operatorname{Ker}(f_{\tilde{c}})$ is generated by the central element $\tilde{c}-1$. Consider a nonzero element $\tilde{b}\cdot(\tilde{c} - 1) \in \tilde{A}_{\tilde{c}}$. Let (18.3.16) be the homogeneous component decomposition of the nonzero element $\tilde{b}$. Then the homogeneous component decomposition of $\tilde{b}\cdot(\tilde{c}-1)$ is $\tilde{b}\cdot(\tilde{c}-1) = -\tilde{b}_1 + \cdots + \tilde{b}_r \cdot \tilde{c}$. This is not a homogeneous element, and hence it cannot lie inside $(\tilde{A}_{\tilde{c}})_0$. From formula (18.3.17) we conclude that $\operatorname{Ker}(g) = 0$. Thus g is a ring isomorphism.

(2) Let

$$h : A \xrightarrow{\simeq} (\tilde{A}_{\tilde{c}})_0 \tag{18.3.18}$$

be the inverse of g. It extends to a graded ring homomorphism $\tilde{h} : A[t, t^{-1}] \to \tilde{A}_{\tilde{c}}$, $t \mapsto \tilde{c}$. This is an isomorphism, because $\tilde{A}_{\tilde{c}}$ is a graded ring of Laurent type. The isomorphism $\tilde{g}$ is the inverse of $\tilde{h}$. □

Exercise 18.3.19 Show that the functor $\operatorname{Ind}_f : \mathbf{M}(\tilde{A}, \mathrm{gr}) \to \mathbf{M}(A)$, $\operatorname{Ind}_f = A \otimes_{\tilde{A}} (-)$, is exact.

Here is the second main result of the section. Connected graded $\mathbb{K}$-rings were defined in Definition 15.2.18.

Theorem 18.3.20 *Let $\tilde{A}$ be a noetherian connected graded ring, let $\tilde{c} \in \tilde{A}$ be a homogeneous central element of degree* 1, *and let $A := \tilde{A}/(\tilde{c} - 1)$. Then:*

(1) *The ring A is noetherian.*
(2) *There is an isomorphism $A^{\mathrm{en}} \otimes^{\mathrm{L}}_{\tilde{A}^{\mathrm{en}}} \tilde{A} \cong A \oplus A[1]$ in $\mathbf{D}(A^{\mathrm{en}})$.*
(3) *The bimodule A is a derived pseudo-finite complex over the ring A^{en}.*

Proof

(1) By Theorem 15.1.28 the ring $\operatorname{Ungr}(\tilde{A})$ is noetherian. Since there is a surjection of rings $\operatorname{Ungr}(\tilde{A}) \to A$, it follows that A is noetherian.

(2) We have these isomorphisms

$$\begin{aligned} A^{\mathrm{en}} \otimes^{\mathrm{L}}_{\tilde{A}^{\mathrm{en}}} \tilde{A} &\cong^1 A \otimes^{\mathrm{L}}_{\tilde{A}} \tilde{A} \otimes^{\mathrm{L}}_{\tilde{A}} A \\ &\cong^2 A \otimes^{\mathrm{L}}_{\tilde{A}} A \cong^3 A \otimes^{\mathrm{L}}_{\tilde{A}} \tilde{A}_{\tilde{c}} \otimes^{\mathrm{L}}_{\tilde{A}_{\tilde{c}}} A \cong^4 A \otimes^{\mathrm{L}}_{\tilde{A}_{\tilde{c}}} A \end{aligned}$$

in $\mathbf{D}(A^{\mathrm{en}})$. The isomorphism $\cong^1$ is a rearrangement of the derived tensor factors. The isomorphism $\cong^2$ is from the left unitor isomorphism $\tilde{A} \otimes^{\mathrm{L}}_{\tilde{A}} A \cong A$ in $\mathbf{D}(\tilde{A} \otimes A^{\mathrm{op}})$. The isomorphism $\cong^3$ is because the ring homomorphism $\tilde{A} \to A$ factors through $\tilde{A}_{\tilde{c}}$, and $A \cong \tilde{A}_{\tilde{c}} \otimes^{\mathrm{L}}_{\tilde{A}_{\tilde{c}}} A$ in $\mathbf{D}(\tilde{A} \otimes A^{\mathrm{op}})$. And, finally, the isomorphism $\cong^4$ is because $A \otimes^{\mathrm{L}}_{\tilde{A}} \tilde{A}_{\tilde{c}} \cong A$ in $\mathbf{D}(\tilde{A} \otimes \tilde{A}^{\mathrm{op}}_{\tilde{c}})$

Now by Lemma 18.3.8 we know that $\tilde{c} - 1$ is a regular central element in $\tilde{A}_{\tilde{c}}$. So there is a short exact sequence

$$0 \to \tilde{A}_{\tilde{c}} \xrightarrow{(\tilde{c}-1)\cdot(-)} \tilde{A}_{\tilde{c}} \xrightarrow{f_{\tilde{c}}} A \to 0 \tag{18.3.21}$$

in $\mathbf{M}(\tilde{A}_{\tilde{c}} \otimes A^{\mathrm{op}})$. Here $\tilde{A}_{\tilde{c}}$ is an A^{op}-module via the ring homomorphism $h : A \to \tilde{A}_{\tilde{c}}$ from formula (18.3.18). We view the short exact sequence (18.3.21) as a quasi-isomorphism $\tilde{P} \to A$ in $\mathbf{C}_{\mathrm{str}}(\tilde{A}_{\tilde{c}} \otimes A^{\mathrm{op}})$, where

$$\tilde{P} := (\cdots \to 0 \to \tilde{A}_{\tilde{c}} \xrightarrow{(\tilde{c}-1)\cdot(-)} \tilde{A}_{\tilde{c}} \to 0 \to \cdots)$$

is a complex concentrated in cohomological degrees $-1, 0$. Because $\tilde{P}$ is semi-free over $\tilde{A}_{\tilde{c}}$, this allows us to calculate

$$A \otimes^{\mathrm{L}}_{\tilde{A}_{\tilde{c}}} A \cong A \otimes_{\tilde{A}_{\tilde{c}}} \tilde{P} \cong (\cdots \to 0 \to A \xrightarrow{0} A \to 0 \to \cdots) \cong A[1] \oplus A$$

in $\mathbf{D}(A^{\mathrm{en}})$.

(3) By Proposition 15.3.24 there is a quasi-isomorphism $\tilde{Q} \to \tilde{A}$ in $\mathbf{C}_{\mathrm{str}}(\tilde{A}^{\mathrm{en}}, \mathrm{gr})$ from a nonpositive complex $\tilde{Q}$ of finite graded-free $\tilde{A}^{\mathrm{en}}$-modules. We can forget the grading now, and just view $\tilde{Q} \to \tilde{A}$ as a free resolution of the module $\tilde{A}$ over the ring $\tilde{A}^{\mathrm{en}}$. We get $P := A^{\mathrm{en}} \otimes_{\tilde{A}^{\mathrm{en}}} \tilde{Q} \cong A^{\mathrm{en}} \otimes^{\mathrm{L}}_{\tilde{A}^{\mathrm{en}}} \tilde{A}$ in $\mathbf{D}(A^{\mathrm{en}})$. The complex P is pseudo-finite semi-free over A^{en}. By item (2) the complex A is a direct summand of P in $\mathbf{D}(A^{\mathrm{en}})$, and hence it is a derived pseudo-finite complex over A^{en}. □

18.4 Graded Rigid NC DC

Here we make a bridge between *balanced dualizing complexes* and *rigid dualizing complexes*. This is done by introducing an intermediate kind of object: the *graded rigid dualizing complex*. All is in the noncommutative setting of course.

The results of this section were originally in [153], with a few improvements later in [183]. Here we give a much more detailed discussion, and some corrections.

We continue with Convention 18.1.1. In particular $\mathbb{K}$ is a base field, and all rings are by default central over $\mathbb{K}$. Throughout this section we assume the following setup:

Setup 18.4.1 We are given a noetherian connected graded central $\mathbb{K}$-ring $\tilde{A}$ and a central homogeneous element $t \in \tilde{A}$ of degree 1. We define the central $\mathbb{K}$-ring $A := \tilde{A}/(t-1)$.

Note that the ring A is not graded. Also the rings $\tilde{A}^{\mathrm{en}}$ and A^{en} need not be noetherian. However:

Proposition 18.4.2 *Under Setup* 18.4.1, *the ring A is noetherian, and A is a derived pseudo-finite complex over A^{en}.*

Proof Use Theorem 18.3.20. □

The concepts of outside and inside actions from the beginning of Section 18.2 make sense in the graded setting. Thus we have a connected graded ring $\tilde{A}^{\mathrm{four}} = \tilde{A}^{\mathrm{en,out}} \otimes \tilde{A}^{\mathrm{en,in}}$. For a complex $\tilde{M} \in \mathbf{D}(\tilde{A}^{\mathrm{en}}, \mathrm{gr})$, its tensor product with itself $\tilde{M} \otimes \tilde{M}$ is an object of $\mathbf{D}(\tilde{A}^{\mathrm{four}}, \mathrm{gr})$, and

$$\mathrm{RHom}_{\tilde{A}^{\mathrm{en,out}}}(\tilde{A}, \tilde{M} \otimes \tilde{M}) \in \mathbf{D}(\tilde{A}^{\mathrm{en,in}}, \mathrm{gr}) = \mathbf{D}(\tilde{A}^{\mathrm{en}}, \mathrm{gr}). \tag{18.4.3}$$

Recall that $\mathbf{D}_{(\mathrm{f},..)}(\tilde{A}^{\mathrm{en}}, \mathrm{gr})$ (resp. $\mathbf{D}_{(..,\mathrm{f})}(\tilde{A}^{\mathrm{en}}, \mathrm{gr})$) is the full subcategory of $\mathbf{D}(\tilde{A}^{\mathrm{en}}, \mathrm{gr})$ on the complexes whose cohomology bimodules are finite over $\tilde{A}$ (resp. $\tilde{A}^{\mathrm{op}}$), and

$$\mathbf{D}_{(\mathrm{f},\mathrm{f})}(\tilde{A}^{\mathrm{en}}, \mathrm{gr}) = \mathbf{D}_{(\mathrm{f},..)}(\tilde{A}^{\mathrm{en}}, \mathrm{gr}) \cap \mathbf{D}_{(..,\mathrm{f})}(\tilde{A}^{\mathrm{en}}, \mathrm{gr}).$$

Definition 18.4.4 Under Setup 18.4.1, a *graded rigid NC dualizing complex* over $\tilde{A}$ is a pair $(\tilde{R}, \tilde{\rho})$, where $\tilde{R} \in \mathbf{D}^{\mathrm{b}}_{(\mathrm{f},\mathrm{f})}(\tilde{A}^{\mathrm{en}}, \mathrm{gr})$ is a graded NC dualizing complex over $\tilde{A}$, in the sense of Definition 17.1.4; and

$$\tilde{\rho} : \tilde{R} \xrightarrow{\simeq} \mathrm{RHom}_{\tilde{A}^{\mathrm{en,out}}}(\tilde{A}, \tilde{R} \otimes \tilde{R})$$

is an isomorphism in $\mathbf{D}(\tilde{A}^{\mathrm{en}}, \mathrm{gr})$, see (18.4.3) with $\tilde{M} := \tilde{R}$.

Balanced dualizing complexes were introduced in Definition 17.2.2.

Theorem 18.4.5 *Under Setup* 18.4.1, *if $\tilde{R}$ is a balanced NC dualizing complex over $\tilde{A}$, then $\tilde{R}$ is a graded rigid NC dualizing complex over $\tilde{A}$.*

We are a bit sloppy in stating the theorem; the proper way to state it is this: if $(\tilde{R}, \tilde{\beta})$ is a balanced dualizing complex over $\tilde{A}$, then there exists an isomorphism $\tilde{\rho}$ such that $(\tilde{R}, \tilde{\rho})$ is a graded rigid NC dualizing complex over $\tilde{A}$.

We need a lemma first. Recall (from Definition 15.2.27) that a graded $\tilde{A}$-module $\tilde{M} = \bigoplus_i \tilde{M}_i$ is called degreewise finite over $\mathbb{K}$ if each homogeneous component $\tilde{M}_i$ is a finite $\mathbb{K}$-module. We denoted by $\mathbf{M}_{\mathrm{dwf}}(\tilde{A}, \mathrm{gr})$ the full subcategory of $\mathbf{M}(\tilde{A}, \mathrm{gr})$ on the degreewise finite graded modules. This is a thick abelian subcategory of $\mathbf{M}(\tilde{A}, \mathrm{gr})$, closed under subobjects and quotients. Then we denoted by $\mathbf{D}_{\mathrm{dwf}}(\tilde{A}, \mathrm{gr})$ the full subcategory of $\mathbf{D}(\tilde{A}, \mathrm{gr})$ on the complexes $\tilde{M}$ such that $\mathrm{H}^q(\tilde{M}) \in \mathbf{M}_{\mathrm{dwf}}(\tilde{A}, \mathrm{gr})$ for all q. This is a full triangulated subcategory.

Lemma 18.4.6 *Let* $\tilde{M}, \tilde{N} \in \mathbf{D}^{\mathrm{b}}_{(\mathrm{f},..)}(\tilde{A}^{\mathrm{en}}, \mathrm{gr})$. *Then*:

(1) *The complexes $\tilde{M}$ and $\tilde{M} \otimes \tilde{N}$ satisfy $\tilde{M} \in \mathbf{D}^{\mathrm{b}}_{\mathrm{dwf}}(\tilde{A}^{\mathrm{en}}, \mathrm{gr})$ and $\tilde{M} \otimes \tilde{N} \in \mathbf{D}^{\mathrm{b}}_{\mathrm{dwf}}(\tilde{A}^{\mathrm{four}}, \mathrm{gr})$.*

(2) *The canonical morphism $\theta : \tilde{M}^* \otimes \tilde{N}^* \to (\tilde{M} \otimes \tilde{N})^*$ in $\mathbf{D}^{\mathrm{b}}_{\mathrm{dwf}}(\tilde{A}^{\mathrm{four}}, \mathrm{gr})$ is an isomorphism.*

Proof

(1) For every q the $\tilde{A}$-module $\mathrm{H}^q(\tilde{M})$ is finite, so it is the image of a finite direct sum of algebraic degree shifts of $\tilde{A}$. Since $\tilde{A} \in \mathbf{M}_{\mathrm{dwf}}(\mathbb{K}, \mathrm{gr})$, and since

the subcategory $\mathbf{M}_{\mathrm{dwf}}(\mathbb{K}, \mathrm{gr})$ is closed under quotients inside $\mathbf{M}(\mathbb{K}, \mathrm{gr})$, we see that $\mathrm{H}^q(\tilde{M}) \in \mathbf{M}_{\mathrm{dwf}}(\mathbb{K}, \mathrm{gr})$. We conclude that

$$\mathrm{H}^q(\tilde{M}) \in \mathbf{M}_{\mathrm{dwf}}(\mathbb{K}, \mathrm{gr}) \cap \mathbf{M}(\tilde{A}^{\mathrm{en}}, \mathrm{gr}) = \mathbf{M}_{\mathrm{dwf}}(\tilde{A}^{\mathrm{en}}, \mathrm{gr}).$$

Let $[q_0, q_1]$ be a finite integer interval that contains $\mathrm{con}(\mathrm{H}(\tilde{M}))$. For every $p \in \mathbb{Z}$ there is an isomorphism

$$\mathrm{H}^p(\tilde{M} \otimes \tilde{N}) \cong \bigoplus_{q \in [q_0, q_1]} (\mathrm{H}^q(\tilde{M}) \otimes \mathrm{H}^{p-q}(\tilde{N}))$$

in $\mathbf{M}(\mathbb{K}, \mathrm{gr})$. Each $\mathrm{H}^q(\tilde{M})$ and $\mathrm{H}^{p-q}(\tilde{N})$ is a finite graded $\tilde{A}$-module, so it is degreewise finite and bounded below (in algebraic degree). Therefore

$$\mathrm{H}^q(\tilde{M}) \otimes \mathrm{H}^{p-q}(\tilde{N}) \in \mathbf{M}_{\mathrm{dwf}}(\mathbb{K}, \mathrm{gr}),$$

and hence so is $\mathrm{H}^p(\tilde{M} \otimes \tilde{N})$. This means that $\tilde{M} \otimes \tilde{N} \in \mathbf{D}^{\mathrm{b}}_{\mathrm{dwf}}(\tilde{A}^{\mathrm{four}}, \mathrm{gr})$.

(2) We start by giving an explicit formula for the canonical morphism θ. There is a canonical homomorphism

$$\tilde{\theta} : \tilde{M}^* \otimes \tilde{N}^* \to (\tilde{M} \otimes \tilde{N})^* \tag{18.4.7}$$

in $\mathbf{C}_{\mathrm{str}}(A^{\mathrm{four}}, \mathrm{gr})$; it is

$$\tilde{\theta}(\phi \otimes \psi)(m \otimes n) := (-1)^{j \cdot k} \cdot \phi(m) \otimes \psi(n) \in \mathbb{K}$$

for $\phi \in (\tilde{M}^*)^i$, $\psi \in (\tilde{N}^*)^j$, $m \in \tilde{M}^k$ and $n \in \tilde{N}^l$. Then $\theta = \mathrm{Q}(\tilde{\theta})$ in $\mathbf{D}(A^{\mathrm{four}}, \mathrm{gr})$. We need to show that under the given finiteness and boundedness conditions, θ is an isomorphism. For that we can forget the A^{four}-module structures, and view θ as a morphism in $\mathbf{D}(\mathbb{K}, \mathrm{gr})$.

But in $\mathbf{D}(\mathbb{K}, \mathrm{gr})$ there are canonical isomorphisms $\tilde{M} \cong \mathrm{H}(\tilde{M})$ and $\tilde{N} \cong \mathrm{H}(\tilde{N})$. This means that we can assume that the differentials of the complexes $\tilde{M}$ and $\tilde{N}$ are zero. We know that for every $\left[\begin{smallmatrix} p \\ i \end{smallmatrix}\right] \in \left[\begin{smallmatrix} \mathbb{Z} \\ \mathbb{Z} \end{smallmatrix}\right] = \mathbb{Z}^2$, in the notation from Section 15.1, the $\mathbb{K}$-modules $\tilde{M}^p_i$ and $\tilde{N}^p_i$ are finite. Also there are uniform bounds on vanishing: there are integers p_0, p_1, i_0 such that $\tilde{M}^p_i = \tilde{N}^p_i = 0$ unless $p_0 \leq p \leq p_1$ and $i_0 \leq i$. An easy calculation shows that for these objects $\tilde{M}$ and $\tilde{N}$, the homomorphism $\tilde{\theta}$ of (18.4.7) is bijective. □

Proof of Theorem 18.4.5 We need to produce a graded rigidifying isomorphism $\tilde{\rho}$ for the graded NC dualizing complex $\tilde{R}$.

Consider the complex $\tilde{R} \in \mathbf{D}^{\mathrm{b}}_{(\mathrm{f},\mathrm{f})}(\tilde{A}^{\mathrm{en}}, \mathrm{gr})$. According to Lemma 18.4.6(1), we know that

$$\tilde{R} \otimes \tilde{R} \in \mathbf{D}^{\mathrm{b}}_{\mathrm{dwf}}(\tilde{A}^{\mathrm{four}}, \mathrm{gr}) \tag{18.4.8}$$

and

$$\tilde{A} \in \mathbf{D}^{\mathrm{b}}_{\mathrm{dwf}}(\tilde{A}^{\mathrm{en}}, \mathrm{gr}). \tag{18.4.9}$$

There is a sequence of isomorphisms in $\mathsf{D}(\tilde{A}^{\mathrm{en}}, \mathrm{gr})$:

$$\begin{aligned} \mathrm{RHom}_{\tilde{A}^{\mathrm{en}},\mathrm{out}}(\tilde{A}, \tilde{R} \otimes \tilde{R}) &\cong^1 \mathrm{RHom}_{\tilde{A}^{\mathrm{en}},\mathrm{out}}((\tilde{R} \otimes \tilde{R})^*, \tilde{A}^*) \\ &\cong^2 \mathrm{RHom}_{\tilde{A}^{\mathrm{en}},\mathrm{in}}(\tilde{R}^* \otimes \tilde{R}^*, \tilde{A}^*) \\ &\cong^3 \mathrm{RHom}_{\tilde{A}^{\mathrm{en}},\mathrm{in}}(\tilde{P} \otimes \tilde{P}, \tilde{A}^*) \\ &\cong^4 \mathrm{RHom}_{\tilde{A}}(\tilde{P}, \mathrm{RHom}_{\tilde{A}^{\mathrm{op}}}(\tilde{P}, \tilde{A}^*)) \\ &\cong^5 \mathrm{RHom}_{\tilde{A}}(\tilde{P}, \tilde{R}) \cong^6 \tilde{R}. \end{aligned} \tag{18.4.10}$$

The explanations for the isomorphisms are:

$\cong^1$: By Theorem 15.3.32, which applies by formulas (18.4.8) and (18.4.9).

$\cong^2$: This is by Lemma 18.4.6(2). Note that the outside action changes to an inside action.

$\cong^3$: Here $\tilde{P} := \mathrm{R}\Gamma_{\tilde{\mathfrak{m}}}(\tilde{A})$, where $\tilde{\mathfrak{m}}$ is the augmentation ideal of $\tilde{A}$. According to Theorems 17.3.19 and 17.2.4, there is an isomorphism $\tilde{R} \cong \tilde{P}^*$ in $\mathsf{D}(\tilde{A}^{\mathrm{en}}, \mathrm{gr})$; dualizing we get an isomorphism $\tilde{R}^* \cong \tilde{P}$ in $\mathsf{D}(\tilde{A}^{\mathrm{en}}, \mathrm{gr})$.

$\cong^4$: This is one of the standard Hom-tensor identities.

$\cong^5$: This is another instance of the isomorphism $\tilde{P}^* \cong \tilde{R}$.

$\cong^6$: By Lemma 17.3.31, the restriction of $\tilde{R}$ to $\mathsf{D}(\tilde{A}, \mathrm{gr})$ satisfies $\tilde{R} \in \mathsf{D}_{\mathrm{f}}(\tilde{A}, \mathrm{gr}) \subseteq \mathsf{D}(\tilde{A}, \mathrm{gr})_{\mathrm{com}}$, so $\tilde{R}$ is isomorphic, in $\mathsf{D}(\tilde{A}^{\mathrm{en}}, \mathrm{gr})$, to its abstract derived completion $\mathrm{ADC}_{\tilde{\mathfrak{m}}}(\tilde{R}) = \mathrm{RHom}_{\tilde{A}}(\tilde{P}, \tilde{R})$. See Section 16.6.

The composition of the isomorphisms in (18.4.10) is the graded rigidifying isomorphism $\tilde{\rho}$. □

The localization homomorphism $\tilde{A} \to \tilde{A}_t$ induces a flat homomorphism of graded rings $\tilde{A}^{\mathrm{en}} \to (\tilde{A}_t)^{\mathrm{en}} = \tilde{A}_t \otimes \tilde{A}_t^{\mathrm{op}}$. This homomorphism can also be viewed as a localization of $\tilde{A}^{\mathrm{en}}$ with respect to the degree 1 central elements $t \otimes 1$ and $1 \otimes t$. There is a corresponding induction functor

$$\mathrm{Ind}_{(\tilde{A}_t)^{\mathrm{en}}} : \mathsf{D}(\tilde{A}^{\mathrm{en}}, \mathrm{gr}) \to \mathsf{D}((\tilde{A}_t)^{\mathrm{en}}, \mathrm{gr}). \tag{18.4.11}$$

Here is what it does on objects: given a complex $\tilde{M} \in \mathsf{D}(\tilde{A}^{\mathrm{en}}, \mathrm{gr})$, we have

$$\mathrm{Ind}_{\tilde{A}^{\mathrm{en}}}(\tilde{M}) = (\tilde{A}_t)^{\mathrm{en}} \otimes_{\tilde{A}^{\mathrm{en}}} \tilde{M} \cong \tilde{A}_t \otimes_{\tilde{A}} \tilde{M} \otimes_{\tilde{A}} \tilde{A}_t .$$

There is a ring homomorphism

$$\mathrm{inc} : (\tilde{A}_t)^{\mathrm{en}}_0 = \mathrm{Deg}_0((\tilde{A}_t)^{\mathrm{en}}) \to (\tilde{A}_t)^{\mathrm{en}},$$

the inclusion of the degree 0 component. And there is a functor

$$\mathrm{Deg}_0 : \mathsf{D}((\tilde{A}_t)^{\mathrm{en}}, \mathrm{gr}) \to \mathsf{D}((\tilde{A}_t)^{\mathrm{en}}_0). \tag{18.4.12}$$

We have a ring isomorphism $h : A \xrightarrow{\simeq} (\tilde{A}_t)_0$; see formula (18.3.18). It gives rise to ring homomorphisms

$$A^{\mathrm{en}} \xrightarrow[\cong]{h^{\mathrm{en}}} (\tilde{A}_t)_0 \otimes (\tilde{A}_t^{\mathrm{op}})_0 \xrightarrow[\subseteq]{\mathrm{inc}} (\tilde{A}_t \otimes \tilde{A}_t^{\mathrm{op}})_0 = (\tilde{A}_t)^{\mathrm{en}}_0 .$$

Relative to this composed ring homomorphism we have the restriction functor

$$\mathrm{Rest}_{A^{\mathrm{en}}} : \mathbf{D}((\tilde{A}_t)_0^{\mathrm{en}}) \to \mathbf{D}(A^{\mathrm{en}}). \tag{18.4.13}$$

Composing the functors from (18.4.11), (18.4.12) and (18.4.13) we get a triangulated functor

$$\mathrm{Rest}_{A^{\mathrm{en}}} \circ \mathrm{Deg}_0 \circ \mathrm{Ind}_{(\tilde{A}_t)^{\mathrm{en}}} : \mathbf{D}(\tilde{A}^{\mathrm{en}}, \mathrm{gr}) \to \mathbf{D}(A^{\mathrm{en}}). \tag{18.4.14}$$

Theorem 18.4.15 *Under Setup* 18.4.1, *let* $\tilde{R} \in \mathbf{D}(\tilde{A}^{\mathrm{en}}, \mathrm{gr})$ *be a graded rigid NC dualizing complex. Define the complex*

$$R := (\mathrm{Rest}_{A^{\mathrm{en}}} \circ \mathrm{Deg}_0 \circ \mathrm{Ind}_{(\tilde{A}_t)^{\mathrm{en}}})(\tilde{R})[-1] \in \mathbf{D}(A^{\mathrm{en}}).$$

Then R *is a rigid NC dualizing complex over* A.

Once more we were a bit sloppy. The proper statement is this: we are given a graded rigid NC dualizing complex $(\tilde{R}, \tilde{\rho})$; then the complex R defined above admits a rigidifying isomorphism ρ, such that (R, ρ) is a rigid NC dualizing complex over A. The proof will come after two lemmas.

Lemma 18.4.16 *Assume that* $\tilde{R}$ *is a graded rigid dualizing complex over* $\tilde{A}$. *Then for every* p *the graded bimodule* $\mathrm{H}^p(\tilde{R})$ *is central over* $\mathrm{Cent}(\tilde{A})$.

Proof We know that the central homotheties

$$\mathrm{chm}_{\tilde{A},\tilde{A}}, \mathrm{chm}_{\tilde{A},\tilde{A}^{\mathrm{op}}} : \mathrm{Cent}(\tilde{A}) \to \mathrm{End}_{\mathbf{M}(\tilde{A}^{\mathrm{en}},\mathrm{gr})}(\tilde{A})$$

are equal graded ring isomorphisms. Therefore the graded version of Theorem 18.2.23 – which is proved the same way – tells us that the ring homomorphisms

$$\mathrm{chm}^{\mathbf{D}}_{R,\tilde{A}}, \mathrm{chm}^{\mathbf{D}}_{R,\tilde{A}^{\mathrm{op}}} : \mathrm{Cent}(\tilde{A}) \to \mathrm{End}_{\mathbf{D}(\tilde{A}^{\mathrm{en}},\mathrm{gr})}(\tilde{R})$$

are equal isomorphisms.

The left $\mathrm{Cent}(\tilde{A})$-module structure on $\mathrm{H}^p(\tilde{R})$ coincides with the categorical action $\mathrm{chm}^{\mathbf{D}}_{R,\tilde{A}}$; and likewise from the right. We see that the left and right $\mathrm{Cent}(\tilde{A})$-module structure on $\mathrm{H}^p(\tilde{R})$ are the same, so it is a central bimodule. □

Lemma 18.4.17 *Assume that* $\tilde{R}$ *is a graded rigid dualizing complex over* $\tilde{A}$. *Then the homomorphisms*

$$\tilde{A}_t \otimes_{\tilde{A}} \tilde{R} \to \tilde{A}_t \otimes_{\tilde{A}} \tilde{R} \otimes_{\tilde{A}} \tilde{A}_t \tag{l}$$

and

$$\tilde{R} \otimes_{\tilde{A}} \tilde{A}_t \to \tilde{A}_t \otimes_{\tilde{A}} \tilde{R} \otimes_{\tilde{A}} \tilde{A}_t \tag{r}$$

in $\mathbf{C}_{\mathrm{str}}(\tilde{A}^{\mathrm{en}}, \mathrm{gr})$ *are quasi-isomorphisms.*

Proof We shall only treat (l); the homomorphism (r) is treated similarly. Let $\tilde{C} := \mathrm{Cent}(\tilde{A})$. Then $\tilde{A}_t = \tilde{A} \otimes_{\tilde{C}} \tilde{C}_t = \tilde{C}_t \otimes_{\tilde{C}} \tilde{A}$. Hence (l) is true if (and only if) the homomorphism $\tilde{C}_t \otimes_{\tilde{C}} \tilde{R} \to \tilde{C}_t \otimes_{\tilde{C}} \tilde{R} \otimes_{\tilde{C}} \tilde{C}_t$ is a quasi-isomorphism.

Because of flatness of localization, for every p, there is a commutative diagram

$$\begin{array}{ccc} \mathrm{H}^p(\tilde{C}_t \otimes_{\tilde{C}} \tilde{R}) & \longrightarrow & \mathrm{H}^p(\tilde{C}_t \otimes_{\tilde{C}} \tilde{R} \otimes_{\tilde{C}} \tilde{C}_t) \\ \cong \downarrow & & \cong \downarrow \\ \tilde{C}_t \otimes_{\tilde{C}} \mathrm{H}^p(\tilde{R}) & \longrightarrow & \tilde{C}_t \otimes_{\tilde{C}} \mathrm{H}^p(\tilde{R}) \otimes_{\tilde{C}} \tilde{C}_t \end{array} \tag{18.4.18}$$

in $\mathbf{M}(\tilde{A}^{\mathrm{en}}, \mathrm{gr})$, with vertical isomorphisms. By Lemma 18.4.16, $\mathrm{H}^p(\tilde{R})$ is a central $\tilde{C}$ bimodule. And localization satisfies $\tilde{C}_t = \tilde{C}_t \otimes_{\tilde{C}} \tilde{C}_t$. This implies that the bottom arrow in diagram (18.4.18) is bijective. Hence the top arrow is also bijective. □

Proof of Theorem 18.4.15 The proof is divided into several steps.

Step 1. This is the easiest step: we prove that for every q the bimodule $\mathrm{H}^q(R)$ is a finite module over A and over A^{op}. In fact, we are going to treat the left module structure of $\mathrm{H}^q(R)$ only; the right module structure is treated the same way, by replacing A with A^{op}.

Let us write

$$\tilde{R}_t := \mathrm{Ind}_{(\tilde{A}_t)^{\mathrm{en}}}(\tilde{R}) \cong \tilde{A}_t \otimes_{\tilde{A}} \tilde{R} \otimes_{\tilde{A}} \tilde{A}_t \in \mathbf{D}((\tilde{A}_t)^{\mathrm{en}}, \mathrm{gr}).$$

According to Lemma 18.4.17, there is an isomorphism $\tilde{R}_t \cong \tilde{A}_t \otimes_{\tilde{A}} \tilde{R}$ in $\mathbf{D}(\tilde{A}_t, \mathrm{gr})$. Since $\tilde{R} \in \mathbf{D}_{\mathrm{f}}(\tilde{A}, \mathrm{gr})$, it follows that $\tilde{R}_t \in \mathbf{D}_{\mathrm{f}}(\tilde{A}_t, \mathrm{gr})$. We identify $(\tilde{A}_t)_0$ with A using the canonical isomorphism h from (18.3.18). By Lemma 18.3.5(2) we get $\mathrm{Deg}_0(\tilde{R}_t) \in \mathbf{D}_{\mathrm{f}}((\tilde{A}_t)_0) = \mathbf{D}_{\mathrm{f}}(A)$. Hence

$$R = \mathrm{Deg}_0(\tilde{R}_t)[-1] \in \mathbf{D}_{\mathrm{f}}(A). \tag{18.4.19}$$

We see that $\mathrm{H}^q(R)$ is a finite A-module for all q.

Step 2. Here we prove that R has finite injective dimension over A and over A^{op}. Again, we only examine the left side; the right side is done similarly, just replacing A with A^{op}.

Because A is noetherian, it suffices to find a uniform bound for the cohomology of $\mathrm{RHom}_A(M, R)$, for all $M \in \mathbf{M}_{\mathrm{f}}(A)$. Take such an A-module M. By Lemma 18.3.4(3) there exists a graded module $\tilde{M} \in \mathbf{M}_{\mathrm{f}}(\tilde{A}, \mathrm{gr})$ such that, letting $\tilde{M}_t := \tilde{A}_t \otimes_{\tilde{A}} M$, we have

$$M \cong \mathrm{Deg}_0(\tilde{M}_t) \tag{18.4.20}$$

in $\mathbf{M}_{\mathrm{f}}(A)$. There are isomorphisms

$$\begin{aligned} &\mathrm{RHom}_{\tilde{A}}(\tilde{M}, \tilde{R}) \otimes_{\tilde{A}} \tilde{A}_t \cong^1 \mathrm{RHom}_{\tilde{A}}(\tilde{M}, \tilde{R} \otimes_{\tilde{A}} \tilde{A}_t) \\ &\quad \cong^2 \mathrm{RHom}_{\tilde{A}}(\tilde{M}, \tilde{R}_t) \cong^3 \mathrm{RHom}_{\tilde{A}_t}(\tilde{M}_t, \tilde{R}_t) \end{aligned} \tag{18.4.21}$$

in $\mathbf{D}(\tilde{A}_t^{\mathrm{op}}, \mathrm{gr})$. The isomorphism $\cong^1$ is an instance of the graded tensor-evaluation isomorphism (Theorem 15.3.27), which is valid because $\tilde{A}$ is noetherian, $\tilde{M} \in \mathbf{M}_{\mathrm{f}}(\tilde{A}, \mathrm{gr})$, $\tilde{R} \in \mathbf{D}^{\mathrm{b}}(\tilde{A}^{\mathrm{en}}, \mathrm{gr})$, and $\tilde{A}_t$ is flat over $\tilde{A}$. The isomorphism $\cong^2$ is by Lemma 18.4.17. And the isomorphism $\cong^3$ is adjunction for the ring homomorphism $\tilde{A} \to \tilde{A}_t$.

Next, according to Lemma 18.3.7, and using the isomorphisms (18.4.20) and (18.4.19), there is an isomorphism

$$\begin{aligned} \mathrm{Deg}_0(\mathrm{RHom}_{\tilde{A}_t}(\tilde{M}_t, \tilde{R}_t)) &\cong \mathrm{RHom}_{\mathrm{Deg}_0(\tilde{A}_t)}(\mathrm{Deg}_0(\tilde{M}_t), \mathrm{Deg}_0(\tilde{R}_t)) \\ &\cong \mathrm{RHom}_A(M, R)[1] \end{aligned}$$

in $\mathbf{D}(A^{\mathrm{op}})$. Combining this with formula (18.4.21), we see that the injective dimension of R over A is at most one more than the graded-injective dimension of $\tilde{R}$ over $\tilde{A}$; and this is known to be finite.

Step 3. In this step we prove that R has the derived Morita property on both sides. As before, we only prove this property on the A side; the A^{op} side is done similarly.

The isomorphisms (18.4.21), but for the complex of bimodules $\tilde{R}$ instead of for the module $\tilde{M}$, become isomorphisms

$$\begin{aligned} &\mathrm{RHom}_{\tilde{A}}(\tilde{R}, \tilde{R}) \otimes_{\tilde{A}} \tilde{A}_t \cong^1 \mathrm{RHom}_{\tilde{A}}(\tilde{R}, \tilde{R} \otimes_{\tilde{A}} \tilde{A}_t) \\ &\quad \cong^2 \mathrm{RHom}_{\tilde{A}}(\tilde{R}, \tilde{R}_t) \cong^3 \mathrm{RHom}_{\tilde{A}_t}(\tilde{A}_t \otimes_{\tilde{A}} \tilde{R}, \tilde{R}_t) \\ &\quad \cong^4 \mathrm{RHom}_{\tilde{A}_t}(\tilde{R}_t, \tilde{R}_t) \end{aligned} \tag{18.4.22}$$

in $\mathbf{D}(\tilde{A}_t^{\mathrm{op}}, \mathrm{gr})$. The last isomorphism $\cong^4$ is another case of Lemma 18.4.17. Define $R' := \mathrm{Deg}_0(\tilde{R}_t) \in \mathbf{D}_{\mathrm{f}}(A)$; so $R = R'[-1]$. We now apply Deg_0 to the last object in (18.4.22), obtaining the isomorphisms

$$\begin{aligned} &\mathrm{Deg}_0(\mathrm{RHom}_{\tilde{A}_t}(\tilde{R}_t, \tilde{R}_t)) \\ &\quad \cong^{(\mathrm{i})} \mathrm{RHom}_{\mathrm{Deg}_0(\tilde{A}_t)}(\mathrm{Deg}_0(\tilde{R}_t), \mathrm{Deg}_0(\tilde{R}_t)) \\ &\quad \cong^{(\mathrm{ii})} \mathrm{RHom}_A(R', R') \cong^{(\mathrm{iii})} \mathrm{RHom}_A(R, R) \end{aligned} \tag{18.4.23}$$

in $\mathbf{D}(A^{\mathrm{op}})$. The isomorphism $\cong^{(\mathrm{i})}$ is due to Lemma 18.3.7, with $\tilde{B} = \tilde{C} := \tilde{A}_t$. The isomorphism $\cong^{(\mathrm{ii})}$ is simply the definition of R', and the isomorphism $\cong^{(\mathrm{iii})}$ is because $R = R'[-1]$.

We know that $\tilde{R}$ has the graded derived NC Morita property on the $\tilde{A}$ side. Using Lemma 18.1.21 (which is true also in the graded situation) we see that $\mathrm{H}^q(\mathrm{RHom}_{\tilde{A}}(\tilde{R}, \tilde{R})) = 0$ for all $q \neq 0$. Because the functors H^q and Deg_0

commute with each other, the isomorphisms (18.4.22) and (18.4.23) tell us that

$$\mathrm{H}^q(\mathrm{RHom}_A(R, R)) = 0 \tag{18.4.24}$$

for all $q \neq 0$.

For $q = 0$ we know that $\mathrm{H}^0(\mathrm{RHom}_{\tilde{A}}(\tilde{R}, \tilde{R}))$ is a graded-free $\tilde{A}^{\mathrm{op}}$-module with basis $\mathrm{id}_{\tilde{R}}$. This is by Lemma 18.1.21, applied in the graded situation. The isomorphisms (18.4.22) send the element $\mathrm{id}_{\tilde{R}} \otimes 1$ to $\mathrm{id}_{\tilde{R}_t}$; and therefore $\mathrm{H}^0(\mathrm{RHom}_{\tilde{A}_t}(\tilde{R}_t, \tilde{R}_t))$ is a graded-free $\tilde{A}_t^{\mathrm{op}}$-module with basis $\mathrm{id}_{\tilde{R}_t}$. The element $\mathrm{id}_{\tilde{R}_t}$ has degree 0, and hence

$$\mathrm{H}^0(\mathrm{Deg}_0(\mathrm{RHom}_{\tilde{A}_t}(\tilde{R}_t, \tilde{R}_t))) \cong \mathrm{Deg}_0(\mathrm{H}^0(\mathrm{RHom}_{\tilde{A}_t}(\tilde{R}_t, \tilde{R}_t)))$$

is a free A^{op}-module with basis $\mathrm{id}_{\tilde{R}_t}$. The isomorphisms (18.4.23) send the element $\mathrm{id}_{\tilde{R}_t}$ to id_R; so $\mathrm{H}^0(\mathrm{RHom}_A(R, R))$ is a free A^{op}-module with basis id_R. Again invoking Lemma 18.1.21, this fact, with formula (18.4.24), says that R has the derived NC Morita property on the A side.

Step 4. Here we produce a rigidifying isomorphism ρ for R, starting with a given graded rigidifying isomorphism

$$\tilde{\rho} : \tilde{R} \xrightarrow{\simeq} \mathrm{RHom}_{\tilde{A}^{\mathrm{en}},\mathrm{out}}(\tilde{A}, \tilde{R} \otimes \tilde{R})$$

in $\mathbf{D}(\tilde{A}^{\mathrm{en}}, \mathrm{gr})$. Here is a list of isomorphisms in $\mathbf{D}((\tilde{A}_t)^{\mathrm{en}}, \mathrm{gr})$.

$$\begin{aligned} \tilde{R}_t &\cong^1 \mathrm{RHom}_{\tilde{A}^{\mathrm{en}},\mathrm{out}}(\tilde{A}, \tilde{R} \otimes \tilde{R}) \otimes_{\tilde{A}^{\mathrm{en}}} (\tilde{A}_t)^{\mathrm{en}} \\ &\cong^2 \mathrm{RHom}_{\tilde{A}^{\mathrm{en}},\mathrm{out}}(\tilde{A}, (\tilde{R} \otimes \tilde{R}) \otimes_{\tilde{A}^{\mathrm{en}}} (\tilde{A}_t)^{\mathrm{en}}) \\ &\cong^3 \mathrm{RHom}_{\tilde{A}^{\mathrm{en}},\mathrm{out}}(\tilde{A}, (\tilde{R} \otimes_{\tilde{A}} \tilde{A}_t) \otimes (\tilde{A}_t \otimes_{\tilde{A}} \tilde{R})) \\ &\cong^4 \mathrm{RHom}_{\tilde{A}^{\mathrm{en}},\mathrm{out}}(\tilde{A}, \tilde{R}_t \otimes \tilde{R}_t) \\ &\cong^5 \mathrm{RHom}_{(\tilde{A}_t)^{\mathrm{en}},\mathrm{out}}(\tilde{A}_t, \tilde{R}_t \otimes \tilde{R}_t). \end{aligned} \tag{18.4.25}$$

The justifications are:

- $\cong^1$: Here we apply the functor $\mathrm{Ind}_{(\tilde{A}_t)^{\mathrm{en}}}$ to the isomorphism $\tilde{\rho}$. The inside action of $\tilde{A}^{\mathrm{en}}$ on $\tilde{R} \otimes \tilde{R}$ is treated as a right action.
- $\cong^2$: This is by the graded tensor-evaluation isomorphism (Theorem 15.3.27). It holds because $\tilde{A}$ is a derived graded-pseudo-finite complex over $\tilde{A}^{\mathrm{en}}$, see Corollary 15.3.26; $\tilde{R} \otimes \tilde{R} \in \mathbf{D}^{\mathrm{b}}(\tilde{A}^{\mathrm{four}}, \mathrm{gr})$; and $(\tilde{A}_t)^{\mathrm{en}}$ is flat over $\tilde{A}^{\mathrm{en}}$.
- $\cong^3$: This is a rearrangement of derived tensor factors.
- $\cong^4$: By Lemma 18.4.17.
- $\cong^5$: This is an adjunction for the graded ring homomorphism $\tilde{A}^{\mathrm{en}} \to (\tilde{A}_t)^{\mathrm{en}}$, with the fact that $\tilde{A}_t = \tilde{A}_t \otimes_{\tilde{A}} \tilde{A} \otimes_{\tilde{A}} \tilde{A}_t$ as graded rings.

Now we apply Deg_0 to (18.4.25), and we obtain the first isomorphism below in $\mathbf{D}(A^{\mathrm{en}})$.

$$\begin{aligned} R' = \mathrm{Deg}_0(\tilde{R}_t) &\cong^{(\mathrm{i})} \mathrm{Deg}_0(\mathrm{RHom}_{(\tilde{A}_t)^{\mathrm{en}},\mathrm{out}}(\tilde{A}_t, \tilde{R}_t \otimes \tilde{R}_t)) \\ &\cong^{(\mathrm{ii})} \mathrm{RHom}_{A^{\mathrm{en}},\mathrm{out}}(A, R' \otimes R')[-1]. \end{aligned} \tag{18.4.26}$$

The isomorphism $\cong^{(\text{ii})}$ is by Theorem 18.3.10, with $\tilde{B} := \tilde{A}_t$, which we are allowed to use because A is a derived pseudo-finite complex over A^{en}; see Theorem 18.3.20. Plugging in $R = R'[-1]$ we get $R \cong \operatorname{RHom}_{A^{\text{en}},\text{out}}(A, R \otimes R)$ in $\mathbf{D}(A^{\text{en}})$. This is the rigidifying isomorphism ρ. □

18.5 Filtered Rings and Existence of Rigid NC DC

In this section we give M. Van den Bergh's proof of the existence of NC rigid dualizing complexes (Theorem 18.5.4). This proof appeared in his seminal paper [153], and then there were some improvements in [183]. The proof here is much more detailed.

We continue with Convention 18.1.1. Specifically, $\mathbb{K}$ is a base field, and all rings are central over $\mathbb{K}$.

Earlier in the book (in Chapter 11) we encountered filtrations of DG modules. Here we are interested in filtrations of rings. By a *filtration of a ring A* we mean a collection $\{F_j(A)\}_{j \geq -1}$ of $\mathbb{K}$-submodules

$$F_{-1}(A) \subseteq F_0(A) \subseteq F_1(A) \subseteq \cdots \subseteq A$$

such that $F_{-1}(A) = 0$, $\bigcup_j F_j(A) = A$, $1_A \in F_0(A)$ and $F_j(A) \cdot F_k(A) \subseteq F_{j+k}(A)$ for all j, k.

Given a filtration $F = \{F_j(A)\}_{j \geq -1}$ of the ring A, we write $\operatorname{Gr}^F_j(A) := F_j(A)/F_{j-1}(A)$ for $j \geq 0$. The *associated graded ring* $\operatorname{Gr}^F(A) := \bigoplus_{j \geq 0} \operatorname{Gr}^F_j(A)$ is a nonnegative algebraically graded ring.

A filtration F on the ring A gives rise to another graded ring: it is the *Rees ring* $\operatorname{Rees}^F(A) := \bigoplus_{j \geq 0} F_j(A) \cdot t^j \subseteq A[t]$. Here t is a central variable of degree 1, which we call the Rees parameter. There is a ring isomorphism $A \cong \operatorname{Rees}^F(A)/(t-1)$ and a graded ring isomorphism $\operatorname{Gr}^F(A) \cong \operatorname{Rees}^F(A)/(t)$. We often use the abbreviations $\bar{A} := \operatorname{Gr}^F(A)$ and $\tilde{A} := \operatorname{Rees}^F(A)$. This is consistent with the notation used earlier in the chapter: $\tilde{A}$ is a graded ring, t is a degree 1 central regular element in it, and $A = \tilde{A}/(t-1)$.

For an element $c \in \mathbb{K}$ we write $\operatorname{pr}_c : \tilde{A} \to \tilde{A}/(t-c)$ for the canonical surjective ring homomorphism. The homomorphisms pr_0 and pr_1 are displayed in this diagram of rings:

$$\begin{array}{ccc} & \tilde{A} = \operatorname{Rees}^F(A) & \\ {\scriptstyle \operatorname{pr}_1} \swarrow & & \searrow {\scriptstyle \operatorname{pr}_0} \\ A & & \bar{A} = \operatorname{Gr}^F(A) \end{array}$$

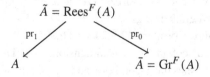

The filtration F can be recovered from the Rees ring as follows:

$$F_j(A) = \mathrm{pr}_1(\tilde{A}_j) = \mathrm{pr}_1\Big(\bigoplus\nolimits_{i=0}^{j} \tilde{A}_i\Big) \subseteq A. \tag{18.5.1}$$

Definition 18.5.2 ([183]) A filtration F of the ring A is called a *noetherian connected filtration* if the graded ring $\mathrm{Gr}^F(A)$ is a noetherian connected graded ring (Definitions 15.2.18 and 15.1.23).

Let $\tilde{B}$ be a connected graded ring. Recall from Definition 15.3.22 that a complex $M \in \mathbf{D}(\tilde{B}, \mathrm{gr})$ is said to be derived graded-pseudo-finite if it belongs to the saturated full triangulated subcategory of $\mathbf{D}(\tilde{B}, \mathrm{gr})$ generated by the bounded above complexes P such that each P^i is a finite graded-free $\tilde{B}$-module.

Proposition 18.5.3 *Assume the ring A admits a noetherian connected filtration F. Then:*

(1) *The Rees ring $\tilde{A} := \mathrm{Rees}^F(A)$ is a noetherian connected graded ring, and $\tilde{A}$ is a derived graded-pseudo-finite complex over $\tilde{A}^{\mathrm{en}}$.*

(2) *The ring A is noetherian, and A is a derived pseudo-finite complex over A^{en}.*

Proof

(1) Let $\bar{A} := \mathrm{Gr}^F(A)$, which is a noetherian connected graded ring. Since $\tilde{A}_0 = \mathrm{Rees}^F_0(A) = F_0(A) = \mathrm{Gr}^F_0(A) = \bar{A}_0 = \mathbb{K}$ and $\tilde{A}_i = \mathrm{Rees}^F_i(A) \cong F_i(A) \cong \bigoplus_{0 \le j \le i} \bar{A}_j$ as ungraded $\mathbb{K}$-modules, we see that $\tilde{A}$ is connected graded. Next, because $\bar{A} = \tilde{A}/(t)$, by Theorem 15.1.35 we deduce that $\tilde{A}$ is noetherian. By Proposition 15.3.24, $\tilde{A}$ is a derived graded-pseudo-finite complex over $\tilde{A}^{\mathrm{en}}$.

(2) By item (1) the graded ring $\tilde{A}$ is connected and noetherian. Since $A = \tilde{A}/(t-1)$, it follows that A is noetherian. And according to Theorem 18.3.20(3), A is a derived pseudo-finite complex over A^{en}. □

Theorem 18.5.4 (Van den Bergh Existence, [153]) *Let A be a ring. Assume A admits a noetherian connected filtration F, such that the connected graded ring $\mathrm{Gr}^F(A)$ has a balanced NC dualizing complex. Then the ring A has a rigid NC dualizing complex (R, ρ). Moreover, (R, ρ) is unique up to a unique rigid isomorphism.*

Proof As before we write $\tilde{A} := \mathrm{Rees}^F(A)$ and $\bar{A} := \mathrm{Gr}^F(A)$. These are noetherian connected graded rings, and $\bar{A} = \tilde{A}/(t)$. Since $\bar{A}$ admits a balanced NC dualizing complex, Corollary 17.3.24 says that $\bar{A}$ satisfies the χ condition and it has finite local cohomological dimension. By Theorem 17.4.35 the graded ring $\tilde{A}$ also satisfies the χ condition and it has finite local cohomological dimension. Using Corollary 17.3.24 for $\tilde{A}$, we conclude that it has a balanced NC dualizing complex $\tilde{R}$.

According to Theorem 18.4.5, $\tilde{R}$ is a graded rigid NC dualizing complex over $\tilde{A}$. Theorem 18.4.15 then says that

$$R := (\mathrm{Rest}_{A^{\mathrm{en}}} \circ \mathrm{Deg}_0 \circ \mathrm{Ind}_{(\tilde{A}_t)^{\mathrm{en}}})(\tilde{R})[-1] \in \mathbf{D}(A^{\mathrm{en}})$$

is a rigid NC dualizing complex over A; i.e. there exists a rigidifying isomorphism ρ for R.

By Proposition 18.5.3(2), A is a derived pseudo-finite complex over A^{en}. Then, according to Theorem 18.2.28, the rigid dualizing complex (R, ρ) is unique up to a unique rigid isomorphism. □

Here is a result we won't prove here. It is needed for the example that comes after it.

Proposition 18.5.5 ([183, Proposition 6.13]) *Let A be a commutative ring, and let $f : A \to B$ be a finite central ring homomorphism. If A has a noetherian connected filtration F, then there exists a noetherian connected filtration G of B, such that $f(F_j(A)) \subseteq G_j(B)$ for all j, and the graded ring homomorphism $\mathrm{Gr}(f) : \mathrm{Gr}^F(A) \to \mathrm{Gr}^G(B)$ is finite.*

Example 18.5.6 Suppose the ring B is finite over its center $\mathrm{Cent}(B)$, and $\mathrm{Cent}(B)$ is a finitely generated $\mathbb{K}$-ring. Then we can find a finite central homomorphism $f : A \to B$, where $A = \mathbb{K}[t_1, .., t_n]$ is a commutative polynomial ring. The grading on A with $\deg(t_i) = 1$ gives rise to a noetherian connected filtration F, by the formula $F_j(A) := \bigoplus_{k \leq j} A_k$. Of course, $\mathrm{Gr}^F(A) \cong A$.

Now we can use Proposition 18.5.5 to conclude that B has a noetherian connected filtration G such that $\mathrm{Gr}^F(A) \to \mathrm{Gr}^G(B)$ is finite. According to Example 17.3.16 and Corollary 17.4.18, the graded ring $\mathrm{Gr}^G(B)$ has a balanced dualizing complex. Therefore, by Theorem 18.5.4, B has a rigid NC dualizing complex.

Sometimes rings come with filtrations of the following type.

Definition 18.5.7 A *central filtration of finite type* of a ring A is a filtration $G = \{G_j(A)\}_{j \geq -1}$ of A, such that the graded ring $\mathrm{Gr}^G(A) = \bigoplus_{j \geq 0} \mathrm{Gr}^G_j(A)$ is finite over its center $C = \bigoplus_{j \geq 0} C_j := \mathrm{Cent}(\mathrm{Gr}^G(A))$, and the commutative graded ring C is finitely generated over $\mathbb{K}$.

In [186] such a filtration was called a "differential filtration of finite type," but this name sounds too confusing when we also talk about DG rings. The reason for the name "differential" will be apparent in Example 18.5.10 below.

It is not hard to see, using Theorem 15.1.35, that if A admits a central filtration of finite type, then it is noetherian. However, much more is true, as the next theorem shows. This is a result from [186] that we present without a proof. A slightly weaker version of this theorem was in the paper [107] of J.C. McConnell and J.T. Stafford, which the authors attributed to J. Bernstein.

Theorem 18.5.8 (Two Filtrations, [186, Theorem 3.1], [107]) *Let A be a ring that admits a central filtration of finite type G. Then A admits a noetherian connected filtration F, such that the noetherian connected graded ring* $\operatorname{Gr}^F(A)$ *is a commutative finitely generated $\mathbb{K}$-ring.*

As explained in Example 17.3.17, the ring $\operatorname{Gr}^F(A)$ has a balanced dualizing complex. Therefore, using Theorems 18.5.8 and 18.5.4, we conclude that:

Corollary 18.5.9 *If A admits a central filtration of finite type, then A has a rigid NC dualizing complex (R, ρ), and it is unique up to a unique rigid isomorphism.*

Example 18.5.10 Assume $\mathbb{K}$ has characteristic 0. Let C be a smooth commutative $\mathbb{K}$-ring of pure dimension n, and let $A := \mathcal{D}(C)$, the ring of differential operators. Consider the filtration $G = \{G_j(A)\}_{j \geq -1}$ of A by order of operators. For this filtration we have $G_0(A) = C$ and $G_1(A) = C \oplus \mathcal{T}(C)$, where $\mathcal{T}(C)$ is the module of derivations. Then $\operatorname{Gr}^G(A)$ is a commutative ring, and there is a canonical graded ring isomorphism $\operatorname{Gr}^G(A) \cong \operatorname{Sym}_C(\mathcal{T}(C))$, the symmetric tensor ring on the finitely generated projective C-module $\mathcal{T}(C)$. This is a commutative finite type $\mathbb{K}$-ring, so G is a central filtration of finite type. Corollary 18.5.9 says that A has a rigid NC dualizing complex (R, ρ). In [166] it was shown that $R \cong A[2n]$.

We end the section with three remarks.

Remark 18.5.11 Here is a brief history of the early days NC dualizing complexes (during the 1990s). The first paper on this topic was [161] by A. Yekutieli from 1992, in which dualizing complexes over NC connected graded rings (always over a base field $\mathbb{K}$) were introduced (see Definition 17.1.4). This material was part of the Ph.D. thesis of Yekutieli, and was inspired by ideas of his advisor M. Artin on a noncommutative analogue of Serre Duality. The paper [161] also contained the definition of balanced dualizing complexes (Definition 17.2.2), a proof of NC Local Duality (Theorem 17.2.7), a proof of the existence of balanced dualizing complexes (Theorem 17.3.5), and some particular calculations (Example 17.3.15).

The paper [182] by Yekutieli and J.J. Zhang from 1997 used the results of [161] to establish NC Serre Duality on noncommutative projective schemes, in the sense of Artin and Zhang [10]. In the same paper we also find Corollary 17.2.13, saying that the existence of a balanced dualizing complex over the noetherian connected ring A implies that A satisfies the χ condition and has finite local cohomological dimension.

The paper [153] by M. Van den Bergh from 1997 begins by establishing the important converse to Corollary 17.2.13, namely that a noetherian connected ring A, which satisfies the χ condition and has finite local cohomological dimension, admits a balanced dualizing complex (Theorem 17.3.19). Thus we have Corollary 17.3.24 on the "two equivalent properties." Van den Bergh went on to define rigid dualizing complexes, to prove their uniqueness (Theorem 18.2.28), and to prove existence of rigid dualizing complexes, in the presence of suitable filtrations and balanced dualizing complexes (Theorem 18.5.4). The subsequent paper [154] by Van den Bergh utilized the rigidity formalism to deduce what is now called Van den Bergh Duality (Theorem 18.6.14 below).

At about the same time that Van den Bergh wrote [153], Yekutieli worked on the paper [165], in which the derived Picard group $\mathrm{DPic}(A)$ of a NC ring A was introduced. The fact that any two NC dualizing complexes R, R' over A are related by $R' \cong T \otimes^{\mathrm{L}}_{A} R$ for some tilting complex T (Theorem 18.1.25, proved in [165]) was used by Van den Bergh in his proof of the uniqueness of rigid dualizing complexes in [153].

The paper [183] by Yekutieli and Zhang from 1999 built on Van den Bergh's ideas. The uniqueness results of Van den Bergh were sharpened, and were upgraded to functoriality (see Remark 18.5.12). The Auslander condition on dualizing complexes (see Remark 18.5.13) was also introduced and studied in [183].

It is hard to track the progress of the theory of NC dualizing complexes beyond the year 2000. This theory has merged with, or has been absorbed into, the more active research on Calabi–Yau rings and categories (see Section 18.6), with its ties to birational algebraic geometry and mathematical physics.

Remark 18.5.12 This remark is about the functorial properties of NC rigid dualizing complexes. In Remark 13.5.17 we talked about the rigid trace homomorphism and the rigid localization homomorphism, between rigid residue complexes over commutative rings. These are actual homomorphisms of complexes. For NC rings there are usually no residue complexes (because of the pathological nature of injective modules in the NC setting). Still we can consider rigid trace morphisms and rigid localization morphisms between NC rigid dualizing complexes. These are morphisms in derived categories. Below is an outline.

Suppose $f : A \to B$ is a finite ring homomorphism, and the rigid dualizing complexes (R_A, ρ_A) and (R_B, ρ_B) exist. A *rigid trace morphism* $\mathrm{tr}_f : (R_B, \rho_B) \to (R_A, \rho_A)$ is a backward morphism $\mathrm{tr}_f : R_B \to R_A$ in $\mathbf{D}(A^{\mathrm{en}})$, which is rigid (the obvious variant of Definition 18.2.13), and is nondegenerate on both sides (see Definition 17.4.20). In [183] the existence and uniqueness of the rigid trace morphism exists was proved, when $f : A \to B$ is finite centralizing, and A admits a noetherian connected filtration F such that $\mathrm{Gr}^F(A)$ has a balanced dualizing complex. The proof is along the same lines as the proof of Theorem 18.5.4 above, combined with the existence of the balanced trace morphism, established in Theorem 17.4.22. Among the applications of the rigid trace: (1) It was used in [166] to calculate the rigid dualizing complex of $A = \mathrm{U}(\mathfrak{g})$, see Example 18.6.13 below. (2) In [184] it was used to calculate multiplicities of indecomposable injectives in minimal injective resolutions.

Now consider a localization ring homomorphism $g : A \to A'$. Assume the rigid dualizing complexes (R_A, ρ_A) and $(R_{A'}, \rho_{A'})$ exist. A *rigid localization morphism* $\mathrm{q}_g : (R_A, \rho_A) \to (R_{A'}, \rho_{A'})$ is a forward morphism $\mathrm{q}_g : R_A \to R_{A'}$ in $\mathbf{D}(A^{\mathrm{en}})$, which is rigid (it respects the two rigidifying isomorphisms), and is nondegenerate on both sides. Existence and uniqueness of the rigid localization morphism was proved in [186, Theorem 6.2]; it is related to Ore conditions. One application of the rigid localization morphism is in the concept of *homological transcendence degree* [185] – the ring A is an integral domain which has a rigid dualizing complex (say, because of the existence of a good filtration on it), and $D = A'$ is the division ring of fractions of A. The homological transcendence degree of D is the unique number n such that the rigid dualizing complex R_D satisfies $\mathrm{H}^{-n}(R_D) \neq 0$.

Remark 18.5.13 In this remark we discuss the *Auslander property* of NC dualizing complexes. Around 1990, J.E. Björk [22] and T. Levasseur [95] had studied NC regular noetherian rings with the Auslander property. This property was later generalized in the papers [163] and [183] by A. Yekutieli and J.J. Zhang. Below is a quick explanation of the meaning and applications of this concept.

Consider a NC noetherian ring A with a NC dualizing complex R. Given a module $M \in \mathbf{M}_{\mathrm{f}}(A)$, its grade with respect to R is the number

$$\mathrm{grade}_R(M) := \inf(\{q \mid \mathrm{Ext}^q_A(M, R) \neq 0\}) \in \mathbb{Z} \cup \{\infty\}.$$

Note that $\mathrm{Ext}^q_A(M, R) \in \mathbf{M}_{\mathrm{f}}(A^{\mathrm{op}})$. The number $\mathrm{grade}_R(N)$ for $N \in \mathbf{M}_{\mathrm{f}}(A^{\mathrm{op}})$ is defined similarly. The complex R is called *Auslander* if for every $M \in \mathbf{M}_{\mathrm{f}}(A)$, every q, and every A^{op}-submodule $N \subseteq \mathrm{Ext}^q_A(M, R)$, the inequality $\mathrm{grade}_R(N) \geq q$ holds; and the same inequality holds with A and A^{op} exchanged.

The original definition used by Björk and Levasseur, for a regular ring A, was for the particular dualizing complex $R := A$.

In [183] it was proved that in many situations the rigid dualizing complex R from Theorem 18.5.4 has the Auslander property. For instance, [183, Theorem 0.7] implies that if A has a noetherian connected filtration F such that $\mathrm{Gr}^F(A)$ is commutative, then the rigid dualizing complex R of A is Auslander.

Having an Auslander rigid dualizing complex gives rise to the *canonical dimension* of A-modules. For $M \in \mathbf{M}_{\mathrm{f}}(A)$ its canonical dimension is $\mathrm{Cdim}(M) := -\mathrm{grade}_R(M)$. This can be considered as a NC version of the rigid dimension from Chapter 13. Indeed, if A is a finite type commutative $\mathbb{K}$-ring, with commutative rigid dualizing complex R, then for a prime ideal $\mathfrak{p} \subseteq A$ and the A-module $M := A/\mathfrak{p}$ we have $\mathrm{Cdim}(M) = \mathrm{rig.dim}_{\mathbb{K}}(\mathfrak{p}) = \dim(A/\mathfrak{p})$, where the second number is from Definition 13.5.10, and the third number is the Krull dimension of the ring $A/\mathfrak{p}$.

As shown in [183], several structural results for NC rings can be deduced from the presence of an Auslander rigid dualizing complex; for instance, the catenarity of $\mathrm{Spec}(A)$, see [183, Theorem 0.1]. Moreover, in [186, Section 7] a link was made between Auslander rigid dualizing complexes and perverse t-structures: the canonical dimension of modules defines a perverse t-structure on $\mathbf{D}^{\mathrm{b}}_{\mathrm{f}}(A)$, and under the duality functor $D := \mathrm{RHom}_A(-, R)$ this t-structure goes to the standard t-structure on $\mathbf{D}^{\mathrm{b}}_{\mathrm{f}}(A^{\mathrm{op}})$.

18.6 Twisted Calabi–Yau Rings

In this final section we study a class of regular rings that is of great importance in recent ring theory. We continue with Convention 18.1.1. Specifically, $\mathbb{K}$ is a base field, and all rings are central over $\mathbb{K}$. For a ring A, its enveloping ring is $A^{\mathrm{en}} = A \otimes A^{\mathrm{op}}$.

Definition 18.6.1 A ring A is called *homologically smooth over* $\mathbb{K}$ if A is a perfect complex over the ring A^{en}.

This definition was proposed by M. Kontsevich, around the year 2000, cf. [154, Erratum]. Clearly if A is homologically smooth over $\mathbb{K}$, then A is a derived pseudo-finite complex over A^{en}. Recall (from Example 18.1.14) that the ring A is called regular if it has finite global cohomological dimension.

Proposition 18.6.2 *If A is homologically smooth over $\mathbb{K}$, then A is a regular ring.*

Proof By Theorem 14.1.26 there is a quasi-isomorphism $\rho : P \to A$ in $\mathbf{C}_{\mathrm{str}}(A^{\mathrm{en}})$, where P is a bounded complex of finite projective A^{en}-modules. Let d be the amplitude of P. Given an A-module M, define $P_M := P \otimes_A M$, and let $\rho_M : P_M \to M$ be the homomorphism in $\mathbf{C}_{\mathrm{str}}(A)$ induced by ρ. Since ρ is a homotopy equivalence in $\mathbf{C}_{\mathrm{str}}(A^{\mathrm{op}})$, it follows that ρ_M is a quasi-isomorphism. Since P^i is a projective A^{en}-module and M is a free $\mathbb{K}$-module, we see that $P^i_M = P^i \otimes_A M$ is a projective A-module (often infinitely generated; cf. Example 18.2.8). The conclusion is that the projective dimension of M is at most d.

A similar consideration shows that every $M' \in \mathbf{M}(A^{\mathrm{op}})$ has projective dimension at most d. So the global homological dimension of A is at most d. □

Remark 18.6.3 Suppose A is a finite type commutative $\mathbb{K}$-ring. It can be shown that A is homologically smooth over $\mathbb{K}$ if and only if it is smooth over $\mathbb{K}$, in the sense of [57]. Cf. [108, Proposition 3.24].

Given a $\mathbb{K}$-ring automorphism μ of A, we define the *μ-twisted bimodule* $A(\mu)$ as follows: as a left A-module, $A(\mu)$ is the free A-module A; and the right A-module action is twisted by μ. To be explicit, let us denote by $a \cdot b$ the multiplication of the ring A. Then the left action of a ring element $a \in A$ on a bimodule element $b \in A(\mu)$ is $a \cdot^{\mu} b := a \cdot b \in A(\mu)$; and the right action is $b \cdot^{\mu} a := b \cdot \mu(a) \in A(\mu)$. In fancier notation, a ring element $a_1 \otimes \mathrm{op}(a_2) \in A^{\mathrm{en}}$ acts on a bimodule element $b \in A(\mu)$ from the left like this: $(a_1 \otimes \mathrm{op}(a_2)) \cdot^{\mu} b = a_1 \cdot b \cdot \mu(a_2) \in A(\mu)$. The graded version of this twisting already appeared in Definition 17.1.7. If ν is another automorphism of A, then $A(\mu) \otimes_A A(\nu) \cong A(\mu \circ \nu)$ in $\mathbf{M}(A^{\mathrm{en}})$.

It is possible to twist other bimodules: given $M \in \mathbf{M}(A^{\mathrm{en}})$ we define

$$M(\mu) := M \otimes_A A(\mu) \in \mathbf{M}(A^{\mathrm{en}}). \tag{18.6.4}$$

Let us make explicit the left and right actions of the ring A^{en} on the module $M(\mu)$, using formula (18.1.8). The original left and right actions of an element $a \in A$ on an element $m \in M$ are denoted by $a \cdot m$ and $m \cdot a$. Then an element $(a_1 \otimes \mathrm{op}(a_2)) \in A^{\mathrm{en}}$ acts on an element $m \in M(\mu)$ from the left and the right by

$$\begin{aligned} (a_1 \otimes \mathrm{op}(a_2)) \cdot^{\mu} m &= a_1 \cdot m \cdot \mu(a_2), \\ m \cdot^{\mu} (a_1 \otimes \mathrm{op}(a_2)) &= a_2 \cdot m \cdot \mu(a_1). \end{aligned} \tag{18.6.5}$$

Definition 18.6.6 A ring A is called a *twisted CY ring* if it has these two properties:

(i) The ring A is homologically smooth over $\mathbb{K}$.

(ii) There is a natural number n, called the *dimension of* A, and a ring automorphism ν of A, called the *Nakayama automorphism*, such that

$$\mathrm{Ext}^i_{A^{\mathrm{en}}}(A, A^{\mathrm{en}}) \cong \begin{cases} A(\nu) & \text{if } i = n \\ 0 & \text{if } i \neq n \end{cases}$$

as A-bimodules.

The ring A is called a *CY ring* if the third property also holds:

(iii) The automorphism ν is inner, so that $A(\nu) \cong A$ as A-bimodules.

In property (ii) the A^{en}-module $\mathrm{Ext}^i_{A^{\mathrm{en}}}(A, A^{\mathrm{en}})$ should be viewed as a special case of $\mathrm{Ext}^i_B(M, N) \in \mathbf{M}(C^{\mathrm{op}})$ for $M \in \mathbf{M}(B)$ and $N \in \mathbf{M}(B \otimes C^{\mathrm{op}})$, where we take $B, C := A^{\mathrm{en}}$, $M := A$ and $N := A^{\mathrm{en}}$, and we use the isomorphism (18.1.5) to switch from a right A^{en}-module to a left A^{en}-module.

The letters "CY" are an abbreviation for "Calabi–Yau." Most texts use the term *CY algebra* of course. See Remark 18.6.20 for a discussion of the background of this definition. Clearly this is a relative notion: it pertains to A as a central $\mathbb{K}$-ring. It is easy to see that the Nakayama automorphism ν is unique up to composition with an inner automorphism.

Here is another definition, taken from [166].

Definition 18.6.7 Let A be a noetherian ring with rigid dualizing complex R. If $R \cong A(\mu)[n]$ in $\mathbf{D}(A^{\mathrm{en}})$ for some automorphism μ and some integer n, then μ is called the *dualizing automorphism* of A.

Like the Nakayama automorphism, the dualizing automorphism is unique up to composition with an inner automorphism.

Proposition 18.6.8 *Let A be a noetherian ring which is homologically smooth over $\mathbb{K}$, let n be a natural number, and let ν be an automorphism of A. The following two conditions are equivalent.*

(i) *A is a twisted CY ring of dimension n with Nakayama automorphism ν.*

(ii) *The complex $R := A(\nu^{-1})[n]$ is a rigid dualizing complex over A.*

Thus the dualizing automorphism is the inverse of the Nakayama automorphism.

Proof The key formula is this: for $M, N \in \mathbf{D}(A^{\mathrm{en}})$ there are isomorphisms

$$\begin{aligned} \mathrm{RHom}_{A^{\mathrm{en}}}(A, M \otimes N) &\cong^1 \mathrm{RHom}_{A^{\mathrm{en}}}(A, A^{\mathrm{en}} \otimes^{\mathrm{L}}_{A^{\mathrm{en}}} (M \otimes N)) \\ &\cong^2 \mathrm{RHom}_{A^{\mathrm{en}}}(A, A^{\mathrm{en}}) \otimes^{\mathrm{L}}_{A^{\mathrm{en}}} (M \otimes N) \qquad (18.6.9) \\ &\cong^3 N \otimes^{\mathrm{L}}_A \mathrm{RHom}_{A^{\mathrm{en}}}(A, A^{\mathrm{en}}) \otimes^{\mathrm{L}}_A M \end{aligned}$$

in $\mathbf{D}(A^{\mathrm{en}})$. The isomorphism $\cong^1$ is the left unitor. The isomorphism $\cong^2$ is by derived tensor-evaluation, and it exists because A is a perfect complex over A^{en};

see Theorem 14.1.22. The isomorphism $\cong^3$ is because the left A^{en} action on $M \otimes N$ is the outside action, cf. formula (18.2.1).

(i) $\Rightarrow$ (ii): By Example 18.1.14 the complex R is dualizing. What is needed is a rigidifying isomorphism for it. We are given that $\mathrm{RHom}_{A^{\mathrm{en}}}(A, A^{\mathrm{en}}) \cong A(\nu)[-n]$ in $\mathsf{D}(A^{\mathrm{en}})$. Plugging this and $M = N := R$ in (18.6.9) gives

$$\begin{aligned}\mathrm{RHom}_{A^{\mathrm{en}}}(A, R \otimes R) &\cong A(\nu^{-1})[n] \otimes^{\mathrm{L}}_A A(\nu)[-n] \otimes^{\mathrm{L}}_A A(\nu^{-1})[n] \\ &\cong A(\nu^{-1} \circ \nu \circ \nu^{-1})[n - n + n] = A(\nu^{-1})[n] = R,\end{aligned}$$

as required.

(ii) $\Rightarrow$ (i): Now we are given that $\mathrm{RHom}_{A^{\mathrm{en}}}(A, R \otimes R) \cong R$ in $\mathsf{D}(A^{\mathrm{en}})$. Substituting $M = N := R$ in (18.6.9) gives

$$A(\nu^{-1})[n] \cong A(\nu^{-1})[n] \otimes^{\mathrm{L}}_A \mathrm{RHom}_{A^{\mathrm{en}}}(A, A^{\mathrm{en}}) \otimes^{\mathrm{L}}_A A(\nu^{-1})[n]$$

in $\mathsf{D}(A^{\mathrm{en}})$. Therefore $A(\nu)[-n] \cong \mathrm{RHom}_{A^{\mathrm{en}}}(A, A^{\mathrm{en}})$ in $\mathsf{D}(A^{\mathrm{en}})$. This is property (iii) of Definition 18.6.6. □

Example 18.6.10 This is a continuation of Example 18.5.10: $\mathbb{K}$ has characteristic 0, C is a smooth commutative $\mathbb{K}$-ring of pure dimension n, and $A = \mathcal{D}(C)$, the ring of differential operators. It is known that A is homologically smooth – in fact, in this case A^{en} is a regular noetherian ring, of dimension $4n$. According to [166, Theorem 2.6], the rigid dualizing complex of A is $R = A[2n]$. Hence A is a CY ring of dimension $2n$.

AS regular graded rings were introduced in Definition 15.4.8.

Theorem 18.6.11 *Let A be a ring that admits some noetherian connected filtration F, such that the graded ring $\mathrm{Gr}^F(A)$ is AS regular of dimension n. Then A is a noetherian twisted CY ring of dimension n, with a Nakayama automorphism ν that respects the filtration F.*

Proof Let $\tilde{A} := \mathrm{Rees}^F(A)$ and $\bar{A} := \mathrm{Gr}^F(A)$. We are given that $\bar{A}$ is AS regular of dimension n. According to Theorem 15.4.13(2), $\tilde{A}$ is AS regular of dimension $n + 1$.

Consider the minimal graded-free resolution $\tilde{Q} \to \tilde{A}$ of $\tilde{A}$ over $\tilde{A}^{\mathrm{en}}$. We know, from Theorem 15.4.2 and Proposition 15.3.24, that each $\tilde{Q}^i$ is a finite graded-free $\tilde{A}^{\mathrm{en}}$-module, and $\tilde{Q}^i = 0$ for $i < -n - 1$. Therefore the complex

$$P := A^{\mathrm{en}} \otimes_{\tilde{A}^{\mathrm{en}}} \tilde{Q} \cong A^{\mathrm{en}} \otimes^{\mathrm{L}}_{\tilde{A}^{\mathrm{en}}} \tilde{A} \in \mathsf{D}(A^{\mathrm{en}})$$

is perfect. But according to Theorem 18.3.20(2), A is a direct summand of P in $\mathsf{D}(A^{\mathrm{en}})$. Thus A is a perfect complex over A^{en}. This is property (ii) of Definition 18.6.6.

By Corollary 17.3.14 the graded ring $\tilde{A}$ has a balanced dualizing complex $\tilde{R} := \tilde{A}(\tilde{\mu}, l)[n+1]$, where $\tilde{\mu}$ is a graded ring automorphism of $\tilde{A}$, and l is an integer. Theorem 18.4.5 says that $\tilde{R}$ is a graded rigid dualizing complex over $\tilde{A}$.

Lemma 18.4.16 says that for every p the $\tilde{A}$-bimodule $\mathrm{H}^p(\tilde{R})$ is a central bimodule over $\mathrm{Cent}(\tilde{A})$. But for $p = -n-1$ we have $\mathrm{H}^{-n-1}(\tilde{R}) \cong \tilde{A}(\tilde{\mu}, l)$, and this implies that $\tilde{\mu}$ acts trivially on the element $t \in \mathrm{Cent}(\tilde{A})$. Hence $\tilde{\mu}$ extends to an automorphism of $\tilde{A}_t$, and there is an isomorphism

$$\mathrm{Ind}_{(\tilde{A}_t)^{\mathrm{en}}}(\tilde{R}) \cong \tilde{A}_t(\tilde{\mu}, l)[n+1] \tag{18.6.12}$$

in $\mathbf{D}((\tilde{A}_t)^{\mathrm{en}}, \mathrm{gr})$. Also there is an induced automorphism μ of the ring A, and formula (18.5.1) shows that μ respects the filtration F.

According to Theorem 18.4.15, the noetherian ring A has a rigid NC dualizing complex

$$R := (\mathrm{Rest}_{A^{\mathrm{en}}} \circ \mathrm{Deg}_0 \circ \mathrm{Ind}_{(\tilde{A}_t)^{\mathrm{en}}})(\tilde{R})[-1] \in \mathbf{D}(A^{\mathrm{en}}).$$

From (18.6.12) we see that $R \cong A(\mu)[n]$ in $\mathbf{D}(A^{\mathrm{en}})$. Now use Proposition 18.6.8 with $\nu := \mu^{-1}$. □

Example 18.6.13 Let $\mathfrak{g}$ be a finite Lie algebra over $\mathbb{K}$, with $\mathrm{rank}_{\mathbb{K}}(\mathfrak{g}) = n$, and let $A := \mathrm{U}(\mathfrak{g})$ be its universal enveloping algebra. The ring A is noetherian twisted CY of dimension n. This was worked out in the paper [166], and is outlined below.

The ring A has a standard noetherian connected filtration F, such that $\bar{A} = \mathrm{Gr}^F(A) \cong \mathbb{K}[t_1, \ldots, t_n]$, the commutative polynomial ring in n variables, all of degree 1. The ring $\bar{A}$ is AS regular of dimension n, as explained in Example 17.3.16. The Rees ring here is the homogeneous universal enveloping ring from Example 15.2.22.

Theorem 18.6.11 says that A is twisted CY of dimension n, with a Nakayama automorphism ν that respects the filtration F. In case $\mathfrak{g}$ is an abelian Lie algebra (i.e. the Lie bracket is zero), then ν is the identity automorphism. When $\mathfrak{g}$ is semi-simple, Van den Bergh [154] proved that ν is trivial too. So in these extreme cases the ring $A = \mathrm{U}(\mathfrak{g})$ is CY (untwisted).

The general case was done in [166]. Consider the character (i.e. rank 1 representation) $\bigwedge^n(\mathfrak{g})$ of $\mathfrak{g}$, the nth exterior power of the adjoint representation. This can be seen as a Lie algebra homomorphism $\epsilon : \mathfrak{g} \to \mathbb{K}$. Then the dualizing automorphism μ is the unique ring automorphism of $A = \mathrm{U}(\mathfrak{g})$ such that $\mu(u) = u - \epsilon(u) \cdot 1_A$ for $u \in \mathfrak{g} \subseteq F_1(A) \subseteq A$. The Nakayama automorphism is $\nu = \mu^{-1}$.

The comultiplication of A makes the tensor product of A-bimodules into a bimodule. This allows us to express $A(\mu)$ like this: $A(\mu) \cong A \otimes \bigwedge^n(\mathfrak{g})$, where the left action of A on $\bigwedge^n(\mathfrak{g})$ is trivial, and the right action is the coadjoint action.

Given an A-bimodule M, its *ith Hochschild homology* is the $\mathbb{K}$-module $\mathrm{HH}_i(A, M) := \mathrm{Tor}_i^{A^{\mathrm{en}}}(A, M)$, and its *ith Hochschild cohomology* is $\mathrm{HH}^i(A, M) := \mathrm{Ext}^i_{A^{\mathrm{en}}}(A, M)$.

Theorem 18.6.14 (Van den Bergh Duality, [154]) *Let A be an n-dimensional twisted CY ring, with Nakayama automorphism ν. For every $M \in \mathbf{M}(A^{\mathrm{en}})$ and $0 \leq i \leq n$ there is an isomorphism*

$$\mathrm{HH}^i(A, M) \cong \mathrm{HH}_{n-i}(A, M(\nu))$$

in $\mathbf{M}(\mathbb{K})$, and it is functorial in M.

The theorem does not require the rings A nor A^{en} to be noetherian. We need a lemma first.

Lemma 18.6.15 *For a complex $M \in \mathbf{D}(A^{\mathrm{en}})$, and an automorphism μ of A, there is an isomorphism $A \otimes^{\mathrm{L}}_{A^{\mathrm{en}}} M(\mu) \cong A(\mu) \otimes^{\mathrm{L}}_{A^{\mathrm{en}}} M$ in $\mathbf{D}(\mathbb{K})$. This isomorphism is functorial in M.*

Proof It suffices to produce a functorial isomorphism $\psi : A \otimes_{A^{\mathrm{en}}} M(\mu) \xrightarrow{\simeq} A(\mu) \otimes_{A^{\mathrm{en}}} M$ for $M \in \mathbf{M}(A^{\mathrm{en}})$. We define ψ on elements $b \in A$ and $m \in M$ by the formula $\psi(b \otimes m) := \mu(b) \otimes m$. It must be verified that the relations coming from the action of the ring A^{en} are respected by ψ. Take a ring element $a_1 \otimes \mathrm{op}(a_2) \in A^{\mathrm{en}}$. Then in the module $A \otimes_{A^{\mathrm{en}}} M(\mu)$ we have equalities

$$\begin{aligned}(a_2 \cdot b \cdot a_1) \otimes m &= (b \cdot (a_1 \otimes \mathrm{op}(a_2))) \otimes m \\ &= b \otimes ((a_1 \otimes \mathrm{op}(a_2)) \cdot^{\mu} m) = b \otimes (a_1 \cdot m \cdot \mu(a_2)).\end{aligned} \tag{18.6.16}$$

See formula (18.6.5). In the module $A(\mu) \otimes_{A^{\mathrm{en}}} M$ we have equalities

$$\begin{aligned}\mu(a_2 \cdot b \cdot a_1) \otimes m &= (\mu(a_2) \cdot \mu(b) \cdot \mu(a_1)) \otimes m \\ &= (\mu(b) \cdot^{\mu} (a_1 \otimes \mathrm{op}(\mu(a_2)))) \otimes m \\ &= \mu(b) \otimes ((a_1 \otimes \mathrm{op}(\mu(a_2))) \cdot m) \\ &= \mu(b) \otimes (a_1 \cdot m \cdot \mu(a_2)).\end{aligned} \tag{18.6.17}$$

The isomorphism ψ sends the first (resp. last) term in (18.6.16) to the first (resp. last) term in (18.6.17). □

Proof of Theorem 18.6.14 We are given an isomorphism

$$A(\nu)[-n] \cong \mathrm{RHom}_{A^{\mathrm{en}}}(A, A^{\mathrm{en}}) \tag{18.6.18}$$

in $\mathsf{D}(A^{\mathrm{en}})$. Consider these isomorphisms

$$\begin{aligned} A \otimes^{\mathrm{L}}_{A^{\mathrm{en}}} M(\nu)[-n] &\cong^1 A(\nu)[-n] \otimes^{\mathrm{L}}_{A^{\mathrm{en}}} M \\ &\cong^2 \mathrm{RHom}_{A^{\mathrm{en}}}(A, A^{\mathrm{en}}) \otimes^{\mathrm{L}}_{A^{\mathrm{en}}} M \\ &\cong^3 \mathrm{RHom}_{A^{\mathrm{en}}}(A, A^{\mathrm{en}} \otimes^{\mathrm{L}}_{A^{\mathrm{en}}} M) \cong^4 \mathrm{RHom}_{A^{\mathrm{en}}}(A, M) \end{aligned} \tag{18.6.19}$$

in $\mathsf{D}(\mathbb{K})$. The first isomorphism is by Lemma 18.6.15. The isomorphism $\cong^2$ comes from (18.6.18). The isomorphism $\cong^3$ is by derived tensor-evaluation, and it holds because A is perfect over A^{en}, see Theorem 14.1.22. The isomorphism $\cong^4$ is clear.

Taking $\mathrm{H}^i(-)$ in (18.6.19) we obtain an isomorphism

$$\mathrm{Tor}^{A^{\mathrm{en}}}_{n-i}(A, M(\nu)) \cong \mathrm{Ext}^i_{A^{\mathrm{en}}}(A, M)$$

in $\mathsf{M}(\mathbb{K})$. Functoriality is clear. □

Remark 18.6.20 Here is a quick discussion of the evolution of the *Calabi–Yau* concept. In algebraic geometry, an n-dimensional smooth proper scheme X over a field $\mathbb{K}$ has the *canonical sheaf* $\omega_X := \Omega^n_{X/\mathbb{K}}$; this is a rank 1 locally free $\mathcal{O}_X$-module. In the paper [24] from 1990, A.I. Bondal and M.M. Kapranov introduced the *Serre functor*

$$\mathrm{S} := \omega_X[n] \otimes_{\mathcal{O}_X} (-). \tag{18.6.21}$$

This is an autoequivalence of the derived category $\mathsf{D}^{\mathrm{b}}_{\mathrm{c}}(X) = \mathsf{D}^{\mathrm{b}}_{\mathrm{c}}(\mathsf{Mod}\,\mathcal{O}_X)$, which admits a bifunctorial isomorphism

$$\mathrm{Hom}_{\mathsf{D}^{\mathrm{b}}_{\mathrm{c}}(X)}(\mathcal{M}, \mathcal{N}) \cong \mathrm{Hom}_{\mathsf{D}^{\mathrm{b}}_{\mathrm{c}}(X)}(\mathcal{N}, \mathrm{S}(\mathcal{M}))^*. \tag{18.6.22}$$

Here $(-)^* := \mathrm{Hom}_{\mathbb{K}}(-, \mathbb{K})$. This is an upgrade of the familiar Serre Duality for X, which is the special case where $\mathcal{M} = \mathcal{O}_X$ and $\mathcal{N}$ is the translation of a locally free $\mathcal{O}_X$-module.

The Serre functor makes sense on any $\mathbb{K}$-linear triangulated category D that has suitable finiteness properties ("smooth" and "proper"). It is a triangulated autoequivalence S of D admitting a bifunctorial isomorphism

$$\mathrm{Hom}_{\mathsf{D}}(M, N) \cong \mathrm{Hom}_{\mathsf{D}}(N, \mathrm{S}(M))^* \tag{18.6.23}$$

in $\mathsf{M}(\mathbb{K})$ for $M, N \in \mathsf{D}$. In [24] it was proved that if a Serre functor exists, then it is unique. In many situations (where this makes sense) the Serre functor is realized by tensoring with the rigid dualizing complex. This is true in the geometric context above, in which the rigid dualizing complex of the scheme X is $\omega_X[n]$; see formula (18.6.21). It is also true for a regular finite $\mathbb{K}$-ring A, as explained in Example 18.6.24 below.

Back in algebraic geometry, a smooth proper scheme X is called Calabi–Yau if the canonical sheaf ω_X is trivial. (Examples are abelian varieties and K3 surfaces.) The CY condition implies that the Serre functor S is isomorphic to T^n, the nth power of the translation functor of $\mathsf{D} = \mathbf{D}^{\mathrm{b}}_{\mathrm{c}}(X)$. Thus the dimension of X is precisely the ratio between the autoequivalences S and T.

Around the year 1994, M. Konstevich began discussing *Homological Mirror Symmetry*, which is a vast and deep program (still not fully understood) relating the algebraic geometry of a CY scheme X (as above) with the *symplectic geometry* of its mirror partner manifold Y. See [88] for more information on this program.

Among the ideas arising in the discussion by Kontsevich was that of a *Calabi–Yau category*. This is a smooth proper $\mathbb{K}$-linear triangulated category D, whose translation functor T and Serre functor S satisfy the numerical relation $S \cong T^n$ for some integer n, called the *CY dimension* of A. More generally, if the relation $S^m \cong T^n$ holds for some $m, n \in \mathbb{Z}$, then D is called *fractionally CY* of dimension (n, m). By abuse of notation, such a ring A is sometimes called fractionally CY of dimension n/m.

The concept of CY triangulated categories later showed up in various mathematical areas, including:

- Algebraic geometry (see [91]).
- Conformal field theories (see [42]).
- Representation theory of algebras and cluster categories (see [36], [33], [82], [84], [92] and Example 18.6.24 below).

In the preprint [52] from 2006, V. Ginzburg defined *Calabi–Yau algebras*. At about the same time K. Brown and J.J. Zhang [32] introduced *rigid Gorenstein algebras*. These are both precisely the CY rings from Definition 18.6.6 above. Eventually the CY terminology was adopted by ring theorists.

In case A is a noetherian connected graded ring, then it is twisted CY in the graded sense if and only if it is AS regular; see [123]. More results on graded CY rings can be found in [124] and its references.

We finish the section (and the book) with an example of a fractionally CY category.

Example 18.6.24 Suppose A is a finite $\mathbb{K}$-ring. (Traditional texts would say that A is a finite dimensional $\mathbb{K}$-algebra.) We claim that the rigid dualizing complex of A is $R := A^*$, the $\mathbb{K}$-linear dual of A. That $R \in \mathbf{D}(A^{\mathrm{en}})$ is a NC dualizing complex over A is very easy to see (cf. Definition 18.1.12). As for rigidity: in this case it is just linear algebra. Indeed, $R \otimes R = A^* \otimes A^* \cong (A^{\mathrm{en}})^*$ in $\mathbf{M}(A^{\mathrm{four}})$, and hence there are canonical isomorphisms

$$\mathrm{Sq}_A(R) = \mathrm{RHom}_{A^{\mathrm{en}}}(A, R \otimes R) \cong \mathrm{Hom}_{A^{\mathrm{en}}}(A, (A^{\mathrm{en}})^*) \cong A^* = R$$

in $\mathbf{D}(A^{\mathrm{en}})$.

Now let's assume that A is regular, i.e. it has finite global cohomological dimension. (If the base field $\mathbb{K}$ is perfect, one can show that A is homologically smooth over it; see [60] and [35].) Then, according to Theorems 14.1.33 and 14.1.22, every object of $\mathbf{D}^{\mathrm{b}}_{\mathrm{f}}(A)$ is a compact object of $\mathbf{D}(A)$. This includes R. So R satisfies condition (iii) of Theorem 14.4.16, and therefore it is a tilting complex over A. We see that the triangulated functor

$$\mathrm{S} := R \otimes^{\mathrm{L}}_A (-) : \mathbf{D}^{\mathrm{b}}_{\mathrm{f}}(A) \to \mathbf{D}^{\mathrm{b}}_{\mathrm{f}}(A)$$

is an autoequivalence.

It turns out that S is a Serre functor of $\mathbf{D}^{\mathrm{b}}_{\mathrm{f}}(A)$. Let us prove this. For $M, N \in \mathbf{D}^{\mathrm{b}}_{\mathrm{f}}(A)$ we have canonical isomorphisms

$$\begin{aligned} \mathrm{RHom}_A(N, \mathrm{S}(M)) &= \mathrm{RHom}_A(N, A^* \otimes^{\mathrm{L}}_A M) \\ &\cong^{\dagger} \mathrm{RHom}_A(N, A^*) \otimes^{\mathrm{L}}_A M \cong N^* \otimes^{\mathrm{L}}_A M \end{aligned}$$

in $\mathbf{D}(\mathbb{K})$. The isomorphism $\cong^{\dagger}$ is by derived tensor-evaluation, and it holds because N is a compact object of $\mathbf{D}(A)$. Therefore we get functorial isomorphisms

$$\begin{aligned} \mathrm{RHom}_A(N, \mathrm{S}(M))^* &\cong (N^* \otimes^{\mathrm{L}}_A M)^* \cong \mathrm{RHom}_{\mathbb{K}}(N^* \otimes^{\mathrm{L}}_A M, \mathbb{K}) \\ &\cong^{\heartsuit} \mathrm{RHom}_A(M, \mathrm{RHom}_{\mathbb{K}}(N^*, \mathbb{K})) \cong \mathrm{RHom}_A(M, N) \end{aligned} \tag{18.6.25}$$

in $\mathbf{D}(\mathbb{K})$. The isomorphism $\cong^{\heartsuit}$ is by adjunction for the ring homomorphism $\mathbb{K} \to A$. Taking H^0 in (18.6.25) we obtain a functorial isomorphism

$$\mathrm{Hom}_{\mathbf{D}^{\mathrm{b}}_{\mathrm{f}}(A)}(N, \mathrm{S}(M))^* \cong \mathrm{Hom}_{\mathbf{D}^{\mathrm{b}}_{\mathrm{f}}(A)}(M, N)$$

in $\mathbf{M}(\mathbb{K})$.

Finally we specialize to the setting of Example 14.5.30. So A is the ring of upper triangular $n \times n$ matrices over $\mathbb{K}$ for some $n \geq 2$. As calculated by A. Yekutieli and E. Kreines in [165], there is an isomorphism

$$\underbrace{R \otimes^{\mathrm{L}}_A \cdots \otimes^{\mathrm{L}}_A R}_{n+1} \simeq A[n \quad 1]$$

in $\mathbf{D}(A^{\mathrm{en}})$. Since the translation functor of $\mathbf{D}^{\mathrm{b}}_{\mathrm{f}}(A)$ is $\mathrm{T} \cong A[1] \otimes^{\mathrm{L}}_A (-)$, we obtain an isomorphism of functors $\mathrm{S}^{n+1} \cong \mathrm{T}^{n-1}$. The conclusion is that the category $\mathbf{D}^{\mathrm{b}}_{\mathrm{f}}(A)$ is fractionally CY of dimension $(n-1, n+1)$.

References

[1] L. Alonso, A. Jeremias and J. Lipman, Local homology and cohomology on schemes, *Ann. Sci. ENS* **30** (1997), 1–39. Correction www.math.purdue.edu/~lipman.

[2] A. Altman and S. Kleiman, *A term of commutative algebra*, http://web.mit.edu/18.705/www.

[3] L. Angeleri Hügel, D. Happel and H. Krause (eds.), *Handbook of Tilting Theory*, London Math. Soc. Lecture Note Ser. **332** (2006), 147–173.

[4] D. Arinkin and R. Bezrukavnikov, Perverse coherent sheaves, *Mosc. Math. J.* **10** (2010), 3–29.

[5] M. Artin, A. Grothendieck and J.-L. Verdier, (eds.), *Séminaire de Géométrie Algébrique du Bois Marie – 1963-64 – Théorie des Topos et Cohomologie Étale des Schémas*, SGA 4, vol. 1, Lecture Notes in Mathematics **269**, Springer, 1972.

[6] M. Artin and W. Schelter, Graded algebras of dimension 3, *Adv. Math.* **66** (1987), 172–216.

[7] M. Artin, L. W. Small and J. J. Zhang, Generic flatness for strongly noetherian algebras, *J. Algebra* **221** (1999), 579–610.

[8] M. Artin, J. Tate and M. Van den Bergh, Some algebras associated to automorphisms of elliptic curves, in *The Grothendieck Festschrift*, Vol. I, Birkhäuser, 1990, pp. 33–85.

[9] M. Artin, J. Tate and M. Van den Bergh, Modules over regular algebras of dimension 3, *Invent. Math.* **106** (1991), 335–388.

[10] M. Artin and J. J. Zhang, Noncommutative projective schemes, *Adv. Math.* **109** (1994), 228–287.

[11] M. Auslander, M. Platzeck and I. Reiten, Coxeter functors without diagrams, *Trans. Amer. Math. Soc.* **250** (1979), 1–46.

[12] L. L. Avramov, Infinite free resolutions, in *Six Lectures on Commutative Algebra*, J. Elias et al., (eds.), Birkhäuser, 1998, pp. 1–118.

[13] L. L. Avramov, S. B. Iyengar and J. Lipman, Reflexivity and rigidity

for complexes, I: Commutative rings, *Algebra Number Theory* **4** (2010), 47–86.

[14] L. L. Avramov, S. B. Iyengar, J. Lipman and S. Nayak, Reduction of derived Hochschild functors over commutative algebras and schemes, *Adv. Math.* **223** (2010) 735–772.

[15] A. A. Beilinson, Coherent sheaves on $\mathbf{P}^n$ and problems in linear algebra, *Funktsional Anal. i Prilozhen.* **12** (1978), 68–69 (Russian). English translation in *Functional Anal. Appl.* **12** (1978), 214–216.

[16] A. A. Beilinson, J. Bernstein and P. Deligne, *Faisceaux pervers*, Astérisque **100**, 1980.

[17] R. Berger, A Koszul sign map, arXiv:1708.01430 (2017).

[18] G. M. Bergman, A note on abelian categories – translating element-chasing proofs, and exact embedding in abelian groups, `http://math.berkeley.edu/~gbergman/papers/unpub/elem-chase.pdf`.

[19] I. N. Bernstein, I. M. Gelfand and V. A. Ponomarev, Coxeter functors and Gabriel's theorem, *Uspekhi Mat. Nauk* **28** (1973), 19–23, Trans.: *Russian Math. Surveys* **28** (1973), 17–32.

[20] J. Bernstein and V. Lunts, *Equivariant Sheaves and Functors*, Lecture Notes in Mathematics **1578**, Springer, 1994.

[21] P. Berthelot, A. Grothendieck and L. Illusie, eds., *Séminaire de Géométrie Algébrique du Bois Marie – 1966-67 – Théorie des intersections et théorème de Riemann-Roch*, SGA 6, Lecture Notes in Mathematics **225**, Springer, 1971.

[22] J.E. Björk, The Auslander condition on noetherian rings, in *Séminaire Dubreil-Malliavin, 1987–1988*, Lecture Notes in Mathematics **1404**, Springer, 1989, pp. 137–173.

[23] M. Bokstedt and A. Neeman, Homotopy limits in triangulated categories, *Compositio Math.* **86** (1993), 209–234.

[24] A. I. Bondal and M. M. Kapranov, Representable functors, Serre functors, and mutations, *Math. USSR Izvestia* **35** (1990), 519–541.

[25] A. I. Bondal and M. M. Kapranov, Enhanced triangulated categories, *Math. USSR Sbornik*, **70** (1991), 93–107.

[26] A. Bondal and M. Van den Bergh, Generators and representability of functors in commutative and noncommutative geometry, *Moscow Math. J.* **3**, Number 1 (2003), 1–36.

[27] A. Borel, *Algebraic D-Modules*, Academic Press, 1987.

[28] N. Bourbaki, *Commutative Algebra*, Hermann, Paris, 1972.

[29] S. Brenner and M. C. R. Butler, Generalizations of the Bernstein–Gelfand–Ponomarev reflection functors, in *Proceedings ICRA II*, Lecture Notes in Mathematics **832**, Springer, 1980, pp. 103–169.

[30] M. P. Brodmann and R. Y. Sharp, *Local Cohomology: An Algebraic Introduction with Geometric Applications*, Cambridge Studies in

Advanced Mathematics **136**, 2nd edition, Cambridge University Press, 2013.

[31] E. Brown, Cohomology theories, *Ann. Math.* **75** (1962), 467–484.

[32] K. A. Brown and J. J. Zhang, Dualising complexes and twisted Hochschild (co)homology for noetherian Hopf algebras, *J. Algebra* **320**, No. 5 (2008), 1814–1850.

[33] A. B. Buan, R. Marsh, M. Reineke, I. Reiten and G. Todorov, Tilting theory and cluster combinatorics, *Adv. Math.* **204** (2006), 572–618.

[34] R.-O. Buchweitz, Maximal Cohen–Macaulay modules and Tate cohomology over Gorenstein rings (1986), 155 pp., `http:hdl.handle.net/1807/16682`.

[35] R.-O. Buchweitz, E. L. Green, D. Madsen and O. Solberg. Finite Hochschild cohomology without finite global dimension, *Math. Res. Lett.* **12** (2005), 805–816.

[36] P. Caldero, F. Chapoton and R. Schiffler, Quivers with relations and cluster tilted algebras, *Algebr. Represent. Theory* **9** (2006), 359–376.

[37] A. Canonaco, M. Ornaghi and P. Stellari, Localizations of the category of A_∞ categories and internal Homs, arXiv:1811.07830 (2018).

[38] A. Canonaco and P. Stellari, A tour about existence and uniqueness of dg enhancements and lifts, *J. Geom. Phys.* **122** (2017), 28–52.

[39] D. Chan, Q. S. Wu and J. J. Zhang, Pre-balanced dualizing complexes, *Israel J. Math.* **132** (2002), 285–314.

[40] X.W. Chen, A note on standard equivalences, *Bulletin LMS* **48** (2016), 797–801.

[41] E. Cline, B. Parshall and L. Scott, Derived categories and Morita theory, *J. Algebra* **104** (1986), 397–409.

[42] K. J. Costello, Topological conformal field theories and Calabi–Yau categories, *Adv. Math.* **210** (2007), 165–214.

[43] D. A. Craven and R. Rouquier, Perverse equivalences and Broué's conjecture, *Adv. Math.* **248** (2013), 1–58.

[44] A. Dold and D. Puppe, Homologie nicht-additiver Funktoren, *Annales de l'Institut Fourier* **11** (1961), 201–312.

[45] D. Dugger and B. Shipley, Topological equivalences of differential graded algebras, *Adv. Math.* **212** (2007), 37–61.

[46] D. Eisenbud, *Commutative Algebra*, Springer, 1994.

[47] P. Freyd, *Abelian Categories*, Harper, 1966.

[48] K. Fukaya, Morse homotopy, A_∞-category and Floer homologies, MSRI preprint No. 020-94, 1993.

[49] P. Gabriel and M. Zisman, *Calculus of Fractions and Homotopy Theory*, Ergebnisse der Mathematik und ihrer Grenzgebiete, Springer, 1967.

[50] D. Gaitsgory, Recent progress in geometric Langlands theory, `www.math.harvard.edu/~gaitsgde/GL/Bourb.pdf`.

[51] S. I. Gelfand and Y. I. Manin, *Methods of Homological Algebra*, Springer, 2002.

[52] V. Ginzburg, Calabi–Yau algebras, eprint arXiv:math/0612139 (2007).

[53] A. Grothendieck, Sur quelques points d'algèbre homologique, *Tôhoku Math. J.* **9** (1957), 119–221.

[54] A. Grothendieck, *Local Cohomology (Lecture Notes by R. Hartshorne)*, Lecture Notes in Mathematics **41**, Springer, 1967.

[55] A. Grothendieck and J. Dieudonné, *Éléments de géometrie algébrique*, collective reference for the whole series.

[56] A. Grothendieck and J. Dieudonné, *Éléments de géometrie algébrique*, Chapitre III, Première partie, Publ. Math. IHES **11**, 1961.

[57] A. Grothendieck and J. Dieudonné, *Éléments de géometrie algébrique*, Chapitre IV, Quatrième partie, Publ. Math. IHES **32**, 1967.

[58] D. Happel, On the derived category of a finite-dimensional algebra, *Commentarii Mathematici Helvetici* **62** (1987), 339–389.

[59] D. Happel, *Triangulated Categories in the Representation of Finite Dimensional Algebras*, Cambridge University Press, 1988.

[60] D. Happel, Hochschild cohomology of finite-dimensional algebras, in *Séminaire d'Algèbre Paul Dubreil et Marie-Paul Malliavin*, Lecture Notes in Mathematics **1404**, Springer, 1989, pp. 108–126.

[61] D. Happel and C. M. Ringel, Tilted algebras, *Trans. Amer. Math. Soc.* **274** (1982), 399–443.

[62] R. Hartshorne, *Residues and Duality*, Lecture Notes in Mathematics **20**, Springer, 1966.

[63] R. Hartshorne, *Algebraic Geometry*, Springer, 1977.

[64] L. Hille and M. Van den Bergh, Fourier-Mukai transforms, in *Handbook of Tilting Theory,* London Math. Soc. Lecture Note Ser. **332** (2006), pp. 147–173.

[65] P. J. Hilton and U. Stammbach, *A Course in Homological Algebra*, Springer, 1971.

[66] V. Hinich, Homological algebra of homotopy algebras, *Comm. Algebra* **25** (1997), 3291–3323. Erratum arXiv:math/0309453.

[67] V. Hinich, Lectures on infinity categories, arXiv:1709.06271 (2018).

[68] M. Hovey, Model categories, *Math. Surv. Monogr.* **63**, AMS, 1999.

[69] B. Huisgen-Zimmermann and M. Saorin, Geometry of chain complexes and outer automorphism groups under derived equivalence, *Trans. Amer. Math. Soc.* **353** (2001), 4757–4777.

[70] D. Huybrechts, *Fourier-Mukai Transforms in Algebraic Geometry*, Oxford Mathematical Monographs, 2006.

[71] N. Jacobson, *Basic Algebra I*, Freeman, 1974.

[72] P. Jørgensen, Local cohomology for non-commutative graded algebras, *Comm. Algebra* **25** (1997), 575–591.

[73] T. V. Kadeisvili, On the theory of homology of fiber spaces, *Uspekhi*

Mat. Nauk **35** (1980), 183–188.

[74] M. Kashiwara, *D-Modules and Microlocal Calculus*, AMS, 2003.

[75] M. Kashiwara and P. Schapira, *Sheaves on Manifolds*, Springer, 1990.

[76] M. Kashiwara and P. Schapira, *Categories and Sheaves*, Springer, 2005.

[77] M. Kashiwara and P. Schapira, *DQ Modules*, Astérisque **345**, 2012.

[78] B. Keller, Chain complexes and stable categories, *Manus. Math.* **67** (1990), 379–417.

[79] B. Keller, Deriving DG categories, *Ann. Sci. Ecole Norm. Sup.* **27** (1994) 63–102.

[80] B. Keller, Introduction to A-infinity algebras and modules, *Homology Homotopy App.*, **3** (2001), 1–35. Addendum: *Homology Homotopy Appl.* **4** (2002), 25–28.

[81] B. Keller, Hochschild cohomology and derived Picard groups, *J. Pure Appl. Algebra* **190** (2004), 177–196.

[82] B. Keller, On triangulated orbit categories, *Doc. Math.* **10** (2005), 551–581.

[83] B. Keller, A-infinity algebras, modules and functor categories, in *Trends in Representation Theory of Algebras and Related Topics*, Contemp. Math. **406**, AMS, 2006, pp. 67–93.

[84] B. Keller, Cluster algebras and derived categories, in *Derived Categories in Algebraic Geometry*, EMS, 2013, pp. 123–184.

[85] G. M. Kelly, Chain maps inducing zero homology maps, *Proc. Cambridge Philos. Soc.* **61** (1965), 847–854.

[86] M. Kontsevich, Homological algebra of mirror symmetry, in *Proceedings of the International Congress of Mathematicians, Zurich, Switzerland 1994*, Vol. 1, Birkhauser, 1995, pp. 120–139.

[87] M. Kontsevich and Y. Soibelman, Homological mirror symmetry and torus fibrations, in *Symplectic Geometry and Mirror Symmetry*, World Sci. Publishing, 2001, pp. 203–263.

[88] M. Kontsevich et al., Simons collaboration on homological mirror symmetry, `https://schms.math.berkeley.edu`.

[89] J. L. Koszul, Sur un type d'algébres différentielles en rapport avec la transgression, in *Colloque de topologie (espaces fibrés)*, Masson, 1951, pp. 73–81.

[90] H. Krause, Localization theory for triangulated Categories, in *Triangulated Categories*, London Math. Soc. Lecture Note Ser. **375**, 2010, pp. 161–253.

[91] A. Kuznetsov, Calabi–Yau and fractional Calabi–Yau categories, *J. Reine Angew. Math.* **753** (2019), 239–267.

[92] S. Ladkani, 2-CY-tilted algebras that are not Jacobian, arXiv:1403.6814 (2014).

[93] T.-Y. Lam, *Lectures on Modules and Rings*, Springer, 1999.

[94] K. Lefèvre-Hasegawa, Sur les A_∞-categories, Ph.D. thesis (2003), arXiv:0310337v1.

[95] T. Levasseur, Some properties of noncommutative regular rings, *Glasgow Math. J.* **34** (1992), 277–300.

[96] J. Lipman, S. Nayak and P. Sastry, Pseudofunctorial behavior of Cousin complexes on formal schemes, in *Variance and Duality for Cousin Complexes on Formal Schemes,* Contemp. Math. **375**, AMS, 2005, pp. 3–133.

[97] J. Lipman, *Dualizing Sheaves, Differentials and Residues on Algebraic Varieties,* Astérisque **117**, 1984.

[98] J. Lipman, Notes on derived functors and Grothendieck duality, in *Foundations of Grothendieck Duality for Diagrams of Schemes*, Lecture Notes in Mathematics **1960**, Springer, 2009.

[99] J. Lurie, Books and papers on derived algebraic geometry, `www.math.harvard.edu/~lurie`.

[100] J. Lurie, *Higher algebra*, `www.math.harvard.edu/~lurie/papers/HA.pdf`.

[101] S. Maclane, *Homology*, Springer, 1994 (reprint).

[102] S. Maclane, *Categories for the Working Mathematician*, Springer, 1978.

[103] E. Macri and P. Stellari, Lectures on non-commutative K3 surfaces, Bridgeland stability, and moduli spaces, arXiv:1807.06169 (2019).

[104] E. Matlis, Injective modules over Noetherian rings, *Pacific J. Math.* **8**, 3 (1958), 511–528.

[105] H. Matsumura, *Commutative Ring Theory*, Cambridge University Press, 1986.

[106] J. C. McConnell and J. C. Robson, *Noncommutative Noetherian Rings*, Wiley, 1987.

[107] J. C. McConnell and J. T. Stafford, Gelfand-Kirillov dimension and associated graded modules, *J. Algebra* **125** (1989), 197–214.

[108] J. S. Milne, *Étale Cohomology*, Princeton University Press, 1980.

[109] J.-I. Miyachi and A. Yekutieli, Derived Picard groups of finite dimensional hereditary algebras, *Compositio Math.* **129** (2001), 341–368.

[110] D. Nadler and E. Zaslow, Constructible sheaves and the Fukaya category, *J. Amer. Math. Soc.* **22** (2009), 233–286.

[111] C. Năstăsescu and F. Van Oystaeyen, *Graded and Filtered Rings and Modules*, Lecture Notes in Mathematics **758**, Springer, 1979.

[112] A. Neeman, The connection between the K-theory localization theorem of Thomason, Trobaugh and Yao and the smashing subcategories of Bousfield and Ravenel, *Ann. Sci. École Norm. Sup.* **25** (1992), 547–566.

[113] A. Neeman, *Triangulated Categories*, Princeton University Press, 2001.

[114] A. Neeman, The Grothendieck duality theorem via Bousfield's techniques and Brown representability, *J. AMS* **9**, (1996), 205–236.

[115] The nLab, an online source for mathematics etc., `http://ncatlab.org`.

[116] M. Olsson, *Algebraic Spaces and Stacks*, AMS Colloquium Publications **62**, 2016.

[117] D. O. Orlov, Triangulated categories of singularities and D-branes in Landau-Ginzburg models, *Tr. Mat. Inst. Steklova* **246** (2004), 240–262.

[118] M. Porta, L. Shaul and A. Yekutieli, On the homology of completion and torsion, *Algebr. Repesent. Theory* **17** (2014), 31–67. Erratum: *Algebr. Represent. Theory*, **18** (2015), 1401–1405.

[119] M. Porta, L. Shaul and A. Yekutieli, Completion by derived double centralizer, *Algebr. Represent. Theory* **17** (2014), 481–494.

[120] M. Porta, L. Shaul and A. Yekutieli, Cohomologically cofinite complexes, *Comm. Algebra* **43** (2015), 597–615.

[121] L. Positselski, Dedualizing complexes of bicomodules and MGM duality over coalgebras, *Algebr. Represent. Theor.* **21** (2018), 737–767.

[122] D. Quillen, Higher algebraic K-theory I, in *Higher K-Theories*, Lecture Notes in Mathematics **341**, Springer, 1973, pp. 85–147.

[123] M. Reyes, D. Rogalski and J. J. Zhang, Skew Calabi–Yau triangulated categories and Frobenius Ext-algebras, *Trans. Amer. Math. Soc.* **369** (2017), 309–340.

[124] M. Reyes and D. Rogalski, A twisted Calabi–Yau toolkit, eprint arXiv:1807.10249 (2018).

[125] J. Rickard, Derived categories and stable equivalence, *J. Pure Appl. Algebra* **61** (1989), 303–317.

[126] J. Rickard, Morita theory for derived categories, *J. London Math. Soc.* **39** (1989), 436–456.

[127] J. Rickard, Derived equivalences as derived functors, *J. London Math. Soc.* **43** (1991), 37–48.

[128] D. Rogalski, Idealizer rings and noncommutative projective geometry, *J. Algebra* **279** (2004), 791–809.

[129] J. Rotman, *An Introduction to Homological Algebra*, Academic Press, 1979.

[130] L. R. Rowen, *Ring Theory* (Student Edition), Academic Press, 1991.

[131] R. Rouquier, Automorphismes, graduations et catégories triangulées, *J. Inst. Math. Jussieu*, **10** (2011), 713–751.

[132] R. Rouquier and A. Zimmermann, Picard groups for derived module categories, *Proc. London Math. Soc.* **87** (2003), 197–225.

[133] M. Saorin, Dg algebras with enough idempotents, their dg modules and their derived categories, *Algebra Discrete Math.* **23** (2017), 62–137.

[134] M. Sato, T. Kawai and M. Kashiwara, Microfunctions and pseudo-differential equations, in *Hyperfunctions and Pseudo-Differential Equations*, Lecture Notes in Mathematics **287**, Springer, 1973, pp. 265–529.

[135] S. Schwede, Algebraic versus topological triangulated categories, in *Proceedings of Conference on Triangulated Categories (Leeds 2006)*, London Math. Soc. Lecture Note Ser. **375** (2006).

[136] S. Schwede and B. Shipley, Stable model categories are categories of modules, *Topology* **42** (2003), 103–153.

[137] L. Shaul, Hochschild cohomology commutes with adic completion, *Algebra Number Theory* **10** (2016), 1001–1029.

[138] L. Shaul, Reduction of Hochschild cohomology over algebras finite over their center, *J. Pure Appl. Algebra* **219** (2015), 4368–4377.

[139] L. Shaul, Relations between derived Hochschild functors via twisting, *Comm. Algebra* **44** (2016), 2898–2907.

[140] B. Shipley, Morita theory in stable homotopy theory, in *Handbook of Tilting Theory*, London Math. Soc. Lecture Note Ser. **332** (2006), 393–409.

[141] P. Seidel, *Fukaya Categories and Picard–Lefschetz Theory*, Zürich Lectures in Advanced Mathematics, EMS, Zürich, 2008.

[142] A. Solotar and P. Zadunaisky, Change of grading, injective dimension and dualizing complexes, *Comm. Algebra* **46** (2018), 4414–4425.

[143] N. Spaltenstein, Resolutions of unbounded complexes, *Compositio Math.* **65** (1988), 121–154.

[144] The Stacks Project, an online reference, J. A. de Jong (ed.), `http://stacks.math.columbia.edu`.

[145] J. T. Stafford and M. Van den Bergh, Noncommutative curves and noncommutative surfaces, *Bull. Amer. Math. Soc.* **38** (2001), 171–216.

[146] J. D. Stasheff, Homotopy associativity of H-spaces, I, *Trans. AMS* **108** (1963), 275–292.

[147] J. D. Stasheff, Homotopy associativity of H-spaces, II, *Trans. AMS* **108** (1963), 293–312.

[148] B. Stenström, *Rings of Quotients*, Springer, 1975.

[149] D. Tamarkin, Microlocal conditions for non-displaceability, arXiv:0809.1584 (2008).

[150] R. W. Thomason and T. Trobaugh, Higher algebraic K-theory of schemes and of derived categories, in *The Grothendieck Festschrift*, Progress in Mathematics **88**, Birkhäuser, 1990, pp. 247–435.

[151] B. Toën, Lectures on DG-categories, in *Topics in Algebraic and Topo logical K-Theory*, Lecture Notes in Mathematics **2008**, Springer, 2011.

[152] B. Toën, Derived algebraic geometry, *EMS Surv. Math. Sci.* **1** (2014), 153–240.

[153] M. Van den Bergh, Existence theorems for dualizing complexes over non-commutative graded and filtered rings, *J. Algebra* **195** (1997), 662–679.

[154] M. Van den Bergh, A relation between Hochschild homology and cohomology for Gorenstein rings, *Proc. Amer. Math. Soc.* **126** (1998), 1345–1348. Erratum: *Proc. Amer. Math. Soc.* **130** (2002), 2809–2810.

[155] J.-L. Verdier, *Catégories Dérivées, état 0*, Lecture Notes in Mathematics **569**, Springer, 1977.

[156] J.-L. Verdier, *Des Catégories Dérivées des Catégories Abéliennes*, Astérisque **239**, 1996.

[157] R. Vyas and A. Yekutieli, Weak proregularity, weak stability, and the noncommutative MGM equivalence, *J. Algebra* **513** (2018), 265–325.

[158] C. Weibel, *An Introduction to Homological Algebra*, Cambridge Studies in Advanced Mathematics **38**, 1994.

[159] Q. S. Wu and J. J. Zhang, Some homological invariants of local PI algebras, *J. Algebra* **225** (2000), 904–935.

[160] Q.S. Wu and J. J. Zhang, Dualizing complexes over noncommutative local rings, *J. Algebra* **239** (2001), 513–548.

[161] A. Yekutieli, Dualizing complexes over noncommutative graded algebras, *J. Algebra* **153** (1992), 41–84.

[162] A. Yekutieli, *An Explicit Construction of the Grothendieck Residue Complex,* Astérisque **208**, 1992.

[163] A. Yekutieli, The residue complex of a noncommutative graded algebra, *J. Algebra* **186** (1996), 522–543.

[164] A. Yekutieli, Residues and differential operators on schemes, *Duke Math. J.* **95** (1998), 305–341.

[165] A. Yekutieli, Dualizing complexes, Morita equivalence and the derived Picard group of a ring, with appendix by E. Kreines, *J. London Math. Soc.* **60** (1999), 723–746.

[166] A. Yekutieli, The rigid dualizing complex of a universal enveloping algebra, *J. Pure Appl. Algebra* **150** (2000), 85–93.

[167] A. Yekutieli, The derived Picard group is a locally algebraic group, *Algebras Representation Theory* **7** (2004), 53–57.

[168] A. Yekutieli, Rigid dualizing complexes via differential graded algebras (survey), in *Proceedings of Conference on Triangulated Categories (Leeds 2006)*, London Math. Soc. Lecture Note Ser. **375** (2006).

[169] A. Yekutieli, Continuous and twisted $L\infty$ morphisms, *J. Pure Appl. Algebra* **207** (2006), 575–606.

[170] A. Yekutieli, *A Course on Derived Categories*, arXiv:1206.6632v2, (2015).

[171] A. Yekutieli, Central extensions of Gerbes, *Adv. Math.* **225** (2010), 445–486.

[172] A. Yekutieli, Duality and tilting for commutative DG rings, arXiv:1312.6411v4 (2016).

[173] A. Yekutieli, The squaring operation for commutative DG rings, *J. Algebra* **449** (2016), 50–107.

[174] A. Yekutieli, Residues and duality for schemes and stacks, lecture notes (2013), `www.math.bgu.ac.il/~amyekut/lectures/resid-stacks/handout.pdf`.

[175] A. Yekutieli, Rigidity, residues and duality for commutative rings, in preparation (2019).

[176] A. Yekutieli, Rigidity, residues and duality for schemes, in preparation (2019).

[177] A. Yekutieli, Rigidity, residues and duality for DM stacks, in preparation (2019).

[178] A. Yekutieli, Derived categories of bimodules, in preparation (2019).

[179] A. Yekutieli, The derived category of sheaves of commutative DG rings, lecture notes (2017), www.math.bgu.ac.il/~amyekut/lectures/shvs-dgrings/notes.pdf.

[180] A. Yekutieli, Derived categories of bimodules, lecture notes (2016), www.math.bgu.ac.il/~amyekut/lectures/der-cat-bimodules/abstract.html.

[181] A. Yekutieli, Another proof of a theorem of Van den Bergh about graded-injective modules, arXiv:1407.5916 (2014).

[182] A. Yekutieli and J. J. Zhang, Serre duality for noncommutative projective schemes, *Proc. Amer. Math. Soc.* **125** (1997), 697–707.

[183] A. Yekutieli and J. J. Zhang, Rings with Auslander dualizing complexes, *J. Algebra* **213** (1999), 1–51.

[184] A. Yekutieli and J. Zhang, Multiplicities of indecomposable injectives, *J. London Math. Soc.* **71** (2005), 100–120.

[185] A. Yekutieli and J. Zhang, Homological transcendence degree, *Proc. London Math. Soc.* **93** (2006) 105–137.

[186] A. Yekutieli and J.J. Zhang, Dualizing complexes and perverse modules over differential algebras, *Compositio Math.* **141** (2005), 620–654.

[187] A. Yekutieli and J. Zhang, Dualizing complexes and perverse sheaves on noncommutative ringed schemes, *Selecta Math.* **12** (2006), 137–177.

[188] A. Yekutieli and J. J. Zhang, Rigid complexes via DG algebras, *Trans. AMS* **360** (2008), 3211–3248.

[189] A. Yekutieli and J. J. Zhang, Rigid dualizing complexes over commutative rings, *Algebras Representation Theory* **12** (2009), 19–52.

[190] A. Yekutieli and J.J. Zhang, Residue complexes over noncommutative rings, *J. Algebra* **259** (2003), 451–493.

Index

Printed in the United States
by Baker & Taylor Publisher Services